Therapeutic and Nutritional Uses of Algae

Therapeutic and Nutritional Uses of Algae

Leonel Pereira
MARE—Marine and Environmental Sciences Centre
and
IMAR—Institute of Marine Research
Department of Life Sciences
University of Coimbra
Coimbra
Portugal

CRC Press
Taylor & Francis Group
Boca Raton London New York

CRC Press is an imprint of the
Taylor & Francis Group, an **informa** business
A SCIENCE PUBLISHERS BOOK

CRC Press
Taylor & Francis Group
6000 Broken Sound Parkway NW, Suite 300
Boca Raton, FL 33487-2742

First issued in paperback 2021

Version Date: 20171011

ISBN 13: 978-0-367-78148-4 (pbk)
ISBN 13: 978-1-4987-5538-2 (hbk)

Library of Congress Cataloging-in-Publication Data

Names: Pereira, Leonel, author.
Title: Therapeutic and nutritional uses of algae / Leonel Pereira, MARE, Marine and Environmental Sciences Centre, and IMAR, Institute of Marine Research, Department of Life Sciences, University of Coimbra, Coimbra, Portugal.
Description: Boca Raton, FL : CRC Press, 2018. | "A science publishers book." | Includes bibliographical references and index of algae species.
Identifiers: LCCN 2017040101 | ISBN 9781498755382 (hardback : alk. paper)
Subjects: LCSH: Marine algae as food. | Marine plants--Nutrition. | Marine algae--Therapeutic use. | Plants, Edible.
Classification: LCC TX402 .P475 2018 | DDC 641.6--dc23
LC record available at https://lccn.loc.gov/2017040101

Visit the Taylor & Francis Web site at
http://www.taylorandfrancis.com

and the CRC Press Web site at
http://www.crcpress.com

Preface

Marine algae (seaweeds and marine microalgae) are one of the largest producers of biomass in the marine environment. They produce a wide variety of chemically active metabolites in their surroundings, potentially as an aid to protect themselves against the environmental stress and aggressive organisms. These active metabolites, such as halogenated compounds, alcohols, aldehydes, terpenoids, polysaccharides and fatty acids derivatives, among other compounds, that are produced by several species of seaweeds and microalgae have antibacterial, antifungal, antiviral, vermifuges, neuroprotective, antitumoral, anti-inflammatory, anti-allergic, antithrombotic, hypocholesterolemic, and hypoglycemic properties, which are effective in the prevention of several diseases and have potential uses as therapeutic drugs. Numerous studies have concentrated on the contribution of marine organisms, including seaweeds and marine microorganisms, in the search for new drugs from natural products.

One important approach to drug development involves assaying folk remedies for active ingredients, and several seaweed species are used as traditional medicines, foods and health care products in various regions of the world. The use of seaweed species to treat fever, lumps, and swelling is recorded in the Oriental medical textbook Donguibogam, published in 1613. Many seaweed species have also been used as herbal medicines in China over several centuries.

The first industrial interest in studying algae started with the aquaculture industry for both microalgae and macroalgae (seaweed). The demand for seaweed as a direct food resource and the demand for fish production needed the application of aquaculture techniques. In the case of seaweed, this was the only choice for sustainable production. The production of microalgae was necessary for the feeding phase of fish larvae to ensure the survival of newly born juveniles and for the feeding of zooplankton. For development of pharmaceutical compounds from a marine source, supply issues will always be a critical problem, and a major obstacle to drug development is the lack of sufficient material. Among marine organisms, seaweed is a promising candidate for drug production because it's relatively easy to obtain adequate, reliable and, the most important, renewable supplies by aquaculture. Thus, such an abundant species with immense aquaculture potential will have a better chance of being developed on a large scale for commercial food and drug production.

Leonel Pereira

Acknowledgements and Credits of Images

This work had the support of "Fundação para a Ciência e Tecnologia" (FCT), through the strategic project UID/MAR/04292/2013 granted to MARE. The author also thanks Alan Critchley (Phycologist) the revision of some chapters of this book.

Credits of Images

Images on Public Domain

Fig. 2.2: Edgar 2016
Fig. 2.3: Eklof 2012
Fig. 2.5: Tike 2016
Fig. 2.6: Yikrazuul 2009
Fig. 5.1: Häggström 2014
Fig. 10.2: Beards 2014
Fig. 10.3: PubChem 2017
Fig. 10.4: Edgar 2017

Originals

Leonel Pereira

Figs. 1.1, 1.2, 1.3, 1.4, 1.5, 1.6, 1.7, 1.8, 1.9, 2.1, 2.4, 3.1, 3.2, 3.3, 10.1, 11.1, 12.1, 12.2

Contents

Preface **v**

Acknowledgements and Credits of Images **vii**

Abbreviations and Acronyms **xi**

Glossary **xxiii**

1. Biodiversity and Description of the Main Algae with Bioactive Properties 1
2. Nutritional Composition of the Main Edible Algae 65
3. Therapeutic Uses of Phycocolloids 128
4. The Cardio-protective Activity of Edible Seaweeds and their Extracts 143
5. Antiviral Activity of Seaweeds and their Extracts 175
6. Antitumor Activity of Seaweeds and their Extracts 212
7. Antifungal Activity of Seaweeds and their Extracts 311
8. Antibacterial Activity of Seaweeds and their Extracts 347
9. Antiparasitic, Insecticidal, and Larvicidal Activities of Seaweeds and their Extracts 428
10. Anti-inflammatory, Anti-allergic, Antipyretic, Antinociceptive, Antithrombotic, and Anti-coagulant Activities of Seaweeds and their Extracts 450
11. Neurological Activities of Seaweeds and their Extracts 485
12. Thalassotherapy and Marine Cosmeceuticals 503

References 523

Index of Algae Species **627**

Abbreviations and Acronyms

1301	: Human T-cell leukemia
13762	: MAT Rat mammary adenocarcinoma
3LL	: Lewis lung carcinoma
4T1	: Mouse breast cancer cells
5-FU	: 5-fluorouracil
αTHR	: Antithrombin
A20	: Murine reticulum cell sarcoma (Mouse B lymphoma cells-leukemia cells)
A2780	: Human ovarian cancer cell line
A549	: Human lung adenocarcinoma
AA	: Arachidonic acid
Aβ	: Beta amyloid protein
AAACa	: Active absorbable algal calcium
AAPH	: 2,2-Azobis(2-amidinopropane) dihydrochloride
ACA	: Active cutaneous anaphylaxis
ACE	: Angiotensin-converting enzyme
ACE-I	: ACE-Inhibitory
ACh	: Acetylcholine
AChE	: Acetylcholinesterase enzyme
ACV	: Acyclovir
A.D.	: After death (see also B.C.)
AD	: Aujeszky's disease
AdV	: Adenovirus
AE	: Aqueous extract
Ag	: Silver
AG	: Arabinogalactan
AGS	: Gastric cancer cell lines
AGC	: Protein kinases includes more than 60 protein kinases in the human genome, classified into 14 families
AgNPs	: Silver nanoparticles of aqueous extract
AHLs	: Acyl-homoserine lactones
AIDS	: Acquired immune deficiency syndrome
AIF	: Apoptosis inducing factor
Akt	: A serine/threonine kinase
ALAT	: Alanine amino-transferase
ALP	: Alkaline-phosphatase
AMV	: Avian myeloblastosis virus (A species of avian type C retroviruses)
AMVN	: 2,2-Azobis(2,4-dimethylvaleronitrile)
AOSC	: Acidic oligosaccharide sugar chain
AP-1	: Activator protein-1
APC	: Activated protein C
APr	: Phosphonoacetic acid

aPTT	:	Activated partial thromboplastin time
ARBs	:	Angiotensin II receptor blockers
ARIs	:	Acute respiratory infections
ASAT	:	Aspartate aminotransferase
ASE	:	Accelerated solvent extraction
ASFV	:	African swine fever virus
ASK1	:	Apoptosis signal-regulating kinase 1
B-16	:	Murine melanoma
B16-BL6	:	Murine melanoma
B16-F10	:	Murine melanoma
BACE	:	Beta-secretases
BAD	:	Bcl2 antagonist of cell death
Bak	:	Pro-apoptotic protein Bak
BAL	:	Bronchoalveolar lavage
BALB/c	:	Is an albino, laboratory-bred strain of the House Mouse from which a number of common sub-strains are derived
BAS(s)	:	Biological Active Substance(s); substance(s) having pronounced physiological activity and producing either stimulatory or inhibitory impact on *in vivo* or *in vitro* biological processes
BAX	:	Apoptosis regulator, also known as bcl-2-like protein 4
BBB	:	Blood-brain barrier
B.C.	:	Before Christ (see also A.D.)
BC	:	Breast cancer
Bcl-xL	:	B-cell lymphoma-extra large
BCNU	:	Carmustine
BDDE	:	Bis(2,3-dibromo-4,5-dihydroxybenzyl) ether
BEL-7402	:	Human hepatocellular carcinoma
BeWo	:	Choriocarcinoma cells
bFGF	:	Basic fibroblast growth factor
BHK	:	Baby Hamster Kidney fibroblasts; BHK-21 cells are susceptible to human adenovirus D, reovirus 3, and vesicular stomatitis virus (Indiana strain); BHK-21 cells are resistant to poliovirus 2; the cells are negative for reverse transcriptase, which means that they lack integral retrovirus genomes
Bid	:	Pro-apoptotic protein Bid
BIP	:	Butyl-isobutyl-phthalate
BMDCs	:	Bone marrow-derived DCs
BoHV-1	:	Bovine herpesvirus type 1
BoHV-5	:	Bovine herpesvirus type 5
BSIs	:	Bloodstream infections
BSR	:	Basal stem rot
BSC	:	Monkey kidney cells
BSE	:	Bovine spongiform encephalitis
BuChE	:	Butyrylcholinesterase
BV-2	:	Murine microglial cell line BV-2
BVDV	:	Bovine viral diarrhea virus
BW	:	Body weight
BXPC3	:	Human primary pancreatic adenocarcinoma
C6	:	Rat brain tumor cells
C6/36	:	Mosquito (*Aedes albopictus*) C6/36 cells
C32	:	Human melanoma cells
Ca	:	Calcium
Caco-2	:	Human epithelial colorectal adenocarcinoma

CaMKII	: Ca^{2+}\calmodulin-dependent kinase II
Caov-3	: Human ovarian cancer cell line
CAT	: Catalase
CC_{50}	: Viable cells concentration; or Cytotoxic concentration
CCL39	: Chinese hamster fibroblasts
CCT	: Cytotoxic chemotherapy
CD4	: Cluster of differentiation 4; is a glycoprotein found on the surface of immune cells such as T helper cells, monocytes, macrophages, and dendritic cells
CD14	: Cluster of differentiation 14, lipopolysaccharide receptor
CD31	: Platelet endothelial cell adhesion molecule-1, PECAM-1/CD31
Cdk	: Cyclin-dependent kinases
CDV	: Canine distemper virus
CEA	: Carcinoembryonic antigen
CEM cell	: a cell line derived from human T cells
CFAV	: Cell fusing agent virus
c-Jun	: Is a protein that in humans is encoded by the JUN gene
CHO	: Chinese hamster ovary
CL1-5	: Human lung carcinoma cells
CLP	: Caulerpin
CLSI	: Clinical and laboratory standards institute
CMV	: Cytomegalovirus
CNS	: Central nervous system
Colo-320DM	: Human colon adenocarcinoma
Colo-357	: Human cell line of metastatic pancreatic adenocarcinoma
Conc	: Concentration
COX-1	: Cyclooxygenase-1
COX-2	: Cyclooxygenase-2
CPT-11	: Irinotecan
CT-26	: Murine colon cancer
CTC	: Cytotoxic T cell
Cu	: Copper
CVB3	: Coxsackie virus B3
CVD	: Cardiovascular diseases
Cx	: Connexin
Da	: Dalton
Daudi	: Human Burkitt's lymphoma
DCM	: Dichloromethane
DCs	: Dendritic cells
DDBT	: 2-(4-(3,5-dihydroxyphenoxy)-3,5-dihydroxyphenoxy) benzene-1,3,5-triol
ddY	: Deutschland, Denken, and Yoken
DENV	: Dengue virus
DF	: Dengue fever
DHA	: Docosahexaenoic acid
DHF	: Dengue hemorrhagic disease
DHOD	: Dihydroorotate dehydrogenase
DLD-1	: Human colorectal adenocarcinoma
DMBA	: 7,12-dimethylbenz[a]anthracene
DMF	: Dimethylformamide $(CH_3)_2NC(O)H$
DMSO	: Dimethyl sulfoxide
DNA	: Deoxyribonucleic acid
DNMT3B	: DNA methyltransferase 3B, which silences tumor suppressors

DON	:	Deoxynivalenol
DPHC	:	Diphlorethohydroxycarmalol
DPP-4	:	Dipeptidyl-peptidase-4
DR5	:	Death receptor 5
DS	:	Double stranded
DSS	:	Dengue shock syndrome
DU-145	:	Human prostate cancer
DW	:	Dry weight
E2F	:	Transcription factors E2Fs
EAC	:	Ehrlich ascites carcinoma
EAE	:	Enzyme assisted extraction
EAT	:	Ehrlich ascites tumor
EBOV	:	Ebola vírus
EC_{50}	:	Effective dose that gives 50% of studied activity
ECM	:	Extracellular matrix
ED-40515(−)	:	Human leukemia Tcell line
EGF	:	Epidermal growth factor
EGFR	:	Epidermal growth factor receptor
EID_{50}	:	Value of egg infective dose fifty
ELISA	:	Enzyme-linked immunosorbent assay
EMCV	:	Encephalomyocarditis virus
EMT	:	Epithelial to mesenchymal transition
ENN	:	N-Ethyl-N′-nitro-N-nitrosoguanidine
EPA	:	Eicosapentaenoic acid
EPO	:	Eosinophil peroxidase
ER	:	Endoplasmic reticulum
ERCC1	:	Excision repair cross complementation 1
ERK	:	Extracellular signal-regulated kinases
ERK1/2	:	Extracellular signal-regulated protein kinases 1 and 2
FA	:	Fat acids
Fas	:	A member of death receptor family
FBHE	:	Fetal bovine heart endothelial
FCV	:	Feline calicivirus
FDA	:	Food and drug administration
Fe	:	Iron
Fem-x	:	Malignant melanoma
FG	:	Female gametophytes
FGF	:	Fibroblast growth factor
FHC	:	Human colon epithelial
FHs	:	74 Int Human normal intestinal
FIV	:	Feline immunodeficiency virus
FluV	:	Influenza virus
FMLP	:	Formyl-Met-Leu-Phe
FOLFOX	:	Oxaliplatin plus 5-fluorouracil/leucovorin
GADD45A	:	Growth arrest and DNA-damage-inducible protein
GAE	:	Gallic acid equivalents
gB	:	Glycoprotein B
gC	:	Glycoprotein C
GC-MS	:	Gas chromatography-mass-spectrometry
GCSF	:	Granulocyte colony-stimulating factor
GFS	:	Galactofucan sulfate
GIP	:	Glucose-dependent insulinotrophic polypeptide

GLP-1	:	Glucagon-like peptide-1
GOTO	:	Human neuroblastoma cells
$Gram^+$	:	Gram-positive
$Gram^-$	:	Gram-negative
GRAS	:	Generally recognized as safe
GRP78	:	Glucose regulated protein 78
GSH	:	Glutathione
GSH-PX	:	Glutathione peroxidase
GST	:	Glutathione-S-transferase
H (H)	:	Hour(s)
H1299	:	Non-small cell lung cancer cells
H22	:	Murine hepatoma
H_2O_2	:	Hydrogen peroxide
HA	:	Hyaluronic acid
HAART	:	Highly active antiretroviral therapy
HAase	:	Hyaluronidase
HaCat	:	Human keratinocytes
HAT	:	Human African trypanosomiasis
HAV	:	Hepatitis A virus
HBV	:	Hepatitis B virus
HBoV	:	Human bocavirus
Hca-F	:	Mouse hepatocarcinoma cells
HCC	:	Hepatocellular carcinoma
HCoV	:	Emerging human coronaviruses
HCT	:	Human colon cancer cells
HCT-8	:	Human colon cancer cells
HCT-15	:	Human colorectal adenocarcinoma cells
HCT-116	:	Human colon cancer cells
HCV	:	Hepatitis C virus
HCMV	:	Human cytomegalo virus
HDF	:	Normal human dermal fibroblast cells
HDL	:	High-density lipoprotein
HEK-293	:	Human embryonic kidney 293 cells, also often referred to as HEK 293, HEK-293, 293 cells, or less precisely as HEK cells, are a specific cell line originally derived from human embryonic kidney cells grown in tissue culture; HEK 293 cells have been widely used in cell biology research for many years, because of their reliable growth and propensity for transfection; they are also used by the biotechnology industry to produce therapeutic proteins and viruses for gene therapy
HeLa	:	Human cervical cancer cells
HEp-2	:	Epidermoid carcinoma cells
HepG2	:	Liver cancer cells
HIF-1α	:	Hypoxia-inducible factor-1α
HFD	:	High-fat diet
HFD-NSK	:	HFD containing 3% NSK
HGF	:	Hepatocyte growth factor
HIV	:	Human Immunodeficiency virus
HIV-RT	:	HIV-reverse transcriptase
HL-60	:	Human promyelocytic leukemia
HMEC-1	:	Human microvascular endothelial cells
HMPV	:	Human metapneumo virus
HMWK	:	High molecular weight kininogen (Fitzgerald factor)

HOS	:	Human osteosarcoma cells
HPIV	:	Human parainfluenza virus
HPV	:	Human papilloma virus
HRV	:	Human rhino virus
HS	:	Heparin sulfate
HS-Sultan	:	Cells human lymphoma
Hs 677.st	:	Human stomach fibroblasts
HSV	:	Herpes simplex virus
HT-29	:	Human colon adenocarcinoma cells
HT1080	:	Human fibrosarcoma cells
HTC-116	:	Colon carcinoma
HTLV-1	:	Human Tcell leukemia virus type 1
HTNV	:	Hantaan virus
HU	:	Human
Huh-6	:	Human liver cancer cells
Huh7	:	Hepatoma cells
HUT-102	:	Human cutaneous T lymphocytes
HUVEC	:	Human umbilical vein endothelial cells
I	:	Iodine
IAV	:	Influenza A virus (H1N1)
IARC	:	International agency for research on cancer
IBV	:	Influenza B virus
IC_{50}	:	Concentration of 50% inhibiting effects
ICR	:	Imprinting control region mice
IEC-6	:	Rat normal intestinal epithelial cells
IgE	:	Allergen-specific immunoglobulin E
IGF-IR	:	Insulin-like, growth Factor-I receptor
IHD	:	Ischemic heart disease
IκBα	:	Inhibitor of NF-κB
IL-1β	:	Interleukin-1β
IL-2	:	Interleukin-2
IL-4	:	Interleukin-4
IL-6	:	Interleukin-6
IL-8	:	Interleukin-8
IL-10	:	Interleukin-10
IL-12	:	Interleukin-12
ILS	:	Increase in life span
IMTA	:	Integrated multi-trophic aquaculture
IFN-β	:	Interferon-beta
IFN-γ	:	Interferon-gamma
iNOS	:	Inducible nitric oxide synthase
IP	:	Intraperitoneal
ISA	:	Infectious salmon anemia
ISFs	:	Insect-specific flaviviruses
IU	:	inhibitory unit; or international units
JAK/STAT	:	Janus kinase/signal transducer and activator of transcription
JB6 Cl41	:	Normal murine epidermis
JEV	:	Japanese encephalitis virus
JNKs	:	The c-Jun N-terminal kinases
Jurkat	:	Human Tcell leukemia
K	:	Potassium
K562	:	Human chronic myelogenous leukemia

KA3IT	:	Virally transformed form cancer cells
KB	:	Human nasopharynx carcinoma
KELLY	:	Cell Line human neuroblastoma
KF	:	Kahalalide F
KK-Ay	:	Mice homozygous for the yellow spontaneous mutation; animal model for the treatment of Human type 2 diabetic nephropathy
KMC-1	:	Cholangiocarcinoma cells
KMG-C	:	Gallbladder carcinoma cells
L929	:	Mouse fibroblast cell line
L-1210	:	Mouse lymphocytic leukemia
LC	:	Lethal concentration
LC_{50}	:	Lethal concentration for 50% mortality
LC_{90}	:	Lethal concentration for 90% mortality
LD_{50}	:	Is the amount of a material, given all at once, which causes the death of 50% (one half) of a group of test organisms; the LD_{50} is one way to measure the short-term poisoning potential (acute toxicity) of a material
LDH	:	Lactase dehydrogenase
LDL	:	Low-density lipoprotein
LLC	:	Lewis lung carcinoma
LLC-MK2 cells	:	Monkey (*Macaca mulatta*) kidney cell strains llc-mk2
LNCaP	:	Human prostate adenocarcinoma
LOVO	:	Human colorectal adenocarcinoma
LPS	:	Lipopolysaccharides
LS-174	:	Human colonic adenocarcinoma cells
LuxR	:	Luminescence transcriptional activator
Lux*S*	:	*S*-ribosylhomocysteine lyase
M	:	molar; a solution with a concentration of 1 mol L^{-1} is equivalent to 1 molar (1 M)
MAAs	:	Mycosporine-like amino acids
MAE	:	Microwave assisted extraction
MAP	:	Microtubule-associated protein
MAPK	:	Mitogen-activated protein kinase
MBC	:	minimum bactericidal concentration
MBP	:	Mean blood pressure
MC	:	Melanoma B16 cells
MCF-7	:	Human breast adenocarcinoma
MCF-10A	:	Non-tumorigenic epithelial cell lines
MCF-12A	:	Immortalized non-cancerous cells
Mcl-1	:	Anti-apoptotic protein Mcl-1
MCMV	:	Murine cytomegalovirus
MCP-1	:	Monocyte chemotaxis protein
MDA	:	Malondialdehyde, an oxidative stress marker
MDA-MB-231	:	Human mammary adenocarcinoma
MDA-MB-435	:	Human mammary carcinoma
MDCK	:	Madine–Darby canine kidney
MDSC	:	Myeloid-derived suppressor cells
ME	:	Methanolic extract
MEK	:	Methyl ethyl ketone (= Butanone)
MEKK1	:	MAPK kinase 1
MEL-28	:	Human melanoma
Meth-A	:	Meth A sarcoma cells
MetS	:	Metabolic syndrome

MeV	:	Measles virus
MFC	:	Minimum fungicidal concentration
Mg	:	Magnesium
MG-63	:	Human osteosarcoma
MGC-803	:	Human gastric cancer
MiaPaCa-2	:	Human pancreatic carcinoma
MIC	:	Minimum inhibitory concentration
MIQ	:	Minimum inhibitory quantity
Min	:	Minimum
min	:	minute(s)
miR-29	:	MicroRNAs
MKN-45	:	Human gastric adenocarcinoma
MLC	:	Minimum lethal concentration
MMP	:	Matrix metalloproteinase
MMP-9	:	Matrix metalloproteinase-9
Mn	:	Manganese
MNNG	:	N-Methyl-N′-nitro-N-nitrosoguanidine
MNV	:	Murine norovirus
MOLT-4	:	Human lymphoblastic leukemia
MPO	:	Myeloperoxidase
MRC-5	:	Human lung fibroblast
mRNA	:	Messenger ribonucleic acid
MRSA	:	Methicillin-resistant *Staphylococcus aureus*
MT-2	:	Human lymphocyte infected with HTLV-1
MT-4	:	Human Tcell leukemia
Mtss1	:	Metastasis suppressor 1
MTT	:	Methyl thiazolyl tetrazolium (colorimetric assay for assessing cell metabolic activity)
MPTP	:	1-methyl-4-phenyl-1,2,3,6-tetrahydropyridine
MU	:	Murine
MUFAs	:	Monounsaturated fatty acids
Na	:	Sodium
NAMALWA	:	Human Burkitt lymphoma cell line
NB4	:	Human promyelocytic cells
NBMS	:	Nutrient-broth micro-dilution bioassay
NCCLS	:	National Committee for Clinical Laboratory Standards
NCI-H292	:	Human lung mucoepidermoid carcinoma cells
NCI-N87	:	Human gastric cancer cells
ND	:	Normal diet
NDV	:	Newcastle disease virus
NF-κB	:	Nuclear factor κB
NIH-3T3	:	Mouse embryonic fibroblast cells
NK	:	Natural killer (cells)
NGF	:	Nerve growth factor
NMF	:	Natural moisturizing factor
NMR	:	Nuclear magnetic resonance
NSCLC-N6	:	Human non-small cell bronchopulmonary carcinoma line
NSK	:	Non-shaved Kombu
NO	:	Nitric oxide
NPs	:	Nanoparticles
NSAID's	:	Non-steroidal anti-inflammatory drugs
NTDs	:	Neglected tropical diseases

OE19	:	Gastro-esophageal adenocarcinoma cells
OE33	:	Human Caucasian esophageal carcinoma cells
OVA	:	Ovalbumin
P	:	Phosphorus
p21	:	CDK inhibitor p21
P-388	:	Murine leukemic cells
p90RSK	:	An AGC kinase of the RSK family. Phosphorylated and activated by Erk1 and -2 in response to many growth factors, polypeptide hormones and neurotransmitters
Pa	:	Chlorophyll-related compound pheophorbide a
PAI-1	:	Plasminogen activator inhibitor-1
Panc-1	:	Pancreatic cancer cells
PancTu1	:	Pancreatic cancer cells
Panc-3.27	:	Human pancreatic cancer
PANC-89	:	Pancreatic cancer cells
PARP	:	Poly-ADP-ribose polymerase
PaRV	:	Parramatta River virus
PBMC	:	Peripheral blood mononuclear cells
PBS	:	Phosphate-buffered saline
PC-3	:	Human prostate cancer
PCA	:	Passive cutaneous anaphylaxis
PCNA	:	Proliferating cell nuclear antigen
PBEE	:	Prepared by enzymatic digestion
PCV	:	Palm Creek virus
PE	:	Protein extract
PEPK\P-eIF2α\CHOP	:	Protein kinase RNA-like ER kinase\phosphorylation of eukaryotic initiation factor2α\enhancer binding protein homologous protein
PGE_2	:	Prostaglandin E_2 (dinoprostone)
PGE_3	:	Prostaglandin E_3
PIVs	:	Parainfluenza virus
PLA_2	:	Phospholipase A_2
PLGF	:	Placenta growth factor
Polio-2	:	Poliovirus type 2
PMA	:	Phorbol 12-myristate 13-acetate
PMN	:	Polymorphonuclear neutrophils
PP	:	Percentage points
PRA	:	Plasma renin activity
pRB	:	Retinoblastoma protein
PSN1	:	Pancreatic adenocarcinoma
PT	:	Prolongation of prothrombin time
PTA	:	Plasma thromboplastin antecedent
PTC	:	Plasma thromboplastin component
PTK	:	Protein tyrosine kinase
PtK1	:	*Potorous tridactylis* normal kidney cells
PUFAs	:	Polyunsaturated fatty acids
RABV	:	Rabies virus infection
Raji	:	Human cell line from hematopoietic origin (Burkitt's lymphoma)
RAS	:	Renin-angiotensin system
RAW264.7	:	Mouse leukemic monocyte macrophage cell line
RBL	:	Rat basophilic leukemia cells
RIF-1	:	Radiation-induced fibrosarcoma
R-MuLV	:	Rauscher routine leukemia virus

ROS	:	Reactive oxygen species
RPMI-7951	:	Human malignant melanoma obtained from the lymph node
RSV	:	Respiratory syncytial virus
RT	:	Reverse transcriptase
RT-PCR	:	Reverse transcription-polymerase chain reaction
RV	:	Rhinovirus
RVFV	:	Reft vally fever virus
S	:	Sulfur
s	:	second
S-180	:	Sarcoma 180
SAPK/JNK	:	Stress-activated protein kinase/Junamino-terminal kinase
SARS-CoV	:	Severe acute respiratory syndrome coronavirus
SBP	:	Systolic blood pressure
SC	:	Subcutaneously
SC-CO_2	:	Supercritical CO_2
SD	:	Sprague Dawley® rat
SF	:	Seaweed fiber
SF-295	:	Human glioblastoma
SFE	:	Supercritical fluid extraction
SFV	:	Semliki forest virus
SGTP	:	Serum glutamic-pyruvic transaminase
SGOT	:	Serum glutamate-oxaloacetate transaminase
SHR	:	Spontaneously hypertensive rat
SH-SY5Y	:	Human neuroblastoma cell line
SK-Hep1	:	Human hepatocarcinoma cells
SK-N-SH	:	Human neuroblastoma cells
SK-OV-3	:	Ovarian carcinoma cells
Smurf2	:	Smad ubiquitination regulatory factor 2
SOCS-3	:	Suppressor of cytokine signaling-3
SOD	:	Superoxide dismutase
SH-SY5Y	:	Is a human derived neural cell line used in scientific research
SI	:	Selectivity index
SiHa	:	Cervix carcinoma
SINV	:	Sindbis virus
SIV	:	Simian immunodeficiency virus
SK Hep-1	:	Human hepatocellular carcinoma
SK-LU-1	:	Human lung carcinoma cell line
SK-MEL-5	:	Human malignant melanoma
SK-MEL-28	:	Human malignant melanoma
SLs	:	sulfolipids
SMMC-7721	:	Human hepatocellular carcinoma
SOCS-3	:	Suppressor of cytokine signaling-3
SOD	:	Superoxide dismutase
SP	:	Sulfated polysaccharides
SPMG	:	Sulfated polymannuroguluronate
SPS	:	Sulfated polysaccharides
SQDG	:	Sulfoquinovosylmonoacylglyceride
SS	:	Single stranded
SSVNIH3T3	:	Virally transformed form cancer cell line
STZ	:	Streptozotocin
SuHV-1	:	Suid herpesvirus type 1
SW-480	:	Human colon carcinoma

T	: Tetrasporophytes
T24	: Human bladder cancer cells
T-47D	: Human breast cancer cells
T98G	: Caucasian human glioblastoma cells
TAM	: Tamoxifen
TB	: Pulmonary tuberculosis
TC	: Total cholesterol
Tc1	: Type 1 cytotoxic T lymphocytes
TCRV	: Tacaribe virus
TDB	: 4,5-dihydroxybenzyl
T2DM	: Type 2 diabetes mellitus
TF	: Tissue factor
TFPI	: Tissue pathway inhibitor
TG	: Triglyceride
Th1	: Murine type 1 helper T cells
THP 1	: Human leukemia
TIMP-1	: Tissue inhibitor metalloproteinase-1
TJ	: Tight junction (proteins)
TK	: Tororokombu
TK^-	: Thymidine kinase deficient
TKd	: Thymidine kinase deficient
TLC	: Thin layer chromatography
TMV	: Tobacco mosaic virus
TNF-α	: Tumor necrosis factor-alpha
TNV	: Tobacco necrotic virus
TOPK	: Oncogenic kinase TOPK
TPA	: 12-*O*-Tetradecanoylphorbol-13-acetate
TRAIL	: TNF-related apoptosis-inducing ligand
TT	: Thrombin time assays
TUNEL	: Transferase-catalyzed deoxyuridine phosphate-nick end labeling
U87MG	: Glioblastoma cell line (Uppsala 87 malignant glioma)
U-937	: Human leukemic monocyte lymphoma
UCP1	: Uncoupling protein 1
UAE	: Ultrasound assisted extraction
u-PA	: Urokinase-type plasminogen activator
uPAR	: Urinary human urokinase-type plasminogen activator receptor
UV	: Ultraviolet
V79-4	: Chinese hamster lung fibroblasts
VACV	: Vaccinia virus
VEGF	: Vascular endothelial growth factor
Vero	: African green monkey kidney
VLDL	: Very low density lipoprotein
VREF	: Vancomycin-resistant *Enterococcus faecium*
VSV	: Vesicular stomatitis virus
WAT	: White adipose tissue
WHCO1	: Human esophageal cancer
WI-38	: Human diploid fibroblast cells
WiDr	: Colon adenocarcinoma cell
WNV	: West Nile virus
WSSV	: White spot syndrome virus
X	: Stuart-Prower factor
XI	: Antihemophilic factor

XIIa	:	Activated Hageman factor
XC	:	Rat sarcoma
XF 488	:	Human CNS tumor cell lines
YAC-1	:	Murine lymphoma
YFV	:	Yellow fever virus
ZIKV	:	Zika virus
Zn	:	Zinc
α	:	Alpha (carrageenan)
β	:	Beta (carrageenan)
ξ	:	Xi (carrageenan)
θ	:	Theta (carrageenan)
λ	:	Lambda (carrageenan)
κ	:	Kappa (carrageenan)
ι	:	Iota (carrageenan)
ν	:	Nu (carrageenan)
wk	:	week(s)
μ	:	Mu (carrageenan)

Glossary

Active metabolite	:	Compound (drug) with therapeutic activity like the parent compound, which must be considered in therapeutic pharmacokinetics
Adultoid	:	A premature adult form of an insect
Acuminate	:	provided with sharp points
Acute	:	sharp at the end; ending in a point
Aerocyst	:	gas-filled bladder
Agar	:	phycocolloid extracted from various genera and species of red seaweed (Gigartinales and Gelidiales) which consists of a heterogeneous mixture of two polysaccharides, agarose and agaropectin
Agaroid	:	a compound like agar in properties that is obtained from certain red algae (as of the genus *Phyllophora*)
Agarose	:	one of two separable components of agar found in some red algae; has a remarkable gelling property
Aglycemia	:	Absence of sugar from blood
Air vesicle/air-bladder	:	a small bladder containing a variety of gases (nitrogen, oxygen and carbon dioxide) and aiding in the buoyancy of certain seaweeds (e.g., brown algae)
Alginate	:	phycocolloid extract from brown algae (Phaeophyceae); are linear macromolecules made up of two types of monomers, the manuronic acid (M) and guluronic acid (G)
Algotherapy	:	The therapeutic use of seaweeds or algae
Alternate (branching)	:	the branches, leaves, etc. are placed singly at different heights on the axis on opposite sides, or at definite angular distances from one another
Apiculum (pl. Apiculae)	:	a short, slender, flexible point
Anthelmintic	:	acting to expel or destroy parasitic intestinal worms
Antheridial	:	relative to Antheridium
Antiangiogenic	:	Of or relating to a naturally occurring substance, drug, or other compound that can destroy or interfere with the fine network of blood vessels needed by tumors to grow and metastasize
Antibiosis	:	A relationship between an organism and an antibiotic produced by another
Anticarcinogen	:	A chemical that acts against cancer
Antifibrotic	:	An agent that blocks or prevents tissue scarring
Antinociceptive	:	Reducing sensitivity to painful stimuli
Antioxidant	:	a molecule capable of inhibiting the oxidation of other molecules; may be of natural origin or laboratory-synthesized

Antithrombotic	:	Preventing or interfering with the formation of a thrombus or blood clotting
Antiproliferative	:	Used or tending to inhibit cell growth; for example, antiproliferative effects on tumor cells
Antithrombic	:	Of, pertaining to, or resembling the action of antithrombin
Anthropophilic	:	Human-seeking or human-preferring, especially regarding: (1) bloodsucking arthropods, denoting the preference of a parasite for the human host as a source of blood or tissues over an animal host; and (2) dermatophytic fungi that grow preferentially on humans rather than other animals
Anxiolytic	:	Drug or substance used to treat anxiety
Apex	:	the tip or summit
Apical	:	located at the tip or highest point
Appressorium (plural Appressoria)	:	A flattened and thickened tip of a hyphal branch, formed by some parasitic fungi, that facilitates penetration of the host plant
Arcuate	:	having the form of a bow; curved
Artemether	:	Is a medication used for the treatment of malaria. The injectable form is specifically used for severe malaria rather than quinine; it may not be as effective as artesunate
Aspergilloma	:	Also known as a mycetoma or fungus ball, is a clump of mold which exists in a body cavity such as a paranasal sinus or an organ such as the lung. It is caused by fungi of the genus *Aspergillus*
Assimilatory filaments	:	pigmented or photosynthetic filaments
Ataxia	:	is a neurological sign consisting of lack of voluntary coordination of muscle movements that includes gait abnormality
Attenuated, attenuating	:	to make thin; make slender or fine
Axial cell	:	the central cell of an axis, sometimes being distinguishable among medullary cells in transverse section
Axil	:	the angle between the upper side of a branchlet (or stem or leaf) and the supporting stem or branch
Axis (pl., axes)	:	a stem-like stalk on which parts or structures are arranged, a line or point forming the center of an object
Bifarious	:	in two vertical rows
Bifurcate	:	divided or forked into two branches
Bilateral	:	with two corresponding or complementary opposite sides
Biomedicinally	:	in terms of, or by means of, biomedicine
Blade	:	a broad, thin flat part of the thallus
Blastoconidia	:	the unit of asexual reproduction produced by budding; seen in yeasts such as *Candida* and *Cryptococcus* spp.; called also blastospore
Blebbing	:	Blebs are protrusions of the cell membrane
Branchlet	:	a small secondary or higher-order branch, usually the ultimate branch in a system of branching (ramulus)
Bulbous	:	bulging, enlarged
Caespitose	:	forming dense tufts or clumps
Carrageenan	:	carrageenans are a special type of polysaccharide extracted from red algae (Rhodophyta, Gigartinales), consisting of galactose units and variable substitution of sulfate groups

Cartilaginous (in phycology)	:	seaweed with cartilage-like consistency; firm
Cervicorn	:	resembling a deer's horn
Clavate	:	club-shaped
Clonogenic	:	giving rise to a clone of cells
Coenocyte (Cenocyte)	:	an organism made up of a multinucleate, continuous mass of protoplasm enclosed by one cell wall, as in some algae and fungi
Coenocytic (Cenocytic)	:	thallus formed of a coenocyte (cenocyte)
Columella	:	In fungi, a sterile invagination of a sporangium
Complanate	:	structures arranged in one plane, uniform flattened
Compressed	:	dorsi-ventrally flattened
Conceptacle	:	one of many specialized hollow chambers containing reproductive structures that appear as dark, dotlike bodies on the surface of receptacles in certain algae and fungi
Conidiophore	:	Specialized hyphal branch of some fungi that produces conidia
Conidia	:	Refers to asexual reproduction in fungi, mitospore or conidiospore may be motile (zoospore, as in Chytridiomycota) or aplanospore (in Zygomycota)
Conidioma (plural Conidiomata)	:	A specialized macroscopic fruiting structure containing masses of conidia
Cortex	:	the outer tissue of a thallus (with or without an enclosing epidermis) that contains photosynthetic pigments
Constriction	:	strangulation
Corticated	:	having a cortex or a similar specialized outer layer
Cruciate	:	shaped like, or resembling, a cross
Crustose	:	having a crust-like appearance
Cryptostomata	:	are like conceptacles but differ having only hairs and are sterile
Cuneate	:	a leaf shape
Cystocarp	:	a reproductive body in red algae (Rhodophyta), developed after fertilization and consisting of filaments bearing carpospores
Decumbent	:	reclining on the substrate
Dermatophyte (adj. Dermatophytic)	:	Any of various parasitic fungi that cause infections of the skin, hair, or nails
Dichotomous (branching)	:	branches that regularly divides in two
Dioecious	:	having the male and female reproductive organs in separate plants
Discoidal	:	having a flat, circular form; disk-shaped
Disk	:	the structure flat, disk-shaped, which ensures the fixation of algae to the substrate
Distichous	:	arranged in two rows along opposite sides of the axis
Divaricate	:	widely spreading branching
Ecostate	:	having no mid-vein
Emarginate	:	having a notched tip or edge
Epilithic	:	growing on the surface of rocks
Epiphyte	:	growing on another plant but not parasitic
Erect	:	forming a frond; not prostrate

Essential amino acids	: amino acids that cannot be synthesized by human or other animal organisms
Factor X	: Also known by the eponym Stuart-Prower factor, is an enzyme of the coagulation cascade; it is a serine endopeptidase; factor Xa is inactivated by protein Z-dependent protease inhibitor (ZPI), a serine protease inhibitor (serpin)
Fastigiate	: having erect, clustered, almost parallel branches
Favus	: Also termed Tinea Favosa, is a chronic inflammatory dermatophytic infection usually caused by *Trichophyton schoenleinii*; rarely, Favus is caused by *Trichophyton violaceum*, *Trichophyton mentagrophytes* var. *quinckeanum*, or *Microsporum gypseum*. Favus typically affects scalp hair but also may infect glabrous skin and nails. The causative agent of mouse Favus is *T. mentagrophytes* var. *quinckeanum*, also termed *Trichophyton quinckeanum*, which can cause Favus in humans, although rarely
Filament	: a plant or branch composed of a linear group of cells joined at their walls, also a chain of cells forming a hair-like strand
Floating	: lying passively on the surface of the water
Frond	: upright part of a macroalgae
Fucoidan	: sulfated polysaccharide found mainly in various species of brown algae (Phaeophyceae)
Fungistatic	: The lowest concentration of the agent that results in the maintenance or reduction of the inoculum of fungus
Furcate	: to divide into two parts; fork
Fusiform	: elongated and tapering at both ends; spindle-shaped
Gametes	: the sexual male and female cells, each gamete is haploid (n), has only single set of chromosomes
Gametophyte	: which produces gametes; the haploid phase of algae that undergo alternation of generations, with each of its cells containing only a single set of chromosomes
Ganglioid	: having or referring to a ganglion-like structure
Granulocytopenia	: a marked decrease in the number of granulocytes; granulocytes are a type of white blood cell filled with microscopic granules that are little sacs containing enzymes that digest microorganisms; granulocytopenia can be genetic and inherited or it can be acquired as, for example, an aspect of leukemia
Habitat	: the local environment or the place in which a plant or animal lives; for marine environments, it is defined per geographical location, physiographic features and the physical and chemical environment
Hair	: a colorless, typically elongate, unicellular or multicellular structure; also, a unicellular or multicellular filament growing from the surface of a thallus, often deciduous
Haplostichous	: present pseudo-parenchymatous constructions
Haptera	: usually cylindrical, multi-branched anchoring structures (usually in Laminariales – Phaeophyceae), more massive than rhizoids
Hemagglutinating	: to cause agglutination of red blood cells
Hepatoprotective	: any drug or substance that prevents the liver damage
Heteromorphic	: having different forms at different periods of the life cycle

Holdfast	:	basal attachment organ of seaweeds
Hyphae	:	thread-like filaments that make up a mycelium of fungus and release enzymes to absorb nutrients from food sources
Immunostimulation	:	Stimulation of the immune response
Inrolled	:	incurved or rolled inwards
Insecticidal	:	capable of killing insects or controlling their growth
Internodal	:	a section or part between two nodes, as of a nerve or stem
Involute	:	having the margins rolled inward
Isodiametric	:	having equal diameters or axes
Laciniate	:	slashed into narrow pointed lobes
Lanceolate	:	narrow and tapering toward the apex
Larvicidal	:	an insecticide or active substance designed to kill larval pests
Lectins	:	carbohydrate-binding proteins (i.e., glycoproteins) found in seaweeds and a wide variety of other life-forms; they are of potential economic importance as topical drug delivery systems
Leishmaniosis (or Leishmaniasis)	:	*Leishmania* infection
Life cycle, or life history	:	a period involving all different generations of a species succeeding each other through means of reproduction, whether through asexual reproduction or sexual reproduction
Linear	:	long and narrow, with parallel margins; also, slightly broader than filiform
Macroalgae	:	multicellular algae, in general marine and sometimes with considerable size (reaching 50 m in length, or greater), who's thalli exhibit a high degree of morphological complexity
Mariculture	:	cultivation of marine organisms in their natural habitats, usually for commercial purposes
Medullary	:	the central portion of a thallus in certain red or brown algae
Mekabu	:	sporophyte phase of *Undaria pinnatifida*
Membranous	:	the consistency of a membrane; laminar thalli, thin, sometimes transparent or semitransparent
Meristoderm	:	the outer layer of the stipe, which resembles meristem in that its cells divide continually to replace tissue damaged by abrasion against rocks
Mitosporic	:	Fungi with asexual reproduction, or by mitosis rather than meiosis
Metastasis	:	the spread of cancer cells from the place where they first formed to another part of the body; in metastasis, cancer cells break away from the original (primary) tumor, travel through the blood or lymph system, and form a new tumor in other organs or tissues of the body; the new, metastatic tumor is the same type of cancer as the primary tumor; for example, if breast cancer spreads to the lung, the cancer cells in the lung are breast cancer cells, not lung cancer cells; the plural form of metastasis is metastases
Micronization	:	the process of reducing the average diameter of a solid material's particles; traditional techniques for micronization focus on mechanical means, such as milling and grinding; modern techniques make use of the properties of supercritical fluids and manipulate the principles of solubility
Minamata	:	disease, sometimes referred to as Chisso-Minamata disease, is a neurological syndrome caused by severe mercury poisoning

Mitogenic	: an agent that induces mitosis
Moniliform	: resembling a string of beads, as the roots of certain plants
Monostromatic	: single cell layered
Mucilaginous	: containing mucilage; which is consistent
Mucoid	: resembling mucus
Mucronate	: terminating in a sharp point
Multiaxial	: having more than one axis; developing in more than a single line or plain; opposed to monoaxial
Mycelium (plural Mycelia)	: the mass of branched, tubular filaments (hyphae) of fungi. The mycelium makes up the undifferentiated body, of a typical fungus
Nociceptive	: causing or reacting to pain
Node	: the site on an axis from which blades and/or branches arise, with rings being formed at the junction of each successive axial cell
Oblanceolate	: lance-shaped, with the thin end at the base
Obtuse	: having a blunt or rounded tip
Obovate	: inversely ovate
Ocellate	: having an ocellus or ocelli
Ocellus (pl. ocelli)	: an enlarged discolored cell in a leaf
Opposite	: referring to a leaf or/and branch arrangement where two leaves or branches arise at the same position at a single node but on reverse sides of the stem; also, referring to locations in the same transverse line but removed by 180° on an axis
Okara	: soymilk residue
Osteoblastogenesis	: the production of osteoblasts
Osteoclastogenesis	: the development of osteoclasts
Ovate	: twice or less as long as broad
Panacea	: a remedy for all disease or ills; cure-all
Parenchymatous	: tissue-like structure of uniform, thin-walled, tightly knit cells forming a tissue resembling the parenchyma of vascular plants
Paronychia	: is a skin infection that occurs around the nails
Pedunculate	: having a peduncle
Percurrent	: extending through the entire length of a structure (e.g., from base to apex of a thallus)
Perennial	: a plant living for three or more years
Pericentral	: surrounding the center
Periclinal	: used to describe a type of cell division in a layer of cells that occurs parallel to an adjacent layer of cells
Phaeophycean	: related to Phaeophyceae class (brown algae)
Phialide	: in fungi, a conidiogenous cell in which the meristematic end remains unchanged as successive conidia are extruded out to form chains
Phycocolloid	: colloid (gel) extracted from algae; the phycocolloids are molecules of great size, consisting of simple sugars, which form part of the cell walls and intercellular spaces of many algae, mainly brown and red

Phylloid	: ribbon-shaped blade
Phytohormones	: any of various hormones produced by plants that control or regulate germination, growth, metabolism, or other physiological activities
Pinnate	: having branchlets set closely together on opposite sides of the main axis, arranged like the plumes of a feather
Pinnately	: resembling a feather in having parts or branches arranged on each side of a common axis
Plurilocular	: having several cells or *loculi*; many-celled sporangia, each cell containing a single spore, as in many algae
Pneumatocyst	: gas-filled bladder
Polysiphonous	: resembling, belonging to, or characteristic of the genus *Polysiphonia* (Rhodophyta)
Proliferous	: producing many side branches or offshoots and normally reproducing vegetatively by buds
Prostrate	: lying flat on the ground
Pseudodichotomous	: forming two unequal branches, or having two equal branches at branch points but with one being derived from a lateral branch
Pyrenoid	: a proteinaceous structure found within the chloroplast of certain algae
Pyriform	: pear-shaped
Racemose	: resembling or borne in a raceme
Radially	: having or characterized by parts so arranged or so radiating
Ramuli (sing., Ramulus)	: determinate branchlets
Receptacle	: structure with conceptacles, present in some brown algae
Retuse	: having a rounded apex and a central depression
Rhizoid	: cell or filament responsible for fixing the thalli to the substrate
Rhodoplast	: Plastid of red algae (Rhodophyta)
Rostrate	: having a beak
Saccate	: in the form of a sac; pouched
Seaweed	: a macroscopic marine alga (non-vascular plant)
Septate	: divided by a septum or septa
Serrate	: having flat small teeth that are projected forward
Schistosome	: any of various chiefly tropical trematode worms of the family Schistosomatidae and especially the genus *Schistosoma*, many of which are parasitic in the blood of Humans and other mammals; also, called "bilharzia", "blood fluke"
Schistosomulum (plural Schistosomula)	: the immature form of a parasitic schistosome after it has entered the blood vessels of its host
Sclerotium (plural Sclerotia)	: is a compact mass of hardened fungal mycelium containing food reserves; also, called "resting bodies"
Scutula	: a yellow, saucer-shaped crust, the characteristic lesion of Favus, consisting of a mass of hyphae, pus, and scales
Smith degradation	: the sequential subjection of polysaccharides to periodate oxidation, reduction with borohydride, and acid hydrolysis
Sorus (pl., Sori)	: is a cluster of sporangia (structures producing and containing spores)

Soxhlet (extractor)	:	is a piece of laboratory apparatus invented in 1879 by Franz von Soxhlet; it was originally designed for the extraction of a lipid from a solid material; typically, a Soxhlet extraction is used when the desired compound has a limited solubility in a solvent, and the impurity is insoluble in that solvent. It allows for unmonitored and unmanaged operation while efficiently recycling a small amount of solvent to dissolve a larger amount of material
Spatulate	:	having a narrow base and a broad rounded apex
Spermatangia	:	an organ or a cell in which gametes are produced
Spinose	:	bearing many spines
Spinous	:	covered with or having spines; thorny
Splenocyte	:	can be any one of the different white blood cell types if it is situated in the spleen or purified from splenic tissue
Sporodochia	:	a small, compact stroma (mass of hyphae) usually formed on host plants parasitized by mitosporic fungi; it bears the conidiophores on which the asexual spores or conidia are formed
Sporophyll	:	a modified leaf-like structure (especially in Laminariales, Phaeophyceae), bearing the spore-producing sporocysts
Sporophyte	:	spore-producing phase (usually diploid) in the life cycle of a plant (or algae) with alternation of generations
Stipe	:	portion located in the base of the algae, situated between the rhizoids and the blade
Stipitate	:	supported on or having a stipe
Stolons	:	axes lying or crawling on the substrate and giving rise to upright branches at intervals
Stratum corneum (latin for 'horny layer')	:	Is the outermost layer of the epidermis, consisting of dead cells (corneocytes). This layer is composed of 15–20 layers of flattened cells with no nuclei and cell organelles. Their cytoplasm shows filamentous keratin. These corneocytes are embedded in a lipid matrix composed of ceramides, cholesterol, and fatty acids
Sublittoral	:	of or relating to the biogeographic region of the ocean bottom between the littoral and bathyal zones, from the low water line to the edge of the continental shelf, or to a depth of approximately 200 m
Terete	:	cylindrical in cross section
Tetrasporangia	:	a sporangium containing four asexual spores
Thalli (sing., Thallus)	:	the plant body of macroalgae
Tomentose	:	covered with short, dense, matted hairs
Tophule	:	reserve structures, typically ovoid, located in base of the branches of some species of the genus *Cystoseira*
Trichotomous	:	having three angles or corners
Trichotomously	:	division into three parts or elements
Tristichously	:	having leaves growing in three ranks
Trypanocidal	:	destructive to trypanosomes
Trypomastigotes	:	Term to replace the older term, "trypanosome stage", which was often confused with the flagellate genus *Trypanosoma*. It denotes the stage (ineffective stage for South American trypanosomiasis and African trypanosomiasis, and the only stage found in Humans in the latter illness)

Tumoricidal	:	Denoting an agent destructive to tumors
Uncalcified	:	not hard, calcareous, or calcified
Unilocular	:	having or consisting of a single chamber or cavity
Uronic (acid)	:	any of a group of organic acids, such as glucuronic acid, derived from oxidation of aldose sugars and occurring in urine
Vasoregulation	:	The regulation of vascular tension
Zonate	:	marked with, divided into, or arranged in zones

CHAPTER 1

Biodiversity and Description of the Main Algae with Bioactive Properties

1.1 Introduction

Marine macroalgae or seaweeds have been used as food, especially in China and Japan, and crude drugs for treatment of many diseases such as iodine deficiency (goiter, Basedow's disease, and hyperthyroidism). Some seaweeds have also been used as a source of additional vitamins and other nutrients (see Chapters 2 and 3), treatment of various intestinal disorders, as vermifuges (Chapter 9), as antiviral (Chapter 5), as antibacterial (Chapter 8), as antifungal (Chapter 7), as antitumoral (Chapter 6), as anti-inflammatory, anti-allergic, and antithrombotic (Chapter 10), as neuroprotective (Chapter 11), as hypocholesterolemic and hypoglycemic agents (Chapter 4), and in thalassotherapy treatments (Chapter 12). Seaweeds have been also employed as dressings, ointments, and in gynecology (Trease and Evanes 2009).

Since always, the coastal communities have used algae in the preparation of home remedies that were later used in the treatment of various health problems. These applications are the product of empirical knowledge of many generations and, in most cases, their mechanism of action is unknown, i.e., few scientific studies have been reported beyond the simple collection and ethnographic record. However, recent investigations conducted in order to analyze the components and causes that alter the functioning and balance of our body are already bearing its first fruits. Thus, we now know that the good results obtained in the treatment of goiter based on the use of *Laminaria* (Ochrophyta, Phaeophyceae) are due to the fact that the origin of this disease is directly related to a food diet low in iodine, which is thus enriched by the intake of these algae, in which the iodine is present in very significant amounts. Apart from these brown algae (kelps), popular culture is yet to know a large number of species that are used because of their therapeutic properties or functional utility in medical practices (Pereira 2011, Pereira 2016, Pereira and Correia 2015).

In recent decades, there has been an increasing interest from researchers and the pharmaceutical industry in seaweed. This fact is due to their enormous potential as a source of molecules and bioactive substances, which may be used in the development of new drugs. Today, in addition to the conventional wisdom, it is known undoubtedly that many of the resulting compounds of algae from secondary metabolism act as stimulators of the primary metabolism of the person who ingests it, stimulating the activity of certain endocrine glands, blood circulation, exchanges of mineral elements, and the physiological elimination of toxins. Due to recent studies in this area of knowledge, several of these compounds were characterized, and their respective pharmacological properties were determined—cholesterol reducers, anticoagulants (important in the prevention of stroke), antimicrobials, antitumor, antiviral, anthelmintics, anti-inflammatory, antacids, growth regulators, immunoregulatory, etc.

1.2 Algotherapy—Herbal Medicine and Algae-based Phytochemistry

Algotherapy is the special branch of herbal medicine that uses marine plants (or rather the bioactive compounds the said active principles are extracted from) with medicinal value, or therapeutic applications that result in attenuation, combat, and/or prevention of diseases and their symptoms.

Marine algae are literally sea vegetables (just as carrots, beans, cabbage, and lettuce are among the terrestrial plants), and have been used for centuries as food by people like the Chinese (with an ancestral tradition of using herbal medicine), Japanese, Scottish, Irish, Icelanders, Scandinavians, Germans, and the Indians of Latin America, who used its potential healing power of various diseases in the form of drugs. Nowadays, the acceptance of the therapeutic potential of these bioactive substances in addition to simple intake of seaweed itself (direct or indirect) is a generally accepted fact, and its extracts are used in drug production, preparation of cosmetics, and therapeutic baths (Pereira and Correia 2015).

Once algae extracts are capable of industrial production by certified technicians (chemical, pharmaceutical, and/or other areas of animal or human health), they assume the status of herbal medicines, i.e., there are medicines made from bioactive vegetable raw materials (without the inclusion of synthetic chemicals), whose effectiveness and safety are validated by ethno-pharmacological surveys, technical and scientific documentation, or clinical evidence replicable.

All algotherapy preparations recourse to curative power of active principles of plants and algae, but also the synergistic effects of cumulative use of more than one species (other algae, and this with terrestrial plants, aquatic and/or fungi), and of inorganic elements, which may incorporate the final composition of the herbal medicine.

1.3 Taxonomy and Description of Some Algae with Bioactive Activity

Seaweeds (or macroalgae) are aquatic photosynthetic organisms belonging to the domain Eukaryota and the kingdoms Plantae (Green and Red algae) and Chromista (Brown algae), respectively.

Although classification systems have changed over time, and according to the authors, it is generally accepted that—(a) the green algae are included in the phylum Chlorophyta and their pigmentation is identical to that of terrestrial plants (chlorophylls a, b and carotenoids), (b) the red algae belong to the phylum Rhodophyta and their photosynthetic pigments are chlorophyll a, phycobilins (R-phycocyanin and R-phycoerythrin), and carotenoids, mostly β-carotene, lutein, and zeaxanthin, and (c) the brown algae are included in the phylum Ochrophyta (or Heterokontophyta), class Phaeophyceae, and their pigments enclose chlorophylls a, c, and carotenoids, dominated by fucoxanthin (Pereira 2009, Pereira 2010a, Pereira 2012).

1.3.1 Chlorophyta (green algae)

Anadyomene stellata (Wulfen) C. Agardh (Fig. 1.1-A)

Common name: see Table 2.1

Description: *A. stellata* consists of erect, bright green, delicate blades with ruffled edges forming densely packed clumps; blades one cell thick, up to 10 cm in height, formed by veins of filaments radiating from base in fan-like branching pattern. Cells between veins arranged in parallel rows. Margins smoothly rounded, formed by small spherical cells; similar to *A. saldanhae* which has cells between veins in random arrangement. Also, similar to *A. lacerata* which has lacerated margins formed by elongated vein cells (Braune and Guiry 2011).

Distribution: see Table 2.1

Uses and bioactivities: Extracts have anticoagulant and antifungal activity (Pereira 2015).

Figure 1.1 Marine Green macroalgae: A–*Anadyomene stellata*, B–*Bryopsis plumosa*, C–*Caulerpa cylindracea*, D–*Caulerpa prolifera*, E–*Caulerpa racemosa*, F–*Caulerpa sertularioides*, G–*Chaetomorpha aerea*, H–*Chaetomorpha linum*, I–*Cladophora prolifera*, J–*Codium adharens*, K–*Codium bursa*, L–*Codium tomentosum*, M–*Halimeda tuna*, N–*Ulva clathrata*, O–*Ulva compressa*.

Bryopsis hypnoides J.V. Lamouroux

Common names: Variously Branched Mossy Feather Weed, Obana-Hanemo

Description: Plants in filamentous tufts, up to 10 cm tall, branching in irregular, scattered pattern. Primary axes highly branched. Fronds decrease in diameter with each successive division; branchlets form irregularly, undifferentiated from axes, constricted at base. Apices rounded. Rhizoidal system fibrous, tightly woven. Color is dull or dark green (Pereira 2017a).

Distribution: Adriatic Sea, Black Sea, NE and NW Atlantic, Atlantic Islands (Azores, Bermuda, Canary Islands, Madeira, Selvage Island), Caribbean Islands, SE and SW Atlantic, Asia and SE Asia, and Pacific Islands.

Uses and bioactivities: The primary structure of bryohealin and of lectin from *B. hypnoides* had little similarity with any known plant lectin, but rather resembled animal lectins with fucolectin domains (Pereira 2015). Extracts of this species possess antibacterial, antifungal, and antiviral activity.

Bryopsis pennata J.V. Lamouroux

Description: Thallus filamentous, bushy, in tuft-like mats, up to 10 cm high. Fronds feather-like, 8–15 mm wide, pinnately branched; lateral branches of uniform length, constricted at base where joined to main axes. Branchlets in two opposite rows on upper half of branch, lower half of branch is bare. Rhizoidal system is fibrous, tightly interwoven. Color is glossy dark green, often with light blue iridescence (Pereira 2015, Pereira 2017b).

Distribution: Atlantic Ocean, the Mediterranean, the Caribbean, Indian and Pacific Oceans, and Australia.

Uses and bioactivities: The extracts of this species have antibacterial, antifungal, antitumor, larvicidal, and cardiotonic activity (Pereira 2015).

Bryopsis plumosa (Hudson) C. Agardh (Fig. 1.1-B)

Common names: see Table 2.1

Description: Thallus bushy, heteromorphic, richly branched, light to dark green, single cell, tubular, erect fronds; branched filaments 2–15 cm long; primary axis and main branches up to 1.5 mm in diameter; lower part of axis bare, upper part pinnately branched with constrictions at base; branches may have thin filamentous (less than 1 mm in diameter), pinnate branchlets with constrictions at base; length of branches decreases towards the apices making thallus pyramid-shaped; apical cells rounded at tip (Braune and Guiry 2011).

Distribution: see Table 2.1

Uses and bioactivities: Produces lectins with biological activity; extracts also possess antioxidant and antimicrobial activity, medicinal and pharmaceutical uses (Pereira 2015).

Capsosiphon fulvescens (C. Agardh) Setchell and N.L. Gardner

Common names: see Table 2.1

Description: This species is rare. They are small plants, green-yellow in color. Upper thalli are in the form of tubes, while the base consists of only one or two rows of cells. The cells are arranged in very pronounced longitudinal cross series loosely bound, sometimes producing false branches. Cell walls are gelatinous (Pereira 2016).

Distribution: see Table 2.1

Uses and bioactivities: Used as food in Korea, in boiled soup with oyster (Pereira 2016). This species has cardio-protective activity.

Caulerpa chemnitzia (Esper) J.V. Lamouroux

Synonym: *Caulerpa peltata* J.V. Lamouroux

Common names: see Table 2.1

Description: Thalli with stolons producing descending rhizoidal branches and ascending branches bearing several ramuli consisting of a short pedicel ending in a disc-like head. The disc-like head differentiates this species from the spherical ramuli of *Caulerpa racemosa* (Domingo and Corrales 2001).

Distribution: see Table 2.1

Uses and bioactivities: This seaweed was reported to be edible and used as medicine for its antifungal properties and has the ability to lower blood pressure. It is also reported to have larvicidal, antiplasmodial, cytotoxic, immunomodulatory, antioxidant, antimicrobial, and anticoagulant properties (Pereira 2015).

Caulerpa cupressoides (Vahl) C. Agardh

Common names: see Table 2.1

Description: Erect uprights up to 25 cm high, green, departing from every 1–10 cm interval on creeping stolon 3 mm in diameter; rhizoids colorless and numerous, each branched irregularly bearing four longitudinal parallel series of branchlets. Branchlets stiff, closely clustered, and conical, more long than wide, 0.3–0.4 mm in diameter, with pointed apices. Fine trabeculae evident in cross sections throughout plant (Pereira 2017c).

Distribution: see Table 2.1

Uses and bioactivities: This seaweed is reported to be edible, and to have antibacterial, antifungal, antioxidant, anticoagulant, analgesic, anti-inflammatory, antithrombotic properties, and is used to treat high blood pressure. However, some *Caulerpa* species produce toxins to protect themselves from browsing fish (Pereira 2015).

Caulerpa cylindracea Sonder (Fig. 1.1-C)

Synonym: *Caulerpa racemosa* var. *cylindracea* (Sonder) Verlaque, Huisman & Boudouresque

Common names: Green Caviar, Invasive Caulerpa racemosa, Sea Grape

Description: A green macroalgae with slender thallus, lacking large rhizoidal pillars, basal part of the upright axes slightly inflated immediately above the attachment to the stolon, clavate branchlets, uncrowded, and radially to distichously disposed (Pereira 2017d).

Distribution: Mediterranean Sea, Atlantic Islands (Canary Islands), SW Asia, Australia, and Pacific Islands.

Uses and bioactivities: This species has antibacterial activity.

Caulerpa mexicana Sonder ex Kützing

Common names: Feather Algae, Fern Algae

Description: *C. mexicana* consists of erect fronds with flattened branchlets arising from a creeping stolon with rhizoidal branches attaching it to the substrate. Fronds 15–25 cm in height, branchlets 2–4 mm in width. Color generally grass-green. Midrib of frond is flat with branchlets oppositely arranged along its axis; branchlets slightly overlapping and pointed upward with a general feather-like appearance. Similar in structure to *C. taxifolia* which has a compressed midrib and branchlets that become constricted at the base (Pereira 2017e).

Distribution: Gulf of Mexico, Caribbean Sea, Western Tropical Atlantic, Asia, SE and SW Asia, Indian Ocean, Australia, and Pacific Islands.

Uses and bioactivities: Extracts have antinociceptive and anti-inflammatory activity.

Caulerpa prolifera (Forsskål) J.V. Lamouroux

Description: Fronds leafy, flat, with a distinct slender stipe, and entire margins, at times undulating, oval to longitudinal, spatula-shaped or linear oblong, tapering toward the base, at the apex also narrowing but bluntly rounded. Occasionally, more of these dark-green photosynthetic organs emerge from the stipes as well as from the margins or surfaces of the blades. The sparsely branching, relatively thin (1–2 mm) but wiry stolons are widely spread out, often extensive areas colonizing in this way; rhizoids in intervals of 0.5–2 cm (Braune and Guiry 2011).

Distribution: NE Atlantic (S Spain to Canary Islands), Caribbean Sea, the Mediterranean, Indian Ocean, and SW Pacific (the Philippines).

Uses and bioactivities: Commonly used in aquaria, to which it is well adapted, this species produces Caulerpenyne, an acetylenic sesquiterpene, which is specific to *Caulerpa* species. This seaweed is reported to have antibacterial, antialgal, antifungal, antifouling, antiproliferative, antioxidant, anticoagulant, and larvicidal activity (Pereira 2015).

Caulerpa racemosa (Forsskål) J.V. Lamouroux (Fig. 1.1-E)

Common names: see Table 2.1

Description: It is a bright green seaweed that resembles long skinny vertical bunches of tiny grapes. It can be very similar in appearance to *C. lentillifera*, though the latter tends to produce denser bunches (though this line can be smudged when *C. racemosa* grows in wave-exposed waters and develops shorter, stronger branches than normal). *C. racemosa* is quite variable in morphology and has many different growth forms that have been identified and named. A horizontal stolon which is attached to the sediment (usually sand) by descending rhizomes gives rise to erect branches at every few centimeters. These branches can reach as much as 30 cm in height and produce many stalked branchlets which vary in shape from spherical to ovate to disc-shaped, sometimes flattening on top or forming ice-cream-cone-type shapes. These plants are coenocytic, which means that the entire plant is made up of one giant cell with many nuclei and no cross-walls. It is mainly due to this characteristic, that any part of a *C. racemosa* plant that is fragmented, even tiny bits of tissue, can regenerate to form entirely new plants (Elbanna and Hegazi 2011).

Distribution: see Table 2.1

Uses and bioactivities: *C. racemosa* is grown commercially in the South Pacific, and harvested wild in other areas; edible seaweed, eaten as salad in Polynesia and the Marquesas Island. *C. racemosa* also contains compounds which function as mild anesthetics, which gives the seaweed clinical value. Traditional medicine of the Philippines uses *C. racemosa* to lower blood pressure and to treat rheumatism. This seaweed is reported to have antibacterial, antiviral, antifungal, insecticidal, antioxidant, anti-inflammatory, anticoagulant, analgesic, hypolipidemic, hypoglycemic, and antitumor activity (Pereira 2015).

Caulerpa scalpelliformis (R. Brown ex Turner) C. Agardh

Common names: see Table 2.1

Description: Stolon slender, 0.5–1 mm in diameter, in small forms of young plants; robust, 1.5–3 mm in diameter, in large, rough-water plants, cartilaginous, naked, and epilithic. Erect fronds medium to dark green, simple to occasionally branched, from 4–10 cm high and 3–6 mm broad in slender forms, to 20 cm high and 2–3 cm broad in robust plants, terete for the basal 1–3 cm, then strongly compressed with an axis 2–3 mm broad in slender plants to 4–10 mm broad in robust plants (Womersley 2003).

Distribution: see Table 2.1

Uses and bioactivities: Considered an edible species in Tasmania, and potential food use in India (Pereira 2016). Extracts of this species have antiviral, antifungal, antibacterial, larvicidal, nymphicidal, and ovicidal activity.

Caulerpa sertularioides (S.G. Gmelin) M. Howe (Fig. 1.1-F)

Common names: see Table 2.1

Description: Thallus up to 6 cm tall, with terete stolons 0.25–1.0 mm in diameter, bearing sparse erect branches which are simple or occasionally dichotomously divided, naked, or branched at the base and bearing plumose, pinnate, undivided branchlets; branchlets cylindrical throughout, not constricted at the base, up to 8 mm long and 200 μ in diameter, up-curved, with pointed tips (N'Yeurt 1999).

Distribution: See Table 2.1

Uses and bioactivities: *C. sertularioides* is an edible crop consumed by humans in some parts of the world, namely in the Caribbean and Central America (Pereira 2016). In the Philippines, this alga is used for dietary and medicinal purposes, and its extracts have antitumor activity.

Caulerpa taxifolia (M. Vahl) C. Agardh

Common names: see Table 2.1

Description: Branches, feather-like, flattened, and upright, 3–10 cm high, rising from a creeping stolon (runner), 1–2 mm in diameter, anchored by rhizoids to the substrate. Branchlets oppositely attached to midrib, flattened, slightly curved upward, tapered at both base and tip, and constricted at point of attachment. Midrib is slightly flattened, appearing oval in cross-section. Dark green to light green (Pereira 2017f).

Distribution: see Table 2.1

Uses and bioactivities: This green seaweed is reported to be edible, to have antioxidant (Box et al. 2008), antibacterial, antiviral, nematicidal, and antifungal properties, and is used to treat tuberculosis and high blood pressure; caulerpenyne from *C. taxifolia* is cytotoxic toward several human cell lines and has anticancer, antitumor, and antiproliferative properties (Pereira 2015).

Caulerpa veravalensis Thivy and V.D. Chauhan

Common names: see Table 2.1

Description: Plants stoloniferous, the branching stolon rather stout, 1.5–2.0 mm in diameter, with thick descending rhizoid-bearing branches at intervals of a few millimeters and erect branches at intervals of 1.0–2.5 cm; thalli of foliar branches 1.0–2.5 mm long, cylindrical, bulbous at the base; blade flat, linear to lanceolate, 4.2–21.5 mm wide, simple or occasionally forked, pinnately divided with a flat mid-rib of 1.0–2.0 mm width; ramuli flat, opposite to alternate, ascending, oblong, slightly arcuate, 1.0–1.5 mm wide, 5–7 mm long, with round apex, with sides parallel throughout or with base a little narrowed, 1.0–2.0 mm apart, occasionally having bifurcate apex (Thivy and Chauhan 1963).

Distribution: see Table 2.1

Uses and bioactivities: This species is reported to be edible and its extracts have antiviral, antifungal, antibacterial, nymphicidal, and ovicidal activity.

Chaetomorpha aerea (Dillwyn) Kützing (Fig. 1.1-G)

Common names: see Table 2.1

Description: Filament unbranched, growing in flocks or communities attached to sand covered or bare rocky substratum, 2–3 cm in height, yellowish-green in color. Cells cylindrical, apical cells 111.2–111.5 μm in length and 110.5–111.6 μm in breadth, middle cells 54.5–62.2 μm in length and 88.8–89.2 μm in breadth, basal cell much longer than the middle and upper one, showing its length to be 133.32–144.5 μm and breadth 55.50–58.8 μm, sheath 17–22.5 μm in thickness, sometimes lamellate in mature portion (Ghosh and Keshri 2010).

Distribution: see Table 2.1

Uses and bioactivities: This species is reported to be edible and its extracts have antifungal, antibacterial, and anti-inflammatory activity; this species is a source of lectin, a specific protein important for regulation of immunity system and cancer diagnosis (Milchakova 2011).

Chaetomorpha antennina (Bory) Kützing

Common names: see Table 2.1

Description: Green plants growing in dense shaped brush tufts between 3–5 cm, but may reach 10–16 cm or more, formed by numerous densely juxtaposed mono-seriate filaments. Non-branched filaments comprised of large multinucleated cells and a single cross-linked chloroplast and many pyrenoids (Pereira 2016).

Distribution: see Table 2.1

Uses and bioactivities: This species is reported to be edible and its extracts have antiviral, antibacterial, nematicidal, antiplasmodial, and larvicidal activity.

Chaetomorpha linoides Kützing

Common names: see Table 2.1

Description: Plants bright green to yellowish-green; composed of unbranched filaments; plants twist together to form clumps or tangles; tangles remain quite rigid when removed from water (Kaliaperumal et al. 1995).

Distribution: see Table 2.1

Uses and bioactivities: This species is reported to be edible and its extracts have antibacterial activity.

Chaetomorpha linum (O.F. Müller) Kützing

Common names: see Table 2.1

Description: Delicate green seaweed; also known as spaghetti algae, it grows as a filamentous loosely entangled mass. Usually free-floating, it may also be attached to rocks and shells. The filaments themselves are unbranched and usually between 5 cm and 30 cm in length. The unattached filaments are wiry, stiff, and curled in appearance. It is bright light to dark green in color (Barnes 2008).

Distribution: see Table 2.1

Uses and bioactivities: This species is reported to be edible and its extracts have antipyretics and antiseptic properties; used in treatment of asthma and cough and as parts of traditional cosmetics (refreshing liquid, skin powder, pulp form for sunlight protection for skin) in the Pacific Islands (Milchakova 2011).

Cladophora glomerata (Linnaeus) Kützing

Common names: see Table 2.1

Description: Description: Plants are up to 20 cm high and light to dark green in color. The texture of thallus is soft and slightly mucilage. Plants usually formed in dense tufts well-branched. Rhizoids are primary and adventitious and descend from the bases of thallus or from the lower segments of the fronds. Primary branches fused together or not and branched in dichotomous manner (Pereira 2015).

Distribution: see Table 2.1

Uses and bioactivities: Edible *Cladophora*, known as "Kai" in northern Thailand, is an economically and ecologically important green alga; its extracts have antibacterial, antiprotozoal, antioxidant, and antifungal activity (Pereira 2015).

Cladophora prolifera (Roth) Kützing (Fig. 1.1- I)

Common names: see Table 2.1

Description: Unattached or basally attached coarse filaments those are usually less than 0.5 mm wide and 3–5 cm long. The filaments are formed of a single row of often swollen cells; if attached then by a discoid base or by rhizoidal outgrowths (Pereira 2015).

Distribution: see Table 2.1

Uses and bioactivities: Extracts have antimicrobial, antifungal, antiviral, antioxidant, and anticoagulant activity (Pereira 2015).

Cladophora rupestris (Linnaeus) Kützing

Common names: see Table 2.1

Description: The bundles, dark green, made up of rough, straight threads have a coarse and rather rigid feel (Braune and Guiry 2011).

Distribution: see Table 2.1

Uses and bioactivities: Extracts have antimicrobial, antifungal, antiviral, antioxidant, and anticoagulant activity (Pereira 2015).

Chaetomorpha spiralis Okamura

Synonym: *Chaetomorpha torta* (Farlow ex Collins) Yendo

Common names: Spaghetti Algae, Green Hair Algae

Description: Filaments rigid, attached when young, soon loosened and entangled among other algae, much coiled and contorted, 20–60 cm long, 0.75–1.25 mm in diameter; cells moniliform to nearly cylindrical; thallus appearing iridescent bluish-green because of thick, opaque cell walls; usually floating (Abbott and Hollenberg 1976).

Distribution: Indian Ocean, Asia and SE Asia, Australia, Pacific Islands, Pacific California, and Mexico.

Uses and bioactivities: Extracts have antiviral, antifungal, antibacterial, antiamoebic, and antiplasmodial activity.

Cladophora vagabunda (Linnaeus) Hoek

Synonym: *Cladophora fascicularis* (Mertens ex C. Agardh) Kützing

Common names: see Table 2.1

Description: Thallus filamentous, spongy, soft tufts, anywhere from 5–50 cm in length. Branches mostly on one side, at times, strongly re-branched and claw-like, maximum number of branches at joints one to four, rarely five; rhizoids fine, often connecting to adjacent filaments by hapteroid-like rhizoids. *C. vagabunda* grows from 4 cm in diameter on wave-swept habitats, to 30 cm high in protected habitats; pale green to grass green (Russell and Balazs 2000).

Distribution: see Table 2.1

Uses and bioactivities: Considered to be an edible species, *C. vagabunda* is the source of valuable protein content; used in pharmaceutical medicines; 5% aquatic extract improves blood formula under adaptation to stress (Pereira 2016, Milchakova 2011).

Codium adhaerens C. Agardh (Fig. 1.1-J)

Common names: see Table 2.1

Description: Spongy thallus, green light, prostrate, irregularly-shaped, and with the appearance of a plane carpet firmly fixed to the substratum. Consists of entangled coenocytic filaments, ending on the surface by narrow and elongated utricles that are difficult to separate. Firm, gelatinous texture, and smooth to the touch (Pereira 2016).

Distribution: see Table 2.1

Uses and bioactivities: Used as food in Hawaii and its extracts have vermifuge, antidiabetic, antibacterial, and antiviral properties (Pereira 2015).

Codium bursa (Linnaeus) C. Agardh (Fig. 1.1-K)

Common names: Green Sponge Ball

Description: Thallus a spongy sphere with a velvety soft, shiny surface; becomes more flattened when increases in size. The internal branched filamentous network becomes looser with increasing size of the sphere; the space fills with water and the surface becomes indented; anchored to substratum by felted filaments (Braune and Guiry 2011).

Distribution: NE Atlantic (Ireland to Canary Islands), and the Mediterranean.

Uses and bioactivities: Have antifungal, antibacterial, and antiplasmodial activity (Pereira 2015).

Codium capitatum P.C. Silva

Description: Thalli dark green, erect, dichotomously branched, cylindrical, up to 20 cm long, slender, smooth (no hairs), attached by weft of branched filaments. Erect branches uniform in diameter (2–3.5 mm) throughout. Utricles cylindrical, 350–700 µm long, 100–200 (250) µm in diameter, usually with distinct constriction below the rounded tip (i.e., capitate); cell wall thickened near end of utricle; hairs seldom present (Anderson et al. 2016).

Distribution: World distribution, recorded from South Africa, Mozambique, Kenya, and Madagascar.

Uses and bioactivities: Extracts have antifungal and antibacterial activity.

Codium decorticatum (Woodward) M. Howe

Common name: Dead Man's Fingers

Description: It is a little branched subtidal species, which can reach 1 m long. The thallus is flattened at the points where the branches are formed; utricles not mucronate and larger than those of other *Codium* species (Pereira 2016).

Distribution: see Table 2.1

Uses and bioactivities: Used as food in Yucatan (Pereira 2016). Extracts have antibacterial, antioxidant, antiviral, larvicidal, antifungal, and antitumor activity (Pereira 2015).

Codium duthieae P.C. Silva

Common names: Flat-forked Velvet-Weed, Forked Codium

Description: Plants yellow-green, up to 600 mm tall, forked, velvety or felty in texture; cylindrical branches that are 3–10 mm in diameter are usually flatter, about 20 mm broad where they fork; tiny bottle-shaped outer parts (utricles) are just visible to the unaided eye (Edgar 2008).

Distribution: W Australia to Victoria and the North coast of Tasmania, Asia, S Africa.

Uses and bioactivities: Extracts have antibacterial activity.

Codium fragile (Suringar) Hariot

Common names: see Table 2.1

Description: Appears as a fuzzy patch of repeatedly branching tubular fingers; these formations hang down from rocks during low tide, which serve as the inspiration for a few of its common names. The color of *C. fragile* ranges from medium green to dark green to blackish green. The entire thallus is velvety and spongy in texture, relatively soft, and is sometimes tomentose with profuse hairs. The erect, dichotomously branched fronds, or fingers, are up to 1 cm wide, and can extend to lengths of over 30 cm. The branches of *C. fragile* are round in cross-section, but may be flattened beneath dichotomies of branches. The fronds usually arise from a spongy disc-shaped holdfast, which resembles a small, broad cushion. A pungent odor is often associated with this species, which is less common for seaweeds. The utricles hairs end with a characteristic peak, and the apical mucronate is visible with a magnifying glass (Pereira 2016).

Distribution: see Table 2.1

Uses and bioactivities: *C. fragile* is used for skincare and antiaging products. It is reported to be a regenerating and anti-free radical ingredient, and have rebalancing and energizing properties. Its wealth of macroelements gives it a re-mineralizing property. One seller claims that *C. fragile* is the ideal ingredient to boost tired and mature skin. It is also used as food in eastern Asia and Chile. This species has also antioxidant, antibacterial, anti-inflammatory, immunostimulating, antitumor, antiangiogenic, and antifouling properties (Pereira 2015).

Codium indicum S.C. Dixit

Synonym: *Codium iyengarii* Børgesen

Description: Holdfast discoid 0.6–2 cm in diameter; stipe erect up to 2 cm high, terete 5–7 mm in diameter, distally flattened, expanded 1–4 cm broad, bifurcate. Thallus erect up to 40 cm high, throughout flat, expanded up to 5 cm broad, dichotomously divided (to 5 orders); dichotomies flattened up to 4 cm broad, at 3 cm distance; axils broadly rounded; distally or laterally strongly proliferous, forming coxcomb-like habit. Proliferations simple or dichotomous (2–5 orders) of variable sizes, flat up to 3 cm broad and 20 cm long, dichotomies flattened; thallus dissecting out into individual utricles (Pereira 2016).

Distribution: see Table 2.1

Uses and bioactivities: Considered to be an edible species in Pakistan; extracts have antiviral, antifungal, antibacterial, and antiparasitic activity.

Codium isthmocladum Vickers

Description: Thalli with green color and 8–10 cm in height, erect, spongy consistency, branched dichotomically four to six times, fixed to the substratum through more or less extended basal portion; made up abundantly branched filaments coenocytic and densely interwoven; upright cylindrical portions, 3–4 mm in diameter (Pereira 2016).

Distribution: see Table 2.1

Uses and bioactivities: Used as food in Yucatan; extracts have antitumor and antiproliferative activity.

Codium tomentosum Stackhouse (Fig. 1.1-L)

Common names: see Table 2.1

Description: A small green alga (up to 30–50 cm long), with a dichotomously branched, cylindrical frond; the frond is solid and spongy with a felt-like touch and has many colorless hairs which can be seen when the plant is immersed in water. The holdfast is disc-like and formed from many fine threads (Pereira 2010a).

Distribution: see Table 2.1

Uses and bioactivities: *C. tomentosum* is used in products from the United States, Germany, France, Italy, and the UK. Some of these products are repair and restoration moisturizers, hydration serums, leg and body creams, muds and butters, bath and shower creams, day creams, night creams, eye creams, antiaging creams, masks, scrubs, lip balms, and lotions. *C. tomentosum* is a popular food in some parts of Asia; this species has anthelminthic and antiprotozoal, antioxidant, antigenotoxic, antitumor, anticoagulant, and antibacterial activity (Pereira 2015).

Dictyosphaeria cavernosa (Forsskål) Børgesen

Common name: Green Bubble Weed

Description: Thallus up to 12 cm in diameter; saclike, hollow, spherical when young, becomes convoluted, ruptured, and irregularly lobed when old. Firm, tough texture consisting of large bubble-shaped cells that are easily seen by the eye. Rhizoids are short, branched, or unbranched. Daughter segments formed as occasional segments become inflated, forming large monostromatic bladders attached to the parent plant.

They may remain attached to the thallus or break away and become independent plants. Grass is green, but sometimes bluish in color (Preskitt 2001).

Distribution: Eastern Atlantic, the Caribbean, Red Sea, Arabian Sea, Indian Ocean, and Pacific Islands (Hawaii).

Uses and bioactivities: Extracts have antiviral, antifungal, and antibacterial activity.

Dictyosphaeria versluysii Weber-van Bosse

Common name: Button Weed

Description: Thallus up to 5 cm in diameter, spherical when young, somewhat flattened solid cushion when mature. Firm, tough texture consisting of large bubble-shaped cells that are easily seen by the eye. Rhizoids are short, generally unbranched. Grass is green, but sometimes bluish in color. Can be easily confused with *D. cavernosa*. *D. cavernosa* forms hollow sacks which are often ruptured and convoluted (Preskitt 2001b).

Distribution: Eastern Atlantic, the Caribbean, Red Sea, Indian Ocean, and Pacific Islands (Hawaii).

Uses and bioactivities: Extracts have antifungal and antibacterial activity.

Halimeda macroloba Decaisne

Common names: Large Leaf Coralline Algae, Erect Sea Cactus, Small Leafed Halimeda, Monnaie de Poseidon, Tuna del Mar

Description: They can grow up to 19 cm tall extending the long, bulbous holdfast, which may extend to 5 cm. Basal segment is somewhat cuneate (wedged shaped, broad above and tapering to the base), or rectangular with slightly expanded anterior portion supporting two or more separate segments, each of which support several younger segments all together forming fan-shaped unit; segments are generally flabellate with dull surface measuring 1.2–3.6 cm broad, 1.5–2 cm tall, and 1–2.3 mm thick; upper margin is entire undulate (wavy) to deeply lobed (Pereira 2017g).

Distribution: Indian Ocean, Asia, pantropical in distribution and widespread in Southeast Asia, Australia, and Pacific Islands.

Uses and bioactivities: Extracts have cardio-protective, antitumor, antifungal, antibacterial, and anti-inflammatory activity.

Halimeda opuntia (Linnaeus) J.V. Lamouroux

Common names: see Table 2.1

Description: Thick, profusely branched clumps of rounded three-lobed or ribbed leaf-like segments, between 10 cm and 25 cm in height. The branches are numerous and are in different planes, rather than nearly in a single plane as some other species are. This alga can cover larger areas with a dense mat so that individual plants are indistinguishable (Braune and Guiry 2011).

Distribution: see Table 2.1

Uses and bioactivities: Seaweed with a high concentration of calcium that can serve as food supplements in the Caribbean Islands (Pereira 2016); extracts have cardio-protective, antitumor, antifungal, antibacterial, and anti-inflammatory activity.

Halimeda tuna (J. Ellis and Solander) J.V. Lamouroux (Fig. 1.1-M)

Common names: Cactus Algae, Sea Cactus, Calcareous Green Seaweed

Description: Thallus calcified, dark green, distinctly segmented with initial branching in one plane; segments disc-like to triangular, up to 2 cm wide. Internodal siphons uncalcified, united in twos or threes, and terminating in pseudo-dichotomous laterals. Surface cells oppressed to one another in a honeycomb pattern, 25–75 μm in diameter (Braune and Guiry 2011).

Distribution: Globally in warmer seas: NE Atlantic (Morocco, Azores, Canary Islands), the Mediterranean, the Caribbean, NW Pacific (Japan), SE Pacific (Chile), SW Pacific, Pacific Islands, Indian Ocean, and Australia.

Uses and bioactivities: This species has antibacterial, antioxidant, anticoagulant, pesticide, antifungal, antitrypanosomal, cytotoxic, and antiproliferative activity (Pereira 2015).

Monostroma nitidum Wittrock

Common names: see Table 2.1

Description: The genus name *Monostroma* means "single layer", which is apt due to the single layer of cells that make up the blades of this alga. Bright green in color, with single cell layer blades (in contrast to Ulva); slender at the holdfast and growing wider toward the apex, often with a slight funnel shape that has splits down the side (Pereira 2016).

Distribution: see Table 2.1

Uses and bioactivities: Besides consumption in China, this species is also exported to Indonesia and the Philippines. Among the green algae, *M. nitidum* is the most popular, as the plant is tender and flavorful. It is also used as a condiment (the dried blades are crushed into a dish, and other ingredients, such as cooked vegetables or pieces of cooked fish, are dipped into the fine green mixture (Pereira 2016); extracts of this species have hypocholesterolemic, cardio-protective, antiviral, anticoagulant, and anti-inflammatory activity.

Penicillus capitatus Lamarck

Common names: Shaving Brush Alga, Mermaid's Shaving Brush

Description: Thallus externally resembling a shaving brush—the simple, cylindrical stipe has a smooth, calcified, rigid surface, and merges, widening, into the filamentous head-tassel, which terminates hemispherically flattened to an inverted pear-shape at the front. The head-filaments are more delicate than in other species, but distinctly calcified (Braune and Guiry 2011).

Distribution: Caribbean Sea, the Bahamas, Florida and Bermuda, and Mediterranean Sea.

Uses and bioactivities: Extracts of this species have antiviral and antifungal activity.

Udotea indica A. Gepp and E.S. Gepp

Common names: see Table 2.1

Description: The thalli are up to 4 cm long, very broad, and slightly calcified. The root-mass forms a small tuft. The terete stipe is up to 1.2 cm long and 1 mm thick. The fronds are green, somewhat rounded, flabellate, orbicular, and sometimes broadly proliferated above; the base is cuneate, distinctly zonate. The blade margins are entire, lobed, or lacerated (Nizamuddin 1963).

Distribution: see Table 2.1

Uses and bioactivities: Extracts of this species have antiviral and antibacterial activity.

Ulva australis Areschoug

Synonyms: *Ulva pertusa* Kjellman

Common names: Bunched Sea Lettuce, Southern Sea Lettuce, Lacy Sea Lettuce, Ana-awosa, Awosa

Description: Plants light green above, darker below; 40–200 mm tall, several sheet-like blades branched near the plant base; blades only slightly longer than broad, edges without microscopic teeth, plant base with thick mass of rhizoids (Edgar 2009).

Distribution: Asia, SE Asia, Indian Ocean, Australia, and Tasmania.

Uses and bioactivities: Used for foodstuff in the Mediterranean Basin and Korea; used to treat fever, heat stroke, urinary problems, lymphatic swellings, antipyretics, goiter, high blood pressure, dropsy, wounds, burn treatment, and as animal fodder; also with antimicrobial, anticoagulant, and anthelminthic activity (Pereira 2016).

Rhizoclonium riparium (Roth) Harvey

Synonym: *R. implexum*

Common name: Rooting Green Thread Weed

Description: Filaments long, slender, decumbent, pale-green, forming wide strata, flaccid, entangled, angularly bent, furnished at the angles with short, root-like processes, which sometimes, but rarely, lengthen into very patent branches, and often attach themselves to neighboring filaments (Harvey 1849).

Distribution: Atlantic, Adriatic Sea, Baltic Sea, Mediterranean Sea, Asia (China, Japan, Russia), SE Asia (Myanmar, the Philippines, Vietnam), SW Asia (Arabian Gulf, Bangladesh, Goa, India, Kuwait, Sri Lanka), Australia and New Zealand, Pacific Islands (Polynesia, Fiji, Micronesia, Mariana Islands, Hawaii, Marshall Islands), North America (Alaska, Baffin Island, Baja California Sur, Florida, California, Connecticut), Central and South America (Baja California, Belize, Costa Rica, El Salvador, Mexico, Argentina, Chile, Peru, Uruguay, Venezuela), Caribbean Sea, Tropical and Subtropical Atlantic.

Uses and bioactivities: Used in wound treatment (Hoppe 1979); also, it has high protein levels in its nutritional composition (Chakraborty and Santra 2008) (see Table 2.5, and Chapter 2).

Ulva clathrata (Roth) C. Agardh (Fig. 1.1-N)

Synonyms: *Enteromorpha clathrata* (Roth) Greville, *Enteromorpha muscoides* (Clemente) Cremades

Common names: see Table 2.1

Description: Plants less than 30 cm long and soft; thallus repeatedly branched in all directions, cylindrical or compressed with narrow branchlets as much as 40 cm long, light green in color; plants grow first attached to the substratum, but later become free floating; cells of the thallus more or less quadrangular in shape with a single cup-shaped chloroplast; three to four pyrenoids. This species forms tufts, bright green, composed of branched axes, which can reach several centimeters long (20–30 cm). The main axis and branches are covered with conical branchlets, which are very characteristic (Pereira 2010a).

Distribution: see Table 2.1

Uses and bioactivities: Extracts have antibacterial, anticoagulant, antifungal, antiprotozoal, larvicidal, analgesic, anticancer, and antiviral activity.

Ulva compressa Linnaeus (Fig. 1.1-O)

Synonyms: *Enteromorpha compressa* (Linnaeus) Nees

Common names: see Table 2.1

Description: Plants attached, light or bright-green in color; adult plants usually tubular; more or less compressed, dilated towards the apex, tapering below, giving several branches from the gradually contracted stalk-like base; branches similar to the main frond; fronds up to 1.5 cm high (Kaliaperumal et al. 1995).

Distribution: see Table 2.1

Uses and bioactivities: Many benefits have been associated with consumption of *U. compressa*, such as cytotoxic, antimicrobial, antiviral, and antioxidant properties; extracts of *U. compressa* are also added to cosmetics products for a soothing quality that reduces skin itchiness and tautness (Pereira 2015).

Ulva fasciata Delile

Common names: see Table 2.1

Description: Thalli thin, sheet-like, up to 50 cm long, consisting of wide blades, 10 cm to 15 cm wide at base, tapering upward to less than 2–5 cm wide at the tip. Basally broadened, but the upper portions divided deeply into many ribbon-like segments; margins smooth, often undulate. Holdfast is small without dark rhizoids. Bright grass-green to dark green, gold at margins when reproductive, and may be colorless when stressed (Braune and Guiry 2011).

Distribution: see Table 2.1

Uses and bioactivities: *U. fasciata* has a very fine texture and lovely fresh taste and is often chopped into salads or used as a relish, though it can also be cooked and used in soups; extracts are also very nutrient rich and are a wonderful addition to natural cosmetic products; have antiviral, algicidal, antifungal, larvicidal, cytotoxic, antifouling, antibacterial, and antioxidant activity (Pereira 2015).

Ulva flexuosa Wulfen

Synonyms: *Enteromorpha flexuosa* (Wulfen) J. Agardh, *Enteromorpha tubulosa* (Kützing) Kützing, *Enteromorpha prolifera* var. *tubulosa* (Kützing) Batters

Common names: see Table 2.1

Description: Plants soft, green, fading to a browner color; 30–60 mm tall branching near the base; branches hollow, about 8 mm broad (Edgar 2009b).

Distribution: see Table 2.1

Uses and bioactivities: Extracts of this species have antioxidant, antibacterial, antivirus, and fungicide (Farasat et al. 2014) activity; they are also used in cosmetics (Milchakova 2011).

Ulva intestinalis Linnaeus (Fig. 1.2-A)

Synonyms: *Enteromorpha intestinalis* (Linnaeus) Nees

Common names: see Table 2.1

Description: *U. intestinalis* is a conspicuous bright grass-green seaweed, consisting of inflated irregularly constricted, tubular fronds that grow from a small discoid base. Fronds are typically unbranched. Fronds may be 10–30 cm or more in length, and 6–18 mm in diameter, the tips of which are usually rounded. Like other members of the genus, *U. intestinalis* is a summer annual, decaying and forming masses of bleached white fronds towards the end of the season (Pereira 2010a).

Distribution: see Table 2.1

Uses and bioactivities: Used for food and animal feed; potential use to produce biofuels; also have antibacterial, antifouling, antioxidant, antifungal, antitumor, and immunomodulatory activity (Pereira 2015).

Ulva lactuca Linnaeus (Fig. 1.2-B)

Common names: see Table 2.1

Description: Thalli thin, sheet-like, up to 50 cm long, consisting of wide blades, 10 cm to 15 cm wide at base, tapering upward to less than 2–5 cm wide at the tip. Basally broadened, but the upper portions divided deeply into many ribbon-like segments; margins smooth, often undulate. Holdfast is small without dark rhizoids. Bright grass-green to dark green, gold at margins when reproductive, and may be colorless when stressed (Braune and Guiry 2011, Pereira 2015).

Distribution: see Table 2.1

Uses and bioactivities: This species have antioxidant, antibacterial, antitumor, anti-inflammatory, antifouling, antifungal, antiviral, and antialgal activity (Pereira 2015).

Ulva linza Linnaeus (Fig. 1.2-C)

Synonym: *Enteromorpha linza* (Linnaeus) J. Agardh

Common names: see Table 2.1

Description: *U. linza* is a large, ribbon-like species of green seaweed that may reach up to 30 cm in length. The thalli are unbranched and often have a frilled margin. The thalli taper into a distinct stipe below and are highly compressed. The width of the thallus is greater in the middle than at the base, and may reach 5 cm in width. *U. linza* is bright light to dark green in coloration (Braune and Guiry 2011).

Distribution: see Table 2.1

Figure 1.2 Marine Green macroalgae: A–*Ulva intestinalis*, B–*Ulva lactuca*, C–*Ulva linza*, D–*Ulva reticulata*, E–*Ulva rigida*, F–*Valonia utricularis*.

Uses and bioactivities: *U. linza* is used as an edible seaweed in many cultures for its high nutrient content and silky texture; extracts are also very nutrient rich and make a beneficial addition to natural cosmetic products; extracts have antibacterial, anti-inflammatory, and antiviral activity (Pereira 2015).

Ulva prolifera O.F. Müller

Synonym: *Enteromorpha prolifera* (O.F. Müller) J. Agardh

Common names: see Table 2.1

Description: The fronds are tubular, though often more or less flattened, with few to many branches. The arrangement of the cells, in longitudinal and transverse rows in the central part of the frond, is characteristic of this species, as are the cylindrical chloroplasts seeming to fill the cell and the usually single, central pyrenoids (Pereira 2015).

Distribution: see Table 2.1

Uses and bioactivities: Used for food and animal feed; potential use to produce biofuels. Its extracts have hypocholesterolemic action; they are used in the treatment of aphthae, back pain, paronychia, lymphatic swellings, goiter, cough, bronchitis, antipyretics, sunstroke treatment, tonsillitis, asthma, nosebleeds, fulvescens, and sore-hand (Pereira 2016); extracts also have antibacterial, antifouling, antioxidant, antifungal, antitumor, immunomodulatory activity (Pereira 2015).

Ulva reticulata Forsskål (Fig. 1.2-D)

Common names: Ribbon Sea Lettuce, Sea Lettuce (see also Table 2.1)

Description: Plants of irregular shape, forming tangled masses of blades from a few centimeters to several meters across. Pale to dark green in color. Young plants initially attached with a small holdfast, but most older plants detached and merely tangled on other algae, rocks, or corals. Reproduction occurring in small patches in the middle of blades, during or after spore release; these patches fall out of the blade and leave a small hole. These holes become larger and form the characteristic pattern or holes in the blades. Blades are two cells thick, cells more tall than wide, or of a more equal dimension.

Distribution: Malaysia, Indonesia, Hawaiian Islands, Okinawa, Japan, Formosa, the Philippines, Indian Ocean, Red Sea (Saudi Arabia) (see also Table 2.1).

Uses and bioactivities: Used as food in salads in the Philippines and Vietnam (Pereira 2016), and occasionally used as animal feed; its extracts have antifungal, antibacterial, antileishmanial, and antiplasmodial activity.

Ulva rigida C. Agardh (Fig. 1.2-E)

Common names: see Table 2.1

Description: Thallus a flat cellular membrane or frond variable in shape and size that may form tufts 2 cm tall, although individuals can grow sheets up to 15 cm tall. The thallus is firm in texture, two cell layers thick, and arises from a small stipe and discoid holdfast. The margin of the frond may have small denticulations near the base (Neto et al. 2005).

Distribution: see Table 2.1

Uses and bioactivities: *U. rigida* is often utilized as a fresh sea vegetable by many island cultures for its high nutrient content and fresh taste; this species is used for animal feed; Green algae extracts are also very nutrient-rich and make a beneficial addition to natural cosmetic products; the polysaccharide ulvan is easily extracted from *U. rigida*; its extracts have antigenotoxicity, antihyperglycemic, immunomodulating, antibacterial, antioxidant, and antileishmanial activity (Pereira 2015, Pereira 2016).

Valonia aegagropila C. Agardh

Common names: see Table 2.1

Description: Thallus green, composed of compressed or loosely interwoven siphonous cells forming hemispherical domes or irregular cushions of indeterminate size; 2–4 cm tall and sometimes reach 15 cm or more in diameter. Sub-dichotomous or irregular branching to several orders; septate only at branch points; cells elongate, 2–4 mm broad, 3–20 mm long; basal cells in contact with substratum serving as rhizoidal cells; cells multinucleate; chloroplasts numerous per cell and discoid, each with single pyrenoid.

Distribution: see Table 2.1

Uses and bioactivities: Used in the Caribbean Islands as food and its extracts have antifungal and antibacterial activity.

Valonia utricularis (Roth) C. Agardh (Fig. 1.2-F)

Common name: Cystic Cysts, Limu Lipuu-Puu

Description: Thallus, translucent light to dark green, primarily consisting of a large (up to 5 mm thick and 20 mm long) bladder- or club- to hose-like cell, branching at the base rhizoidally. Later, due to outgrowths of this cell, cylindrical-clavate branches, often contorted and almost gapless densely packed, thus forming intertwined erect stands (Braune and Guiry 2011).

Distribution: Warm NE Atlantic (Portugal to Canary Islands), Mediterranean Sea, Caribbean Sea, Indian Ocean, NW Pacific (Japan, China), Indo-Pacific (the Philippines, Vietnam), Pacific Islands (Hawaii), and Australia.

Uses and bioactivities: Considered to be an edible alga in the Hawaiian Islands (Pereira 2016); its extracts show antifungal, antiviral, cytotoxic, and antimitotic activity (Pereira 2015).

Valoniopsis pachynema (G. Martens) Børgesen

Common name: Astro-Turf Algae

Description: Filamentous green algae forming stiff, spongy mats of tangled filaments on intertidal rocks, dead corals, or hard substrates. It covers the substrate completely, giving it a ball-like appearance. If the substrate is a dead reef, they tend to cover the entire area as small, green, hairy clumps due to its thick turf-like appearance.

Distribution: Warm NE Atlantic (Portugal to Canary Islands), Mediterranean Sea, Caribbean Sea, Indian Ocean, NW Pacific (Japan, China), Indo-Pacific (the Philippines, Vietnam), Pacific Islands, and Australia.

Uses and bioactivities: A study carried out by Kumar et al. (2009) revealed that high Calcium (Ca) levels were found in *V. pachynema* (476.67%) when compared to nine important seaweed species screened for macronutrients along Okha coast, Gujarat. Sulfur content was 104.02%, calorific content was 12.9%, and ash content was 37.0%. This study was carried out by the research team to identify potential seaweed as food supplements or as a spice to improve the nutritive value in the diet. Venkatesalu et al. (2012) studied the seasonal variation in the fatty acid composition of *V. pachynema* along with 51 algal species in the Gulf of Mannar Biosphere Reserve. Its extracts have antiviral, antifungal, and antibacterial activity.

1.3.2 Ochrophyta, Phaeophyceae (brown algae)

Adenocystis utricularis (Bory de Saint-Vincent) Skottsberg

Description: Thallus medium to dark brown, with a small discoid holdfast 1–3 mm across, bearing one to several clavate to pyriform, clustered, saccate, mucoid vesicles, each with a short solid stipe, 4–6 cm long and 0.5–2 cm in diameter, epilithic or on crustose coralline algae; structure multiaxial and haplostichous, with an apical pit containing a tuft of phaeophycean hairs surrounded by laterally adjacent filaments with periclinal cell divisions, differentiating into a pseudo-parenchymatous cortex (Womersley 1987).

Distribution: see Table 2.1

Uses and bioactivities: Used for foodstuff, and its extracts have antiviral activity (Pereira 2016).

Alaria esculenta (Linnaeus) Greville (Fig. 1.3-A)

Synonym: Alaria macroptera (Ruprecht) Yendo

Common names: see Table 2.1

Description: Plants with olive or yellow-brown fronds up to 4 m long and 25 cm wide. Attached by a root-like holdfast at the base, from which a narrow flexible stipe arises, which continues into the leafy part of the plant as a distinct mid-rib. The reproductive structures, apparent as dark-brown areas, are confined to unbranched leafy appendages borne on the stipe, usually in two rows. This is the only kelp-like plant in Ireland and Britain with a distinct midrib, and is the only one with sporangia borne at the base of the frond in special leaflets called sporophylls (Pereira 2017h).

Distribution: see Table 2.1

Uses and bioactivities: *A. esculenta* can be used for a variety of purposes—from human consumption and alginate production to fodder and body-care products. It is rapidly gaining popularity in the natural foods market. It can be ordered from many sellers as whole, flaked, milled, or powdered. It is used for antiaging body creams, foot creams, bath soaks, body and face masks, body polish, UV-protecting facial moisturizers, self-tanning lotions, lip balm, night-creams, and nutritional supplements, to name a few (Pereira 2015). Extracts of this species also have cardio-protective, antitumor, and antibacterial activity.

Ascophyllum nodosum (Linnaeus) Le Jolis (Fig. 1.3-B)

Common names: see Table 2.1

Description: *A. nodosum* is a large and common brown alga. The fronds are olive-brown. It is a species of the N Atlantic Ocean, also known as Norwegian kelp, knotted Kelp, or knotted wrack. It is common on the north-western coast of Europe (from Svalbard to Portugal), including east Greenland and the NE coast of N America. It has long fronds with large egg-shaped airbladders. The fronds can reach 2 m in length. When ripe they are yellow, and as the tide goes out they can form huge piles of seaweed. It lives up to 15 years and is a dominant species of the middle shore (Pereira 2010a).

Distribution: see Table 2.1

Figure 1.3 Marine Brown macroalgae: A–*Alaria esculenta*, B–*Ascophyllum nodosum*, C–*Bifurcaria bifurcata*, D–*Canistrocarpus cervicornis*, E–*Colpomenia sinuosa*, F–*Cystoseira brachycarpa*, G–*Cystoseira abies-marina*, H–*Cystoseira compresssa*, I–*Cystoseira humilis*, J–*Cystoseira mediterranea*, K–*Cystoseira tamariscifolia*, L–*Dictyopteris polypodiodes*, M–*Dictyota bartayresiana*, N–*Dictyota dichotoma*, O–*Dictyota mertensii*.

Uses and bioactivities: *A. nodosum* is very effective at accumulating nutrients and minerals from the surrounding seawater, and this is what makes them a valuable resource for human enterprise. This species is harvested for use in items such as food, fertilizer, soil conditioners, animal feed, skin and hair care products, cleaners, degreasers, equestrian products, and nutritional supplements. It is also popular in cosmetology and thalassotherapy (see Chapter 12). The industry has more than 200 types of products from over 100 companies which include *A. nodosum* as an ingredient (Pereira 2015). Extracts of this species have anticoagulant, antiviral, and anti-inflammatory, antibacterial, antioxidant, nematicidal, agricultural bio-stimulant, antitumor, antifouling, and phytobiotic activity.

Bifurcaria bifurcata R. Ross (Fig. 1.3-C)

Common names: see Table 2.1

Description: Up to 30 cm in length; olive-yellow in color, but much darker when dry; holdfast expanded and knobby; frond cylindrical, unbranched near base then branching dichotomously. Elongate reproductive bodies present at ends of branches. Rounded air bladders sometimes present (Pereira 2015).

Distribution: see Table 2.1

Uses and bioactivities: A linear cytotoxic diterpene bifurcadiol was isolated from the brown alga *B. bifurcata* by Di Guardia et al. (1999) which exhibit cytotoxicity against cultured human tumor cell lines. Extracts also have antifouling, antibacterial, antiprotozoal, antioxidant, and antitumoral activity (Pereira 2015).

Canistrocarpus cervicornis (Kützing) De Paula and De Clerck (Fig. 1.3-D)

Synonym: *Dictyota cervicornis* Kützing

Common names: see Table 2.1

Description: Thallus tufted, yellow-brown, erect, somewhat intertwined; branches screw-like twisted, narrow ribbons, mostly forking, but asymmetrically branching (one fork longer than the other), widening at the fork-base (up to 4 mm), outer thallus parts narrow (1–2 mm wide), slightly tapering upward, tips acute, margins smooth; anchored to right substrata by a discoidal holdfast (Pereira 2017i).

Distribution: see Table 2.1

Uses and bioactivities: Used for foodstuff; extracts have antiviral, anti-snake venom, antifouling, antifungal, antiproliferative, antioxidant, and anticoagulant activity (Pereira 2015, Pereira 2016).

Cladosiphon novae-caledoniae Kylin

Common name: Mozuku

Description: Frond erect, caespitose from a small discoid base, cylindrical, lubricious, up to 30 cm high, 1 mm to 1.5 mm in diameter, moderately, irregularly alternately branched; central axis sympodial, polysiphonous, composed of cylindrical large cells (medullary layer cells), length-wise, loosely, more or less parenchymatously arranged; subcortical layer very thin, consisting of one to two cells, divaricate transforming into the assimilatory filaments; assimilatory filaments of the cortical layer, 160–300 μm high, 10–30 cells long, the lower cells cylindrical, 7–8 μm in diameter, the upper cells swollen, 8–10 μm in diameter, strongly constricted at dissepiments, curved in the apical portion; hairs hyaline, 8–10 μm in diameter, uniseriate with cylindrical cells covered with basal sheaths; unilocular sporangia borne on the basal part of the assimilatory filaments, elliptical-obovate, 50–55 μm in length, 25–30 μm in width; plurilocular sporangia transformed from the upper segments of the assimilatory filaments, seriate, at maturity with unilateral openings; both sporangia on the same individual; frond brownish in color, closely adhering to paper when dried (Ajisaka 1991).

Distribution: Pacific Islands (New Caledonia).

Uses and bioactivities: Enzyme-digested fucoidan extracts derived from seaweed Mozuku of *C. novae-caledoniae* inhibit invasion and angiogenesis of tumor cells (see Chapter 6) (Ye et al. 2005, Zhang et al. 2013).

Cladosiphon okamuranus Tokida

Common names: see Table 2.1

Description: This sea vegetable is slender, pale brown, sparsely branched, slippery, floppy (not stiff), and jelly-like. Branches are 1–1.5 mm wide and the plant can reach a length of at least 30 cm. The branching pattern may be alternate (first to one side, then to the other) or irregular (Novaczek 2001).

Distribution: see Table 2.1

Uses and bioactivities: This sea vegetable is highly prized in Japan, and consumed with soy sauce and vinegar (Pereira 2016), and its extracts have cardio-protective, antiviral, anticoagulant, and anti-inflammatory activity.

Colpomenia sinuosa (Mertens ex Roth) Derbès and Solier (Fig. 1.3-E)

Common names: see Table 2.1

Description: Thallus bladder-like, smooth, slick, hollow, crisp, spherical to sac-like, irregularly expanded or somewhat lobed, up to 30 cm diameter, 10 cm high, golden-brown; often covered with fine colorless hairs; reproductive sori as dark raised patches on surface. Membrane 300–500 μm thick, of 4–6 cell layers; medullary cells up to 240 μm diameter; cortex 1–2 cells thick; surface cells darkly pigmented; surface-phaeophycean-hairs transparent, in scattered clusters; holdfast not apparent, attachment at many points (Pereira 2010a).

Distribution: see Table 2.1

Uses and bioactivities: Edible, used as food, fertilizer, and source of alginic acid. Extracts have antibacterial, antifungal, antioxidant, antitumor, antileukemic, antiprotozoal, and hypolipidemic activity (Pereira 2015, Pereira 2016).

Cystoseira abies-marina (S.G. Gmelin) (Fig. 1.3-F)

Common names: see Table 2.1

Description: At the base a branching, gnarled stem with teeth-like appendages. Upper thallus tufted, forking, or lateral branching at more or less the same length, single branches thin, covered with scattered dark dots, and conspicuous bilateral saw-teeth-like; conceptacles as wart-like swellings in the upper parts of the thallus; without swim bladders (Braune and Guiry 2011).

Distribution: see Table 2.1

Uses and bioactivities: This species is used directly for food and as fertilizer. Extracts have antioxidant, antimicrobial, and cytotoxic activity (Pereira 2015).

Cystoseira barbata (Stackhouse) C. Agardh

Common names: see Table 2.1

Description: Thallus large, bushy, richly branched, dark-brown to light-olive, 20–170 cm long, 4–12 mm wide, holdfast conical, unattached form is also known; stem very rough, bearing main and additional branches; main branches 10–40 cm long, cylindrical, arranged alternately or quasi irregularly; branchlets often form panicles close to apices of main branches; additional branches 5–10 cm long; air vesicles present on branches during winter and spring, oblong, 7–15 mm long, 2–5 mm wide, solitary or moniliform, 2–10 per branch; receptacles develop on apices of branches, 2–20 mm long, 1–3 mm wide, oval or lanceolate, spinules absent, surface smooth and slightly sinuous; apices with a sterile spike; unattached plants prostrate, thin, 20–80 cm long; branches very thin and long, with sparse lateral branchlets; receptacles nearly always absent; air vesicles reduced (Milchakova 2011).

Distribution: see Table 2.1

Uses and bioactivities. *C. barbata* is a source of alginic acid salts, iodine-containing amino acids, PUFA, antimicrobial, antimitotic, antiviral, antibacterial and antitumor BASs; micro- and macroelements (e.g., iodine, bromine, potassium, calcium, magnesium, chlorine, sulfur, and selenium) (Milchakova 2011, Pereira 2015).

Cystoseira brachycarpa J. Agardh (Fig. 1.3-G)

Synonym: *Cystoseira balearica* Sauvageau

Description: Plant caespitose, up to 20–25 cm in height, attached to the substratum by a more or less compact discoid base formed of haptera; axes numerous, two to six with 8 cm high; apices of the axes not very prominent, flattened, and smooth; tophules absent. Primary branches cylindrical, seasonally either with smooth bases or covered with small spinose; conical appendages, some of which can give rise to smaller branches; secondary and tertiary branches also cylindrical and covered with spinose appendages (Pereira 2017j).

Distribution: Mediterranean Sea.

Uses and bioactivities: The linear diterpenes eleganolone and elegandiol, isolated from *C. brachycarpa*, inhibit contractile activities of acetylcholine and histamine on ileum musculature of guinea pigs. Extracts have antimicrobial and antiviral, antifungal, cytotoxic, and antimitotic activity (Pereira 2015).

Cystoseira compressa (Esper) Gerloff and Nizamuddin (Fig. 1.3-H)

Description: *C. compressa* has a discoid base with several spined axes; the axes have denticulate margins. It is irregularly branched with compressed-primary ramifications, and compact crawling receptacles; the axes have a height of 10–100 mm, and a thickness of 2–5 mm. This alga lacks any leaves, tophules, and air vesicles; *C. compressa* has an olive-brown coloration (Pereira 2017k).

Distribution: NE Atlantic (Azores, Canary Islands) and Mediterranean Sea.

Uses and bioactivities: Extracts have anti-inflammatory and antiproliferative, antimicrobial, antiviral, and cytotoxic activity (Pereira 2015).

Cystoseira crinita Duby

Description: Thallus large, bushy, 10–120 cm high, dark-brown; holdfast giving rise to as many as 20 shoots; richly branched; stem 5–80 cm long; 2–4 mm wide, resilient, flexible and smooth; main branches 6–18 cm long; additional branches few, 3–10 cm long; air vesicles large, 5–8 mm long, 4–5 mm width, triangular, inflated, solitary; receptacles forms on surface of air vesicles, cylindrical, abortive process absent; apices blunt (Milchakova 2011).

Distribution: NE Atlantic (Azores, Canary Islands) and Mediterranean Sea.

Uses and bioactivities: *C. crinita* is a source of alginic acid salts, PUFA, as well as micro- and macroelements (e.g., iodine, bromine, potassium, calcium, magnesium, chlorine, sulfur, and selenium), and its extracts have anti-inflammatory and antiproliferative, antimicrobial, antitumor, antiviral, and cytotoxic activity (Milchakova 2011, Pereira 2015).

Cystoseira humilis Schousboe ex Kützing (Fig. 1.3-I)

Description: Characterized by highly differentiated basal and apical regions and the presence of catenate pneumatocysts (air vesicles). Old plants have an elongated main axis, and in time the primary laterals become proportionally elongated. Their lower parts are strongly flattened into foliar expansions or basal leaves. Fertile regions which bear conceptacles are known as receptacles. These are normally found at the tips of the branches. Their basal and apical regions are highly differentiated. They have catenated pneumatocysts (air vesicles). The aerocysts or air vesicles keep the organism erect, by causing it to float in strong currents (Pereira 2017l).

Distribution: One of the most widely distributed genera of the Fucales order and provides an essential habitat for many epiphytes, invertebrates, and fish. Found mostly in temperate regions of the Northern Hemisphere, such as the Atlantic, the Mediterranean, Indian and Pacific Oceans.

Uses and bioactivities: Extracts have antibacterial activity (Pereira 2015).

Cystoseira mediterranea Sauvageau (Fig. 1.3-J)

Description: *C. mediterranea* has an upright thallus, with a single cylindrical main axis; irregular ramifications are present with the alga reaching a height of 400 mm. The receptacles are short and cylindrical, and whilst it has gas-filled vesicles, it is lacking in tophules. The main axis has a relatively soft texture, whilst the outer extremities are rough, almost spiny. It is a brown to olive green color and displays a bluish green iridescence when submerged (Pereira 2017m).

Distribution: Mediterranean Sea.

Uses and bioactivities: Extracts have antitumor, antimicrobial, and antifungal activity (Pereira 2015).

Cystoseira nodicaulis (Withering) M. Roberts

Common name: see Table 2.1

Description: Thallus up to 1 m long, usually solitary, attached by an irregular conical disc. Axis cylindrical usually branched, with smooth, rounded apex immersed between bases or tophules of developing laterals. Lateral branch systems (below) radial or distichous, with greenish-blue iridescence when first formed, about 50 cm long, repeatedly branched in a pinnate manner, either regularly or irregularly, with infrequent cryptostomata and bearing spine-like appendages; deciduous in summer; first-formed laterals of the season with tophules, later without; tophules ovoid, up to 15 mm long, smooth or covered with small tubercles, persistent on axis after rest of lateral has been shed. Receptacles formed in the ultimate branchlets, simple or branched, nodose, usually bearing spine-like appendages; air vesicles inconspicuous, dilations of ultimate branchlets, solitary, in series or confluent; sometimes absent (Braune and Guiry 2011).

Distribution: see Table 2.1

Uses and bioactivities: Considered to be an edible species; extracts have antileukemia and neurological activity.

Cystoseira tamariscifolia (Hudson) Papenfuss (Fig. 1.3-K)

Common names: see Table 2.1

Description: *C. tamariscifolia* is bushy seaweed, up to 60 cm in length but usually 30–45 cm. It has a cylindrical frond and branches irregularly. The reproductive bodies on the end of branches are long, oval, and spiny. Small air bladders are usually found below the reproductive bodies. *C. tamariscifolia* is olive-green in color, almost black when dry. When the plant is seen underwater, it has a blue-green iridescence. Found in rockpools and on the lower shore; grows on both rocky shores and gravelly flats (Pereira 2015, Pereira 2016).

Distribution: see Table 2.1

Uses and bioactivities: Extracts have antibacterial and antifungal, antioxidant and antitumor, and cytotoxic activity (Pereira 2015).

Costaria costata (C. Agardh) De A. Saunders

Common names: Five-Ribbed Kelp, Seersucker Kelp, Short Kelp, Sujime

Description: Thallus of this light to medium brown kelp has a branched holdfast (haptera), a somewhat flattened, finely grooved stipe, and an elliptical blade up to 2 m long and 35 cm wide, with five parallel ribs running its length. Three ribs project on one side of the blade and two ribs on the other. Between the ribs the blade is profusely wrinkled or puckered (Lindberg and Lindstrom 2017).

Distribution: N America (Alaska to S California), S America (Chile, Argentina), and Asia (Japan).

Uses and bioactivities: Used for foodstuff in NE Pacific, especially for the manufacture of potash salts (Tokida 1954). It has anti-inflammatory activity (see Table 10.1 in Chapter 10).

Dictyopteris delicatula J.V. Lamouroux

Description: This species has erect, light-brown, strap-shaped blades attached to the substratum at basal holdfast or to adjacent branches, creating a tangled mass 2–8 cm in height. Dichotomous to irregularly branch bi-layered blades, 0.5–5.0 mm wide, have a distinctly raised midrib that may be several cells thick. Cells of blades are arranged in parallel rows at acute angles to midrib. Scattered clusters of hyaline hairs in dense tufts arise along midrib on only one surface of midrib (Schmitt and Mamoozadeh 2014).

Distribution: see Table 2.1

Uses and bioactivities: Used for foodstuff; the heterofucans of this species have anticoagulant, antioxidant, and antitumor activities (Pereira 2016).

Dictyopteris polypodioides (A.P. De Candolle) J.V. Lamouroux (Fig. 1.3-L)

Synonym: *Dictyopteris membranacea* Batters

Common names: see Table 2.1

Description: Thallus flat and leaf-like, up to 300 mm long and 20–30 mm broad; fronds olive to yellow-brown, translucent, and ± regularly dichotomously forked with a prominent midrib extending to the apices. Margins sometimes split to the midrib. Initially with an unpleasant smell shortly after collection, and degenerating quickly (Pereira 2010a).

Distribution: see Table 2.1

Uses and bioactivities: Extracts have antitumor, antifungal, antibacterial, antileishmanial, and anticoagulant activity.

Dictyota bartayresiana J.V. Lamouroux (Fig. 1.3-M)

Synonym: *Dictyota bartayresii* J.V. Lamouroux

Common names: see Table 2.1

Description: Thallus erect, iridescent blue and green in the water, or light brown, often with dark olive-brown bands, 9–14 cm high, erect, not entangled, a little harsh to the touch, attached to the substratum by irregularly shaped holdfast with rhizoids, thallus branched dichotomously; segments without midrib, 1–1.5 cm long, 2–4 mm broad above a fork, broadening to 6–10 mm below the next fork; margin entire, tips are pointed except in young branches (Pereira 2015).

Distribution: see Table 2.1

Uses and bioactivities: Extracts have antibacterial activity.

Dictyota caribaea Hörnig and Schnetter

Description: It is an erect olive-brown alga with strap-shaped fronds that measure up to 32 mm long in the studied area. The fronds have unequal dichotomous divisions that range between 1–2 mm wide, 90–140 mm thick, and have non-constricted branches with sharp apexes (De Oliveira et al. 2009).

Distribution: see Table 2.1

Uses and bioactivities: Used in the Caribbean Islands as food and its extracts have antitumor and antibacterial activity.

Dictyota dichotoma (Hudson) J.V. Lamouroux (Fig. 1.3-N)

Common names: see Table 2.1

Description: Thallus flat, homogenous yellow-brown to darker brown, with fairly regular dichotomous branches with parallel sides up to 30 cm long, the tips usually bifid; branches 3 mm to 12 mm wide, membranous, without a mid-rib (Pereira 2010a).

Distribution: see Table 2.1

Uses and bioactivities: Extracts of this seaweed also have anticoagulant, cytotoxic, antitumor, anti-inflammatory, antifungal, larvicidal, antimicrobial, and antifouling activity; extracts have also been used as a liquid fertilizer; it is also used in treatments for goiter and scrofula; it is also used as a preventive medicine for heart disease and stroke (Milchakova 2011, Pereira 2015).

Dictyota dichotoma var. *intricata* (C. Agardh) Greville

Synonym: *Dictyota linearis* Trevisan

Common names: see Table 2.1

Description: Thalli medium-brown of flat blades branching regularly into twos (dichotomous), a narrow variety, var. *intricata*, with relatively long distances between branching, is common in sheltered rocky substrata (Santhanam 2015).

Distribution: see Table 2.1

Uses and bioactivities: Extracts have antiviral, antitumor, antibacterial, and larvicidal activity.

Dictyota implexa (Desfontaines) J.V. Lamouroux

Synonyms: *Dictyota dichotoma* var. *implexa* (Desfontaines) S.F. Gray; *Dictyota linearis* (C. Agardh) Greville

Description: Thallus bushy, brown, consisting of intertwined, erect fascicles; thallus regularly forking, width at the base 2–3 mm, abruptly narrowing toward the tip, then filamentous (< 0.5 mm) (Braune and Guiry 2011).

Distribution: Atlantic Islands (Azores and Canary Islands) and the Mediterranean.

Uses and bioactivities: Extracts have antibacterial, antifungal, and antioxidant activity (Pereira 2015).

Dictyota mertensii (C. Martius) Kützing (Fig. 1.3-O)

Description: Thallus bushy, brown, often iridescent blue-green under water, erect, robust, and stately; the flattened fronds show a branching pattern atypical for the genus—distinct main axes branch repeated-alternating, terminal forkings of the latitudinal axes turn into spur-like, 1–2 mm long, pointed or rounded tips (Braune and Guiry 2011).

Distribution: see Table 2.1

Uses and bioactivities: Used in the Caribbean Islands as food and its extracts have antibacterial activity (Pereira 2015, Pereira 2016).

Ecklonia arborea (Areschoug) M.D. Rothman, Mattio and J.J. Bolton

Synonym: *Eisenia arborea* Areschoug

Common names: see Table 2.1

Description: It has a relatively short (up to 1 m), broad stipe running from the holdfast to the cluster of strap-like blades that grow off at a split at the top of the plant. The entire plant reaches about 1.5 m tall and the two clusters of blades at the top make the plant resemble a mini palm tree or a cheerleader's pom-pom (Pereira 2016).

Distribution: see Table 2.1

Uses and bioactivities: *E. arborea* is often a part of Asian cuisine and is sold as a dried whole or crushed leaves under the name "Arame" and is also used as feed for aquaculture of abalone. This seaweed contains up to 20 times the levels of elements found in land plants. Their mineral content can include calcium, copper, iron, magnesium, potassium, and zinc. They are also rich in vitamins. They are highly nutritious; containing beta-carotene (a potent antioxidant) and "Arame" contains particularly high levels of iodine. In Peru, it has been used as folk medicine with anti-allergic properties. Due to this and the naturally moisturizing properties of *E. arborea*, it is often used in cosmetics, soaps, and skin care products (Pereira 2016).

Ecklonia bicyclis Kjellman

Synonym: *Eisenia bicyclis* (Kjellman) Setchell

Common names: see Table 2.1

Description: *E. bicyclis* is a rather small kelp with a stiff, woody stipe up to 1 m tall and two flattened, oval fronds or lobes, with many lateral blades. Each year fronds are shed and new ones develop, creating a branched and feathery plant. This kelp is native to warm-temperate waters of the Pacific, specifically around Japan, where it displays a distinct seasonality of growth and reproduction (Smith 1904).

Distribution: see Table 2.1

Uses and bioactivities: Used for preparation of miso soup, nori-jam, salads, and seaweed powder for various foods in the Japanese cuisine (Smith 1904). It contains a fair amount of calcium, which is present in chelated form, that is, bound to an organic or amino acid, which permits the body to absorb more of the mineral (Pereira 2011). The extracts of *E. bicyclis* have antihypertensive, cardio-protective, antiviral, antitumor, antifungal, antibacterial, anti-inflammatory, anti-allergic, antithrombotic, and anticoagulant activity.

Ecklonia cava Kjellman

Common names: see Table 2.1

Description: *E. cava* is a large brown algae and an important kelp, forming vast underwater forests with plants growing up to 3 m in length. The large strong holdfast is composed of many branched haptera which give rise to a single plant with a long cylindrical stipe of 1 m to 2 m. Many long, smooth, leathery blades emerge from the stipe, forming a clump at the top of the plant, very reminiscent of a palm tree. *E. cava* grows exclusively in subtidal areas of deep pools, where they are anchored directly to the rocky substratum (Pereira 2016).

Distribution: see Table 2.1

Uses and bioactivities: Most of the harvest of *E. cava* is used for food in China, Japan, and Korea, where it is boiled with soy sauce. This species has antihypertensive action, and is used for alginates extraction (Pereira 2016). Its extracts have cardio-protective, antiviral, antibacterial, anti-inflammatory, anti-allergic, anti-coagulant, and neurological activity.

Ectocarpus siliculosus (Dillwyn) Lyngbye (Fig. 1.4-B)

Description: Plants tufted, often only one to a few centimeters tall, but in exceptional cases up to 20 cm. Axes freely branched, main axis not distinguishable; filaments up to 30 μm in diameter, tapering toward the apices. Sometimes forming terminal pseudo-hairs, forms soft beards on larger plants or other firm substrata (Pereira 2015).

Distribution: NE Atlantic (Greenland to Canary Islands, North Sea, Baltic), the Mediterranean, NW Atlantic (Canada, USA); SE and SW Atlantic, the Caribbean, NW Pacific (Japan, China), NE Pacific, Australia and New Zealand, and Sub-Antarctica.

Uses and bioactivities: Extracts have antioxidant and antibacterial activity.

Fucus ceranoides Linnaeus (Fig. 1.4-A)

Common names: see Table 2.1

Description: Large brown intertidal seaweed, restricted to growing in estuaries or near freshwater streams on the shore. *F. ceranoides* does not have air bladders, but the side of the fronds is often inflated. Frond thin with smooth margin, fan shaped with prominent midrib, without air bladders but frond on either side may be inflated, reproductive bodies narrow, pointed fronds at ends of branches (Pereira 2010a).

Distribution: see Table 2.1

Uses and bioactivities: Extracts have antioxidant and antitumor activity.

Fucus serratus Linnaeus (Fig. 1.4-C)

Common names: see Table 2.1

Description: *F. serratus* is a robust, olive-brown shrubby seaweed. It can grow in high densities low on the shore, forming dense mats of long ribbons up to 1 meter long and 2–5 cm across. It attaches to rocks

Figure 1.4 Marine Brown macroalgae: A–*Fucus ceranoides*, B–*Ectocarpus siliculosus*, C–*Fucus serratus*, D–*Fucus spiralis*, E–*Halopteris scoparia*, F–*Himanthalia elongata*, G–*Laminaria hyperborea*, H–*Laminaria ochroleuca*, I–*Leathesia marina*, J–*Padina gymnospora*, K–*Padina pavonica*, L–*Pelvetia canaliculata*, M–*Petalonia fascia*, N–*Saccharina latissima*, O–*Saccorhiza polyschides*.

via a discoid holdfast about 3 cm in diameter. Though technically a brown alga, it can vary in color from olive green through reddish brown (though it often has a greenish tint). It typically grows up to 70 cm, but has been recorded at over 2 m in length in very sheltered environments. The flat, strap-like fronds have a forward-pointing serrated edge, a distinct midrib, and grow from a short stipe. The fronds are bifurcating (splitting in two repeatedly) frond on either side may be inflated, reproductive bodies narrow, pointed fronds at ends of branches (Pereira 2009).

Distribution: see Table 2.1

Uses and bioactivities: *F. serratus* survives by filtering the ocean for nutrients and thus, amasses a huge number of minerals and vitamins. Used for hundreds of years in seaweed baths, the oils from this seaweed have positive effects on skin, hair, and body. This seaweed is used as a food, is harvested for cosmetics, and is harvested to make fertilizer. Its edible properties are very similar to that of bladderwrack and interest in this plant is growing, as being a thyroid stimulant, it might counter obesity by increasing the metabolic rate. It is also known to help women with abnormal menstrual cycling patterns and/or menstrual-related disease histories. It can be stored, dried to make a nutritious tea, and be used in soups and stews as a flavoring (Pereira 2015).

Fucus spiralis Linnaeus (Fig. 1.4-D)

Common names: see Table 2.1

Description: Well-grown fronds are usually easily recognizable by the flattened, twisted, dichotomously branched thallus, lacking bladders, and the large, oval receptacles at the frond tips, each receptacle being surrounded by a narrow rim of vegetative frond. Nevertheless, younger plants are not always so easy to identify, and even mature plants can be confused with *F. ceranoides* or with bladderless forms of *F. vesiculosus*. Both species, however, have narrower, more pointed, rimless receptacles (Pereira 2009, Pereira 2010a).

Distribution: see Table 2.1

Uses and bioactivities: *F. spiralis* has been used historically for treatment of obesity, gout, goiter, and corns, and also in weight reducing and revitalizing bath treatments. It has been used for cattle feed, and as an organic manure. This alga, regularly exposed to sun radiation and its oxidative consequences, has developed optimal bioelectronics characters. It has a high concentration of phloroglucinol derivatives, including phenol acid, and in turn has been used in products from companies in France and the UK, such as nutritional supplements, skin serum, body lotion, and compounds and extracts used as ingredients in other skin and hair products. Extracts of this species also have antifouling, antimicrobial, antioxidant, anticoagulant, and antiproliferative activity (Pereira 2015).

Fucus vesiculosus Linnaeus

Common name: Bladderwrack

Description: *F. vesiculosus* is generally larger and has a lighter color than *F. spiralis*, and has air bladders (aerocysts) arranged on both sides of the middle rib (Pereira 2010a). It can be found in high densities, living for about four to five years. Under sheltered conditions, the fronds have been known to grow up to 2 m in Maine, N America (Pereira 2016).

Distribution: N and NE Atlantic, from Greenland to Canary Islands, NW Atlantic, from Canadian Arctic to Caribbean Sea, and W Mediterranean.

Uses and bioactivities: This species is commonly used as a food in Japan, though less so in Europe and North America (Alaska); consumed in W Europe, Spain, Portugal, Scotland, and Ireland (Pereira 2016). Primary chemical constituents of *F. vesiculosus* include mucilage, alginic acid, mannitol, β-carotene, zeaxanthin, iodine and iodine salts, bromine, potassium, volatile oils, and many other minerals, as well as polysaccharides (see Chapter 2). When used in hot seawater baths or steamed, the plants are said to release certain substances that promote good skin, lower the blood pressure, and ease arthritic and rheumatic pains. *F. vesiculosus* has been shown to help women with abnormal menstrual cycling patterns and menstrual-related disease histories. A popular use of *F. vesiculosus* in herbal medicine is as a source of iodine (it was the original source of iodine, discovered in 1811), an essential nutrient for the thyroid gland; it can be used in the treatment of underactive thyroid glands (hypothyroidism) and goiter, a swelling of the thyroid gland related to iodine deficiency (Pereira 2010a); in Alaska it is made into tea (Chapman and Chapman 1980). It can be stored, dried, and added to soups and stews in flakes or powder form for flavor (Pereira 2016b).

Halidrys siliquosa (Linnaeus) Lyngbye

Common names: Sea Oak, Pod Weed

Description: A large sturdy brown alga 0.3–1 m in length (occasionally up to 2 m) rising from a strong, flattened, cone-shaped holdfast. The main stem is flattened and branches alternate to give a distinct zigzag appearance. The stem bears a few, flattened ribbon-like leafy fronds. The ends of some branches bear characteristic pod-shaped air bladders (about 0.5 cm wide by 1–4 cm long) that are divided by transverse septa into 10 or 12 compartments. The branches also bear reproductive bodies that appear similar to the bladders, but lack the septa. Young plants are olive-green in color, while older specimens are dark brown and leathery (Pereira 2015).

Distribution: NE Atlantic (from Faroe Islands to Portugal, North Sea, Baltic).

Uses and bioactivities: Source of alginic acid. Extracts have antibiotic, antifouling, antioxidant, and antitumor activity (Pereira 2015).

Halopteris scoparia (Linnaeus) Sauvageau (Fig. 1.4-E)

Synonym: *Stypocaulon scoparium* (Linnaeus) Kützing

Common names: see Table 2.1

Description: Dark brown algae that forms beautiful fluffy clumps in shallow rocky-bottomed water. Growing only up to 15 cm in length, *H. scoparium* has a main axis with alternate plumed branches which are more or less fan shaped when flat, though when buoyed up by water, they form inverted cone-shaped tufts with a very delicate appearance due to the many filamentous branches. Plants are usually attached to rocks with small to extensive discs often obscured by many matted rhizoids, though these are lacking in free-living plants. These plants are characterized by pure, sheltered waters with high light levels. *H. scoparium* often forms clumps on cobble in the shallow sublittoral zone, though they are just as likely to be found in shallow tide pools or sandy-bottomed areas (Pereira 2009, Pereira 2010a).

Distribution: see Table 2.1

Uses and bioactivities: *H. scoparium* is known to be an ingredient in compounds used in personal care products; it contains growth substances (phytohormones) that include auxins, gibberellins, cytokinin's, abscisic acid, and betaines. Extracts of this species have antiprotozoal, antifungal and antimitotic, antioxidant, antileukemia, and antimicrobial activity (Pereira 2015).

Himanthalia elongata (Linnaeus) S.F. Gray (Fig. 1.4-F)

Common names: see Table 2.1

Description: Common brown seaweed with two stage morphology. Small button-like thalli are first produced, from which long strap-like reproductive fronds (receptacles) are formed in autumn. The strap-like reproductive fronds grow quickly between February and May, reaching a length of up to 3 m. The plant releases gametes from June until the winter, when it starts to decay. Thalli commonly live for 2–3 years and reproduce once before dying (Pereira 2009, Pereira 2010a).

Distribution: see Table 2.1

Uses and bioactivities: *H. elongata* is known to provide high levels of vitamins A, C, and E along with essential amino acids and other natural minerals. It has several properties which make it attractive as an ingredient in personal care and cosmetic products, for instance, it is absorbent, viscosity controlling, skin protecting, and can be used as a binding agent (Pereira 2015). *H. elongata* is used in shampoos and hair treatments, facial cleansers and skin care products, and also as a fertilizer. Sea spaghetti is rich in phosphorus, a mineral known to enhance brain function, helping to preserve memory, concentration, and mental agility. Extracts of this species have hypoglycemic, antibiotic, neuropharmacological, antimicrobial, and antioxidant activity (Pereira 2015, Pereira 2016).

Hydroclathrus clathratus (C. Agardh) M.A. Howe

Common names: see Table 2.1

Description: *H. clathratus* is a very interesting brown seaweed, appearing as a 6 cm to 10 cm yellow-brown clump of very porous, chain-like tissue. The plant has a very open, sponge-like structure with a complex series of holes perforating narrow, fleshy strips. Its name *clathratus*, which means latticed, is an apt description. *H. clathratus* is found worldwide in warm seas, but is an uncommon species on hard reefs. *H. clathratus* is more often found on in calm, shallow areas where it is anchored in bare sand (Braune and Guiry 2011).

Distribution: see Table 2.1

Uses and bioactivities: *H. clathratus* has been used for centuries in traditional cuisine, and medicine of island cultures, such as Hawaii. *H. clathratus* is known to possess anticancer, anti-herpetic, anti-inflammatory, and anticoagulant properties, and is now used as a mineral supplement in cosmetics and as a soil-additive (fertilizer) for its high concentration of micronutrients. Extracts have antiviral, antitumor, cytotoxic, antiviral, and antimicrobial activity (Pereira 2015).

Ishige okamurae Yendo

Common names: see Table 2.1

Description: This species has leathery branched narrow fronds consisting of cylindrical hairs, uniseriate plurilocular sporangia lacking sterile terminal cells, apical growth, pyrenoid-less discoid plastids, and an isomorphic life history (Yendo 1907).

Distribution: see Table 2.1

Uses and bioactivities: Used as food in China and Korea; the simplest way involves powdering the dried blades, then mixing with condiments; or the dried thallus may be steamed, then boiled, each for about 15 minutes, mixed with soy sauce, and eaten. Their extracts have anti-inflammatory, anti-diabetic, and antioxidant action. Some fraction extracts (phlorotannins) exhibited inhibitory activity against acetylcholinesterase that could be used as potential functional food ingredients or nutraceuticals for preventing Alzheimer's disease (see Chapter 11) (Pereira 2015).

Iyengaria stellata (Børgesen) Børgesen

Description: Thallus brown, a hemispherical cushion, irregularly grooved, with a wrinkled surface of bulbous outgrowths, cartilaginous. Solid at first, later hollow; closely related to *Colpomenia* (Braune and Guiry 2011).

Distribution: SE Atlantic and NW Indian Ocean.

Uses and bioactivities: Extracts have antibacterial activity.

Jolyna laminarioides S.M. Guimarães

Description: The plants consist of one to several erect blades arising from a discoid holdfast. Anatomical features include a meristoderm, an outer and inner cortex, a transition region, a medulla that is composed of elongated cells (fiber-like cells), and a ground tissue made of a dense, irregular weft of interconnected filaments (Guimarães et al. 1986).

Distribution: Western tropical Atlantic (Brazil) and SW Asia (Oman, Pakistan, Yemen).

Uses and bioactivities: Extracts have cardio-protective and antibacterial activity.

Laminaria digitata (Hudson) J.V. Lamouroux

Common names: Tangle, Sea girdles, Sea wand, Sea ware, Tangle tail, Wheelbangs, Sea Tangle, Horsetail Kelp, Kelp, Strap wrack, Oarweed, Oar weed, Horsetail tangle, Sea Girdle

Description: *L. digitata* is a dark brown alga with a smooth, flexible stipe and can reach 3 m to 4 m in length. *L. digitata* can grow to be 4 to 5 years old. Fixed with a root-like holdfast, this kelp grows on rocky bottoms in the upper subtidal zone in sheltered to moderately exposed areas from 1 m to 25 m in depth (Cabioc'h et al. 2014).

Distribution: N Atlantic, Atlantic Islands (Canary Islands, Greenland, Iceland), and Baltic Sea.

Uses and bioactivities: Used as food in Ireland and Iceland, in W Europe and UK, and potential use in Norway, *L. digitata* is imported in Japan and China for making dashi, a soup stock, and for other culinary purposes, such as accelerating the cooking time of vegetables, such as beans and lentils (Pereira 2016). Extracts of this species have antibacterial and antioxidant activity.

Laminaria hyperborea (Gunnerus) Foslie (Fig. 1.4-G)

Common names: see Table 2.1

Description: *L. hyperborea* is often difficult to distinguish from *L. digitata*, particularly when plants are young. However, the stipe of *L. digitata* is darker and usually oval in cross section instead of cylindrical, is not thicker at the base, does not snap easily, and does not have epiphytes on the stipe. Also, the belt of *L. hyperborea* is nearly always below (in deeper water) any belt of *L. digitata*. *L. ochroleuca* is a similar species, but has a smooth stipe (not rough like *L. hyperborea*), and its fronds are typically more golden or yellow. *Saccorhiza polyschides*, another commercially utilized kelp, appears similar to *L. hyperborea* from the frond, but has a twisted stipe near a bulbous haptera, which is very different from *L. hyperborea*'s bird claw holdfast (Pereira 2010a, Pereira 2015).

Distribution: see Table 2.1

Uses and bioactivities: *L. hyperborea* is one of the two kelp species commercially exploited by the hydrocolloid industry; *L. hyperborea* is also utilized by the cosmetic and agrochemical industries and for biotechnological applications, and by the food industry for emulsifiers and gelling agents; drift kelp has long been collected as an agricultural fertilizer and soil conditioner. *L. hyperborea* is still harvested and used in popular kelp meal fertilizer products; extracts have antifungal, anticoagulant, antibacterial, antioxidant, and anticoagulant activity (Pereira 2015).

Laminaria ochroleuca Bachelot de la Pylaie (Fig. 1.4-H)

Common names: see Table 2.1

Description: *L. ochroleuca* is a glossy, yellow-brown kelp that is prevalent along the intertidal zones. This kelp is quite conspicuous as it grows quite large under the right conditions. The maximum length recorded is 4 m long, but this length is rarely attained and occurs only in specific areas. Under normal conditions *L. ochroleuca* is more likely to reach a maximum length of about 2 m. It has a large heavy holdfast made up of thick haptera (up to 18 cm in diameter) that support the plant and anchor it to the rock. This holdfast gives rise to a fairly long, rigid, round, epiphyte-free stipe that tapers somewhat as it approaches the blade. This stipe is so strong and stiff that it stands erect when the plant is out of the water. The blade of this kelp is large, flat, and leathery, and is divided into 5–20 strap-like digits. This kelp is easily distinguished by the distinct yellow area at the junction of the stipe and the blade. The entire plant actually has a very lovely yellowish hue to its smooth, bright, and glossy tissue. *L. ochroleuca* is a perennial kelp that retains its stipe and holdfast yearlong, but regenerates a new blade each year (Pereira 2009, Pereira 2010a).

Distribution: see Table 2.1

Uses and bioactivities: Extracts of *L. ochroleuca* have been found to act as a central nervous system depressant with a slight analgesic activity. It is also able to effectively guard DNA against UV rays and premature aging. With these properties, it is becoming more widely used in cosmetics and in natural therapeutic medicine. Applied topically, *L. ochroleuca* helps reduce inflammation and, like many forms of seaweed, has some amount of moisture-binding properties due to its sterol content. It is used in products from Spain, Germany, France, and the United Kingdom; these include rescue balms, repairing and antiaging moisturizers, as well as dried and packaged foods. Extracts also have antimicrobial, antifungal, and antialgal activity (Pereira 2015).

Leathesia marina (Lyngbye) Decaisne (Fig. 1.4-I)

Synonym: *Leathesia difformis* Areschoug

Common names: Sea cauliflower, Sea balls, Punctured Ball Weed, Nebarimo

Description: Thallus round in young stages, light brown, firm-fleshy and slimy-smooth, later hollow, and with an irregularly convoluted surface (Braune and Guiry 2011).

Distribution: NE Atlantic (Iceland to Canary Islands), the Mediterranean, NW Atlantic, SE Atlantic (Namibia, South Africa), Indian Ocean (South Africa), NW Pacific (Japan, China), NE Pacific (Alaska to California), SW Pacific (the Philippines), Australia, New Zealand, Antarctica, and Sub-Antarctica.

Uses and bioactivities: Extracts have agricultural bio-control and bio-stimulating, antifouling, antiviral, antioxidant, and antitumor activity (Pereira 2015).

Lobophora variegata (J.V. Lamouroux) Womersley ex E.C. Oliveira

Common names: Encrusting Fan-Leaf Alga, Leathery Lobeweeds

Description: These algae occur in orange brown to dark brown colors and bear a leathery feel. The blades may appear fan-shaped, flabellate, or as crusts. They are often tightly adhered to the substrate (Pereira 2015).

Distribution: see Table 2.1

Uses and bioactivities: Extracts have antifungal, antiprotozoal, antimicrobial, anti-inflammatory, hypoglycemic, antifungal, antifouling, anti-inflammatory, antioxidant, and anticoagulant activity (Pereira 2015).

Padina australis Hauck

Common names: see Table 2.1

Description: Thallus is yellow-brown in color, and does not change the original color when dried. Thallus is upright with rhizoid holdfast, fan shape, membrane-like texture, 6–18 cm in height, and 5–15 cm in width. Concentric hair bands are too narrow which make it difficult to recognize, and then fertile (1.0–2.0 mm) in width, has reproductive organs and sterility zones (2.0–2.3 mm) in width, lacks reproductive organs alternating arranged on the outer surface of thallus. Proliferations are formed on the hair bands at the older part of thallus. Slight calcification on the inner rather than outer surface of the thallus (Wang 2014, Santhanam 2015).

Distribution: see Table 2.1

Uses and bioactivities: Extracts have antiviral and antibacterial activity.

Padina boergesenii Allender and Kraft

Common names: see Table 2.1

Description: Thalli light brown to tan, moderately ventrally calcified, 4–6 cm long and wide; blades broadly or narrowly lobed, short-stipitate from a bulbous fibrous holdfast (Kraft 2009).

Distribution: see Table 2.1

Uses and bioactivities: Extracts of this brown seaweed have hepatoprotective, anti-diabetic, and antioxidant activity.

Padina gymnospora (Kützing) Sonder

Common names: see Table 2.1

Description: Thallus erect, slender fan-shaped, funnel-like rolled up and narrowing towards the base, divided at the short, stipe-like base, proliferating, upper margin split; lamina with delicate concentric lines, tightly rolled at the front margin. Delicate, only slightly calcified (Braune and Guiry 2011).

Distribution: see Table 2.1

Uses and bioactivities: Extracts of this species have anti-inflammatory, hemagglutinating and anticancer, neuroprotective, antioxidant, antifungal, antiviral, and antibacterial properties (Pereira 2015, Pereira 2016).

Padina pavonica (Linnaeus) Thivy (Fig. 1.4-K)

Common name: see Table 2.1

Description: The fronds are thin and leafy, flattish and entire when young, but often concave, or almost funnel-shaped in mature specimens, with a laciniate or irregularly lobed margin. The inner (or upper) surface is covered in a thin coating of slime, and the outer (or lower) surface is banded with zones of light brown, dark brown, and olive green. Small, fine hairs form concentric lines, 3–5 mm apart, from the outer margin continuing down the outer (colored) surface of the fronds (Braune and Guiry 2011).

Distribution: see Table 2.1

Uses and bioactivities: *P. pavonica* collected from the Buleji coast (Pakistan) can be utilized as food, medicines, and fodder; this species have cosmetic uses for skin antiaging; high anticoagulant activity has been reported for purified fucan sulfates from *P. pavonica* (xylofucomannoglucuronan); extracts have also allelopathic, antifungal, antibacterial, anticoagulant, antitumor, antiviral, cytotoxic, and antimitotic activity (Pereira 2015).

Pelvetia canaliculata (Linnaeus) Decaisne and Thuret (Fig. 1.4-L)

Common names: see Table 2.1

Description: *P. canaliculata*, often called channeled wrack, is a very common brown alga (Phaeophyceae) of Europe. It is the only species remaining in the monotypic genus *Pelvetia*. It is relatively small, not growing longer than 15 cm. When viewed underwater, its color can be light olive-green and sometimes yellowish brown. *P. canaliculata*'s color can range from dark brown to dark olive to very dark or blackish green when dried. Its extremities can appear swollen and orange during spring and summer. These bumpy, irregularly v-shaped swellings with forked tips at the ends of the fronds are its reproductive structures. It appears bushy and grows in dense tufts. Each frond is curled longitudinally (rolled lengthwise) to form a distinct channel. It is irregularly dichotomously branched; each tough and thick branch is of uniform width up to 1 cm, lacks a midrib, and lacks air vesicles or bladders. *P. canaliculata* is a perennial species; it is at least two years old before it reaches maturity, and has a life span of up to 4 or 5 years, growing 3 cm to 4 cm per year (Pereira 2009, Pereira 2010a).

Distribution: see Table 2.1

Uses and bioactivities: According to one leading cosmetics producer (see Chapter 12), *P. canaliculata* stimulates the synthesis of collagens and proteoglycans, which are responsible for giving connective tissue its elastic properties. Using it as a compound can increase the skin's firmness and reduce the appearance of lines and wrinkles. It is reported from another leading cosmetic company that it increases microcirculation and can help reduce fat and cellulite. It is a source of alginic acid and fucoidan, and the extracts of this species have antioxidant, anticoagulant, antifungal, antifouling, and antibiotic activity (Pereira 2015, Pereira 2016).

Petalonia fascia (O.F. Müller) Kuntze

Common names: see Table 2.1

Description: Thallus consists of erect, dorsoventrally-flattened lamina, arising from a holdfast singly or in clusters; linear or broadly lanceolate to almost elliptical, abruptly or gradually narrowing towards the base and merging into a short stipe, only slightly narrowing at the tip, rounded, and often frayed. Thallus undivided, membranous, and thin when young, later leathery-tough with smooth, sometimes undulating, margins (Braune and Guiry 2015).

Distribution: see Table 2.1

Uses and bioactivities: Used as directly food, and the extracts have antimicrobial and antioxidant activity (Pereira 2015).

Saccharina latissima (Linnaeus) C.E. Lane, C. Mayes, Druehl and G.W. Saunders (Fig. 1.4-N)

Common names: see Table 2.1

Description: Just like most kelps, *S. latissima* has blades (lamina), stipes, and holdfasts, which is attached to substrates. The sporophytes of *S. latissima* have a rich medium-brown color, a long undivided frond without a midrib, and a profusely branched holdfast. Mucilage ducts are absent from stipe; the blade often has two rows of bullations formed in two longitudinal rows parallel to the central axis. The frond of *S. latissima* has a distinctive frilly undulating margin. The stipe of *S. latissima* may be up to 50 cm long (Pereira 2009, Pereira 2010a).

Distribution: see Table 2.1

Uses and bioactivities: Extracts of this species have anticoagulant, antifouling, antimicrobial, and antioxidant activity (Pereira 2015).

Saccharina angustata (Kjellman) C.E. Lane, C. Mayes

Synonym: *Laminaria angustata* Kjellman

Common names: see Table 2.1

Description: Grows as one very long, dark, linear frond arising from a very short stipe; the holdfast is rather small and composed of many branched haptera. The blade itself is over a meter long with wavy edges (Pereira 2016).

Distribution: see Table 2.1

Uses and bioactivities: Used as food in Japan and Russia (Pereira 2016), and its extracts have antitumor, antibacterial, and anticoagulant activity.

Saccharina japonica (Areschoug) C.E. Lane, C. Mayes, Druehl and G.W. Saunders

Synonym: *Laminaria japonica* Areschoug

Common names: see Table 2.1

Description: This species grows as a single blade (reaching 10 m in length) with a short stipe. The holdfast is rather small compared to the overall size of the plant it supports, and composed of irregular haptera. Often multiple plants will grow together from an entangled mass of holdfasts. The blade is entire, tapering towards the tip and rounding out at the base, with the widest portion about 1/3 of the way up the blade. The various subspecies of *S. japonica* vary in blade form specifics (rippled edges or not, etc.), range in color from golden yellow to olive brown, and are even reported to differ in taste (Braune and Guiry 2011).

Distribution: see Table 2.1

Uses and bioactivities: Extracts have cardio-protective, antitumor, antibacterial, anti-allergic, anti-inflammatory, and anticoagulant activity.

Saccharina longicruris (Bachelot de la Pylaie) Kuntze

Common names: see Table 2.1

Description: The hollow, cylindrical stipe itself can reach up to 10 m long, plus the frond which adds another 1 m to 2 m; the large, branched holdfast grips firmly to the rocky substratum allowing a single long, thin, olive-brown, leafy blade to float near the surface. The midsection of these blades is somewhat thicker, but the edges spread and thin and become wide and ruffled (Pereira 2015).

Distribution: see Table 2.1

Uses and bioactivities: *S. longicruris* is an edible species of kelp that is related to traditional Japanese Kombu, but is thinner, more tender, and cooked quickly. It is high in minerals and micronutrients, and is particularly delicious as it contains naturally occurring monosodium glutamate—a little known feature of many kelp species (Pereira 2015); its extracts have antibacterial activity.

Saccharina longissima (Miyabe) C.E. Lane, C. Mayes, Druehl and G.W. Saunders

Synonym: *Laminaria angustata* var. *longissima* (Miyabe) Miyabe

Common names: see Table 2.1

Description: Large, flexible thalli with a rubbery texture; size can range from less than 1 m to several. Color ranges from light brown to dark brown (Pereira 2016).

Distribution: see Table 2.1

Uses and bioactivities: Extracts have antitumor and anticoagulant activity.

Saccorhiza polyschides (Lightfoot) Batters (Fig. 1.4-O)

Common names: see Table 2.1

Description: *S. polyschides* is a kelp species with a distinctive large, warty holdfast and a flattened stipe with a frilly margin. The stipe is twisted at the base and widens to form a large flat lamina, which is divided into ribbon-like sections. The species is an annual, and very fast growing. It is opportunistic and colonizes available hard substrata in the sublittoral (Pereira 2009, Pereira 2010a).

Distribution: see Table 2.1

Uses and bioactivities: Source of alginic acid, and possible source of biofuel. Extracts of this species have hypoglycemic, anti-settlement, cytotoxic, and antiplasmodial activity (Pereira 2015).

Sargassum aquifolium (Turner) C. Agardh

Synonym: *Sargassum binderi* Sonder ex J. Agardh

Common names: see Table 2.1

Description: This large, dark brown plant with golden-brown tips has a long central stem and branches with blades that resemble oak tree leaves. The plant may be a meter or more in length. Blades have toothed edges and the larger ones are rather tough. Leaves at the bottom are large, compared with the ones near the top. The stem also has many small grape-like bladders attached to it, which are filled with air and hold the plant up in the water (Novaczek 2001).

Distribution: see Table 2.1

Uses and bioactivities: It contains high levels of iodine, which prevents goiter. It also has alginic acid, fucoidan, and laminarin substances which act as a preventative medicine for heart disease and stroke. Alginates can help remove poisonous metals such as lead and radioactivity from one's body. Basal parts, rich in alginic acid, can be dried for use as a dressing for cuts and burns (Novaczek 2001, Pereira 2016).

Sargassum filipendula C. Agardh (Fig. 1.5-D)

Common names: see Table 2.1

Description: The habit of *S. filipendula* is so similar to that of other species which have been described that it needs but slight attention. This species grows attached to rocks below low water mark, and therefore, unlike *Fucus* and *Ascophyllum*, is never exposed to the air. Vegetative plants and reproductive plants bearing all stages of conceptacles are plentiful in summer. Sporelings are also abundant and easily collected, for the discharged eggs and their products. The sporelings remain attached for some time by mucilage to the surface of reproductive branches near the parent conceptacles. The stem arises from a small disc-shaped holdfast and passes into long cylindrical branches which bear spirally arranged leaves, berry-like floats, which seem to be modified portions of leaves, as generally stated, and short reproductive branches. This form may attain a height of 60 cm, but is commonly shorter (Pereira 2017n).

Figure 1.5 Marine Brown macroalgae: A–*Sargassum muticum*, B–*Sargassum vulgare*, C–*Taonia atomaria*, D–*Sargassum filipendula*, E–*Undaria pinnatifida*, F–*Zonaria tournefortii*.

Distribution: see Table 2.1

Uses and bioactivities: *S. filipendula* is used in traditional cuisines of much of South America and Asia. The nutrient-rich extracts of *S. filipendula* are also used in cosmetic products from lotions to face masks (Pereira 2016); extracts of this species have antioxidant and anti-proliferative activity.

Sargassum fusiforme (Harvey) Setchell

Synonyms: *Cystophyllum fusiforme* Harvey, *Hizikia fusiformis* (Harvey) Okamura

Common names: see Table 2.1

Description: Several to many stipes arising from a common holdfast area. These form the main axes of the thalli around which short, terete branchlets of variable size grow in whorls; when wet, *S. fusiforme* has a yellow-brown color, but dries to nearly black when removed from water making the once supple, hydrated tissues thick and tough.

Distribution: see Table 2.1

Uses and bioactivities: Japanese people consider Hiziki to be a special health food (Arasaki and Arasaki 1983). In Japan, the young fronds are collected and dried in the sun, in which condition they are a well-known article of merchandise; cooked in soy it is eaten by the peasantry, but not by the better classes (Yendo 1902, Smith 1904). Extracts of this species have cardio-protective, antitumor, and antibacterial activity.

Sargassum hemiphyllum (Turner) C. Agardh

Common name: see Table 2.1

Description: Fronds 80–100 cm high, up to 130 cm. Holdfast composed of irregularly ramifying filamentous rhizoidal, some of which develop as stolon and give rise to a new shoot. These filamentous rhizoidals are 1.5–4 cm long, 1–5 mm wide; main axis terete, smooth, 3–5 cm long, and about 2 mm in diameter. Primary branches giving rise from main axis, terete, slightly compressed and narrower toward apices, 1–2 mm in diameter. Secondary branches spirally giving rise from primary branches, terete to compressed, sometimes with spiny processes, more slender than that of primary branches, less than 1 mm in diameter, alternate, up to 13 cm long, at intervals of 2–4 cm, beset with leaves, vesicles, and receptacles. Leaves coarse, lower leaves oblong-elliptical to oblong-obovate or lanceolate, 1–3 cm long, 5–8 cm wide, margin entire or undulate, bases cuneate and asymmetrical, midrib inconspicuous, absent or not apparent, cryptostomata small and inconspicuous, irregularly scattered on both sides of midrib; upper leaves elongated elliptical, lanceolate or bat-shaped, 1.5–3.5 cm long, 1.5–4 mm wide, apex obtuse or acute, margin entire or undulate, bases cuneate and asymmetrical, midrib inconspicuous, vanishing near the middle part of leaves, one row of small and inconspicuous cryptostomata scattered on both sides of midrib. Vesicles spherical, obovate or elongated elliptical, 3–10 mm long, 2–7 mm in diameter, apex rounded, with sharp tip and (or) auricular tip, margin with or without ridge, pedicel flattened, 3–20 mm long, cryptostomata rarely scattered on the vesicles and pedicels. Plant is dioecious. Female receptacle terete or slightly compressed, 2–3 mm long, 0.8 mm in diameter, margin and upper part, solitary or racemose, stalked (Pereira 2016).

Distribution: see Table 2.1

Uses and bioactivities: Extracts of this species have anti-inflammatory, antioxidant, and immune-stimulating activity.

Sargassum horneri (Turner) C. Agardh

Common name: see Table 2.1

Description: Immature specimens have flat, symmetrical, fern-like blades with notched tips. As the alga grows, it becomes loosely branched in a zigzag pattern, develops small air bladders, and may reach lengths of more than 6 m. Very thick, nearly impenetrable forests may form (Bushing 2014).

Distribution: see Table 2.1

Uses and bioactivities: Used for foodstuff in China, Japan, and Korea, to treat goiter, for animal fodder, and to produce alginates. Their extracts have antiviral, osteoblastogenesis, and osteoclastogenesis activity (Pereira 2016).

Sargassum ilicifolium (Turner) C. Agardh

Common names: see Table 2.1

Description: Thallus axis robust, smooth, brown. Species with particularly large, wide, oval leaflets (phylloides), margins toothed, midrib conspicuous but not reaching the tip of the leaf, hair pits (darker dots) conspicuous, scattered on both sides of the midrib, relatively large; float bladders almost spherical, stiped; receptacles compressed, short wedge-shaped, with terminal teeth (Braune and Guiry 2011).

Distribution: see Table 2.1

Uses and bioactivities: Extracts have cardio-protective, antibacterial, and antifungal activity.

Sargassum johnstonii Setchell and N.L. Gardner

Description: Holdfast crustose or conical (with few haptera) and terete, with smooth primary axes, usually continuous from the base. Secondary branches bifarious in some thalli. Blades narrowly lanceolate and linear or aciculate. Some blades flat, 1.0–2.5 cm in length and 1.0–2.5 mm wide, expanded, ecostate, or rarely faintly midrib, with denticulate to smooth margins. Cryptostomata present, vesicles elliptical with smooth margins. Pedicels shorter than the vesicle length, the apical part of the vesicle (with apiculae or crowns) is blade-shaped. Receptacles cylindrical and branched several times, with few denticulations (thorns). Thalli reach up to 1–2 m in length (Andrade-Sorcia et al. 2014).

Distribution: NE and E Pacific (Baja California), SW Asia (India).

Uses and bioactivities: Extracts have cardio-protective, antibacterial, and antifungal activity.

Sargassum miyabei Yendo

Synonym: *Sargassum kjellmanianum* Yendo

Common names: see Table 2.1

Description: Thallus to 10–20 cm or more in length, with one to a few simple, terete to compressed, stipes 1–20 cm long arising from a discoid-conical holdfast; stipes bearing radially or distichously borne, long primary branches, produced seasonally from the stipe apices and subsequently deciduous, leaving scars or other residues on the stipe. Primary branches 10 cm to 20 cm or more long, distichously, tristichously or radially branched with terete, angular, compressed or three-side axes; the receptacles are distinctly longer than the female ones, often measuring 10–15 mm, or sometimes up to 30 mm, in length, while the receptacles of the female plant measure 3–8 mm in length (Tokida 1954).

Distribution: see Table 2.1

Uses and bioactivities: Used for foodstuff in Japan, especially for the manufacture of potash salts (Tokida 1954), and its extracts have antitumor activity.

Sargassum muticum (Yendo) Fensholt (Fig. 1.5-A)

Common names: Wireweed, Japanese Weed, Jap Weed, Strangle Weed, Tamahahakimoku

Description: *S. muticum* is a large brown seaweed, varying in color from dark brown to pale, yellowish-brown depending on the season and the growing conditions. *S. muticum* has regularly alternating lateral shoots or branches, on a central perennial stem. It attaches to the substrate with a disc-shaped holdfast. It has numerous small 2–3 mm round or pear-shaped air-bladders which sit on small stems and cause the alga to stand upright in the water or float if parts of the alga are detached from the basal stem. *S. muticum* has a frond which may be 75–120 cm long in its native range, but normally reaches a length of 1.5–2 m in Swedish waters, 6–7 m in French waters, and up to 8.5 m in Norwegian waters. Lateral branches detach in the summer or autumn, leaving a short perennial basal stem over winter. During the summer, cigar-shaped reproductive receptacles develop in the areas where the annual shoot or branch attaches to the stem, but may also sit on top of the branch (Pereira 2010a, Pereira 2015).

Distribution: Highly invasive seaweeds, originally from Japan, now colonize large parts of the NE Atlantic (Norway to Portugal, North Sea), the W Mediterranean, N Pacific, NW Pacific (Japan), and NE Pacific (Alaska to Mexico).

Uses and bioactivities: This species is ecologically and economically important seaweed in Asia, which acts as spawning, nursery, and feeding grounds for fishes, shellfishes, and other marine organisms (Tsukidate 1984). On the European and American west coasts, seasonal harvesting has been suggested as a control strategy in shellfish areas and in channels frequented by small motor boats, since eradication by hand removal, physical and chemical methods were unsuccessful (Kraan 2008).

In the last decade, some *Sargassum* species have been used in alginate production, biosorption of toxic heavy metals and successfully tested for pharmacological experiments, encapsulation, and elicitation of plant growth (Larsen et al. 2003, Davis et al. 2003, Mao et al. 2004, Torres et al. 2007, Yabur et al. 2007). In this context and due to its large canopies in Morocco, *S. muticum* could offer a promising opportunity to make use of this biological material for industrial applications while controlling the invasion and preserving autochthonous species (El Atouani et al. 2016).

Source of Laminaran (Pereira 2016c, see also Chapter 2), and extracts have antitumor (Villarreal-Gómez et al. 2010), antibiotic (Glombitza et al. 1982), antioxidant (Le Lann et al. 2008), antialgal (Hellio et al. 2002), antifungal (Peres et al. 2012), and antifouling (Plouguerné et al. 2008, Bazes et al. 2009, Plouguerné et al. 2010, Silkina et al. 2012) activity.

Sargassum natans (Linnaeus) Gaillon

Common names: see Table 2.1

Description: As the common name suggests, Common Gulfweed is the most common *Sargassum* species found in the Sargasso Sea and washed up on Bermuda's beaches. *S. natans* is bushy seaweed with narrow leaf blades which are golden brown with toothed edges. The rubbery-textured leaves range from 2–6 mm wide and 2–10 cm long. The gas-filled floats are less than 6 mm in diameter and are held on short stalks along the stems among the leaves. The floats of *S. natans* have a single protruding spine 2–5 mm long. *S. natans* does not have a single main stem; instead it grows in many directions forming clumps that can reach 60 cm long. It is these clumps that form together into much larger mats (Pereira 2016).

Distribution: see Table 2.1

Uses and bioactivities: Extracts have antiprotozoal, trypanocidal, and leishmanicidal activity.

Sargassum oligocystum Montagne

Common names: see Table 2.1

Description: Fronds 10–60 cm. Holdfast discoid 5–10 mm in diameter; main axis terete, with warty processes, 0.5–1 cm long, 1.5–2 mm in diameter, bearing one to six primary branches. Primary branches flattened, 10–60 cm long, opposite or alternate, 1.5–3.5 mm wide, giving rise to secondary branches from laterals. Secondary branches alternate, 5–20 cm long, shorter lower and upper parts, longer on middle parts, 1.5–3 mm wide. Branchlets giving rise from the axils of secondary branches, alternately arranged, similar to that of primary branches, but more slender, less than 1.5 mm in diameter, 2–4 cm long; leaves coarse, lanceolate or elongated-elliptical, 3–6 cm long, 6–10 mm wide, margins dentate, apices obtuse, bases cuneate, sessile, midrib vanishing below the apex (Pereira 2016).

Distribution: see Table 2.1

Uses and bioactivities: Used as food in Thailand, and its extracts have antitumor, trypanocidal, and leishmanicidal activity.

Sargassum polycystum C. Agardh

Synonym: *Sargassum myriocystum* J. Agardh

Common names: see Table 2.1

Description: Thallus 35 cm tall, with yellowish-brown color, attached with discoid holdfast; main axis cylindrical and rough due to the presence of large outgrowth, supporting alternately arranged branches bearing leaves and vesicles; in young thalli leaves are longer and broader, measuring 13–42 mm long including the stalk and 2.5–11.5 mm wide; leaves are generally oblong slightly tapered, retuse (slightly rounded) or emarginate at the tip finally serrated throughout the margin; mature thalli fewer leaves smaller, 7–15 mm long including stalk and 17–40 mm wide, oblanceolate, oblong with tapered bases, the apices are rounded, obtuse to acute outer margin is coarsely serrate; prominent midrib at a short distance from apex of the leaves; cryptostomates are scattered on the surface of the blade. Pedunculate vesicles are ovate or spherical with a diameter of 1.5–3 mm; these are tipped with spinose or thin leaf-like extensions and with few cryptostomates; vesicles may be solitary or may form clusters attached to the primary or secondary branches, and are more numerous but smaller in mature thalli (Guiry and Guiry 2017).

Distribution: see Table 2.1

Uses and bioactivities: Used for alginate production, animal feed, fertilizer, and for medical purposes; extracts of this species have antioxidant, hypoglycemic, and antibacterial activity.

Sargassum siliquastrum (Mertens ex Turner) C. Agardh

Synonym: *Sargassum tortile* (C. Agardh) C. Agardh

Common names: see Table 2.1

Description: Thallus up to 10–20 cm or more in length, with one to a few simple, terete to compressed, stipes 1–20 cm long arising from a discoid-conical holdfast; stipes bearing radially or distichously borne, long primary branches, produced seasonally from the stipe apices and subsequently deciduous, leaving scars or other residues on the stipe. Primary branches 10 cm to 20 cm or more long, distichously, tristichously or radially branched with a terete, angular, compressed or three-sides axes; basal laterals simple or branched, compressed and relatively narrow (in most species) leaf-like, 3–15 mm broad, entire or with dentate margins; upper laterals usually branched, with slender, compressed to terete, ramuli (Womersley 1987).

Distribution: see Table 2.1

Uses and bioactivities: In Korea it is used for foodstuff, and for goiter treatment; extracts have antitumor and cytoprotective activity.

Sargassum swartzii C. Agardh

Synonym: *Sargassum wightii* Greville ex J. Agardh

Common name: see Table 2.1

Description: The plants have smooth, flat branches, leaves with sparse serrations. The vesicles are oval to elliptical without spines. This seaweed is 1–1.5 m high, and found on rocks in subtidal zones along shorelines with moderate wave activity (Pereira 2016).

Distribution: see Table 2.1

Uses and bioactivities: Extracts of this species have antiviral, larvicidal, antioxidant, and anti-cholinesterase (Pereira 2016) activity.

Sargassum thunbergii (Mertens ex Roth) Kuntze

Common names: see Table 2.1

Description: Thallus upright, cartilaginous-firm, fronds with a conical outline, the smooth axes are laterally covered with the genus-characteristic cone-shaped phylloid-complexes; which may have embedded float bladders. This plant shows a high degree of morphological differentiation, having a perennial holdfast, a stipe, branches, leaves, and vesicles. Different from other species belonging to the genus *Sargassum*, *S. thunbergii* shows a distinctive feature in the branch systems. Every primary branch develops to an individual-like main axis (Braune and Guiry 2011).

Distribution: see Table 2.1

Uses and bioactivities: In Japan and Korea it is used for foodstuff, especially for the manufacture of potash salts (Tokida 1954, Kang 1968), as animal fodder, and manure (Arasaki and Arasaki 1983). Extracts of this species have vermifuge, antibacterial, antitumor, anti-allergic, anti-inflammatory, and hypocholesterolemic activity.

Sargassum vulgare C. Agardh (Fig. 1.5-B)

Common names: see Table 2.1

Description: *S. vulgare* has a bush-like thallus reaching 150–700 mm high; the fronds are oval, flattened, olive-green to brown, and possess a central rib and undulate edge. The base of the fronds has hollow, spherical vesicles, of 5–8 mm, and clusters of reproductive bodies; these are held in place by a pedicle. The alga is attached to the substrate with irregular rhizoidal branches. *S. vulgare* has an olive brown to dark brown coloration (Pereira 2017o).

Distribution: see Table 2.1

Uses and bioactivities: Source of alginic acid, and the extracts of this species have anthelminthic, antimicrobial, antifungal, antitumor, antilipemic, antifouling, antimicrobial, anticoagulant, antithrombotic, and anti-inflammatory activity (Pereira 2015).

Scytosiphon lomentaria (Lyngbye) Link

Common names: see Table 2.1

Description: Light to dark brown, sometimes hollow, cylindrical seaweed, that grows up to 33 cm tall. Fronds may occur in groups or singly, are unbranched to 2.3 mm wide, tapering at both ends and arising from a short stipe and small attachment disc (Neto et al. 2005).

Distribution: see Table 2.1

Uses and bioactivities: Used directly as food, and the extracts have antioxidant, antiviral, antimicrobial, antitumor, and antifouling activity (Pereira 2015).

Spatoglossum asperum J. Agardh

Description: Plants are dark to dirty green in color with indistinct holdfast. Thallus is 20–35 cm or more in height. The branches are strap-like, dichotomously divided with large and small lobes. The lobes are elongate, linear lanceolate, attenuated towards base. Apex is acute or rounded, margin sinuate, irregularly dentate with larger or smaller proliferation. Surface is smooth and its cells are arranged in more or less distinct rows. Thallus has many epiphytes among which *Melobesia* is common. This seaweed is found on rocks in the subtidal zone of shores exposed to strong wave action (Sahoo 2010).

Distribution: W Atlantic, Tropical and Subtropical W Atlantic, Indian Ocean, Asia, SE Asia and SW Asia, Australia, and Pacific Islands.

Uses and bioactivities: Extracts have cardio-protective, antitumor, antibacterial, antifungal, and nematicidal activity.

Spatoglossum schroederi (C. Agardh) Kützing

Common name: see Table 2.1

Description: Thalli erect, arising from a matted rhizoidal holdfast, up to 80 cm long, complanate, divided into sub-dichotomous to sub-palmate segments, 0.5–5 cm broad, with undulate to dentate margins, lacking midrib or veins. Growth initiated from a short row of apical cells (Pereira 2016).

Distribution: see Table 2.1

Uses and bioactivities: Extracts have antitumor, antithrombotic, and anticoagulant activity.

Stoechospermum polypodioides (Lamouroux) J. Agardh

Synonym: *Stoechospermum marginatum* (C. Agardh) Kützing

Description: Thalli are erect, racemose, repeatedly dichotomously branched, yellowish-brown, 4.1–28 cm in height, and attached to the substratum by means of a fibrous holdfast. Branches are obtriangular, loosely twisted, apical margins involute, internodal segments 1–5 cm long and 0.2–1.5 cm broad, margins entire, without rhizoids, angles between the branches wider in the lower portion (about 110°) and abruptly narrower in the upper portion (about 40°). Plants possess marginal growth (Kyaw et al. 2009).

Distribution: Indian Ocean, South Africa, Tanzania, Kenya, Madagascar, Somalia, Ethiopia and Egypt, Red Sea, Yemen, Oman, Pakistan, and SE Arabian coast, India, Ceylon (Sri Lanka), Mauritius, Myanmar, Indo-Pacific Oceans (Malaysia), Pacific Ocean (California), and Australia.

Uses and bioactivities: Extracts have antiviral, antibacterial, antifungal, nematicidal, and anticoagulant activity.

Stypopodium zonale (Lamouroux) Papenfuss

Common name: Leafy Flat-Blade Alga

Description: Bushy plant formed by flat, squared-off blades that are irregularly branched and split; up to 30 cm high. Concentrically banded in a wide range of colors, including shades of yellow, green, and brown. Often with tints of iridescent green and/or blue (De Kluijver et al. 2017).

Distribution: Caribbean Islands, Atlantic Islands (Bermuda, Canary Islands, Cape Verde Islands, Selvage Islands), W Atlantic, Tropical W Atlantic, SW and SE Asia, Australia, and Pacific Islands.

Uses and bioactivities: Extracts have antiviral, antibacterial, and antifungal activity.

Taonia atomaria (Woodward) J. Agardh (Fig. 1.5-C)

Synonym: *Dictyota ciliata* Lamouroux

Common name: see Table 2.1

Description: Thalli erect, attached by matted, branched rhizoids, up to 30 cm long, complanate, flabellate or lacerate with many elongate, cuneate branches, 0.5–6 cm broad, often tapering to the apex. Growth initiated from a marginal row of apical cells. Thallus two cells thick near apex, increasing to 5–7 cells thick towards the thallus base, not arranged in rows in transverse section. Hairs in concentric lines across the thallus. Sporangia solitary, in irregularly scattered groups or in short concentric zones adjacent to the hair lines, partially embedded in or completely external to the frond, without a stalk cell or with 1–3 stalk cells, 80–140 mm high when mature, producing four spores. Gametophytes dioecious. Antheridial sori whitish, irregularly shaped, scattered over both thallus surfaces, 100–800 mm across, bounded by 1–2 layers of sterile cells. Antheridia with 15–20 tiers of locules, with one stalk cell (Pereira 2017p).

Distribution: see Table 2.1

Uses and bioactivities: Used in food supplements and cosmetics, and its extracts have antioxidant and antibacterial activity.

Turbinaria conoides (J. Agardh) Kützing

Common names: see Table 2.1

Description: The holdfast is branched in this species, and there are numerous secondary branches. The thin triangular leaves are concave at the center and have single margins with large sharp serrations. The plants grow on rocks in subtidal zones, along relatively calm shorelines with water movement (Santhanam 2015).

Distribution: see Table 2.1

Uses and bioactivities: Extracts have cardio-protective, antiviral, antitumor, antifungal, antibacterial, anti-inflammatory, larvicidal, and pupicidal activity.

Turbinaria decurrens Bory

Common names: see Table 2.1

Description: Plants yellowish-brown or brown in color; 20–25 cm in height; possesses elongated cylindrical branches; leaves possessing angular margins (Kaliaperumal et al. 1995).

Distribution: see Table 2.1

Uses and bioactivities: Extracts have cardio-protective, antiviral, antifungal, and antibacterial activity.

Undaria pinnatifida (Harvey) Suringar (Fig. 1.5-E)

Common names: see Table 2.1

Description: *U. pinnatifida* or Wakame is a large brown kelp with a branched holdfast giving rise to a stipe. Just above the holdfast, the stipe has very wavy edges, giving it a corrugated appearance. The stipe gives rise to a blade that is broad, flattened, and lanceolate. It has a distinct midrib. The margins of the blade are wavy. Plants can reach an overall length of 1 m to 3 m. *U. pinnatifida* is an annual species with two separate life stages (Pereira 2015).

Distribution: see Table 2.1

Uses and bioactivities: *U. pinnatifida* is economically important as a food crop, next to Nori, on the Japanese menu, and is eaten both dried and fresh. In East Asian countries, the seaweed is known as Wakame and is treated as a delicacy, often added to miso soup. Extracts of this species have antihypertensive, immunomodulating, antidiabetic, antiviral, cytotoxic, antioxidant, antitumor, antiedema, antiplasmodial, anti-osteoporotic, anti-inflammatory, anti-obesity, antihypertensive, and antithrombotic activity (Pereira 2015).

Zonaria tournefortii (J.V. Lamouroux) Montagne (Fig. 1.5-F)

Description: Thallus erect, with flat lobed sections divided into wedge-shaped segments, often incised and proliferating; lamina inconspicuously banded by distant, concentric rows of hairs which are parallel to the distal margin; additionally, delicate lines run radially from the base to the upper margin; margins not inrolled, lamina not calcified; basally thickened, stipe-like, branching, stipe extending like a midrib into the segments; cushion-like rhizoidal network (Braune and Guiry 2011).

Distribution: Warm NE Atlantic (Madeira, Canary Islands, W Africa), the Mediterranean, the Caribbean, SW Atlantic (Brazil), and SW Indian Ocean (South Africa).

Uses and bioactivities: Extracts have antibacterial activity.

1.3.3 Rhodophyta (red algae)

Acanthophora spicifera (M. Vahl) Børgesen

Common names: see Table 2.1

Description: *A. spicifera* has a large, irregularly-shaped holdfast for attachment to hard bottoms. From the holdfast, erect fronds begin to branch out. The main branches have short, determinate branchlets that are irregularly-shaped and spinose. Branchlets are hook-like, brittle and fragment easily under heavy wave action. Color is highly variable, and can be shades of red, purple, or brown (Littler et al. 1989). *A. spicifera* grows upright to approximately 25 cm.

Distribution: see Table 2.1

Uses and bioactivities: Extracts have cardio-protective, antiviral, antitumor, antifungal, and antibacterial activity.

Amphiroa fragilissima (Linnaeus) Lamouroux

Common names: see Table 2.1

Description: Thalli with dense cushion-like tufted growth, very brittle to strong calcification; branches cylindrical, thin, segmented, rather regularly forking, sometimes also trichotomous, the angles between two fork branches usually rather wide (broadly Y-shaped); the segments are slightly swollen at the endings; yellowish-red to whitish-pink. *A. fragilissima* is extremely fragile, as the name implies. The calcified branches will often crack and break upon collection and handling. This species could be confused with the similarly sized *Jania* spp. (Braune and Guiry 2011).

Distribution: see Table 2.1

Uses and bioactivities: Used on functional foods, and the extracts have antiviral, antibacterial, cytotoxic and antioxidant, oxytocic, and espamogenic activity (Pereira 2015).

Asparagopsis armata Harvey (Fig. 1.6-A)

Common names: see Table 2.1

Description: In north-eastern Europe, gametophyte plants occurring from June or July–August or September (sometimes overwintering), pale purplish-red, quickly degenerating when removed from the water and becoming distinctly orange; fronds bushy, with a cylindrical axis up to 1 mm wide and 200 mm long, arising from bare, creeping stolons; irregularly branched, with four rows of branchlets, simple, short, branchlets alternating with longer ones with four rows of simple filamentous ramuli. Lower branchlets unbranched, long, tapered, with harpoon-like barbs. Tetrapsorophyte ("*Falkenbergia*-phase") occurring all year round, but most obvious in October–March, brownish-red, much branched, filamentous, in dense cotton-wool-like tufts up to 15 mm in diameter (Pereira 2017q).

Distribution: see Table 2.1

Uses and bioactivities: *A. armata* extract is a powerful antioxidant with antibacterial qualities and is a valued ingredient in many cosmetic products. Presents strong cytotoxicity against human cancer cell lines. *A. armata* is also harvested or grown to produce phycocolloid. Extracts have anti-Leishmania, antioxidant, antiviral, antifungal, and antimicrobial activity (Pereira 2015).

Asparagopsis taxiformis (Delile) Trevisan (Fig. 1.6-B)

Common names: see Table 2.1

Description: Thallus fluffy, fine, filamentous creeping mats or tufts, up to 4 cm high, pale red to gray-pink; branching irregular to alternate. Branches cylindrical, occasionally moniliform (with segments swollen or bead-like), 30–80 µm in diameter, central axial filament surrounded by three pericentral cells; cells commonly pointed at tips, twice as long as broad, each set rotated approximately 60°; apex with single prominent apical cell cutting off lens-shaped cell basally; holdfast initially disc-like, later becoming

Figure 1.6 Marine Red macroalgae: A–*Asparagopsis armata*, B–*Asparagopsis taxiformis*, C–*Boergeseniella thuyoides*, D–*Calliblepharis jubata*, E–*Callophyllis laciniata*, F–*Ceramium virgatum*, G–*Champia parvula*, H–*Chondrcanthus acicularis*, I–*Chondracanthus teedei* var. *lusitanicus*, J–*Chondria dasyphylla*, K–*Chondrus crispus*, L–*Cryptopleura ramosa*.

branched, tangled, creeping, forming filamentous mass. Tetrasporangia solitary on outer filaments, not in groups or series, formed from one pericentral cell of segment. Fluffy appearance and shaped like a Christmas tree. Grows 3–15 cm high (Braune and Guiry 2011).

Distribution: see Table 2.1

Uses and bioactivities: Extracts have antifouling, anti-cyanobacterial, antifungal, anticoagulant, and antibacterial activity (Pereira 2015).

Boergeseniella thuyoides (Harvey) Kylin (Fig. 1.6-C)

Common name: see Table 2.1

Description: Cylindrical, cartilaginous, tufted, deep brownish-purple fronds, up to 150 mm high, from creeping rhizoidal base; fronds distichously bi-tripinnate, patent, short, of nearly uniform length giving branches a linear appearance; ramuli short, spine-like; polysiphonous, central siphon with 8–12 pericentral siphons and outer cortication of small, colored cells; articulations as broad as long, barely visible (Pereira 2015).

Distribution: see Table 2.1

Uses and bioactivities: Extracts of this species have antiviral and antibacterial activity (Pereira 2015).

Bonnemaisonia hamifera Hariot

Common name: Bonnemaison's Hook Weed, Pink Cotton Wool

Description: Gametophyte plants occurring from March–June, brownish-red, fronds feathery, with a slightly flattened axis up to 1 mm wide and 350 mm long, attached to *Cytstoseira* and other algae by crosier-shaped, hook-like modified branches. Tetrasporophyte (*Trailliella* phase) plants occurring all year, but most obvious in October–March, brownish-red, much branched, filamentous, in dense cotton-wool-like tufts up to 25 mm in diameter (Pereira 2015).

Distribution: NE Atlantic (Scandinavia to Canary Islands), SE Atlantic (South Africa), the Mediterranean, NW Pacific (Russia, Japan), and NE Pacific (California, Mexico).

Uses and bioactivities: Extracts of this species have antibacterial, antioxidant, and antihypertension activity (Pereira 2015).

Brongniartella byssoides (Goodenough and Woodward) F. Schmitz

Description: Soft, tufted, deep purplish red fronds, up to 300 mm long; main axis well-defined, bi- or tripinnate, bearing alternate, distichous branches. Branches and branchlets articulated, with central siphon and 5–7 pericentral siphons, clothed with short, slender, repeatedly dichotomously branched, monosiphonous ramuli. Its color is light red brown to almost black when dry (Braune and Guiry 2011).

Distribution: NE Atlantic (Scandinavia to Portugal, North Sea, E Baltic Sea) and Mediterranean Sea.

Uses and bioactivities: Great antioxidant potential; strong cytotoxicity against human cancer cell lines (Pereira 2015).

Bryothamnion triquetrum (S.G. Gmelin) M. Howe

Common name: see Table 2.1

Description: Thallus coarse and bushy, up to 25 cm high; dark brown to red; branches numerous, irregularly alternate; branchlets stiff up to 3 mm long, in three vertical rows, creating triangular branches; apices of branches incurved and pointed; axes polysiphonous with 7–9 pericentral cells, heavily corticated, and generally triangular in transverse section (Pereira 2016).

Distribution: see Table 2.1

Uses and bioactivities: Extracts have antitumor, antibacterial, and neurological activity.

Calliblepharis jubata (Goodenough and Woodward) Kützing (Fig. 1.6-D)

Description: *C. jubata* is brownish-red in color. It has a thallus consisting of a branched holdfast that gives rise to an erect frond that expands into a dichotomous or irregularly divided blade. The outline of the frond is variable but it commonly has a cylindrical or very slightly compressed stipe. Its blades are about 6 mm broad and 30 cm long with narrow branches. The branches appear long and tendril-like. Long branchlets (proliferations) arise from the blade surface and margins of the branches (Pereira 2009, Pereira 2010a).

Distribution: NE Atlantic (Ireland to Mauritania) and W Mediterranean.

Uses and bioactivities: Source of carrageenan (Pereira et al. 2009, Pereira and van de Velde 2011, Pereira 2013), and its extracts have hemagglutinins, antimycobacterial, anticoagulant, antiprotozoal, and cytotoxic activity (Pereira 2015).

Callophycus serratus (Harvey ex Kützing) P.C. Silva

Common name: see Table 2.1

Description: This plant is named by the author as large wire weed because it is stiff and rubbery. The branching is opposite, like on a feather and the dark red branches are flattened. It grows to be quite large (10–20 cm tall) (Novaczek 2001).

Distribution: see Table 2.1

Uses and bioactivities: Extracts have antifungal, antibacterial, antioxidant, antiplasmodial, and neurological activity.

Callophyllis laciniata (Hudson) Kutzing (Fig. 1.6-E)

Common names: Table 2.1

Description: A short soft, fleshy thallus arises and expands immediately from a small round (discoid) holdfast. The stipe is inconspicuous or absent and the crimson, brownish or purple-red fronds are much divided spreading out into wedge-shaped divisions of about 1–3 cm broad. The whole blade can be up to 15 cm long and fan-shaped with overlapping sections. Frond tips are rounded and end in tiny marginal leaflets giving a fringed appearance (Pereira 2009, Pereira 2010a).

Distribution: Table 2.1

Uses and bioactivities: Extracts have antioxidant and cytotoxic activity (Zubia et al. 2009a).

Callophyllis variegata (Bory) Kützing

Common names: see Table 2.1

Description: Anchored to rocky substrates by a discoid holdfast. The thallus, typically made up of flat, sometimes deeply and irregularly divided branches and glossy or semi-glossy blades (leaf-like structures), extends upward from the holdfast, sometimes supported on a short stipe (stem-like stalk). The blades, which lack midribs and veins, have large cells in the medulla (central region) and ragged or smooth margins. *Callophyllis* have fleshy, flexible branches and blades, though some (particularly intertidal specimens) may be brittle. Most species grow between 5 cm and 30 cm in height, though some are slightly smaller or larger (Pereira 2016).

Distribution: see Table 2.1

Uses and bioactivities: *C. variegata*, harvested off the southern coast of Chile, is popular edible seaweed; consumed in cooked or dried and rehydrated form. In Chile, the demand for edible seaweeds has increased and *C. variegata* (Carola) is one of the most popular (McHugh 2003, Pereira 2016). Extracts have antiviral (see Chapter 5) activity.

Centroceras clavulatum (C. Agardh) Montagne

Common names: see Table 2.1

Description: Thalli are filamentous turfs, usually in entangled mats, 3–5 cm in height, and bright red to dark brown in color. Erect branches are un-branched or sub-dichotomously branched with straight, incurved apices. Axial cells are completely corticated with covering axial cells, 15–18 μm in diameter arranged in one layer, rectangular in shape, between successive segments. The covering cortical cells at nodes are usually in two layers and with spines. Tetrasporangia are borne on outer nodes, 45–55 μm in diameter in whorls (Pereira 2016).

Distribution: see Table 2.1

Uses and bioactivities: Extracts have antiviral, antitumor, antifungal, antibacterial, nematicidal, and anti-*Leishmania* activity.

Ceramium virgatum Roth

Synonym: *Ceramium rubrum* C. Agardh

Common names: see Table 2.1

Description: Small red seaweed growing up to 30 cm tall. It has a filamentous frond that is irregularly and dichotomously branched, with the branches narrowing toward pincer-like tips. The holdfast is a minute conical disc that extends into a dense mass of rhizoidal filaments. The plant is reddish-brown to purple in color and has a banded appearance when viewed closely (Pereira 2009, Pereira 2010a).

Distribution: see Table 2.1

Uses and bioactivities: *C. virgatum* is used as an extract for cosmetic products, produces an agar-type polysaccharide, and has antiviral, antibacterial, antioxidant, antialgal, antiprotozoal, antimycobacterial, and cytotoxic activity (Pereira 2015).

Champia parvula (C. Agardh) Harvey

Common names: see Table 2.1

Description: Soft, gelatinous, pinkish red, much branched fronds, densely matted, with blunt apices, up to 100 mm high. Axes segmented, with nodal diaphragms, segments about broad as long, filled with watery mucilage (Braune and Guiry 2011).

Distribution: see Table 2.1

Uses and bioactivities: Extracts have anticoagulant and antiherpetic activity (Pereira 2015).

Chondracanthus acicularis (Roth) Fredericq (Fig. 1.6-H)

Synonym: *Gigartina acicularis* (Roth) J.V. Lamouroux

Common names: see Table 2.1

Description: Cartilaginous, cylindrical, or compressed, purple-red or blackish fronds, sometimes with greenish or whitish spots, up to 100 mm long, irregularly bipinnately branched, branches curved, sharply pointed (Pereira 2009, Pereira 2010a).

Distribution: see Table 2.1

Uses and bioactivities: Source of carrageenan (Pereira et al. 2009, Pereira and van de Velde 2011, Pereira 2013). Extracts have antiviral, antibacterial, antioxidant, antifungal, and anticoagulant activity (Pereira 2015).

Chondracanthus teedei var. *lusitanicus* (Rodrigues) Bárbara and Cremades (Fig. 1.6-I)

Synonym: *Gigartina teedei* var. *lusitanica* J.E. De Mesquita Rodrigues

Common names: Table 2.1

Description: The fronds of this alga are cartilaginous-membranous, with purple-violet color that darkens by desiccation, becoming greenish-yellow with decay. The main axes of the fronds, as their ramifications are wide, reaching 1 cm in the older portions. This species is confused sometimes with *Calliblepharis jubata* (Pereira 2009, Pereira 2010a).

Distribution: Table 2.1

Uses and bioactivities: Source of carrageenan (Pereira et al. 2009, Pereira and van de Velde 2011, Pereira 2013), and its extracts have antifungal activity.

Chondria dasyphylla (Woodward) C. Agardh (Fig. 1.6-J)

Common name: Table 2.1

Description: Thalli of cylindrical erect axes or decumbent tufts, brownish-red or yellowish, 8–15 cm high, up to 1 mm diameter, tubular, coarse-fibrous, attached by basal holdfast; main axes distinct, richly branched, branching sparsely at irregular intervals; branchlets multiple, 3–20 mm long, 0.5–1 mm in diameter, simple or brachiated, often growing in small clusters (Milchakova 2011).

Distribution: Table 2.1

Uses and bioactivities: Extracts of this seaweed have larvicidal, antibacterial, antifungal, antitumor, antiprotozoal, antiviral, antifertility, and hypoglycemic activity (Milchakova 2011, Pereira 2015, Pereira 2016).

Chondrophycus brandenii (Saito and Womersley) Nam

Synonym: *Laurencia brandenii* Saito and Womersley

Description: Thallus red to red-brown, soft, drying adherent to paper, 4–8 cm high, with one to several axes bearing laterals irregularly radially for 3–4 orders, branches terete to larger ones slightly compressed; axes 3–4 mm in diameter, laterals 1.5–2 mm and ultimate ramuli 0.8–1.5 mm in diameter. Discoid holdfast with 1–3 mm across. Species epilithic, fixed on shells, or epiphytic on Amphiboles. Internal structure: Epidermal cells isodiametric and rounded, 15–25 µm across near apices, without secondary pit-connections and corps; in section, obconical to rounded, 20–35 µm across and long, with spaces below epidermal and between cortical cells which lack lenticular thickenings. Cells uninucleate, larger multinucleate; rhodoplasts discoid, ribbon-like in larger cells (Womersley 2003).

Distribution: SW Asia and Australia.

Uses and bioactivities: Extracts have antifungal, antibacterial, and insecticidal activity.

Chondrus crispus Stackhouse (Fig. 1.6-K)

Common names: see Table 2.1

Description: Highly variable (polymorphous) thalli may reach 15 cm long, cartilaginous consistency and reddish-pink or brown color and iridescent in water. These algae are fixed by a disc that starts as an unbranched stipe gradually expanding into fan-like blade, repeatedly dichotomously divided, with ends rounded or truncated. On the surface of the blades may appear small dilations (2–3 mm in diameter), which are the reproductive structures (Pereira 2009, Pereira 2010a).

Distribution: see Table 2.1

Uses and bioactivities: The gelling and thickening properties of carrageenan are used widely in the cosmetics, food, and pharmaceutical industries; boiled with milk, sugar, and spices as a pudding or healthy drink (Pereira 2016). Examples of applications include making ice cream and air fresheners, beer clarification, and treatment for coughs and diarrhea (Pereira 2015).

Chondrus ocellatus Holmes

Common names: see Table 2.1

Description: This species is about 5–8 cm high, twice or thrice-forked from near the base and the segments are divergent at a wide angle, varying from 0.6–0.8 cm in diameter, with rounded axils. At the base the frond tapers to a point, and the branches are slightly constricted at intervals. The cystocarps are surrounded with a raised ring, giving them an ocellate appearance (Holmes 1895).

Distribution: see Table 2.1

Uses and bioactivities: Used for production of carrageenan, food tranquilizer, homogenizer, and to treat intestinal disorders (Pereira 2016).

Coccotylus brodiei (Turner) Kützing

Synonym: *Phyllophora brodiaei* (Turner) Endlicher

Description: The repeatedly branched leafy bushes are usually up to 15 cm high, but can reach 33 cm. They are red-brown, lightly colored in new parts and darker in older parts. The cylindrical stem is as a rule relatively long and robust. Branches, in the form of stipitate leaves, spring from all parts of the plants. Branches springing from the leaves usually grow from the rim and most characteristically from the apex, but at times also from the flat side of the blade. Leaf growth is arrested at the end of the growing season and new growth resumes as stipitate leaves. Often the stipitate leaves occur in a row. Occasionally the stem or stipes of long branches are merely intermittently flattened. The leaf growth rarely continues without a stipe. The transition from stipe to blade is gradual with a wedge-shaped leaf widening from the stipe, or the basal part can be band-shaped. The blades are typically dichotomously divided into two lobes attenuated to a pointed apex. However, often three or more lobes are present (Lundsteen and Nielsen 2015).

Distribution: Artic (White Sea), NE Atlantic, and Atlantic Islands (Greenland).

Uses and bioactivities: Extracts have antifungal, antibacterial, and anticoagulant activity.

Corallina officinalis Linnaeus

Common names: see Table 2.1

Description: Whitish pink to lilac, calcified, articulated fronds, 60–120 mm high, axis cylindrical to compressed, repeatedly pinnate from and expanded discoid base, branching often irregular. Growth form very variable often stunted. In unfavorable habitats, erect system vestigial, but extensive base may be present (Pereira 2009, Pereira 2010a).

Distribution: see Table 2.1

Uses and bioactivities: *C. officinalis* is a very popular ingredient among cosmetics, health, and personal care companies. There are known sellers of *C. officinalis*-based products in the United States, China, Italy, France, Switzerland, and Germany. Extracts have anthelmintic, antibacterial, and antioxidant activity (Milchakova 2011, Pereira 2015).

Coronaphycus elatus (C. Agardh) Metti

Synonym: *Laurencia elata* (C. Agardh) J.D. Hooker and Harvey

Description: Plants large and robust, up to 40 cm in height, branching alternate or subopposite and in one plane, axes evenly compressed, cartilaginous, denuded in lower branches, often branching in triads at < 45-degree angles between ultimate branchlets and the supporting branch (Metti 2012).

Distribution: Australia and New Zealand.

Uses and bioactivities: Extracts have antiviral and antibacterial activity.

Cryptonemia crenulata (J. Agardh) J. Agardh

Description: Foliose, linear-lanceolate to broad and lobed, entire, proliferous or dissected into numerous cuneate, lanceolate or obovate lobes, occasionally laciniate, usually with distinct stalk or stipe, continuing as a midrib in some species, branched or unbranched, and with small discoid holdfast (Pereira 2016).

Distribution: see Table 2.1

Uses and bioactivities: Extracts have antiviral activity.

Cryptopleura ramosa (Hudson) L. Newton (Fig. 1.6-L)

Description: A short thallus with a stout midrib arises from a discoid holdfast, widening as it branches into flattened red-brown or red-purple fronds to a height of 20 cm. The fronds are thin and membranous, around 2.5 cm in width and may have a slight blue iridescence under water. Repeated branching gives it a bushy tangled appearance with the branches tapering to rounded tips. Morphology is variable and blades can be either erect or prostrate and broadly wedged or strap-shaped. Margins may be smooth, undulating, and denticulate or hooked, frequently with a blue iridescence underwater (Hughes 2017).

Distribution: NE Atlantic (Scandinavia to Canary Islands), SW Atlantic (Brazil, Uruguay), and Mediterranean Sea.

Uses and bioactivities: Extracts have antifouling, antiviral, algicidal, and anti-*Trypanosoma* activity (Pereira 2015).

Cystoclonium purpureum (Hudson) Batters (Fig. 1.7-A)

Common name: see Table 2.1

Description: Rather soft, cylindrical, dull-purplish pink fronds, 3 mm wide, up to 600 mm long. Branches numerous, alternate, branchlets tapered at both ends; branches sometimes drawn out into long twisting tendrils; multiaxial, medulla a cordlike strand of loosely interwoven, narrow filaments, surrounded by large, rounded cells, with outer layer of small, angular, assimilatory cells; said to have an onion-like smell shortly after collection (Pereira 2015).

Distribution: see Table 2.1

Uses and bioactivities: Source of carrageenan; its extracts have nematocidal, antibacterial, and anticoagulant activity (Braune and Guiry 2011, Pereira 2015).

Delesseria sanguinea (Hudson) J.V. Lamouroux (Fig. 1.7-B)

Common name: Sebeech, Sea Beech

Description: A conspicuous crimson seaweed up to 30 cm in length. Blades are oval or lanceolate, leaf like and reminiscent of beech leaves. The membranous lamina has a wavy margin and is supported by a conspicuous midrib with opposite pairs of lateral veins. The irregularly shaped, thickened holdfast (about 0.5 cm in diameter) gives rise to a short cylindrical stipe about 1 cm long. The stipe branches sparingly giving rise to spirally arranged blades (about 1.5–4 cm wide). The leaves may be pointed in young specimens. In autumn, the membranous lamina is lost so that only the midrib remains (Tyler-Walters 2017).

Distribution: NE Atlantic (Scandinavia to Portugal, North Sea, Baltic Sea) and the Mediterranean (sporadically).

Uses and bioactivities: *D. sanguinea* is used in the cosmetics industry for its anticoagulant properties and vitamin K content; the active principle being termed Delesserine; also have anti-inflammatory and anti-aging skin activity (Pereira 2015).

Dilsea carnosa (Schmidel) Kuntze (Fig. 1.7-C)

Common name: see Table 2.1

Description: *D. carnosa* has large and fleshy fronds with no veins, easily identifiable by its 15 cm to 30 cm long and 5 cm to 20 cm wide spoon-shaped blades. It presents a beautiful dark reddish blood-color with its fronds normally describing a perfect single tear. *D. carnosa* settles to substrate through an ovoid-shaped disc (approximately 30 mm to 40 mm length and 20 mm to 30 mm wide) from which emerge several fronds through small cylindrical stipes [30, 59, 60, 120]. Upon touch its fronds resemble a hard, flat, and thick leather (*carnosum*, from the Latin: fleshy, thick), as well as a similar odor out of water. In their juvenile stages *D. carnosa* thalli has a well-defined ligular shape. At older ages *D. carnosa* may present vertical slits or circular cutting sections on its fronds (rarely deviating from its original form). *D. carnosa* may also present yellowing tonalities in the apical end of its thallus, also characteristic of late stages of development (Marques and Pereira 2016).

Distribution: see Table 2.1

Uses and bioactivities: Produces a sulfated polysaccharide of the lambda-carrageenan family; extracts have anti-settlement, antiprotozoal, antifungal, and antifouling activity (Pereira 2015, Marques and Pereira 2016).

Figure 1.7 Marine Red macroalgae: A–*Cystoclonium purpureum*, B–*Delesseria sanguinea*, C–*Dilsea carnosa*, D–*Ellisolandia elongata*, E–*Dumontia contorta*, F–*Eucheuna denticulatum*, G–*Furcellaria lumbricalis*, H– *Gelidium pulchellum*, I–*Gelidium spinosum*, J–*Gigartina pistillata*, K–*Grateloupia filicina*, L–*Gracilaria gracilis*.

Dumontia contorta (S.G. Gmelin) Ruprecht (Fig. 1.7-E)

Synonym: *Dumontia incrassata* (O.F. Müller) J.V. Lamouroux

Common names: see Table 2.1

Description: Erect thallus, cylindrical when young, compressed when older, tubular-hollow, brownish-red to crimson-purple, often yellow-brown at the tips. Thallus with irregular simple lateral branches, tapering at both ends, sometimes slightly swollen, often twisted around the longitudinal axis; soft, gelatinous. Holdfast a small, persistent disc, expanding with age (Braune and Guiry 2011).

Distribution: see Table 2.1

Uses and bioactivities: Extracts have antiviral and antioxidant activity (Pereira 2015).

Ellisolandia elongata (J. Ellis and Solander) K.R. Hind and G.W. Saunders (Fig. 1.7-D)

Synonyms: *Coralina elongata* J. Ellis and Solander, *Corallina mediterranea* Areschoug

Description: Whitish pink to reddish lilac, calcified, articulated fronds, fish-bone-like arrangement, up to 50 mm high, axis compressed, repeatedly pinnate from discoid base, more abundantly and regularly branched than *Corallina officinalis*; articulations small (Pereira 2010a, Pereira 2015).

Distribution: see Table 2.1

Uses and bioactivities: Used for R-phycoerythrin extraction and for functional foods. Extracts have antimicrobial, antifungal, and antiviral activity (Pereira 2015).

Euchema denticulatum (N.L. Burman) Collins and Hervey (Fig. 1.7-F)

Synonyms: *Eucheuma spinosum* J. Agardh

Common names: Spinosum, Agar Gésér, Agar Poeloe, Agar-Agar Kasar, Agar-Agar Haloes, East Indian Carrageen, Kirinsai, Ryukyu-Tsunomata

Description: Thallus pale-yellowish, cartilaginous-thought, knotty-gnarled to roundish in cross-section; with irregular loose branching; branches loosely covered with outgrowths, papillae and bent spikes (Braune and Guiry 2011).

Distribution: Indian Ocean, W Pacific, and Pacific Islands.

Uses and bioactivities: Currently one of the most important sources of raw materials for carrageenan extraction (iota variant). Extracts have antiviral, antitumor, antibacterial, and anticoagulant activity (Pereira et al. 2009b).

Furcellaria lumbricalis (Hudson) J.V. Lamouroux (Fig. 1.7-G)

Synonym: *Furcellaria fastigiata* (Turner) J.V. Lamouroux

Common names: see Table 2.1

Description: Cartilaginous, cylindrical, brownish-black fronds, repeatedly dichotomously branched, fastigiate, up to 2 mm diameter and 300 mm long, with acute apices; attached by much-branched rhizoids. Multiaxial, medulla of cylindrical cells interspersed with rhizoids, cortex of irregular filaments, inner cells elliptical, outer cells narrow, elongated, in radial rows (Pereira 2015).

Distribution: see Table 2.1

Uses and bioactivities: Source of hybrid carrageenan (Furcellaran). The sulfated polysaccharides of this species have immunostimulatory activity (Pereira 2015).

Gelidiella acerosa (Forsskål) Feldmann and G. Hamel

Common names: see Table 2.1

Description: *G. acerosa* is a red alga with yellowish brown, tufted, entangled, erect, cylindrical thalli reaching 6 cm tall. The ends of the fronds are pinnately divided, giving it a feathered appearance. Branch tips of *G. acerosa* terminate in a single apical cell. Short, thick branches attached to the substratum by stoloniferous rhizoids form dense mats along shallow reefs. *G. acerosa* is found on surf-exposed and moderately wave sheltered rocks and reefs in the lower mid-littoral and the sublittoral zone, and in tide pools at higher levels on the shore (Pereira 2017r).

Distribution: see Table 2.1

Uses and bioactivities: *G. acerosa* is an important commercial species for agar production. It has also been used traditionally for the preparation of agar-forming hard jellies, eaten fresh and prepared as a salad vegetable, or cooked and eaten mixed with rice (Pereira 2015, Pereira 2016). Extracts have antifungal and antibacterial activity.

Gelidium pulchellum (Turner) Kützing (Fig. 1.7-H)

Description: Cartilaginous, regularly or irregularly bipinnate, dark red-brown fronds, 50–100 mm high, arise from a creeping base. Main axes narrow, cylindrical, somewhat flattened above. Ultimate branches short, pointed at first, later more or less spatulate, particularly when reproductive; appearance variable with habitat and time of year (Pereira 2015).

Distribution: NE Atlantic (Ireland and Britain to Portugal and Morocco) and Australia.

Uses and bioactivities: Source of agar, and its extracts have antibacterial and antiviral activity (Pereira 2015).

Gelidium spinosum (S.G. Gmelin) P.C. Silva (Fig. 1.7-I).

Synonym: *Gelidium latifolium* Bornet ex Hauck

Common names: see Table 2.1

Description: Small alga, cartilaginous, crimson to purplish-red, 20–60 mm long. Main axes distinctly flattened, often narrower at base, ultimate branches short, often opposite, spine-like or spatulate (Braune and Guiry 2011).

Distribution: see Table 2.1

Uses and bioactivities: Source of Agar. Extracts have antiviral, cytotoxic, and antibacterial activity (Pereira 2015).

Gigartina pistillata (S.G. Gmelin) Stackhouse (Fig. 1.7-J).

Common names: see Table 2.1

Description: *G. pistillata* is the type species of the genus *Gigartina* and their thalli are erect, up to 20 cm tall, dark red or red-brown, cartilaginous, elastic, dichotomously branched, attached to the substrate through a small disc (Pereira 2009, Pereira 2010a).

Distribution: see Table 2.1

Uses and bioactivities: Edible seaweed, carrageenan producer, and the sulfated polysaccharides of this species have antitumor, antiviral, anti-inflammatory, and antioxidant activity (Pereira et al. 2009b, Pereira 2013, Pereira 2015, Pereira 2016).

Gigartina skottsbergii Setchell and N.L. Gardner

Common names: see Table 2.1

Description: Thick, leathery, brick red thallus that is generally circular to ellipsoidal in shape. *G. skottsbergii* attaches to rocky substratum by a short strong stipe and holdfast (Pereira 2016).

Distribution: see Table 2.1

Uses and bioactivities: It is one of the most important and highly valued carrageenophyte resources growing in cold temperate water, and its extracts have antiviral (Pereira et al. 2009b, Pereira 2016) activity.

Gloiopeltis furcata (Postels and Ruprecht) J. Agardh

Common names: see Table 2.1

Description: Thallus is rusty red to golden yellow, up to 5 cm tall. The smooth, narrow cylindrical branches fork infrequently and are more or less dichotomous; they lack spines and have a rubbery to slippery texture. The annual thallus grows from a perennial basal crust each spring (Lindberg and Lindstrom 2017).

Distribution: see Table 2.1

Uses and bioactivities: *G. furcata* has long been utilized as a food source in Asia (Japan, Vietnam) where it is also used as a sizing material in silk and other textile industries. *G. furcata* is also a raw material for textile binding. Studies now show that extracts of *G. furcata* inhibit the growth of several human cancer cell lines, and are able to significantly lower blood glucose levels (Pereira 2016).

Gracilaria canaliculata Sonder

Synonym: *Gracilaria crassa* Harvey ex J. Agardh

Common names: see Table 2.1

Description: Thallus upright to partially decumbent, up to 12 cm in height, red to purple (often with a blotchy iridescence underwater), cartilaginous; sub-dichotomously branched; branches terete or slightly compressed, 2–5.0 mm in diameter. Structure apparently uniaxial but central axis obscure, pseudo-parenchymatous, with large medullary cells grading abruptly into a smaller-celled cortex (Huisman and Parker 2016).

Distribution: see Table 2.1

Uses and bioactivities: Used for agar extraction in India, Hawaii, and SE Asia (Pereira 2016). Extracts have antiviral, antifungal, and larvicidal activity.

Gracilaria changii (B.M. Xia and I.A. Abbott) I.A. Abbott, J. Zhang and B.M. Xia

Common names: Sarer, Sarai Kao Kwang

Description: Thalli 6–20 cm tall, robust, purplish-brown to dark brown when dry, with one to many axes, 1.0–3.5 mm in diameter, arising from a disc-like holdfast or from a percurrent axis; branching of two to four orders, irregular, alternate or second; branches turgid, cylindrical, 0.3–2.5 mm in diameter, abruptly constricted at the base forming a slender stipe, slightly swollen distally, tapering towards the tip (Phang and Lewmanomont 2016).

Distribution: see Table 2.1

Uses and bioactivities: Coastal communities, especially in Kelantan and Terengganu (eastern coast Peninsular Malaysia) and Selangor (western coast Peninsular Malaysia), collect *G. changii* for food. After washing, the seaweed is blanched in boiling water and served together with onions, chili, grated coconut, and lime juice, as an appetizer. In Thailand, large quantities are collected from the eastern coast of the Gulf of Thailand for food, agar extraction, and abalone feed (Phang and Lewmanomont 2016). Extracts of this species have cardio-protective, antifungal, and antibacterial activity.

Gracilaria edulis (S.G. Gmelin) P.C. Silva

Common names: see Table 2.1

Description: Thalli up to 27 cm tall, brownish-red, each arising from a discoid holdfast; branching dense and fastigiated, divaricate, dichotomous to trichotomous, up to seven orders and with long branch intervals; branches 1–1.5 mm in diameter, cartilaginous, flexuous, with or without a constriction at their bases or with only a slight constriction, cylindrical, ending in pointed apices (Phang and Lewmanomont 2016b).

Distribution: see Table 2.1

Uses and bioactivities: Extracts have antifungal and antibacterial activity.

Gracilaria gracilis (Stackhouse) Steentoft, L.M. Irvine and Farnham (Fig. 1.7-L)

Common names: see Table 2.1

Description: Cartilaginous, cylindrical, dull-purple fronds, up to 500 mm long, one or several are arising from small, fleshy, perennial discoid holdfast. Branching very irregular, sparse or profuse, branches up to 2 mm diameter, apices pointed; intertidal tissue of large thin-walled cells with narrow outer cortical zone of small colorless cells (Pereira 2009, Pereira 2016).

Distribution: see Table 2.1

Uses and bioactivities: Source of agar and used directly for animal feed. Extracts have antimicrobial and antioxidant activity (Pereira 2015).

Gracilaria vermiculophylla (Ohmi) Papenfuss

Common names: see Table 2.1

Description: *G. vermiculophylla* is a red macroalga that is cartilaginous, cylindrical, and up to 50 cm long. It is coarsely branched, often profusely so. It can be found as loose-lying thalli or attached to small stones or shells. Red algae are often found in the vegetative state, and the characterization of reproductive structures is often necessary for correct identification of *Gracilaria* species (Pereira 2015).

Distribution: see Table 2.1

Uses and bioactivities: *G. vermiculophylla* is widely collected to produce agar, which is used extensively in the pharmaceutical and food industries. Extracts have antibacterial and antioxidant activity (Pereira 2015).

Gracilariopsis longissima (S.G. Gmelin) M. Steentoft, L.M. Irvine and W.F. Farnham

Common names: see Table 2.1

Description: Thalli are from almost simple to profuse and irregularly branched, with the slender cylindrical axis throughout the plant. Cystocarps are scattered throughout the thallus, protruding from the thallus surface. Spermatia are formed near the surface of the thallus. Tetrasporangia are cruciate and are scattered in the cortex (Pereira 2015).

Distribution: see Table 2.1

Uses and bioactivities: Source of agar, this species has been used in traditional medicine to treat pulmonary tuberculosis, stomach disorders, urinary diseases, dropsy, and goiter (Pereira 2015, Pereira 2016). Extracts have antibacterial activity (Pereira 2015).

Grateloupia filicina (J.V. Lamouroux) C. Agardh (Fig. 1.7-K)

Common names: see Table 2.1

Description: Compressed, tufted, dark purplish brown fronds, up to 120 mm high, main axis 1–4 mm broad. Once or twice pinnate, axes and branchlets tapered at base and apex. This is a soft limp and slippery smooth seaweed with somewhat flattened branches that can be red green brown or almost black. Size and shape vary greatly from 0.5–5 mm wide to 2–30 cm long, with either few or many branches (Pereira 2015).

Distribution: see Table 2.1

Uses and bioactivities: Used as food and source of carrageenan. Extracts have antioxidant, antimicrobial, antiviral, and anticoagulation activity (Pereira 2015).

Grateloupia turuturu Yamada (Fig. 1.8-A)

Common names: see Table 2.1

Description: Thallus flat, membranous, with short stipe, the single fronds linear to broad-lanceolate, undivided or irregularly dividing from the base, narrowing towards the base as well as the stipe; sometimes proliferating on the margins and the surface; consistency gelatinous-slippery but firm; discoid holdfast; violet to crimson-red, often greenish at the top thallus (Braune and Guiry 2011, Pereira 2015).

Figure 1.8 Marine Red macroalgae: A–*Grateloupia turuturu*, B–*Gymnogongrus griffithsiae*, C–*Halurus equisetifolius*, D–*Heterosiphonia plumosa*, E–*Hypnea musciformis*, F–*Jania adhaerens*, G–*Jania rubens*, H–*Laurencia obtusa* , I–*Mastocarpus stellatus*, J–*Osmundea hybrida*, K–*Osmundea pinnatifida*, L–*Palisada perforata*, M–*Palmaria palmata*, N–*Plocamium cartilagineum*.

Distribution: see Table 2.1

Uses and bioactivities: Extracts have antifouling, antibacterial, anticoagulant, antioxidant, and antiviral activity (Pereira 2015).

Gymnogongrus griffithsiae (Turner) Martius (Fig. 1.8-B)

Common name: see Table 2.1

Description: Thallus stiff-erect, wiry-cartilaginous, cylindrical to compressed, brown-red to blackish purple fronds, up to 75 mm high, from an expanded discoid base; repeatedly dichotomous, fastigiated, with rounded, somewhat flattened apices (Braune and Guiry 2011, Pereira 2015).

Distribution: see Table 2.1

Uses and bioactivities: Produces sulfated galactans (carrageenans) with antioxidant and antiviral activity (Pereira 2015).

Halopithys incurva (Hudson) Batters

Description: Tough, cylindrical, cartilaginous, shaggy, dark red fronds, 250 mm long. Main branches alternate or sub-dichotomous, simple or pectinate in lower parts, much branched above, often curved and hooked. Branches with usually double row of short, pointed ramuli on the upper side, ramuli straight, curved or hooked, slightly narrowed at base; axis of one central, five pericentral siphons, with several layers of cortical cells, outermost small, colored; articulations visible, more short than broad (Braune and Guiry 2011).

Distribution: Warmer NE Atlantic (Ireland to Canary Islands) and Mediterranean Sea.

Uses and bioactivities: Extracts have antibacterial, antioxidant, antitumor, and antiviral activity (Pereira 2015).

Halurus equisetifolius (Lightfoot) Kützing (Fig. 1.8-C)

Common names: see Table 2.1

Description: Thallus densely tufted, pale ore bright crimson, spongy, firm, main axes loosely or densely irregularly laterally branched, uniseriate, very densely covered on all sides with whorls of short, repeatedly dichotomously branched lateral branchlets; basal disc of condensed filaments (Braune and Guiry 2011).

Distribution: see Table 2.1

Uses and bioactivities: Extracts have antifungal and anti-trichomonal activity.

Heterosiphonia plumosa (J. Ellis) Batters (Fig. 1.8-D)

Description: *H. plumosa* is a red to deep crimson seaweed which appears black when dried. This species has a flattened, fern-like appearance with a hairy thallus growing from a discoid holdfast. The fronds are flat or slightly cylindrical up to 20 cm in length and 0.5 cm in diameter at the base, tapering toward the apex. The primary branching from the main frond occurs in a single plane, and is alternately, yet irregularly spaced with up to 1 cm between each branch. The secondary branches are progressively shorter toward the apex, and each branch bears numerous pointed branchlets giving an overall tufted and feather-like appearance (Rowley 2008).

Distribution: NE Atlantic (Scandinavia to Portugal).

Uses and bioactivities: Extracts have antioxidant and antitumor activity.

Hypnea musciformis (Wulfen) J.V. Lamouroux (Fig. 1.8-E)

Common names: see Table 2.1

Description: Clumps or masses of loosely intertwined, cylindrical branches, 10–20 cm tall, 0.5 1.0 cm diameter, that become progressively slenderer towards tips. Firm, cartilaginous, highly branched. Branching is variable and irregular, often tendril-like and twisted around axes of other algae. The ends of many axes and branches are flattened with broad hooks. Holdfasts are small, inconspicuous, or lacking. Usually red, but can be yellowish brown in high light environments or nutrient-poor waters (Abbott 1999).

Distribution: see Table 2.1

Uses and bioactivities: In some areas, *H. musciformis* is grown for harvest of kappa-carrageenan. It also contains a high amount of naturally-occurring antioxidants, and is nourishing to the skin. *H. musciformis* is used in over 100 hair color and hair care products, such as shampoos, conditioners, and styling gels. It is used in more than 20 sunless tanning products, and in many antiaging creams and applications. *H. musciformis* is used in facial treatments, toners, and moisturizers, makeup and cosmetics, and eye treatments (Pereira 2015).

Hypnea valentiae (Turner) Montagne

Common names: see Table 2.1

Description: Thallus medium to dark red-brown, 5–30 cm high, usually with a main percurrent axis and long percurrent lateral branches, much branched with irregular radial laterals becoming gradually shorter, main branches with numerous short, spinous, branchlets; all branches and branchlets directed upward and not at right angles to parent branches; hamate branches absent. All branches terete, axes 1–2 mm in diameter below, decreasing gradually to lesser branches 200–300 μm in diameter, branchlets basally 150–200 μm in diameter, tapering from their base to an acute apex. Attachment by small discoid haptera to stones and shells (Womersley 1987).

Distribution: see Table 2.1

Uses and bioactivities: Used as food and for carrageenan production; in India, Bangladesh, and Vietnam this species is used as food (Pereira 2016). Extracts of this species have antifungal, antivenom, anti-inflammatory, larvicidal, and antibacterial activity.

Jania adhaerens J.V. Lamouroux (Fig. 1.8-F)

Common name: Wide Horned Algae, Finely Forked Coraline, Kabkhunpai Bon Khonha

Description: This species is small (4–5 mm high), heavily calcified, and forms intricately entwined clumps. The branches are pinkish-red in color, regularly dichotomous, terete or slightly compressed, and slightly curve downward. Terminal segments are acuminate. This is an epiphytic species, decumbent on other seaweeds, such as the species of Sargassaceae, growing in the sublittoral zones along shorelines moderately exposed to water movement (Pereira 2015).

Distribution: Atlantic Islands, the Mediterranean, SW and SE Atlantic, the Caribbean, Indo-Pacific Oceans, and Australia.

Uses and bioactivities: Extracts have antigenotoxic, antifungal, antiviral, and antimicrobial activity (Pereira 2015).

Jania rubens (Linnaeus) J.V. Lamouroux (Fig. 1.8-G)

Common names: Slender-Beaded Coral Weed, Fine Coral Moss

Description: Slender, rose-pink, articulated, calcified fronds, up to 50 mm high; repeatedly dichotomously branched, luxuriant specimens secondarily pinnate. Segments cylindrical, up to 120 μm in diameter, those bearing branches somewhat compressed, up to 180 μm in diameter. Fixed by small conical disc, but spreading vegetatively by developing attachment discs from branches in contact with solid substratum (Pereira 2015).

Distribution: Baltic Sea, NE Atlantic (Norway to Portugal), Senegal, East Africa, Atlantic Islands (Azores, Canary Islands), Brazil, Mediterranean Sea, Indian Ocean, Black Sea, and China Sea.

Uses and bioactivities: Extracts of this species have antitumor, bio-insecticide, antimicrobial, anthelmintic and cytotoxic, antifouling, and antifungal activity (Pereira 2015).

Kappaphycus alvarezii (Doty) Doty ex P.C. Silva

Common names: see Table 2.1

Description: Algae tough, fleshy, firm; up to 2 m tall. Thalli coarse, with axes and branches 1–2 cm in diameter; heavy, with major axes relatively straight, lacking secondary branches near apices. Frequently and irregularly branched, most branches primary, secondary branches intercalated between primary branches or mostly lacking. Shiny green to yellow orange (Abbott 1999).

Distribution: see Table 2.1

Uses and bioactivities: *K. alvarezii* is predominantly utilized as raw material for kappa-carrageenan extraction, although it is also used directly as a fresh, whole food source (eaten in seaweed salads and used in other recipes) in the tropical areas where it is harvested, namely in Indonesia, the Philippines, Malaysia, and Vietnam (Pereira 2016). Extracts have antitumor, antifungal, antibacterial, and anticoagulant activity.

Laurencia dendroidea J. Agardh

Synonyms: *Laurencia majuscula* (Harvey) A.H.S. Lucas, *Laurencia scoparia* J. Agardh

Description: *L. dendroidea* possesses all the characters that are typical of the genus, such as the production of the first pericentral cell underneath the basal cell of the trichoblast, the production of tetrasporangia from particular pericentral cells, without the formation of additional fertile pericentral cells, spermatangial branches that are produced from one of two laterals on the suprabasal cell of the trichoblasts, and a procarp-bearing segment with five pericentral cells. The species is recognizable primarily by its thallus, which is usually densely branched from the base to the upper portions, pyramidal in outline, and contains one fertile pericentral cell (the fourth) in its tetrasporangial segments (Cassano et al. 2012).

Distribution: Atlantic Islands (Bermuda), Caribbean Sea, and Brazil.

Uses and bioactivities: Extracts have antitumor, antiviral, antifungal, antibacterial, anthelmintic, nematicidal, antileishmanial, acaricidal, insect repellent, and antiprotozoal activity.

Laurencia microcladia Kützing

Description: Thallus erect, flexible, soft texture, forming dense tufts up to 15 cm in length. The thallus has greenish-yellow and pinkish apices easily visible. Flaccid consistency does not adhere completely to the herbarium sheet when dry; thalli fully cylindrical with 385–410 µm in diameter; branching irregularly, alternating, dense in the upper two-thirds of the plant, to three branching orders. Adhered to the substrate by rhizoids, from which branches emerge, ending in small secondary locking discs (Aylagas et al. 2010).

Distribution: NE Atlantic (from France to Mauritania), Atlantic Islands (Azores, Canary Islands), the Mediterranean, NW Atlantic (USA), the Caribbean, and SW Atlantic (Venezuela, Brazil).

Uses and bioactivities: Have antimitotic, cytotoxic, antiparasitic, and antifungal activity (Pereira 2015).

Laurencia obtusa (Hudson) J.V. Lamouroux (Fig. 1.8-H)

Common names: see Table 2.1

Description: Tufted thalli, light orange to brownish-orange, with cylindrical main axes from which emerge spirally to almost oppositely, repeatedly in the same manner branched lateral axes; these shorten towards the apex, outline therefore distinctly pyramidal; terminal branches short-cylindrical to club-shaped; texture crisp, brittle. Offshoot-like holdfasts (Braune and Guiry 2011).

Distribution: see Table 2.1

Uses and bioactivities: Extracts have antibacterial, antimalarial, antitumor, antioxidant, and antifouling activity (Pereira 2015).

Mastocarpus stellatus (Stackhouse) Guiry (Fig. 1.8-I)

Common names: see Table 2.1

Description: A small red alga (up to 17 cm in length), the fronds are channeled with a thickened edge and widen from a narrow stipe with disc-like holdfast. The channeling is often slight and is most noticeable at the base of the frond. Mature plants have conspicuous growths of short, stout papillae (reproductive bodies) on the fronds. The plant is dark reddish-brown to purple in color and may be bleached. The common name false Irish moss is used as it may be confused with *C. crispus* (Irish moss); the main features separating the two species are the channeled frond and appearance of reproductive bodies on mature plants (Pereira 2009, Pereira 2010a).

Distribution: see Table 2.1

Uses and bioactivities: Source of carrageenan (Pereira et al. 2009, Pereira 2013). Extracts have antioxidant, antibacterial, anti-Trypanosoma, and antiprotozoal activity.

Nothogenia fastigiata (Bory) P.G. Parkinson

Description: Thallus (gametophyte) grey-red to red-purple, mucilaginous, 1–8 cm high, erect, 1–4 times furcate from near the base, tapering above in larger plants, apices rounded to contracted, branches terete to compressed, 2–5 mm broad. Holdfast discoid, 1–2 mm across; epilithic. Structure multiaxial, with a loosely filamentous medulla with cells 1–2 μm in diameter, becoming hollow, surrounded by a cortex of anticlinal filaments 2–4 cells long where unbranched, cells 2–3 μm in diameter. Tetrasporophyte crustose, discoid, with a basal layer bearing erect filaments 10–20 cells long (Delépine et al. 1979).

Distribution: Antarctic and the sub-Antarctic Islands (Fuegia), Falkland Islands (Islas Malvinas), Auckland Islands, Australia, and New Zealand.

Uses and bioactivities: Extracts have antiviral and anticoagulant activity.

Osmundea hybrida (A.P. de Candolle) K.W. Nam (Fig. 1.8-J)

Common name: False Pepper Dulse

Description: Cylindrical, cartilaginous, tufted, dark purple to greenish-yellow fronds, 150 mm long; main axis with repeatedly pinnate branching, branches mostly alternate, shorter toward apex giving pyramidal outline; ultimate ramuli short, patent, truncate; axis monosiphonus with elongated pericentral cells and 1–2 outer layers of rounded colored cells; apex concave, with ephemerally colorless dichotomous hairs surrounding apical cell (Pereira 2009, Pereira 2015).

Distribution: NE Atlantic (Britain to Morocco) and SW Atlantic (Brazil).

Uses and bioactivities: Extracts have antifungal (Pereira 2015) activity.

Osmundea pinnatifida (Hudson) Stackhouse (Fig. 1.8-K)

Synonym: *Laurencia pinnatifida* (Hudson) J.V. Lamouroux

Common names: see Table 2.1

Description: A small red alga (up to 8 cm in length), it is tough and cartilaginous with flattened fronds. Branching is alternate and occurs in one plane only, with branches becoming shorter toward their apex and broadly rounded. The plant is highly variable in size and coloration, depending upon its location on the shore. Higher shore plants are generally dwarfed and yellow green in color, owing to exposure to high levels of sunshine while on the lower shore they are reddish-brown (Pereira 2009, Pereira 2010a).

Distribution: see Table 2.1

Uses and bioactivities: This aromatic seaweed is dried and used as a pepper- or curry-flavored spice in Scotland, Ireland, and Portugal (Azores Islands) (Pereira 2015, Pereira 2016). Extracts of this species have antibacterial, antioxidant, antileishmanial, anticancer, antifouling, insecticidal, and antifungal activity (Pereira 2015).

Palisada perforata (Bory) K.W. Nam (Fig. 1.8-L)

Synonyms: *Chondrophycus papillosus* (C. Agardh) D.J. Garbary and J.T. Harper, *Laurencia papillosa* (C. Agardh) Greville

Common names: see Table 2.1

Description: This species presents all typical features of the genus *Palisada*, being characterized by turf-like growth, the presence of arcuate and decumbent branches with erect branches disposed unilaterally (Pereira 2017s).

Distribution: see Table 2.1

Uses and bioactivities: *P. perforata* is the source of antimicrobial, fungicidal and antitumor BASs, PUFAs, phycocolloids, agaroids and phenol-bromides, rare microelements; also, provide natural fertilizer (Milchakova 2011, Pereira 2016).

Palisada poiteaui (J.V. Lamouroux) K.W. Nam

Synonym: *Laurencia poiteaui* (J.V. Lamouroux) M. Howe

Description: Thallus cartilaginous and bushy. Main axes cylindrical, 0.6 mm in diameter, medullary cells 75–100 µm in diameter. Branchlets short (0.8–1.3 mm long), alternate, cylindrical, 0.4 mm in diameter, alternating in pairs, blunt, wart-like and often swollen. Tetrasporangia tetrahedrally divided, 100 µm in diameter when mature, located near branchlet tips (Nam 2007).

Distribution: Atlantic Islands (Bermuda, Cape Verde Islands), North America (Florida, Mexico, North Carolina, Texas), Central America (Belize), Caribbean Islands (Bahamas, Barbados, Caicos Islands, Cuba, Hispaniola, Jamaica, Lesser Antilles, Netherlands Antilles, Puerto Rico, Virgin Islands), South America (Brazil, Colombia, Venezuela), Africa (Kenya, Tanzania), Indian Ocean Islands (Réunion), SW Asia (India, Sri Lanka), SE Asia (the Philippines), Australia, and New Zealand.

Uses and bioactivities: Extracts have antiviral and antifungal activity.

Palmaria palmata (Linnaeus) Weber and Mohr (Fig. 1.8-M)

Common names: see Table 2.1

Description: A foliose red algae with a tough flat frond usually between 20 cm and 50 cm in length, but sometimes up to 1 m. The algae grow directly from a small discoid holdfast gradually widening and subdividing. The stipe is inconspicuous, rarely up to 5 mm long. Older parts may have small "leaflets" along the margin especially where damaged; dark red, with purple tints under water (Pereira 2010a, Pereira 2015).

Distribution: see Table 2.1

Uses and bioactivities: Extracts of this species have cardio-protective, antibacterial antioxidant, and antitumor activity (Pereira 2015).

Plocamium cartilagineum (Linnaeus) P.S. Dixon (Fig. 1.8-N)

Common names: see Table 2.1

Description: Bright scarlet seaweed up to 30 cm in length with branching fronds (Fig. 1.8-N). The branching occurs alternately along the fronds and becomes more frequent toward the tips. The general appearance of these seaweeds can greatly vary from very compact and closely branched to a much broader appearance with widely separated branching. The tips are incurving and ultimate branching occurs only to one side, giving a distinctive feathery or comb-like appearance (Pereira 2015).

Distribution: see Table 2.1

Uses and bioactivities: *P. cartilagineum* extracts are also high in lipolytic (lipid-digesting) sterols and as such are useful as an additive in slimming applications such as creams and massage products where they can provoke the release of fatty acids and eliminate surface fat, acting as a skin-firmer (Pereira 2015). Extracts have antimicrobial, insecticidal, cytotoxic and antitumor, antioxidant, and antiviral activity.

Polysiphonia denudata (Dillwyn) Greville ex Harvey (Fig. 1.9-A)

Common name: Wide-branched Siphon Weed

Description: Small fronds reddish-brown in color; composed thallus of cylindrical, uncrossed axes that branch in dichotomous form giving rise to secondary branches bearing cystocarps (Pereira 2017t).

Distribution: Caribbean Islands, SW and SE Atlantic, the Mediterranean, SW Asia, and Australia.

Uses and bioactivities: Extracts have antiviral and antitumor activity.

Polysiphonia mollis J.D. Hooker and Harvey

Common names: see Table 2.1

Description: Thallus red-brown, usually 4–20 cm high, with a single, erect, basal axis (occasionally a slight prostrate part) and profusely branched above (often denuded below in older plants) sub-dichotomously to laterally to form dense, fastigiated to spreading, soft tufts; holdfast discoid, small; commonly epiphytic on *Posidonia*, *Heterozostera*, *Halophila* (marine vascular plants) or larger algae (Womersley 2003, Pereira 2016).

Distribution: see Table 2.1

Uses and bioactivities: Used as food in Hawaii (Reed 1906, MacCaughey 1918), and have high protein levels (Chakraborty and Santra 2008) (see Table 2.5 and Chapter 2).

Porphyra linearis Greville (Fig. 1.9-B)

Common name: see Table 2.1

Description: Delicate, linear, membranous, purple-brown fronds, 20–40 mm (200 mm) long and 5–10 (25 mm) broad, usually simple with short stipe from basal holdfast; orange patches when reproductive. Narrow stem attaches the base and appears mainly in winter (Pereira 2009, Pereira 2010a).

Distribution: see Table 2.1

Uses and bioactivities: Used for food and for aquaculture. Extracts have antimicrobial activity (Pereira 2015).

Figure 1.9 Marine Red macroalgae: A–*Polysiphonia denudata*, B–*Porphyra linearis*, C–*Pterocladiella capillacea*, D– *Pyropia leucosticta*, E–*Solieria chordalis*, F–*Sphaerococcus coronopifolius*.

Portieria hornemannii (Lyngbye) P.C. Silva

Synonyms: *Chondrococcus hornemannii* (Lyngbye) F. Schmitz

Common name: see Table 2.1

Description: Thalli up to 20 cm tall, usually smaller, bright orange to red, composed of several erect, overlapping flattened branches arising from small discoid holdfasts. Branching in one plane, irregularly pinnate-alternate, in up to 5(7) orders, forming rounded axils; diameter of primary branches not exceeding 7 mm; terminal branches at distal portion of thalli with slightly expanded curved or enrolled tips; lower lateral branchlets with simple acute teeth (Trono 2016).

Distribution: see Table 2.1

Uses and bioactivities: *P. hornemannii* is a source of carrageenan related to lambda-carrageenan (Trono 2016). Extracts have antifungal and antibacterial activity.

Pterocladiella capillacea (S.G. Gmelin) Santelices and Hommersand (Fig. 1.9-C)

Synonyms: *Gelidium capillaceum* (S.G. Gmelin) Meneghini, *Pterocladia capillacea* (S.G. Gmelin) Bornet

Common name: see Table 2.1

Description: Thallus dark brownish red, commonly grows in dense tufts about 4 cm high, and composed of prostrate axes that give rise to the flattened erect axes: branching pinnate to irregularly opposite with uniaxial growth from single apical cells; inner structure pseudoparenchymatous, with long, slender, thick-walled rhizine cells filling the spaces between elongated medullary cells that are surrounded by a three-layered cortex of smaller pigmented cells (Pereira 2015).

Distribution: see Table 2.1

Uses and bioactivities: Source of agar. Extracts have antioxidant, antibacterial, antinociceptive and anti-inflammatory, anticoagulant, and antitumor activity (Pereira 2015).

Pyropia leucosticta (Thuret) Neefus and J. Brodie (Fig. 1.9-D)

Synonyms: *Porphyra leucosticta* Thuret

Common name: see Table 2.1

Description: Delicate membranous monostromatic reddish-brown fronds, becoming pink on drying, to 150 mm long, with very short stipe from basal holdfast (Braune and Guiry 2011).

Distribution: see Table 2.1

Uses and bioactivities: Produces high percentage of vitamin C and natural carotenoids, and therefore *P. leucosticta* could become a valuable source of raw material to obtain such compounds, which have multiple uses in the pharmaceutical field, cosmetic, and food industry; used for food, and for IMTA aquaculture. Extracts have high antioxidant, and low antiprotozoal, antimycobacterial, and cytotoxic activity (Pereira 2015).

Solieria chordalis (C. Agardh) J. Agardh (Fig. 1.9-E)

Description: Thallus tufted, vividly red, axes cylindrical, thin, only sparsely branching at first, later characteristically with relatively long, unilaterally inserted appendices, height up to 20 cm (Pereira 2015).

Distribution: NE Atlantic (from N France to S Morocco) and the Mediterranean.

Uses and bioactivities: Source of carrageenan. Extracts have hemagglutinating and Immunostimulation activity (Pereira 2015).

Sphaerococcus coronopifolius Stackhouse (Fig. 1.9-F)

Common name: see Table 2.1

Description: Narrow, compressed, two-edged, cartilaginous, scarlet fronds, main axes dark brownish-red, up to 300 mm long; branching abundant, distichous, sub-dichotomous or alternate, terminal branchlets acute, fringed with short marginal proliferations. Tetrasporophyte is *Haematocelis fissurata*, a thick crust with oily fissures, with which it is sometimes found (Pereira 2015).

Distribution: see Table 2.1

Uses and bioactivities: Extracts have antifouling, cytotoxic and antimitotic, antiviral, antibacterial, and antifungal activity (Pereira 2015).

Vidalia colensoi (J.D. Hooker and Harvey) J. Agardh

Synonym: *Osmundaria colensoi* (J.D. Hooker and Harvey) R.E. Norris

Common name: Fretsaw weed

Description: This species presents typical ramifications that remind teeth alternately arranged, especially in coastal areas undergoing intense wave action. It does not feel slimy but rough and can grow to dense patches (Anthoni 2007).

Distribution: Australia and New Zealand.

Uses and bioactivities: This species has antitumor activity (see Chapter 6).

CHAPTER 2

Nutritional Composition of the Main Edible Algae

2.1 Introduction

Seaweeds, sometimes referred to as "sea vegetables", are daily food items among the countries of East Asia, especially Japan, Korea, and China, the island nations of the Pacific, and the Celtic cultures of Europe (i.e., Ireland, Scotland, and Brittany). These resources form the basis for a multi-billion USD business employing sea farmers and processors (Pereira 2016).

Edible algae are those that can be eaten directly or used in the preparation of other foods (as gelling or emulsifying agents). These photosynthetic organisms belong to one of several groups of multicellular algae, most conveniently segregated into green, red, or brown forms. Collectively, algae, i.e., both microscopic unicellular forms and macroalgae are used throughout the world in coastal areas and are particularly important in the cuisines of China, Japan, and Korea since pre-historic times (see Table 2.1).

Seaweeds have been consumed as vegetables since the beginning of the fourth century in Japan, and the sixth century in China. On an average, the Japanese eat 1.4 kg seaweed per person per year. This ancient tradition and everyday habit has made possible a large number of epidemiological studies which have demonstrated the many health benefits linked to seaweed consumption (Johnston 1966, MacArtain et al. 2007, Mouritsen 2013). The spread of Japanese and Chinese cuisine and of public perceptions on the values of "health foods" throughout the world brought new attention to sea vegetables as an underutilized resource for the kitchen. Although, of the recent interests, there remains a tremendous untapped potential for increasing the consumption of "sea veggies" as everyday food items, dietary supplements, food texturizers, flavoring (umami components), coloring, and condiments (list not exhaustive)—especially amongst mainstream markets. For example, France was one of the first European countries to establish a specific regulation concerning the use of seaweeds for human consumption as non-traditional food substances.

In general terms, it can be said that a varied diet that includes a proportion of seaweed product, i.e., up to 10% as in Japan, promotes wellness. This is due principally to the relatively high concentration of important minerals and vitamins—sea vegetables have been described as "nutrient dense" in this regard. These minerals in seaweeds are in what are known as chelated and colloidal forms, which enhance their bioavailability in the body. Seaweeds are also a good source of protein and essential amino acids. In addition, marine algae have much greater fiber content than traditional/terrestrial vegetables and fruits, as they are largely composed of both soluble and insoluble dietary fiber. Because dietary fiber is indigestible, it contributes to the diet with no calories, and enhances digestive function by absorbing water, and thereby easing the passage of food through the gastro-intestinal system. Soluble fiber, particularly, slows the rate at which polysaccharides are absorbed, which helps to lower blood sugar levels, a potential advantage for diabetics (Pereira 2011, Mouritsen 2013).

2.2 Nutritional Ingredients from Selected Seaweeds and their Properties

2.2.1 Carbohydrates

The chemical composition and abundance of carbohydrates vary amongst seaweed species. Red seaweeds varieties consist of different typical carbohydrates including: floridean starch (α-1,4-binding glucan), cellulose, xylan, and mannan. Moreover, their water-soluble fiber fraction is formed by sufur-containing galactans, e.g., agar and carrageenan (see Chapter 3; Jimenez-Escrig and Sanchez-Muniz 2000).

The typical carbohydrates in brown seaweeds varieties consist of fucoidan, laminaran (β-1,3-glucan), cellulose, alginates, and mannitol. Brown seaweeds fibers are mainly cellulosic with insoluble alginates. In contrast, the amorphous, slimy fraction of fibers consists mainly of water-soluble alginates and/or fucoidan (Kolb et al. 1999, El-Said and El-Sikaily 2013).

The typical carbohydrates of edible seaweeds in general are not digestible by the human gastro-intestinal tract and, therefore, they are referred to as insoluble dietary fiber. The content of total dietary fiber ranges from 33–50 g 100 g DW (see Table 2.2; Lahaye 1991, Jiménez-Escrig and Cambrodón 1999, Ruperez and Saura-Calixto 2001, Pereira 2011).

Sulfated polysaccharides

As previously described, some edible seaweeds may be rich sources of sulfated polysaccharides, including some that have become valuable additives in the food industry because of their rheological properties as gelling and thickening agents. In addition, sulfated polysaccharides are recognized to possess a number of biological activities, including anti-coagulant, anti-fungal, anti-viral, and immuno-inflammatory activities, which might find further and future relevance in nutraceutical/functional food properties, cosmetic/cosmeceutical, and pharmaceutical applications (Jiao et al. 2011, Soares et al. 2016).

Fucoidan

Fucans are sulfated polysaccharides which comprise a fucose backbone. One of the best studied fucans from the brown algae is fucoidan (see Table 2.2), which was first isolated by Kylin (Kylin 1913). The fucoidan from *Fucus vesiculosus* has been available commercially for decades (Sigma-Aldrich Chemical Company, St. Louis, MO, US). Early work on its structure showed that it contained primarily (1→2) linked 4-O-sulfated fucopyranose residues (Fig. 2.1). However, 3-linked fucose with 4-sulfated groups was subsequently reported to be present on some of the fucose residues. Additionally, it was determined to contain branches every 2–3 fucose residues. Subsequently, Chevolot and colleagues reported that the fucoidan from *Fucus vesiculosus* and *Ascophyllum nodosum* contained a predominant disaccharide motif with sulfate at the 2-position of the 3-linked fucose and sulfate groups on the 2- and 3-positions of the 4-linked fucose (Chevolot et al. 2001, Pereira et al. 2013).

Fucoidan can be easily "cooked" out of most edible brown algae by simmering for 20–40 minutes in water. When consumed, it seemed to reduce the intensity of the inflammatory response and promote more rapid tissue healing after wound or surgical trauma. This led to recommendations that brown seaweed broth be administered after auto-collision, sports injuries, bruising falls, muscle and joint damage, and deep tissue cuts, including surgery (Fitton 2011).

However, the commercial importance of fucoidans is presently much lower than that of seaweed hydrocolloids; these polysaccharides are now attracting considerable attention because of the growing market for them as "bioactive polysaccharides" in a wide arena of diverse applications (Lorbeer et al. 2013). More recently, anti-coagulant and anti-thrombotic activities are amongst the most studied biological effects of fucoidans. Commonly, the anti-coagulant activity of fucoidans is mediated through the activation of thrombin inhibitors, although direct thrombin inhibition and competitive binding of fibrinogen to block the actions of thrombin are also possible (Kuznetsova et al. 2003, Lorbeer et al. 2013, Wu et al. 2016).

Table 2.1 Seaweed used as food and foodstuff Worldwide (adapted from Pereira 2016).

Species	Common names	Countries
Acanthopeltis japonica (R)	Toriashi, Yuikiri	Japan, Korea
Acanthophora spicifera (R)	Spiny Seaweed, Spiny Sea Plant	Caribbean, Central America, India, China, Indonesia, Philippines, Vietnam, Thailand, Tahiti, Fiji
Acetabularia major (G)	Umbrella Alga	Malaysia
Acrothrix pacifica (B)	Nise-Mozuku	Japan
Adenocystis utricularis (B)	Adenocistie, Adenocisto	Argentina
Aegagropila linnaei (G)	Lake Balls, Marimo, Kai	Thailand
Agardhiella subulata (R)	Gulaman-Dagat	Philippines
Agarum clathratum (B)	Aname	Japan, Korea
Ahnfeltia plicata (R)	Bushy Ahnfelt's Seaweed, Black Scour Weed, Landlady's Wing	Japan
Ahnfeltiopsis concinna (R)	Tufted Seaweed, Saimi, Limu 'Aki'aki	Hawaii
Ahnfeltiopsis devoniensis (R)	Devonshire Fan Weed	NE Atlantic
Ahnfeltiopsis gigartinoides (R)	Sea Nibbles, Itanigusa Saimi	Russia, Japan, Hawaii
Ahnfeltiopsis paradoxa (R)	Harigane	Japan
Ahnfeltiopsis vermicularis (R)	Limu Vavaloli	Hawaii
Alaria crassifolia (B)	Lesser Kelp, Chigaiso	China, Korea, Japan
Alaria esculenta (B)	Daberlocks, Bladderlocks	Norway, Iceland, Scotland, Ireland
Alaria marginata (B)	Winged Kelp, Pacific Wakame	Alaska, British Columbia (Canada)
Alaria ochotensis (B)	Karafuto-Wakame	Japan
Alaria pylaiei (B)	Suvdluistsit, Me'cgomei	Iceland, Alaska, Siberia
Anadyomene stellata (G)	Repollo de Mar	Caribbean, Costa Rica
Analipus japonicus (B)	Bottlebrush Seaweed, Far Needle, Sea Fir Needle	China, Japan
Antrocentrum nigrescens (R)	Red Threads	China
Arthrothamnus bifidus (B)	Nekoashi-Kombu	Japan

Table 2.1 contd. ...

Table 2.1 contd. ...

Species	**Common names**	**Countries**
Arthrothamnus kurilensis (B)	Chishima-nekoashikombu	Japan
Ascophyllum nodosum (B)	Wrack, Yellow Tang, Asco	Iceland, Greenland, Scotland, Ireland
Asparagopsis armata (R)	Harpoon Weed, Harpoon-Alga	NE Atlantic, Australia, New Zealand
Asparagopsis taxiformis (R)	Supreme-Limu, Kagikenori, Limu Kohu	China, Korea, Indonesia, Hawaii
Bangia atropurpurea (R)	Purple Sea Scum	Korea
Bangia fuscopurpurea (R)	Velvet Thread Weed, Ushike-Nori	Taiwan, Japan, Thailand, Hawaii
Betaphycus gelatinus (R)	Macassar Agar Agar, Kirinsai	Vietnam, Philippines, Indonesia, Malaysia, Japan
Betaphycus philippinensis (R)	Kanutkanot	Philippines
Betaphycus speciosus (R)		Tasmania (Australia)
Bifurcaria bifurcata (B)	Brown Tuning Fork Weed	NE Atlantic
Boergeseniella thuyoides (R)	Tufted Conifer Weed	NE Atlantic
Boodlea composita (G)		India
Bostrychia radicans (R)		Myanmar, Thailand, Indonesia
Bostrychia tenella (R)	Pakpako	Pakistan, Philippines
Brassicophycus brassicaeformis (B)	Hanging Wrack	South Africa
Bryothamnion triquetrum (R)	Pelo de Cabra	Costa Rica
Callophycus serratus (R)	Large Wire Weed	Pacific Islands
Callophyllis adnata (R)	Lablabig	Philippines
Callophyllis laciniata (R)	Beautiful Fan Weed	NE Atlantic
Callophyllis variegata (R)	Large Wire Weed, Carola	Chile
Caloglossa bengalensis (R)		Myanmar, Indonesia
Caloglossa leprieurii (R)	Zhegucai	Myanmar, Thailand, Korea, Indonesia
Campylaephora hypnaeoides (R)	Ego Nori	Korea, Japan
Capsosiphon fulvescens (G)	Maesaengui	Korea
Catenella caespitosa (R)	Creeping Chain Weed	Pakistan
Catenella impudica (R)	Burmese Moss	Asia
Catenella nipae (R)	Kolongkong	Myanmar, Indonesia, Thailand

Caulerpa bikinensis (G)		Pacific Islands
Caulerpa brachypus (G)	Mini Caulerpa, Sea Mustard	Pacific Islands
Caulerpa brownii (G)	Sea Rimu	Tasmania (Australia), New Zealand
Caulerpa cactoides (G)		Tasmania (Australia)
Caulerpa chemnitzia (G)	Big Parasol Green Seaweed, Flattop Seagrape	Indonesia, Philippines, Malaysia, Pacific Islands
Caulerpa corynephora (G)		Thailand
Caulerpa cupressoides (G)	Cactus Tree Alga, Toothed Stolon, Kaka, Byakushinzuta	Japan, Pacific Islands
Caulerpa cupressoides var. *lycopodium* (G)	Mamanga	French Polynesia
Caulerpa fergusonii (G)		Malaysia
Caulerpa flexilis (G)	Feathery Caulerpa, Fern Caulerpa, Sea Rimu	Asia (Japan), Tasmania (Australia)
Caulerpa geminata (G)		Tasmania (Australia)
Caulerpa hodkinsoniae (G)		Tasmania (Australia)
Caulerpa lamourouxii (G)		Philippines, Australia
Caulerpa lentillifera (G)	Green Caviar, Sea Grapes, Small Seagrape	Pakistan, Philippines, Malaysia, Hawaii Is
Caulerpa longifolia (G)	Feather Caulerpa, Long-Filament Caulerpa	Tasmania (Australia)
Caulerpa macrodisca (G)		Thailand, Philippines
Caulerpa microphysa (G)	Ar-Arusip	Philippines
Caulerpa obscura (G)		Tasmania (Australia)
Caulerpa okamurae (G)		Korea, Japan
Caulerpa racemosa (G)	Sea Grapes, Mouse Plant, Green Sea Feathers, Coarse Seagrape, Rimu	Caribbean, Central America, Bangladesh, Malaysia, New Zealand, Tonga, Hawaii, Cook Island, Fiji
Caulerpa scalpelliformis (G)		India, Tasmania (Australia)
Caulerpa serrulata (G)	Serrated Green Seaweed, Toothed Spiral, Yorezuta	Thailand, Malaysia, Singapore, Philippines, Tonga
Caulerpa sertularioides (G)	Green Feather Alga, Delicate Feathery Green Seaweed	India, Bangladesh, Indonesia, Thailand, Tonga
Caulerpa taxifolia (G)	Green Sea Palm, Lukay-Lukay	Philippines
Caulerpa trifaria (G)		Tasmania (Australia)
Caulerpa veravalensis (G)		India
Caulerpa vesiculifera (G)	Beaded Caulerpa	Tasmania (Australia)

Table 2.1 contd. ...

Table 2.1 contd. ...

Species	**Common names**	**Countries**
Centroceras clavulatum (R)	Red Panache, Limu Hulu	Nigeria, India, Hawaii
Ceramium boydenii (R)	Fairy Herb, Amikusa	China, Japan
Ceramium deslongchampsii (R)	Estuary Banded Picer Weed	Asia
Ceramium kondoi (R)	Igisu	China, Korea, Japan
Ceramium virgatum (R)	Fa-Tsai	Japan
Chaetomorpha aerea (G)	Tarugata-Juzumo	Tasmania (Australia)
Chaetomorpha antennina (G)	Riprippiis, Limu Manu	Asia, Hawaii
Chaetomorpha coliformis (G)	Mermaids Necklace	Tasmania (Australia)
Chaetomorpha crassa (G)	Hair-Shaped Green Algae	China, Taiwan, Korea, Thailand, Philippines
Chaetomorpha intestinalis (G)		Caribbean
Chaetomorpha javanica (G)	Lumut-Laut	Indonesia, Malaysia
Chaetomorpha linoides (G)	Nool Pasi	India
Chaetomorpha linum (G)	Flax Brick Weed, Spaghetti Algae,Warakuzumo	India, Philippines; Pacific Islands
Chaetomorpha valida (G)		Tasmania (Australia)
Champia compressa (R)	Limu Oolu	Hawaii
Champia feldmannii (R)		W and SW Atlantic
Champia parvula (R)	Little Fat Sausage Weed	Hawaii
Chnoospora implexa (R)		Philippines, Hawaii
Chnoospora minima (B)	Hornwort, Coontail	India, Indochina, Vietnam, French Polynesia, Guam, Hawaii
Chondracanthus acicularis (R)	Creephorn	Portugal, W Africa
Chondracanthus chamissoi (R)	Chicorea de Mar, Suginori	Peru, Chile
Chondracanthus exasperatus (R)	Turkish Towel	E Pacific
Chondracanthus glomeratus (R)	Cochayuyo	Peru
Chondracanthus intermedius (R)	Kosugi Algae	China, Taiwan, Korea
Chondracanthus teedei (R)	Cata-Nori, Shikin-Nori	Japan, Portugal
Chondracanthus tenellus (R)	Matsuba-Gusa, Suginori	China, Korea, Japan, Taiwan, Philippines, Pacific Islands
Chondria capillaris (R)	Limu O-olu	Hawaii

Chondria coerulescens (R)	Iridescent Cartilage Weed	NE Atlantic
Chondria crassicaulis (R)	Seushil, Yuna	Korea
Chondria dasyphylla (R)	Diamond Cartilage Weed	Korea
Chondrophycus cartilaginous (R)	Kulot	Philippines
Chondrophycus dotyi (R)	Limu Lipe-Epe-e	Hawaii
Chondrophycus succisus (R)	Limu Lipe-Epe-e	Hawaii
Chondrophycus undulatus (R)	Kohu-Sozo, It-Ittip	Philippines
Chondrus armatus (R)	Togetsunomata	Japan
Chondrus crispus (R)	Irish-Moss, Carragheen Moss	NE Atlantic
Chondrus elatus (R)	Dee-horn Vegetable, Kotoji-Tsunomata, Tsunomata	China, Taiwan
Chondrus ocellatus (R)	Japanese-Moss, Hosokenomimi, Tsunomata	China, Korea, Japan, Hawaii
Chondrus pinnulatus (R)	Kotojitsunomata, Hirakotiji, Hirasaimi	Korea, Japan, Russia
Chondrus yendoi (R)	Ezo-Tsunomata, Kuroha-Ginnanso	Japan
Chorda filum (B)	Dead Men's Ropes, Sea Lace	Scotland, China, Korea, Japan
Chordaria flagelliformis (R)	Slimy Whip Weed, Naga-Matsumo	Korea, Japan
Chrysymenia enteromorpha (R)		Bangladesh
Chylocladia rigens (R)	Limu akuila, Limu kihe	Hawaii
Cladophora glomerata (G)	Kai	Thailand
Cladophora laetevirens (G)	Pelillo, Espinaca de Mar	Caribbean, Thailand
Cladophora patentiramea (G)	Imu ouoho	French Polynesia
Cladophora prolifera (G)		Morocco
Cladophora rugulosa (G)	Kulkulasisi	Philippines
Cladophora rupestris (G)	Slobán, Common Green Branched Weed	Norway, Phillippines
Cladophora sericea (G)	Graceful Green Hair	Philippines, Hawaii
Cladophora vagabunda (G)		India, Peru
Cladosiphon okamuranus (B)	Slender Slippery Weed	Okinawa-Mozuku
Codium adhaerens (G)	Haimiru, Limu Aalaula	Japan, Hawaii
Codium arabicum (G)	Rat-Faces Algae, Imu Tutae, Bagaba	Philippines, French Polynesia

Table 2.1 contd. ...

Table 2.1 contd. ...

Species	Common names	Countries
Codium australicum (G)		Australia (Tasmania, Victoria), New Zealand
Codium bartlettii (G)	Green Sea Antler	Philippines
Codium bursa (G)	Green Sponge Ball	Turkey
Codium cylindricum (G)	Long Songzao	Korea
Codium decorticatum (G)	Dead Man's Fingers	Indi, Caribbean
Codium dichotomum (G)	Imose-Miru, Yezo-Miru	Korea, Japan, Argentina
Codium dimorphum (G)		Tasmania (Australia)
Codium duthieae (G)	Duthie's Upright Codium	Tasmania (Australia)
Codium dwarkense (G)		India
Codium edule (G)	Pukpuklo, Limu, Miru	Philippines, Hawaii, French Polynesia
Codium fragile (G)	Sponge Seaweed, Sponge Weed, Green Sea-Velvet, Green Sponge Fingers, Fragile Green Sponge Fingers, Miru	Denmark, Korea, China, Japan, Philippines, Tasmania (Australia)
Codium galeatum (G)	Dead Man's Fingers	Tasmania (Australia)
Codium geppiorum (G)	Sagati, Totoyava, Pocpoclo	Philippines, Fiji, Cook Island, Polynesia
Codium harveyi (G)		Tasmania (Australia)
Codium indicum (G)		Pakistan
Codium intricatum (G)	Finger Algae, Mosure-Miru	Pakistan, Philippines
Codium isthmocladum (G)		Yucatan (Mexico)
Codium muelleri (G)	Pokpoklo, Siling Siling, Limu Wawae-Moa	Hawaii
Codium papillatum (G)	Popoklo	Philippines
Codium platylobium (G)	Miru	Japan
Codium reediae (G)	Dead Man's Fingers, Antler Seaweed, Limu Wawae'iole	Philippines, French Polynesia, Hawaii
Codium repens (G)	Pukpuklo	Philippines
Codium subtubulosum (G)	Sagati	Korea, Japan
Codium taylorii (G)		Caribbean
Codium tenue (G)	Papu-Lo, Puk-Puklo, Pupulo, Itomiru	Indonesia, Thailand, Philippines
Codium tomentosum (G)	Miru, Susu-Lopek, Laur-Laur, Chorão-do-Mar	Spain, Portugal, India, Thailand, Japan, Indonesia, Malaysia, Hawaii

Codium vermilara (G)	Alga Candelabro Vermiforme	Japan
Colpomenia peregrina (B)	Oyster Thief	NE Atlantic
Colpomenia sinuosa (B)	Papery Sea Bubble, Silver-Ballon, Sea Ballon, Sea Potato, Fukuronori, Hukuronori	Persian Gulf, India, Philippines
Cordylecladia erecta (R)	Erect Clublet	NE Atlantic
Costaria costata (B)	Ribbed Kelp, Seersucker Kelp, Short Kelp, Sujime	Korea
Cryptonemia crenulata (R)		Caribbean
Cryptonemia obovata (R)		Chile
Cymathaere triplicata (B)	Three-Ribbed Kelp	Alaska
Cystoclonium purpureum (R)	Purple Claw Weed	N Atlantic
Cystoseira abies-marina (B)		Gulf of Mannar
Cystoseira baccata (B)	Bushy Berry Wrack	N Atlantic
Cystoseira barbata (B)	Cistoseira	Bulgaria, Ukraine
Cystoseira crinita (B)	Cistoseira	Bulgaria, Ukraine
Cystoseira nodicaulis (B)	Bushy Noduled Wrack	N Atlantic
Cystoseira tamariscifolia (B)	Rainbow Bladderweed	N Atlantic
Dermocorynus dichotomus (R)		Asian Countries
Dermonema pulvinatum (R)		China, Taiwan
Dermonema virens (R)		Caribbean, China, Taiwan, Vietnam
Desmarestia dudresnayi subsp. *tabacoides* (B)		Korea
Devaleraea ramentacea (R)		Nova Scotia, Canada
Dictyopteris australis (B)	Limu Lipoa	Hawaii, India
Dictyopteris delicatula (B)		Caribbean
Dictyopteris plagiogramma (B)	Limu Lipoa	Hawaii
Dictyopteris polypodioides (B)	Netted Wing Weed, Uraboshiyahazu	NE Atlantic, India
Dictyopteris repens (B)	Auke	Easter Is
Dictyota acutiloba (B)	Brown Ribbon Weed, Limu Alani	Hawaii, Indonesia, Malaysia, Thailand
Dictyota bartayresiana (B)	Alani	Hawaii

Table 2.1 contd. ...

Table 2.1 contd. ...

Species	**Common names**	**Countries**
Dictyota caribaea (B)		Caribbean
Dictyota ciliolata (B)		Caribbean
Dictyota dichotoma (B)	Divided Net Weed, Brown Fan Weed, Amijigusa	Worldwide distribution, India, Indonesia, Malaysia, Thailand
Dictyota dichotoma var. *intricata* (B)	Common Forked Weed	Indian Ocean
Dictyota friabilis (B)		Polynesia
Dictyota menstrualis (B)		Caribbean
Dictyota mertensii (B)		Caribbean
Dictyota sandvicensis (B)	Limu Alani	Hawaii
Dictyota spiralis (B)		NE Atlantic
Dictyota stolonifera (B)		Caribbean
Digenea simplex (R)	Makuri, Kaijinso	Japan, SE Asia
Dilsea carnosa (R)	Red Rags	Britain, Iceland, Ireland, Scotland
Dumontia contorta (R)	Dumont's Tubular Weed, Ryumonso	Japan
Durvillaea antarctica (B)	Bull Kelp, Conchajugo	Chile, Chiloe Is, Peru
Durvillaea potatorum (B)	Bull Kelp	Australia
Ecklonia arborea (B)	Southern Sea Palm, Stiped Kelp, Arame	Asia
Ecklonia bicyclis (B)	Arame, Kajimi, Sagarame	Japan
Ecklonia cava (B)	Paddle Weed, Kajime	China, Japan, Korea
Ecklonia kurome (B)	Kunbu, Miangichai, Kurome	Japan
Ecklonia maxima (B)	Sea Bamboo	Namibia, South Africa
Ecklonia radiata (B)	Brown Kelp	New Zealand
Ecklonia stolonifera (B)	Kompi, Kizame Arame	Japan, Korea
Eckloniopsis radicosa (B)		Japan
Egregia menziesii (B)	Feather Boa, Boa Kelp	British Columbia, Canada
Eualaria fistulosa (B)	Dragon Kelp, Kausam, Kauan, Oni-Wakame	Alaska, Japan
Eucheuma amakusaense (R)		Japan
Eucheuma arnoldii (R)	Kanutkanot	Philippines

Eucheuma cartilagineum (R)		Japan
Eucheuma denticulatum (R)	Macassar, East Indian Carrageen, Spinosum, Unicorn Seaweed, Kanutkanot, Tambalang	China, Japan, Malaysia, Korea, Philippines, Hawaii
Eucheuma edule (R)	Agar-Agar-Besar, Hai Ts'Ai Mu, Makuri, Kaijinso	China, Indonesia
Eucheuma edule f. *majus* (R)	Agar-Agar-Besar	Indonesia, Pacific Is (New Caledonia)
Eucheuma horridum (R)		Indonesia, Malaysia, Philippines
Eucheuma isiforme (R)	Sea Moss, Seamoss	Jamaica, Trinidad
Eucheuma nudum (R)		Bermuda, Florida (US)
Eucheuma serra (R)	Bulung, Bulung Lipan, Bulung Djukut Lelipan	Japan, Indonesia, Malaysia, Borneo, Philippines
Eudesme virescens (B)	Yezo-Mozuku, Nise-Futomozuku	Japan
Feldmannia indica (B)	Limu Akaakoa, Limu Huluilio	Hawaii
Fucus ceranoides (B)	Horned Wrack, Estuary wrack	NE Atlantic
Fucus distichus (B)	Two-Headed Wrack, Popweed, Bladderwrack, Rockweed	NE Pacific
Fucus distichus subsp. *evanescens*	Popweed, Rockweed, Arctic Wrack, Hibatsunomata, Hibamata	Canada, Japan
Fucus serratus (B)	Serrated wrack, Saw wrack, Toothed wrack	NE Atlantic
Fucus spiralis (B)	Jelly Bags, Spiral Wrack, Flat Wrack, Spiraled Wrack	Azores (Portugal)
Fucus vesiculosus (B)	Paddy Tang, Paddy-Tang, Red Fucus, Dyers Fucus, Swine Tang, Sea Ware, Bladder, Rockweed, Bladderwrack, Popping Wrack, Wrack, Bladder Wrack	Japan, N America (Alaska), W Europe, Spain, Portugal, Scotland, Ireland
Furcellaria lumbricalis (R)	Black Carrageen, Clawed Fork Weed	Baltic Sea Region
Ganonema farinosum (R)	Baris-Baris	Philippines
Gayralia brasiliensis (G)		Brazil, Japan
Gayralia oxysperma (G)		Caribbean, Central America
Gelidiella acerosa (R)	Little Wire Weed, Shimatengusa	India, China, Indonesia, Malaysia, Vietnam, Philippines
Gelidiella indica (R)		India
Gelidiophycus divaricatus (R)	Shihua, Tanmae, Hime Tengusa	China, Japan, Korea
Gelidium amansii (R)	Ceylon Moss, Tengusa, Makusa, Genso	China, Japan, Korea, East Asia
Gelidium attenuatum (R)	Limo Loloa	Hawaii

Table 2.1 contd. ...

Table 2.1 contd. ...

Species	Common names	Countries
Gelidium corneum (R)	Atlantic Agar, Tengusa	NE Atlantic, Hawaii
Gelidium elegans (R)	Makusa	Asia
Gelidium filicinum (R)	Limu Loloa, Limu Ekahakaha	Hawaii
Gelidium japonicum (R)	Onigusa, Oyakusa	China
Gelidium micropterum (R)		Hawaii
Gelidium pacificum (R)	Obusa	Korea
Gelidium pulvinatum (R)		Hawaii
Gelidium pusillum (R)	Hai-Tengusa	Bangladesh, India, Hawaii
Gelidium serrulatum (R)	Seamoss	Caribbean
Gelidium spinosum (R)	Spiny Straggle Weed, Limu Loloa	Indonesia, Borneo, Philippines, Hawaii
Gelidium vagum (R)	Yore-Kusa	Japan, Korea, China
Gelidium vittatum (R)	Red Ribbons	South Africa
Gigartina papillata (R)	Grapestone, Tar Spot, Turkish Washcloth	Iceland, NE Pacific
Gigartina pistillata (R)	Pestle Weed	Iberian Peninsula
Gigartina polycarpa (R)	Tongue weed	South Africa
Gigartina skottsbergii (R)	Red Marine Algae, Luga Roja	Chile
Gloiopeltis complanata (R)	Hana-Fu-Nori	Japan, Korea
Gloiopeltis furcata (R)	Cretan Sea Plant, Chi Tsai, Chiao Tsai, Kaau Tsoi, Hung Tsai, Kita-Funori, Fukuro-Funori, Funori, Jelly Moss	Bering Sea and Aleutian Is, Alaska to Baja California, Mexico, China, Taiwan, Vietnam, Korea, Japan, Russia
Gloiopeltis tenax (R)	Glueweed, Jelly Moss, Funori, Yanagi-Funori, Hai Lo	China, Taiwan, Korea
Gracilaria arcuata (R)	Kanutkanot	Indonesia
Gracilaria articulata (R)		China
Gracilaria birdiae (R)	Macarrão do Mar	Brazil
Gracilaria blodgettii (R)	Guraman	Indonesia, tropical zones of W Atlantic
Gracilaria bursa-pastoris (R)	Shepherd's Purse Wart Weed, Limu Manauea, Ogo	Caribbean, England, Wales, Hawaii
Gracilaria canaliculata (R)	Susueldot-Baybay, Taiwan-Ogonori	Vietnam, Philippines, Hawaii
Gracilaria changii (R)		Malaysia

Gracilaria chilensis (R)	Pelillo	Japanese, Philippines, Hawaii
Gracilaria cornea (R)		Caribbean
Gracilaria coronopifolia (R)	Limu-Manauea, Limu Manauea, Cao-Caoyan, Gargararao	SE Asia, Hawaii
Gracilaria corticata (R)		SE Asia
Gracilaria damaecornis (R)		Caribbean
Gracilaria debilis (R)	Limu Koelen	Hawaii
Gracilaria domingensis (R)		Brazil, Caribbean, Central America
Gracilaria edulis (R)	Ceylon Moss	India, Indonesia, Philippines, Japan, Malaysia, Hawaii
Gracilaria ephemera (R)	Limu Aau	Samoa
Gracilaria foliifera (R)	Limo-Folha	Cuba, Gulf of Mexico, India
Gracilaria gigas (R)		Indonesia
Gracilaria gracilis (R)	Slender Wart Weed	NE Atlantic, South Africa, Vietnam
Gracilaria hainanensis (R)		China
Gracilaria heteroclada (R)		Philippines, Vietnam
Gracilaria incurvata (R)	Marakawayan	Philippines
Gracilaria maramae (R)		Polynesia, Fiji
Gracilaria minor (R)	Boeloeng, Bulung Buku	Indonesia, Taiwan, China, Philippines, Micronesia
Gracilaria multipartita (R)	Cleaved Wart Weed	NE Atlantic, Mediterranean
Gracilaria pacifica (R)	Red Ogo Seaweed, California Limu, Red Spaghetti	Alaska to California
Gracilaria parvispora (R)	Limu Ogo, Long Ogo, Ogo	Baja California, Japan, Korea, Polynesia, Hawaiian Is
Gracilaria pudumadamensis (R)		India, Pakistan
Gracilaria salicornia (R)	Boeloeng, Gorilla Ogo, Ogo, Robusta	Indian Ocean, China, Japan, Malaysia, Philippines, Taiwan, Thailand, Singapore, Vietnam, Australia, Pacific Is
Gracilaria tenuistipitata (R)		China, Japan, Indonesia, Malaysia, Singapore, Thailand, Vietnam
Gracilaria tenuistipitata var. *liui* (R)	Lumot	China, Taiwan, Philippines, Thailand, Vietnam
Gracilaria textorii (R)	Lablabig, Kabonori	Caribbean Is, Indian Ocean, NE and E Pacific, India, Sri Lanka, Yemen, China, Japan, Korea, Russia, Philippines, Taiwan, Vietnam, Australia
Gracilaria tikvahiae (R)	Graceful Red Weed, Ogo	Tropical and Subtropical W Atlantic, Caribbean Sea, SW Atlantic, Hawaii

Table 2.1 contd. ...

Table 2.1 contd. ...

Species	Common names	Countries
Gracilaria vermiculophylla (R)	Ogonori, Komulkosiraegi	N Atlantic, Tropical and Subtropical W Atlantic, China, Japan, Korea, Russia, Vietnam
Gracilariopsis andersonii (R)	Sea Spaghetti, Ogo	Caribbean Sea, Tropical W Atlantic, China, British Columbia to Mexico
Gracilariopsis lemaneiformis (R)	Red Wiry Weed	Tropical and Subtropical W Atlantic, E Pacific, China, Japan, Philippines, Hawaiian Is
Gracilariopsis longissima (R)	Thin Dragon Beard Plant, Ceylon Moss	NE Atlantic, Baltic Sea, Mediterranean, SW and SE Atlantic, Indian Ocean, Iran, Israel, Sri Lanka, Vietnam, Australia, Hawaiian Is, Polynesia, Samoan Is
Gracilariopsis tenuifrons (R)		Brazil, Venezuela, Cuba and Guadalupe
Grateloupia asiatica (R)		China, Japan, Korea, Vietnam
Grateloupia divaricata (R)	Katanori	China, Japan, Korea, Russia, Philippines, Vietnam, Australia
Grateloupia doryphora (R)	Cochayuyo	Adriatic Sea, Britain, France, Spain, Brazil, Venezuela, S Africa, Canary Is, Caribbean Is, Indian Ocean, China, Philippines, New Zealand
Grateloupia elliptica (R)		Japan, Korea
Grateloupia filicina (R)	Chop-Chop, Limu Pakaele-awa'a, Limu Hulu-hulu-waen, Limu Hula Hula, Mukade-Nori, Ratanho, Ratenho	NE Atlantic, Azores, Canary Is, Cape Verde Is, Madeira, Adriatic Sea, Mediterranean Sea, Tropical and Subtropical W and E Atlantic, Indian Ocean, China, Japan, Korea, Taiwan, Indonesia, Philippines, Singapore, Vietnam, Australia, Polynesia, Micronesia, Fiji, Hawaiian Is, Mariana Is, Marshall Is, Antarctic Is (Fuegia)
Grateloupia hawaiiana (R)		Hawaiian Is
Grateloupia indica (R)		Indian Ocean (India)
Grateloupia lanceolata (R)	Fudaraku	NE Atlantic (France), NE Pacific (California), NW Pacific (China, Japan)
Grateloupia livida (R)	Hai-Ts'ai, Tongue Centipede Algae	China, Japan, Korea, Taiwan, Vietnam
Grateloupia turuturu (R)	Devil's Tongue Weed, Red Lettuce, Tsurutsuru	Chile, Peru, native to Japan, China and Korea, but has spread to the NE Atlantic, Mediterranean, S America, Australia and New Zealand
Griffithsia corallinoides (R)	Mrs Griffiths's Coral Weed	NE Atlantic, Atlantic Is (Azores, Canary Is), Mediterranean, Indian Ocean, Asia (Japan, Korea, Vietnam)
Griffithsia ovalis (R)	Limu Moopunaa-ka-lipoa, Limu Moo-puna, Limu Ka-lipoa, Limu Au-pupu	S Atlantic, Indian Ocean, Philippines, Singapore, Australia, Pacific Is (Polynesia, Micronesia, Fiji, Mariana Is, Marshall Is, Hawaiian Is)

Gymnogongrus disciplinalis (R)	Limu Awikiwiki, Limu Nei, Limu Vavaloli	SE Pacific, Pacific Is (Hawaiian Is)
Gymnogongrus griffithsiae (R)	Ito-Okitsunori	NE and E Atlantic, Mediterranean, NW and SW Atlantic, Caribbean, Japan, Australia
Halimeda discoidea (G)	Money Plant, Uchiwasabotengusa	Atlantic Is (Bermuda, Canary Is, Cape Verde Is), Tropical and subtropical W Atlantic, India Ocean, Asia (China, Japan, Taiwan, Indonesia, the Philippines, Singapore, Thailand, Vietnam), Australia (Papua New Guinea), Pacific Is (Samoa, Polynesia, Micronesia, Fiji, Hawaiian Is, Mariana Is, Marshall Is, Palau, Solomon Is)
Halimeda incrassata (G)	Mitsudesabotengusa	Tropical and Subtropical W Atlantic, Caribbean, Indian Ocean, Asia (China, Japan, Taiwan), SE Asia (Indonesia, Philippines, Singapore, Thailand, Vietnam), Papua New Guinea, Australia
Halimeda opuntia (G)		Tropical and subtropical W Atlantic; Caribbean, Indian Ocean, Asia (China, Japan, Taiwan), SE Asia (Indonesia, Malaysia, Myanmar, the Philippines, Singapore, Thailand, Vietnam), Papua New Guinea, Australia; Pacific Is (Samoa, Polynesia, Micronesia, Fiji, Hawaiian Is, Mariana Is, Marshall Is, Republic of Palau, Solomon Is)
Halopteris scoparia (B)	Sea Flax Weed, Taggtofs, Hake-Kashirazaki, Yezokashirazaki	Norway to Cape Verde (Atlantic), Mediterranean, Adriatic, Black Sea
Halosaccion glandiforme (R)	Sea Sac, Dead Man's Fingers, Salt Sacs, Benifukuronori	Asia (Commander Is, Russia), NE Pacific (Alaska to California)
Halurus equisetifolius (R)	Sea Tail, Sea Horsetail	NE Atlantic, Atlantic Is (Azores, Canary Is), Mediterranean, Asia (Japan)
Halurus flosculosus (R)	Mrs. Griffith's Little Flower	NE Atlantic, Adriatic Sea, Atlantic Is (Azores, Canary Is, Cape Verde Is), Mediterranean, Auckland Is
Halymenia dilatata (R)	Gayunggayong	Indian Ocean, China, Japan, Korea, Taiwan, Indonesia, Malaysia, the Philippines, Singapore, Thailand, Vietnam, Australia, Pacific Is (Micronesia, Fiji, Mariana Is, Palau, Solomon Is)
Halymenia durvillei (R)	Aragan-Ilek, Gayunggayong, Gayong-Gayong, Limu Lepeahina, Limu Mumu, Limu A'au	Indian Ocean, Indonesia, Malaysia, Philippines, Singapore, Thailand, Australia, Pacific Is (Samoa, Polynesia, Micronesia, Fiji, Guam, Mariana Is, Palau, Solomon Is)
Halymenia flooresii (R)	Red Sea Lettuce, Dragons Tongue	Warm and temperate Atlantic, Indian Ocean, China, Japan, Taiwan, Indonesia, Malaysia, Philippines, Singapore, Australia, Pacific Is (Micronesia, Fiji, Mariana Is)
Halymenia floridana (R)	Red Sea Lettuce, Dragons Tongue	Atlantic Is (Bermuda, Selvage Is), Tropical and Subtropical W Atlantic, Caribbean, Indian Ocean

Table 2.1 contd. ...

Table 2.1 contd. ...

Species	Common names	Countries
Halymenia formosa (R)	Limu Lepe-o-Hina, Limu Lepe'Ula'Ula, Limu Lepeahina	Indian Ocean, Oman, Japan, Korea, Taiwan, Indonesia, Malaysia, Philippines, Singapore, Australia, Pacific Is (Hawaiian Is, Samoan Is)
Halymenia harveyana (R)	Gayunggayong	Africa, SE Asia (Indonesia, Philippines), Australia
Halymenia maculata (R)	Gayunggayong	China, Indonesia, Malaysia, Philippines, Singapore, Vietnam, Thailand, Australia (Papua New Guinea), Pacific Is (Fiji, Palau)
Himanthalia elongata (B)	Sea Thong, Thongweed, Buttonweed, Sea Haricots, Sea Spaghetti, Thong Weed	NE Atlantic (from Norway to Portugal), Baltic Sea
Hormophysa cuneiformis (B)	Tahalib, Wedgeshaped, Chainweed, Aragan, Irrakkai Pasi	Indian Ocean, Japan, Malaysia, Taiwan, Thailand, Philippines, Singapore, Vietnam, Australia, Pacific Is (Polynesia, Micronesia, Mariana Is, Palau, Fiji, New Caledonia, Solomon Is)
Hormosira banksii (B)	Neptune's Necklace	Australia and New Zealand
Hydroclathrus clathratus (B)	South Sea Colander, Sponge Seaweed, Kagomenori	Widely distributed in warm, subtropical and tropical seas
Hydropuntia caudata (R)		Tropical and Subtropical W Atlantic, Caribbean
Hydropuntia cornea (R)		Caribbean and S Atlantic (Brazil)
Hydropuntia crassissima (R)	Wild Seamoss	Atlantic Is (Bermuda), Tropical and Subtropical W Atlantic, Asia (Japan)
Hydropuntia edulis (R)	Ceylon Moss, Jafna Moss, Agar-Agar, Atjar, Ceylon, Doejoeng	Indian Ocean, SE Asia (Indonesia, Malaysia, Philippines), Australia, Pacific Is (Hawaiian Is)
Hydropuntia eucheumatoides (R)	Anggapang, Ambaang, Cauot-Cauot, Kauatkauat, Cavot-Cavot	Indian Ocean, China, Japan, Taiwan, Indonesia, Philippines, Thailand, Vietnam, Australia, Pacific Is (Micronesia, Fiji, Guam, Polynesia, Palau, Solomon Is)
Hydropuntia fisheri (R)		China, Malaysia, Myanmar, Philippines, Singapore, Thailand, Vietnam
Hydropuntia secunda (R)		Gulf of Mexico, Tropical W Atlantic, Caribbean Sea
Hypnea aspera (R)		Indian Ocean, China, Korea, Taiwan, Indonesia, Philippines, Vietnam, Australia
Hypnea cenomyce (R)		Atlantic Is (Cape Verde Is), Tropical and Subtropical W Atlantic, Indian Ocean, SW Asia (Bangladesh, Sri Lanka); Asia (Japan, Taiwan), SE Asia (Indonesia, Philippines, Vietnam), Australia, Pacific Is (Easter Is)
Hypnea charoides (R)	Spiny Red Weed, Kulot Ti Pusa, Kabutsu, Limu Vai, Ibaranori	W Indian Ocean (E Africa, Madagascar, Mauritius), China, Japan, Indonesia, Thailand, Philippines, Pacific Is and Australia

Hypnea chordacea (R)	Limu Huna	Asia (China, Japan, Taiwan, Indonesia), Pacific Is (Hawaiian Is)
Hypnea divaricata (R)	Arien	Atlantic Is (Cape Verde Is), Indian Ocean, India, Indonesia, Myanmar, Philippines, Australia, Pacific Is (Polynesia, Fiji, Samoan Is)
Hypnea esperi (R)	Ragutirit	Indian Ocean, Japan, Taiwan, Philippines, Singapore, Vietnam, Australia (Papua New Guinea), Pacific Is (Polynesia, Easter Is, Micronesia, Fiji, Polynesia, Marshall Is, Hawaiian Is), Antarctic Is (Fuegia); SE Pacific (Chile)
Hypnea hamulosa (R)		SW Atlantic (Brazil), Indian Ocean, Asia (Taiwan, Indonesia, Thailand), Australia; Pacific Is (Polynesia)
Hypnea japonica (R)	Japanese Red Algae	China, Japan, Korea, Russia, Taiwan, Pacific Is (Central Polynesia)
Hypnea musciformis (R)	Su-Wei-Tung, Crozier Weeds	China, Japan, Korea, Russia, Taiwan, Pacific Is (Central Polynesia)
Hypnea nidifica (R)	Limu Huna, Wane'One'O	Tropical and Subtropical Atlantic, Indian Ocean; SE Asia (Indonesia), New Zealand, NE Pacific (Alaska to California), Pacific Is (Hawaiian Is)
Hypnea pannosa (R)	Blue Hypnea, Gulot, Lumi Cevata, Limu 'ava, Tattered Sea Moss, Kulot	Widely distributed throughout the tropics and subtropics
Hypnea saidana (R)	Saidaibara	Indian Ocean, Japan, Korea, Taiwan, Philippines, Australia, Pacific Is (Polynesia, New Caledonia, Hawaiian Is)
Hypnea spicifera (R)	Agar Agar, Limu Huna	SE Atlantic, Indian Ocean, Pacific Is (Hawaiian Is), SE Pacific
Hypnea spinella (R)	Sa Ts'ai, Limu Huna, Boeloeng	Atlantic Is (Ascension, Bermuda, Canary Is, Cape Verde Is, Selvage Is), Tropical and Subtropical W and E Atlantic, Mediterranean, Persian Gulf, Indian Ocean, SW Asia (Arabian Gulf, Cyprus, India, Iran, Oman, Sri Lanka, Turkey), Asia (China, Japan, Korea, Taiwan), SE Asia (Indonesia, Myanmar, Singapore, Thailand, Vietnam), Australia, Pacific Is (Samoa, Polynesia, Micronesia, Hawaiian Is, Mariana Is, Marshall Is)
Iridaea cordata (R)	Rainbow Kelp, Rainbow Seaweed	Antarctic Is, NE Pacific (Alaska to California), SE Pacific (Argentina, Chile), Auckland Is
Iridophycus subdichotomus (R)	Chishima-Ginnan	Asia (Japan, Russia)
Ishige foliacea (B)	Neolppae	Asia (Japan, Korea), NE Pacific (Gulf of California)
Ishige okamurae (B)	Tieding Cai, Pae	Asia (China, Japan, Korea, Taiwan)
Ishige sinicola (B)	Iroro	Asia (China, Japan, Korea and Taiwan), W Pacific (California and Mexico)

Table 2.1 contd. ...

Table 2.1 contd. ...

Species	Common names	Countries
Kappaphycus alvarezii (R)	Adik Goma, Adik Kalas, Agal Agal, Agal Agal Besar, Agar Agar Besar, Agar Agar Palau, Agar Agar Seru Laut, Agar-Agar, Chilin-t' Sai, Cottonii, Elkhorn Sea Moss, Eucheuma, Eucheuman, Guso, Kab Kab, Kappa, Kirinsai, Purdoy, Tambalang, Tambalang Milo, Vanguarda, Goso	Native to the Indo-pacific (Malaysia and Philippines) but widely introduced and cultivated in the W Pacific, SW Atlantic, and Indian Oceans
Kappaphycus cottonii (R)	Guso, Kanot-Kanot, Kanutkanot	Indian Ocean, Asia (China, Japan, Taiwan), SE Asia (Indonesia, Malaysia, Philippines, Singapore, Vietnam), Pacific Is (Micronesia, Fiji, Guam, Mariana Is)
Kappaphycus striatus (R)	Kanot-Kanot, Kanutkanot, Kapkap	Asia (China, Japan), SE Asia (Indonesia, Malaysia, Philippines, Singapore), Pacific Is (Micronesia, Fiji, Hawaiian Is, Palau)
Laminaria abyssalis (B)		Tropical and Subtropical W and E Atlantic, Caribbean Sea
Laminaria digitata (B)	Tangle, Sea Girdles, Tangle Tail, Wheelbangs, Sea Wand, Sea Ware, Sea Tangle, Horsetail Kelp, Kelp, Strap Wrack, Oarweed, Oar Weed, Horsetail Tangle, Sea Girdle, Kombu Breton	N Atlantic, Atlantic Is (Canary Is, Greenland, Iceland), Baltic Sea
Laminaria hyperborea (B)	Mirkle, Kelpie, Liver Weed, Pennant Weed, Strapwrack, Cuvie, Tangle, Split Whip Wrack, Tangleweed, Cuvie, May-Weed, Sea Rods, Forest Kelp, Northern Kelp	European N Atlantic cold-temperate species which does not extend into areas influenced by Arctic waters; its range is the NE Atlantic Ocean, from Scandinavia south to Spain and the Canary Is, the Baltic Sea and the N Sea
Laminaria ochroleuca (B)	Golden Kelp, Atlantic Kombu	Warm-temperate species of kelp, and is most common in the NE Atlantic from the British Isles to the Sahara and the Atlantic Is, Mediterranean
Laminaria pallida (B)	Split-Fan Kelp, Split Fan	Atlantic Is (Canary Is, Tristan da Cunha Is), SE Atlantic (Namibia, S Africa), Indian Ocean
Laminaria setchellii (B)	Southern Stiff-Stiped Kelp, Split Blade Kelp, Split Kelp, Stiff-Stiped Kelp, Oar Weed	From the Aleutian Is in Alaska to Baja California in Mexico
Laminaria yezoensis (B)		Asia (Japan, Russia), NE Pacific (Alaska, British Columbia)
Laurencia botryoides (R)		Indian Ocean, SE Asia (Malaysia, Singapore), Australia, Pacific Is (Fiji)
Laurencia composita (R)	Kulot	Tropical and Subtropical W Atlantic, Asia (China, Japan, Korea, Philippines)
Laurencia coronopus (R)		Black Sea, SE Asia

Laurencia flexilis (R)		Atlantic Is (Canary Is, Madeira, Selvage Is), Indian Ocean, Asia (China, Japan), SE Asia (Indonesia, Philippines), Australia, Pacific Is (Polynesia, Micronesia, Tahiti)
Laurencia glomerata (R)	Limu Maneoneo, Limu Lipuupuu	SE Atlantic, India Ocean, SE Asia (Indonesia), Pacific Is (Hawaiian Is)
Laurencia mcdermidiae (R)		Hawaiian endemic
Laurencia nidifica (R)	Mustard Limu, Limu Mane'one'o, Limu Ape'ape'e	Atlantic Is (Cape Verde Is, Madeira), SE Atlantic, Indian Ocean, Asia (Japan, Korea), SE Asia (Indonesia, Philippines, Thailand, Vietnam), Australia, Pacific Is (Polynesia, Micronesia, Hawaiian Is, Samoan Is
Laurencia obtusa (R)	Rounded Brittle Fern Weed, Corsican Moss, Limu Ho`onunu, Sangan, Sanga	Worldwide in warm temperate to tropical seas
Laurencia pinnata (R)	Kulot	Asia (China, Japan, Korea, Russia, Indonesia, Malaysia, Philippines), Pacific Is (Polynesia)
Laurencia tropica (R)		China, Japan, Korea, Taiwan, Philippines, Vietnam, Pacific Is (Micronesia, Mariana Is)
Laurencia viridis (R)	Erva-Malagueta	Atlantic Is (Azores, Canary Is, Cape Verde Is, Madeira, Selvage Is)
Lemanea fluviatilis (R)		Indian Ocean, India
Lessonia corrugata (B)	Tasmanian Kombu, Strap Weed	Tasmania
Lessonia flavicans (B)		S America (Argentina, Chile, Falkland/Malvinas Is), Antarctic and the Sub-Antarctic Is
Lessoniopsis littoralis (B)	Short Kelp, Ocean Ribbons, Flat Pompom Kelp, Ocean Ribbon, Strap Kelp	Pacific coast of N America, from Kodiak Island, Alaska to central California
Liagora albicans (R)	Limu Pu-aki, Limu puaki, Limu Puak	Warm seas worldwide
Lobophora variegata (B)		Atlantic Is (Canary Is, Cape Verde Is), Pacific Is (Easter Is), S America (Brazil), Africa (Mauritius), Asia (Sri Lanka, Taiwan)
Lomentaria catenata (R)	Fushitsunagi	Asia (China, Japan, Korea), Australia
Macrocystis pyrifera (B)	Huiro, Giant Kelp, Sea Ivy, Giant Pacific Kelp, Small Perennial Kelp, Giant Perennial Kelp, Kelp Gigante	From the Kodiak Archipelago, Alaska, to northern Baja California, Mexico, South of Monterey, California. In the S hemisphere, S America, from Valparaíso, Chile southward to Tierra del Fuego, Argentina, S Africa, S Australia, Tasmania, New Zealand, and several Sub-Antarctic Is

Table 2.1 contd. ...

Table 2.1 contd. ...

Species	Common names	Countries
Mastocarpus pacificus (R)	Hoso-Ibonori	Asia (Japan, Russia, Taiwan), NE Pacific (Alaska)
Mastocarpus papillatus (R)	Grapestone, Tar Spot, Turkish Washcloth	N Atlantic (Iceland), NE Pacific (Alaska to California, Mexico)
Mastocarpus stellatus (R)	Grape Pip Weed, Cluimhin Cait, Carragheen, False Irish Moss	NE Atlantic (Scandinavia to Mauretania, N Sea), NW Atlantic (Newfoundland and Nova Scotia to Rhode Island)
Mazzaella denticulata (R)		SE Pacific (Chile, Peru)
Mazzaella japonica (R)	Akabaginnansou	Asia (China, Japan, Russia)
Mazzaella laminarioides (R)	Kuroba-Ginnansô, Atsuba-Ginnansô, Luga, Luga Corta, Luga Cuchara	Atlantic Is (Gough Is), Mediterranean, Asia (Japan, Russia), NE Pacific (Alaska to California), SE Pacific (Chile), Antarctic Is
Mazzaella splendens (R)	Rainbow Seaweed, Rainbow Leaf Seaweed, Splendid Iridescent Seaweed	NE Pacific (Alaska to California)
Melanamansia glomerata (R)	Limu Li-pepe-iao, Limu Pepe-iao, Amansia, Lipepeiao, Limu Ha'ula	Tropical Indian Ocean, SE Asia, Japan, Tropical Pacific Ocean
Meristotheca papulosa (R)	Rosy Pudding Plant, Tosakanori, Keikansai, Tosaka-Nori, Lumu Mie'ta	Asia (China, Japan, Korea, Taiwan), SE Asia (Indonesia, Philippines), Australia, Pacific Is (Polynesia)
Meristotheca procumbens (R)	Lum Mi'a	Australia, Pacific Is (Cook Is, Fiji, Polynesia, Samoa)
Mesophyllum lichenoides (R)	Pink plates	NE Atlantic (Ireland to Mauritania), Mediterranean
Monostroma angicava (G)	Yezo-Hitoegusa	China, Japan, Korea
Monostroma antarcticum (G)		Japan, New Zealand
Monostroma crassidermum (G)	Atsukawa-Hitoe	Japan, Korea
Monostroma grevillei (G)	Green Laver, Hotparae, Strutsallat	NE Atlantic, Canadian Arctic, Atlantic Is (Azores, Greenland, Iceland), Asia (Japan, Korea, Russia), NE Pacific (Alaska, California), Antarctic Is
Monostroma kuroshiense (G)	Shimanto Nori	Japan
Monostroma latissimum (G)	Jade Nori, Hai Tsai, Hai-Cai, Awo-Nori, Aonori, Hitoegusa, Aonoriko	Adriatic Sea, NE Atlantic, SW Atlantic (Brazil), Asia (Japan, Korea, Taiwan, Philippines), Australia and New Zealand
Monostroma nitidum (G)	Zi-Cai, Aonori, Aonoriko, Tsukushi-Amanori, Hitoegusa	Asia (China, Japan, Korea, Taiwan, Thailand, Philippines, Vietnam), New Zealand, Pacific Is (Fiji)
Monostroma quaternarium (G)		NE Atlantic (Britain, Spain), NE Pacific (Alaska to Baja California, Costa Rica, Peru)

Nemacystus decipiens (B)	Haida, Hai Yun, Mozuku, Mozuka, Tanga'u	India Ocean, SW Asia (Arabian Gulf, Bahrain, India, Kuwait, Pakistan, Saudi Arabia), Asia (China, Japan, Korea), Australia, Pacific Is (Hawaiian Is, Tonga Is)
Nemalion elminthoides (R)	Threadweed, Sea Noodles, Sea Noodle, Umi-Zomen, Umisomen, Tsukomo Nori, Esparguete-da-Costa	NE Atlantic (Scandinavia to Canary Is), SE Atlantic, SW Atlantic (Brazil, Uruguay), Mediterranean, NW Pacific (Japan), NE Pacific (Alaska to Mexico), Australia and New Zealand
Nemalion multifidum (R)	Tsukomo-Nori	N Atlantic, Baltic Sea, Asia (Japan)
Nemalion vermiculare (R)	Crop of Threads, Sea Noodles, Threadweed, Somen-Nori, Tsukomo-Nori, Umi-Somen, Umisomen, Guksunamul	Asia (China, Japan, Korea, Russia)
Neodilsea yendoana (R)	Akaba, Akahata	Asia (Japan, Russia)
Neohypophyllum middendorffii (R)	Chikaputsuro, Setakemaa	Asia (Japan, Russia), NE Pacific (Alaska)
Neorhodomela larix (R)	Black Pine, Fujimatsumo	Asia (Bearing Sea, Japan, Korea, Russia), NE Pacific (Alaska to California)
Nereocystis luetkeana (B)	Ribbon Kelp, Giant Kelp, Bull Whip Kelp, Bull Kelp, Sea Whip, Horsetail Kelp, Bladder Kelp, Sea Otter's Cabbage, Serpent Kelp, Giant Bull Kelp, Bullwhip Kelp	NE Pacific (Alaska to Central California), NW Pacific (Commander Is)
Nitophyllum adhaerens (R)	Limu Haula	Tropical and Subtropical W Atlantic, Caribbean Sea, SE Asia (Vietnam), Pacific Is (Polynesia, Micronesia, Guam, Hawaiian Is)
Odonthalia corymbifera (R)	Hakesaki-Nokogirihiba	Asia (Japan, Korea, Russia)
Odonthalia floccosa (R)	Bewildering Brush, Sea Brush, Fusa-Nokogirihiba	Asia (Bearing Sea, Japan); NE Pacific (Alaska to California)
Odonthalia kamtschatica (R)	Kamchakka-Nokogirihiba	Asia (Japan, Russia), NE Pacific (Alaska, British Columbia)
Osmundea osmunda (R)	Pepper Dulse, Royal Fern-Weed	NE Atlantic, Atlantic Is (Madeira), Tropical and Subtropical W and E Atlantic, Asia (Korea)
Osmundea pinnatifida (R)	Limu Maneoneo, Limu Olipeepee, Limu Lipee, Argacinho-das-Lapas, Botelho-Preto, Pele-de-Lapa, Erva-Malagueta	N Atlantic, Atlantic Is, N Sea, English Channel, Mediterranean, Pacific Is (Hawaiian Is)
Osmundea truncata (R)		NE Atlantic, Atlantic Is (Canary Is, Selvage Is), Adriatic Sea, Baltic Sea, Black Sea, Mediterranean
Padina antillarum (B)		Atlantic Is (Selvage Is), Tropical and subtropical W and E Atlantic, SW Asia (India, Sri Lanka), SE Asia (Indonesia, Philippines, Singapore)
Padina arborescens (B)	Umiuchiwa	Asia (China, Korea, Japan, Taiwan), SE Asia (Philippines)

Table 2.1 contd. ...

Table 2.1 contd. ...

Species	Common names	Countries
Padina australis (B)	Fan-Leaf Seaweed, Dunggan-Dunggan, Agar-agar Daun Besar	SE Atlantic, Indian Ocean, SW Asia (Arabian Gulf, Bangladesh, India, Iran, Kuwait), Asia (China, Japan, Korea, Taiwan), SE Asia (Indonesia, Myanmar, the Philippines, Singapore, Thailand, Vietnam), Australia, Pacific Is (Polynesia, Easter Is, Micronesia, Fiji, Hawaiian Is, Palau, Solomon Is)
Padina boergesenii (B)	Leafy Rolled-Blade Alga	Tropical, subtropical and warm-temperate seas
Padina boryana (B)	Ear-Like Seaweed, Limu Lautaliga	Adriatic Sea, Mediterranean, Equatorial E Atlantic (S. Tomé and Príncipe), Indian Ocean, Red Sea, Asia (China, Japan, Taiwan, Indonesia, Malaysia, Thailand, Singapore, Vietnam), Australia, Pacific Is (Samoa, Micronesia, Mariana Is, Polynesia, Marshall Is, Fiji, Hawaiian Is, Solomon Is)
Padina crispata (B)		Tropical W Atlantic, Caribbean Sea, Tropical E Pacific
Padina tetrastromatica (B)		SE Atlantic, Indian Ocean, SW Asia (Arabian Gulf, India, Iran, Oman, Sri Lanka), Asia (China), SE Asia (Indonesia, Malaysia, Philippines, Thailand), Australia
Palisada intermedia (R)	Kulot	Atlantic Is (Cape Verde Is), Tropical and Subtropical W and E Atlantic, Asia (China, Korea, Taiwan, Indonesia, Philippines), Australia (Papua New Guinea)
Palisada perforata (R)	Culot, Kulot, Culot-Tumeng, Limu Maneoneo, Limu Lipeepee, Papirasozo	Atlantic Is (Ascension, Canary Is, Cape Verde Is), Tropical and Subtropical W and E Atlantic, Adriatic Sea, Mediterranean, Persian Gulf, Indian Ocean, Asia (Indonesia, Japan, Taiwan, Korea, Philippines, Thailand, Vietnam), Australia, Pacific Is (Fiji, Micronesia, Mariana Is, New Caledonia, Hawaiian Is)
Palmaria hecatensis (R)	Leathery Dulse, Stiff Red Ribbon	NE Pacific (Alaska to Oregon, Aleutian Is)
Palmaria mollis (R)	Dulse, Ribbon Seaweed, Pacific Dulse, Red Ribbon, Red Kale	NE Pacific (Alaska to Oregon, Aleutian Is)
Palmaria palmata (R)	Dillisk, Dillesk, Crannogh, Water Leaf, Sheep Dulse, Dried Dulse, Dulse, Shelldulse	North coasts of the Atlantic and Pacific Oceans, as far north as Arctic Canada and Russia, and as far south as Portugal in Europe, and New Jersey and California in the United States (US). In the W Pacific, the southern range of *P. palmata* includes Japan and Korea
Papenfussiella kuromo (B)	Kuromo	Atlantic Is (Azores, Canary Is, Selvage Is), Asia (China, Japan, Korea, Russia)

Parviphycus pannosus (R)		Tropical and subtropical W Atlantic, Atlantic Is (Azores, Canary Is, Cape Verde Is, Madeira, Selvage Is), SW Mediterranean (Balearic Is, Corsica, France, Greece, Italy, Malta, Sardinia, Spain, Tunisia), Indian Ocean, SW Asia (Bangladesh, India, Israel, Turkey), SE Asia (Indonesia), Australia, Pacific Is (Polynesia, Micronesia, Fiji, Mariana Is, Samoa), Asia (China, Japan, Korea, Russia)
Pelvetia canaliculata (B)	Cow Tang, Channeled Wrack, Channel Wrack, Botelho-Bravo	NE Atlantic, from the Arctic Ocean to the Iberian Peninsula, in the English Channel, and in the N Sea. It is common on the Atlantic shores of Europe from Iceland to Spain and Portugal
Petalonia binghamiae (B)	Habonori, Miyeoksoi	Atlantic Is (Azores), Tropical and subtropical W and E Atlantic, Asia (China, Japan, Korea, Taiwan), Australia and New Zealand, Pacific Is (Hawaiian Is), NE Pacific (California)
Petalonia fascia (B)	Sea Petals, Broad Leaf Weed, Haba-Nori, Seiyohabanori, Hondawara	N Atlantic (Greenland to Canary Is), Mediterranean, NW Atlantic (Arctic to New Jersey), SE Atlantic (Senegal, Namibia, S Africa), SW Atlantic (Brazil, Uruguay), Indian Ocean (Pakistan, S Africa), NW Pacific (Japan, China), NE Pacific (Alaska to California); SE Pacific (Chile), Australia, New Zealand, Antarctica, Sub-Antarctica
Phymatolithon calcareum (R)	Calcified Seaweed, Mäerl	NE Atlantic, Atlantic Is (Faroe Is, Iceland)
Pithophora roettleri (G)	Limu Palawai, Limu Lipalawai	NE Atlantic, Tropical and Subtropical W Atlantic, Asia (Japan), SE Asia (Singapore, Vietnam), Pacific Is (Hawaiian Is)
Pleurophycus gardneri (B)	Kelp, Tender Kelp, Sea Spatula	NE Pacific (Alaska to California)
Plocamium cartilagineum (R)	Cockscomb, Cock's Comb, Branched Cock's Comb, Red Comb Weed	NE Atlantic (Scandinavia to Senegal, North Sea), SE Atlantic (Namibia), Mediterranean, Indian Ocean (Pakistan, Mauritius), NW Pacific (Japan), Pacific Is; NE Pacific (Alaska to California), SE Pacific (Chile), Australia, New Zealand, Antarctica
Polyneura latissima (R)	Crisscross Network	NE Pacific (Alaska to California, Mexico), Asia (Russia, Philippines)
Polyopes affinis (R)	Limu pala-wai, Limu Li-pala Wai, Come-Nori, Kome-Nori, Matsunori	Asia (China, Japan, Korea, Indonesia, Philippines)
Polyopes lancifolius (R)		N Atlantic (Channel Is), Asia (China, Japan, Korea, Taiwan)
Polyopes prolifer (R)	Kome-Nori	Asia (China, Japan, Korea, Philippines, Taiwan)
Polysiphonia fucoides (R)	Black Siphon Weed	Arctic, White Sea, N Atlantic, Adriatic Sea, Black Sea, Atlantic Is (Azores, Canary Is, Greenland, Iceland, Madeira, Selvage Is), Tropical and Subtropical Atlantic

Table 2.1 contd. ...

Table 2.1 contd. ...

Species	Common names	Countries
Polysiphonia mollis (R)	Limu Pualu, Limu Hawane, Limu Pepe-iao	Indian Ocean, SE Asia (Indonesia, Philippines), Australia, Pacific Is (Polynesia, Fiji, Hawaiian Is), NE Pacific (Alaska to Baja California), SE Pacific (Chile)
Polysiphonia morrowii (R)	Moroitogusa	Mediterranean, Asia (China, Japan, Korea, Russia), New Zealand, SE Pacific (Chile)
Porphyra akasakae (R)	Murone-Amanori	Asia (Japan, Korea)
Porphyra angusta (R)	Kosuji-Nori	Asia (Japan, Korea, Taiwan)
Porphyra atropurpurea (R)	Limu Luau, Lipahee, Garnet	Mediterranean (Italy, Turkey), SW Atlantic (Brazil), SE Asia (Indonesia, Philippines)
Porphyra capensis (R)	Cape Laver, Purple Laver	SE Atlantic (Angola, Namibia, S Africa), India Ocean, Antarctic Is, SE Pacific (Chile)
Porphyra indica (R)	Laver, Nori	Indian Ocean
Porphyra kanyakumariensis (R)		Indian Ocean (India)
Porphyra laciniata (R)	Laver Sloke, Laver, Laver Slack, Red Laver, Purple Laver, Hoshinori, Asakusa-Nori, Asakusa Nori	Indian Ocean (India)
Porphyra linearis (R)	Winter Laver, Erva-do-Calhau, Erva-Patinha	NE and NW Atlantic, Atlantic Is, Mediterranean, Adriatic Sea, Baltic Sea, NE Pacific (Alaska), Asia (Japan)
Porphyra marcosii (R)	Gamet	SE Asia (Philippines), Australia
Porphyra marginata (R)		Asia (China)
Porphyra monosporangia (R)		Asia (China)
Porphyra ochotensis (R)	Ana-Amanori	Asia (Japan, Russia), NE Pacific (Alaska)
Porphyra okamurae (R)	Kumo-Nori	Asia (Japan, Russia), NE Pacific (Alaska)
Porphyra oligospermatangia (R)		Asia (China)
Porphyra purpurea (R)	Chi Choy, Purple Laver, Purple Vegetable, Red Laver, Purple Laver, Rose Nori, Red Nori, Amanori, Asakusa Nori, Hoshinori, Nori	Currently growing across most the Northern hemisphere, likely due to its preference for low temperature water
Porphyra umbilicalis (R)	Laver-Bread, Purple Laver, Sloak, Slook, Laver, Tough Laver, Laitue Rouge, Chishima-Kuronori, Erva-Patinha, Folhuda, Nori-Atlântico	NE Atlantic (Norway to Portugal), W Mediterranean., NW Atlantic (Labrador in Canada to the mid-Atlantic coast of the US), Asia (Japan)

Portieria hornemannii (R)	Hosoba-Naminohana	Indian Ocean, Asia (China, Japan, Korea, Taiwan), SE Asia (Indonesia, Philippines, Singapore, Sri Lanka, Vietnam), Australia, Papua New Guinea, Pacific Is (Fiji, Hawaiian Is, Mariana Is, Micronesia, Polynesia)
Postelsia palmaeformis (R)	Sea Palm, Sea Palm Kelp, See-Palme, Gaye	NE Pacific (from Hope Is, British Columbia South to the Southern central coast of California)
Prasiola japonica (G)	Kawa-Nori, Datyagawa-Nori, Nikko-Nori, Kawanori	Asia (China, Japan)
Protomonostroma undulatum (G)	Hida-Hitoe	N Atlantic, Atlantic Is (Greenland, Iceland), Tropical and Subtropical W Atlantic, Baltic Sea, Asia (China, Japan, Korea, Russia), Argentina, Antarctic Is, NE Pacific (Alaska)
Pseudochorda nagaii (B)	Nise-Tsurumo	N Atlantic, Atlantic Is (Greenland, Iceland), Tropical and Subtropical W Atlantic, Baltic Sea, Asia (China, Japan, Korea, Russia), Argentina, Antarctic Is, NE Pacific (Alaska)
Pterocladia lucida (R)	Agar Weed	Atlantic Is (St. Helena), SE Asia (Indonesia), Indian Ocean, Australia and New Zealand
Pterocladiella capillacea (R)	Yimaocai, Limu Loloa, Obakusa, Kata-Obakusa, Branched Wing Weed, Kaeumu, Musgo	NE Atlantic, Atlantic Is (Azores, Canary Is, Madeira), Adriatic Sea, SE and SW Atlantic, Asia (China, Japan, Korea, Taiwan), Australia and New Zealand; Pacific Is (Hawaiian Is)
Pterosiphonia bipinnata (R)	Black Tassel, Itoyanagi	Asia (Japan, Russia), NE Pacific (Alaska to California)
Pterygophora californica (B)	Pompom Kelp	NE Pacific (Alaska to California, Mexico)
Pyropia abbottiae (R)	Black Seaweed, Summer Seaweed	NE Pacific (Alaska to British Columbia), Asia (Commander Is)
Pyropia acanthophora (R)		Atlantic Is (Canary Is), Tropical and subtropical W Atlantic
Pyropia columbina (R)	Common Porphyra, Purple Laver, Maori, Karengo, Luche, Luche Rojo, Yuyo, Cochayuyo	Chile, Argentina, Falkland Is, New Zealand
Pyropia dentata (R)	Oni-Amonari	Asia (China, Japan, Taiwan)
Pyropia fallax (R)	False Laver	NE Pacific (Alaska to Oregon)
Pyropia haitanensis (R)	Zicai	Asia (China)
Pyropia kuniedae (R)	Maruba-Nori	Asia (Japan, Korea)
Pyropia leucosticta (R)	Pale Patch Laver, Limu Lua'u Limu Lipahee, Limu luan, Limu Lua'u, Erva-do-Calhau, Erva-Patinha, Cocha	NE Atlantic, Atlantic Is, N Sea, Baltic Sea, Mediterranean, Black Sea, NW Atlantic (Maine to Florida), Antarctic and the Sub-Antarctic Is
Pyropia nereocystis (R)	Chi Choy, Bull Kelp Laver, Red Laver, Purple Laver, Rose Nori, Red Nori, Nori	N America (Alaska, British Columbia, California, Washington)

Table 2.1 contd. ...

Table 2.1 contd. ...

Species	Common names	Countries
Pyropia onoi (R)		Asia (Japan, Russia)
Pyropia perforata (R)	Chi Choy, Black Seaweed, Nori, Laver, Wild Nori, Red Laver	NE Pacific (Alaska to Oregon), Asia (Japan, Malaysia), Antarctic and Sub-Antarctic Is
Pyropia pseudolanceolata (R)	Laver Seaweed	NE Pacific (Alaska to California), E Pacific (Peru)
Pyropia pseudolinearis (R)	Uppurui-Nori	NE Pacific (Alaska), Asia (Japan)
Pyropia seriata (R)	Ichimatsu-Nori	Asia (China, Japan, Korea, Russia)
Pyropia spiralis (R)		Tropical and subtropical W Atlantic)
Pyropia suborbiculata (R)	Tsz Tsai, Tsu Tsoi, Tsu Tsai, Chi Tsai, Hung Tsoi, Hung Tsai, Chi Choy, Hai Tsai, Hai Tso, Zi-Cai, Red Laver, Gamet, Mambiama, Maruba-Amanori, Iwanori, Kim	Tropical and subtropical W Atlantic, Asia (China, Japan, Korea, Taiwan, Philippines, Vietnam), Australia and New Zealand, NE Pacific
Pyropia tenera (R)	Tsz Tsai, Tsu Tsoi, Tsu Tsai, Chi Tsai, Hung Tsoi, Hung Tsai, Chi Choy, Hai Tsai, Tai Tso, Zicai, Asakusa Nori, Amanori, Hoshi-Nori, Kuro-Nori, Sushi Nori, Chishima Kuro-Nori, Tisima, Kim, Nuru	Indian Ocean, NW Pacific (China, Japan, Korea)
Pyropia torta (R)	Teal Nori, Winter Seaweed, Winter Black Seaweed	N America (Alaska, British Columbia, Washington), Asia (Commander Is)
Pyropia vietnamensis (R)	Limu Pahe'e	Indian Ocean; Asia (China, Indonesia, Taiwan, Thailand, Vietnam); Pacific Is (Hawaiian Is)
Pyropia yezoensis (R)	Zicai, Open Sea Nori, Susab-Nori, Susabi-Nori, Amanori, Kim	Asia (China, Japan, Korea, Russia)
Rhodoglossum pulchrum (R)	Akaba-Ginnansô, Usaba-Ginnansô	Asia (Japan), NE Pacific (Alaska)
Rhodomela sachalinensis (R)	Niretsu-Fujimatsu	Asia (Japan, Russia)
Rhodothamniella floridula (R)	Sand binder	NE Atlantic (Ireland and Britain to Portugal), SE Atlantic (Namibia, S Africa), SW Atlantic (Argentina)
Rhodymenia corallina (R)		Falkland Is/Islas Malvinas, Indian Ocean, Australia (Papua New Guinea), SE Asia (Indonesia), Pacific Is (Polynesia), E Pacific (Chile, Peru)
Rhodymenia pseudopalmata (R)	Dulse, Rosy Fan Weed	NE Atlantic, Adriatic Sea, Atlantic Is (Azores, Canary Is, Cape Verde Is, Madeira, Selvage Is), Gulf of Mexico, Caribbean Sea, Tropical and subtropical Atlantic, E Pacific (Galapagos Is)

Rosenvingea intricata (B)	Slippery Cushion, Samsamit	N and S Atlantic (temperate and tropical Atlantic), Indian Ocean Is, SW Asia (Arabian Gulf, Bangladesh, India, Yemen), Asia (China, Japan), SE Asia (Indonesia, Philippines, Vietnam), Australia, Pacific Is (Polynesia, Micronesia, Fiji, Hawaiian Is, Mariana Is, Marshall Is, New Caledonia, Samoan Archipelago)
Saccharina angustata (B)	Hai Dai, Hai Tai, Kunpu, Tender Kombu, Shreaded Kombu, Kizami-Kombu, Kombu, Sopaushi, Shiohoshi-Kombu, Urakawa-Kombu, Shamani Kombu, Tokachi-Kombu, Dashi-Kombu, Mizu-Kombu	Asia (Japan, Russia)
Saccharina bongardiana (B)	Split Kelp	NE Pacific (Alaska to California), NW Pacific (Kamchatka, Japan and Russia)
Saccharina cichorioides (B)	Chiimi-Kombu	Asia (Japan, Korea, Russia)
Saccharina cichorioides f. *coriacea* (B)		Asia (China, Japan, Russia)
Saccharina dentigera (B)	Mendocino Coast Kombu, Kumade-Kombu	NE Pacific (Alaska to California), Asia (Japan, Russia)
Saccharina diabolica (B)	Thin Snow Kombu, Black Kombu, Cloudy Kombu, White-Pulpy Kombu, Black Kombu, Cloudy Kombu, Oni-Kombu, Kuro-Tororo Kombu, Oboro Kombu	Asia (Japan, Russia)
Saccharina groenlandica (B)	Short Kelp, Kombu, Split Kelp, Sugar Wrack	Arctic (Canada), Atlantic Is (Greenland), NW Atlantic, NE Pacific (Alaska to California), Asia (Russia)
Saccharina gyrata (B)	Sea Banner, Tororo Kombu	Asia (Japan, Russia)
Saccharina japonica (B)	Hai Dai, Hai Tai, Kunpu, Royal Kombu, Makombu, Shinori-Kombu, Hababiro-Kombu, Oki-Kombu, Uchi Kombu, Moto-Kombu, Minmaya-Kombu, Ebisume, Kombu, Hirome, Umiyama-Kombu, Rishiri-Kombu, Para-Kompo, Dashi Kombu, Menashi-Komou, Birodo-Kombu, Teshio-Kombu, Kuro-Kombu, Koteshio, Hosome-Kombu, Shio-Kombu, Hoiro-Kombu, Hae tae, Tasima, Dasima	Asia (China, Japan, Korea)
Saccharina japonica f. *longipes* (B)	Enaga-Onikombu	Asia (Japan, Russia)
Saccharina latissima (B)	Sea Belt, Poor Man's Weather Glass, Sweet Wrack, Sugar wrack, Sugar Tang, Oarweed, Tangle, Kelp, Sugar Sea Belt, Sweet Tangle, Sugarwrack, Zuckertang, Royal or Sweet Kombu, Laminaire Sucrée, See-Palme, Karafuto Kombu, Karafuto Tororo Kombu, Kan-Hoa, Rabeiro, Kombu-Real	N and NE Atlantic (Greenland to Portugal, N Sea, Baltic), NW Atlantic (Canadian Arctic to Massachusetts), NE Pacific (Alaska to California)

Table 2.1 contd. ...

Table 2.1 contd. ...

Species	Common names	Countries
Saccharina longicruris (B)	Kelp, Oarweed, Atlantic Kombu, Atlantic Kelp	Arctic (Canada), N Atlantic (New England, US)
Saccharina longipedalis (B)		Asia (Japan)
Saccharina longissima (B)	Hai Dai, Hai Tai, Kunpu, Naga-Kombu, Ma Kombu, Nimotsu-Kombu, Gimberi-Kombu, Kimberi-Kombu, Mizu-Kombu, Shima-Kombu, Wakaoi, Barner Kombu	Asia (China, Japan, Russia)
Saccharina religiosa (B)	Hosame-Kombu, Saimatsu-Kombu, Aegidasima	Asia (Japan, Korea, Russia)
Saccharina sculpera (B)	Gagome	Asia (Japan, Korea)
Saccharina sessilis (B)	Bubbly Kelp, Sea Cabbage, Sweet Kombu	Asia (Japan, Korea)
Saccorhiza polyschides (B)	Furbelows, Furbellows, Caixeira, Carocha, Cintas, Golfe, Golfo, Limo-Correia, Limo-Corriola	NE Atlantic, Atlantic Is (Canary Is)
Sarcodia dentata (R)	Tosaka, Tosaka Nori	Indian Ocean, Asia (Japan), SE Asia (Indonesia), Auckland Is
Sarcodia montagneana (R)	Bebiroe, Bibiru, Atsuba-Nori	Indian Ocean, SW Asia (India, Oman, Sri Lanka, Yemen), Asia (China, Japan, Taiwan), SE Asia (Indonesia, Philippines), Australia, Antarctica and sub-Antarctic Is
Sarcodiotheca gaudichaudii (R)	Red String Seaweed	NE Atlantic (Britain), NE and E Pacific (Alaska to California, Chile, Galápagos Is, Peru)
Sarconema filiforme (R)		Indian Ocean, Asia (China), SE Asia (the Philippines), Australia, Pacific Is (Polynesia, Samoa)
Sarcothalia crispata (R)	Luga Negra, Negra, Crespa	SE Pacific (Chile)
Sargassum aquifolium (B)	Qunbar Al-ma, Binder Sargassum Weed, Sea Oak, Limu Kala, Limu-Kala, Limu Honu, Holly Limu, Kala-Launui, Kala-Lauli'i, Rimu Akau, Arien Wari, Limu Vaova	Asia (China, Japan, Taiwan), SE Asia (Indonesia, Philippines, Singapore, Vietnam, Malaysia), SW Asia (Arabian Gulf, India, Iran, Kuwait), Indian Ocean Is, Australia, Pacific Is (Samoa, Polynesia, Fiji, Hawaiian Is, Solomon Is)
Sargassum cinctum (B)	Aragan, Pinong Samo	Indian Ocean, Asia (China, Indonesia, Philippines, Vietnam), Australia, Pacific Is (New Caledonia, Samoan Is)
Sargassum confusum (B)	Aragan, Fushisuji-Moku, Hushisujimoku	Asia (China, Japan, Korea, Philippines, Vietnam)
Sargassum corderoi (B)		SE Asia (Philippines), Australia
Sargassum coreanum (B)		Asia (Japan, Korea)

Sargassum cymosum (B)	Limu Kala, Sargazo	Atlantic Is (Azores, Bermuda, Canary Is, Cape Verde Is, Selvage Is), Tropical and Subtropical W and E Atlantic, Caribbean, Indian Ocean, Asia (China, Japan, Taiwan), SE Asia (Vietnam), Pacific Is (Hawaiian Is)
Sargassum fulvellum (B)	Gulf Weed, Hondawara, Mojaban, Wu Lei Ma Wei Zao	Asia (China, Japan, Korea and Taiwan)
Sargassum fusiforme (B)	Hai Tso, Chiau Tsai, Hai Ti Tun, Hai Toe Din, Hai Tsao, Hoi Tsou, Chu-Chiau Ts'ai, Hijiki, Hiziki, Nongmichae, Tot, Yang Xi Cai	Asia (China, Japan, Korea)
Sargassum hemiphyllum (B)	Aragan, Ban Ye Ma Wei Zao	Asia (China, Japan, Korea, Philippines, Taiwan)
Sargassum henslowianum (B)	Heng Shi Ma Wei Zao	Asia (China, Hong Kong, Japan, Taiwan), SE Asia (Vietnam)
Sargassum horneri (B)	Devil Weed, Aka-Moku, Akamoku, Kwaengsaegi-Mojaban, Tong Zao	Asia (China, Japan, Korea, Taiwan, Philippines), NE Pacific (California to Mexico)
Sargassum muticum (B)	Wireweed, Jap Weed, Tamahahakimoku, Hai Shu Zi	Native to waters around Japan, Russia, Korea, and China, but has subsequently spread via ballast water and oyster shells to such far distant places as Europe, Mediterranean to the Adriatic, and N America from Alaska to Baja California
Sargassum obtusifolium (B)	Limu Kala	SE Asia, Pacific Is, S America
Sargassum oligocystum (B)		Indian Ocean, Asia (China, Japan, Taiwan, Indonesia, Malaysia, Philippines, Singapore, Thailand, Vietnam), Australia, Pacific Is (New Caledonia, Palau, Samoa, Solomon Is)
Sargassum pallidum (B)	Hai Hao Zi, Da Hao Zi, Hai Gen Cai, Hai Cao, Hai Zao	Asia (China, Japan, Korea, Russia), SE Asia (Indonesia)
Sargassum platycarpum (B)	Sargazo	Tropical and subtropical Atlantic, Caribbean Is
Sargassum polyceratium (B)	Sargazo	Tropical and Subtropical W Atlantic, Caribbean, SE Asia (Indonesia, Philippines)
Sargassum polycystum (B)	Agar-Agar Koepan, Agar-Agar Koepan, Arien Harulu, Kattaikkorai, Pu Zhi Ma Wei Zao	Caribbean Sea, Indian Ocean, Asia (China, Taiwan, Xisha Is, Indonesia, Malaysia, Myanmar, Moluccas, Philippines, Singapore, Thailand, Vietnam), Australia, Pacific Is (Polynesia, Micronesia, Fiji, Mariana Is, New Caledonia, Solomon Is, Tonga)
Sargassum polyphyllum (B)		Indian Ocean, Australia, Pacific Is (Polynesia, Fiji, Hawaiian Is, New Caledonia)
Sargassum sagamianum (B)		Asia (Japan, Korea, Philippines)
Sargassum serratifolium (B)	Nokogiri-Moku	Asia (China, Hong Kong, Japan, Korea, Taiwan, Philippines), New Zealand

Table 2.1 contd. ...

Table 2.1 contd. ...

Species	Common names	Countries
Sargassum siliquastrum (B)	Lie Ye Ma Wei Zao	Asia (Hong Kong, Japan, Korea, and Vietnam), Pacific Islands (New Caledonia)
Sargassum siliquosum (B)	Aragan, Yore-Moku	Indian Ocean, Asia (China, Japan, Taiwan), SE Asia (Indonesia, Malaysia, Philippines, Singapore, Vietnam), Australia
Sargassum swartzii (B)	Kattaikkorai	W Indian Ocean (Bangladesh, India, Kenya, Pakistan, Sri Lanka, Tanzania), Asia (China, Japan, Korea, Taiwan, Vietnam), SW Asia (Indonesia, Malaysia, Singapore), Pacific Is (New Macedonia)
Sargassum tenerrimum (B)	Kattaikkorai	Indian Ocean, Asia (China, Hong Kong, Taiwan), SE Asia (Malaysia, Philippines, Vietnam), Australia, Pacific Is (Micronesia, Polynesia, Samoa)
Sargassum thunbergii (B)	Mouse Tail Algae, Earth Worm, Shuweizhao, Shu Wei Zao, Djichungi, Umitoranoo	Atlantic Is (Canary Is), Asia (China, Japan, Korea, Taiwan)
Sargassum vulgare (B)	Gulf Weed, Beerentang, Sargasso	NE Atlantic, Mediterranean, Caribbean Sea, SE Atlantic, Indian Ocean, W Pacific (Philippines)
Saundersella simplex (B)	Golden Bottlebrush Epiphyte, Motsukichasômen	Arctic, Atlantic Is (Greenland), Asia (Japan), NE Pacific (Alaska, British Columbia)
Schizymenia apoda (R)	Orange Sheets	Atlantic Is (Azores, Madeira, Tristan da Cunha Is), SE Atlantic (Namibia, Somalia, S Africa), Asia (China, Korea, Taiwan)
Schizymenia binderi (R)		S America (Chile, Peru); Antarctic and Sub-Antarctic Is (Fuegia)
Scinaia aborealis (R)	Glassweed	Tropical and Subtropical W Atlantic; SE Asia (Philippines), Australia, Pacific Is (Fiji, New Caledonia)
Scinaia hatei (R)		SE Asia (India, Oman, Pakistan, Yemen)
Scinaia hormoides (R)	Gargarnatis	Asia (Japan, Indonesia, Philippines), Australia (Papua New Guinea), Pacific Is (Hawaiian Is)
Scinaia moniliformis (R)	Juzuhusanori	Indian Ocean, Asia (China, Japan, Taiwan, Philippines), Australia
Scytosiphon lomentaria (B)	Leather Tube, Chipolata Weed, Soda Straws, Whip Tube, Kayamonori	Cosmopolitan in temperate and cold seas
Sebdenia flabellata (R)		SW Atlantic, Asia (Japan, Korea, Taiwan), SE Pacific, Australia
Silvetia babingtonii (B)	Yezo-Ishige	Asia (Japan, Korea, Russia)

Silvetia siliquosa (B)	Lijao Kai, Ttumbugi	Asia (China and Korea)
Sirophysalis trinodis (B)		Arabian Gulf, Indian Ocean, SE Asia (Indonesia, Philippines), Australia, Pacific Is (New Caledonia)
Solieria filiformis (R)		N and S Atlantic, Mediterranean
Solieria robusta (R)	Blubber Weed, Tender Golden Weed, Tajuk Bau'no, Lumitamana	Asia (China, Japan, Taiwan), Indian Ocean, SE Asia (Indonesia, Philippines, Singapore), SW Asia (Arabian Golf, India, Iran, Kuwait, Pakistan, Sri Lanka, Yemen), Australia, Pacific Is (Fiji)
Spatoglossum schroederi (B)	Espatogloso	Atlantic Is (Azores, Bermuda, Canary Is), Tropical and Subtropical W and E Atlantic, Caribbean Sea, Mediterranean, Indian Ocean, Australia, Pacific Is (Micronesia), Tropical E Pacific
Sphacelaria indica (B)		SE Asia (Singapore)
Sphaerococcus coronopifolius (R)	Berry Wart Cress	NE Atlantic, Atlantic Is (Azores), Mediterranean, SW Asia (Turkey)
Sphaerotrichia divaricata (B)	Kusa-Mozuku	N Atlantic, Atlantic Is (Azores), Baltic Sea, Asia (China, Japan), Australia; NE Pacific (British Columbia)
Sphaerotrichia firma (B)	Ishi-Mozuku	N Atlantic, Atlantic Is (Azores), Baltic Sea, Asia (China, Japan), Australia, NE Pacific (British Columbia)
Splachnidium rugosum (B)	Dead-Man's Finger, Gummy Weed	Atlantic Is (Tristan da Cunha), E Pacific (Chile), Indian Ocean, Australia and New Zealand
Spyridia filamentosa (R)	Hairy Basket Weed, Limu Hulupuaa, Hulu Pua'a	Adriatic Sea, NE Atlantic, Mediterranean, Atlantic Is (Azores, Bermuda, Canary Is, Cape Verde Is, Madeira, Selvage Is), Tropical and subtropical Atlantic, Caribbean Is (Bahamas, Barbados, Cayucos Is, Cuba, Jamaica, Martinique, Puerto Rico, Trinidad), Indian Ocean, Asia (China, Japan, Korea, Taiwan), SE Asia (Indonesia, Philippines, Vietnam, Singapore), Australia and New Zealand, Pacific Is (Samoa, Polynesia, Micronesia, Fiji, Hawaiian Is, Mariana Is, Marshall Is, Solomon Is), E Pacific (California, Mexico)
Spyridia fusiformis (R)		Indian Ocean
Spyridia griffithsiana (R)		NE Atlantic, Mediterranean Sea
Stepharocystis crassipes (B)	Nebuto-Moku	Asia (Japan, Russia)
Stepharocystis geminata (B)	Chain Bladder, Yezo-Moku	Asia (Japan), NE Pacific (Alaska, British Columbia, Aleutian Is)
Stepharocystis hakodatensis (B)	Uga-no-Moku, Uganomoku	Asia (Japan, Korea), SE Asia (Philippines)

Table 2.1 contd. ...

Table 2.1 contd. ...

Species	Common names	Countries
Stephanocystis osmundacea (B)	Woody Chain Bladder, Bladder Chain Kelp	Pacific Ocean (N Oregon through N Mexico)
Taonia atomaria (B)	Dotted Peacock Weed	NE Atlantic (Britain to Mauretania), Atlantic Is (Azores, Canary Is, Madeira, Selvage Is), Adriatic Sea; Mediterranean, Persian Gulf
Tinocladia crassa (B)	Futo-Mozuku, Somen-Nori	Asia (China, Japan, Korea, Russia), NE Pacific (California)
Titanophora weberae (R)	Aragan Elik	Asia (Japan, Taiwan, Indonesia, Philippines, Vietnam), Australia, Pacific Is (Samoa, Polynesia, Micronesia, Fiji, Palau, Solomon Is)
Trichogloea requienii (R)	Barisbaris	Atlantic Is (Cape Verde Is), Tropical and Subtropical W Atlantic, Indian Ocean, Asia (China, Japan, Korea, Taiwan, Indonesia, Philippines), Australia; Pacific Is (Fiji, Polynesia, Hawaiian Is, Mariana Is)
Turbinaria conoides (B)	Samo, Labi-Labi, Agar-Agar Lesong, Pakkoda Pasi	Indian Ocean, SW Asia (Arabian Gulf, India, Iran, Saudi Arabia, Sri Lanka), Asia (China, Japan, Taiwan), SE Asia (Indonesia, Malaysia, Philippines, Singapore, Thailand, Vietnam), Australia, Pacific Is (Polynesia, Micronesia, Palau, Samoa, Solomon Is)
Turbinaria decurrens (B)	Pakkoda Pasi	Indian Ocean, Asia (China), SE Asia (Indonesia, Philippines, Singapore, Thailand, Vietnam), Australia, Papua New Guinea, Pacific Is (Micronesia, Palau)
Turbinaria ornata (B)	Ornate Seaweed, Spiny-Leaf Seaweed, Spiny Tops, Tubol, Rappamoku, Limu Lautalatala, Rimu Taratara, Pakkoda Pasi	Indian Ocean, Asia (China, Japan, Korea, Taiwan, Indonesia, Malaysia, Myanmar, Philippines, Singapore, Thailand, Vietnam), Australia, Pacific Is (Samoa, Polynesia, Cook Is, Micronesia, Fiji, Hawaiian Is, Mariana Is, Palau, Marshall Is, Solomon Is)
Turbinaria turbinata (B)	Turbinweed	Caribbean, Indian Ocean, W Pacific (Philippines)
Turnerella mertensiana (R)	Red Sea-Cabbage, Yezo-Nameshi, Oba-Sô	Asia (Japan, Korea, Russia), NE Pacific (Alaska, British Columbia)
Udotea indica (G)		Indian Ocean, SE Asia (Philippines), Pacific Is (Micronesia, Marshall Is)
Ulothrix flacca (G)	Hoso-Hibimidoro	NE Atlantic, Adriatic Sea, Baltic Sea, Arctic (Canada), Atlantic Is (Azores, Canary Is, Greenland, Iceland), Tropical and Subtropical W Atlantic, Mediterranean, Asia (China, Japan, Korea, Russia, Taiwan, Vietnam), New Zealand, Antarctic Is, NE Pacific (Alaska to California)
Ulva australis (G)	Lacy Sea Lettuce, Ana-Awosa, Awosa	Indo-Pacific (Indonesia through Australia and New Zealand), but has been widely introduced and is now found throughout Asia, East Africa, Mediterranean, and both shores of North America
Ulva clathrata (G)	Tahalib, Tai Tyau, Taitiao, Spiky Tendrils, Stone Hair, Bright Green nori, Aonori, Parae	Widely distributed in warm seas from the W Pacific through the Atlantic Ocean, Gulf of Mexico, Mediterranean, Indian Ocean and SE Pacific

Ulva compressa (G)	Thread Weed, Tape Weed, Bagisbagis, Lumot, Limu Ele-ele, Awo-Nori, Parae	Cosmopolitan
Ulva conglobata (G)		SW Asia (India), Asia (China, Japan, Korea, and Taiwan), SE Asia (Malaysia, Vietnam)
Ulva fasciata (G)	Limu Papahapapa, Pahapaha, Limu Palahalaha, Pakaiea, Papahapaha	Adriatic Sea, NE Atlantic, Atlantic Is (Azores, Bermuda, Canary Is, Cape Verde Is), Mediterranean, Florida, Mexico, Caribbean Is, Tropical and Sub-Tropical W Atlantic, S Atlantic, India Ocean, SE Asia, Australia and New Zealand, Pacific Is (Hawaii Is)
Ulva flexuosa (G)	Tahalib, Winding Nori, Limu Ele-ele, Limu Pipilani, Imu Vai (Freshwater Algae), Imutapaa (Ripe Algae), Imu Ketaha (Encroaching Algae), Imu Ouohu (Hair Algae)	N and S Atlantic, SW Asia (Abu Dhabi, Arabian Gulf, Cyprus, Israel, Kuwait, Turkey), Asia (China, Japan, Korea, Taiwan), SE Asia (Indonesia, Singapore, Vietnam), Australia, Pacific Is (Samoa, Polynesia, Hawaiian Is)
Ulva flexuosa subsp. *paradoxa* (G)	Lumot, Limu Ele'ele, Limu Pipilani, Watage-Awonori	NE Atlantic, Tropical and Subtropical Atlantic, Caribbean Sea, Atlantic Is (Bermuda, Canary Is), Adriatic Sea, Baltic Sea, Black Sea, Indian Ocean, Asia (Japan), SE Asia (Indonesia, Philippines, Singapore), Australia, Pacific Is (Fiji, Samoa, Hawaiian Is)
Ulva intestinalis (G)	Gut Weed, Grass Kelp, Gut Weed, Hollow Green Nori, Hollow Green wee, Tubular Sea Lettuce, Sea Grass, Boaonori, Erva-Patinha, Erva-Patinha Verde, Erva-do-Calhau	Occurs worldwide, except in polar waters
Ulva lactuca (G)	Sea Lettuce, Lettuce Laver, Green Laver, Sea Grass, Thin Stone Brick, Chicory Sea Lettuce, Limu Papahapapa, Pahapaha, Limu Pahapaha, Limu Palahalaha, Pakaiea, Papahapaha, Limu Paha-paha, Limu Pala-haloha, Aosa, Imu Kokuu, Kokuu, Imu Sarata (Salad or Lettuce Algae)	Worldwide distribution in temperate and tropical waters
Ulva linza (G)	Slender Sea Lettuce, Doubled Ribbon Weed, Green String Lettuce, Hai-Cai, Limu Ele-ele, Awo-nori, Usuba-Awonori, Usaba-Aonori, Parae, Ipparae	NE Atlantic, Atlantic Is (Azores Is, Bermuda, Canary Is, Iceland, Madeira, Selvage Is), Tropical and Subtropical W and E Atlantic, NE Pacific (Alaska), SW Asia (Arabian Gulf, Cyprus, Israel, Turkey), Asia (China, Japan, Korea, Russia, Taiwan), SE Asia (Indonesia), Australia and New Zealand; Pacific Is (Hawaiian Is)
Ulva prolifera (G)	Tai-Tiao, Limu Ele-ele, Hulu'ilio, Suji-awnori, Suji-Aonori, Parae	Arctic (Canada, Svalbard, White Sea), Adriatic Sea, Black Sea, NE Atlantic, Atlantic Is (Azores, Bermuda, Canary Is, Greenland, Iceland, Madeira), Tropical and Subtropical Atlantic, W Mediterranean, Asia (China, Japan, Korea, Taiwan, Indonesia, Philippines), New Zealand, Pacific Is (Samoa, Polynesia, Hawaiian Is), NE Pacific (Alaska)

Table 2.1 contd. ...

Table 2.1 contd. ...

Species	Common names	Countries
Ulva reticulata (G)	Seda-Seda, Pattu Pasi	Tropical and Subtropical W Atlantic, Indian Ocean Is, SW Asia (Arabian Gulf, Bahrain, India, Kuwait, Pakistan, Persian Gulf, Saudi Arabia, Sri Lanka, Yemen), Asia (China, Japan, Korea, Taiwan), SE Asia (Indonesia, Malaysia, Philippines, Singapore, Thailand, Singapore, Vietnam), Australia, Pacific Is (Hawaiian Is), Antarctic Is
Ulva rigida (G)	Green Laver, Lattuga Marina, Lechuga de Mar, Meersalat, Sea Lettuce, Stijve Zeesla, Ulve Rigide, Alface-do-Mar	Worldwide distribution in temperate and warm seas
Ulva rotundata (G)		NE Atlantic, Atlantic Is (Canary Is), Tropical and subtropical W Atlantic, Adriatic Sea
Ulva stenophylla (G)		Indian Ocean, Australia and New Zealand, NE Pacific (British Columbia to California)
Ulvaria splendens (G)	Kuro-Hitoegusa, O-Hitoegusa	NE Atlantic, Atlantic Is (Greenland), Baltic sea, Asia (Japan, Korea, Russia), NE Pacific (Alaska, British Columbia, California), SE Pacific (Chile)
Umbraulva japonica (G)		Asia (Japan, Korea, Taiwan)
Undaria peterseniana (B)		Asia (Japan and Korea)
Undaria pinnatifida (B)	Qun Dai Cai, Asia Kelp, Apron-Ribbon Vegetable, Sea Mustard, Precious Sea Grass, Wakame, Miyok, Miyeouk	Kelp indigenous to the NW Pacific Ocean and the cold temperate coastal regions of Japan, China, Korea, and Southeast Russia. *U. pinnatifida* has been spread around the world by international shipping and mariculture, and has extended its range to include four continents since 1980's. It is now growing in the temperate Pacific Ocean, the SE temperate Indian Ocean, the Mediterranean, and the temperate N and S Atlantic. *U. pinnatifida* was grown in the French Bretagne as food, which increased exposure of this seaweed to Europeans
Undaria undarioides (B)	Wakame	Asia (Japan, Korea)
Valonia aegagropila (G)	Bolitas de Mar, Bio-Bolas	Adriatic Sea, Mediterranean, Atlantic Is (Canary Is, Selvage Is), Tropical and Subtropical W Atlantic, Caribbean, Mexican Gulf, Arabian Gulf, Indian Ocean, Asia (China, Japan, Taiwan), SE Asia (Indonesia, Malaysia, Philippines, Singapore, Vietnam), Australia, Pacific Is (Samoa, Polynesia, Micronesia, Fiji, Hawaiian Is, Mariana Is, Marshall Is, Solomon Is)

Valonia macrophysa (G)	Tamago Valonia, Bolitas de Mar, Bio-Bolas	NE Atlantic, Adriatic Sea, Mediterranean, Atlantic Is (Ascension, Azores, Bermuda, Canary Is, Madeira, Selvage Is), Tropical and Subtropical W Atlantic, Caribbean, Indian Ocean, Asia (Japan), SE Asia (Indonesia, Philippines, Vietnam), Australia, Pacific Is (Samoa, Polynesia, Fiji)
Valonia utricularis (G)	Limu Lipuu-puu	NE Atlantic (Portugal to Canary Is), Mediterranean, Caribbean, Indian Ocean, Asia (Japan, China), SE Asia (Philippines, Vietnam), Pacific Is, Australia
Vertebrata lanosa (R)	Wrack Siphon Weed, Mæhre	N Atlantic, Atlantic Is (Iceland)
Wildemania amplissima (R)	Northern Pink Laver, Red Cellophane	Arctic (Norway), White Sea, NE Atlantic (Scandinavia), Asia (Japan), NE Pacific (Alaska to California)
Wildemania miniata (R)	Red Nori, Nori	N Atlantic, Atlantic Is (Greenland, Iceland), Baltic Sea, NE Pacific (Alaska to California, Chile), Asia (Russia)
Yonagunia formosana (R)	Limu	Caribbean Is, Indian Ocean, Asia (China, Japan, Taiwan, Philippines, Vietnam), Pacific Is (Samoa, Polynesia, Fiji)
Zonaria antillarum (B)		Atlantic Is (Selvage Is), Tropical and Subtropical W and E Atlantic, SW Asia (India, Sri Lanka), SE Asia (Indonesia, Philippines, Singapore)

G – green algae (Chlorophyta), B – brown algae (Phaeophyceae), R – red algae (Rhodophyta)

Table 2.2 Composition of some fucoidans extracted* from brown seaweed (adapted from Ale 2012, Ale and Meyer 2013, Li et al. 2008, Cardoso et al. 2014, Cuong et al. 2015).

Species	Order	Chemical composition
Adenocystis utricularis	Ectocarpales	fucose, galactose, mannose, xylose, GlcA, sulfate
Ascophyllum nodosum	Fucales	fucose (49%), xylose (10%), GlcA (11%), sulfate
Bifurcaria bifurcata	Fucales	fucose, xylose, mannose, glucose, galactose, sulfate
Chorda filum	Laminariales	fucose, xylose, mannose, glucose, galactose, uronic acid, sulfate
Cladosiphon okamuranus	Ectocarpales	fucose, glucose, uronic acid, sulfate
Dictyota menstrualis	Dictyotales	fucose/xylose/galactose/sulfate (1/0.5/2/2)
Ecklonia kurome	Laminariales	fucose, galactose, mannose, GlcA, glucose, xylose, sulfate
Fucus distichus	Fucales	fucose/sulfate/acetate (1/1.21/0.08)
F. evanescens	Fucales	fucose/sulfate/acetate (1/1.23/0.36)
F. serratus	Fucales	fucose/sulfate/acetate (1/1/0.1)
F. spiralis	Fucales	fucose, xylose, mannose, glucose, galactose, uronic acid, sulfate
F. vesiculosus	Fucales	fucose,sulfate
Himanthalia elongata	Fucales	fucose, xylose, GlcA, sulfate
Laminaria hyperborea (as *L. cloustonii*)	Laminariales	fucose, galactose, xylose, uronic acid, sulfate
L. digitata	Laminariales	fucose, xylose, mannose, glucose, galactose, uronic acid, sulfate
Lessonia flavicans (as *L. vadosa*)	Laminariales	fucose/sulfate (1/1.12)
Macrocystis pyrifera	Laminariales	fucose/galactose (18/1), sulfate
Padina pavonica	Dictyotales	fucose/galactose, sulfate (9/1/9)
Saccharina angustata (as *L. angustata*)	Laminariales	fucose, galactose, mannose, xylose, GlcA, sulfate
S. religiosa (formerly *L. religiosa*)	Laminariales	fucose, xylose, mannose, glucose, rahmnose, uronic acid, sulfate
Sargassum acinarium (as *S. linifolium*)	Fucales	fucose, mannose, galactose, xylose, uronic acid
S. fusiforme (formerly *Hizikia fusiformis)*	Fucales	fucose/xylose/uronic acid/galactose/sulfate (1/0.8/0.7/0.8/0.4) and (1/0.3/0.4/1.5/1.3)
S. henslowianum	Fucales	α (1→3)-linked L-fucopyranose backbone and sulfate groups occupied mostly at C-2, C-4 and sometimes at C-3 position of fucose residues
S. henslowianum	Fucales	α (1→3)-linked L-fucopyranose backbone and sulfate groups occupied mostly at C-2, C-4 and sometimes at C-3 position of fucose residues
S. stenophyllum	Fucales	fucose, galactose, mannose, sulfate
Silvetia babingtonii (as *Pelvetia wrightii*)	Fucales	fucose/galactose (10/1), sulfate
Undaria pinnatifida	Laminariales	fucose, mannose, xylose, rhamnose, galactose, glucose, sulfate
U. pinnatifida (Mekabu)	Laminariales	fucose/galactose (1/1.1), sulfate

*Fucoidans were obtained by acidified or alkali/water solutions, followed by precipitation, mostly with ethanol.

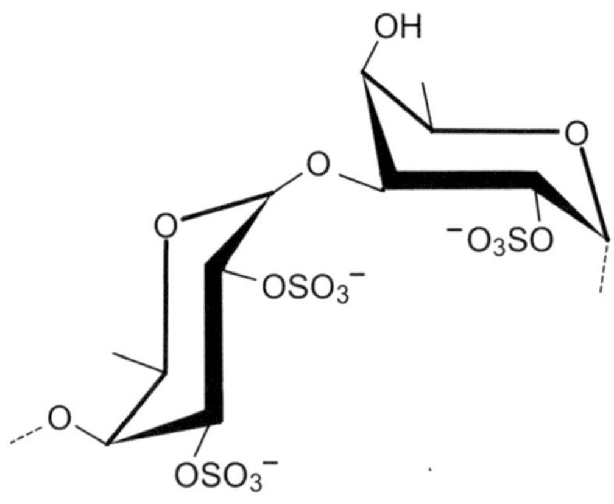

Figure 2.1 Idealized structure of the chemical units of Fucoidan.

Several studies reported that the anti-coagulant functionalities of fucoidans extracted from *F. vesiculosus* and *Eklonia cava* were due to thrombin-inhibition mediated via plasma anti-thrombin-III; their anti-coagulant activity was similar to that of heparin (Li et al. 2008, Holtkamp et al. 2009, Jiao et al. 2011). Moreover, fucoidans from *Saccharina longissima*, *S. latissima*, *Laminaria digitata*, *Fucus serratus*, *F. distichus*, *F. evanescens*, and *Ascophyllum nodosum* were also described to possess strong anti-coagulant activities both *in vitro* and *in vivo* models (see also Chapter 10) (Li et al. 2008, Holtkamp et al. 2009).

In general, structural-bioactive studies suggested that the anti-coagulant/anti-thrombin activities of fucoidans were mainly dependent on the content and/or positioning of sulfate groups, as well as the molecular weight of the polymers (Li et al. 2008). Moreover, their monomeric composition, types of linkages, and branching might also exert moderate modulation on the biological properties of fucoidans (Jiao et al. 2011, Thanh-Sang and Kim 2013, Lorbeer et al. 2013). In this context, it is possible that the greater anticoagulant/antithrombin activities exhibited by longer fucoidans are due to the higher content of fucose and sulfate groups (Li et al. 2008, Fitton 2011, Jiao et al. 2011, Thanh-Sang and Kim 2013, Fitton et al. 2015), though this is still under debate (Cardoso et al. 2014, Cardoso et al. 2015).

Fucoidan isolated from the brown seaweed *Fucus vesiculosus* (Béress et al. 1993) shows inhibitory effect on the replication of DNA viruses: Herpes virus (HSV-1, HSV-2) and HCMV (Baba et al. 1988b, Béress et al. 1993, Moen and Clark 1993) (see Table 3.1). These compounds are also active against RNA viruses: VSV, Sinbis virus (SINV), and HIV-1 (Baba et al. 1988b). However, this compound is not active against Coxsackievirus, Poliovirus, and Parainfluenza virus. A water-soluble, non-carbohydrate component of fucoidan isolated from *F. vesiculosus* can inhibit HIV RT *in vitro* at μg mL^{-1} (Moen and Clark 1993). Pre-incubation of cell-free virus to 200 μg mL^{-1} causes 100% reduction in the amount of HIV-1 p24 antigen released. Tests show that these effects are not due to the killing of target cells. In fact, fucoidan produced no adverse effects on cell proliferation and protein metabolism. The pre-incubation of target cells with fucoidan actually protects them from HIV-1 infection (Moen and Clark 1993). In addition to its antiviral activity, fucoidan possesses low anticoagulation properties (see also Chapter 5) (Baba et al. 1988, Béress et al. 1993, Moen and Clark 1993, Yasuhara-Bell and Lu 2010, Wu et al. 2016).

Fucoidans are also reported to inhibit the replication of several enveloped viruses (see Table 3.2) such as human immune-deficiency and human cytomegalovirus, among others (Li et al. 2008, Jiao et al. 2011). The mechanisms for such activity are thought to occur via inhibition of cell infection by viral sorption, or hampering of viral-induced, syncytium formation (Jiao et al. 2011, Thanh-Sang and Kim 2013).

Fucoidans from *Saccharina japonica*, *Cladosiphon okamuranus*, *Adenocystis utricularis*, *Stoechospermum marginatum*, *Cystoseira indica*, *Dictyota mertensii*, *Lobophora variegata*, *Fucus vesiculosus*, *Spatoglossum schroederi*, and *Undaria pinnatifida* showed impressive, positive results in both *in vitro* and *in vivo* models of infection by poliovirus III, adenovirus III, ECHO6 virus, coxsackie B3 virus, coxsackie A16, Newcastle disease virus (NDV), HSV-1, HSV-2, HIV, and avian reverse transcriptase (Li et al. 2008, Thanh-Sang and Kim 2013, Wu et al. 2016).

Anti-tumor activities of fucoidans include the inhibition of tumor proliferation, the stimulation of tumor cell apoptosis, blocking of tumor cell metastasis, and enhancement of various immune responses (Koyanagi et al. 2003). In this context, fucoidans from several macroalgal species (e.g., *Saccharina japonica*, *S. latissima*, *Laminaria digitata*, *Fucus serratus*, *F. distichus*, and *F. vesiculosus*) proved to be useful and were regarded as good candidates for future cancer therapy (Li et al. 2008, Thanh-Sang and Kim 2013). Besides those, the commercial fucoidans branded "Tokida" (from cultured *Cladosiphon okamuranus*) and that from the Korean cultured sporophyte (Miyeokgui) of *U. pinnatifida* also revealed promising anti-tumoral activities, as tested in *in vitro* models (Thanh-Sang and Kim 2013).

Other important biological activities demonstrated by fucoidans included anti-oxidant, anti-inflammatory, and anti-allergic, although others cannot be forgotten, e.g., hepato-protection, cardio-protection, stomach protection and anti-obesity (Li et al. 2008, Jiao et al. 2011, Thanh-Sang and Kim 2013). Examples of fucoidans showing promising anti-oxidant activities in *in vitro* models included those obtained from *S. japonica*, *Canistrocarpus cervicornis*, *F. vesiculosus*, *Dictyota cervicornis*, *Sargassum filipendula*, *Dictyopteris delicatula*, and *S. japonica* (Li et al. 2008, Jiao et al. 2011, Thanh-Sang and Kim 2013), while the anti-inflammatory activity of several fucoidans (*L. digitata*, *F. evanescens*, *F. serratus*, *F. distichus*, *F. spiralis*, *A. nodosum*, *C. okamuranus*, *Padina gymnospora*, and *S. latissima*) have been described through inhibition of leucocyte recruitment, in an inflammation model in rats (Cumashi et al. 2007). Moreover, inhibition of the expression of inducible nitric oxide synthase (iNOS) has also been demonstrated for fucoidans, such as that from the Sigma-Aldrich Chemical Co. (from *F. vesiculosus*). Furthermore, commercial fucoidans (from Mekabu and Sigma-Aldrich Chemical Co.) together with those isolated from *A. nodosum*, *F. evanescens*, *C. okamuranus*, and from other several members of the Order Laminariales (Phaeophyceae) were described to exhibit anti-complementary activities, rendering their potential as anti-allergen (Li et al. 2008, Jiao et al. 2011, Thanh-Sang and Kim 2013, Wu et al. 2016).

Ascophyllan

Ascophyllan isolated from the brown alga *Ascophyllum nodosum* (Phaeophyceae) is a fucose-containing sulfated polysaccharide, which has a similar but distinct characteristic monosaccharide composition and chemical structure to fucoidan (Jiang et al. 2011). Ascophyllan (xylofucoglycuronan) has similar but obviously distinct composition characteristics from those of fucoidans isolated from *A. nodosum* and *Fucus vesiculosus* (Larsen et al. 1970, Kloareg et al. 1986). Specifically, ascophyllan has fucose and xylose in about equimolecular proportion, whereas fucoidans have a much higher ratio of fucose than of xylose. The sulfate levels of ascophyllan and fucoidans isolated from *A. nodosum* and *F. vesiculosus* were 9.6, 19.4, and 22.6%, respectively (Nakayasu et al. 2009). In addition, the apparent molecular mass of ascophyllan estimated by gel filtration chromatography was about 390 kDa, which is much higher than that of fucoidan (Wang et al. 2013).

Mohsin et al. (2014) characterize and evaluate the antioxidant activity of a sulfated polysaccharide ascophyllan isolated from marine brown algae *Padina tetrastromatica*. The results showed that the ascophyllan fraction AF3 showed stronger free-radical-scavenging abilities and had good antioxidant effect. Available data obtained by *in vitro* models suggest that there is a correlation between the sulfate content and antioxidant activity.

In work carried out by Jiang et al. (2014), ascophyllan and crude extract administered via the oral route showed greater antitumor effects than via IP (Intraperitoneal) route, and the tumor sizes in mice treated with ascopyllan and crude extract were reduced by a mean of 68.7 ± 6.8% and 42.4 ± 24.8% by the oral route, and 41.4 ± 16.1% and 13.6 ± 20.6% by IP route, compared to control mice. Splenic natural killer cell activity in the mice treated with ascophyllan and crude extract by IP route was significantly enhanced, while only a slight increase of this activity was observed in orally-treated mice. Furthermore, increase in spleen weight of tumor-bearing mice was slightly suppressed by oral administration of ascophyllan, whereas IP administration resulted in further enlargement. Analysis of serum cytokines revealed that oral treatment with ascophyllan resulted in significant increase of tumor necrosis factor-α and interleukin-12 levels. Since ascophyllan showed no direct cytotoxic effect on sarcoma-180 cells, orally-administered ascophyllan is suggested to exhibit its antitumor activity through the activation of the host immune system (Jiang et al. 2014).

Zhang et al. (2014) investigated the effect of ascophyllan, a sulfated polysaccharide purified from *A. nodosum*, on the maturation of mouse dendritic cells (DCs) *in vitro* and *in vivo*. Ascophyllan induced up-regulation of co-stimulatory molecules and production of pro-inflammatory cytokines in bone marrow-derived DCs (BMDCs). Interestingly, ascophyllan induced a higher degree of co-stimulatory molecule up-regulation and pro-inflammatory cytokine production than fucoidan, a marine-derived polysaccharide with well-defined effect for promoting DCs maturation. Ascophyllan also promoted the generation of IFNγ (Interferon gamma)-producing Th1 (murine Type 1 helper T cells) and Tc1 (Type 1 cytotoxic T lymphocytes) cells in the presence of DCs in an IL (cytokines)-12-dependent manner. Finally, myeloid differentiation primary response 88 (MyD88) signaling pathway was essential for DCs maturation induced by ascophyllan. Taken together, these results demonstrate that ascophyllan induces DCs maturation, and consequently enhances Th1 and Tc1 responses *in vivo*. This knowledge could facilitate the development of novel therapeutic strategies to combat infectious diseases and cancer (Zhang et al. 2014).

Laminaran

Laminarans are a category of small glucans present in either soluble or insoluble form. The first form is characterized by its complete solubility in cold water, while the other is only soluble in hot water (Kylin 1913, Chevolot et al. 2001). This polysaccharide is composed of D-glucose with β-(1,3) linkages, with β-(1,6) intra-chain branching (see Fig. 2.2; Kadam et al. 2015a, Pereira and Ribeiro-Claro 2015). Laminaran, also known as "laminarin" or "leucosin", was first isolated from members of the Laminariaceae (Phaeophyceae) by Schmiedeberg (1885). Laminaran is a food reserve of brown algae. It is located in vacuoles present in cells. Laminaran is found in the fronds of *Laminaria* and *Saccharina* species. The main sources of laminaran are outlined in Table 2.3.

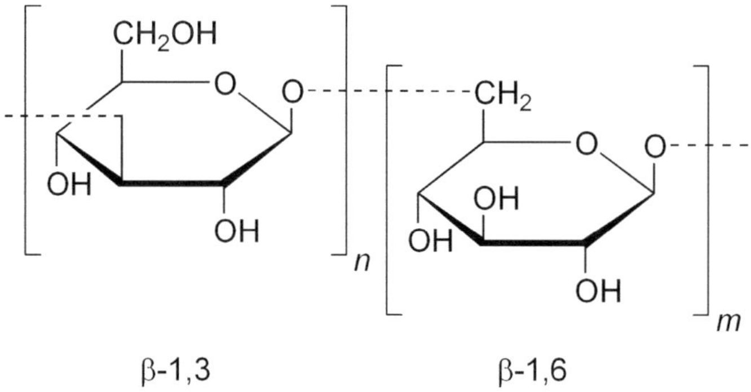

Figure 2.2 Chemical structure of β-1,3-1,6-glucan (Laminarin, Laminaran) (after Edgar 2016).

Table 2.3 Main sources of Laminaran.

Species	Content (% DW)	Reference
Alaria esculenta	11.10–18.30	Schiener et al. 2015
Ascophyllum nodosum	15.02–35.62	Kadam et al. 2015b
Fucus distichus	10–20	Dethier and Williams 2009
F. serratus	0.1–1.3	Kremer 1975
Laminaria digitata	0–35	Devillé et al. 2004 MacArtain et al. 2007
L. hyperborea	36.97–91.76	Kadam et al. 2015b
Saccharina latissima	0–33.00	Haug and Jensen 1954
Sargassum acinarium (as *S. linifolium*)	4.20–10.98	Abdelfat and Hussein 1973
S. muticum	0.3	Gorham and Lewey 1984
Undaria pinnatifida	3	Je et al. 2009

Laminarans are basically a class of low-molecular weight storage β-glucans, which are composed of (1,3)-β-D-glucan (Rioux et al. 2007). They consist of (1,3)-β-D-glucopyranose residues with some 6-O-branching in the main chain, and some β-(1,6)-intra-chain links are also present. For example, laminaran from *Eisenia bicyclis* is composed of a linear chain of (1,3) and (1,6) linkage in a ratio of 2:1 (1,3)-β-D-glucans (Shin et al. 2009). The molecular weight of laminaran is approximately 5 kDa and is dependent upon the degree of polymerization, which is in the range of 20–25 glucose moieties (Nelson and Lewis 1974, Alderkamp et al. 2007). The molecular weight of laminaran from *Saccharina longicruris* is reported to be in the range of 2.89–3.32 kDa, depending on the extraction conditions (Rioux et al. 2010), including the solvent type used (Kadam et al. 2015b). So, not all laminarans have the same characteristics. Different species of raw material are present in different structures of molecules, and therefore have different biological properties.

The laminaran prepared from kelp by water extraction were reported to be able to effectively inhibit the adsorption of HIV on lymphocytes and the activity of HIV reverse transcriptase at the concentration of 50 μg mL^{-1}, which suggest that laminaran polysaccharides possess good inhibitory effect on HIV replication (Muto et al. 1988).

Laminaran has been extensively investigated for potential bio-functional activities. Algae polysaccharides have numerous biological activities because they enhance macrophage immune responses (Schepetkin and Quinn 2006). Similarly, laminaran can be used to achieve the activation of macrophages leading to immune-stimulatory, anti-tumor, and wound-healing activities (Ibrahim et al. 2005, Lee et al. 2012). Thus, it is also termed as a biological response modifier (Bohn and BeMiller 1995). The biological activities of laminarans can be enhanced or modified using various techniques including irradiation, sulfation, reduction, and oxidation. During irradiation, the molecular weight of the laminarans investigated was reported to be significantly reduced with the formation of carbonyl groups. This is suggested to lead to enhanced anti-oxidant activities (Choi et al. 2011).

Porphyran

Porphyran is a sulfated polysaccharide isolated from selected (red) algae of the Order Bangiales, Phylum Rhodophyta, especially from the genus *Porphyra*/*Pyropia* (Villarroel and Zanlungo 1980, Bhatia et al. 2008). The chemical structure of porphyran comprises of a linear backbone of alternating 3-linked β-D-galactose and 4-linked α-L-galactose-6-sulfate or 3,6-anhydro-α-L-galactose units (Fig. 2.3) (Zhang et al. 2005).

Porphyran is effective against contact hypersensitivity induced by 2, 4, 6 tri-nitrochlorobenzene. The mechanism is suggested to operate by suppressing the serum levels of IgE and IFN-γ. Porphyran has

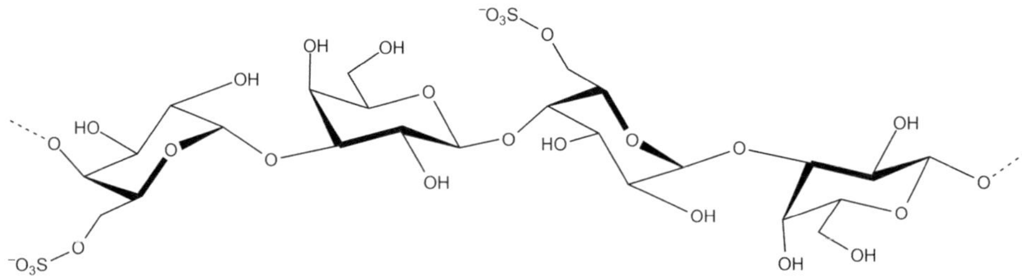

Figure 2.3 Idealized structure of the chemical units of Porphyran (after Eklof 2012).

excellent inhibitory activities against hyaluronidase, responsible for the release of histamine from mast cells (Ishihara et al. 2005).

Porphyran produces hypocholesterolemic and hypolipidemic effects due to reduced cholesterol absorption in the gut, with increased fecal cholesterol content, and a hypoglycemic response (see also Chapter 4). It is also reported to reduce total cholesterol, free cholesterol, and the levels of triglycerides and phospholipids in the liver. These substances are likely to be exploited by "nutraceutical" companies that market them as "health products" (Kiriyama et al. 1968, Ito and Tsuchida 1972, Lamela et al. 1989, Nishide et al. 1993, Renn et al. 1994a, Renn et al. 1994b, Dumelod et al. 1999, Ara et al. 2002, Nishide and Uchida 2003, Panlasigui et al. 2003, Tsuge et al. 2004).

Some sulfated polysaccharides from red algae also showed anti-viral activities towards viruses responsible for human infectious diseases (see also Chapters 3 and 6). Porphyran is reported to inhibit HIV reverse transcriptase *in vitro*. It has minimal effects on human DNA and RNA polymerase activity. Some agaroids, such as high molecular weight galactan sulfate, also have anti-viral properties against the *Herpes simplex* virus, human cytomegalo virus (HCMV), dengue virus (DENV), respiratory syncytial virus (RSV), and influenza virus due to the inhibition of the initial viral attachment to host cells (Nakashima et al. 1987b, Mazumder et al. 2002, Zhu et al. 2003a, Zhu et al. 2003b, Smit 2004).

Porphyran displays anti-tumor properties, in particular the ester sulfates of the polysaccharides may be related to the anti-tumor activities reported (see also Chapters 3 and 6). The degree of polyanionic properties may be intimately related to this activity. Porphyran reportedly induced apoptosis by caspase-3 or caspase-9 activation, while decreasing the IGF-1R phosphorylation and down regulated Akt phosphorylation. The caspase-induce apoptosis were activated in a sequential cascade of cleavages from their inactive forms. Activation of caspase-3 leads to the cleavage of a number of proteins, one of which is poly (ADP-ribose) polymerase (PARP) (Kwon and Nam 2006). The cleavage of the poly (ADP-ribose) polymerase is the hallmark of apoptosis. The expression level of anti-apoptotic molecules, such as Bcl-2, gradually decreased, whereas the pro-apoptotic molecule Bad opposed the action of Bcl-2, which increased in response to levels of exposure to porphyran. Porphyran decreased the expression levels in AGS gastric cancer cells by negatively regulating the insulin-like, growth Factor-I receptor (IGF-IR) phosphorylation, which is a potent mitogen and growth stimulatory factor for several kinds of cells (over-expression and enhanced activation of which is frequently observed in human cancers) (Kwon and Nam 2006). Thus, Insulin-like, growth Factor-I receptor (IGF-IR) phosphorylation was decreased in porphyran-treated AGS cells, which correlated with Akt activation. Porphyran down regulates Akt phosphorylation. The PI3-kinase/Akt pathway is viewed as a key player to cell survival in different systems. The IGF-I stimulation induced Akt activation decreased in cells treated with porphyran, which suggested that inhibition of Akt phosphorylation may be an important mechanism for porphyran-induced apoptosis (Yamamoto et al. 1974, Ito and Sugiura 1976, Usui et al. 1980, Yamamoto et al. 1982, Yamamoto et al. 1984a, Yamamoto et al. 1984b, Yamamoto et al. 1986, Yamamoto and Maruyama 1985, Baserga et al. 1994, LeRoith et al. 1995, Parrizas et al. 1997, Baserga 1999, Kandel and Hay 1999).

Ulvan

Ulvan (Fig. 2.4) is a sulfated, water-soluble polysaccharide which has physio-chemical and biological features of great potential in food, pharmaceutical, agricultural, and chemical applications (Lahaye and Robic 2007, Jiao et al. 2011).

Ulvan may represent 8–29% of algal dry weight, and is produced by some species belonging to the Phylum Chlorophyta (green algae), mostly belonging to the Class Ulvophyceae (Robic et al. 2009). It is mainly made up of disaccharide repeating sequences composed of sulfated rhamnose and glucuronic acid, iduronic acid, or xylose (Percival and McDowell 1967, Quemener et al. 1997).

The two major repeating disaccharides are aldo-biuronic acids designated as: Type A, ulvano-biuronic acid 3-sulfate (A3s) and Type B, ulvano-biuronic acid 3-sulfate (B3s) (Fig. 2.4). Partially sulfated xylose residues at O-2 can also occur in place of uronic acids. Low proportions of galactose, glucose, mannose, and protein were also generally described as components of ulvan. Additionally, minor repeating units were being reported to contain sulfated xylose, replacing the iduronic acid or glucuronic acid components (Lahaye and Robic 2007, Jiao et al. 2011, Pereira and Ribeiro-Claro 2015).

Ulvan have exhibited strong anti-oxidant (Qi et al. 2005, Qi et al. 2006), anti-tumor (Kaffer et al. 1999), immune-stimulatory (Leiro et al. 2007), anti-inflammatory (Leiro et al. 2007, Chiellini and Morelli 2011), anti-hypercholesterolemic (Rizk et al. 2016), anti-hyperlipidemic (Taboada et al. 2010) (see also Chapter 4), anticoagulant/antithrombotic (Zhang et al. 2008), anti-viral (Ivanova et al. 1994, Karnjanapratum et al. 2012), anti-bacterial (Choi et al. 2013), anti-protozoal, hyperplasia prevention, gastrointestinal, regenerative, and nano-medicine applications (Patel 2012).

Alves et al. (2013b) determined the cytotoxicity of ulvan via MTT test, using fibroblast-like cell, where cells were incubated with ulvan, *in vitro*. Quantified protein and total DNA stands data were directly correlated with hyaluronic acid use as control (non-cytotoxic); the studies conducted by these researchers (Alves et al. 2013b) of ulvan showed encouraging results in terms of cytotoxicity.

Meanwhile, cytotoxicity is one of the most important factors to determine the use of a bio-material for biomedical purposes, as in tissue engineering and wound healing (Venkatesan et al. 2015). Ulvan can be used in biomedical applications, especially in tissue engineering. Bio-functionalized ulvan hydrogels with ALP (Alkaline Phosphatase) enzyme served as inducers of mineralization in osteoblastic differentiation (Toskas et al. 2012, Venkatesan et al. 2015) and the cytocompatibility of ulvan, as well as its relative membrane are promising features for (Toskas et al. 2012) scaffold development (Alves et al. 2013a, Rossi and van Griensven 2014). The inter-molecular charges of ulvan and chitosan (i.e., anionic and cationic charges, respectively) produced stable supra-molecular structures via electrostatic interactions and could also form stabilized membranes. Ulvan/chitosan poly-electrolytes were found to mimic the extra-cellular matrix structure providing points of attachment for osteoblasts, believed to be enhanced by the nano-fibrous structure of this construct. It is suggested that this construct could be a suitable scaffold for applications in tissue engineering (Toskas et al. 2012, Venkatesan et al. 2015).

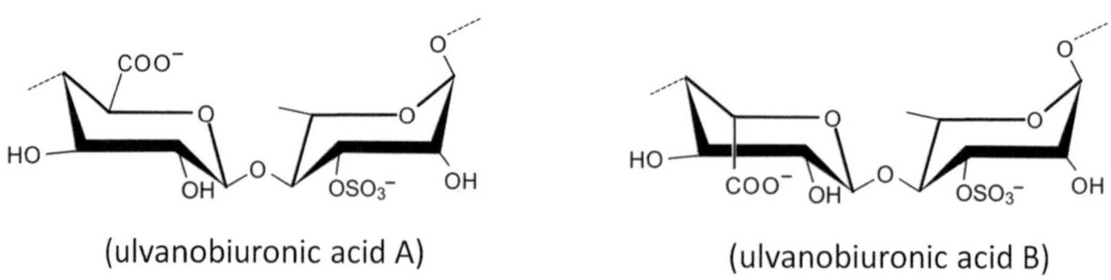

Figure 2.4 Idealized structure of the chemical units of Ulvan.

2.2.2 Minerals

All essential minerals are available in dietary seaweeds. No land plant even remotely approaches seaweeds as sources of metabolically-required minerals. Seaweeds can provide minerals often absent from freshwater algae and terrestrial crops grown on mineral-depleted soils (Drum 2013).

As an example, Japanese people consume more than 1.6 kg seaweed DW per year per capita (Fleurence 1999). Moreover, because of the minerals present in seaweed (e.g., Sodium, Potassium, Calcium, Magnesium, Iron, Zinc, Manganese, etc.), they are needed for human nutrition. However, this wide range in mineral content (8–40%) is not found in edible land plants, due to many factors, such as phylum/class of the seaweeds, geographical origin, seasonal, environmental and physiological variations (Rupérez 2002, Nisizawa 2006). Certain seaweed are also one of the most important vegetable sources of Ca, which may be as high as 7% of the dry weight, and up to 25–34% in the (calcareous) chalky seaweed *Lithothamnion* (El-Said and El-Sikaily 2013). The dry weight of the mineral present in macroalgae was obtained by burning off organic material and weighing the remaining ash (Kazutosi 2002). Other abundant elements in seaweeds include: K, Na, Mg, Zn, etc. (Drum 2013) (see Table 2.4).

Examples of fundamental minerals for human beings are Iron (Fe) for hemoglobin and Iodine (I) for thyroxin (Soetan et al. 2010). Dietary I is essential for the production of thyroid hormones, thyroxine, and triiodothyronine, which regulate many important physiological processes (Haldimann et al. 2005). I deficiency has effects on growth and development, because of inadequate production of thyroid hormones. The health consequences of I deficiency are referred to as goiter, with increased occurrence of hypo-thyroidism in moderate-to-severe cases of iodine deficiency, or decreased occurrence of hypo-thyroidism in mild, iodine deficiency, and increased susceptibility of the thyroid gland to the effects of exposure to radiation. The consequences may be abortion, still-birth, congenital anomalies, perinatal and infant mortality, or endemic cretinism, which may occur in neonates. Iodine deficiency during childhood and adolescence may cause delays of physical development and impairment of mental function or iodine-induced hyper-thyroidism in adults (Kapil 2007, Zimmermann and Crill 2010). In cases of severe iodine deficiency, hypo-thyroidism and developmental brain damage may be the dominating disorders (Laurberg et al. 2010, Bernal 2015). However, conversely, excess dietary iodine may lead to thyro-toxicosis, and could be connected with hyper-thyroidism, eu-thyroidism, hypo-thyroidism, or auto-immune thyroid disease (Alzahrani et al. 2005, Burgi 2010, Laurberg et al. 2010). However, thyroid possesses the adaptation mechanisms which regulate thyroid hormones synthesis and secretion and protect from thyrotoxicosis (Wolff 1989, Mišurcová et al. 2011a).

Zinc, iron, manganese, and copper are essential metals for human beings, as they are involved in a variety of biological activities as functions of enzymes. Implications for dietary deficiencies are well-known to cause human abnormalities (Anuradha et al. 2015). Iron is the most essential metal as it is an integral part of many proteins and enzymes which maintain good health in humans (Dallman 1986). The availability of dietary iron (proportion of the total iron content of a food that the gastro-intestinal tract and organism are able to utilize) has added a most important point of view in the study of iron metabolism (Sherman et al. 1934).

Manganese (Mn) is an essential trace element for both animals and plants; deficiencies result in severe skeletal and reproductive abnormalities in mammals (Sivaperumal et al. 2007). The highest Mn levels are concentrated in tissues with high-energy demands, such as the brain, the retina, and dark skins (with a high content of melatonin pigment). Further, bone, liver, pancreas, and the kidney contain high Mn concentrations (Aschner and Aschner 2005). Mn is involved in the metabolism of proteins, lipids, and carbohydrates; it performs as a co-factor for various enzymes. Mn is required for normal immune function, regulation of blood sugars and cellular energy, reproduction, digestion, and bone growth Manganese also aids in defense mechanisms against free radicals, and together with vitamin K, supports blood clotting and hemostasis. Finally, it is essential for the development and function of the brain (Aschner and Aschner 2005, Takeda 2003). A large portion of Mn is bound to manganese metallo-proteins. Approximately 3–5%

Table 2.4 Mineral composition of some edible seaweeds (mg 100 g^{-1} DW).

Species	Na	K	P	Ca	Mg	Fe	Zn	Mn	Cu	I	References
Chlorophyta (Green seaweed)											
Capsosiphon fulvescens	2048	1210	–	445	629	30	1.16	9.3	0.54	–	Hwang et al. 2008
Caulerpa lentillifera	8917	970–1142	1030	780–1874	630–1028	9.3–21.4	2.6–3.5	7.9	0.11– 2.2	1.424	Ratana-Arporn and Chirapart 2006 Matanjun et al. 2009
C. racemosa	2574–6990	3180–3920	29.71	1852–5970	384–1610	30–81	1–7	4.91–5.78	0.47–0.8	–	Santoso et al. 2006 Kumar et al. 2011a
C. scalpelifprmis	6610	4300	–	3820	1390	23.56	1.31	1.43	0.33	–	Kumar et al. 2011a
C. veravalensis	7230	3220	–	2320	860	17.44	1.76	1.66	0.45	–	Kumar et al. 2011a
Cladophora rupestris	–	–	87	2900	1200	1000	3	2.4	1.7	6.3	Maehre et al. 2014
Codium fragile	–	–	–	–	–	942.5	–	–	–	15.38	Hou and Yan 1998
C. tomentosum	1179	429–3729	180	513–550	83–1046	28.3	1.8	1.9	0.6	19.3	El-Said and El-Sikaily 2013 Rodrigues et al. 2015 Lloréns et al. 2016
Ulva spp.	–	–	–	–	–	–	–	–	–	16.25	MacArtain et al. 2007
U. australis (as *U. pertusa*)	1670	2040	–	170	1620	578.75	–	–	–	3.38	Hou and Yan 1998
U. clathrata	–	–	–	–	–	417.5	–	–	–	–	Peña-Rodríguez et al. 2011
Ulva fasciata	4910–14350	3125–4340	3125	3300–3940	2180–3478	5.38–44.54	2.81–4.52	2.30–8.13	0.62–1.16	12.88	Mageswaran et al. 1985 Pise and Sabale 2010 Kumar et al. 2011b
U. intestinalis	2221	5890–8650	120	550	1500	600	2.1–2.5	1.3	0.43–0.49	11.5–13	Hou and Yan 1998 Rohani-Ghadikolaei et al. 2012 El-Said and El-Sikaily 2013 Maehre et al. 2014

U. lactuca	1805–2790	1771–2414	50–140	350–840	2540–3318	21–147	0.8–2.3	1.1–15.4	0.6–1.3	2.1–5.38	Castro-Gonzalez et al. 1996 Hou and Yan 1998 Tabarsa et al. 2012a Cavaco 2017 Maehre et al. 2014
U. prolifera	573–846	1210	–	530–1139	717–930	13–67	1.47–1.91	2.27–9.48	0.66–1.11	–	Hwang et al. 2008
U. reticulata	4050	1540–3680	180	4700	2330	32.62	3.88	1.28	0.29	1.124	Ratana-Arporn and Chirapart 2006 Kumar et al. 2011b
U. rigida	1595–4100	1561–4710	210	524–1770	1770–2094	28.3–30.44	0.6–4.03	1.6–6.3	0.5–1.08	6.5	Taboada et al. 2009 Kumar et al. 2011b Romarís-Hortas et al. 2011
U. stenophylla	–	–	–	–	–	–	–	–	–	2.75	Smith et al. 2010
Phaeophyceae (Brown seaweed)											
Alaria esculenta	3625	4850	230–434	800–1000	870–900	8.7–18	3.4–4.9	0.56–1.0	0.24	19–22	Maehre et al. 2014 Lloréns et al. 2016
Ascophyllum nodosum	3700–4695	2300–3060	200	2300–4200	600–900	50–58.75	1.0	2.3	3.75	72.5–150	Smith et al. 2010 Astorga-España et al. 2015 Lloréns et al. 2016
Chnoospora minima	–	–	–	–	–	–	–	–	–	18.25	Mageswaran et al. 1985
Colpomenia sinuosa	423.83–2451	919–3510	–	377–5227	114–781	1581	1.9	9.11	0.94	–	Rohani-Ghadikolaei et al. 2012 Tabarsa et al. 2012b El-Said and El-Sikaily 2013
Cystoseira corniculata	2655	13153	453	1161	2710	–	3.65	0.78	–	–	Polat and Ozogul 2009
Dictyota dichotoma	633.85	3417	–	3257	–	1196	–	8.5	1.29	–	Tabarsa et al. 2012b
Durvillaea antarctica	–	–	–	–	–	–	–	–	–	29.13	Smith et al. 2010
Ecklonia radiata	–	–	–	–	–	–	–	–	–	372–399	Smith et al. 2010
Fucus spiralis	1429	976	–	118	163	–	–	–	–	–	Paiva et al. 2014
F. vesiculosus	2450–5469	2500–4322	315	725–938	670–994	4–11	3.71	5.50	< 0.5	14.5	Ruperez 2002, Sáa 2002

Table 2.4 contd. ...

Table 2.4 contd. ...

Species	Na	K	P	Ca	Mg	Fe	Zn	Mn	Cu	I	References
Himanthalia elongata	4030–4100	7830–8250	205–240	715–720	90–724	30.08–59	1.7–4.5	2.3	0.1	10.7–15	Sáa 2002 Maehre et al. 2014 Lloréns et al. 2016
Hormosira banksii	–	–	–	–	–	–	–	–	–	104.1	Smith et al. 2010
Laminaria digitata	3818	11579	120	1005	659	3.29	1.77	< 0.5	< 0.5	70	Rupérez 2002 Maehre et al. 2014
L. hyperborea	–	–	160	800	640	12	2.2	0.65	0.17	350	Maehre et al. 2014
L. ochroleuca	2542	6031	228	959	658	9.4	4.2	0.4	–	305	Lloréns et al. 2016
Lobophota veriagata	2556	3443	147	135.4	2.343	0.119	5.653	1.960	0.119	3.110	Thennarasan and Murugesan 2015
Macrocystis pyrifera	3370	8510	360	810	770	8.27	–	–	0.117	211.6	Smith et al. 2010 Astorga-España et al. 2014
Padina gymnospora	–	–	–	–	–	19.67–29.3	4.26	0.63	0.293–0.324	–	Robledo and Pelegrin 1997 Amado-Filho et al. 2008
P. pavonica	926.97	2970	113	319	–	250	3.05	2.03–10.33	1.21	–	Polat and Ozogul 2009 Tabarsa et al. 2012b
P. tetrastromatica	1190	2490	–	665	1140	34.24	17.09	12.94	0.85	–	Kumar et al. 2011b
Pelvetia canaliculata	–	–	73	830	960	13	3.1	0.86	0.26	21	Maehre et al. 2014
Saccharina japonica	2532–3260	4350–5951	150–300	225–910	550–757	1.19–43	0.89–1.63	0.13–0.65	0.25–0.4	130–690	Funaki et al. 2001 Kolb et al. 2004
S. latissima	2620–3133	4330–5843	165–173	750–810	324–715	11.1	1.8	0.8	–	15.9–323	Sáa 2002 Lloréns et al. 2016
Saccorhiza polyschides	–	7654	232	911	797	7.9	6.5	0.8	0.3	–	Rodrigues et al. 2015
Sargassum fusiforme	–	–	–	1860	687	88.6	1.35	–	–	–	Sugawa-Katayama and Katayama 2009
S. henslowianum	–	–	–	–	–	–	–	16.6	–	520	Hou and Yan 1998 Mišurcová et al. 2011a

S. ilicifolium	9140	4640–8766	3500	4320	817–3348	18.34–25.4	2.2–6.40	1.6–9.62	0.28–1.20	–	Pise and Sabale 2010 Rohani-Ghadikolaei et al. 2012
S. miyabei (formerly *S. kjellmanianum*)	–	–	–	–	–	182.5	–	–	–	20.38	Hou and Yan 1998
S. muticum	–	5756	228	918	1504	19	2.5	1.1	0.5	–	Rodrigues et al. 2015
S. naozhouense	3250	4170	120	66.98	–	14.7	9.08	5.84	0.36	–	Peng et al. 2013
S. swartzii	5710	7710	–	1860	700	28.26	5.28	3.84	0.58	–	Kumar et al. 2011b
S. tenerrimum	5090	8360	–	2510	820	19.88	3.57	4.31	0.40	–	Kumar et al. 2011
S. thunbergii	–	–	–	–	–	297.5	–	–	–	–	Hou and Yan 1998
Scytosiphon lomentaria	–	–	–	–	–	223.75	–	–	–	–	Hou and Yan 1998
Undaria pinnatifida	4880–6494	4475–6810	235–450	680–1380	405–680	1.54–858	0.944–1.8	0.332–1.9	0.185	22–30	Sáa 2002 Kolb et al. 2004 Smith et al. 2010 Lloréns et al. 2016
Rhodophyta (Red seaweed)											
Ceramium boydenii	–	–	–	–	–	–	–	–	–	7.13	Hou and Yan 1998
Chondrus crispus	1200–4270	1350–3184	135–592	403–1120	600 –752	4–47.7	7.14–11	1.3	< 0.5	6.1–26	Ruperez 2002, Saá 2002 Maehre et al. 2014 Lloréns et al. 2016
Corynomorpha prismatica	–	–	–	–	–	–	–	–	–	29.75	Mageswaran et al. 1985
Ellisolandia elongata (as *Corallina mediterranea*)	2086	328	–	45075	4977	27.70	3.33	6.27	0.69	–	El-Din and El-Ahwany 2016
Gelidiella acerosa	7720	13930	–	600	490	28.68	4.31	3.07	0.66	–	Kumar et al. 2011b
Gelidium amansii	–	–	–	–	–	253.75	–	–	–	20.37	Hou and Yan 1998
G. micropterum	4480	8740	–	614	890	19.98	6.12	3.96	0.48	–	Kumar et al. 2011b
Gracilaria sp.	5465	3417	–	402	565	3.65	4.35	–	–	–	Krishnaiah et al. 2008
G. bursa-pastoris	2908	4460	–	333	173	40.97	–	–	–	–	El-Said and El-Sikaily 2013

Table 2.4 contd. ...

Table 2.4 contd. ...

Species	Na	K	P	Ca	Mg	Fe	Zn	Mn	Cu	I	References
Rhodophyta (Red seaweed)											
G. canaliculata (as *G. crassa*)	4105	11170	–	255.63	438.5	29.7	6.33	8.33	0.886	–	Baghel et al. 2014
G. changii	–	–	–	651	–	95.6	13.8	–	0.8	–	Norziah and Ching 2000
G. corticata	5360–9080	2818–10860	5500	1300–4337	630–3568	13.76–27.04	0.96–5.8	1.72–3.64	0.41–0.92	18.63	Mageswaran et al. 1985 Pise and Sabale 2010 Kumar et al. 2011b
G. debilis	5720	15920	–	560	300	10.23	1.81	5.26	0.29	–	Kumar et al. 2011b
G. dura	5580	15070	–	350	290	7.88	4.26	3.27	0.45	–	Kumar et al. 2011b
G. edulis (as *G. lemaneiformis*)	–	–	–	–	–	–	–	–	–	426	Wen et al. 2006
G. fergusonii	4860	15460	–	811	410	8.93	7.29	2.46	0.42	–	Kumar et al. 2011b
G. gracilis	–	6510	226	344	175	9	2.5	2	0.4	–	Rodrigues et al. 2015
G. salicornia	2890–1035	9680–11380	–	948–1340	510	36–67	5.38	4.08–4.16	0.53–0.57	–	Kumar et al. 2011b Tabarsa et al. 2012a
Gracilariopsis longissima (as *Gracilaria confervoides*)	2280	817	–	94	29	113	–	–	35.25	–	Hou and Yan 1998 El-Said and El-Sikaily 2013
Grateloupia turuturu	–	1628	228	281	695	5	6.9	2.5	0.3	–	Rodrigues et al. 2015
Hypnea musciformis	2422	822		379	115	–	–	–	–	–	El-Said and El-Sikaily 2013
H. valentiae	–	7460	–	–	387	80.3	3.1	3.7	0.39	–	Rohani-Ghadikolaei et al. 2012
Jania rubens	2086	327	–	4234	2987	47.5	3.02	9.53	0.36	–	El-Din and El-Ahwany 2016
Kappaphycus alvarezii	2230	4100	120	840	740	66	2.76	1.10	0.47	–	Kumar et al. 2015a
Laurencia okamurae	–	–	–	–	–	–	–	–	–	22.5	Hou and Yan 1998
Osmundea pinnatifida	–	2610	173	541	480	37	5.8	1.2	0.5	–	Rodrigues et al. 2015
Palisada perforata (as *Laurencia papillosa*)	2529	3633	145	795	446	–	2.56	4.93	–	–	Polat and Ozogul 2009

Palmaria palmata	1595–1766	7310–7608	235–308	148–560	97.6–610	12.8–50	2.86–3.4	1.14–4.5	0.376–0.4	7.75–73.75	Sáa 2002 MacArtain et al. 2007 Romarís-Hortas et al. 2011 Maehre et al. 2014
Pterocladiella capillacea	684–3966	837–5090	–	610–744	77.09–221	18–22.7	0.19–4.21	3.33	0.43–0.54	–	El-Said and El-Sikaily 2013 Khairy and El-Sheikh 2015 El-Din and El-Ahwany 2016
Pyropia columbina	414	1444	380	444	492	22	1.46	–	0.51	–	Cian et al. 2014
P. tenera	3627	3500	–	390	565	10.3	2.21	2.72	< 0.5	–	Rupérez 2002
Porphyra umbilicalis	940–1050	2030–2650	235	300–330	370–950	23–46	0.7	–	0.875	11.8–17.3	Sáa 2002 Maehre et al. 2014
Pyropia vietnamensis (as *Porphyra vietnamensis*)	2450–6560	176–319	–	140–612	400–590	33–298	0.93–3.27	4.22–10.00	0.54–1.05	–	Rao et al. 2007
P. yezoensis	570	2400	–	440	650	13	10	2	1.47	–	Noda 1993
Sarcodia ceylanica	–	–	–	–	–	–	–	–	–	17.88	Mageswaran et al. 1985
Sarconema filiforme	3900	9080	–	1580	480	14.14	6.09	1.52	0.48	–	Kumar et al. 2011b

– Not determined

of ingested Mn is absorbed and is cleared from the blood by the liver, and excreted in bile (Mergler 1999). The absorption of Mn is known to be influenced by the presence of other trace elements, phytate, and ascorbic acid (Aschner and Aschner 2005).

Mn deficiencies can lead to several diseases including osteoporosis, epilepsy, impaired growth, poor bone formation and skeletal defects, abnormal glucose tolerance, and altered lipid and carbohydrate metabolism (Aschner and Aschner 2005, Nkwenkeu et al. 2002). Manganese toxicity is associated with damaged ganglia structures and leads to neuro-psychiatric symptoms and behavioral dysfunctions reminiscent of Parkinson's disease, which is the most common form of Parkinsonism, and is caused by neuro-degenerative disease, drugs, toxicants, and infections (Cersosimo and Koller 2006, Nkwenkeu et al. 2002, Ordoñez-Librado et al. 2010). High liver Mn content has been reported in alcoholics, causing liver disease, and may affect hepatic fibrogenesis (Rodríguez-Moreno et al. 1997).

As stated above, Iron (Fe) is an essential element for humans because of its participation in fundamental cell functions. Iron is the most abundant transition metal in the body, which takes part in the utilization of oxygen, and as a component of numerous enzymes, it affects many important metabolic processes, including oxygen transport, DNA synthesis, and electron transport (Lieu et al. 2001, Puntarulo 2005). The main part, 60–70% of Fe, is bound to hemoglobin in circulating erythrocytes; 10% of Fe is present in the form of myoglobin, cytochromes, and iron-containing enzymes, with 20–30% of surplus Fe being stored as ferritins and hemosiderin (Lieu et al. 2001). Fe is stored in the liver, spleen, and bone marrow associated with specific proteins (Puntarulo 2005). Fe deficiency is considered to be the most common nutritional disorder worldwide, anemia (Deegan et al. 2005, Puntarulo 2005), but can be induced by the plant-based diets of vegans, which contains less bio-available Fe (Martínez-Navarrete et al. 2002). Fe deficiency adversely affects the cognitive performance, behavior, physical growth, immune status, and morbidity from infections of all age groups. Iron-deficient humans have impaired gastro-intestinal functions, and altered patterns of hormone production and metabolism (Walker 1998, WHO 2001).

Homeostatic mechanisms are very important for the prevention of accumulation of excess Fe, which is believed to generate oxidative stress by catalysis of a variety of chemical reactions involving free radicals, which could result in cell damage (Pietrangelo 2002, Puntarulo 2005). Excess Fe accumulation has been suggested to promote cancer and increase cardio-vascular risks (Martínez-Navarrete et al. 2002). Iron overload may be observed in some cases, including an excessive dietary Fe intake, inherited diseases, e.g., idiopathic haemochromatosis, congenital atransferrinemia, or the medical treatment of thalassemia (Fontecave and Pierre 1993, Shamsian et al. 2009).

Zinc is an essential micronutrient for both animals and humans. It is described as a co-factor in nearly 300 enzymes, and is responsible for certain physiological functions, such as protein synthesis and energy metabolism, for which a relatively high level of Zn is said to be required for maintenance. Zn is necessary for growth and development; it is a structural ion of biological membranes; it has roles in gene expression and endocrine function, DNA synthesis, RNA synthesis, and cell division (O'Dell 2000, Salgueiro et al. 2002). Zn alone is an antioxidant; it regulates immune response and has a role in vitamin A metabolism (Rink and Haase 2007, Salgueiro et al. 2000). Zinc interacts with important hormones involved in bone growth and enhances the effects of vitamin D in bone metabolism (Salgueiro et al. 2002). The majority (85%) of Zn in the whole (human) body is deposited in muscles and bones, 11% is in the skin and liver; the remainder is in other tissues. Zinc constitutes about 33 $\mu g\ g^{-1}$ of an adult body mass. A high level of Zn is present in the brain (Tapiero and Tew 2003). Disturbances of Zn homeostasis have been associated with several diseases, including diabetes mellitus, and the alteration of Zn homeostasis in the brain may be associated with the manifestation of epileptic seizures (Chausmer 1998, Takeda 2000).

Copper is an essential part of several enzymes and necessary for the synthesis of haemoglobin. It is an integral metal for biological functions (Anuradha et al. 2015). Copper (Cu) and Zinc (Zn) are reported as essential nutrients for humans, stimulating fundamental metabolic protein synthesis; however, both can be toxic depending on the concentration (Underwood 1977, Onianwa et al. 1999). An adult human body contains about 1.5 $\mu g\ g^{-1}$ of copper (Nascentes et al. 2004), which is essential as a constituent of some metallo-enzymes and in the catalysis of metabolic oxidation (Underwood 1977).

Seaweed salt

The word "salary" was derived from the word "salt". Salt was highly valued and its production was legally restricted in ancient times, so it was historically used as a method of trade and currency. The word "salad" also originated from "salt", and began with the early Romans salting their leafy greens and vegetables. Undeniably, the history of salt is both broad-ranging and unique, leaving its indelible mark in cultures across the globe (Mouritsen 2013, ASSC 2016).

In earlier times, salt (NaCl) was a precious commodity, which was sometimes referred to as "white gold" and was used by the Vikings as a medium of commercial exchange. As far back as 6050 BC, salt was an important and integral part of the world's history, as it has been inter-woven into the daily lives of countless historic civilizations. Salt was used as a part of Egyptian religious offerings, and was a valuable trade item between the Phoenicians and their Mediterranean empire. Salt and history have been inextricably intertwined for millennia, with great importance placed on salt by many different races and cultures of people. Even today, the history of salt touches our daily lives (Mouritsen 2013, ASSC 2016).

Today, most ordinary, "white table salt" is derived from underground salt domes, formed millions of years ago. Also known as "rock salt", it is almost pure sodium chloride (NaCl), approx. 99.3%, and has a smattering of other salts and minerals. Consequently, the major source of salt throughout the Middle Ages was the sea. Coastal peoples were always able to obtain what is known as "grey salt". In warm climates, this was done by allowing seawater to evaporate in the sun, and in colder areas by boiling the water over an open fire. This type of salt consists of 90–96% sodium chloride and Ca, 4–7% water, as well as varying amounts of other salts and minerals, depending on how the salt was produced (Mouritsen 2013, Pereira and Correia 2015).

Sodium (Na) is vital for certain functions in the body. However, it is very easy to consume far more than is required, as salt is also often "hidden" in commodity/processed foods. There is a large body of evidence to show that too much sodium causes high blood pressure, which is one of the most significant contributors to both heart disease and stroke. Too much dietary salt raises the risk of developing eye problems and kidney disease. A natural alternative to salt is a very good way to cut down sodium intake. Seaweed has, on average, between 9–12% sodium (Na), versus table and gourmet salts, which have as much as 98% sodium (Na). However, what about all that salty "flavor" that "tastes so good"? A healthy alternative to the consumption of sodium (Na) is the consumption of algae naturally rich in potassium (K) (see Table 2.4). All living cells need potassium (K) all the time to function and stay alive; there are no exceptions (Krotkiewski et al. 1991, Dhemla and Varma 2015, Jakobsen 2016).

A nonpharmacological approach in the treatment of mild hypertension is often advocated. In an attempt to decrease sodium (Na) and increase potassium (K) intake, 62 middle-aged patients with mild hypertension were given a potassium (K) loaded ion-exchanging Na-adsorbing K-releasing seaweed preparation (seaweed fiber, SF). The mean blood pressure (MBP), evaluated in a double-blind crossover manner with four weeks' familiarization and wash-out periods, showed a significant decrease after four weeks on 12 and 24 g/day SF, but not on 6 g/day or placebo treatment. Systolic blood pressure during submaximal exercise decreased on all three SF doses. The decrease in MBP appeared to be significantly higher in Na-sensitive (11.2 mmHg, P less than .001) than in Na-insensitive (5.7 mmHg, P less than .05) patients, and was in salt-sensitive patients significantly correlated to the increase in plasma renin activity (PRA). The urinary Na excretion decreased, the K urinary increased, and the Na/K urinary excretion ratio decreased, indicating that the decrease of MBP was dependent on the decreased intestinal absorption of sodium (Na) and increased absorption of K released from the seaweed preparation. A Na-K ion-exchanging seaweed preparation is an effective means of decreasing Na and increasing K intake, and may be used for antihypertensive treatment in mild hypertension (Krotkiewski et al. 1991).

The human tongue, just as the average beginner analytical chemistry students, seems to have difficulty distinguishing potassium from sodium: both taste "salty". In equal amounts, potassium (K) is up to 8 times "saltier" than sodium (Na). The use of edible seaweeds is a delicious high-potassium, salt-replacement in most foods, e.g., on popcorn (Drum 2013, Pereira 2016). Some examples of seaweeds species used as condiments are *Porphyra*/*Pyropia* spp. (Nori), *Osmundea pinnatifida* (Pepper Dulse), among others (Braune and Guiry 2011, Guiry 2016, Pereira 2016).

2.2.3 Proteins and amino acids

Seaweed protein is a source of all amino acids, especially glycine, alanine, arginine, proline, glutamic and aspartic acids (see Table 2.5). Algal Essential Amino Acids (EAAs) represent almost a half of total amino acids and their protein profile is close to the profile of egg protein. In case of non-EAAs, all three groups of macroalgae (i.e., green, brown, and red seaweeds) contain the similar amounts. Certain red seaweeds (*Porphyra*/*Pyropia* spp.) seem to be a good source of protein because the average value may reach 47% in some species (see Table 2.5). The issue of protein malnutrition supports the trend to find a new and alternative source of protein. Some selected seaweeds could play an important role in the above-mentioned challenge because of a relatively high content of nitrogen compounds. Algae may be used in the food industry as a source of ingredients with high nutritional quality (Cerna 2011, Pereira 2011).

In general, higher protein contents are found in green and red seaweeds (10–47% of dry weight—DW) as compared to brown seaweeds (5–24% DW) (Burtin 2003, Matanjun et al. 2009, Polat and Ozogul 2009). In these certain instances, their protein levels may be comparable with soybean (35% DW). As important to amount of protein is the "quality"—most seaweed proteins contain all of the essential amino acids at levels close to that recommended by FAO/WHO (Wong and Cheung 2000, Matanjun et al. 2009). Some seaweed such as *Sargassum vulgare* contains high level of methionine (1.7%), which is not generally available in other seaweeds (Barbarino and Lourenço 2005). Others, such as *Ulva* spp., have a range reported as 26–32% of aspartic and glutamic acid of the total amino acid, respectively (Fleurence 1999), which explains their properties as flavor enhancers (the "umami" effect). The protein contents of sub-tropical edible seaweeds, e.g., *Kappaphycus alvarezii* (formerly *Eucheuma cottonii*) (Rhodophyta; 9.76% DW) and *Caulerpa lentillifera* (Chlorophyta; 10.41% DW) were higher than *S. polycystum* (Phaeophyceae; 5.4% DW), and their protein chemical scores were between 20% and 67% (Matanjun et al. 2009), some of which may be comparable to certain animal proteins. Some other seaweeds, such as *Capsosiphon fulvescens*, *Chaetomorpha ligustica* (formerly *Lola capillaris*), *Rhizoclonium riparium* (as *Rhizoclonium implexum*), *Ulva compressa*, *U. intestinalis*, *U. lactuca*, *U. prolifera* (Chlorophyta), *Cystoseira abies-marina*, *Dictyota ceylanica*, *Fucus spiralis* (Phaeophyceae), *Catenella caespitosa* (formerly *C. repens*), *Gelidiella acerosa*, *Gelidium microdon*, *Osmundea pinnatifida*, *Polysiphonia mollis*, *Porphyra* sp., *Pterocladiella capillacea*, and *Sphaerococcus coronopifolius* (Rhodophyta; see Chakraborty and Santra 2008, Hwang et al. 2008, Patarra et al. 2011 for further details) also have reportedly significant levels of protein (see Table 2.5). The highest protein content was found in those red algae collected in the summer (i.e., a range 4.8–12.8% DW) which was significantly reduced during winter (Renaud and Luong-Van 2006). Phycobiliproteins were demonstrated to have antioxidant properties which are beneficial for the prevention and treatment of neuro-degenerative diseases, cancers, and gastric ulcers (Gonzalez et al. 1999).

2.2.4 Lipids

Seaweeds are often referred to as "low-energy" food. Despite their low lipid content, ω-3 and ω-6 polyunsaturated fatty acids (PUFAs) may form a significant part of the profile of seaweed lipids. As such, PUFAs are important components of all cell membranes and precursors of eicosanoids, which are known as essential bio-regulators of many cellular processes. PUFAs effectively reduce the risk of cardiovascular disease, cancer, osteoporosis, and diabetes. Since there is frequent use of seaweeds in Asia and an increasing utilization as food in other parts of the world, there is potential for them to contribute to the enhancement of a low level of ω-3 PUFAs, especially in the Western diet. The major commercial sources of ω-3 PUFAs are fish, but their wide usage as food additives is limited because of the typical "fishy smell", unpleasant taste, and lack of stability (since they are powerful antioxidants). Nevertheless, the growing requirements

Table 2.5 Nutrient composition of selected edible seaweed (% dry weight).

Species	Protein	Ash	Dietary fiber	Carbo-hydrate	Lipid	References
Chlorophyta (green seaweed)						
Capsosiphon fulvescens	29.50	15.89	31.19	45.27	0.91	Hwang et al. 2008
Caulerpa lentillifera	10–13	24–37	33	38–59	0.86–1.11	Pattama and Chirapart 2006 Matanjun et al. 2009
C. racemosa	17.8–18.4	7–19	64.9	33–41	9.8	El-Sarraf and El-Shaarawy 1994 Akhtar and Sultana 2002 Santoso et al. 2006, Kumar et al. 2011b
C. taxifolia	12.44	–	–	23.86	0.32	Kokilam and Vasuki 2014
Chaetomorpha ligustica (as *Lola capillaris*)	40.87	–	–	22.32	4.05	Chakraborty and Santra 2008
Codium sp.	10	–	5	53	1.0	Lloréns et al. 2016
C. fragile	8–11	21–39	5.1	39–67	0.5–1.5	Ortiz et al. 2009 Guerra-Rivas et al. 2011
Gayralia oxysperma	6.03–10.02	16.83–31.22	3.06–11.73	46.5–72.28	0.53–3.28	Pádua et al. 2004
Rhizoclonium riparium	21.09	–	–	15.34	3.37	Chakraborty and Santra 2008
Ulva compressa	21–27	18.6	33–45	48.2	0.3	Burtin 2003, Mamatha et al. 2007 Patarra et al. 2011
U. fasciata	3.45–17.08	17.75–20.61	9.32–10.85	5.92–19.68	0.51–6.3	Pádua et al. 2004 Kokilam and Vasuki 2014
U. intestinalis (as *Enteromorpha intestinalis*)	6.15	–	–	30.58	7.13	Chakraborty and Santra 2008
U. lactuca	10–25	12.9–36.8	10.28–38	36–43	0.6–3.9	Fleurence 1999 Morrissey et al. 2001 Manivannan 2008, Kumar et al. 2011b Cavaco 2017
U. pertusa	20–26	–	–	47.0	–	Fleurence 1999, Pengzhan et al. 2003
U. prolifera	24.90–27.40	10.93–14.15	36.66–45.94	50.67–51.14	1.09–1.49	Hwang et al. 2008
U. rigida	18–19	28.6	38–41	43–56	0.9–2.0	Santoso et al. 2006 Taboada et al. 2010, Kumar et al. 2011b
U. reticulata	17–20	–	65.7	50–58	1.7–2.3	Shanmugam and Palpandi 2008 Kumar et al. 2011b

Table 2.5 contd....

Table 2.5 contd....

Species	Protein	Ash	Dietary fiber	Carbo-hydrate	Lipid	References
Phaeophyceae (brown seaweed)						
Alaria esculenta	9–15	–	42.86–43	46–52	1–2	Applegate and Gray 1995 Morrissey et al. 2001 Lloréns et al. 2016
Ascophyllum nodosum	10	–	50	52	1.7	Lloréns et al. 2016
Bifurcaria bifurcata	8.57	30.15	–	–	5.81	Alves et al. 2016b
Cystoseira abies-marina	6.81	–	56.34	–	–	Patarra et al. 2011
C. corniculata	30.43	28.97	–	–	5.45	Polat and Ozogul 2009
Canistrocarpus cervicornis (as Dictyota cervicornis)	18.82	20.48	–	–	–	Pádua et al. 2004
Dictyota ceylanica	18.52	–	–	3.33	2.61	Chakraborty and Santra 2008
Ecklonia bicyclis Kjellman (as *Eisenia bicyclis*)	7.5	9.72	10–12	60.6	0.1	Mitchell 2000, Mišurcová et al. 2010
Fucus spiralis	10.77	–	63.88	–	–	Patarra et al. 2011
F. vesiculosus	3–14	14–30	45–59	46.8	1.9	Applegate and Gray 1995 Fleurence 1999, Saá 2002 Truus et al. 2001, Ruperez 2002 Díaz-Rubio et al. 2008
Himanthalia elongata	5–17	30–36	9.48–37	44–61	0.5–3.1	Morrissey et al. 2001 Saá 2002, Burtin 2003 López-López et al. 2009 Gómez-Ordóñez et al. 2010 Lloréns et al. 2016
Laminaria digitata	8–15	37.59	37.3	48	1.0	Fleurence 1999, Morrissey et al. 2001 Ruperez 2002, Burtin 2003
L. ochroleuca	9–17.2	38.0	11.7–36	43.0–52	1.5–1.8	Cavaco 2017 Lloréns et al. 2016
Lobophora variegata	23.13	–	–	19.34	0.27	Thennarasan and Murugesan 2015
Padina pavonica	35.40	44.92	–	–	5.89	Polat and Ozogul 2009
Saccharina japonica	4.1–7.5	9.1–26.63	10–22.8	51.9–84.4	1.0–2.4	Mitchell 2000, Dawczynski 2007 Mišurcová et al. 2010, Cavaco 2017
S. latissima	6–26	34.78–37.3	9.89–30	44.1–61	0.5–1.8	Morrissey et al. 2001, Saá 2002 Gómez-Ordóñez et al. 2010 Cavaco 2017 Lloréns et al. 2016

Table 2.5 contd....

Table 2.5 contd....

Species	Protein	Ash	Dietary fiber	Carbo-hydrate	Lipid	References
Sargassum filipendula	8.72	44.29	6.57	3.73	–	Robledo and Freile-Pelegrin 1997
S. fusiforme	11.6	19.77	17–62	30.6	1.4	Mitchell 2000, Dawczynski 2007 Mišurcová et al. 2010
S. ilicifolium	3.86	–	–	6.72	5.7	Pise and Sabale 2010
S. vulgare	9.2–19.9	13.01–30.35	4.8–10.5	52.6–68.5	0.15–0.79	Marinho-Soriano et al. 2006
Undaria pinnatifida	12–23	26–41.2	8.81–46	30.1–51	1.5–10.1	Yamada et al. 1991 Mitchell 2000, Ruperez 2002 Sáa 2002, Burtin 2003 Dawczynski 2007 López-López et al. 2009 Mišurcová et al. 2010 Lloréns et al. 2016
Rhodophyta (red seaweed)						
Acanthophora spicifera	17.27	16.58	–	–	–	Pádua et al. 2004
Catenella caespitosa (as *C. repens*)	8.42	–	–	28.96	5.29	Chakraborty and Santra 2008
Chondrus crispus	11–21	21.08	10–34	55–68	0.9–3.0	Indergaard and Minsaas 1991 Morrissey et al. 2001 Ruperez 2002, Saá 2002 Lloréns et al. 2016
Gelidium microdon	15.18	–	57.37	–	–	Patarra et al. 2011
G. pristoides	11.80	14	–	43.10	0.90	Foster and Hodgson 1998
Gelidiella acerosa	9.18	–	–	14.34	3.83	Chakraborty and Santra 2008
Gracilaria canaliculata (as *G. crassa*)	5.18	43.18	–	42.0	1.30	Baghel et al. 2014
G. cervicornis	14.29–22.71	8.07–13.11	4.87–7.67	57.72–68.29	0.33–0.51	Marinho-Soriano et al. 2006
G. changgi	6.9	22.7	24.7	–	3.3	Norziah and Ching 2000
G. chilensis	13.7	18.9	–	66.1	1.3	Ortiz et al. 2009
G. cornea	5.47	29.06	5.21	36.29	–	Robledo and Freile-Pelegrin 1997
G. corticata	3.3	–	–	4.2	4.2	Pise and Sabale 2010
Hypnea cervicornis	18.72	21.96	–	–	–	Pádua et al. 2004
H. japonica	19	–	–	4.28	1.42	Wong and Cheung 2000
Osmundea pinnatifida	20.64	–	33.82	–	–	Patarra et al. 2011
Palisada perforata (as *Laurencia papillosa*)	34.65	44.05	–	–	6.73	Polat and Ozogul 2009

Table 2.5 contd....

Table 2.5 contd....

Species	Protein	Ash	Dietary fiber	Carbo-hydrate	Lipid	References
Palmaria palmata	8–35	15–30	28.57–38	46–56	0.7–3	Indergaard and Minsaas 1991 Fleurence 1999 Morrissey et al. 2001, Saá 2002 Lloréns et al. 2016
Polysiphonia mollis	16.59	–	–	25.81	5.79	Chakraborty and Santra 2008
Porphyra sp.	25.80	–	40.98	–	–	Patarra et al. 2011
Porphyra umbilicalis	29–39	12	29–35	43	0.3	Sáa 2002, López-López et al. 2009
Pterocladiella capillacea	20.52	–	52.08	–	–	Patarra et al. 2011
Pyropia spiralis (as *Porphyra spiralis*)	16.56	12.17	–	–	–	Pádua et al. 2004
P. tenera	7.9–47	17.5–20.5	12–35	44.3–72.3	0.7–2.3	Fleurence 1999, Mitchell 2000 Ruperez 2002, Burtin 2003 Mišurcová et al. 2010, Cavaco 2017
P. yezoensis	31–44	7.8	48.6	44.4	2.1	Indergaard and Minsaas 1991 Noda 1993, Dawczynski 2007
Sphaerococcus coronopifolius	19.56	–	41.25	–	–	Patarra et al. 2011

of healthy functional foods led to the production of PUFAs as nutraceuticals in controlled, batch culture of marine microalgae, especially *Thraustochytrium* and *Schizochytrium* strains (Mišurcová et al. 2011b).

The term "a balanced diet" of a heterotrophic organism implies the intake of essential nutrients for growth and reproduction. Some of the essential nutrients may be used for reconstruction decomposed and/or used in the production of new metabolites essential to the primary metabolism. However, there are others which cannot be produced, causing them to be obtained externally through ingestion. Among these are polyunsaturated fats, of which the most familiar are the fatty acids of omega type. These fatty acids control the cholesterol that binds to lipoproteins (carriers of these fats in blood plasma), that is, the balance between HDL / High Density Lipoprotein, or good cholesterol, and LDL / Low Density Lipoprotein, the bad cholesterol. The first should be kept at a high rate, the second at a low rate. HDL carries excess cholesterol into the bloodstream to the liver, where it is catabolized, while LDL does reverse transport, thus promoting its accumulation in tissues and organs (Pereira and Correia 2015).

EPA (eicosapentaenoic acid) and DHA (docosahexaenoic acid) (see Fig. 10.4, Chapter 10) are carboxylic acids of omega-3 type, which are considered the most important polyunsaturated fatty acids for human health (220 mg daily), because the human body is unable to produce them, and only obtains them through the intake of foods that contain them. Amongst the main sources of this type of unsaturated fats are, in addition to algae, the fish from the cold deep waters (sardines, salmon, trout, tuna, and mackerels, etc.). There is also another family of polyunsaturated fatty acids, the omega-6 group, whose primary sources are, in addition to the algae, the vegetable oils (soy, corn, sunflower, etc.), and fats and eggs. Curiously, in the human body, both groups of these polyunsaturated fatty acids are reported to interact, i.e., so that omega-3 can act effectively and give rise to all its potential benefits, there needs to be a balance between the consumption of omega-3 and omega-6 in the diet (Pereira and Correia 2015).

EPA (Omega-3) (see Fig. 10.4, Chapter 10) is the only fatty acid which can lower triglycerides (i.e., decreases hepatic synthesis of these fats) and hence the conversion of cholesterol (LDL) in the liver—which is thus available to be transported to various tissues, which can then be deposited, thereby increasing the likelihood of some diseases. Accordingly, high levels of eicosapentaenoic acid may play an important role in the prevention of thrombotic strokes, and may also lower the risk of artherosclerosis, ischemic heart disease or angina pectoris, and the risk of myocardial infarction (Aragão 2009, Pereira and Correia 2015).

For example, the red seaweed Dulse (*Palmaria palmata*), beyond the polyphenols, vitamins A, B_{12} and C, high levels of fiber, protein, minerals, and arginine, have low concentration of saturated fatty acids, and various polyunsaturated fatty acids—such as linoleic and arachidonic of omega-6 family, or EPA omega-3 family. It is therefore a good seaweed for restorative in anemia states, asthenia (weakness), and for postoperative processes. It also strengthens the vision (has high levels of vitamin A) and is recommended for the treatment of gastric and intestinal disorders, and for the regeneration of mucosa (respiratory, gastric, and vaginal). Due to the high content of vitamin B_{12} this red seaweed is suggested to provide a protection against cardiovascular diseases (Aragão 2009); and because this vitamin also reduces the homocysteine levels in blood, when high amounts are deposited in blood vessels (Pereira and Correia 2015).

2.2.5 Vitamins

Vitamins can be divided into those that are either water or fat-soluble. Water-soluble vitamins include both B-complex vitamins and vitamin C. The B-complex vitamins are the largest group, and have roles associated with metabolism, muscle tone, cell growth, and the nervous system. For example, Nori (*Porphyra/Pyropia* spp.) (red algae) and Sea Lettuce (*Ulva* spp.) (green algae) are good sources of vitamin B_{12}, which has an important role in DNA synthesis. Vitamin C is a water-soluble vitamin that is important for gum health; iron absorption, and resistance to infection (see Table 2.6).

Fat-soluble vitamins include vitamins A, D, E, and K. Vitamin A (retinol) plays an important role in bone growth, tooth development, reproduction, and cell division. Vitamin D, another fat-soluble vitamin, is important for bone growth and maintenance. Vitamins E and K also have several biological functions, including antioxidant activity and blood clotting. In addition to their biochemical functions and antioxidant activity, seaweed-derived vitamins have been demonstrated to have other health benefits, such as reducing hypertension, preventing cardiovascular disease, and reducing the risk of cancer (LTS 2005, Škrovánková 2011, Mills 2012).

Although some seaweeds contain both water and fat-soluble vitamins, their vitamin composition is variable and depends on several factors. For example, evidence exists of seasonal variation in the vitamin content of the seaweed *Eisenia arborea*, where fat-soluble vitamins follow a different pattern to those that are water-soluble. Another factor affecting seaweed vitamin content is light exposure, as plants growing in bright light can contain higher levels of some vitamins (Mills 2012).

The genus and species of seaweed are also critical factors which can affect vitamin composition. For example, the level of niacin (vitamin B_3) in some brown seaweed (e.g., *Laminaria* spp.) is approximately one tenth of the level found in the red seaweed, *Pyropia tenera* (as *Porphyra tenera*). Other factors that can influence vitamin content include season, geographical location, salinity, and sea temperature. Vitamin content can also be affected by processing, as both heat and dehydration can have a significant effect on the vitamin levels (Škrovánková 2011, Mills 2012).

2.2.6 Pigments

In agreement with what was stated above, seaweeds are classified taxonomically according to their pigment composition, which includes chlorophylls, carotenoids, and phycobilins. Many of these compounds have been commercialized for many years for coloring purposes, but importantly, the interest in their commercial applications has significantly increased in recent decades, as promising applications in human health are being established. In this context, fucoxanthin is probably the main macroalgal pigment under the spotlight of several industries (Pereira 2009, Pereira 2016).

Table 2.6 Vitamin content of some edible seaweed (mg 100 g edible portion).

Species	A (Retinol)	B_1 (Thiamin)	B_2 (Riboflavin)	B_3 (Niacin)	B_5 (Panththenic Acid)	B_6 (Pyridoxine)	B_8 (Biotin)	B_{12} (Cobalamin)	C (Ascorbic Acid)	D (Cholecalciferol)	E	Folic Acid	References
Chlorophyta (green seaweed)													
Capsosiphon fulvescens	0.278	–	–	–	–	–	–	0.61	28.2	–	–	–	Hwang et al. 2008
Caulerpa lentillifera	–	0.05	0.02	1.09	–	–	–	–	1.00	–	2.22	–	Ratana-Arporn and Chirapart 2006
Codium sp.	0.5	–	0.4	–	–	–	–	–	–	–	3.1	–	Lloréns et al. 2016
C. fragile	0.527	0.223	0.559	–	–	–	–	–	< 0.223	–	–	–	García et al. 1993
Ulva fasciata			0.1	6.6					22	–			
U. lactuca	0.017	< 0.024	0.533	98*	–	–	–	6*	< 0.242	–	–	–	García et al. 1993 Morrissey et al. 2001
U. pertusa	–	–	–	–	–	–	–	–	30–241**		–	–	Tsuchiya 1950
U. prolifera	0.126–0.195	–	–	–	–	–	–	0.31–0.99	8.9–10.5	–	–	–	Hwang et al. 2008
U. reticulata	–	0.01	0.13	–	–	–	–	–	–	–	–	–	Ratana-Arporn and Chirapart 2006
U. rigida	9581	0.47	0.199	< 0.5	1.70	< 0.1	0.012	6	9.42	–	19.70	0.108	Taboada et al. 2009
Phaeophyceae (brown seaweed)													
Alaria esculenta	–	0.6	0.3–2.7	11	–	6	–	5	100–500*	–	–	–	Morrissey et al. 2001 Lloréns et al. 2016
Ascophyllum nodosum	–	1.5	0.6	1.9	–	< 0.1	–	< 0.1	52	–	–	15	Lloréns et al. 2016

Fucus vesiculosus	0.307	0.02	0.035	–	–	–	–	–	14.124	–	–	–	García et al. 1993 Sáa 2002
Himanthalia elongata	0.079–0.3	0.020–0.3	0.020–4.5	–	–	–	–	–	28.56–66	–	5.8	0. 176–0.258	García et al. 1993 Sáa 2002 Quirós et al. 2004 Lloréns et al. 2016
Laminaria digitata	–	0.3–1.250	0.138–0.8	2.6–6.12	–	6.41	6.41	0.0005	16–35.5	–	3.43–4.7	–	MacArtain et al. 2007 Lloréns et al. 2016
L. ochroleuca	0.041	0.058	0.212	–	–	–	–	–	0.353	–	–	0.479	García et al. 1993 Quirós et al. 2004
Lobophora variegata	10.340	0.3771	0.3491	4.0162	1.36	0.3040	–	0.119	23.430	0.6442	2.13	1.983	Thennarasan and Murugesan 2015
Saccharina japonica	0.481	0.2	0.85	1.58		0.09	–	–	–		–	–	Kolb et al. 2004
S. latissima	0.04–0.4	0.05–0.2	0.21–0.4	1.7	–	0.2	–	0.0003–0.2	0.35	18	1.6		Sáa 2002 Lloréns et al. 2016
Undaria pinnatifida	0.04–0.22	0.17–0.30	0.23–1.4	2.56	–	0.18	–	0.0036	5.29		1.4–2.5	0.479	García et al. 1993, Saá 2002 Kolb et al. 2004 Quirós et al. 2004
Rhodophyta (red seaweed)													
Chondrus crispus	< 0.1	< 0.1	2.5	3.2	–	0.4	–	0.6–4*	10–13*	16	–	4.7	Morrissey et al. 2001 Rupérez 2002 Lloréns et al. 2016
Gracilaria spp.	–	–	–	–	–	–	–	–	16–149**	–	–	–	Tsuchiya 1950

Table 2.6 contd. ...

Table 2.6 contd. ...

Species	A (Retinol)	B_1 (Thiamin)	B_2 (Riboflavin)	B_3 (Niacin)	B_5 (Panththenic Acid)	B_6 (Pyridoxine)	B_8 (Biotin)	B_{12} (Cobalamin)	C (Ascorbic Acid)	D (Cholecalciferol)	E	Folic Acid	References
Palmaria palmata	1.59–3.7	0.073–1.56	0.51–1.91	1.89–2.6	–	6.8–8.99	–	0.009–3.5	6.34–34.5	–	2.2–13.9	0.267–3.5	Morrissey et al. 2001 Sáa 2002, Quirós et al. 2004 Lloréns et al. 2016
Porphyra umbilicalis	3.65	0.144	0.36	–	–	–	–	0.029	4.214	60	–	0.363	García et al. 1993, Saá 2002 Quirós et al. 2004
Pyropia yezoensis	16000***	0.129	0.382	11.0	–	–	–	0.052	–	–	–	–	Noda 1993, Watanabe et al. 2000

* expressed as *ppm*; ** expressed as *mg%*; *** expressed as *I.U.*

Phycobiliproteins (Fig. 2.5)

Red algae are rich in phycobiliproteins, i.e., water soluble pigments found in the cytoplasm or in the stroma of the chloroplasts, which are formed by complexes of phycobilins with co-valently bound proteins. Chemically, phycobilins are open-chain tetrapyrrole chromophores bearing A, B, C, and D rings. These chromophores link to the polypeptide chain at conserved positions, either by one cysteinyl thioester linkage through the vinyl substituent on the pyrrole ring A, or occasionally, by two cysteinyl thioester linkages through the vinyl substituent on both A and D pyrrole rings (Cai et al. 2012). The phycobilins are the main component determining the color of the phycobiliproteins. Based on their absorption properties, they can be blue (phycocyanobilin), red (phycoerythrobilin), yellow (phycourobilin), or purple (phycobiliviolin). Molecular pigments are organized in supra-molecular complexes (i.e., phycobilisomes) and they exert a fundamental role in the photosynthetic process of the red algae (Pereira 2009). Phycoerythrin (Fig. 2.5) is the most common phycobiliprotein in many red algae, with levels, on a dry weight basis, of approximately 0.2% for *Polysiphonia stricta* and *Pyropia* (as *Porphyra*) *yezoensis*, 0.5% for *Palmaria palmata* and *Gracilaria gracilis*, and 12% for *G. tikvahiae* (Romay et al. 2003, Wang 2002, Jespersen et al. 2005, Sekar and Chandramohan 2008, Kim et al. 2013). R-phycoerythrin, together with other phycobiliproteins, have been used for decades as natural colorants in foods (e.g., chewing gum, ice creams, soft drinks, fermented milk products, milk shakes, desserts, jellies, and coated sweet cakes, cosmetic, and pharmaceutical products) (Jespersen et al. 2005, Sekar and Chandramohan 2008). In general, the colors are very stable and tolerate high temperatures, pH changes, and light (Sekar and Chandramohan 2008). Moreover, R-phycoerythrin has specialized applications in analytical techniques, such as flow cytometry, cell sorting, and histo-chemistry (Lorbeer et al. 2013). C-phycocyanin, R- and B-phycoerythrin are currently used in the cosmetic industry for production of lipsticks, eye-liners, and other high value cosmetics (Kim et al. 2013).

The biological properties of phycoerythrin and/or phycobiliproteins include anti-oxidant, anti-inflammatory, neuro-protective, immune-modulator, anti-viral, anti-tumor, cardio-vascular, and liver protection (Romay et al. 2003, Sekar and Chandramohan 2008, Chang et al. 2011, Kim et al. 2013). Due to their biological properties, many patents have been established towards applications of these pigments for nutritional supplements and therapeutic agents (Sekar and Chandramohan 2008).

Typically, extraction of phycobiliproteins comprises the disruption of cells and a primary isolation from the alga by chemical and physical techniques. The extraction yield can be improved by the addition of other processes such as freezing, sonication, and homogenization, or the use of enzymes, e.g., lyzozymes (Romay et al. 2003). Phycobiliproteins are then purified, usually by chromatographic methods (Sekar and Chandramohan 2008), or using novel techniques such as immuno-absorption and genetic recombination.

From the reported data, *Ellisolandia elongata* (formerly *Corallina elongata*), *Gracilaria gracilis*, *Grateloupia turuturu*, and *Palmaria palmata* are present in several places around the world and hence, can be considered potential candidates for the extraction and applications of phycobiliproteins (Pereira 2009, Pereira 2015).

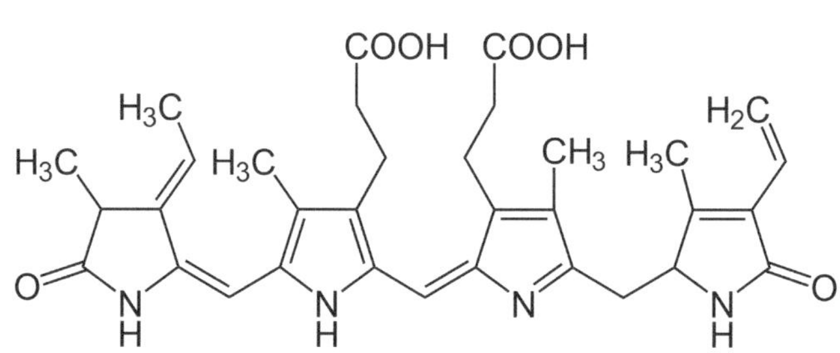

Figure 2.5 Phycoerythrobilin is the typical chromophore in phycoerythrin; it is similar to porphyrin of chlorophyll for example, but tetrapyrrole is linear, not closed into ring with metal ion in the middle (after Tiker 2016).

Fucoxanthin (Fig. 2.6)

Fucoxanthin is an orange-colored accessory pigment, belonging to the xanthophylls (carotenoids) and one of the most abundant (Dembitsky and Maoka 2007, Hosokawa et al. 2009, Kim et al. 2011, Kadam et al. 2013, Rajauria and Abu-Ghannam 2013). It is the major carotenoid of edible brown seaweeds (Phaeophyceae), where it binds to several proteins, together with chlorophyll a in the thylakoid of chloroplasts. Fucoxanthin was first isolated from the marine brown seaweeds: *Fucus*, *Dictyota*, and *Laminaria* by Willstätter and Page (1914), and its complete structure was elucidated by Englert et al. (1990). From the structural point of view, this compound is unique, with an unusual allenic bond and some oxygenic functional groups such as epoxide, hydroxyl, carbonyl, and carboxyl moieties (Fig. 2.6) (Yan et al. 1999, Cardoso et al. 2014).

Due to the presence of double bonds in the polyene chain of the carotenoid, fucoxanthin might exist in trans and/or cis configurations. As for carotenes in general, the trans forms of fucoxanthin are thermodynamically more stable than the cis counterparts due to reduced steric hindrance (Nakazawa et al. 2009). Accordingly, all-trans fucoxanthin has been isolated from several seaweed sources and, in particular, it has been shown to account for 88% of the total fucoxanthin in fresh *Undaria pinnatifida* (Holdt and Kraan 2011).

Fucoxanthin is extremely vulnerable to degradation, which mainly occurs by oxidative cleavage and/or epoxidation of the backbone. Degradation might be triggered by diverse external agents, such as high temperature, high pressure, light, and the presence of acid or oxygen. In this sense, storage and processing conditions can compromise the stability of fucoxanthin, resulting in oxidative degradation and isomerization (Holdt and Kraan 2011). Indeed, the levels of fucoxanthin were reported to significantly decrease after drying (Miyashita and Hosokawa 2008). Moreover, light and pH decreases were reported to degrade the pigment, possibly related to trans-cis isomerization reactions (Yip et al. 2014). These modifications currently limit the use of pure fucoxanthin as an ingredient of functional food preparations (Fung et al. 2013). It is noteworthy that the stability of fucoxanthin might be improved in the presence of other organic ingredients such as polyphenols (Fung et al. 2013).

The content of fucoxanthin in seaweeds differs greatly between species, with reported contents between 0.022 mg g^{-1} and 3.7 mg g^{-1} of dry weight (Mori et al. 2004, Kanazawa et al. 2008, Terasaki et al. 2009, Holdt and Kraan 2011). Also, the fucoxanthin content can be highly variable during the season and life cycle of the seaweed. In general, the levels of fucoxanthin increase from winter to spring (mature phase of the sporophyte), and decrease during summer (during the senescence phase) (Terasaki et al. 2009, Nomura et al. 2013).

Several *in vitro* and *in vivo* experiments suggested that fucoxanthin exerts important health-promoting activities, mainly due to its anti-oxidant properties (Piovan et al. 2013). The pigment has been shown to possess a strong ability to scavenge or quench DPPH• radicals, nitrobenzene with linoleic acids radical adduct (NB-L) and 12-doxyl-steric acid (12-DS) (Sachindra et al. 2007).

Besides being a good antioxidant, bioactivities reported for fucoxanthin also include anti-obesity, anti-diabetic, anti-inflammatory, anti-malarial, anti-aging, anti-tumural, and protective effects on liver, brain, bones, skin, and eyes (Holdt and Kraan 2011, Peng et al. 2012, Fung et al. 2013, Lorbeer et al. 2013). Because of its claimed health-associated properties, the pigment is being evaluated for further use as a food supplement, as a therapeutic agent in the treatment of obesity, metabolic syndrome, diabetes, and wrinkle formation (Lorbeer et al. 2013).

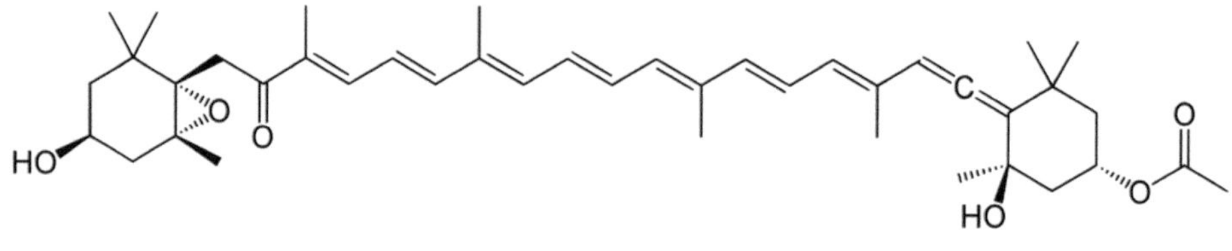

Figure 2.6 Chemical structure of Fucoxanthin (after Yikrazuul 2009).

Fucoxanthin, as other algal carotenoids, is commonly extracted with hexane and other non-polar solvents, by liquid solvent extraction. The solvent disrupts the cell membranes and dissolves lipids, lipoproteins, and the membranes of chloroplasts (Humphrey 2004, Hosikian et al. 2010). Special care must be taken in the extraction procedure (e.g., low temperature and being kept in the dark) due to the high instability of fucoxanthin. Alternative methods such as the enzyme-assisted and microwave-assisted extractions and pressurized liquid extraction techniques have been used in an attempt to minimize the degradation of the fucoxanthin (Dembitsky and Maoka 2007, Pasquet et al. 2011, Billakanti et al. 2013, Piovan et al. 2013). As reported, the species *Undaria pinnatifida*, *Fucus vesiculosus*, *Sargassum siliquastrum*, *S. fulvellum*, *S. fusiforme*, *Himanthalia elongata*, *Eisenia bicyclis*, *Laminaria digitata*, *Saccharina japonica*, and *Ascophyllum nodosum* (Phaeophyceae) can be candidates for fucoxanthin extraction (Buggeln and Carige 1973).

CHAPTER 3

Therapeutic Uses of Phycocolloids

3.1 The Main Types of Phycocolloids

Colloids are extracted compounds that form colloidal solutions, an intermediate state between a solution and a suspension; they can be used commercially as thickeners, gelling agents, and stabilizers for suspensions and emulsions. Hydrocolloids are carbohydrates, which form viscous solutions when dissolved in water.

Many seaweeds produce hydrocolloids, associated with the cell wall and intercellular spaces. Members of the red algae (Rhodophyta) produce galactans (e.g., carrageenans and agars) and the brown algae (Ochrophyta, Phaeophyceae) produce uronates (alginates) (Pereira et al. 2009, Bixler and Porse 2011, Pereira and van de Velde 2011, Pereira et al. 2013b).

Sulfated galactans (e.g., agars and carrageenans) can be obtained from red algae and alginates, and other sulfated polysaccharides (e.g., laminaran and fucoidan) are obtained from selected brown algae.

Phycocolloids (the collective term used for colloids extracted from seaweeds) are used in global food industries as natural additives and have a number of different European codes: E400 (alginic acid), E401 (sodium alginate), E402 (potassium alginate), E403 (ammonium alginate), E404 (calcium alginate), E405 (propylene glycol alginate), E406 (agar), E407 (carrageenan), and E407a (semi-refined carrageenan or "processed *Eucheuma* seaweed") (Pereira et al. 2013). Agar, alginates, and carrageenans are the colloids with the highest economic and commercial significance, since these polysaccharides exhibit high molecular weights, high viscosity, and excellent gelling, stabilizing, and emulsifying properties. All polysaccharides are water soluble and can be extracted with hot water or alkaline solutions (Minghou 1990).

3.1.1 Agar

Agar (Fig. 3.1) is the colloquial Malay name for red algae, also called agar-agar and kanten. It is an extract of certain red algae, sold in granular or powder form, or as flakes or long strips. This colloid is widely used as a gelling agent and is rarely eaten on its own (see recipe at the end of this book). In its purest form, agar is a tasteless and odorless polysaccharide, which normally contains proteins, vitamins, and minerals, like the red algae from which it is derived (Pereira 2010b, Mouritsen 2013).

Kanten was discovered by accident. For more than a thousand years, the Japanese have eaten a dish called "Tokoroten", which is prepared from the red alga *Gelidium amansii* (called "Tengusa" in Japan), and then letting the mixture stiffen. At some point towards the end of the 17th century, leftover "Tokoroten" was thrown (or put) away outside on a freezing cold day. When it was found, it had become a dry, whitish solid. The result was giving the name Kanten (Shimamura 2010, Mouritsen 2013).

Agar is light in color, semi-transparent, and very brittle when dry. When soaked, it absorbs water and, in contrast to gelatin (which is prepared from animal origin), it must be allowed to completely swell before the water is warmed to a temperature above its melting point of 85°C. Agar can be used as a gelling agent as it cools (see Table 3.1) (Pereira 2011, Mouritsen 2013).

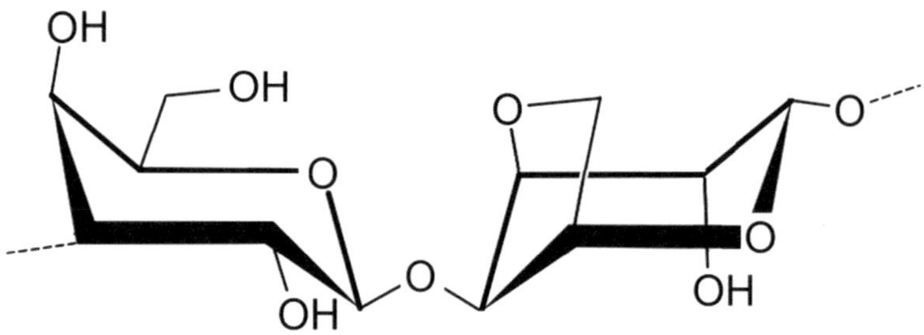

Figure 3.1 Idealized structure of the chemical units of Agar.

Agar consists of a mixture of at least two polysaccharides, i.e., agarose and agaropectin (Armisen and Galatas 2000). Typically, agarose is the predominant fraction of agar (in the range of 50–90%) (Araki 1937, Nussinovitich 1997), and is also responsible for its gelling properties (Nussinovitich 1997). Agarose typically consists of a high molecular weight polysaccharide composed of repeating units of (1→3)-β-D-galactopyranosyl-(1→4)-3,6-anhydro-α-L galactopyranose (see Fig. 3.1). Some variations can occur, depending on factors such as the species of seaweed from which the extract is made, as well as environmental and seasonal conditions affecting the raw material (Armisen and Galatas 2000). In turn, agaropectin is less clearly defined; it is a more complex polysaccharide of lower molecular weight than agarose, and it also has thickening properties (Armisen and Galatas 2000, Pereira 2011). Its structure is essentially made up of alternating (1→3)-β-D-galactopyranose, and (1→4)-3,6-anhydro-α-L-galactopyranose residues (Armisen and Galatas 2000, Qi et al. 2008).

3.1.1.a Biological properties of agar

Agar has medicinal uses, including pharmaceutical, as well as industrial applications, including use as a suspending agent for radiological solutions (i.e., barium sulfate), as a bulk laxative, as it gives a smooth and non-irritating hydrated bulk in the digestive tract, and as a formulation ingredient for tablets and capsules used to carry and control the rate of release of the drugs included. Pharmaceutical grade agar has a viscous consistency. For microbiology, agar is the medium of choice for culturing bacteria on solid substrates. Agar also has some applications in molecular microbiology, where it can be used to obtain DNA information (Dumitriu 2004). More recently, agar was used in a newly-developed medium, i.e., combined de-activators-supplemented agar medium (CDSAM), to evaluate the viability of dermatophytes in skin scales (Adachi and Watanabe 2007). The experimental data from this clinical study indicated that CDSAM was more useful than the standard media which are normally used in accurately evaluating the efficacy of anti-fungal drugs.

The possibility to use agar and agarose beads for sustained (time) release of water-soluble drugs was investigated by Nakano et al. (1979). Agarose had a significantly lower sulfate content, better optical clarity, and increased gel strength with respect to agar, but it was considerably more expensive (Kojima et al. 1978). Agar beads containing phenobarbitone sodium, as a water soluble and hypnotic drug, were prepared by El-Raheem et al. (1988). The encapsulation procedure consisted of dissolving the drug in a hot (around 70°C) agar aqueous solution and then dropping that solution in a cold bath containing a non-solvent for agar (i.e., acetone or ethyl acetate). It was reported that agar beads were instantaneously formed by gelification (not unlike alginate beads in a Ca bath—see later). The results of dissolution and release studies indicated that agar beads could be useful for the preparation of sustained-release dosage, although not many further studies have been developed till date (Laurienzo 2010).

Anti-tumor activity was found in an agar-type polysaccharide from a cold-water extraction process—as *Gracilaria* species and hydrolysates of agar resulted in agaro-oligosaccharides with activity against

Table 3.1 Applications of seaweed phycocolloids (adapted from van de Velde and De Ruiter 2002, Dhargalkar and Pereira 2005, Pereira 2004, 2008, 2011, Pereira and Ribeiro-Claro 2015).

Use/Food additives	Phycocolloid	Function
Baked food	Agar Kappa, Iota, Lambda	Improving quality and controlling moisture
Beer and wine	Alginate Kappa	Promotes flocculation and sedimentation of suspended solids
Canned and processed meat	Alginate Kappa	Hold the liquid inside the meat and texturing
Cheese	Kappa	Texturing
Chocolate milk	Kappa, Lambda	Keep the cocoa in suspension
Cold preparation puddings	Kappa, Iota, Lambda	Thicken and gelling
Condensed milk	Iota, Lambda	Emulsify
Dairy creams	Kappa, Iota	Stabilize the emulsion
Fillings for pies and cakes	Kappa	Give body and texture
Frozen fish	Alginate	Adhesion and moisture retention
Gelled water-based desserts	Kappa + Iota Kappa + Iota + CF	Gelling
Gums and sweets	Agar Iota	Gelling, texturing
Hot preparation flans	Kappa, Kappa + Iota	Gelling and improving taste
Jelly tarts	Kappa	Gelling
Juices	Agar Kappa, Lambda	Viscosity, emulsifier
Low calorie gelatins	Kappa + Iota	Gelling
Milk ice-cream	Kappa + GG, CF, X	Stabilize the emulsion and prevent ice crystals formation
Milkshakes	Lambda	Stabilize the emulsion
Salad dressings	Iota	Stabilize the suspension
Sauces and condiments	Agar Kappa	Thicken
Soymilk	Kappa + Iota	Stabilize the emulsion and improve taste
Cosmetics		
Shampoos	Alginate	Vitalization interface
Toothpaste	Carrageenan	Increase viscosity
Lotions	Alginate	Emulsification, elasticity and skin firmness
Lipstick	Alginate	Elasticity, viscosity
Medicinal and Pharmaceutical uses		
Dental mould	Alginate	Form retention
Laxatives	Alginate Carrageenan	Indigestibility and lubrification
Tablets	Alginate Carrageenan	Encapsulation
Metal poisoning	Carrageenan	Binds metal
HSV	Alginate	Inhibit virus

Table 3.1 contd....

Table 3.1 contd....

Use/Food additives	Phycocolloid	Function
Industrial and Lab uses		
Paints	Alginate	Viscosity and suspension glazing
Textiles	Agar, Carrageenan	Sizing and glazing
Paper making	Alginate, Agar, Carrageenan	Viscosity and thickening
Analytical separation	Alginate Carrageenan	Gelling
Bacteriological media	Agar	Gelling
Electrophoresis gel	Agar, Carrageenan	Gelling

Non-seaweed colloids: CF – Carob Flour; GG – Guar Gum; X – Xanthan.

α-glucosidase, and also exhibited antioxidant properties (Fernandez et al. 1989, Chen et al. 2005). It has also been reported that administration of agar-agar led to decreased concentrations of blood glucose and exerted an anti-aggregation effect on red blood cells (Kraan 2012).

In an extraction of *Gelidium* sp., agarose could be separated from agar with a yield of about 42%, and the agar content varied seasonally from 26–42% in *Gelidium* sp. (Mouradi-Givernaud et al. 1992, Jeon et al. 2005). Agaro-oligosaccharides have also been shown to suppress the production of a pro-inflammatory cytokine and an enzyme associated with the production of nitric oxide (Enoki et al. 2003).

3.1.2 Alginates

"Alginate" is the term usually used for the salts of alginic acid, but can also refer to all the derivatives of alginic acid and alginic acid itself; in some publications, the term "algin" is used instead of alginate. Chemically, alginates (Fig. 3.2) are linear co-polymers of β-D-mannuronic acid (M) and α-L-guluronic acid (G) (1-4)-linked residues, arranged either in hetero-polymeric (MG), and/or homo-polymeric (M or G) blocks (Larsen et al. 2003, Pereira et al. 2003, Leal et al. 2008).

Alginic acid is present in the cell walls of brown seaweeds, where it is partially responsible for their flexibility (McHugh 2003). Alginic acid was discovered in 1883 by E.C.C. Stanford, a British pharmacist, who called it algin. Alginic acid is extracted as a mixed salt of sodium and/or potassium, calcium, and magnesium. Since Stanford discovered algin, the name has been applied to a number of substances, e.g., alginic acid and all alginates derived from alginic acid. The extraction process is based on the conversion of an insoluble mixture of alginic acid salts of the cell wall in a soluble salt (alginate) which is appropriate for the water extraction (Lobban et al. 1988, Lahaye 2001).

While any brown seaweed could be used as a source of alginate, the actual chemical structure of alginate varies from one genus to another, and similar variability is found in the properties of the alginate that is extracted from the seaweed. Since the main applications of alginate are in thickening aqueous solutions and forming gels, its quality is judged on how well it performs in these uses (McHugh 2003).

25 to 30 years ago, almost all extraction of alginates took place in Europe, USA, and Japan. The major change in the alginate industry over the past few decades has been the emergence of producers in China in the 1980s. Initially, production was limited to low-cost, low-quality alginate for the internal, industrial markets, as produced from the locally cultivated *Saccharina (Laminaria) japonica*. By the 1990s, Chinese producers were competing in western industrial markets to sell alginates, primarily based on low prices (Pereira 2011).

A high-quality alginate forms strong gels and gives thick, aqueous solutions. A good raw material for alginate extraction should also give a high yield of alginate. Brown seaweeds that fulfill the above criteria are species of: *Ascophyllum*, *Durvillaea*, *Ecklonia*, *Fucus*, *Saccharina (Laminaria)*, *Lessonia*, *Macrocystis*,

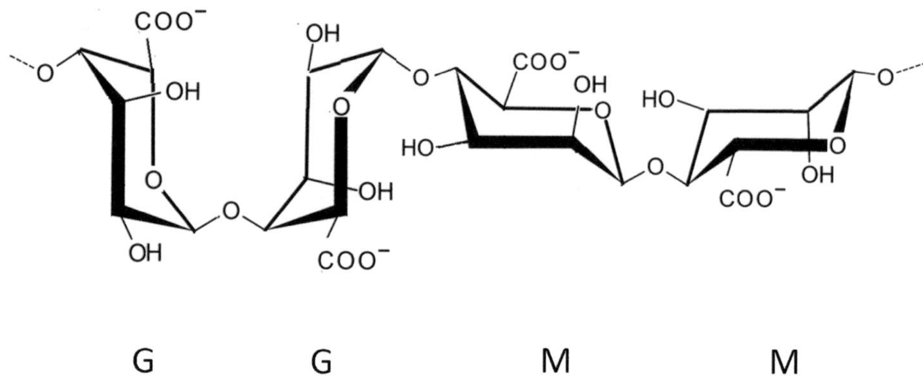

Figure 3.2 Idealized structure of the chemical units of alginic acid.

and *Sargassum*. However, *Sargassum* is only used when nothing else is available—its alginate is usually of borderline quality, and the yield is usually low (Pereira 2008).

The goal of the extraction process is to obtain dry, powdered, sodium alginate. The Ca and Mg salts do not dissolve in water; the sodium salt does. The rationale behind the extraction of alginate from the seaweed is to convert all the alginate salts to the sodium salt, dissolve them in water, and remove the seaweed residue by filtration (McHugh 2003).

Water-in-oil emulsions such as mayonnaise and salad dressings are less likely to separate into their original oil and water phases if thickened with alginate. Sodium alginate is not useful when the emulsion is acidic, due to insoluble alginic acid forms. For these applications, propylene glycol alginate (PGA) is used since this is stable in mild, acidic conditions. Alginate improves the texture, body, and sheen of yoghurt, but PGA is also used in the stabilization of milk proteins under acidic conditions, as found in some yoghurts. Some fruit drinks have fruit pulp added, and it is preferable to keep this in suspension. The addition of sodium alginate, or PGA in acidic conditions can prevent sedimentation of the pulp and create foam (hence it's used in creating the "head" on certain beers). In chocolate milk, the cocoa can be kept in suspension by an alginate/phosphate mixture, although in this application, it faces strong competition from another seaweed gel, i.e., carrageenan. Small amounts of alginate can thicken and stabilize whipped cream (Nussinovitich 1997, Onsoyen 1997).

Alginates have several commercial applications based on their thickening, gelling, emulsifying, and stabilizing abilities. They are used in the food industry for improving the textural quality of numerous products, such as salad dressing, ice-cream, beer, jelly, and lactic drinks, but also in cosmetics, pharmaceuticals, textiles, and painting industries (Murata and Nakazoe 2001, Kim and Lee 2008, Pereira 2011).

3.1.2a Biological properties of alginates

Alginate molecules may be playing vital roles in the fields of pharmacy and medicine. Due to its ability to retain water, it is mainly used as a gelling and/or viscosifying agent. Alginates can also be used in controlled drug delivery systems. In this case, the rate of drug release depends on the type and molecular weight of the alginates used. Alginates have been mostly used marine derived polymers in the preparation of drug delivery particles with the shape of spheres of different sizes (Tonnesen and Karlsen 2002). Alginate instantaneously forms gel-spheres at pH > 6 by ionotropic gelation with divalent cations such as Ca^{2+} (Russo et al. 2007), Ba^{2+}, or Zn^{2+} and for this, it is widely used for micro-encapsulation of drugs. On the other hand, at low pH, hydration of alginic acid leads to the formation of a high-viscosity "acid gel".

The ability of alginate to form two types of gel dependent on pH, i.e., an acid gel and an ion-tropic gel, gives the polymer unique properties as compared to neutral macromolecules, and it can be tailor-made for several applications (Laurienzo 2010).

The micro-encapsulation technique has been developed particularly for the oral delivery of proteins, as they are quickly denatured and degraded in the hostile environment of the stomach. The protein is encapsulated in a core material that, in turn, is coated with a biocompatible, semi permeable membrane, which controls the release rate of the protein while protecting it from biodegradation. Due to its mild gelation conditions at neutral pH, an alginate gel can act as core material in this application, while poly(ethylene glycol) (PEG), which exhibits properties such as protein resistance, low toxicity, and immunogenicity (Merrill and Salzman 1983), together with the ability of preserving the biological properties of proteins (Han et al. 1989, Tu et al. 1993), can act as a coating membrane. A chitosan/PEG-alginate micro-encapsulation process (Chen et al. 1998), applied to biological macromolecules such as albumin or hirudin (Chandy et al. 1998), was reported to be a good candidate for oral delivery of bioactive peptides.

Several examples of alginate-encapsulated drugs, other than proteins, can also be found in the literature, i.e., Qurrat-ul-Ain et al. (2003) reported that alginate micro-particles showed better drug bioavailability and reduction of systemic side effects as compared to free drugs in the treatment of tuberculosis. Polyelectrolyte coating of alginate microspheres showed to be a promising tool to achieve release systems characterized by approximately zero-order release kinetics, release of up to 100% of entrapped drugs (dexamethasone) within one month, and improved biocompatibility (Jayant et al. 2009). Composites technology has been applied to alginate for drug delivery purposes. As an example, montmorillonite-alginate nanocomposites have been recently proposed as a system for sustained release of B_1 and B_6 vitamins (Kevadiya et al. 2010). The vitamins are interspersed in the nanocrystals of the inorganic phase, and successively the hybrid B_1/B_6 montmorillonite (MMT) is further used for the synthesis of B_1/B_6-MMT-alginate Nano-composite.

Drug release is controlled by the formation of a hydrated viscous layer around the tablet, which acts as a barrier to drug diffusion and water penetration. Water-soluble drugs are mainly released by diffusion of dissolved drug molecules across the gel layer, while poorly-soluble drugs are mainly released by erosion mechanisms. Modulation of the drug release rate has been achieved by incorporating pH-independent hydrocolloids as gelling agents, or adding poly-cationic hydrocolloids such as chitosan (Miyazaki et al. 1994, Tapia et al. 2004). Several muco-adhesive systems based on alginate have been developed (Gavini et al. 2002, El-Kamel et al. 2002). The main shortcoming of alginates is their rapid erosion at neutral pH; furthermore, adhesion to mucosal tissues is reduced when cross-linked with divalent cations. Alginates have been extensively used to modify the performances of other polysaccharides, such as chitosan, through the realization of alginate-coated, chitosan micro-spheres (Strand et al. 2002). In the literature, it is also possible to find acrylic modified polysaccharides developed with the aim to obtain finer control over release rate, or to improve their adhesive properties (Mumper et al. 1994, Shojaei et al. 2000).

Due to the large variety of possible chemical compositions and molecular weights of alginate, various preparations result in different effects on biological systems. A biological effect of alginate initially was hinted at in the transplantation trails of encapsulated Islets of Langerhan for diabetes control. Overgrowth of alginate capsule by phagocytes and fibroblasts resembling foreign body/inflammatory reaction was reported. The main hurdles to the widespread use of islet transplantation for the treatment of Type 1 diabetes continues to be the insufficient number of appropriate donors and the need for immune-suppression. Micro-encapsulation has been proposed as a means to protect transplanted islets from the host's immune system. In a study by Qi et al. (2012), the function of human pancreatic islets encapsulated in Ca^{2+}/Ba^{2+}– alginate micro-beads transplanted intra-peritoneal in diabetic Balb/c mice was investigated. The study concluded that the Ca^{2+}/Ba^{2+}– alginate micro-beads could protect human islets from xenogeneic rejection in immune competent mice, without immune-suppression. However, as the grafts ultimately failed, this was likely secondary to a macrophage-mediated foreign body reaction (Nalamothu et al. 2014).

Alginate forms weak gels by the addition of the divalent cation Ca^{2+} and have been extensively studied and applied as a biomaterial in wound healing, tissue engineering (Lee and Mooney 2012), orthopedics, and dental implant surgery because of its low toxicity, relatively low cost, good bio-compatibility, and osteo-conductivity (Lin et al. 2009, Venkatesan et al. 2015a). The thickening, gel-forming, and stabilizing properties of alginates results in their being amongst the most widely used biopolymer with the broadest range of applications, including tissue engineering, drug delivery, biosensors, and wound dressings (Valente et al. 2012).

Composite materials of alginate with ceramics play a major role in increasing the mechanical strength of the extracellular matrix. Injectable calcium phosphate-alginate hydrogel-umbilical cord mesenchymal stem cell (UCMSC) paste, as reported by Zhao et al. (2010) showed remarkable osteogenic differentiation with alkaline phosphatase activity (ALP), osteocalcin (OC), collagen I, and mineralization expressions. Bouhadir et al. (2001) introduced a hydrogel containing chondrocytes in to the dorsal region of mice. After seven weeks, the mice were dissected; the hydrogel-containing chondrocyte scaffolds were uncovered and removed. A white opalescence was observed in which the appearance of native cartilage was confirmed by standard trichrome blue staining. A reduction in weight of the hydrogel implant within a span of seven weeks was attributed to the degradation of the hydrogel, and release of oxidized alginate was much faster than *in vitro* studies on the same construct. These results were in contrast with the unmodified alginate and chondrocytes in which small cartilage-like tissues were surrounded by huge amount of residues from the alginate. This injectable scaffold could find future applications in regeneration of cartilage-like tissues. Moshaverinia et al. developed an injectable and biodegradable, oxidized alginate microbead encapsulation to periodontal ligaments and gingival mesenchymal stem cells (GMSCs), and in an *in vitro* study, showed high level osteo-differentiation and adipo-differentiation (Moshaverinia et al. 2012, Venkatesan et al. 2015b).

Alginates, fucoidans, and laminaran are all reported to also have anti-bacterial effects (see Chapter 2). Extracts were tested against nine bacteria, including *Escerichia coli*, *Salmonella*, *Staphylococcus*, and *Listeria*. They appeared to be effective against *E. coli* and *Staphylococcus*. Sodium alginate seemed to demonstrate strong anti-bacterial properties; it not only bound, but also killed the bacteria. Other studies conducted on seaweed extracts found that fucoidan appeared to function as a good prebiotic (a substance that encourages the growth of beneficial bacteria in the intestines). An anti-inflammatory effect was also reported from some of the extracts. Thus far, no toxic effects have emerged in use for human health applications (Hennequart 2007, Gupta and Abu-Ghannam 2011, Nalamothu et al. 2014).

A prominent marine polysaccharide "drug" named "911" was reported to be derived from alginate polysaccharide and was found to exhibit promising activity against HIV-1 at both chronic infection of the H9 cells and acute infection of MT4 cells *in vitro* and *in vivo* (Table 3.2) (Ahmadi et al. 2015). These special effects revealed that the "911" drug inhibited the viral replication of HIV via significantly decreasing the activity of reverse transcriptase (RTase), thereby discontinuing adsorption of the virus and improving the defense mechanisms of the host cells (Xin et al. 2000a, b, Xianliang et al. 2000). Additional inhibitory results were also reported for the hepatitis B virus (HBV), stating that "911" drug inhibited viral replication by suppressing the activity of DNA polymerase (Jiang et al. 2003). Wang et al. (2012) discovered that sulfated poly-mannuroguluronate (the sulfated form of alginate) was a promising anti-AIDS drug candidate, as it caused the inhibition of HIV-1 infection mainly through the robust attachment of the viral gp120 protein together with CD4 molecules on the surface of T-cells. Moreover, there was a correlation between the size of sulfated polymannuroguluronate (SPMG) oligosaccharides and their inhibitory significance, such that the octa-saccharide would be the minimally active fragment preventing syncytium formation and reducing the p24 core antigen level in HIV-IIIB-infected CEM cells (Meiyu et al. 2003, Liu et al. 2005).

Moreover, it has been published that alginates with molecular weights greater than, or equal to, 50 kDa could prevent obesity, hypo-cholesterolemia, and diabetes (Kimura et al. 1996, Lordan et al. 2011, Harden et al. 2012, El Khoury et al. 2014, Nalamothu et al. 2014).

Clinical observations of volunteers who were 25–30% overweight showed that a drug containing alginic acid significantly decreased body weight (Zee et al. 1991, Georg et al. 2012). Interestingly it was found that cooking alginates into the bread did not reduce the molecular size of the alginate or affect its inhibitory properties. These data demonstrated the robustness of the alginate to lipase inhibition despite the cooking process and further digestive process when consumed. Therefore, adding alginate to bread may be one vehicle with the potential for application in treatments for obesity (Houghton et al. 2015).

Table 3.2 Antiviral activities of selected seaweed polysaccharides.

Seaweed group	Polysaccharide	Antiviral effect	References
Chlorophyta (green seaweeds)	Sulfated Galactans	Anti-HSV	Ohta et al. 2009
	Sulfated Heterorhamnans	Anti-HSV	Cassolato et al. 2008
	Ulvans	Anti-IAV	Ivanova et al. 1994 Karnjanapratum et al. 2012
Phaeophyceae (brown seaweeds)	Alginates and SPMG	Anti-HBV, HIV, IAV	Xin et al. 2000a, b Geng et al. 2003 Jiang et al. 2003 Miao et al. 2004 Liu et al. 2005 Hui et al. 2008 Wang et al. 2012a Ahmadi et al. 2015
	Laminarins	Anti-HIV	Muto et al. 1988 Ahmadi et al. 2015
	Sulfated Fucans and Fucoidans	Anti-AdV, CMV, Coxsackievirus, DENV, HBV, HIV, HSV, IAV, NDV, Poliovirus, Reovirus, SINV, VACV, VSV	Baba et al. 1988b Venkateswaran et al. 1989 Akamatsu et al. 2003 Hidari et al. 2008 Li et al. 2008 Queiroz et al. 2008 Trinchero et al. 2009 Karmakar et al. 2010 Jiao et al. 2011 Thanh-Sang and Kim 2013 Rabanal et al. 2014 Ahmadi et al. 2015
Rhodophyta (red seaweeds)	Agars and Agaroids	Anti-DENV, HCMV, HSV, Influenza virus, RSV	Nakashima et al. 1987a, b Mazumder et al. 2002 Zhu et al. 2003a, b Smit 2004
	ι-family Carrageenans	Anti-DENV, HAV, HPV, HRV, HSV, IAV	Gonzalez et al. 1987 Girond et al. 1991 Carlucci et al. 1997a Carlucci et al. 2004 Talarico et al. 2005 Buck et al. 2006 Talarico and Damonte 2007 Grassauer et al. 2008 Leibbrandt et al. 2010 Stephanie et al. 2010 Talarico et al. 2011 Ahmadi et al. 2015 Morokutti-Kurz et al. 2015
	κ-family Carrageenans	Anti-AdV, CMV, Coxsackievirus, Enterovirus, HIV, HPV, HSV, IAV, Poliovirus, Reovirus, SINV, VACV, VSV	Baba et al. 1988b Yamada et al. 1997 Yamada et al. 2000 Carlucci et al. 2004 De SF-Tischer et al. 2006 Chiu et al. 2012 Wang et al. 2012b Tang et al. 2013

Table 3.2 contd....

Table 3.2 contd....

Seaweed group	Polysaccharide	Antiviral effect	References
Rhodophyta (red seaweeds)	λ-family Carrageenans	Anti-AdV, CMV, Coxsackievirus, DENV, Enterovirus, HAV, HIV, HPV, HSV, Poliovirus, RABV, Reovirus, RVFV, SINV, VACV, VSV	Baba et al. 1988b Girond et al. 1991 Yamada et al. 1997 Carlucci et al. 1999 Yamada et al. 2000 Carlucci et al. 2002 Carlucci et al. 2004 Buck et al. 2006 Talarico and Damonte 2007 Gomaa and Elshoubaky 2015 Luo et al. 2015
	Porphyrans	Anti-HIV	Bhatia et al. 2008
	Sulfated Galactans	Anti-DENV, HAV, HIV, HMPV, HSV	Matsuhiro et al. 2005 Chattopadhyay et al. 2007 Bouhlal et al. 2011 Mendes et al. 2014 Ahmadi et al. 2015
	Sulfated Xylomannans and Xylogalactans	HCMV, HPIV, HSV, IAV, IBV, Junin and Tacaribe virus (Arenavirus), RSV, SIV	Damonte et al. 1994 Ghosh et al. 2009 Perez Recalde et al. 2012 Ray et al. 2015

In Type II diabetes treatment, taking 5 g of sodium alginate every morning was found to prevent a post-prandial (after eating) increase of glucose, insulin, and C-peptide levels, and slowed down gastric transit (Torsdottir et al. 1991). A meal supplemented with 5% kelp-derived alginate decreased the glucose absorption balance over an 8 hour period in pigs. Similar studies have been made on rats and humans (Vaugelade et al. 2000, El Khoury et al. 2015).

Another health effect is that the binding property of alginic acid to divalent metallic ions is correlated to the degree of the gelation or precipitation in the range of Ba < Pb < Cu < Sr < Cd < Ca < Zn < Ni < Co < Mn < Fe < Mg. No intestinal enzymes can digest alginic acid. This means that heavy metals taken into the human body are absorbed or rendered insoluble by alginic acid in the intestines and cannot be absorbed into the body tissue (Arasaki and Arasaki 1983). In several countries, such as Germany, Japan, Belgium, USA, and Canada, the use of alginic acid and its derivatives for the treatment of gastritis and gastro-duodenal ulcers, as well as the use of alginates as anti-ulcer remedies, is protected by several patents (see Sheth 1967, Bogentoff 1981, Nicolai et al. 1981, Borgo 1984, Cardoso et al. 2014, Nalamothu et al. 2014). Several products comprising drugs containing alginates have been shown to effectively suppress post-prandial and acidic refluxes, by the binding of bile acids and treatment of duodenal ulcers in humans (Nicolai et al. 1981). Examples are "Gaviscon™" (i.e., sodium alginate, sodium bicarbonate, and calcium carbonate), "Algitec™" (sodium alginate and cimetidine, an H2 antagonist) and "Gastralgin™" (alginic acid, sodium alginate, aluminum hydroxide, magnesium hydroxide, and calcium carbonate; Brailski and Dimitrov 1987, Washington and Denton 1995, Khotimchenko et al. 2001, Kwiatek et al. 2011, Nalamothu et al. 2014).

Clinical trials showed that orally administered sodium alginate promoted regeneration of the mucous membrane in the stomach, suppressed inflammation, while also eradicating colonies of *Helicobacter pylori* in the mucous membrane and normalized non-specific resistance of the latter in 4–15-year-old children. It also promoted restoration of the intestinal biocenosis (see Kraan 2012, Nalamothu et al. 2014, De Jesus Raposo et al. 2015 for details). Other studies showed the positive dietary effects of alginates on fecal microbial fauna, changes in concentrations of compounds and acids, and prebiotic properties that promote health (Terada et al. 1995, Wang et al. 2006, Kraan 2012).

Furthermore, the algins are reported to have anti-cancer properties (Murata and Nakazoe 2001, Hu et al. 2004, Laurienzo 2010) and a bioactive food additive "Detoxal™", containing calcium alginate, has reported anti-toxic effects on hepatitis. This nutritional supplement decreased the content of lipid peroxidation products and normalized the concentrations of lipids and glycogen in the liver (Khotimchenko et al. 2001, Kraan 2012).

3.1.3 Carrageenans

For several hundred years, carrageenan has been used as a thickening and stabilizing agent in food in Europe and in Asia. In Europe, the use of carrageenan started more than 600 years ago in Ireland, in the village of Carraghen, on the south Irish coast. There, flans were made by cooking the so-called "Irish Moss" (a common red seaweed species named *Chondrus crispus*) in milk. The name "carrageenan", as it was first called, was used for the first time in 1862 to name the extract from *C. crispus*, and was dedicated to this village (Tseng 1945). The name was later changed to carrageenan, so as to comply with the "-an" suffix for the names of polysaccharides (Pereira et al. 2009, Pereira 2011). The extraction method was first described by Smith in 1844 (van de Velde and de Ruiter 2002).

Irish Moss has been used in industry since the 19th century, initially in the clarification of beer (Therkelsen 1993). However, the industrial extraction of carrageenan had its beginnings in 1930, in New-England, using *Chondrus crispus* and *Mastocarpus stellatus* thalli, for the preparation of a superior suspension agent for cocoa on chocolate milk. However, the interruption of agar imports from Japan during World War II led to its replacement by carrageenan. This situation was the starting point of a booming industry (Ribier and Godineau 1984).

Fractionation of crude, carrageenan extracts started in the early 1950s (Smith and Cook 1954, Smith et al. 1955), resulting in the characterization of the different carrageenan types. A Greek prefix was introduced to identify the different notional types of carrageenan. In the same period, the molecular structures of carrageenans were determined (O'Neill 1955a, b). The structure of 3,6-anhydro-D-galactose in kappa (κ) carrageenan, as well as the type of linkages between galactose and anhydrogalactose rings, was determined.

Today, the industrial manufacture of carrageenan is no longer limited to using just *C. crispus* (Irish Moss) as a raw material; numerous red seaweed species (within the Gigartinales, Rhodophyta) are currently used. For a long period of time, these seaweeds were harvested from naturally occurring populations. Commercial carrageenans (Fig. 3.3) are usually classified into κ (kappa), ι (iota), and λ (lambda) carrageenans (Pereira and van de Velde 2011).

Carrageenans represent one of the major texturizing ingredients used by the food industry (see Table 3.1); they are natural ingredients, which have been used for decades in food applications and are **G**enerally **R**egarded **A**s **S**afe (GRAS). Carrageenans are commercially important hydrophilic colloids which occur as a matrix in numerous species of red seaweed (Stanley 1987). They are the third most important hydrocolloid in the food industry, after gelatin (from animal origins) and starch (plant origin) (Pereira and van de Velde 2011).

The modern industry of industrial extraction and marketing of carrageenans dates from the 1940s, where it was found to be the ideal stabilizer for the suspension of cocoa in chocolate milk. In the last decades, due to its functional physical properties (its rheology), i.e., gelling, thickening, emulsifying, and stabilizing abilities, carrageenans have been employed in the processed food industry to improve the texture of many convenience foods, e.g., cottage cheese, puddings, and dairy desserts, and also the hydrogel properties have been used in the manufacture of sausages, patties, and low-fat hamburgers for retention of water, flavors, and preservatives (van de Velde and de Ruiter 2002, van de Velde et al. 2004, Pereira et al. 2009, Pereira and van de Velde 2011, Li et al. 2014).

The most commonly used commercial carrageenans are extracted from open-watered sub-tropically farmed *Kappaphycus alvarezii, Eucheuma denticulatum* and some minor domesticated strains of

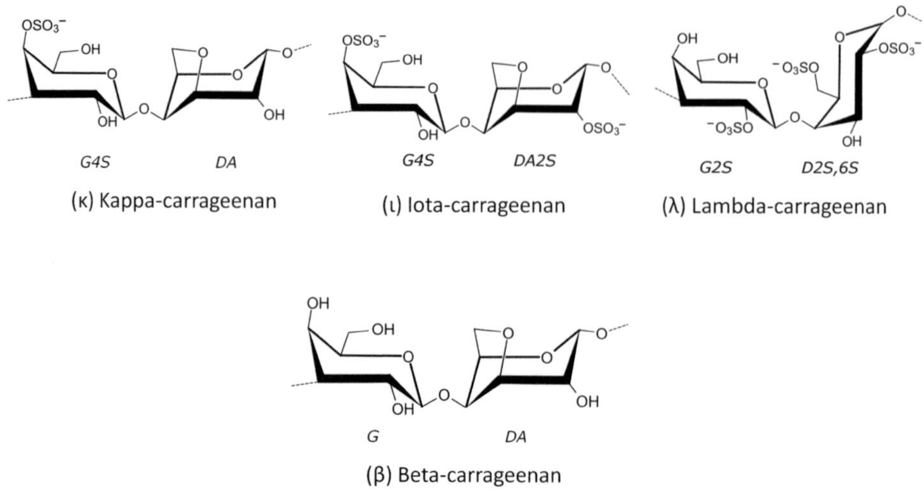

Figure 3.3 Idealized structure of the chemical units of the main types of Carrageenan.

eucheumatoids from the SE Asian triangle (McHugh 2003, Hurtado et al. 2016). Primarily wild-harvested genera from "cold water" habitats, such as *Chondrus*, *Furcellaria*, *Gigartina*, *Chondracanthus*, *Sarcothalia*, *Mazzaella*, *Iridaea*, *Mastocarpus*, and *Tichocarpus* are also raw materials of carrageenan (Pereira et al. 2013b).

The original source of a mix of "carrageenans" was from the red seaweed *Chondrus crispus*, which continues to be used, but in limited quantities. *Betaphycus gelatinus* is used for the extraction of beta (β) carrageenan (Fig. 3.3). Some South American red algae which were used previously only in minor quantities have, more recently, received attention from carrageenan producers, as they seek to increase diversification of raw materials in order to provide for the extraction of new carrageenan types with different physical (rheological) functionalities (Pereira et al. 2009b).

Large carrageenan processors fueled the development of *Kappaphycus alvarezii* (which goes by the name "cottonii" to the trade) and *Eucheuma denticulatum* (commonly referred to as "spinosum" in the trade) farming in several countries including the Philippines, Indonesia, Malaysia, Tanzania, Kiribati, Fiji, Kenya, and Madagascar (McHugh 2003, Pereira 2011, Hurtado et al. 2016). Indonesia has recently overtaken the Philippines as the world's largest producer of dried carrageenophytes biomass—which is entirely produced through open water cultivation on floating rope and raft structures (Pereira 2011, Hurtado et al. 2016).

3.1.3.a Biological properties of carrageenans

From a human health perspective, it has been reported that carrageenans may have anti-cancer, anti-hyperlipidemic, anti-coagulant, immune-modulatory, and anti-viral properties (Cáceres et al. 2000, Vlieghe et al. 2002, Panlasigui et al. 2003, Zhou et al. 2004, Skoler-Karpoff et al. 2008, Campo et al. 2009, Wijesekara et al. 2011). Carrageenans are also classically used as agents for the induction of experimental inflammation and inflammatory pain (Morris 2003).

Historically, Irish Moss or Carrageen (this is an unspecified mix of naturally co-occurring *Chondrus crispus* and *Mastocarpus stellatus*) has a large number of medical applications, some of which date from the 1830s. Indeed, it is still used in Ireland to make traditional medicinal teas and cough medicines to combat colds, bronchitis, and chronic coughs. It is said to be particularly useful for dislodging mucus and has anti-viral properties. Carrageenans are also used as suspension agents and stabilizers in other drugs, lotions, and medicinal creams. Other medical applications are as an anti-coagulant in blood-products, and for the treatment of bowel problems such as diarrhea, constipation, and dysentery. They are also used to make internal poultices to control stomach ulcers (Morrissey et al. 2001, Kraan 2012, Pereira and Correia 2015).

Previous studies conducted by Panlasigui et al. (1997) and by Dumelod et al. (1999), showed that carrageenan, incorporated into common Philippine foods such as "fishballs" and "arroz caldo", has hypoglycemic effects in normal subjects. A study done on another seaweed extract showed that when incorporated into "puto", "siomai", and a meal composed of rice and meatballs with sweet-and-sour sauce, agar had a glucose lowering effect in normal people (Panlasigui et al. 1998). The study conducted by Panlasigui et al. in 2003 indicated that regular inclusion of carrageenan in the diet may result in reduced blood cholesterol and lipid levels in human beings.

Many reports exist of anticoagulant activity and inhibited platelet aggregation of carrageenan (Hawkins and Leonard 1962, Hawkins and Leonard 1963, Kindness et al. 1979, Güven et al. 1991). Among the carrageenan types, λ-carrageenan (primarily from *C. crispus* and other cold water carrageenophytes) has approximately twice the "activity" of unfractionated carrageenan and four times the activity of κ-carrageenan (*Kappaphycus alvarezii*—formerly *Eucheuma cottoni*, and *E. denticulatum*—formerly *E. spinosum*). The most active carrageenan has approximately 1/15th the activity of heparin (Hawkins and Leonard 1962), but the sulfated galactan from *Grateloupia indica* collected from Indian waters, exhibited anti-coagulant activity as potent as heparin (Sen et al. 1994). The principal basis of the anti-coagulant activity of carrageenan appeared to be an anti-thrombotic property. λ-carrageenan showed greater anti-thrombotic activity than κ-carrageenan, probably due to its higher sulfate content, whereas the activity of the unfractionated material remained between the two.

It was found that the toxicity of carrageenans depended on the molecular weight and not the sulfate content. Similar results were obtained with λ-carrageenan of *Coccotylus brodiei* (formerly *Phyllophora brodiaei*), which gave the highest blood anti-coagulant activity (Sen et al. 1994).

The anti-microbial activities of numerous algal species have been tested and reported, presenting an extended spectrum of action against bacteria and fungi (Pereira et al. 2017). Carrageenans have proved to have effects against some bacterial strains such as *Salmonella enteritidis*, *S. typhimurium*, *Vibrio mimicus*, *Aeromonas hydrophila*, *Escherichia coli*, *Listeria monocytogenes*, and *Staphylococcus aureus*. The growth of all the bacterial strains except *L. monocytogenes* was significantly inhibited by them, particularly by the ι-carrageenan. A growth inhibition experiment using *S. enteritidis* showed that the inhibitory effect of the carrageenans was not bactericidal but bacteriostatic. Removal of the sulfate residues eliminated the bacteriostatic effect of ι-carrageenan, suggesting that the sulfate residues in carrageenan play an essential role in this effect (Venugopal 2008). In 2014, Sebaaly et al. reported that carrageenans isolated from the red alga *Corallina* sp. exhibited anti-bacterial activity against *Staphylococcus epidermis*. Infra-red spectroscopy (IR) showed that the isolated carrageenan was of λ-type.

Carrageenans are selective inhibitors of several enveloped and non-enveloped viruses, and act predominantly by inhibiting the binding or internalization of virus into the host cells (Table 3.2) (Buck et al. 2006, Grassauer et al. 2008). Carrageenans are exceptionally potent inhibitors of (Human Papilloma Virus) HPV *in vitro*, by inhibiting the initial stage of infection (Buck et al. 2006). Notably, various carrageenans are also extremely effective against a range of sexually transmitted HPV-types that lead to cervical cancer and genital warts (Gonzalez et al. 1987, Zeitlin et al. 1997). Several *in vitro* studies suggested that carrageenans may also have other antiviral properties, inhibiting the replication of herpes and hepatitis A virus (Gonzalez et al. 1987, Girond et al. 1991, Marchetti et al. 1995, Carlucci et al. 1999). More recently, Schiller and co-workers demonstrated that carrageenan was an extremely potent infection inhibitor of a broad range of genital HPVs and there were indications that carrageenan-based, sexual lubricant gels may offer protection against HPV transmission (Buck et al. 2006, Roberts et al. 2007, Rodriguez et al. 2014). The carrageenans extracted from *Acanthophora spicifera* can inhibit both herpes simplex (HSV-1) and rift valley fever virus (RVFV) and had low cytotoxicity to Vero cells (Gomaa and Elshoubaky 2015).

Researchers from the Population Council, a non-profit, research organization in the USA, conducted a clinical trial investigating whether a carrageenan-based, sexual lubricant was effective as a topical microbicide by blocking HIV infection in women. The study was conducted from March 2004–2007 in

South Africa, and enrolled more than 6,000 volunteers. The results demonstrated that the experimental microbicide gel was safe for use, but did not conclusively provide protection against HIV in the level 3 trials (Skoler-Karpoff et al. 2008, Campo et al. 2009, Population Council 2016). These results were inconsistent with *in vitro* assays, since the carrageenan was active against HIV, but at concentrations about 1,000 times greater than those required to inhibit papillomaviruses (Buck et al. 2006). However, carrageenans may serve as models for designing novel, anti-HIV agents, improving their therapeutic properties through chemical modifications. Carrageenans continue to be promoted for "female and male sexual health" with regard to lubrication of sex aids and toys.

Carrageenans from red marine algae are known to also be a potent inflammatory agent in rodents; they are reported to produce the tumor necrosis factor-α (TNF-α) in response to bacterial lipo-polysaccharide (Ogata et al. 1999) in prime mice leucocytes. Moreover, some types of carrageenan induce potent macrophage activation (Nacife et al. 2000, Nacife et al. 2004), while some carrageenans and fucoidan appear to inhibit macrophage functions (Van Rooijen and Sanders 1997, Yang et al. 2006). However, sulfated polysaccharides may have potential biomedical applications in stimulating the immune system or in controlling macrophage activity to reduce associated negative effects (Leiro et al. 2007).

Carlucci et al. (2002) found that λ-type carrageenan was active against the replication of HSV upon its firm interaction that leads to inactivation of HSV virion. They also discovered that λ-carrageenan and moderately cyclized μ/ι-carrageenan (see Pereira et al. 2003, Pereira 2004, Pereira et al. 2009) isolated from *Gigartina skottsbergii* exerted promising anti-viral activities towards diverse strains of *Herpes simplex* virus HSV-1 and HSV-2, during virus attachment stage (Carlucci et al. 1997b, 2004). Surprisingly, similar results were reported by a different group of researchers, who analyzed the chemical structure and anti-viral activity of selected carrageenans (i.e., λ, κ, and ι) against HSV-2 infection (Zacharopoulos and Phillips 1997, De SF-Tischer et al. 2006). A recent *in vitro* study conducted by Grassauer et al. (2008) reported the inhibitory effects of ι-carrageenan against human rhino-virus (HRV) proliferation by preventing the primary phases of virus replication. They suggested that this effect was possibly attributed to the suppression of the allosteric activity of virus particles during their entry (Grassauer et al. 2008, Ahmadi et al. 2015). Carrageenan nasal spray appeared as an effective treatment of common cold in children and adults, and is now a commercial product.

A nasal spray containing only ι-carrageenan, or together with "Zanamivir", provides an easy-to-apply treatment of upper respiratory tract infections in patients under suspicion of infection by influenza A (H1N1) (Leibbrandt et al. 2010, Morokutti-Kurz et al. 2015). Patients were found to benefit from the fast and efficient treatment of an uncomplicated influenza of the upper respiratory tract. Due to faster clearance of the influenza virus from the upper respiratory tract and the independent anti-viral mechanism of carrageenan and "Zanamivir", the likelihood to develop escape mutations against "Zanamivir" will be reduced. Both individual compounds can reduce severity and/or duration of the influenza illness and a combination is expected to work similarly. Additionally, due to the broad antiviral effectiveness of carrageenan, patients would receive, in parallel, a treatment of concomitant viral infections. Therefore, patients would benefit from a decreased probability for the development of complications. In consideration of the complications known to accompany an influenza virus illness, this combinational therapy meets an urgent medical need (Leibbrandt et al. 2010, Morokutti-Kurz et al. 2015). The pharmaceutical company Boehringer Ingelheim sells a nasal spray called Bisolviral®, which uses this bioactivity of ι-carrageenan (Eccles et al. 2010, Koenighofer et al. 2014, Eccles et al. 2015).

Additionally, ι-carrageenan was proven to be effective against dengue virus replication in mosquito and mammalian cells; however, the mode of anti-viral action of ι-carrageenan in both cell types was interestingly distinctive. In a Vero cell line, the inhibitory activity was exerted at the early stage of virus adhesion, probably due to some primary receptors, whereas in the mosquito cell it affected cell proliferation and protein synthesis (Talarico and Damonte 2007, Talarico et al. 2011).

Until 1935, pneumonia was the leading recorded cause of human death in the USA. 100 years ago, five of the top ten causes of death in men were respiratory diseases. Today, asthma is the leading cause of juvenile school absenteeism and is increasingly an affliction of adults. Red algae (carrageenophytes) containing carrageenan have been used for centuries as treatments for respiratory ailments, especially intractable sinus infections and lingering pneumonias. Asthma was not separated out as such in the old literature (Drum 2016).

A recent *in vivo* study in mice, revealed that low molecular weight carrageenans (i.e., 3, 5, and 10 kDa), as well as acetylated and sulfated derivatives, had substantial inhibitory effects against the influenza virus. Furthermore, the smallest κ-carrageenan with appropriate degrees of sulfation and acetylation was the greatest anti-viral candidate against influenza virus (Tang et al. 2013).

Acute viral, upper-respiratory tract infection, also known as the common cold, is the most frequently observed infectious disease in human beings. Children may get four to eight upper respiratory infections per year, and adults suffer from two to four episodes per year (Goldmann 2001). In the majority of cases, the common cold is caused by respiratory virus such as Rhinovirus, Coronavirus, Parainfluenza, Influenza, Respiratory Syncytial Virus, Adenovirus, Enterovirus, and Metapneumovirus (Monto and Sullivan 1993, Makela et al. 1998, Monto et al. 2001). According to the work of Koenighofer et al. (2014), the administration of carrageenan nasal spray in children as well as in adults suffering from virus-confirmed common cold reduced duration of disease, increased viral clearance, and reduced relapses of the symptoms.

Rabies, caused by rabies virus (RABV), is an acute, fatal encephalitic disease that affects many warm-blooded mammals. Currently, post-exposure prophylaxis regimens are effective for most rabies cases, but once the clinical signs of the disease appear, current treatment options become ineffective. Recently, works of Luo et al. (2015) indicate that λ-CGP32 is a promising agent that can inhibit RABV infection mainly by inhibiting viral internalization and glycoprotein-mediated cell fusion, and can be used for the development of novel anti-RABV drugs.

Marine algae collectively are prolific sources of a multitude of sulfated polysaccharides, which may explain the low incidence of certain cancers in countries that traditionally consume marine foods including a seaweed diet. Breast cancer is one of the most common types of non-skin cancer in females. In the study made by Murad et al. (2015), sulfated carrageenan, predominantly consisting of ι-carrageenan, extracted from the red alga *Palisada perforata* (formerly *Laurencia papillosa*), demonstrated potential cytotoxic activity against the MDA-MB-231 cancer cell line. These findings suggested that sulfated carrageenan may serve as a potential therapeutic agent to target breast cancer via prompting apoptosis (Murad et al. 2015).

λ-Carrageenan is a seaweed polysaccharide which has been generally used as pro-inflammatory agent in basic research. However, the way the immune-modulating activity of λ-carrageenan affects tumor micro-environment remains unknown. Luo et al. (2015) found that intra-tumoral injection of λ-carrageenan could inhibit tumor growth in B_{16}-F_{10} and $_4T_1$-bearing mice, and also enhance the tumor-immune response by increasing the number of tumor-infiltrating M_1 macrophages, dendritic cells (DCs), and more activated $CD_4^+CD_8^+$ T lymphocytes in the spleen. In addition, λ-carrageenan enhanced the secretion of $IL_{17}A$ in the spleen and significantly increased the level of TNF-α in the tumor, most of which was secreted by infiltrating macrophages. Moreover, λ-carrageenan exhibited an efficient adjuvant effect in ovalbumin (OVA)-based preventative and therapeutic vaccine for cancer treatment, which significantly enhanced the production of the anti-OVA antibody. The toxicity analysis suggested that λ-carrageenan had a good safety profile. Thus, λ-carrageenan might be used both as a potent anti-tumor agent and an efficient adjuvant in cancer immunotherapy.

Exposure of human (and other mammalian) cells to ultraviolet (UV) light induces various deleterious responses. Damage to cells by UV-A is thought to involve reactive oxygen species (ROS), including singlet oxygen, the superoxide and hydroxyl radicals, and hydrogen peroxide. Some of the major harmful effects are DNA damage, systemic immune suppression, and accelerated aging (Didier et al. 2001, Chignell and Sik 2003). When thymic lymphocytes were exposed to a dose of 90 mJ cm^2 UV-A, their viability was decreased by about 60%. However, the viability of the lymphocytes was significantly increased, as compared to the control group when carrageenan oligosaccharides and their derivatives were pre-administered to cells before UV-A radiation. This phenomenon showed that carrageenan oligosaccharides and their derivatives can protect the lymphocytes against UVA injury. Like a H_2O_2 induced oxidative stress model, the chemical modification of carrageenan oligosaccharides also had no significant effect on their protective effect. The studies of Yuan et al. (2006) suggested that carrageenan oligosaccharides and their derivatives showed relevant anti-oxidant activity, both *in vitro* and in a cell system.

According to Sokolova et al. (2011), the anti-oxidant actions of λ-, ξ-, κ-, κ/β-, and κ/ι-carrageenans against reactive oxygen/nitrogen species depended on the polysaccharide concentration and such structural characteristics as the presence of hydrophobic 3,6-anhydrogalactose units, the amount and position of the sulfate groups, and an oxidant agent, on which sample antioxidant action is directed. The results of Sokova et al. (2011) indicated that carrageenans possess an anti-oxidant capacity *in vitro*, and this action notably depended on the structure of the polysaccharide itself, rather than just the reducing capacity of the polysaccharides.

CHAPTER 4

The Cardio-protective Activity of Edible Seaweeds and their Extracts

4.1 Introduction

Worldwide, the burden of chronic diseases, such as cardiovascular disease (CVD), cancer, diabetes, and obesity is increasing rapidly. In 2001, chronic diseases contributed to approximately 59% of the 56.5 million total reported deaths in the world, and 46% of the global burden of disease. Cardiovascular disease is the term given for the cluster of disorders afflicting the heart and blood vessels, including hypertension (high blood pressure), coronary heart disease (heart attack), cerebrovascular disease (stroke), heart failure, and peripheral vascular disease. In 1999, CVD alone contributed to a third of global deaths, and by 2010 it would become the leading cause of death in developing countries. The majority of CVD cases are controllable and preventable. It has been reported that low intake of fruits and vegetables is also associated with a high mortality rate from CVD afflictions (Rissanen et al. 2003, Temple and Gladwin 2003). More recently, many research studies have identified a protective role for a diet rich in selected seaweeds and their derived nutraceuticals against CVD (Mayakrishnan et al. 2013, Cornish et al. 2015).

Although the most basic causal agents which contribute directly to obesity are dietary intakes which are high in calories, saturated fats, sugars, and salt, all of which are also often coupled with physical inactivity, it is prudent to also acknowledge the impact of socio-economic and environmental influences. For example, children of obese parents run a greater risk of becoming obese, and this consequence is not necessarily of genetic origin, but related more to family environment and conditioning (Birch and Fisher 1998). Other studies found that gender inequality, parental education, physical and social environment, and socio-economic indices can all affect the eating habits of children and adults (Vereecken and Maes 2010, Pabayo et al. 2012, Wells et al. 2012). Additional risk factors for CVD include diabetes mellitus (Schoenaker et al. 2012), sedentary lifestyle (Healy et al. 2011), and hypertension (Parati et al. 2012), all of which are major components of a cluster of metabolic abnormalities known as the "Metabolic Syndrome" (MetS), which also contribute to CVD pathologies (Monteiro and Azevedo 2010, Babio et al. 2012, Barona et al. 2012). Primary hypertension occurs when there is a sustained increase in blood pressure (Tierney et al. 2010), a condition subject to a world-wide, epidemic growth, and is influenced by increased dietary intakes of sodium chloride (also see Chapter 2) (Bibbins-Domingo et al. 2010, He and MacGregor 2010). Parati et al. (2012) reported that highly fluctuating but increased blood pressure level is associated with even greater CVD risks and this is also affected by obesity. Obesity in children has been repeatedly correlated to endothelial dysfunction, in that the associated over-production of reactive oxygen species (ROS) contributes markedly to that condition by reducing the bioactivity of nitric oxide (NO) within the vascular system. Such inactivation of NO, a key regulator of endothelial function, has been closely correlated to the development of CVD (Forstermann 2010, Montero et al. 2011, Lantorno et al. 2014, Cornish et al. 2015).

Marine algae in general can be a source of several functional materials, including polyunsaturated fatty acids (PUFAs), polysaccharides, essential minerals and vitamins, antioxidants, enzymes, and bioactive peptides (Pomponi 1999, Kim and Wijesekara 2010, Pereira 2011). Among marine organisms, algae may be a rich source of structurally diverse bioactive compounds with various biological activities. Recently, their importance as a source of novel bioactive substances has been growing rapidly, and research has revealed that compounds of marine algal origin exhibit effective cardio-protective activity (Barrow and Shahidi 2008, Mayakrishnan et al. 2013). Though it is a focus for a narrow audience to look for products focusing only on heart health, products with low saturated fats and cholesterol have found a high value niche in the market and have a high level of acceptance. In the same arena, omega-3 fatty acids still maintain their recognition amongst consumers, having identified effects, including protection against CVD, various inflammatory and auto-immune conditions, and enhanced cognitive health (Mayakrishnan et al. 2013, Cornish et al. 2015).

4.2 Anti-hyperlipidemic and Hypo-cholesterolemic Activities of Dietary Seaweeds and their Extracts

Abnormally high concentrations of cholesterol can occur in the body as a result of various factors. These include genetic disorders, ingestion of certain drugs, or dietary imbalances caused by the consumption of excessive calories, saturated fatty acids, and excess cholesterol (Cox and Garcia-Palmieri 1990). Unhealthy plasma cholesterol levels are a result of increases in low-density lipoprotein (LDL), or a decrease in high-density lipoprotein (HDL) (Salter 2013). Excess triglycerides in the diet, the main constituents of fats and oils, have been implicated when in combination with low levels of HDL as increasing the risk of heart attack and stroke (Bulliyya 2000, Center for Disease Control and Prevention 2016). Although the evidence remains controversial (Siri-Tarino et al. 2010, de Souza et al. 2015), excess dietary saturated fats from the over-consumption of red meats and certain dairy products, for example, play a principal role in the prevalence and incidence of MetS and central obesity (Babio et al. 2012). Trans fats found in a wide variety of baked goods and fast food products have been identified as critical elements in the development and manifestation of CVD in humans (Mann 1994, Cercaci et al. 2006, Ganguly and Pierce 2012, Cardoso et al. 2015, Cornish et al. 2015).

In recent years, there have been several important studies reporting the positive effects on lipid metabolism as a result of dietary supplementation with extracts of macroalgae. Ruqqia et al. (2015) showed that amongst the ethanol extracts of 13 seaweed species, those of *Jolyna laminarioides* and *Sargassum aquifolium* (formerly *S. binderi*), Phaeophyceae, exhibited comparable hypo-lipidemic potential to common hypolipidemic drugs such as Bezafibrate and Fenofibrate (through decreased total cholesterol –TC, triglyceride–TG, and low-density lipoprotein–LDL) in diet-induced hyper-lipidemic rats and in triton-induced hyper-lipidemic rats. The extract from *Melanothamnus afaqhusainii* (Rhodophyta) was also moderately active in lowering the levels of TC, TG, and LDL in triton-induced hyper-lipidemic rats. Liver and cardiac enzymes such as lactate dehydrogenase, alkaline phosphatase, and aspartate alanine aminotransferase were not adversely affected by administration of these three extracts (Ruqqia et al. 2015).

In addition, the ethanolic extract of *Iyengaria stellata*, Phaeophyceae (10 mg 200 g^{-1} body weight to rabbits for 30 days), showed an overall decrease in total plasmatic lipid levels, although an increase in the contents of the liver enzymes alkaline phosphatase, glutamic-pyruvic transaminase (SGPT), glutamic oxaloacetic transaminase and γ-transaminase (except for SGPT) were also registered. It is noted that as SGPT is a more specific indicator of liver injury, the overall results gathered by the authors suggested that the intake of *I. stellata* extract as a hypo-lipidemic agent should be followed by the monitoring of liver enzymes to ensure liver safety (Riaz et al. 2014).

In a pharmacological screening made by Ara et al. (2002), ethanol extracts of several seaweeds (i.e., Chlorophyta–*Caulerpa racemosa*, Rhodophyta–*Solieria robusta*, Phaeophyceae–*Colpomenia sinuosa*, *Iyengaria stellata*, *Spatoglossum asperum*) were tested for their effect on the lipid profile in normal, triton-

Table 4.0 Seaweed species associated with cardiovascular diseases (CVD) betterment factors.

Species	CVD betterment factors	Delivery format	Test systems	References
Chlorophyta (Green seaweed)				
Derbesia tenuissima	Decreased plasma triglycerides and total cholesterol	Supplement (insoluble fiber)	Rat model of human MetS	Kumar et al. 2015b
Ulva linza	Reduction of cholesterol and serum triglycerides	Whole foods	Rats	Ramirez-Higuera et al. 2014
Ulva ohnoi	Decreased plasma triglycerides and total cholesterol	Supplement (insoluble fiber)	Rat model of human MetS	Kumar et al. 2015b
Phaeophyceae (Brown seaweed)				
Ascophyllum nodosum	Fermentable fiber-short-chain fatty-acid production	Extract	Human gut microbes	Ramnani et al. 2012
Dictyopteris undulata	Inhibition of adipocyte differentiation	Extract	Adipocyte cell lines	Lee et al. 2011
Ecklonia bicyclis	Vascular oxidative stress tolerance	Extract	Human (males)	Kang et al. 2003
E. cava	Reduction in body fat ROS inhibition	Mice Adipocyte cell lines	Extract Extract	Park et al. 2012 Lee et al. 2011
Himanthalia elongata	Decreased plasma cholesterol antioxidant activities	Whole food	Rats	Moreira et al. 2011
Ishige okamurae	ROS inhibition	Extract	Adipocyte cell lines	Lee et al. 2011
Lessonia trabeculata	Reduction in cholesterol and serum triglycerides	Whole food	Rats	Ramirez-Higuera et al. 2014
Saccharina japonica (formerly *Laminaria japonica*)	Reduction in cholesterol and serum triglycerides	Extract mixture	Mice	Shin and Yoon 2012
Sargassum fusiforme (as *Hizikia fusiformis*)	Nitric oxide inhibition	Extract	Mouse macrophage cell line	Yang et al. 2010
S. fulvellum	Suppression of triglyceride absorption	Extract	Human macrophages	Matsumoto et al. 2010
S. polycystum	Decreased plasma total cholesterol and triglycerides	Whole food	Rats	Awang et al. 2014
S. thunbergii	Nitric oxide inhibition	Extract	Rats	Yang et al. 2010
Undaria pinnatifida	Suppression of triglyceride absorption Reduction in plasma cholesterol levels Inflammation inhibition	Extract Whole foods Extract	Rats Wistar rats Adipocyte cell lines	Matsumoto et al. 2010 Moreira et al. 2010 Kim and Lee 2012

Table 4.0 contd....

Table 4.0 contd....

Species	CVD betterment factors	Delivery format	Test systems	References
Rhodophyta (Red seaweed)				
Callophyllis japonica	Reduction in total cholesterol and serum triglycerides	Extract	Mice	Kang et al. 2012
Chondrus crispus	Oxidative stress tolerance n-3 PUFA; inflammation inhibition Nitric oxide inhibition	Extract Extract Extract	*Caenorhabditis elegans* Human macrophages Macrophage RAW264.7 cells	Sangha et al. 2013 Robertson et al. 2015 Banskota et al. 2013
C. ocellatus	Inhibition of adipocyte differentiation	Extract	Adipocyte cell lines	Lee et al. 2011
Chondrophycus intermedius (formerly *Laurencia intermedia*)	ROS inhibition	Extract	Adipocyte cell lines	Lee et al. 2011
Gelidium corneum (formerly *Gelidium sesquipedale*)	Fermentable fiber-short-chain fatty-acid production	Extract	Human gut microbes	Ramnani et al. 2012
Gloiopeltis furcata	Nitric oxide inhibition	Extract	Mouse macrophage cell line	Yang et al. 2010
Gracilaria sp.	Fermentable fiber-short-chain fatty-acid production	Extract	Human gut microbes	Ramnani et al. 2012
Grateloupia elliptica	Nitric oxide inhibition	Extract	Mouse macrophage cell line	Yang et al. 2010
Laurencia okamurae	Nitric oxide inhibition	Extract	Mouse macrophage cell line	Yang et al. 2010
Mastocarpus stellatus	Reduction in triglycerides and total cholesterol	Whole food	Wistar rats	Gómez-Ordóñez et al. 2012
Plamaria palmata	n-3 PUFA; inflammation inhibition	Extract	Human macrophages	Robertson et al. 2015
Porphyra dioica	n-3 PUFA; inflammation inhibition	Extract	Rats	Robertson et al. 2015
P. umbilicalis	Reduction in plasma cholesterol levels	Whole food	Human macrophages	Moreira et al. 2010

induced and high fat diet-induced hyper-lipidemic rats. In a similar screening made by Hetta et al. (2009), sulfated polysaccharides (SP) extracted from *Acanthophora spicifera* (Rhodophyta) lowered the level of total serum lipids, total cholesterol (TC), triglycerides (TG), and low-density lipoproteins (LDL) by 48%, 49.6%, 63%, and 80.6%, respectively. For SP extracted from *Sirophysalis trinodis* (formerly *Cystoseira trinodis*, Phaeophyceae), total lipids, TC, TG, and LDL were decreased by 25.5%, 49%, 51%, and 91%, respectively (Hetta et al. 2009). Most of the seaweeds tested demonstrated promising results in lowering the serum cholesterol, triglyceride, and LDL-cholesterol levels, followed by an increase in HDL-cholesterol, as has been observed with other natural hypo-cholesterolemic agents (Singer 1981, Tsi et al. 1995). The protective role of HDL-cholesterol in atherogenesis was shown by Frick et al. (1987), emphasizing that an 11% increase in HDL-cholesterol concentration could reduce the risk of myocardial infarction by 34%.

Triton-induced and diet-induced hyper-lipidemic rats showed a significant decrease in serum cholesterol, triglyceride, and LDL-cholesterol levels when treated with the ethanolic extract of *Caulerpa racemosa*, *C. sinuosa*, *Solieria robusta*, *Iyengaria stellata*, and *Spatoglossum asperum*. It has been shown that seaweeds (i.e., Wakame) increased the activity of the hepatic enzymes responsible for fatty acid oxidation (Murata et al. 1999). These effects may be due to a direct effect on cholesterol biosynthesis (Vazquez-Freire et al. 1996). It was shown that the hypo-lipidemic action of other natural products was linked to an increased fecal bile acid excretion and with an associated inhibition of cholesterol biosynthesis (Singh et al. 1990). The reduced LDL-cholesterol level in hyper-lipidemic rats possibly may be due to receptor-mediated catabolism of LDL and enhanced hepatic biosynthesis of protein and nucleic acid (Singh et al. 1992a, b). For a pharmacological agent to be effective in combating hyper-lipidemia, it should act to reduce the high LDL-cholesterol level, because low density lipoprotein transport accounts for approximately 70% of plasma cholesterol in humans (Spilman et al. 1992). High fat diet-induced hyper-lipidemic rats treated with an ethanol extract of *S. asperum* showed a significant decrease in serum enzyme concentrations of low-density lipoprotein (LDH), alkaline-phosphatase (ALP), serum glutamate-oxaloacetate transaminase (SGOT)/aspartate aminotransferase (ASAT), compared to the control of the high fat diet rats (Ara et al. 2002).

When supplementing the diets of hyper-cholesterolemic Wistar rats (WI) with 21% of *Himanthalia elongata* (Phaeophyceae) or 23% of *Gigartina pistillata* (Rhodophyta), Villanueva and coworkers (2014) showed that *Himanthalia*-treated rats presented a reduction in the plasmatic levels of TG (28%), while increasing those of HDL (20%). In turn, the *Gigartina*-supplemented diet produced a significant decrease of 31% in TG, 18% in total cholesterol (TC), and 16% in LDL. An identical effect was observed in studies using diets supplemented with tropical green seaweed (i.e., 5% of dried *Derbesia tenuissima*, Chlorophyta). Consumption of *D. tenuissima* did not change the total body fat mass, but it was reported to decrease the plasma levels of TG by 38% and TC by 17% (Kumar et al. 2015b).

Rats fed with a HF (hight-fat) diet supplemented with 5% freeze dried *Gracilaria changii* (Rhodophyta) powder had significantly reduced plasma TC (−39.19%), LDL-C (−36.36%), and triglycerides (TG) content (− 25.45%). Meanwhile, 10% seaweed powder significantly lowered the plasma TC, LDL-C and TG content by −40.34, −35.95 and −30.91% respectively, compared to the HF group. As suggested by the authors, one of the potential mechanisms of action explaining the hypo-lipidemic effects of *G. changii* might be due to its high dietary fiber (i.e., 61.29%) (unpublished). Supplementation of *G. changii* to normal diet rats showed a less significant effect on the plasma TC, HDL-C, LDL-C, and TG levels (Chan et al. 2014). These differences suggested that administration of dietary *G. changii* was probably more effective for hyper-lipidemia treatment, instead of preventive ones. In addition, *Ecklonia cava* (Phaeophyceae) supplementation (methanol extracted) dose dependently suppressed TG, TC, and LDL concentrations in the serum in both normal and Streptozotocin (STZ)-diabetic mice (supplementation of 5% *E. cava* in diabetic mice caused a decreased serum levels of 72, 53, and 78%, respectively), but failed to affect the HDL concentrations in normal mice (Kim and Kim 2012).

In a study conducted by Najam et al. (2010), methanolic extracts of *Hypnea musciformis* (Rhodophyta) were tested for their pharmacological activities on rabbits and mice. *H. musciformis* extracts significantly decreased the serum total cholesterol, triglyceride, and low-density lipoprotein cholesterol levels of the rabbits tested. This is an important finding, since decreased levels of cholesterol and total lipids together minimize the incidence of many cardiovascular problems (Najam et al. 2010). Later, Ara et al. (2005), using oil fractions of *Spatoglossum asperum* (Phaeophyceae), showed hypolipidemic activity in normal, triton-induced and high fat diet-induced hyper-lipidemic rats without producing any ill effects on cardiac and liver enzymes.

In an animal model fed on a cholesterol-rich diet, a seaweed mixture comprising *Ecklonia bicyclis* (formerly *Eisenia bicyclis*), *Sargassum fusiforme* (formerly *Hizikia fusiformis*), *Undaria pinnatifida* sporophytes, Phaeophyceae and *Pyropia yezoensis* (formerly *Porphyra yezoensis*), Rhodophyta exerted anti-hyper-lipidemic activity by lowering triglyceride, serum total cholesterol, LDL-cholesterol, and free cholesterol levels (Amano et al. 2005).

Besides the evidence provided from the dietary supplementation of seaweeds and/or crude extracts on lipid metabolism, several authors have also described positive effects for purified fractions and/or isolated compounds. Amongst them, the majority of the research has focused on two major groups of compounds, namely sulfated polysaccharides and lipids (Cardoso et al. 2015). Some recent examples from the literature are described below.

A reduction of the risk of cardiovascular diseases through the dietary consumption (or supplementation) by various beneficial seaweeds is suggested due to their modifying effects on the gastrointestinal tract (GIT), such as emulsification of bile acid and interfering with lipid micelle formation, dilution of lipase concentration, binding with cholesterol, and slowing down lipid absorption (Rajapakse and Kim 2011). Studies carried out using rats reported that alginic acid (a component of brown algal structure) leads to decreased concentration of cholesterol, and is often coupled with increased faecal cholesterol content and a hypo-cholesterolemic response (Dumelod et al. 1999). Furthermore, porphyran (a red algal polysaccharide; see also Chapter 2 and Fig. 2.3) significantly lowered the artificially enhanced level of hypertension and blood cholesterol in rats, thereby conserving cardiac health (Noda 1993).

Thomesa et al. (2010) reported results investigating whether a fucoidan extract of *Cladosiphon okamuranus* (Phaeophyceae) might affect isoproterenol-induced myocardial infarction in rats. Isoproterenol induction was marked by elevations in the serum levels of creatine phosphor-kinase, lactate dehydrogenase, and transaminases in animals. On the contrary, fucoidan treatment protected the structural and functional integrity of the myocardial membrane, as was evidenced from the significant reduction in elevated levels of these serum marker enzymes in rats treated with fucoidan, when compared to the group of isoproterenol-treated control animals. Increases in the activities of these enzymes were reported as diagnostic indicators of myocardial infarction, indicative of cellular damage, and loss of functional integrity of cell membranes (Bhakta et al. 1999). The reversal of these enzyme activities by pre-treatment with fucoidan indicated its therapeutic potential against myocardial infarction.

Dietary fibers are known to support reduced cholesterol levels, and recent studies have shown that those with an ion-exchange capacity contain more potent effects on the lowering of blood cholesterol (Guillon and Champ 2000). For example, low density lipo-protein (LDL) cholesterol was significantly lower in rats fed on a diet containing dried *Ulva rigida* (Taboada et al. 2010). Ulvan, a polysaccharide from green algae (see also Chapter 2), may also modulate lipid metabolism in rats and mice. A decrease of serum high-density lipoprotein cholesterol (HDL-cholesterol) and an increase of low-density lipoprotein cholesterol (LDL-cholesterol) and triglyceride are considered to be significant risk factors in cardiovascular diseases. Ulvan (see Fig. 2.4, Chapter 2), or ulvan-derived oligosaccharides, significantly lowered the level of serum total cholesterol, LDL-cholesterol and reduced triglyceride, while they increased the levels of serum HDL-cholesterol (Pengzhan et al. 2003).

Yu et al. (2003a, b) reported that ulvan and its sub-fractions had significantly different and even opposing effects, on lipid parameters. Significant reductions in serum total and LDL-cholesterol concentrations and elevations of daily bile acid excretion in ulvan-fed rats suggested a mechanism of cholesterol breakdown into bile acid, by which ulvan effectively lowered serum cholesterol levels. When a diet supplemented with ulvan was consumed, other nutritional components were more easily digested and absorbed from the small intestine, and ulvan became a major component in the gut lumen. Ulvan was observed to gel with calcium ions in the body and increased the viscosity of the lumen's contents, thereby interfering with absorption of bile acid from the ileum. LDL-cholesterol was removed from the blood and converted into bile acids by the liver to replace the bile acids lost in the stool, and consequently, serum LDL-cholesterol also decreased simultaneously (Marlett et al. 2002, Lahaye and Robic 2007).

A relevant study focusing the effects of ulvans in lipid metabolism was performed by Hassan et al. (2011). These authors described that the intra-gastric administration of *Ulva lactuca* (Chlorophyta) sulfated polysaccharides to dietary-induced hyper-cholesterolemic rats could cause a more apparent effect in the increment of HDL-C level (i.e., by 180%) when compared to that induced by its oral administration. These observations paved the way for discussions of the effects of the administration mode and how to take differences in outcome (Hassan et al. 2011).

When ulvan was degraded into smaller fractions (i.e., its oligomeric groups), the viscosity decreased and its ability to interfere with bile acids might be expected to reduce and even disappear. As a result, the cholesterol-lowering action of ulvan was observed to degrade; this postulation was in agreement with the results for total and LDL-cholesterol concentrations in the fractionated ulvan group. However, one confusing problem was that bile acid excretion in various ulvan-fed rats showed no significant differences.

To elucidate the mechanism of action played by ulvan in lowering cholesterol, further studies on various enzymes, such as HMG-CoA reductase or 7-alpha hydroxylase are required (Mayakrishnan et al. 2013).

Borai et al. (2015) reported that the oral administration of sulfated polysaccharides (SP) from *Ulva fasciata* (Chlorophyta) to induced-hyper-cholesterolemic rats for four consecutive weeks did not exert any side effects, and simultaneously caused a significant decrease in the serum lipid profile, by reducing the plasma TC, TG, LDL, and also very low density lipo-protein-cholesterol (VLDL). Notably, the scores for the effects of administered ulvans from *Ulva fasciata* were better than those observed for the reference drug: Fluvastatin. Additionally, Hoang et al. (2015) reported that two types of SP isolated from the green alga *Monostroma nitidum* showed the ability to reduce cellular lipid concentrations in lipid-loaded hepatocytes, when compared to control; this reduction was accompanied by a reduced expression of cholesterol synthesis genes, and increased expression of genes dictating cholesterol degradation, LDL uptake, and peroxisomal β-oxidation.

Fucoidan (Fig. 2.1, Chapter 2) is an SP from brown seaweed, and its use in dietary treatments also reduced lipid peroxidation activity and increased enzymatic and non-enzymatic antioxidant levels in rats, indicating a protective role for the SP against isoproterenol-induced myocardial infarction. The formation of free radicals and accumulation of lipid peroxides is one possible biochemical mechanism for the myocardial damage caused by this catecholamine. Free radical scavenging enzymes such as catalase, superoxide dismutase, and glutathione peroxidase are the first-line, cellular defense mechanism against oxidative injury, decomposing O_2 and H_2O_2 before their interaction to form the more reactive hydroxyl radical (OH·). The equilibrium between these enzymes is therefore essential for the effective removal of oxygen stress in intracellular organelles. The second-line of defense consists of non-enzymatic scavengers, viz., α-tocopherol and ascorbic acid, which scavenge residual free radicals, escaping decomposition by antioxidant enzymes. Improved antioxidant levels (enzymatic- SOD, CAT, GPx, GSH, GST; non-enzymatic-reduced glutathione, α-tocopherol, ascorbic acid) in the hearts of animals treated with fucoidan indicated the potential of fucoidan in the treatment of myocardial infarction. Isoproterenol administration raised the LDL-cholesterol and decreased HDL cholesterol levels in the serum. Interestingly, treatment with fucoidan reversed the effects of isoproterenol. Increased total cholesterol and LDL-cholesterol and decreased HDL cholesterol are associated with higher risks for myocardial infarction. High levels of circulating cholesterol and triglycerides and their accumulation in heart tissue are associated with cardiovascular damage. Hyper-triglyceridemic patients at risk of cardiovascular disease often develop a lipoprotein profile characterized by elevated triglyceride, dense LDL, and low HDL cholesterol, which causes myocardial membrane damage (Kelly 1999). Hyper-triglyceridemia has been seen in isoproterenol-treated rats in cases of ischemic heart disease. The report of anti-hyper-triglyceridemia activity of fucoidan signified that the myocardial membrane was protected against isoproterenol-induced damage. Moreover, histo-pathological findings confirmed the induction of myocardial infarction by isoproterenol and the protection rendered by the fucoidan treatment to the cardiac muscle (Mayakrishnan et al. 2013).

Although fucoidans are high-molecular-weight polysaccharides, they exert biological effects after oral administration in experimental animals. Fucoidans are reported to activate lymphocytes and macrophages to render immune protection to the host. They may exert their effects through gut-associated immunity before absorption, which is then transferred to the systemic immune response via lymph nodes and peripheral blood (Wu et al. 1998, Sithranga and Kathiresan 2010).

Recent findings suggested that polysaccharides administered orally might also be partially absorbed intact via the hepatic portal vein and central lacteals into the general blood circulatory system. They have been shown to accumulate in the mesenteric lymph nodes, Peyer's patches, spleen, liver, and kidneys. Therefore, fucoidan have the potential to modulate the defense activity of the body, both at the mucous membrane and systemically.

Laminaran (Fig. 2.2, Chapter 2) are a grouping of sulfated polysaccharides found in brown seaweeds and are commercially generally derived from kelps or *Laminaria* and *Saccharina* spp. (Phaeophyceae). Two forms of laminaran are recognized; they are referred to as soluble and insoluble laminaran. Microsomal polysaccharide of laminaran has been noted to produce hyper-cholesterolemic and hypo-lipidemic responses due to reduced cholesterol absorption in the gut (Panlasigui et al. 2003, Mayakrishnan et al. 2013). The observation was often coupled to an increase in the faecal cholesterol content and a hypoglycemic response

(Dumelod et al. 1999). A similar result was obtained with the fucoidan fraction extracted from the brown seaweed *Sargassum henslowianum* collected in Vietnam, having lowered cholesterol, triglyceride, and LDL-cholesterol levels in obese mice (Cuong et al. 2015).

Collectively, Thomesa et al. (2010) showed that fucoidan was non-toxic and could protect cardiac cells from isoproterenol-induced myocardial infarction. The hypothesis is that the protection of the myocardium by fucoidan was likely due to the detoxification of isoproterenol through the antioxidant defense system and alterations in the lipid profile. Further studies are needed to determine the molecular mechanism by which fucoidan(s) of various sources can act on the myocardium to beneficially affect the cardiovascular system. Studies along this line can open new avenues for novel seaweed compounds in the treatment of cardiovascular dysfunction (Mayakrishnan et al. 2013).

Additionally, Kim and co-workers (2014b) recently demonstrated that the dietary supplementation of a commercial fucoidan (Haewon Biotech, Inc., Seoul, Korea) induced a significant decrease on the plasma levels of TG, total cholesterol, and of LDL-C. As reported by the authors, the beneficial effects of the fucoidan were probably partially associated to the down-regulation of the adipogenic transcription factor.

The findings of Panlasigui et al. (2003) showed that use of carrageenan (see also Chapter 3) as the main source of dietary fiber may have brought about hypo-cholesterolemic effects due to its ability to bind bile acids and cholesterol in the lower small intestine. By binding up bile acids, less cholesterol can be absorbed. Since bile is made from cholesterol, circulating cholesterol is further decreased, since it is used to compensate for the lost bile. Reductions in serum cholesterol and lipids may also be due to changes in the physical properties of the intestinal contents related to the ability of fibers to provide bulk, volume, and viscosity to the contents, all of which are important in slowing the rate of digestion and absorption of nutrients. Dietary fibers are known to provide bulk since they are not digested, so they remain intact during transit of food through the small intestines. The water-retaining capacity of carrageenans results in an increased volume of intestinal contents. On the other hand, the viscosity of the intestinal contents increases due to the presence of viscous polysaccharides, such as carrageenan (Mayakrishnan et al. 2013).

Carrageenans were also recently used in a clinical trial study. Patients with ischemic heart disease (IHD) were reported to exhibit a significant effect on lipid profile by a short-term dietary carrageenan supplement. The prophylactic administration of the carrageenan food supplement in the complex therapy of IHD patients significantly decreased the plasma TC levels by 16.5% and LDL-C by 33.5%, as compared to the baseline measurements (background therapy control and experimental groups) (Sokolova et al. 2014).

The free amino-acid fraction of seaweed is a mixture of amino acids, and mainly composed of taurine, alanine, amino butyric acid, omithine, citrulline, and hydroxyproline (Holdt and Kraan 2011). Taurine is an amino acid present in high concentrations in some red algae; it acts as an antioxidant and protects against toxicity of various heavy metals, such as lead and cadmium, by preventing their absorption in the stomach. Taurine has been shown to be effective in reducing the secretion of serum lipids and apolipoprotein B100, a structural component of low density lipoproteins, thereby reducing the risk of atherosclerosis and coronary heart disease. These finding were supported by several research reports, concluding that taurine supplementation exerted a hypo-cholesterolemic effect in young, overweight adults. Taurine has also been shown to relieve complications in people with congestive heart failure by increasing the force and effectiveness of heart–muscle contractions (Mochizuki et al. 1999, Lourenço and Camilo 2002, Mendis and Kim 2011).

Amongst the various studies on macroalgae, lipids, fucoxanthin (see Fig. 2.6, Chapter 2), and fatty acids are widely mentioned when it comes to the regulation of dyslipidemia. Notably, the beneficial effects of fucoxanthin (pigment) on cardiovascular diseases were recently reviewed by Gammone et al. (2015). In agreement with other reported studies, these authors highlighted that fucoxanthin metabolites (i.e., amarouciaxanthin A and fucoxanthinol) were the active *in vivo* forms of fucoxanthin, and that their ability in regulating lipids plasmatic levels were ascribed mainly due to their antioxidant activity. In addition, fucoxanthin and its metabolite fucoxanthinol are reported to exhibit extra-metabolic beneficial effects, including the regulation of polyunsaturated fatty acid (PUFAs) biosynthesis, by the promotion of the up-regulation of enzymatic activities related to the bioconversion of omega-3 PUFA and omega-6 PUFA to docosahexaenoic acid (DHA) and arachidonic acid (AA), respectively (Tsukui et al. 2007, 2009, Miyashita et al. 2011, Cardoso et al. 2015). Furthermore, it has been demonstrated that both of these compounds could induce a decreased content of eicosapentaenoic acid (EPA), as shown in cultured rat hepatoma BRL-3A, thus suggesting a down-regulation of certain metabolic enzyme pathways such as fatty acid desaturase and elongase (Aki et al. 2014). In rodents fed with fucoxanthin supplement, it was found to promote the synthesis of DHA in the liver (Tsukui et al. 2007), resulting in the improvement of the lipid profile, since this acid inhibited the synthesis of thromboxane A2 and enhanced the production of prostacyclin, a prostaglandin that produced vasodilation and less "sticky" blood platelets (see Riccioni et al. 2011 for details).

Fatty acids from seaweeds by themselves may have beneficial properties relevant to cardiovascular diseases, since they are good sources of long-chain omega-3 PUFAs, EPA, and DHA acids (Cardoso et al. 2015). As demonstrated in KK-Ay mouse, DHA and AA levels were significantly increased by the feeding of lipids from *Sargassum horneri* and *Stephanocystis hakodatensis* (formerly *Cystoseira hakodatensis*), both members of the Phaeophyceae, but not by those obtained from another brown kelp—*Undaria pinnatifida* (Airanthi et al. 2011). As suggested by these authors, this difference could be due to the higher fucoxanthin content in the first two seaweeds (hence not all seaweeds are the same, and one should be very specific in terms of the seaweed tested, its extracts, and the model in which the efficacy is demonstrated)—this is not dissimilar to land plants. The authors additionally reported that the plasma levels of TC, HDL, and phospholipid of the mice fed with these seaweed lipid fractions were significantly increased, whilst those of hepatic cholesterol and triacylglycerol were decreased, as compared to the control group (Airanthi et al. 2011).

4.3 Hypoglycemic, Anti-diabetic, and Anti-obesity Activities Reported in Selected Seaweeds

In recent times, the global tendency towards reduced physical activity of humans associated with augmented dietary intake of fats, sugars, and calories is leading to a burgeoning issue of overweight, obesity, and lifestyle-related diseases, e.g., diabetes, hypertension, dyslipidemia, and metabolic syndrome. In particular, obesity, which is characterized as a state of low-level inflammation, is a powerful determinant both in the development of insulin resistance and in the progression to Type 2 diabetes (Gammone and D'Orazio 2015). Diabetes mellitus is a chronic metabolic disease affecting an estimated 382 million adults globally, and this figure is projected to increase to approximately 592 million by 2035. Type 2 diabetes mellitus (T2DM) accounts for approximately 90% of the cases of diabetes, and is quickly becoming a global epidemic of the 21st century (International Diabetes Federation 2013).

The publication of Lamela et al. (1989), showed the effects of several extracts of selected seaweeds on hypoglycemic activity in rabbits. Ethanolic extracts of *Laminaria ochroleuca*, *Saccorhiza polyschides*, and *Fucus vesiculosus* (Phaeophyceae) were administered orally to normal animals, and their effects on glycemia and tri-glyceridemia were evaluated. Crude polysaccharides and protein solutions from *Himanthalia elongata* (Phaeophyceae) and *Codium tomentosum* (Chlorophyta) were also assayed. Polysaccharides and proteins from *H. elongata* caused a significant reduction in blood glucose eight hours after intravenous administration. The study showed that 5 mg kg^{-1} of crude polysaccharide lowered glycemia by about 18% in normal rabbits, and by about 50% in alloxan-diabetic animals, while the protein solution lowered glycemia in diabetic rabbits by about 30%.

In other published data, Nwosu et al. (2011) showed that phenolic-rich extracts from *Palmaria palmata* (Rhodophyta), *Ascophyllum nodosum*, and *Alaria esculenta* (Phaeophyceae) inhibited α-amylase activity to some extent, but *Ascophyllum* extracts were very effective, with an IC_{50} of ≈0.1 μg mL^{-1} GAE (gallic acid equivalents). The *Ascophyllum* extracts also inhibited α-glucosidase (see Table 5.1), the other key enzyme involved in starch digestion and blood glucose regulation, at relatively low levels (e.g., IC_{50} ≈20 μg mL^{-1} GAE). After fractionation on Sephadex LH-20, the inhibitory activity from *Ascophyllum* was concentrated to the fraction which, from mass spectrometric evidence, was found to be enriched in phlorotannins. These components are described as having the capacity to inhibit α-amylase and α-glucosidase activities at μM levels, which are easily achievable in the gut. This may explain the anti-diabetic properties associated with algal extracts and algal phenolics in various *in vivo* studies.

Table 4.1 Glucose levels reduction through inhibition of α-amylase and α-glucosidase activities.

Species	Active agents	Activity	Test systems	References
Chlorophyta (Green seaweed)				
Caulerpa racemosa	Acetone crude extract	α-amylase inhibition, IC_{50} = 0.09 mg/mL	*In vitro* assay	Teixeira et al. 2007
C. racemosa	Methanolic extract Acetate extract	• α-amylase inhibition (300 μg/mL) = 35% • α-amylase inhibition (300 μg/mL) = 68%	*In vitro* assay	Shanmugaasokan et al. 2013
Chaetomorpha aerea	Chloroform extract	α-amylase inhibition, IC_{50} = 408.9 mg/mL	*In vitro* assay	Unnikrishnan et al. 2016
Chlorodesmis sp.	Methanolic extract	α-amylase inhibition, IC_{50} = 147.6 mg/mL	*In vitro* assay	Unnikrishnan et al. 2016
Cladophora rupestris	Methanolic extract	α-amylase inhibition (500 μg/mL) = 14%	*In vitro* assay	Unnikrishnan et al. 2016
Halimeda macroloba	Water extract	α-glucosidase inhibition, IC_{50} = 6.388 mg/mL	*In vitro* assay	Chin et al. 2015
Ulva intestinalis	Methanolic extract	α-amylase inhibition (500 μg/mL) = 59%	*In vitro* assay	Unnikrishnan et al. 2016
U. lactuca	Water extract	α-amylase inhibition, IC_{50} = 67 μg/mL α-glucosidase inhibition, IC_{50} = 53 μg/mL	*In vitro* assay	SenthilKumar and Sudha 2012
U. lactuca	Methanolic extract	α-amylase inhibition, IC_{50} = 1.00 mg/mL α-glucosidase inhibition, IC_{50} = 1.14 mg/mL	*In vitro* assay	Cavaco 2017
Phaeophyceae (Brown seaweed)				
Alaria marginata	Ethyl acetate extract	α-amylase inhibition, IC_{50} = 63.28 μg/mL α-glucosidase inhibition, IC_{50} = 15.66 μg/mL	*In vitro* assay	Kellogg et al. 2014
Ascophyllum nodosum	Aqueous ethanolic extract	α-glucosidase inhibition, IC_{50} = 77 μg/mL	*In vitro* assay	Zhang et al. 2007
A. nodosum	Water extract	α-amylase inhibition, IC_{50} = 1.34 μg phenolics α-glucosidase inhibition, IC_{50} = 0.24 μg phenolics	*In vitro* assay	Apostolidis and Lee 2010
A. nodosum	Phlorotannin-rich extract	α-amylase inhibition, IC_{50} ~ 0.1 μg/mL GAE α-glucosidase inhibition, IC_{50} ~ 20 μg/mL GAE	*In vitro* assay	Nwosu et al. 2011
A. nodosum	Cold water and ethanol extracts	α-amylase inhibition (water), IC_{50} = 53.6 μg/mL α-amylase inhibition (ethanol), IC_{50} = 44.7 μg/mL	*In vitro* assay	Lordan et al. 2013

Table 4.1 contd....

Table 4.1 contd....

Species	Active agents	Activity	Test systems	References
A. nodosum	Fucoidan	• α-amylase inhibition, IC_{50} = 0.12–4.64 mg/mL • α-glucosidase inhibition, IC_{50} = 0.013–0.047 mg/mL	*In vitro* assay	Kim et al. 2014c
Ecklonia bicyclis	Phloroglucinol derivatives	α-amylase inhibition at 1 mM by: • Dieckol, inhibition = 97.5% • 1-(3′,5′-dihydroxyphenoxy)-7-(2″,4″,6″-trihydroxyphenoxy)-2,4,9-trihydroxydibenzo-1,4-Dioxin, inhibition = 89.5% • Eckol, inhibition = 87.5%	*In vitro* assay	Okada et al. 2004
E. bicyclis	• Phlorofucofuroeckol-A • Dieckol • 7-Phloroeckol • Eckol • Dioxinodehydroeckol • Phloroglucinol	α-glucosidase inhibition by: • Phlorofucofuroeckol-A, IC_{50} = 1.37 μM • Dieckol, IC_{50} = 1.61 μM • 7-Phloroeckol, IC_{50} = 6.13 μM • Eckol, IC_{50} = 22.78 μM • Dioxinodehydroeckol, IC_{50} = 34.60 μM • Phloroglucinol, IC_{50} = 141.18 μM	*In vitro* assay	Moon et al. 2011
E. bicyclis	Fucofuroeckol A	• α-amylase inhibition, IC_{50} = 42.91 μM • α-glucosidase inhibition, IC_{50} = 131.34 nM	*In vitro* assay	Eom et al. 2012
	Dioxinodehydroeckol	• α-amylase inhibition, IC_{50} = 472.70 μM • α-glucosidase inhibition, IC_{50} = 93.33 nM		
E. cava	• Dieckol • Fucodiphloroethol G • Phlorofucofuroeckol A • 6,6′-bieckol • 7-Phloroeckol	α-glucosidase inhibition by: • Dieckol, IC_{50} = 10.8 μM • Fucodiphloroethol G, IC_{50} = 19.5 μM • Phlorofucofuroeckol A, IC_{50} = 19.7 μM • 6,6′-bieckol, IC_{50} = 22.2 μM • 7-Phloroeckol, IC_{50} = 49.5 μM	*In vitro* assay	Lee et al. 2009
	• Dieckol • 7-Phloroeckol • 6,6′-bieckol • Fucodiphloroethol G • Phlorofucofuroeckol A	α-amylase inhibition by: • Dieckol, IC_{50} = 124.9 μM • 7-Phloroeckol, IC_{50} = 250.0 μM • 6,6′-bieckol, IC_{50} > 500 μM • Fucodiphloroethol G, IC_{50} > 500 μM • Phlorofucofuroeckol A, IC_{50} > 500 μM		

Table 4.1 contd....

Table 4.1 contd....

Species	Active agents	Activity	Test systems	References
E. cava	Dieckol	• α-amylase inhibition, IC_{50} = 0.66 mM • α-glucosidase inhibition, IC_{50} = 0.24 mM	*In vitro* assay	Lee et al. 2010
E. maxima	• Eckol • Dibenzo [1,4] dioxine-2,4,7,9-tetraol • Phloroglucinol	α-glucosidase inhibition by: • Eckol, IC_{50} = 11.16 μM • Dibenzo [1,4] dioxine-2,4,7,9-tetraol, IC_{50} = 33.69 μM • Phloroglucinol, IC_{50} = 1991 μM	*In vitro* assay	Rengasamy et al. 2013
E. stolonifera	Water extract	α-glucosidase inhibition against: • α-glucosidase (Saccharomyces), IC_{50} = 0.026 mg/mL • Rat intestinal maltase, IC_{50} = 4.213 mg/mL • Rat intestinal sucrose, IC_{50} = 10.10 mg/mL • Rat intestinal isomaltase, IC_{50} > 100 mg/mL • Rat intestinal glucoamylase, IC_{50} > 100 mg/mL	*In vitro* assay	Iwai 2008
	Methanolic extract	α-glucosidase inhibition against: • α-glucosidase (Saccharomyces), IC_{50} = 0.022 mg/mL • Rat intestinal maltase, IC_{50} = 0.772 mg/mL • Rat intestinal sucrose, IC_{50} = 4.056 mg/mL • Rat intestinal isomaltase, IC_{50} > 100 mg/mL • Rat intestinal glucoamylase, IC_{50} = 5.851 mg/mL		
E. stolonifera	• Phlorofucofuroeckol-A • Dieckol • 7-Phloroeckol • Eckol • Dioxinodehydroeckol • Phloroglucinol	α-glucosidase inhibition by: • Phlorofucofuroeckol-A, IC_{50} = 1.37 μM • Dieckol, IC_{50} = 1.61 μM • 7-Phloroeckol, IC_{50} = 6.13 μM • Eckol, IC_{50} = 22.78 μM • Dioxinodehydroeckol, IC_{50} = 34.60 μM • Phloroglucinol, IC_{50} = 141.18 μM	*In vitro* assay	Moon et al. 2011

Table 4.1 contd....

Table 4.1 contd....

Species	Active agents	Activity	Test systems	References
Fucus distichus	Phlorotannin	• α-amylase inhibition, IC_{50} = 13.9 μg/mL • α-glucosidase inhibition, IC_{50} = 0.89 μg/mL	*In vitro* assay	Kellogg et al. 2014
F. distichus	Ethyl acetate extract	α-amylase inhibition, IC_{50} = 13.98 μg/mL α-glucosidase inhibition, IC_{50} = 0.89 μg/mL	*In vitro* assay	Kellogg et al. 2014
F. vesiculosus	Cold water and ethanol extracts	• α-glucosidase inhibition (water), IC_{50} = 0.32 μg/mL • α-glucosidase inhibition (ethanol), IC_{50} = 0.49 μg/mL	*In vitro* assay	Lordan et al. 2013
F. vesiculosus	Fucoidan	α-glucosidase inhibition, IC_{50} = 0.049 mg/mL	*In vitro* assay	Kim et al. 2014c
Ishige okamurae	DPHC	• α-amylase inhibition, IC_{50} = 0.53 mM • α-glucosidase inhibition, IC_{50} = 0.16 mM	*In vitro* assay	Heo et al. 2009
Padina arborescens	Methanolic extract	• α-amylase inhibition, IC_{50} = 0.23 mg/mL • α-glucosidase inhibition, IC_{50} = 0.26 mg/mL	*In vitro* assay	Park and Han 2012
P. gymnospora	Methanolic extract Acetate extract	• α-amylase inhibition (300 μg/mL) = 58% • α-amylase inhibition (300 μg/mL) = 71%	*In vitro* assay	Shanmugaasokan et al. 2013
P. pavonica	Polyphenols	• α-amylase inhibition, IC_{50} = 0.36 mg/mL • α-glucosidase inhibition, IC_{50} = 26.57 mg/mL	*In vitro* assay	Husni et al. 2014
	Phlorotannins	• α-amylase inhibition, IC_{50} = 0.25 mg/mL • α-glucosidase inhibition, IC_{50} = 34.40 mg/mL		
P. sulcata	Water/Ethanolic extract	Against DPP-4 activity with IC_{50} = 2.306 mg/mL	*In vitro* assay	Chin et al. 2015
Saccharina japonica	BIP	α-glucosidase inhibition, IC_{50} = 38.00 μM	*In vitro* assay and hypoglycemic effect in Streptozotocin-induced diabetic mice	Bu et al. 2010
Sargassum aquifolium (as *Sargassum binderi*)	Water/Ethanolic extract	Against DPP-4 activity with IC_{50} = 2.194 mg/mL	*In vitro* assay	Chin et al. 2015

Table 4.1 contd....

Table 4.1 contd....

Species	Active Agents	Activity	Test Systems	References
S. glaucescens	Methanolic extract	α-amylase inhibition, IC_{50} = 8.9 mg/mL	*In vitro* assay	Payghami et al. 2015
S. hemiphyllum	Acetone extract	• α-amylase inhibition, IC_{50} = 0.35 mg/mL • Sucrose inhibition, IC_{50} = 1.89 mg/mL • Maltase inhibition, IC_{50} = 0.09 mg/mL • α-amylase inhibition (300 μg/mL) = 57%	*In vitro* assay	Hwang et al. 2015
S. ilicifolium (as *S. duplicatum*)	Methanolic extract Acetate extract	• α-amylase inhibition (300 μg/mL) = 65% • α-amylase inhibition (300 μg/mL) = 68%	*In vitro* assay	Shanmugaasokan et al. 2013
S. patens	DDBT	• α-amylase inhibition, IC_{50} = 3.2 μg/mL • α-glucosidase inhibition against: • Rat intestinal maltase, IC_{50} = 114.0 μg/mL • Rat intestinal sucrose, IC_{50} = 25.4 μg/mL	*In vitro* assay	Kawamura-Konishi et al. 2012
S. polycystum	Water extract	• α-amylase inhibition, IC_{50} = 60 μg/mL • α-glucosidase inhibition, IC_{50} = 50 μg/mL	*In vitro* assay	SenthilKumar and Sudha 2012
S. polycystum	Ethanolic extract	α-amylase inhibition (10 mg/mL) = 46%	*In vitro* assay	Balasubramaniam et al. 2016
S. ringgoldianum	Methanolic extract	• α-amylase inhibition, IC_{50} = 0.18 mg/mL • α-glucosidase inhibition, IC_{50} = 0.12 mg/mL	*In vitro* assay	Lee and Han 2012
S. swartzii (as *S. wightii*)	Methanolic extract	α-amylase inhibition (300 μg mL^{-1}) = 53%	*In vitro* assay	Shanmugaasokan et al. 2013
S. swartzii (as *S. wightii*)	Fucoidan	α-glucosidase inhibition, IC_{50} = 132 μg	*In vitro* assay	Kumar et al. 2015c
S. swartzii (as *S. wightii*)	Ethanol extraction and $CaCl_2$ precipitation method (Fucoidan)	α-amylase inhibition, IC_{50} = 103.83 μg/mL	*In vitro* assay	Lakshmana-Senthil et al. 2015
S. tenerrimum	Methanolic extract	α-amylase inhibition (300 μg/mL) = 65%	*In vitro* assay	Shanmugaasokan et al. 2013
Spatoglossum asperum	Methanolic extract	• α-amylase inhibition, IC_{50} = 55 μg/mL • α-glucosidase inhibition, IC_{50} = 61 μg/mL	*In vitro* assay	
Spatoglossum schroederi	Acetone crude extract	α-amylase inhibition, IC_{50} = 0.58 mg/mL	*In vitro* assay	Teixeira et al. 2007

Table 4.1 contd....

Table 4.1 contd....

Species	Active agents	Activity	Test systems	References
Turbinaria conoides	Water/Ethanolic extracts	Against DPP-4 activity with IC_{50} = 3.594 mg/mL	*In vitro* assay	Chin et al. 2015
T. ornata	Methanolic extract	• α-amylase inhibition (300 µg/mL) = 45%	*In vitro* assay	Shanmugaasokan et al. 2013
Rhodophyta (Red seaweed)				
Eucheuma denticulatum	Ethanolic extract	α-amylase inhibition (10 mg/mL) = 67%	*In vitro* assay	Balasubramaniam et al. 2016
Gracilaria corticata	Water extract	• α-amylase inhibition, IC_{50} = 82 µg/mL • α-glucosidase inhibition, IC_{50} = 87 µg/mL	*In vitro* assay	SenthilKumar and Sudha 2012
G. edulis	Water extract	• α-amylase inhibition, IC_{50} = 83 µg/mL • α-glucosidase inhibition, IC_{50} = 46 µg/mL	*In vitro* assay	SenthilKumar and Sudha 2012
G. edulis	Methanolic extract	• α-amylase inhibition (300 µg/mL) = 57%	*In vitro* assay	Shanmugaasokan et al. 2013
	Acetone extract	• α-amylase inhibition (300 µg/mL) = 53%		
	Acetate extract	• α-amylase inhibition (300 µg/mL) = 60%		
G. gacilis	Methanolic extract	α-amylase inhibition (300 µg/mL) = 68%	*In vitro* assay	Shanmugaasokan et al. 2013
G. opuntia	Water extract (sulfated polygalactans)	• α-amylase inhibition, IC_{50} = 0.04 mg/mL • α-glucosidase inhibition, IC_{50} = 0.09 mg/mL	*In vitro* assay	Makkar and Chakraborty 2016
Grateloupia elliptica	2,4,6-Tribromophenol	α-glucosidase inhibition against: • *B. stearothermophilus*, IC_{50} = 130.3 µM • *S. cerevisiae*, IC_{50} = 60.3 µM • Rat intestinal maltase, IC_{50} = 5.0 mM • Rat intestinal sucrose, IC_{50} = 4.2 mM	*In vitro* assay	Kim et al. 2008
	2,4-Dibromophenol	α-glucosidase inhibition against: • *B. stearothermophilus*, IC_{50} = 230.3 µM • *S. cerevisiae*, IC_{50} = 110.4 µM • Rat intestinal maltase, IC_{50} = 4.8 mM • Rat intestinal sucrose, IC_{50} = 3.6 mM		

Table 4.1 contd....

Table 4.1 contd....

Species	Active agents	Activity	Test systems	References
Kappaphycus alvarezii	Water extract (sulfated polygalactans)	• α-amylase inhibition, IC_{50} = 0.15 mg/mL • α-glucosidase inhibition, IC_{50} = 0.09 mg/mL	*In vitro* assay	Makkar and Chakraborty 2016
Odonthalia corymbifera	Bromophenols	α-glucosidase (*S. cerevisiae*) inhibition by: • BDDE, IC_{50} = 0.098 μM • 4-Bromo-2,3-dihydroxy-6-hydroxymethylphenyl 2,5-dibromo-6-hydroxy-3-hydroxymethylphenyl ether, IC_{50} = 25.0 μM • 4-Bromo-2,3-dihydroxy-6-methoxymethylphenyl 2,5-dibromo-6-hydroxy-3-methoxymethylphenyl ether, IC_{50} = 53.0 μM • 2,3-dibromo-4,5-dihydroxybenzyl alcohol, IC_{50} = 89.0 μM	*In vitro* assay	Kurihara et al. 1999
Polyopes lancifolius	BDDE	α-glucosidases inhibition against: • *B. stearothermophilus*, IC_{50} = 0.12 μM • *S. cerevisiae*, IC_{50} = 0.098 μM • Rat intestinal maltase, IC_{50} = 1.20 mM • Rat intestinal sucrose, IC_{50} = 1.00 mM	*In vitro* assay	Kim et al. 2010b
Portieria hornemannii (as *Chondrococcus hornemannii*)	Methanolic extract Acetate extract	• α-amylase inhibition (300 μg/mL) = 61% • α-amylase inhibition (300 μg/mL) = 94%	*In vitro* assay	Shanmugaasokan et al. 2013
Symphyocladia latiuscula	Bromophenols	α-glucosidase (yeast) inhibition by: • BDDE, IC_{50} = 0.03 μM • 2,3,6-Tribromo-4,5-dihydroxybenzyl Alcohol, IC_{50} = 11.0 μM BDDE: • Sucrose inhibition, IC_{50} = 2.4 mM • Maltase inhibition, IC_{50} = 3.2 mM 2,3,6-Tribromo-4,5-dihydroxybenzyl Alcohol: • Sucrose inhibition, IC_{50} = 4.2 mM • Maltase inhibition, IC_{50} > 5.0 mM	*In vitro* assay	Kurihara et al. 1999

Table 4.2 Glucose levels reduction by seaweed via miscellaneous mechanisms.

Species	Active agents	Activity	Test systems	References
Chlorophyta (Green seaweed)				
Capsosiphon fulvescens	• Capsofulvesin A • Capsofulvesin B • Chalinasterol	Aldose reductase inhibition by: • Capsofulvesin A, IC_{50} = 52.53 μM • Capsofulvesin B, IC_{50} = 101.92 μM • Chalinasterol, IC_{50} = 345.27	Rat lens aldose reductase and advanced glycation end-products inhibition assays *in vitro*	Islam et al. 2014
Ulva rigida	Ethanolic extract	• Reduction of post-prandial blood glucose level • Antioxidant activity	Wistar diabetic rats	Celikler et al. 2009
Phaeophyceae (Brown seaweed)				
Ascophyllum nodosum	Aqueous ethanolic extract	Stimulation of basal glucose uptake into cells	3T3-L1 adipocytes *in vitro*	Zhang et al. 2007
A. nodosum	Polyphenolic extracts	• Improved fasting blood glucose level • Decreased blood total cholesterol and glycated serum protein levels	Streptozotocin-induced diabetic mice	Zhang et al. 2007
Ecklonia bicyclis	Phloroglucinol	AGE formation inhibition at 1 mM: • Eckol, inhibition = 96.2% • 1-(3′,5′-dihydroxyphenoxy)-7-(2″,4″,6″-trihydroxyphenoxy)-2,4,9-trihydroxydibenzo-1,4-Dioxin, inhibition = 91.1% • Dieckol, inhibition = 86.7%	*In vitro* assay	Okada et al. 2004
E. bicyclis	• Phlorofucofuroeckol-A • Dieckol • 7-Phloroeckol • Eckol • Dioxinodehydroeckol • Phloroglucinol	PTP1B inhibition by: • Phlorofucofuroeckol-A, IC_{50} = 0.56 μM • Dieckol, IC_{50} = 1.18 μM • 7-Phloroeckol, IC_{50} = 2.09 μM • Eckol, IC_{50} = 2.64 μM • Dioxinodehydroeckol, IC_{50} = 29.97 μM • Phloroglucinol, IC_{50} = 55.48 μM	*In vitro* assay	Moon et al. 2011
Ecklonia cava	Methanolic extract	Reduction of post-prandial blood glucose level partially attributed to the AMP-activated protein kinase/ACC and PI-3K/Akt cellular signal pathways	C_2C_{12} myoblast cells and streptozotocin-induced diabetic mice	Kang et al. 2008
E. cava	Dieckol	Reduction of post-prandial blood glucose level and delayed absorption of dietary carbohydrates	Streptozotocin-induced diabetic mice	Lee et al. 2010

Table 4.2 contd....

Table 4.2 contd....

Species	Active agents	Activity	Test systems	References
E. cava	Methanolic extract	• Decreased blood glucose concentration • Prevented the loss of β-cell mass hence improved insulin secretion	Streptozotocin-induced diabetic mice	Kim and Kim 2012
E. cava	Dieckol	• Reduction of blood glucose, glycosylated hemoglobin levels • Reduction of hepatic lipids concentration and also improvement of impaired glucose tolerance • Reduction of glucose-6-phosphatase and phosphoenolpyruvate carboxykinase enzymes activities • Increased glucokinase activity	C57BL/KsJ-db/db diabetic mice	Lee et al. 2012b
Ecklonia stolonifera	• Phlorofucofuroeckol-A • Dieckol • 7-Phloroeckol • Eckol • Dioxinodehydroeckol • Phloroglucinol	PTP1B inhibition by: • Phlorofucofuroeckol-A, IC_{50} = 0.56 μM • Dieckol, IC_{50} = 1.18 μM • 7-Phloroeckol, IC_{50} = 2.09 μM • Eckol, IC_{50} = 2.64 μM • Dioxinodehydroeckol, IC_{50} = 29.97 μM • Phloroglucinol, IC_{50} = 55.48 μM	*In vitro* assay	Moon et al. 2011
Himanthalia elongata	Crude polysaccharides	Reduction of post-prandial blood glucose level	Alloxan-induced diabetic rabbits	Lamela et al. 1989
H. elongata	Fucan	Reduction of post-prandial blood glucose level	Alloxan-induced diabetic rabbits	Lamela et al. 1993
Ishige okamurae	DPHC	Reduction of post-prandial blood glucose level and delayed absorption of dietary carbohydrates	Streptozotocin-induced diabetic mice	Heo et al. 2009
Padina arborescens	Methanolic extract	Reduction of post-prandial blood glucose level delayed absorption of dietary carbohydrates	Streptozotocin-induced diabetic mice	Park and Han 2012
P. sulcata	Butanol fraction	Stimulation of GLP-1 secretion of 40.67 pM GIP per million cells per h at 5.0 mg/mL	pGIP/neo STC-1 cells *in vitro*	Chin et al. 2015
Petalonia binghamiae	Extract	• Increased up-regulation of GLUT4 mRNA, PPARγ and terminal marker protein aP2 up-regulation • Stimulation of 3T3-L1 adipocytes differentiation and expression of IRS-1 with increased uptake of glucose	3T3-L1 adipocytes *in vitro*	Kang et al. 2008

Table 4.2 contd....

Table 4.2 contd....

Species	Active agents	Activity	Test systems	References
Saccharina japonica (formerly *Laminaria japonica*)	Porphyrin derivatives	AGE formation inhibition by: • Pheophorbide a, IC_{50} = 49.43 μM • Pheophytin a, IC_{50} = 228.71 μM	*In vitro* assay	Son et al. 2011
S. japonica (formerly *L. japonica*)	Porphyrin derivatives	Aldose reductase inhibition by: • Pheophorbide a, IC_{50} = 12.31 μM • Pheophytin a, IC_{50} > 100 μM	Rat lens aldose reductase assay *in vitro*	Son et al. 2011
Sargassum aquifolium (as *Sargassum binderi*)	Ethanolic precipitates	DPP-4 inhibition, IC_{50} = 2.194 mg/mL	*In vitro* assay	Chin et al. 2015
S. aquifolium (as *S. binderi*)	Water extracts	Stimulation of GIP secretion of 5.46 pM GIP per million cells per h at 2.5 mg/mL	pGIP/neo STC-1 cells *in vitro*	Chin et al. 2015
S. aquifolium (as *S. binderi*)	Butanol fraction	Stimulation of GLP-1 secretion of 56.38 pM GIP per million cells per h at 5.0 mg/mL	pGIP/neo STC-1 cells *in vitro*	Chin et al. 2015
Sargassum polycystum	Ethanolic and water extracts	• 150 and 300 mg/kg of ethanolic extract and 300 mg/kg of water extract significantly reduced blood glucose and HbA1C levels • Significant reduction of serum total cholesterol, triglyceride levels	Streptozotocin-induced diabetic rat given high-sugar, high-fat diet	Motshakeri et al. 2013
Sargassum ringgoldianum	Methanolic (80%) extract	Reduction of post-prandial blood glucose level and delayed absorption of dietary carbohydrates	Streptozotocin-induced diabetic mice	Lee and Han 2012
Turbinaria conoides	Ethanolic precipitates	DPP-4 inhibition, IC_{50} = 3.594 mg/mL	*In vitro* assay	Chin et al. 2015
T. conoides	Water extracts	Stimulation of GIP secretion of 5.00 pM GIP per million cells per h at 2.5 mg/mL	pGIP/neo STC-1 cells *in vitro*	Chin et al. 2015
Undaria pinnatifida	Fucoxanthin	Promotion of Adrb3 and GLUT4 mRNA expressions in skeletal muscle tissues	High-fat diet mice	Maeda et al. 2009
Rhodophyta (Red seaweed)				
Laurencia similis	Bromophenols	PTP1B inhibition by: • 2′,5′,6′,5,6-pentabromo-3′,4′,3,4-tetramethoxybenzophenone, IC_{50} = 2.66 μg/mL • 3′,5′,6′,6-tetrabromo-2,4-dimethyldiphenyl ether, IC_{50} = 2.97 μg/mL	*In vitro* assay	Qin et al. 2010

Table 4.2 contd....

Table 4.2 contd....

Species	Active agents	Activity	Test systems	References
		• 2,5,8-tribromo-3-bromoamino-7-bromomethylnaphthalene, IC_{50} = 65.30 μg/mL • 2,5,6-tribromo-3-bromoamino-7-bromomethylnaphthalene, IC_{50} = 69.80 μg/mL		
Rhodomela confervoides	Bromophenol derivatives	PTP1B inhibition by: • 2,2′,3-tribromo-3′,4,4′,5-tetrahydroxy-6′-ethyloxy-methyldiphenylmethane, IC_{50} = 0.84 μM • Bis(2,3-dibromo-4,5-dihydroxybenzyl) ether, IC_{50} = 1.5 μM • 3-bromo-4,5-bis(2,3-dibromo-4,5-dihydroxybenzyl) pyrocatechol, IC_{50} = 1.7 μM • 2,2′,3,3′-tetrabromo-4,4′,5,5′-tetra-hydroxydiphenyl methane, IC_{50} = 2.4 μM	PTP1B and hypoglycemic effect in streptozotocin-induced diabetic Wistar rats	Shi et al. 2008
Symphyocladia latiuscula	Bromophenols	PTP1B inhibition by: • 1,2-bis(2,3,6-tribromo-4,5-dihydroxyphenyl)-ethane, IC_{50} = 3.5 μM • 2,3,6-tribromo-4,5-dihydroxybenzyl methyl ether, IC_{50} = 3.9 μM • Bis(2,3,6-tribromo-4,5-dihydroxyphenyl)-ethane, IC_{50} = 4.3 μM • 2,3,6-tribromo-4,5-dihydroxybenzaldehyde, IC_{50} = 19.40 μM	*In vitro* assay	Liu et al. 2011

Dilution and slowing down the action of carbohydrases in the gut by seaweed fibers would have a positive impact on the regulation of blood glucose levels. Taken together, control of starch digestion in the diet could help to control blood glucose in Type 2 diabetes. It was suggested that 5 g of sodium alginate administered daily to Type 2 diabetic patients prevented a post-prandial increase of glucose and insulin and slowed down gastric transit (Torsdottir et al. 1991). Hydrolysates of agar (from selected red seaweeds) resulted in agaro-oligosaccharides possessing an activity against α-glucosidase (Chen et al. 2005). Moreover, *Ascophyllum nodosum* extracts (fibers) at 50 mg mL^{-1} completely inhibited amylase activity. A meal supplemented with 5% alginates derived from key brown seaweeds decreased the glucose absorption balance over eight hours in pigs, and similar studies have been completed on rats and humans (Vaugelade et al. 2000). Taken together, these findings suggested that various seaweed fibers could have a positive effect by influencing the inhibition of starch digestive enzymes (amylase) at very low concentrations, and thereby maintained glycemic control *in vivo* (Rajapakse and Kim 2011).

According to the experimental results obtained by Korukanti et al. (2013), *Fucus vesiculosus* (Phaeophyceae) was found to be beneficial when tested on two different animal models, i.e., diet-induced and chemical-induced models. Diet-induced animal model typically mimics the obesity gained through high-calorie diet in human beings. *F. vesiculosus* prevented the rats from becoming obese, when given along with high-calorie diet. Biochemical and physical parameters were also maintained at normal levels.

Chemical-induced animal model typically represents hyperlipidemia in human beings without weight gain. When hyperlipidemic animals were treated with *F. vesiculosus*, the levels of HDL, LDL, and VLDL were brought back to normal. So, *F. vesiculosus* can be taken up for further research on human subjects (Korukanti et al. 2013).

The anti-obesity and anti-diabetic effects of some allenic compounds including fucoxanthin (see also Chapter 2) were reported by Miyashita and Hosokawa (2008), Miyashita et al. (2011), and Gammone and D'Orazio (2015). A few molecular targets offer hope for anti-obesity therapeutics. One of the keys to success could be the induction of uncoupling protein 1 (UCP1) in abdominal white adipose tissue (WAT), and the regulation of cytokine secretions from both abdominal adipose cells and macrophage cells infiltrated into adipose tissue. Anti-obesity effects of fucoxanthin (see also Chapter 2), a characteristic carotenoid exactly belonging to xanthophylls, have been reported. Nutrigenomic studies reveal that fucoxanthin induces UCP1 in abdominal WAT mitochondria, leading to the oxidation of fatty acids and heat production in WAT. Fucoxanthin improves insulin resistance and decreases blood glucose levels through the regulation of cytokine secretions from WAT. The key structure of anti-obesity effect is suggested to be the carotenoid end of the polyene chromophore, which contains an allenic bond and two hydroxyl groups. Fucoxanthin, which can be isolated from edible brown seaweeds, recently displayed its many physiological functions and biological properties. We reviewed recent studies, and this article aims to explain the essential background of fucoxanthin, focusing on its promising potential anti-obesity effects. In this respect, fucoxanthin can be developed into promising marine drugs and nutritional products in order to become a helpful functional food. These compounds improved insulin resistance and decreased blood glucose levels through the regulation of cytokine secretions from WAT by inducing UCP1. The key structures of these activities were thought to be an allenic bond and two hydroxyl groups.

Some reports have focused on the alginate (see also Chapter 3) contained in seaweed. Sodium alginate from the brown seaweed *Laminaria digitata* is currently marketed as a weight-loss supplement, but its effects on gastric motor functions and satiation are unknown. Odunsi et al. (2010) clinically investigated the effects of 10 days of treatment with alginate or placebo on gastric function, satiation, appetite, and gut hormones associated with satiety in overweight or obese adults. They found that treatment with alginate for 10 days had no effect on any of the above parameters. These results suggested that the daily continuous intake of alginates may be required to prevent obesity.

Tororokombu (TK) is a traditional Japanese food that is made by shaving *Saccharina japonica* [formerly *Laminaria japonica* and *L. ochotensis* (Kombu)] very thinly. Miyata et al. (2009) first investigated the effects of NSK (non-shaved Kombu) and TK on the absorption of TGs in the intestine by an oil-loading test in female SD rats (seven weeks old). One feature of this study is the improved efficiency of dissolution of the active component. SD rats were first divided into three groups: distilled water-treated, NSK-treated, and TK-treated groups. Next, corn oil (5 mL/kg) was administered orally. The elevation of the serum TG level in the NSK- and TK-treated groups was significantly lower than that in the normal rat group.

The anti-obesity effects of NSK and TK were investigated by a long-term experiment on obese female ddY mice (four weeks old) induced by a HFD for 63 days. Mice were divided into four groups: ND (normal diet), HFD, HFD-NSK (HFD containing 3% NSK), and HFD-TK (HFD containing 3% TK) groups. The body weights on the 63rd day after the treatment started, and the serum TC levels in both the HFD-NSK and the HFD-TK groups, were significantly lower than those in the HFD group. The parauterus adipose tissue weight, and hepatic TG, serum TG, and TC levels in the HFD-TK group were significantly less than those in the HFD-NSK group. The fecal TG and TC levels in the HFD-TK group were significantly higher than those in the HFD group, and fecal TG in the HFD-TK-group was significantly higher than that in the HFD-NSK group. Consequently, it was demonstrated that TK consumption reduced the accumulation of visceral fat caused by HFD, and this effect of TK was more powerful than that of NSK, due to TG and cholesterol excretion in the feces. This report concluded that alginate may be one of the active components in *Laminaria* spp. and *Saccharina* spp. In previous reports, alginate has been reported to have hypoglycemic and cholesterol-lowering effects by acting as a viscous soluble dietary fiber (Kimura et al. 1996, Pasquier et al. 1996, Paxman et al. 2008).

Petalonia binghamiae (Phaeophyceae) is an edible brown alga, and is consumed as a traditional food in fishery areas of northeast Asia (Pereira 2016). Galactosyldiacylglycerol (Mizushina et al. 2001) and fucoxanthin-related compounds (Mori et al. 2004) have been reported as bioactive compounds from *P. binghamiae.* Its potent anti-obesity and anti-diabetic activities were reported in several studies. The water-soluble extract of *P. binghamiae*, prepared by enzymatic digestion (PBEE), suppressed adipocyte differentiation and adipogenesis via inhibition of adipogenic specific gene expression and insulin-stimulated uptake of glucose in an *in vitro* study. In rat model with high-fat diet (HFD)-induced obesity, PBEE exhibited a potent anti-obesity effect. PBEE supplementation reduced body weight, fat storage, serum levels of glutamic pyruvic and glutamic oxaloacetic transaminases, but increased the serum level of HDL-cholesterol (Rang et al. 2010). Moreover, the dietary administration of its ethanol extract decreased hyperglycemia, and improved glucose tolerance in STZ-induced diabetic mice, in which PBEE exerts an *in vivo* anti-diabetic effect by mediating both insulin-like and insulin-sensitizing actions in adipocytes (Kang et al. 2008).

After Zhan et al. (2007), an aqueous ethanolic extract of *Ascophyllum nodosum* (Phaeophyceae) was found to be active in both assays, inhibiting rat intestinal α-glucosidase (IC_{50} = 77 mg mL^{-1}), and stimulating basal glucose uptake into 3T3-L1 adipocytes during a 20-minute incubation three-fold (at 400 mg mL^{-1} extract). Bioassay-guided fractionation of the *A. nodosum* extract showed that α-glucosidase inhibition was associated with polyphenolic components in the extract. These polyphenolics, along with other constituents, appeared to be responsible for the stimulatory activity on glucose uptake.

The inhibitory activity of α-glucosidase of two brown seaweeds [*Sargassum ilicifolium* (formerly *S. duplicatum*) and *Turbinaria decurrens*] as laminaran faction is highest, followed by the fraction of fucoidan, but the fraction of alginate had no inhibitory activity of α-glucosidase. The fraction of these brown seaweeds that is potentially as anti-diabetic type 2 is as laminaran of *S. ilicifolium* (IC_{50} = 36.13 ppm), laminaran of *T. decurrens* (IC_{50} = 43.65 ppm), fucoidan of *T. decurrens* (IC_{50} = 63.39 ppm), and fucoidan of *S. ilicifolium* (IC_{50} = 75.10 ppm) (Hardoko et al. 2014).

As claimed by Senthilkumar et al. (2014), *Padina boergesenii* (Phaeophyceae) extracts showed a high-quality anti-diabetic activity in STZ induced diabetic rats. The effective dose of *P. boergesenii* extract was found to be 400 mg/kg body weight. The action of this species was comparable to anti-diabetic drug Glibenclamide. Results of this experimental study indicated that *P. boergesenii* has potent anti-diabetic activity in STZ-induced experimental diabetes in rats.

According to the work of Mohapatra et al. (2016), it was demonstrated that ethyl acetate extracts of *Sargassum swartzii* [formerly *S. wightii* (Phaeophyceae)] and *Ulva fasciata* (Chlorophyta) influence glycemic control, and are effective in lowering blood lipids. Furthermore, these extracts are also helpful in improving antioxidant enzyme activities and glycogen content in liver and skeletal muscle. Consequently, the findings of this study suggest that the use of these seaweeds may minimize the chances of developing serious complications in type 2 diabetic individuals. But further studies are necessary to corroborate these results, and to make dietary recommendations of these seaweeds for patients with type 2 diabetes.

In another study (Unnikrishnan et al. 2015), the edible seaweeds *Sargassum polycystum* and *S. swartzii* (formerly *S. wightii*), were investigated for their anti-diabetic potential using *in vitro* enzyme inhibitory assays. Among the various extracts, petroleum ether and ethyl acetate extracts of *S. swartzii* showed significant inhibitory effects against α-amylase (IC_{50} 378.3 μg mL^{-1}) and α-glucosidase (IC_{50} 314.8 μg mL^{-1}). Methanol extract of *S. swartzii* showed the highest inhibition against dipeptidyl peptidase-IV (DPP-IV) (IC_{50} 38.27 μg mL^{-1}), and moderate antioxidant activity was observed in acetone extract of *S. swartzii* (44%). Similarly, ethyl acetate extract of *S. polycystum* showed the highest inhibition against α-amylase (IC_{50} 438.5 μg mL^{-1}), and methanol extract of *S. polycystum* showed maximum inhibition against α-glucosidase (IC_{50} 289.7 μg mL^{-1}) and DPP-IV (36.94 μg mL^{-1}). The antioxidant activity was poor (22%). The extracts were investigated for *in vitro* cytotoxicity, DNA fragmentation in macrophages, and hemolytic activity against erythrocytes, but no notable toxicity was observed with any of the tested extracts. Gas chromatography–mass spectrometry revealed the presence of the anti-diabetic compound fucosterol, and other major bioactive compounds, giving an insight into the anti-diabetic and antioxidant properties of these algae. This study reveals the possible mechanisms of anti-diabetic action *in vitro*, and these two seaweeds may also have an anti-diabetic action *in vivo*.

As stated by Chin et al. (2015), crude water extracts of *Halimeda macroloba* (Chlorophyta), *Padina sulcata*, *Sargassum aquifolium* (formerly *S. binderi*), and *Turbinaria conoides* (Phaeophyceae) possessed potent inhibitory activities against α-glucosidase and DPP-4. The highest inhibitory activity against α-glucosidase was found in water extracts of the green seaweed species *H. macroloba* with an IC_{50} value of 6.388 mg mL^{-1}. Crude water extracts of the brown seaweeds studied, namely *P. sulcata*, *S. aquifolium*, and *T. conoides*, exhibited potent DPP-4 inhibition compared to the green seaweed *H. macroloba* (see Table 5.1). The brown seaweed also stimulates secretion of glucose-dependent insulinotrophic polypeptide (GIP) and glucagon-like peptide-1 (GLP-1) from pGIP neo STC-1 cells *in vitro*. *H. macroloba* stimulated GLP-1 secretion, but not secretion of GIP.

Therefore, the studies made by Krish and Das (2014) determined that *Cladophora rupestris* (Chlorophyta), methanol, and ethyl acetate extracts showed microbiological and α-glucosidase inhibition. Inhibition of α-amylase and α-glucosidase, enzymes involved in the digestion of carbohydrate, can significantly decrease the postprandial increase of blood glucose after a mixed carbohydrate diet, and can therefore be an important strategy in the management of blood glucose level in type 2 diabetic patients (Ali et al. 2006, Krish and Das 2014).

4.4 Hypotensive Activities of Seaweeds

Hypertension is a major risk factor for stroke (ischemic and hemorrhagic), myocardial infarction, heart failure, chronic kidney disease, peripheral vascular disease, cognitive decline, and premature death (see also Chapter 2). Untreated hypertension is associated with a progressive rise in blood pressure, often culminating in a treatment-resistant state due to associated vascular and renal damage (NCGC 2011).

Blood pressure is quantified as diastolic and systolic pressures measured in millimeters of mercury (mmHg). The diastolic pressure represents the pressure during ventricular relaxation in diastole, whereas the systolic pressure represents the peak pressure due to ventricular contraction during systole. Either or both pressures have specified upper limits of normal and elevation, and either or both pressures are used to define hypertension (NCGC 2011).

Blood pressure is normally distributed in the population, and there is no natural cut-point above which "hypertension" definitively exists and below which it does not. Epidemiological studies demonstrate that the aforementioned disease risk associated with blood pressure is a continuous relationship, and above blood pressures of 115/70 mmHg, the risk of cardiovascular events doubles for every 20/10 mmHg rise in blood pressure. The threshold blood pressure determining the presence of hypertension is defined as the level of blood pressure above which treatment has been shown to reduce the development or progression of disease. Primary hypertension was previously termed "essential hypertension" because of a long-standing view that high blood pressure was sometimes "essential" to perfuse diseased and sclerotic arteries. It is now recognized that the diseased and sclerotic arteries were most often the consequence of the hypertension, and thus the term "essential hypertension" is redundant, and the term "primary hypertension" is preferred. Primary hypertension refers to the majority of people with sustained high blood pressure (approximately 90%) encountered in clinical practice, for which there is no obvious, identifiable cause. The remaining 10% are termed "secondary hypertension", for which specific causes for the blood pressure elevation can be determined (for example, Conn's adenoma, renovascular disease, or phaeochromocytoma) (NCGC 2011).

A vast epidemiological literature describes an apparent relationship between raised blood pressure and lifestyle choices and habits. For example, observational studies have shown that people with raised blood pressure tend to also have low dietary calcium (Ca). Does inadequate intake of dietary calcium (Ca) promote raised blood pressure, or is the relationship a spurious one, arising from inadequate adjustment for other hard-to-measure influences (a common problem in observational studies). There is similar controversy about the role of diet, exercise, alcohol, caffeine, potassium (K) and magnesium (Mg) supplements, sodium (Na) (table) salt, and relaxation therapies (see also Chapter 2). Cause and effect can only be established by repeated and methodologically sound randomized controlled trials, supported by evidence of a plausible biological mechanism, particularly when the potential benefit is small (NCGC 2011).

Our systolic (and to a lesser extent our diastolic) blood pressure tends to increase as we grow older. It is unhelpful to think of a single threshold above which we suddenly have problematically high blood pressure, although such thresholds can be useful to spur us into action. A review of our lifestyle helps us to identify changes we can make which may reduce our blood pressure and thus delay, reduce, or remove the need for long term drug therapy, as well as lead a healthier life. The cumulative trial evidence suggests that individuals who develop improved habits of regular exercise, sensible diet, and relaxation can reduce their blood pressure. Forming these habits will take determination and support. Health care professionals can provide advice, encouragement, and materials, but ultimately have limited scope to influence poor dietary habits and inadequate exercise, which result partly from the busy and stressful pace of life, and partly from personal choice. Much of the research evidence for lifestyle change uses regular time spent together in groups for support and encouragement. Patient and healthcare organizations may be able to help provide patients with, or point them to local groups which encourage a lifestyle change, particularly those promoting healthy eating and regular exercise (NCGC 2011).

Angiotensin-converting enzyme (ACE, peptidyldipeptide hydrolase EC 3.4.15.1) is a zinc-containing metalloenzyme, which has been determined to perform an important physiological function in the pathogenesis of cardiovascular and renal diseases, and in the regulation of blood pressure. Various cardiometabolic syndromes are known to be operant in hypertension. Several metabolic actions of ACE in the rennin–angiotensin system have been shown to result in an increase in blood pressure by the hydrolysis of the rennin-induced decapeptide, angiotensin I aspartic acid–arginine–valine–tyrosine–isoleucine–histidine–proline–phenylalanine–histidine–leucine (Asp-Arg-Val-Tyr-Ile-His-Pro-PheHis-Leu) to octapeptide angiotensin II (Asp-ArgVal-Tyr-Ile-His-Pro-Phe), a potent vasoconstrictor. This reaction stimulates aldosterone secretion in adrenal/cardiovascular tissue, promotes sodium and water retention, and causes an increase in rennin generation in the kidneys. ACE also catalyzes the degradation of bradykinin, a vasodilator in the kallikrein–kinin system. Further, ACE is implicated in cell oxidative stress, augmenting the generation of reactive oxygen species (ROS) and peroxynitrite, and also in thrombosis, during which ACE induces platelet activation, aggregation, and adhesion (McFarlane et al. 2003, Jung et al. 2006). Considering a series of metabolic actions of ACE, the inhibition of ACE may prevent hypertension, other cardiovascular and renal diseases, and oxidative stress-associated diseases. For the treatment of cardiovascular diseases in the arteries, heart, brain, and kidney, the synthetic ACE inhibitors captopril, benazepril, enalapril, and lisinopril are usually prescribed. However, it has been reported that these synthetic drugs show undesirable side-effects like coughing, taste disturbances, hyperpotassemia, and skin rashes (Cleland et al. 1998). Due to the side-effects of synthetic ACE inhibitors, and considering the preventive potential discovered in foods, herbs, and seaweeds as bioactive sources, there is great interest in the possibility of deriving ACE inhibitors from natural products (Ueno et al. 1998, Sato et al. 2005). Several classes of ACE-inhibitory compounds have been discovered, including tannins, flavonoids, xanthones, terpenoids, peptides, and caffeoylquinic acid derivatives (Table 4.3) (Liu et al. 2003, Jung et al. 2006). Most ACE-inhibitory peptides have been isolated from the hydrolysates of seaweeds (Sato et al. 2005).

Edible seaweeds have been considered over the past few decades as promising organisms for providing both novel biologically active substances and essential compounds for human nutrition (MacArtain et al. 2007). However, till date, scarce work on the potential ACE-inhibitory (ACE-I) compounds such as biopeptides (Suetsuna and Nakano 2000, Sato et al. 2002a, b) or phlorotannins (Jung et al. 2006) on seaweeds has been done. As mentioned before, inhibition of ACE is a well-established approach in the treatment of hypertension and, because of that, many authors have screened the potential of seaweeds in inhibiting this enzyme, either using crude extracts, purified fractions, and/or isolated components (Table 4.3).

Cha et al. (2006b) screened the *in vitro* ACE-inhibitory activity of methanol and aqueous extracts from 26 red Korean (Jeju island) algae, obtained at 20°C or at 70°C (Table 4.3). The authors have found several potential extracts, with IC_{50} values for ACE-inhibitory activity in the range of 12.21–124.69 μg mL^{-1}, being the lowest value found for the aqueous extract of *Fushitsunagia catenata*—formerly *Lomentaria catenata* (Rhodophyta) at 20°C, and *Lithophyllum okamurae* recorded the second highest activity. From MeOH extract at 20°C (20 ME) *Ahnfeltiopsis flabelliformis* possessed the strongest ACE-inhibitory activity. Remarkable activities from MeOH extracts at 70°C (70 ME) were observed in *Grateloupia filicina*, *Polyopes lancifolius* (formerly *Sinkoraena lancifolia*), and *Pachymeniopsis lanceolata* (formerly *Grateloupia lanceolata*). Those of the 70 ME from *Bonnemaisonia hamifera*, *Gracilaria vermiculophylla*, *Grateloupia filicina*, *Lithophyllum okamurae*, *Pachymeniopsis lanceolata*, and *Polyopes lancifolius* ranged from 25.82 to 124.69 μg mL^{-1}.

Table 4.3 Recent studies reporting inhibitory abilities of seaweeds extracts and isolates on the renin-angiotensin system (RAS).

Species	**Extraction**	**Inhibition**	**References**
	ACE-I inhibition of extracts		
Chlorophyta (Green seaweeds)			
Capsosiphon fulvescens	Ethanolic extract	15.51% ACE inhibition activity	Jung et al. 2006
Chaetomorpha linum	Methanolic and aqueous extracts, obtained at 20°C and 70°C	20.54–40.27% ACE inhibition activity	Cha et al. 2006a
Codium contractum	Methanolic and aqueous extracts, obtained at 20°C and 70°C	22.54–34.08% ACE inhibition activity	Cha et al. 2006a
C. fragile	Methanolic and aqueous extracts, obtained at 20°C and 70°C	3.37–29.90% ACE inhibition activity	Cha et al. 2006a
Monostroma nitidum	Methanolic and aqueous extracts, obtained at 20°C and 70°C	16.02–47.79% ACE inhibition activity	Cha et al. 2006a
Ulva australis (as *Ulva pertusa*)	Methanolic and aqueous extracts, obtained at 20°C and 70°C	19.20–29.90% ACE inhibition activity	Cha et al. 2006a
U. compressa (formerly *Enteromorpha compressa*)	Methanolic and aqueous extracts, obtained at 20°C and 70°C	18.87–52.98% ACE inhibition activity	Cha et al. 2006a
U. conglobata	Methanolic and aqueous extracts, obtained at 20°C and 70°C	13.51–34.08% ACE inhibition activity	Cha et al. 2006a
U. intestinalis (formerly *Enteromorpha intestinalis*)	Methanolic and aqueous extracts, obtained at 20°C and 70°C	16.19–33.75% ACE inhibition activity	Cha et al. 2006a
U. linza (formerly *Enteromorpha linza*)	Methanolic and aqueous extracts, obtained at 20°C and 70°C	7.83–52.84% ACE inhibition activity	Cha et al. 2006a
Phaeophyceae (Brown seaweeds)			
Ecklonia cava	Aqueous extracts, obtained at 70°C	More than 50% ACE inhibition activity	Cha et al. 2006a
E. cava	Proteolytic (Kojizyme, Flavourzyme, Neutrase, Alcalase, and Protamex) digest of aqueous extract, at 70°C	About 90% ACE inhibition activity	Cha et al. 2006a
E. cava	Ethanolic extract	Above 50% ACE inhibition activity	Jung et al. 2006
E. stolonifera	Ethanolic extract	64.86% ACE inhibition activity	Jung et al. 2006
Ishige foliacea	Methanolic and aqueous extracts, obtained at 20°C and 70°C	Weak ACE inhibition activity	Cha et al. 2006a
I. sinicola	Methanolic extracts, obtained at 70°C	More than 50% ACE inhibition activity	Cha et al. 2006a

Table 4.3 contd....

Table 4.3 contd....

Species	Extraction	Inhibition	References
I. okamurae	Methanolic and aqueous extracts, obtained at 20°C and 70°C	3.81–30.40% ACE inhibition activity	Cha et al. 2006a
Leathesia marina (as *L. difformis*)	Methanolic and aqueous extracts, obtained at 20°C and 70°C	11.24–37.59% ACE inhibition activity	Cha et al. 2006a
Myagropsis myagroides	Methanolic extracts, obtained at 20°C	More than 50% ACE inhibition activity	Cha et al. 2006a
Myelophycus simplex	Methanolic and aqueous extracts, obtained at 20°C and 70°C	3.65–48.63% ACE inhibition activity	Cha et al. 2006a
Padina arborescens	Methanolic and aqueous extracts, obtained at 20°C and 70°C	3.81–38.09% ACE inhibition activity	Cha et al. 2006a
Petalonia binghamiae (as *Endarachne binghamiae*)	Methanolic and aqueous extracts, obtained at 20°C and 70°C	15.18–45.12% ACE inhibition activity	Cha et al. 2006a
Petrospongium rugosum	Methanolic extracts, obtained at 20°C and 70°C	More than 50% ACE inhibition activity	Cha et al. 2006a
Saccharina japonica (as *Laminaria ochotensis*)	Methanolic extracts, obtained at 70°C	More than 50% ACE inhibition activity	Cha et al. 2006a
Sargassum coreanum	Methanolic and aqueous extracts, obtained at 20°C and 70°C	7.32–32.58% ACE inhibition activity	Cha et al. 2006a
S. fulvellum	Methanolic and aqueous extracts, obtained at 20°C and 70°C	8.50–29.23% ACE inhibition activity	Cha et al. 2006a
S. fusiforme (as *Hizikia fusiformis*)	Methanolic extracts, obtained at 70°C	About 87% ACE inhibition activity	Cha et al. 2006a
S. fusiforme (as *Hizikia fusiformis*)	Ethanolic extract	25.68% ACE inhibition activity	Jung et al. 2006
S. horneri	Methanolic and aqueous extracts, obtained at 20°C and 70°C	17.69–70.70% ACE inhibition activity	Cha et al. 2006a
S. horneri	Methanolic extracts, obtained at 70°C	More than 50% ACE inhibition activity	Cha et al. 2006a
S. piluliferum	Methanolic and aqueous extracts, obtained at 20°C and 70°C	7.99–36.59% ACE inhibition activity	Cha et al. 2006a
S. siliquastrum	Methanolic and aqueous extracts, obtained at 20°C and 70°C	15.35–31.57% ACE inhibition activity	Cha et al. 2006a
S. thunbergii	Methanolic and aqueous extracts, obtained at 20°C and 70°C	0.30–32.24% ACE inhibition activity	Cha et al. 2006a
Scytosiphon lomentaria	Methanolic and aqueous extracts, obtained at 20°C and 70°C	4.55–38.60% ACE inhibition activity	Cha et al. 2006a

Table 4.3 contd....

Table 4.3 contd....

Species	Extraction	Inhibition	References
Undaria pinnatifida	Methanolic extracts, obtained at 70°C	More than 50% ACE inhibition activity	Cha et al. 2006a
U. pinnatifida	Ethanolic extract	53.15% ACE inhibition activity	Jung et al. 2006
Rhodophyta (Red seaweeds)			
Ahnfeltiopsis flabelliformis	Methanolic and aqueous extracts (tested concentration: 200 μg mL^{-1})	6.91–97.59% ACE inhibition activity	Cha et al. 2006b
Bonnemaisonia hamifera	Methanolic and aqueous extracts (tested concentration: 200 μ g mL^{-1})	8.05–71.36% ACE inhibition activity	Cha et al. 2006b
Chondracanthus tenellus (formerly *Gigartina tenella*)	Ethanolic extract	18.92% ACE inhibition activity	Jung et al. 2006
Chondria crassicaulis	Methanolic and aqueous extracts (tested concentration: 200 μ g mL^{-1})	18.13–58.43% ACE inhibition activity	Cha et al. 2006b
Chondrophycus undulatus	Methanolic and aqueous extracts (tested concentration: 200 μ g mL^{-1})	2.73–16.79% ACE inhibition activity	Cha et al. 2006b
Chondrus crispus	Methanolic and aqueous extracts (tested concentration: 200 μ g mL^{-1})	12.42–27.63% ACE inhibition activity	Cha et al. 2006b
Fushitsunagia catenata (as *Lomentaria catenata*)	Methanolic and aqueous extracts (tested concentration: 200 μ g mL^{-1})	6.15–98.92% ACE inhibition activity	Cha et al. 2006b
Gelidium amansii	Methanolic and aqueous extracts (tested concentration: 200 μ g mL^{-1})	9–17.93% ACE inhibition activity	Cha et al. 2006b
G. amansii	Ethanolic extract	58.11% ACE inhibition activity	Jung et al. 2006
Gloiopeltis furcata	Methanolic and aqueous extracts (tested concentration: 200 μ g mL^{-1})	7.29–31.05% ACE inhibition activity	Cha et al. 2006b
Gracilaria textorii	Methanolic and aqueous extracts (tested concentration: 200 μ g mL^{-1})	18.13–65.40% ACE inhibition activity	Cha et al. 2006b
Gracilariopsis longissima (formerly *Gracilaria verrucosa*)	Methanolic and aqueous extracts (tested concentration: 200 μ g mL^{-1})	7.29–74.02% ACE inhibition activity	Cha et al. 2006b
Grateloupia cornea (as *Prionitis cornea*)	Methanolic and aqueous extracts (tested concentration: 200 μ g mL^{-1})	1.59–25.54% ACE inhibition activity	Cha et al. 2006b
G. elliptica	Methanolic and aqueous extracts (tested concentration: 200 μ g mL^{-1})	9.19–68.13% ACE inhibition activity	Cha et al. 2006b
G. filicina	Methanolic and aqueous extracts (tested concentration: 200 μ g mL^{-1})	7.48–83.13% ACE inhibition activity	Cha et al. 2006b

Table 4.3 contd....

Table 4.3 contd....

Species	Extraction	Inhibition	References
Halymenia dilatata	Methanolic and aqueous extracts (tested concentration: 200 μ g mL^{-1})	12.99–25.35% ACE inhibition activity	Cha et al. 2006b
Laurencia okamurae	Methanolic and aqueous extracts (tested concentration: 200 μ g mL^{-1})	15.08–78.01% ACE inhibition activity	Cha et al. 2006b
Lithophyllum okamurae	Methanolic and aqueous extracts (tested concentration: 200 μ g mL^{-1})	8.05–89.23% ACE inhibition activity	Cha et al. 2006b
Martensia denticulata	Methanolic and aqueous extracts (tested concentration: 200 μ g mL^{-1})	13.18–23.35% ACE inhibition activity	Cha et al. 2006b
Neosiphonia japonica (as *Polysiphonia japonica*)	Methanolic and aqueous extracts (tested concentration: 200 μ g mL^{-1})	6.15–19.47% ACE inhibition activity	Cha et al. 2006b
Pachymeniopsis lanceolata (as *Grateloupia lanceolata*)	Methanolic and aqueous extracts (tested concentration: 200 μ g mL^{-1})	4.25–89.04% ACE inhibition activity	Cha et al. 2006b
Phacelocarpus sp.	Methanolic and aqueous extracts (tested concentration: 200 μ g mL^{-1})	6.72–22.50% ACE inhibition activity	Cha et al. 2006b
Polyopes affinis (as *Carpopeltis affinis*)	Methanolic and aqueous extracts (tested concentration: 200 μ g mL^{-1})	3.30–59.95% ACE inhibition activity	Cha et al. 2006b
P. lancifolius (as *Sinkoraena lancifolia*)	Methanolic and aqueous extracts (tested concentration: 200 μ g mL^{-1})	3.68–80.86% ACE inhibition activity	Cha et al. 2006b
Pterocladiella capillacea	Methanolic and aqueous extracts (tested concentration: 200 μ g mL^{-1})	4.25–17.17% ACE inhibition activity	Cha et al. 2006b
Pyropia tenera (formerly *Porphyra tenera*)	Methanolic and aqueous extracts (tested concentration: 200 μ g mL^{-1})	0.82–25.59% ACE inhibition activity	Cha et al. 2006b
P. tenera (formerly *P. tenera*)	Ethanolic extract	15.77% ACE inhibition activity	Jung et al. 2006
Schizymenia dubyi	Methanolic and aqueous extracts (tested concentration: 200 μ g mL^{-1})	3.87–26.11% ACE inhibition activity	Cha et al. 2006b
Scinaia okamurae	Methanolic and aqueous extracts (tested concentration: 200 μ g mL^{-1})	12.04–42.84% ACE inhibition activity	Cha et al. 2006b
Silvetia siliquosa (formerly *Pelvetia siliquosa*)	Ethanolic extract	45.95% ACE inhibition activity	Jung et al. 2006
	ACE-I or Renin inhibition associated with peptides		
Chlorophyta (Green seaweeds)			
Caulerpa microphysa	Pepsin, alcalase, flavourzyme (enzyme-digested extracts)	ACE-I IC_{50} (mg/L): pepsin = 0.20; flavourzyme = 29.74; alcalase = 31.71	Lin et al. 2012

Table 4.3 contd....

Table 4.3 contd....

Species	Extraction	Inhibition	References
Phaeophyceae (Brown seaweeds)			
Undaria pinnatifida	Aqueous hot extraction dialysis, and chromatography	ACE-I IC_{50} (μM): Tyr-His = 5.1; Lys-Trp = 10.8; Lys-Tyr = 7.7; Lys-Phe = 28.3; Phe-Tyr = 3.7; Val-Trp = 10.8; Val-Phe = 43.7; Ile-Tyr = 2.7; Ile-Trp = 12.4; Val-Tyr = 11.3	Suetsuna et al. 2004
U. pinnatifida	Pepsin-digested extracts	ACE-I IC_{50} (μM): Ala-Ile-Tyr-Lys = 213; Tyr-Lys-Tyr-Tyr = 64.2; Lys-Phe-Tyr-Gly = 90.5; Tyr-Asn-Lys-Leu = 21	Suetsuna and Nakano 2000
Rhodophyta (Red seaweeds)			
Palmaria palmata	Papain-digested extracts	Ile-Arg-Leu-Ile-Ile-Val-Leu-Met-Pro-Ile-Leu-Met-Ala Renin inhibitory bioassay: decrement of renin activities by 58.97% at 1 mg/mL	Fitzgerald et al. 2012
P. palmata	Chymotrypsin (ChTr) or trysin (Tr) extraction	< 10 kDa fractions of PP: hydrolyzed with ChTr (ACE IC_{50} 460.05 mg/mL)	Bondu et al. 2014
Pyropia columbina (formerly *Porphyra columbina*)	Aqueous followed by enzymatic extraction	23.9–38.1% ACE inhibition activity	Cian et al. 2012
P. columbina (formerly *P. columbina*)	Enzyme proteolysis enhanced extraction	ACE-I IC_{50} (g/L): 1.01	Cian et al. 2013
P. columbina (formerly *P. columbina*)	Enzymatic in thermostatic reactor extraction	ACE-I IC_{50} (g/L): 1.2–1.7	Cian et al. 2014b
P. columbina	Enzymatic hydrolysis of aqueous extracted proteins	Protein extract ACE-I IC_{50} 12.4 g/L and its hydrolysates ACE-I IC_{50} 35.3–44.8 g/L	Cian et al. 2015
Pyropia yezoensis (formerly *Porphyra yezoensis*)	pH and enzymatic extraction	ACE-I IC_{50} (g/L): 1.6	Qu et al. 2010
Solieria chordalis	Chymotrypsin (ChTr) or trysin (Tr) extraction	< 10 kDa fractions: hydrolyzed with ChTr (IC_{50} ACE 3.50 mg/mL) or Tr (IC_{50} ACE 20.34 mg/mL)	Bondu et al. 2014
	ACE-I inhibition associated with antioxidants		
Phaeophyceae (Brown seaweeds)			
Ecklonia cava	Purified phlorotannins	IC_{50} (mM): phloroglucinol = 2.57 ± 0.09; eckol = 2.27 ± 0.08; triphlorethol-A = 2.01 ± 0.36; dieckol = 1.47 ± 0.04; eckstolonol = 2.95 ± 0.28	Wijesinghe et al. 2011
E. stolonifera	Purified phlorotannins	Best inhibition recorded for eckol, dieckol and phlorofucofuroeckol. IC_{50} (μM): eckol = 70.82; phlorofucofuroeckol A = 12.74; dieckol = 34.25	Jung et al. 2006
Saccharina japonica	Supercritical CO_2 *versus* acetone and methanolic extracts	IC_{50} (μg/mL): supercritical CO_2 extraction = 0.89 ± 0.07; acetone/MeOH extraction = 1.05 ± 0.14	Sivagnanam et al. 2015
Sargassum horneri	Supercritical CO_2 *versus* acetone and methanolic extracts	IC_{50} (μg/mL): supercritical CO_2 extraction = 0.97 ± 0.11; acetone/MeOH extraction = 1.28 ± 0.50	Sivagnanam et al. 2015

On the other hand, Nishide and Uchida (2003) have reported lowering of systolic blood pressure (antihypertensive responses) and lower levels of total cholesterol, free cholesterol, triglyceride, and phospholipids in the liver, in rats fed with *Ulva* (Chlorophyta) powder.

Biologically active peptides are food-derived peptides that can exhibit diverse activities, including opiate-like, mineral binding, immunomodulatory, antimicrobial, antioxidant, antithrombotic, hypocholesterolemic, and blood pressure-lowering actions (Erdmann et al. 2008). Bioactive peptides have been detected in different animal and vegetable protein sources, milk peptides being by far the most commonly known source (Jiménez-Escrig et al. 2010, Pihlanto et al. 2008).

The isolation of protein from seaweeds is a difficult task due to the link between polysaccharides and protein within the seaweed matrix. It is described that the extraction of proteins from the tissues of Laminariales (Nagai et al. 2008), namely *Saccharina japonica* (Kim et al. 2011b), is difficult due to high levels of non-protein interfering compounds, mainly viscous polysaccharides. As a consequence, isoelectric point (Ma et al. 1996) or ammonium sulfate saturation (Hernández-Mireles and Rito-Palomares 2006) or trichloro acetic acid (Barbarino and Lourenço 2005) approaches, which are commonly used for protein precipitation, are not completely useful for seaweeds. Thus, to solve this task, different approaches are proposed, such as proteolytic treatment of the whole seaweeds, followed by filtration and dialysis (Suetsuna and Nakano 2000) or treatment of seaweed matrix with alginate lyase S to obtain an enriched-protein precipitate which is recovered by centrifugation (Sato et al. 2002a, b).

The identification of ACE-inhibitory peptides derived from *Undaria pinnatifida* (Wakame), and hypotensive action of orally administered peptides on spontaneously hypertensive rats (SHRs) is described. These studies are based on the previous evidence that dietary ingestion of whole Wakame, one of the most widely eaten brown seaweeds in Japan, has been shown to decrease blood pressure in humans. Specifically, the systolic blood pressure (SBP) of patients decreased significantly after daily oral administration of 3.3 g of dried Wakame after four weeks (Nakano et al. 1998). In the work of Suetsuna and Nakano (2000), Wakame powder was digested using pepsin. Then, the filtrate of enzymatic digestion was dialyzed, the outer solution was applied sequentially to a Dowex 50W column Hþ form, and peptides were eluted with ammonium solution. After concentration under vacuum, the residue was fractionated on a SP-Sephadex C-25 column, and a peptide power was obtained. The fractions having a molecular weight of 300–1000 kDa were collected and concentrated to dryness. The total yield of the peptide powder from 23.6 g of seaweed powder was 3.7 g. The peptides on the most ACE-inhibitory potent fraction were purified further by HPLC with an ODS-5 column. Although approximately 100 peaks were detected by this chromatography, potent inhibitory peptides were obtained in four peaks. Afterward, using protein sequencing, primary structures of the individual peptides were identified. The amino acid sequences of the peptides were Ala-Ile-Tyr-Lys, Tyr-Lys-Tyr-Tyr, Lys-Phe-Tyr-Gly, and Tyr-Asn-Lys-Leu. All of the active peptides had a tyrosine and lysine residue in the structure. Apart from this research, some peptides with potent ACE-inhibitory activity *in vitro* or intravenously are inactive in oral administration. Thus, hypotensive activity of each tetrapeptide is evaluated by measuring the SBP on SHR after oral administration of chemically synthesized tetrapeptides [50 mg/kg of body weight (BW)] (Suetsuna and Nakano 2000). SBP did not change in control rats during the study period (six hours). Captopril (10 mg/kg BW) lowered SBP significantly. A single dose (50 mg/kg BW) of the tetrapeptides significantly reduced SBP in SHR. This work firstly isolated the bioactive peptide, and then evaluated the activity of each synthesized peptide in a rat model.

The ACE-inhibitory and antihypertensive activities of Wakame hydrolysates have been investigated in another study, with a different research design (Sato et al. 2002a, b). To obtain an isolated protein residue, Wakame was treated with alginate lyase S at 45°C for 18 hours, and an enriched protein precipitate (46.3% dry matter) was recovered by centrifugation. Then Wakame was hydrolyzed using 17 kinds of proteases at different pH and temperature conditions, and ultrafiltered hydrolysates were tested for the inhibitory activity of the ACE. Among the proteases used in this study, Wakame hydrolysates of pepsin, protease S and N Amano, and proleather FG-F were able to produce potent ACE inhibitors *in vitro*. The yield of the different enzymes used ranged from 115 to 239 mg, as the weight of the solid contents obtained from 1 g of dried Wakame. In a second step, in order to evaluate the antihypertensive activity *in vivo* of hydrolysates produced by the four selected proteases, single oral administrations of hydrolysates were given to SHR (n¼ 6) at dosages of 100 and 1000 mg protein kg^{-1} BW. All the Wakame hydrolysates used in this test decreased the SBP in SHR, especially hydrolysates from protease S Amano or proleather FG-F. Digestion stability was evaluated by the change in IC_{50} values of hydrolysates before and after treatment with gastrointestinal proteases (pepsin, trypsin, and chymotrypsin) to simulate *in vivo* resistance to digestion. In addition, a long-term feeding of hydrolysates was assayed on SHR. Seven-week-old SHR were fed a diet containing 0%, 0.01%, 0.1%, and 1.0% of the protease S Amano hydrolysate for 10 weeks. The SBP in the Wakame hydrolysate group tended to be lower than in the control group. Summarizing, there is no correlation between the *in vitro* and *in vivo* studies. These results indicated that *in vivo* experiments —single oral administration test and long-term feeding test—are important for the final evaluation of the antihypertensive effects of peptides. Among 17 proteolytic enzymes tested *in vitro*, it has been found that hypertension in SHR was suppressed by the Wakame protease S Amano hydrolysates.

Lately, distinct studies applying proteolytic enzymatic digestion to seaweeds have led to the detection of a number of renin- or ACE-I bioactive peptides. Examples of potential peptides have been described for red macroalgae (e.g., from *Pyropia columbina*, *Pyropia yezoensis*, and *Palmaria palmata*), as well as for green algae (*Caulerpa microphysa*) and brown algae (*Undaria pinnafita*) (see Table 4.3). As seaweed hydrolysates consist of a complex mixture of constituents and the amino acid sequence of bioactive peptides in ACE-active hydrolysates has not been commonly determined, more experimental data should be gathered in order to allow solid conclusions on structural-active relations. However, according to Suetsuna et al. (2004), the presence of Tyr residues in dipeptides seems to improve their ability of targeting ACE.

Moreover, a study of isolation of potential antihypertensive agents (fucosterol and polyphenols) has been derived from seaweeds—Phaeophyceae [*Ecklonia cava*, *E. stolonifera*, *Sargassum fusiforme* (formerly *Hizikia fusiformis*), *Silvetia siliquosa* (formerly *Pelvetia siliquosa*), and *Undaria pinnatifida*], Rhodophyta [*Chondracanthus tenellus* (formerly *Gigartina tenella*), *Chondria crassicaulis*, *Gelidium amansii*, and *Pyropia tenera* (formerly *Porphyra tenera*)], and Chlorophyta [*Capsosiphon fulvescens*] (Jung et al. 2006). The study includes the crude extracts of selected edible Korean seaweeds which were screened for ACE-inhibitory activity. Seaweed bioactive constituents are extracted with ethanol followed by partitioning with organic solvents: *n*-hexane, dichloromethane, ethyl acetate, and n-butanol. Then the fractions extracted are chromatographed over a silica gel column yielding different sub-fractions and evaluated. In the case of the extract containing phloroglucinol, purification over an RP-18 column is used. Among the tested seaweeds, the ethanol extracts at a concentration of 163.93 mg/mL of *E. stolonifera* and *E. cava* appeared to be the most active, with inhibition of 64.86 ± 0.58% and 166.67 ± 4.20%, respectively. With the notable exception of *S. fusiforme*, the other brown algae *S. siliquosa*, and *U. pinnatifida* also exhibited favorable ACE-inhibitory activity, between 46% and 53%. Among the red algae tested, only *G. amansii* exhibited significant ACE-inhibitory effects, with an inhibition of 58.11 ± 1.73%. Column chromatography of the n-hexane and ethyl acetate fractions led to the isolation of fucosterol and 6 phlorotannins, as phloroglucinol, and its oligomers eckstolonol, eckol, phlorofucofuroeckol A (a pentamer), dieckol (a hexamer), and triphlorethol A (a trimer) from the *Ecklonia* and *Eisenia* species of brown algae. The ACE-inhibitory properties of phlorofucofuroeckol A, dieckol, and eckol ranked high, with IC_{50} values of 12.74 ± 0.15, 34.25 ± 3.56, and 70.82 ± 0.25 μM, respectively. Summarizing, other bioactive compounds, besides peptides, may be responsible for the antihypertensive capacity of seaweeds (Jiménez-Escrig and Cambrodón 1999, Jiménez-Escrig et al. 2010, 2011).

Heart diseases, such as arteriosclerosis, coronary heart disease, stroke, peripheral arterial disease, and heart failure, may be caused by hypertension or blood pressure greater than 140 mmHg systolic and/or 90 mmHg diastolic pressures (Lo and Li-Chan 2005). The ACE (dipeptidyl carboxypeptidase, EC 3.4.15.1) performs an important physiological function in the pathogenesis of cardiovascular and renal diseases through blood pressure regulation. In the renin–angiotensin system, ACE catalyzes the conversion of the inactive decapeptide angiotensin I (Asp-Arg-Val-Tyr-Ile-His-Pro-Phe-His-Leu) to the potent vasoconstrictor, the octapeptide angiotensin II (Asp-Arg-Val-Tyr-Ile-His-Pro-Phe), by hydrolytic removal of the histidyl leucine group from the C-terminal (Ondetti and Cushman 1982). Further, ACE is implicated in cell oxidative stress, through the generation of reactive oxygen/nitrogen species (Jung et al. 2006).

Certain biologically active peptides may act as ACE-inhibitory peptides, and thus, may prevent hypertension and its pathological consequences. ACE-inhibitory peptides from foods are less active than synthetic drugs such as captopril; however, their significance lies in the fact that they meet the need for naturalness and safety (Wu and Ding 2002).

4.5 Oxidative Stress and the Risk of Cardiovascular Diseases

The chronic oxidative stress associated with obesity occurs because the reactive oxygen species (ROS) production is increased. ROS production is accelerated because excess adipocytes are constantly being formed, increasing both in size and number (Monteiro and Azevedo 2010, Lee et al. 2011). In a well-documented review, Cornish et al. (2015), reported high intake of dietary antioxidants in vitamin (see Table 2.6 in Chapter 2) form to be associated with a reduced risk of CVD.

CHAPTER 5

Antiviral Activity of Seaweeds and their Extracts

5.1 Introduction

Viruses can be considered obligate intracellular parasites, constituted by either a RNA or DNA genome surrounded by a protective protein coat (Hans 1996). Viruses are grouped on the basis of size and shape, chemical composition and structure of the genome, and mode of replication. Helical morphology is seen in nucleocapsids of many filamentous and pleomorphic viruses. Icosahedral morphology is characteristic of the nucleocapsids of many "spherical" viruses. Many viruses also have an outer envelope (Gelderblom 1991).

The genome of a virus may consist of DNA or RNA, which may be single stranded (SS) or double stranded (DS), linear or circular. The entire genome may occupy either one nucleic acid molecule (monopartite genome) or several nucleic acid segments (multipartite genome). The different types of genome necessitate different replication strategies (Gelderblom 1991).

The viruses that infect humans are currently grouped into 21 families, reflecting only a small part of the spectrum of the multitude of different viruses whose host ranges extend from vertebrates to protozoa and from plants and fungi to bacteria (Gelderblom 1991). Therefore, enteroviruses and lentiviruses can cause a large number of human diseases (Fig. 5.1) (Soares 2015).

5.1.1 Caliciviruses

Caliciviruses are positive-sense, single stranded RNA viruses containing four recognized genera —Lagovirus, Norovirus, Sapovirus, and Vesivirus. They are ubiquitous in the environment and are a major cause of disease in humans and many animals. Examples include Norwalk virus, a norovirus, thought to be responsible for roughly 90% of epidemic, non-bacterial outbreaks of gastroenteritis in humans around the world. Lack of a suitable cell culture system for human caliciviruses limited studies in previous decades, however the recent application of modern genomic technologies has revolutionized the field, leading to an explosion in calicivirus publications (Hansman et al. 2010).

Norovirus is a group of non-enveloped viruses that have a single-stranded, positive sense RNA genome (Atmar 2010). Norovirus infection can usually be caused by contaminated water or food, and it spreads via human contact with infected materials through the fecal-oral route (Choi et al. 2014). This infection has been recognized as a leading cause of epidemics with the symptoms of vomiting, diarrhea, mild fever, abdominal cramping, and nausea in the community (La Rosa et al. 2013). According to the Centers for Disease Control and Prevention (2011), over 70% of water-related and over 50% of food-consumption gastroenteritis patients resulted from norovirus infection in USA. Likewise, norovirus causative issues have been a great concern around the world during the recent years. Norovirus has characteristics such as low infectious dose, prolonged shedding period, strong stability, great diversity, and frequent genome mutations (Lee et al. 2015d). Norovirus could be effectively reduced by using disinfectants such as alcohols

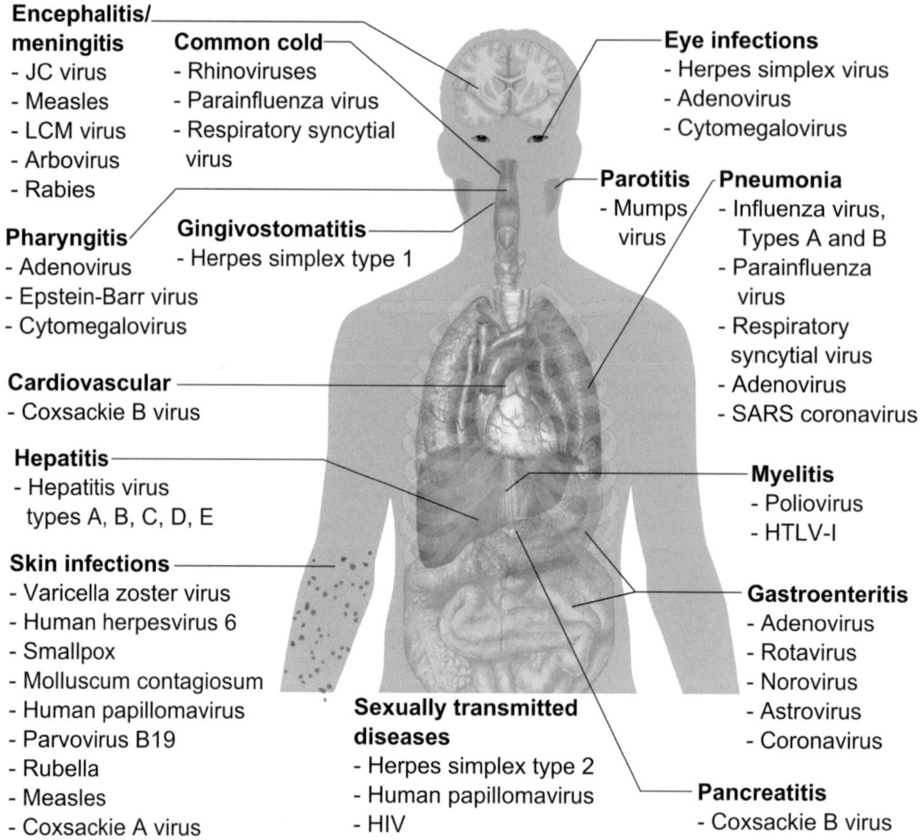

Figure 5.1 Overview of viral human infections (adapted from Häggström 2014).

and povidone idodines (Belliot et al. 2008). However, chemical sanitizers can cause various side effects in humans, such as fever and itching (Choi et al. 2014). Recently, attempts have been made to find sustainable solutions from medicinal plants and marine organisms (Balunas and Kinghorn 2005, Jain et al. 2008, Choi et al. 2014). Among marine organisms, chitosan and chitosan oligosaccharides were reported to be able to effectively reduce the infectivity of human enteric viral surrogates (Feline calicivirus –FCV) (Su et al. 2009, Davis et al. 2012). However, there are limitations in the progress of this study due to inefficient cell culture amplification process for norovirus (Guix et al. 2007, Lay et al. 2010). Recent studies revealed that the structure and genetic relatedness of murine norovirus (MNV) to human norovirus makes this virus a promising and relevant surrogate for studying the environmental survivability of human norovirus (Cannon et al. 2006, Zhang et al. 2012).

5.1.2 Enteroviruses

Enteroviruses are a genus of the family Picornaviridae, a group of single stranded RNA viruses, which includes polioviruses, echoviruses, group A and B coxsackieviruses (CVB), and numbered enteroviruses (Rueckert 1996). They have been implicated in a large variety of human diseases, ranging from mild illnesses to severe clinical diseases such as myocarditis, meningitis, encephalitis, and paralysis (Melnick 1996, Pallansch and Roos 2001).

5.1.3 Flaviviruses

Flavivirus is a genus of the family Flaviviridae composed of nearly 80 members. Many flaviviruses are arthropod-borne viruses that cause important human diseases, including Yellow fever virus (YFV), Dengue virus (DENV), West Nile virus (WNV), and Japanese encephalitis virus (JEV) (Lindenbach and Rice 2001). This genus also includes the Zika virus (ZIKV) and several other viruses which may cause encephalitis (Shi 2012), as well as insect-specific flaviviruses (ISFs), such as Cell fusing agent virus (CFAV), Palm Creek virus (PCV), and Parramatta River virus (PaRV) (McLean et al. 2015).

Flaviviruses are named from the Yellow fever virus (YFV), the type virus for the family; the word flavus means "yellow" in Latin. The name yellow fever in turn originated from its propensity to cause yellow jaundice in victims (Mitchell 1744).

Flaviviruses share several common aspects—common size (40–65 nm), symmetry (enveloped, icosahedral nucleocapsid), nucleic acid (positive-sense, single-stranded RNA around 10,000–11,000 bases), and appearance in the electron microscope (Shi 2012).

Most of these viruses are transmitted by the bite from an infected arthropod (mosquito or tick) and hence, classified as arboviruses. Human infections with these viruses are typically incidental, as humans are usually unable to replicate the virus to high enough titers to re-infect the arthropods needed to continue the virus lifecycle—man is then a dead-end host. The exceptions to this are the Yellow fever, Dengue, and Zika viruses, which still require mosquito vectors, but are so well-adapted to humans as to not necessarily depend upon animal hosts (although they continue to have important animal transmission routes, as well) (Shi 2012).

In particular, DENV has re-emerged in recent years as an increasingly important public health threat affecting more than 100 countries worldwide, with nearly 50 million infections each year and over 2.5 billion people at risk (Gubler 2002). DENV circulates in nature as four serotypes (DENV-1 to DENV-4), which are transmitted to humans by two species of mosquitoes, *Aedes aegypti* and *Aedes albopictus* (Arthropoda). Infection with DENV produces a wide spectrum of clinical illness ranging from silent infection to either a mild febrile, self-limited acute syndrome known as Dengue fever (DF) or the severe and often fatal Dengue hemorrhagic disease (DHF) and Dengue shock syndrome (DSS). Instead of the increasing global incidence of DF and DHF occurring in the last decades, there is neither specific chemotherapy nor a vaccine for treatment and prevention of DENV infection. Supportive medical care and symptomatic treatment through hydration are the most important aids to patients and to improve survival in the severe forms of the disease. Consequently, new approaches for the control of DENV infection are urgently needed (Talarico et al. 2005).

5.1.4 Lentiviruses

Lentiviruses are a subfamily of retroviruses, single-stranded RNA virus with a viral reverse transcriptase, that are characterized by long incubation periods between infection of the host and the manifestation of clinical disease. Human immunodeficiency virus type 1, the causative agent of AIDS, is the most widely studied lentivirus. However, the lentiviruses that infect sheep, goats, and horses were identified and studied prior to the emergence of human immunodeficiency virus type 1 (Clements and Zink 1996).

Acute respiratory infections (ARIs) are the major cause of childhood morbidity and mortality worldwide. Viruses account for the majority of ARIs in young children and most infections are attributed to respiratory syncytial virus (RSV), parainfluenza virus (PIVs), influenza virus (FluV), rhinovirus (RV), and adenovirus (AdV) (Monto 2002, Mackay et al. 2003, Weigl et al. 2007, Canducci et al. 2008). Over the past few years, newly described viruses were associated to respiratory infection, such as human metapneumovirus (HMPV), the emerging human coronaviruses (HCoV), human bocavirus (HBoV), and the new human papilloma viruses KIPyV and WUPyV (Kahn 2007, Sloots et al. 2008).

Human metapneumovirus is a species in the genus Metapneumovirus, family Paramyxoviridae, and was isolated in 2001 by Van Den Hoogen and collaborators, in previously virus-negative nasopharyngeal aspirates from children with respiratory tract infections (Van Den Hoogen et al. 2001). Although HMPV infections have been diagnosed in all age groups, this virus has its greatest effect in children (Deffrasnes et al. 2007). Several studies demonstrated that HMPV accounts for a major proportion of hospitalizations for lower respiratory tract infections in infants and young children (Nicholson et al. 2006, Van Den Hoogen et al. 2003, Williams et al. 2004). The most frequent diagnoses in hospitalized children are bronchiolitis and pneumonia, but occasionally HMPV may also cause severe illnesses that require treatment at intensive care units (Schildgen et al. 2005).

5.1.5 Main targets for antiviral drugs

Specific events in virus replication identified as targets for antiviral agents are viral adsorption, penetration, uncoating, and viral nucleic acid synthesis, as well as viral protein synthesis. Specificity for infected cells may occur when virus-specified enzymes (e.g., thymidine kinase-induced by *Herpes simplex* virus or *Varicella-zoster* virus) activate drugs (e.g., Acyclovir) (Gelderblom 1996).

5.1.6 Antiviral activity of marine algal polysaccharides and other metabolites

In recent years, the constant outbreak of some emerging or remerging viral diseases has caused serious harm to human health. During the last decades, the number of antiviral products approved for clinical use has been increased from five to more than 30 drugs. The potential antiviral activity of marine algal polysaccharides was first shown by Gerber et al. (1958), describing that the polysaccharides extracted from *Gelidium robustum* (formerly *Gelidium cartilagenium*) (Rhodophyta) protected the embryonic eggs against Influenza B or Mumps virus. Many species of marine algae contain significant quantities of complex structural sulfated polysaccharides that have been shown to inhibit the replication of enveloped viruses including members of the Flavivirus, Togavirus, Arenavirus, Rhabdovirus, Orthopoxvirus, and Herpes Virus families (Witvrouw and De Clercq 1997). Polysaccharides extracted from Rhodophyta have been shown to exhibit antiviral activity against a wide spectrum of viruses, including important human pathogenic agents, such as Human Immunodeficiency virus (HIV), Herpes simplex virus (HSV), Vesicular stomatitis virus (VSV), and Cytomegalovirus (CMV) (Witvrouw and De Clercq 1997). The chemical structure including the degree of sulfation, molecular weight, constituent sugars, conformation, and dynamic stereochemistry are caused to determine the antiviral activity of algal sulfated polysaccharides (Lüscher-Mattli 2000, Damonte et al. 2004, Adhikari et al. 2006). In addition, both the degree of sulfation and the distribution of sulfate groups on the constituent polysaccharides play an important role in the antiviral activity of these sulfated polysaccharides. Algal polysaccharides with low degrees of sulfation are generally inactive against viruses (Damonte et al. 2004).

Marine polysaccharides can either inhibit the replication of the virus by interfering with the viral life cycle or improve the host antiviral immune responses to accelerate the process of viral clearance. The life cycle of viruses differs greatly between species, but there are six basic stages in the life cycle of viruses-viral adsorption, viral penetration, uncoating of capsids, biosynthesis, viral assembly, and viral release. Marine polysaccharides can inhibit viral life cycle at different stages or directly inactivate virions before virus infection. Specific antiviral mechanism of marine polysaccharides is commonly related to specific structure features of the polysaccharides and specific viral serotypes (Damonte et al. 2004). Carrageenan might inhibit virus infection via direct actions on the virus surface by its negative charge (Wang et al. 2008). Several studies showed that carrageenan has a direct virucidal action on some enveloped viruses, which makes the viruses lose the ability to infect cells, thus effectively reducing the virus multiplication. Carlucci et al. (2002) found that lambda-type carrageenan could firmly bind to the *Herpes simplex* virus (HSV), leading to the inactivation of the HSV virion, thus inhibiting the replication of HSV. Their studies further suggest that carrageenan changes the structure of the glycoproteins gB and gC of HSV (Carlucci et al. 1999, Carlucci et al. 2002). Moreover, Harden et al. (2009) reported that carrageenan polysaccharides derived from red algae could directly inactivate HSV-2 at low concentrations. The virucidal activities increase with increased molecular weight of carrageenan polysaccharide up to 100 kDa, after which the virucidal activities level off. The direct virucidal actions of carrageenan may be due to the formation of a stable virion–carrageenan complex, where binding is not reversible and hence the sites on the viral envelope required for virus attachment to host cells are occupied by the sulfated polysaccharide, which renders the virus unable to complete the subsequent infection process (Damonte et al. 2004). Several studies have shown that carrageenan can mask the positive charge of host cell surfaces by the negative charge of its sulfate groups, so as to interfere with the adsorption process of viruses. Mazumder et al. (2002) obtained a high molecular weight sulfated galactan from red algae, and showed its antiviral activities against *Herpes simplex* virus 1 and 2 in bioassays, which is likely due to an inhibition of the initial viral attachment to the host cells. Carlucci et al. (1997a, 1999) noted that lambda carrageenan and partially cyclized mu/iota-carrageenan from *Gigartina skottsbergii* have potent antiviral effects against different strains of HSV types 1 and 2 during the virus adsorption stage. They subsequently confirmed the firm binding of carrageenan to virus receptors on the host cell surface. Their studies demonstrate that lambda carrageenan interferes with the adsorption process of the virus to the host cell surfaces. Buck et al. (2006) found that carrageenan could directly bind to the HPV capsid, so as to inhibit not only the viral adsorption process but also the subsequent entry and uncoating process of the virus. They also found that the inhibition actions of carrageenan against HPV might be related to a mechanism that is independent of the heparan sulfate after viral adsorption (Buck et al. 2006). Moreover, Talarico and coworkers reported that lambda and iota-carrageenans could interfere with both DENV-2 adsorption and internalization into host cells, and they are only effective if added together with the virus or shortly after infection (Talarico et al. 2007). The mechanism of this inhibition action may be due to the fact that although DENV virus can enter into host cell in the presence of carrageenans, their subsequent uncoating and release from endosomes may be interfered by the carrageenans. The inhibitory action of iota-carrageenan on the uncoating process of dengue virus may be attributed to the direct interaction of carrageenans with the virus membrane glycoprotein E (gE) (Talarico et al. 2005, Talarico et al. 2007, Talarico et al. 2007b). Talarico and co-workers (Talarico et al. 2007, Talarico et al. 2011), also reported that iota-carrageenan could inhibit dengue virus (DENV) replication in mammalian and mosquito cells, and the mode of action of iota-carrageenan in both cell types is strikingly different. In conclusion, the antiviral activities of carrageenans are very broad, which can suppress the replication of both enveloped and non-enveloped viruses. The antiviral effects of carrageenans are closely related to the molecular weights and the degree of their sulfation. Moreover, the inhibitory actions of carrageenans on different viruses are usually different, which are associated with the types of carrageenans, the virus serotypes, and the host cell itself (Damonte et al. 2004, Talarico et al. 2005, Soares 2015).

Wang et al. (2007) isolated a sulfoquinovosylmonoacylglyceride (SQDG) from the n-butanol fraction of the invasive *Caulerpa racemosa* collected from the South China Sea. The SQDG compound was characterized using spectroscopic methods as (2S)-1,2-di-O-palmitoyl-3-O-(6'-sulfo-α-D-quinovopyranosyl) glycerol, and was active against HSV-2, with a 50% inhibitory concentration (IC_{50}) of 15.6 mg mL^{-1} against both standard and clinical strains of HSV-2.

Since the discovery of the human immunodeficiency virus (HIV), many drugs have been developed in an attempt to inhibit its replication. However, HIV is resistant to treatment with known drugs (Kuritzkes 2007, Hirsch et al. 2008, Manhanzva et al. 2015). This is because HIV has a high mutation rate and does not have an effective mechanism for error correction during replication. One of the strategies researchers adopted was the combination of two or more drugs known as Highly Active Antiretroviral Therapy (HAART) (Marrazzo et al. 2014). Such treatment may reduce viral load to undetectable levels in the blood and provide long-lasting clinical benefit. However, some patients do not respond to this treatment, making the search for new molecules with anti-HIV activity an urgent need. Seaweeds are a source of many diverse chemicals. Several extracts, fractions, and natural products isolated from seaweeds have demonstrated effective anti-HIV activity (Vo and Kim 2010), making it an interesting base from which to develop new medicines. Of the seaweeds, red seaweeds produce metabolites such as acetogenins, sesquiterpenes, monoterpenes, bromophenols, and sulfated polysaccharides that can be used for anti-HIV drug development. *Acanthophora spicifera* (Rhodophyta) is an excellent model for studies of biological activity in Brazil because it has a large biomass on the Rio de Janeiro coast, is easily identified (Perrone et al. 2006), has an experimental cultivation (Kaliaperumal et al. 1986), and is a food source to other species, indicating a low toxicity. Furthermore, fractions rich in sulfated polysaccharides from the red seaweed *Acanthophora* are an effective antiviral against HSV-1 and HSV-2 strains (Duarte et al. 2004) (see also Section 5.4 below).

5.2 Antiviral Activity of Chlorophyta (Green Seaweeds)

In 1997, a wide-ranging study on methanol extracts of seaweed species and their antiviral activity was performed. In that work, several species were tested for anti-HSV-1 activity. Among the green algae, five were demonstrated to have antiviral activity—*Acrosiphonia coalita*, *Codium fragile*, *Ulva linza* (formerly *Enteromorpha linza*), *Ulva australis* (formerly *Ulva pertusa*), and *Ulva* sp. (Kim et al. 1997). Hudson et al. (1999) showed anti-HSV activity of methanol extracts of the species *C. fragile*, *U. linza*, and *U. australis*, as well as the virucidal activity of these algae. In 1999, Santos et al. showed the antiviral effect of aqueous cold extracts from *Ulva fasciata* and *Codium decorticatum* against Acyclovir (ACV)-resistant HSV-1, with 92.9 and 98.5% of inhibition at the 1.2% concentration, respectively.

The communication made by Lakshmi et al. (2006) deals with the biological activities of the extracts of 48 marine florae. The biological screening includes tests for antibacterial (see Chapter 8), antiviral, among others. From among green algae, the crude extracts from *Bryopsis hypnoides*, *Bryopsis plumosa*, *Caulerpa racemosa*, *Caulerpa veravalensis*, *Chaetomorpha* sp., *Codium dwarkense*, *Codium decorticatum* (formerly *Codium elongatum*), *Ulva clathrata* (formerly *Enteromorpha clathrata*), *Dictyosphaeria cavernosa*, *Udotea indica*, and *Valoniopsis pachynema* were active against Ranikhet disease virus (NDV), Semliki forest virus (SFV), Encephalomyocarditis virus (EMCV), Herpes simplex virus (HSV), Influenza 'A' virus (IAV), and Japanese B encephalitis virus (JEV).

C. fragile was subjected to enzymatic hydrolysis in the studies made by Kulshreshtha et al. (2015), and the extracts produced were tested for their antiviral activity. *C. fragile* had a higher amount of neutral sugar and ash. Moreover, extracts obtained from enzyme-assisted hydrolysis exhibited antiviral activity against HSV-1, with IC_{50} values in the range of 36–52 μg mL^{-1} (Kulshreshtha et al. 2015).

According to Mendes et al. (2010), three extracts of *Ulva fasciata* collected at Rasa beach (Brazil) presented excellent inhibitory activity against human metapneumo virus (HMPV), despite the method used for its preparation (maceration, decoctions, and maceration of the decoction), while only the extract prepared by maceration from the algae collected at Forno beach (Brazil) presented similar activity. The other two extracts prepared by decoction and maceration of decoction showed only partial inhibition of the viral replication.

Ivanova et al. (1994, 1994b) isolated a biologically active polysaccharide and a hetero-saccharide from *Ulva lactuca*. These compounds exhibited antiviral activity *in vitro* against a number of human and avian influenza. They also stimulated macrophages, B-cells, and T-cells in mice.

Extracts of algae have been shown to possess a broad range of biological activities, including antivirus activity. In this study, we identified that the sulfated polysaccharide extracts from *Ulva lactuca* can inhibit Japanese encephalitis virus (JEV) infection in Vero cells. Mechanistic studies further revealed that the *Ulva* sulfated polysaccharide extracts can block virus adsorption and thus make the virus unable to enter cells. The *Ulva* sulfated polysaccharide extracts also effectively decrease the production of pro-inflammatory cytokines in the JEV-infected primary mixed glia cells. The results showed that the *Ulva* extracts have an excellent anti-JEV activity in preventing the virus from entering into cells, which is likely to be a result of forming virus-polysaccharide complexes. Moreover, in mice essays, the *Ulva* extracts were shown to possess many advantages, such as low production cost, broad spectrum of antivirus, low toxicity, high safety, broad acceptability, and novel mode of action. At the current stage, the polysaccharide extracts of *U. lactuca* may be suitable to serve as a healthy food supplement for daily antiviral prevention, particularly anti-flavivirus (Chiu et al. 2012b).

Dichloromethane extracts from 12 species of green macroalgae—*Avrainvillea elliottii*, *Bryopsis* sp., *Chaetomorpha antennina*, *Cladophora prolifera*, *Codium decorticatum*, *Codium spongiosum*, *Caulerpa racemosa*, *Halimeda opuntia*, *Halimeda tuna*, *Penicillus capitatus*, *Udotea flabellum*, and *Ulva fasciata*, were tested against HSV-1 Acyclovir-resistant strains, and all, except for *A. elliottii*, showed a percentage inhibition (PI) of the virus between 55.3% and 99.9%. In fact, four species, namely *C. prolifera*, *C. decorticatum*, *P. capitatus*, *U. flabellum*, and *U. fasciata*, exhibited an inhibition percentage ≥ 90%, which is considered a strong anti-HSV-1 activity. It was also shown that the major components of the extracts of *C. decorticatum*, *P. capitatus*, and *U. fasciata* were triglycerides and fatty acids (Soares et al. 2012).

Lee et al. (2004) showed the anti-HSV-1 activity of ten natural sulfated polysaccharides extracted from nine green algae—*Caulerpa brachypus*, *Caulerpa okamurae*, *Caulerpa scalpelliformis*, *Chaetomorpha linum* (formerly *Chaetomorpha crassa*), *Chaetomorpha spiralis*, *Codium adhaerens*, *Codium fragile*, *Codium latum*, and *Monostroma nitidum*. The polysaccharides showed low cytotoxicity and potent anti-HSV-1 activity, with antiviral activity of 50% (IC_{50}) at concentrations ranging from 0.38 μg mL^{-1} to 8.5 μg mL^{-1} in experiments in which the samples were added at the same time as viral infection. However, the polysaccharides were shown to be less potent when added after infection. The polysaccharides isolated from the algae *M. nitidum*, *C. brachypus*, *C. okamurai*, and *C. latum* presented the best results for inhibition of virus replication. *Ulva compressa* (formerly *Enteromorpha compressa*) was also tested, but showed no activity against HSV-1 (De Souza-Barros et al. 2015).

C. racemosa also has a sulfated polysaccharide with antiviral activity against three different strains of HSV-1, with IC_{50} values ranging from 2.2 μg mL^{-1} to 4.2 μg mL^{-1}, and a low cytotoxic effect (Ghosh et al. 2004). In fact, polysaccharides derived from marine organisms have shown a variety of antiviral activities (Wang et al. 2012a). Macedo et al. (2012) have demonstrated that caulerpin isolated from *C. racemosa* shows low cytotoxicity and anti-HSV-1 activity with an IC_{50} value of 1.29 μg mL^{-1}, while ACV showed an IC_{50} value of 1.09 μg mL^{-1} in the referred study. However, caulerpin has a selectivity index (SI) greater than ACV given its low toxicity. The SI represents the degree of security for the use of the molecule. In addition to having low cytotoxicity and potent antiviral activity, caulerpin is present in many species of seaweed and has actually been isolated from more than 12 species of *Caulerpa*, from *Codium decorticatum* and from *Halimeda incrassata* (Teixeira 2012). Hence, this molecule may be a good candidate for utilization in antiviral drugs (HSV-1) (Teixeira 2012, De Souza-Barros et al. 2015).

Some green seaweed also shows activity against Herpes simplex virus type 2 (HSV-2). Some examples are *A. elliottii*, *Bryopsis* sp., *C. antennina*, *C. fragile*, *C. racemosa*, *C. spongiosum*, *H. opuntia*, *H. tuna*, *U. flabellum*, and *P. capitatus* (Ghosh et al. 2004, Ohta et al. 2009, Soares et al. 2012, De Souza-Barros et al. 2015).

5.3 Antiviral Activity of Phaeophyceae (Brown Algae)

The brown algae (Phaeophyceae class) are recognized by their large size, serving as a food source for many marine animals such as *Macrocystis pyrifera*, which is responsible for creating authentic forests in the Pacific Ocean, and may grow 30 cm per day, and grow up to 65 meters or more in length (Pereira and Correia 2015). One of the examples of antiviral activity of brown algae is *Laminaria abyssalis*, whose extract showed activity in the fight against *Herpes simplex* virus (Santos et al. 1999).

Kim et al. (1997) found anti-HSV-1 activity in methanol extracts of seven brown algae—*Analipus japonicus*, *Colpomenia bullosa*, *Sargassum thunbergii*, *Scytosiphon lomentaria*, *Undaria pinnatifida*, *Nereocystis luetkeana*, and *Postelsia palmaeformis*. Hudson et al. (1999) confirmed the antiviral activity (HSV-1) of *C. bullosa*, *S. lomentaria*, and *U. pinnatifida*.

Santos et al. (1999) have shown an antiviral effect of aqueous cold extracts from *Padina gymnospora*, *Laminaria abyssalis*, and *Sargassum vulgare* against ACV-resistant HSV-1, with 99%, 100%, and 92% inhibition at the concentration of 1.2%, respectively.

Galactofucan sulfate (GFS) purified from aqueous extract of *Undaria pinnatifida* was evaluated for antiviral activity against 14 clinical strains of HSV-1. Four of these strains were resistant to Acyclovir, while the other 10 were susceptible. The median IC_{50} value of GFS for the 14 strains was 32 μg mL^{-1}, ranging from 1.0 μg mL^{-1} to 128 μg mL^{-1}. However, no significant difference in the anti-HSV activity of galactofucan sulfate was found between ACV-susceptible and ACV-resistant strains (Thompson and Dragar 2004). Hemmingson et al. (2006) also showed anti-HSV-1 activity of a galactofucan sulfate extract from *U. pinnatifida*. Its fractions had an IC_{50} value ranging from 1.1 μg mL^{-1} to 4.6 μg mL^{-1} (De Souza-Barros et al. 2015).

Six Hong Kong seaweed species, namely *Colpomenia sinuosa* (IC_{50}=22.1 μg mL^{-1}), *Dictyota dichotoma* (IC_{50}=24.3 μg mL^{-1}), *Hydroclathrus clathratus* (IC_{50}=6.25 μg mL^{-1}), *Lobophora varicgata* (IC_{50}=18.5 μg mL^{-1}), *Padina australis* (IC_{50}=58.9 μg mL^{-1}), and *Sargassum hemiphyllum* (IC_{50}=19.1 μg mL^{-1}), were extracted with hot water and tested for antiviral activity against the standard strains of

HSV-1. The extract of *H. clathratus* had the highest antiviral effect on HSV-1, followed by the extract of *L. variegata*. The extracts of *H. clathratus* and *L. variegata* also showed inhibitory activity against ACV-resistant HSV-1 and clinical strains. Additionally, four fractions obtained from the water extract of *H. clathratus* were tested and showed an IC_{50} range of 1.6 μg mL^{-1} to 6.5 μg mL^{-1} (Wang et al. 2008).

In the works of Premnathan et al. (1992) the brown algae *Sargassum tenerrimun* and the red algae *Acanthophora spicifera* (see Table 5.1), which were earlier reported to be ineffective against Newcastle disease virus (NDV) (Naqvi et al. 1980), present some variations on their antiviral activity that can be attributed to the ecological factors prevailing at the time of collection and the growth stage of the algae (Premnathan et al. 1992).

Queiroz et al. (2008) assessed the activity of fucans isolated from *F. vesiculosus* (from the coast of Natal, Brazil) as inhibitors of HIV from reverse transcriptase (RT). These fucans had a pronounced inhibitory effect *in vitro* on the avian-RT at a concentration of 0.5–1.0 mg mL^{-1}. The alginic acid (1.0 mg mL^{-1}) inhibited the RT activity by 51.1% using activated DNA. The authors attributed the inhibitory to the fucans to the presence of sulfate groups, as desulfation resulted in the loss of this effect. Furthermore, it was suggested that fucan activity was not only dependent on the ionic changes but also on the sugar rings that act to spatially orientate the charges in a configuration that recognizes the enzyme, thus determining the specificity of the binding (Queiroz et al. 2008).

Furthermore, recent studies demonstrated that seaweed derived sulfated polysaccharides could be used as vaginal antiviral formulations without disturbing essential functions of the vaginal epithelial cells and normal bacterial flora (Béress et al. 1993). It will be a continuous challenge to select the most promising drug candidates among the wide array of available sulfated polysaccharides compounds. There are numerous advantages over other classes of antiviral drugs, such as relatively low production costs, broad spectrum of antiviral properties, low cytotoxicity, safety, wide acceptability, and novel modes of action. These suggest marine algal sulfated polysaccharides as promising drug candidates in the near future, and further studies are needed with clinical trials for these antiviral sulfated polysaccharides.

Hemmingson et al. (2006) studied the anti-viral activity of a galactofucan sulfate extracted from *Undaria pinnatifida*, collected from the east coast of Tasmania, Australia. It was found to be a potent inhibitor of the Herpes virus HSV-1, HSV-2, and human cytomegalo virus (HCMV), with IC_{50} values of 1.1 mg mL^{-1}, 0.2 mg mL^{-1}, and 0.5 mg mL^{-1}, respectively.

The study done by Mendes and collaborators (2011) demonstrated that the crude extract and the meroditerpenoids epitaondiol, atomaric acid, and its methyl ester derivative isolated from the marine seaweed *Stypopodium zonale* possess strong anti-HMPV activity. Unlike atomaric acid and its methyl ester derivative, the meroditerpene epitaondiol was capable of inhibiting the penetration stage into cells of the viral replication, even though it has shown higher cytotoxicity, probably due to particular molecular recognition events associated to its biological mechanism of action that may be further explored. Taken together, these studies confirm the enormous potential of Brazilian marine algae as a source for biomolecules and also suggest that bio-guided assays are a valuable tool in order to assist the identification of potent bioactive compounds (Mendes et al. 2011).

A fucan from *Cladosiphon okamuranus* (see Table 5.1) composed of glucuronic acid and sulfated fucose units potently inhibited infection of BHK-21 cells with Dengue virus type 2 (DENV-2), but showed little effect on the other three serotypes of the virus (Hidari et al. 2008). Sulfation of the fucan was necessary for this activity; surprisingly, carboxyl-reduction of the glucuronic acids to glucose units also abolished the fucans' antiviral properties. Analysis of the structure of the envelope glycoproteins from the four serotypes of dengue virus suggested that arginine-323 in DENV-2, which is proximal to the putative heparin binding site, was critical for the interaction with the fucan (Jiao et al. 2011).

The low cytotoxicity, as well as the potent antiviral activity of dolabellanodienotriol isolated from the alga *Dictyota friabilis* (formerly *Dictyota pfaffii*) against HIV-1 reverse transcriptase (Barbosa et al. 2004, Cirne-Santos et al. 2008, Abrantes et al. 2010) motivated studies on the toxicity of this substance. These studies demonstrated the low toxicity of this compound in BALB/c mice (Garrido et al. 2011). Other diterpenes isolated from *D. dichotoma* (isopachydictyolal) and *Dictyota dichotoma* var. *intricata* (formerly *Dictyota linearis*) were tested. However, they did not show significant antiviral activity against HSV-1 in concentrations below the maximum non-toxic dose and proved to be toxic for Vero cells in

Table 5.1 Seaweed species and seaweed extracts with antiviral activity.

Species	Tested product, fraction or extract	Virus	Observed results	References
Chlorophyta (Green seaweed)				
Acrosiphonia coalita	Extract	HSV-1	Inhibition at 200 μg mL^{-1}	Kim et al. 1997
A. orientalis	Polysaccharide	WSSV	Survival rate (88%) of shrimp	Manilal et al. 2009
Avrainvillea elliottii	Extract	HSV-2	Inhibition percent = 60.2%	Soares et al. 2012
Bryopsis sp.	Extract	HSV-1 HSV-2	Inhibition percent = 82.2% Inhibition percent = 87.4%	Soares et al. 2012
Caulerpa brachypus	Sulfated polysaccharides (ramnose and galactose derivatives)	HSV-1	IC_{50} = 1.9 and 0.65 μg mL^{-1}	Lee et al. 2004 Wang et al. 2014
C. okamurae f. *oligophylla*	Sulfated polysaccharides	HSV-1	IC_{50} = 0.55 μg mL^{-1}	Lee et al. 2004 Wang et al. 2014
C. prolifera	Crude organic extracts	HSV-1 VSV	Inhibition zone to 1 mm	Ballesteros et al. 1992
C. racemosa	Extract, sulfated polysaccharide, and caulerpin	HSV-1	2.2 to 4.2 μg mL^{-1} (against reference strains and TK^{-}) IC_{50} = 1.29 μg mL^{-1} Inhibition percent = 57.3%	Ghosh et al. 2004 Macedo et al. 2012 Soares et al. 2012
C. racemosa	Glycolipids Dichloromethane and methanol crude extracts	HSV-2 DENV	IC_{50} = 15.6 mg mL^{-1} IC_{50} = 62.5 mg mL^{-1}	Wang et al. 2007 Koishi et al. 2012
C. racemosa	Ethanol extract Caulerpin	NDV BVDV	0.1 mL ethanol extract = 2.7 log 10 EID_{50} reduction in virus titer IC_{50} = 2 μM	Ibraheem et al. 2012 Pinto et al. 2012
C. scalpelliformis	Extracts Sulfated polysaccharides	EMCV SFV VACV HSV-1	1–24% activity 50–74% activity 50–74% activity IC_{50} = 1.6 μg mL^{-1}	Premnathan et al. 1992 Lee et al. 2004
C. taxifolia	Crude extract	FIV	Extract concentration of 10^{-3} mg mL^{-1} almost fully abolished the viral activity	Nicoletti et al. 1999
	Toluene extract	TNV	A concentration of 1000 μg mL^{-1} reduced the number of lesion formed by TNV by nearly 64%	Sethi 2015
Chaetomorpha antennina	Aqueous extract Dichloromethane and methanol crude extracts	HSV-1 HSV-2 DENV	Inhibition percent = 53.3% Inhibition percent = 85.9% IC_{50} = 31.25 μg mL^{-1}	Soares et al. 2012 Koishi et al. 2012
C. linum	Extracts	HBV	50–74% activity	Premnathan et al. 1992
C. linum (as *C. crassa*)	Sulfated polysaccharides	HSV-1	IC_{50} = 8.5 μg mL^{-1}	Lee et al. 2004
C. spiralis	Sulfated polysaccharides	HSV-1	IC_{50} = 1.9 μg mL^{-1}	Lee et al. 2004

Table 5.1 contd....

Table 5.1 contd....

Species	Tested product, fraction or extract	Virus	Observed results	References
Cladophora prolifera	Extract	HSV-1	Inhibition percent = 90%	Soares et al. 2012
C. vagabunda	Lipophilic extract	FluV	Minimum inhibitory concentration at 0.8 mg mL^{-1}, inhibition percent = 50%	Serkedjieva et al. 2000
Chlorella vulgaris var. *autotrophica** (as *Chlorella autotrophica*)	Sulfated soluble exopolysaccharide	ASFV	Inhibition percent = 15.3%	Fabregas et al. 1999
Codium adhaerens	Sulfated polysaccharides	HSV-1	IC_{50} = 1.0 μg mL^{-1}	Lee et al. 2004
C. decorticatum (as *C. elongatum*)	Methanol extract	SFV VACV	Inhibition percent = 50% Inhibition percent = 45%	Kamat et al. 1992
C. decorticatum	Extract Extract (major components are triglycerides and fatty acids)	HSV-1	IC_{50} = 1.0 μg mL^{-1} Inhibition percent = 98% Inhibition percent = 99.9%	Santos et al. 1999 Lee et al. 2004 Soares et al. 2012
C. fragile	Methanol extract	HSV SINV Poliovirus	Min. antiviral conc. = 8 μg mL^{-1} Min. antiviral conc. = 4 μg mL^{-1} Min. antiviral conc. = 100 μg mL^{-1}	Hudson et al. 1998
C. fragile	Enzymatic hydrolysis extract	HSV-1	IC_{50} = 36–52 μg mL^{-1}	Kulshreshtha et al. 2015
C. fragile subsp. *fragile*	Extract Extract Sulfated polysaccharides	HSV-1	Inhibition at 50 μg mL^{-1} Minimum inhibitory concentration at 8 μg mL^{-1} (under light) and virucidal IC_{50} = 0.86 μg mL^{-1}	Kim et al. 1997 Hudson et al. 1998 Lee et al. 2004
C. latum	Sulfated polysaccharides	HSV-1	IC_{50} = 0.38 μg mL^{-1}	Lee et al. 2004 Wang et al. 2014
C. spongiosum	Extract	HSV-1 HSV-2	Inhibition percent = 55.3% Inhibition percent = 55.3%	Soares et al. 2012
Dictyosphaeria cavernosa	Dichloromethane and methanol crude extracts	DENV	IC_{50} = 31.25 μg mL^{-1}	Koishi et al. 2012
Flabellia petiolata	Crude organic extracts	HSV-1 VSV	Inhibition zone to 1 mm	Ballesteros et al. 1992
Gayralia oxysperma	Sulfated heterorhamnan	HSV-1 HSV-2	IC_{50} = 0.27–0.30 μg mL^{-1} IC_{50} = 0.036–0.054 μg mL^{-1}	Cassolato et al. 2008
Halimeda opuntia	Aqueous extract	HSV-1 HSV-2	Inhibition percent = 73.1% Inhibition percent = 68.4%	Soares et al. 2012
H. tuna	Diterpene aldehyde (Halitunal) Crude organic extracts Aqueous extract	Murine coronavirus HSV-1 and VSV HSV-1 HSV-2	IC_{50} = 020 μg mL^{-1} Inhibition zone to 1 mm Inhibition percent = 84.1% Inhibition percent = 82.2%	Koehn et al. 1991 Ballesteros et al. 1992 Soares et al. 2012

Table 5.1 contd....

Table 5.1 contd....

Species	Tested product, fraction or extract	Virus	Observed results	References
Monostroma grevillei	Methanolic extract	HIV-1 *In vitro*	Inhibition of HIV-1 (a) integrase, (b) reverse transcriptase a: 200 mg mL^{-1}, b: ≤ 35%	Ahn et al. 2002
M. nitidum	Sulfated polysaccharides	HSV-1	IC_{50} = 0.4–3.7 μg mL^{-1}	Lee et al. 2004 Wang et al. 2014
Palmophyllum crassum	Crude organic extract	HSV-1 and VSV	Inhibition zone to 1 mm	Ballesteros et al. 1992
Penicillus capitatus	Extract (major components are triglycerides and fatty acids)	HSV-1 HSV-2	Inhibition percent = 93.0% Inhibition percent = 96%	Soares et al. 2012
Udotea flabellum	Aqueous extract	HSV-1 HSV-2	Inhibition percent = 90% Inhibition percent = 75%	Soares et al. 2012
Ulva australis (as *Ulva pertusa*)	Extract Extract	HSV-1	Inhibition at 100 μg mL^{-1} Minimum inhibitory concentration at 4 μg mL^{-1} (under light) and virucidal	Kim et al. 1997 Hudson et al. 1998
U. clathrata	Ethanol extract Sulfated polysaccharide (ulvan)	VACV NDV	Tested IC_{50} = 0.1 μg mL^{-1}	Lakshmi et al. 2006 Aguilar-Briseño et al. 2015
U. compressa (formerly *Enteromorpha compressa*)	Sulfated polysaccharides	HSV-1	IC_{50} = 49–58 μg mL^{-1}	Wang et al. 2014
U. flexuosa (formerly *Enteromorpha flexuosa*)	Phenolic compound 2-(2-hydroxyphenoxy)-1-phenylethanol	WSSV	Effectively controlled multiplication = 95%	Velmurugan et al. 2015
U. intestinalis	Sulfated polysaccharide (ulvan)	MeV	IC_{50} = 3.6 μg mL^{-1}	Morán-Santibañez et al. 2016
U. lactuca	Extract Ethanol extract Polysaccharides (heteroglycuronans)	EMCV NDV IAV(H1N1)	50–74% activity 0.1 mL = 2.7 log 10 EID_{50} reduction in virus titer Inhibition percent = 20–50%	Premnathan et al. 1992 Ibraheem et al. 2012 Jiao et al. 2012
U. lactuca	Sulfated polysaccharide extracts	JEV	Can inhibit JEV infection	Chiu et al. 2012b
U. linza (formerly *Enteromorpha linza*)	Extract Methanol extract	HSV-1 SINV	Inhibition at 200 μg mL^{-1} Min. antiviral conc. = 8 μg mL^{-1}, (under light) and virucidal	Kim et al. 1997 Hudson et al. 1998
U. linza (as *U. fasciata*)	2-N-palmitoyl-4,5-dihydro-1,3,4,5 tetrahydoxy sphingosine Methanol extract	SFV	Inhibition percent = 80% Inhibition percent = 75%	Mithlesh et al. 1988 Kamat et al. 1992

Table 5.1 contd....

Table 5.1 contd....

Species	Tested product, fraction or extract	Virus	Observed results	References
U. linza (as *U. fasciata*)	Extract Extract (major components are triglycerides and fatty acids) Sulfolipids	HSV-1 *In vitro*	Inhibition percent = 92.9% Inhibition percent = 99.9% Concentrations = 10.0–20.0 μg mL^{-1} Inhibition = 18.75–46.87%	Santos et al. 1999 Soares et al. 2012 El Baz et al. 2013
U. linza (as *U. fasciata*)	Extraction methods: Maceration Decoction Maceration of the decoction	HMPV	Inhibition at: 25 μg mL^{-1} 100 μg mL^{-1} 12.5 μg mL^{-1}	Mendes et al. 2010
U. rigida	Water extract	FluV	Minimum inhibitory concentration at 0.7 mg mL^{-1}, inhibition percent > 9%	Serkedjieva et al. 2000
U. rigida (as *U. armoricana*)	Enzyme-assisted extracts and polysaccharide fractions (ulvans)	HSV-1	IC_{50} = 320.9–373.0 μg mL^{-1}	Hardouin et al. 2016
Valonia utricularis	Crude organic extract	HSV-1 VSV	Inhibition zone to 1 mm	Ballesteros et al. 1992
Phaeophyceae (Brown seaweed)				
Adenocystis utricularis	Galactofucan	HSV-1 and 2 HIV-1	IC_{50} = 0.28 μg mL^{-1} IC_{50} = 0.6 μg mL^{-1}	Ponce et al. 2003 Trinchero et al. 2009
Analipus japonicus	Extract	HSV-1	Inhibition at 25 μg mL^{-1}	Kim et al. 1997
Ascophyllum nodosum	Polysaccharides (sulfated fucans)	IAV (H1N1)	IC_{50} = 100.7–191.4 μg mL^{-1} Inhibition percent = 52.7–69.3%	Jiao et al. 2012
Canistrocarpus cervicornis (formerly *Dictyota cervicornis*)	Diterpenes Dichloromethane and methanol crude extracts	HSV-1 DENV	Inhibition percent = 90.0% Inhibition percent = 99.0% IC_{50} = 31.25 μg mL^{-1}	Vallim et al. 2010 Koishi et al. 2012
Canistrocarpus cervicornis	Extract ointment	HSV-1 *In vivo* (BALB/c mice)	2% or 0.4 mg cm^{-2} $dose^{-1}$	De Souza-Barros et al. 2017
Cladosiphon okamuranus	Fucoidan Fucoidan Sulfated polysaccharide	DENV-2 NDV NDV	IC_{50} = 4.7 μg mL^{-1} IC_{50} = 0.75 μg mL^{-1} IC_{50} = 0.01 μg mL^{-1}	Hidari et al. 2008 Elizondo-Gonzalez et al. 2012 Aguilar-Briseño et al. 2015
C. okamuranus	Fucoidan	CDV *In vitro*	Fucoidan actively inhibited CDV replication in Vero cells at a 50% inhibitory concentration (IC_{50}) of 0.1 μg mL^{-1}. The derived selectivity index (SI_{50}) was > 20,000	Trejo-Avila et al. 2014

Table 5.1 contd....

Table 5.1 contd....

Species	Tested product, fraction or extract	Virus	Observed results	References
Colpomenia bullosa	Extract Ethanol extract	HSV-1 SINV	Inhibition at 100 μg mL^{-1} Minimum inhibitory concentration at 8 μg mL^{-1} (under light) and virucidal	Kim et al. 1997 Hudson et al. 1998
C. sinuosa	Extract	HSV-1	IC_{50} = 22.1 μg mL^{-1}	Wang et al. 2008
Cystoseira barbata	Lipophilic extract Water extract	FluV	Minimum inhibitory concentration at 5 mg mL^{-1}, inhibition percent = 50% Minimum inhibitory concentration at 1–2 mg mL^{-1}, inhibition percent = 90%	Serkedjieva et al. 2000
C. brachycarpa (as *C. balearica*)	Crude organic extracts	HSV-1 VSV	Inhibition zone to 1–2 mm	Ballesteros et al. 1992
C. compressa	Ethanol extract	NDV	0.1 mL = 2.9 log 10 EID_{50} reduction in virus titer	Ibraheem et al. 2012
C. crinita	Water extract	FluV HSV-1	Minimum inhibitory concentration at .25 mg mL^{-1}, inhibition percent = 50% IC_{50} = 0.3 mg mL^{-1}	Serkedjieva et al. 2000 Kamenarska et al. 2009
Dictyopteris delicatula	Dichloromethane and methanol crude extracts Aqueous extract	DENV HSV-1 HSV-2	IC_{50} = 125 μg mL^{-1} Inhibition percent = 82.2% Inhibition percent = 77.6%	Koishi et al. 2012 Soares et al. 2012
Dictyota dichotoma	Extract Sulfated fucogalactans	HSV-1 HSV-1 CVB3	IC_{50} = 24.3 μg mL^{-1} IC_{50} = 12.5–15.6 μg mL^{-1} IC_{50} = 12.5 μg mL^{-1}	Wang et al. 2008 Rabanal et al. 2014
D. dichotoma var. *intricata*	Crude organic extracts	HSV-1 VSV	Inhibition zone to 1 mm	Ballesteros et al. 1992
D. fastigiata (formerly *Dilophus fasciculatus*)	Sulfolipids	HSV-1 *In vitro*	IC_{50} = 18.75–70.2 μg mL^{-1}	El Baz et al. 2013
D. friabilis (formerly *D. pfaffii*)	Dolabelladienetriol Diterpens	HSV-1 BVDV	IC_{50} = 1.2 μM and may inhibit early events of virus replication IC_{50} = 2–2.3 μM	Cirne-Santos et al. 2008 Abrantes et al. 2010 Pinto et al. 2012
D. friabilis	Dolabelladienetriol	HIV-1	Inhibition percent = 20–99% IC_{50} = 0.15–14.4 μM	Stephens et al. 2017
D. menstrualis	Diterpenes Aqueous extract	HSV-1 HIV-1	IC_{50} = 1.6 μM and may inhibit early events of virus replication IC_{50} = 10 and 35 μM Inhibition percent = 20.6%	Abrantes et al. 2010 Pereira et al. 2004 Saores et al. 2012
D. menstrualis	Diterpenes	BVDV	IC_{50} = 2.8 μM	Pinto et al. 2012
D. mertensii	Sulfated fucans Dichloromethane and methanol crude extracts	HIV-1 DENV	Reverse transcriptase inhibition IC_{50} = 31.25 μg mL^{-1}	Queiroz et al. 2008 Koishi et al. 2012

Table 5.1 contd....

Table 5.1 contd....

Species	Tested product, fraction or extract	Virus	Observed results	References
Ecklonia arborea (formerly *Eisenia arborea*)	Sulfated polysaccharide (fucoidan and alginic acid)	MeV	IC_{50} = 0.275 μg mL^{-1}	Morán-Santibañez et al. 2016
E. bicyclis (formerly *Eisenia bicyclis*)	Methanolic extract Dieckol Phlorofucofuroeckol-A	FCV MNV	IC_{50} = 80 μg mL^{-1} IC_{50} = 0.9 μM	Choi et al. 2014 Eom et al. 2015
E. cava	Methanolic extract	HIV-1 *In vitro*	Inhibition of HIV-1 (a) integrase, (b) reverse transcriptase a: 200 mg mL^{-1}, b: 25 mg mL^{-1}	Ahn et al. 2002
E. cava	Phlorotannins Methanolic extract	HIV-1 FCV	IC_{50} = 0.51–81.5 μM IC_{50} = 1.07 μM IC_{50} = 210 μg mL^{-1}	Ahn et al. 2004 Artan et al. 2008 Choi et al. 2014
Ecklonia stolonifera	Methanolic extract	FCV	IC_{50} = 340 μg mL^{-1}	Choi et al. 2014
Fucus distichus subsp. *evanescens*	Fucoidan	HTNV	Reduces the number of infected cells	Pavliga et al. 2016
F. vesiculosus	Fucoidan Sulfated fucans	HIV-1	Inhibit HIV *in vitro* and was synergistic with AZT IC_{50} = 50 μg mL^{-1} Reverse transcriptase inhibition	Sugawara et al. 1989 Moen and Clark 1993 Queiroz et al. 2008
F. vesiculosus	Polysaccharides, fucoidan, polyphenols	HIV-1 *In vitro*	HIV-induced syncytium formation and reverse transcriptase inhibition	Béress et al. 1993
F. vesiculosus	Fucoidans	IAV (H1N1) HSV-1	IC_{50} = 74.8–180.5 μg mL^{-1}, inhibition percent = 47.6–77.9% IC_{50} = 2.41–5.69 μg mL^{-1}	Jiao et al. 2012 Zayed et al. 2016
Hydroclathrus clathratus	Extract and fractions Sulfated polysaccharide	HSV-1 HSV-1 RVFV	IC_{50} = 22.1 μg mL^{-1}, IC_{50} = 1.6 to 6.5 μg mL^{-1} IC_{50} = 100.5 μg mL^{-1} IC_{50} = 95.2 μg mL^{-1}	Wang et al. 2008 Gomaa and Elshoubaky 2015
H. clathratus	Ethanol extract	NDV	0.1 mL = 2.7 log 10 EID_{50} reduction in virus titer	Ibraheem et al. 2012
Ishige okamurae	Ethanol extract Phlorotannins	SINV HIV-1	Minimum inhibitory concentration at 8 μg mL^{-1} (under light) and virucidal IC_{50} = 9.1–25.2 μM	Hudson et al. 1998 Ahn et al. 2006
I. okamurae	Methanolic extract	HIV-1 *In vitro*	Inhibition of HIV-1 (a) integrase, (b) reverse transcriptase a: 200 mg mL^{-1}, b: 25 mg mL^{-1}	Ahn et al. 2002
Laminaria abyssalis	Extract	HSV-1	Inhibition percent = 100%	Santos et al. 1999
Leathesia marina (formerly)	Sulfated fucan	HSV-1	IC_{50} = 0.7 μg mL^{-1}	Feldman et al. 1999
Lobophora variegata (formerly Zonaria *variegata*)	Extract	HBV EMCV NDV SFV	50–74% activity 50–74% activity 1–24% activity 25–49% activity	Premnathan et al. 1992

Table 5.1 contd....

Table 5.1 contd....

Species	Tested product, fraction or extract	Virus	Observed results	References
L. variegata	Sulfated fucans Extract Aqueous extract	HIV-1 HSV-1 HSV-1	Reverse transcriptase inhibition Inhibition percent = 92%, IC_{50} = 18.5 µg mL^{-1} Inhibition percent = 92%	Queiroz et al. 2008 Moreira et al. 2011 Soares et al. 2012
L. variegata	Dichloromethane and methanol crude extracts	DENV	IC_{50} = 62.5 µg mL^{-1}	Koishi et al. 2012
Macrocystis pyrifera	Sulfated polysaccharide	MeV	IC_{50} = 1.00 µg mL^{-1}	Morán-Santibañez et al. 2016
Marginariella boryana (vegetative thalli)	Sulfated-fucan extracts	HSV-1	IC_{50} = 7.57 µg mL^{-1}	Wozniak et al. 2015
Myelophycus simplex	Methanolic extract	HIV-1 *In vitro*	Inhibition of HIV-1 (a) integrase, (b) reverse transcriptase a: 200 mg mL^{-1}, b: ≤ 35%	Ahn et al. 2002
Nereocystis luetkeana	Extract	HSV-1	Inhibition at 200 µg mL^{-1}	Kim et al. 1997
Padina australis	Extract	HSV-1	IC_{50} = 58.9 µg mL^{-1}	Wang et al. 2008
P. gymnospora	Extract Aqueous extract	NDV SFV VACV HSV-1 HSV-2	1–24% activity 25–49% activity 25–49% activity IC_{50} = 58.9 µg mL^{-1}, Inhibition percent = 99% Inhibition percent = 85.9% Inhibition percent = 43.8%	Premnathan et al. 1992 Santos et al. 1999 Soares et al. 2012
P. gymnospora (as *P. crassa*)	Methanolic extract	HIV-1 *In vitro*	Inhibition of HIV-1 (a) integrase, (b) reverse transcriptase a: 200 mg mL^{-1}, b: ≤ 35%	Ahn et al. 2002
P. gymnospora	Dichloromethane and methanol crude extracts	DENV	IC_{50} = 125 µg mL^{-1}	Koishi et al. 2012
P. pavonica	Ethanol extract Sulfated polysaccharide	NDV HSV HAV	0.1 mL = 2.8 log 10 EID_{50} reduction in virus titer Inhibition = 72.3% at 20 µg mL^{-1} Inhibition = 73.3% at 20 µg mL^{-1}	Ibraheem et al. 2012 Mohamed and Agili 2013
P. tetrastromatica	Extract	HBV NDV VACV	25–49% activity 1–24% activity 25–49% activity	Premnathan et al. 1992
P. tetrastromatica	Fucoidan derivatives	HSV-1 HSV-2	IC_{50} = 0.74–1.05 µg mL^{-1} IC_{50} = 0.30–0.39 µg mL^{-1}	Karmakar et al. 2010
Papenfussiella lutea	Sulfated-fucan extracts	HSV-1	IC_{50} = 0.98 µg mL^{-1}	Wozniak et al. 2015
Polycladia indica (formerly *Cystoseira indica*)	Sulfated fucan	HSV-1	IC_{50} = 2.8 µg mL^{-1}	Mandal et al. 2007

Table 5.1 contd....

Table 5.1 contd....

Species	Tested product, fraction or extract	Virus	Observed results	References
P. myrica (formerly *Cystoseira myrica*)	Water extract Sulfated polysaccharide	HSV-1 HSV HAV	IC_{50} = 99–125 μg mL^{-1} Inhibition = 62.2% at 20 μg mL^{-1} Inhibition = 66.15% at 20 μg mL^{-1}	Zandi et al. 2007 Mohamed et al. 2015
P. myrica (formerly *Cystoseira myrica*)	Ethanol extract	NDV	0.1 mL = 2.3 log 10 EID_{50} reduction in virus titer	Ibraheem et al. 2012
Postelsia palmaeformis	Extract	HSV-1	Inhibition at 200 μg mL^{-1}	Kim et al. 1997
Punctaria latifolia	Butanol extract Chloroform extract Water extract	HSV-1	IC_{50} = 0.07 mg mL^{-1} IC_{50} = 0.02 mg mL^{-1} IC_{50} = 0.75 mg mL^{-1}	Kamenarska et al. 2009
Saccharina cichorioides (formerly *Laminaria cichorioides*)	Fucoidan	HTNV	Reduces the number of infected cells	Pavliga et al. 2016
S. japonica	Sulfated polymannuro-guluronate (SPMG) Sulfated fucans Methanolic extract	HIV-1 FCV	Inhibition of infection IC_{50} = 100 μg mL^{-1}	Lu et al. 2007 Hui et al. 2008 Choi et al. 2014
S. japonica	Fucoidan	HTNV	Reduces the number of infected cells	Pavliga et al. 2016
Saccharina sculpera (as *Kjellmaniella crassifolia*)	Fucoidan	IAV *In vitro*	IC_{50} = 34.4 μg mL^{-1} CC_{50} = 2752.6 μg mL^{-1}	Wang et al. 2017
Sargassum confusum	Methanolic extract	HIV-1 *In vitro*	Inhibition of HIV-1 (a) integrase, (b) reverse transcriptase a: 200 mg mL^{-1}, b: 10, 25, 100 mg mL^{-1}	Ahn et al. 2002
S. cymosum	Extract (major compounds are phenolic compounds)	HSV-1 HSV-2	Inhibition percent = 98.2% Inhibition percent = 90%	Soares et al. 2012
S. fulvellum	Methanolic extract	FCV	IC_{50} = 100 μg mL^{-1}	Choi et al. 2014
S. hemiphyllum	Methanol extracts	*In vitro*	Used human cell lines: HDL, WI-38, HL-60, MG63, MOLT-4 and WIL-2NS Promote production of interferon β (IFN-β) in MG-63 cells	Nakano et al. 1997
S. hemiphyllum	Sulfated fucan	HSV-1 *In vitro*	Selectivity index = 11000 and 7100 (depending on the time of addition)	Preeprame et al. 2001
S. hemiphyllum	Methanolic extract	HIV-1 *In vitro*	Inhibition of HIV-1 (a) integrase, (b) reverse transcriptase a: 200 mg mL^{-1}, b: 100 mg mL^{-1}	Ahn et al. 2002
S. hemiphyllum	Extract	Aujeszky's disease (AD) virus *In vivo*	Antiviral activity by modulating host immunodefense system	Nakano and Kamei 2005

Table 5.1 contd....

Table 5.1 contd....

Species	Tested product, fraction or extract	Virus	Observed results	References
S. hemiphyllum var. *chinense*	Powder and extract	WSSV	Enhanced the immunity and resistance	Huynh et al. 2011
S. horneri	Sulfated fucan	CMV, HIV-1, HSV-1 *In vitro*	IC_{50} = 1 mL^{-1}	Hoshino et al. 1998
S. horneri	Extract	HSV-1 *In vitro*	IC_{50} = 19.1 μg mL^{-1}	Wang et al. 2008
S. ilicifolium	Aqueous extract	HSV	Inhibition of infection	Thenmozhi et al. 2013
S. johnstonii	Extract	HBV VACV	50–74% activity 25–49% activity	Premnathan et al. 1992
S. latifolium	Sulfated polysaccharide fractions	HSV-1 HAV	Inhibition percent = 25 to 81% Inhibition percent = 28 to 63%	Mohsen et al. 2007
S. mcclurei	Fuidan	HIV-1	IC_{50} = 0.33–0.7 μg mL^{-1}	Thuy et al. 2015
S. micracanthum	Plastoquinones	HSV-1 HSV-2 HCMV MMR MeV Adeno virus FluV Poliovirus CMV	IC_{50} = 1.8–24 μM	Iwashima et al. 2005
S. muticum	Methanolic extract	HIV-1 *In vitro*	Inhibition of HIV-1 (a) integrase, (b) reverse transcriptase a: 200 mg mL^{-1} (45–55%), b: ≤ 35%	Ahn et al. 2002
S. muticum	Enzymatic extracts	HSV-1 *In vitro*	EC_{50} = 225.1–430.1 μg mL^{-1}	Puspita et al. 2017
S. naozhouense	Alginates and fucoidan	HSV-1	IC_{50} = 8.92 μg mL^{-1}	Peng et al. 2013
S. patens	Sulfated polysaccharide	HSV-1	IC_{50} = 5.5 μg mL^{-1}	Zhu et al. 2006
S. polyceratium	Extract	HSV-1	Inhibition percent = 86.8%	Soares et al. 2012
S. polycystum (formerly *S. myriocystum*)	Ethanol extract	VACV	Inhibition percent = 75%	Kamat et al. 1992
S. polycystum	Fucoidan	HIV-1	IC_{50} = 0.33–0.7 μg mL^{-1}	Thuy et al. 2015
S. ringgoldianum	Methanolic extract	HIV-1 *In vitro*	Inhibition of HIV-1 (a) integrase, (b) reverse transcriptase a: ≤ 35%, b: 100 mg mL^{-1} (55–65%)	Ahn et al. 2002
S. sagamianum	Ethanol extract	SINV	Minimum inhibitory concentration at 47 μg mL^{-1} (under light) and virucidal	Hudson et al. 1998
S. swartzii (formerly *S. wightii*)	Ethanol extract	VACV	Inhibition percent = 50%	Kamat et al. 1992

Table 5.1 contd....

Table 5.1 contd....

Species	Tested product, fraction or extract	Virus	Observed results	References
S. swartzii (formerly *S. wightii*)	Extract Fucoidan	HBV NDV SFC VACV HIV-1	25–49% activity 1–24% activity 50–74% activity 50–74% activity IC_{50} = 1.56–6.25 μg mL^{-1}	Premnathan et al. 1992 Dinesha et al. 2013
S. tenerrimum	Extract	EMCV NDV SFC VACV	50–74% activity 50–74% activity 1–24% activity 25–49% activity	Premnathan et al. 1992
S. thunbergii	Extract	HSV-1	Inhibition at 200 μg mL^{-1}	Kim et al. 1997
S. thunbergii	Methanolic extract	HIV-1 *In vitro*	Inhibition of HIV-1 (a) integrase, (b) reverse transcriptase a: 200 mg mL^{-1} (45–55%), b: ≤ 35%	Ahn et al. 2002
S. trichophyllum	Sulfated polysaccharide: fucose (79.1 mol%), galactose (19.9 mol%), and sulfate content (25.5%)	HSV-2	IC_{50} = 18–410 μg mL^{-1}	Lee et al. 2011
S. vulgare	Water extract	HTLV-1 *In vitro*	78.8% syncytium inhibition at 5% concentration	Romanos et al. 2002
S. vulgare	Extract Extract Glycolipids	HSV-1 HSV-2	Inhibition percent = 99.8% Inhibition percent = 76% Inhibition percent = 39.7% Inhibition percent = 96%	Santos et al. 1999 Soares et al. 2012 Plouguerné et al. 2013
S. vulgare	Dichloromethane and methanol crude extracts	DENV	IC_{50} = 62.5 μg mL^{-1}	Koishi et al. 2012
S. vulgare var. *nanum*	Dichloromethane and methanol crude extracts	DENV	IC_{50} = 62.5 μg mL^{-1}	Koishi et al. 2012
Scytosiphon lomentaria	Extract Ethanol extract Water extract	HSV-1 SINV HSV-1	Inhibition at 100 μg mL^{-1} Minimum inhibitory concentration at 8 μg mL^{-1} (under light) and virucidal IC_{50} = 0.4 mg mL^{-1}	Kim et al. 1997 Hudson et al. 1998 Kamenarska et al. 2009
Scytothamnus australis	Sulfated-fucan extracts	HSV-1	IC_{50} > 20 μg mL^{-1}	Wozniak et al. 2015
Silvetia compressa (formerly *Pelvetia compressa*)	Sulfated polysaccharide	MeV	IC_{50} = 1.00 μg mL^{-1}	Morán-Santibañez et al. 2016
Spatoglossum schroederi	Sulfated fucan	HIV-1	Reverse transcriptase inhibition	Queiroz et al. 2008
Splachnidium rugosum	Sulfated fucan	HSV-1 HSV-2	IC_{50} = 7.9–29 μg mL^{-1} IC_{50} = 5.3–6.6 μg mL^{-1}	Harden et al. 2009
Sphacelaria indica	Xylogalactofucan	HSV-1	IC_{50} = 1.3 μg mL^{-1}	Bandyopadhyay et al. 2011
Splachnidium rugosum	Sulfated-fucan extracts	HSV-1	IC_{50} = 1.18 μg mL^{-1}	Wozniak et al. 2015

Table 5.1 contd....

Table 5.1 contd....

Species	Tested product, fraction or extract	Virus	Observed results	References
Stilophora tenella	Butanol extract Water extract	HSV-1	$IC_{50} = 0.06$ mg mL^{-1} $IC_{50} = 0.5$ mg mL^{-1}	Kamenarska et al. 2009
Stoechospermum polypodioides (formerly *Stoechospermum marginatum*)	Sulfated fucan	HSV-1	$IC_{50} = 3.6$ µg mL^{-1}	Adhikari et al. 2006
Taonia atomaria	Sulfolipids	HSV-1 *In vitro*	$IC_{50} = 18.75–70.2$ µg mL^{-1}	El Baz et al. 2013
Stypopodium zonale	Extract (meroditerpenes) Aqueous extract	HSV-1 HIV-1 HMPV HSV-1 HSV-2	Inhibition percent = 96.8%, $IC_{50} = 1.28$ µM Inhibition concentration at 2 µg mL^{-1} $IC_{50} = 1.34$ µM, $IC_{50} = 2.38$ µM Inhibition percent = 96.8% Inhibition percent = 95.8%	Soares et al. 2007 Mendes et al. 2011 Soares et al. 2012
Turbinaria ornata	Ethanol extract Fuidan	NDV HIV-1	0.1 mL = 1 log 10 EID_{50} reduction in virus titer $IC_{50} = 0.33–0.7$ µg mL^{-1}	Ibraheem et al. 2012 Thuy et al. 2015
Undaria pinnatifida	Extract Ethanol extract Fucoidan Galactofucan sulfate Galactofucan sulfate extract and its fractions Sulfated-fucan extracts	HSV-1	Inhibition at 100 µg mL^{-1} Minimum inhibitory concentration at 4 µg mL^{-1} (under light) and virucidal $IC_{50} = 2.5$ µg mL^{-1} $IC_{50} = 1$ to 128 µg mL^{-1} $IC_{50} = 1.1$ to 4.6 µg mL^{-1} $IC_{50} = 0.77$ µg mL^{-1}	Kim et al. 1997 Hudson et al. 1998 Lee et al. 2004 Thompson and Dragar 2004 Hemmingson et al. 2006 Wozniak et al. 2015
U. pinnatifida	Fucoidan	IAV *In vitro*	Influenza A infection by different subtypes (H5N3 and H7N2) was modulated by oral ingestion of a low molecular weight *U. pinnatifida* fucoidan fraction (9 kDa) in mouse models	Hayashi et al. 2013
U. pinnatifida	Mekabu fucoidan	IAV	Favorable effects in the control of not only seasonal influenza virus (FluV) but also avian influenza virus infections (IAV)	Synytsya et al. 2014
Zanardinia typus (formerly *Zanardinia prototypus*)	Crude organic extracts Chloroform extract	HSV-1 VSV HSV-1	Inhibition zone to 1 mm $IC_{50} = 0.05$ mg mL^{-1}	Ballesteros et al. 1992 Kamenarska et al. 2009
Rhodophyta (Red seaweed)				
Acanthophora spicifera	Extract Ethanol extract	EMCV HBV NDV NDV	75–100% activity 25–49% activity 50–74% activity 0.1 mL = 1.6 log 10 EID_{50} reduction in virus titer	Premnathan et al. 1992 Ibraheem et al. 2012

Table 5.1 contd....

Table 5.1 contd....

Species	Tested product, fraction or extract	Virus	Observed results	References
A. spicifera	Sulfated agaran Sulfated polysaccharide	HSV-1 HSV-1 RVFV	IC_{50} = 0.6 to > 50 μg mL^{-1} IC_{50} = 80.5 μg mL^{-1} IC_{50} = 75.8 μg mL^{-1}	Duarte et al. 2004 Gomaa and Elshoubaky 2015
A. spicifera	*In vitro* Dichloromethane, ethyl acetate, acetone, methanol extracts	HIV-1	CC_{50} values of 31, 45, 38, and 179 μg mL^{-1}, respectively Ethyl acetate extract inhibited approximately 60% of the viral load IC_{50} value of 33.17 μg mL^{-1} for the reverse transcriptase	Nogueira et al. 2016
Acrosorium sp.	Methanolic extract	HIV-1 *In vitro*	Inhibition of HIV-1 (a) integrase, (b) reverse transcriptase a: 200 mg mL^{-1} (45–55%), b: ≤ 35%	Ahn et al. 2002
Agardhiella subulata (as *Agardhiella tenera*)	Sulfated galactan	HIV-1 HIV-2	IC_{50} = 0.6 μg mL^{-1} IC_{50} = 0.5 μg mL^{-1}	Witvrouw et al. 1994
Alsidium corallinum	Methanol, chloroform-methanol, dichlomethane, and water extracts	HSV-1	IC_{50} = 36.6–71.5 μg mL^{-1}	Bouhlal et al. 2010
Asparagopsis armata	Sulfated galactans Extract	HIV-1 HSV-1	Inhibition of an early step of infection IC_{50}< 2.5 μg mL^{-1}	Haslin et al. 2001 Bouhlal et al. 2010
Bangia fuscopurpurea	Water extract	HSV-1	IC_{50} = 0.18 mg mL^{-1}	Kamenarska et al. 2009
Boergeseniella thuyoides	Extract Sulfated polysaccharides	HSV-1	IC_{50} = 12.6 μg mL^{-1} IC_{50} = 17.2 μg mL^{-1}	Bouhlal et al. 2010 Bouhlal et al. 2011
Bossiella sp.	Methanol extract	HIV-1	EC_{50} = 49.2 μg mL^{-1}	Ahn et al. 2002
Bostrychia montagnei	Sulfated polysaccharides	HSV-1	IC_{50} = 12.9 to more than 50 μg mL^{-1}	Duarte et al. 2001
B. radicans	Extract	HSV-1	Inhibition percent = 86.5%	Soares et al. 2012
Callophyllis variegata	Galactans	HSV-1	IC_{50} = 0.16 to 1.55 μg mL^{-1}	Rodríguez et al. 2005
Ceramium virgatum (formerly *Ceramium rubrum*)	Water extract CH_3Cl extract	FluV	Minimum inhibitory concentration at 0.12–1 mg mL^{-1}, Inhibition percent > 99% Minimum inhibitory concentration at 0.05 mg mL^{-1}, Inhibition percent = 50%	Serkedjieva et al. 2000

Table 5.1 contd....

Table 5.1 contd....

Species	Tested product, fraction or extract	Virus	Observed results	References
C. virgatum (formerly *C. rubrum*)	Extract Extract	HSV-1	IC_{50} = 0.8 μg mL^{-1} (strain Kupka) IC_{50} = 0.5 μg mL^{-1} (strain KOS) IC_{50} = 12.4 μg mL^{-1}	Serkedjieva 2004 Bouhlal et al. 2010
Chondracanthus acicularis (formerly *Gigartina acicularis*)	λ-carrageenan Extract	HIV-2 HSV-2 HSV-1	IC_{50} = 1.9 μg mL^{-1} Inhibition percent = 92.4% Inhibition percent = 68.4%	Witvrouw and De Clercq 1997 Soares et al. 2012
C. acicularis (formerly *G. acicularis*)	κ and ι-carrageenan's	HSV-1 Poliovirus	IC_{50} = 0.00275 mg mL^{-1} IC_{50} = 0.047 mg mL^{-1}	Montanha et al. 2009
C. tenellus (formerly *Gigartina tenella*)	Sulfolipid	HIV-1	The inhibition was dose-dependent, and complete (more than 90%) inhibition of DNA polymerase α, DNA polymerase β and HIV-reverse transcriptase type 1 (HIV-RT) was observed at concentrations of 5, 10, and 30 μM, respectively	Ohta et al. 1998
Chondria armata	Methanol extract	SFV	Inhibition percent = 75%	Kamat et al. 1992
C. atropurpurea	Chondriamide A	HSV-2	Antiviral activity	Palermo et al. 1992
C. crassicaulis	Methanol extract	HIV-1	EC_{50} = 19.9 μg mL^{-1}	Ahn et al. 2002
Chondrus armatus	κ-carrageenan (low molecular weight derivatives) Carrageenans	TMV HTNV	Antiviral activity = 40–88% Reduces the number of infected cells	Kalitnik et al. 2013 Pavliga et al. 2016
C. crispus	Enzymatic hydrolysis extract	HSV-1	IC_{50} = 77.6–129.8 μg mL^{-1}	Kulshreshtha et al. 2015
C. ocellatus	Ethanol extract	Poliovirus	Min. antiviral conc. = 200 μg mL^{-1}	Hudson et al. 1998
Constantinea simplex	Extract	HSV-1 HSV-2	Prophylactic effect	Richards et al. 1978
Corallina officinalis	Water extract	FluV	Minimum inhibitory concentration at 2.5 mg mL^{-1}, Inhibition percent = 50%	Serkedjieva et al. 2000
C. panizzoi	Extract	HSV-1	Inhibition percent = 68.4%	Soares et al. 2012
C. pilulifera	Extract Extract	HSV-1	Inhibition at 100 μg mL^{-1} Minimum inhibitory concentration at 8 μg mL^{-1} (under light) and virucidal	Kim et al. 1997 Hudson et al. 1998
C. vancouveriensis	Extract	HSV-1	Inhibition at 200 μg mL^{-1}	Kim et al. 1997
Cryptonemia crenulata	Sulfated galactans (major components) DL-hybrid galactan	HSV-1 DENV-2 DENV-2	IC_{50} = 0.5 to 2.8 μg mL^{-1} IC_{50} ≈ 1 μg mL^{-1} IC_{50} = 0.8–16 μg mL^{-1}	Talarico et al. 2004 Talarico et al. 2005 Talarico et al. 2007b

Table 5.1 contd....

Table 5.1 contd....

Species	Tested product, fraction or extract	Virus	Observed results	References
Cryptopleura ramosa	Sulfated polysaccharides	HSV-1	IC_{50} = 1.6 to 4.2 μg mL^{-1}	Carlucci et al. 1997b
Cryptosiphonia woodii	Extract	HSV-1 HSV-2	Inhibition	Deig et al. 1974
Digenea simplex	Dichloromethane and methanol crude extracts	DENV	IC_{50} = 62.5 μg mL^{-1}	Koishi et al. 2012
Dumontia contorta	Sulfated polysaccharides	HSV DENV	Blocking of virus adsorption	Magill et al. 2009
D. simplex	Methanolic extract	HIV-1 *In vitro*	Inhibition of HIV-1 (a) integrase, (b) reverse transcriptase a: 200 mg mL^{-1} (45–55%), b: ≤ 35%	Ahn et al. 2002
Ellisolandia elongata (formerly *Corallina elongata*)	Crude organic extracts	HSV-1 VSV	Inhibition zone to 1 mm	Ballesteros et al. 1992
Euchema denticulatum	ι-carrageenan	HSV-1 Poliovirus	IC_{50} = 0.026 mg mL^{-1} IC_{50} = 0.115 mg mL^{-1}	Montanha et al. 2009
E. serra	Lectin	IAV IBV	IC_{50} = 12.4 nM IC_{50} = 20.4 nM	Sato et al. 2015 Liao et al. 2003
Farlowia mollis	Extract	HSV-1 HSV-2	Prophylactic effect	Deig et al. 1974 Richards et al. 1978
Furcellaria lumbricalis	Polysaccharides (carrageenan)	IAV (H1N1)	Inhibition percent = 20–50%	Jiao et al. 2012
Galaxaura rugosa (formerly *G. elongata*)	Ethanol extract	VACV NDV	Inhibition percent = 20% 0.1 mL = 2.4 log 10 EID_{50} reduction in virus titer	Kamat et al. 1992 Ibraheem et al. 2012
Ganonema farinosum (formerly *Liagora farinosa*)	Ethanol extract	NDV	0.1 mL = 1.6 log 10 EID_{50} reduction in virus titer	Ibraheem et al. 2012
Gelidiella acerosa	Extract	HBV VACV	50–74% activity 1–12% activity	Premnathan et al. 1992
Gelidium attenuatum	Methanol, chloroform-methanol, dichlomethane, and water extracts	HSV-1	IC_{50} = 23.8–36.9 μg mL^{-1}	Bouhlal et al. 2010
G. microdon (as *G. spinulosum*)	Methanol and dichloromethane extracts	HSV-1	IC_{50} = 40.1–197.0 μg mL^{-1}	Bouhlal et al. 2010
G. corneum (formerly *G. sesquipedale*)	Dichlomethane extract	HSV-1	IC_{50} = 71.6 μg mL^{-1}	Bouhlal et al. 2010
G. pulchellum	Methanol, chloroform-methanol, dichlomethane, and water extracts	HSV-1	IC_{50} = 15.0–100 μg mL^{-1}	Bouhlal et al. 2010

Table 5.1 contd....

Table 5.1 contd....

Species	Tested product, fraction or extract	Virus	Observed results	References
G. pusillum	Methanol, chloroform-methanol, dichlomethane, and water extracts	HSV-1	IC_{50} = 35.8 > 250 µg mL^{-1}	Bouhlal et al. 2010
G. spinosum (formerly *G. latifolium*)	Lipophilic extract Water extract Water extract Methanol extract	FluV HSV-1 HSV-1	Minimum inhibitory concentration at 2.5 mg mL^{-1}, Inhibition percent = 90% Minimum inhibitory concentration at 0.25 mg mL^{-1}, Inhibition percent = 50% IC_{50} = 1.9 mg mL^{-1} IC_{50} = 2.5 mg mL^{-1}	Serkedjieva et al. 2000 Kamenarska et al. 2009 Bouhlal et al. 2010
Gigartina pistillata	λ-carrageenan	HIV-2	IC_{50} = 2.9 µg mL^{-1}	Witvrouw and De Clercq 1997
G. skottsbergii	Extract	HSV-1	IC_{50} < 1 to 4.1 µg ml^{-1} IC_{50} = 0.4 to 3.3 µg mL^{-1} and virucidal	Carlucci et al. 1997a Carlucci et al. 1999
G. skottsbergii	λ-carrageenan	BoHV-1 SuHV-1	IC_{50} = 0.52 µg mL^{-1} IC_{50} = 10.42 µg mL^{-1}	Diogo et al. 2015
Gloiopeltis furcata	Methanolic extract	HIV-1 *In vitro*	Inhibition of HIV-1 (a) integrase, (b) reverse transcriptase a: 200 mg mL^{-1} (45–55%), b: ≤ 35%	Ahn et al. 2002
Gracilaria sp.	Dichloromethane and methanol crude extracts	DENV	IC_{50} = 125 µg mL^{-1}	Koishi et al. 2012
G. chilensis	Included in salmon diet	ISA	Antiviral activity: 25.55–68.02%	Lozano et al. 2016
Gracilaria corticata	Extract	EMCV HBV NDV SFV	25–49% activity 50–74% activity 50–74% activity 1–24% activity	Premnathan et al. 1992
G. corticata	Sulfated galactan	HSV-1 HSV-2	IC_{50} = 0.19 µg mL^{-1} IC_{50} = 0.24 µg mL^{-1}	Mazumder et al. 2002
G. domingensis	Aqueous extract	HSV-2	Inhibition percent = 43.8%	Soares et al. 2012
G. fisheri	Sulfated galactans	WSSV	IC_{50} = 62.25 µg mL^{-1}	Rudtanatip et al. 2014
G. foliifera	Extract	NDV VACV	25–49% activity 25–49% activity	Premnathan et al. 1992
G. pacifica	Methanol extract	SINV	IC_{50} = 200.0 µg mL^{-1}	Kamat et al. 1992
G. tenuistipitata	Aqueous extract	HCV	IC_{50} = 325 µg mL^{-1}	Chen et al. 2013
Grateloupia indica	Sulfated galactans	HSV-1	IC_{50} = 0.12 to 1.06 µg mL^{-1}	Chattopadhyay et al. 2007
G. filicina	Sulfated galactans	HIV-1	IC_{50} = 0.0006 µM	Wang et al. 2007
G. longifolia	Sulfated galactans	HIV-1	IC_{50} = 0.003–0.010 µg mL^{-1}	Wang et al. 2007

Table 5.1 contd....

Table 5.1 contd....

Species	Tested product, fraction or extract	Virus	Observed results	References
G. turuturu	Methanol extract	SINV	Min. antiviral conc. = 24 μg mL^{-1} (under light)	Hudson et al. 1998
Griffithsia sp.	Lectin (Griffithsin)	HIV-1	IC_{50} = 0.043–0.63 nM	Mori et al. 2005
Griffithsia sp.	Lectin (Griffithsin)	EBOV HCV HIV-1 HIV-2 SARS-CoV	Potential antiviral therapeutic agent against	Barton et al. 2014
Gymnogongrus griffithsiae	Sulfated galactans (major components) kappa/iota/nu carrageenan	HSV-1 DENV-2	IC_{50} = 1.0 to 5.6 μg mL^{-1} $IC_{50} \approx$ 1 μg mL^{-1}	Talarico et al. 2004 Talarico et al. 2005
G. griffithsiae	Sulfated polysaccharide	DENV-2	IC_{50} = 0.9 μg mL^{-1}	Talarico et al. 2005
G. torulosus	Sulfated polysaccharide (DL-galactan hybrids)	DENV-2 HSV-2	IC_{50} = 0.19–1.7 μg mL^{-1} IC_{50} = 0.6–16 μg mL^{-1}	Pujol et al. 2002
Halopithys incurva	Methanol, chloroform-methanol, dichlomethane, and water extracts	HSV-1	IC_{50} = 64.8–86.6 μg mL^{-1}	Bouhlal et al. 2010
Hypnea cervicornis	Sulfated polysaccharide	VACV	Inhibition percent = 40–100%	Kamat et al. 1992
H. japonica	Methanolic extract	HIV-1 *In vitro*	Inhibition of HIV-1 (a) integrase, (b) reverse transcriptase a: 200 mg mL^{-1} (45–55%), b: ≤ 35%	Ahn et al. 2002
H. musciformis	Sulfated polysaccharide	SFV VACV	Inhibition percent = 50–100% Inhibition percent = 50%	Kamat et al. 1992
H. musciformis	Methanol, chloroform-methanol, dichlomethane, and water extracts	HSV-1	IC_{50} = 32.94–100.18 μg mL^{-1}	Bouhlal et al. 2010
H. musciformis	Extract Extract Extract	HSV-1	Inhibition percent = 99.9% Inhibition percent = 57.3% IC_{50} = 23.5 μg mL^{-1} IC_{50} = 11.9 μg mL^{-1} (natural), IC_{50} = 0.67 μg mL^{-1} (phytohormone) and virucidal (both)	Santos et al. 1999 Soares et al. 2012 Bouhlal et al. 2010 Mendes et al. 2012
H. musciformis	Aqueous extract	HSV-2	Inhibition percent = 74.9%	Soares et al. 2012
H. musciformis	Dichloromethane and methanol crude extracts	DENV	IC_{50} = 31.25 μg mL^{-1}	Koishi et al. 2012
H. spinella	Carrageenans	HSV-1	Inhibition percent = 92%	Soares et al. 2012
Jania crassa	Extract	HSV-1 HSV-2	Inhibition percent = 43.8%	Soares et al. 2012

Table 5.1 contd....

Table 5.1 contd....

Species	Tested product, fraction or extract	Virus	Observed results	References
J. spectabilis (as *Cheilosporum spectabile*)	Extract	EMCV HBV NDV SFV VACV	75–100% activity 25–49% activity 50–74% activity 75–100% activity 1–12% activity	Premnathan et al. 1992
Kallymenia requienii	Crude organic extracts	HSV-1 VSV	Inhibition zone to 1 mm	Ballesteros et al. 1992
Kappaphycus alvarezii (formerly *Eucheuma cottonii*)	κ-carrageenan Lectin	HIV-1	IC_{50} = 12 μg mL^{-1} IC_{50} = 7.3–12.9 nM	Witvrouw and De Clercq 1997 Hirayama et al. 2016
K. alvarezii	κ-carrageenan (low molecular weight derivatives)	TMV	Antiviral activity = 28–85%	Kalitnik et al. 2013
Kappaphycus cottonii	κ-carrageenan	HSV-1 Poliovirus	IC_{50} = 0.068 mg mL^{-1} IC_{50} = 0.227 mg mL^{-1}	Montanha et al. 2009
Laurencia coronopus	Water extract	HSV-1	IC_{50} = 0.1 mg mL^{-1}	Kamenarska et al. 2009
L. dendroidea	Dichloromethane and methanol crude extracts Extract	DENV HSV-1 HSV-2	IC_{50}< 0.5 μg mL^{-1} Inhibition percent = 97.5% Inhibition percent = 43.8%	Koishi et al. 2012 Soares et al. 2012
Lithothamnion muelleri	Sulfated polysaccharides	HSV-1	IC_{50} = 7.39–125.79 μg mL^{-1}	Malagolia et al. 2014
Mazzaella parksii (as *Mazzaella cornucopiae*)	Extract	HSV-1	Inhibition at 200 μg mL^{-1}	Kim et al. 1997
Meristotheca gelidium (as *Meristiella gelidium*)	Carrageenans	HSV-2 DENV-2	IC_{50} = 0.04 to 0.11 μg mL^{-1} IC_{50} = 0.14 to 1.6 μg mL^{-1}	SF-Tischer et al. 2006
Mesophyllum expansum (as *Lithophyllum expansum*)	Crude organic extracts	HSV-1 VSV	Inhibition zone to 2–4 mm	Ballesteros et al. 1992
Nemalion elminthoides	Sulfated polysaccharides: mainly mannose	HSV-1 *In vitro*	EC_{50} = 5.4 μg mL^{-1}	Recalde et al. 2009
N. elminthoides	Sulfated xylomannans and modifications	HSV-1 HSV-2 DENV-2	IC_{50} = 189 to 26.3 μg mL^{-1} IC_{50} = 0.70 to 16.8 μg mL^{-1} IC_{50}< 1.25 to 16.1 μg mL^{-1}	Recalde et al. 2012
Neorhodomela aculeata	Extracts	HRV	IC_{50} = 15.5 to 18.27 μg mL^{-1}	Park et al. 2012b
Neorhodomela munita	Methanolic extract	HIV-1 *In vitro*	Inhibition of HIV-1 (a) integrase, (b) reverse transcriptase a: 200 mg mL^{-1} (45–55%), b: ≤ 35%	Ahn et al. 2002

Table 5.1 contd....

Table 5.1 contd....

Species	Tested product, fraction or extract	Virus	Observed results	References
Nothogenia fastigiata	Sulfated xylomannan	HSV-1 (F strain) HSV-1 (KOS strain) HSV-2 HCMV (AD169 strain) HCMV (Davis strain) Poliovirus IAV IBV HPIV-3 RSV Juanin virus TCRV SIV	IC_{50} = 0.6 μg mL^{-1} IC_{50} = 1.5 μg mL^{-1} IC_{50} = 2.5 μg mL^{-1} IC_{50} = 10 μg mL^{-1} IC_{50} = 2.8 μg mL^{-1} IC_{50}> 50 μg mL^{-1} IC_{50} = 0.2 μg mL^{-1} IC_{50} = 20 μg mL^{-1} IC_{50}> 100 μg mL^{-1} IC_{50} = 0.9 μg mL^{-1} IC_{50} = 10 μg mL^{-1} IC_{50} = 7.8 μg mL^{-1} IC_{50} = 0.4 μg mL^{-1}	Damonte et al. 1994
N. fastigiata	Sulfated polysaccharides	HSV-1	IC_{50} = 0.6 to more than 100 μg mL^{-1}	Kolender et al. 1997
Osmundaria obtusiloba (formerly *Vidalia obtusiloba*)	Extract Extract Extract Sulfoglicolipids	HSV-1	Inhibition percent = 55.3% Inhibition percent = 90% Inhibition percent = 82.2 to 99.9% Inhibition percent = 75%	Santos et al. 1999 Mattos et al. 2011 Soares et al. 2012 De Souza et al. 2012
Palisada perforata	Dichloromethane and methanol crude extracts	DENV	IC_{50} = 125 μg mL^{-1}	Koishi et al. 2012
P. perforata (formerly *Laurencia papillosa*)	Butanol extract Chloroform extract Sulfolipids	HSV-1 *In vitro*	IC_{50} = 0.06 mg mL^{-1} IC_{50} = 0.4 mg mL^{-1} IC_{50} = 18.75–70.2 μg mL^{-1}	Kamenarska et al. 2009 El Baz et al. 2013
Palmaria palmata	Polysaccharides (xylans)	IAV (H1N1)	Inhibition percent = 20–50%	Jiao et al. 2012
Peyssonelia sp.	Sesquiterpene hydroquinones, peyssonol A and peyssonol B	HIV-1 HIV-2	HIV-1 RT IC_{50} 34.5–38.7 μM HIV-2 RT IC_{50} 23.7–28.0 μM	Loya et al. 1995
Phyllophora crispa (as *Phyllophora nervosa*)	Lipophilic extract Water extract	FluV	Minimum inhibitory concentration at 0.25 mg mL^{-1}, Inhibition percent = 50% Minimum inhibitory concentration at 0.25–1.8 mg mL^{-1}, Inhibition percent > 99%	Serkedjieva et al. 2000
Plocamium brasiliense	Extract Extract	HSV-1 HSV-2 BoHV-5	Inhibition percent = 43.8% Inhibition percent = 77.6% Inhibition of attachment = 0.3–1.5 μg	Soares et al. 2012 Pinto et al. 2014

Table 5.1 contd....

Table 5.1 contd....

Species	Tested product, fraction or extract	Virus	Observed results	References
P. cartilagineum	Complex sulfated galactan Methanol, chloroform-methanol, and water extracts	HSV-1 HSV-2 HSV-1	IC_{50} = 5.4 µg mL^{-1} IC_{50} = 2.4 µg mL^{-1} IC_{50} = 91.9–100.04 µg mL^{-1}	Harden et al. 2009 Bouhlal et al. 2010
Polysiphonia denudata	Lipophilic extract Water extract	FluV	Minimum inhibitory concentration at 0.5–1.6 mg mL^{-1}, Inhibition percent = 90% Minimum inhibitory concentration at 0.03–0.25 mg mL^{-1}, Inhibition percent > 99%	Serkedjieva et al. 2000
P. denudata	Extract Butanol extract Water extract	HSV-1 HSV-2 HSV-1	IC_{50} = 8.7 to 47.7 mg mL^{-1} IC_{50} = 0.02 mg mL^{-1} IC_{50} = 0.75 mg mL^{-1}	Serkedjieva 2000b Kamenarska et al. 2009
Polyopes affinis (formerly *Carpopeltis affinis*)	Ethanol extract	SINV	Min. antiviral conc. = 6 µg mL^{-1} (under light)	Hudson et al. 1998
*Porphyridium purpureum** (formerly *P. cruentum*)	Sulfated soluble exopolysaccharide	ASFV	Inhibition percent = 6.1%	Fabregas et al. 1999
Portieria hornemannii (formerly *Chondrococcus hornemannii*)	Ethanol extract	VACV	Inhibition percent = 30%	Kamat et al. 1992
Pterocladiella capillacea (formerly *Pterocladia capillacea*)	Dichlomethane and water extracts	HSV-1	IC_{50} = 21.7–42.3 µg mL^{-1}	Bouhlal et al. 2010
P. capillacea	Extract Extract Sulfated galactan	HSV-1	IC_{50} = 3.2 to 6.1 µg mL^{-1} Inhibition percent = 99.9% Inhibition percent = 68.4%	Pujol et al. 1996 Santos et al. 1999 Soares et al. 2012
Pterosiphonia complanata	Extract	HSV-1	IC_{50} = 10.4 µg mL^{-1}	Bouhlal et al. 2010
Pyropia columbina	Included in salmon diet	ISA	Antiviral activity = 20.29–46.54%	Lozano et al. 2016
Sarcothalia atropurpurea (formerly *Gigartina atropurpurea*)	FG (κ-2 carrageenan) T (unusual λ-carrageenan)	HSV-1 HSV-2 HSV-1 HSV-2	IC_{50} = 36 µg mL^{-1} IC_{50} = 2.4 µg mL^{-1} IC_{50} = 1.5 µg mL^{-1} IC_{50} = 0.7 µg mL^{-1}	Harden et al. 2009
Schizymenia binderi	DL-hybrid sulfated galactan	HSV-1 HSV-1	IC_{50} = 0.8 µg mL^{-1} IC_{50} = 0.6 µg mL^{-1}	Matsuhiro et al. 2005
S. dubyi	Sulfated heteropolysaccharide	HSV-1, HSV-2, VSV and Pol *In vitro*	Syncytial formation was completely suppressed at 20 µg mL^{-1} and delayed at 10 µg mL^{-1} No toxicity was found at these concentrations	Bourgougnon et al. 1993

Table 5.1 contd....

Table 5.1 contd....

Species	Tested product, fraction or extract	Virus	Observed results	References
S. dubyi	Sulfated glucuronogalactan	HIV-1	Inhibition concentration of 5 μg mL^{-1}	Bourgougnon et al. 1996
S. dubyi	Methanolic extract	HIV-1 *In vitro*	Inhibition of HIV-1 (a) integrase, (b) reverse transcriptase a: 200 mg mL^{-1} (45–55%), b: ≤ 35%	Ahn et al. 2002
S. pacifica	Aqueous extract Sulfated polysaccharides	R-MuLV HIV	Inhibition percent = 87% IC_{50} = 9.5 × 103 inhibitory units (IU) mL^{-1}	Nakashima et al. 1987a, b
Scinaia hatei	Sulfated xylomannans	HSV-1 and HSV-2	IC_{50} = 0.5–1.4 μg mL^{-1} IC_{50} = 0.5 to 4.6 μg mL^{-1}	Lu et al. 2007 Mandal et al. 2008
Sebdenia flabellata (as *Sebdenia polydactyla*)	Sulfated xylomannans	HSV-1	IC_{50} = 0.35–2.8 μg mL^{-1}	Ghosh et al. 2009
Solieria chordalis	Carraguard Hydrolysate	HIV-1 HSV-1	Inhibition IC_{50} = 23–101 μg mL^{-1}	Skoler-Karpoff et al. 2008 Hardouin et al. 2014
S. filiformis	Sulfated polysaccharide (ι-carrageenan)	MeV	IC_{50} = 0.985 μg mL^{-1}	Morán-Santibañez et al. 2016
Sphaerococcus coronopifolius	Extract Sulfated polysaccharides	HSV-1	IC_{50} = 4.4 μg mL^{-1} IC_{50} = 4.1 μg mL^{-1}	Bouhlal et al. 2010, 2011
Spyridia clavata	Extract	HSV-1 HSV-2	Inhibition percent = 85.9% Inhibition percent = 20.6%	Soares et al. 2012
S. fusiformis	Ethanol extract	VACV	Inhibition percent = 20%	Kamat et al. 1992
S. hypnoides (as *S. insignis*)	Methanol extract	VACV	Inhibition percent = 50%	Kamat et al. 1992
Stenogramma interruptum (as *Stenogramme interrupta*)	Carrageenans	HSV-1 HSV-2	IC_{50} = 4.3–9.3 μg mL^{-1} IC_{50} = 0.9–2.9 μg mL^{-1}	Cáceres et al. 2000
Symphyocladia latiuscula	Extract Extract Extract and three isolated compounds	HSV-1	Inhibition at 100 μg mL^{-1} Minimum inhibitory concentration at 16 μg mL^{-1} (under light) and virucidal IC_{50} = 0.91–3.02 μM IC_{50} = 7.2–11.2 μM IC_{50} = 4.1 μM Extract and TDB showed *in vivo* antiviral	Kim et al. 1997 Hudson et al. 1998 Park et al. 2005
S. marchantioides	Extract	HSV-1	Inhibition at 100 μg mL^{-1} Minimum inhibitory concentration at 12 μg mL^{-1} (under light) and virucidal	Kim et al. 1997 Hudson et al. 1998

Table 5.1 contd....

Table 5.1 contd....

Species	Tested product, fraction or extract	Virus	Observed results	References
Tichocarpus crinitus	κ/β-carrageenan κ/β-carrageenan (low molecular weight derivatives)	TMV	With 1 mg mL^{-1} the infection reduction was 87% Antiviral activity = 32–77%	Nagorskaia et al. 2008 Kalitnik et al. 2013
Tricleocarpa cylindrica (formerly *Galaxaura cylindrica*)	Extract Sulfolipids	HSV-1 *In vitro*	Inhibition percent = 83.4% IC_{50} = 18.75–70.2 μg mL^{-1}	Soares et al. 2012 El Baz et al. 2013
T. fragilis	Aqueous extract	IAV (H1N1)	IC_{50} = 47.1 μg mL^{-1}	Pérez-Riverol et al. 2014
Vertebrata lanosa (formerly *Polysiphonia lanosa*)	Polysaccharides	IAV (H1N1)	Inhibition percent = 20–70%	Jiao et al. 2012
Vidalia obtusiloba	Extract	HSV-1	Inhibition percent = 93.0%	Soares et al. 2012

*Marine microalgae

different dilutions. All concentrations were tested in duplicate, and the highest assayed concentration was 100 μg mL^{-1} (Siamopoulou et al. 2004).

Dichloromethane extracts of the algae *Dictyopteris delicatula*, *Dictyota menstrualis*, *Lobophora variegata*, *Padina gymnospora*, *Sargassum cymosum*, *Sargassum polyceratium*, *Sargassum vulgare*, and *Stypopodium zonale* collected in Rio de Janeiro (Brazil) were tested against HSV-1. All species showed antiviral effects with the percentages of inhibition ranging from 20.6% to 98.2%. The extracts *L. variegata*, *S. zonale*, and *S. cymosum* had inhibitory percentages greater than 90%. The major compounds from *S. zonale* and *S. cymosum* extracts, respectively atomaric acid and phenolic compounds, were identified by means of nuclear magnetic resonance techniques (NMR) (Soares et al. 2012). Soares et al. (2007) tested three meroditerpenes isolated from *S. zonale* against HSV-1 and determined IC_{50} values of 1.28 μM, 1.34 μM, and 2.38 μM, respectively (Soares et al. 2007).

Likewise, sulfoquinovosyldiacylglycerols purified from a chloroform/methanol extract of *Sargassum vulgare* showed potent activity against HSV-1, with a percentage of inhibition ranging from 96.9% to 99% (Plouguerné et al. 2013). That work once again confirmed the activities described in 1999 and 2012 (Santos et al. 1999, Soares et al. 2012).

According to Soares et al. (2012), the extract of *Dictyota menstrualis* was less active against HSV-1, with merely 20.6% inhibition. However, a compound isolated from *D. menstrualis* has shown a potent inhibitory activity against HSV-1, with an IC_{50} value of 1.6 μM (Abrantes et al. 2010). In the same work, Abrantes et al. (2010) demonstrated the activity of a dolabelladienotriol isolated from the alga *D. friabilis* (formerly *D. pfaffii*), with an IC_{50} value of 1.2 μM. It has also been suggested that these compounds may inhibit early events during replication of HSV-1.

Several diterpenes were isolated from the brown alga *Canistrocarpus cervicornis* (formerly *Dictyota cervicornis*) (Teixeira and Kelecom 1988, Teixeira et al. 1986a, b, Fleury et al. 1994), and two of these diterpenes were shown to be active against HSV-1 *in vitro*. One of the diterpenes showed 90% inhibition, while other diterpene achieved a 99% inhibition of virus replication (see Vallim et al. 2010).

SGDGs were identified in fractions obtained after the purification of the organic extract of the Brazilian macroalga *Sargassum vulgare*. The main SQDG responsible for the anti-HSV-1 and anti-HSV-2 activities highlighted was characterized as 1, 2-di-O-palmitoyl-3-O-(6-sulfo-α-D-quinovopyranosyl)-glycerol (Plouguerné et al. 2013).

A research was conducted by Eom and collaborators (2015), whose main objective was to develop effective and safe marine-derived antiviral compounds against Norovirus (MNV). The ethyl acetate

(EtOAc)-extract from *Ecklonia bicyclis* (formerly *Eisenia bicyclis*) exhibited strong antiviral activity against murine norovirus (MNV) as a norovirus surrogate. Among the phlorotannins from *E. bicyclis*, dieckol (DE) and phlorofucofuroeckol-A (PFF) were known to possess the strongest antibacterial activity. In this study, DE and PFF were evaluated for antiviral activity against MNV. DE and PFF exhibited strong anti-MNV activity with 50% effective concentration (IC_{50}) of 0.9 μM.

Sulfated polysaccharides, such as fucoidan, have been isolated from various types of algae and have been shown to possess a broad antiviral spectrum. Fucoidan can be isolated from many brown algae (Hayashi et al. 2008), including *Undaria pinnatifida* (Lee et al. 2004). *In vitro* tests showed anti-HSV-1 activity of fucoidan extracted from *U. pinnatifida*, with an IC_{50} value of 2.5 μg mL^{-1} (Lee et al. 2004). *In vivo* tests also have demonstrated the ability of fucoidan to inhibit the replication of HSV-1 (Hayashi et al. 2008).

The cultivated brown alga *Sargassum naozhouense* has shown low cytotoxicity and potent antiviral activity against HSV-1 (IC_{50}=8.92 μg mL^{-1}). The fact that it contains 21% of water-soluble polysaccharides, alginates, and, probably, fucoidan has motivated the cultivation of its polysaccharides for future development of potential anti-HSV-1 drug (Peng et al. 2013). However, further studies regarding the compounds isolated from these polysaccharides are needed to identify precisely which of them are responsible for the activity described.

Three fractions of plastiquinones were isolated from *Sargassum micracanthum*, for the inhibition of Measles virus (MeV) and Cytomegalovirus (CMV) (Iwashima et al. 2005). It was suggested that it can be used as a lead compound in anti-human cytomegalovirus (HCMV) drug.

The study done by Lakshmi et al. (2006) deals with the biological activities of the extracts of 48 marine florae. The biological screening includes tests for antibacterial, antifungal, and antiviral, among others. From among brown algae, the crude extracts from *Colpomenia sinuosa*, *Polycladia indica* (formerly *Cystoseira indica*), *Dictyopteris woodwardia*, *Sargassum johnstonii*, *Spatoglossum variabile*, *Stoechospermum polypodioides* (formerly *Stoechospermum marginatum*), *Dictyota dichotoma*, *Turbinaria condensata*, *Turbinaria conoides*, and *Turbinaria decurrens* were active against Ranikhet disease virus (NDV), Semliki forest virus (SFV), Encephalomyocarditis virus (EMCV), Herpes simplex virus (HSV), Influenza 'A' virus (IAV), and Japanese B encephalitis virus (JEV).

One sulfated polysaccharide isolated from *Sargassum patens* has been found to have anti-HSV-1 activity, with an IC_{50} value of 5.5 μg mL^{-1} (Zhu et al. 2006). Similarly, a fraction extracted from *Polycladia indica* (formerly *Cystoseira indica*), containing mainly sulfated fucan, also showed antiviral activity *in vitro*, with an IC_{50} of 2.8 μg mL^{-1} (Mandal et al. 2007). In addition, Ponce et al. (2003) have shown the great antiviral potential of fractions obtained from *Adenocystis utricularis*. The most active fraction (IC_{50}=0.28 μg mL^{-1}) was rich in fucose and galactose. The sulfated fucans isolated from *Leathesia marina* (formerly *Leathesia difformis*) and *Sargassum horneri* also have shown anti-HSV-1 activity *in vitro*, with IC_{50} values of 0.7 μg mL^{-1} and 1.0 μg mL^{-1}, respectively (Feldman et al. 1999, Hoshino et al. 1998). *S. horneri* had another sulfated fucan tested by the same group in 2001. When the sulfated fucan was evaluated for antiviral activity based on the selectivity index (SI), it was found to be a potent inhibitor of the replication of HSV-1. The selectivity index of this sulfated fucan was 11,000 when it was added to the medium at the same time as the viral infection and throughout the incubation thereafter, and 7,100 when it was added to the medium immediately after the viral infection (Preeprame et al. 2001). Its potency was similar to that of the sulfated fucan tested in 1998 (Hoshino et al. 1998).

The aqueous extract and a fraction with sulfated fucan isolated from *Stoechospermum polypodioides* (formerly *Stoechospermum marginatum*) showed potent antiviral activity against HSV-1, with IC_{50} values of 1.15 μg mL^{-1} and 3.55 μg mL^{-1}, respectively. These extracts were also active against strains of TK-deficient HSV-1. However, the desulfation of the extract (IC_{50}=31.8 μg mL^{-1}) and the fraction rich in sulfated fucans (IC_{50}=50 μg mL^{-1}) was followed by a significant decrease in antiviral activity (Adhikari et al. 2006), which indicates that the antiviral action may be attributed to sulfated polysaccharides.

Taking these points into account, it is possible to assume that brown seaweed polysaccharides and their derivatives, in particular alginates and fucans, have good antiviral activity. In addition, alginate polysaccharides may potentially be used in the development of new types of antiviral drugs with high efficiency and low toxicity (Wang et al. 2012a).

In the works of Synytsya et al. (2014), the defensive effects of Mekabu fucoidan (obtained from *Undaria pinnatifida*) were evaluated on mice infected with avian influenza A virus (H5N3 and H7N2 subtypes); its efficacy was determined in reducing viral replication and increasing antibody production. Oral administration of Mekabu Fucoidan resulted in suppressing virus yields. In addition, the production of neutralizing antibodies and mucosal IgA in the animals inoculated with the avian influenza A viruses (IAV) were significantly increased. These works suggested that Mekabu fucoidan could be used for the prevention of viral infection (see Table 5.1).

The brown algae *Colpomenia sinuosa*, *Dictyopteris delicatula*, *Dictyota dichotoma*, *Hydroclathrus clathratus*, *Lobophora variegata*, *Padina australis*, *Padina gymnospora*, *Sargassum hemiphyllum*, *Sargassum cymosum*, *Sargassum vulgare*, *Stypopodium zonale*, and *Undaria pinnatifida*, also show anti-HSV-2 activity (Thompson and Dragar 2004, Wang et al. 2008, Soares et al. 2012).

Marine brown algae accumulate a variety of phloroglucinol-based polyphenols, as phlorotannins of low, intermediate, and high molecular weight containing both phenyl and phenoxy units. Furthermore, phlorotannins consist of phloroglucinol units linked to each other in various ways, and are of wide occurrence amongst marine organisms, especially brown and red algae (Glombitza and Li 1991, Singh and Bharate 2006). Firstly, Ahn et al. (2004) reported that 8,8'-bieckol and 8,4'''-dieckol, which were isolated from the brown algae *Ecklonia cava* (see Table 5.1), show an inhibitory effect on HIV-1 reverse transcriptase and protease. The inhibition against reverse transcriptase of 8,8'-bieckol with a biaryl linkage (IC_{50}, 0.5 μM) was 10-fold higher than that of 8,4'''-dieckol with a diphenyl ether linkage (IC_{50}, 5.3 μM), although these two phlorotannis are dimers of eckol. The authors suggested that the steric hindrance of the hydroxyl and aryl groups near the biaryl linkage of 8,8'-bieckol caused the potent inhibitory activity. Moreover, 8,8'-bieckol selectively inhibited reverse transcriptase over protease, and the inhibitory effect was comparable to the positive control nevirapine (IC_{50}, 0.28 μM). It is clear that the 8,8'-bieckol possessed higher inhibitory activity than 8,4'''-dieckol. In the next work, Ahn and collaborators (2006) showed that *Ishige okamurae*-derived diphlorethohydroxycarmalol also has an inhibitory effect on HIV-1. This compound exhibited inhibitory effects on the HIV-1 reverse transcriptase and integrase with IC_{50} values of 9.1 μM and 25.2 μM, respectively. However, diphlorethohydroxycarmalol did not show inhibitory activity against the HIV-1 protease. Furthermore, 6,6'-bieckol, one of the main phloroglucinol derivative naturally occurring in *Ecklonia cava*, has a potent inhibition against HIV-1 induced syncytia formation, lytic effects, and viral p24 antigen production Artan et al. (2008). Besides, 6,6'-bieckol selectively inhibited the activity of HIV-1 reverse transcriptase enzyme with an IC_{50} of 1.07 μM, as well as the inhibition of HIV-1 entry. Additionally, it exhibited no cytotoxicity at a concentration where it inhibited HIV-1 replication almost completely. Therefore, 6,6'-bieckol can be employed for the development of new generation therapeutic agents against HIV (Vo and Kim 2010).

Pereira et al. (2004) studied the effect of two diterpenes ((6R)-6-hydroxydichotoma-3,14-diene-1,17-dial, named Da-1, and (6R)-6-acetoxidichotoma-3,14-diene-1,17-dial, named AcDa-1) isolated from Brazilian seaweed, *Dictyota menstrualis*, on HIV-1 replication (see Table 5.1). The compounds were reported to have an effect on an early step of the virus replicative cycle or during virus adsorption/penetration. The isolated compounds were shown to inhibit the RNA-dependent DNA-polymerase activity of HIV-1 reverse transcriptase (RT) in a dose-dependent manner with an IC_{50} of 40 mM and 70 mM. However, the diterpenes were not as strong as the well-known non-nucleoside inhibitor of the HIV-1 RT nevirapine (IC_{50} 40 nM).

According to Thuy et al. (2015), fucoidans extracted from the three species (*Sargassum mcclurei*, *S. polycystum*, and *Turbinaria ornata*) from Vietnam, displayed similar antiviral activities, against HIV-1, with a mean IC_{50} ranging from 0.33 μg mL^{-1} to 0.7 μg mL^{-1} while displaying no cell toxicity.

5.4 Antiviral Activity of Rhodophyta (Red Algae)

The inhibitory effects of polysaccharides from marine algae on viral replication were reported almost six decades ago. In 1958, Gerber et al. reported that algal polysaccharides extracted from *Plocamium cartilagineum* (formerly *Gelidium cartilagineum*) exhibited antiviral activity toward Mumps virus (MuV) and influenza B virus (IBV), protecting chicken embryos against the infection. In 1977, Ehresmann et al. associated the inhibition of Herpes simplex (HSV) and other viruses with polysaccharide fractions from extracts of 10 red algae; similar observations were made by Richards et al. (1978). Previously, in 1987, Nakashima et al. (1987a, b) reported inhibition of HIV reverse transcriptase by sulfated polysaccharides from the red alga *Schizymenia pacifica*. The polysaccharides with antiviral activity were shown to be highly sulfated (Huheihel et al. 2002).

A potent HIV-inactivating protein, griffithsin, was isolated from the red alga *Griffithsia* sp. (see Table 5.1). Griffithsin is a new type of lectin that displays potent antiviral activity against laboratory strains and primary isolates of HIV-1 (IC_{50} = 0.043–0.63 nM) (Mori et al. 2005). This activity requires binding to viral glycoproteins (e.g., gp120, gp41 and gp160) in a monosaccharide-dependent manner (Yasuhara-Bell and Lu 2010).

In the works of Richards et al. (1978), extracts of two species of red algae, *Constantinea simplex* and *Farlowia mollis*, were tested for antiviral activity in tissue culture and in experimental infections of mice (see Table 5.1). Treatment of confluent mouse embryo fibroblast cell monolayers with either compound before viral inoculation was effective in inhibiting the replication of Herpes simplex virus type 1 and type 2 (HSV-1, HSV-2), Vaccinia virus (VACV or VV), and Vesicular stomatitis virus (VSV), but not Encephalomyocarditis virus (EMCV), Semliki Forest virus (SFV), or Murine cytomegalovirus (MCMV). Prophylactic administration of these extracts was effective in reducing final mortality or prolonging the mean day of death of animals inoculated by the intraperitoneal, intracerebral, or intranasal routes with HSV-2 (Richards et al. 1978).

In a comparative evaluation of diverse sulfated polysaccharides, Baba et al. (1988a, b) found that many, but not all, had anti-HIV activity. More recently, fucoidan, a complex sulfated polysaccharide from the alga *Fucus vesiculosus*, was found to inhibit HIV *in vitro* and was synergistic with AZT (Sugawara et al. 1989) (see also Section 5.3 and Table 5.1). This activity presumably resulted from a direct interaction of the polysaccharide with the HIV binding site of the target cells.

With respect to the human immunodeficiency virus (HIV), the agent of acquired immunodeficiency syndrome (AIDS), with about 34 million people infected worldwide (Talarico et al. 2016), a λ-carrageenan purified from *Schizymenia pacifica* (Nakashima et al. 1987a), as well as commercial λ-, ι-, and κ-carrageenans (Baba et al. 1988a, b, Lynch et al. 1994) and low-molecular weight derivatives (Yamada et al. 1997, 2000) prevented *in vitro* HIV-1 infection of lymphocytes and T cell lines and also virus transmission to epithelial cells (Pearce-Pratt and Phillips 1996).

Sulfated xylomannans from the red seaweed *Sebdenia polydactyla* inhibited the propagation of HSV-1 in Vero cells (Gosh et al. 2009). The activity was abolished by desulfation of the xylomannan and, conversely, over-sulfated derivatives exhibited enhanced potency. Mohsen reported that sulfated polysaccharide fractions isolated from *Sargassum latifolium* (brown algae) (see Section 5.3 and Table 5.1) inhibited HSV-1 in the plaque assay with the most effective fraction having greater sulfate ester content and molecular weight compared to the other fractions studied (Mohsen et al. 2007). It has been generally observed that antiviral activity of sulfated polysaccharides increases with their molecular weight (Witvrouw and De Clerq 1997).

The sulfated glucuronogalactan from red algae *Schizymenia dubyi* was reported to possess anti-HIV activity. Bourgougnonl et al. (1996) determined the antiviral activity with HIV-1 by measuring the protective effect of sulfated glucuronogalactan against the virus-induced cytopathogenicity in MT4 cells over eight days. As shown in their study, the syncitial formation was completely suppressed with 5 μg mL^{-1} of this polysaccharide. Furthermore, the HIV-1 reverse transcriptase was inhibited at concentrations as low as 5 μg mL^{-1}, without cytotoxicity to MT4 cells. They suggested that the mechanism of action of this polysaccharide *in vitro* can be been mainly attributed to the inhibition of virus-host cell attachment or an early step of HIV infection (see Table 5.1).

The red algae *Caloglossa leprieurii*, *Chondria armata*, *Hypnea musciformis*, and *Laurencia dendroidea* (formerly *Laurencia majuscula*), according to Teixeira (2012), possess antiviral activity. Additionally, a sulfated polysaccharide (sulfated xylomannan) from *Nothogenia fastigiata* inhibits efficiently the replication of herpes simplex virus type 1 (HSV-1), type 2 (HSV-2), human cytomegalovirus (HCMV), respiratory syncytial virus (RSV), influenza A (IAV) and B virus (IBV), Juanin (Argentine hemorrhagic fever) and Tacaribe virus (TCRV), and simian immunodeficiency virus (SIV) (see Table 5.1) (Damonte et al. 1994).

The study done by Lakshmi et al. (2006) deals with the biological activities of the extracts of 48 marine florae. The biological screening includes tests for antibacterial, antifungal (see Chapter 7), and antiviral, among others. From among red algae, the crude extracts from *Chondria dasyphylla*, *Gracilaria corticata*, *Gracilaria canaliculata* (formerly *Gracilaria crassa*), *Halymenia porphyroides*, *Heterosiphonia muelleri*, *Hypnea musciformis*, *Laurencia obtusa*, *Palisada poiteaui* (formerly *Laurencia poiteaui*), *Scinaia moniliformis* (formerly *Scinaia indica*), and *Solieria robusta* were active against Ranikhet disease virus (NDV), Semliki forest virus (SFV), Encephalomyocarditis virus (EMCV), Herpes simplex virus (HSV), Influenza A virus (IAV), and Japanese B encephalitis virus (JEV).

Most studies indicate that sulfated polysaccharides prevent attachment of enveloped viruses (including herpes-viruses) to cells. *In vitro* experiments by Montanha et al. (2009) have shown that the carrageenans from red seaweeds *Chondracanthus acicularis* (formerly *Gigartina acicularis*) and *Euchema denticulatum* inhibited adhesion and replication of HSV-1 by inhibiting viral DNA synthesis, but were ineffective against non-enveloped Poliovirus. For example, the IC_{50} value for *C. acicularis* carrageenan against HSV-1 was 0.0027 mg mL^{-1}, while SI (selectivity index) > 545 and > 32, respectively (see Table 5.1). The authors have shown that the antiviral activity of carrageenan was determined by seaweed species. IC_{50} towards HSV-1 was 0.0027 mg mL^{-1} for *C. acicularis* carrageenan, while the same parameter for carrageenan from the *E. denticulatum* seaweed was 0.026 mg mL^{-1}, and SI > 545 and > 71, respectively, compared with the control. Reduction of the virus titer (MOI—multiplicity of infection of 0.01) induced by carrageenan from the *C. acicularis* seaweed was 0.9 ± 0.25 log10 in the case of virus pretreatment and 1.9 ± 0.25 log10 after one cycle of virus replication as compared with control (Montanha et al. 2009, Besednova et al. 2016).

Methanol extracts from the red algae *Corallina pilulifera*, *Corallina vancouveriensis*, *Symphyocladia latiuscula*, *Symphyocladia marchantioides*, and *Mazzaella parksii* (formerly *Mazzaella cornucopiae*) have been found to inhibit HSV-1, though the latter appears to have some cytotoxicity (Kim et al. 1997). The antiviral activity (HSV-1) of *C. pilulifera*, *S. latiuscula*, and *S. marchantioides* was confirmed later by Hudson et al. (1999). Park et al. (2005) also studied the antiviral activity of *S. latiuscula*. The methanol extract and three isolated compounds were tested against HSV-1 strains resistant to acyclovir and phosphonoacetic acid (APr), thymidine kinase deficient (TKd), and an *in vitro* wild-type strain. The extracts and the compounds were active against all strains, except for compound 11, where the activity was determined only for the wild-type strain. The compound 2,3,6-tribromo-methyl ether of 4,5-dihydroxybenzyl (TDB) has shown the most potent activity, with IC_{50} values of 3.02 µg mL^{-1} for the wild-type strain, 0.91 µg mL^{-1} for the APr strain, and 1.41 µg mL^{-1} for the TKd strain. The therapeutic efficacy of the methanol extract and the TDB compound was further examined in BALB/c mice subcutaneously infected with HSV-1. Three daily oral administrations of the extract and the TDB were able to delay the onset of skin lesions (local vesicles) and limit the development of new lesions in the mild zoster, although no signs of toxicity were noticed in comparison with controls. Furthermore, TDB has reduced the amount of virus in brain and skin and may, therefore, be a promising anti-HSV-1 agent (Park et al. 2005).

The aqueous extract of *Polysiphonia denudata* has also been shown to reduce viral infectivity and cytopathic effect of HSV-1 (Serkedjieva 2000). In addition, Serkedjieva (2004) has shown antiviral activity of an extract of *Ceramium virgatum* (formerly *Ceramium rubrum*) against Kupka, and KOS strains, with IC_{50} values of 0.8 mg mL^{-1} and 0.5 mg mL^{-1}, respectively. A 1.2% cold aqueous extract of *Pterocladiella capillacea* (formerly *Pterocladia capillacea*), *Hypnea musciformis*, and *Osmundaria obtusiloba* (formerly *Vidalia obtusiloba*) showed antiviral activity against ACV-resistant HSV-1, with 99.6%, 95.5%, and 55.3% inhibition, respectively. At 2.5% concentration, the *O. obtusiloba* extract had better results with 95.5% inhibition.

De Souza et al. (2012) isolated an anti-HSV glycolipid-enriched fraction from the Brazilian seaweed *O. obtusiloba*. The major compound of the active fraction was identified as the SQDG1,2-di-O-acyl-3-O-(6-deoxy-6-sulfo-α-D-glucopyranosyl)-sn-glycerol (see Table 5.1).

A sulfated galactan found in *P. capillacea* was described showing anti-HSV-1 activity, with an IC_{50} value ranging from 3.2 µg mL^{-1} to 6.1 µg mL^{-1} (Pujol et al. 1996). Other anti-HSV-1 sulfated polysaccharides included sulfated galactans found in *Bostrychia montagnei* (IC_{50} = 12.9 to more than 50.0 µg mL^{-1}) (Duarte et al. 2001), *Cryptopleura ramosa* (IC_{50} = 1.6 to 4.2 µg mL^{-1}) (Carlucci et al. 1997b), and *Nothogenia fastigiata* (IC_{50} = 0.6 to more than 100 µg mL^{-1}) (Kolender et al. 1997).

Commercial λ- and ι-carrageenans (extracted from *Gigartina* spp. and *Eucheuma denticulatum*—formerly *Eucheuma spinosum*, respectively), were in that order the most active Dengue virus agents, and particularly DENV-2 presented the highest inhibition with IC_{50} values lower than 1 μg mL^{-1}. The susceptibility of DENV-4 and DENV-1 was irregular to carrageenans, with a high inhibitory effect of λ- and ι-carrageenans against DENV-4 and a low or undetected inhibitory action against DENV-1 up to 50 μg mL^{-1} of polysaccharide. The variation in DENV serotype susceptibility to carrageenans was not attributed to differential growth ability in Vero cells (Talarico et al. 2005). Since CGNs lacked cytotoxicity in Vero cells up to 1,000 μg mL^{-1}, the inhibition of λ- and ι-carrageenans against DENV-2 and DENV-3 was highly specific with SI in the range 2,500–6,666 for DENV-2, and 244–500 for DENV-3, and was also independent of the input multiplicity of infection (Talarico and Damonte 2007).

As reported with Vero cells, λ- and ι-carrageenans were selective inhibitors of DENV-2 and DENV-3 multiplication in human hepatoma HepG2 cells with IC_{50} values lower than 1 μg mL^{-1} (see Table 5.1). In addition to the homogeneous carrageenans, the κ/ι/ν-carrageenan extracted from south American red seaweed *Gymnogongrus griffithsiae* exhibited antiviral effect against DENV-2 and DENV-3 multiplication in monkey Vero cell as well as in human hepatoma HepG2 and foreskin PH cells, with selectivity index (SI) values in the range of 556–3.226 for DENV-2 and 96–192 for DENV-3 (Talarico et al. 2005). Furthermore, DL-galactan fractions extracted from red seaweed *Callophyllis variegata*, mainly composed of hybrid carrageenan, showed antiviral effectiveness against DENV-2 infection of Vero cells with SI values around 2.400 (Rodríguez et al. 2005).

Mechanistic studies suggested that the target for commercial λ-CGN in DENV-2 in Vero cells was extended not only to adsorption, as observed for HSV and HIV, but also to a post-adsorption event blocking the viral nucleocapsid internalization into the cytoplasm. The virions may enter the cell in the presence of CGN, but the probable association of CGN with the E virion glycoprotein may block the uncoating of nucleocapsid and escape from endosomes (Talarico et al. 2011). Furthermore, the λ-CGN was able to produce a very weak direct inactivating effect on DENV-2 virions, indicating that the blockade observed in DENV-2 multiplication in mammalian cells can be totally due to an interference with the viral entry process (Talarico et al. 2016).

Regarding the host cell, the antiviral activity against DENV-2 of carrageenans showed a differential susceptibility and mode of action for mammalian and mosquito cells, depending on the type of carrageenan. The ι-, λ- and κ-carrageenans blocked DENV-2 infection in mammalian Vero cells, but only ι-carrageenans were DENV inhibitors in mosquito C6/36 cells (Talarico et al. 2011). The lack of antiviral activity against DENV-2 of λ- and κ-carrageenans in mosquito cells was in agreement with previous studies of polysaccharides (including heparin and sulfated galactans), which showed effective inhibition of DENV infection in diverse mammalian cells, but were totally inactive in mosquito cells (Talarico et al. 2005, Thaisomboonsuk et al. 2005). The inhibition of DENV-2 multiplication in Vero cells was exerted by interference with the initial interaction of glycoprotein E (gE) with heparin sulfate (HS) in the adsorption and internalization processes (Talarico et al. 2007b, 2011), but the inhibition in C6/36 cells was independent of HS (Talarico et al. 2011). Iota-carrageenans, extracted from *Eucheuma denticulatum* (formerly *Eucheuma spinosum*) and *Meristotheca gelidium* (formerly *Meristiella gelidium*) (De SF-Tischer et al. 2006), failed to affect DENV-2 adsorption to C6/36 cells and did not show any virucidal effect against DENV-2 virions. The inhibitory activity of ι-carrageenans against DENV-2 in mosquito cells was exerted by cell-pretreatment before infection and may be mainly due to a long lasting effect produced by any cellular factor triggered as a response of the cell to the presence of the compound, since the proliferation activity of C6/36 cells was affected in the presence of this type of carrageenan (Talarico et al. 2011). Differences in ionic and hydrophobic zones in carrageenan structures may allow specific interactions that probably account for the differential effects among carrageenans observed in mosquito cells.

The antiviral activity of carrageenans against DENV *in vitro* infection was assessed and confirmed by different approaches which can be used in drug discovery—via real-time monitoring the cellular oxygen consumption rates which reflect cellular metabolic activity in BHK-21 cells (Huang et al. 2014) and by a high-content cell-based immunofluorescence assay in HEK-293 cells (Shum et al. 2010, Talarico et al. 2016).

Hybrid carrageenan has also shown antiviral activity against respiratory viruses, such as HMPV, a member of the family Paramyxoviridae and an important agent of respiratory tract disease worldwide, especially in the pediatric and elderly populations (Collins and Ruth 2013). Mendes et al. (2014) described that sulfated DL-galactans extracted from the red seaweed *Cryptonemia seminervis* showed inhibitory activity against human metapneumo virus (HMPV) in LLC-MK2 cells, by either exerting a virucidal effect on viral particles or interfering with viral adsorption to target cells. The anti-HMPV activity was dependent on the molecular weight of polysaccharides, showing lower inhibitory effect in depolymerized fractions (Mendes et al. 2014).

Dichloromethane extracts from 16 species of red macroalgae—*Corallina panizzoi*, *Jania adhaerens*, *Bostrychia radicans*, *Centroceras clavulatum*, *Chondracanthus acicularis*, *Cryptonemia seminervis*, *Gracilaria domingensis*, *Hypnea musciformis*, *Hypnea spinella*, *Jania crassa*, *Laurencia dendroidea*, *Osmundaria obtusiloba*, *Plocamium brasiliense*, *Pterocladiella capillacea*, *Spyridia clavata*, and *Tricleocarpa cylindrica*, were tested against HSV-1 Acyclovir-resistant strains. All extracts, except for those extracted from *C. clavulatum*, *C. seminervis*, *G. domingensis*, *J. adhaerens*, *C. clavulatum*, and *G. domingensis* have shown an inhibition percentage ranging from 43.8% to 97.5% (Soares et al. 2012). *L. dendroidea* had the best inhibitory activity, followed by *H. spinella* (inhibition = 92%) and *O. obtusiloba* (inhibition = 90%). Most of the compounds of *L. dendroidea* extract have been identified by nuclear magnetic resonance techniques (NMR) (Soares et al. 2012). Elatol was first isolated by Sims et al. (1974) from *Coronaphycus elatus* (formerly *Laurencia elata*), and several species of *Laurencia* have this sesquiterpene as a major secondary metabolite (Santos et al. 2010). Although Soares et al. (2012) reported the anti-HSV-1 activity for *L. dendroidea* collected in the coast of Rio de Janeiro, this species presented a high cytotoxicity against VERO cells and showed no antiviral activity in the work of Bianco et al. (2013). However, the authors did not specify which species were really tested against HSV-1.

The fractions obtained with the acetone and the methanol lipid extracts from *O. obtusiloba* were evaluated, and the results showed that both have potent activity against HSV-1, with a percentage of inhibition ranging from 82.2% to 99.9% (Mattos et al. 2011). De Souza et al. (2012), tested a purified fraction containing sulfoglycolipids from seaweed in order to identify the molecules responsible for this activity. The results have shown a slightly less potent activity against HSV-1, yielding a 75% inhibition. This fact may be related to the lack of other polar lipids present in the fractions previously tested, which probably act in synergism with sulfoglycolipids (De Souza et al. 2012).

Bouhlal et al. (2010) have demonstrated that the aqueous extract of *Asparagopsis armata* (IC_{50} < 2.5 μg mL^{-1}) was the most efficient extract as an antiviral, with a selectivity index greater than 250 μg mL^{-1}. Apart from that extremely potent extract, the extracts from *Sphaerococcus coronopifolius* (IC_{50} = 4.4 μg mL^{-1}), *Plocamium cartilagineum* (IC_{50} = 6.4 μg mL^{-1}), *Pterosiphonia complanata* (IC_{50} = 10.4 μg mL^{-1}), *Ceramium virgatum* (formerly *C. rubrum*) (IC_{50} = 12.4 μg mL^{-1}), *Boergeseniella thuyoides* (IC_{50} = 12.6 μg mL^{-1}), and *H. musciformis* (IC_{50} = 23.5 μg mL^{-1}) have shown the greatest efficiency. The results have also demonstrated that the aqueous extracts were more effective in inhibiting HSV-1 than methanol, dichloromethane, and methanol-dichloromethane extracts.

According to Faral-Tello et al. (2012), the extract from *H. musciformis* has no antiviral activity (HSV-1). Nevertheless, this alga has been mentioned in three previous studies as showing anti-HSV-1 activity (Santos et al. 1999, Bouhlal et al. 2010, Soares et al. 2012). This result may be related to the extraction, which is performed with different solvents in the studies. Mendes et al. (2012) tested extracts from *H. musciformis* cultured with different phytohormones, revealing the possibility of enhancing the anti-HSV-1 activity by growing the seaweed in the presence of phytohormones.

Three fractions containing galactans were isolated from the alga *Callophyllis variegata*, revealing a potent anti-HSV-1 activity with IC_{50} value ranging from 0.16 μg mL^{-1} to 1.55 μg mL^{-1} (Rodríguez et al. 2005). The aqueous extracts from *Gymnogongrus griffithsiae* and *Cryptonemia crenulata* and their fractions exhibited antiviral activity (HSV-1), with IC_{50} values ranging from 0.5 μg mL^{-1} to 5.6 μg mL^{-1} and sulfated galactans as major compounds (Talarico et al. 2004). Sulfated galactans with activity against HSV-1 were also found in algae *Grateloupia indica* (IC_{50} = 0.12 to 1.06 μg mL^{-1}) and *Schizymenia binderi* (IC_{50} = 0.18 to 0.76 μg mL^{-1}) (Chattopadhyay et al. 2007, Matsuhiro et al. 2005).

The sulfated polysaccharides from *C. crenulata*, i.e., galactan (see Table 5.1), were selective inhibitors of DENV-2 multiplication in Vero cells with IC_{50} values of 1.0 μg mL^{-1}, where the IC_{50} values for the reference polysaccharides heparin and DS8000 were 1.9 μg mL^{-1} and 0.9 μg mL^{-1}, respectively (Talarico et al. 2005). However, the compound has lower antiviral effect against DENV-3 and DENV-4, and was totally inactive against DENV-1. The inhibitory effect of C2S-3 against DENV-2 was slightly higher when treatment was by adsorption (IC_{50} = 2.5 ± 0.1 μg mL^{-1}) with respect to treatment only during internalization (IC_{50} = 5.5 μg mL^{-1}) (Talarico et al. 2007b). Thus, the inhibitory effect was increased when C2S-3 was included at both stages of adsorption and internalization (Kadir et al. 2013).

The inhibitory properties against DENV-2 of the sulfated polysaccharide from *Gymnogongrus griffithsiae* (see Table 5.1), kappa carrageenan was evaluated in Vero cells (Talarico et al. 2005). The compound effectively inhibits DENV-2 multiplication at the IC_{50} value of 0.9 μg mL^{-1}, which is the same as the IC_{50} value for the commercial polysaccharides DS8000. However, the compound has lower antiviral effect against DENV-3 and DENV-4, and was totally inactive against DENV-1 (Kadir et al. 2013).

Sulfated galactans are the major extracellular polysaccharides produced by red algae (De SF-Tischer et al. 2006), and aqueous extracts are known to be rich in polysaccharides. Several sulfated carbohydrates (see also Chapter 2) from seaweed, cyanobacteria, and animal sources have been described as showing potent inhibitory activity against several human and animal viruses (Bouhlal et al. 2011).

The sulfated polysaccharides are primarily classified as carrageenan and agarans based on their stereochemistry. Galactans of the L-series are termed agarans, while those of the D-series are called carrageenans (see also Chapter 3) (Knutsen et al. 1994).

Sulfated agarans from *Acanthophora spicifera* have been isolated, and some of the fractions containing these agarans showed potent anti-HSV-1 activity. This result may be related to the fact that agarans from *A. spicifera* are structurally analogous to the minimal binding sequence required for the interaction of heparan sulfate (HS) with the glycoprotein C of HSV-1. The average standard of sulfation allows the possibility of these disaccharide units having a minimal structure similar to sequence HS, which would be responsible for the activity described (Duarte et al. 2004).

The inhibitory effects of sulfated polysaccharides appear to be based primarily on their ability to interfere with the initial attachment of the virus to the target cell, leading to a block of virus entry. The initial contact between the virus envelope and the host cell occurs by ionic interaction between the positively charged regions of viral glycoproteins and regions of the surface of the negatively charged cell. The result of this interaction is the binding of the glycoprotein C (gC) and, in some cases, glycoprotein B (gB) from the virus, with HS (WuDunn and Spear 1989, Lycke et al. 1991, Shieh et al. 1992).

Carrageenans isolated from *Gigartina skottsbergii* also have been identified as potent inhibitors of HSV-1 (including acyclovir-resistant variants and clinical isolates), with IC_{50} values in the range of < 1.0 μg mL^{-1} to 4.1 μg mL^{-1} (Carlucci et al. 1997a). In 1999, the same group showed that carrageenans isolated from this alga are potent and selective inhibitors of HSV-1. The antiviral IC_{50} values, which were determined by virus yield reduction assays in different cell lines, ranged from 0.4 μg mL^{-1} to 3.3 μg mL^{-1} with no cytotoxic effects. Time of addition and attachment studies suggested that the main target for antiviral action of the carrageenans is virus adsorption, whereas no effect on virus internalization was detected on either early or late protein synthesis. As a matter of fact, some carrageenans have already shown virucidal activity (Carlucci et al. 1999). Capsules containing *Gigartina* are already on sale in some countries. Carrageenans also have been extracted from Chilean samples of *Stenogramme interrupta*, which might have promising antiherpetic activity (Cáceres et al. 2000).

Gymnogongrus torulosus was investigated for its *in vitro* antiviral properties against DENV-2 in Vero cells. Galactan (DL-galactan hybrids) extracted from this plant was active against DENV-2, with IC_{50} values in the range of 0.19–1.7 μg mL^{-1} (Pujol et al. 2002).

Water-soluble sulfated polysaccharides isolated from two red algae, namely *S. coronopifolius* and *Boergeseniella thuyoides*, collected on the coast of Morocco were capable of inhibiting the *in vitro* replication of HSV-1 with IC_{50} values of 4.1 μg mL^{-1} and 17.2 μg mL^{-1}, respectively (Bouhlal et al. 2011).

The data obtained in the work of Soares (2015) revealed that the viral infection by Lentivirus was reduced upon exposure to a pre-treatment with extracts from female gametophytes (FG) and tetrasporophytes (T) of *Chondracnthus teedei* var. *lusitanicus*. Although the inhibitions were not statistically significant, FG (producers of kappa/iota hybrid carrageenan) and T (producers of xi-carrageenan) of *C. teedei* var. *lusitanicus* extracts was able to reduce 14% of the virus infection, and the Tetra extract was able to reduce, approximately, 35% of the virus infection. Damonte et al. (2004) reported that marine polysaccharides can either inhibit the replication of virus by interfering with its viral life cycle (viral adsorption, viral penetration, uncoating of capsids, biosynthesis, viral assembly, and viral release) at different stages or directly inactivate virions before virus infection. The results relative to the effects of the pre-treatment with FG and T extracts might be explained by the interference of these extracts with the adsorption process of the virus. In fact, Carlucci et al. (1997a, b, 1999) noted that lambda-carrageenan (which is the same carrageenan family found in T extract), and partially cyclized mu/iota-carrageenan from *Gigartina skottsbergii* (Gigartinales) showed potent antiviral effects against different strains of HSV types 1 and 2 during the virus adsorption stage (see also Chapter 3). They subsequently confirmed the firm binding of carrageenan to virus receptors on the host cell surface, and demonstrated that lambda-carrageenan interferes with the adsorption process of the virus to the host cell surfaces. Furthermore, Mazumder et al. (2002) described that a sulfated galactan extracted from a red alga showed antiviral activity against *Herpes simplex* virus 1 and 2, which was likely due to an inhibition of the initial viral attachment to the host cells. Moreover, Buck et al. (2006) found that carrageenan could directly bind to the HPV capsid, so as to inhibit not only the viral adsorption process but also the subsequent entry and uncoating process of the virus (Soares 2015). The data obtained in this study showed that the FG extract proved to have a virucidal effect on Lentivirus. These results might be explained by the interference of these extracts in the multiplication of the virus. In fact, Damonte et al. (2004) reported that the direct virucidal actions of carrageenan may be due to the formation of a stable virion-carrageenan complex where binding is not reversible and, therefore, the sites on the viral envelope required for virus attachment to host cells are occupied by the sulfated polysaccharide. This results in an inability of the virus to complete the subsequent infection process (Soares 2015).

Chondrus crispus was subjected to enzymatic hydrolysis in the studies done by Kulshreshtha et al. (2015), and the extracts produced were tested for their antiviral activity. This species was characterized by higher levels of protein and sulfate. The commercial enzymes tested showed higher yields of bioactive extract. Since enzyme extraction is solvent-free, the hydrolysates obtained were of food grade and could be easily utilized in the cosmetics and food industries. Furthermore, extracts obtained from enzyme-assisted hydrolysis exhibited antiviral activity, against HSV-1, with IC_{50} values in the range of 77.6–129.8 $\mu g\ mL^{-1}$ (Kulshreshtha et al. 2015).

Like other seaweeds, red algae also have some species with anti-HSV-2 activity, such as algae—*Acanthophora spicifera*, *Bostrychia montagnei*, *Callophyllis variegata*, *Ceramium virgatum*, *Chondracanthus acicularis*, *Cryptonemia crenulata*, *Cryptopleura ramosa*, *Gigartina skottsbergii*, *Gracilaria domingensis*, *Gymnogongrus griffithsiae*, *Hypnea musciformis*, *Jania crassa*, *Laurencia dendroidea*, *Nothogenia fastigiata*, *Plocamium brasiliense*, *Pterocladiella capillacea*, *Polysiphonia denudata*, *Schizymenia binderi*, and *Spyridia clavata* (Pujol et al. 1996, Carlucci et al. 1997a, Kolender et al. 1997, Serkedjieva 2000, 2004, Duarte et al. 2001, 2004, Talarico et al. 2004, Matsuhiro et al. 2005, Rodríguez et al. 2005, Mendes et al. 2012, Soares et al. 2012).

Hepatitis C virus (HCV) is an important human pathogen leading to hepatocellular carcinoma. Using an *in vitro* cell-based HCV replicon and JFH-1 infection system, Chen and collaborators (2013) demonstrated that an aqueous extract of the seaweed *Gracilaria tenuistipitata* concentration-dependently inhibited HCV replication at nontoxic concentrations (see Table 5.1). The aqueous extract of *G. tenuistipitata* synergistically enhanced interferon-α (IFN-α) anti-HCV activity in a combination treatment.

CHAPTER 6

Antitumor Activity of Seaweeds and their Extracts

6.1 Introduction

Cancer figures among the leading cause of morbidity and mortality worldwide. According to the estimates from the International Agency for Research on Cancer (IARC), 14 million new cancer cases and 8.2 million cancer deaths were accounted in 2012 (Ferlay et al. 2013, 2015). The global cancer burden is growing at an alarming pace, projected to reach about 21.7 million new cancer cases and 13 million cancer deaths by 2030 (American Cancer Society 2015). Cancer is the generic term for a group of diseases, also known as malignant tumors that can affect any part of the body (WHO 2002).

The most commonly diagnosed cancers were lung (1.82 million), breast (1.67 million), and colorectal (1.36 million); the most common causes of cancer deaths were lung cancer (1.6 million deaths), liver cancer (745,000 deaths), and stomach cancer (723,000 deaths) (Ferlay et al. 2015).

Most cancers are named by the cells that give rise to them. Breast cancer, for example, is the name attributed to a malignant tumor whose origin is breast cells (Silva 2015). Malignant tumors of the brain, lung, breast, prostate, skin, and colon are among the diseases known as cancer. There are different causes, evolution, and treatments for each type, but there is a common feature to all of them—the uncontrolled division and growth of cells (NIH 2016). The process of cell turnover is normally well controlled throughout life by basic biological mechanisms. In cancer, however, the control mechanisms are disrupted. Cells in the affected part of the body grow beyond their usual boundaries, invade adjoining tissues, and may spread to secondary organs or tissues as metastases (WHO 2002). Several approaches on targeted therapies have significantly changed the treatment of cancer over the last few years (Brannon-Peppas and Blanchette 2004, Kreitman 2006, Chari 2008, Mathew and Verma 2009). Cisplatin, carboplatin, and oxaliplatin are still involved in nearly 50% of all anticancer therapies worldwide (Liu et al. 2014). However, problems of platinum resistance and undesirable side effects are limiting their future use, thereby emphasizing the urgent need for new anticancer agents (Silva 2015).

6.1.1 Approaches to cancer treatment

The initial presentation of a tumor is the result of genetic changes which occur in critical genes within a cell that control its normal growth and development. This leads to unregulated proliferation of the cell and its progeny, which results, except in the case of most hematological malignancies, in the development of a solid tumor. The increase in tumor size depends on a large number of interacting factors, including a reduction in cell death (particularly through the apoptosis pathway), an increase in proliferation, and the generation of new blood vessels to provide oxygen and nutrients to support the growing mass (King and

Robins 2006). Tumors are classified in relation to their tissue of origin to include squamous cell carcinomas, adenocarcinomas, sarcomas, lymphomas, and neuro-ectodermal cancers; however, even within each group, there is much heterogeneity (Weinberg 2013). Six main characteristics of tumors have been proposed—the ability to proliferate without reliance on external growth signals, insensitivity to anti-growth signals, resistance to apoptosis, limitless replicative potential, the ability to encourage angiogenesis, and the ability to metastasize (Hanahan and Weinberg 2011). In many instances, this final characteristic is the one that results in the death of the individual, since there are very few effective treatments of metastatic cancer. As tumors grow, the genome becomes increasingly unstable with the acquisition of further deleterious mutations that promote malignant progression (Murphy et al. 2014).

The complexity of tumor pathology makes treatment very challenging. When a patient has a single solid tumor, it may be possible to remove it through surgical excision or by sterilization with external beam radiotherapy; however, in many cases, this is not possible. Even when surgery or radiotherapy is used as a first-line treatment, in many patients this approach is not curative. This can be for a number of reasons, e.g., lack of accessibility for complete excision, limitations in the dose of radiation used due to the risk of damage to critical normal structures in the radiation field, etc. When surgery or radiotherapy is not possible or limited, then cytotoxic chemotherapy (CCT) is the main treatment option. CCT can be used as a first-line treatment or as an adjunct to other therapeutic approaches. The aim of CCT is to destroy all cancer cells within the body, including those that have spread to distant tissues, i.e., metastasized (Kintzios 2004). Conventional CCT drugs are primarily designed to act on rapidly proliferating cells (King and Robins 2006). Unfortunately, many healthy cells in the body with high rates of proliferation can also be affected, e.g., cells in the bone marrow, intestinal villi, and hair follicles. Side effects are consequently moderate to severe, and include nausea, anemia, weakening of the immune system, hair loss, diarrhea, and vomiting. Due to the heterogeneous nature of tumors and their inherent genetic instability, tumors may develop resistance to CCT through selection for resistant sub-clones. The majority of the most common cancers are therefore not curable with a single CCT drug, and treatment approaches frequently include surgery and/or radiotherapy, where practical, and/or a combination of several drugs (Kintzios 2004).

More recently, research into CCT has focused on improving the specificity of cytotoxic agents by targeting specific features of the tumor micro-environment, including the poorly formed vasculature (Wu and Staton 2012), hypoxic tumor cell sub-populations (McKeown et al. 2007), and infiltrating inflammatory cells (Joyce 2005). In addition, considerable efforts are being made to identify molecular targets specific to tumor cells (Kaufman et al. 2008). However, till date, there have only been a few successes with this approach, and most novel agents still require combination with more conventional cytotoxic drugs and/or surgery/radiotherapy. One approach to identify new CCT agents is to mine the natural environment (Murphy et al. 2014).

6.2 Seaweeds as Sources of Anticancer Compounds

Most of the anticancer drugs currently used in chemotherapy are cytotoxic to normal cells and cause immunotoxicity (Harada et al. 1997). Many studies have focused on water soluble antitumor active substances from marine algae. Due to the biological properties of seaweeds, viz., reduction of plasma cholesterol, binding of biliary steroids, inhibition of carcinogenic fecal flora, binding of pollutants, stimulation of the immune system, and the protective effects of beta-sitosterols, seaweed is suggested as a breast cancer anticarcinogen (Teas et al. 2013). Certain algae have long been used in traditional Chinese herbal medicine in the treatment of cancer (Yamamoto et al. 1984). The habit of consuming seaweeds among the Japanese could be an important factor for lower breast cancer rates reported in Japan (Mariya and Ravindran 2013, Murphy et al. 2014).

In the past three decades, many researchers have worked on the antitumor activity of seaweeds. Extracts or even isolated compounds from green, brown, and red seaweeds have exhibited antitumor properties, especially *in vitro*, against human tumor cells (Sithranga and Kathiresan 2010, Murphy et al. 2014).

One way to determine if and how an extract can kill cancer cells is to measure inhibitors of kinase enzymes which are crucial to the survival of cancer cells. Kinase inhibitors provide a new target for anticancer agents that are more specific, efficacious, and with less toxic side effects, and there are several examples already in clinical trials (Dancey and Sausville 2003). In their project, Winberg et al. (2011), conducted assays on extracts from 12 seaweed taxa to determine if there was cytotoxic activity expressed as kinase inhibition (Table 6.1).

Regarding the red seaweeds (Rhodophyta) (see Table 6.1 and Section 6.5), extracts from the edible *Palmaria palmata* inhibited epithelial cancer cell (HeLa) proliferation *in vitro* (Yuan et al. 2005). The compound chondriamide-A from *Chondria* sp. exhibits cytotoxicity towards human nasopharyngeal and colorectal cancer cells. The alcoholic extract of the *Acanthophora spicifera* exhibited tumoricidal activity on Ehrlich's ascites carcinoma cells developed in mice (at 20 mg kg^{-1}), comparable to the standard drug 5-fluorouracil; which was further evidenced by increase in the mean survival time, decrease in tumor volume, and viable cell count (Sithranga and Kathiresan 2010). It is important to emphasize that *Chondria* sp. and *Acanthophora spicifera* are phylogenetically closed to *Osmundea pinnatifida*, since they belong to the family Rhodomelaceae (Silva 2015).

Sulfated polysaccharides, including fucoidans and carrageenans, inhibit tumor metastasis in rat test systems by inhibiting the action of the tumor cell-derived heparanases involved in membrane crossing (Parish and Snowden 1985, Coombe et al. 1987, Parish et al. 1987).

Sulfated polysaccharides from *Sargassum miyabei* (formerly *Sargassum kjellmanianum*) (see Table 6.1 and Section 6.4) (Iizima-Mizui et al. 1985) inhibited mouse S-180 tumor growth and carrageenan, while not active alone, significantly potentiated the effect of mitomycin against leukemia 1210 ascites tumor in mice (Matsumoto et al. 1988). Carrageenan has also been found to stimulate lectin dependent cell mediated cytotoxicity against HEp-2 human epipharynx carcinoma cells (Perl et al. 1983). Various brown algae (Phaeophyceae) viz., *Lessonia nigrescens*, *Saccharina japonica* (formerly *Laminaria japonica*), *Sargassum ringgoldianum*, *Scytosiphon lomentaria*, the red algae (Rhodophyta), *Pyropia yezoensis* (formerly *Porphyra yezoensis*), and *Betaphycus gelatinus* (formerly *Eucheuma gelatinae*), and the green alga (Chlorophyta, see Table 6.1 and Section 6.3), *Ulva prolifera* (formerly *Enteromorpha prolifera*) have shown antitumor activity against Meth-A fibrosarcoma (Noda et al. 1990).

Fucoidans isolated from the brown algae *Sargassum thunbergii* and *Sargassum miyabei* (formerly *S. kjellmanianum*) have shown antitumor activity (Itoh et al. 1995, Zhuang et al. 1995). Ulvan (see also Chapters 2 and 3) extracted from *Ulva lactuca* (Chlorophyta) has shown cytotoxicity against human colon cell line (Kaeffer et al. 1999). Fucosterols of *Turbinaria conoides* (Phaeophyceae) have displayed cytotoxicity against Murine (MU) and Human (HU) cell lines (Sheu et al. 1999). Caulerpenyne of *Caulerpa taxifolia* (Chlorophyta) has exhibited antitumor activity against HU neurobloma cell line by inhibiting microtubule assembly and tubulin aggregation (Barbier et al. 2001). Compounds of dihydroxysargaquinone and sargatriol from *Sargassum siliquastrum* (formerly *S. tortile*) and diterpene from *Sargassum notarisii* (formerly *S. crispum*) are known for their cytotoxic activities (Numata et al. 1991, Ayyad et al. 2001). Cytotoxicity of *Sargassum polycystum* against some human cancer cell lines (Ly et al. 2005) *in vitro*, antitumor and antiproliferative activity of *Hydroclathrus clathratus* (Ayyad et al. 1991), antitumor activity of the red algae *Gracilaria corticata* against Jurkat and molt-4 human cancer cell lines (Keivan et al. 2010), and *Spyridia filamentosa* against human prostate carcinoma epithelium like cell lines DU-145 (Taskin et al. 2010). The brown algae *Colpomenia sinuosa*, *Polycladia myrica* (formerly *Cystoseira myrica*), and *Sargassum swartzii* against colon carcinoma (HT-29), colorectal adenocarcinoma (Caco-2), breast ductal carcinoma (T47D), tamoxifen resistant breast ductal carcinoma (T47D-T. R), and estrogen independent breast carcinoma (MDAMB468) cell lines has been reported (Khanavi et al. 2010). The blue-green algae (Cyanobacteria) *Aphanizomenon flosaquae*, *Arthrospira platensis*, and the green microalgae (Chlorophyta) *Dunaliella salina* and *Haematococcus pluvialis* against the growth of human leukemia cell line HL-60 and the biphenotypic B myelomonocytic leukemia cell line MV-4-11 has been reported (Bechelli et al. 2011).

Table 6.1 Seaweed species and seaweed extracts with anticancer activity.

Species	Tested product, fraction or extract	Study model	Observed results	References
Chlorophyta (Green seaweed)				
Avrainvillea digitata	DCM-methanol (7:3) and water extracts	Tumor cells: Hep-2, HeLa, KB Normal cells: MDCK *In vivo*	Anti-proliferative activity of methanolic extract against KB cells; IC_{50} = 73.8 μg mL^{-1}	Moo-Puc et al. 2009
Bryopsis pennata	5-OHKF (a kahalalide derivative)	Tumor cells: SK-N-SH Normal cells *In vivo*	5-OHKF showed no anti-tumor effects	Gao et al. 2009
Bryopsis spp.	Kahalalide F	Tumor cells: CCL131 Normal cells *In vivo*	IC_{50} values against A-549, HT-29 and LOVO are 2.5, 0.25 and < 1.0 μg mL^{-1}, respectively; against P-388 and KB, 10 and > 10 μg mL^{-1}. Active against CV-1 cells with an IC_{50} of 0.25 μg mL^{-1}	Hamann and Scheuer 1993
Bryopsis spp.	Kahalalide F	Tumor cells: A-549, HT-29, LOVO, P-388, KB Normal cells: CV-1 *In vivo*	Showed no cytotoxic effect against neuroblastoma cells at concentration of 100 μg mL^{-1}	Kan et al. 1999
Bryopsis spp.	Kahalalide O	Tumor cells: P-388, A549, HT29, MEL28 Normal cells *In vivo*	Did not inhibit growth of cell lines at concentration of 10 μg mL^{-1}	Horgen et al. 2000
Bryopsis spp.	Kahalalide F (synthesised)	Tumor cells Normal cells *In vivo*: phase I clinical trial, 38 patients with advanced solid tumors. Dose 266–1,200 μg m^{-2}, 1 h weekly infusion	Recommended dose 650 μg m^{-2}, good safety profile	Pardo et al. 2008
Bryopsis spp.	Kahalalide F (synthesised)	Tumor cells Normal cells *In vivo*: phase II clinical trial, 24 patients with advanced malignant melanoma. Dose 650 μg m^{-2}, 1 h weekly infusion	Good safety profile, but no anti-tumor response. Trial closed	Martín-Algarra et al. 2009

Table 6.1 contd....

Table 6.1 contd....

Species	Tested product, fraction or extract	Study model	Observed results	References
Bryopsis spp.	Kahalalide F	Tumor cells Normal cells *In vivo*: phase I clinical trial, 106 patients with advanced solid tumors. Dose: up to 1,200 µg m^{-2}, 3–24 h weekly infusion	Aimed to determine recommended dose for further phase II studies of a prolonged weekly intravenous infusions of KF. 3 and 24 h infusion times were found to have an acceptable safety profile	Salazar et al. 2013
Capsosiphon fulvescens	Glycoprotein	Tumor cells: AGS Normal cells: IEC-6 *In vivo*	Reduced cell viability to 68% in AGS (dose 1–3 µg mL^{-1}), no effect on IEC-6. Induced apoptosis and sub-G1 arrest. Increased levels of Fas, FADD, cleaved caspases 3, 8 and 9, cleaved PARP, Bcl-2 family proteins, cytochrome C and Apaf-1	Kim et al. 2012b
C. fulvescens	Glycoprotein	Tumor cells: AGS Normal cells *In vivo*	Showed dose-dependent inhibition of growth and cell invasion. Expression of TJ proteins, MMP-2 and MMP-9 decreased	Kim et al. 2013b
Caulerpa chemnitzia (as *Caulerpa racemosa* var. *laetevirens*)	Methanolic extract	Tumor cells: L-1210 and P-388 Normal cells: NIH-3T3 *In vitro*	Extract exhibited high cytotoxicity to tumor cells and no cytotoxicity to normal cells	Harada et al. 1997
C. cupressoides	Methanolic extract	Tumor cells: L-1210 and P-388 Normal cells: NIH-3T3 *In vitro*	Extract exhibited high cytotoxicity to tumor cells and to normal cells	Harada et al. 1997
C. cupressoides	Sulfated polysaccharides	Tumor cells: HeLa Normal cells *In vivo*	38.4% of cell proliferation inhibition at 2.0 mg mL^{-1}	Costa et al. 2010
C. cupressoides	Methanolic seaweed extracts	Tumor cells: HeLa, SiHa	Extract displayed the highest inhibition of HeLa cells, approximately 32%. However, like the other tested extracts it also had just a time-dependent effect	Gomes et al. 2015
C. cylindracea (as *C. racemosa* var. *cylindracea*)	Caulerpenyne	Tumor cells: SHSY5Y, Kelly Normal cells *In vivo*	Cytotoxic to both cell lines, induces apoptosis	Cavas et al. 2006

C. cylindracea	Caulerpin and caulerpin acid	Tumor cells: A2780 *In vitro*	Both compounds were found to selectively inhibit respiratory complex II activity, while complexes I, III, and IV remained functional. This provided novel insight toward the potential use of metabolites from invasive *Caulerpa* species for the treatment of human ovarian carcinoma cisplatin-resistant cells	Ferramosca et al. 2016
C. microphysa	Pepsin-digested extracts	Tumor cells: WEHI-3, HL-60 Normal cells: RAW 264.7 *In vivo*	Cytotoxic effect seen at ≥ 25 μg mL^{-1} for both tumor cell lines. Little effect on RAW 264.7. Increased DNA damage in tumor cell lines	Lin et al. 2012
C. prolifera	Sulfated polysaccharides	Tumor cells: HeLa Normal cells *In vivo*	57.1% of cell proliferation inhibition at 0.1 mg mL^{-1}	Costa et al. 2010
C. prolifera	Methanolic seaweed extracts	Tumor cells: HeLa, SiHa	The inhibition did not surpass the value of 35% under any of the evaluated conditions. Nevertheless, they showed a dose-dependent and time-dependent inhibitory activity tendency The average inhibitory activity was 20%, however, this effect was not dose-dependent	Gomes et al. 2015
C. racemosa	Alkaloid: caulerpin	Tumor cells: C32 Normal cells: FEK4 *In vivo*	Caulerpin was cytotoxic to FEK4 cells	Rocha et al. 2007
C. racemosa	Sulfated polysaccharide with 3.9–7.9% uronic acid and protein	Tumor cells: K562, H22 *In vitro* and *in vivo*	Inhibition rate of K562 cells *in vitro* at the concentration of 6.0–10.0 mg mL^{-1} and of H22 tumor transplanted in mice at a dose of 100 mg kg^{-1} day were 59.5–83.8% (48 h) and 53.9% (14 days), respectively. Moreover, at a lower dose (0.05–0.2 mg mL^{-1}) and longer time (72 h), extract exhibited stronger inhibition effect on K562 cells	Ji et al. 2008b
C. racemosa	Methanol extract	Tumor cells: A549, B16F10, HL-60 Normal cells: Vero *In vitro*	IC_{50} = 30.17 μg mL^{-1} for HL-60 cells	Lakmal et al. 2014
C. serrulata f. *lata*	Methanolic extract	Tumor cells: L-1210 and P-388 Normal cells: NIH-3T3 *In vitro*	Extract exhibited cytotoxicity to tumor cells and no cytotoxicity to normal cells	Harada et al. 1997

Table 6.1 contd....

Table 6.1 contd....

Species	Tested product, fraction or extract	Study model	Observed results	References
C. sertularioides	Crude methanol, water and PBS extracts	Tumor cells: MOLT-4, K562, HeLa, KB Normal cells *In vivo*	The methanolic extract strongly inhibited telomerase activity when added to MOLT-4 cell culture at a level of 1.25% (v/v)	Kanegawa et al. 2000
C. sertularioides	Crude extract (ethanol or acetone)	Tumor cells: Caco-2 Normal cells *In vivo*	Both ethanol and acetone extracts decreased Caco-2 viability when cells were treated with 600 μM hydrogen peroxide	Gunji et al. 2007
C. sertularioides	Sulfated polysaccharides	Tumor cells: HeLa Normal cells *In vivo*	36.4% of cell proliferation inhibition at 2.0 mg mL^{-1}	Costa et al. 2010
C. sertularioides	Methanolic seaweed extracts	Tumor cells: HeLa, SiHa	The inhibition did not surpass the value of 35% under any of the evaluated conditions. Nevertheless, they showed a dose-dependent and time-dependent inhibitory activity tendency	Gomes et al. 2015
C. sertularioides f. *longipes*	Methanolic extract	Tumor cells: L-1210 and P-388 Normal cells: NIH-3T3 *In vitro*	Extract exhibited high cytotoxicity to tumor cells and low cytotoxicity to normal cells	Harada et al. 1997
Caulerpa spp.	Alkaloid: caulerpin	Tumor cells: T47D, MCF-7, MDAMB-231, DU145, PC-3 Normal cells: HMEC *In vivo*	Cytotoxic to all cell lines, PC-3 most sensitive, DU145 least. Studied HIF-1 inhibition: under hypoxic conditions, caulerpin may disrupt mitochondrial ROS-regulated HIF-1 activation and HIF-1 downstream target gene expression through inhibition of the transport or delivery of electrons to mitochondrial complex III	Liu et al. 2009b
C. taxifolia	Crude water and methanol (plus 10,11-epoxycaulerpenyne, taxifolial A and D, caulerpenyne)	Tumor cells Normal cells: BHK 21/C13 *In vivo*: Swiss, female	*In vitro*: seasonal differences found, IC_{50}: water/winter (800 μg mL^{-1}), methanol/winter (250 μg mL^{-1}), methanol/summer (150 μg mL^{-1}). *In vivo*: toxicity on mice was tested. Water extracts in summer/autumn was not toxic at 2 g kg^{-1}. In winter/spring, they were lethal at 1 g kg^{-1}. Methanol extracts from summer were lethal at 1 g/kg but not lethal at 1.5 g kg^{-1} in winter	Lemée et al. 1993

C. taxifolia	(1) 10,11-epoxycaulerpenyne, (2) taxifolial A, (3) taxifolial D, (4) caulerpenyne	Tumor cells Normal cells: BHK 21/C13 *In vivo*: Swiss mice	IC_{50} in BHK21/C13 cells: (1) 11 μg mL^{-1}, (4) 15 μg mL^{-1}, (2) and (3) both non-toxic at 20 μg mL^{-1}. *In vivo*: lethality: (1) 75 mg kg^{-1}, (2) and (4) both nontoxic at 100 mg kg^{-1}, (3) not determined	Lemée et al. 1993
C. taxifolia	Caulerpenyne	Tumor cells: several colorectal tumor cell lines and CAL 27 Normal cells *In vivo*	IC_{50}: colorectal cancer cells (approximately 7 μM) Cells exhibited early shift into S phase followed by a blockade in G_2/M	Fischel et al. 1995
C. taxifolia	Caulerpenyne	Tumor cells: SK-N-SH *In vitro*	IC_{50} of 10 ± 2 μM after 2 h of incubation	Barbier et al. 2001
C. taxifolia	Caulerpal A and B	Tumor cells: HL-60, MCF-7 Normal cells *In vivo*	No significant cytotoxicity was found against these cell lines	Mao et al. 2006
Chaetomorpha antennina	Aqueous extract	Tumor cells: MCF-7 *In vitro*	IC_{50} = 1020 μg mL^{-1}	Arulvasu et al. 2014
C. linum (as *C. crassa*)	Methanol extract	Tumor cells: A549, B16F10, HL-60 Normal cells: Vero *In vitro*	IC_{50} = 179.88 μg mL^{-1} for HL-60 cells	Lakmal et al. 2014
Cladophora herpestica (as *Cladophoropsis zollingeri*)	Crude methanol, water and PBS extracts	Tumor cells: MOLT-4, K562, HeLa, KB Normal cells *In vivo*	Methanol extract showed telomerase inhibiting activity	Kanegawa et al. 2000
Cladophoropsis vaucheriiformis	Methanol extract	Tumor cells: L-1210 and P-388 Normal cells: NIH-3T3 *In vitro*	Extract exhibited high viability (86%) to normal cells, showing selective cytotoxicity to tumor cells. Extract acted strongly only against leukemic cell lines L-1210 and P-388	Harada et al. 1997
C. vaucheriiformis	Methanol Phosphate-buffered saline (PBS)	Tumor cells: HDF, HL-60, L-1210, MOLT-4, NIH-3T3 *In vitro*	Extracts with most selective cytotoxic activity, did not show cytotoxicity to any human leukemic cell lines tested at 50 μg mL^{-1} However, this extracts showed strong cytotoxicity to two human leukemic cell lines and NIH-3T3 at 100 μg mL^{-1}	Harada and Kamei 1997
Codium adhaerens	Phosphate buffered saline extract	Tumor cells: L-1210 and P-388 Normal cells: NIH-3T3 *In vitro*	Extract exhibited cytotoxicity to tumor cells and no cytotoxicity to normal cells	Harada et al. 1997

Table 6.1 contd....

Table 6.1 contd....

Species	Tested product, fraction or extract	Study model	Observed results	References
C. coactum (formerly *C. coarctatum*)	Methanolic extract	Tumor cells: L-1210 and P-388 Normal cells: NIH-3T3 *In vitro*	Extract exhibited cytotoxicity to tumor cells and high cytotoxicity to normal cells	Harada et al. 1997
C. decorticatum	Glycoprotein (GLP) (about 48 kDa)	Tumor cells: MCF-7, Siha, A549 *In vivo*	The time-dependent inhibitory concentrations (IC_{50}) of GLP: 24 h–60 μg mL^{-1} (MCF-7), 75 μg mL^{-1} (SiHa), 55 μg mL^{-1} (A549) 48 h–45 μg mL^{-1} (MCF-7), 50 μg mL^{-1} (SiHa), 40 μg mL^{-1} (A549)	Senthilkumar and Jayanthi 2016
C. isthmocladum	Sulfated polysaccharides	Tumor cells: HeLa Normal cells *In vivo*	42.1% of cell proliferation inhibition at 2.0 mg mL^{-1}	Costa et al. 2010
C. isthmocladum	Methanolic seaweed extracts	Tumor cells: HeLa, SiHa	*C. isthmocladum* extract, after 72 h treatment, also showed dose-independent inhibitory activity of around 20%, however different from *C. racemosa* extract, which also exhibited inhibitory activity (~ 22%) after 48 h under the experimental conditions used	Gomes et al. 2015
C. fragile	Seaweed powder	Tumor cells: Ehrlich carcinoma *In vivo*	Oral administration of 1600 mg seaweed powder per kg mouse per day for 28 days. Inhibition rate: 43.34	Noda et al. 1990
C. fragile	Methanolic extract	Tumor cells: L-1210 and P-388 Normal cells: NIH-3T3 *In vitro*	Extract exhibited high cytotoxicity to tumor cells and high cytotoxicity to normal cells	Harada et al. 1997
C. fragile	Methanolic extract	Tumor cells: HT-29, KB Normal cells: NIH-3T3 *In vitro*	IC_{50} > 50 (KB), > 50 (HT-29), > 50 (NIH-3T3) μg mL^{-1}	Xu et al. 2004b
C. fragile	Siphonaxanthin	Tumor cells: HL-60 Normal cells *In vivo*	Siphonaxanthin induces apoptosis through caspase-3 activation (dose range 5–20 μM), but siphonaxanthin was cytotoxic. It increased expression of GADD45α and DR5 and suppressed Bcl-2	Ganesan et al. 2011

C. intricatum	Phosphate buffered saline extract	Tumor cells: L-1210 and P-388 Normal cells: NIH-3T3 *In vitro*	Extract exhibited cytotoxicity to tumor cells and no cytotoxicity to normal cells	Harada et al. 1997
C. pugniforme	Water and ethanol extract	Tumor cells: EAC, S-180 *In vitro*	EAC: Inhibition = 41.6–53.9% (0.5–1.0 mg/mouse/day) S-180: Inhibition = 37.2% (0.6–4.2 mg/mouse/day)	Nakazawa et al. 1974
Codium sp.	Ethanol extract 100 μg mL^{-1}	Protein kinase A inhibition *In vitro*	1–25% inhibition	Winberg et al. 2011
C. tenue	Methanolic extract	Tumor cells: L-1210 and P-388 Normal cells: NIH-3T3 *In vitro*	Extract exhibited cytotoxicity to tumor cells and high cytotoxicity to normal cells	Harada et al. 1997
Enteromorpha attenuata	Methanolic extract	Tumor cells: HepG2, MCF-7 Normal cells: VERO cell lines *In vivo*	The average inhibitory activity was 87.43 and 64.75%, respectively, using 500 μg mL^{-1} of extract Only at a high concentration of 1 mg mL^{-1}, 60% cytoxicity was seen in normal VERO cell lines	Narasimhan et al. 2013
Gayralia oxysperma	Sulfated hetero-rhamnans	Tumor cells: U87MG *In vitro* (MTT assay)	The homogeneous products and the crude extracts containing the sulfated hetero-rhamnans showed cytotoxic effect against U87MG cells	Ropellato et al. 2015
Halicoryne wrightii	Crude methanol, water and PBS extracts	Tumor cells: MOLT-4, K562, HeLa, KB Normal cells *In vivo*	Methanol extract showed telomerase inhibiting activity	Kanegawa et al. 2000
Halimeda discoidea	Methanolic extract	Tumor cells: L-1210 and P-388 Normal cells: NIH-3T3 *In vitro*	Extract exhibited cytotoxicity to tumor cells and no cytotoxicity to normal cells	Harada et al. 1997
H. discoidea	Ethyl acetate extract	Tumor cells: U937, HL-60, HuH-7 Normal cells *In vivo*	Concentrations: 25–200 mg mL^{-1} Inhibition effects on U937 cells: 31.8–87.3% Inhibition effects on HL-60 cells: 15.4–98.1%	Huang et al. 2005
H. incrassata	Phosphate buffered saline extract	Tumor cells: L-1210 and P-388 Normal cells: NIH-3T3 *In vitro*	Extract exhibited high cytotoxicity to tumor cells and high cytotoxicity to normal cells	Harada et al. 1997
H. macroloba	Methanolic extract	Tumor cells: L-1210 and P-388 Normal cells: NIH-3T3 *In vitro*	Extract exhibited high cytotoxicity to tumor cells and no cytotoxicity to normal cells	Harada et al. 1997

Table 6.1 contd....

Table 6.1 contd....

Species	Tested product, fraction or extract	Study model	Observed results	References
H. macroloba	Crude extract (ethanol or acetone)	Tumor cells: Caco-2 Normal cells *In vivo*	Both ethanol and acetone extracts decreased Caco-2 viability when cells were treated with 600 μM hydrogen peroxide	Gunji et al. 2007
*Mougeotia nummuloides**	Methanol extract	Tumor cells: C6, HeLa Normal cells: Vero *In vitro*	Extract significantly inhibited the proliferation of Vero, HeLa and C6 cancer cell lines with IC_{50} and IC_{75} values. *M. nummuloides* extract exhibited higher activity than 5-FU and cisplatin on Vero and C6 cells at high concentrations	Erenler et al. 2016
Monostroma nitidum	Seaweed powder	Tumor cells: Ehrlich carcinoma; Meth-A fibrosarcoma *In vivo*	Oral administration of 1600 mg seaweed powder per kg mouse per day for 28 days. Inhibition rate: 24.3% Intraperitoneal administration of 50 mg kg^{-1} d^{-1} for seven days. Inhibition rate: 12.3%	Noda et al. 1990
M. nitidum	Sulfated polysaccharides	Tumor cells: Meth A fibrosarcoma *In vivo*	Intraperitoneal administration of 40 mg kg^{-1} for seven days. Inhibition rate: 26.5%	Noda et al. 1990
Penicillus dumetosus	DCM-methanol (7:3) and water extracts	Tumor cells: Hep-2, HeLa, KB Normal cells: MDCK *In vivo*	Anti-proliferative activity of methanolic extract against KB cells; IC_{50} = 88.2 μg mL^{-1}	Moo-Puc et al. 2009
P. dumetosus	DCM-methanol crude extract	Tumor cells: Hep-2, HeLa, SiHa, KB Normal cells: MDCK *In vivo*	Effect of light on extract IC_{50} was studied. In general, increased light exposure led to decreased IC_{50} values. Light treatment also increased cytotoxicity in normal cells, although the extract was less cytotoxic to normal cells	Moo-Puc et al. 2011b
Rhipocephalus phoenix	DCM-methanol (7:3) and water extracts	Tumor cells: Hep-2, HeLa, KB Normal cells: MDCK *In vivo*	Anti-proliferative activity of methanolic extract against KB cells; IC_{50} = 95.3 μg mL^{-1}	Moo-Puc et al. 2009
Rhizoclonium riparium	Methanolic extract	Tumor cells: HeLa *In vitro*	IC_{50} = 506.081 μg mL^{-1}	Paul and Kundu 2013
Tydemania expeditionis	3 sulfated cycloartanes	–	IC_{50} values of 100, 32 and 39 μM. Activity mediated through inhibition of protein kinase	Govindan et al. 1994

T. expeditionis	Unsaturated fatty acids	Tumor cells: BT-549, DU4475, MDAMB-468, MDA-MB-231, HCT116, NCI-H446, SHP-77, PC-3, LNCaPFGC, DU145, A2780/DDP-S, CCRF-CEM Normal cells *In vivo*	IC_{50} values from 1.3 to 14.4 μM	Jiang et al. 2008
Udotea conglutinata	DCM-methanol (7:3) and water extracts	Tumor cells: Hep-2, HeLa, KB Normal cells: MDCK *In vivo*	Anti-proliferative activity of methanolic extract against HeLa and KB cells; IC_{50} = 60.8 and 66.3 μg mL^{-1}, respectively	Moo-Puc et al. 2009
U. flabellum	DCM-methanol (7:3) and water extracts	Tumor cells: Hep-2, HeLa, KB Normal cells: MDCK *In vivo*	Anti-proliferative activity of methanolic extract against HeLa and KB cells; IC_{50} = 45.5 and 47.5 μg mL^{-1}, respectively	Moo-Puc et al. 2009
U. flabellum	Crude extract. PP content 1.7–3.4%	Tumor cells: Hep-2, HeLa, SiHa and KB Normal cells: MDCK *In vivo*	Cytotoxic to cancerous cells, much less cytotoxic to normal cells. Authors do not link anti-cancer activity and PP content, but discuss the possibility that purified PP extracts may have anti-cancer activity. Also found link between increasing light exposure and increasing cytotoxicity	Moo-Puc et al. 2011a
Ulva australis (as *Ulva pertusa*)	Seaweed powder	Tumor cells: Ehrlich carcinoma; Meth-A fibrosarcoma *In vivo*	Oral administration of 1600 mg seaweed powder per kg mouse per day for 28 days. Inhibition rate: 32.6% Intraperitoneal administration of 50 mg kg^{-1} d^{-1} for seven days. Inhibition rate: 32.6%	Noda et al. 1990
U. australis (as *U. pertusa*)	Methanolic extract	Tumor cells: HT-29, KB Normal cells: NIH-3T3 *In vitro*	IC_{50} = 29.9 (KB), 14.9 (HT-29), 23.8 (NIH-3T3) μg mL^{-1}	Xu et al. 2004b
U. australis	Ethanol extract 100 μg mL^{-1}	Protein kinase A inhibition *In vitro*	0% inhibition	Winberg et al. 2011
U. flexuosa	Ethanol extract 100 μg mL^{-1}	Protein kinase A inhibition *In vitro*	1–25% inhibition	Winberg et al. 2011

Table 6.1 contd....

Table 6.1 contd....

Species	Tested product, fraction or extract	Study model	Observed results	References
U. flexuosa	Ethylacetate and methanol extracts	Tumor cells: MCF-7, HeLa Normal cells: Vero *In vitro*	Ethylacetate fraction exhibited cytotoxicity to MCF-7, HeLa and Vero (IC_{50} = 100 µg ml^{-1}). Methanol fraction also exhibited cytotoxic activity in MCF7, Vero (IC_{50} > 100 µg mL^{-1}) and HeLa (IC_{50} = 100 µg mL^{-1})	Mashjoor et al. 2016
U. intestinalis (formerly *Enteromorpha intestinalis*)	Methanolic extract	Tumor cells: L-1210 and P-388 Normal cells: NIH-3T3 *In vitro*	Extract exhibited high cytotoxicity to tumor cells and high cytotoxicity to normal cells	Harada et al. 1997
U. intestinalis (formerly *E. intestinalis*)	Polysaccharides	Tumor cells: S-180 *In vitro* and *in vivo*	The tumor inhibition was 61.17%, 67.65% and 70.59% at doses of 100, 200 and 400 mg kg^{-1}, respectively. However, no direct cytotoxicity was detected. At dose of 100, 200 and 400 mg kg^{-1}, a significant increase ($P < 0.01$) in relative spleen and thymus weight, and production of tumor necrosis factor α (TNF-α) were observed in extracted polysaccharides treated groups	Jiao et al. 2009
U. intestinalis (formerly *E. intestinalis*)	Methanolic extract	Tumor cells: HeLa *In vitro*	IC_{50} = 309.048 µg mL^{-1}	Paul and Kundu 2013
U. intestinalis (formerly *E. intestinalis*)	Acetone extract	Tumor cells: Fem-x, A549, LS174 and K562 cell lines	Extract of *U. intestinalis* expressed the stronger cytotoxic activity towards all tested cell lines with IC_{50} values ranging from 74.73 to 155.39 µg mL^{-1}	Kosanić et al. 2015
U. lactuca	Water soluble from methanol extract	Tumor cells: U937, RAW 264.7 Normal cells: splenocytes *In vivo*	Inhibited tumor cell growth whilst stimulating splenocytes (dose 25–100 µg mL^{-1}); extract increased NO production in RAW 264.7 cells	Lee et al. 2004b
U. lactuca	Methanol extract	Tumor cells: EAT *In vivo*	Group that received extract alone showed slight changes in serum tumor marker levels in comparison with the control group	Ahmed et al. 2011
U. lactuca	Ethanol extract 100 µg mL^{-1}	Protein kinase A inhibition *In vitro*	1–25% inhibition	Winberg et al. 2011
U. lactuca	Crude extract. Extract standardized to percentage PP, PP content of 0-50 µg mL^{-1} tested	Tumor cells: Caco-2 Normal cells *In vivo*	Extracts reduced viability of Caco-2 cells	Nwosu et al. 2011

U. lactuca	Dichloromethane, chloroform, methanol, ethanol, water extracts, hexane and chloroform fractions	Tumor cells: HEp-2, K562, NCI-H292 *In vitro*	Water extract (48.5 μg mL^{-1}) was active against HEp-2	Guedes et al. 2013
U. lactuca	Acetone extract	Tumor cells: Fem-x, A549, LS174 and K562 cell lines	Extracts of *U. lactuca* expressed very weak cytotoxic activity towards Fem-x and K562 cell lines with IC_{50} values range 93.31 and 169.54 μg mL^{-1}, while cytotoxic activity against A569 and LS174 cells was not detected (IC_{50} > 200 mg mL^{-1})	Kosanić et al. 2015
U. lactuca	Ulvan	Tumor cells: MCF-7 *In vivo*	50 mg kg^{-1} body weight every other day for 10 weeks were administered orally to the Wistar rats	Abd-Ellatef et al. 2017
U. linza (formerly *Enteromorpha linza*)	Methanolic extract	Tumor cells: HT-29, KB Normal cells: NIH-3T3 *In vitro*	IC_{50} > 50 (KB), > 50 (HT-29), > 50 (NIH-3T3) μg mL^{-1}	Xu et al. 2004b
U. linza (formerly *E. linza*)	Methanolic extract	Tumor cells: HepG2, MCF-7 Normal cells: VERO cell lines *In vivo*	The average inhibitory activity was 71.44 and 84.77%, respectively, using 500 μg mL^{-1} of extract Only at a high concentration of 1 mg mL^{-1}, 60% cytoxicity was seen in normal VERO cell lines	Narasimhan et al. 2013
U. linza (formerly *U. fasciata*)	Sulfolipids	Tumor cells: MCF7, HepG2 *In vitro*	Concentrations = 1.0–10.0 μg mL^{-1} Growth inhibition = 65.40–79.48% (MCF7), 48.15–72.97% (HepG2)	El Baz et al. 2013
U. linza (formerly *U. fasciata*)	Methanolic extract	Tumor cells: A549 Normal cells: VERO *In vitro*	Cytotoxic effect against human A549 lung adenocarcinoma cancer cell lines in a concentration dependent manner, with IC_{50} = 12 μg mL^{-1}	Rani et al. 2013
U. linza (formerly *U. fasciata*)	Aqueous extract	Tumor cells: MCF-7 *In vitro*	IC_{50} = 300 μg mL^{-1}	Arulvasu et al. 2014
U. linza (formerly *U. fasciata*)	Aqueous extract	Tumor cells: EAC Silver Nanoparticle's of aqueous extract (AgNPs) *In vitro*	Cytotoxic effect (62–94%) of different AgNPs concentrations (42–98 μg mL^{-1})	Khalifa et al. 2016
U. prolifera (formerly *Enteromorpha prolifera*)	Seaweed powder	Tumor cells: Ehrlich carcinoma *In vivo*	Oral administration of 1600 mg seaweed powder per kg mouse per day for 28 days Inhibition rate: 51.7%	Noda et al. 1990

Table 6.1 contd....

Table 6.1 contd....

Species	Tested product, fraction or extract	Study model	Observed results	References
U. prolifera (formerly *E. prolifera*)	Methanol/acetone extraction	Tumor cells Normal cells *In vivo*: ICR mice with DMBA-induced skin tumors; after 1 week TPA applied twice-weekly for 20 weeks	Skin application of extract significantly reduced tumorigenesis. 4 treatment combinations were tested; the largest reduction was seen when extract was applied before each application of both DMBA and TPA	Higashi-Okai et al. 1999
U. reticulata	Crude extract (ethanol or acetone)	Tumor cells: Caco-2 Normal cells *In vivo*	Both ethanol and acetone extracts decreased Caco-2 viability when cells were treated with 600 µM hydrogen peroxide	Gunji et al. 2007
U. rigida	Methanol extract	Tumor cells: EAC *In vitro*	The maximal protective effects against hepatic lipid peroxidation (64.8 ± 4.5 and 69.6 ± 13.3 nmol g^{-1} of tissue) were observed when solid tumor bearing mice treated with 80 and 100 µg mL^{-1} of methanol extracts of *Ulva rigida*, respectively	Salem and Ibrahim 2011
Phaeophyceae (Brown seaweed)				
Alaria esculenta	Crude (DCM/MeOH) extract. PP content 1.0–10.9%	Tumor cells: Daudi, Jurkat, K562 Normal cells *In vivo*	Crude extract concentrations: 50–500 mg L^{-1}; Inhibition: 20.40–76.64%	Zubia et al. 2009b
A. esculenta	Crude extract. Extract standardized to percentage PP, PP content of 0-50 µg mL^{-1} tested	Tumor cells: Caco-2 Normal cells *In vivo*	Extracts reduced viability of Caco-2 cells	Nwosu et al. 2011
A. esculenta	Seaweed in capsules	Insulin-like growth factor 1 (IGF-1) is an anabolic hormone important for growth and development. However, high-circulating serum concentrations in adults are associated with increased risk of postmenopausal breast cancer	Participants consumed 5 g/day placebo or seaweed in capsules for 7 wk. During the 7th wk, a high-soy protein isolate powder was added (2 mg kg^{-1} body weight aglycone equivalent isoflavones). Overnight fasting blood samples were collected after each intervention period. Soy significantly increased serum IGF-1 concentrations compared to the placebo (21.2 nmol L^{-1} for soy vs. 16.9 nmol L^{-1} for placebo; P = 0.0001). The combination of seaweed and soy significantly reduced this increase by about 40% (21.2 nmol L^{-1} for soy alone vs. 19.4 nmol L^{-1}; P = 0.01). Concurrent seaweed and soy consumption may be important in modifying the effect of soy on IGF-1 serum concentrations	Teas et al. 2011

Alaria sp.	Fucoidan	Tumor cells: T-47D and RPMI-7951 *In vitro*	Inhibiting proliferation and colony formation	Vishchuk et al. 2012
Ascophyllum nodosum	Fucoidan	Tumor cells: MDA-MB-231 Normal cells: HUVECs *In vivo*	Fucoidan showed 66% inhibition of the tumor cell adhesion to human platelets	Cumashi et al. 2007
A. nodosum	Unfractionated fucoidan	Tumor cells: HCT116 *In vitro*	Fucoidan treatment of HCT116 cells induced activation of caspases-9 and -3 and the cleavage of PARP, led to apoptotic morphological changes, and altered mitochondrial membrane permeability	Foley et al. 2011
A. nodosum	Crude extract. PP content 0.2–0.5%	Tumor cells: Caco-2 Normal cells *In vivo*	Antioxidant and antiproliferative activity of the extracts was examined. Extracts had antiproliferative effects (dose 0.55–5.5 mg mL^{-1} tested). Good anti-oxidant activity *in vitro*. Caco-2 cells used as a model for antioxidant effect rather than anticancer model	O'Sullivan et al. 2011
A. nodosum	Sulfated polysaccharide (ascophyllan)	Tumor cells: B16 melanoma	Anti-metastatic effects	Abu et al. 2015
A. nodosum	Fucoidan	Tumor cells: YAC-1 *In vivo*	Fucoidan of *A. nodosum* delayed human neutrophil apoptosis (50%) at higher concentration, prevented apoptosis at concentration of 50–100 μg mL^{-1}	Zhang et al. 2015
Asperococcus bullosus	Crude (DCM/MeOH) extract. PP content 1.0–10.9%	Tumor cells: Daudi, Jurkat, K562 Normal cells *In vivo*	Crude extract concentrations: 50–500 mg L^{-1}; Inhibition: 8.43–44.61%	Zubia et al. 2009b
Bifurcaria bifurcata	Crude chloroform/methanol extract	Tumor cells: NSCLC-N6 Normal cells *In vivo*	The extract has an IC_{50} of 4 μg mL^{-1} after 72 h of treatment on this particularly Chemo-resistant cell line and induced G_1 cell cycle arrest	Moreau et al. 2006
B. bifurcata	Crude (DCM/MeOH) extract. PP content 1.0–10.9%	Tumor cells: Daudi, Jurkat, K562 Normal cells *In vivo*	Crude extract concentrations: 50–500 mg L^{-1}; Inhibition: 54.57–90.97%	Zubia et al. 2009b
B. bifurcata	Methanolic and dichloromethane extracts	Tumor cells: Caco-2 cells, HepG-2 cells *In vitro*	IC_{50} of cell proliferation inhibition and cytotoxicity of the methanolic extract against Caco-2 (82.31 μg mL^{-1}, 90.09 μg mL^{-1}, respectively), and against HepG-2 cells (95.63 μg mL^{-1}, 123.9 μg mL^{-1}, respectively)	Alves et al. 2016b

Table 6.1 contd....

Table 6.1 contd....

Species	Tested product, fraction or extract	Study model	Observed results	References
Canistrocarpus cervicornis (formerly *Dictyota cervicornis*)	Sulfated polysaccharides	Tumor cells: HeLa Normal cells *In vivo*	About 32.0% of cell proliferation inhibition at 2.0 mg mL^{-1}	Costa et al. 2010
Chorda filum	Methanolic extract	Tumor cells: HT-29, KB Normal cells: NIH-3T3 *In vitro*	IC_{50} > 50 (KB), > 50 (HT-29), > 50 (NIH-3T3) µg mL^{-1}	Xu et al. 2004b
Chordaria flagelliformis	Phosphate buffered saline extract	Tumor cells: L-1210 and P-388 Normal cells: NIH-3T3 *In vitro*	Extract exhibited showing selective cytotoxicity to tumor cells. Extract acted strongly only against leukemic cell lines L-1210 and is less toxic to normal cells	Harada et al. 1997
Cladosiphon novae-caledoniae	Fucoidan Low mol weight fucoidan (< 500 Da)	Tumor cells: MDA-MB-231, MCF-7, HU fibro-sarcoma HT1080 cells; HeLa cells Normal cells *In vitro* and *in vivo*	HeLa cells treated with fucoidan extract (fucoidan ex.–HeLa) significantly suppressed all parameters of tubules formation assay Synergistic, dose-dependent reduction in cytotoxicity towards both cell lines in combination treatment with cisplatin, tamoxifen or paclitaxel. Induction of apoptosis, reduced expression of Bcl-xLandMcl-1 (both cell lines), increased expression of Bax (MCF7only). Modulation of ERK and Akt phosphorylation in certain combinations. Enhanced ROS and reduced GSH	Ye et al. 2005 Zhang et al. 2013
C. novae-caledoniae	Fucoidan	Tumor cells: MCF-7, MDA-MB-231, HeLa, HT1080 *In vitro*	Inducing apoptosis of MCF-7 cells and inhibiting growth of MCF-7, MDA-MB-231, HeLa, HT1080 cell lines	Zhang et al. 2011
C. okamuranus	Seaweed powder	Tumor cells: Ehrlich carcinoma; Meth-A fibrosarcoma *In vivo*	Oral administration of 1600 mg seaweed powder per kg mouse per day for 28 days. Inhibition rate: 47.8% Intraperitoneal administration of 50 mg kg^{-1} d^{-1} for seven days. Inhibition rate: 24.8%	Noda et al. 1990
C. okamuranus	Fucoidan	Tumor cells: Human T cell leukemia *In vitro* and *in vivo*	Partial inhibition of growth of tumors of an HTLV-1-infected T-cell line transplanted subcutaneously in severe combined immune deficient mice	Haneji et al. 2005

C. okamuranus (commercially cultured)	Fucoidan and its sulfate derivatives	Tumor cells: U-937 *In vivo*	Oversulfated fucoidan is derived from native fucoidan from *C. okamuranus* by chemical modification. The oversulfated fucoidan exhibited anti-proliferative activity in U937 cells, but the activity of native fucoidan was weak. The activity of oversulfated fucoidan is related to apoptosis via caspase-3 and -7 activation-dependent pathway	
C. okamuranus	Fucoidan	Tumor cells: HCC, KMG-C, KMC-1 *In vitro*	Cell proliferation was suppressed in 13 cell lines in a time- and/or dose-dependent manner; this suppression was marked in the hepatocellular carcinoma, cholangio carcinoma and gallbladder carcinoma cell lines. In contrast, proliferation of the neuroblastoma and 1 of the 2 ovarian carcinoma cell lines was not affected	Fukahori et al. 2008
C. okamuranus	Fucoidan	Tumor cells: HCC, HuH-6 *In vivo*	Inhibiting growth by stimulating macrophages suppressing cell growth	Hayakawa and Nagamine 2009
C. okamuranus	Fucoidan	Tumor cells: Huh7 and HepG2 *In vitro*	Fucoidan inhibited the growth of Huh7 cells and HepG2 cells in a dose-dependent manner, with a 50% inhibition of cell growth (IC_{50}) of 2.0 and 4.0 mg mL^{-1}, respectively. α-fetoprotein levels in medium collected from fucoidan-treated cells were significantly decreased in Huh7 cells but not in HepG2 cells	Nagamine et al. 2009
C. okamuranus (commercially cultured)	Acetyl fucoidan	Tumor cells: RAW 264.7 *In vitro*	Acetyl fucoidan induced activation of macrophages leading to NO, TNF-α and IL-6 production. Such effects were not due to endotoxin contamination, but to acetyl fucoidan itself. Consequently, acetyl fucoidan has potent effects on macrophage activation. Acetyl fucoidan-induced NO production was significantly decreased by desulfation and deacetylation of acetyl fucoidan. The results suggested that sulfate and acetyl groups of acetyl fucoidan act as active sites, or contribute to maintain the conformation of active form of acetyl fucoidan on acetyl fucoidan-induced NO production	Teruya et al. 2009

Table 6.1 contd....

Table 6.1 contd....

Species	Tested product, fraction or extract	Study model	Observed results	References
C. okamuranus	Fucoxanthin, fucoxanthinol	Tumor cells: BCBL-1, HeLa, TY-1 infected with human herpes virus 8 Normal cells: PBMC *In vivo*: Female C.B-17/Icr-SCID mice injected i.p. with BCBL-1 cells	*In vitro*: both caused cell cycle arrest in G_1 and caspase-dependent apoptosis; also reduced NF-κB, AP-1 and Akt and down-regulated anti-apoptotic proteins and cell cycle regulators. Proteasome degradation was responsible for the low levels of proteins after fucoxanthinol treatment. *In vivo*: reduction in tumors in mice treated with fucoxanthin	Yamamoto et al. 2011
C. okamuranus	Fucoidan	Tumor cells: Colon 26 *In vivo*	Mice fed fucoidan were fed about 5 g kg^{-1} day of each fucoidan Natural killer (NK) cell-mediated Showed significantly increased survival times compared with that observed in the control group	Azuma et al. 2012
C. okamuranus	Fucoidan	Tumor cells: Sarcoma-180 *In vitro*	The anti-tumor activity of fucoidan on S-180 cells is mediated through increased NO production by fucoidan-stimulated macrophages via nuclear factor-κB-dependent signaling pathway	Takeda et al. 2012
C. okamuranus	Fucoidan	Tumor cells: MSN45 *In vitro*	Fucoidan impeded the MKN45 cell cycle by approximately 50%, and inhibited cell proliferation. LDH assays showed no immediate cytotoxic effects of fucoidan at 24 h exposure, however longer time courses revealed cell growth inhibition at 4 days in a dose-dependent manner	Yoshimoto et al. 2015
Coccophora langsdorfii	Phosphate buffered saline extract	Tumor cells: L-1210 and P-388 Normal cells: NIH-3T3 *In vitro*	Extract exhibited highly cytotoxicity to tumor cells and less cytotoxicity to normal cells	Harada et al. 1997
C. langsdorfii	Fucoidan	Tumor cells: SK-MEL-5, SK-MEL-28, MDA-MB-231 *In vitro*	Fucoidan inhibited colony formation of SK-MEL-5 and SK-MEL-28 melanoma cells (the percentage of inhibition was 28 and 76, respectively) and weakly inhibited colony formation of breast adenocarcinoma cells MDA-MB-231 (the percentage of inhibition was about 5)	Imbs et al. 2016

Colpomenia bullosa	Seaweed powder	Tumor cells: Ehrlich carcinoma *In vivo*	Oral administration of 1600 mg seaweed powder per kg mouse per day for 28 days. Inhibition rate: 18.7%	Noda et al. 1990
C. sinuosa	Phosphate buffered saline extract	Tumor cells: L-1210 and P-388 Normal cells: NIH-3T3 *In vitro*	Extract exhibited cytotoxicity to tumor cells and no cytotoxicity to normal cells	Harada et al. 1997
C. sinuosa	Methanolic extract	Tumor cells: HT-29, KB Normal cells: NIH-3T3 *In vitro*	IC_{50} > 50 (KB), > 50 (HT-29), > 50 (NIH-3T3) μg mL^{-1}	Xu et al. 2004b
C. sinuosa	Ethyl acetate extract	Tumor cells: U937, HL-60, HuH-7 Normal cells *In vivo*	Concentrations: 25–200 mg mL^{-1} Inhibition effects on U937 cells: 16.4–96.1% Inhibition effects on HL-60 cells: 39.2–99.8%	Huang et al. 2005
C. sinuosa	Ethanol/water extract	Tumor cells: HL-60, HT-29 Also tested: B-16, A-549 Normal cells: HaCaT *In vivo*	Concentration of extract: 100 μg mL^{-1} HL-60 inhibition = 86.5% HT-29 inhibition = 34.3% B16 inhibition = 37.6% A549 inhibition = 1.1% HaCat inhibition = 17.3%	Kim et al. 2009
Colpomenia sp.	Ethanol extract 100 μg mL^{-1}	Protein kinase A inhibition *In vitro*	26–50% inhibition	Winberg et al. 2011
Costaria costata	Water-ethanol extracts	Tumor cells: DLD-1, HT-29 Normal cells *In vivo*	Extracts obtained from *Costaria* also varied in their inhibitory effect: extracts of May specimens suppressed colony growth by 38 and 31%; those of July specimens, by 50 and 44%, for DLD-1 and HT-29 cells, respectively	Imbs et al. 2009
C. costata	Fucoidan	Tumor cells: SK-MEL-28, DLD-1 *In vitro*	Inhibiting colony formation	Ermakova et al. 2011
Cystophora moniliformis	Linear and cyclic C18 terpenoids	Tumor cells: P-388 Normal cells *In vivo*	Most compounds (7) showed little antitumor activity (IC_{50s} > 40 μM). An inseparable mixture of 2 compounds (3:1 ratio) displayed moderate antitumor activity (IC_{50} of 45 μM)	Reddy and Urban 2008

Table 6.1 contd....

Table 6.1 contd....

Species	Tested product, fraction or extract	Study model	Observed results	References
Cystoseira amentacea var. *stricta* (as *Cystoseira stricta*)	Ethanol extract	Tumor cells: MCF-7, LNCaP, PC-3 Normal cells: MCF-10A *In vitro*	Extract shows cytotoxic effects on the cancer cells (LNCa, PC-3 and MCF-7) but not on the non-tumorigenic cell line MCF-10, suggesting that these extracts might be interesting for further studies regarding their potential anticancer effects. Extract showed significant effects on human breast adeno-carcinoma MCF-7 cells (87% inhibition of growth)	Montalvão et al. 2016
C. barbata	Ethanol extract	Tumor cells: MCF-7, LNCaP, PC-3 Normal cells: MCF-10A *In vitro*	Extract shows cytotoxic effects on the cancer cells (LNCa, PC-3 and MCF-7) but not on the non-tumorigenic cell line MCF-10, suggesting that these extracts might be interesting for further studies regarding their potential anticancer effects	Montalvão et al. 2016
C. compressa	Methanol and chloroform, ethyl acetate and methanol fractions	Tumor cells: A549, HCT15, MCF7 Normal cells *In vivo*	IC_{50} for chloroform fraction (78–80 µg mL^{-1}), ethyl acetate fraction (27–50 µg mL^{-1}) and methanol fraction (110–130 µg mL^{-1})	Mhadhebi et al. 2012
C. compressa	Water extract	Tumor cells: A549 (lung cell carcinoma), HCT15 (colon cell carcinoma), MCF7 (breast adenocarcinoma) Normal cells: MDCK (Mardin–Darby canine kidney) and rat fibroblast *In vitro*	IC_{50} = 90.3, 20.3, 29.5, 510.5, 450.3 µg mL^{-1}, respectively	Mhadhebi et al. 2014
C. crinita	Water extract	Tumor cells: A549 (lung cell carcinoma), HCT15 (colon cell carcinoma), MCF7 (breast adenocarcinoma) Normal cells: MDCK (Mardin–Darby canine kidney) and rat fibroblast *In vitro*	IC_{50} = 49.5, 26.4, 17.9, 192.4, 182.6 µg mL^{-1}, respectively	Mhadhebi et al. 2014

C. mediterranea	Mediterraneol A (hydroquinone diterpene)	Tumor cells Normal cells *In vivo*: P388 mouse leukemia screen	Range of novel structures reported. They inhibited mitotic division of fertilized urchin eggs with an ED_{50} in the range 2 μg mL^{-1} and also showed *in vivo* activity. Detailed results not given as compounds were screened by NCI, presumably unpublished. Treatment/control of 128% at dose of 32 mg kg^{-1}	Francisco et al. 1985, 1986
C. mediterranea	Methanol extract	Tumor cells: DU-145, LNCaP, MCF-7, PC-3 *In vitro*	Extract in 100 g mL^{-1} concentration showed 55% cell inhibition against the MCF-7 cells	Taskin et al. 2010
C. sedoides	Water extract	Tumor cells: A549 (lung cell carcinoma), HCT15 (colon cell carcinoma), MCF7 (breast adenocarcinoma) Normal cells: MDCK (Mardin–Darby canine kidney) and rat fibroblast *In vitro*	IC_{50} = 42.5, 26.4, 17.9, 192.4, 182.6 μg mL^{-1}, respectively	Mhadhebi et al. 2014
C. tamariscifolia	Hexane, ethyl ether and dichloromethane fractions	Cytotoxity essay: Brine shrimp (*Artemia salina*)	Highest cytotoxic activity was determined by CH_2CI_2 fraction, in 20 g mL^{-1} concentration of extract (30% cell inhibition)	Abourriche et al. 1999
C. tamariscifolia	Crude (DCM/MeOH) extract. PP content 1.0–10.9%	Tumor cells: Daudi, Jurkat, K562 Normal cells *In vivo*	Crude extract concentrations: 50–500 mg L^{-1}; Inhibition: 60.05–91.83%	Zubia et al. 2009b
C. usneoides	Chloroform extract Meroterpenes: usneoidone E and usneoidone Z	Tumor cells: L-1210 *In vitro*	Showed antitumor properties against L-1210 cell line	Urones et al. 1992
Desmarestia ligulata	Crude (DCM/MeOH) extract. PP content 1.0–10.9%	Tumor cells: Daudi, Jurkat, K562 Normal cells *In vivo*	Crude extract concentrations: 50–500 mg L^{-1}; Inhibition: 16.26–63.02%	Zubia et al. 2009b
D. viridis	Methanolic extract	Tumor cells: HT-29, KB Normal cells: NIH-3T3 *In vitro*	IC_{50} > 50 (KB), > 50 (HT-29), > 50 (NIH-3T3) μg mL^{-1}	Xu et al. 2004b

Table 6.1 contd....

Table 6.1 contd....

Species	Tested product, fraction or extract	Study model	Observed results	References
Dictyopteris delicatula	Sulfated polysaccharides	Tumor cells: HeLa Normal cells *In vivo*	61.0% of cell proliferation inhibition at 2.0 mg mL^{-1}	Costa et al. 2010
D. delicatula	7 characterized fucoidan fractions	Tumor cells: HeLa Normal cells *In vivo*	Screened for bioactivity. Proliferation assay: best fractions (2 mg mL^{-1}) inhibited tumor cell growth by 60–90%	Magalhães et al. 2011
D. delicatula	Methanolic extract	Tumor cells: HeLa, SiHa	Extract exhibited an inhibition rate of around 22% after 48 h	Gomes et al. 2015
D. divaricata	Methanolic extract	Tumor cells: HT-29, KB Normal cells: NIH-3T3 *In vitro*	IC_{50} > 50 (KB), > 50 (HT-29), > 50 (NIH-3T3) µg mL^{-1}	Xu et al. 2004b
D. divaricata	Ethanol/water extract	Tumor cells: HL-60, HT-29 Also tested: B-16, A-549 Normal cells: HaCaT *In vivo*	Concentration of extract: 100 µg mL^{-1} HL-60 Inhibition = 85.3% HT-29 Inhibition = 32.2% B16 Inhibition = 29.3% A549 Inhibition = 6.5% HaCat = Inhibition = 19.2%	Kim et al. 2009
D. polypodioides (formerly *D. membranacea*)	Aqueous, ethanolic and chloroformic extracts	Tumor cells: KB *In vitro*	*D. polypodioides* significantly inhibits the growth of the cell cultures in all the three different crude extracts, was quantitatively screened in different periods of the year for the presence of cytostatic compound(s)	Kosovel et al. 1988
D. polypodioides (formerly *D. membranacea*)	Methanolic extract	Tumor cells: L-1210 and P-388 Normal cells: NIH-3T3 *In vitro*	Extract exhibited highly cytotoxicity to tumor cells and to normal cells	Harada et al. 1997
D. polypodioides (formerly *D. membranacea*)	Ethanol extract	Tumor cells: MCF-7, LNCaP, PC-3 Normal cells: MCF-10A *In vitro*	Extract displayed a general cytotoxicity against all four cell lines, with inhibition of growth in the range of 60–96% with concentration of 50 mg mL^{-1}	Montalvão et al. 2016

D. prolifera	Ethanol/water extract	Tumor cells: HL-60, HT-29 Also tested: B-16, A-549 Normal cells: HaCaT *In vivo*	Concentration of extract: 100 μg mL^{-1} HL-60 Inhibition = 53.3% HT-29 Inhibition = 1.5% B16 Inhibition = 17.9% A549 Inhibition = 4.3% HaCat = Inhibition = 14.2%	Kim et al. 2009
D. undulata	Methanolic extract	Tumor cells: L-1210 and P-388 Normal cells: NIH-3T3 *In vitro*	Extract exhibited highly cytotoxicity to tumor cells and to normal cells	Harada et al. 1997
Dictyota ciliolata	Methanolic seaweed extracts	Tumor cells: HeLa, SiHa	Seaweed extracts showed a dose-dependent and time-dependent inhibitory activity tendency. Inhibition rate of around 50% after 72 h of experiment (at 0.2 mg mL^{-1})	Gomes et al. 2015
D. coriacea (as *Pachydictyon coriaceum*)	Methanolic extract	Tumor cells: L-1210 and P-388 Normal cells: NIH-3T3 *In vitro*	Extract exhibited highly cytotoxicity to tumor cells and to normal cells	Harada et al. 1997
D. dichotoma	Methanolic extract	Tumor cells: L-1210 and P-388 Normal cells: NIH-3T3 *In vitro*	Extract exhibited highly cytotoxicity to tumor cells and to normal cells	Harada et al. 1997
D. dichotoma	Crude (DCM/MeOH) extract. PP content 1.0–10.9%	Tumor cells: Daudi, Jurkat, K562 Normal cells *In vivo*	Crude extract concentrations: 50–500 mg L^{-1}; Inhibition: 21.95–61.95%	Zubia et al. 2009b
D. dichotoma	Hexane, dichloromethane-, ethyl acetate and methane fractions	Tumor cells: MiaPaCa-2, Panc-1, BXPC-3, Panc-3.27 Normal cells *In vivo*	Dose-dependent inhibition of cell growth, induction of apoptosis. Inhibition of NF-κB. Modulated EGFR phosphorylation, kRas, AurKb and Stat3 (see paper for details on extract/cell line efficacy). Little characterization carried out; authors recommend more to aid discovery of active compounds	Aravindan et al. 2013
D. dichotoma	Dichloromethane, chloroform, methanol, ethanol, water extracts, hexane and chloroform fractions	Tumor cells: HEp-2, K562, NCI-H292 *In vitro*	Dichloromethane extracts (16.3 μg mL^{-1}) was more active against HEp-2	Guedes et al. 2013
D. fasciola (formerly *Dilophus fasciola*)	Sulfolipids	Tumor cells: MCF7, HepG2 *In vitro*	Concentrations = 1.0–10.0 μg mL^{-1} Growth inhibition = 71.12–79.34% (MCF7), 57.36–80.29% (HepG2)	El Baz et al. 2013

Table 6.1 contd....

Table 6.1 contd....

Species	Tested product, fraction or extract	Study model	Observed results	References
D. menstrualis	Sulfated polysaccharides	Tumor cells: HeLa Normal cells *In vivo*	58.4% of cell proliferation inhibition at 2.0 mg mL^{-1}	Costa et al. 2010
D. menstrualis	Methanolic seaweed extracts	Tumor cells: HeLa, SiHa	Seaweed extracts showed a dose-dependent and time-dependent inhibitory activity tendency. Maximum inhibition (approximately 80%) with 0.2 mg mL^{-1} after 48 h exposure	Gomes et al. 2015
D. mertensii	Sulfated polysaccharides	Tumor cells: HeLa Normal cells *In vivo*	About 45.0% of cell proliferation inhibition at 2.0 mg mL^{-1}	Costa et al. 2010
D. spinulosa	Methanolic extract	Tumor cells: L-1210 and P-388 Normal cells: NIH-3T3 *In vitro*	Extract exhibited highly cytotoxicity to tumor cells and to normal cells	Harada et al. 1997
Ecklonia arborea (formerly *Eisenia arborea*)	Methanol-water extract	Tumor cells: RBL-2H3 Normal cells *In vivo*	Inhibition effect: 74.3% (concentration: 1 mg mL^{-1})	Sugiura et al. 2006
E. bicyclis (formerly *Eisenia bicyclis*)	Seaweed powder	Tumor cells: Ehrlich carcinoma *In vivo*	Oral administration of 1600 mg seaweed powder per kg mouse per day for 28 days. Inhibition rate: 37.5%	Noda et al. 1990
E. bicyclis (formerly *Eisenia bicyclis*)	Neutral lipid fractions	Tumor cells: Meth-A fibrosarcoma *In vivo*	Intraperitoneal administration of 40 mg kg^{-1} for seven days. Inhibition rate: 27.0%	Noda et al. 1990
E. bicyclis (formerly *Eisenia bicyclis*)	Phosphate buffered saline extract	Tumor cells: L-1210 and P-388 Normal cells: NIH-3T3 *In vitro*	Extract exhibited highly cytotoxicity to tumor cells and no cytotoxicity to normal cells	Harada et al. 1997
E. cava	Seaweed powder	Tumor cells: Ehrlich carcinoma; Meth-A fibrosarcoma *In vivo*	Oral administration of 1600 mg seaweed powder per kg mouse per day for 28 days. Inhibition rate: 25.0% Intraperitoneal administration of 50 mg kg^{-1} d^{-1} for seven days. Inhibition rate: 35.9%	Noda et al. 1990
E. cava	Phosphate buffered saline extract	Tumor cells: L-1210 and P-388 Normal cells: NIH-3T3 *In vitro*	Extract exhibited highly cytotoxicity to tumor cells and no cytotoxicity to normal cells	Harada et al. 1997

E. cava	Crude extract	Tumor cells: CT-26, THP 1, B-16, U-937 Normal cells: V79-4 *In vivo*	Toxic against all cancer cell lines tested, low cytotoxicity against the normal cell line, V79-4. Extract induced apoptosis in CT-26 cells and had good antioxidant activity. More toxic than polysaccharide fraction	Athukorala et al. 2006
E. cava	Crude extract (ethanol) 58% polyphenols	Tumor cells: HT1080 Normal cells: HDF *In vivo*	No cytotoxic effect on HT1080 or HDF cells (doses up to 100 μg mL^{-1}). Inhibited MMP-2 and MMP-9 activity. Link proposed between polyphenols and anticancer activity	Kim et al. 2006
E. cava	Methanol-water extract	Tumor cells: RBL-2H3 Normal cells *In vivo*	Inhibition effect: 91.9% (concentration: 1 mg mL^{-1})	Sugiura et al. 2006
E. cava	Ethanol/water extract	Tumor cells: HL-60, HT-29 Also tested: B-16, A-549 Normal cells: HaCaT *In vivo*	Concentration of extract: 100 μg mL^{-1} HL-60 Inhibition = 6.7–15.4% HT-29 Inhibition = 8.1% B16 Inhibition = 9.3% A549 Inhibition = 2.8% HaCat = Inhibition = –21.7%	Kim et al. 2009
E. cava	Phlorotannins: (1) dioxinodehydroeckol, (2) complex dioxin analogue structure provided	Tumor cells: MCF-7, MDA-MB-231 Normal cells *In vivo*	Compound (1) had the greatest cytotoxic effect on both MCF-7 and MDA-MB-231 cells. It induced apoptosis, increased Bax, p53, caspases 3 and 9 and PARP cleavage. Bcl2 and NF-κB were down-regulated	Kong et al. 2009
E. cava	6,6′-Bieckol, a phlorotannin	Tumor cells: HT1080 Normal cells *In vivo*	Cell viability in HT1080 cells not affected (dose: up to 250 μM). Reduced tumor invasion and migration by matrigel assay. Inhibited expression of MMP-2 and MMP-9, which are both expressed in malignant tumors. No clear effect on AP-1 but NF-κB was reduced	Zhang et al. 2010
E. cava	Fucoidan	Tumor cells: SK-MEL-28, DLD-1 *In vitro*	Inhibiting colony formation	Ermakova et al. 2011
E. cava	Crude ethyl acetate extract. PP content 30%	Tumor cells: A549 Normal cells *In vivo*	Extract showed a significant inhibition of migration and invasion of A549 cells in a concentration-dependent manner. It strongly inhibited MMP-2, Akt and p38, but not JNK and ERK. No significant cytotoxicity was observed. Link between polyphenol content and anticancer activity could not be made	Lee et al. 2011b

Table 6.1 contd. …

Table 6.1 contd....

Species	Tested product, fraction or extract	Study model	Observed results	References
E. cava	Fucodiphlorethol G, eckol, dieckol, phlorofucofuroeckol A and a complex 1,4-dioxin analogue	Tumor cells: HeLa, HT1080, A549 and HT-29 Normal cells: MRC-5 *In vivo*	IC_{50} values of compounds against cancer cell lines were in the range of 180–360 μM. The extracts were less cytotoxic to MRC-5 cells (IC_{50} not specified but above 400 μM)	Li et al. 2011
E. cava	Phlorotannin: Dieckol	Tumor cells: SK Hep-1 Normal cells *In vivo*	TPA induced cell motility was decreased by Dieckol. Decreased AP-1 in MAPK signaling pathways and MMP-9 activity	Oh et al. 2011
E. kurome	Phosphate buffered saline extract	Tumor cells: L-1210 and P-388 Normal cells: NIH-3T3 *In vitro*	Extract exhibited highly cytotoxicity to tumor cells and less cytotoxicity to normal cells	Harada et al. 1997
E. radiata	Ethanol extract 100 μg mL^{-1}	Tumor cells Normal cells *In vivo*	Inhibition of kinase A was assayed Showed the highest levels of inhibition = 76–100%	Winberg et al. 2011
Ectocarpus confervoides	Methanolic extract	Tumor cells: HT-29, KB Normal cells: NIH-3T3 *In vitro*	IC_{50} > 50 (KB), > 50 (HT-29), > 50 (NIH-3T3) μg mL^{-1}	Xu et al. 2004b
Fucus ceranoides	Crude (DCM/MeOH) extract. PP content 1.0–10.9%	Tumor cells: Daudi, Jurkat, K562 Normal cells *In vivo*	Crude extract concentrations: 50–500 mg L^{-1}; Inhibition: 32.09–84.68%	Zubia et al. 2009b
F. distichus	Fucoidan	Tumor cells: MDA-MB-231 Normal cells: HUVECs *In vitro*	Fucoidan showed 80% inhibition of the tumor cell adhesion to human platelets	Cumashi et al. 2007
F. evanescens	Fucoidan	Tumor cells: LLC Normal cells *In vivo*: LLC cells implanted in thigh of C57Bl/6 mice	At 10 mg kg^{-1}, there was moderate antitumor and antimetastatic effects; also enhanced anti-metastatic, but not antitumor effect, of cyclophosphamide. 10 mg kg^{-1} was well tolerated, but 25 mg kg^{-1} fucoidan, both alone and with cyclophosphamide, was toxic causing some deaths	Alekseyenko et al. 2007
F. evanescens	Fucoidan	Tumor cells: MDA-MB-231 Normal cells: HUVECs *In vitro*	Fucoidan showed 78% inhibition of the tumor cell adhesion to human platelets	Cumashi et al. 2007

F. evanescens	Fucoidan	Tumor cells: MT-4, NAMALWA *In vivo*	500 μg mL^{-1} enhanced etoposide-induced caspase dependent cell death pathway in MT-4 but not NAMALWA cell line	Philchenkov et al. 2007
F. evanescens	Water-ethanol extracts	Tumor cells: DLD-1, HT-29 Normal cells *In vivo*	Extract (concentration of 50 μg mL^{-1} suppressed the growth of cell colonies by 67 and 63%, correspondingly for DLD-1 and HT-29 cells	Imbs et al. 2009
F. evanescens	Fucoidan	Tumor cells: HCT-116 Normal cells: JB6 Cl41 *In vitro* and *ex vivo*	Fucoidan effectively prevented EGF-induced neoplastic cell transformation through inhibition of TOPK/ERK1/2/MSK 1 signaling axis. *In vitro* studies showed that the fucoidan attenuated mitogen-activated protein kinases downstream signaling in a colon cancer cells with different expression level of TOPK, resulting in growth inhibition	Vishchuk et al. 2016
F. serratus	Fucoidan	Tumor cells: MDA-MB-231 Normal cells: HUVECs *In vitro*	Fucoidan showed 80% inhibition of the tumor cell adhesion to human platelets	Cumashi et al. 2007
F. serratus	Crude (DCM/MeOH) extract. PP content 1.0–10.9%	Tumor cells: Daudi, Jurkat, K562 Normal cells *In vivo*	Crude extract concentrations: 50–500 mg L^{-1}; Inhibition: 24.86–69.19%	Zubia et al. 2009b
F. serratus	Crude extract. PP content 0.2–0.5%	Tumor cells: Caco-2 Normal cells *In vivo*	Anti-oxidant and antiproliferative activity of the extracts was examined. Extracts had antiproliferative effects (dose 0.55–5.5 mg mL^{-1} tested). Good anti-oxidant activity *in vitro*. Caco-2 cells used as a model for antioxidant effect rather than anticancer model	O'Sullivan et al. 2011
F. spiralis	Methanol and dichloromethane extracts	Tumor cells: HepG-2 *In vitro*	Anti-proliferative activity IC_{50} values obtained for methanolic extract: 1039.0 μg mL^{-1}	Alves et al. 2016
F. vesiculosus	Fucoidan	Tumor cells: Lung metastases of 13762 MAT *In vivo*	Inhibiting metastases	Coombe et al. 1987
F. vesiculosus	Fucoidan	Tumor cells: MC *In vitro*	Increasing the tumoricidal activity of macrophages	Choi et al. 2005

Table 6.1 contd....

Table 6.1 contd....

Species	Tested product, fraction or extract	Study model	Observed results	References
F. vesiculosus	Fucoidan	Tumor cells: *In vitro*	Mouse spleen lymphocytes became cytotoxic to tumor cells after culture with arabinogalactan (AG) and fucoidan at concentrations of 10–100 μg mL^{-1}. Also, AG and fucoidan were mitogenic in spleen lymphocytes and peripheral macrophages. Macrophages treated with AG and fucoidan (10–100 μg mL^{-1}) exhibited induced tumoricidal activity and increased phagocytosis, lysosomal enzyme activity, and production of nitrite, H_2O_2, tumor necrosis factor (TNF)-α, and interleukin (IL)-6	Kasai et al. 2005
F. vesiculosus	Fucoidan	Tumor cells: MDA-MB-231 Normal cells: HUVECs *In vivo*	Fucoidan showed 80% inhibition of the tumor cell adhesion to human platelets	Cumashi et al. 2007
F. vesiculosus	Fucoidan	Tumor cells: HCT-15 Normal cells *In vitro*	Inhibited growth (0–100 μg mL^{-1}) with increased DNA fragmentation, apoptosis, Bax and caspases 3 and 9 expression; Bcl-2 and Akt were decreased	Hyun et al. 2009
F. vesiculosus	Fucoidan	Tumor cells: HL-60, THP-1, NB4 *In vitro*	Fucoidan induced apoptosis of HL-60, NB4, and THP-1 cells, but not K562 cells	Jin et al. 2010
F. vesiculosus	Fucoidan	Tumor cells: HT-29, HCT116 Normal cells: FHC *In vivo*	Dose-dependent inhibition of tumor cell growth; no effect on normal cells (doses 0–20 μg mL^{-1}). Induced apoptosis via the death receptor and mitochondrial pathways	Kim et al. 2010a
F. vesiculosus	Fucoidan	Tumor cells: LLC, B16 Normal cells	*In vitro*: both fucoidans reduced the viability of LLC and B16 cells by inducing apoptosis *In vivo*: fucoidan enhanced NK cell activity (no tumor cells implanted)	Ale et al. 2011a
F. vesiculosus	Fucoidan	*In vivo*: male C57BL/6J mice injected IP with fucoidan (50 mg kg^{-1}) Tumor cells: B16 Normal cells	Dose-dependent inhibition (0.1–1 mg mL^{-1}); induction of apoptosis and increased caspase-3 activity	Ale et al. 2011b

F. vesiculosus	Crude extract. PP content 0.2–0.5%	Tumor cells: Caco-2 Normal cells *In vivo*	Anti-oxidant and antiproliferative activity of the extracts was examined. Extracts had antiproliferative effects (dose 0.55–5.5 mg mL^{-1} tested). Good anti-oxidant activity *in vitro*. Caco-2 cells used as a model for antioxidant effect rather than anticancer model	O'Sullivan et al. 2011
F. vesiculosus	Fucoidan	Tumor cells Normal cells: DC *In vivo*	DC cells were exposed to 5-FU. Fucoidan improved cyto-protection—increased cell viability and size compared to cells treated only with 5-FU; it affected apoptosis-related proteins (cIAP-1, cIAP-2, Bcl-xL and Bax) and also increased immune-related surface markers of DC cells	Jeong et al. 2012
F. vesiculosus	Fucoidan	Tumor cells: A549 *In vitro*	Fucoidan exhibits anti-metastatic effect on A549 lung cancer cells via the down-regulation of ERK1/2 and Akt-mTOR as well as NF-kB signaling pathways	Lee et al. 2012c
F. vesiculosus	Fucoidan	Tumor cells: 4T1 *In vitro* and *in vivo*	Inducing apoptosis, antiangiogenesis and inhibiting lung metastasis of breast cancer suppressing tumor growth and migration, antiangiogenesis, arresting G_1-phase of cell cycle and inducing apoptosis	Xue et al. 2012 Hsu et al. 2013
F. vesiculosus	Hydrophilic extract	Tumor cells: Colo-357, Panc-1, Panc-89, PancTu1 *In vitro*	Inhibited the growth of different tumor cell lines significantly. The EC_{50} (effective half maximal concentration) values of extract range between 17.35 µg mL^{-1} for PancTU1 (95% CI: 16.74–17.99), 17.5 µg mL^{-1} for Panc89 (95% CI: 17.24–17.77), 19.23 µg mL^{-1} for Panc1 (95% CI: 18.52–19.98) and 28.9 µg mL^{-1} for Colo-357 (95% CI: 22.71–32.11)	Geisen et al. 2015
F. vesiculosus	Fucoidan	Tumor cells: 4T1 *In vitro* and *in vivo*	Fucoidan significantly inhibited cell growth, increased cell death, and induced G_1 cell cycle arrest in 4T1 cells. Fucoidan also reduced β-catenin expression and T cell factor/lymphoid-enhancing factor reporter activity	Xue et al. 2013

Table 6.1 contd.…

Table 6.1 contd....

Species	Tested product, fraction or extract	Study model	Observed results	References
F. vesiculosus	Fucoidan	Tumor cells: MDA-MB-231, HCT 116 *In vitro*	Fucoidan exerts its anti-tumor function by modulating ER stress cascades. Contribution of ER stress to the fucoidan-induced cell apoptosis augments our understanding of the molecular mechanisms underlying its antitumor activity and provides evidence for the therapeutic application of fucoidan in cancer	Chen et al. 2014
F. vesiculosus	Fucoidan	Tumor cells: LLC1 *In vitro* and *in vivo*	Fucoidan reduces tumor size in LLC1-xenograft male C57BL/6 mice. Moreover, we found that LLC1-bearing mice continuously fed fucoidan showed greater antitumor activity than mice with discontinuous feeding. Fucoidan inhibited the *in vitro* growth of lung cancer cells	Hsu et al. 2014
F. vesiculosus	Fucoidan	Tumor cells: HepG2 *In vitro*	Arresting the G_0/G_1 cell cycle and inducing apoptosis	Min et al. 2014
F. vesiculosus	Fucoidan	Tumor cells: HT29 *In vitro* and *in vivo*	Inducing apoptosis, antiangiogenesis and inhibiting lung metastasis of breast cancer suppressing tumor growth and migration, antiangiogenesis, arresting G_1-phase of cell cycle and inducing apoptosis	Han et al. 2015a, 2015b
F. vesiculosus	Low-molecular weight fucoidan derivatives	Tumor cells: HeLa, MCF-7 *In vitro*	Has anti-proliferative activities and apoptosis-inducing activities against human breast cancer (MCF-7) and human cervical epithelioid carcinoma (HeLa) cells	Kasai et al. 2015
F. vesiculosus	Fucoidan	Tumor cells: YAC-1 *In vivo*	Fucoidan of *F. vesiculosus* delayed human neutrophil apoptosis (50%) at higher concentration, prevented apoptosis at concentration of 50–100 µg mL^{-1}	Zhang et al. 2015
Halidrys siliquosa	Crude (DCM/MeOH) extract. PP content 1.0–10.9%	Tumor cells: Daudi, Jurkat, K562 Normal cells *In vivo*	Crude extract concentrations: 50–500 mg L^{-1}; Inhibition: 72.40–94.90%	Zubia et al. 2009b

Himantothallus grandifolius	Ethanolic extract	Tumor cells: A375, A549, Hep-2, HeLa Normal cells: Hek-293 *In vitro*	Extract showed selectivity to the non-tumor line with enhanced cytotoxicity in tumor cells according to the concentration and exposure time After 72 h treatment, the HeLa strain was more susceptible to the extract, followed by lines Hep2, A375 and A549	Gambato et al. 2014
Hormophysa cuneiformis (formerly *H. triquetra*)	Hexane, dichloromethane-, ethyl acetate and methane fractions	Tumor cells: MiaPaCa-2, Panc-1, BXPC-3, Panc-3.27 Normal cells *In vivo*	Dose-dependent inhibition of cell growth, induction of apoptosis. Inhibition of NF-κB. Modulated EGFR phosphorylation, kRas, AurKb and Stat3. Little characterization carried out; authors recommend more to aid discovery of active compounds	Aravindan et al. 2013
Hydroclathrus clathratus	Seaweed powder	Tumor cells: Ehrlich carcinoma *In vivo*	Oral administration of 1600 mg seaweed powder per kg mouse per day for 28 days. Inhibition rate: 26.7%	Noda et al. 1990
H. clathratus	Ethyl acetate portion of ethanol extract	Tumor cells: HL-60 Normal cells	Induced apoptosis, activated caspases 3 and 9, up-regulated Bax and downregulated Bcl-xL, increased ROS	Kim et al. 2012a
Ishige foliacea	Seaweed powder	Tumor cells: Ehrlich carcinoma *In vivo*	Oral administration of 1600 mg seaweed powder per kg mouse per day for 28 days. Inhibition rate: 36.8%	Noda et al. 1990
I. foliacea	Methanol-water extract	Tumor cells: RBL-2H3 Normal cells *In vivo*	Inhibition effect: 63.3% (concentration: 1 mg mL^{-1})	Sugiura et al. 2006
I. okamurae	Seaweed powder	Tumor cells: Ehrlich carcinoma; Meth-A fibrosarcoma *In vivo*	Oral administration of 1600 mg seaweed powder per kg mouse per day for 28 days. Inhibition rate: 20.4% Intraperitoneal administration of 50 mg kg^{-1} d^{-1} for seven days. Inhibition rate: 19.5%	Noda et al. 1990
I. okamurae	Methanol-water extract	Tumor cells: RBL-2H3 Normal cells *In vivo*	Inhibition effect: 63.2% (concentration: 1 mg mL^{-1})	Sugiura et al. 2006

Table 6.1 contd....

Table 6.1 contd....

Species	Tested product, fraction or extract	Study model	Observed results	References
I. okamurae	Ethanol/water extract	Tumor cells: HL-60, HT-29 Also tested: B-16, A-549 Normal cells: HaCaT *In vivo*	Concentration of extract: 100 μg mL^{-1} HL-60 Inhibition = 36.3% HT-29 Inhibition = 3.2% B16 Inhibition = 4.1% A549 Inhibition = 1.5% HaCat = Inhibition = –6.7%	Kim et al. 2009
I. sinicola	Methanolic extract	Tumor cells: L-1210 and P-388 Normal cells: NIH-3T3 *In vitro*	Extract exhibited highly cytotoxicity to tumor cells and to normal cells	Harada et al. 1997
I. sinicola	Crude methanol, water and PBS extracts	Tumor cells: MOLT-4, K562, HeLa, KB Normal cells *In vivo*	PBS extract showed telomerase inhibiting activity	Kanegawa et al. 2000
Laminaria brasiliensis	Alginic acid	Tumor cells: HeLa *In vitro*	Alginic acid (160.0 μg mL^{-1}) promoted atypical mitoses in HeLa cells	Stevan et al. 2001
L. digitata	Fucoidan	Tumor cells: MDA-MB-231 Normal cells: HUVECs *In vivo*	Fucoidan showed 80% inhibition of the tumor cell adhesion to human platelets	Cumashi et al. 2007
L. hyperborea	Crude extract. PP content 0.2–0.5%	Tumor cells: Caco-2 Normal cells *In vivo*	Anti-oxidant and antiproliferative activity of the extracts was examined. Extracts had antiproliferative effects (dose 0.55–5.5 mg mL^{-1} tested). Good anti-oxidant activity *in vitro*. Caco-2 cells used as a model for antioxidant effect rather than anticancer model	O'Sullivan et al. 2011
L. setchellii	Crude extract, PP content 0.2–1.3%	Tumor cells: HeLa Normal cells *In vivo*	HeLa cell proliferation was inhibited between 0% and 55% at 0.5–5 mg mL^{-1} algal extract	Yuan and Walsh 2006

Landsburgia quercifolia	Deoxylapachol, 2-(3-methyl-2-butenyl)-2,3-epoxy-1,4-naphthalenedione	Tumor cells: P-388 Normal cells: BSC *In vivo*	IC_{50}: deoxylapachol, P388 cells (0.6 μg mL^{-1}), BSC cells (10 μg mL^{-1}). 2-(3-methyl-2-butenyl)-2,3-epoxy-1,4-naphthalenedione, P388 cells (0.8 μg mL^{-1}). BSC cells (10 μg mL^{-1}). Other molecules tested were found to be much less cytotoxic than the 2 cited	Perry et al. 1991
Leathesia marina (formerly *Leathesia difformis*)	Phosphate buffered saline extract	Tumor cells: L-1210 and P-388 Normal cells: NIH-3T3 *In vitro*	Extract exhibited cytotoxicity to tumor cells and less cytotoxicity to normal cells	Harada et al. 1997
L. marina (formerly *L. nana*)	Bromophenols	Tumor cells: A549, BGC-823, MCF-7, Bel7402, HCT-8 Normal cells *In vivo*	Four compounds toxic to cell lines, with IC_{50} values in range 0.001–0.020 μM mL^{-1}	Xu et al. 2004
L. marina (formerly *L. difformis*)	Methanolic extract	Tumor cells: HT-29, KB Normal cells: NIH-3T3 *In vitro*	IC_{50} = 12.65 (KB), = 40.60 (HT-29), > 50 (NIH-3T3) μg mL^{-1}	Xu et al. 2004b
L. marina (formerly *L. nana*)	Ethanol extract and Bromophenols	Tumor cells: A549, BGC-823, MCF-7, B16-BL6, HT-1080, A2780, Bel7402, HCT-8 Normal cells *In vivo*: S-180 cells inoculated SC in the right axilla of Kunming mice	Most compounds had IC_{50} values below 10 μg mL^{-1} against the cell lines tested. Caused a decrease in PTK and increase in c-kit expression. *In vivo*: crude extract reduced tumor growth	Shi et al. 2009
L. marina (formerly *L. nana*)	bis(2,3-Dibromo-4,5-dihydroxybenzyl) ether	Tumor cells: A549, HCT-116, HeLa, HCT- 8, K562, SMMC-7721 Normal cells: MCF-10A *In vivo*	IC_{50}, approximately 14–36 μg mL^{-1} for tumor cells, 50 μg mL^{-1} for normal cells. K562 most sensitive (IC_{50}, 14 μg mL^{-1}) and further work carried out on these cells. Induced apoptosis by a mitochondrial mediated pathway. Induced ROS generation and arrested the cell cycle in the S phase. Inhibits topoisomerase I activity. Modelling showed that it may bind to the minor groove of DNA	Liu et al. 2012b
Lessonia nigrescens	Seaweed powder	Tumor cells: Ehrlich carcinoma *In vivo*	Oral administration of 1600 mg seaweed powder per kg mouse per day for 28 days. Inhibition rate: 60.0%	Noda et al. 1990
Lobophora variegata	Acetone/dichloromethane/chloroform crude extracts	Tumor cells: C32 Normal cells: FEK4 *In vivo*	IC_{50} values for C32 cells were > 100 μg mL^{-1}	Rocha et al. 2007

Table 6.1 contd....

Table 6.1 contd....

Species	Tested product, fraction or extract	Study model	Observed results	References
L. variegata	DCM-methanol (7:3) and water extracts	Tumor cells: Hep-2, HeLa, KB Normal cells: MDCK *In vivo*	Anti-proliferative activity of methanolic extract against KB cells IC_{50} = 68.4 μg mL^{-1}	Moo-Puc et al. 2009
Macrocystis pyrifera (as *Macrocystis integrifolia*)	Crude extract, PP content 0.2–1.3%	Tumor cells: HeLa Normal cells *In vivo*	HeLa cell proliferation was inhibited between 0% and 69% at 0.5–5 mg mL^{-1} algal extract	Yuan and Walsh 2006
M. pyrifera	Fucoidan	Tumor cells: MC, LLC, C57BL/6J mice *In vitro* and *in vivo*	Fucoidan from *M. pyrifera* promoted NK cell activation and cytotoxic activity against YAC-1 cells. In addition, *M. pyrifera* fucoidan induced the strongest activation of spleen DCs and T cells and ovalbumin (OVA) specific immune responses compared to other fucoidans	Zhang et al. 2015
Myagropsis myagroides	Methanolic extract	Tumor cells: L-1210 and P-388 Normal cells: NIH-3T3 *In vitro*	Extract exhibited highly cytotoxicity to tumor cells and to normal cells	Harada et al. 1997
Myelophycus simplex	Methanolic extract	Tumor cells: HT-29, KB Normal cells: NIH-3T3 *In vitro*	IC_{50} > 50 (KB), > 50 (HT-29), > 50 (NIH-3T3) μg mL^{-1}	Xu et al. 2004b
M. simplex	Ethanol/water extract	Tumor cells: HL-60, HT-29 Also tested: B-16, A-549 Normal cells: HaCaT *In vivo*	Concentration of extract: 100 μg mL^{-1} HL-60 Inhibition = 88.1% HT-29 Inhibition = 34.6% B16 Inhibition = 47.4% A549 Inhibition = 6.4% HaCat = Inhibition = 12.2%	Kim et al. 2009
Nemacystis decipiens	Phosphate buffered saline extract	Tumor cells: L-1210 and P-388 Normal cells: NIH-3T3 *In vitro*	Extract exhibited highly cytotoxicity to tumor cells and to normal cells	Harada et al. 1997
Nereocystis luetkeana	Crude extract, PP content 0.2–1.3%	Tumor cells: HeLa Normal cells *In vivo*	HeLa cell proliferation was inhibited between 0% and 69% at 0.5–5 mg mL^{-1} algal extract	Yuan and Walsh 2006

Padina antillarum	Ethylacetate and methanol extracts	Tumor cells: MCF-7, HeLa, Vero *In vitro*	Ethylacetate fraction exhibited cytotoxicity to MCF-7, HeLa and Vero (IC_{50} = 100 µg ml^{-1}). Methanol fraction also exhibited cytotoxic activity in MCF7, Vero (IC_{50} > 100 µg mL^{-1}) and HeLa (IC_{50} = 100 µg mL^{-1})	Mashjoor et al. 2016
P. arborescens	Seaweed powder	Tumor cells: Ehrlich carcinoma *In vivo*	Oral administration of 1600 mg seaweed powder per kg mouse per day for 28 days. Inhibition rate: 16.1%	Noda et al. 1990
P. arborescens	Phosphate buffered saline extract	Tumor cells: L-1210 and P-388 Normal cells: NIH-3T3 *In vitro*	Extract exhibited cytotoxicity to tumor cells and less cytotoxicity to normal cells	Harada et al. 1997
P. australis	Crude extract (ethanol or acetone)	Tumor cells: Caco-2 Normal cells *In vivo*	Both ethanol and acetone extracts decreased Caco-2 viability when cells were treated with 600 µM hydrogen peroxide. However, when Caco-2 cells were treated with 700 or 800 µM hydrogen peroxide, the ethanol and acetone extracts from *P. australis* increased cell viability significantly more than those from the other seaweeds	Gunji et al. 2007
P. australis	Fucoxanthin	Tumor cells: H1299 Normal cells *In vivo*	Data obtained from the methyl thiazolyl tetrazolium (MTT) assay indicated that fucoxanthin reduced the viability of H1299 cell lines, showing an IC_{50} value of 2.45 mM	Jaswir et al. 2011
P. boergesenii	Methanol extract	Tumor cells: HepG2 *In vitro*	IC_{50} = 1.67 mg mL^{-1}	Kanagarajjeevitha et al. 2014
P. boergesenii	Ethylacetate and methanol extracts	Tumor cells: MCF-7, HeLa Normal cells: Vero *In vitro*	Ethylacetate fraction exhibited cytotoxicity to MCF-7, HeLa and Vero (IC_{50} = 100 µg ml^{-1}). Methanol extract of *P. boergesenii* demonstrated cytotoxic effect against all cell lines (IC_{50} > 100 µg mL^{-1})	Mashjoor et al. 2016
P. gymnospora	Dichloromethane, chloroform, methanol, ethanol, water extracts, hexane and chloroform fractions	Tumor cells: HEp-2, K562, NCI-H292 *In vitro*	Chloroform fraction (8.2 µg mL^{-1}) was more active against HEp-2 as well as ethanol extracts (15.9 µg mL^{-1})	Guedes et al. 2013

Table 6.1 contd....

Table 6.1 contd....

Species	Tested product, fraction or extract	Study model	Observed results	References
P. pavonica	Dichloromethane Methanol extracts	Tumor cells: KB *In vitro*	Dichloromethane extract showed high cytotoxic activity against the human buccal epidermal carcinoma (KB) cells in 10 g mL^{-1} concentration, while dichloromethane/methanol extract showed moderate inhibition	Ktari and Guyot 1999
P. pavonica	Methanol extract	Tumor cells: DU-145, LNCaP, MCF-7, PC-3 *In vitro*	Extract showed good cytotoxic activities with 200 than 100 g mL^{-1} concentrations in 24 h (higher than 80% inhibition)	Taskin et al. 2010
P. pavonica (formerly *P. pavonia*)	Methanol extract	Tumor cells: MDA-MB-453, HeLa Normal cells: MRC-5 *In vitro*	IC_{50} values 86.45 μg mL^{-1} related to HeLa cell and 74.59 μg mL^{-1} related to MDA-MB453 cell The extracts did not exert any significant cytotoxicity toward normal human fetal lung fibroblast cells (MRC-5)	Stanojković et al. 2013
P. tetrastromatica	Hexane, dichloromethane-, ethyl acetate and methane fractions	Tumor cells: MiaPaCa-2, Panc-1, BXPC-3, Panc-3.27 Normal cells *In vivo*	Dose-dependent inhibition of cell growth, induction of apoptosis. Inhibition of NF-κB. Modulated EGFR phosphorylation, kRas, AurKb and Stat3 (see paper for details on extract/cell line efficacy). Little characterization carried out; authors recommend more to aid discovery of active compounds	Aravindan et al. 2013
P. tetrastromatica	Methanolic extract	Tumor cells: A549 Normal cells: VERO *In vitro*	Cytotoxic effect against human A549 lung adenocarcinoma cancer cell lines in a concentration dependent manner, with $IC_{50} = 222.78$ μg mL^{-1}	Rani et al. 2013
P. tetrastromatica	Methanolic extract	Tumor cells: MCF-7 Normal cells: 3T3 *In vitro*	The methanolic extract of *P. tetrastromatica* showed considerable antioxidant activity through inhibition of DPPH and hydroxyl radicals, with median inhibitory concentration (IC_{50}) values of 45.57 ± 1.63 and 36.58 ± 2.13 μg mL^{-1}, respectively. Treatment of the human breast adenocarcinoma cell line MCF-7 with the methanolic extract, at a concentration range of 0–500 μg mL^{-1}, showed considerable antiproliferative effects, with an IC_{50} value of 125 ± 2.03 μg mL^{-1}	Chia et al. 2015

Pelvetia canaliculata	Crude extract. PP content 0.2–0.5%	Tumor cells: Caco-2 Normal cells *In vivo*	Anti-oxidant and antiproliferative activity of the extracts was examined. Extracts had antiproliferative effects (dose 0.55–5.5 mg mL^{-1} tested). Good anti-oxidant activity *in vitro*. Caco-2 cells used as a model for antioxidant effect rather than anticancer model	O'Sullivan et al. 2011
Perithalia capillaris	Three different Quinone's	Tumor cells: HL-60 Normal cells *In vivo*	IC_{50} of Quinone (5-(1,1-dimethylprop-2-enyl)-2-(3-methylbut-2-enyl) cyclohexa-2,5-diene-1,4-dione), 0.34 μM. Other compounds less active	Sansom et al. 2007
Petalonia binghamiae	Ethanol/water extract	Tumor cells: HL-60, HT-29 Also tested: B-16, A-549 Normal cells: HaCaT *In vivo*	Concentration of extract: 100 μg mL^{-1} HL-60 Inhibition = 60.6$% HT-29 Inhibition = 37.0% B16 Inhibition = 54.9% A549 Inhibition = 4.1% HaCat = Inhibition = 23.6%	Kim et al. 2009
P. fascia	Ethanol extract 100 μg mL^{-1}	Protein kinase A inhibition *In vitro*	26–50% inhibition	Winberg et al. 2011
Phaeophyceae (brown algae)	Fucoxanthin	Tumor cells: PC-3	Fucoxanthinol now been established as a potent anti-carcinogenic compound and a study was done to evaluate the efficacy of the pure fractioned fucoxanthin and its metabolites produced into the gastrointestinal tract against human prostate cancer cells (PC-3) Fucoxanthin metabolized into fucoxanthinol and finally amarouciaxanthin A Fucoxanthin, fucoxanthinol, and amarouciaxanthin A, exibited 50% cell growth inhibition of PC-3 cells at 3.0, 2.0, and 4.6 μM concentrations respectively	Asai et al. 2004
Phyllospora comosa	Ethanol extract 100 μg mL^{-1}	Protein kinase A inhibition *In vitro*	1–25% inhibition	Winberg et al. 2011
Polycladia myrica (formerly *Cystoseira myrica*)	Ethanolic extract and purified diterpenes	Tumor cells: SSVNIH3T3, KA3IT Normal cells: NIH3T3 *In vitro*	All compounds exhibited moderate cytotoxicity on the cancer cell line KA3IT ($IC_{50} \approx 5$ μg mL^{-1})	Ayyad et al. 2003

Table 6.1 contd....

Table 6.1 contd....

Species	Tested product, fraction or extract	Study model	Observed results	References
P. myrica (formerly *C. myrica*)	(1) 3-Keto-22-epi-28-norcathasterone; (2) cholest-4-ene-3,6-dione	Tumor cells: HEPG-2 and HCT116 Normal cells *In vivo*	IC_{50} for (1): HEPG-2 (2.96 μM) and HCT116 (12.38 μM). IC_{50} for (2): HEPG-2 (5.63 μM) and HCT116 (1.16 μM)	Hamdy et al. 2009
P. myrica (formerly *C. myrica*)	Crude methanol (70%) extract and hexane, chloroform, ethyl acetate and MeOH-H2O fractions of this	Tumor cells: HT-29, Caco-2, T47D, T47D-T. R, MDA-MB468 Normal cells: NIH 3T3 *In vivo*	Hexane fraction had good cytotoxic action and induced apoptosis in Caco-2 and T47D cells. Mechanism was found to be estrogen receptor independent by testing estrogen receptor negative and positive breast cancer cells. Was also cytotoxic to normal cells, but at higher IC_{50}	Khanavi et al. 2010
P. myrica (formerly *C. myrica*)	Aqueous extract	Tumor cells: EAC Silver Nanoparticle's of aqueous extract (AgNPs) *In vitro*	Cytotoxic effect (53–83%) of different AgNPs concentrations (42–98 μg mL^{-1})	Khalifa et al. 2016
Punctaria latifolia	Methanolic extract	Tumor cells: HT-29, KB Normal cells: NIH-3T3 *In vitro*	IC_{50} = 38.99 (KB), = 45.42 (HT-29), > 50 (NIH-3T3) μg mL^{-1}	Xu et al. 2004b
P. plantaginea	Methanolic extract	Tumor cells: HT-29, KB Normal cells: NIH-3T3 *In vitro*	IC_{50} > 50 (KB), > 50 (HT-29), > 50 (NIH-3T3) μg mL^{-1}	Xu et al. 2004b
Rugulopteryx okamurae (formerly *Dilophus okamurae*)	Methanolic extract	Tumor cells: L-1210 and P-388 Normal cells: NIH-3T3 *In vitro*	Extract exhibited highly cytotoxicity to tumor cells and less cytotoxicity to normal cells	Harada et al. 1997
R. okamurae (formerly *D. okamurae*)	Methanol Phosphate-buffered saline (PBS)	Tumor cells: HDF, HL-60, L-1210, MOLT-4, NIH-3T3 *In vitro*	Extracts with weak selective cytotoxic activity to L-1210 cells exhibited not only strong cytotoxicity to L-1210, but also to human leukemic cells, HL-60 and MOLT-4 at 50 μg mL^{-1}	Harada and Kamei 1997
Saccharina angustata (formerly *Laminaria angustata*)	Raw seaweed	*In vivo*: DMBA-induced mammary tumors in female Sprague-Dawley rats	Causes delay in time for tumor to appear and a reduction in the number of histologically confirmed tumors	Teas et al. 1984

S. angustata (formerly *L. angustata*)	Raw seaweed	*In vivo*: AOM-induced intestinal carcinogenesis in male F344 rats	The incidence and multiplicity of intestinal tumors did not vary between control and seaweed groups. The incidence and multiplicity of colon adenomas and the size of colon tumors increased in rats fed the seaweed containing diet compared to control. No effect on fecal bile acids and fecal cholesterol, total neutral sterols decreased in the seaweed group. Authors suggested that dietary seaweed increases the risk for colon tumors	Reddy et al. 1985
S. angustata (formerly *L. angustata*)	Seaweed powder	Tumor cells: Meth-A fibrosarcoma *In vivo*	Intraperitoneal administration of 50 mg kg^{-1} d^{-1} for seven days. Inhibition rate: 36.6%	Noda et al. 1990
S. angustata (formerly *L. angustata*)	Phospholipid fractions	Tumor cells: Meth-A fibrosarcoma *In vivo*	Intraperitoneal administration of 40 mg kg^{-1} for seven days. Inhibition rate: 58.0%	Noda et al. 1990
S. cichorioides (formerly *L. cichorioides*)	Fucoidan	Tumor cells Normal cells: JB6 Cl41 *In vivo*	No cytotoxicity towards JB6 Cl41 cells up to 200 μg mL^{-1}. Decreased the phosphorylation of EGFR in JB6 Cl41 cells (but not EGFR levels). Fucoidan suppressed EGF-induced phosphorylation of MEK, ERK1/2, p90RSK, JNKs and c-Jun	Lee et al. 2008
S. cichorioides (formerly *L. cichorioides*)	Water-ethanol extracts	Tumor cells: DLD-1, HT-29 Normal cells *In vivo*	*L. cichorioides* extracts, the strongest inhibitory effect was shown by the extract of the alga collected in July and September. So, growth of the colonies treated with the *L. cichorioides* extract of a July specimen declined by 64 and 56%, for the DLD-1 and HT-29 cells, respectively, and by 50 and 52%, respectively, when treated with the extract of a September specimen of the alga. Extracts prepared from algae collected in May inhibited colony growth only by 9 and 23%	Imbs et al. 2009
S. cichorioides	Fucoidan	Tumor cells: HCT 116 *In vivo* and *in vitro*	Combination of a fucoidan and resveratrol. Significant inhibition of HCT 116 colony formation was associated with the sensitization of cells to resveratrol by the fucoidan	Vishchuk et al. 2013b
S. gurjanovae	Fucoidan	Tumor cells Normal cells: JB6 Cl41 *In vitro*	Inhibiting EGF promoted neoplastic cell transformation	Lee et al. 2008b

Table 6.1 contd....

Table 6.1 contd....

Species	Tested product, fraction or extract	Study model	Observed results	References
S. gurjanovae	Fucoidan	Tumor cells: DLD-1 *In vitro*	Inhibiting colony formation	Shevchenko et al. 2015
Saccharina japonica (formerly *Laminaria japonica* var. *ochotensis*)	Raw seaweed	*In vivo*: DMBA-induced tumors in rats	*S. japonica* showed an inhibitory effect on tumorigenesis, and tumor incidence was lower. A significant delay in the time to first noticeable tumor	Yamamoto et al. 1987
S. japonica (formerly *L. japonica*)	Seaweed powder	Tumor cells: Ehrlich carcinoma *In vivo*	Oral administration of 1600 mg seaweed powder per kg mouse per day for 28 days. Inhibition rate: 57.6%	Noda et al. 1990
S. japonica (formerly *L. japonica*)	Raw seaweed	*In vivo*: AOM-induced colon tumors in male Sprague-Dawley rats	All seaweed feeding regimens reduced aberrant intestinal crypt formation. Suggests anti-carcinogenic effects mediated through both the blocking of initiation and the suppression of cell proliferation in initiated cells	Lee and Sung 2003
S. japonica (formerly *L. japonica*)	Methanolic extract	Tumor cells: HT-29, KB Normal cells: NIH-3T3 *In vitro*	IC_{50} > 50 (KB), > 50 (HT-29), > 50 (NIH-3T3) $\mu g\ mL^{-1}$	Xu et al. 2004b
S. japonica	Fucoidan	Cells: HUVEC and aortic ring of Wistar rats *In vitro* and *ex vivo*	Fucoidan (30 kDa) had similar effects to native fucoidan; inhibition of HUVEC tube formation and angiogenesis in an *ex vivo* model, although their effects were weaker than native fucoidan. On the other hand, 15–20 kDa fucoidan enhanced HUVEC migration, but did not inhibit HUVEC tube formation. Thus, 15–20 kDa fucoidan would have proangiogenic effect on angiogenesis. These results elucidate that 20–30 kDa would be a critical point to characterize the role of fucoidans on angiogenesis	Matsubara et al. 2005
S. japonica (formerly *L. japonica*)	Crude extract	Tumor cells: IEC-6 Normal cells *In vivo*	Stimulates growth of these normal gastrointestinal cells by activating the epidermal growth factor receptor signaling pathway	Go et al. 2009

S. japonica (formerly *L. japonica*)	Crude extract	Tumor cells: HT-29, AGS, HepG2 Normal cells *In vivo*	Cytotoxic to all cell lines especially HT-29. Induced apoptosis which may be mediated via more than 1 pathway, including the Fas signaling pathway, the mitochondrial pathway and cell cycle arrest	Go et al. 2010
S. japonica (formerly *L. japonica*)	Crude extract. PP content 200 μg mL^{-1} of extract. Fractions of this also studied, no characterization was made	Tumor cells: BEL-7402, P388 Normal cells *In vivo*	Extract inhibited growth of cells and had free radical scavenging effect. Caused apoptosis in 1 of the fractions. Links anticancer effect and PP content, but limited evidence for this claim	Yang et al. 2010b
S. japonica	Fucoidan	Tumor cells: T-47D and RPMI-7951 *In vitro*	Inhibiting proliferation and colony formation	Vishchuk et al. 2012
S. latissima (fomerly *L. saccharina*)	Fucoidan	Tumor cells: MDA-MB-231 Normal cells: HUVECs *In vitro*	Fucoidan showed 80% inhibition of the tumor cell adhesion to human platelets	Cumashi et al. 2007
S. latissima (fomerly *L. saccharina*)	Fucoidan fractions: (1) unfractionated, (2) mostly sulfated fucoidans and (3) *O*-sulfated manno-glucuronofucans	Tumor cells: MDA-MB-231, B16-F10 Normal cells: HUVEC *In vivo*: C57BL/6 (B6) mice were injected SC with B16-F10 cells plus PBS or a fucoidan (100 μg), fucoidan also given every 3 days (50 mg kg^{-1}). Angiogenesis was examined after 7 and 21 days	*In vitro*: 100 μg mL^{-1} of (1) and (2), not (3) reduced HUVEC tubulogenesis, PAI-1 and bFGF-induced pathways. All fractions inhibited adhesion between MDA-MB-231 cells and human platelets; no effects on B16-F10 proliferation. *In vivo*: reduced vascularization in 7-day study (hemoglobin content and CD34+ cells). Reduced tumor angiogenesis, micro-vessel density and tumor weight was observed in 21-day study with (1) or (2), not (3)	Croci et al. 2011
S. latissima	Fucoidan	Tumor cells: Raji *In vitro*	Inhibiting proliferation and migration	Schneider et al. 2015
S. religiosa (formerly *Laminaria religiosa*)	Raw seaweed	*In vivo*: DMBA-induced tumors in rats	*S. religiosa* showed an inhibitory effect on tumorigenesis, and tumor incidence was lower. A significant delay in the time to first noticeable tumor. Tumor weight per rat in each group was significantly lower in the *S. religiosa* fed rats	Yamamoto et al. 1987
S. religiosa (formerly *L. religiosa*)	Phosphate buffered saline extract	Tumor cells: L-1210 and P-388 Normal cells: NIH-3T3 *In vitro*	Extract exhibited highly cytotoxicity to tumor cells and less cytotoxicity to normal cells	Harada et al. 1997

Table 6.1 contd....

Table 6.1 contd....

Species	Tested product, fraction or extract	Study model	Observed results	References
S. sculpera (as *Kjellmaniella crassifolia*)	Seaweed powder	Tumor cells: Ehrlich carcinoma; Meth-A fibrosarcoma *In vivo*	Oral administration of 1600 mg seaweed powder per kg mouse per day for 28 days. Inhibition rate: 36.1% Intraperitoneal administration of 50 mg kg^{-1} d^{-1} for seven days. Inhibition rate: 36.6%	Noda et al. 1990
Saccharina spp. [*S. angustata* (as *L. angustata*), *S. longissima* (as *L. angustata* var. *longissima*), *S. japonica* (as *L. japonica* var. *ochotensis*)]	Powdered seaweed, hot-water extract, non-dialyzable fraction and the residue of hot-water extractions	Tumor cells: Sarcoma-180 Normal cells *In vivo*	Extractions were incorporated into a basic diet gave tumor inhibition ratios of 70.3 to 83.6%. IP injection of 10 preparations from 6 edible seaweeds, including the 3 listed, was also effective with inhibition ratios of 61.9 to 95.2%	Yamamoto et al. 1986
Saccorhiza polyschides	Crude (DCM/MeOH) extract. PP content 1.0–10.9%	Tumor cells: Daudi, Jurkat, K562 Normal cells *In vivo*	Crude extract concentrations: 50–500 mg L^{-1}; Inhibition: 20.00–72.29%	Zubia et al. 2009b
Sargassum angustifolium	Fucosterol (crude fractions also studied)	Tumor cells: HT-29, Caco-2, T47D Normal cells: NIH 3T3 *In vivo*	IC_{50} (μg mL^{-1}): 28 (T47D), > 70 (Caco-2), 70 (HT-29), > 70 (NIH3T3)	Khanavi et al. 2012
S. angustifolium	Methanol extract	Tumor cells: HT-29, HeLa, MCF-7 *In vitro*	Cytotoxic activities IC_{50} against HT-29 (121.8 μg mL^{-1}), HeLa (97.9 μg mL^{-1}), and MCF-7 (67.3 μg mL^{-1})	Mehdinezhad et al. 2016
S. aquifolium (as *S. crassifolium*)	Methanol extract	Tumor cells: A549, B16F10, HL-60 Normal cells: Vero *In vitro*	IC_{50} > 200 μg mL^{-1} for HL-60 cells	Lakmal et al. 2014
S. boveanum	Methanol extract	Tumor cells: HT-29, HeLa, MCF-7 *In vitro*	Cytotoxic activities IC_{50} against HT-29 (125.6 μg mL^{-1}), HeLa (82.3 μg mL^{-1}), and MCF-7 (60.4 μg mL^{-1})	Mehdinezhad et al. 2016

S. carpophyllum	7 different steroids	Tumor cells: HL-60, P-388, MCF-7, HCT-8, 1A9, HOS, PC-3 Normal cells *In vivo*	Fucosterol and 24-ethylcholesta-4,24(28)-dien-3,6-dione have excellent cytotoxicity towards P-388 cells; IC_{50s} of 0.7 and 0.8 mg mL^{-1}, respectively. 24R, 28R- and 24S, 28S-epoxy-24-ethylcholesterol was active against MCF-7, HCT-8, 1A9, HOS and PC-3 with IC_{50s} of 4–10 mg mL^{-1}	Tang et al. 2002
S. cinereum	Fucoidan	Tumor cells: HCT-15 *In vitro*	Fucoidan extract caused about 50% of cell death after 24 h of incubation with 75 g mL^{-1}	Somasundaram et al. 2016
S. confusum	Methanolic extract	Tumor cells: L-1210 and P-388 Normal cells: NIH-3T3 *In vitro*	Extract exhibited highly cytotoxicity to tumor cells and less cytotoxicity to normal cells	Harada et al. 1997
S. fallax	Fallahydroquinone, fallaquinone, fallachromenoic acid, sargaquinone, sargahydroquinoic acid, sargaquinoic acid, sargachromenol	Tumor cells: P388 Normal cells *In vivo*	IC_{50}: sargaquinoic acid (17 μM) and sargahydroquinoic acid (14 μM). Remainder were less active (IC_{50s} > 27–32 μM)	Reddy and Urban 2009
S. filipendula	Sulfated polysaccharides	Tumor cells: HeLa Normal cells *In vivo*	61.1% of cell proliferation inhibition at 0.1 mg mL^{-1}	Costa et al. 2010
S. filipendula	Fucoidan	Tumor cells: HeLa, PC3, HepG2 Normal cells *In vitro*	All fractions tested showed some cytotoxic action at doses of 0.1–2 mg mL^{-1}	Costa et al. 2011a
S. filipendula	Fucoidan	Tumor cells: HeLa Normal cells *In vivo*	Inhibited proliferation and caused apoptosis (0.1 to 2.0 mg mL^{-1}); the greatest effect was seen after 72 h. AIF was released into cytosol with a decrease in Bcl-2 and increase I Bax; no effects were found on caspases or levels of ERK, p38, p53, pAKT and NF-κB	Costa et al. 2011b
S. fluitans	DCM-methanol (7:3) and water extracts	Tumor cells: Hep-2, HeLa, KB Normal cells: MDCK *In vivo*	Anti-proliferative activity of methanolic extract against KB cells IC_{50} = 77.3 μg mL^{-1}	Moo-Puc et al. 2009
S. fulvellum	Hot water extraction	Tumor cells: S-180 Normal cells *In vivo*	10 mg kg^{-1}	Yamamoto et al. 1974

Table 6.1 contd....

Table 6.1 contd....

Species	Tested product, fraction or extract	Study model	Observed results	References
S. fulvellum	Methanolic extract	Tumor cells: L-1210 and P-388 Normal cells: NIH-3T3 *In vitro*	Extract exhibited highly cytotoxicity to tumor cells and less cytotoxicity to normal cells	Harada et al. 1997
S. fusiforme (formerly *Hizikia fusiformis*)	Fucoxanthin	Tumor cells Normal cells *In vivo*: B6C3F1 mice: SC injection of DMH 6× over 3 weeks, then fucoxanthin treatment for 7 weeks	Colonic crypt epithelial cells showed reduced proliferation and number of aberrant crypt foci; no liver or kidney toxicity seen	Kim et al. 1998
S. fusiforme (formerly *H. fusiformis*)	Seaweed powder	Tumor cells: Ehrlich carcinoma *In vivo*	Oral administration of 1600 mg seaweed powder per kg mouse per day for 28 days. Inhibition rate: 49.7%	Noda et al. 1990
S. fusiforme (formerly *H. fusiformis*)	Phosphate buffered saline extract	Tumor cells: L-1210 and P-388 Normal cells: NIH-3T3 *In vitro*	Extract exhibited cytotoxicity to tumor cells and no cytotoxicity to normal cells	Harada et al. 1997
S. fusiforme	Methanolic extract	Tumor cells: HT-29, KB Normal cells: NIH-3T3 *In vitro*	IC_{50} > 50 (KB), > 50 (HT-29), > 50 (NIH-3T3) μg mL^{-1}	Xu et al. 2004b
S. fusiforme (formerly *H. fusiformis*)	Ethanol/water extract	Tumor cells: HL-60, HT-29 Also tested: B-16, A-549 Normal cells: HaCaT *In vivo*	Concentration of extract: 100 μg mL^{-1} HL-60 Inhibition = 86.5% HT-29 Inhibition = 59.2% B16 Inhibition = 37.2% A549 Inhibition = 2.1% HaCat = Inhibition = 64.6%	Kim et al. 2009
S. fusiforme (formerly *H. fusiformis*)	Ethyl alcohol extract	Tumor cells: U937 Normal cells *In vivo*	Doses of 30–50 μg mL^{-1} reduced viability to 50–60%. Treatment with extract increased the cleaved forms of caspases 3, 8 and 9 and PARP and decreased Bcl-2, IAP-1, IAP-2 and XIAP	Kang et al. 2011
S. fusiforme	Sulfated fucoidan FP08S2	Tumor cells: A549 *In vitro* and *in vivo*	Fucoidan inhibited the tube formation, migration and invasion of HMEC-1 cells *in vitro*. Also FP08S2 impaired angiogenesis *in vivo*, resulting in antitumor effects in A549 tumor xenograft model	Chen et al. 2016

S. hemiphyllum	Hedaol A, B and C	Tumor cells: P-388 Normal cells *In vivo*	IC_{50} of 5.1, 2.2 and 50 µg mL^{-1} to P-388 cells for Heladaol A, B and C, respectively	Takada et al. 2001
S. hemiphyllum	Fucoidan with molecular weight of 760 Da	Tumor cells: T24 *In vitro* and *in vivo*	Inhibiting angiogenesis and tumor growth	Chen et al. 2015
S. hemiphyllum	Fucoidan	Tumor cells: SK-Hep1 and HepG2 *In vitro*	Suppressing invasion and migration	Yan et al. 2015
S. henslowianum	Fucoidan	*In vivo*: male C57BL/6J mice injected IP with fucoidan (50 mg kg^{-1}) Tumor cells: B16 Normal cells	Dose-dependent inhibition (0.1–1 mg mL^{-1}); induction of apoptosis and increased caspase-3 activity	Ale et al. 2011b
S. horneri	Seaweed powder	Tumor cells: Ehrlich carcinoma *In vivo*	Oral administration of 1600 mg seaweed powder per kg mouse per day for 28 days. Inhibition rate: 38.6%	Noda et al. 1990
S. horneri	Methanolic extract	Tumor cells: L-1210 and P-388 Normal cells: NIH-3T3 *In vitro*	Extract exhibited highly cytotoxicity to tumor cells and less cytotoxicity to normal cells	Harada et al. 1997
S. horneri	Water extract	Tumor cells: Meth-A, Meth-A tumor bearing BALB/c female mice Normal cells *In vitro* and *in vivo*	0.05% × 105 mL/mouse	Matsuda et al. 2005
S. horneri	Fucoidan	Tumor cells: SK-MEL-28, DLD-1 *In vitro*	Inhibiting colony formation	Ermakova et al. 2011
S. ilicifolium (formerly *S. cristaefolium*)	Fucoidan	Tumor cells: HT-29 *In vitro*	Inhibition of growth in human colon cancer cell line HT-29 by fucoidan = 500 µg mL^{-1} for 48 h	
S. latifolium	Chloroform, ethanol extracts	Tumor cells: AGS, HeLa, PC12 Normal cells: NIH 3T3 *In vitro*	Polysaccharide extracted (ethanolic extract derivative) have been proved to have antiproliferative potential against human lymphoblastic leukemia and the IC_{50} dose was found to be significantly lower (17.18 µg mL^{-1}), the extract not only induced apoptosis but also able to induce the carcinogen detoxification enzyme Glutathione-S-transferases	Gamal-Eldeen et al. 2009

Table 6.1 contd....

Table 6.1 contd....

Species	Tested product, fraction or extract	Study model	Observed results	References
S. miyabei (as *S. kjellmanianum*)	Water extract	Tumor cells: S-180 Normal cells *In vivo*	93.7% for 100 mg mL^{-1}	Yamamoto et al. 1981
S. macrocarpum	Phosphate buffered saline extract	Tumor cells: L-1210 and P-388 Normal cells: NIH-3T3 *In vitro*	Extract exhibited highly cytotoxicity to tumor cells and to normal cells	Harada et al. 1997
S. micracanthum	Methanolic extract	Tumor cells: L-1210 and P-388 Normal cells: NIH-3T3 *In vitro*	Extract exhibited cytotoxicity to tumor cells and highly cytotoxicity to normal cells	Harada et al. 1997
S. mcclurei	Fucoidan	Tumor cells: DLD-1 Normal cells *In vivo*	No cytotoxicity at doses of 1 to 200 μg mL^{-1}. Inhibited colony formation at 100 μg mL^{-1}, no link to sulfation	Thinh et al. 2013
S. micracanthum	Seaweed powder	Tumor cells: Ehrlich carcinoma; Meth-A fibrosarcoma *In vivo*	Oral administration of 1600 mg seaweed powder per kg mouse per day for 28 days. Inhibition rate: 42.6% Intraperitoneal administration of 50 mg kg^{-1} d^{-1} for seven days. Inhibition rate: 26.3%	Noda et al. 1990
S. micracanthum	3 plastoquinones	Tumor cells: HeLa 229, MDCK Normal cells: Vero *In vivo*	IC_{50} values 8–35 μM, with similar cytotoxicity's to normal cells	Iwashima et al. 2005
S. micracanthum	Plastoquinones	Tumor cells: colon 26-L5 Normal cells *In vivo*	IC_{50} values 1.5–17.5 μg mL^{-1}	Mori et al. 2005b
S. micracanthum	Methanol-water extract	Tumor cells: RBL-2H3 Normal cells *In vivo*	Inhibition effect: 20.0% (concentration: 1 mg mL^{-1})	Sugiura et al. 2006
S. miyabei (formerly *S. kjellmanianum*)	Sulfated polysaccharide fraction	Tumor cells: L-1210 *In vivo*	Sulfated fraction was observed to contain nearly 50% more ester sulfate than in normal extract and to be effective against L-1210 leukemia showing an ILS value of 26%	Yamamoto et al. 1984b

S. miyabei (formerly *S. kjellmanianum*)	Methanolic extract	Tumor cells: HT-29, KB Normal cells: NIH-3T3 *In vitro*	$IC_{50} > 50$ (KB), > 50 (HT-29), > 50 (NIH-3T3) μg mL^{-1}	Xu et al. 2004b
S. muticum	Methanolic extract	Tumor cells: MCF-7, MDA-MB-231 Normal cells: Vero *In vitro*	Extracts were cytotoxic against breast cancer cell lines in a dose-dependent manner, with IC_{50} of 22 μg mL^{-1} for MCF-7 and 55 μg mL^{-1} for MDA-MB-231 cell lines. The percentages of apoptotic MCF-7-treated cells increased from 13 to 67% by increasing the concentration of the extracts	Namvar et al. 2013
S. muticum	Extracts	Tumor cells: A549, HTC-116, PSN1, T98G *In vitro*	Colon carcinoma cells were inhibited by more than 50% in the presence of all extracts at 250 and 500 μg mL^{-1}. The IC_{50} values for growth inhibition of lung carcinoma A549 cells, colon carcinoma HCT15 cells and breast adenocarcinoma MCF7 cells ranged from 78–82 mg mL^{-1}, 27–50 mg mL and 110–130 mg mL^{-1}, respectively	Casas et al. 2016
S. muticum	Water-soluble polysaccharides (Fucoidn)	Tumor cells: DLD-1 human colon carcinoma cells	The preparations were nontoxic for DLD-1 human colon cancer cells at concentrations up to 400 μg mL^{-1}	Usol'tseva et al. 2017
S. oligocystum	Aqueous extract	Tumor cells: Daudi, K562 *In vitro* (MTT assay)	The most potent antitumor activity has been shown at concentrations 500 μg mL^{-1} and 400 μg mL^{-1} of the alga extract on Daudi and K562 cell lines, respectively	Zandi et al. 2010
S. oligocystum	Methanolic extract	Tumor cells: L-1210 and P-388 Normal cells: NIH-3T3 *In vitro*	Extract exhibited highly cytotoxicity to tumor cells and to normal cells	Harada et al. 1997
S. oligocystum	Methanol extract	Tumor cells: HT-29, HeLa, MCF-7 *In vitro*	Cytotoxic activities IC_{50} against HT-29 (133.9 μg mL^{-1}), HeLa (96.7 μg mL^{-1}), and MCF-7 (56.9 μg mL^{-1})	Mehdinezhad et al. 2016

Table 6.1 contd....

Table 6.1 contd....

Species	Tested product, fraction or extract	Study model	Observed results	References
S. pallidum	Crude aqueous extract	Tumor cells Normal cells *In vivo*: gastric cancer induced by MNNG in male Wistar rats	Serum IL-2, 4 and 10 increased and IL-6, IL-1β, TNF-α decreased in rats receiving extract. Gastric mucosa and serum MDA decreased, gastric mucosa and serum GSH decreased, antioxidant enzymes (SOD, CAT, GSH-Px) increased	Zhang et al. 2012b
S. piluliferum	Methanolic extract	Tumor cells: L-1210 and P-388 Normal cells: NIH-3T3 *In vitro*	Extract exhibited highly cytotoxicity to tumor cells and less cytotoxicity to normal cells	Harada et al. 1997
S. polycystum	Fucoidan	Tumor cells: MDA-MB-231 *In vitro*	Fucoidan was effective against some types of cancer such as lung, liver, bladder and colon cancers	Ly et al. 2005
S. polycystum	Crude extract (ethanol or acetone)	Tumor cells: Caco-2 Normal cells *In vivo*	Both ethanol and acetone extracts decreased Caco-2 viability when cells were treated with 600 μM hydrogen peroxide	Gunji et al. 2007
S. patens	Seaweed powder	Tumor cells: Meth-A fibrosarcoma *In vivo*	Intraperitoneal administration of 50 mg kg^{-1} d^{-1} for seven days. Inhibition rate: 55.5%	Noda et al. 1990
S. ringgoldianum	Seaweed powder	Tumor cells: Ehrlich carcinoma; Meth-A fibrosarcoma *In vivo*	Oral administration of 1600 mg seaweed powder per kg mouse per day for 28 days. Inhibition rate: 46.5% Intraperitoneal administration of 50 mg kg^{-1} d^{-1} for seven days. Inhibition rate: 39.1%	Noda et al. 1990
S. ringgoldianum	Neutral lipid fractions	Tumor cells: Meth-A fibrosarcoma *In vivo*	Intraperitoneal administration of 40 mg kg^{-1} for seven days. Inhibition rate: 42.6%	Noda et al. 1990

S. ringgoldianum	Sulfated fucans: Fucoidan II A fraction B fraction C fraction	Meth-A fibrosarcoma *In vivo*	Intraperitoneal administration of 40 mg kg^{-1} for seven days. Inhibition rate (%): 78.1 32.0 26.2 34.7	Noda et al. 1990
S. ringgoldianum	Phospholipid fractions	Tumor cells: Meth-A fibrosarcoma *In vivo*	Intraperitoneal administration of 40 mg kg^{-1} for seven days. Inhibition rate: 47.1%	Noda et al. 1990
S. ringgoldianum	Methanolic extract	Tumor cells: L-1210 and P-388 Normal cells: NIH-3T3 *In vitro*	Extract exhibited highly cytotoxicity to tumor cells and less cytotoxicity to normal cells	Harada et al. 1997
S. ringgoldianum	Methanol-water extract	Tumor cells: RBL-2H3 Normal cells *In vivo*	Inhibition effect: 70.0% (concentration: 1 mg mL^{-1})	Sugiura et al. 2006
S. siliquastrum (as *S. tortile*)	Seaweed powder	Tumor cells: Ehrlich carcinoma *In vivo*	Oral administration of 1600 mg seaweed powder per kg mouse per day for 28 days. Inhibition rate: 38.6%	Noda et al. 1990
S. siliquastrum	Methanolic extract	Tumor cells: L-1210 and P-388 Normal cells: NIH-3T3 *In vitro*	Extract exhibited highly cytotoxicity to tumor cells and to normal cells	Harada et al. 1997
S. siliquastrum	Sargachromanol E	Tumor cells: HL-60 Normal cells *In vivo*	Caused caspase 3-mediated apoptosis in HL-60 cells	Heo et al. 2011
S. stenophyllum	Sulfated fucans	Tumor cells: HeLa	Sulfated polysaccharide promoted very accentuated morphologic modifications in HeLa cells at low concentrations (2.5 μg mL^{-1}), caused significant alterations in the cellular morphology and reduction of the growth, such effects being dose dependent	Stevan et al. 2001
S. stenophyllum	Polysaccharide powder fractions	Tumor cells: B16F10 *In vivo* and *in vitro*	Polysaccharide powder from *S. stenophyllum* at doses of 1.5 and 150 μg/animal has the ability to retard the increase in tumor volume by at least 2.5 and 5 days, respectively. Treatment with either dose of polysaccharide powder did not induce any deaths or body weight loss, suggesting little or no toxicity	Dias et al. 2005

Table 6.1 contd....

Table 6.1 contd....

Species	Tested product, fraction or extract	Study model	Observed results	References
Sargassum sp.	Methanolic extract	Tumor cells: L-1210 and P-388 Normal cells: NIH-3T3 *In vitro*	Extract exhibited highly cytotoxicity to tumor cells and less cytotoxicity to normal cells	Harada et al. 1997
Sargassum sp.	Fucoidan	Tumor cells: LLC, B16 Normal cells	*In vitro*: both fucoidans reduced the viability of LLC and B16 cells by inducing apoptosis. *In vivo*: fucoidan enhanced NK cell activity (no tumor cells implanted).	Ale et al. 2011a
Sargassum sp.	Fucoidan	Tumor cells: HepG2 Normal cells *In vivo*: female nu/nu nude mice injected SC with Bel-7402 cells. Treated with fucoidan 20, 200 mg kg^{-1} injected IP *In vitro*: no effect on bFGF, VEGF, IL-8 and heparanase observed in HepG2. *In vivo*: no effect on bFGF and VEGF	Reduction in tumor volume and weight. Reduction in PCNA, no apoptosis found	Zhu et al. 2013
S. subrepandum	Methanol extract	Tumor cells: EAT *In vivo*	Group that received extract alone showed slight changes in serum tumor marker levels in comparison with the control group	Ahmed et al. 2011
S. swartzii	Fucoidan	Tumor cells: MDA-MB-231 *In vitro*	Fucoidan concentration of 0.5 μg mL^{-1}, almost all the breast cancer cells were inactivated	Ly et al. 2005
S. swartzii	Crude methanol (70%) extract and hexane, chloroform, ethyl acetate and MeOH-H2O fractions of this	Tumor cells: HT-29, Caco-2, T47D, T47D-T. R, MDA-MB468 Normal cells: NIH 3T3 *In vivo*	Hexane fraction had good cytotoxic action and induced apoptosis in Caco-2 and T47D cells. Mechanism was found to be estrogen receptor independent by testing estrogen receptor negative and positive breast cancer cells. Was also cytotoxic to normal cells, but at higher IC_{50}	Khanavi et al. 2010
S. swartzii (formerly *S. wightii*)	Chloroform, ethanol extracts	Tumor cells: AGS, HeLa, PC12 Normal cells: NIH 3T3 *In vitro*	IC_{50} values of this fraction against AGS, HeLa, MCF-7 and PC12 cell lines after 24 h were determined, 43.61, 46.92, 99.38, 158.8 μg mL^{-1}, respectively	Manojkumar 2013

S. swartzii (formerly *S. wightii*)	Methanolic extract	Tumor cells: A549 Normal cells: VERO *In vitro*	Cytotoxic effect against human A549 lung adenocarcinoma cancer cell lines in a concentration dependent manner, with IC_{50} = 18 μg mL^{-1}	Rani et al. 2013
S. thunbergii	Seaweed powder	Tumor cells: Ehrlich carcinoma; Meth-A fibrosarcoma *In vivo*	Oral administration of 1600 mg seaweed powder per kg mouse per day for 28 days. Inhibition rate: 41.7% Intraperitoneal administration of 50 mg kg^{-1} d^{-1} for seven days. Inhibition rate: 35.3%	Noda et al. 1990
S. thunbergii	Fucoidan	Tumor cells: EAC Normal cells *In vivo*	Markedly inhibited the growth of Ehrlich ascites carcinoma (EAC) at the dose of 20 mg kg^{-1} per day x10 with no sign of toxicity in mice	Itoh et al. 1993
S. thunbergii	Fucoidan with molecular weight of 19,000	Tumor cells: EAC *In vivo*	Inhibiting growth	Zhuang et al. 1995
S. thunbergii	Methanolic extract	Tumor cells: L-1210 and P-388 Normal cells: NIH-3T3 *In vitro*	Extract exhibited highly cytotoxicity to tumor cells and to normal cells	Harada et al. 1997
S. thunbergii	Methanolic extract	Tumor cells: HT-29, KB Normal cells: NIH-3T3 *In vitro*	IC_{50} > 50 (KB), > 50 (HT-29), > 50 (NIH-3T3) μg mL^{-1}	Xu et al. 2004b
S. thunbergii	Methanol-water extract	Tumor cells: RBL-2H3 Normal cells *In vivo*	Inhibition effect: 40.7% (concentration: 1 mg mL^{-1})	Sugiura et al. 2006
S. thunbergii	Possibly fucosterol	Tumor cells: HL-60, HT-29 Also tested: B-16, A-549 Normal cells: HaCaT *In vivo*	HL_{60}: most seaweeds were toxic at 100 μg mL^{-1}. HT29: *Sargassum fusiforme* (formerly *Hizikia fusiformis*) and *S. thunbergii* were toxic at 100 μg mL^{-1}. Normal cells: mostly low cytotoxicity, *S. fusiforme* inhibited growth up to 65.6%	Kim et al. 2009
S. vestitum	Ethanol extract 100 μg mL^{-1}	Tumor cells Normal cells *In vivo*	Inhibition of kinase A was assayed. Showed the highest levels of inhibition = 76–100%	Winberg et al. 2011
S. vulgare	Dichloromethane, chloroform, methanol, ethanol, water extracts, hexane and chloroform fractions	Tumor cells: HEp-2, K562, NCI-H292 *In vitro*	Water extract (18.7 μg mL^{-1}) was active against HEp-2	Guedes et al. 2013

Table 6.1 contd....

Table 6.1 contd....

Species	Tested product, fraction or extract	Study model	Observed results	References
S. vulgare	Ethanol extract	Tumor cells: MCF-7, LNCaP, PC-3 Normal cells: MCF-10A *In vitro*	Extract shows low to moderate cytotoxic effects on the cancer cells (LNCa, PC-3 and MCF-7) but not on the non-tumorigenic cell line MCF-10. Extract showed moderate effects on human breast adeno-carcinoma MCF-7 cells (45% inhibition of growth)	Montalvão et al. 2016
S. yendoi	Methanolic extract	Tumor cells: L-1210 and P-388 Normal cells: NIH-3T3 *In vitro*	Extract exhibited highly cytotoxicity to tumor cells and no cytotoxicity to normal cells	Harada et al. 1997
Scytosiphon lomentaria	Seaweed powder	Tumor cells: Ehrlich carcinoma *In vivo*	Oral administration of 1600 mg seaweed powder per kg mouse per day for 28 days. Inhibition rate: 69.8%	Noda et al. 1990
S. lomentaria	Phosphate buffered saline extract	Tumor cells: L-1210 and P-388 Normal cells: NIH-3T3 *In vitro*	Extract exhibited high cytotoxicity to tumor cells and no cytotoxicity to normal cells	Harada et al. 1997
S. lomentaria	Methanolic extract	Tumor cells: HT-29, KB Normal cells: NIH-3T3 *In vitro*	IC_{50} = 45.3 (KB), = 32.3 (HT-29), = 48.7 (NIH-3T3) μg mL^{-1}	Xu et al. 2004b
S. lomentaria	Ethanol/water extract	Tumor cells: HL-60, HT-29 Also tested: B-16, A-549 Normal cells: HaCaT *In vivo*	Concentration of extract: 100 μg mL^{-1} HL-60 Inhibition = 79.4% HT-29 Inhibition = 2.4% B16 Inhibition = 23.9% A549 Inhibition = 3.6% HaCat = Inhibition = 22.5%	Kim et al. 2009
S. lomentaria	Methanol extract	Tumor cells: DU-145, LNCaP, MCF-7, PC-3 *In vitro*	Extract in 100 g mL^{-1} concentration showed low toxicities against the cell lines However, 200 g mL^{-1} concentration of extract showed higher toxicity in 48 h (< 40% cell viability)	Taskin et al. 2010

Spatoglossum asperum	Hexane, dichloromethane-, ethyl acetate and methane fractions	Tumor cells: MiaPaCa-2, Panc-1, BXPC-3, Panc-3.27 Normal cells *In vivo*	Dose-dependent inhibition of cell growth, induction of apoptosis. Inhibition of NF-κB. Modulated EGFR phosphorylation, kRas, AurKb and Stat3 (see paper for details on extract/cell line efficacy). Little characterization carried out; authors recommend more to aid discovery of active compounds	Aravindan et al. 2013
S. stipitatum (formerly *zonaria stipitata*)	Methanolic extract	Tumor cells: L-1210 and P-388 Normal cells: NIH-3T3 *In vitro*	Extract exhibited cytotoxicity to tumor cells and less cytotoxicity to normal cells	Harada et al. 1997
S. schroederi	Acetone/dichlorometha-ne/chloroform crude extracts	Tumor cells: C32 Normal cells: FEK4 *In vivo*	IC_{50} values for C32 cells were > 100 μg mL^{-1}	Rocha et al. 2007
S. schroederi	Sulfated polysaccharides	Tumor cells: HeLa Normal cells *In vivo*	About 45.0% of cell proliferation inhibition at 2.0 mg mL^{-1}	Costa et al. 2010
Stoechospermum polypodioides (formerly *Stoechospermum marginatum*)	Hexane, dichloromethane-, ethyl acetate and methane fractions	Tumor cells: MiaPaCa-2, Panc-1, BXPC-3, Panc-3.27 Normal cells *In vivo*	Dose-dependent inhibition of cell growth, induction of apoptosis. Inhibition of NF-κB. Modulated EGFR phosphorylation, kRas, AurKb and Stat3 (see paper for details on extract/cell line efficacy). Little characterization carried out; authors recommend more to aid discovery of active compounds	Aravindan et al. 2013
S. polypodioides (formerly *Stoechospermum marginatum*)	Methanolic extracts	Tumor cells: BeWo, EAT, HUVEC *In vivo*	Extract inhibits *in vitro* proliferation of EAT/BeWo cells in a dose-dependent manner. Inhibition of proliferation was 80.55, 83.54, 90.79, 95.43, and 96.36% on the proliferation in BeWo cells and 4.62, 5.76, 5.85, 6.71, and 7.40% was recorded, while in HEK 293 cells at 0.005, 0.025, 0.050, 0.075, and 0.1 mg mL^{-1} concentrations	Vinayak et al. 2014
Stypopodium flabelliforme	14-keto-stypodiol diacetate	Tumor cells: DU-145 *In vitro*	Extract concentrations of 5 μM decreased cell growth by 14%, while at 45 μM a 61% decrease was found, as compared with control cells incubated with the solvent but in the absence of the drug	Depix et al. 1998
S. flabelliforme	(1) 2β,3α-epitaondiol, (2) flabellinol, (3) flabellinone, (4) stypotriolaldehyde, (5) stypohydroperoxide	Tumor cells: NCI-H460, Neuro-2a Normal cells *In vivo*	(1), (2) and (3) were moderately cytotoxic to neuro-2a cells (LC_{50} ranges from 2 to 11 μM) and to NCIH460 cells (LC_{50} of 9–24 μM)	Sabry et al. 2005

Table 6.1 contd....

Table 6.1 contd....

Species	Tested product, fraction or extract	Study model	Observed results	References
S. flabelliforme	Mero-diterpenoids	Tumor cells: Caco-2, SH-SY5Y, RBL-2H3, RAW.267 Normal cells: V79 *In vivo*	Concentrations tested ≤ 50 μM. Noncancerous V79 cells: 3 compounds had no/little effect, but 3 inhibited proliferation significantly. Cancer cells: some compounds showed selective toxicity, e.g., stypodiol—had little effect on V79 cells but toxic to SH-SY5Y cells	Pereira et al. 2011
S. zonale	Methanolic extract	Tumor cells: L-1210 and P-388 Normal cells: NIH-3T3 *In vitro*	Extract exhibited highly cytotoxicity to tumor cells and to normal cells	Harada et al. 1997
S. zonale	Acetone/dichloromethane/ chloroform crude extracts	Tumor cells: C32 Normal cells: FEK4 *In vivo*	IC_{50} values for C32 cells were > 100 μg mL^{-1}	Rocha et al. 2007
Taonia atomaria	Atomarianone A and B	Tumor cells: NSCLC-N6, A549 Normal cells *In vivo*	IC_{50} values of < 7.35 μM	Abatis et al. 2005
T. atomaria	Sulfolipids	Tumor cells: MCF7, HepG2 *In vitro*	Concentrations = 1.0–10.0 μg mL^{-1} Growth inhibition = 79.46–85.69% (MCF7), 71.19–85.77% (HepG2)	El Baz et al. 2013
Turbinaria conoides	Crude extract (ethanol or acetone)	Tumor cells: Caco-2 Normal cells *In vivo*	Both ethanol and acetone extracts decreased Caco-2 viability when cells were treated with 600 μM hydrogen peroxide	Gunji et al. 2007
T. conoides	Fucoidan	Tumor cells: MiaPaCa-2 and Panc-1 *In vivo*	Exhibited 28.99 and 17.88% cell death in MiaPaCa-2 and Panc-1 cells respectively at a concentration of 100 μg mL^{-1} at 24 h. This drastically increased to 57.25 and 46.32 and 128.54 and 68.67% after 48 and 72 h incubation, respectively	Delma et al. 2015

T. conoides	Fucoidan with 53% fucose and 38% sulfate	Tumor cells: A549 Normal cells: Vero *In vitro*	Inhibited the growth of cancer cells in a dose-dependent manner and potent anticancer activities were 24.9–73.5% in the concentrations of 31.25–500 μg mL^{-1}. The CTC_{50} value against the cancer cell was found to be 45 μg mL^{-1} and the CTC_{50} value of normal Vero cell line is 325 μg mL^{-1}	Marudhupandi et al. 2015
T. conoides	Fucoidan	Tumor cells: Lung cancer A549 *In vitro*	Fucoidan induced a dose-dependent reduction in cell survival of lung cancer A549 cells by MTT assay (IC_{50} = 75 μg mL^{-1})	Alwarsamy et al. 2016
T. ornata	Phosphate buffered saline extract	Tumor cells: L-1210 and P-388 Normal cells: NIH-3T3 *In vitro*	Extract exhibited cytotoxicity to tumor cells and to normal cells	Harada et al. 1997
T. ornata	Methanol, chloroform, ethyl acetate and aqueous extracts. Highest PP content 1.3 μg mL^{-1}, the highest flavonoid content 88 μg mL^{-1}	Tumor cells: A549, HCT-15, MG-63, PC-3 Normal cells *In vivo*	Although some association was found between inhibition of proliferation of A549 and MG-63 cells with higher values of total flavonoid content (see paper for details), authors suggest many bioactive molecules in extract could be responsible for growth inhibition effect	Murugan and Iyer 2014
T. turbinata	DCM-methanol (7:3) and water extracts	Tumor cells: Hep-2, HeLa, KB Normal cells: MDCK *In vivo*	Anti-proliferative activity of methanolic extract against KB, HeLa and Hep-2 cells; IC_{50} = 29.8, 84.8 and 93.2 μg mL^{-1}, respectively	Moo-Puc et al. 2009
T. turbinata	Aqueous extract	Tumor cells: EAC Silver Nanoparticle's of aqueous extract (AgNPs) *In vitro*	Cytotoxic effect (67–99%) of different AgNPs concentrations (42–98 μg mL^{-1})	Khalifa et al. 2016
Undaria pinnatifida	Seaweed powder	Tumor cells: Ehrlich carcinoma; Meth-A fibrosarcoma *In vivo*	Oral administration of 1600 mg seaweed powder per kg mouse per day for 28 days. Inhibition rate: 18.1% Intraperitoneal administration of 50 mg kg^{-1} d^{-1} for seven days. Inhibition rate: 24.3%	Noda et al. 1990

Table 6.1 contd....

Table 6.1 contd....

Species	Tested product, fraction or extract	Study model	Observed results	References
U. pinnatifida	Sulfated fucans: Fucoidan I A fraction B fraction C fraction	Meth-A fibrosarcoma *In vivo*	Intraperitoneal administration of 40 mg kg^{-1} for seven days. Inhibition rate (%): 53.4 45.5 51.4 38.6	Noda et al. 1990
U. pinnatifida	Glycolipid fractions	Tumor cells: Meth-A fibrosarcoma *In vivo*	Intraperitoneal administration of 40 mg kg^{-1} for seven days. Inhibition rate: 55.8%	Noda et al. 1990
U. pinnatifida	Dichloromethane crude extract	Tumor cells Normal cells *In vivo*: skin tumors in ICR mice initiated with DMBA and promoted with TPA	*U. pinnatifida* was most potent and used for *in vivo* testing. 15 weeks of topical treatment with extract: number of mice with tumors and no. of tumors per mouse was reduced	Ohigashi et al. 1992
U. pinnatifida	Phosphate buffered saline extract	Tumor cells: L-1210 and P-388 Normal cells: NIH-3T3 *In vitro*	Extract exhibited highly cytotoxicity to tumor cells and to normal cells	Harada et al. 1997
U. pinnatifida	Raw seaweed	*In vivo*: DMBA-induced mammary tumors in female Sprague-Dawley rats	Weights of mammary tumors were significantly lower and serum total iodine concentration was significantly higher than in control group	Funahashi et al. 1999
U. pinnatifida	Aqueous extract of Mekabu (sporophyte phase)	Tumor cells: MCF-7, T-47D and MDAMB-231 Normal cells: MCF-10A *In vivo*: DMBA-induced mammary tumors in Sprague-Dawley rats	*In vivo*: strong suppressive effect on mammary carcinogenesis (given daily in drinking water, without toxicity). *In vitro*: strongly induced apoptosis in 3 human breast cancer cell lines. No induction of apoptosis was seen in normal human mammary cells	Funahashi et al. 2001
U. pinnatifida	Fucoxanthin	Tumor cells: PC-3, DU 145, LNCaP Normal cells *In vivo*	Fucoxanthin reduced cell viability of PC-3 (> 5 μM) and DU 145 and LNCaPs (> 10 μM); showed DNA fragmentation and reduce cell viability through increased apoptosis	Kotake-Nara et al. 2001
U. pinnatifida	Fucoxanthin	Tumor cells: Caco-2 Normal cells *In vivo*	Showed that fucoxanthin is hydrolyzed to fucoxanthinol during absorption both *in vitro* and *in vivo*	Sugawara et al. 2002

U. pinnatifida	Mekabu fucoidan	Tumor cells: P-388 *In vivo*	The survival of mice was prolonged when Mekabu fucoidan was administered for 4 days before tumor cell inoculation, compared with non-treated mice. Fucoidan significantly enhanced the cytolytic activity of NK cells and increased the amount of IFN-gamma produced by T cells up to about 2-fold compared with non-treated mice. The antitumor effect of Mekabu fucoidan appears to be mediated by IFN-gamma-activated NK cells	Maruyama et al. 2003
U. pinnatifida	Fucoxanthin	Tumor cells: Caco-2, HT-29 and DLD-1 Normal cells *In vivo*	Reduced cell viability (test doses 7.6, 15.2 μM); Bcl-2 expression decreased and apoptosis increased. A greater effect on Caco-2 cell viability was seen when combined with troglitazone	Hosokawa et al. 2004
U. pinnatifida	Methanolic extract	Tumor cells: HT-29, KB Normal cells: NIH-3T3 *In vitro*	IC_{50} > 50 (KB), > 50 (HT-29), > 50 (NIH-3T3) $\mu g\ mL^{-1}$	Xu et al. 2004b
U. pinnatifida	Fucoxanthin	Tumor cells: HL-60 and H2O2 resistant HL-60 variants (HP50-2, HP100-1) Normal cells *In vivo*	Fucoxanthin induced apoptosis through loss of mitochondrial membrane potential. Caspases 3 and 9 were activated, Bcl-2, Bcl-XL or Bax were unchanged, no evidence found for ROS involvement	Kotake-Nara et al. 2005
U. pinnatifida	Aqueous extract of Mekabu (sporophyte phase)	Tumor cells: MDA-MB231 Normal cells *In vivo*	Mekabu extract activates caspases 3, 6, and 8 and contributes to intracellular signaling to induce apoptosis in a human breast cancer cell line	Sekiya et al. 2005
U. pinnatifida	Mekabu fucoidan	Tumor cells: A20 *In vivo*	Diet containing 1% Mekabu fucoidan (0.034g/mouse/day) for 10 days and subcutaneously (SC) inoculated with A20 leukemia cells. Thereafter, the mice were fed with the diet containing fucoidan for 40 days. Mekabu fucoidan inhibited tumors by 65.4%	Maruyama et al. 2006
U. pinnatifida	Fucoxanthin, fucoxanthinol	Tumor cells: HUVEC Normal cells *In vivo: ex vivo* aortic segments from male Wister rats	Fucoxanthin (> 10 μM) suppressed HUVEC proliferation; there was no effect on HUVEC chemotaxis. Microvessel (CD31+ve) formation suppressed (10–20 μM). Both chemicals suppressed microvessel growth in rat aortic rings	Sugawara et al. 2006

Table 6.1 contd....

Table 6.1 contd....

Species	Tested product, fraction or extract	Study model	Observed results	References
U. pinnatifida	Hydrolyzed fucoidans	Tumor cells: A549 Normal cells *In vivo*	Microwave hydrolysis of fucoidan was not as effective as boiling acid hydrolysis in improving cytotoxicity; suggested partial removal of sulfate groups by microwaving decreased fucoidan bioactivity (0–1 mg mL^{-1} tested)	Yang et al. 2008
U. pinnatifida	Fucoidan isolated from sporophyte	Tumor cells: PC-3, HeLa, A549, HepG2 Normal cells: Vero *In vivo*	Dose-dependent cytotoxicity to cancer cell lines; not cytotoxic to normal cell line (0–1 mg mL^{-1})	Synytsya et al. 2010
U. pinnatifida	Edible seaweed	Experimental design: Korean women (n = 3 62) aged 30–65 years old, with histologically confirmed breast cancer, paired with control cases according to age and menopausal status. Seaweed consumption, and other foods, was assessed using a food frequency questionnaire	Case-control study of seaweed consumption and cancer risk in Korean women. Miyeok (*Undaria*) consumption did not have any significant associations with breast cancer	Yang et al. 2010a
U. pinnatifida	Fucoidan	Tumor cells: A549 *In vitro*	Induces apoptosis of A549 human lung cancer cells through down-regulation of p38, PI3K/Akt, and the activation of the ERK1/2 MAPK pathway	Boo et al. 2011
U. pinnatifida	Fucoidan	Tumor cells: AGS Normal cells *In vivo*	Connection shown between increasing cytotoxicity and increasing molecular weight or sulfate content. Dose-dependent cytotoxicity found at 0.2–0.8 mg mL^{-1}	Cho et al. 2011
U. pinnatifida	Fucoxanthin	Tumor cells: HL-60 Normal cells *In vivo*	Fucoxanthin induces apoptosis through caspase-3 activation (dose range 5–20 μM). It increased expression of GADD45α and DR5 and suppressed Bcl-2	Ganesan et al. 2011
U. pinnatifida	Fucoxanthin, fucoxanthinol	Tumor cells: Caco-2, HepG2, Neuro2a Normal cells *In vivo*	Both reduced proliferation of tumor cells in a dose-dependent manner; on combining with cisplatin, neither reduced the effect of cisplatin	Mise and Yasumoto 2011

U. pinnatifida	Fucoxanthin	Tumor cells: MGC-803 Normal cells *In vivo*	Reduced cell viability was dose and time dependent. Increases in DNA ladder, hypodiploid cells and caspase 3 were found. Apoptotic cells were > 93% after 72 h in 20 μM fucoxanthin	Yu et al. 2011
U. pinnatifida	Ethanol extract of Mekabu (sporophyte phase)	Tumor cells: HCT116 Normal cells *In vivo*	Extract induced apoptosis through a different mechanism from 5-FU and CPT-11; this suggests Mekabu could be a useful adjunct to chemotherapy in colorectal cancer	Nishibori et al. 2012
U. pinnatifida	Fucoidan	Tumor cells Normal cells: HUVECs	Inhibited proliferation, migration and tube formation in HUVECs. Suppressed rat aortic ring angiogenesis. Reduced VEGF in HUVECs	Liu et al. 2012a
U. pinnatifida	Fucoidan	Tumor cells Normal cells: HUVECs *In vitro* and *in vivo*	About 40% of cell proliferation and cell migration and 61% of tube formation by HUVECs were inhibited by 400 μg mL^{-1} fucoidan, the maximum concentration tested *Ex vivo* angiogenesis assay demonstrated that at 100 μg mL^{-1}, fucoidan caused significant reduction in microvessel outgrowth	Liu et al. 2012c
U. pinnatifida	Fucoidan	Tumor cells: PC-3 Normal cells *In vitro*	Inhibited growth in dose-dependent manner and induced apoptosis. Results suggested effect through inactivation of PI3K/Akt and the p38 MAPK pathways and activation of ERK1/2 MAPK pathway. G0/G1 phase arrest. Down-regulation of Wnt/β-catenin pathway	Boo et al. 2013
U. pinnatifida	Fucoxanthin	Tumor cells: HepG2 Normal cells *In vivo*	Was found to improve efficacy of cisplatin treatment. Reduced cell viability. Increased Bax/Bcl-2 ratio, probably through inhibition of NF-κB, and inhibited ERCC1 expression through ERK and PI3K/AKT	Liu et al. 2013

Table 6.1 contd....

Table 6.1 contd....

Species	Tested product, fraction or extract	Study model	Observed results	References
U. pinnatifida	Edible seaweed	American healthy postmenopausal women (n = 15), African American and European American, 10 breast cancer (BC) survivors, 5 no history of BC). 3-month single-blinded placebo-controlled clinical trial. Consumed 10 capsules daily (5 g day^{-1}) of placebo for 4 weeks, seaweed for 4 weeks, then placebo for 4 weeks	Blood and urine samples were collected after each treatment period. Reversible reduction in levels of uPAR after treatment period. UPAR is associated with unfavorable prognosis in BC patients	Teas et al. 2013
U. pinnatifida	Fucoidan	Tumor cells: SMMC-7721 Normal cells *In vitro*	Taken together, our findings demonstrate that fucoidan induces apoptosis in SMMC-7721 cells via the ROS-mediated mitochondrial pathway, by increasing ROS production, inducing mitochondrial oxidative damage, MMP depolarization and release of cytochrome c, combined with downregulation of *XIAP* and *Livin* and activation of caspase-3 and caspase-9	Yang et al. 2013
U. pinnatifida	Fucoidan	Tumor cells: Hca-F *In vitro* and *in vivo*	Inhibits the hypoxia-induced expression, nuclear translocation and activity of HIF-1α, the synthesis and secretion of VEGF-C and HGF, cell invasion and lymphatic metastasis in a mouse hepatocarcinoma Hca-F cell line. Fucoidan also suppressed lymphangiogenesis *in vitro* and *in vivo*	Teng et al. 2015
U. pinnatifida	Fucoidan	Tumor cells: YAC-1 *In vivo*	Strongly delayed human neutrophil apoptosis (80%) at low concentration, showed a dose-dependent inhibiting effect on neutrophil apoptosis at concentrations between 5–100 $\mu g\ mL^{-1}$	Zhang et al. 2015
Rhodophyta (Red seaweed)				
Acanthopeltis japonica	Seaweed powder	Tumor cells: Ehrlich carcinoma *In vivo*	Oral administration of 1600 mg seaweed powder per kg mouse per day for 28 days. Inhibition rate: 38.5%	Noda et al. 1990

Acanthophora spicifera	Crude ethanol extract	Tumor cells Normal cells *In vivo*: Ehrlich ascites carcinoma implanted IP in Swiss albino mice	Extract decreased tumor volume and weight and increased survival compared to the saline treated control. Decreased hemoglobin and packed cell volume, increased white blood cell and SOD and CAT levels	Lavakumar et al. 2012
A. spicifera	Methanol, chloroform, ethyl acetate and aqueous extracts. Highest PP content 1.3 μg mL^{-1}, the highest flavonoid content 88 μg mL^{-1}	Tumor cells: A549, HCT-15, MG-63, PC-3 Normal cells *In vivo*	Although some association was found between inhibition of proliferation of A549 and MG-63 cells with higher values of total flavonoid content (see paper for details), authors suggest many bioactive molecules in extract could be responsible for growth inhibition effect	Murugan and Iyer 2014
A. spicifera	Methanolic seaweed extracts	Tumor cells: HeLa, SiHa	With respect to *A. spicifera* extract, a decreased viability of ~ 20% was observed already in the lower concentration tested, however this activity did not increase with increasing concentration or time of exposure to the extract	Gomes et al. 2015
Acrosorium flabellatum	Crude methanol, water and PBS extracts	Tumor cells: MOLT-4, K562, HeLa, KB Normal cells *In vivo*	PBS extract showed telomerase inhibiting activity	Kanegawa et al. 2000
Actinotrichia fragilis	Methanolic extract	Tumor cells: L-1210 and P-388 Normal cells: NIH-3T3 *In vitro*	Extract exhibited cytotoxicity to tumor cells and high cytotoxicity to normal cells	Harada et al. 1997
Aglaothamnion pseudobyssoides	Crude (DCM/MeOH) extract. PP content 1.0–10.9%	Tumor cells: Daudi, Jurkat, K562 Normal cells *In vivo*	Crude extract concentrations: 50–500 mg L^{-1}; Inhibition: 33.85–79.42%	Zubia et al. 2009a
Ahnfeltiopsis flabelliformis (formerly *Gymnogongrus flabelliformis*)	Seaweed powder	Tumor cells: Ehrlich carcinoma *In vivo*	Oral administration of 1600 mg seaweed powder per kg mouse per day for 28 days. Inhibition rate: 25.0%	Noda et al. 1990
A. flabelliformis	Methanolic extract	Tumor cells: L-1210 and P-388 Normal cells: NIH-3T3 *In vitro*	Extract exhibited high cytotoxicity to tumor cells and high cytotoxicity to normal cells	Harada et al. 1997
A. flabelliformis	Methanolic extract	Tumor cells: HT-29, KB Normal cells: NIH-3T3 *In vitro*	IC_{50} > 50 (KB), > 50 (HT-29), > 50 (NIH-3T3) μg mL^{-1}	Xu et al. 2004b

Table 6.1 contd....

Table 6.1 contd....

Species	Tested product, fraction or extract	Study model	Observed results	References
A. paradoxa (formerly *Ahnfeltia paradoxa*)	Crude methanol, water and PBS extracts	Tumor cells: MOLT-4, K562, HeLa, KB Normal cells *In vivo*	Methanol extract showed telomerase inhibiting activity	Kanegawa et al. 2000
Amphiroa anceps (formerly *Amphiroa dilatata*)	Methanolic extract	Tumor cells: L-1210 and P-388 Normal cells: NIH-3T3 *In vitro*	Extract exhibited cytotoxicity to tumor cells and no cytotoxicity to normal cells	Harada et al. 1997
A. beauvoisii (formerly *A. zonata*)	Methanol Phosphate-buffered saline (PBS)	Tumor cells: HDF, HL-60, L-1210, MOLT-4, NIH-3T3 *In vitro*	Strong cytotoxicity to all human leukemic cell lines tested and murine leukemic cells L-1210 at the final concentrations from 15 to 375 $\mu g\ mL^{-1}$	Harada and Kamei 1997
A. beauvoisii (formerly *A. zonata*)	Palmitic acid	Tumor cells: MOLT-4, U-937 Normal cells: HDF *In vitro* and *in vivo*	At concentrations ranging from 12.5 to 50 $\mu g\ mL^{-1}$, palmitic acid shows selective cytotoxicity to human leukemic cells, but no cytotoxicity to normal HDF cells Palmitic acid induces apoptosis in the human leukemic cell line MOLT-4 at 50 $\mu g\ mL^{-1}$, and also shows *in vivo* antitumor activity in mice	Harada et al. 2002
A. fragilissima	Methanol extract	Tumor cells: A549 *In vitro*	The LC_{50} (lethal concentration) of methanol extract was observed at 125 $\mu g\ mL^{-1}$ concentration of the extract. Whereas, only 16.9% cell viability was observed at 1 mg mL^{-1} concentration	Viswanathan et al. 2014
Asparagopsis armata	Crude (DCM/MeOH) extract. PP content 1.0–10.9%	Tumor cells: Daudi, Jurkat, K562 Normal cells *In vivo*	Crude extract concentrations: 50–500 mg L^{-1}; Inhibition: 28.56–65.73%	Zubia et al. 2009a
A. armata	Methanol and dichloromethane extracts	Tumor cells: HepG-2 *In vitro*	The most potent anti-proliferative activity was induced by dichloromethane extract (1000 $\mu g\ mL^{-1}$, 24 h), with 99.6% of cell's proliferation reduction. Cytotoxicity assays with an IC_{50} of 567.9 and 473.1 $\mu g\ mL^{-1}$, respectively	Alves et al. 2016
A. taxiformis	Methanolic extract	Tumor cells: L-1210 and P-388 Normal cells: NIH-3T3 *In vitro*	Extract exhibited cytotoxicity to tumor cells and cytotoxicity to normal cells	Harada et al. 1997

A. taxiformis	Methanolic extracts	Tumor cells: BeWo, EAT, HUVEC *In vivo*	Metanholic extract (100 μg mL^{-1}) of red seaweed *A. taxiformis* inhibits about 40% mouse mammary carcinoma cell (EAT) cell proliferation	Vinayak et al. 2014
Betaphycus gelatinus (formerly *Eucheuma gelatinae*)	Seaweed powder	Tumor cells: Ehrlich carcinoma Meth-A fibrosarcoma *In vivo*	Oral administration of 1600 mg seaweed powder per kg mouse per day for 28 days. Inhibition rate: 52.1% Intraperitoneal administration of 50 mg kg^{-1} d^{-1} for seven days. Inhibition rate: 26.9%	Noda et al. 1990
Bonnemaisonia hamifera	Ethanol/water extract	Tumor cells: HL-60, HT-29 Also tested: B-16, A-549 Normal cells: HaCaT *In vivo*	Concentration of extract: 100 μg mL^{-1} HL-60 Inhibition = 86.5% HT-29 Inhibition = 5.5% B16 Inhibition = 41.4% A549 Inhibition = 9.7% HaCat = Inhibition = 8.1%	Kim et al. 2009
Bostrychia montagnei	Sulfated galactans	Tumor cells: HeLa *In vitro*	At maximum concentration of 80 μg mL^{-1}, these fractions promoted atypical mitoses in HeLa cells, presence of atypical nuclei and blebs	Stevan et al. 2001
Botryocladia occidentalis	Methanolic seaweed extracts	Tumor cells: HeLa, SiHa	The dependency of *B. occidentalis* extract on time and/or dose could not be identified clearly. However, *B. occidentalis* methanolic extract presented inhibitory activity of approximately 10% in 24 h that later tended to rise to nearly 20%	Gomes et al. 2015
Brongniartella byssoides	Crude (DCM/MeOH) extract. PP content 1.0–10.9%	Tumor cells: Daudi, Jurkat, K562 Normal cells *In vivo*	Crude extract concentrations: 50–500 mg L^{-1}; Inhibition: 74.86–94.80%	Zubia et al. 2009a
Bryothamnion triquetrum	DCM-methanol (7:3) and water extracts	Tumor cells: Hep-2, HeLa, KB Normal cells: MDCK *In vivo*	Anti-proliferative activity of methanolic extract against KB cells IC_{50} = 62.9 μg mL^{-1}	Moo-Puc et al. 2009
Calliblepharis jubata	Crude (DCM/MeOH) extract. PP content 1.0–10.9%	Tumor cells: Daudi, Jurkat, K562 Normal cells *In vivo*	Crude extract concentrations: 50–500 mg L^{-1}; Inhibition: 10.51–68.10%	Zubia et al. 2009a
Callithamnion tetragonum	Crude (DCM/MeOH) extract. PP content 1.0–10.9%	Tumor cells: Daudi, Jurkat, K562 Normal cells *In vivo*	Crude extract concentrations: 50–500 mg L^{-1}; Inhibition: 24.71–73.17%	Zubia et al. 2009a

Table 6.1 contd....

Table 6.1 contd....

Species	Tested product, fraction or extract	Study model	Observed results	References
Callophycus serratus	Bromophycolides A and B; debromophycolide A	Tumor cells: BT-549, DU4475, MDAMD-468, NCI-H446, PC-3, SHP-77, LNCaP-FGC, HCT116, MDA-MB-231, A2780/DDP-S, DU145 Normal cells *In vivo*	Mean anticancer activity across the cell lines studied was 6.9, 27.7 and > 76 μM for bromophycolides A, B and debromophycolide A, respectively. Bromophycolides A caused apoptosis in A2780 with data indicating cell cycle arrest in G_1	Kubanek et al. 2005
C. serratus	Bromophycolides C-I	Tumor cells: cell panel as above Normal cells *In vivo*	Range of IC_{50} in the 11 cell lines was 9 to 42.6 μM	Kubanek et al. 2006
C. serratus	Callophycoic acids A–H, brominated diterpene-benzoic acids and phenols	Tumor cells: Panel of 11 cancer cell lines Normal cells *In vivo*	IC_{50} values of > 25 μM for 7 of the 11 cell lines; IC_{50} for other 4 had range 20.6–24.5 μM	Lane et al. 2007
C. serratus	Bromophycolides J-Q	Tumor cells: BT-549, DU4475, MDAMD-468, NCI-H446, PC-3, SHP-77, LNCaP-FGC, HCT116, MDA-MB-231, A2780/DDP-S, DU145 Normal cells *In vivo*	Range of IC_{50} in the 11 cell lines was 3.1–10 μM	Lane et al. 2009
C. serratus	Bromophycolide R, S, T, U	Tumor cells: BT-549, DU4475, MDAMD-468, NCI-H446, PC-3, SHP-77, LNCaP-FGC, HCT116, MDA-MB-231, A2780/DDP-S, DU145 Normal cells *In vivo*	Mean IC_{50} of 16–19 μM across the cell lines studied	Lin et al. 2010
Callophyllis crispata	Ethanol/water extract	Tumor cells: HL-60, HT-29 Also tested: B-16, A549 Normal cells: HaCaT *In vivo*	Concentration of extract: 100 μg mL^{-1} HL-60 Inhibition = 88.6% HT-29 Inhibition = 28.3% B16 Inhibition = 7.7% A549 Inhibition = 8.7% HaCat = Inhibition = 1.3%	Kim et al. 2009

C. laciniata	Crude (DCM/MeOH) extract. PP content 1.0–10.9%	Tumor cells: Daudi, Jurkat, K562 Normal cells *In vivo*	Crude extract concentrations: 50–500 mg L^{-1}; Inhibition: 19.26–68.93%	Zubia et al. 2009a
Caulacanthus ustulatus	Crude (DCM/MeOH) extract. PP content 1.0–10.9%	Tumor cells: Daudi, Jurkat, K562 Normal cells *In vivo*	Crude extract concentrations: 50–500 mg L^{-1}; Inhibition: 17.40–66.73%	Zubia et al. 2009a
Ceramium boydenii	Methanolic extract	Tumor cells: HT-29, KB Normal cells: NIH-3T3 *In vitro*	IC_{50} > 50 (KB), > 50 (HT-29), > 50 (NIH-3T3) μg mL^{-1}	Xu et al. 2004b
C. ciliatum	Crude (DCM/MeOH) extract. PP content 1.0–10.9%	Tumor cells: Daudi, Jurkat, K562 Normal cells *In vivo*	Crude extract concentrations: 50–500 mg L^{-1}; Inhibition: 18.75–66.15%	Zubia et al. 2009a
C. kondoi	Methanolic extract	Tumor cells: HT-29, KB Normal cells: NIH-3T3 *In vitro*	IC_{50}: > 50 (KB), 49.3 (HT-29), > 50 (NIH-3T3) μg mL^{-1}	Xu et al. 2004b
C. japonicum	Methanolic extract	Tumor cells: HT-29, KB Normal cells: NIH-3T3 *In vitro*	IC_{50}: > 50 (KB), > 50 (HT-29), > 50 (NIH-3T3) μg mL^{-1}	Xu et al. 2004b
Ceratodictyon spongiosum	Ceratodictyols A–F	Tumor cells: HeLa Normal cells *In vivo*	IC_{50} value of 67 μM each	Akiyama et al. 2009
Champia feldmannii	Sulfated polysaccharides	Tumor cells: S-180 *In vitro* and *in vivo*	The inhibition rates of sarcoma 180 tumor development were 48.62 and 48.16% at the doses of 10 and 25 mg kg^{-1}, respectively In addition, sulfated polysaccharide was also able to increase the response elicited by 5-fluorouracil (5-FU) from 48.66 to 68.32%	Lins et al. 2008
Chondracanthus intermedius (formerly *Gigartina intermedia*)	Seaweed powder	Tumor cells: Ehrlich carcinoma *In vivo*	Oral administration of 1600 mg seaweed powder per kg mouse per day for 28 days. Inhibition rate: 12.5%	Noda et al. 1990
Chondria capillaris (formerly *Chondria tenuissima*)	Methanolic extract	Tumor cells: HT-29, KB Normal cells: NIH-3T3 *In vitro*	IC_{50} = 42.7 (KB), > 50 (HT-29), > 50 (NIH-3T3) μg mL^{-1}	Xu et al. 2004b

Table 6.1 contd....

Table 6.1 contd....

Species	Tested product, fraction or extract	Study model	Observed results	References
C. crassicaulis	Methanolic extract	Tumor cells: L-1210 and P-388 Normal cells: NIH-3T3 *In vitro*	Extract exhibited high cytotoxicity to tumor cells and no cytotoxicity to normal cells	Harada et al. 1997
C. crassicaulis	Ethanol/water extract	Tumor cells: HL-60, HT-29 Also tested: B-16, A-549 Normal cells: HaCaT *In vivo*	Concentration of extract: 100 μg mL^{-1} HL-60 Inhibition = 86.8% HT-29 Inhibition = 2.9% B16 Inhibition = 38.0% A549 Inhibition = 7.7% HaCat = Inhibition = 5.0%	Kim et al. 2009
Chondrophycus ceylanicus	Methanol extract	Tumor cells: A549, B16F10, HL-60 Normal cells: Vero *In vitro*	IC_{50} = 102.17 μg mL^{-1} for HL-60 cells	Lakmal et al. 2014
Chondrus crispus	Ethanol/water extract	Tumor cells: HL-60, HT-29 Also tested: B-16, A-549 Normal cells: HaCaT *In vivo*	Concentration of extract: 100 μg mL^{-1} HL-60 Inhibition = 87.0% HT-29 Inhibition = 39.8% B16 Inhibition = 25.4% A549 Inhibition = 1.1% HaCat = Inhibition = 10.6%	Kim et al. 2009
C. crispus (wild and cultivated)	Methanol extract	Tumor cells: HeLa, U-937 *In vitro*	Cultivated *C. crispus* extract exhibited the least efficacy at the low dose concentrations (0.125–1.0 mg mL^{-1}), but high efficacy at the highest dose concentrations (2.0–4.0 mg mL^{-1}) compared to the other red macroalgal extracts. Extract induced marked changes in HeLa cell morphology with increasing concentrations of the extracts. HeLa cell morphology was altered from the typical rhomboid–tetrahedral shape observed with the untreated control cells into progressively more contracted and rounded cells characteristic of apoptosis with lower cell density and ultimately, membrane blebbing and apoptotic bodies at the 1.0 and 4.0 mg mL^{-1} concentrations	Athukorala et al. 2016

C. ocellatus f. *crispus*	Seaweed powder	Tumor cells: Meth-A fibrosarcoma *In vivo*	Intraperitoneal administration of 50 mg kg^{-1} d^{-1} for seven days. Inhibition rate: 34.7%	Noda et al. 1990
C. ocellatus	Phosphate buffered saline extract	Tumor cells: L-1210 and P-388 Normal cells: NIH-3T3 *In vitro*	Extract exhibited high cytotoxicity to tumor cells and high cytotoxicity to normal cells	Harada et al. 1997
C. ocellatus	Methanolic extract	Tumor cells: HT-29, KB Normal cells: NIH-3T3 *In vitro*	IC_{50} > 50 (KB), > 50 (HT-29), > 50 (NIH-3T3) µg mL^{-1}	Xu et al. 2004b
C. ocellatus	Ethanol/water extract	Tumor cells: HL-60, HT-29 Also tested: B-16, A-549 Normal cells: HaCaT *In vivo*	Concentration of extract: 100 µg mL^{-1} HL-60 Inhibition = 54.9% HT-29 Inhibition = 8.4% B16 Inhibition = 12.2% A549 Inhibition = 2.6% HaCat Inhibition = 2.4%	Kim et al. 2009
Corallina crassisima (as *Marginisporum crassissimum*)	Water extract	Tumor cells: B16-BL6, JYG-B, KPL-1 Normal cells *In vivo*: female C57BL/6J mice inoculated in the tail vein with B16-BL6 cells	*In vitro*: extracts inhibited growth of all tumor cell lines and also invasion of B16-BL6 cells. *In vivo*: lung metastasis of B16-BL6 cells was inhibited by IP administration of extract; also increase survival time	Hiroishi et al. 2001
C. officinalis	Methanolic extract	Tumor cells: HT-29, KB Normal cells: NIH-3T3 *In vitro*	IC_{50} > 50 (KB), > 50 (HT-29), > 50 (NIH-3T3) µg mL^{-1}	Xu et al. 2004b
C. pilulifera	Seaweed powder	Tumor cells: Ehrlich carcinoma *In vivo*	Oral administration of 1600 mg seaweed powder per kg mouse per day for 28 days. Inhibition rate: 38.4%	Noda et al. 1990
C. pilulifera	Methanolic extract	Tumor cells: L-1210 and P-388 Normal cells: NIH-3T3 *In vitro*	Extract exhibited cytotoxicity to tumor cells and to normal cells	Harada et al. 1997
Chrysymenia wrightii	Methanolic extract	Tumor cells: HT-29, KB Normal cells: NIH-3T3 *In vitro*	IC_{50} > 50 (KB), > 50 (HT-29), > 50 (NIH-3T3) µg mL^{-1}	Xu et al. 2004b
Cystocionium purpureum	Crude (DCM/MeOH) extract. PP content 1.0–10.9%	Tumor cells: Daudi, Jurkat, K562 Normal cells *In vivo*	Crude extract concentrations: 50–500 mg L^{-1}; Inhibition: 15.74–62.23%	Zubia et al. 2009a

Table 6.1 contd....

Table 6.1 contd....

Species	Tested product, fraction or extract	Study model	Observed results	References
Dasya baillouviana (as *Dasya pedicellata*)	Methanol extract	Tumor cells Normal cells: L6 *In vivo*	IC_{50} = 14.7 μg mL^{-1}	Süzgeç-Selçuk et al. 2011
Dasysiphonia japonica (formerly *Heterosiphonia japonica*)	Methanolic extract	Tumor cells: HT-29, KB Normal cells: NIH-3T3 *In vitro*	IC_{50} > 50 (KB), > 50 (HT-29), > 50 (NIH-3T3) μg mL^{-1}	Xu et al. 2004b
Dichotomaria falcata (formerly *Galaxaura falcata*)	Crude methanol, water and PBS extracts	Tumor cells: MOLT-4, K562, HeLa, KB Normal cells *In vivo*	Methanol extract showed telomerase inhibiting activity	Kanegawa et al. 2000
D. marginata (formerly *G. marginata*)	Ethylacetate extract (Desmosterols)	Tumor cells: A549, HeLA, P338 *In vitro*	Cytotoxicity with effective doses 0.28 μg mL^{-1} (P338), 1.00 μg mL^{-1} (A549), and 0.40 μg mL^{-1} (HeLa)	Sheu et al. 1997
D. obtusata (formerly *G. robusta*)	Crude methanol, water and PBS extracts	Tumor cells: MOLT-4, K562, HeLa, KB Normal cells *In vivo*	Methanol extract showed telomerase inhibiting activity	Kanegawa et al. 2000
Digenea simplex	Methanolic extract	Tumor cells: L-1210 and P-388 Normal cells: NIH-3T3 *In vitro*	Extract exhibited cytotoxicity to tumor cells and no cytotoxicity to normal cells	Harada et al. 1997
D. simplex	Dichloromethane, chloroform, methanol, ethanol, water extracts, hexane and chloroform fractions	Tumor cells: HEp-2, K562, NCI-H292 *In vitro*	Ethanol extract (32.2 μg mL^{-1}) was effective against HEp-2 and methanol extract (33.8 mL^{-1}) against K562	Guedes et al. 2013
Dilsea carnosa	Crude (DCM/MeOH) extract. PP content 1.0–10.9%	Tumor cells: Daudi, Jurkat, K562 Normal cells *In vivo*	Crude extract concentrations: 50–500 mg L^{-1}; Inhibition: 27.00–70.68%	Zubia et al. 2009a
Dumontia contorta	Crude (DCM/MeOH) extract. PP content 1.0–10.9%	Tumor cells: Daudi, Jurkat, K562 Normal cells *In vivo*	Crude extract concentrations: 50–500 mg L^{-1}; Inhibition: 19.28–65.78%	Zubia et al. 2009a
Ellisolandia elongata (formerly *Corallina elongata*)	Aqueous extract	Tumor cells: EAC Silver Nanoparticle's of aqueous extract (AgNPs) *In vitro*	Cytotoxic effect (41–74%) of different AgNPs concentrations (42–98 μg mL^{-1})	Khalifa et al. 2016

Eucheuma amakusaense	Seaweed powder	Tumor cells: Ehrlich carcinoma *In vivo*	Oral administration of 1600 mg seaweed powder per kg mouse per day for 28 days. Inhibition rate: 38.7%	Noda et al. 1990
E. denticulatum (formerly *E. spinosum*)	Seaweed powder	Tumor cells: Ehrlich carcinoma; Meth-A fibrosarcoma *In vivo*	Oral administration of 1600 mg seaweed powder per kg mouse per day for 28 days. Inhibition rate: 17.0–30.1% Intraperitoneal administration of 50 mg kg^{-1} d^{-1} for seven days. Inhibition rate: 24.3%	Noda et al. 1990
E. denticulatum (formerly *E. spinosum*)	Sulfated polysaccharide: ι-carrageenan	Tumor cells: Meth A fibrosarcoma *In vivo*	Intraperitoneal administration of 40 mg kg^{-1} for five days. Inhibition rate: 40.1%	Noda et al. 1990
E. serra	Lectin: agglutinin	Tumor cells: Colon26 Normal cells *In vivo*: Colon26: mouse colon adenocarcinoma cells in BALB/c mice	*In vitro*: induced cell death; the increased expression of caspase 3 and translocation of phosphatidylserine in lectin-treated Colon26 cells suggested cell death was induced through apoptosis. *In vivo*: intravenous injection of extract significantly inhibited growth of tumors; DNA fragmentation, indicating apoptosis, was detected in tumor cells following treatment. Small decrease in body weight in treated animals	Fukuda et al. 2006
E. serra	Lectin: agglutinin	Tumor cells: Colon201, HeLa, MCF-7, HB4C5 Normal cells: MCF10-2A *In vivo*	Cytotoxic to Colon201 and HeLa, not to malignant (MCF-7) or nontumourigenic (MCF10-2A) breast cancer cells. Induced apoptosis by caspase 3 activation. Lipid vesicles as a drug delivery system were investigated, with promising results	Sugahara et al. 2001
Furcellaria lumbricalis (formerly *Furcellaria fastigiata*)	Crude (DCM/MeOH) extract. PP content 1.0–10.9%	Tumor cells: Daudi, Jurkat, K562 Normal cells *In vivo*	Crude extract concentrations: 50–500 mg L^{-1}; Inhibition: 24.23–75.19%	Zubia et al. 2009a
Galaxaura filamentosa	Galaxamide	Tumor cells: GRC-1, HepG2 Normal cells *In vivo*	Antitumor effects were reported with IC_{50} values of 4.26 μg mL^{-1} (GRC-1) and 4.63 μg mL^{-1} (HepG2)	Xu et al. 2008
Ganonema farinosum (formerly *Liagora farinosa*)	Phosphate buffered saline extract	Tumor cells: L-1210 and P-388 Normal cells: NIH-3T3 *In vitro*	Extract exhibited cytotoxicity to tumor cells and no cytotoxicity to normal cells	Harada et al. 1997
Gastroclonium ovatum	Crude (DCM/MeOH) extract. PP content 1.0–10.9%	Tumor cells: Daudi, Jurkat, K562 Normal cells *In vivo*	Crude extract concentrations: 50–500 mg L^{-1}; Inhibition: 11.11–52.05%	Zubia et al. 2009a

Table 6.1 contd.…

Table 6.1 contd....

Species	Tested product, fraction or extract	Study model	Observed results	References
Gelidiella acerosa	Methanolic extract	Tumor cells: L-1210 and P-388 Normal cells: NIH-3T3 *In vitro*	Extract exhibited cytotoxicity to tumor cells and no cytotoxicity to normal cells	Harada et al. 1997
G. acerosa	Methanol extract	Tumor cells: A549, B16F10, HL-60 Normal cells: Vero *In vitro*	IC_{50} = 104.43 μg mL^{-1} for HL-60 cells	Lakmal et al. 2014
Gelidium amansii	Seaweed powder	Tumor cells: Ehrlich carcinoma *In vivo*	Oral administration of 1600 mg seaweed powder per kg mouse per day for 28 days. Inhibition rate: 37.2%	Noda et al. 1990
G. amansii	Methanolic extract	Tumor cells: L-1210 and P-388 Normal cells: NIH-3T3 *In vitro*	Extract exhibited high cytotoxicity to tumor cells and high cytotoxicity to normal cells	Harada et al. 1997
G. amansii	Methanolic extract	Tumor cells: HT-29, KB Normal cells: NIH-3T3 *In vitro*	IC_{50} > 50 (KB), > 50 (HT-29), > 50 (NIH-3T3) μg mL^{-1}	Xu et al. 2004b
G. crinale	Aqueous extract	Tumor cells: EAC Silver Nanoparticle's of aqueous extract (AgNPs) *In vitro*	Cytotoxic effect (46–61%) of different AgNPs concentrations (42–98 μg mL^{-1})	Khalifa et al. 2016
G. japonicum	Seaweed powder	Tumor cells: Ehrlich carcinoma *In vivo*	Oral administration of 1600 mg seaweed powder per kg mouse per day for 28 days. Inhibition rate: 15.0%	Noda et al. 1990
Gloiopeltis furcata	Seaweed powder	Tumor cells: Ehrlich carcinoma *In vivo*	Oral administration of 1600 mg seaweed powder per kg mouse per day for 28 days. Inhibition rate: 37.7%	Noda et al. 1990
G. furcata	Methanolic extract	Tumor cells: HT-29, KB Normal cells: NIH-3T3 *In vitro*	IC_{50} = 35.7 (KB), 38.2 (HT-29), > 50 (NIH-3T3) μg mL^{-1}	Xu et al. 2004b

G. furcata	Methanolic extract	Tumor cells: HepG2 *In vivo*	Markedly inhibited human hepatocellular carcinoma (HepG2) cell proliferation and induced the G2/M arrest of the cell cycle in a dose-dependent manner (from 10 to 500 μg mL^{-1})	Bae and Choi 2007
G. furcata	Ethanol/water extract	Tumor cells: HL-60, HT-29 Also tested: B-16, A-549 Normal cells: HaCaT *In vivo*	Concentration of extract: 100 μg mL^{-1} HL-60 Inhibition = 89.3% HT-29 Inhibition = 37.9% B16 Inhibition = 27.2% A549 Inhibition = 2.0% HaCat = Inhibition = 31.7%	Kim et al. 2009
Gracilaria caudata	DCM-methanol (7:3) and water extracts	Tumor cells: Hep-2, HeLa, KB Normal cells: MDCK *In vivo*	Anti-proliferative activity of methanolic extract against Hep-2 cells IC_{50} = 87.4 μg mL^{-1}	Moo-Puc et al. 2009
G. caudata	Sulfated polysaccharides	Tumor cells: HeLa Normal cells *In vivo*	About 40.0% of cell proliferation inhibition at 2.0 mg mL^{-1}	Costa et al. 2010
G. cervicornis	DCM-methanol (7:3) and water extracts	Tumor cells: Hep-2, HeLa, KB Normal cells: MDCK *In vivo*	Anti-proliferative activity of methanolic extract against KB and HeLa cells IC_{50} = 68.2 and 75.5 μg mL^{-1}, respectively	Moo-Puc et al. 2009
G. cornea (formerly *Hydropuntia cornea*)	DCM-methanol (7:3) and water extracts	Tumor cells: Hep-2, HeLa, KB Normal cells: MDCK *In vivo*	Anti-proliferative activity of methanolic extract against Hep-2 cells IC_{50} = 74.5 μg mL^{-1}	Moo-Puc et al. 2009
G. corticata	Water crude extract	Tumor cells: Jurkat, MOLT-4 *In vitro*	9.336 and 9.726 μg μL^{-1} of algal extract were the most effective concentrations against Jurkat and molt-4 cells, respectively	Zandi et al. 2010b
G. corticata	Methanol, chloroform, ethyl acetate and aqueous extracts. Highest PP content 1.3 μg mL^{-1}, the highest flavonoid content 88 μg mL^{-1}	Tumor cells: A549, HCT-15, MG-63, PC-3 Normal cells *In vivo*	Although some association was found between inhibition of proliferation of A549 and MG-63 cells with higher values of total flavonoid content (see paper for details), authors suggest many bioactive molecules in extract could be responsible for growth inhibition effect	Murugan and Iyer 2014
G. corticata	Methanolic extract	Tumor cells: HepG2, MCF-7 Normal cells: VERO *In vivo*	The average inhibitory activity was 91% and 93%, respectively, using 500 μg mL^{-1} of extract. Only at a high concentration of 1 mg mL^{-1}, 60% cytoxicity was seen in normal VERO cell lines	Narasimhan et al. 2013

Table 6.1 contd....

Table 6.1 contd....

Species	Tested product, fraction or extract	Study model	Observed results	References
G. corticata	Aqueous extract	Tumor cells: MCF-7 *In vitro*	IC_{50} = 200 μg mL^{-1}	Arulvasu et al. 2014
G. corticata	Methanol extract	Tumor cells: A549, B16F10, HL-60 Normal cells: Vero *In vitro*	IC_{50} > 200 μg mL^{-1} for HL-60 cells	Lakmal et al. 2014
G. edulis	Methanolic extract	Tumor cells: A549 Normal cells: VERO *In vitro*	Cytotoxic effect against human A549 lung adenocarcinoma cancer cell lines in a concentration dependent manner, with IC_{50} = 118.98 μg mL^{-1}	Rani et al. 2013
G. edulis	GE ethyl acetate extract (GEEA)	Tumor cells: A549 *In vitro* and *in vivo*	100 μg mL^{-1} treated group showed the maximum growth inhibition of A549 at 48 h. The IC_{50} value was found to be 24.5 ± 19.1 μg mL^{-1} at 48 h.	Sakthivel et al. 2016
G. gracilis	Crude (DCM/MeOH) extract. PP content 1.0–10.9%	Tumor cells: Daudi, Jurkat, K562 Normal cells *In vivo*	Crude extract concentrations: 50–500 mg L^{-1}; Inhibition: 18.32–62.78%	Zubia et al. 2009a
G. lemaneiformis	Polysaccharide	Tumor cells: Sarcoma-180 *In vivo*	Ingredient with marked antitumor activity and an ideal potential nontoxic preventive agent	Chen et al. 2008
G. lemaneiformis	Polysaccharide	Tumor cells: CL1-5 *In vitro*	Could significantly inhibit the proliferative activity and alter the cell morphology of lung tumor cells	Kang et al. 2015
G. lemaneiformis	Polysaccharide	Tumor cells: MKN45, NSCLC, A549, HeLa *In vitro*	Extracts demonstrated growth inhibitory effects on HeLa, MKN45, and A549 cells in a dose-dependent manner, with the most remarkable effect on A549 cells. The IC_{50} was approximately 50 μg mL^{-1} at 48 h in A549 cells	Kang et al. 2016
G. manilaensis	Dichloromethane extract Acetone and ethyl acetate extracts	Tumor cells: Caov-3, MDA-MB-231, MCF-7, HeLa Normal cells: L929, MDCK *In vitro*	Strongest cytotoxic on MDA-MB-231 (53.90 μg mL^{-1}) and HeLa (95.50 μg mL^{-1}) Cytotoxic on MCF-7 (66.07 μg mL^{-1}) and Caov-3 (69.67 μg/mL^{-1})	Abdullah et al. 2013

G. tenuistipitata	Crude aqueous extract	Tumor cells: H1299 Normal cells *In vivo*	The extract reduced H_2O_2-induced oxidative damage of DNA in H1299 and protected the cells against H_2O_2-induced cytotoxicity. The extract itself was not found to be cytotoxic. H_2O_2-induced G2/M cell cycle arrest was reduced when cells were co-treated with the extract (1–4 mg mL^{-1} tested)	Yang et al. 2012
G. tenuistipitata	Methanol extract	Tumor cells: Ca9-22 Normal cells *In vivo*	IC_{50} 0.326 mg mL^{-1} Caused apoptosis, increased ROS and DNA damage	Yeh et al. 2012
G. vermiculophylla (formerly *G. asiatica*)	Gracilarioside, gracilamides	Tumor cells: A375-S2 Normal cells *In vivo*	Gracilarioside induced 18.2% cell death at 20.0 μg mL^{-1}. Two gracilamides tested showed weak toxicity (11.7% cell death at 30.0 μg mL^{-1})	Sun et al. 2006
Gracilariopsis andersonii (formerly *Gracilaria sjoestedtii*)	Methanolic extract	Tumor cells: HT-29, KB Normal cells: NIH-3T3 *In vitro*	IC_{50} > 50 (KB), > 50 (HT-29), > 50 (NIH-3T3) μg mL^{-1}	Xu et al. 2004b
G. longissima (formerly *Gracilaria verrucosa*)	Seaweed powder	Tumor cells: Ehrlich carcinoma Meth-A fibrosarcoma *In vivo*	Oral administration of 1600 mg seaweed powder per kg mouse per day for 28 days. Inhibition rate: 19.3% Intraperitoneal administration of 50 mg kg^{-1} d^{-1} for seven days. Inhibition rate: 24.9%	Noda et al. 1990
G. longissima (formerly *G. verrucosa*)	Methanolic extract	Tumor cells: HT-29, KB Normal cells: NIH-3T3 *In vitro*	IC_{50} > 50 (KB), > 50 (HT-29), > 50 (NIH-3T3) μg mL^{-1}	Xu et al. 2004b
Grateloupia acuminata	Phosphate buffered saline extract	Tumor cells: L-1210 and P-388 Normal cells: NIH-3T3 *In vitro*	Extract exhibited high cytotoxicity to tumor cells and no cytotoxicity to normal cells	Harada et al. 1997
G. cornea (formerly *Prionitis cornea*)	Ethanol/water extract	Tumor cells: HL-60, HT-29 Also tested: B-16, A-549 Normal cells: HaCaT *In vivo*	Concentration of extract: 100 μg mL^{-1} HL-60 Inhibition = 79.5% HT-29 Inhibition = 9.8% B16 Inhibition = 9.9% A549 Inhibition = 9.9% HaCat = Inhibition = 22.7%	Kim et al. 2009

Table 6.1 contd....

Table 6.1 contd....

Species	Tested product, fraction or extract	Study model	Observed results	References
G. elliptica	Ethanol/water extract	Tumor cells: HL-60, HT-29 Also tested: B-16, A-549 Normal cells: HaCaT *In vivo*	Concentration of extract: 100 μg mL^{-1} HL-60 Inhibition = 88.3% HT-29 Inhibition = 44.6% B16 Inhibition = 20.7% A549 Inhibition = 5.1% HaCat = Inhibition = 33.1%	Kim et al. 2009
G. elliptica	Chlorophyll-related compound pheophorbide a (Pa)	Tumor cells: HeLa, U87MG, SK-OV-3 Normal cells: HUVEC *In vitro*	Strongest anticancer activity of Pa exhibited on U87MG cells with IC_{50} values of 2.8 μg mL^{-1}	Cho et al. 2014
G. filicina	Seaweed powder	Tumor cells: Meth-A fibrosarcoma *In vivo*	Intraperitoneal administration of 50 mg kg^{-1} d^{-1} for seven days. Inhibition rate: 14.0%	Noda et al. 1990
G. filicina	Phosphate buffered saline extract	Tumor cells: L-1210 and P-388 Normal cells: NIH-3T3 *In vitro*	Extract exhibited high cytotoxicity to tumor cells and no cytotoxicity to normal cells	Harada et al. 1997
G. filicina	Methanolic extract	Tumor cells: HT-29, KB Normal cells: NIH-3T3 *In vitro*	IC_{50} > 50 (KB), > 50 (HT-29), > 50 (NIH-3T3) μg mL^{-1}	Xu et al. 2004b
G. filicina	Crude (DCM/MeOH) extract. PP content 1.0–10.9%	Tumor cells: Daudi, Jurkat, K562 Normal cells *In vivo*	Crude extract concentrations: 50–500 mg L^{-1}; Inhibition: 2.40–51.47%	Zubia et al. 2009a
G. imbricata	Seaweed powder	Tumor cells: Ehrlich carcinoma *In vivo*	Oral administration of 1600 mg seaweed powder per kg mouse per day for 28 days. Inhibition rate: 25.0%	Noda et al. 1990
G. livida	Seaweed powder	Tumor cells: Ehrlich carcinoma *In vivo*	Oral administration of 1600 mg seaweed powder per kg mouse per day for 28 days. Inhibition rate: 29.8%	Noda et al. 1990
G. sparsa	Phosphate buffered saline extract	Tumor cells: L-1210 and P-388 Normal cells: NIH-3T3 *In vitro*	Extract exhibited cytotoxicity to tumor cells and no cytotoxicity to normal cells	Harada et al. 1997
G. turuturu	Phosphate buffered saline extract	Tumor cells: L-1210 and P-388 Normal cells: NIH-3T3 *In vitro*	Extract exhibited cytotoxicity to tumor cells and no cytotoxicity to normal cells	Harada et al. 1997

G. turuturu	Methanolic extract	Tumor cells: HT-29, KB Normal cells: NIH-3T3 *In vitro*	$IC_{50} > 50$ (KB), > 50 (HT-29), > 50 (NIH-3T3) $\mu g\ mL^{-1}$	Xu et al. 2004b
Heterosiphonia gibbesii	DCM-methanol (7:3) and water extracts	Tumor cells: Hep-2, HeLa, KB Normal cells: MDCK *In vivo*	Anti-proliferative activity of methanolic extract against KB cells $IC_{50} = 99.2\ \mu g\ mL^{-1}$	Moo-Puc et al. 2009
H. plumosa	Crude (DCM/MeOH) extract. PP content 1.0–10.9%	Tumor cells: Daudi, Jurkat, K562 Normal cells *In vivo*	Crude extract concentrations: 50–500 mg L^{-1}; Inhibition: 28.12–75.15%	Zubia et al. 2009a
Hyalosiphonia caespitosa	Methanolic extract	Tumor cells: HT-29, KB Normal cells: NIH-3T3 *In vitro*	$IC_{50} > 50$ (KB), > 50 (HT-29), > 50 (NIH-3T3) $\mu g\ mL^{-1}$	Xu et al. 2004b
Hydrolithon boergesenii (formerly *Hydrolithon reinboldii*)	Unsaturated fatty acids	Tumor cells: BT-549, DU4475, MDAMB-468, MDA-MB-231, HCT116, NCI-H446, SHP-77, PC-3, LNCaPFGC, DU145, A2780/DDP-S, CCRF-CEM Normal cells *In vivo*	IC_{50} values from 1.3 to 14.4 μM	Jiang et al. 2008
Hypnea charoides	Methanolic extract	Tumor cells: L-1210 and P-388 Normal cells: NIH-3T3 *In vitro*	Extract exhibited cytotoxicity to tumor cells and no cytotoxicity to normal cells	Harada et al. 1997
Hypnea japonica	Seaweed powder	Tumor cells: Ehrlich carcinoma *In vivo*	Oral administration of 1600 mg seaweed powder per kg mouse per day for 28 days. Inhibition rate: 17.7%	Noda et al. 1990
H. musciformis	Dichloromethane, chloroform, methanol, ethanol, water extracts, hexane and chloroform fractions	Tumor cells: HEp-2, K562, NCI-H292 *In vitro*	Dichloromethane extract and chloroform fraction showed the best cytotoxic activity against K562 (3.8 $\mu g\ mL^{-1}$ and 6.4 $\mu g\ mL^{-1}$, respectively) Chloroform fraction (6.0 $\mu g\ mL^{-1}$) was more active against HEp-2 and chloroform fraction (15.0 $\mu g\ mL^{-1}$) against the cell NCI-H292 Water extract (43.6 $\mu g\ mL^{-1}$) was active against HEp-2	Guedes et al. 2013

Table 6.1 contd....

Table 6.1 contd....

Species	Tested product, fraction or extract	Study model	Observed results	References
H. musciformis	Ethanol extract	Tumor cells: MCF-7, LNCaP, PC-3 Normal cells: MCF-10A *In vitro*	Extract displayed a general cytotoxicity against all four cell lines, with inhibition of growth in the range of 60–96% with concentration of 50 mg mL^{-1} Extract also showed selective antiproliferative effect at the lower test concentration of 5 mg mL^{-1}	Montalvão et al. 2016
H. saidana	Phosphate buffered saline extract	Tumor cells: L-1210 and P-388 Normal cells: NIH-3T3 *In vitro*	Extract exhibited cytotoxicity to tumor cells and no cytotoxicity to normal cells	Harada et al. 1997
Iridaea cordata	$CaCl_2$ and ethanol precipitations following diluted acid extraction at room temperature Carrageenan-like sulfated galactan	Tumor cells: PC-3, HeLa, HT-29 Normal cells: Vero cell *In vitro*	Weak antitumor activity against PC-3 (prostate cancer), HeLa (cervical cancer), and HT-29 (human colon adenocarcinoma)	Kim et al. 2016
Jania adhaerens	Methanol extract	Tumor cells: MCF-7, HepG-2, A-549, HT-29 and MDBK *In vitro* (MTT assay)	IC_{50} = 85.03 μg mL^{-1} against MCF-7 cells	Mosaddegh et al. 2014
J. rubens	16β-Hydroxy-5α-cholestane-3,6-dione	Tumor cells: KB Normal cells *In vivo*	IC_{50} value of 0.5 μg mL^{-1}	Ktari et al. 2000
J. rubens	Methanol extract	Tumor cells: EAT *In vivo*	The group that received extract showed slight changes in both serum tumor marker levels in comparison with the control group	Ahmed et al. 2011
Kallymenia perforata	Phosphate buffered saline extract	Tumor cells: L-1210 and P-388 Normal cells: NIH-3T3 *In vitro*	Extract exhibited cytotoxicity to tumor cells and no cytotoxicity to normal cells	Harada et al. 1997
Kallymenia reniformis	Crude (DCM/MeOH) extract. PP content 1.0–10.9%	Tumor cells: Daudi, Jurkat, K562 Normal cells *In vivo*	Crude extract concentrations: 50–500 mg L^{-1}; Inhibition: 1.69–42.75%	Zubia et al. 2009a

Kappaphycus alvarezii (formerly *Kappaphycus cottonii*)	Seaweed powder	Tumor cells: Ehrlich carcinoma; Meth-A fibrosarcoma *In vivo*	Oral administration of 1600 mg seaweed powder per kg mouse per day for 28 days. Inhibition rate: 4.3% Intraperitoneal administration of 50 mg kg^{-1} d^{-1} for seven days. Inhibition rate: 39.8%	Noda et al. 1990
K. alvarezii (formerly *K. cottonii*)	Sulfated polysaccharide: κ-carrageenan	Tumor cells: Meth A fibrosarcoma *In vivo*	Intraperitoneal administration of 40 mg kg^{-1} for five days. Inhibition rate: 54.0%	Noda et al. 1990
K. alvarezii (formerly *Eucheuma cottonii*)	Crude extract, PP content 2%. Identification attempt using HPLC; catechin, rutin and quercetin were found. No quantitative information given	Tumor cells: MCF-7, MB-MDA-431 Normal cells: Vero *In vivo*: female Sprague-Dawley rats inoculated SC with LA7 cells into the mammary fat pad	*In vitro*: MCF-7: IC_{50} 20 μg mL^{-1}, MB-MDA-431: IC_{50} 42 μg mL^{-1}. No toxicity to Vero cells. Apoptosis was observed in cancer cells. *In vivo*: tumor incidence and volume was reduced by extract. Tumors showed apoptosis and rats had better anti-oxidative status. Authors propose that the anti-cancer effect may be related to a number of extract components	Namvar et al. 2012
K. alvarezii (formerly *E. cottonii*)	Ethanol extract	Tumor cells Normal cells *In vivo*: female Sprague-Dawley rats injected SC with LA-7 cells	Oral supplementation of powdered extract (100 mg kg^{-1}), or oral treatment with TAM (10 mg kg^{-1}) for 28 days. Suppressed tumor growth more effectively than TAM, no visible side effects for extract, slight liver and kidney lesions for TAM-treated. Extract decreased MDA, increased GSH	Shamsabadi et al. 2013
K. alvarezii	Methanol, chloroform, ethyl acetate and aqueous extracts. Highest PP content 1.3 μg mL^{-1}, the highest flavonoid content 88 μg mL^{-1}	Tumor cells: A549, HCT-15, MG-63, PC-3 Normal cells *In vivo*	Although some association was found between inhibition of proliferation of A549 and MG-63 cells with higher values of total flavonoid content (see paper for details), authors suggest many bioactive molecules in extract could be responsible for growth inhibition effect	Murugan and Iyer 2014
K. alvarezii (formerly *E. cottonii*)	Crude extract	Tumor cells: HeLa, HCT-116, SKLU-1 Normal cells: Vero *In vitro*	Fibroblast - Extract: 0.5–20.0 mg mL^{-1}, Cells viability percentage: 59.7–68.5% HeLa - Extract: 0.5–20.0 mg mL^{-1}, Cells viability percentage: 11–52.25% SK-LU-1 - Extract: 0.5–20.0 mg mL^{-1}, Cells viability percentage: 12–63.7% HCT-116 - Extract: 0.5–20.0 mg mL^{-1}, Cells viability percentage: 5–40.5%	Lee et al. 2015

Table 6.1 contd....

Table 6.1 contd....

Species	Tested product, fraction or extract	Study model	Observed results	References
Laurencia catarinensis	Hexane extract	Tumor cells: MES-SA and the doxorubicin-resistant mutant, MES-SA/Dx5, HeLa Normal cells *In vivo*	$\geq$ 80% mortality of MES-SA cells	Stein et al. 2011
L. dendroidea (formerly *L. majuscula*)	Majapolene A, B, majapolone, majapols A–D	Tumor cells: NCI 60-cell *in vitro* tumor panel Normal cells *In vivo*	Majapolene A displayed modest cytotoxic activity in NCI panel screen	Erickson et al. 1995
L. dendroidea	Hexane extract	Tumor cells: MES-SA and the doxorubicin-resistant mutant, MES-SA/Dx5, HeLa Normal cells *In vivo*	$IC_{50} = 91 \pm 13$ μg mL^{-1}	Stein et al. 2011
Laurencia intricata	Ethanol/water extract	Tumor cells: HL-60, HT-29 Also tested: B-16, A-549 Normal cells: HaCaT *In vivo*	Concentration of extract: 100 μg mL^{-1} HL-60 Inhibition = 87.3% HT-29 Inhibition = 2.0% B16 Inhibition = 34.5% A549 Inhibition = 2.1% HaCat = Inhibition = 17.1%	Kim et al. 2009
L. glandulifera	Tetrahydrofuran acetogenins	Tumor cells: HT-29, MCF-7, PC-3, HeLa and A431 Normal cells *In vivo*	No significant cytotoxicity was found against the 5 cell lines at 10 μM	Kladi et al. 2009
L. mariannensis	Laurenmariannol, (21α)-21-hydroxythyrsiferol and thyrsiferol	Tumor cells: P-388 Normal cells *In vivo*	IC_{50}: laurenmariannol (0.60 μg mL^{-1}), (21α)-21-hydroxythyrsiferol (6.60 μg mL^{-1}) and thyrsiferol (0.30 μg mL^{-1})	Ji et al. 2008
L. microcladia	12 novel and known sesquiterpenes	Tumor cells: K562, MCF7, PC3, HeLa, A431, A549, NSCLC-N6 Normal cells: CHO *In vivo*	Cytotoxic activity ranged between 15.8 and 320 μM. Compounds were nearly always more cytotoxic to cancer than normal cells	Kladi et al. 2006

L. microcladia	Sesquiterpenes	Tumor cells: HT29, MCF7, PC3, HeLa, A431 Normal cells *In vivo*	Bulkiest compound (a dimer) had no cytotoxicity (IC_{50} > 300 μM). Others had IC_{50} ranging from 75 to 288 μM	Kladi et al. 2007
L. microcladia	DCM-methanol (7:3) and water extracts	Tumor cells: Hep-2, HeLa, KB Normal cells: MDCK *In vivo*	Anti-proliferative activity of methanolic extract against KB and Hep-2 cells IC_{50} = 47.3 and 97.2 μg mL^{-1}, respectively	Moo-Puc et al. 2009
L. microcladia	Elatol	Tumor cells: B16F10, A549, DU145, MCF-7, L929 Normal cells: L929 *In vivo*: C57BL6 mice were injected SC with B16F10 cells	*In vitro*: reduced viability of all cell lines. B16F10 cells: G1 and sub-G1 cell cycle arrest with reduction in cyclin D1, cyclin E, cdk2 and cdk4. Increased apoptosis was accompanied by a decrease in Bcl-xl and increase in Bak, caspase 9 and p53. *In vivo*: both oral (3, 10 and 30 mg kg^{-1}) and IP (1, 3 and 10 mg kg^{-1}) administration reduced tumor growth	Campos et al. 2012
L. obtusa	12 novel and known sesquiterpenes	Tumor cells: K562, MCF7, PC3, HeLa, A431, A549, NSCLC-N6 Normal cells: CHO *In vivo*	Cytotoxic activity ranged between 15.8 and 320 μM. Compounds were nearly always more cytotoxic to cancer than normal cells	Kladi et al. 2006
L. obtusa	Phenolic content of methanol extract and its fractions (F2 and F3)	Tumor cells: A549, HCT15 and MCF7 Normal cells *In vivo*	Higher phenolic and flavonoid contents found in fraction than crude extract exhibited a cytotoxic effect, 50% inhibition of cell growth (IC_{50}) was obtained at concentrations of 110, 175 and 75 μg mL^{-1}, respectively against human tumor cell lines. IC_{50} were obtained at concentrations of 50, 35.95 and 20.5 μg mL^{-1}, respectively against human tumor cell lines tested. The fraction F2 revealed the lowest activity with IC_{50} of 375, 405 and 309 μg mL^{-1}, respectively against human tumor cell lines tested. We then evaluated the clonogenic inhibition of methanol extract and F3 against the same three human tumor cell lines (50–1000 μg mL^{-1}). Methanol extract showed a significant clonogenic inhibition at concentration-related manner, and IC_{50} were obtained at concentrations of 437.4, 406.25 and 357 μg mL^{-1}, respectively against human tumor cell lines tested. Similarly, F3 produced significant clonogenic inhibition too. IC_{50} are 100, 131 and 68.75 μg mL^{-1}, respectively	Dellai et al. 2013

Table 6.1 contd....

Table 6.1 contd....

Species	Tested product, fraction or extract	Study model	Observed results	References
L. obtusa	Aqueous extract	Tumor cells: EAC Silver Nanoparticle's of aqueous extract (AgNPs) *In vitro*	Cytotoxic effect (55–86%) of different AgNPs concentrations (42–98 μg mL^{-1})	Khalifa et al. 2016
L. okamurae	Methanolic extract	Tumor cells: L-1210 and P-388 Normal cells: NIH-3T3 *In vitro*	Extract exhibited high cytotoxicity to tumor cells and to normal cells	Harada et al. 1997
Laurencia sp.	Methanolic extract	Tumor cells: L-1210 and P-388 Normal cells: NIH-3T3 *In vitro*	Extract exhibited high cytotoxicity to tumor cells and no cytotoxicity to normal cells	Harada et al. 1997
L. translucida	Hexane extract	Tumor cells: MES-SA and the doxorubicin-resistant mutant, MES-SA/Dx5, HeLa Normal cells *In vivo*	IC_{50} = 16 ± 7 μg mL^{-1}	Stein et al. 2011
L. viridis	Polyether triterpenoids	Tumor cells: Jurkat, MM144, HeLa, CADO-ES1 Normal cells *In vivo*	IC_{50} values of 2.0–34.5 μM depending on compound and cell line. Jurkat and CADO-ES1 cells particularly sensitive	Pacheco et al. 2011
Lomentaria articulata	Crude (DCM/MeOH) extract. PP content 1.0–10.9%	Tumor cells: Daudi, Jurkat, K562 Normal cells *In vivo*	Crude extract concentrations: 50–500 mg L^{-1}; Inhibition: 9.98–50.89%	Zubia et al. 2009a
L. catenata	Phosphate buffered saline extract	Tumor cells: L-1210 and P-388 Normal cells: NIH-3T3 *In vitro*	Extract exhibited high cytotoxicity to tumor cells and some cytotoxicity to normal cells	Harada et al. 1997
L. clavellosa	Crude (DCM/MeOH) extract. PP content 1.0–10.9%	Tumor cells: Daudi, Jurkat, K562 Normal cells *In vivo*	Crude extract concentrations: 50–500 mg L^{-1}; Inhibition: 16.08–64.21%	Zubia et al. 2009a

L. hakodatensis	Ethanol/water extract	Tumor cells: HL-60, HT-29 Also tested: B-16, A-549 Normal cells: HaCaT *In vivo*	Concentration of extract: 100 μg mL^{-1} HL-60 Inhibition = 87.5% HT-29 Inhibition = 25.7% B16 Inhibition = 42.2% A549 Inhibition = 4.6% HaCat = Inhibition = 14.6%	Kim et al. 2009
Lophocladia sp.	Lophocladines A and B	Tumor cells: NCI-H460, neuro-2a, MDA-MB-435 Normal cells *In vivo*	IC_{50} values: lophocladine A NCI-H460 and neuro-2a (> 45 μM), MDA-MB-435 (> 450 μM). Lophocladine B, MDA-MB-435 (3.1 μM), NCIH460 (64.6 μM), neuro-2a cells (> 45 μM) Lophocladine B caused microtubule depolymerization and cell cycle arrest in G_2/M	Gross et al. 2006
Mertensia denticulata	Methanolic extract	Tumor cells: L-1210 and P-388 Normal cells: NIH-3T3 *In vitro*	Extract exhibited high cytotoxicity to tumor cells and to normal cells	Harada et al. 1997
Mastocarpus stellatus	Methanol extract	Tumor cells: HeLa, U-937 *In vitro*	HeLa cell caspase activity was influenced by the wild-harvested and cultivated red macroalgal extracts in a dose-dependent manner; the induction of caspase activity in HeLa cells was observed with macroalgal extracts at the lower dose of 0.25 mg mL^{-1} and the wild-*harvested M. stellatus*	Athukorala et al. 2016
Mazzaella japonica	Phosphate buffered saline extract	Tumor cells: L-1210 and P-388 Normal cells: NIH-3T3 *In vitro*	Extract exhibited cytotoxicity to tumor cells and no cytotoxicity to normal cells	Harada et al. 1997
Meristotheca coacta	Crude methanol, water and PBS extracts	Tumor cells: MOLT-4, K562, HeLa, KB Normal cells *In vivo*	Methanol extract showed telomerase inhibiting activity	Kanegawa et al. 2000
M. papulosa	Seaweed powder	Tumor cells: Ehrlich carcinoma *In vivo*	Oral administration of 1600 mg seaweed powder per kg mouse per day for 28 days. Inhibition rate: 48.8%	Noda et al. 1990
Neurymenia fraxinifolia	Neurymenolide A and B	Tumor cells: DU4475 Normal cells *In vivo*	IC_{50}: neurymenolide A (3.9 μM) and neurymenolide B (19.0 μM)	Stout et al. 2009

Table 6.1 contd....

Table 6.1 contd....

Species	Tested product, fraction or extract	Study model	Observed results	References
Osmundea pinnatifida	Dichloromethane extract	Tumor cells: HeLa Normal cells: Vero *In vitro*	IC_{50} = 129.3 µg mL^{-1}	Silva 2015
Otohimella japonica (formerly *Liagora japonica*)	Phosphate buffered saline extract	Tumor cells: L-1210 and P-388 Normal cells: NIH-3T3 *In vitro*	Extract exhibited cytotoxicity to tumor cells and no cytotoxicity to normal cells	Harada et al. 1997
Pachymeniopsis lanceolata (formerly *Grateloupia lanceolata*)	Ethanol/water extract	Tumor cells: HL-60, HT-29 Also tested: B-16, A-549 Normal cells: HaCaT *In vivo*	Concentration of extract: 100 µg mL^{-1} HL-60 Inhibition = 87.0% HT-29 Inhibition = 0.9% B16 Inhibition = 35.4% A549 Inhibition = 5.5% HaCat = Inhibition = 18.7%	Kim et al. 2009
Palisada perforata (formerly *Laurencia papillosa*)	Methanolic extract	Tumor cells: L-1210 and P-388 Normal cells: NIH-3T3 *In vitro*	Extract exhibited high cytotoxicity to tumor cells and to normal cells	Harada et al. 1997
P. perforata (formerly *L. papillosa*)	Crude methanol, water and phosphate buffered saline extracts	Tumor cells: MOLT-4, K562, HeLa, KB Normal cells *In vivo*	Methanol extract showed telomerase inhibiting activity	Kanegawa et al. 2000
P. perforata (formerly *L. papillosa*)	Sulfolipids	Tumor cells: MCF7, HepG2 *In vitro*	Concentrations = 1.0–10.0 µg mL^{-1} Growth inhibition = 76.93–94.19% (MCF7), 37.35–79.89% (HepG2)	El Baz et al. 2013
P. perforata (formerly *L. papillosa*)	Ethanol extract	Tumor cells: MCF-7, LNCaP, PC-3 Normal cells: MCF-10A *In vitro*	Extract shows cytotoxic effects on the cancer cells (LNCa, PC-3 and MCF-7) but not on the non-tumorigenic cell line MCF-10, suggesting that these extracts might be interesting for further studies regarding their potential anticancer effects	Montalvão et al. 2016
P. yamadana (formerly *L. yamadana*)	Crude methanol, water and PBS extracts	Tumor cells: MOLT-4, K562, HeLa, KB Normal cells *In vivo*	Methanol extract showed telomerase inhibiting activity	Kanegawa et al. 2000

Palmaria palmata	Crude extract, PP content 1.3%	Tumor cells: HeLa Normal cells *In vitro*	The grade 1 and 2 Dulse extract inhibition of HeLa cell proliferation was dose-dependent over 0.5–5.0 mg mL^{-1} and maximal at 48 and 72 h incubation	Yuan et al. 2005
P. palmata	Crude extract, PP content 0.2–1.3%	Tumor cells: HeLa Normal cells *In vivo*	HeLa cell proliferation was inhibited between 0% and 78% at 0.5–5 mg mL^{-1} algal extract	Yuan and Walsh 2006
P. palmata	Crude extract. Extract standardized to percentage PP, PP content of 0–50 μg mL^{-1} tested	Tumor cells: Caco-2 Normal cells *In vivo*	Extracts reduced viability of Caco-2 cells	Nwosu et al. 2011
P. palmata	Methanol extract	Tumor cells: HeLa, U-937 *In vitro*	HeLa cell caspase activity was influenced by the wild-harvested and cultivated red macroalgal extracts in a dose-dependent manner; the greatest induction of caspase activity in HeLa cells was observed with 2.00 mg mL^{-1} of the wild *P. palmata*	Athukorala et al. 2016
Phymatolithon calcareum (formerly *Lithothamnion calcareum*)	Mineral extract	Tumor cells Normal cells *In vivo*	Extract was found to be effective against colon cancer	Dame et al. 2011
Plocamium cartilagineum	Furoplocamioid C, prefuroplocamioid, pirene, others including halogenated cyclohexanes mertensene, violacene and lindane	Tumor cells: CT26, SW480, HeLa, SkMel28 Normal cells: CHO *In vivo*	Some compounds had good inhibitory potential; for most compounds, the cytotoxic effect was reversible	De Ines et al. 2004
P. cartilagineum	Crude (DCM/MeOH) extract. PP content 1.0–10.9%	Tumor cells: Daudi, Jurkat, K562 Normal cells *In vivo*	Crude extract concentrations: 50–500 mg L^{-1}; Inhibition: 30.07–70.85%	Zubia et al. 2009a
P. cartilagineum	Methanol and dichloromethane extracts	Tumor cells: HepG-2 *In vitro*	The most potent anti-proliferative activity was induced by dichloromethane extract (1000 μg mL^{-1}, 24 h), with 85.13% of cell's proliferation reduction	Alves et al. 2016
P. corallorhiza	Plocoralides A-C	Tumor cells: WHCO1 Normal cells *In vivo*	IC_{50} range 9.3–34.8 μM	Knott et al. 2005
P. corallorhiza	Methanol and dichloromethane extract Polyhalogenated monoterpene	Tumor cells: MCF-7, MDA-MB-231 Normal cells: MCF-12A *In vitro*	Extract was highly toxic to both the MCF-7 and MCF-12A cell lines (IC_{50} of 3.6 and 3.5 μM, respectively)	De La Mare et al. 2012

Table 6.1 contd....

Table 6.1 contd....

Species	Tested product, fraction or extract	Study model	Observed results	References
P. cornutum	Halogenated monoterpenes	Tumor cells: WHCO1 Normal cells *In vivo*	IC_{50} values range 6.6–87.6 µM	Antunes et al. 2011
P. cornutum	Methanol and dichloromethane extract Polyhalogenated monoterpene	Tumor cells: MCF-7, MDA-MB-231 Normal cells: MCF-12A *In vitro*	Extract was highly toxic to both the MCF-7 and MCF-12A cell lines (IC_{50} of 3.6 and 3.5 µM, respectively)	De La Mare et al. 2012
P. suhrii	Halogenated monoterpenes	Tumor cells: WHCO1 Normal cells *In vivo*	IC_{50} values range 6.6–87.6 µM	Antunes et al. 2011
P. telfairiae	Methanolic extract Phosphate buffered saline extract	Tumor cells: L-1210 and P-388 Normal cells: NIH-3T3 *In vitro*	Extract exhibited high cytotoxicity to tumor cells and no cytotoxicity to normal cells Extract exhibited cytotoxicity to tumor cells and high cytotoxicity to normal cells	Harada et al. 1997
P. telfairiae	Methanolic extract	Tumor cells: HT-29 *In vitro*	The treatment of HT-29 cells with various extract concentrations resulted in the inhibition of growth and the induction of apoptosis in a dose-dependent manner	Kim et al. 2007
Polyneura bonnemaisonii	Crude (DCM/MeOH) extract. PP content 1.0–10.9%	Tumor cells: Daudi, Jurkat, K562 Normal cells *In vivo*	Crude extract concentrations: 50–500 mg L^{-1}; Inhibition: 8.72–38.97%	Zubia et al. 2009a
Polyopes affinis (formerly *Carpopeltis affinis*)	Seaweed powder	Tumor cells: Ehrlich carcinoma *In vivo*	Oral administration of 1600 mg seaweed powder per kg mouse per day for 28 days. Inhibition rate: 20.0%	Noda et al. 1990
P. affinis (formerly *C. affinis*)	Ethanol/water extract	Tumor cells: HL-60, HT-29 Also tested: B-16, A-549 Normal cells: HaCaT *In vivo*	Concentration of extract: 100 µg mL^{-1} HL-60 Inhibition = 86.3% HT-29 Inhibition = 57.9% B16 Inhibition = 45.0% A549 Inhibition = 6.5% HaCat = Inhibition = 17.6%	Kim et al. 2009
P. lancifolius	Methanol extract	Tumor cells: T24 Normal cells *In vivo*	Doses of up to 150 µg mL^{-1} did not affect cell viability. Decreased MMP-9 and decreased cell invasion by matrigel assay	Jayasooriya et al. 2012

Polysiphonia stricta (as *Polysiphonia urceolata*)	Methanolic extract	Tumor cells: HT-29, KB Normal cells: NIH-3T3 *In vitro*	IC_{50} = 40.0 (KB), 26.0 (HT-29), 42.36 (NIH-3T3) µg mL^{-1}	Xu et al. 2004b
Porphyra purpurea	Crude (DCM/MeOH) extract. PP content 1.0–10.9%	Tumor cells: Daudi, Jurkat, K562 Normal cells *In vivo*	Crude extract concentrations: 50–500 mg L^{-1}; Inhibition: 8.72–62.59%	Zubia et al. 2009a
Porphyra/Pyropia sp.	Porphyran	Tumor cells: AGS *In vitro*	The addition of 0.1% porphyran also reduced DNA synthesis after 24 h of exposure, suggesting that porphyran inhibits cancer cell growth by both decreasing cell proliferation and inducing apoptosis	
Porphyra/Pyropia sp.	Ethanol extract 100 µg mL^{-1}	Protein kinase A inhibition *In vitro*	76–100% inhibition	Winberg et al. 2011
Porphyra/Pyropia sp.	Edible seaweed	Experimental design: Korean women (*n* = 362) aged 30–65 years old, with histologically confirmed breast cancer, paired with control cases according to age and menopausal status. Seaweed consumption, and other foods, was assessed using a food frequency questionnaire	Case-control study of seaweed consumption and cancer risk in Korean women. Inverse associations were found between Gim (*Porphyra*) intake and the risk of breast cancer in pre and post-menopausal women	Yang et al. 2010a
Porphyra/Pyropia spp.	Ethanol extract 100 µg mL^{-1}	Tumor cells Normal cells *In vivo*	Inhibition of kinase A was assayed. Showed the highest levels of inhibition = 76–100%	Winberg et al. 2011
Portieria hornemannii	(3S,6R)-6-Bromo-3-(bromomethyl)-2,3,7-trichloro-7-methyloct-1-ene (holomon)	Tumor cells: NCI *in vitro* tumor panel Normal cells *In vivo*	Cytotoxic to chemoresistant cell lines. Selected for preclinical drug development by NCI; this was limited due to lack of supply (Fuller et al. 1994); synthesis of the compound has since been published	Fuller et al. 1992
P. hornemannii	Halogenated monoterpenes	Tumor cells: NCI *in vitro* tumor panel Normal cells *In vivo*	Structure activity study. Results expressed as panel average cytotoxicity. 4 compounds had good cytotoxicity's (GI_{50} of 0.7–1.3 µM)	Fuller et al. 1994
P. hornemannii (formerly *Chondrococcus hornemannii*)	Methanolic extract	Tumor cells: L-1210 and P-388 Normal cells: NIH-3T3 *In vitro*	Extract exhibited high cytotoxicity to tumor cells and to normal cells	Harada et al. 1997

Table 6.1 contd....

Table 6.1 contd....

Species	Tested product, fraction or extract	Study model	Observed results	References
Pyropia columbina (formerly *Porphyra columbina*)	Sulfated galactans	Tumor cells: HeLa *In vitro*	At maximum concentration of 80 μg mL^{-1}, these fractions promoted atypical mitoses in HeLa cells, presence of atypical nuclei and blebs	Stevan et al. 2001
P. dentata (formerly *Porphyra dentata*)	Sterol fraction (cholesterol-15%, β-sitosterol-55%, campesterol-30%) (crude fractions also studied)	Tumor cells: 4T1 Normal cells *In vivo*: female BALB/c mice injected with 4T1 cells SC in mammary fat pads. Injected IP with 20 μL of sterol at dosages of 5, 10 or 25 mg kg^{-1} day^{-1} every 3 days for 18 consecutive days	*In vitro*: reduced cell viability and induced apoptosis. *In vivo*: reduced tumorigenesis and increased survival. Did not inhibit MDSC. Decrease in ROS levels in MDSC	Kazłowska et al. 2013
P. leucosticta (formerly *Porphyra leucosticta*)	Crude (DCM/MeOH) extract. PP content 1.0–10.9%	Tumor cells: Daudi, Jurkat, K562 Normal cells *In vivo*	Crude extract concentrations: 50–500 mg L^{-1}; Inhibition: 16.18–61.00%	Zubia et al. 2009a
P. tenera (formely *Porphyra tenera*)	Raw seaweed	*In vivo*: DMBA-induced tumors in rats	*P. tenera* showed an inhibitory effect on tumorigenesis, and tumor incidence was lower	Yamamoto et al. 1987
P. yezoensis (formerly *Porphyra yezoensis*)	Seaweed powder	Tumor cells: Ehrlich carcinoma; Meth-A fibrosarcoma *In vivo*	Oral administration of 1600 mg seaweed powder per kg mouse per day for 28 days. Inhibition rate: 53.2% Intraperitoneal administration of 50 mg kg^{-1} d^{-1} for seven days. Inhibition rate: 24.4%	Noda et al. 1990
P. yezoensis (formerly *Porphyra yezoensis*)	Sulfated polysaccharide: Porphyran	Tumor cells: Meth A fibrosarcoma *In vivo*	Intraperitoneal administration of 40 mg kg^{-1} for seven days. Inhibition rate: 58.4%	Noda et al. 1990
P. yezoensis (formerly *Porphyra yezoensis*	Phospholipid fractions	Tumor cells: Meth-A fibrosarcoma *In vivo*	Intraperitoneal administration of 40 mg kg^{-1} for seven days. Inhibition rate: 64.0%	Noda et al. 1990
P. yezoensis (formerly *Porphyra yezoensis*)	Sulfoquinovosyl-diacylglycerol with acyl chains of different lengths and degrees of unsaturation	Tumor cells Normal cells *In vivo*	100 μg L^{-1} caused cytotoxicity; 5–50 μg L^{-1} did not. Telomerase inhibition was found at IC_{50} of 22 μM	Eitsuka et al. 2004
P. yezoensis (formerly *Porphyra yezoensis*)	Methanolic extract	Tumor cells: HT-29, KB Normal cells: NIH-3T3 *In vitro*	IC_{50} > 50 (KB), > 50 (HT-29), > 50 (NIH-3T3) μg mL^{-1}	Xu et al. 2004b

Rhodomela confervoides	Methanolic extract	Tumor cells: HT-29, KB Normal cells: NIH-3T3 *In vitro*	IC_{50} = 37.0 (KB), 40.5 (HT-29), > 50 (NIH-3T3) μg mL^{-1}	Xu et al. 2004b
R. confervoides	Bromophenols	Tumor cells: A549, BGC-823, MCF-7, Bel7402, HCT-8 Normal cells *In vivo*	No cytotoxicity at 10 μg mL^{-1}	Zhao et al. 2005
R. confervoides	Bromophenols	Tumor cells: HCT-8, Bel7402, BGC-823, A549, A2780 Normal cells *In vivo*	Analysis of bromophenols from red algae. Pilot cytotoxicity study showed 4 out of 8 novel bromopolyphenols had moderate cytotoxicity compared to 5FU	Ma et al. 2006
Schizymenia dubyi	Sulfated heteropolysaccharide	Tumor cells: NSCLC-N6 *In vitro*	With 150 μg mL^{-1} tumor cells appeared to be inhibited in the G_1 phase of the cell cycle	Bourgougnon et al. 1994
Scinaia moniliformis	Phosphate buffered saline extract	Tumor cells: L-1210 and P-388 Normal cells: NIH-3T3 *In vitro*	Extract exhibited cytotoxicity to tumor cells and cytotoxicity to normal cells	Harada et al. 1997
Sphaerococcus coronopifolius	Dichloromethane extract	Tumor cells: HepG-2 *In vitro*	Cytotoxicity and anti-proliferation assays with an IC_{50} of 14.1 and 32.3 μg mL^{-1}, respectively	Alves et al. 2016
Spyridia filamentosa	Methanol extract	Tumor cells: DU-145, LNCaP, MCF-7, PC-3 *In vitro*	The highest cytotoxic activity among all extracts was shown by extract in 100 g mL^{-1} concentration with higher than 90% cell inhibition against the DU-145 cell line and about 80% against the MCF-7 cell line Extracts were assessed against four tumor cell lines and the most sensitive cell line to the extract was found as DU-145, while the LN-CAP cell line was the most resistant to the extract; however, in both concentrations (100 and 200 g mL^{-1}) of extracts, viabilities of cell lines did not exceed 50%	Taskin et al. 2010
Symphyocladia latiuscula	Methanolic extract	Tumor cells: HT-29, KB Normal cells: NIH-3T3 *In vitro*	IC_{50} = 36.2 (KB), 43.3 (HT-29), > 50 (NIH-3T3) μg mL^{-1}	Xu et al. 2004b
Trichogloeopsis mucosissima	Phosphate buffered saline extract	Tumor cells: L-1210 and P-388 Normal cells: NIH-3T3 *In vitro*	Extract exhibited low cytotoxicity to tumor cells and low cytotoxicity to normal cells	Harada et al. 1997

Table 6.1 contd....

Table 6.1 contd....

Species	Tested product, fraction or extract	Study model	Observed results	References
Tricleocarpa cylindrica (formerly *Galaxaura fastigiata*)	Methanolic extract Phosphate buffered saline extract	Tumor cells: L-1210 and P-388 Normal cells: NIH-3T3 *In vitro*	Extracts exhibited cytotoxicity to tumor cells and no cytotoxicity to normal cells	Harada et al. 1997
T. cylindrica (formerly *Galaxaura cylindrica*)	Sulfolipids	Tumor cells: MCF7, HepG2 *In vitro*	Concentrations = 1.0–10.0 μg mL^{-1} Growth inhibition = 66.37–82.46% (MCF7), 28.31–72.27% (HepG2)	El Baz et al. 2013
T. fragilis (formerly *G. oblongata*)	Ethyl acetate extract	Tumor cells: U937, HL-60, HuH-7 Normal cells *In vivo*	Concentrations: 25–200 mg mL–1 Inhibition effects on U937 cells: 6.3–95.9% Inhibition effects on HL-60 cells: 32.3–97.0%	Huang et al. 2005
Vertebrata lanosa (formerly *Polysiphonia lanosa*)	Bromophenols	Tumor cells: DLD-1, HCT-116 Normal cells *In vivo*	Several compounds were tested but lanosol n-propyl ether was the most active compound from the seaweed—IC_{50} 12.4 μM (DLD-1) and 1.32 μM (HCT-116). Other synthesized isomers were also tested and reported	Shoeib et al. 2004
V. lanosa (formerly *P. lanosa*)	Crude (DCM/MeOH) extract. PP content 1.0–10.9%	Tumor cells: Daudi, Jurkat, K562 Normal cells *In vivo*	Crude extract concentrations: 50–500 mg L^{-1}; Inhibition: 38.16–80.88%	Zubia et al. 2009a
Vidalia colensoi (formerly *Osmundaria colensoi*)	Bromophenols	Tumor cells: HL-60 Normal cells *In vivo*	One of the isolated compounds, lanosol butenone, had moderate cytotoxic activity against leukemia cells (IC_{50} = 8 μM)	Popplewell and Northcote 2009

6.3 Antitumor Activity of Chlorophyta (Green Seaweeds)

Powdered tissues from four species of air-dried marine green algae were screened for antitumor activity (Noda et al. 1990). Significant activity against Ehrlich carcinoma (EAC) was found in the green algae *Codium fragile* (43.4% inhibition) and *Ulva prolifera* (formerly *Enteromorpha prolifera*) (51.7%) (see Table 6.1).

In the studies done by Harada et al. (1997), it was found that most algae extracts showed cytotoxic activity not only to mouse lymphocytic leukemia (L-1210) cells, but also to mouse embryonic fibroblast (NIH-3T3) normal cells. However, among the active extracts, methanolic extracts from green algae, *Caulerpa chemnitzia* (formerly *Caulerpa racemosa* var. *laetevirens*), *Cladophoropsis vaucheriaeformis*, *Halimeda discoidea*, and *Halimeda macroloba* showed selective cytotoxic activity to L-1210 tumor cells. In particular, *C. vaucheriaeformis* exhibited markedly selective cytotoxicity to these cells. The results of the time course study of L-1210 cell growth in *C. vaucheriaeformis* extract-added culture showed that the cytotoxicity had a cytostatic effect only on the tumor cells. The positive control, mitomycin C, on the other hand, which inhibits DNA synthesis of cells, strongly inhibited the growth of both L-1210 and NIH-3T3 cells, indicating a non-selective cytotoxicity. In the test of the *in vitro* cytotoxic spectrum, the green algae extract from *C. vaucheriaeformis* also had a strong effect against murine leukemic P-388 cells (see Table 6.1). It appears that the active substance in this extract may act specifically against leukemic cell lines. Similarly, extracts from the other green alga, *H. discoidea* also had a strong effect on mouse leukemic cell lines (Harada et al. 1997).

When the toxicity of crude water and methanol extracts from *Caulerpa taxifolia* was studied on Swiss mice, seasonality was found to have an effect. Water extracts were found to be more toxic when isolated in the winter and spring, and methanol extracts were more toxic in summer (Lemée et al. 1993). The development of chemically induced skin tumors was reduced in ICR mice by a methanol/acetone extract from the green seaweed *Ulva prolifera* (formerly *Enteromorpha prolifera*) (Higashi-Okai et al. 1999).

In the studies done by (Moo-Puc et al. 2009), excluding *Halimeda tuna*, all the species of Chlorophyta tested showed some cytotoxic and antiproliferative activity against tumor cells (see Table 6.1).

The first published reports on the cytotoxic activity of extracts of green algae *Ulva intestinalis* (formerly *Enteromorpha intestinalis*) and *Ulva lactuca* (see Table 6.1) on Human colonic adenocarcinoma (LS174), Human lung adenocarcinoma (A549), Malignant melanoma (Fem-x), and Human chronic myelogenous leukemia (K562) cell lines were made by Kosanić et al. (2015). Exceptionally, it has been shown that the polysaccharide from *U. intestinalis*, although not directly demonstrating cytotoxic activity *in vitro*, may be associated with its potent immunostimulating effect and increased antitumor response in mice (Jiao et al. 2009). Furthermore, other studies and findings reported that algal components are responsible for anticancer activities of algae (Nakajima et al. 2009, Go et al. 2010, Ermakova et al. 2011). However, it is difficult to determine the contribution of individual components for the overall anticancer effects. Often, the activities of the extracts may be the result of synergistic or antagonistic effects of several compounds (Kosanić et al. 2015).

A marine carotenoid Siphonaxanthin derived from green growth has as of late been seen to exhibit less outflow of Bcl-2 and an astoundingly upgraded initiation of caspase-3 alongside up-controlled representation of Gadd45α and Dr5 in human leukemia (HL-60) cells. It has been additionally reported that siphonaxanthin inferred from *Codium fragile* are stronger development inhibitors against HL-60 cells than fucoxanthin (Ganesan et al. 2011, Farooqi et al. 2012, Sharif et al. 2014).

In the literature (see Table 6.1) there is some data on anticancer activity of extracts from *U. lactuca* and *U. intestinalis*, but anticancer activity of some other algal extracts were studied by other researchers. Salem and Ibrahim (2011) found anticancer activity for different extracts of the *Ulva rigida* on the Ehrlich ascites carcinoma (EAC) cell line.

In the works of Costa and collaborators (2010), the viability of HeLa cells treated with sulfated polysaccharides for 72 hours was determined using a colorimetric MTT-based assay. All sulfated polysaccharides showed antiproliferative activity in a dose-dependent manner. Among the Chlorophyta, sulfated polysaccharides from *Caulerpa prolifera* were the most active with 57.1% of cell proliferation inhibition at 0.1 mg mL^{-1}. The sulfated polysaccharides from *Codium isthmocladum*, *Caulerpa cupressoides*, and *Caulerpa sertularioides* have also shown high antiproliferative activity of 42.1%, 38.4%, and 36.4%, respectively, but only at 2.0 mg mL^{-1} (see Table 6.1).

Different studies report that sulfated polysaccharides with antitumor activity display lower or absent cytotoxicity to normal cells (Vishchuk et al. 2013). The sulfated hetero-rhamnans from *Gayralia oxysperma* showed no cytotoxicity for Vero cells (Cassolato et al. 2008). In this context, the sulfated polysaccharides obtained from seaweeds could be a good antitumor candidate, since the majority of chemotherapeutics agents cytotoxic to cancer cells are also cytotoxic to normal cells (Ropellato et al. 2015). Sulfated hetero-rhamnans produced by *G. oxysperma* were utilized for the preparation of two homogeneous and highly sulfated Smith-degraded products. The homogeneous products and the crude extracts containing the sulfated hetero-rhamnans showed cytotoxic effect against U87MG cells (Ropellato et al. 2015).

In the studies done by Lakmal et al. (2014), sample cytotoxicity effect against Vero cells was determined by the MTT assay. The methanol extract of marine algae samples was incubated at 50, 100, and 200 µg mL^{-1} concentrations for 24 hours with Vero cells and the cell viability (%) was determined *in vitro*. All samples at 50 and 100 µg mL^{-1} concentrations showed a high cell viability (%) compared to the controls. Among the samples, *Chaetomorpha linum* (formerly *Chaetomorpha crassa*) indicated the least cytotoxicity on Vero cells at 200 µg mL^{-1} concentration. The methanol extracts of *Caulerpa racemosa* showed significant cytotoxic effects at 200 µg mL^{-1} concentration. The cell viability (%) of the cancer cells HL-60, B16-F10, and A549 were determined by MTT assay with incubation of different concentrations (50, 100, and 200 µg mL^{-1}) of methanol extract of seaweeds for 24 hours. Among the cancer cells, HL-60 cells were significantly suppressed ($p < 0.05$) with the *C. racemosa* extract dose-dependently compared to the control. In particular, the highest growth inhibitory activity on HL-60 cells was observed at 200 µg mL^{-1} of *C. racemosa*, which was approximately 95%. The same concentration of *C. racemosa* showed a marked growth inhibitory activity against B16F10 and A459 cancer cells. The calculated IC_{50} values for the growth inhibitory activity of the cancer HL-60 cells are listed in Table 6.1. According to the results presented by Lakmal et al. (2014), the lowest cancer cell growth inhibitory activity (IC_{50} value 30.17 µg mL^{-1}) was reported in *C. racemosa*. Therefore, the methanol extract of *C. racemosa* was used to determine the mechanism of anticancer activity against HL-60 cancer cells, and the apoptotic body formation and accumulation of DNA content percentage in the sub-G_1 phase of the cell cycle of HL-60 were evaluated (Lakmal et al. 2014).

Methanolic extract of green algae were also able to decrease the rate of HeLa cell viability (Gomes et al. 2015). However, these inhibitions did not surpass the value of 35% under any of the evaluated conditions. Moreover, the inhibition pattern differed among extracts. The methanolic extracts of *Caulerpa prolifera* and *Caulerpa sertularioides* seaweeds showed inhibitory effects under most of the conditions tested. However, in many cases, this effect presented no considerable difference in time and concentration. Nevertheless, they showed a dose-dependent and time-dependent inhibitory activity tendency. Methanolic extract of *Caulerpa racemosa*, after 24 and 48 hour treatments, produced no inhibitory activity, but after 72 hours of exposure, the average inhibitory activity was 20%. However, this effect was not dose-dependent. *Codium isthmocladum* extract, after 72 hours of treatment, also showed dose-independent inhibitory activity of around 20%. This was, however, different from *C. racemosa* extract, which also exhibited inhibitory activity (~ 22%) after 48 hours under the experimental conditions used. *Caulerpa cupressoides* extract displayed the highest inhibition of HeLa cells—approximately 32%. However, like the two previous MEs it also had just a time-dependent effect (Gomes et al. 2015).

Another study with MEs of *Caulerpa chemnitzia* (formerly *Caulerpa peltata*), *C. racemosa*, *Caulerpa taxifolia*, and *Codium decorticatum* (formerly *Codium elongatum*) showed that they are weak antiproliferative agents (Vinayak et al. 2014). These data suggest that seaweeds from these genera do not synthesize antiproliferative compounds with high activity. However, studies with these seaweeds are scarce and more data is needed to confirm this fact (see Table 6.1).

The effect of different concentrations of silver nanoparticles of aqueous extract (AgNPs), synthesized by the marine green algae *Ulva fasciata* was investigated on Ehrlich Ascites Carcinoma (EAC) (Khalifa et al. 2016). In general, results showed a decrease in EAC cells viability by increasing AgNPs concentration; the maximum percentage of inhibition of EAC cells was 94% with 98 µg mL^{-1} AgNPs synthesized by *Ulva linza* (Khalifa et al. 2016).

In the work by Mashjoor et al. (2016), MTT assay was used as an indirect measure to determine the viability of breast adenocarcinoma (MCF7) and monkey kidney (Vero) cells, as well as HeLa cells exposed to the ethyl acetate and methanol extract of marine macroalgae. The viability of *Ulva flexuosa* was reduced in a dose-related manner after 24 or 48 hours. The results are given. Both ethyl acetate and methanol extracts caused cell death in a concentration dependent manner (see also Table 6.1). Cytotoxicity observed against the three cells (MCF7, HeLa, and Vero), at all concentrations, was tested using *U. flexuosa* extracts. The treatment of cells with *U. flexuosa*, ethyl acetate, and methanol extracts at high concentrations showed a great decrease in viability. Ethyl acetate fraction exhibited cytotoxicity to MCF-7, HeLa, and Vero (IC_{50} = 100 μg ml^{-1}). Methanol fraction also exhibited cytotoxic activity in MCF7, Vero (IC_{50} > 100 μg mL^{-1}), and HeLa (IC_{50} = 100 μg mL^{-1}) (Mashjoor et al. 2016).

6.4 Antitumor Activity of Phaeophyceae (Brown Seaweeds)

Antitumor activity in aqueous extracts of seaweed was first demonstrated by Nakazawa et al. (1974). Ito and Sugiura (1976) obtained a polysaccharide fraction from *Sargassum thunbergii* with antitumor activity. The fraction showed a great ILS value (Increase in Life Span) against Ehrlich carcinoma (ascite form). An aqueous extract from *Sargassum miyabei* (formerly *Sargassum kjellmanianum*) also had a high antitumor activity against sarcoma 180 (Yamamoto et al. 1984b). The aqueous extracts were assumed to contain polysaccharides (Yamamoto et al. 1977). Based on these findings, Noda et al. (1989, 1990) surveyed various seaweeds that showed antitumor activity against mice implanted with Ehrlich carcinoma or Meth-A fibrosarcoma (see Table 6.1).

Brown algal polysaccharides from *Ecklonia*, *Sargassum*, and *Undaria* species display antitumor properties (Yamamoto et al. 1974, Ito and Sigiura 1976, Takahashi 1983, Usui et al. 1980, Yamamoto et al. 1982, 1984a, 1984b, Noda et al. 1989, 1990). In particular, the ester sulfates of the polysaccharides may be related to the antitumor activity (Jolles et al. 1962, Yamamoto et al. 1984b, Yamamoto and Maruyama 1985). This assumption could be supported, in part, by the results obtained by Noda et al. (1990), but it does not explain the low but positive effect of alginates. Another reasonable assumption is that the degree of polyanionic properties may be intimately related to this activity. However, G- and M-alginate fragments showed no activity despite their polyanionic properties, so the polymerization degree of such polysaccharides and/or the ultrastructure of these polyanions may hold the key to their activity (Noda et al. 1989, 1990) (see Table 6.1).

Compared to polysaccharides, studies on the antitumor effects of lipids are confined in number—fatty acids (Ando et al. 1969, Ito et al. 1982, Tolnai and Morgan 1962, Townsend et al. 1960), glyceryl ethers (Ando et al. 1972), unsaturated fatty acids and their ester derivatives (Nishikawa et al. 1976), amino acid-fatty acid salts (Tolnai and Morgan 1966), and polyunsaturated fatty acids (Mertin and Hunt 1976). However, none of these substances was highly effective. Ito et al. (1982) suggested that fatty acids induce a change in the lipid composition of tumor cells, resulting in damage to the tumor cells. We found some lipid fractions to be highly effective against Meth-A fibrosarcoma. These effective lipids, therefore, hold promise of future utilization, but there is a need for precise structural studies. In this connection, it should be added that glycoproteins obtained from some brown algae showed antitumor activity (Yamamoto et al. 1974, Nakazawa et al. 1976, Noda et al. 1989) (see Table 6.1).

Suzuki et al. (1980) reported that administration of partially purified polysaccharide from *Saccharina longissima* (formerly *Laminaria angustata* var. *longissima*) to mice implanted with ascite form of L-1210 leukemia, Meth A fibrosarcoma, and B-16 melanoma gave increase in Life Span (ILS) of 49%, 100%, and 92%, respectively. Yamamoto et al. (1977, 1981) also reported that extracts from *Sargassum miyabei* (formerly *Sargassum kjellmanianum*) and *S. fulvellum* against Sarcoma 180 (solid form) showed high inhibition rates such as 93.7% and 91.5%, respectively. Similarly, Takahashi (1983) obtained results from *in vitro* experiments for ddY mice and nude mice that crude fucoidan from *Ecklonia bicyclis* (formerly *Eisenia bicyclis*) showed inhibition rates of 86.6% and 68.9% against Sarcoma 180 (solid form), respectively. In this experiment, the increase of phagocytosis in the reticuloendothelial system of the mouse was observed, indicating the enhancement of the immunological activity.

The edible alga, *Undaria pinnatifida*, exhibits antitumor activity against implanted Lewis lung carcinoma (3LL), and the active substance is thought to be a polysaccharide (Ohigashi et al. 1992). In the study done by Harada et al. (1997) in an *in vitro* assay, this alga showed weak cytotoxic activity against mouse lymphocytic leukemia (L-1210) cells. Crude or partially purified polysaccharides from various brown algae showed antitumor activity against experimental tumor. A partially purified fucoidan from *Ecklonia bicyclis* (formerly *Eisenia bicyclis*) exhibits antitumor activity against L-1210 leukemia (Yamamoto et al. 1984). In the study done by Harada et al. (1997), phosphate-buffered saline (PBS) extract from *E. bicyclis* (formerly *E. bicyclis*) also showed activity against L-1210 cells and relatively low cytotoxicity to normal NIH-3T3 cells. In the report by Noda et al. (1990), the brown algae, *Sargassum ringgoldianum* and *Scytosiphon lomentaria* showed antitumor activity against implanted Ehrlich carcinoma (EAC).

In the study done by Harada et al. (1997), 1446 samples of marine algae were collected from 79 points along Japan's coastline. The total number of species was 306; out of 306 seaweed species, 81 exhibited cytotoxic activities against mouse lymphocytic leukemia L-1210 cells (36 brown, 30 red, 14 green algae). Among the three types of sample algae, the highest activity against L-1210 was found in brown algae (see Table 6.1).

Chronic exposure of the skin to ultraviolet B (UVB) radiation induces oxidative stress, which plays a crucial role in the induction of skin cancer. In the study done by Hwang et al. (2006), the effect of dietary feeding and topical application of brown algae polyphenols on UVB radiation-induced skin carcinogenesis in SKH-1 mice was investigated. SKH-1 hairless mice were randomly divided into nine groups, including control, UVB control, and treatment groups. They were treated orally (0.1% and 0.5% with AIN-76 diet, w/w) and topically (3 and 6 mg 0.2 mL^{-1} of vehicle) with brown algae polyphenols and irradiated with UVB for 26 weeks. Dietary feeding (0.1% and 0.5%) of brown algae polyphenols significantly reduced tumor multiplicity (45% and 56%) and tumor volume (54% and 65%), and topical administration (3 and 6 mg) significantly decreased tumor multiplicity (60% and 46%) and tumor volume (66% and 57%), respectively, per tumor-bearing mouse. Dietary feeding and topical administration of the polyphenols also inhibited tumor incidence by 6% and 21%, respectively, but the results were not significant. Dietary and topical administration of the polyphenols markedly inhibited cyclooxygenase-2 activity and cell proliferation. These observations showed that brown algae polyphenols have an antiphotocarcinogenic effect which may be associated with the prevention of UVB-induced oxidative stress, inflammation, and cell proliferation in the skin (Hwang et al. 2006).

Noda et al. (1990) reported that administration of neutral lipid fractions, those from *Ecklonia bicyclis* (formerly *Eisenia bicyclis*) and *Sargassum ringgoldianum* showed significant inhibition of Meth-A fibrosarcoma (see Table 6.1); the glycolipid fractions of *Undaria pinnatifida* and *S. ringgoldianum* showed significant inhibition, and the phospholipid fractions from *Saccharina angustata* (formerly *Laminaria angustata*) and *S. ringgoldianum* showed significant activity.

Testing several algae species in China with potential antitumor activity, Xu et al. (2004b) found that ethanolic extracts from *Scytosiphon lomentaria* and hexane from *Dictyopteris divaricata* showed cytotoxic activity against human oral epidermoid carcinoma (KB). Substances isolated from algae, such as fucoidan, laminaran (see also Chapter 2), and terpenoids show activity against cancer cell strains (Gerwick and Bernart 1993, Synytsya et al. 2010) and the search for new drugs from these organisms is growing. In order to be considered effective in treating cancer, it is necessary that a drug have a selective antitumor activity without side effects (Xu et al. 2004b). Jolles et al. (1963) were the first researchers to report the influence of a sulfated degraded laminarin obtained from seaweed extract in inhibiting the growth of tumor cells.

In the studies done by Moo-Puc et al. (2009), three of the five species of Phaeophyceae tested (*Turbinaria turbinata*, *Lobophora variegata*, and *Dictyota caribaea*) showed some important activity against the human nasopharynx carcinoma (KB) cell line. The extract from *T. turbinata* exhibited high cytotoxic and antiproliferative activity against KB cells (IC_{50} = 23.94–29.84 μg mL^{-1}), but also showed cytotoxic activity against normal cells (see Table 6.1).

In the study done by Kim et al. (2009), the cytotoxic effects of 23 extracts of seaweeds, collected from Jeju Island (South Korea), were tested *in vitro* on human promyelocytic leukemia (HL-60), human colon carcinoma (HT-29), murine melanoma (B16), and human lung cancer (A549) cells. These cells were subjected to 72 hours of exposure at either a 25 or 100 mg mL^{-1} concentration (see Table 6.1). The percentage inhibition of seaweed extracts (25 mg mL^{-1}) from *Colpomenia sinuosa*, *Dictyopteris divaricata*, *Dictyopteris prolifera*, *Sargassum fusiforme* (formerly *Hizikia fusiformis*), and *Sargassum thunbergii*, on HL-60 cell lines was over 40%. Of the 23 seaweeds, *D. divaricata* and *S. thunbergii* displayed the strongest cytotoxic effects on human HL-60. Furthermore, *S. thunbergii* and *S. fusiforme* also demonstrated cytotoxic effects greater than 50% on HT-29 cell lines. Were compared the cytotoxic effects of seaweed extracts on cancer cells with the effects on normal cells (human keratinocytes, HaCaT) (Kim et al. 2009).

In the studies by Costa et al. (2010), the viability of HeLa cells treated with sulfated polysaccharides for 72 hours was determined using a colorimetric MTT-based assay. The brown algae *Canistrocarpus cervicornis* (formerly *Dictyota cervicornis*), *Dictyota mertensii*, and *Spatoglossum schroederi* showed antiproliferative effect, with about 32.0%, 45.0%, and 45.0%, respectively, but only at 2.0 mg mL^{-1}. Interestingly, sulfated polysaccharides from *Sargassum filipendula* were the most active with 61.1% of cell proliferation inhibition at 0.1 mg mL^{-1}, while sulfated polysaccharides from *Dictyopteris delicatula* and *Dictyota menstrualis* also presented an excellent antiproliferative activity in high concentration (2.0 mg mL^{-1}) with 61.0% and 58.4% of inhibition ratio, respectively.

Some brown seaweeds produce sulfated homo-heterofucans, which have shown effective antitumor activities with a wide range of mechanisms, which stands as a valuable source (Xin et al. 2000b). Sulfated polysaccharides, such as heparin, cellulose sulfate, and dextran sulfate, were reported to block the infectivity of papillomaviruses (Spoden et al. 2013). Polysaccharide fraction (sulfated fucans) isolated from the brown seaweed *Sargassum stenophyllum* was reported to promote morphological modifications in HeLa cells at low (2.5 μg mL^{-1}) concentrations (Stevan et al. 2001). Polysaccharides and terpenoids from brown algae are considered as promising bioactive molecules with anticancer activity by Devery et al. (2001) and Taskin et al. (2010). Polysaccharide-rich extract from *Sargassum filipendula* showed antiproliferative effect on HeLa cell (human uterine adenocarcinoma cell) proliferation (Costa et al. 2010) (see Table 6.1).

Results obtained in one of the earliest works demonstrated that purified fucoidan isolated from the alga *Ecklonia bicyclis* (formerly *Eisenia bicyclis*) inhibited the growth of the sarcoma 180 tumor (Usui 1980). Sulfated polysaccharides extracted from *Sargassum miyabei* (formerly *Sargassum kjellmanianum*) inhibited the growth of implanted sarcoma 180 cells, but had no effect on L-1210 leukemia cells. However, when the SKCF polysaccharide fraction was further sulfated, it acquired anti-leukemic properties (Yamamoto et al. 1984b). Fucoidan significantly reduced lung metastases resulting from the intravenous injection of cells from the rat mammary adenocarcinoma 13762 MAT (Coombe et al. 1987). In the experiments of Zhuang et al. (1995), 31 neutral and acidic polysaccharide fractions were obtained from *Sargassum thunbergii* and tested for antitumor activity in mice with Ehrlich carcinoma transplanted intraperitoneally. Only two fractions of these 31, which were identified as L-fucans with molecular weights of 19 and 13.5 kDa, had antitumor activity (Table 6.1).

Meroterpenoids, Sargol, Sargol-I, and sargol-II were isolated from the brown alga *Sargassum siliquastrum* (formerly *Sargassum tortile*), and showed cytotoxic activity (Numata et al. 1991). A linear cytotoxic diterpene bifurcadiol was isolated from the brown alga *Bifurcaria bifurcata* by Guardia et al. (1999), which exhibited cytotoxicity against cultured human tumor cell lines (A549, SK-OV-3, XF 498 and HCT).

The hydroquinone diterpene (Mediterraneol A) from *Cystoseira mediterranea* has shown inhibitory effects on mitotic cell division (Francisco et al. 1985). Fractions of the alcohol extract of *Polycladia myrica* (formerly *Cystoseira myrica*) have demonstrated four new cytotoxic hydroazulene diterpenes. All compounds exhibited moderate cytotoxicity on the cancer cell line KA3IT ($IC_{50} \approx 5$ μg mL^{-1}), and showed reduced cytotoxicity towards the normal NIH3T3 cells. The total alcohol extract showed more cytotoxic activity against the normal cell line than on the other virally transformed forms (Ayyad et al. 2003).

In the works of Roszkowsky et al. (1989), fucoidan was injected intraperitoneally (10 mg kg^{-1}) and exerted moderate antitumor and anti-metastatic effects, and it increased the anti-metastatic, but not the antitumor action of cyclophosphan, by more than twice the amount. *In vivo*, repeated administration of fucoidan significantly inhibited the settling of metastatic sarcoma L-1 cells in the lungs of Balb/C-Mice (Roszkowsky et al. 1989). Similar effects of fucoidan action were described for several experimental models. For instance, a fucoidan extract from *Ascophyllum nodosum* exerted *in vitro* antiproliferative activity on a cell line derived from a non-small-cell human bronchopulmonary carcinoma (NSCLC-N6) with a block observed in the G_1 phase of the cell cycle. Experimentations conducted *in vivo* showed that this extract had antitumor activity at sub-toxic doses (Riou et al. 1996). Fucoidan from *Saccharina japonica* (formerly *Laminaria japonica*) inhibited *in vitro* the cells of hepatoma QGY7703 at their logarithmic growth phase (Shi et al. 2000). Commercially produced fucoidans inhibited proliferation and induced apoptosis in human lymphoma HS-Sultan cell lines (Aisa et al. 2005). Alekseyenko et al. (2007) tested the effects of *Fucus distichus* subsp. *evanescens* fucoidans with molecular weights of 20–40 kDa on the pulmonary adenocarcinoma C57B/6 cell line of mice. Fucoidans extracted from *Fucus distichus*, *Fucus serratus*, *Fucus vesiculosus*, and *Saccharina latissima* (formerly *Laminaria saccharina*) blocked MDA-MB-231 breast carcinoma cell adhesion to platelets (Cumashi et al. 2007) (see also Table 6.1).

Low molecular weight fucans extracted from *Ascophyllum nodosum* inhibited human colon adenocarcinoma Colo320DM proliferation. The active concentration of the sulfated polysaccharide was about 100 μg mL^{-1} (Ellouali et al. 1993). Commercially obtained *Fucus vesiculosus* homofucan decreased *in vitro* protein content in a medium with myeloleukemia HL60 cells, which indirectly suggested an antiproliferative effect; the IC_{50} was approximately 30 μg mL^{-1} (Queiroz et al. 2006).

In vitro, fucoidan extract prepared from the brown alga *Cladosiphon novae-caledoniae* inhibited the invasive ability of human fibrosarcoma HT1080 cells, possibly via suppressing matrix metalloproteinase activities. It also blocked angiogenesis of carcinoma HeLa cells by suppressing the expression and secretion of vascular endothelial growth factor (VEGF). The intracellular H_2O_2 level and H_2O_2 released from tumor cells were both repressed upon treatment with fucoidan extract. The authors believe that these effects might result, at least partially, from the anti-oxidative potential of fucoidan (Ye et al. 2005).

In vitro model experiments with mammary breast adenocarcinoma MDA-MB231 cells and Lewis lung 3LL carcinoma cells demonstrated that fucoidan inhibited tumor cell invasion through the matrigel basement membrane matrix, and inhibited their adhesion to laminin. These results suggest that tumor cell adhesion to the extracellular matrix is blocked in the presence of fucoidan, which is caused by the suppression of the laminin-induced increase in extracellular urokinase plasminogen activator (u-PA) (Soeda et al. 1994, Haroun-Bouhedja et al. 2002). Native fucoidans extracted from *Undaria pinnatifida* displayed an anticancer activity of 37.6%. When hydrolyzed in boiling water with HCl for 5 minutes, fucoidans (490 kDa) expressively increased anticancer activity to 75.9% (Yang et al. 2008).

The level of the algal flora from the North Atlantic (French coasts) highlights that the crude extracts of *Bifurcaria bifurcata*, *Cystoseira tamariscifolia*, *Desmarestia ligulata*, *Dictyota dichotoma*, and *Halidrys siliquosa* presented significant cytotoxic activities against Daudi (Human Burkitt's lymphoma), Jurkat (Human leukemic T cell lymphoblast), and K562 (Human chronic myelogenous leukemia) cell lines. One should note that the crude extract from *Alaria esculenta* also shows a significant reduction of the cell viability with Daudi (73.74%) and Jurkat (61.01%) cell lines, conversely to the crude extract from *Fucus ceranoides*, which exhibits a significant increase in Daudi cell viability (128.79%) (Zubia et al. 2009b).

The ability of fucoidan from *Fucus distichus* subsp. *evanescens* to increase the apoptosis induced by etoposide (inhibitor of DNA topoisomerase II) was investigated. It has been shown that the incubation of MT-4, but not Namalwa cells, in the presence of a highly purified preparation of fucoidan (500 μg mL^{-1}) increases the sensitivity of these cells to etoposide with the subsequent induction of caspase-3-independent pathways of apoptosis. The results suggested a new role for fucoidan—the combination of fucoidan with approved anticancer drugs for a synergistic effect (Philchenkov et al. 2006, 2007). Commercially available crude fucoidan from the brown seaweed *Fucus vesiculosus* (Sigma-Aldrich) became very attractive, in spite of the absence of sufficient data on its exact structure. A number of researchers used this product without any purification. Thus, it was demonstrated fucoidan from *F. vesiculosus* induced apoptosis of human lymphoma HS-Sultan cells (Aisa et al. 2005), myeloid leukemia U937 cells (Park et al. 2013), and HL-60, NB4, THP-1 cells (Jin et al. 2010) via the activation of caspase-9 and -3 accompanied by down-regulation of Bcl-2 and Bax expression, as well as changes in the phosphorylation of ERK, JNK, p38, and Akt kinases.

The antitumor and antimetastatic activities of fucoidan from *F. distichus* subsp. *evanescens* were investigated *in vivo* using mice with transplanted Lewis lung adenocarcinoma (C57Bl/6 mice). Fucoidan (10 mg kg^{-1}) alone possesses moderate antitumor and antimetastatic effects (see also Table 6.1). In addition, this fucoidan potentiates the antimetastatic but not antitumor activities of cyclophosphamide (drug used for chemotherapy) and, at 25 mg kg^{-1}, increases the toxic effect of cyclophosphamide (Alekseyenko et al. 2007).

Several authors investigated the effect of fucoidans from *F. vesiculosus* on colorectal cancer growth. The fucoidan was shown to suppress growth of human colon carcinoma HCT-15 cells on 62% at concentration 100 μg mL^{-1} (Hyun et al. 2009), and to induce substantial reductions in viable cell numbers of HT-29 and HCT116 at a dose of 20 μg mL^{-1} (Kim et al. 2010a). The *in vitro* anticancer activity (soft agar model) of fucoidans derived from nine brown algae species was studied. The fucoidan from *F. evanescens* (200 μg mL^{-1}) was nontoxic to DLD-1 and HT-29 cells and inhibited their colony formation (DLD-1 at 50% and HT-29 at 30%) (Vischuk et al. 2009). On the other hand, Han and co-workers (2015a) examined the growth inhibiting effect of the same fucoidan on HT-29 cells. The cell growth was found to be significantly decreased following treatment with polysaccharide (200 μg mL^{-1}). Because of the heterogeneity in structural characteristics within seaweed, differing extraction conditions used by researchers can give rise to the isolation of distinct fucan forms (Li et al. 2008). This fact can explain the difference in doses of fucoidan used for the treatment of the same cell lines. This fucoidan (400 μg mL^{-1}) was also nontoxic to the melanoma cell lines SK-Mel-5 and SK-Mel-28 and inhibited the cell proliferation (48 hours) of these cells in a dose-dependent manner. The fucoidan from *F. evanescens* (800 μg mL^{-1}) inhibited the colony formation of SK-MEL-5 (63%) and SK-MEL-28 (70%) cells, and the content of α-1,4-fucose residues in the fucoidan molecule is important for its inhibiting activity (Anastyuk et al. 2012). The cancer-preventive efficacy of the fucoidan from *F. evanescens in vitro* and *ex vivo* was investigated. Fucoidan participates in the prevention of neoplastic cell transformation and in the progression of colon carcinomas through lymphokine-activated killer T-cell-originated protein kinase (TOPK) (Vishchuk et al. 2016).

It was shown that fucoidan from *Saccharina gurjanovae* (formerly *Laminaria gurjanovae*) suppressed neoplastic cell transformation by inhibiting the phosphorylation of epidermal growth factor (EGF) receptor in mouse epidermal JB6 Cl41 cells and suppressing the extracellular signal-regulated kinase or c-jun *N*-terminal kinases. Moreover, EGF-induced c-fos and c-jun transcriptional activities were inhibited by fucoidan, resulting in the suppression of AP-1 activity (Lee et al. 2008b). The same authors also found that fucoidan from *Saccharina cichorioides* (formerly *Laminaria cichorioides*) inhibited the EGF or 12-*O*-tetradecanoylphorbol-13-acetate-induced neo-plastic cell transformation in epidermal cells of mice. The *in vitro* binding assay revealed that fucoidan directly interacted with EGF, suggesting that the antitumor promoting effect of fucoidan might be due to preventing the binding of EGF to its epidermal cell surface receptor. Oral administration of fucoidan extracted from *Undaria pinnatifida* inhibited the tumor growth of A20 leukemia cells and P-388 tumors in mice (see Table 6.1). Fucoidan significantly augmented natural killer (NK) cells, enhanced their cytolytic activity, and increased the amount of IFN-γ produced by T-cells (Maruyama et al. 2003, 2006).

An *in vitro* model of HT-29 and DLD-1 cancer cells from human intestine was used to study the properties of fucoidans from nine brown algae. The highest antitumor activity was found in the fucoidans from *Saccharina cichorioides* (formerly *Laminaria cichorioides*). Its main chain consisted of α-1,3-linked fucose, and the content of sulfates was 36.0%. The fucoidans from *Fucus distichus* subsp. *evanescens* and *Undaria pinnatifida* had lower activities. The main chain of the first polysaccharide is composed of 1,3- and 1,4-linked α-L-fucopyranose units; in the second polysaccharide it consists of the alternating units of 1,3- and 1,4-linked α-L-fucopyranose and galactose; sulfates make up 31% and 36%, respectively. The fucoidans from *Sargassum denticarpum*, *Sargassum mcclurei*, and *Sargassum swartzii* have a structure similar to that from *F. evanescens*, but with sulfate contents of 11.2%, 2.8%, and 5.9%, respectively, and with negligible antitumor activities (Vischuk et al. 2009).

Aqueous extracts of brown alga *Sargassum oligocystum*, gathered from the Persian Gulf seashore, have showed antitumor activity against K-562 and Daudi human cancer cell lines. The most potent antitumor activity has been shown at concentrations of 500 μg mL^{-1} and 400 μg mL^{-1} of the alga extract on Daudi and K562 cell lines, respectively. In this study, the effective concentration is more than its counterpart in other studies and it seems that the main reason for this difference is due to the use of crude extract in this study (Zandi et al. 2010). Antitumor activity has also been observed with the macroalga *Sargassum stenophyllum in vivo*. The extracted polysaccharide powder fraction at doses of 1.5 and 150 μg/animal has the ability to retard the increase in tumor volume by at least 2.5 and 5 days, respectively. Treatment with either dose of extracted polysaccharide did not induce any deaths or body weight loss, suggesting little or no toxicity (Dias et al. 2005). Also, the methanol extract of *Sargassum swartzii* collected from the Persian

Gulf has demonstrated cytotoxic effect against T-47D cells (IC_{50} < 100 μg mL^{-1}). In the same study, the species of *Sargassum* showed no cytotoxicity (see also Table 6.1) (Khanavi et al. 2010).

Brown seaweed with low molecular weight fucoidan mediated the broad-spectrum growth inhibition of human carcinoma cells, including HeLa cervix adenocarcinoma, HT1080 fibrosarcoma, K562 leukemia, U-937 lymphoma, A549 lung adenocarcinoma, and HL-60 (Zhang et al. 2011). Fucoidan produced from brown algae was reported to inhibit HPV pseudo virus infection *in vitro* with the IC_{50} value of 1.1 μg mL^{-1} (Buck et al. 2006). Brown algae could be a relevant source of anticancer compound (Manivannan et al. 2008). Heterofucans from *Sargassum filipendula* exhibited antiproliferative effects on cervical cells (Costa et al. 2011a, b). Fucoxanthin has been shown to induce apoptosis in human cervical cancer HeLa cells (Hou et al. 2013, Ye et al. 2014) (see also Table 6.1).

In another work, Vinayak et al. (2014) showed that methanolic extracts (0.1 mg mL^{-1}) of *Spatoglossum asperum*, *Spatoglossum variabile*, and *Sargassum marginatum* exhibited inhibition rates of around 82%, 80%, and 40%, respectively, against Ehrlich ascites tumor (EAT) cells. In addition, these methanolic extracts did not affect the viability of normal cells. These results indicate that methanolic extracts of these two genera of brown algae may be more effective against breast cancer cell lines (Vinayak et al. 2014).

Methanolic extracts of the brown seaweeds *Spatoglossum schroederi* and *Sargassum filipendula* presented no antiproliferative activity (see Table 6.1) under any of the tested conditions (Gomes et al. 2015). *Dictyota mertensii* methanolic extract showed significant activity only after 72 hours and at the two highest concentrations evaluated. *Dictyopteris delicatula* extract exhibited an inhibition rate of around 22% after 48 hours. However, this rate did not change after 72 hours. A dose-dependent effect was not observed. The most effective extracts were those of *Dictyota cilliolata* and *Dictyota menstrualis*. *D. cilliolata* extract presented a time-dependent effect and achieved a cell viability inhibition rate of around 50% after 72 hours of experiment (at 0.2 mg mL^{-1}). *D. menstrualis* extract presented a dose-dependent and time-dependent effect, reaching its maximum inhibition (approximately 80% of HeLa and SiHa tumor cells) with 0.2 mg mL^{-1} after 48 hours of exposure (Gomes et al. 2015).

In the work by Alves et al. (2016b), the results obtained by dichloromethane extract of *Bifurcaria bifurcata* extract are particularly interesting when compared to the IC_{50} obtained for cisplatin and tamoxifen (commercial drugs) in cytotoxicity and antiproliferative tests. In the specific case of cisplatin (92.00 and 80.11 μg mL^{-1}), the values of IC_{50} obtained on Caco-2 cells were very similar to the values obtained by dichloromethane fraction (90.09 and 82.31 μg mL^{-1}) in cytotoxicity and antiproliferative tests, respectively. Similar effect was also verified on HepG-2 cells cytotoxicity assay. As this fraction is a crude extract, these results suggest that this extract has molecules with high antitumor potential (see Table 6.1).

The effect of different concentrations of silver nanoparticles of aqueous extract (AgNPs), synthesized by the marine brown algae *Turbinaria turbinata*, was investigated on Ehrlich Ascites Carcinoma (EAC) (see also Table 6.1). Overall, the results showed a reduction in EAC cells viability by increasing AgNPs concentration; the maximum percentage of inhibition of EAC cells was 99%, with 98 μg mL^{-1} AgNPs synthesized by *Turbinaria turbinata* (Khalifa et al. 2016).

In the study done by Motalvão et al. (2016), the extracts of reproductive *Dictyopteris polypodioides* (formerly *Dictyopteris membranacea*) displayed a general cytotoxicity against all four cell lines, with inhibition of growth in the range of 60–96%, with concentration of 50 mg mL^{-1}. However, extracts from the *Cystoseira* species, and of microalgae *Nitzschia thermalis* (Bacillariophyta), displayed cytotoxic effects on the cancer cells (LNCa, PC-3, and MCF-7), but not on the non-tumorigenic cell line MCF-10, suggesting that these extracts might be interesting for further studies regarding their potential anticancer effects (Motalvão et al. 2016).

6.5 Antitumor Activity of Rhodophyta (Red Seaweeds)

Significant activity against Ehrlich carcinoma (EAC) was found in the red algae *Pyropia yezoensis* (formerly *Porphyra yezoensis*) (53.2%) and *Betaphycus gelatinus* (formerly *Eucheuma gelatinae*) (52.1%). Four red algae showed appreciable antitumor activity against Meth-A fibrosarcoma (see Table 6.1) (Noda et al. 1990).

The results obtained by Noda et al. (1990) show that red algal polysaccharides, viz., porphyran from *Porphyra/Pyropia* and carrageenans from *Eucheuma denticulatum* (ι-carrageenan) and *Kappaphycus alvarezii* (κ-carrageenan) (Pereira et al. 2009b), display antitumor properties (Noda et al. 1989) (see Table 6.1). The results of the studies by Zhou et al. (2006) indicated that the degraded lambda-carrageenan (extracted from *Chondrus ocelatus*) could add the antitumor activities of 5-Fu and improve the immunocompetence damaged by 5-Fu.

In the studies done by Moo-Puc et al. (2009), two of the 14 species of Rhodophyta tested (*Bryothamnion triquetrum* and *Laurencia microcladia*) were selectively cytotoxic against cancer cells (see Table 6.1). The *B. triquetrum* extract showed the highest cytotoxic and selective activity against Epidermoid carcinoma (Hep-2) cells (IC_{50} = 8.29 μg mL^{-1}, SI = 12.04). It also showed medium activity against Human nasopharynx carcinoma (KB) and Human cervical cancer cells (HeLa cells) (IC_{50} = 32.57 and 48.45 μg mL^{-1}, respectively), but low antiproliferative activity against KB cells (IC_{50} = 62.98 μg mL^{-1}). On the other hand, *Gracilaria cervicornis* showed high cytotoxic activity against KB cells (IC_{50} = 19.23 μg mL^{-1}), as well as cytotoxic activity against normal cells (IC_{50} = 48.64 μg mL^{-1}). The antiproliferative activity was low against KB and HeLa cells (IC_{50} = 68.28 and 75.56 μg mL^{-1}, respectively).

Villarreal-Gómez et al. (2010) evaluated the anticancer activities of extracts from several seaweeds. *Centroceras clavulatum* showed anticancer activity with IC_{50} value of 6.492 μg mL^{-1} against HCT-116 colon cancer cells. Some red algae also showed interesting results on liver cancer (HepG2) cells. Among these, *Asparagopsis armata*, *Plocamium cartilagineum*, and *Sphaerococcus coronopifolius* had the highest effects on HepG2 cells. Several studies have reported the antimicrobial activity of *A. armata* against different types of microorganisms, showing that this alga produces compounds with pharmacological potential (Genovese et al. 2009, Paul et al. 2005, 2006, Zbakh et al. 2012, Alves et al. 2016); however, the cytotoxicity effects of these algae in cells lines were only reported by Zubia et al. (2009a). On the other hand, the activity demonstrated by *P. cartilagineum* may be due to the presence of halogenated monoterpenes, since these molecules were already isolated from this alga and revealed high activity on SW480 cells (Inés et al. 2004). Nevertheless, the best activity in both studies was exhibited by the dichloromethane extract of *S. coronopifolius* that exhibited the lowest IC_{50} on cell viability and cell proliferation studies. This potential is not surprising, since previous studies isolated diterpes with great interest from *S. coronopifolius* collected in the Mediterranean Sea (Piazza et al. 2011, Smyrniotopoulos et al. 2009). These molecules showed interesting results against several cells lines and were able to overcome the natural resistance of certain tumor cells to apoptosis (Smyrniotopoulos et al. 2008, 2010).

The potential of *S. coronopifolius* gets even more interesting when compared to the standard drugs, cisplatin and tamoxifen. In the anti-proliferative tests, the IC_{50} exhibited by the *S. coronopifolius* dichloromethane extract was similar to the IC_{50} exhibited by these antitumor drugs. Like this, in the cytotoxicity test, the dichloromethane extract of *S. coronopifolius* displayed an IC_{50} smaller than those drugs (Alves et al. 2016). On the other hand, the environmental conditions to which algae are subjected can result in the production of different molecules, and the possibility of the presence of new molecules involved in the cytotoxicity and anti-proliferative effects presented in this work must be considered. Moreover, several authors have associated the production of compounds with toxicity and anti-proliferative activities to temporal-space variations, depending on the community or the season, pointing to the important role of biotic and abiotic factors (Marti et al. 2004, Osman et al. 2010, Taskin et al. 2010).

Among all types of cancer, breast cancer is the leading cause of death among women. Consumption of food containing antioxidants was proven to be effective in reducing cancer incidence caused by oxidative damages. In a study carried out by Namvar et al. (2012), *Kappaphycus alvarezii* (formerly *Eucheuma cottonii*) samples were obtained from the north coast of Sabah in Malaysia. *K. alvarezii* polyphenol-rich extract was discovered to exhibit anti-proliferative and apoptotic effect on the estrogen-dependent MCF-7 and estrogen-independent MB-MDA-431 breast cancer cell lines. Besides, *K. alvarezii* polyphenol-rich extract was found to be more potent towards estrogen-dependent MCF-7, as compared to estrogen-independent MB-MDA-431. The mechanisms of the antitumor activity against MCF-7 breast cancer are via hormone modulation and apoptosis induction without cell cycle arrest. During hormonal regulation, the biosynthesis of estrogen in the cancer cells was downregulated, thus giving it its anti-estrogenic effect. Most importantly, *K. alvarezii* polyphenol-rich extract does not have any cytotoxic effect on normal Vero cells (Namvar et al. 2012).

At the level of the flora of the North Atlantic (French Coasts), crude red algal extracts of *Asparagopsis armata*, *Brongniartella byssoides*, and *Heterosiphonia plumosa* had strong cytotoxic activities against Daudi and Jurkat cells (see Table 6.1) (Zubia et al. 2009a).

In the studies by Costa et al. (2010), the viability of HeLa cells treated with sulfated polysaccharides for 72 hours was determined using a colorimetric MTT-based assay. The red algae *Gracilaria caudata* showed high antiproliferative activity, with about 40.0% cell proliferation inhibition at 2.0 mg mL^{-1}.

Methanolic extract (100 µg mL^{-1}) of red seaweed *Asparagopsis taxiformis* inhibits about 40% of cell proliferation (mammary carcinoma cell-EAT) in mouse (Vinayak et al. 2014). Another study showed that a methanolic extract of *Gloiopeltis furcata* markedly inhibited human hepatocellular carcinoma (HepG2) cell proliferation, and induced the G_2/M arrest of the cell cycle in a dose-dependent manner (from 10 to 500 µg mL^{-1}) (Bae and Choi 2007). In addition, methanolic extract of *Gracilaria corticata* was used against HepG2 and human breast adenocarcinoma (MCF-7) cells. The average inhibitory activity was 91% and 93%, respectively, using 500 µg mL^{-1} of extract (Narasimhan et al. 2013).

Mosaddegh et al. (2014) reported that the methanol extract of marine algae *Jania adhaerens* from the Chabahar coast of Oman Sea exhibited cytotoxic effects on MCF-7 and HT-29 cell lines, but the chloroform fraction demonstrated higher cytotoxicity than the methanol extract to HT-29, MCF7, HepG-2, A-549, and MDBK cell lines (see Table 6.1).

According to Gomes et al. (2015) the methanolic extracts of some red seaweed promoted a modest inhibition (10% to 20%) of the HeLa cell viability. The dependency of *Botryocladia occidentalis* extract on time and/or dose could not be identified clearly. However, *B. occidentalis* extract presented inhibitory activity of approximately 10% in 24 hours, that later tended to rise to nearly 20%. With respect to *Acanthophora spicifera* extract, a decreased viability of $\approx$ 20% was observed already in the lower concentration tested; however this activity did not increase with increasing concentration or time of exposure to the extract (see Table 6.1) (Gomes et al. 2015).

In the studies by Montalvão et al. (2016), *Hypnea musciformis* displayed a general cytotoxicity against all four cell lines, with inhibition of growth in the range of 60–96%, with concentration of 50 mg mL^{-1}. However, extracts of the red algae *Palisada perforata* (formerly *Laurencia papillosa*) displayed cytotoxic effects on the cancer cells (LNCa, PC-3, and MCF-7), but not on the non-tumorigenic cell line MCF-10, suggesting that these extracts might be interesting for further studies regarding their potential anticancer effects. Interestingly, *Hypnea musciformis* extract also showed this selective antiproliferative effect at the lower test concentration of 5 mg mL^{-1} (Montalvão et al. 2016).

CHAPTER 7

Antifungal Activity of Seaweeds and their Extracts

7.1 Introduction

Invasive fungal infections are a serious threat to human health and, although the current antifungal therapies have been significantly improved, the outcome is still far from satisfactory (Zhai and Lin 2011). On the one hand, the number of therapeutic options for the treatment of invasive fungal infections is quite limited when compared with those available to treat bacterial infections. In fact, two of the three classes of antifungal drugs (azoles and polyenes) that are currently used had been introduced in clinical practice by 1980, and the third class, the echinocandins, had been discovered more recently (Roemer and Krysan 2014). Additionally, proper and early diagnosis is quite challenging, conditioning the future treatment. The choice of an appropriate antifungal therapy for invasive candidiasis depends upon specific clinical circumstances that includes, among others, the condition of the patient, relevant comorbidities, severity of the illness, histories of recent azole exposure, and intolerance to antifungal agents (Pappas et al. 2009). Generally, amphotericin B-based preparations, the echinocandins antifungal agents, and the azole antifungal agents play a role in treatment of invasive candidiasis, while therapy for mucosal infections is dominated by the latter (Pappas et al. 2004). In turn, the antifungal compounds approved by the Food and Drug Administration (FDA) that have *in vitro*, *in vivo*, and clinical activity against *Aspergillus* species and are licensed for treatment of invasive aspergillosis are—the polyene amphotericin B deoxycolate (D-AMB) and its lipid formulations (AMB lipid complex [ABLC], L-AMB, and AMB colloidal dispersion [ABCD]); the azoles itraconazole, voriconazole, posaconazole; and the echinocandins caspofungin (Walsh et al. 2008). Although there are no standardized therapies for infections caused by dematiaceous fungi, the azoles voriconazole, posaconazole and itraconazole, and, in some cases, AMB, have demonstrated the most consistent *in vitro* activity against this group of fungi (Chowdhary et al. 2014).

On the other hand, the development of resistance to current antifungals is a real and worrying burden, being reported either in *Candida albicans* (Cannon et al. 2007) or in *Aspergilus fumigatus*. Azole resistance in *A. fumigatus* is an emerging problem (Chowdhary et al. 2013, Howard et al. 2009, Snelders et al. 2008, van der Linden et al. 2015), causing resistant invasive and noninvasive aspergillosis and, therefore, compromising severely its clinical use. The increasing need for novel antifungal agents is therefore evident (Silva 2015).

As both fungi and mammals are eukaryotic organisms, the antifungal drugs should exploit differences between their cells to kill only the fungi without damaging the mammalian host. Due to its specific composition, the fungal cell wall and the underlying plasma membrane are unique targets for the development of antifungal drugs (Tada et al. 2013).

Seaweeds are directly exposed in the various oceanic environment conditions and are supposed to be susceptible to ambient microorganisms. However, they have incredible survival ability because they possess an inherently available chemical defense mechanism. Accordingly, many bioactive compounds (e.g., terpenes, phlorotannins, etc.) could be present (Peng et al. 2015). Currently the search for new antifungal agents is a growing need, owing to the increment of fungal infections, but also because of the increase in resistance to antifungal agents (Richardson and Warnock 2012). Consequently, large attention has been given to natural products with antifungal properties.

7.2 Pathogenic Fungi

Fungal kingdom is constituted by a wide variety of eukaryotic organisms with a diverse range of forms and functions. Among other reasons, these organisms are grouped in a different kingdom due to the presence of chitin in their cell walls (Rai and Bridge 2009). Based on the characteristics of their morphologic, physiologic, and reproductive structures, fungi are classified into seven phyla—Microsporidia, Chytridiomycota, Blastocladiomycota, Neocallimastigomycota, Glomeromycota, Basidiomycota, and Ascomycota (James et al. 2006, Hibbett et al. 2007, Margulis and Chapman 2009). Of these, Ascomycota constitutes the largest, most diverse, and ubiquitous phylum, occurring in a wide variety of ecosystems. Species belonging to this phylum possess a specialized sac-like structure (ascus) in which meiotic spores (ascospores) are produced (Pereira 2011b). However, a great number of Ascomycota occur as a single-celled yeast that reproduces by budding or binary fission (Schoch et al. 2009). Among the wide variety of species belonging to this phylum, some filamentous fungi and yeasts have a special importance from the health care point of view, since these can cause opportunistic infections and skin and mucosa diseases, respectively (Richardson and Warnock 2012).

The frequency of invasive mycoses due to opportunistic fungal pathogens has increased significantly over the past two decades (Rees et al. 1998, Trick et al. 2002, Ostrosky-Zeichner et al. 2003, Hajjeh et al. 2004, Pfaller and Diekema 2004, Walsh et al. 2004). Serious life-threatening infections are being reported with an ever-increasing array of pathogens, including the well-known opportunists *Candida albicans*, *Cryptococcus neoformans*, and *Aspergillus fumigatus* (Mirza et al. 2003, Hajjeh et al. 2004, Pfaller and Diekema 2004). Although *A. fumigatus* heads the list of these opportunistic molds (Denning et al. 1998, Lin et al. 2001, Diekema et al. 2003), infections due to less common but antifungal-resistant species such as *Aspergillus terreus* and dematiaceous filamentous fungi (e.g., *Bipolaris*, *Cladophialophora*, and *Alternaria*) are being reported with greater frequency (Iwen et al. 1998, Flemming et al. 2002, Baddley et al. 2003, Walsh et al. 2004).

7.2.1 Filamentous fungi

Alternaria alternata

The mold *Alternaria* is a well-recognized allergy-causing fungus. *Alternaria* spores can be detected from spring through late fall in most temperate areas, and can reach levels of thousands of spores per cubic meter of air. *Alternaria* spores can be at their highest concentrations during dry, windy conditions that are ideal for the spores to become airborne. *Alternaria* is currently comprised of about 40–50 species. It is commonly isolated from plants, soil, food, and indoor air. One of the species, *A. alternata*, has been isolated from numerous kinds of organic materials in damp situations, including textiles, stored food, canvas, cardboard and paper, electric cables, polyurethane, jet fuel, sewage and effluents; *A. alternata* is recognized as an important allergen with airborne spores and mycelial fragments being responsible for the allergic symptoms in individuals with rhinitis or bronchial asthma. *Alternaria* sensitivity can also lead to severe and potentially fatal asthma. Studies have shown that up to 70% of mold-allergic patients have skin test reactivity to *Alternaria*. It has also been shown that prolonged heavy exposure to *A. alternata* spores and mycelial fragments mimics that of other allergens, such as cat dander and dust mites. It has also been recorded as an opportunistic pathogen causing skin diseases particularly in immunocompromised patients, such as bone marrow transplant patients (Kung'u 2016).

Alternaria brassicola

This fungus can affect host species at all stages of growth, including seeds. In seedlings, symptoms include dark stem lesions immediately after germination that can result in damping-off, or stunted seedlings.

A. brassicola produces black stripes or dark brown, sharp-edged lesions on the hypocotyl of the seedling. It grows in the vascular system and rapidly infects the entire seedling. This pathogen affects most cruciferous crops, including broccoli and cauliflower (*Brassica oleracea* var. *botrytis*), field mustard and turnip (*B. rapa*, formerly *B. campestris*), leaf or Chinese mustard (*B. juncea*), Chinese or celery cabbage (*B. pekinensis*), cabbage (*B. oleracea* var. *capitata*), rape (*B. campestris*), and radish (*Raphanus sativus*) (Valkonen and Koponen 1990).

Alternaria dauci

It is a plant pathogen; the English name of the disease it incites is "carrot leaf blight". "*Alternaria* Leaf Blight" is a foliar disease of carrots caused by the fungus *A. dauci*. *A. dauci* is included in the porri species group of *Alternaria*, which is classified for having large conidium and a long, slender filiform beak (Farrar et al. 2004).

Alternaria infectoria

Alternaria species are increasingly found as etiologic agents of human disease, due to the growing number of immunocompromised patients (de Hoog et al. 2001). *A. infectoria* is a rare opportunistic agent of phaeohyphomycosis (Dubois et al. 2005, Gilaberte et al. 2005), a human infection that usually affects the sub-cutaneous tissue, in particular, the nervous system (Li and de Hoog 2009).

Alternaria longipes

It is a plant pathogen; the English name of the disease it incites is "brown spot". The first symptoms of brown spot disease are small, circular, dark-brown lesions on the lower leaves of field plants, and sometimes on old seedlings or the senescing leaves of transplants. The lesions enlarge as the leaves mature and are surrounded by an irregular, yellow halo. The halo is caused by the secretion of toxins. Often there are numerous lesions, about 25 mm in diameter, which may coalesce and affect more than 80% of the leaf area. They are also common allergens in humans, growing indoors and causing hay fever or hypersensitivity reactions (Nowicki et al. 2012).

Aspergillus flavus

This fungus has a worldwide distribution and normally occurs as a saprophyte in soil and on many kinds of decaying organic matter. However, it is also a recognized pathogen of humans and animals. It is a causative agent of otitis, keratitis, acute and chronic invasive sinusitis, and pulmonary and systemic infections in immunocompromised patients. *A. flavus* is second only to *A. fumigatus* as the cause of human invasive aspergillosis (Hedayati et al. 2007). Colonies of this species are granular, flat, often with radial grooves, yellow at first but quickly becoming bright to dark yellow-green with age. Conidial heads are typically radiate, later splitting to form loose columns (mostly 300–400 μm in diameter), biseriate but having some heads with phialides borne directly on the vesicle (uniseriate). Conidiophore stipes are hyaline and coarsely roughened, often more noticeable near the vesicle. Conidia are globose to sub-globose (3–6 μm in diameter), pale green, and conspicuously echinulate. Some strains produce brownish sclerotia (Ellis 2016d).

Aspergillus fumigatus

A. fumigatus is one of the most ubiquitous of the airborne saprophytic fungi (Vanden Bossche et al. 1988, Pitt 1994, Haines 1995). Until recently, *A. fumigatus* was viewed as a weak pathogen responsible for allergic forms of the disease, such as farmer's lung, a clinical condition observed among individuals exposed repeatedly to conidia, or aspergilloma, an overgrowth of the fungus on the surface of preexisting cavities in the lungs of patients treated successfully for tuberculosis (Pennington 1994, Kwon-Chung et al. 1992). Over the past 10 years, *A. fumigatus* has become the most prevalent airborne fungal pathogen, causing severe and usually fatal invasive infections in immunocompromised hosts in developed countries (Bodey and Vartivarian 1989, Andriole 1993, Beck-Sagué and Jarvis 1993, Dixon et al. 1996, Groll et al. 1996, Denning 1998).

Aspergillus niger

A. niger is the most abundant species of Aspergillus in nature, as it can grow on a large variety of substances. *A. niger* can even grow in environments with very little nutrients available. In houses, it is often found growing on damp walls. Of the *Aspergillus* species, *A. niger* infects humans the third most often. A fungal ball in the lungs is eventually created by *A. niger* after it infects a person's lungs and begins to grow. The health effects of *A. niger* include hearing problems and even hearing loss. The group of diseases caused by *Aspergillus* exposure is known as Aspergillosis. The main diseases of Aspergillosis are—allergic bronchopulmonary aspergillosis, acute invasive aspergillosis, and disseminated invasive aspergillosis (Blackmold 2016).

Aspergillus ochraceus

This fungus is known to produce the toxin ochratoxin A, one of the most abundant food-contaminating mycotoxins, and citrinin. It also produces the dihydroisocoumarin mellein. It is a filamentous fungus in nature and has characteristic biseriate conidiophores. Traditionally a soil fungus, it has now begun to adapt to varied ecological niches, like agricultural commodities, farmed animals, and marine species. In humans and animals, the consumption of this fungus produces chronic neurotoxic, immunosuppressive, genotoxic, carcinogenic, and teratogenic effects. Its airborne spores are one of the potential causes of asthma in children and lung diseases in humans. The pig and chicken populations in the farms are the most affected by this fungus and its mycotoxins (Bennet and Klich 2003).

Aspergillus terreus

A. terreus complex includes species with biseriate, columnar conidial heads in shades of buff to brown. Colonies are typically suede-like and cinnamon-buff to sand-brown in color with a yellow to deep dirty-brown reverse. Conidial heads are compact, columnar (up to 30–50 μm in diameter) and biseriate. Conidiophore stipes are hyaline and smooth-walled. Conidia are globose to ellipsoidal (1.5–2.5 μm in diameter), hyaline to slightly yellow and smooth-walled. *A. terreus* occurs commonly in soil and is occasionally reported as a pathogen of humans and animals (Raper and Fennell 1965).

Botryotrichum piluliferum

With expanding colonies, they become brownish due to pigmented hyphae and erect, verrucose hairs; "aleurioconidia" numerous in cultures, often in clusters, spherical or nearly so, thick-walled, colorless, 13–18 μm, formed on short hyphal branches; phialoconidia obovate, colorless, 3–4 × 2–2.5 μm, formed basipetally on clustered ampulliform cells, aggregated in droplets (Abdel-Azeem 2017).

Botrytis cinerea

It is the agent of grey mold in cyclamen, and attacks plants at all stages of their development. Climatic conditions for cyclamen growing 15°C to 20°C are ideal conditions for the spread of this fungus. The grey mold which afflicts ornamental plants is everywhere, and can attack cyclamen along with many other species. It is particularly dangerous in glasshouses, for it is here that it finds environmental conditions which favor its development. *B. cinerea* is an airborne fungus. It keeps quite well in the soil in the form of small hard nodules, dark in color, made of intertwined hyphae—these nodules are known as sclerotia or "resting bodies". In this form, the life functions of the fungus are in slow motion and it can wait for a number of years for favorable conditions to return (Cyclamen 2016).

Cladosporium cladosporioides

Cladosporium species are ubiquitous worldwide, and commonly isolated from soil and organic matter. They represent the most frequently isolated airborne fungi. The genus has undergone several revisions. The well-known thermotolerant true human-pathogenic species, is one of the most common fungi in outdoor air, where its spores are important in seasonal allergic diseases. While this species rarely causes invasive disease in animals, it is an important agent of plant disease, attacking both the leaves and fruits of many plants. Colonies are slow growing, mostly olivaceous-brown to blackish-brown but also sometimes grey, buff or brown, suede-like to floccose, often becoming powdery due to the production of abundant conidia (Ellis 2016m).

Cladosporium sphaerospermum

It is common on wet building material, such as gypsum board, ceiling board, windowsills, insulation material, acrylic and oil painted walls, painted wood, and wallpaper. *C. sphaerospermum* may cause allergies in sensitive individuals (Kung'u 2016b).

Colletotrichum orbiculare (formerly *Colletotrichum lagenarium*)

Colletotrichum spp. causes anthracnose diseases in many economically important crops worldwide. The classification of species within this genus traditionally has been based on conidial shape and size, presence of sclerotia, appressoria production, and, often, on host origin. *C. orbiculare* causes anthracnose and fruit rot. Leaf spots are often large (> 10 mm in diameter) and pale brown to grey with distinct margins. Lesions on fruit appear as brownish discolorations, often growing to 20–30 mm in diameter, which become sunken, wrinkled, and dark, with concentric rings of conidiomata (Liu et al. 2007).

Colletotrichum capsici

It is a plant pathogen which causes leaf blight on *Chlorophytum borivilianum* (safed musli), basil, chickpea, and pepper, as well as dieback in pigeon pea and anthracnose in poinsettia (Nayaka et al. 2009).

Colletotrichum gloeosporioides

It is known to infect a wide variety of hosts. However, this summary deals specifically with its effects on papaya. Information about the host range of this fungus may be found in other summaries discussing this organism. This fungus produces hyaline, one-celled, ovoid to oblong, slightly curved or dumbbell shaped conidia, 10–15 μm in length and 5–7 μm in width. Masses of conidia appear pink or salmon-colored (Dickman 1993).

Colletotrichum lindemuthianum

It is a fungus which causes anthracnose, or black spot disease, of the common bean plant (*Phaseolus vulgaris*). It is considered a hemibiotrophic pathogen because it spends part of its infection cycle as a biotroph, living off the host but not harming it, and the other part as a necrotrophic, killing and obtaining nutrients from the host tissues. Bean anthracnose, caused by the fungus *C. lindemuthianum*, is a major disease of beans (*P. vulgaris*), causing serious crop loss in many parts of the world. Under moist conditions, small, pink masses of spores are produced in the lesions. Spores produced on cotyledon and stem lesions may spread to the leaves. Symptoms generally occur on the underside of the leaves as linear, dark brick-red to black lesions on the leaf veins. As the disease progresses, the discoloration appears on the upper leaf surface. Leaf symptoms often are not obvious and may be overlooked when examining bean fields (Dillard 1988).

Colletotrichum musae

It is a plant pathogen primarily affecting the genus Musa, which includes bananas and plantains. It is best known as a cause of anthracnose (the black and brown spots), indicating ripeness in bananas (Zakaria et al. 2009).

Cunninghamella bertholletiae

Synonymy: *Cunninghamella elegans*

It is characterized from white to grey, rapidly growing colonies, producing erect, straight, branching sporangiophores. These sporangiophores end in globose or pyriform-shaped vesicles from which several one-celled, globose to ovoid, echinulate or smooth-walled sporangiola develop on swollen denticles. Chlamydospores and zygospores may also be present. The genus now contains seven species with

C. bertholletiae—the only known species to cause disease in humans and animals, often in association with trauma and immunosuppression (Ellis 2016t).

Drechslera oryzae

This fungus produces oval, eye-shaped spots with a conspicuous dark-brown spot in the center and light brown margin in rice (brown leaf spot).

Epidermophyton floccosum

It is an anthropophilic dermatophyte with a worldwide distribution which often causes tinea pedis, tinea cruris, tinea corporis, and onychomycosis. It is not known to invade hair *in vivo* and no specific growth requirements have been reported. Colonies are usually slow-growing, greenish-brown or khaki-colored with a suede-like surface, raised and folded in the center, with a flat periphery and submerged fringe of growth. Older cultures may develop white pleomorphic tufts of mycelium. A deep yellowish-brown reverse pigment is usually present. Microscopic morphology shows characteristic smooth, thin-walled macroconidia which are often produced in clusters growing directly from the hyphae (Ellis 2016g).

Epicoccum nigrum

Synonym: *Epicoccum purpurascens*

It is a cosmopolitan saprophyte of worldwide distribution which is occasionally isolated as a contaminant from clinical specimens like skin. Colonies are fast growing, suede-like to downy, with a strong yellow to orange-brown diffusible pigment. When sporulating, numerous black sporodochia (aggregates of conidiophores) are visible. Conidia are formed singly on densely compacted, non-specialised, determinant, slightly pigmented conidiophores. Conidia are globose to pyriform, mostly 15–25 μm in diameter with a funnel-shaped base and broad attachment scar, often seceding with a protuberant basal cell, i.e., aleuric or rhexolytic dehiscence of conidia. Conidia become multicellular (dictyoconidia), darkly pigmented, and have a verrucose external surface (Ellis 2016l).

Erysiphe polygonion

Anamorph: *Oidium anacardii*, *O. balsamii*, *O. erysiphoides*, and *O. mangiferae*

It causes the powdery mildew. Conidia are spread mainly by wind. They require 90% to 100% relative humidity and temperature between 26°C to 28°C for germination. High humid conditions, without precipitation or precipitation with temperatures between 23ºC to 32ºC, favor the development of this pathogen. This fungus has a white powdery appearance. It appears on leaves in the summer time. Infection normally begins on older leaves, typically close to the junction between the lamina and petiole, and it develops on both ab- and adaxial surfaces (Reddy et al. 1994).

Fusarium dimerum complex

The *F. dimerum* complex contains 12 phylogenetically distinct species, including *F. delphinoides*, *F. penzigii*, and *F. dimerum*. These are regarded as cosmopolitan saprotrophs in soil and on plant materials. They have also been isolated from human corneal ulcers after trauma and from disseminated or localized infections in immunocompromised patients. Colonies grow slowly; the surface is usually orange to deep apricot due to confluent conidial slime; aerial mycelium is sometimes floccose and whitish (Guarro 2013).

Fusarium graminearum

Teleomorph: *Gibberella zeae*

It is of worldwide importance on small grain cereals and corn, occurring under a wide range of soil and environmental conditions. *F. graminearum* has become one of the most significant cereal diseases. The fungus can produce several mycotoxins, including deoxynivalenol (DON) and zearalenone (Turkington et al. 2014).

Fusarium incarnatum

This fungus occasionally causes infections in humans and animals. Colonies grow rapidly; aerial mycelium floccose, at first whitish, they later become avellaneous to buff-brown; the reverse is pale, and then become peach-colored. Conidiophores scattered in the aerial mycelium, loosely-branched; polyblastic conidiogenous cells abundant (Guarro 2013).

Fusarium oxysporum f. *lycopersici*

F. oxysporum f. *lycopersici* is a soil-borne plant pathogen in the class Sordariomycetes, and causes *Fusarium* wilt specifically in tomato. This disease was first described in England in 1895. It is of worldwide importance, where at least 32 countries had reported the disease, which is particularly severe in countries with warm climate. However, the development and use of resistant cultivars have nearly eliminated the concern over this disease (Wong 2003).

Fusarium sambucinum var. *sambucinum*

Synonym: *Fusarium roseum*

Most *Fusarium* species are soil fungi and have a worldwide distribution. Some are plant pathogens, causing root and stem rot, vascular wilt, or fruit rot. Several species have emerged as important opportunistic pathogens in humans causing hyalohyphomycosis (especially in burn victims and bone marrow transplant patients), mycotic keratitis, and onychomycosis (Guarro 2013). Other species cause storage rot and are important mycotoxin producers. Colonies are usually fast-growing, pink bright-colored with a cottony aerial mycelium (Pereira 2011b).

Fusarium solani

The *F. solani* complex contains at least 60 species and accounts for about 50% of human infections caused by fusaria. All are ubiquitous soil-borne pathogens responsible for vascular wilts, rots, and damping-off diseases of a broad range of plants. Colonies grow rapidly, 4.5 cm in four days, aerial mycelium white to cream, and become bluish-brown when sporodochia are present. Macroconidia are formed after 4–7 days from short multiple branched conidiophores which may form sporodochia (Ellis 2016e).

Fusarium udum

It's a soil-borne disease, but is also transmitted on roots. Cultures pale sulphureous to rose buff, becoming salmon-orange or somewhat purple; growth rate medium, covering the surface of a petri dish after 10–14 days at 25°C; aerial mycelium felted. Conidia initially produced on simple or verticillately branched conidiophores and later from pionnotal or small sporodochia when they form a salmon-colored mass; variable in size, with a strongly curved apex; there is no clear distinction between microconidia and macroconidia (Bensch 2016).

Fusarium verticillioides

Synonyms: *Fusarium moniliforme*, *Gibberella moniliformis*

It is a fungal plant pathogen, and causes a disease in rice called bakanae, which is Japanese and means "foolish seedlings". The afflicted plants are at best infertile with empty panicles, producing no edible grains; at worst, they are incapable of supporting their own weight, topple over, and die (hence "foolish seedling"). In humans with normal immune systems, fusarial infections may occur in the nails and in the eye. In humans whose immune systems are weakened in a particular way, aggressive fusarial infections penetrating the entire body and bloodstream may be caused by members of the Fusarium genera. The most significant economic impact of *F. verticillioides* is its ability to produce fumonisin mycotoxins. Various diseases caused by fumonisins have been reported in animals, such as liver and kidney cancer (Ensembl Fungi 2016).

Ganoderma sp.

Basal Stem Rot (BSR) disease caused by *Ganoderma boninense* is the most destructive disease in oil palm, especially in Indonesia and Malaysia. The available control measures for BSR disease, such as cultural practices and mechanical and chemical treatment have not proved satisfactory due to the fact that *Ganoderma* has various resting stages, such as melanised mycelium, basidiospores, and pseudosclerotia (Susanto et al. 2005).

Geotrichum spp.

The genus *Geotrichum* and related species have undergone extensive taxonomic revision. The three species of prime interest to medical mycology are *Geotrichum candidum* (formerly *Galactomyces candidus*), *Magnusiomyces capitatus* (previously known as *Geotrichum capitatum*), and *Saprochaete clavata* (previously known as *Geotrichum clavatum*). Colonies are fast growing, flat, white to cream, dry and finely suede-like with no reverse pigment. Hyphae are hyaline, septate, branched and break up into chains of hyaline, smooth, one-celled, sub-globose to cylindrical arthroconidia (Ellis 2016p).

Gibberella spp.

It is a fungal plant pathogen, and causes bakanae disease in rice seedlings by overloading them with the phytohormone gibberellin as its own metabolic product. In 1926, Japanese scientists observed that rice plants infected with *Gibberella* had abnormally long stems ("foolish seedling disease"). Gibberellin is a plant hormone that promotes cell elongation, flower formation, and seedling growth (Yazdani et al. 2011).

Humicola insolens

Soft-rot fungus

Lindra thalassiae

It is a scolecosporous pyrenomycete infecting leaves of turtle grass (*Thalassia testudinum*). *L. thalassiae* shows a euryhaline growth response in the mycelial stage, but requires higher concentrations of seawater for maximal reproduction (Meyers and Hoyo 1966).

Macrophomina phaseolina

This fungus has a wide host range and is responsible for causing losses on more than 500 cultivated and wild plant species. It is a fungus that is a plant pathogen that causes charcoal rot on many plant species, including *Zea mays* and *Pinus* spp. This fungus causes seedling blight, root rot, and charcoal rot of more than 500 crop and non-crop species (Smith and Carvil 1977). It has a very wide distribution covering most of the tropics and subtropics, extending well into temperate zones, having occurrence as far north as the United Kingdom and as far south as New Zealand (Songa 1995). It is an important pathogen of crops, particularly where high temperatures and water stress occurs during the growing season. The fungus is reported to be soil, seed and stubble borne. The evidence suggests that it is primarily a root inhibiting fungus and produces tuber or cushion shaped black sclerotia 1–8 mm in diameter. These sclerotia serve as a primary means of survival (Khan 2007).

Microsporum canis

It is a zoophilic dermatophyte of worldwide distribution and is a frequent cause of ringworm in humans, especially children. Invades hair, skin and rarely nails. Cats and dogs are the main sources of infection. Invaded hairs show an ectothrix infection and fluoresce a bright greenish-yellow under Wood's ultra-violet light. Colonies are flat, spreading, white to cream-colored, with a dense cottony surface which may show some radial grooves. Colonies usually have a bright golden yellow to brownish yellow reverse pigment, but non-pigmented strains may also occur (Ellis 2016c).

Microsporum gypseum

Teleomorphs: *Arthroderma gypsea* and *Arthroderma incurvatum*

It is a geophilic fungus with a worldwide distribution which may cause infections in animals and humans (a Mycosis called Dermatophytosis), particularly children and rural workers during warm humid weather. Usually produces a single inflammatory skin or scalp lesion. colonies are usually flat, spreading, suede-like to granular, with a deep cream to tawny-buff to pale cinnamon colored red surface (Ellis 2016r). *M. gypseum* is a filamentous fungus dermatophyte. Capillary parasitism is of the ectotrix type (Carvalho and Lopes 2016a).

Mucor spp.

The genus *Mucor* contains about 50 recognized *taxa*, many of which have widespread occurrence and are of considerable economic importance. However, only a few thermotolerant species are of medical importance and human infections are only rarely reported. Colonies are very fast growing, cottony to fluffy, white to yellow, becoming dark-grey, with the development of sporangia. Sporangiophores are erect, simple or branched, forming large (60–300 µm in diameter), terminal, globose to spherical, multispored sporangia (Mendoza et al. 2015, Pereira 2011b).

Mucor hiemalis

Colonies are very fast growing, cottony to fluffy, white to yellow, becoming dark-grey, with the development of sporangia. Sporangiophores are erect, simple or branched, forming large (60–300 µm in diameter), terminal, globose to spherical, multispored sporangia, without apophyses and with well-developed subtending columella. They have been reported as infectious agents, although their inability to grow at temperatures above 32ºC raises doubt to their validity as human pathogens, and their pathogenic role may be limited to cutaneous infections (De Hoog et al. 2001, Pereira 2011b).

Mucor indicus

This species differs from others of *Mucor* by its characteristic deep-yellow colony color, growth at over 40ºC, assimilating ethanol, but not nitrate, and thiamine dependence. Colonies are characteristically deep-yellow, aromatic, and have a maximum growth temperature of 42ºC. Sporangiophores are hyaline to yellowish, erect or rarely circinate and repeatedly sympodially branched, with long branches (De Hoog et al. 2001).

Penicillium spp.

Penicillium is a very large and ubiquitous genus which currently contains 354 accepted species. Many species are common contaminants on various substrates and are known as potential mycotoxin producers. Correct identification is therefore important when studying possible *Penicillium* contamination of food. Human pathogenic species are rare, however opportunistic infections leading to mycotic keratitis, otomycosis, and endocarditis. Colonies are usually fast growing, in shades of green, sometimes white, mostly consisting of a dense felt of conidiophores. Microscopically, chains of single-celled conidia are produced in basipetal succession from a specialized conidiogenous cell called a phialide (Ellis 2016i).

Penicillium citrinum

It is an anamorph, mesophilic fungus species of the genus of *Penicillium* which produces tanzawaic acid, mevastatin, quinocitrinine A and B, and nephrotoxic citrinin. *P. citrinum* is often found on moldy citrus fruits and occasionally occurs in tropical spices and cereals. Due to its mesophilic character, *P. citrinum* occurs worldwide (Houbraken et al. 2010).

Penicilium expansum

It is a fungus with mono-verticillate branched septate conidiophores with sporulating cells (phialides) that forms a circular colony, white-green or grey, with smooth edges (entire margin). It is the disease agent in vines that may impart an off-flavor to the wine. Small amounts of infected berries have been reported to contribute to a moldy taste in wine (Delage et al. 2003).

Penicillium funiculosum

This fungus is a plant pathogen infecting pineapples. *P. funiculosum* occurs in tropical areas as well as temperate ones. It is also used as a source of the enzymes xylanase and beta-glucanase, which are a non-starch polysaccharide hydrolyzing enzymes. In culture, it presents white-colored colonies, sometimes of a pale green-grey, slightly slimy and shiny with erected funiculi at the center, filamentous and lighter towards the edges (Pitt and Hocking 1999).

Penicillium simplicissimum

Synonym: *Penicillium janthinellum*

It is an anamorph species of the genus of *Penicillium* which can promote plant growth; this species occurs on food and its primary habitat is in decaying vegetation (Hossain et al. 2007).

Pythium aphanidermatum

It is a fungus-like pathogen which has never been reported as a cause of human infection. However, Calvano et al. (2011) reported a case of *P. aphanidermatum* invasive wound infection in a 21-year-old male injured during combat operations in Afghanistan. It is a soil-borne plant pathogen, and their cell walls are made of cellulose instead of chitin, they are diploid in their vegetative state, and they contain coenocytic hyphae (Calvano et al. 2011, Parker 2017). *P. aphanidermatum* has a wide host range, and can have an economic impact on the cultivation of soybeans, beets, peppers, chrysanthemum, cucurbits, cotton, and turf-grasses; however, because *P. aphanidermatum* requires warmer temperatures, it is often seen in greenhouses and has a large impact in poinsettia production. It is a major cause of root rot in papaya production in subtropical areas (Calvano et al. 2011, Parker 2017).

Phytophthora infestans

This fungus causes serious losses of potato crops worldwide and is probably the most important pathogen of potato and tomato today. The disease, late blight, is famous for the destruction of the potato crops in Ireland in the 1840s and the resulting famine and death of over a million people. Today, epidemics recur in disease conducive environments.

P. infestans produces microscopic, asexual spores called sporangia. These sporangia are hyaline (clear), lemon-shaped and 20–40 µm long. When placed in water or in very high relative humidity, the cytoplasm in the sporangia divide and many swimming zoospores emerge from each sporangium. Sporangia are formed on specialized branches called sporangiophores. The branched sporangiophore, with swellings at the points where sporangia were attached are distinctive for *P. infestans* and useful for identification of this pathogen. In the absence of sufficient water or with temperatures above 24°C, no zoospores form. However, sporangia germinate by producing germ tubes that penetrate the host (Uchida 2016).

Pseudallescheria boydii

Typically found in stagnant and polluted water, it has been implicated in the infection of immunocompromised and near-drowned pneumonia patients. Its asexual (anamorphic) form is *Scedosporium apiospermum*. Treatment of infections with *P. boydii* is complicated by its resistance to many of the standard antifungal agents normally used to treat infections by filamentous fungi (Wiederhold and Lewis 2009).

Pyricularia oryzae

Teleomorph: *Magnaporthe grisea*

It is also known as rice blast fungus, rice rotten neck, rice seedling blight, blast of rice, oval leaf spot of graminea, pitting disease, ryegrass blast, and Johnson spot. It is a plant-pathogenic fungus that causes a serious disease affecting rice (Talbot 2003).

Rhizopus spp.

The genus *Rhizopus* is characterized by the presence of stolons and pigmented rhizoids, the formation of sporangiophores, singly or in groups from nodes directly above the rhizoids, and apophysate, columellate, multispored, generally globose sporangia. After spore release the apophyses and columella often collapse to form an umbrella-like structure (Pereira 2011b). Sporangiospores are globose to ovoid, one-celled, hyaline to brown, and striate in many species (Ellis 2016j).

Saprolegnia parasitica

Water moulds (oomycetes) of the order Saprolegniales, such as *Saprolegnia* and *Aphanomyces* species, are responsible for devastating infections on fish in aquaculture, fish farms, and hobby fish tanks. Members of the genus *Saprolegnia* cause Saprolegniosis, a disease that is characterised by visible white or grey patches of filamentous mycelium on the body or fins of freshwater fish (van West 2006).

Trichoderma asperellum

Conidiophores are highly branched and thus difficult to define or measure, loosely or compactly tufted, often formed in distinct concentric rings or borne along the scant aerial hyphae. Main branches of the conidiophores produce lateral side branches that may be paired or not, the longest branches distant from the tip and often phialides arising directly from the main axis near the tip. The branches may re-branch, with the secondary branches often paired and longest secondary branches being closest to the main axis. The typical *Trichoderma* conidiophore, with paired branches assumes a pyramidal aspect (Samuels 2006).

Trichoderma reesei

Anamorph: *Hypocrea jecorina*

It is a mesophilic and filamentous fungus. *T. reesei* has the capacity to secrete large amounts of cellulolytic enzymes (cellulases and hemicellulases). Microbial cellulases have industrial application in the conversion of cellulose, a major component of plant biomass, into glucose (Kumar et al. 2008).

Trichoderma viride

It is an antagonistic fungus which prevents the crops from diseases such as root rots, wilts, brown rot, damping off, charcoal rot and other soil-borne diseases in crops. It is suitable for sugarcane, pulses, oilseeds, cotton, vegetables, banana, coconut, oil palm, chilies, lime, coffee and tea, areca nut and rubber, flower crops, and spices. *T. viride* is a mold which produces spores asexually, by mitosis. It is the anamorph of *Hypocrea rufa*, its teleomorph, which is the sexual reproductive stage of the fungus and produces a typical fungal fruiting body. The mycelium of *T. viride* can produce a variety of enzymes, including cellulases and chitinases, which can degrade cellulose and chitin, respectively (Volk 2010b).

Trichophyton mentagrophytes

It is a zoophilic fungus with a worldwide distribution and a wide range of animal hosts including mice, guinea-pigs, kangaroos, cats, horses, sheep, and rabbits. Produces inflammatory skin or scalp lesions in humans, particularly in rural workers. Kerion of the scalp and beard may occur. Invaded hairs show an ectothrix infection but do not fluoresce under Wood's ultra-violet light. Distribution is worldwide (Ellis 2016b). Colonies are generally flat, white to cream in color, with a powdery to granular surface. Some cultures show central folding or develop raised central tufts or pleomorphic suede-like to downy areas. Reverse pigmentation is usually a yellow-brown to reddish-brown color. Numerous single-celled microconidia are formed, often in dense clusters. Microconidia are hyaline, smooth-walled, and are predominantly spherical to sub-spherical in shape (Ellis 2016b).

Trichophyton rubrum

It is an anthropophilic fungus that has become the most widely distributed dermatophyte of humans (Ellis 2016h). It frequently causes chronic infections of skin, nails and rarely scalp. Granulomatous lesions may sometimes occur. Infected hairs do not fluoresce under Wood's ultraviolet light, and microscopically may show endothrix or ectothrix type of invasion. Morphologically *T. rubrum* exhibits a spectrum of overlapping characters; for example culture surface texture may vary from downy to suede-like; culture surface pigmentation may vary from white to cream to deep red; culture reverse pigmentation may vary from colorless to yellowish to yellow-brown to wine red; numbers of microconidia range from none to scanty to many; shape of microconidia vary from slender clavate to pyriform; numbers of macroconidia range from none to scanty to many, and may or may not have terminal projections (Ellis 2016h).

Trichophyton schoenleinii

It is an anthropophilic fungus causing favus in humans. Favus is a chronic, scarring form of tinea capitis characterized by saucer-shaped crusted lesions or scutula and permanent hair loss. Invaded hairs remain intact and fluoresce a pale greenish-yellow under Wood's ultra-violet light. Favus was once common in Eurasia and North Africa, however its incidence is now in decline. Colonies are slow growing, waxy or suede-like with a deeply folded honey-comb-like thallus and some subsurface growth. The thallus is cream-colored to yellow to orange brown. Cultures are difficult to maintain in their typical convoluted form, and rapidly become flat and downy. No reverse pigmentation is present. No macroconidia and microconidia are seen in routine cultures, and numerous chlamydospores may be present in older cultures (Ellis 2016j).

Trichophyton simii

This fungus is characterized morphologically by the development of both smooth-walled macro- and microconidia. Macroconidia are mostly borne laterally directly on the hyphae or on short pedicels, and are thin- or thick-walled, clavate to fusiform, and range from 4–8 × 8–50 μm in size. Macroconidia are few or absent in many species. Microconidia are spherical, pyriform to clavate, or of irregular shape and range from 2–3 × 2–4 μm in size. This fungus can infect human skin, hair, and nails (Ellis 2016k).

Trichophyton tonsurans

It is an anthropophilic fungus with a worldwide distribution which causes inflammatory or chronic non-inflammatory finely scaling lesions of skin, nails, and scalp. It is a common cause of tinea capitis in Australian Aborigines and African Americans. Invaded hairs show an endothrix infection and do not fluoresce under Wood's ultra-violet light. Colonies show considerable variation in texture and color. They may be suede-like to powdery, flat with a raised center or folded, often with radial grooves. The color may vary from pale-buff to yellow (Ellis 2016n).

Trichophyton verrucosum

It is a zoophilic fungus causing ringworm in cattle. Infections in humans result from direct contact with cattle or infected fomites, and are usually highly inflammatory involving the scalp, beard, or exposed areas of the body. Invaded hairs show an ectothrix infection, and fluorescence under Wood's ultra-violet light has been noted in cattle but not in humans. Colonies are slow growing, small, button or disc-shaped, white to cream-colored, with a suede-like to velvety surface, a raised center, and flat periphery with some submerged growth. Reverse pigment may vary from non-pigmented to yellow. Broad, irregular hyphae with many terminal and intercalary chlamydospores. Chlamydospores are often in chains (Ellis 2016v).

Trichophyton violaceum

Synonymy: *Trichophyton yaoundei*

It is an anthropophilic fungus causing inflammatory or chronic non-inflammatory finely scaling lesions of skin, nails, beard, and scalp, producing the so-called "black dot" tinea capitis. Distribution is worldwide, particularly in the near East, Eastern Europe, Russia, and North Africa. Invaded hairs show an endothrix infection, and do not fluoresce under Wood's ultra-violet light. Colonies are very slow growing, glabrous or waxy, heaped and folded, and deep violet in color. Cultures often become pleomorphic, forming white sectors. Occasional non-pigmented strains may occur. Hyphae are relatively broad, tortuous, much branched, and distorted. Young hyphae usually stain well in lactophenol cotton blue, whereas older hyphae stain poorly, and show small central fat globules and granules. Typically, no conidia are present, although occasional pyriform microconidia have been observed on enriched media. Numerous chlamydospores are usually present, especially in older cultures (Ellis 2016u).

Thanatephorus cucumeris

The most widely recognized species of this genus was originally described by Julius Kühn in a potato in 1858; it is a basidiomycete fungus that does not produce any asexual spores (called conidia) and only occasionally will the fungus produce sexual spores (basidiospores). In nature, this fungus reproduces asexually and exists primarily as vegetative mycelium and/or sclerotia (Ceresini 1999).

Ulocladium spp.

It resembles to the genus *Alternaria*, in which it has already included species of this genus contain both plant pathogens and food spoilage agents; other species contain enzymes that are biological control agents; some species of the genus can invade homes and are a sign of moisture because the mold requires water to thrive; they can cause plant diseases or hay fever and more serious infections in immuno-suppressed individuals.

Verticillium alboatrum

Verticillium wilt is a very destructive fungal disease in cool climates. It affects several hundred species of trees, shrubs, vines, flowers, house plants, vegetables, fruits, field crops, and weeds. The causal agent is the soil-inhabiting fungus *V. alboatrum*. In hot weather, the leaves on one or more branches turn dull green to yellow, wilt, and wither, often from the base upward (Encyclopaedia Britannica 1998).

Verticillium dahliae

It is a soil-borne pathogen, and belongs to a group of fungi which do not have a known sexual stage. *V. dahliae* has a wide host range. Over 300 woody and herbaceous plant species are known to be susceptible to this fungal pathogen. The disease, *Verticillium* wilt, is problematic in temperate areas of the world, especially in irrigated regions. There are no curative measures once a plant is infected (Gómez-Alpízar 2001).

7.2.2 Yeasts

Yeasts are members of the phylum Ascomycota and can be found in plants, animals, soil, water, and atmosphere, and also in extreme environments (e.g., osmophilic and halotolerant yeasts). As non-motile organisms, yeasts rely on aerosols, animal vectors, and human activity for their natural scatter (Walker 1998b).

Candida albicans

Yeasts from the genus *Candida*, particularly *C. albicans*, are the most important agents causing yeast infection. This opportunistic pathogen is normally associated with mucosal, cutaneous, and nail infections (normally known as candidiasis), but can cause acute or chronic invasive infections in immunocompromised or debilitated individuals. This yeast is part of human flora and, in most cases, infections are derived from the individual's own reservoir in mouth, gastrointestinal tract, lower genital tract, or skin. When there are serious impairments of host defenses, life-threatening invasive infections can occur (Richardson and Warnock 2012).

Candida glabrata

On Sabouraud's dextrose agar colonies are white to cream colored, smooth, and glabrous yeast-like in appearance. Microscopic morphology shows numerous ovoid, budding yeast-like cells or blastoconidia, 2.0–4.0 × 3.0–5.5 µm in size (Ellis 2016a). *C. glabrata* is a highly opportunistic pathogen of the urogenital tract, and of the bloodstream (Candidemia); it is especially prevalent in the elderly, HIV positive, and other immunocompromised patients.

Candida kefyr

Synonym: Candida pseudotropicalis

It is a rare cause of candidiasis, and is usually associated with superficial cutaneous manifestations rather than systemic disease. It has been isolated from nails and lung infections. Environmental isolations have been made from cheese and dairy products. Colonies are white to cream colored, smooth, glabrous, and yeast-like in appearance. Microscopic morphology shows numerous short-ovoid to long-ovoid, budding yeast-like cells or blastoconidia, 3.0–6.5 × 5.5–11.0 µm, sometimes becoming elongate (up to 16.0 µm) in size (Ellis 2016s).

Candida krusei

On Sabouraud's dextrose agar colonies are white to cream colored, smooth, glabrous, yeast-like colonies. Microscopic morphology shows predominantly small, elongated to ovoid budding yeast-like cells or blastoconidia, 2.0–5.5 × 4.0–15.0 µm in size. *C. krusei* is regularly associated with some forms of infant diarrhea and occasionally with systemic disease. It has also been reported to colonize the gastrointestinal, respiratory, and urinary tracts of patients with granulocytopenia. Environmental isolations have been made from beer, milk products, skin, feces of animals and birds, and pickle brine (Ellis 2016f).

Candida parapsilosis

It is an opportunistic human pathogen which may cause both superficial cutaneous infections, especially of the nail and systemic disease, especially endocarditis. Other clinical manifestations include endophthalmitis and fungemia. Environmental isolations have been made from intertidal and oceanic waters, pickle brine, cured meats, olives, skin, and feces. Colonies are white to cream colored, smooth, glabrous, and yeast-like in appearance. Microscopic morphology shows predominantly small, globose to ovoid budding yeast-like cells or blastoconidia, 2.0–3.5 × 3.0–4.5 um in size, with some larger elongated forms present (Ellis 2016q).

Candida tropicalis

It is a major cause of septicemia and disseminated candidiasis. It is also found as part of the normal human mucocutaneous flora and environmental isolations that have been made from feces, shrimp, kefir, and soil. Colonies white to cream-colored smooth, glabrous, and yeast-like (Ellis 2016o).

Saccharomyces cerevisiae

When translated, it means "sugar fungus". That is what this yeast uses for food (Baker's yeast). They are found in the wild, growing on the skins of grapes and other fruits (Volk 2010).

7.3 Methods of Studying the Antifungal Effect of Extracts

Most authors use agar diffusion assays to determine the antifungal activity of algal extracts (Stein et al. 2011b). The technique works adequately with well-defined inhibitors (Hewitt and Vincent 1989), but with extracts that contain unknown components there are problems with false positive and false negative results (Ellof 1998).

The type of agar, the salt concentration, the incubation temperature, and the molecular size of the antimicrobial components can all influence the results obtained with agar diffusion assays (Marsh and Goode 2007). Furthermore, this technique also cannot distinguish between fungicidal (the lowest concentration of the agent that results in no growth) and fungistatic (the lowest concentration of the agent that results in the maintenance or reduction of the inoculum) effects (Hammer et al. 2003), and does not permit determination of the minimum inhibitory concentration (MIC) (Ellof 1998).

Lopes et al. (2015) proposed phlorotannin extracts due to their antifungal activity against several yeast and dermatophyte strains, using a micromethod for the evaluation of the MIC (minimum inhibitory concentration) and the MLC (minimum lethal concentration).

7.4 Antifungal Activity of Chlorophyta (Green Algae)

Several investigations have been carried out worldwide. In this respect, in USA, Kulik (1995) tested extracts from macroalgae by spraying on plants and reported a pronounced reduction in the disease incidence of *Botrytis cinerea* on strawberries, *Erysiphe polygoni* on turnips, and reported that macroalgae produce various biologically active compounds (Kulik 1995).

In a research work done by Mtolera and Semesi (1996), the antifungal activity of extracts from six green algae from Tanzania was screened against the yeast *Candida albicans*, using a disc assay method. Of the six species tested, *Valonia aegagropila* extract was most active against *C. albicans*. After 24 months' storage, the extract of *V. aegagropila* exhibited halved antifungal activity. The extracts of *Halimeda opuntia* and *Halimeda tuna* showed mild activity, and extract of *Ulva australis* (formerly *Ulva pertusa*) was not active against *C. albicans*. The antifungal activity shown by *Caulerpa racemosa* in the study may be attributed to caulerpin (see Fig. 10.3, Chapter 10) or caulerpein (Paul et al. 1987), or flexin and trifarin (Blackman and Wells 1978), or by caulerpanyene (Amico et al. 1978). The inactivity shown by an extract of *Caulerpa mexicana* suggests that the species might be lacking the substances mentioned.

Of the 35 seaweeds collected along the coast of Sri Lanka and screened against the human pathogenic bacteria (see Chapter 8) and fungi, 26 species exhibited antibacterial and/or antifungal activity (Bandara et al. 1988). The extracts of *Cladophora* sp., *Caulerpa chemnitzia*, *Caulerpa racemosa*, *Halimeda macroloba*, *Microdictyon umbilicatum* (formerly *Microdictyon agardhianum*), *Ulva fasciata*, and *Valoniopsis pachynema* showed antifungal activity, particularly active against *Candida albicans*, *Cryptococcus neoformans*, and *Cladosporium cladosporioides* (Bandara et al. 1988).

In France, Hellio et al. (2000) tested antimicrobial potentiality of many seaweed extracts and observed a conspicuous decrease in the development of the fungi tested. In China, Yi et al. (2001) used ethanol, acetone, and methanol-toluene to extract antibiotics from several species of marine algae belonging to the Chlorophyta phylum and revealed that the strongest antifungal activities were exhibited by the ethanol extract, but the least were by the methanol-toluene extract.

Some species of marine benthic algae collected from different coastal areas of Karachi (Pakistan) were investigated for their antifungal activities (Rizvi and Shameel 2004). *Codium indicum* (formerly *Codium iyengarii*) displayed a good inhibition activity against *Aspergillus niger* (46%), *Trichophyton schoenleinii* (50%), *Microsporum canis* (50%), *Pseudallescheria boydii* (54%), and *Fusarium solani* (56.3%); while in another observation it showed significant antifungal activity against a variety of pathogens (Ali et al. 2000).

In another study (Val et al. 2001), extracts from several species of seaweed from Canary Islands (Spain) were screened to produce antifungal and antibacterial compounds against a panel of bacteria, mycobacterium, yeasts, and fungi. Regarding antifungal activity, two of the green macroalgae species tested—*Cymopolia barbata* and *Ulva clathrata* (formerly *Enteromorpha muscoides*), presented activity against the filamentous fungi *Aspergillus fumigatus*, and/or the yeasts *Candida albicans* and *Saccharomyces cerevisiae*.

The study done by Lakshmi et al. (2006) deals with the biological activities of the extracts of 48 marine florae. The biological screening includes tests for antibacterial, antifungal, among others. From among green algae, the crude extracts from *Bryopsis hypnoides*, *Bryopsis plumosa*, *Caulerpa racemosa*, *Caulerpa veravalensis*, *Chaetomorpha* sp., *Codium dwarkense*, *Codium decorticatum* (formerly *Codium elongatum*), *Ulva clathrata* (formerly *Enteromorpha clathrata*), *Dictyosphaeria cavernosa*, *Udotea indica*, and *Valoniopsis pachynema* were active against *Candida albicans*, *Cryptococcus neoformans*, *Trichophyton mentagrophytes*, *Aspergillus fumigatus*, and *Sporotrichum schenckii*.

Atlantic marine algae were screened in growth inhibition assays against the pathogenic fungus *Lindra thalassiae*, the saprophytic fungus *Scolecobasidium salinum* (formerly *Dendryphiella salina*), the oomycetes *Phytophthora spinosa* (formerly *Halophytophthora spinosa*), and the labyrinthulomycetes (formerly slime moulds) *Schizochytrium aggregatum* (Puglisi et al. 2006). Overall, 95% of all species surveyed in this study yielded either hydrophilic or lipophilic extracts that were active against one or more assay microorganism. Of the crude extracts, 77% were active against two or more assay microbes. Broad-spectrum activity against three or four assay microorganisms was observed in the extracts from 50% to 21% of all species, respectively. The green alga *Bryopsis pennata* was one of the species to yield extracts active against all assay microorganisms. Overall, 76% of all green algae yielded extracts that were active against *P. spinosa*, and 62% yielded extracts that were active against *S. salinum*. While most of these extracts were active against multiple assay microorganisms, extracts from a few species were selectively active against only one assay microorganism. For example, the extracts from *Caulerpa cupressoides* and *Dictyosphaeria versluysii* were only active against *S. aggregatum*. Selective activity against *P. spinosa* was observed in extracts from *Caulerpa ambigua*, *Udotea argentea*, *Boodlea composita*, and *Ulva clathrata* (formerly *Enteromorpha clathrata*). While none of the extracts were selectively active against either *L. thalassiae* or *S. salinum*, the extract from *Caulerpa racemosa* was the only green alga with selective activity against both fungi. Among extracts with broad-spectrum activity, the extract from *Bryopsis pennata* inhibited the growth of all assay microorganisms. Broad-spectrum activity against fungi was observed in the extracts from *Caulerpa serrulata*, *Codium geppiorum*, *Neomeris annulata*, and most species of algae within the family Udoteaceae. Among members of the Udoteaceae, *Avrainvillea obscura* was the only species to yield extracts active against *P. spinosa* and *S. aggregatum*. Further, the extracts from *Halimeda macroloba*, *Halimeda opuntia*, and *Tydemania expeditionis* were active against one or both fungi (Puglisi et al. 2007).

The goal of the study done by Engel et al. (2006) was to systematically screen extracts from marine plants for antimicrobial effects against marine pathogens and saprophytes. Lipophilic and hydrophilic extracts from species of 49 marine algae and three seagrasses collected in the tropical Atlantic were screened for antimicrobial activity against five ecologically relevant marine microorganisms from three separate kingdoms. These assay microbes consisted of the pathogenic fungus *Lindra thalassiae*, the saprophytic fungus *Dendryphiella salina*, and the saprophytic stramenopiles, *Halophytophthora spinosa* and *Schizochytrium aggregatum*. The green algae *Halimeda copiosa* and *Penicillus capitatus* were the only species to yield extracts active against all assay microorganisms. Among all assay microorganisms, both fungi were the most resistant to the extracts tested, with less than 21% of all extracts inhibiting the growth of either *L. thalassiae* or *D. salina*. In contrast, over half of all lipophylic extracts were active against the stramenopiles *H. spinosa* and *S. aggregatum* (Engel et al. 2006).

In a study done by Stirk et al. (2007), seasonal variation in antifungal activity in seven South African seaweeds (methanolic extracts) was evaluated. No seasonal variation was observed in antifungal activity tested against *C. albicans*. All species had a consistent MIC of 6.25 mg mL^{-1}, with the two exceptions being *Codium capitatum* collected on 22/03/2004 and *Halimeda cuneata* collected on 23/05/2004, which had MIC values of 3.125 mg mL^{-1}.

Paulert et al. (2007) studied the antifungal activity of cell-wall polysaccharides and crude extracts from the seaweed *Ulva fasciata* against filamentous fungi and yeast. The antifungal activity was assessed by agar diffusion assay and by means of the broth dilution method estimating the minimal inhibitory concentration (MIC). MIC was determined for the fungi *Colletotrichum lindemuthianum* (plant pathogen), *Trichophyton mentagrophytes*, and *Microsporum canis* (dermatophyte pathogens). The methanol insoluble extract inhibited the growth of *T. mentagrophytes* at a concentration of 2 mg mL^{-1}. In contrast, ulvans did not show any *in vitro* activity towards all test organisms. In the study by Kolanjinathan and Stella (2011), the minimum inhibitory concentration (MIC) value of *Ulva reticulata* and *Ulva lactuca* against fungi was ranged between 4 mg mL^{-1} to 64 mg mL^{-1}. The lowest MIC (4 mg mL^{-1}) value was recorded against *Candida albicans* and *Candida glabrata* (Kolanjinathan and Stella 2011).

In the works of Wefky et al. (2009), some algal extracts were tested against fish and human pathogenic bacteria—*Aeromonas hydrophila*, *Vibrio anguillarum*, *Pseudomonas fluorescens*, *Staphylococcus aureus*, *Pseudomonas aeruginosa*, and *Salmonella typhimurium*. The methanolic extract of *Ulva lactuca* strongly inhibited the growth of *A. hydrophila* (27 mm), *V. anguillaum* (24 mm), *S. aureus* (20 mm), and *P. fluorescens* (20 mm), but no activity was exerted against *P. aeruginosa* or *S. typhimurium*. The methanolic extract of *Ulva compressa* (formerly *Enteromorpha compressa*) showed good inhibitory activity against *A. hydrophila*, and *P. fluorescens* with inhibition zones of 19 mm and 17 mm, respectively, and a moderate inhibitory activity against *V. anguillarum* (inhibition zone = 13 mm). On the other hand, *Ulva fasciata* showed no antibacterial activity against tested pathogenic of bacteria-fish (*A. hydrophila*, *V. anguillaum*, *P. fluorescens*), human (*S. aureus*, *S. typhimurium*) or human-fish (*P. aeruginosa*). The ethyl acetate extracts of *Ulva lactuca* showed antibacterial activities against *A. hydrophila*, *V. anguillaum*, *P. fluorescens*, and *P. aeruginosa*. The tested pathogens were varied in their response to the antibacterial action of the different extracts. The most susceptible organisms were *A. hydrophila* and *P. fluorescens*. The growth of *A. hydrophila* was strongly inhibited by the methanolic extracts of *U. lactuca* (inhibition zone = 27 mm).

In a study carried out in a medical college and hospital in India (Prabhakar et al. 2008), the algal extracts of *Caulerpa scalpelliformis* have exhibited anticandidial activity at 100 mg mL^{-1}. The zone of inhibition produced by the 70% aqueous ethanol extracts of *C. scalpelliformis* by agar diffusion test against all the 25 isolates of *Candida* species ranged from 7 mm to 22 mm. Fluconazole positive control produced a zone of inhibition of 18 mm or more on all the 25 isolates of *Candida* species tested. 5% aqueous DMSO negative control did not have any inhibitory effect on any of the 25 isolates of *Candida* species tested (Prabhakar et al. 2008).

Pandurangan et al. (2010) has studied the antifungal activities of six important seaweeds, namely the macroalgae like *Cladophora glomerata*, *Ulva lactuca*, and *Ulva reticulata*. These seaweeds were screened against fungal pathogens including *Aspergillus niger*, *Aspergillus flavus*, *Aspergillus fumigatous*, *Saccharomyces cervisiae*, and *Mucor indicus*. The antifungal activity of the plant extract was tested using well diffusion method using different concentrations. The zone of inhibition was calculated by measuring the diameter of the inhibition zone around the well including the diameter.

With a view to explore the finding of new molecules with therapeutic efficacy for human use, the alcoholic extracts of 33 identified species of marine flora, collected from Indian coasts, were prepared and screened for a wide range of biological activities (Lakshmi et al. 2010). From among green algae, the alcoholic extracts from *Caulerpa scalpelliformis*, *Chaetomorpha spiralis* (formerly *Chaetomorpha torta*), and *Ulva reticulata* were active against *Candida albicans*, *Cryptococcus neoformans*, *Trichophyton mentagfrophytes*, *Aspergillus fumigatus*, and *Sporotrichum schenckii*.

Aruna et al. (2011) screened the antifungal activities of six seaweeds, namely the green seaweed *Cladophora glomerata*, *Ulva lactuca*, and *Ulva reticulata*, against fungal pathogens *Aspergillus niger*, *Aspergillus flavus*, *Aspergillus fumigatus*, *Saccharomyces cerevisiae*, and *Mucor indicus*. The zone of inhibition ranged between 56–58 mm in aqueous extract and 54–56 mm in methanolic extract. The maximum activity (56 mm) was recorded from 200 mg of aqueous extract of *U. lactuca* against *A. flavus*; the methanolic extract showing the maximum activity (56 mm) was recorded from 200 mg of *U. lactuca* against *A. niger* and minimum (4 mm) by 50 mg of *U. reticulata* against *A. flavus*.

In the study done by Kolanjinathan and Stella (2011), the antifungal activity of *Ulva lactuca* and *Ulva reticulata* extracts were determined by Disc diffusion method proposed by Bauer et al. (1966). The antifungal activity of marine seaweeds extract *U. reticulata* and *U. lactuca* was investigated against *Aspergillus niger*, *Aspergillus flavus*, *Aspergillus fumigatus*, *Saccharomyces cerevisiae*, *Candida albicans*, and *Candida glabrata*. The zone of inhibition of *U. reticulata* and *U. lactuca* extracts against fungal pathogens ranged between 6 mm to 13 mm at 10 mg mL^{-1}. The methanol extract of *U. reticulata* (10 mg mL^{-1}) showed highest mean zone of inhibition (13 mm) against *A. niger* followed by *A. fumigatus* (10 mm), *C. glabrata* (9 mm), *C. albicans* (9 mm) and *A. flavus* (8 mm). No zone of inhibition was recorded against *S. cerevisiae*. The zone of inhibition obtained from the Hexane extract of seaweed *U. reticulata* against fungal pathogens was comparatively very less when compared to the other solvent extracts. No zone of inhibition was seen in DMSO blind control and the positive control Flucanozole showed zone of inhibition ranging from 12 mm to 15 mm against the test fungal pathogens. The minimum inhibitory concentration (MIC) value of *U. reticulata* against fungi ranged between 4 mg mL^{-1} to 64 mg mL^{-1}. The lowest MIC (4 mg mL^{-1}) value was recorded against *C. albicans* and *C. glabrata* (Kolanjinathan and Stella 2011). The *U. lactuca* methanol extract (10 mg mL^{-1}) showed highest mean zone of inhibition (13 mm) against *C. glabrata* followed by *C. albicans* (12 mm), *A. flavus* (12 mm), *A. fumigatus* (11 mm) and *A. niger* (9 mm). No zone of inhibition was recorded against *S. cerevisiae*. The zone of inhibition obtained from the Hexane extract of seaweed *U. lactuca* against fungal pathogens was comparatively very less when compared to the other solvent extracts. No zone of inhibition was seen in DMSO blind control and the positive control Flucanozole showed zone of inhibition ranging from 10 mm to 15 mm against the test fungal pathogens. The minimum inhibitory concentration (MIC) value of *U. lactuca* against fungi was ranged between 4 mg mL^{-1} to 32 mg mL^{-1}. The lowest MIC (4 mg mL^{-1}) value was recorded against *C. albicans* and *C. glabrata* (Kolanjinathan and Stella 2011).

In the work by Galal et al. (2011), crude methanolic extracts of *Codium fragile* exhibited strong activity against most of the tested fungi (*Alternaria*, *Botryotrichum*, *Fusarium*, and *Ulocladium* species). All the tested fungi were sensitive to Nestatin (standard antibiotic), except *F. oxysporium* (the most resistant fungi), where the dry weight recorded from 68% to 69%, protein recorded 69% to 52%, pectinase and cellulase activity 62% to 58% of control. The most sensitive fungi were *U. botrytis*, where the dry weight, protein content, pectinase, and cellulase activities all were completely inhibited in cellulose and pectin media. The most active algae are the methanolic extract of *C. fragile* inhibited pectinase and cellulase enzymes activities for all the tested fungi except *A. brassicicola* and *F. oxysporum*. This study confirms the broad antifungal effect of *C. fragile* using the methanolic extracts (Galal et al. 2011).

In the studies done by Lavanya and Veerappan (2012), antifungal activity of two species of marine green macroalgae *Codium decorticatum* and *Caulerpa scalpelliformis* using different solvents acetone, methanol, chloroform, diethyl ether, ethyl acetate, hexane, and aqueous were evaluated against *Fusarium oxysporum*, *Fusarium udum*, *Fusarium solani*, *Thanatephorus cucumeris* (formerly *Rhizoctonia solani*), *Alternaria alternata*, *Botrytis cinerea*, *Candida albicans*, *Candida krusei*, *Aspergillus niger*, and *Aspergillus flavus*.

In a study carried out in a hospital in Libya (El-Fatimy and Abdel-Moneim 2011), methanolic extract of green alga *Caulerpa prolifera* was tested against four types of dermatophytes (Tinea)—*Trichophyton tonsurans* (Tinea capitis or scalp ringworm), *Microsporum canis* (Tinea corporis or ringworm), *Trichophyton mentagrophytes* (Tinea pedis or athlete's foot), and *Trichophyton rubrum* (Tinea unguin or onychomycosis). At high concentration of algal extract (100 µm) nails and hair Tinea were obviously affected (25.36 mm and 21.65 mm, respectively), where foot and skin Tinea were not affected (9 mm). Meanwhile, the lower concentration (50 µm) has no clear effect on the four types of Tinea tested (Fatemi and Said 2011).

Lavanya and Veerappan (2012) reported that the aqueous extract of *Codium decorticatum* and *Caulerpa scalpelliformis* showed no activity against *Aspergillus flavus*. The highest activities of *Codium decorticatum* were recorded in ethyl acetate extract against *Fusarium udum* (10 mm) and methanol extract against *Candida krusei* (10 mm) and chloroform extract against *Candida albicans* (10 mm). The minimum activities (2 mm) were observed in acetone extract against *Fusarium oxysporum*; methanol extract against *Fusarium solani*, *C. albicans*, *A. flavus*, diethyl ether extract against *Trichophyton solani*, *C. krusei*; chloroform extract against *Fusarium udum*, *Alternaria alternata*, *C. krusei*; ethyl acetate extract against *A. flavus*. There was no activity in hexane extract against all tested pathogens. The highest activities of *Caulerpa scalpelliformis* were recorded in acetone extract against *F. udum* (14 mm), followed by *T. solani* (8 mm), chloroform extract (8 mm), and water extract (6 mm) against *B. cinerea*. The minimum (2 mm) activities were observed in acetone extract against *C. krusei*, methanol extract against *T. solani* and *C. albicans*, diethyl ether extract against *A. flavus*. No activities in hexane and ethyl acetate extract against all tested pathogens (Lavanya and Veerappan 2012).

Halimeda tuna was examined for antifungal activity *in vitro* using the well diffusion method (Indira et al. 2013). The activity was against nine fungal strains (*Aspergillus niger*, *Aspergillus flavus*, *Alternaria alternaria*, *Candida albicans*, *Epidermophyton floccossum*, *Trichophyton mentagrophytes*, *Trichophyton rubrum*, *Pencillium* sp., and *Rhizopus* sp.). The range of 15.62 µg mL^{-1} to 250 µg mL^{-1} were obtained for the methanol extract with the fungi agents. Moreover, 15.62 µg mL^{-1} to 500 µg mL^{-1} and 31.25 µg mL^{-1} to 500 µg mL^{-1} obtained for the ethanol extract and chloroform extract was recorded against the fungal isolates. The MIC values obtained in the antifungal assays were 62.5 µg mL^{-1} to 250 µg mL^{-1}. All the extracts exhibited a minimum fungicidal concentration (MFC) at a concentration of 500 µg mL^{-1}, while aqueous extracts had a MFC value ranged between 250 µg mL^{-1} to 500 µg mL^{-1} (Indira et al. 2013).

In India, Indira et al. (2013) displayed the antifungal property of seaweed *Halimeda tuna* against nine strains (*Aspergillus niger*, *Aspergillus flavus*, *Alternaria alternaria*, *Candida albicans*, *Epidermophyton floccossum*, *Trichophyton mentagrophytes*, *Trichophyton rubrum*, *Penicillium* sp., and *Rhizopus* sp.). In this study the methanol extract showed a higher antifungal activity than that of other extracts and aqueous extract.

Two prenylated para-xylenes, named caulerprenylols A and B, isolated from the green alga *Caulerpa racemosa*, collected from the Zhanjiang coastline, China, possess a broad spectrum of antifungal activity against *Candida glabrata*, *Trichophyton rubrum*, and *Cryptococcus neoformans*, with MIC_{80} values between 4 µg mL^{-1} and 64 µg mL^{-1} when compared to Amphotericin B (MIC_{80} values of 2.0 µg mL^{-1}, 1.0 µg mL^{-1}, and 4.0 µg mL^{-1}, respectively) (Liu et al. 2013b).

In a study of *Chaetomorpha linum* harvested in the Gulf of Mannar (India) by Devi et al. (2014), the methanolic extract was tested for its antifungal activity. All the extracts exhibited different degrees of antifungal activity against *Fusarium dimerum* and *Trichoderma reesei*. The growth of *T. reesei* was highly inhibited by all the tested concentrations (5–20%) of methanol extracts compared with control, the corresponding inhibition ranging from 72–81%. The extract showed comparatively very low activity against *F. dimerum*, ranging from 76–86%.

A preliminary research work (Manigandan and Kolanjinathan 2014) was carried out to find out the antifungal activity of *Ulva lactuca* extracts collected from Mandapam coastal regions of Tamil Nadu. The acetone extract of *U. lactuca* showed maximum activity against *Aspergillus niger* of 12.0 mm zone of inhibition, followed by *Aspergillus flavus* (11.0 mm), and minimum zone of inhibition was observed in *Candida albicans* (10.0 mm). The ethyl acetate extracts showed maximum activity against *A. niger* and *A. flavus* of 10.0 mm zone of inhibition, and minimum activity was observed in *C. albicans* (08.0 mm). The chloroform and hexane extracts showed minimum activity of 8.0 mm against *A. niger* and no zone of inhibition was observed against *A. flavus*.

In the study done by Sukumaran and Thevanathan (2014), methanol and *n*-hexane extracts of *Pithophora roettleri* (formerly *Pithophora oedogonia*) were tested for activity against *Colletotrichum lindemuthianum* and *Drechslera oryzae*, two of the economically important pathogenic fungi in India. Visually evident promotion of radial mycelial growth by the methanolic extract of *P. roettleri* was recorded for both pathogens. In contrast *n*-hexane extracts inhibited the radial mycelial growth of *D. oryzae* at concentrations above 100 μg mL^{-1}. At concentrations below 100 μg mL^{-1}, *n*-hexane extracts also promoted the growth of the pathogenic fungi. The morphology of the test fungi was significantly altered by the algal extract to produce dense mats. Inhibitory effect being marginal, the dense branching of the mycelium in response to treatment with extracts of *P. roettleri* at all concentrations should be considered to indicate a inductor effect rather than an inhibitory effect. This kind of inductor effect by the extract of the fresh water alga *P. roettleri* on fungal plant pathogens is not known for other fresh water and marine algae. Compounds that delay conidiation are of importance in management of diseases caused by fungal pathogens. Delayed conidiation will imply prolonged vegetative growth, which means vulnerability of the fungus to disease management techniques and a decreased incidence of infections. Both methanolic and *n*-hexane extracts of *P. roettleri* delayed conidial formation in *C. lindemuthianum*. Methanolic extract of the alga did not affect conidial initiation in *D. oryzae*, while *n*-hexane extract delayed the process by four to six days. A delay in conidiation by three to six days in the presence of *n*-hexane extract at 1000 μg mL^{-1} concentration may be significant for consideration in the management of *D. oryzae* (Sukumaran and Thevanathan 2014).

In the study done by Rajarajan and Selvaraju (2014), the effect of *Caulerpa racemosa* occurring along the coastal regions of south India was investigated against *Fusarium oxysporum*. In the study, the methanol extract shows promising antifungal activity against the fungal strain, and shows maximum activity. The minimum inhibitory concentration of methanol extract was tested against the fungal strains. The 100 μg mL^{-1} concentration showed inhibitory effect against fungal strain. As the concentration increases from 100 μg mL^{-1}, 500 μg mL^{-1}, 1 mg mL^{-1}, 10 mg mL^{-1}, 50 mg mL^{-1}, the inhibitory effect also increases proportionally the maximum inhibitory effect absorbed with 50 mg mL^{-1} of methanol extract. The active principles of algae are responsible for antifungal activity. It is clear that the algae *C. racemosa* has the potential to control the fungal pathogen *F. oxysporum*, which causes the fungal disease to a larger extent.

The antifungal activity of *Caulerpa racemosa* crude extract was studied by Sekar and Kolanjinathan (2015) against two pathogenic yeasts. The crude extract showed zone of inhibition against C. *albicans* (12 mm) and *C. glabrata* (12 mm). Extract of *C. racemosa* showed best MIC at 40 mg mL^{-1}.

The study done by Selim et al. (2015) was undertaken to explore the inhibitory effect of seaweed extracts of *Ulva lactuca* with five solvents (methanol, ethanol, methylene chloride, chloroform, and hexane) and water at different concentrations, along with control against the mycelial growth *of Fusarium solani*, *Rhizoctonia solani*, *Sclerotinia sclerotiorum*, *Alternaria solani*, *Phytophthora infestanse*, and *Botrytis cinerea* by poison food technique. The results revealed that the extract of methanol showed the highest antifungal activity against all the tested fungi, while water extract showed a moderate antifungal activity against all the tested plant pathogenic fungi. *Phytophthora infestans* was more sensitive than the other tested fungi to all solvent extracts of *U. lactuca*. The results also showed that the highest antifungal activity was recorded in winter season, while the lowest one was in summer. Methanol extract of *U. lactuca* showed the highest antifungal activity against all the tested fungi, especially *Phytophthora infestans*, *Sclerotinia sclerotioru*, and *Alternaria solani* with EC_{50} values of 1149 μg mL^{-1}, 1225 μg mL^{-1}, and 1330 μg mL^{-1}, respectively. Methylene chloride extract exhibited the strongest antifungal effect against *Phytophthora infestans* with an EC_{50} value of 891 μg mL^{-1}. *Phytophthora infestans* was most sensitive than the other tested fungi to all solvent extracts of *U. lactuca*. Chloroform extract of *U. lactuca* showed weaker antifungal activity against all the tested fungi. Concerning the effect of ethanol extract, it gave a moderate effect against most of the tested fungi, while it was the most effective one against *Botrytis cinerea*, with an EC_{50} value of 1017.8 μg mL^{-1}.

In a research study done by Chowdhury et al. (2015), antifungal activity of extracts of freshwater and marine algae from Bangladesh was evaluated. The extracts obtained from *Ulva lactuca* showed inhibition of growth of *Candida albicans* (13.0–17.7 mm), and extracts from *Ulva prolifera* (formerly *Enteromorpha prolifera*) also showed inhibition of growth of the yeast studied (13–16.1 mm).

In the works of Pinteus et al. (2015), two green algae, *Ulva compressa* and *Codium adhaerens*, presented high antifungal activity against *Saccharomyces cerevisiae* growth, in the dichloromethane and *n*-hexane fractions with inhibitory effects of 86%, 35% and 70%, 37% respectively, however, no antifungal activity was achieved for green algae with the methanolic extraction (Pinteus et al. 2015).

When the antifungal activity was carried out with the green seaweed *Caulerpa scalpelliformis* by Radhika and Priya (2016), both crude and fractionated extracts showed good activity against all the fungal strains tested. *Alternaria* sp. was inhibited the most (7.5 mm), succeeded by *A. fumigatus*, *Gibberella* sp., *A. terreus*, and *Ganoderma* sp. All the eight different extracts inhibited *Gibberella* sp. Aqueous and methanol extracts had little or no inhibiting factor against *A. terreus* and *Ganoderma* sp. *Gibberella* sp. and *Alternaria* sp. were the most susceptible against all the seaweeds tested and *Ganoderma* sp. was the most resistant, followed by *A. terreus*. Among the solvents used for extraction, ethanol proved to be a good one, having the highest inhibition rate against the fungal pathogens tested (Radhika and Priya 2016).

In the research done by Mashjoor et al. (2016), antifungal and cytotoxic activities of the extracts of marine macroalgae collected in the Persian Gulf (Iran) were evaluated. Methanol extract of *Ulva flexuosa* shows moderate activity against *Candida albicans* (13.0 mm), with a minimum inhibitory concentration (MIC) of 7.5 mg mL^{-1}, and is highly active against *Saccharomyces cerevisiae* (17.0 mm), with a MIC value of 3.75 mg mL^{-1}. Ethyl-acetate extract shows high activity against *C. albicans* (17.0 mm), with a minimum inhibitory concentration (MIC) of 1.87 mg mL^{-1}, and also against *S. cerevisiae* (21.0 mm), with a MIC value of 0.93 mg mL^{-1}.

In a large study of bioprospecting of several algal species from the Aegean Sea, the antifungal activity of macroalgal extracts was evaluated (Montalvão et al. 2016). The most significant results were achieved by *Codium fragile* ethanolic extract with growth inhibition of *Candida albicans* (49% at 100 μg mL^{-1}), and *Chaetomorpha aerea* extract with 30% of growth inhibition.

In the study done by Sahnouni et al. (2016), extracts of two marine green algae (*Ulva rigida* and *Ulva intestinalis*), harvested from Arzew Gulf (Western Algeria) were investigated for their antifungal activity against human pathogens. A significant result of this study was that the methanolic extract of *U. rigida* showed good antifungal activity against *Aspergillus niger* (24.02 mm), *Candida albicans* (21.5 mm), and *Cryptococcus neoformans* (20.5 mm). A similar observation was recorded by Ertürk and Taş (2011). Antifungal activity results of *U. intestinalis* were more significant than those observed by methanolic extract *of U. rigida*. The highest inhibition activity was observed in *C. neoformans* (23.5 mm) and *C. albicans* (21.2 mm). These results are not in agreement with those obtained by Darah and Lim (2015). No activity was exhibited by methanolic extract of *U. intestinalis* (formerly *Enteromorpha intestinalis*) on tested fungi and yeast according to these researchers.

7.5 Antifungal Activity of Phaeophyceae (Brown Algae)

Ethanolic extracts of 17 species of seaweeds were tested against the plant pathogenic root infecting fungi *Marcophomina phaseolina*, *Rhizoctonia solani* (teleomorph: *Thanatephorus cucumeris*), *Fusarium solani*, and *Fusarium oxysporum*. Extracts of *Spatoglossum asperum* and *Spatoglossum* sp. inhibited the radial growth of the fungi such as *M. phaseolina*, *R. solani*, and *Fusarium solani in vitro* when used at 6 mg^{-1} disc. Extracts obtained from the brown alga *Stoechospermum polypodioides* (formerly *Stoechospermum marginatum*) control only the growth of *F. solani* at the same concentration (Ara et al. 1998).

According to Cowan et al. (1999), the brown seaweeds contain high amounts of flavonoid and phenolic compounds, which could be the reason for antifungal activity. Furthermore, a meroditerpenoid metabolite isolated and characterized from the brown alga *Cystoseira tamariscifolia* as Methoxybifurcarenone possess antifungal activity against three tomato pathogenic fungi—*Botrytis cinerea*, *Fusarium oxysporum* f. *lycopersici* (formerly *Fusarium oxysperium* sp. *mycopersici*), and *Verticillium alboatrum* (Bennamara et al. 1999). Long before, Fenical et al. (1973) reported that the two sesquiterpenes-zonarol and isozonarol from *Dictyopteris undulata* (formerly *Dictyopteris zonarioides*) strongly inhibited the growth of 10 species of pathogenic fungi causing diseases in plants. In the study done by Abourriche et al. (1999), the hexane, ethyl ether, and dichloromethane fractions of the brown alga *Cystoseira tamariscifolia* extracts showed interesting antifungal activities against *Botrytis cinerea*, *Fusarium oxysporum*, and *Verticillium alboatrum*. All tested fractions showed antifungal activity, but less than that of amphotericin. However, the *in vivo* antifungal action of ethanol fraction is very interesting in the field of plant diseases and suggests the need of further research extended to other mold species and other plants. A meroditerpenoid metabolite (Methoxybifurcarenone) isolated from the brown alga *C. tamariscifolia* possesses antifungal activity against three tomato pathogenic fungi—*B. cinerea*, *F. oxysporum* f. *lycopersici*, and *V. alboatrum* (Bennamara et al. 1999). Differences in the action on the three mold isolates were observed. The activity of the three concentrations of 0.1 mg mL^{-1}, 0.5 mg mL^{-1}, and 1 mg mL^{-1} on the mycelium of *B. cinerea* were significantly higher than those observed with the other strains. This assay may suggest that the methoxybifurcarenone had a similar antifungal activity on *F. oxysporum* f. *lycopersici* and *V. alboatrum*. All the concentrations tested influenced the mold growth and the inhibition zone diameters expressed in mm are observed with a concentration of 0.1 mg mL^{-1} for all the molds. These diameters were more developed with higher concentrations. The most sensitive strains were *B. cinerea* which showed higher zones than the other strains (Bennamara et al. 1999).

According to Ara et al. (2005), chloroform and methanol fractions of an ethanol extract of *Spatoglossum asperum* showed antifungal activity (inhibition of 3 mm and 4 mm) against the highly destructive plant pathogen *Macrophomina phaseolina*, while the *n*-hexane fraction showed activity (inhibition of 2 mm) against *Thanatephorus cucumeris* (formerly *Rhizoctonia solani*) and *Fusarium solani*. The oily fractions of ethanol extract from *S. asperum* showed growth inhibition of *M. phaseolina*, *F. solani*, *Fusarium oxysporum*, and *T. cucumeris* by producing a zone of inhibition of 6 mm, 13 mm, 6 mm, and 18 mm, respectively.

The study done by Lakshmi et al. (2006) deals with the biological activities of the extracts of 48 marine florae. The biological screening includes tests for antibacterial, antifungal, among others. From among brown algae, the crude extracts from *Colpomenia sinuosa*, *Polycladia indica* (formerly *Cystoseira indica*), *Dictyopteris woodwardia*, *Sargassum johnstonii*, *Spatoglossum variabile*, *Stoechospermum polypodioides* (formerly *Stoechospermum marginatum*), *Dictyota dichotoma*, *Turbinaria condensata*, *Turbinaria conoides*, *Dictyota dichotoma*, and *Turbinaria decurrens* were active against *Candida albicans*, *Cryptococcus neoformans*, *Trichophyton mentagrophytes*, *Aspergillus fumigatus*, and *Sporotrichum schenckii*.

Atlantic marine algae were screened in growth inhibition assays against the saprophytic fungus *Scolecobasidium salinum* (formerly *Dendryphiella salina*), the pathogenic fungus *Lindra thalassiae*, the oomycetes *Phytophthora spinosa* (formerly *Halophytophthora spinosa*), and the labyrinthulomycetes (formerly slime moulds) *Schizochytrium aggregatum* (Puglisi et al. 2007). Overall, 95% of all species surveyed in this study yielded either hydrophilic or lipophilic extracts that were active against one or more assay microorganism. Of the crude extracts, 77% were active against two or more assay microbes. Broad-spectrum activity against three or four assay microorganisms was observed in the extracts from 50% to 21% of all species, respectively. Brown algae comprised 14% of all species surveyed in this study; overall, 88% of all phaeophyceae yielded extracts active against *P. spinosa*, and 63% yielded extracts active against *S. salinum*; only extracts from 38% to 13% were active against *L. thalassiae* and *S. aggregatum*, respectively. Apart from extracts from *Hydroclathrus clathratus*, all other species of brown algae yielded extracts that were active against *P. spinosa* and one or both fungi. Broad-spectrum activity against three or more assay microorganisms was observed in extracts from all species of the families Dictyotaceae and Sargassaceae. Among members of the family Dictyotaceae, including *Dictyota bartayresiana*, *Lobophora* sp., *Padina jonesii*, and *Padina boryana* (formerly *Padina tenuis*), antimicrobial activity was most prevalent among extracts from *D. bartayresiana*, inhibiting the growth of both fungi and stramenopiles. Similarly, the extracts from both *Padina* species were active against one or both fungi and *P. spinosa*, while the extracts from *Lobophora* sp. were also active against *S. salinum* and *P. spinosa*. Similarly, extracts from *Sargassum ilicifolium* (formerly *Sargassum cristaefolium*), *Sargassum polycystum*, and *Turbinaria ornata*, representing the family Sargassaceae, were active against one or both fungi and *P. spinosa* (Puglisi et al. 2007).

In a study carried out in a medical college and hospital in India (Prabhakar et al. 2008), the algal extracts of *Sargassum swartzii* (formerly *Sargassum wightii*) have exhibited anticandidial activity at 10 mg mL^{-1}. The zone of inhibition produced by the 70% aqueous ethanol extracts of *S. swartzii* by agar diffusion test against all the 25 isolates of *Candida* species ranged from 7 mm to 22 mm. Fluconazole positive control produced a zone of inhibition of 18 mm or more on all the 25 isolates of *Candida* species tested. 5% aqueous DMSO negative control did not have any inhibitory effect on any of the 25 isolates of *Candida* species tested (Prabhakar et al. 2008).

In the study done by Cosoveanu et al. (2010), the biological activity of the brown algae *Alaria esculenta*, *Fucus vesiculosus*, *Fucus* sp. (Bioalgua®), *Arthrospira platensis* (formerly *Spirulina platensis*) (Cyanobacteria), and *Ecklonia maxima* (as Kelpak®), methanolic extracts was tested *in vitro* against *Fusarium sambucinum* var. *sambucinum* (formerly *Fusarium roseum*), *F. oxysporum*, *Alternaria alternata*, *A. dauci*, *A. longipes*, *Trichoderma viride*, *Botrytis cinerea*, *Aspergillus niger*, and *Penicillium expansum*. Their potential toxic effects were evaluated on mycelial growth. Results are presented as effective concentration, which inhibits mycelial growth by 50% and 90%. Almost all the algal extracts tested showed antifungal activity as ethanol extracts (Cosoveanu et al. 2010).

In the works conducted by Pandithurai et al. (2015b), the methanolic extract of *Spatoglossum asperum* showed the highest percentage of inhibition against all the pathogens studied, that is: *Candida albicans* 57.14% (14.67 mm), *Candida tropicalis* 54.75% (15.33 mm), *Trichophyton mentagrophytes* 50.85% (16.44 mm). The chloroform and aqueous extract of *S. asperum* showed moderate activities against all the fungal strain were studied. However, ethyl acetate and hexane extract don't show any activity against the above pathogens studied. Lavanya and Veerappan (2012) reported that the methanolic extract of *Sargassum swartzii* and *Turbinaria conoides* exhibited highest activity against *C. albicans*. Similarly, the methanolic extract of *S. asperum* showed the highest activity against *C. albicans*. In the present study, the aqueous extract of *S. asperum* showed moderate activity against *C. tropicalis* and it was correlated with the study of Boonchum et al. (2011). They also reported that the aqueous extract of *T. conoides* showed moderate activity against *C. tropicalis*. Chloroform extract of marine brown algae *S. swartzii* and *T. conoides* exhibited moderate activity (Lavanya and Veerappan 2012) against *C. albicans*, and it was also found to show moderate activity against *C. albicans*. Selvaraj et al. (2006) reported that the chloroform extract of *Stoechospermum polypodioides* (formerly *Stoechospermum marginatum*) showed moderate antifungal activity against *Trichophyton mentagrophytes*, *A. flavus*, and *C. albicans*. In the same way, in the present study it was observed that the chloroform extract of *S. asperum* exhibited moderate activity against *C. albicans*, *C. tropicalis*, and *T. mentagrophytes*, and higher activity against *A. flavus*. Lavanya and Veerappan (2012) reported that ethyl acetate and hexane extracts of *S. wightii* and *T. conoides* showed no activities against *Candida albicans*, whereas, in the present study ethyl acetate extract showed lesser activities and no activity was observed for the hexane extract. The aqueous extract of *S. asperum* against *A. flavus* shows no activity, and it was supported by Lavanya and Veerappan (2012). It is correlated with the present study that the methanolic extract of the brown seaweed *S. asperum* showed the highest activity against *C. albicans*. Many earlier reports have shown the antifungal potential of seaweeds (Selvaraj et al. 2006).

In a study done by Stirk et al. (2007), seasonal variation in antifungal activity in seven South African seaweeds (methanolic extracts) was evaluated. No seasonal variation was observed in antifungal activity tested against *C. albicans*. *Dictyota humifusa* was the most active seaweed tested, with a MIC of 3.125 mg mL^{-1} for all the collections, except the sample collected on 12/09/2004, which had a MIC of 6.26 mg mL^{-1}.

In the research conducted by Galal et al. (2011), the antifungal capacity of the methanol and ethyl acetate crude extracts of seaweeds collected from the Red Sea, Hurghada, Egypt, was tested. The concentration of 50 mg mL^{-1} was tested against *Alternaria alternata*, *Alternaria brassicola*, *Botryotrichum piluliferum*, *Fusarium oxysporum*, and *Ulocladium botrytis*. All the tested fungi were sensitive to Nestatin (standard antibiotic), except *F. oxysporum*, where the dry weight recorded 68% of absolute control in pectin medium. The most sensitive fungi were *U. botrytis* where the dry weight was completely inhibited in both cellulose and pectin media. Methanolic extracts showed much more bioactivity (inhibition of fungal dry weight) than ethyl acetate extracts. *U. botrytis* and *A. brassicicola* were the most sensitive fungi (showed complete inhibition of protein content) when ethyl acetate and methanolic extracts were employed, respectively.

In the study developed by Manivannan et al. (2011) with brown seaweeds collected in Gulf of Mannar (Indian Ocean), the methanol extract of *Turbinaria conoides* showed maximum activity against *Candida albicans* (18.00 mm) and *Penicillium* sp. (18.00 mm), and minimum activity against *Aspergillus niger* (4.66 mm). Chloroform extract showed maximum activity against *Aspergillus terreus* (17.00 mm) and minimum activity against *A. niger* (4.00 mm). The ethanol extracts showed maximum activity against *C. albicans* (18.33 mm) and *Penicillium* sp. (18.33 mm), and minimum activity against *A. niger* (5.33 mm). Diethyl ether extract showed maximum activity against *Candida glabrata* (16.33 mm), and minimum activity against *A. niger* (11.00 mm). Petroleum ether extract showed maximum activity against *Aspergillus flavus* (15.66 mm) and *A. terreus* (15.66 mm), and minimum activity against *A. niger* (7.00 mm). Ethyl acetate extract showed maximum activity against *Penicillium* sp. and *C. albicans* (18.33 mm), and minimum activity against *A. niger* (4.33 mm). Acetone extract showed maximum activity against *Cryptococcus neoformans* (13.00 mm), and minimum activity against *A. niger* (3.00 mm). The antibiotic disc showed maximum activity against *A. flavus* (20.66 mm), and maximum activity against *C. neoformans* (12.66 mm) (Manivannan et al. 2011).

The methanol extract of *Padina gymnospora* recorded maximum activity against *Cryptococcus neoformans* (20.00 mm) and minimum activity against *Aspergillus niger* (12.00 mm). The chloroform extract showed maximum activity against *Aspergillus terreus* (16.33 mm) and minimum activity against *Candida albicans* (13.66 mm). The ethanol extract showed maximum activity against *Aspergillus flavus* (18.66 mm), and minimum activity against *C. neoformans* (12.66 mm). The diethyl ether extract showed maximum activity against *A. flavus* (19.00 mm), and minimum activity against *Penicillium* sp. (12.33 mm). Petroleum ether extract showed maximum activity against *Candida glabrata* (16.66 mm) and minimum activity against *A. niger* (12.66 mm). Ethyl acetate extract showed maximum activity against *A. flavus* (16.33 mm), and minimum activity against *C. albicans* (11.66 mm). Acetone extract had maximum activity against *C. neoformans* (23.00 mm) and minimum activity against *A. niger* (13.00 mm). The antibiotic disc showed maximum activity against *C. neoformans* (15.66 mm), and minimum activity against *A. terreus* (11.66 mm) and *C. glabrata* (11.66 mm) (Manivannan et al. 2011).

The methanol extract of *Sargassum tenerrimum* showed maximum activity against *Candida albicans* (15.00 mm), and minimum activity against *Aspergillus niger* (5.66 mm). The chloroform extract had maximum activity against *Aspergillus flavus* (14.66 mm), and minimum activity against *Penicillium* sp. (4.66 mm). Ethanol extract had maximum activity against *Penicillium* sp. (18.66 mm), and minimum activity against *A. niger* (5.66 mm). The diethyl ether extract showed maximum activity against *Aspergillus terreus* (18.00 mm), and minimum activity against *Cryptococcus neoformans* (12.00 mm). Petroleum ether extract showed maximum activity against *Penicillium* sp. (16.66 mm) and minimum activity against *A. niger* (12.33 mm). The ethyl acetate extract had maximum activity against *A. niger* (16.33 mm) and minimum activity against *C. albicans* (6.66 mm). The antibiotic disc showed maximum activity against *C. albicans* (21.33 mm) and minimum activity against *A. flavus* (10.66 mm) (Manivannan et al. 2011).

In a study made in Algeria, the highest minimum inhibitory concentration (MIC) was obtained with methanolic extract of *Cystoseira tamariscifolia* against the three fungi strains—*Aspergillus niger*, *Candida albicans*, and *Mucor ramaniannus* (Saidani et al. 2012). Extracts of *Padina pavonica* were more efficient against the three fungi (*C. albicans*, *A. niger*, and *M. ramaniannus*), because MIC values, in this case, were very low compared to that of *C. tamarisciflora*, except in the case of *A. niger*. Nystatin used as a positive control presented the lowest MIC value against all the strains tested in this study. Polyphenol standards except quercetin showed low MIC values against *C. Albicans* and *M. ramaniannus*. However, the MIC value was high concerning *A. niger*, which presented resistance against all algae methanolic extracts and polyphenol standards (Saidani et al. 2012). However, the highest inhibiting effect was noted for *Padina pavonica* against *C. albicans* (diameter of inhibition zone—26 mm). *A. niger* showed resistance against the majority of the methanolic extracts. The evaluation of minimum inhibitory concentrations showed that extracts of *P. pavonica* were very efficient against *M. ramaniannus* and *C. albicans* (Saidani et al. 2012).

In the investigation conducted by Bibiana et al. (2012), antifungal activity of *Sargassum swartzii* (formerly *Sargassum wightii*) extracts were found to active against all tested fungi. The maximum activity of diethyl ether extract was shown against *Aspergillus flavus* (8 mm) and *Aspergillus fumigatus* (7 mm); of petroleum ether extracts was shown against *A. fumigatus* (8 mm); of acetone extract was shown against *A. flavus* (8 mm); and of acetic acid extracts was shown against most of the tested fungal strains with the maximum activity of 8 mm.

In the work carried out on brown algae harvested on the Lebanese Mediterranean coast (Khaled et al. 2012), no antifungal activity was detected from total extract and fractions of *Sargassum vulgare*, while the ethyl acetate fraction from *Padina pavonica* showed a moderate activity against *Candida glabrata* (diameter of inhibition = 16 mm) and a lesser activity against *Candida krusei* (diameter of inhibition = 14 mm).

Six organic extracts prepared with different solvents (methanol, acetone, hexane, chloroform, and dichloromethane-methanol) and aqueous extract of 27 species of marine algae from the Atlantic coast of Sidi Bouzid (El Jadida, Morocco) were studied for antifungal activities against *Candida tropicalis* and *Cryptococcus neoformans* (Oumaskour et al. 2012). For brown algae, a diameter of inhibition ranging between 10 mm and 15 mm was observed in the dichloromethane/methanol extract of *Laminaria ochroleuca* against *C. neoformans*, and in the methanolic extract of *Colpomenia sinuosa* and *Saccorhiza polyschides* (formerly *Saccorhiza bulbosa*) toward *C. tropicalis*. For the green algae studied, no activity was detected against fungi. Of the 27 species tested, those belonging to Phaeophyceae were the most active in comparison to Chlorophyta. The same result was reported by Caccamese et al. (1980) and Pesando and Carm (1984).

In the works developed by Peres et al. (2012), the extracts of *Stypopodium zonale*, *Ascophyllum nodosum*, *Sargassum muticum*, *Pelvetia canaliculata*, *Fucus spiralis*, *Sargassum filipendula*, *Sargassum stenophyllum*, and *Laminaria hyperborea* were tested for their antifungal activity against *Colletotrichum orbiculare* (formerly *Colletotrichum lagenarium*) (anthracnose fruit rot, plant disease pathogen). *S. zonale*, *L. dendroidea*, *P. canaliculata*, *S. muticum*, *A. nodosum*, and *F. spiralis* extracts significantly inhibited the *C. lagenarium* growth, but did not significantly inhibit the *A. flavus* growth. The presence of terpenes in all of these extracts was confirmed by thin layer chromatography, whereas the presence of phenolic compounds was confirmed only in extracts of *P. canaliculata*, *A. nodosum*, and *S. muticum*.

In a study done by Mhadhebi et al. (2012b), several organic extracts obtained from marine algae—*Cystoseira compressa*, *Cystoseira crinita*, *Cystoseira sedoides*, and *Dictyopteris polypodioides* (formerly *Dictyopteris membranacea*), collected from Tunisian Mediterranean coast, were evaluated for their antifungal activities against five human pathogenic yeasts using the agar disc diffusion assay. The chloroformic and the ethyl acetate extract obtained from *C. crinita* and *C. sedoides* showed a higher antifungal activity against four *Candida* strains. These findings suggest that the chloroformic and ethyl acetate extracts of the brown algae could contain a new antifungal compound(s). The diameters of the inhibition zones of *C. crinita* and *C. sedoides* extracts, against *Candida krusei*, were 23 mm and 33.3 mm, respectively for the chloroformic extract, and 17.6 mm and 24 mm, respectively for the ethyl acetate extract. The diameters of the inhibition zones of *C. crinita* and *C. sedoides* against *Candida kefyr* were 44.3 mm and 42.6 mm, respectively for the chloroformic extract (10 mg/disc), and 20.3 mm and 28.3 mm, respectively for the ethyl acetate extract. The diameters of the inhibition zone produced by the brown algae were greater than those produced by the reference drug, Amphotericin B (10 μg/disc). In addition, the chloroformic and the ethyl acetate extracts of *D. polypodioides* showed a moderate antifungal activity against *C. kefyr* and the diameters of the inhibition zones were in the range of 11 mm. Also, we noted that the petroleum ether extract of *C. sedoides* showed a diameter of the inhibition zone of 14 mm against *C. krusei*, and the diameters of the inhibition zone of the petroleum ether extract of *C. crinita* against *C. kefyr* and *C. albicans* were 15 mm and 17 mm, respectively.

According to Thenmozhi et al. (2013), the antifungal activity of the aqueous extract of *Sargassum ilicifolium* on the fungal species of *Trichoderma asperellum* showed that at higher concentrations of 20% and 25%, it was better than that of the patented medicine, namely, Clotrimazole.

Antifungal activity of seaweed extracts in petroleum ether, benzene, chloroform, ethanol, ethyl acetate, and water extracts was observed by Thinakaran and Sivakumar (2013). All the extracts used in this study exhibited antifungal activity against *Pythium aphanidermatum* and *Colletotrichum capsici*. The lowest inhibition zone effect was noticed for *Sargassum swartzii* (formerly *Sargassum wightii*), followed by *Sargassum ilicifolium*, and *Turbinaria conoides*.

Jassbi et al. (2013) reported antioxidant and antifungal activity in the water extracts of two brown algae—*Cystoseira myrica* and *Sargassum boveanum* from the Persian Gulf. The active substances were identified as free fatty acids, fucosterol, cholesterol, and 22-dehydroxychlosterol.

In the Master's thesis works of Carvalho (2013) and Silva (2014), three distinct samples of *Bifurcaria bifurcata* were used—wild-type sample from Baleal Island (Peniche, Portugal), wild-type sample from Buarcos (Figueira da Foz, Portugal), and cultured samples from a laboratory culture (MARE-UC, Coimbra, Portugal). A broth macrodilution assay was applied for the determination of the minimum inhibitory concentration (MIC) and minimum lethal concentration (MLC) of the methanolic extracts obtained from the *B. bifurcata* samples. From the tested dermatophyte strains, only *Epidermophyton floccosum* demonstrated sensibility for extracts of *B. bifucata* collected in Baleal and cultivated in a laboratory. However, *B. bifurcata* extracts collected in Baleal presented a higher antifungal capacity against *E. floccosum* (100–200 μg mL^{-1} MIC, 200 μg mL^{-1} MLC), and also *Microsporum canis* (400 μg mL^{-1} MIC, $\geq$ 400 μg mL^{-1} MLC), *Trichophyton mentagrophytes* (100 μg mL^{-1} MIC, > 800 μg mL^{-1} MLC), *M. gypseum* (800 μg mL^{-1} MIC, $\geq$ 800 μg mL^{-1} MLC), *T. mentagrophytes* var. *interdigitale* (800 μg mL^{-1} MIC, $\geq$ 800 μg mL^{-1} MLC), and *T. rubrum* (200 μg mL^{-1} MIC, $\geq$ 400 μg mL^{-1} MLC) (Carvalho 2013, Silva 2014).

A preliminary work (Manigandan and Kolanjinathan 2014) was carried out to find the antifungal activity of *Padina gymnospora* extracts collected from Mandapam coastal regions of Tamil Nadu (India). The acetone extract showed maximum activity against *Candida albicans* of 12.0 mm zone of inhibition, followed by *Aspergillus niger* (11.0 mm), and minimum zone of inhibition was observed in *Aspergillus flavus* (9.0 mm). The ethyl acetate extracts showed maximum activity against *C. albicans* and *A. flavus* of 10.0 mm zone of inhibition, and minimum activity was observed in *A. niger* (8.0 mm). The chloroform and hexane extracts showed minimum activity of 8.0 mm.

In a research study on antifungal activity of *Sargassum swartzii* (formerly *Sargassum wightii*) from the Mandapam coast of Tamil Nadu, India (Vengadesan and John 2014), methanol extract of *S. swartzii* showed activity on various fungus, such as *Aspergillus niger* (15 mm), *Candida albicans* (25 mm), *Cunninghamella bertholletiae* (17 mm), *Aspergillus flavus* (10 mm), *Mucor hiemalis* (22 mm), and *Trichophyton violaceum* (19 mm), compared to the standard drugs and studied through zone of inhibition as per the standard method.

In the study done by Ponnanikajamideen et al. (2014), the antifungal activity of various solvent extracts (dimethyl sulfoxide (DMSO), benzene, acetic acid, hexane, diethyl ether, and chloroform) of *Anthophycus longifolius* (formerly *Sargassum longifollium*) was tested against several fungi—*Aspergillus niger*, *Candida albicans*, *Aspergillus fumigatus*, *Aspergillus terreus*, and *Fusarium* sp. The antifungal activity of the extracts against the tested fungi was as follows—*A. niger* (9.06 mm to 14.4 mm); *A. fumigatus* (5.2 mm to 12.2 mm); *A. terreus* (5.8 mm to 11.9 mm); *C. albicans* (6.2 mm to 10.2 mm); *Fusarium* sp. (7.2 mm to 11.2 mm).

Antifungal activity of these species belonging to red, green, and brown seaweeds was explored and extracted in acetone, ethanol, and chloroform (Kumar et al. 2014). The highest inhibiting effect was noted for *Sargassum tenerrimum* and *Turbinaria ornata*, against *Aspergillus niger* and *Penicillium simplicissimum* (formerly *Penicillium janthinellum*) in chloroform extracts and ethanolic extract, which cause opportunistic infection of HIV-infected person, lung disease, aspergillosis, and otomycosis (fungal ear infections) (Kumar et al. 2014).

Ismail et al. (2014) reported the antifungal activities of aqueous extracts and fractions of *Zonaria tournefortii* against the pathogenic fungi *Candida albicans* and *Cryptococcus neoformans*. The minimum inhibitory concentration (MIC) of seaweed extracts against *C. albicans* and *C. neoformans* were relatively low compared to the positive control MIC value (1 mg mL^{-1} against *C. albicans* and 0.5 mg mL^{-1} against *C. neoformans*).

In vitro studies were conducted to evaluate the effect of seaweed liquid extracts of *Sargassum polycystum* (formerly *Sargassum myriocystum*) (Ambika and Sujatha 2014). Among the concentrations, 30% showed a better performance. Significant difference was observed in the seaweed extract of *S. polycystum* (30%), which inhibited the mycelial growth of *Fusarium udum* (formerly *Fusarium oxysporum* f. sp. *udum*), which recorded lowest mycelial growth of 28.1 mm, 33.9 mm, 34.7 mm, 39.4 mm, and 44.3 mm. *S. polycystum* showed inhibited radial growth in the range of 19% to 42% after 24 hours, 20% to 46% after 48 hours, 34% to 58% after 72 hours, 27% to 54% after 96 hours, and 27% to 51% after 108 hours compared to control.

The group of Kim et al. (2014d) investigated the efficacy of an antifungal agent from the marine brown alga *Ecklonia bicyclis* (formerly *Eisenia bicyclis*). The methanolic extract of *E. bicyclis* evinced potential antifungal activity against *Candida* species. The ethyl acetate extract from *E. bicyclis* demonstrated the strongest antifungal activity against *Candida* species among five solvent-soluble extracts. Indeed, the ethyl acetate extract showed minimum inhibitory concentrations (MICs) ranging from 4 mg mL^{-1} to 8 mg mL^{-1}. Furthermore, the ethyl acetate extract considerably reversed the high-level fluconazole resistance of *Candida* species. The MIC values of fluconazole against *Candida* species decreased substantially (from 64 μg mL^{-1} to 4 μg mL^{-1}) in combination with the MIC of the ethyl acetate extract (4 mg mL^{-1}). The fractional inhibitory concentration indices of fluconazole ranged from 0.531 to 0.625 in combination with 4 mg mL^{-1}, 2 mg mL^{-1}, or 1 mg mL^{-1} of the ethyl acetate extract against *Candida* isolates, indicating that these combinations exert a marked synergistic effect against *Candida* isolates (Kim et al. 2014d).

In a study made with seaweeds collected from Mandapam coast of Tamil Nadu, India (Sekar and Kolanjinathan 2015), the antifungal activity of selected marine macroalgae crude extracts was studied against two pathogenic yeasts (*Candida albicans* and *Candida glabrata*). The crude methanolic extract of *Padina gymnospora* showed maximum mean zone of inhibition against *C. albicans* (15 mm), followed by *Turbinaria conoides* (15 mm), and *Sargassum swartzii* (formerly *Sargassum wightii*) (15 mm), at 300 mg mL^{-1}. The crude hexane extract of marine macroalgae crude extracts showed minimum zone of inhibition against *C. albicans* when compared to the other solvent extracts. No zone of inhibition was seen in DMSO negative control, and the positive control Fluconazole (100 units) showed 17 mm inhibition zone against *C. albicans*. The minimum inhibitory concentration (MIC) of marine macroalgae crude extracts against *C. albicans* was ranged between 20 mg mL^{-1} to 640 mg mL^{-1}. The crude methanolic extract of *P. gymnospora*, *T. conoides*, and *S. swartzii* showed best MIC at 20 mg mL^{-1} against *C. albicans*. The crude methanolic extract of *P. gymnospora* showed maximum mean zone of inhibition against *C. glabrata* (16 mm), followed by *T. conoides* (15 mm), and *S. swartzii* (14 mm). The crude hexane extract of marine macroalgae crude extracts showed minimum zone of inhibition against *C. glabrata* when compared to the other solvent extracts. The minimum inhibitory concentration (MIC) of marine macroalgae crude extracts against *C. glabrata* was ranged between 20 mg mL^{-1} to 640 mg mL^{-1}. The crude methanolic extract of *P. gymnospora* showed best MIC at 20 mg mL^{-1} against *C. glabrata*. Extracts of *T. conoides* and *S. swartzii* showed best MIC at 40 mg mL^{-1} (Sekar and Kolanjinathan 2015).

In a research study done by Chowdhury et al. (2015), antifungal activity of extracts of marine algae from Bangladesh was evaluated. The ethanolic extract obtained from *Dictyopteris polypodioides* (formerly *D. membranacea*) showed inhibition of growth of *Candida albicans* (10.2 mm), and extracts from *Sargassum vulgare* also showed inhibition of growth of the yeast studied (10–11.2 mm).

In the study done by Pandithurai et al. (2015b), the antifungal activity of various solvent extracts aqueous/water, methanol, chloroform, ethyl acetate, and hexane of *Spatoglossum asperum* was tested against fungal dermatophytes and non-dermatophyte. The highest inhibition was recorded in chloroform 98.83% (11.86 mm) and methanolic extract 92.91% (11.15 mm) against *Aspergillus flavus*, in the same way Lavanya and Veerappan (2012) reported that the chloroform extract showed 10 mm zone of inhibition against *A. flavus*, and no inhibition was observed by them in the methanolic extract of marine brown alga *Turbinaria conoides*. In the study by Pandithurai et al. (2015b), no inhibition zone was observed in aqueous, ethyl acetate, and hexane extract against *A. flavus*. The methanolic extract showed the highest percentage of inhibition against all the pathogens studied—*Candida albicans* 57.14% (14.67 mm) and *Candida tropicalis* 54.75% (15.33 mm); *Trichophyton mentagrophytes* 50.85% (16.44 mm), chloroform and aqueous extract of *S. asperum* showed moderate activities against the fungal strain studied. However, ethyl acetate and hexane extract don't show any activity against the above pathogens studied. Lavanya and Veerappan (2012) reported that the methanolic extract of *Sargassum swartzii* (formerly *S. wightii*) and *T. conoides* exhibited the highest activity against *C. albicans*, and similarly the methanolic extract of *S. asperum* showed the highest activity against *C. albicans*. In the present study, aqueous extract of *S. asperum* showed moderate activity against *C. tropicalis*, and it was correlated with the study by Boonchum et al. (2011), in which it was reported that the aqueous extract of *T. conoides* showed moderate activity against *C. tropicalis* (Pandithurai et al. 2015b).

Cyclohexane, chloroform, and ethanol extracts of *Sargassum vulgare* displayed higher antifungal activity than acetone and ethyl acetate extracts (Khallil et al. 2015). The highest inhibitory actions of cyclohexanic extract were recorded against *Epicoccum nigrum*, *Fusarium oxysporum*, *Cladosporium cladosporioides*, and *Aspergillus ochraceus*. A moderate inhibitory action was recorded against *Penicillium citrinum*, *Aspergillus niger*, *Alternaria alternata*, and *Aspergillus flavus*. Both *A. flavus* and *A. niger* were not affected by the chloroform extract of *S. vulgare*, whereas another experimented fungal species was slightly or moderately affected. Acetone extract did not display inhibitory action against all tested fungi, except *E. nigrum*, which was slightly affected. Nearly similar inhibitory action was recorded for ethyl acetate extract against *F. oxysporum*, and a moderate inhibitory action was recorded against *E. nigrum*. *A. alternata*, *A. flavus*, *E. nigrum*, and *C. cladosporioides* did not show responses against the ethanol extract of *S. vulgare*, but the other tested fungi were slightly affected (Khallil et al. 2015).

Cyclohexanic extract of *Colpomenia sinuosa* exhibited the highest antifungal activity in comparison to the other applied extracts. The strongest inhibitory action was recorded against *F. oxysporum*, followed by *C. cladosporiodes*. The weakest inhibitory action was recorded against *A. flavus*, *A. niger*, *A. ochraceus*, and *E. nigrum*. A moderate inhibitory action was monitored against *P. cirinum* (Khallil et al. 2015). Chloroform extract did not inhibit both *A. alternata* and *A. niger*, but displayed a slight inhibitory action against the remaining tested fungi. Acetone extract exhibited no inhibitory action against all tested fungi, except for *C. cladosporiodes* which was strongly inhibited and *E. nigrum*, which was slightly inhibited. Ethyl acetate extract exerted the weakest inhibitory action since all the tested fungi were not negatively affected by its application, except for *F. oxysporum* which was slightly inhibited. Ethanol extract showed a slight inhibitory action against all experimented fungi, except for *A. alternata*, *A. flavus*, and *A. niger*, which exhibited no inhibitory action (Khallil et al. 2015).

It was observed that cyclohexane extract of *Cystoseira barbata* was the best organic solvent for extracting the effective antifungal material from the applied algal species and exhibited the highest antifungal potential particularly against *F. oxysporum*, followed by *A. ochraceus* and *E. nigrum* (Khallil et al. 2015). The weakest inhibitory action of cyclohexanic extract was recorded against *A. alternata* and *A. flavus*. A moderate inhibitory action was recorded against *P. citrinum*, *A. niger*, and *C. cladosporiodes* (Khallil et al. 2015). Chloroform and ethanol extracts followed cyclohexanic extract as antifungal activity exhibited low potentiality against most of the tested fungi. There was no activity recorded in chloroform extract against *A. alternata*, *A. niger*, and *E. nigrum*. A slight antifungal activity of chloroform extract was recorded against *F. oxysporum*, followed by *A. flavus*, *A. ochraceus*, *P. citrinum*, and *C. cladosporiodes*. Five of the experimented fungi were not negatively affected by ethanol extract, whereas the other three fungal species were slightly inhibited. Both acetone and ethyl acetate extracts displayed the lowest antifungal activity, since the former did not display antifungal activity against all experimented fungi. Similarly, all experimented fungi except *F. oxysporum* resisted ethyl acetate extract (Khallil et al. 2015).

In general, extracts of *Dictyota dichotoma* displayed low antifungal potential in comparison to the other applied seaweeds (Khallil et al. 2015). Cyclohexanic and ethanolic extracts of *D. dichotoma* exhibited higher antifungal activity than the other three applied extracts. Cyclohexanic extract was more effective than both employed patented medicine reagents. The tested fungi varied in response to cyclohexanic extract, since *A. alternata* and *A. niger* were not retarded. *A. flavus*, *A. ochraceus*, and *C. cladosporiodes* were slightly or moderately inhibited but *P. citrinum*, *E. nigrum*, and *F. oxysporum* were strongly inhibited. Chloroform extract of *Dictyota dichotoma* did not display inhibitory activity against both *A. alternata* and *A. niger*, whereas the remainder of the experimented fungi were slightly inhibited. All tested fungi were not affected by acetone extract of *D. dichotoma*, except *F. oxysporum*, which was slightly inhibited. Similarly, no inhibitory action was recorded against all tested fungi resulting from ethyl acetate application.

Ethanol extract showed different results ranging from no inhibitory action against *A. alternata*, *A. flavus*, *E. nigrum*, and *C. cladosporiodes*, to slight inhibitory action against the remaining three fungal species (Khallil et al. 2015).

The highest antifungal activity was recorded for cycloxanic extract of *Dictyopteris polypodioides* (formerly *Dictyopteris membranacea*), followed by chloroform and ethanol extracts (Khallil et al. 2015). The maximum inhibitory action of cyclohexanic extract was recorded against *F. oxysporun*, and a moderate inhibitory action was recorded against *P. citrinum*, *A. ochraceus*, and *E. nigrum*. A slight inhibitory action was recorded for *A. alternata*, *C. cladosporiodes*, and *A. niger*. Chloroform extract did not display inhibitory action against *A. alternata* and *E. nigrum*, whereas the remaining fungal species were slightly inhibited. Both acetone and ethyl acetate extracts presented the weakest inhibitory action since all the experimented fungi do not inhibit except *F. oxysporum*. A similar slight inhibitory action was recorded for *F. oxysporum* when treated with ethyl acetate extract. No inhibitory action was recorded for ethanol extract against *A. alternata*, *A. flavus*, *A. niger*, *E. nigrum*, and *C. cladosporiodes*, whereas a slight inhibitory action was recorded against the remaining fungal species (Khallil et al. 2015).

In the works by the Sabour research group, methanolic extract of *Dictyopteris polypodioides* has been tested against four yeast strains of *Candida* and five phytopathogenic fungi (Belattmania et al. 2016). The obtained results showed that *D. polypodioides* exhibited strong antifungal activities with diameters of growth inhibition ranging between 7 mm and 20 mm against phytopathogenic fungi (*Verticillium dahliae*, *Fusarium oxysporum*, *Fusarium graminearum*, *Botrytis cinerea*, and *Geotrichum* sp.), and between 12 mm and 32 mm against the tested yeasts (*Candida albicans*, *Candida glabrata*, *Candida krusei*, and *Candida parapsilosis*). The antifungal activity of seaweeds extracts and isolated compounds has not been extensively studied, mainly because in the past few years more attention has been paid to pathogenic bacteria, which is, by far, more explored (Mayer et al. 2013). Nevertheless, the antifungal activity of seaweeds has been demonstrated by some researchers where the results showed that fungal inhibition depends on seaweeds species and solvent used. For example, Guedes et al. (2012) found that methanolic, ethanolic, and dichloromethanolic extracts of seaweeds were most active against dermatophytes and *Candida* spp., compared to other solvents.

In vitro studies were conducted by Ambika and Sujatha (2015) to evaluate the effect of seaweed extracts of *Sargassum myricocystum* against the mycelial growth of *Macrophomina phaseolina* at different concentrations of 10%, 15%, 20%, 25%, and 30%, along with control by poison food technique. The result revealed that extract of *S. myricocystum* showed significant antifungal activity against the pathogen. *S. myricocystum* (30%) extract recorded the lowest mycelial growth (45.2 mm, 50.6 mm, 58.4 mm, and 61.5 mm) at 24 hours, 48 hours, 72 hours, and 96 hours after incubation (Ambika and Sujatha 2015).

The antifungal activity of selected marine macroalgae crude extracts was studied by Sekar and Kolanjinathan (2015) against two pathogenic yeasts. The crude methanolic extract of *Padina gymnospora* showed maximum mean zone of inhibition against *Candida albicans* (15 mm), followed by *Turbinaria conoides* (15 mm), and *Sargassum swartzii* (as *Sargassum wightii*) (15 mm). The crude hexane extract of marine macroalgae crude extracts showed minimum zone of inhibition against *C. albicans* when compared to the other solvent extracts. No zone of inhibition was seen in DMSO negative control, and the positive control Fluconazole (100 units) showed 17 mm inhibition zone against *C. albicans*. The minimum inhibitory concentration (MIC) of marine macroalgae crude extracts against *C. albicans* was ranged between 20 mg mL^{-1} to 640 mg mL^{-1}. The crude methanolic extract of *P. gymnospora*, *T. conoides*, and *S. swartzii* showed best MIC at 20 mg mL^{-1} against *C. albicans*.

The antifungal activity of selected marine macroalgae crude extracts was also studied against *C. glabrata*. The crude methanolic extract of *P. gymnospora* showed maximum mean zone of inhibition against *C. glabrata* (16 mm), followed by *T. conoides* (15 mm), and *S. swartzii*. No zone of inhibition was seen in DMSO negative control and the positive control Fluconazole (100 units) showed zone of inhibition of 18 mm against the *C. glabrata*. The minimum inhibitory concentration (MIC) of marine macroalgae crude extracts against *C. glabrata* was ranged between 20 mg mL^{-1} to 640 mg mL^{-1} (Sekar and Kolanjinathan 2015).

In the study on *Padina tetrastomatica* done by Radhika and Priya (2016), this species had the highest inhibition against *Gibberella* sp. (7.5 mm) and the second best against *Alternaria* sp. (6.5 mm). *A. fumigatus* was also moderately inhibited by both crude and fractions of the seaweed extract. Comparing the crude and fraction, the ethanol fraction of *P. tetrastomatica* had the highest activity (7.5 mm) followed by ethanol crude extract. Acetone and methanol extracts also showed good antifungal activity. Aqueous extract had the least activity as it showed no inhibition against four of the five pathogens tested with the exception of *A. fumigatus*. On comparing the inhibitory capacity of the crude and fraction extracts, the crude extracts had lesser antifungal activity except for acetone extracts where the crude was more successful in curbing the pathogens. Methanol extracts also showed a similar pattern against *A. terreus* and *A. fumigatus* (Radhika and Priya 2016).

In the study done by Taherpour et al. (2016), none of the examined crude extracts of *Padina* sp. showed toxicity against *Aspergillus flavus* and *Candida albicans*. This result was previously obtained by Padmakumar and Ayyakkannu (1997), which confirmed resistance of *A. flavus* against seaweed extracts. Therefore, the results of the present study were similar to Tüney et al. (2006), whose investigation evaluated antifungal activities of methanol, acetone, and diethyl ether extracts of *Padina pavonica*.

In the Master's thesis works of Pires (2016), the antifungal activity of the methanolic extracts of *Saccharina latissima* against *Candida glabrata* was assessed through micro-dilution assay. The extracts were tested at decreasing concentrations. Initially 500 μg mL^{-1} > 250 μg mL^{-1} > 100 μg mL > 10 μg mL^{-1}, these concentrations would allow a rapid approach to MIC values because they are widely dispersed. The minimum inhibitory concentration (MIC) for the extracts was 25 μg mL^{-1}. In the same work, some assays growth inhibition of *Candida krusei* treated with methanolic extracts of *Laminaria ochroleuca* was observed, however, these results were not clear. As such, the author considers there had been only a vestigial activity (Pires 2016).

In the study done by Mashjoor et al. (2016), antifungal and cytotoxic activities of the extracts of marine macroalgae collected in the Persian Gulf (Iran) were evaluated. All extracts exhibited moderate antifungal activity, except for one resistant antifungal strain—*Aspergillus niger*. Methanol extract of *Padina antillarum* show moderate activity against *Candida albicans* (13.0 mm), with a minimum inhibitory concentration (MIC) of 16 mg mL^{-1}, and is highly active against *Saccharomyces cerevisiae* (16.0 mm), with a MIC value of 7.5 mg mL^{-1}. Ethylacetate extract shows moderate activity against *S. cerevisiae* (7.5 mm), with a minimum inhibitory concentration (MIC) of 7.5 mg mL^{-1}, and is highly active against *C. albicans* (18.0 mm), with a MIC value of 7.5 mg mL^{-1}. Methanol extract of *Padina boergeseni* shows moderate activity against *Candida albicans* (12.0 mm), with a minimum inhibitory concentration (MIC) of 15 mg mL^{-1}, and also against *Saccharomyces cerevisiae* (14.0 mm), with a MIC value of 7.5 mg mL^{-1}. Ethylacetate extract shows moderate activity against *C. albicans* (12.0 mm), with a minimum inhibitory concentration (MIC) of 15 mg mL^{-1}, and is highly active against *S. cerevisiae* (15.0 mm), with a MIC value of 7.5 mg mL^{-1}.

7.6 Antifungal Activity of Rhodophyta (Red Algae)

There are numerous studies testing inhibition against bacteria, some viruses, and marine fungi, but there are few studies of the activity of compounds or extracts of *Laurencia* complex against human pathogenic fungi. Ballantine et al. (1987) collected *Laurencia obtusa*, *Yuzurua poiteaui* (as *Laurencia poitei*, formerly *Palisada poitei*), and *Laurencia* sp. from Puerto Rico, and evaluated the antifungal properties of chloroform/methanol (2:1) extracts against *Candida albicans*. The *L. obtusa* extract showed a 0.5 mm inhibition using the disc-agar method.

Tariq (1991) reported that extracts of *Dilsea carnosa*, *Osmundea pinnatifida* (formerly *Laurencia pinnatifida*), *Odonthalia dentata*, and *Vertebrata lanosa* (formerly *Polysiphonia lanosa*) reduced the rate of colony extension in *Microsporum canis* and *Trichophyton verrucosum*, with seasonal variations in the levels of inhibitory activity.

In the work made by Carg (1993), the highly volatile fractions extracted from the genera *Asparagopsis*, *Bonnemaisonia*, and *Ptilonia* (family Bonnemaisoniaceae) have been investigated. A great variety of halogenated alkanes, saturated and unsaturated ketones, aldehyde, alcohols, epoxides, and halogenated derivatives of acetic and acrylic acids have shown antifungal activity against *Fusarium* sp. (Garg 1993).

In another study (Val et al. 2001), extracts from 44 species of seaweed from Canary Islands (Spain) were screened to produce antifungal and antibacterial compounds against a panel of $Gram^-$ and $Gram^+$

bacteria, mycobacterium, yeasts, and fungi. *Asparagopsis taxiformis* and *Osmundea hybrida* presented activity against the filamentous fungi *Aspergillus fumigatus*, and/or the yeasts *Candida albicans* and *Saccharomyces cerevisiae*. *A. taxiformis* was the species with the strongest activities against the broadest spectrum of target microorganisms. *A. taxiformis* and *O. hybrida* were active against mycobacterium too. Only one species – *A. taxiformis* – showed activity against the whole panel of nine target microorganisms, including *C. albicans*, *S. cerevisiae*, and *A. fumigatus* (Val et al. 2001).

Some species of marine benthic algae collected from different coastal areas of Karachi (Pakistan) were investigated for their antifungal activities (Rizvi and Shameel 2004). *Botryocladia leptopoda* exhibited the greatest antifungal inhibition activity against *Microsporum conis* (50%), *Pseudallescheria boydii* (52%), *Trichophyton mentagrophytes* (55%), *Trichophyton simii* (60%), *Fusarium solani* (60%), and *Fusarium verticillioides* (formerly *Fusarium moniliforme*) (66.6%). The other red algae with good antifungal activity were *Champia compressa*, *Hypnea valentiae*, and *Sarconema filiforme* (formerly *Sarconema furcellatum*). The ethanolic extracts derived from seven seaweed species showed no detectable antifungal activity against *Epidermophyton floccosum*, *Microsporum canis*, and *Trichophyton rubrum* (Alam et al. 1994). It appears that different seaweeds behave variably against a variety of fungal species (Rizvi and Shameel 2004).

Solieria robusta inhibited growth of root infecting fungi *Macrophomina phaseolina*, *Thanatephorus cucumeris* (formerly *Rhizoctonia solani*) (Sultana et al. 2005), of *Fusarium solani*, the human pathogens *Pseudallescheria boydii* and *Trichophyton schoenleinii*, and the animal pathogenic fungi *Microsporum canis* and *Trichophyton simii* (Rizvi and Shameel 2004).

Atlantic marine algae were screened in growth inhibition assays against the saprophytic fungus *Scolecobasidium salinum* (formerly *Dendryphiella salina*), the pathogenic fungus *Lindra thalassiae*, the oomycetes *Phytophthora spinosa* (formerly *Halophytophthora spinosa*), and the labyrinthulomycetes (formerly slime moulds) *Schizochytrium aggregatum* (Puglisi et al. 2007). Overall, 80% of all red algae yielded extracts active against *P. spinosa*, and 56% against *S. salinum*. In comparison, the extracts from 36% of all Rhodophyta were active against *L. thalassiae*, and only 16% yielded extracts active against *S. aggregatum*. While the extracts from *Laurencia* sp. and *Lithophyllum kotschyanum* were only active against *P. spinosa*, and extracts from *Hypnea pannosa* were selectively active against *S. salinum*, all other extracts from red algae were active against multiple assay microorganisms. Among these, *Portieria hornemannii* was the only species to yield extracts active against all assay microorganisms. Broad-spectrum activity against three or four assay microorganisms was observed in extracts from *Dasyphila plumarioides*, *Gibsmithia hawaiiensis*, *Rhodymenia divaricata*, and most species belonging to the families Rhodomelaceae and Corallinaceae. Among the members of the family Rhodomelaceae, including *Acanthophora spicifera*, *Tolypiocladia glomerulata*, and three species of the genus *Laurencia*, the extracts from *A. spicifera* and all *Laurencia* species were active against only one or two assay microbes, and the extracts from *T. glomerulata* were active against fungi and stramenopiles. Among *Amphiroa fragilissima*, *Lithophyllum kotschyanum*, *Lithophyllum pygmaeum* (formerly *Lithophyllum moluccense*), and *Mastophora rosea*, only the extracts from *M. rosea* were active against fungi and stramenopiles. While the extracts from *L. pygmaeum* were active against *L. thalassiae*, *S. salinum*, and *P. spinosa*, the extracts from *L. kotschyanum* were only active against *P. spinosa*. The extracts from *A. fragilissima* were only active against *S. salinum* (Puglisi et al. 2007).

Cordeiro et al. (2006) obtained a protein fraction rich in lectin from the red seaweed *Hypnea musciformis* by precipitation with ammonium sulfate (F40/70). It was screened for chitinase and β-1,3-glucanase activity, and assessed for antifungal potential against the human pathogen yeasts *Candida albicans* and *Blastodendrion arztii* (formerly *Candida guilliermondii*). The F40/70 fraction showed chitinase and β-1,3-glucanase enzymes with specific activities of 276.43 units' mg^{-1} protein and 1880.7 units' mg^{-1} protein, respectively. It can inhibit the growth of *B. arztii* at the concentrations of 45 µg protein mL^{-1}, 100 µg protein mL^{-1}, and 450 µg protein mL^{-1}, but it showed only a discrete inhibition against *C. albicans*, irrespective of the tested concentrations. The inhibitory action was shown to be fungistatic.

The study done by Lakshmi et al. (2006) deals with the biological activities of the extracts of 48 marine florae. The biological screening includes tests for antibacterial (see Chapter 8), antifungal, among others. From among red algae, the crude extracts from *Chondria dasyphylla*, *Gracilaria corticata*, *Gracilaria canaliculata* (formerly *Gracilaria crassa*), *Halymenia porphyroides*, *Heterosiphonia muelleri*, *Hypnea musciformis*, *Laurencia obtusa*, *Palisada poiteaui* (formerly *Laurencia poiteaui*), *Scinaia moniliformis* (formerly *Scinaia indica*), and *Solieria robusta* were active against *Candida albicans*, *Cryptococcus neoformans*, *Trichophyton mentagrophytes*, *Aspergillus fumigatus*, and *Sporotrichum schenckii*.

In Pakistan, Khanzada et al. (2007) screened various fractions of ethanolic extract of *Solieria robusta* for antifungal activity against five fruit spoiling fungi (*Aspergillus flavus*, *A. niger*, *A. ochraceus*, *Penicillium funiculosum*, and *Phytophthora infestans*) isolated from fruits, and reported that all fractions were able to inhibit fungal growth. All five fractions of ethanolic extract of *S. robusta* showed activity against fruit spoiling fungi. The quantum of activity exhibited in the fractions of seaweed varied from mild in ethanol to significant activity in aqueous fractions. Aqueous fraction inhibited 99% growth of *A. niger* with 20 mg mL^{-1} concentration. Activity continued to the lower concentration of 0.02 mg mL^{-1}. Antifungal activity of aqueous fraction was the highest, whereas methanol, ethyl acetate, and chloroform fractions showed moderate inhibitions with 20 mg mL^{-1} concentration. All the test fungi, more or less, were inhibited by all fractions of *S. robusta* extract. It was observed that the aqueous fraction retained the highest inhibition ratios at the lower concentrations. During the bioassays of minimum concentration (0.02 mg mL^{-1}), a significant inhibition of *A. niger* (89%), *P. infestans* (10%), and *P. funiculosum* (9%) was observed in aqueous fraction. At the minimal concentration (0.02 mg mL^{-1}) 18%, 14%, 10%, and 6% inhibition of *P. infestans* was observed in chloroform, ethyl acetate, and ethanol fractions, respectively. While the ethanol fraction was the only fraction it could not inhibit the growth of *P. infestans*, the methanol fraction (1%) inhibited *P. funiculosum*. Inhibition ratios increased at lower concentrations of ethyl acetate fraction against *P. funiculosum* and at higher concentrations inhibition ratio decreased. Aqueous fraction was the dominant and the most active fraction of the ethanolic extract of seaweed with antifungal activity against all test fungi (Khanzada et al. 2007).

In 2007, Salvador et al. evaluated the antifungal and antibacterial activity of 82 Iberian macroalgae (18 Chlorophyta, 25 Phaeophyceae, and 39 Rhodophyta) against three $Gram^{+}$ bacteria, two $Gram^{-}$ bacteria (see Chapter 8), and one yeast (*Candida albicans*). The bioactivity was analyzed from crude extracts of fresh and lyophilized samples. Of the seaweeds analyzed, 67% were active against at least one of the six test microorganisms. The highest percentage of active *taxa* was found in Phaeophyceae (84%), followed by Rhodophyta (67%), and Chlorophyta (44%). Nevertheless, red algae had both the highest values and the broadest spectrum of bioactivity. *Bonnemaisonia asparagoides*, *Bonnemaisonia hamifera*, *Asparagopsis armata* (and *Falkenbergia rufolanosa* phase) (Bonnemaisoniales) were the most active *taxa*. In this study, Ceramiales and Gigartinales had noteworthy antimicrobial activity, and Bonnemaisoniales was the order that had the highest bioactivity.

According to Lane et al. (2009b), bromophycolides and callophycoic acids from *Callophycus serratus* extracts inhibited growth of *Lindra thalassiae*, a marine fungal pathogen, and represent the largest group of algal antifungal chemical defenses reported to date.

Aruna et al. (2010) reported that among the seaweeds tested, the highest rate of antifungal activity was noticed in the red alga *Kappaphycus alvarezii*. The methanolic extract of red seaweeds *Asparagopsis taxiformis*, *Chondrophycus brandenii* (as *Laurencia brandenii*), *Chondrophycus ceylanicus* (as *Laurencia ceylanica*), and *Hypnea valentiae* showed higher activity against *Candida albicans*.

Specimens of the red alga *Bostrychia tenella* collected from São Paulo coast (Brazil) presented high activity in an antifungal assay with the phytopathogenic fungi *Cladosporium cladosporioides* and *Cladosporium sphaerospermum* (Felício et al. 2010).

Elsie and Rajan (2010) had investigated the antifungal activity of three different solvents extracts of *Gelidiella acerosa*. The *G. acerosa* extracts are obtained from three different solvents like methanol, ethanol, and acetone and various antifungal activities were observed. In methanolic extracts the maximum activity is seen in *Candida albicans* (5 mm) and *Aspergillus flavus* (5 mm). No activity is seen in *Candida tropicalis*, *Aspergillus fumigatus*, and *Aspegillus niger*. In ethanol extract, the maximum activity is seen in *C. albicans* (7 mm) and *C. tropicalis* (7 mm), and minimum activity is seen in *A. flavus* (5 mm), and no activity in *A. fumigatus*. The extracts obtained induced inhibition in *C. albicans* (6 mm), but showed minimal activity against the pathogens *C. tropicalis*, *A. fumigatus*, and *A. niger*.

With a view to explore the finding of new molecules with therapeutic efficacy for human use, the alcoholic extracts of 33 identified species of marine flora, collected from Indian coasts, were prepared and screened for a wide range of biological activities (Lakshmi et al. 2010). From among red algae, the alcoholic extracts from *Gelidiella acerosa*, *Hypnea musciformis*, and *Portieria hornemannii* (formerly *Chondrococcus hornemanii*) were active against *Candida albicans*, *Cryptococcus neoformans*, *Trichophyton mentagfrophytes*, *Aspergillus fumigatus*, and *Sporotrichum schenckii*.

In the research done by Elsie et al. (2011), antimicrobial activities of ethanolic and acetone extracts of *Gelidiella acerosa* against human pathogens like *Aspergillus flavus*, *Aspergillus niger*, *Aspergillus fumigatus*, *Candida albicans*, and *Candida tropicalis* were tested. Ethanol extracts of *G. acerosa* showed the maximum activity against *C. albicans* (7 mm), *C. tropicals* (7 mm), *A. niger* (7 mm), and minimum zone formation in *A. flavus* (5 mm), and no zone was observed in *A. fumigatus*. The referred results show that *G. acerosa* exhibit high antifungal activity (5 mm and 7 mm) in ethanol extract when compared to acetone extracts.

In an initial report of the antifungal activities of the secondary metabolites of five species of *Laurencia* collected in the state of Espírito Santo, Brazil (Stein et al. 2011), against three strains of pathogenic fungi—*Candida albicans*, *Candida parapsilosis*, and *Cryptococcus neoformans*, the minimum inhibitory concentrations (MIC) of the algal extracts were determined by serial dilution method. All results showing maintenance or reduction of the inoculum were defined as fungistatic, while fungicidal action observed no growth in the 10 μL fungistatic samples sub-cultured in Sabouraud Agar. The MIC of the chloroform and methanol extracts of *Laurencia dendroidea* were < 31.25 μg mL^{-1} against *C. albicans* with a fungistatic effect; the chloroform extract of *Laurencia catarinensis* showed the same results against *C. parapsilosis*. A fungistatic effect was observed for the methanol extract of *Laurencia aldingensis* against *C. parapsilosis* (MIC < 31.25 μg mL^{-1}), *C. neoformans* (33.9 μg mL^{-1}), and *C. albicans* (65.2 μg mL^{-1}). The fungicidal effects of the hexane and chloroform extracts of *L. aldingensis* against *C. parapsilosis* were indicated by MIC values of 49.0 μg mL^{-1} and 57.8 μg mL^{-1}, respectively; the same extracts were fungistatic against *C. albicans* and *C. neoformans* between the concentrations of 85.3 μg mL^{-1} and 130.8 μg mL^{-1}. Moreover, the chloroform extract of *L. catarinensis* was fungicidal against *C. albicans* and *C. neoformans* at 303.8 μg mL^{-1} and 600.4 μg mL^{-1}, respectively. Of all the extracts of *Laurencia intricata*, the chloroform extract was the most active, with a MIC higher than 220 μg mL^{-1}. The *Laurencia translucida* extracts showed no activity against *C. albicans*, while between 177 μg mL^{-1} and 743 μg mL^{-1} were required to observe a fungistatic effect against *C. parapsilosis* or *C. neoformans*. Comparison of the three different fungal strains indicates that *C. albicans* was the most susceptible to maintenance or reduction of growth by the extracts, while *C. parapsilosis* was the most susceptible to the fungicidal effects of the algal extracts. *L. aldingensis* appears to be a particularly interesting alga, showing activity against all three fungal strains tested. In general, the results show that the aqueous extracts stimulate cells growth, except for the crude water extracts of *L. aldingensis* and *L. dendroidea*, which show a reasonable percentage of inhibition (110 μg mL^{-1} and 100 μg mL^{-1}, respectively) against *C. neoformans*. The MIC values of the positive control Fluconazole were in agreement with those reported by the National Committee Clinical Laboratory Standards (NCCLS) (MIC ≤ 2 μg mL^{-1}) for the reference antifungal agent.

The synthesis, characterization, and application of biologically synthesized nanomaterials are an important aspect in nanotechnology. The study done by Vivek et al. (2011) deals with the synthesis of silver nanoparticles (Ag-NPs) using the aqueous extract of red seaweed *Gelidiella acerosa* as the reducing agent to study the antifungal activity. The formation of Ag-NPs by reduction of the aqueous Ag^+ during exposure to the aqueous extract of *G. acerosa* showed reddish-brown color, which suggested the formation of Ag-NPs in the solution. Further the nanoparticles synthesis by green route by using *G. acerosa* extract was found highly active against tested fungal species at a concentration of 50 μL of synthesized Ag nanoparticles. The results showed higher antifungal activity against *Mucor indicus* (22.3 vs. 21.3) and *Trichoderma reesei* (17.2 vs. 14.3), whereas moderate activity was revealed against *Fusarium dimerum* (13.15 vs. 13.0), *Humicola insolens* (12.2 vs. 12.1) when compared to standard antifungal agent Clotrimazole.

The antifungal efficiency of various solvent extracts of algae *Acanthaphora spicifera* was evaluated by Pandian et al. (2011). The antifungal activity of various extract of *A. spicifera* were tested against various microorganism. The methanol extracts showed maximum antifungal activity than petroleum ether and chloroform extracts. The zone of inhibition were ranged between 15.1 mm to 2.8 mm against fungal strains. Methanol extract shows promising antifungal activity against all fungal strains. This extract presented maximum activity against *Microsporum gypseum* (15.1 mm), *Aspergillus niger* (13.4 mm), and *Candida albicans* (11.2 mm). The further study was carried out only with methanolic extract. The minimum inhibitory concentration of methanol extract was tested against fungal strains. The 100 μg mL^{-1} concentration showed inhibitory effect against fungal strains. As the concentration increases from 100 μg mL^{-1}, 500 μg mL^{-1}, 1 mg mL^{-1}, 10 mg mL^{-1}, 50 mg mL^{-1}, the inhibitory effect also increases proportionally the maximum inhibitory effect absorbed with 50 mg mL^{-1} of methanol extract as compared to standard drug. From the study it's observed that if the crude extract is purified it could show similar activity when compared to the standard antifungal agent amphotericin (Pandian et al. 2011).

In a study done by Mhadhebi et al. (2012b), several organic extracts obtained from marine algae-*Halurus equisetifolius* and *Gelidium spinosum* (formerly *Gelidium latifolium*) collected from Tunisian Mediterranean coast, were evaluated for their antifungal activities against five human pathogenic yeasts using the agar disc diffusion assay. The petroleum ether extract of *H. equisetifolius* and *G. spinosum* (formerly *G. latifolium*) presented antifungal activity with a diameter of inhibition zone in the range of 11 mm against *Candida glabrata*.

Lavanya and Veerappan (2012) reported that in a study with two red macroalgae extracts, the highest inhibition zones (8 mm) were recorded in acetone *Acanthophora spicifera* extract against *Fusarium udum*, followed by methanol extract against *Candida albicans* (6 mm) and ethyl acetate extract against *Alternaria alternata* (6 mm). The lowest inhibition zones (2 mm) were measured in acetone extracts of *A. alternata*, *Trichophyton solani*, *C. albicans*, *Aspergillus niger*, and *Aspergillus flavus* in chloroform extract against *Candida krusei* and *A. flavus*. There was no activity in hexane extract against all tested pathogens. The maximum inhibition zones (4 mm) were noted in acetone extract of *Gracilaria canaliculata* (formerly *Gracilaria crassa*) against *F. udum*, *C. albicans*; diethyl ether extract against *A. flavus* chloroform extract against *A. alternata*, *T. solani*; water extract against *T. solani*, *B. cinerea*. The minimum activities (2 mm) were measured in acetone extract against *R. solani*, *C. krusei*; in methanol extract against *C. albicans*; in diethyl ether extract against *C. krusei*; in chloroform extract against *Fusarium solani*. Hexane and ethyl acetate extract showed no activity against all tested pathogens (Lavanya and Veerappan 2012).

In the research conducted by Bibiana et al. (2012), antifungal activity of *Kappaphyus alwarezii* extracts were found to be present against all tested fungi. The maximum activity of diethyl ether extract was shown against *Trichophyton rubrum* (8 mm), and *Microsporum gypseum* (7 mm); of petroleum ether extracts was shown against *Aspergillus niger* and *A. flavus* (8 mm); of acetone extract was shown against *M. gypseum* and *T. rubrum* (8 mm); and of acetic acid extracts was shown against most of the tested fungal strains with the maximum activity of 8 mm.

In the study done by Saidani et al. (2012), antifungal activity of four species of marine algae of Bejaia coast (Algeria) was explored. The highest inhibiting effect was noted for *Rhodomela confervoides*, against *Candida albicans* (diameter of inhibition zone was 24 mm) and *Mucor ramaniannus* (diameter of inhibition zone was 26 mm). *Aspergillus niger* showed resistance against majority of methanolic extracts.

Genovese et al. (2013) confirmed the antifungal activity of *Asparagopsis taxiformis*, from the Straits of Messina (Italy), against *Aspergillus fumigatus*, *Aspergillus terreus* and *Aspergillus flavus*. The lowest MIC observed were < 0.15 mg mL^{-1}, and the highest were > 5 mg mL^{-1}.

In a study developed by Zhang et al. (2013b), the common seaweeds from Dalian coastline of the North Yellow Sea (China) were found to yield considerable biomass and some natural constituents with remarkable antifungal activity. Methanolic extract of *Symphyocladia latiuscula* and *Rhodomela confervoides* showed inhibition zones against *Pyricularia oryzae* with a diameter of 35 mm and 20 mm, respectively.

Jassbi et al. (2013) reported antioxidant and antifungal activity in the water extracts of a red algae, *Hypnea flagelliformis* from the Persian Gulf. The active substances were identified as free fatty acids, fucosterol, cholesterol, and 22-dehydroxychlosterol.

A preliminary research (Manigandan and Kolanjinathan 2014) was carried out to find out the antifungal activity of *Gracilaria corticata* and *Hypnea valentiae* extracts collected from Mandapam coastal regions of Tamil Nadu (India). The acetone extract of *G. corticata* showed maximum activity against *Aspergillus flavus* of 12.0 mm zone of inhibition, followed by *Aspergillus niger* (12.0 mm), and minimum zone of inhibition was observed in *Candida albicans* (9.0 mm). The ethyl acetate extracts showed maximum activity against *A. niger* of 10.0 mm zone of inhibition, followed by *C. albicans* and *A. flavus* (8.0 mm). The chloroform and hexane extracts showed minimum activity of 8.0 mm. The acetone extract of *H. valentiae* showed maximum activity against *A. niger* of 15.0 mm zone of inhibition, followed by *C. albicans* (11.0 mm), and minimum zone of inhibition was observed in *A. flavus* (10.0 mm). The ethyl acetate extracts showed maximum activity against *A. flavus* (10.0 mm) followed by *C. albicans* of 9.0 mm zone of inhibition, and minimum activity was observed in *A. niger* (8.0 mm). In chloroform extract, *C. albicans* showed maximum zone of inhibition (10.0 mm) and hexane extracts showed minimum activity of 8.0 mm against *A. flavus*, followed by *C. albicans*, and no zone of inhibition was observed against *A. niger*.

According to Aswathi and Jamila (2014), the acetone extract of *Gracilaria edulis* showed highest antifungal activity against *Aspergillus flavus* (18 mm), and it was followed by *Aspergillus terreus* (17 mm) and *Fusarium incarnatum* (formerly *Fusarium semitectum*) (17 mm). Moderate activity was observed in the ethanol extract showing inhibition zone of 13 mm against *Aspergillus niger*, *Mucor* sp., and *Trichoderma viride* (Aswathi and Jamila 2014).

Saprolegniasis, which is caused by the oomycete *Saprolegnia parasitica*, is an important illness that affects the salmonid farming. The study carried out by Cortés et al. (2014) focused on the antifungal activity of a lipophilic extract of *Ceramium virgatum* (formerly *Ceramium rubrum*) on *S. parasitica*. The maximum inhibition activity for dichloromethane and for stearic acid (> 17.6% of inhibition) occurred at 250 μg mL^{-1}. These results are consistent with other studies of antifungal inhibition in Rhodophyta seaweeds (Cortés et al. 2014).

Banana and papaya are among the most important crops in the tropics, with a value amounting to millions of dollars per year. However, these fruits suffer significant losses due to anthracnose, a fungal disease. It is well known that certain seaweed extracts possess antifungal activity, but no published data appears to exist on the practical application of this property. In the study done by Machado et al. (2014), five Brazilian seaweed extracts were screened for their activity against banana and papaya anthracnose fungi (*Colletotrichum gloeosporioides* and *Colletotrichum musae*). Strong fungus-inhibitory effects of the seaweeds *Ochtodes secundiramea* and *Laurencia dendroidea* extracts were observed on both papaya (100% and 98% respectively) and banana (89% and 78% respectively). This impressive activity could be associated with halogenated terpenes, the major components of both extracts (Machado et al. 2014).

In a research work done by Machado et al. (2014b), one of the main objectives was to determine whether the species retain their antifungal potential after the period of laboratory cultivation, correlating the changes in physiology and those from bioactivity. Extracts from the macroalgae, *Ochtodes secundiramea* and *Palisada flagellifera*, after laboratory cultivation showed a significant increase in antifungal potential which could be positively correlated with the change in concentration and ratios of chlorophyll-*a* and accessory pigments. The results for antifungal activity corroborate those of Machado et al. (2011) who found inhibition of mycelial growth in *Colletotrichum gloeosporioides* of approximately 90% for *O. secundiramea* and around 50% for *Hypnea musciformis*.

In the work done by Kausalya and Rao (2015), the ethanolic extract of *Gelidium pusillum* showed maximum zone of inhibition against several fungal strains, i.e., *Rhizopus stolonifer* (17 mm), *Saccharomyces cerevisiae* (16 mm), *Candida albicans* (15 mm), *Aspergillus niger* (13 mm), *Thanatephorus cucumeris* (formerly *Rhizoctonia solani*) (13 mm), and *Mucor racemosus* (11 mm), with a concentration of 500 mg mL^{-1}. The methanolic extract of *G. pusillum* showed maximum zone of inhibition against fungal strains *C. albicans* (15 mm), *T. cucumeris* (formerly *R. solani*) (14 mm), *S. cerevisiae* (14 mm), *M. racemosus* (12 mm), *R. stolonifer* (11 mm), *A. niger* (12 mm), with a concentration of 500 mg mL^{-1}. The chloroform extract of *G. pusillum* showed maximum zone of inhibition against fungal strains, *R. stolonifer* (15 mm), *A. niger* (15 mm), *S. cerevisiae* (14 mm), *C. albicans* (13 mm), *T. cucumeris* (formerly *R. solani*) (11 mm), *M. racemosus* (11 mm), with a concentration of 500 mg mL^{-1}. Water extracts of *Gelidium pusillum* showed maximum zone of inhibition against fungal strains, *Thanatephorus cucumeris* (formerly *Rhizoctonia solani*) (13 mm), *C. albicans* (13 mm), *S. cerevisiae* (12 mm), *R. stolonifer* (12 mm), *M. racemosus* (11 mm), *A. niger* (11 mm), with a concentration of 500 mg mL^{-1}. Minimum inhibitory concentration of (MIC) values of *G. pusillum* against fungus ranged between 35 mg mL^{-1} to 85 mg mL^{-1}. The lowest MIC (35 mg mL^{-1}) value of ethanol extract was against *R. stolonifera* (Kausalya and Rao 2015).

The ethanol extract of *Centroceras clavulatum* showed maximum zone of inhibition against fungal strains such as *Aspergillus niger* (12 mm), *Rhizopus stolonifer* (10 mm), *Saccharomyces cerevisiae*

(11 mm), with a concentration of 500 mg mL^{-1}. No zone of inhibition was noticed against *Thanatephorus cucumeris* (formerly *Rhizoctonia solani*) with a concentration of 100 mg mL^{-1}, 300 mg mL^{-1}, and 500 mg mL^{-1}). The minimum inhibitory concentration (MIC) value of *C. clavulatum* against fungus was ranged between 45 mg mL^{-1} to 85 mg mL^{-1}. The lowest MIC (50 mg mL^{-1}) value of ethanol extract was against *A. niger* (Kausalya and Rao 2015).

Pinteus et al. (2015) evaluated the methanol, *n*-hexane, and dichloromethane extracts of macroalgae from Peniche coast (Portugal) for antifungal activity. All the red algae (Rhodophyta) presented high antifungal activity in the three fractions, with the only exception of *Plocamium cartilagineum* in the methanolic fraction. The red algae *Sphaerococcus coronopifolius*, *Asparagopsis armata*, and *P. cartilagineum* dichloromethane extracts (1 mg mL^{-1}) induced an inhibitory activity against *S. cerevisiae* of 86%, 96%, 76%, and 86%, respectively. *S. coronopifolius* revealed high antifungal potential for *n*-hexane (IC_{50} = 40.2 µg mL^{-1}), dichloromethane (IC_{50} = 78.9 µg mL^{-1}), and methanolic (IC_{50} = 55.18 µg mL^{-1}) extracts against *Saccharomyces cerevisiae* growth. The antifungal potency of the *S. coronopifolius* extracts was similar to the standard amphotericin B.

In the Master's thesis work (Silva 2015), the antifungal activity of the extracts against *Candida albicans* was assessed through two different approaches according to the Clinical and Laboratory Standards Institute (CLSI) standard procedures—disc-based testing and broth microdilution testing. In these works, the antifungal activity of *Osmundea pinnatifida* extracts were also tested against *Aspergillus fumigatus* and *Alternaria infectoria*. Until the execution of this work, studies in which the evaluation of *C. albicans* susceptibility to seaweed extracts was carried out using the standardized disc diffusion and broth microdilution methods are scarce. When concerning *O. pinnatifida* extracts, this may even be the first time. Nevertheless, Rizvi and Shameel (2005), using a different methodology, found that *O. pinnatifida* crude methanol extract inhibited the *C. albicans* growth by 3.43% at 400 µg mL^{-1}. On the other hand, Hellio et al. (2000) evaluated, through a National Committee for Clinical Laboratory Standards (NCCLS) standardized macrodilution method, the antifungal capacity of ethanol and dichloromethane fractions of *O. pinnatifida* (formerly *Laurencia pinnatifida*) against *C. albicans*. The results were negative (did not inhibit the growth of the yeast) at the concentrations tested, from 4 µg mL^{-1} to 96 µg mL^{-1}. Despite the different methodologies and yeast strains used, as well as the distinct biogeography of the samples, the results obtained in the cited works agree with that presented herein, suggesting that extracts of this nature don't have the potential to be used as a natural antifungal agent against *C. albicans*. Furthermore, the results revealed that the great majority of *O. pinnatifida* extracts promoted the fungal growth of *A. infectoria*, as compared to the negative control, regardless of the concentration tested (10 µg mL^{-1} and 100 µg mL^{-1}). An effect promoting fungal growth was also verified by Barreto et al. (2002), which found that *Osmundaria serrata* ethanol extract, at 31.25 µg mL^{-1}, stimulated *Colletotrichum gloeosporioides* growth, which is a common plant-pathogenic fungi that can also cause infection in humans (Cho et al. 2015). Nevertheless, methanol extract (10 µg mL^{-1}) and *n*-hexane extract of the IMTA-cultivated seaweed (10 µg mL^{-1} and 100 µg mL^{-1}) inhibited poorly the radial growth (3.7%, 1.79%, and 3.7%, respectively) as compared with control, after incubation for eight days. Although all treatments have been performed in duplicate, in some of them (namely 10 µg mL^{-1} *n*-hexane extract of the IMTA-cultivated seaweed, 100 µg mL^{-1} dichloromethane extract of wild sample, and 100 µg mL^{-1} *n*-hexane extract of the wild seaweed) a uniform radial growth was not observed, invalidating one of the tests. On the other hand, all the tested extracts were found to be effective against *Aspergillus fumigatus*, reducing the radial growth as compared to control. The extracts tested at 10 µg mL^{-1} show a percentage of radial growth inhibition that range from 2.94% to 9.28%, and from 7.84% to 19.83% for 100 µg mL^{-1} of extract. The extracts tested at the highest concentration showed higher values of radial growth inhibition than those tested at the lowest concentration, which seems to indicate a dose-dependent effect. Among the extracts in study, the hexane fractions stand out from the remaining as the most effective against *Aspergillus fumigatus*. Curiously, the fraction obtained from IMTA-cultivated seaweed showed a stronger effect than the wild sample, inhibiting almost 20% of the mycelial growth as compared to control, after four days of exposure (Silva 2015).

In the study on *Acanthophora spicifera* made by Radhika and Priya (2016), the ethanol extracts showed good antifungal activity (6 mm) followed by acetone and then methanol extracts. Aqueous extract had the least activity. *Aspergillus fumigatus* was the highest inhibited fungal species, followed by *Alternaria* sp. and *Aspergillus terreus*. *Ganoderma* sp. showed the least activity against all the pathogens tested. The fractions had a slightly higher rate of inhibition when compared to the crude extracts, with the exception of ethanol extract, where the crude showed a higher rate of inhibition against *A. terreus* and *A. fumigatus* and in acetone extract in which the crude extracts showed promising antifungal activity more than the fraction in fighting against the pathogen *Gibberline* sp. Aqueous extract had no activity against *A. terreus* and *Ganoderma* sp. (Radhika and Priya 2016).

Carrageenans extracted from *Chondracanthus teedei* var. *lusitanicus* were studied in order to determine their potential antifungal activity (Soares 2015, Soares et al. 2016). FTIR-ATR and FT-Raman spectroscopic analysis confirmed the presence of a hybrid kappa/iota carrageenan belonging to the gametophyte phase and a hybrid xi/theta carrageenan in the tetrasporophyte phase. Kappa/iota and xi/theta carrageenan induced the formation of swollen hyphal segments in *Alternaria infectoria*, upon exposure to 125 μg mL^{-1} and 60 μg mL^{-1}, respectively. The observed phenotype was similar to those induced by antifungals targeting the fungal cell wall. When exposed to 87.5 μg mL^{-1} of kappa/iota carrageenan, *Aspergillus fumigatus* hyphae became shortened and highly branched, a phenotype commonly observed in response to antifungals. These morphological alterations were associated with a decrease of the β-glucan content in *A. infectoria* after exposure to 150 μg mL^{-1} of kappa/iota and to 100 μg mL^{-1} of xi/theta carrageenan. On the other hand, the chitin cell wall content of *A. fumigatus* decreased significantly upon exposure to 150 μg mL^{-1} of both extracts, which triggered an increase in the content of β-glucan. Overall, the present work shows that carrageenans extracted from *C. teedei* var. *lusitanicus* cause alterations on the *A. fumigatus* and *A. infectoria* cell walls, indicating a marked antifungal activity (Soares 2015, Soares et al. 2016).

CHAPTER 8

Antibacterial Activity of Seaweeds and their Extracts

8.1 Introduction

Bacteria comprise a large domain of unicellular prokaryotic organisms, therefore being morphologically characterized for the absence of a membrane-bound nucleus and other intracellular organelles, such as mitochondria. Prokaryotes are the ancestors of all life forms, having developed an impressive adaptability that allows them to thrive in almost all available ecological habitats. The bacteria are an extremely diverse group of organisms differing in size, shape, habitat, and metabolism (Rogers 2011). Species of bacteria can be divided into two major groups, called Gram-positive (Gram$^+$) and Gram-negative (Gram$^-$), based on the Gram stain reaction. The differences in cell wall structure are at the heart of the Gram stain: Gram$^+$ bacteria have very thick cell walls consisting primarily of peptidoglycan (as much as 90% of the cell wall), whereas Gram$^-$ bacteria cell wall is chemically complex and consists of at least two layers—a thinner layer of peptidoglycan (about 10% of the total cell wall) and an outer membrane that also contains polysaccharides linked (Madigan et al. 2014). Some bacteria live in symbiotic relationships with plants and animals, where they carry out important functions for the host, such as nitrogen fixation and cellulose breakdown. Additionally, they play a vital role in several ecosystems, contributing to the degradation of organic matter in soil, increasing its fertility and thus, sustaining higher life forms. Some bacteria can cause disease in plants, animals, and humans, while others are harmless (Rogers 2011). This is the case of normal microbiota, which comprises a great variety of microorganisms, including bacteria and unicellular fungi that colonize the skin and mucous membranes (e.g., oral cavity, respiratory tract, gastrointestinal tract, and urogenital tract) of every healthy human. These organisms are estimated to outnumber human cells by a factor of 10. It has been proven that normal microbiota provides a first line of defense against microbial pathogens, assists in digestion, is involved in toxin degradation, and contributes to the maturation of the immune system. However, shifts in the normal microbiota or stimulation of inflammation by these commensals may cause diseases such as inflammatory bowel disease (Carroll and Hobden 2016). On the other hand, bacteria have been the cause of some of the deadliest diseases and widespread epidemics of humans, including plague, cholera, dysentery, diphtheria, typhoid fever, typhus, pneumonia, and tuberculosis. Water purification, immunization (vaccination), and antibiotic treatment have reduced the morbidity and the mortality of bacterial diseases at least in the developed countries (Isbary and Stolz 2012).

Mankind's discovery of antibiotics ushered in a new age of medicine during the 19th century, an age wherein many predicted an end to diseases that had plagued mankind for centuries, with the appearance of penicillin during World War II as the first miracle drug (Wainwright 1991). From 1940s to almost 1980s, many classes of antibiotics discovered have helped tame many of the terrors of human health. The use of these "wonder drugs", combined with improvements in sanitation, housing, nutrition, and the advent of widespread immunization programs, led to a dramatic drop in deaths from diseases that were previously widespread, untreatable, and frequently fatal. Over the years, antimicrobials have saved the lives and eased the suffering of millions of people (Chanda et al. 2010).

In fact, antibiotics have revolutionized medicine in many respects (Davies and Davies 2010). Antibiotics are synthetic, natural, or semisynthetic molecules and can be classified according to the bacterial cellular component or system they affect, in addition to whether they induce cell death (bactericidal drugs), or merely inhibit cell growth (bacteriostatic drugs) (Walsh 2003, Kohanski et al. 2010). Most current bactericidal antibiotics are natural products or semisynthetic derivatives that inhibit DNA (quinolones), RNA (rifamycins), cell wall (β-lactans and glycopeptides), or protein synthesis. Drugs that inhibit protein synthesis are among the broadest classes of antibiotics and can be divided into two subclasses according to the ribosomal subunit they target–50S inhibitors (macrolides, lincosamides, streptogramins, amphenicols, and oxazolidinones) and 30S inhibitors (tetracyclines and amino cyclitols) (Kohanski et al. 2010). However, their effectiveness and ease of access led to a serious threat: the rise of antibiotic resistance in hospitals, communities, and the environment concomitant with their use (Davies and Davies 2010).

The extraordinary genetic capacities of microbes, namely their ability to exchange small packages of genetic information on plasmids, have benefited from man's overuse of antibiotics to exploit every source of resistance genes and every means of horizontal gene transmission (i.e., from one cell to another) to develop multiple mechanisms of resistance for each and every antibiotic introduced into practice clinically, agriculturally, or otherwise (Davies and Davies 2010, Morse and Meitzner 2016). In May 2015, a global action plan was endorsed by World Health Assembly to tackle antimicrobial resistance (including antibiotic resistance). The increase of the investment in new medicines is part of the strategic objectives delineated to achieve the action plan goal, which is to ensure continuity of successful treatment and prevention of infectious diseases with effective and safe medicines (WHO 2015).

The use of and search for drugs and dietary supplements derived from plants have accelerated in recent years. Ethnopharmacologists, botanists, microbiologists, phycologists, and natural-products chemists are combing the Earth for phytochemicals and "leads" which could be developed for treatment of infectious diseases. While 25 to 50% of current pharmaceuticals are derived from plants, including algae, none are even used as antimicrobials. Traditional healers have long used plants to prevent or cure infectious conditions; western medicine is trying to duplicate their successes. Plants are rich in a wide variety of secondary metabolites, such as tannins, terpenoids, alkaloids, and flavonoids, which have been found *in vitro* to have antimicrobial properties (Cowan et al. 1999).

Marine algae have received a lot of attention as potential sources of compounds possessing a wide range of biological activities, including antimicrobial properties. The antimicrobial activities of numerous algae species have been tested and reported, presenting an extended spectrum of action against bacteria and fungi (Shalaby 2011, Guedes et al. 2012).

8.2 Some of the Bacteria that Promote Infection

8.2.1 *Escherichia coli* and *Staphylococcus aureus*—from commensalism to lethal infections

A high number of bacteria are harmless to humans and inhabit the human body without causing adverse effects. The great majority is found in skin and gut and most of them are beneficial. Nevertheless, there are also a large number of infectious bacteria capable of causing morbidity and mortality, especially in countries with ineffective health care conditions and in immunocompromised patients. Others, which seem less harmful, are largely responsible for serious infections, including nosocomial infections. Among these are, for example, some serotypes of *Escherichia coli* (Spicer 2007, Paterson 2006) and *Staphylococcus aureus* (Archer 1998).

Escherichia coli is a rod-shaped Gram$^-$ bacteria that colonizes the gastrointestinal tract of human infants within a few hours after birth, being the most abundant facultative anaerobe of the human intestinal microflora. Usually, *E. coli* coexists with its human host in good health and with mutual benefit for decades, rarely causing disease except in immunocompromised hosts or where the normal gastrointestinal barriers are breached (as in peritonitis, for example). The niche of commensal *E. coli* strains is the mucous layer of the mammalian colon, where they are a highly successful competitor (Kaper et al. 2004). This could be due to its ability to utilize gluconate in the colon in a more efficient way than other resident species, allowing it to occupy a highly specific metabolic niche (Sweeney et al. 1996). Nevertheless, several highly adapted *E. coli* clones have acquired specific virulence attributes, which confers an increased capacity to adapt to new niches. Only the most successful combinations of virulence factors have persisted to become specific pathotypes of *E. coli* that are capable of causing disease in healthy individuals, such as enteric/diarrhea disease, urinary tract infections, or sepsis/meningitis (Kaper et al. 2004).

Staphylococcus aureus is a round-shape, facultative anaerobic Gram$^+$ bacteria with the ability to asymptomatically colonize healthy individuals. The ecological niches of *S. aureus* strains are the anterior nares. Over time, three patterns of carriage can be distinguished in healthy subjects: about 20% of people are persistent carriers (carrying almost always one type of strain), 60% are intermittent carriers, and approximately 20% almost never carry *S. aureus* (Kluytmans et al. 1997). However, *S. aureus* also constitutes a versatile and dangerous pathogen in humans (Lowy 1998). Although it is naturally susceptible to almost every antibiotic that has ever been developed, the treatment of infections caused by *S. aureus* has become more difficult due to its notorious capacity to become resistant to antibiotics (Chambers and DeLeo 2009).

E. coli and *S. aureus* are the main causes of bloodstream infections (BSIs) in humans. A remarkable increase in the number of BSIs caused by *E. coli* along with an alarming increase of antimicrobial multi-resistance were observed in Europe from 2002 to 2009 (Gagliotti et al. 2011). In turn, of all the resistance traits that *S. aureus* has acquired since the introduction of antimicrobial chemotherapy, methicillin resistance is clinically the most important, since a single genetic element confers resistance to the most commonly prescribed class of antimicrobials, the β-lactam antibiotics (Grundmann et al. 2006). The antimicrobial therapy used to treat infections caused by these microorganisms has therefore become problematic, emphasizing the need of a rational use along with a pharmaceutical investment in antibiotic research and development (Silva 2015).

8.2.2 Oral bacteria

Dental caries are among the major dental infections caused by oral microorganisms. Some of these microorganisms participate in plaque formation, the accumulation of which results in dental caries or periodontal disease, or both. *Streptococcus mutans* multiplies in plaque and then produces organic acids, such as lactic acid, which induce dental decay. Tooth brushing and flossing are good methods for preventing dental caries. However, these activities may be difficult for individuals who are physically handicapped. Using a mouthwash containing chemotherapeutic agents would be a simple way to prevent dental caries by controlling the numbers of these microorganisms (Murata et al. 1989). The presence of nutrients, epithelial debris, and secretions makes the oral cavity a favorable habitat for a great variety of oral bacteria like *Streptococci*, *Lactobacilli*, *Staphylococci*, *Corynebacteria*, and with a great number of anaerobes, especially Bacteroides. The mouth presents a floral succession with age and corresponding ecological changes in the oral cavity. These bacteria colonizing the dental surface and gingiva have coevolved with their host to establish a highly sophisticated relationship between pathogenic and mutualistic bacteria coexisting in homeostasis. Plaque is a biofilm on the surfaces of the teeth. The accumulation of plaque results in dental caries, and leads to gingivitis or periodontal diseases. Some research has shown that oral bacteria may contribute to increased risk of heart attacks, strokes, and lung disease and may be associated with premature childbirth in some women (Chung et al. 2006).

8.2.3 Diarrhea causing bacteria

Diarrhea is one the most common complaints faced by internists and primary care physicians, and accounts for many referrals to gastroenterologists. Acute infective diarrhea contributes to the high rates of morbidity and mortality worldwide. Diarrhea constitutes 70% of food borne diseases (Guandalini and Vaziri 2011).

The Centers for Disease Control and Prevention (CDC.GOV) estimates that each year, roughly 1 in 6 Americans (or 48 million people) get sick, 128,000 are hospitalized, and 3,000 die of foodborne diseases. A recent World Health Organization (WHO) report estimates that about 1.4 million people died from diarrhea diseases in 2015 (WHO 2017c). Most of the fatalities and morbidities occur in children below the age of 5 years. Use of certain drugs, chemotherapeutic agents or toxins can also be associated with watery diarrhea, and should always be considered in any chronic diarrhea (Ratnaike and Jones 1998).

Diarrhea may be classified as acute or chronic and it is the eighth highest killer disease in the world (WHO 2017c), and the third in developing countries (Khan 2016). Diarrhea is the condition of having three or more loose motions within 24 hours. An estimated 7.5 million people, including 2.5 million children, die as a result of diarrhea every year. *Escherichia coli* is one important pathogen which causes diarrhea and the other bacteria, *Salmonella typhi*, causes diarrhea and typhoid fever (Salem et al. 2011).

Diarrhea and typhoid fever are common diseases and a major cause of morbidity and mortality in the third world countries. In developed countries, diarrhea and typhoid fever have almost been eliminated due to good sewage and water treatment facilities. Although in New York (USA), alone, 3,500 people became ill and 639 of them died with typhoid fever in 1906, and in the early 1900s in the United States, typhoid fever was common and a deadly disease (Ray 2002, Caper 2011).

8.3 Assessment of Antimicrobial Activity

Several methods are widely used by researchers to detect and to measure the antimicrobial activity of algal extracts or their metabolites. Authors refer most often to *in vitro* and sometimes *in vivo* assays, but different algal extracts quantities and microorganisms are tested, making it difficult to unify results. In some cases, an initial *in vitro* screening is followed by an *in vivo* study, but most studies on the antimicrobial effects of seaweed are either only *in vitro* or only *in vivo* (Pérez et al. 2016).

8.3.1 In vitro assays

MIC represents the lowest concentration of crude or purified algal extracts that inhibit the bacterial or fungal growth (Silva 2015). The concentration serial broth (micro) dilution assay has been used in several studies (Hellio et al. 2001, Bazes et al. 2009, Boisvert et al. 2015). Minimum inhibitory concentration for the extracts of algae are usually determined by broth microdilution method, according to procedures described in the reference document M07-A10 of Clinical and Laboratory Standards Institute (CLSI 2015), with some modifications. This approved guideline addresses methods for dilution of antimicrobial susceptibility testing for bacteria that grow aerobically (Silva 2015).

García-Bueno et al. (2014) tested the antibacterial activity of water-soluble seaweed extracts in 96-well plates (growth inhibition assay). Bacterial growth in the presence of algal extracts was monitored by measuring optical density at 490 nm every 30 min for 24 hours. After incubation, the intensity of growth in the presence of the tested compounds and controls was compared. Dubber and Harder (2008) used a highly sensitive growth inhibition assay that recorded the fluorescence intensity of stained DNA, and Cox et al. (2014) assessed the antibacterial activity of varying concentrations of a hydrophilic extract from *Himanthalia elongata* (brown algae) in carbohydrate and protein model food systems.

8.3.2 In vivo assays

In the literature related to the assessment of antimicrobial activity of the crude extracts or fractions of algae, the *in vivo* assays are less numerous. *In vivo* assays depend critically on the target organism or substrate, varying between one another. Vatsos and Rebours (2015) refer to pharmacodynamics, pharmacokinetic studies, and artificial challenges like survival rate, progress of disease, and severity of signs or fewer ions in the studies applied to aquaculture. For possible use in humans, they are much more restricted, so the detection assays *in vitro* antimicrobial activity will be fundamental. In this section, some of the *in vivo* assays based on published studies are collected (Pérez et al. 2016).

8.4 Antibacterial Activity of Chlorophyta (Green Algae)

Table 8.1 presents some examples of the microorganisms inhibited by solvent extracts from green, brown, and red seaweed species.

Bactericidal and bacteriostatic compounds were first isolated from algae when chloroform and benzene fatty acid extracts of chlorellin, from *Chlorella vulgaris* (microalga), were found to inhibit *Pseudomonas aeruginosa*, *Staphylococcus aureus*, *Streptococcus pyogenes*, and *Bacillus subtilis* (Pratt et al. 1944). Chlorellin was not suitable for large scale commercial use, however it heralded further research into algal antimicrobial inhibitors in genera such as *Scenedesmus* sp. (microalga) (Willis 2007). As discussed, several chemical functional groups in algae, such as phlorotannins, fatty acids, peptides, terpenes, polysaccharides, polyacetylenes, sterols, indole alkaloids, aromatic organic acids, shikimic acid, polyketides, hydroquinones, alcohols, aldehydes, ketones, and halogenated furanones have been reported as bacterial inhibitors. The mechanism of pharmacological action for some remains uncertain, however, methods of bacterial inhibition employed by the following functional groups have been proposed.

Antibacterial activity has been widely reported from marine algae and has been reviewed by Sieburth (1964) and Burkholder (1973). Subsequent surveys of antibacterial activity in algae have been reported from the Mediterranean (Caccamese et al. 1979, 1980, 1981, Moreau et al. 1984), Puerto Rico (Burkholder et al. 1960, Welch 1962, Almodovar 1964, Ballantine and Almodovar 1977), India (Rao and Parekh 1981), China (Ma and Tan 1984), Australia (Reichelt and Borowitzka 1984), Argentina (Espeche et al. 1984), and Atlantic Canada (Ragan 1984). The survey reported here was undertaken to assess the distribution of lipid-soluble bioactive materials among these macroalgal species, and to evaluate more thoroughly this source of potentially useful biomedicinally active substances. A number of secondary algal metabolites with biological activity are known (Norris and Fenical 1985), and these authors have stated that approximately 90% of known algal secondary products are lipid-soluble (Ballantine et al. 1987).

The crude extracts obtained from the green alga *Chlorococcum infusionum* (formerly *Chlorococcum humicola*) (a freshwater/terrestrial microalga species) showed bioactivity against the pathogenic bacteria such as *Cryptococcus neoformans*, *Bacillus pumillus*, *Escherichia coli*, *Sarcina lutea*, *Bacillus subtilis*, and *Staphylococcus aureus* (Pande and Gupta 1977).

Rao and Parekh (1981) found that the crude extracts obtained from the green seaweeds such as *Caulerpa taxifolia*, *Caulerpa scalpelliformis*, *Halimeda tuna*, and *Ulva intestinalis* (formerly *Enteromorpha intestinalis*), showed considerable bioactive potential. Rao and Karmarkar (1986) reported that ethanol extract of *Ulva* sp. (formerly *Enteromorpha* sp.) showed high activity against *Staphylococcus aureus* than the extracts of acetone and *n*-butanol.

The greatest degree of activity within the Chlorophyta was seen in the order Caulerpales, in which 76% of species were active. High activity among Caulerpales is not surprising as previous studies have documented bioactive compounds within the order (Nakatsu et al. 1981, Norris and Fenical 1982, Paul and Fenical 1984, Fenical and Paul 1984). Seven of the nine Puerto Rican species of *Caulerpa* assayed were also seen to be active. Bioactivity in this genus is well known (Aguilar-Santos and Doty 1968, Amico et al. 1978, Blackman and Wells 1978, Capon et al. 1981, Norris and Fenical 1982, Ballantine et al. 1987).

Studies were conducted on 30 species of seaweeds collected along the coast of Mandapam, Tamil Nadu for their hemolytic and antimicrobial potential. Results indicated that extracts obtained from the seaweeds such as *Ulva compressa* (formerly *Enteromorpha compressa*) and *Acrocladus herpesticus* (formerly *Cladophoropsis zollingeri*), showed antibacterial activity against the Gram$^-$ bacteria and Gram$^+$ cultures of *Bacillus*. A strong hemolytic activity was shown by the extract of *C. herpestica* (Rao et al. 1991).

Several species of marine macrophytes found along the coast of Central Mediterranean were screened for the production of antibacterial potential. The maximum level of activity was found among the Chlorophyta and some members of the Bryopsidales (*Flabellia petiolata*, *Caulerpa prolifera*, *Halimeda tuna*) were the most active species (Ballesteros et al. 1992).

Antimicrobial potential of six marine green algae found along the coast of Tanzania were screened against three bacterial species, viz., *Staphylococcus aureus*, *Bacillus subtilis*, *Escherichia coli*, and a yeast,

Candida albicans using a disc assay method. A brine shrimp bio-assay using newly hatched *Artemia salina* larvae was used for the cytotoxicity study of crude extracts from three algal species. Of the six species tested, the extract of *Valonia aegagropila* was the most active against all the pathogens tested, and its extract was even more active against the test bacteria than Penicillin G at a concentration of 2.5 μg mol^{-1}. The extracts of *Halimeda opuntia* and *Halimeda tuna* showed mild activity against all pathogens tested. The extract of *Ulva australis* (formerly *Ulva pertusa*) was more active against the bacteria such as *S. aureus* and *B. subtilis*, but less active against *E. coli*, and was not active against the fungus *C. albicans* (see Chapter 7). The extract of *Caulerpa mexicana* was inactive against all the pathogens tested. Occasional development of antimicrobial resistance colonies within the inhibition zones were observed from the extracts of *H. opuntia* and *H. tuna* when they were assayed against *C. albicans* and *E. coli* (Mtolera and Semesi 1996).

Furthermore, compounds synthesized by *Scenedesmus costatum* (microalga), and partially purified from its organic extract, exhibited activity against aquaculture bacteria because of their fatty acids longer than 10 carbon atoms in chain length—which apparently induce lysis of bacterial protoplasts. The ability of fatty acids at large to interfere with bacterial growth and survival has been known for quite some time, but recent structure-function relationship studies suggest that said ability depends on both their chain length and degree of unsaturation. Such compounds as cholesterol can antagonize antimicrobial features (Lampe et al. 1998), so both composition and concentration of free lipids should be taken into account (Benkendorff et al. 2005).

Extracts from several southern African seaweeds were screened against 12 bacteria (Vlachos et al. 1997). The Chlorophyta extracts showed activity against the $Gram^+$ bacteria, but limited activity against the $Gram^-$ bacteria, a trend similar to that observed for the Rhodophyta extracts. *Codium capitatum*, *Codium platylobium*, and *Codium tenue* extracts inhibited five of the six, and *Codium duthieae* inhibited four of the six $Gram^+$ bacteria tested. Different species of the genus *Codium* thus showed similar ranges and degrees of activity against similar $Gram^+$ bacteria. *Bacillus subtilis* was inhibited by all the seaweed methanolic extracts studied.

Lyengaroside A was isolated from the alga *Codium indicum* (formerly *Codium iyengarii*) and displayed a moderate antibacterial activity (Ali et al. 2002). Extract of *Caulerpa prolifera* exhibited moderate to significant activity against unidentified strains of marine bacteria (Smyrniotopoulos 2003).

According to Ravikumar et al. (2002), among the seaweed extracts, the acetone extract of *Caulerpa cupressoides* gives maximum inhibitory activity (9 mm) against the bacterium *Escherichia coli*; and acetone extracts of *C. cupressoides* and *Chaetomorpha linoides* gives the maximum inhibitory activity (8 mm) against *Streptococcus pyogens*. Bioactivity of extracts obtained from different regions of the thallus (apical, basal, and stolon) of *Caulerpa* spp. (*Caulerpa ashmeadii*, *Caulerpa paspaloides* and *Caulerpa prolifera*) was evaluated; it was observed that the stolon of *Caulerpa* have the highest antibacterial potential; only three $Gram^+$ bacteria (*Bacillus subtilis*, *Streptococcus faecalis*, and *Micrococcus luteus*) were inhibited by the extracts; growth in different bacteria was inhibited by 10 Chlorophyta (100%) (Freile-Pelegrin and Morales 2004).

Selvin and Lipton (2004) tested the secondary metabolites of two seaweeds for bio-toxicity potential. Both species showed potent activity in antibacterial, brine shrimp cytotoxicity, larvicidal, antifouling, and ichthyotoxicity assays. The green alga *Ulva fasciata* exhibited broad-spectrum antibacterial activity.

Atlantic marine algae were screened in growth inhibition assays against the pathogenic bacterium *Pseudoaltermonas bacteriolytica* (Puglisi et al. 2007). Among the green algae, only 29% yielded extracts that were active against *P. bacteriolytica*. For example, the extracts from *Caulerpa cupressoides* and *Dictyosphaeria versluysii* were only active against this pathogenic bacterium. Among members of the Udoteaceae, *Avrainvillea obscura* was the only species to yield extracts active against *P. bacteriolytica*. Further, the extracts from *Halimeda macroloba*, *Halimeda opuntia*, and *Tydemania expeditionis* were active against the tested bacterium (Puglisi et al. 2007).

The methanol, dichloromethane, hexane, chloroform, and the volatile components of *Ulva linza* (formerly *Enteromorpha linza*) were tested *in vitro* for their antimicrobial activity against five $Gram^+$, four $Gram^-$ bacteria, and *Candida albicans* (Sukatar et al. 2006). Gas chromatography-mass-spectrometry

Table 8.1 Antibacterial activity of different solvent extracts from seaweed.

Species	Solvent and/or extract	Bacteria	References
Chlorophyta (green seaweed)			
Acrosiphonia orientalis	Methanol:toluene (3:1)	10 Human pathogen bacteria	Shanmughapriya et al. 2008b
A. orientalis (formerly *Spongomorpha indica*)	Methanol	*Actinomyces viscosus, Streptococcus mitis, Streptococcus mutans*	Sujatha et al. 2012
Anadyomene stellata	Hexane, ethyl acetate, methanol	*E. coli, L. monocytogene, S. enterica, Agrobacterium tumefaciens, P. aerigunosa, S. aureus, M. luteus*	Kolsi et al. 2015
Avrainvillea obscura	Crude extracts	*Pseudoaltermonas bacteriolytica*	Puglisi et al. 2007
Boodlea composita	Methanol	*Vibrio harveyi, V. alginolyticus, V. vulnificus*	Manilal et al. 2010
Bryopsis hypnoides	Crude extracts	*S. aureus, E. coli, S. faecalis, Klebsiella pnuemoniae, P. aeruginosa*	Lakshmi et al. 2006
B. myosuroides	Ethanol	*B. subtilis, L. innocua, L. monocytogenes, Micrococcus* sp., *S. aureus, A. lwoffii, S. enteritidis*	Vlachos et al. 1997
B. pennata	Methanol	*Vibrio parahaemolyticus, V. alcaligenes*	Manilal et al. 2010
B. plumosa	Crude extracts	*S. aureus, E. coli, S. faecalis, Klebsiella pnuemoniae, P. aeruginosa*	Lakshmi et al. 2006
Caulerpa ashmeadii	Crude extracts	*B. subtilis, S. faecalis, M. luteus*	Freile-Pelegrin and Morales 2004
C. chemnitzia	Hexane, chloroform, ethyl acetate, acetone, methanol	*B. subtilis, S. pyogenes, E. coli, K. pneumoniae, P. mirabilis, P. vulgaris, P. aeruginosa, S. typhimurium, S. flexneri, V. cholerae*	Raj et al. 2015
C. corynephora	Water, ethanol, chloroform, petroleum ether, hexane	*E. coli, K. pneumoniae, P. aeruginosa, P. mirabilis, S. marcescens, S. typhi, A. calcoaceticus, Enterobacter cloacae, S. aureus, E. faecalis, S. pyogenes, B. cereus*	Johnson and Raja 2015
C. cylindracea	Crude extracts	*Vibrio* spp.	Ördög et al. 2004 Rizzo et al. 2013
C. cupressoides	Acetone Dichloromethane, methanol, ethyl acetate	*E. coli, S. pyogens* *Pseudoaltermonas bacteriolytica*	Ravikumar et al. 2002 Puglisi et al. 2007
C. lentillifera	Methanol, ethyl acetate	*Escherichia coli, Staphylococcus aureus, Streptococcus* sp., *Salmonella* sp.	Nagappan and Vairappan 2014

Table 8.1 contd....

Table 8.1 contd....

Species	Solvent and/or extract	Bacteria	References
C. mexicana	Crude extracts	*S. aureus, B. subtilis, E. coli*	Mtolera and Semesi 1996
C. parvula	Methanol	*Vibrio parahaemolyticus, V. alcaligenes*	Manilal et al. 2010
C. paspaloides	Crude extracts	*B. subtilis, S. faecalis, M. luteus*	Freile-Pelegrin and Morales 2004
C. prolifera	Methanol/toluene Crude extracts Crude extracts	*E. coli, B. subtilis* *B. subtilis, S. faecalis, Micrococcus luteus* Unidentified strains of marine bacteria	Ballesteros et al. 1992 Ali et al. 2002 Smyrniotopoulos 2003
C. prolifera	Crude extracts	*B. subtilis, S. faecalis, M. luteus* *S. aureus, E. coli, E. faecalis*	Freile-Pelegrin and Morales 2004 Zbakh et al. 2012
C. racemosa	Crude extracts Crude extracts Hexane, diethyl ether, chloroform, ethyl acetate, ethanol, acetone, methanol	*S. aureus, E. coli, S. faecalis, Klebsiella pnuemoniae, P. aeruginosa* *S. aureus* *B. subtilis, K. pneumoniae*	Lakshmi et al. 2006 Kandhasamy and Arunachalam 2008 Jebasingh et al. 2011
C. racemosa	Ethyl acetate Crude extracts Crude extracts Methanol	*S. aureus, B. cereus, E. coli, E. faecalis, Salmonella* sp., *P. aeruginosa* *Mycobacterium tuberculosis* *S. aureus, E. coli, S. faecalis, K. pnuemoniae, P. aeruginosa* *S. aureus, B. subtilis, Bacillus* spp., *S. epidermidis, E. coli, S. typhi, Klebsiella* spp., *P. aeruginosa*	Salem et al. 2011 Süzgeç-Selçuk et al. 2011 Puglisi et al. 2007 Alghazeer et al. 2013
C. racemosa	Ethyl acetate Methanol, diethylether, water	*E. faecalis* and Vancomycin-resistant *E. faecalis* *E. coli, S. aureus, Streptococcus* sp., *Salmonella* sp.	Chandrasekaran et al. 2014 Nagappan and Vairappan 2014
C. scalpelliformis	Acetone and *n*-butanol Ethanol Methanol	*S. aureus* *S. aureus, E. coli, M. tuberculosis, S. faecalis, K. pnuemoniae, P. aeruginosa* *Vibrio parahaemolyticus, Salmonella* sp., *Shewanella* sp., *E. coli, K. pneumoniae, S. pyogenes, S. aureus, E. faecalis, P. aeruginosa, Proteus mirabilis*	Rao and Parekh 1981 Lakshmi et al. 2010 Lavanya and Veerappan 2011
C. scalpelliformis	Hexane, diethyl ether, chloroform, ethyl acetate, ethanol, acetone, methanol Water, ethanol, chloroform, petroleum ether, hexane	*B. subtilis, K. pneumoniae, Escherichia coli, P. aeruginosa* *E. coli, K. pneumoniae, P. aeruginosa, P. mirabilis, S. marcescens, S. typhi, A. calcoaceticus, Enterobacter cloacae, S. aureus, E. faecalis, S. pyogenes, B. cereus*	Jebasingh et al. 2011 Johnson and Raja 2015

C. scalpelliformis	Acetone, ethanol	*E. coli, S. typhi*	Khan 2016
C. taxifolia	Acetone and *n*-butanol	*S. aureus*	Rao and Parekh 1981 Lavanya and Veerappan 2011
C. veravalensis	Crude extracts Crude extracts	*S. aureus, E. coli, S. faecalis, Klebsiella pnuemoniae,* *S. aureus, E. coli, S. faecalis, K. pnuemoniae, P. aeruginosa*	Lakshmi et al. 2006 Puglisi et al. 2007
C. zeyheri	Ethanol	*B. subtilis, L. innocua, L. monocytogenes, Micrococcus* sp., *S. aureus, A. lwoffii, S. enteritidis*	Vlachos et al. 1997
Chaetomorpha aerea	Acetone, methanol, ethanol Water	*E. coli* *Bacilus subtilis, Micrococcus luteus, S. aureus*	Seenivasan et al. 2010 Pierre et al. 2011
C. antennina	Crude extracts Methanol Ethanol	*P. aeroginosa, S. aureus, E. coli, E. aeroginosa and S. paratyphi, K. pneumoniae, B. subtilis, Citrobacter* sp., *Proteus* sp., *S. epidemis* *Actinomyces viscosus, Streptococcus mitis, Streptococcus mutans* *V. parahaemolyticus*	Dhasarathan and Theriappan 2011 Sujatha et al. 2012 Thanigaivel et al. 2014
C. antennina	Water, ethanol, chloroform, petroleum ether, hexane Water Water at 100°C followed by precipitation using ethanol (sulfated polysaccharides)	*E. coli, K. pneumoniae, P. aeruginosa, P. mirabilis, S. marcescens, S. typhi, A. calcoaceticus, Enterobacter cloacae, S. aureus, E. faecalis, S. pyogenes, B. cereus* *E. coli, K. pneumoniae, P. vulgaris, P. aeruginosa, S. aureus, S. pyogenes, E. faecalis* *E. coli, Staphylococcus* sp., *Proteus* sp., *Streptococcus* sp., *Enterococci* sp.	Johnson and Raja 2015 Manchu et al. 2015 Puglisi et al. 2007
C. linum	Chloroform, ethyl acetate Hexane, diethyl ether, chloroform, ethyl acetate, ethanol, acetone, methanol Chloroform:methanol (2:1) Water, methanol	*Shigella flexneri, Vibrio cholerae, E. coli, B. subtilis, Bacillus brevis* *Enterobacter aerogenes, K. pneumoniae, Bacillus subtilis, E. coli, P. aeruginosa* *Vibrio ordalii, V. vulnificus* *Pseudomonas aeruginosa, E. coli*	Patra et al. 2009 Jebasingh et al. 2011 Cavallo et al. 2013 Kumar and Padhi 2016
C. linoides	Acetone Ethanol, methanol, chloroform	*E. coli, S. pyogens* *S. aureus, B. subtilis, Lactobacillus acidophilus, E. coli, P. aeruginosa, P. mirabilis*	Ravikumar et al. 2002 Arunachalam et al. 2014
C. spiralis (formerly *C. torta*)	Ethanol	*S. aureus, E. coli, M. tuberculosis, S. faecalis, K. pneumoniae, P. aeruginosa*	Lakshmi et al. 2010

Table 8.1 contd.…

Table 8.1 contd....

Species	Solvent and/or extract	Bacteria	References
Chaetomorpha sp.	Crude extracts	*S. aureus, E. coli, S. faecalis, K. pnuemoniae, P. aeruginosa*	Lakshmi et al. 2006
*Chlorella vulgaris**	Chloroform, benzene	*P. aeruginosa, S. aureus, S. pyogenes, B. subtilis*	Pratt et al. 1944
Chlorococcum infusionum (formerly *Chlorococcum humicola*)*	Crude extracts	*C. neoformans, B. pumillus, E. coli, Sarcina lutea, B. subtilis, S. aureus*	Pande and Gupta 1977
Cladophora albida	Methanol	*Vibrio harveyi, V. alginolyticus, V. vulnificus, V. parahaemolyticus, V. alcaligenes*	Manilal et al. 2010
C. conglomerata	Water	*E. coli, S. pyogenes, C. conglomerata, P. aeruginosa, S. typhi*	Mansuya et al. 2010
C. fascicularis	Methanol	*A. viscosus, S. mitis, S. mutans*	Sujatha et al. 2012
C. glomerata	Methanol	*Acinetobacter baumannii, V. fischeri, V. vulnificus, V. anguillarum, V. parahaemolyticus*	Yuvaraj et al. 2011
C. herpestica (as *Cladophoropsis zollingeri*)	Crude extracts	*Bacillus* sp.	Rao et al. 1991
C. prolifera	Ethanol	*B. subtilis, L. innocua, L. monocytogenes, Micrococcus* sp., *S. aureus, A. lwoffii, S. enteritidis*	Vlachos et al. 1997
C. rupestris	Crude extracts	*B. subtilis, B. cereus*	Salvador et al. 2007
C. rupestris	Chloroform:methanol (2:1) Methanol, ethyl acetate, ethanol Chloroform:methanol (2:1) Chloroform:methanol (2:1)	*V. metschnikovii, V. ordalii, V. salmonicida, V. vulnificus, V. salmonicida; V. vulnificus, V. ordalii, V. cholera* *E. coli, P. aeruginosa, S. aureus, V. harveyii, V. parahaemolyticus, V. alginolyticus* *S. aureus, V. harveyii, V. parahaemolyticus, V. alginolyticus, Enterococcus* sp., *Streptococcus agalactiae, V. fluvialis*	Cavallo et al. 2013 Krish and Das 2014 Stabili et al. 2014
Cladophora sp.	Ethanol	*S. aureus, S. pyogenes*	Muñoz-Ochoa et al. 2010
C. vagabunda	Oil extracts	*B. cereus, L. monocytogenes, E. coli, S. enteritidis*	Horincar et al. 2014
Codium adherens	Acetone, methanol, chloroform, diethyl ether, ethyl acetate, ethanol, petroleum	*E. coli, Staphylococcus* sp., *Proteus* sp., *Streptococcus* sp., *Enterococci* sp.	Karthikaidevi et al. 2009
C. amplivesiculatum	Ethanol	*S. aureus, S. pyogenes*	Muñoz-Ochoa et al. 2010
C. bursa	Crude extracts	*Mycobacterium tuberculosis*	Süzgeç-Selçuk et al. 2011
C. capitatum	Methanol	Gram^{+} bacteria	Vlachos et al. 1997

C. cuneatum (formerly *Codium simulans*)	Ethanol	*S. aureus, S. pyogenes*	Muñoz-Ochoa et al. 2010
C. decorticatum (formerly *C. elongatum*)	Crude extracts	*S. aureus, E. coli, S. faecalis, Klebsiella pnuemoniae, P. aeruginosa*	Lakshmi et al. 2006
	Crude extracts	*S. aureus, E. coli, S. faecalis, K. pnuemoniae, P. aeruginosa* *Vibrio parahaemolyticus, Salmonella* sp., *Shewanella* sp., *E. coli,*	Puglisi et al. 2007
	Methanol	*K. pneumoniae, S. pyogenes, S. aureus, E. faecalis, P. aeruginosa, Proteus mirabilis*	Lavanya and Veerappan 2011
C. duthieae	Ethanol	*B. subtilis, L. innocua, L. monocytogenes, Micrococcus* sp., *S. aureus*	Vlachos et al. 1997
C. dwarkense	Crude extracts	*S. aureus, E. coli, S. faecalis, K. pnuemoniae, P. aeruginosa*	Lakshmi et al. 2006
	Crude extracts	*S. aureus, E. coli, S. faecalis, K. pnuemoniae, P. aeruginosa*	Puglisi et al. 2007
C. fragile	Ethyl acetate	*S. aureus, B. cereus, E. coli, E. faecalis, Salmonella* sp., *P. aeruginosa*	Salem et al. 2011
	Hexane, ethyl acetate, methanol	*E. coli, L. monocytogenes, S. enterica, Agrobacterium tumefaciens, P. aerigunosa, S. aureus, M. luteus*	Kolsi et al. 2015
C. indicum (formerly *C. iyengarii*)	Methanol, ethyl acetate (Lyengaroside A)	*Corynebacterium diphteriae, E. coli, K. pneumonia, Snigellabdysentri, S. aureus, P. aeruginosa, S. typhi, S. pyogenes*	Ali et al. 2002
	Acetone, ethanol	*E. coli, S. typhi*	Khan 2016
C. platylobium	Ethanol	*B. subtilis, L. innocua, L. monocytogenes, Micrococcus* sp., *S. aureus*	Vlachos et al. 1997
C. shameelii	Acetone, ethanol	*E. coli, S. typhi*	Khan 2016
C. tenue	Ethanol	*B. subtilis, L. innocua, L. monocytogenes, Micrococcus* sp., *S. aureus*	Vlachos et al. 1997
Dictyosphaeria cavernosa	Crude extracts	*S. aureus, E. coli, S. faecalis, K. pnuemoniae, P. aeruginosa*	Lakshmi et al. 2006
	Crude extracts	*S. aureus, E. coli, S. faecalis, K. pnuemoniae, P. aeruginosa*	Puglisi et al. 2007
D. versluysii	Dichloromethane, methanol, ethyl acetate	*Pseudoaltermonas bacteriolytica*	Puglisi et al. 2007
Enteromorpha attenuata	Methanol	9 food borne pathogens	Narasimhan et al. 2013
Flabellia petiolata	Methanol/toluene	*E. coli, B. subtilis*	Ballesteros et al. 1992
	Hexane, ethyl acetate, methanol	*E. coli, L. monocytogene, S. enterica, Agrobacterium tumefaciens, P. aerigunosa, S. aureus, M. luteus*	Kolsi et al. 2015
Halimeda cuneata	Ethanol	*B. subtilis, L. innocua, L. monocytogenes, Micrococcus* sp., *S. aureus, A. lwoffii, S. enteritidis*	Vlachos et al. 1997

Table 8.1 contd....

Table 8.1 contd....

Species	Solvent and/or extract	Bacteria	References
H. gracilis	Acetone	*E. coli, P. aeruginosa, S. aureus, K. pneumonia, E. faecalis*	Kolanjinathan and Stella 2009
H. macroloba	Crude extracts Ethanol, methanol, chloroform Water, ethanol, chloroform, petroleum ether, hexane	*P. bacteriolytica* *S. aureus, B. subtilis, Lactobacillus acidophilus, E. coli, P. aeruginosa, P. mirabilis* *E. coli, K. pneumoniae, P. aeruginosa, P. mirabilis, S. marcescens, S. typhi, A. calcoaceticus, Enterobacter cloacae, S. aureus, E. faecalis, S. pyogenes, B. cereus*	Puglisi et al. 2007 Arunachalam et al. 2014 Johnson and Raja 2015
H. opuntia	Crude extracts Crude extracts Methanol Methanol, ethanol, petroleum ether, dimethyl formamide	*S. aureus, B. subtilis, E. coli* *P. bacteriolytica* *S. aureus* *E. coli, Salmonella typhi, Shigella dysenteriae, K. pneumoniae, Enterobacter aerogenes*	Mtolera and Semesi 1996 Puglisi et al. 2007 Selim 2012 Al-Judaibi 2014
H. tuna	Acetone and *n*-butanol Methanol/toluene Crude extracts	*S. aureus* *E. coli, B. subtilis* *S. aureus, B. subtilis, E. coli*	Rao and Parekh 1981 Ballesteros et al. 1992 Mtolera and Semesi 1996
H. tuna	Acetone, methanol, chloroform, diethyl ether, ethyl acetate, ethanol, petroleum Acetone and *n*-butanol Chloroform, ethanol, methanol	*E. coli, Staphylococcus* sp., *Proteus* sp., *Streptococcus* sp., *Enterococci* sp. *S. aureus* *S. aureus, S. typhimurium, Salmonella paratyphi, Klebsiella oxytoca, E. coli, P. mirabillis, Lactobacillus vulgaris, Pseudomonas* sp., *K. pneumonia and V. cholerae*	Karthikaidevi et al. 2009 Lavanya and Veerappan 2011 Indira et al. 2013
*Haematococcus pluvialis**	Ethanol	*E. coli, S. aureus*	Santoyo et al. 2009
*Planktochlorella nurekis**	Crude extracts	*C. jejuni, E. coli, Salmonella enterica* var. *enteritidis, Salmonella enterica* var. *infantis, Arcobacter butzleri, Lactobacillus johnsonii*	Cermák et al. 2015
Pseudocodium devriesii	Ethanol	*B. subtilis, L. innocua, L. monocytogenes, Micrococcus* sp., *S. aureus, A. lwoffii, S. enteritidis*	Vlachos et al. 1997
*Rhizoclonium hieroglyphicum**	Water, ethanol	*S. aureus*, Methicillin-resistant *S. aureus, Propionibacterium acnes*	Mungmai et al. 2014
*Scenedesmus costatum**	Crude extracts	Aquaculture bacteria	Benkendorff et al. 2005
Scenedesmus sp.*	Crude extract (chlorellin)	Several Bacteria	Willis 2007
Tydemania expeditionis	Crude extracts	*P. bacteriolytica*	Puglisi et al. 2007

Udotea indica	Crude extracts Crude extracts	*S. aureus, E. coli, S. faecalis, K. pnuemoniae, P. aeruginosa* *S. aureus, E. coli, S. faecalis, K. pnuemoniae, P. aeruginosa*	Lakshmi et al. 2006 Puglisi et al. 2007
U. indica	Acetone, ethanol	*E. coli, S. typhi*	Khan 2016
Ulva australis (formerly *Ulva pertusa*)	Crude extracts	*B. cereus, L. monocytogenes, E. coli, S. enteritidis* *S. aureus, B. subtilis, E. coli* *Helicobacter pylori*	Baik and Kang 1986 Mtolera and Semesi 1996 Lee et al. 2009b
U. australis (formerly *U. pertusa*)	Crude extracts Ethanol	*G. vaginalis* *G. vaginalis*	Choi et al. 2014b Ha et al. 2014
U. clathrata (formerly *Enteromorpha ramulosa*)	Crude extracts	Gram$^+$ and Gram$^-$ bacteria	Gonzalez et al. 2001
U. clathrata (formerly *E. clathrata*)	Crude extracts Crude extracts	*S. aureus, E. coli, S. faecalis, K. pnuemoniae, P. aeruginosa* *S. aureus, E. coli, S. faecalis, K. pnuemoniae, P. aeruginosa*	Lakshmi et al. 2006 Puglisi et al. 2007
U. compressa	Ethanol, acetone, methanol	*B. subtilis, S. aureus, E. coli, S. typhi, K. pneumoniae, C. albicans*	Osman et al. 2010
U. compressa (formerly *E. compressa*)	Crude extracts Hexane, diethyl ether, chloroform, ethyl acetate, ethanol, acetone, methanol Methanol Crude extracts (phthalate esters derivatives)	*S. aureus, E. faecalis, K. pneumoniae, E. faecalis* *B. subtilis, K. pneumoniae, Escherichia coli, P. aeruginosa* *E. coli, P. aeruginosa, Proteus vulgaris, K. pneumoniae, S. aureus, B. subtilis* Pathogenic Gram$^+$ and Gram$^-$ bacteria, namely *K. pneumoniae*	Ibtissam et al. 2009 Jebasingh et al. 2011 Elnabris et al. 2013 Osman et al. 2013
U. compressa	Hexane, diethyl ether, chloroform, ethyl acetate, ethanol, acetone, methanol Ethanol Water, ethanol, chloroform, petroleum ether, hexane Acetone, ethanol	*S. aureus, E. coli, E. faecalis* *Gardnerella vaginalis* *E. coli, K. pneumoniae, P. aeruginosa, P. mirabilis, S. marcescens, S. typhi, A. calcoaceticus, Enterobacter cloacae, S. aureus, E. faecalis, S. pyogenes, B. cereus* *E. coli, S. typhi*	Zbakh et al. 2012 Ha et al. 2014 Johnson and Raja 2015 Khan 2016
U. compressa (formerly *E. compressa*)	Water, methanol	*P. aeruginosa, E. coli*	Kumar and Padhi 2016

Table 8.1 contd....

Table 8.1 contd....

Species	Solvent and/or extract	Bacteria	References
U. fasciata	Crude extracts Ethanol, acetone, methanol Methanol	Several bacteria *B. subtilis, S. aureus, E. coli, S. typhi, K. pneumoniae, C. albicans* *E. coli, S. aureus, P. aeruginosa, B. cereus, M. luteus, Xanthomonas campestris, Erwinia carotovora*	Selvin and Lipton 2004 Sukatar et al. 2006 Paulert et al. 2007
U. fasciata	Ethyl acetate/*n*-hexane Ethanol, acetone, methanol	*V. parahaemolyticus, V. alginolyticus* *B. subtilis, S. aureus, E. coli, S. typhi, K. pneumoniae, C. albicans*	Chakraborty et al. 2010 Osman et al. 2010
U. fasciata	Methanol, butanol, water Crude extracts Methanol	*P. aeroginosa, S. aureus, E. coli, E. aeroginosa, S. paratyphi, K. pneumoniae, B. subtilis, Citrobacter* sp., *Proteus* sp., *S. epidemis* *A. hydrophila, P. fluorescens, Proteus* sp., *V. alginolyticus, Enterobacter* sp. *A. viscosus, S. mitis, S. mutans*	Priyadharshini et al. 2011 Dhasarathan and Theriappan 2011 Sujatha et al. 2012
U. fasciata	Crude extracts (phthalate esters derivatives) Ethanol, methanol, hexane, acetone Water, ethanol, chloroform, petroleum ether, hexane	Pathogenic Gram$^+$ and Gram$^-$ bacteria, namely *K. pneumoniae* *Vibrio navarrensis* *E. coli, K. pneumoniae, P. aeruginosa, P. mirabilis, S. marcescens, S. typhi, A. calcoaceticus, Enterobacter cloacae, S. aureus, E. faecalis, S. pyogenes, B. cereus*	Osman et al. 2013 Johnson and Raja 2015
U. fascita	Water at 100°C followed by precipitation using ethanol (sulfated polysaccharides) Acetone, ethanol	*E. coli, Staphylococcus* sp., *Proteus* sp., *Streptococcus* sp., *Enterococci* sp. *E. coli, S. typhi*	Khan 2016
U. flexuosa	Ethyl acetate, methanol	*Staphylococcus epidermidis, E. coli, P. aeruginosa, E. faecalis, K. pneumonia*	Mashjoor et al. 2016
U. intestinalis (formerly *E. intestinalis*)	Acetone, *n*-butanol Acetone, methanol	*S. aureus* *B. cereus, M. flavus, C. freundii, K. pneumoniae, P. testosterone*	Rao and Parekh 1981 Nair et al. 2007
U. intestinalis (formerly *E. intestinalis*)	Hexane, diethyl ether, chloroform, ethyl acetate, ethanol, acetone, methanol Acetone and *n*-butanol	*B. subtilis, K. pneumoniae, Escherichia coli, P. aeruginosa* *S. aureus*	Jebasingh et al. 2011 Lavanya and Veerappan 2011
U. intestinalis	Oil extracts Acetone	*B. cereus, L. monocytogenes, E. coli, S. enteritidis* *Bacillus mycoides, B. subtilis, E. coli. K. pneumonia, S. aureus, A. flavus, A. fumigatus, C. albicans, P. purpurescens, P. verrucosum*	Horincar et al. 2014 Kosanić et al. 2015

U. lactuca	Methanol, dichloromethane, hexane, chloroform Organic extracts Crude extracts	*B. subtilis, M. luteus, S. aureus* *K. pneumoniae, S. aureus* *S. aureus*	Kim et al. 2007b Baky et al. 2008 Kandhasamy and Arunachalam 2008
U. lactuca	Crude extracts Acetone Ethanol, acetone, methanol	*S. aureus, E. faecalis, K. pneumoniae, E. faecalis* *E. coli, P. aeruginosa, S. aureus, K. pneumonia, E. faecalis* *B. subtilis, S. aureus, E. coli, S. typhi, K. pneumoniae, C. albicans*	Ibtissam et al. 2009 Kolanjinathan and Stella 2009 Osman et al. 2010
U. lactuca	Methanol Buthanol Ethanol, methanol, hexane, acetone	*B. subtilis, B. cereus, M. luteus, K. pneumonia, S. aureus, E. coli* *S. aureus* *E. coli, P. aeruginosa, Proteus vulgaris, K. pneumoniae, S. aureus, B. subtilis*	Abirami and Kowsalya 2011 Águila-Ramírez et al. 2012 Elnabris et al. 2013
U. lactuca	Crude extracts (phthalate esters derivatives) Acetone Ethanol, methanol, chloroform	Pathogenic Gram$^+$ and Gram$^-$ bacteria, namely *K. pneumoniae* *Shigella sonnei, B. subtilis, E. feacalis, S. typhimurium, E. coli, S. aureus, S. pyogenes, Staphylococcus epidermis* *S. aureus, B. subtilis, L. acidophilus, E. coli, P. aeruginosa, P. mirabilis*	Osman et al. 2013 Saritha et al. 2013 Arunachalam et al. 2014
U. lactuca	Ethanol Water, ethanol, chloroform, petroleum ether, hexane Acetone	*M. luteus, E. coli, B. thermosphacta* *E. coli, K. pneumoniae, P. aeruginosa, P. mirabilis, S. marcescens, S. typhi, A. calcoaceticus, Enterobacter cloacae, S. aureus, E. faecalis, S. pyogenes, B. cereus* *B. mycoides, B. subtilis, E. coli, K. pneumoniae, S. aureus, A. flavus, A. fumigatus, C. albicans, P. purpurescens, P. verrucosum*	Boisvert et al. 2015 Johnson and Raja 2015 Kosanić et al. 2015
U. lactuca	Ethanol Acetone, ethanol	*K. pneumonia, A. hydrophilla, S. aureus, E. coli, Pseudomonas* sp. *E. coli, S. typhi*	Radhika and Mohaideen 2015 Khan 2016
U. lactuca	Methanol	Gram$^+$ (*B. subtilis, C. diphtheriae* and *S. aureus*) and Gram$^-$ (*E. coli, P. aeruginosa* and *S. paratyphi*)	Alagan et al. 2017
U. linza (formerly *E. linza*)	Methanol, dichloromethane, hexane, chloroform Ethanol, acetone, methanol Crude extracts Methanol	*S. aureus* *B. subtilis, S. aureus, E. coli, S. typhi, K. pneumoniae, C. albicans* *Prevotella intermedia, Porphyromonas gingivalis* 9 food borne pathogens	Sukatar et al. 2006 Osman et al. 2010 Choi et al. 2012 Narasimhan et al. 2013
U. linza (formerly *E. linza*)	Essential oil	*B. cereus, S. aureus*	Patra and Baek 2016
U. linza (formerly *E. linza*)	Methanol:water (4:1) Ethanol Water/dichloromethane	*Prevotella intermedia, Porphyromonas gingivalis* *G. vaginalis* *E. coli, S. typhimurium*	Park et al. 2013b Ha et al. 2014 Patra et al. 2015

Table 8.1 contd....

Table 8.1 contd....

Species	Solvent and/or extract	Bacteria	References
U. prolifera (formerly *E. prolifera*)	Water, methanol	Several bacteria	
	Petroleum ether, diethyl ether, ethyl acetate and methanol	*E. coli, K. pneumoniae,* Methicillin-resistant *Staphylococcus aureus, B. subtilis*	Omar et al. 2012
	Water at 100°C followed by precipitation using ethanol (sulfated polysaccharides)	*E. coli, Staphylococcus* sp., *Proteus* sp., *Streptococcus* sp., *Enterococci* sp.	Kailas and Nair 2016
U. prolifera	Crude extracts	*B. subtilis, S. aureus, K. pneumoniae*	Gonzalez et al. 2001
	Chloroform:Methanol (2:1)	*V. ordali*	Cavallo et al. 2013
U. reticulata	Crude extracts	*S. aureus, P. aeruginosa*	Gonzalez et al. 2001
	Acetone, methanol, chloroform, diethyl ether, ethyl acetate, ethanol, petroleum ether	*E. coli, Staphylococcus* sp., *Proteus* sp., *Streptococcus* sp., *Enterococci* sp.	Devi et al. 2009
	Acetone, methanol, chloroform, diethyl ether, ethyl acetate, ethanol, petroleum	*E. coli, Staphylococcus* sp., *Proteus* sp., *Streptococcus* sp., *Enterococci* sp.	Karthikaidevi et al. 2009
U. reticulata	Ethanol	*S. aureus, E. coli, M. tuberculosis, S. faecalis, K. pnuemoniae, P. aeruginosa*	Lakshmi et al. 2010
	Petroleum ether, diethyl ether, ethyl acetate and methanol	*E. coli, K. pneumoniae,* Methicillin-resistant *Staphylococcus aureus, B. subtilis*	Omar et al. 2012
U. reticulata	Hexane	*E. coli, S. typhi, B. subitils, P. aeruginosa*	Dhanya et al. 2016
U. rigida	Hexane, diethyl ether, chloroform, ethyl acetate, ethanol, acetone, methanol	*S. aureus, E. coli, E. faecalis*	Zbakh et al. 2012
	Hexane, ethyl acetate, methanol	*E. coli, L. monocytogene, S. enterica, Agrobacterium tumefaciens, P. aerigunosa, S. aureus, M. luteus*	Kolsi et al. 2015
Ulva sp. (formerly *Enteromorpha* sp.)	*n*-butanol	*S. aureus*	Rao and Karmarkar 1986
U. taeniata (formerly *U. dactylifera*)	Ethanol	*S. aureus, S. pyogenes*	Muñoz-Ochoa et al. 2010
Valonia aegagropila	Crude extracts	*S. aureus, B. subtilis, E. coli*	Mtolera and Semesi 1996
V. utricularis	Crude organic extract	HSV-1 VSV Inhibition zone to 1 mm	Ballesteros et al. 1992

Valoniopsis pachynema	Crude extracts Crude extracts	*S. aureus, E. coli, S. faecalis, Klebsiella pnuemoniae, P. aeruginosa* *S. aureus, E. coli, S. faecalis, K. pnuemoniae, P. aeruginosa*	Lakshmi et al. 2006 Puglisi et al. 2007
V. pachynema	Acetone, ethanol	*E. coli, S. typhi*	Khan 2016
Phaeophyceae (Brown seaweed)			
Anthophycus longifolius	Ethanol	*B. subtilis, L. innocua, L. monocytogenes, Micrococcus* sp., *S. aureus, Acinetobacter lwoffii*	Vlachos et al. 1997
Ascophyllum nodosum	Phlorotannins	*E. coli* resistant to Ampicillin, Kanamycin, and Nalidixic acid	Wang et al. 2009
Bifurcaria bifurcata	Ethanol (eleganolone) Diethyl-ether Methanol	Several bacteria *Cobetia marina, Pseudoalteromonas haloplanktis* *E. coli, P. aeruginosa, S. aureus*	Biard et al. 1980 Maréchal et al. 2004 Alves et al. 2016b
Bifurcariopsis capensis	Ethanol	*B. subtilis, L. innocua, L. monocytogenes, Micrococcus* sp., *S. aureus*	Vlachos et al. 1997
Chnoospora implexa	Ethanol	*S. aureus, S. pyogenes*	Muñoz-Ochoa et al. 2010
Cladostephus spongiosus f. *verticillatus*	Crude extracts Methanol, dichloromethane, hexane (essential oils)	*S. aureus, M. luteus, E. faecalis, E. coli, E. aerogenes, E. coli* *B. subtilis, S. aureus*	Taskin et al. 2007 Demirel et al. 2009
Colpomenia sinuosa	Crude extracts	*B. subtilis, B. cereus, S. aureus*	Salvador et al. 2007
C. sinuosa	Crude extracts Diethyl ether Methanol, acetone, diethyl ether, ethanol Methanol, dichloromethane, hexane Ethanol	*S. aureus, E. coli, S. faecalis, K. pnuemoniae, P. aeruginosa* *E. faecalis, S. aureus, S. epidermidis, P. aeruginosa, E. coli* *E. faecalis, S. aureus, S. epidermidis, P. aeruginosa, E. coli* *B. subtilis, S. aureus* *S. aureus, S. pyogenes*	Lakshmi et al. 2006 Liao et al. 2003 Tüney et al. 2006 Demirel et al. 2009 Muñoz-Ochoa et al. 2010
C. tuberculata	Ethanol, acetone, methanol	*B. subtilis, S. aureus, E. coli, S. typhi, K. pneumoniae*	Osman et al. 2010
Cystoseira barbata	Crude extracts Crude extracts	*B. subtilis, B. cereus* *S. aureus, M. luteus, E. faecalis, E. coli, E. aerogenes, E. coli*	Salvador et al. 2007 Taskin et al. 2007
C. brachycarpa (formerly *C. balearica*)	Crude extracts	*Bacillus megatherium, S. aureus*	Rao and Parekh 1981
C. compressa	Crude extracts	*B. subtilis, B. cereus, S. aureus*	Salvador et al. 2007
C. mediterranea	Diethyl ether	*E. faecalis, S. aureus, S. epidermidis, P. aeruginosa, E. coli*	Tüney et al. 2006
C. tamariscifolia	Crude extracts	*B. subtilis, B. cereus*	Salvador et al. 2007

Table 8.1 contd....

Table 8.1 contd....

Species	Solvent and/or extract	Bacteria	References
Dictyopteris delicatula	Ethanol	*S. aureus, S. pyogenes*	Muñoz-Ochoa et al. 2010
D. longifolia	Ethanol	*B. subtilis, L. innocua, L. monocytogenes, Micrococcus* sp., *S. aureus, Acinetobacter lwoffii*	Vlachos et al. 1997
D. polypodioides (as *D. membranacea*)	Methanol, acetone, diethyl ether, ethanol Ethanol, acetone	*E. faecalis, S. aureus, S. epidermidis, P. aeruginosa, E. coli* *Enterococcus faecium, Streptococcus agalactiae, B. subtilis*	Tüney et al. 2006 Akremi et al. 2017
D. undulata	Ethanol	*S. aureus, S. pyogenes*	Muñoz-Ochoa et al. 2010
Dictyota bartayresiana (formerly *Dictyota bartayresii*)	Crude extracts (diterpenoid compound)	Several bacteria	Norris and Fenical 1982
D. cervicornis (formerly *D. indica*)	Methanol	*S. aureus, B. subtilis, E. coli, P. aeruginosa*	Muñoz-Ochoa et al. 2010
D. ciliolata (as *D. ciliata*)	Acetone, ethanol	*E. coli, S. typhi*	Khan 2016
D. dichotoma	Crude extracts Crude extracts Crude extracts	*Bacillus megatherium, S. aureus* Several bacteria *S. aureus, E. coli*	Rao and Parekh 1981 Moreau et al. 1984 Hornsey and Hide 1985
D. dichotoma	Crude extracts Crude extracts Methanol, dichloromethane, hexane (essential oils) Methanol	*S. aureus, E. coli, S. faecalis, K. pnuemoniae, P. aeruginosa* *S. aureus, M. luteus, E. faecalis, E. coli, E. aerogenes, E. coli* *B. subtilis, S. aureus* *S. aureus, B. subtilis, E. coli, P. aeruginosa*	Lakshmi et al. 2006 Taskin et al. 2007 Demirel et al. 2009 Muñoz-Ochoa et al. 2010
D. dichotoma	Methanol	*E. coli, L. monocytogene, S. enterica, A. tumefaciens, P. aerigunos, S. aureus, M. luteus*	Kolsi et al. 2015
D. dichotoma var. *intricata* (as *D. linearis*)	Methanol, acetone, diethyl ether, ethanol	*E. faecalis, S. aureus, S. epidermidis, P. aeruginosa, E. coli*	Tüney et al. 2006
D. flabellata	Ethanol Methanol	*S. aureus, S. pyogenes* *S. aureus, S. pyogenes*	Muñoz-Ochoa et al. 2010 Kolsi et al. 2015
D. implexa (as *D. dichotoma* var. *implexa*)	Methanol, dichloromethane, hexane (essential oils)	*B. subtilis, S. aureus*	Demirel et al. 2009
Dictyota sp.	Crude extracts Dichloromethane:methanol (2:1)	*Bacillus megatherium, S. aureus* *S. aureus, E. faecalis, P. aeruginosa*	Rao and Parekh 1981 Bianco et al. 2013
Dictyota spiralis	Crude extracts	*B. subtilis, B. cereus, S. aureus*	Salvador et al. 2007

Ecklonia bicyclis (formerly *Eisenia bicyclis*)	Methanol, acetone Phlorofucofuroeckol Methanol, methanol/hexane, methanol/dichloromethane, methanol/butanol	Methicillin-resistant *S. aureus* and *P. aeruginosa* Methicillin-resistant *S. aureus* *S. aureus, S. epidermidis, Propionibacterium acnes*	Al Hazzani et al. 2014 Eom et al. 2014 Lee et al. 2014
E. cava	Acetone, chloroform, methanol	*Micrococcus* sp., *Shigella fleschneri, Salmonella paratyphi*	Moorthi and Balasubramanian 2015
E. maxima	Ethanol	*B. subtilis, L. innocua, L. monocytogenes, Micrococcus* sp., *S. aureus, A. lwoffii*	Vlachos et al. 1997
E. radiata	Ethanol	*B. subtilis, L. innocua, L. monocytogenes, Micrococcus* sp., *S. aureus, A. lwoffii*	Vlachos et al. 1997
Ectocarpus siliculosus	Diethyl ether	*E. faecalis, S. aureus, S. epidermidis, P. aeruginosa, E. coli*	Tüney et al. 2006
Fucus distichus subsp. *evanescens*	Ethyl acetate	*Hemophilus influenzae, Legionella pneumophila, P. acnes, S. pyogenes, Clostridium difficile,* Methicillin-resistant *S. aureus*	Amiguet et al. 2011
F. spiralis	Methanol Crude extract and 0.5% natural sorbic acid	*S. aureus* *S. aureus, B. cereus, B. subtilis, P. fluorescens, E. coli, K. pneumoniae, Aeromonas hydrophila, V. alginolyticus, V. parahaemolyticus*	Ibtissam et al. 2009 Lück and Jager 2012 Rodríguez-Martínez et al. 2016
F. guiryi (as *F. spiralis* var. *platycarpus*)	Crude extracts	*B. subtilis, B. cereus, S. aureus*	Salvador et al. 2007
F. vesiculosus	0.05% trifluoroacetic acid (polyhydroxylated fucophlorethol) Methanol Methanol (fucoidan)	*E. coli, P. mirabilis, S. aureus, S. epidermidis* *E. coli, S. aureus* *Vibrio alginolyticus*	Sandsdalen et al. 2003 Ibtissam et al. 2009 Nishiguchi et al. 2014
Halopteris filicina	Crude extracts	*S. aureus, M. luteus, E. faecalis, E. coli, E. aerogenes*	Taskin et al. 2007
Halidrys saliquosa	Ethyl acetate	*P. aeruginosa, E. coli, S. aureus*	Le Lann et al. 2016
Hapalospongidion macrocarpum	Crude extracts	*B. subtilis, B. cereus, S. aureus, E. coli, P. aeruginosa*	Salvador et al. 2007
Himanthalia elongata	Acetone, ethanol Crude extracts Hexane, ether, water (pressurized liquid extraction)	*E. faecalis, P. aeruginosa, L. monocytogenes, S. abony* *L. monocytogenes, S. abony, E. faecalis, P. aeruginosay* *E. coli, S. aureus*	Cox et al. 2010 Gupta et al. 2010 Plaza et al. 2010

Table 8.1 contd....

Table 8.1 contd....

Species	Solvent and/or extract	Bacteria	References
H. elongata	Methanol, water Water, methanol and mixtures Diethyl ether, *n*-hexane, chloroform Chloroform, diethylether, hexane, methanol, water	*L. monocytogenes, E. faecalis, P. aeruginosa, S. abony* *L. monocytogenes, E. faecalis, P. aeruginosa, S. abony* *L. monocytogenes* *L. monocytogenes, S. abony*	Rajauria et al. 2012 Rajauria et al. 2013 Rajauria and Abu-Ghannam 2013 Cox et al. 2014
Hydroclathrus clathratus	Dichloromethane, methanol, ethyl acetate Ethanol Petroleum ether, chloroform, methanol	*Pseudoaltermonas bacteriolytica* *S. aureus, S. pyogenes* *S. aureus, S. typhimurium, E. coli, Klebsiella* sp., *Proteus* sp., *Citrobacter* sp., *Pseudomonas* sp.	Puglisi et al. 2007 Muñoz-Ochoa et al. 2010 Singh and Raadha 2015
Iyengaria stellata	Acetone, ethanol	*E. coli, S. typhi*	Khan 2016
Laminaria digitata	Acetone, ethanol Chloroform/methanol (2:1), ethanol	*E. faecalis, P. aeruginosa, L. monocytogenes, S. abony* *L. monocytogenes, S. abony, E. faecalis, P. aeruginosa*	Cox et al. 2010 Gupta et al. 2010
Lobophora variegata	Chloroform/methanol (2:1), ethanol Organic extracts Ethyl acetate:methanol (1:1)	*B. subtilus, S. aureus, P. aeruginosa, E. coli* *P. bacteriolytica* *V. shiloi, Photobacterium eurosenbergii, Alteromonas* sp., *A. macleodii, Pseudoalteromonas* sp., *P. prydzensis, Shewanella* sp., *Photobacterium eurosenbergii, Listonella pelagia, V. sinaloensis, V. harveyi, V. brasiliensis*	Ballantine et al. 1987 Engel et al. 2006 Morrow et al. 2011
L. variegata	Methanol Organic extracts Dichloromethane:methanol (1:1)	*V. alginolyticus, V. vulnificus, V. parahaemolyticus, B. subtilis, B. cereus, M. luteus, S. typhimurium, Aeromonas hydrophila, E. coli* *Klebsilla pneumonia, E. coli, S. typhi, V. cholera* *E. faecalis, E. coli, S. aureus*	Manilal et al. 2010, 2010b, 2012 Sivakumar 2014 Gutiérrez-Cepeda et al. 2015
Padina australis	*n*-hexane, dichloromethane, methanol Methanol, acetone, ethyl acetate, chloroform	*B. cereus* *B. subtilis, S. aureus, P. aeruginosa, E. coli*	Chiao-Wei et al. 2011 Jaswir et al. 2014
P. boergesenii	Butanol, propanol, acetone	*P. aeruginosa, S. aureus*	Rao and Karmarkar 1986
P. concrescens	Methanol Ethanol	*V. alginolyticus, V. vulnificus, V. parahaemolyticus, B. subtilis* *S. aureus, S. pyogenes*	Manilal et al. 2010 Muñoz-Ochoa et al. 2010
P. gymnospora	Methanol, ethyl acetate Hexane	*S. aureus, B. cereus, E. coli, E. feacalis, Salmonella* sp., *P. aeruginosa* *P. aeruginosa, S. typhi, K. pneumonia, B. subtilis*	Salem et al. 2011 Rosaline et al. 2012

P. gymnospora	Ethanol, methanol Water, ethanol	*V. parahaemolyticus, E. coli, S. aureus, P. aeruginosa, S. enterica, V. brasiliensis, V. xuii, V. navarrensis* *P. aeruginosa*	Thanigaivel 2015
P. mexicana	Ethanol	*S. aureus, S. pyogenes*	Muñoz-Ochoa et al. 2010
P. pavonica	Diethyl ether Diethyl ether, methanol Crude extracts	*E. faecalis, S. aureus, S. epidermidis, P. aeruginosa, E. coli* *E. faecalis, S. aureus, S. epidermidis, P. aeruginosa, E. coli* *B. subtilis, B. cereus, S. aureus*	Liao et al. 2003 Tüney et al. 2006 Salvador et al. 2007
P. pavonica	Methanol Methanol	*E. coli* *E. coli, S. aureus*	Muñoz-Ochoa et al. 2010 El-Fatimy and Abdel-Moneim 2011
P. pavonica	Acetone, ethanol	*E. coli, S. typhi*	Khan 2016
P. sanctae-crucis	Methanol, ethanol	*E. coli, S. aureus, P. aeruginosa*	Nogueira et al. 2014
Padina sp.	Ethanol	*L. monocytogenes, B. cereus, S. aureus*	Dussault et al. 2016
P. tetrastromatica	Butanol, propanol, acetone Ethyl acetate, methanol, water	*P. aeruginosa, S. aureus* *B. subtilis, E. coli, Pseudomonas* sp., *S. pyrogenes, S. aureus, P. vulgaris, K. pneumoneae, Serratia marganii*	Rao and Karmarkar 1986 Kotnala et al. 2009
P. tetrastromatica	Methanol Ethanol, chloroform Crude extracts	*V. alginolyticus, V. vulnificus, V. parahaemolyticus, B. subtilis* 6 strains of Gram$^+$ and Gram$^-$ bacterial isolates *P. aeruginosa, K. pneumonia, S. aureus, B. subtilis*	Manilal et al. 2010 Rangaiah et al. 2010 Subba et al. 2010
Petalonia fascia	Methanol, dichloromethane, hexane (essential oils)	*B. subtilis, S. aureus*	Demirel et al. 2009
Polycladia indica (as *Cystoseira indica*)	Acetone, ethanol	*E. coli, S. typhi*	Khan 2016
Polycladia myrica (as *Cystoseira myrica*)	Water, chloroform:methanol (2:1)	*S. aureus, S. viridians, S. flexneri*	El-Sheekh et al. 2014
Rosenvingea intricata	Ethanol	*S. aureus, S. pyogenes*	Muñoz-Ochoa et al. 2010
Saccharina angustata (formerly *Laminaria angustata*)	Methanol Subcritical water hydrolysis	*Bacillus mesentericus* *E. coli, S. typhimurium, S. aureus, B. cereus*	Saito and Nakamura 1951 Meillisa et al. 2013
S. japonica	Subcritical water hydrolysis	*E. coli, S. typhimurium, S. aureus, B. cereus*	Meillisa et al. 2013
S. japonica (formerly *L. japonica*)	Ethanol Seaweed powder (500–900 μm)	*Actinomyces naeslundii, Actinomyces odontolyticus, Streptococcus mutans, Fusobacterium nucleatum, Porphyromonas gingivalis* *E. coli, S. typhimurium, B. cereus, S. aureus*	Kim et al. 2013c Siahaan et al. 2014

Table 8.1 contd. ...

Table 8.1 contd....

Species	**Solvent and/or extract**	**Bacteria**	**References**
S. japonica (formerly *L. japonica*)	Methanol, acetone Water Hexane, ethanol, acetone/methanol	Methicillin-resistant *S. aureus* and *P. aeruginosa* *E. coli, S. aureus, B. cereus* *E. coli, S. aureus, P. aeruginosa, S. typhimurium*	Al Hazzani et al. 2014 Patra et al. 2015 Sivagnanam et al. 2015
S. latissima	Acetone, ethanol Crude extracts	*E. faecalis, P. aeruginosa, L. monocytogenes, S. abony* *L. monocytogenes, S. abony, E. faecalis, P. aeruginosay*	Cox et al. 2010 Gupta et al. 2010
S. longicruris	Enzymatic hydrolysis with trypsin	*S. aureus*	Beaulieu et al. 2015
Sargassum aquifolium (formerly *Sargassum binderi*)	Methanol, acetone, ethyl acetate, chloroform	*B. subtilis, S. aureus, P. aeruginosa, E. coli*	Jaswir et al. 2014
S. boveanum	Acetone, ethanol	*E. coli, S. typhi*	Khan 2016
S. elegans	Ethanol	*B. subtilis, L. innocua, L. monocytogenes, Micrococcus* sp., *S. aureus, A. lwoffii*	Vlachos et al. 1997
S. filipendula	Ethanol and methanol extracts *In vitro* 3.13–6.25 mg mL^{-1}	*S. aureus, B. subtilis, S. agalactiae, E. coli, P. aeruginosa, K. pneumoniae, S. flexneri, C. albicans, S. cerevisae, A. niger, T. mentagrophytes*	Morales et al. 2006
S. fulvellum	Methanol, acetone, ethyl acetate, chloroform	*B. subtilis, S. aureus, P. aeruginosa, E. coli*	Jaswir et al. 2014
S. horneri	Acetone:ethanol (1:1)	*E. coli, L. monocytogenes, B. cereus, S. aureus*	Sivagnanam et al. 2015
S. horridum	Ethanol	*S. aureus, S. pyogenes, E. coli*	Muñoz-Ochoa et al. 2010
S. hystrix	Ethanol and methanol extracts *In vitro* 3.13–6.25 mg mL^{-1}	*S. aureus, B. subtilis, S. agalactiae, E. coli, P. aeruginosa, K. pneumoniae, S. flexneri, C. albicans, S. cerevisae, A. niger, T. mentagrophytes*	Morales et al. 2006
S. ilicifolium (formerly *S. cristaefolium*)	Dichloromethane, methanol, ethyl acetate	*Pseudoaltermonas bacteriolytica*	Puglisi et al. 2007
S. incisifolium	Ethanol	*B. subtilis, L. innocua, L. monocytogenes, Micrococcus* sp., *S. aureus, A. lwoffii*	Vlachos et al. 1997
S. miyabei (as *S. kjellmanianum*)	Ethanol extract *In vitro*	*S. aurens*	Xu et al. 2002
S. plagiophyllum	Methanol	*B. subtilis, S. aureus*	Jaswir et al. 2014

S. polycystum	Dichloromethane, methanol, ethyl acetate	*Pseudoaltermonas bacteriolytica*	Puglisi et al. 2007
S. polycystum (formerly *S. myriocystum*)	Methanol	*S. aureus, B. subtilis, E. coli, P. aeruginosa*	Kavita et al. 2014
S. ramifolium	Water, chloroform:methanol (2:1)	*S. aureus, S. flexneri, S. typhi, E. coli*	El-Sheekh et al. 2014
S. swartzii (as *S. wightii*)	Benzene, chloroform and methanol extract *In vitro*	*S. aureus, E. coli, P. vulgaris, P. aeruginosa, S. paratyphi, S. typhi, S. typhimurium*	Sastry and Rao 1994
S. swartzii (as *S. wightii*)	Acetone, ethanol	*E. coli, S. typhi*	Khan 2016
S. vulgare	Diethyl ether	Gram$^+$ and Gram$^-$ bacteria	Rao et al. 1988
S. vulgare	Ethanol, acetone, methanol Ethyl acetate, methanol Diethyl ether, ethanol	*B. subtilis, S. aureus, E. coli, S. typhi, K. pneumoniae* *E. coli, L. monocytogene, S. enterica, A. tumefaciens, P. aerigunosa, S. aureus, M. luteus* *K. pneumoniae, S. aureus*	Osman et al. 2010 Kolsi et al. 2015 El Shafay et al. 2016
Scytosiphon lomentaria	Methanol, dichloromethane, hexane (essential oils) Methanol	*B. subtilis, S. aureus* *S. aureus, S. typhimurium, E. coli*	Demirel et al. 2009 Taskin et al. 2010
Sirophysalis trinodis (formerly *Cystoseira trinodis*)	Methanol Methanol, ethyl acetate	*B. subtilis, S. aureus, P. aeruginosa* *S. aureus, B. cereus, E. coli, E. feacalis, Salmonella* sp., *P. aeruginosa*	Muñoz-Ochoa et al. 2010 Salem et al. 2011
Spatoglossum variabile	Acetone, ethanol	*E. coli, S. typhi*	Khan 2016
Stepharocystis osmundacea (formerly *Cystoseira osmundacea*)	Ethanol	*S. pyogenes*	Muñoz-Ochoa et al. 2010
Stoechospermum polypodioides (formerly *Stoechospermum marginatum*)	Crude extracts Crude extracts Methanol	*V. cholera, Klebsiella* sp. *S. aureus, E. coli, S. faecalis, K. pnuemoniae, P. aeruginosa* *S. aureus, B. subtilis, E. coli, P. aeruginosa*	Ely et al. 2004 Lakshmi et al. 2006 Kavita et al. 2014
S. polypodioides (formerly *S. marginatum*)	Acetone, ethanol	*E. coli, S. typhi*	Khan 2016
Stypopodium zonale	Ethanol	*B. subtilis, L. innocua, L. monocytogenes, Micrococcus* sp., *S. aureus, A. lwoffii*	Vlachos et al. 1997
Turbinaria condensata	Crude extracts	*S. aureus, E. coli, S. faecalis, K. pnuemoniae, P. aeruginosa*	Lakshmi et al. 2006
T. conoides	Crude extracts	*S. aureus, E. coli, S. faecalis, K. pnuemoniae, P. aeruginosa*	Lakshmi et al. 2006

Table 8.1 contd....

Table 8.1 contd....

Species	Solvent and/or extract	Bacteria	References
T. decurrens	Crude extracts	*S. aureus, E. coli, S. faecalis, K. pnuemoniae, P. aeruginosa*	Lakshmi et al. 2006
T. ornata	Phenolic compounds Methanol	*Aeromonas hydrophila, B. subtilis, E. faecalis, E. coli, K. pneumoniae, P. vulgaris, P. aeruginosa, Shigella flexneri, S. aureus* *S. aureus, B. subtilis, E. coli, P. aeruginosa*	Vijayabaskar and Shiyamala 2011 Kavita et al. 2014
T. triquetra	Crude extracts Chloroform, methanol (fucoxanthin) Methanol, ethanol, petroleum ether, dimethyl formamide	*B. subtilis,* Methicillin-resistant *S. aureus, E. coli, K. pneumoniae, P. aeruginosa* *E. coli, B. cereus, B. subtilis, K. pneumoniae, S. aureus, P. aeruginosa* *E. coli, S. typhi, S. dysenteriae, K. pneumoniae, E. aerogenes*	Omar et al. 2012 Deyab and Abou-Dobara 2013 Al-Judaibi 2014
Zanardinia typus (as *Zanardinia prototypus*)	Crude extracts	*Bacillus megatherium, S. aureus*	Rao and Parekh 1981
Z. typus	Crude extracts	*B. subtilis, B. cereus, S. aureus*	Salvador et al. 2007
Zonaria subarticulata	Ethanol	*B. subtilis, L. innocua, L. monocytogenes, Micrococcus* sp., *S. aureus, A. lwoffii*	Vlachos et al. 1997
Z. tournefortii	Ethanol	*B. subtilis, L. innocua, L. monocytogenes, Micrococcus* sp., *S. aureus, A. lwoffii*	Vlachos et al. 1997
Rhodophyta (Red seaweed)			
Acanthaphora spicifera	Petroleum ether, methanol, chloroform	*E. coli, M. gypseum*	Pandian et al. 2011
Alsidium corallinum	Methanol	*E. coli, K. pneumoniae, S. aureus*	Bouhlal et al. 2010b
Amansia multifida	Hexane, chloroform, ethanol	*E. aerogenes, K. pneumoniae, P. aeruginosa, S. typhi, S. choleraesuis, S. marcescens, V. cholerae, B. subtilis, S. aureus*	Lima-Filho et al. 2002
Amphiroa ephedraea	Ethanol	*B. subtilis, L. innocua, L. monocytogenes, Micrococcus* sp., *S. aureus, A. lwoffii, S. enteritidis*	Vlachos et al. 1997
A. fragilissima	Dichloromethane, methanol, ethyl acetate	*Pseudoaltermonas bacteriolytica*	Puglisi et al. 2007
Arthrocardia carinata	Ethanol	*B. subtilis, L. innocua, L. monocytogenes, Micrococcus* sp., *S. aureus, A. lwoffii, S. enteritidis*	Vlachos et al. 1997
Asparagopsis armata	Crude extracts	*B. subtilis, B. cereus, S. aureus, E. coli, P. aeruginosa*	Salvador et al. 2007
A. armata	Crude extracts Dichloromethane:methanol (50:50) Methanol	*E. coli, S. aureus, E. faecalis* *B. cereus, S. aureus, E. faecalis* *E. coli, B. subtilis*	Zbakh et al. 2012 Oumaskour et al. 2013 Pinteus et al. 2015

A. armata (*Falkenbergia rufolanosa*-phase)	Crude extracts Dichloromethane, methanol, water Crude extracts	3 pathogenic bacteria *Aeromonas salmonicida* subsp. *salmonicida, A. hydrophila* subsp. *hydrophila, Pseudomonas anguilliseptica, Vibrio anguillarum, Yersinia ruckeri* *B. subtilis, B. cereus, S. aureus, E. coli, P. aeruginosa*	Pesando and Caram 1984 Bansemir et al. 2006 Salvador et al. 2007
Asparagopsis sp.	Crude extracts	*Vibrio* spp.	Manilal et al. 2011
A. taxiformis	Crude extracts	6 pathogenic *Vibrio* strains	Manilal et al. 2009b
A. taxiformis	Ethanol, acetone	*E. coli, S. typhi*	Khan 2016
A. taxiformis (*Falkenbergia hillebrandii*-phase)	Chloroform, acetone Methanol	*S. aureus* *B. subtilis, K. pneumonia, P. aeruginosa, S. aureus, S. epidermidis, Vibrio harveyi, V. alginolyticus, V. vulnificus, V. parahaemolyticus, V. alcaligenes*	Olessen et al. 1963 Manilal et al. 2010
Bonnemaisonia asparagoides	Crude extracts	*S. aureus, E. coli, B. subtilis, S. pyogenes, P. morganii*	Hornsey and Hide 1974
B. hamifera	Crude extracts Crude extracts Crude extracts	*S. aureus, E. coli, B. subtilis, S. pyogenes, P. morganii* *S. aureus, E. coli* *B. subtilis, B. cereus, S. aureus, E. coli, P. aeruginosa*	Hornsey and Hide 1974 Hornsey and Hide 1985 Salvador et al. 2007
Calliblepharis fimbriata	Acetone, ethanol	*E. coli, S. typhi*	Khan 2016
Callophycus serratus	Water, methanol, dichloromethane	Methicillin-resistant *S. aureus*, Vancomycin-resistant *E. faecium*	Lane et al. 2009
Centroceras clavulatum	Ethanol	*B. subtilis, M. luteus, S. aureus, S. mutans, S. anginosus, L. acidophilus, E. coli, E. aerogenes, K. pneumonia, P. aeuroginosa, Erwinia caratovora, P. vulgaris*	Kausalya and Rao 2015
Ceramium ciliatum	Methanol, *n*-hexane, dichloromethane	*B. subtilis, E. coli*	Pinteus et al. 2015
C. deslongchampsii	Crude extracts	*B. subtilis, B. cereus, S. aureus, E. coli, P. aeruginosa*	Salvador et al. 2007
C. diaphanum var. *elegans*	Chloroform, water, *n*-butanol	*S. aureus*	Kamenarska et al. 2009
C. nitens	Ethanolic (lipid-soluble extracts)	*B. subtilis, S. faecalis, M. luteus*	Freile-Pelegrín and Morales 2004
C. virgatum (as *C. rubrum*)	Crude extracts	Several bacteria	Chesters and Stott 1956 Roos 1957 Ikawa et al. 1973

Table 8.1 contd....

Table 8.1 contd....

Species	Solvent and/or extract	Bacteria	References
C. virgatum (as *C. rubrum*)	Dichloromethane, methanol, water Hexane, methanol Methanol	*Aeromonas salmonicida* subsp. *salmonicida, A. hydrophila* subsp. *hydrophila, P. anguilliseptica, V. anguillarum, Yersinia ruckeri* *Listonella anguillarum, Pseudomonas anguilliseptica, Aeromonas salmonicida* *E. coli, E. faecalis, S. aureus*	Bansemir et al. 2006 Dubber and Harder 2008 Bouhlal et al. 2010b
C. virgatum (as *C. rubrum*)	Methanol Dichloromethane	*M. tuberculosis* *Yersinia ruckeri*	Süzgeç-Selçuk et al. 2011 Cortés et al. 2014
C. virgatum	Hexane	*S. enteritidis, E. coli, L. monocytogenes, B. cereus*	Horincar et al. 2014
Chondracanthus acicularis	Methanol	*E. coli, K. pneumoniae, E. faecalis, S. aureus*	Bouhlal et al. 2010
C. canaliculatus	Ethanol	*S. aureus, Streptococcus pyogenes*	Muñoz-Ochoa et al. 2010
Chondrophycus brandenii (as *Laurencia brandenii*)	Crude extracts Methanol, petroleum ether:chloroform (6:4)	6 pathogenic *Vibrio* strains *B. subtilis, S. typhi, V. cholerae, S. pneumoniae*	Manilal et al. 2009b Manilal et al. 2011
C. ceylanicus (formerly *L. ceylanica*)	Crude extracts	6 pathogenic *Vibrio* strains	Manilal et al. 2009b
Chondria armata	Methanol extract (Galactoglycerolipids)	*Klebsiella* sp., *Candida albicans*	Al-Fadhli et al. 2006
C. dasyphylla	Crude extracts	*S. aureus, E. coli, S. faecalis, K. pnuemoniae, P. aeruginosa*	Lakshmi et al. 2006
C. littoralis	Crude extracts	*E. coli, P. aeruginosa*	Martinez Nadal et al. 1963
C. oppositiclada	Cycloeudesmol	*S. aureus*	Fenical and Sims 1974
Chondriopsis dasyphylla f. *pyrifera* (as *Laurencia intricata*)	Crude extracts	*B. cereus, S. aureus*	Salvador et al. 2007
Chondrus crispus	Acetone Crude extracts	*S. aureus, E. coli, B. subtilis, S. pyogenes, P. morganii* *S. aureus* *S. aureus, E. coli*	Hornsey and Hide 1974 Hornsey and Hide 1976 Hornsey and Hide 1985
C. crispus	Methanol 95% ethanol Methanol, diethyl acetate	*L. monocytogenes, Salmonella abony, E. faecalis, P. aeruginosa* *Pseudoalteromonas elyakovii, Vibrio aestuarianus, Polaribacter irgensii, Halomonas marina, Shewanella putrefaciens* *S. enteritidis, P. aeruginosa, Listeria innocua, S. aureus*	Cox et al. 2010 Chambers et al. 2011 Mendes et al. 2013
C. crispus	95% ethanol/toluene Water, methanol	*Cobetia marina, Marinobacter hydrocarbonoclasticus* *S. enteritidis*	Salta et al. 2013 Kulshreshtha et al. 2016
Corallina sp.	Crude extracts	*S. epidermidis, S. aureus, E. feacalis, E. coli, P. aeruginosa*	Sebaaly et al. 2014

C. vancouveriensis	Ethanol	*S. aureus, S. pyogenes*	Muñoz-Ochoa et al. 2010
Coronaphycus elatus (as *Laurencia elata*)	Elatol	*S. epidermis, K. pneumonia, Salmonella* sp.	Sims 1974 Vairappan 2003
Dasya scoparia	Ethanol, acetone, methanol-toluene	*P. solancearum, E. coli, B. subtilis, S. aureus*	Yi et al. 2001
Dichotomaria diesingiana (as *Galaxaura diesingiana*)	80% ethanol	*B. subtilis, S. aureus, A. lwofii, E. coli*	Vlachos et al. 2001
Dilsea carnosa	Crude extracts Acetone	*S. aureus, E. coli, B. subtilis, S. pyogenes, P. morganii* *S. aureus*	Hornsey and Hide 1974 Hornsey and Hide 1976
Ellisolandia elongata (formerly *Corallina elongata*)	Water Ethanol, acetone, methanol Methanol, methanol:dichloromethane (50:50)	*S. aureus* *B. subtilis, S. aureus, E. coli, S. typhi, K. pneumoniae* *B. cereus, B. thuringiensis, B. subtilis, C. sporogenes, S. aureus, S. aureus* subsp. *aureus, M. smegmatis, E. faecalis, Bacillus* sp., *E. coli, Pseudomonas* sp.	Kamenarska et al. 2009 Osman et al. 2010 Oumaskour et al. 2013
Enantiocladia prolifera	Methanol, water	*S. typhi, S. aureus, B. subtilis*	Adaikalaraj et al. 2012
Eucheuma denticulatum	Crude extract Methanol	*S. aureus, S. pyogenes* Methicillin-resistant *S. aureus, E. coli, S. pyogenes, V. cholerae, P. aeruginosa, K. pneumoniae*	Al-Haj et al. 2009 Al-Haj et al. 2010
Ganonema farinosum	Ethanol	*S. aureus, S. pyogenes*	Muñoz-Ochoa et al. 2010
Gelidium amansii	Crude extracts	*Mycobacterium tuberculosis, M. avium, M. phlei*	Kamimoto 1956
G. attenuatum	Methanol	*E. coli, K. pneumoniae, E. faecalis, S. aureus*	Bouhlal et al. 2010b
G. abbottiorum	Ethanol	*B. subtilis, L. innocua, L. monocytogenes, Micrococcus* sp., *S. aureus, A. lwoffii, S. enteritidis*	Vlachos et al. 1997
G. corneum (formerly *G. sesquipedale*)	Dichloromethane:methanol (50:50), methanol	*B. cereus, S. aureus, E. faecalis*	Oumaskour et al. 2013
G. filicinum	Crude extracts	*Mycobacterium smegmatis, M. tuberculosis*	Maurer 1965
G. microdon (formerly *G. spinulosum*)	Methanol	*E. coli, E. faecalis, S. aureus*	Bouhlal et al. 2010b
G. micropterum	Methanol	*V. parahaemolyticus, V. alcaligenes*	Manilal et al. 2010
G. pulchellum	Methanol	*E. coli, E. faecalis, S. aureus*	Bouhlal et al. 2010b

Table 8.1 contd....

Table 8.1 contd....

Species	Solvent and/or extract	Bacteria	References
G. pusillum	Methanol	*E. coli, K. pneumonia, E. faecalis, S. aureus*	Bouhlal et al. 2010b
	Methanol	*V. harveyi, V. alginolyticus, V. vulnificus, V. parahaemolyticus, V. alcaligenes*	Manilal et al. 2010
	Methanol, ethanol, chloroform	*B. subtilis, M. luteus, S. aureus, S. mutans, S. anginosus, Lactobacillus acidophilus, E. coli, E. aerogenes, K. pneumonia, P. aeuroginosa, Erwinia caratovora, Proteus vulgaris*	Kausalya and Rao 2015
G. robustum	Ethanol	*S. aureus, S. pyogenes*	Muñoz-Ochoa et al. 2010
G. spinosum	Chloroform, *n*-butanol (1,2-dihydroxy ethane sulfonate), water	*E. coli, S. aureus*	Kamenarska et al. 2009
Gelidium spp.	Methanol, water	*S. typhi, S. aureus, B. subtilis*	Adaikalaraj et al. 2012
Gloiopeltis furcata	Ethanol, acetone, methanol-toluene	*P. solancearum, E. coli, B. subtilis, S. aureus*	Yi et al. 2001
Gloiosiphonia capillaris	Crude extracts	*S. aureus, E. coli, B. subtilis, S. pyogenes, P. morganii*	Hornsey and Hide 1974
Gracilaria canaliculata (formerly *Gracilaria crassa*)	Crude extracts	*S. aureus, E. coli, S. faecalis, K. pnuemoniae, P. aeruginosa*	Lakshmi et al. 2006
	Methanol	*V. parahaemolyticus, Salmonella* sp., *Shewanella* sp., *E. coli, K. pneumoniae, S. pyogenes, S. aureus, E. faecalis, P. aeruginosa, P. mirabilis*	Lavanya and Veerappan 2011
G. cervicornis	95% ethanol	*S. aureus*	Perez et al. 1990
G. changii	Methanol, methanol/chloroform, diethyl ether, ethyl acetate, butanol	*P. aeruginosa, B. subtilis*	Sasidharan et al. 2009
	Methanol	Methicillin-resistant *S. aureus, E. coli, S. pyogenes, V. cholerae, P. aeruginosa, K. pneumoniae*	Al-Haj et al. 2010
G. cornea	Dichloromethane, methanol, water	*Aeromonas salmonicida* subsp. *salmonicida, A. hydrophila* subsp. *hydrophila, Pseudomonas anguilliseptica, Vibrio anguillarum, Yersinia ruckeri*	Bansemir et al. 2006
G. corticata	Ethanol	*B. subtilis, Bacillus megaterium, S. aureus, Streptococcus viridians*	Usmanghani et al. 1984
	Crude extracts	*S. aureus*	Vidyavathi and Sridhar 1991
	Methanol	*Pseudomonas fluorescens, Edwardsiella tarda, V. alginolyticus, P. aeruginosa, A. hydrophila*	Choudhury et al. 2005
G. corticata	Crude extracts	*S. aureus, E. coli, S. faecalis, K. pnuemoniae, P. aeruginosa*	Lakshmi et al. 2006
	Crude extracts	*M. luteus, S. epidermidis, E. faecalis*	Shanmughapriya et al. 2008
	Methanol, acetone, chloroform, hexane, ethyl acetate	*S. aureus, S. pyogenes, S. epidermis, B. subtilis, B. cereus, K. pneumoniae, E. aerogenes*	Kolanjinathan and Stella 2011b
G. corticata	Ethanol	*K. pneumonia, A. hydrophilla, S. aureus, E. coli, Pseudomonas* sp.	Radhika and Mohaideen 2015

G. corticata	Acetone, ethanol	*E. coli, S. typhi*	Khan 2016
G. corticata	Sulfated polysaccharide	*S. aureus, K. pneumoniae, S. typhi, V. cholerae, K. oxytoca, E. coli, S. paratyphi, P. mirabilis, V. parahaemolyticus, S. pyogenes*	Seedevi et al. 2017
G. corticata var. *cylindrica*	Petroleum ether, chloroform, methanol	*S. aureus, S. typhimurium, E. coli, Klebsiella* sp., *Proteus* sp., *Citrobacter* sp., *Pseudomonas* sp.	Singh and Raadha 2015
G. corticata var. *cylindrica*	Petroleum ether, chloroform, methanol Water at 100°C followed by precipitation using ethanol (sulfated sugars)	Several bacteria *E. coli, B. cereus, S. abony*	Singh and Raadha 2015
G. debilis	Ethanol	*S. aureus, M. smegmatis*	Albuquerque et al. 1983
G. debilis (formerly *G. fergusonii*)	Acetone, chloroform, diethyl ether, ethanol, methanol (coumerins, phenols, quinones, steroids and tannins) Methanol, water	*K. pneumoniae, P. aeruginosa, B. subtilis, S. aureus* *S. typhi, S. aureus, B. subtilis*	Bai 2010 Adaikalaraj et al. 2012
G. dendroides	Chloroform, ethanol, petroleum ether	*E. coli, P. aeruginosa, S. aureus, E. faecalis*	Al-Saif et al. 2014
G. domingensis	Methanol, ethanol	*E. coli, S. aureus*	Albuquerque et al. 1983
G. dura	Crude extracts Chloroform:methanol (2:1)	*B. cereus, S. aureus, E. coli* *V. ordalii, V. alginolyticus*	Salvador et al. 2007 Cavallo et al. 2013
G. edulis	Butanol, propanol Crude extracts Chloroform	*K. pneumonia, P. aeruginosa* *S. aureus, V. cholerae, S. dysentriae, S. bodii, S. paratyphi, P. aeruginosa, K. pneumonia* *V. cholera, S. aureus, S. dysenteriae, S. bodii, S. paratyphi, P. aeruginosa, K. pneumonia, E. coli*	Ravikumar et al. 2002 Vallinayagam et al. 2009 Almeida et al. 2011
G. gracilis	Chloroform:methanol (2:1)	*V. salmonicida*	Cavallo et al. 2013
G. ornata	Crude sulfated sugars	*E. coli, B. subtilis, E. aerogens, S. typhi, S. choleraesuis*	Amorim et al. 2012
G. pulvinata (as *G. pygmaea*)	Acetone, ethanol	*E. coli, S. typhi*	Khan 2016
G. salicornia	Ethyl acetate	*S. aureus, E. coli*	Saeidnia et al. 2009
Gracilaria sp.	Hexane	*B. subtilis*	Wong and Cheung 2002
G. subsecundata	Ethanol	*S. aureus, S. pyogenes*	Muñoz-Ochoa et al. 2010
G. vermiculophylla	Ethyl acetate, methanol, diethyl ether	*S. enteritidis, P. aeruginosa, L. innocua, S. aureus*	Mendes et al. 2013
Gracilariopsis longissima	Methanol, chloroform:methanol (2:1) Chloroform:methanol (2:1)	*V. alginolyticus, V. fluvialis, V. vulnificus, V. cholerae* *V. alginolyticus, V. vulnificus, V. ordalii, V. salmonicida*	Stabili et al. 2012 Cavallo et al. 2013

Table 8.1 contd....

Table 8.1 contd....

Species	Solvent and/or extract	Bacteria	References
G. andersonii (formerly *Gracilariopsis sjoestedtii*)	Methanol, ethanol	*E. coli, S. aureus*	Sebaaly et al. 2014
G. longissima (formerly *Gracilaria verrucosa*)	Crude extracts Methanol, water Methanol, ethanol, chloroform, water	*S. aureus, E. coli* *S. typhi, S. aureus, B. subtilis* *S. paratyphi, E. aerogenes, S. epidermidis, S. typhi, S. flexneri*	Hornsey and Hide 1985 Adaikalaraj et al. 2012 Varier et al. 2013
Grateloupia filicina	Ethanol, acetone, methanol-toluene Methanol Methanol Methanol	*P. solancearum, E. coli, B. subtilis, S. aureus* *V. harveyi, V. alginolyticus, V. vulnificus, V. parahaemolyticus, V. alcaligenes* *S. aureus, B. subtilis, E. coli, P. aeruginosa* *E. coli, P. aerugenosa, S. aureus, B. subtilis*	Yi et al. 2001 Manilal et al. 2010 Al Hazzani et al. 2014 Kavita et al. 2014
G. livida	Ethanol, diethyl ether	*S. aureus, E. coli, P. aeruginosa*	Jiang et al. 2013
Halopitys incurva	Dichloromethane, methanol, water Methanol Methanol	*A. salmonicida* subsp. *salmonicida, A. hydrophila* subsp. *hydrophila, P. anguilliseptica, V. anguillarum, Y. ruckeri* *E. coli, K. pneumoniae, E. faecalis, S. aureus* *S. aureus, S. aureus* subsp. *aureus, M. smegmatis, E. faecalis, Bacillus* sp., *E. coli, Pseudomonas* sp.	Bansemir et al. 2006 Bouhlal et al. 2010b Oumaskour et al. 2013
Halymenia porphyroides	Crude extracts	*S. aureus, E. coli, S. faecalis, K. pnuemoniae, P. aeruginosa*	Lakshmi et al. 2006
Heterosiphonia muelleri	Crude extracts	*S. aureus, E. coli, S. faecalis, K. pnuemoniae, P. aeruginosa*	Lakshmi et al. 2006
Hypnea musciformis	Crude extracts Butanol Methanol-dichloromethane	3 pathogenic bacteria *S. pyogens* *M. luteus, B. cereus, B. subtilis, E. coli, A. hydrophila, P. aeruginosa, V. fischeri, V. alginolyticus, V. harveyi, Aeromonas* sp.	Pesando and Caram 1984 Ravikumar et al. 2002 Selvin and Lipton 2004
H. musciformis	Crude extracts Methanol Methanol	*S. aureus, E. coli, S. faecalis, K. pnuemoniae, P. aeruginosa* *K. pneumonia, E. faecalis, S. aureus* *E. coli, K. pneumoniae, E. faecalis, S. aureus*	Lakshmi et al. 2006 Kandhasamy et al. 2008 Bouhlal et al. 2010b
H. musciformis	Alcoholic extracts Methanol Methanol	*S. aureus, E. coli, M. tuberculosis, S. faecalis, K. pnuemoniae, P. aeruginosa* *S. aureus* *S. aureus, S. typhimurium, E. coli, E. faecali*	Lakshmi et al. 2010 Rhimou et al. 2010 Taskin et al. 2010
H. musciformis	Methanol, water Crude extracts Ethanol	*S. typhi, S. aureus, B. subtilis* *E. coli, S. aureus, E. faecalis* *V. parahaemolyticus, E. coli, S. aureus, P. aeruginosa, S. enterica,* antibiotic-resistant *V. brasiliensis, V. xuii, and V. navarrensis*	Adaikalaraj et al. 2012 Zbakh et al. 2012 Silva et al. 2013

H. musciformis	Methanol Acetone, ethanol	*S. paratyphi, E. aerogenes, S. epidermidis, S. typhi, S. flexneri* *E. coli, S. typhi*	Varier et al. 2013 Khan 2016
H. panrosa	Methanol Acetone, ethanol	*S. aureus, B. subtilis, E. coli, P. aeruginosa* *E. coli, S. typhi*	Kavita et al. 2014 Khan 2016
H. valentiae	Crude extracts Ethanol Methanol	6 pathogenic *Vibrio* strains *S. aureus, S. pyogenes* *S. aureus, B. subtilis, E. coli, P. aeruginosa*	Manilal et al. 2009b Muñoz-Ochoa et al. 2010 Kavita et al. 2014
Jania rubens	Water Methanol, dichloromethane, hexane, chloroform and volatile oil extracts Ethanol, acetone, methanol	*B. subtilis, S. aureus* 5 Gram$^+$, 4 Gram$^-$ bacteria *B. subtilis, S. aureus, E. coli, S. typhi, K. pneumoniae, C. albicans*	Soliman et al. 1994 Karabay-Yavasoglu et al. 2007 Osman et al. 2010
J. rubens	Methanol, ethanol, acetone, chloroform	*V. fluvialis*	El-Din and El-Ahwany 2016
J. sagittata (as *Cheilosporum sagittatum*)	Ethanol	*B. subtilis, L. innocua, L. monocytogenes, Micrococcus* sp., *S. aureus, A. lwoffii, S. enteritidis*	Vlachos et al. 1997
J. virgata (formerly *Haliptilon virgatum*)	Crude extracts *n*-butanol	*S. aureus* *S. aureus, E. coli*	Kamenarska et al. 2006a Kamenarska et al. 2009
Kappaphycus alvarezii	Methanol, Benzene, *n*-hexane, ethylacetate, methanol, chloroform	*P. fluorescens, S. aureus and less inhibition on V. cholerae, P. mirabilis*	Rajasulochana et al. 2009
Laurencia complanata	Methanol	*B. subtilis, L. innocua, L. monocytogenes, Micrococcus* sp., *S. aureus*	Vlachos et al. 1997
L. coronopus	*n*-butanol	*S. aureus*	Kamenarska et al. 2009
L. decidua	Crude extracts	*B. subtilis, B. cereus*	Caccamese et al. 1979
L. dendroidea	Acetone	*S. aureus, E. faecalis*	Bianco et al. 2013
L. dendroidea (formerly *L. majuscula*)	Acetone (Elatol, iso-obtusol)	*S. epidermidis, K. pneumoniae, Salmonella* sp.	Vairappan 2003
L. filiformis	Acetone, ethanol	*E. coli, S. typhi*	Khan 2016
L. johnstonii	Butanol Ethanol	*S. aureus* *S. aureus, S. pyogenes*	Águila-Ramírez et al. 2012 Horincar et al. 2014
L. mariannensis	Halogenated sesquiterpenes	Several bacteria	Gonzalez et al. 1982
L. obtusa	Elatol Crude extracts Crude extracts	*K. pneumonia, Salmonella* sp. 3 pathogenic bacteria *S. aureus, E. coli, S. faecalis, K. pnuemoniae, P. aeruginosa*	Gonzalez et al. 1976, 1979 Pesando and Caram 1984 Lakshmi et al. 2006
L. obtusa	Crude extracts	*B. subtilis*	Salvador et al. 2007

Table 8.1 contd....

Table 8.1 contd....

Species	Solvent and/or extract	Bacteria	References
L. okamurae	Ethanol, acetone, methanol-toluene	*P. solancearum, E. coli, B. subtilis, S. aureus*	Yi et al. 2001
L. pacifica	Crude extracts Ethanol	*B. subtilis, B. cereus* *S. aureus, S. pyogenes*	Caccamese et al. 1979 Muñoz-Ochoa et al. 2010
L. snackeyi	Methanol, ethanol, hexane, water, Na_2SO_4 (Snakeol and Snakediol)	*S. typhi, E. coli*	Kamada and Vairappan 2017
Laurencia sp.	Methanol	*P. aeruginosa, S. epidermidis, V. harveyi, V. alginolyticus, V. vulnificus, V. parahaemolyticus*	Manilal et al. 2010
Mastocarpus stellatus (formerly *Gigartina stellata*)	Crude extracts	*S. aureus, E. coli, B. subtilis, S. pyogenes, P. morganii*	Hornsey and Hide 1974
Mazzaella capensis	Ethanol	*B. subtilis, L. innocua, L. monocytogenes, Micrococcus* sp., *S. aureus, A. lwoffii, S. enteritidis*	Vlachos et al. 1997
Melanothamnus afaqhusainii	Petroleum ether, methanol, chloroform	*S. aureus, E. coli, K. pneumoniae, P. mirabilis, P. aeruginosa*	Khan et al. 2011
Myriogramme minuta (formerly *Drachiella minuta*)	Dichloromethane, methanol, water	*A. salmonicida* subsp. *salmonicida, A. hydrophila* subsp. *hydrophila, P. anguilliseptica, V. anguillarum, Y. ruckeri*	Bansemir et al. 2006
Neorhodomela larix	Ethanol	*S. aureus, S. pyogenes*	Muñoz-Ochoa et al. 2010
Neurymenia fraxinifolia	Neurymenolide	Methicillin-resistant *S. aureus*, Vancomycin-resistant *E. faecium*	Stout et al. 2009
Ochtodes secundiramea	Acetone	*S. aureus*	Bianco et al. 2013
Odonthalia corymbifera	Bromophenols	*S. aureus, B. subtilis, M. luteus, P. vulgaris, S. typhimurium*	Oh et al. 2008
Osmundaria colensoi	Lanosol butenone and rhodomelol	*Mycobacterium smegmatis*	Popplewell and Northcote 2009
O. obtusiloba	Dichloromethane:methanol (2:1)	*P. aeruginosa*	Bianco et al. 2013
O. serrata	80% ethanol Lanosol enol ether	*B. subtilis, S. aureus, A. lwoffii, E. coli* *Halomonas* spp., *Marinococcus* sp., *Pseudomonas* spp., *Vibrio* spp., *Bacillus* spp., *Enterobacter cloacae, Enterococcus faecalis, E. coli, K. pneumoniae, Propionibacterium acnes, Salmonella typhimurium, Serratia marcescens, S. aureus*	Vlachos et al. 2001 Barreto and Meyer 2006
Osmundea pinnatifida	Methanol Methanol	*S. aureus* *S. aureus, E. coli*	Rizvi and Shameel 2005 Rizvi 2010
O. pinnatifida (formerly *Laurencia pinnatifida*)	Acetone Crude extracts	*S. aureus* *E. coli, K. pneumoniae, P. aeruginosa*	Hornsey and Hide 1976 Hellio et al. 2000

Osmundea truncata	Crude extracts	*B. subtilis, B. cereus, S. aureus, E. coli, P. aeruginosa*	Salvador et al. 2007
Palisada cruciata (formerly *Laurencia cruciata*)	Acetone, Butanol	*P. aeruginosa, S. pyogens*	Ravikumar et al. 2002
P. perforata (as *Chondrophycus papillosus*)	*n*-butanol	*S. aureus*	Kamenarska et al. 2009
P. perforata (formerly *Laurencia papillosa*)	Methanol	*E. coli, P. aerugenosa, S. aureus, B. subtilis, K. pneumoniae, Shigella flexineri*	Kavita et al. 2014
P. poiteaui (formerly *L. poiteaui*)	Crude extracts	*S. aureus, E. coli, S. faecalis, K. pnuemoniae, P. aeruginosa*	Lakshmi et al. 2006
Peyssonnelia rubra	Crude extracts	*B. subtilis, B. cereus, S. aureus, E. coli, P. aeruginosa*	Salvador et al. 2007
Plocamium cartilagineum	Crude extracts Crude extracts Methanol *n*-hexane, dichloromethane	*E. coli, S. aureus* *B. subtilis, B. cereus, S. aureus, E. coli, P. aeruginosa* *E. coli, E. faecalis, S. aureus* *E. coli, B. subtilis*	Hornsey and Hide 1985 Salvador et al. 2007 Bouhlal et al. 2010 Pinteus et al. 2015
P. rigidum	Ethanol	*B. subtilis, L. innocua, L. monocytogenes, Micrococcus* sp., *S. aureus, A. lwoffii, S. enteritidis*	Vlachos et al. 1997
P. telfairiae	Ethanol, acetone, methanol-toluene	*P. solancearum, E. coli, B. subtilis, S. aureus*	Yi et al. 2001
Polysiphonia tuticorinensis	Methanol	*E. coli, P. aerugenosa, S. aureus, B. subtilis*	Kavita et al. 2014
P. virgata (as *Carradoriella virgata*)	Ethanol	*B. subtilis, L. innocua, L. monocytogenes, Micrococcus* sp., *S. aureus, A. lwoffii, S. enteritidis*	Vlachos et al. 1997
Porphyra dioica	Ethyl acetate, dichloromethane, methanol:water (1:1)	*E. coli, B. cereus, L. brevis. E. faecalis, Candida* sp., *S. aureus, E. faecalis*	Mendes et al. 2013
Portieria hornemannii	Ethanol Methanol Methanol	*B. subtilis, L. innocua, L. monocytogenes, Micrococcus* sp., *S. aureus, A. lwoffii, S. enteritidis* *V. harveyi, V. alginolyticus, V. vulnificus* *Vibrio parahaemolyticus*	Vlachos et al. 1997 Manilal et al. 2010 Roohi Fatima et al. 2016
P. hornemannii (as *Chondrococcus hornemanii*)	Crude extracts Alcoholic extracts Methanol	*S. aureus, E. coli* *S. aureus, E. coli, M. tuberculosis, S. faecalis, K. pnuemoniae, P. aeruginosa* *Aeromonas hydrophila, Vibrio parahaemolyticus*	Bandara et al. 1988 Lakshmi et al. 2010 Ganeshamurthy et al. 2013, 2014
Pterocladiella capillacea	Methanol	*P. fluorescens, V. anguillarum, P. aeruginosa*	Wefky et al. 2009

Table 8.1 contd.…

Table 8.1 contd....

Species	Solvent and/or extract	Bacteria	References
P. capillacea (as *Gelidium capillaceum*)	Crude extracts	*Mycobacterium tuberculosis, Mycobacterium avium, Mycobacterium phlei*	Kamimoto 1956
P. capillacea (formerly *Pterocladia capillacea*)	Ethanol, acetone, methanol Cold water-extracted polysaccharides Methanol, ethanol, acetone, chloroform	*B. subtilis, S. aureus, E. coli, S. typhi, K. pneumoniae, C. albicans* *B. cereus, S. aureus, P. fluorescens, E. coli* *V. fluvialis*	Osman et al. 2010 Abou Zeid et al. 2014 El-Din and El-Ahwany 2016
Pterosiphonia complanata	Methanol Methanol	*E. coli, E. faecalis, S. aureus* *B. cereus, B. thuringiensis, B. subtilis, C. sporogenes, S. aureus, S. aureus* subsp. *aureus, M. smegmatis, E. faecalis, Bacillus* sp., *E. coli, Pseudomonas* sp.	Bouhlal et al. 2010b Oumaskour et al. 2013
Ptilophora pinnatifida (as *Beckerella pinnatifida*)	Ethanol	*B. subtilis, L. innocua, L. monocytogenes, Micrococcus* sp., *S. aureus, A. lwoffii, S. enteritidis*	Vlachos et al. 1997
Pyropia haitanensis (as *Porphyra haitanensis*)	1,8-dihydroxy-anthraquinone	*S. aureus*	Wei et al. 2015
Rhodymenia californica	Ethanol	*S. aureus, S. pyogenes*	Muñoz-Ochoa et al. 2010
Rytiphlaea tinctoria	Crude extracts	*B. subtilis, B. cereus, S. aureus, E. coli, P. aeruginosa*	Salvador et al. 2007
Sarcodiotheca furcata	Water, chloroform:methanol (2:1)	*S. aureus, S. pneumoniae, Shigella flexneri, E. coli, S. typhi, K. pneumoniae, P. mirabilis*	El-Sheekh et al. 2014
Sarconema filiforme	Methanol, dimethyformamide	*P. aeruginosa, E. coli, S. marcescens, S. aureus, E. feacali, B. subtilis, B. cereus*	Selim 2012
Scinaia hatei	Acetone, ethanol	*E. coli, S. typhi*	Khan 2016
S. moniliformis (as *S. indica*)	Crude extracts	*S. aureus, E. coli, S. faecalis, K. pnuemoniae, P. aeruginosa*	Lakshmi et al. 2006
Solieria robusta	Crude extracts	*S. aureus, E. coli, S. faecalis, K. pnuemoniae, P. aeruginosa*	Lakshmi et al. 2006
Sphaerococcus coronopifolius	Bromosphaerone, 12S-hydroxybromosphaerodiol	*S. aureus*	Etahiri et al. 2001
S. coronopifolius	Crude extracts Water and organic extracts *n*-hexane, dichloromethane Methanol	*B. subtilis, B. cereus, S. aureus* *B. cereus, B. thuringiensis, B. subtilis, C. sporogenes, S. aureus, S. aureus* subsp. *aureus, M. smegmatis, E. faecalis, Bacillus* sp., *E. coli, Pseudomonas* sp. *E. coli, B. subtilis* *S. aureus*	Salvador et al. 2007 Oumaskour et al. 2013 Pinteus et al. 2015 Rodrigues et al. 2015b

Sphondylothamnion multifidum	Crude extracts	*S. aureus, E. coli, B. subtilis, S. pyogenes, P. morganii*	Hornsey and Hide 1974
Spyridia filamentosa	Methanol	*S. aureus, E. coli, E. faecalis*	Taskin et al. 2010
Vertebrata lanosa (formerly *Polysiphonia lanosa*)	Crude extracts Crude extracts Water	*S. aureus, E. coli* *E. coli, K. pneumoniae, P. aeruginosa* Methicillin-resistant *S. aureus, S. aureus, E. cloacae, Clostridium perfringens*	Hornsey and Hide 1985 Hellio et al. 2000 Tan et al. 2013

*Marine microalgae

(GC-MS) analysis of the volatile components of *U. linza* resulted in the identification of 35 compounds, which constituted 84.76% of the total compounds. The volatile components of *U. linza* consisted of *n*-tetratriacontane (8.45%), 1-heptadecanamine (6.65%), and docosane (6.46%) as major components. The methanol and chloroform extracts showed more potent antimicrobial activity than hexane and dichloromethane extracts. Generally, when compared with the standard antibiotic, Tobramycin, especially methanol extracts (4 mg/disc) exhibited more antimicrobial activity. For example, 2 mg kg^{-1} of methanol extract of *U. linza* showed same antibacterial activity against *Staphylococcus aureus* (inhibition zone is 15 mm) with tobramycin, while 4 mg kg^{-1} of methanol extract of *U. linza* showed the more potent antibacterial activity against *S. aureus* (inhibition zone is 25 mm) than Tobramycin (inhibition zone is 16 mm). The volatile oils of these algae did not remarkably inhibit the growth of tested microorganisms (Sukatar et al. 2006). Studies investigating disease resistance in marine plants have indicated that secondary metabolites may have important defensive functions against harmful marine microorganisms.

The communication made by Lakshmi et al. (2006) deals with the biological activities of the extracts of 48 marine florae. The biological screening includes tests for antibacterial, antifungal, among others. From among green algae, the crude extracts from *Bryopsis hypnoides*, *Bryopsis plumosa*, *Caulerpa racemosa*, *Caulerpa veravalensis*, *Chaetomorpha* sp., *Codium dwarkense*, *Codium decorticatum* (formerly *Codium elongatum*), *Ulva clathrata* (formerly *Enteromorpha clathrata*), *Dictyosphaeria cavernosa*, *Udotea indica*, and *Valoniopsis pachynema* were active against *Staphylococcus aureus*, *Escherichia coli*, *Streptococcus faecalis*, *Klebsiella pnuemoniae*, and *Pseudomonas aeruginosa*.

In the study made by Kim et al. (2007b), the *in vitro* antimicrobial activity of the marine green algae *Ulva lactuca*, collected from the coastal area of Busan, Korea, was examined against $Gram^{+}$ bacteria. The ethyl-ether extract of algae exhibited a broad-spectrum of antibacterial activity, in particular, the *U. lactuca* extract showed strong activity against the bacterium methicillin-resistant *Staphylococcus aureus*. The minimal inhibitory concentration (MIC) values of the *U. lactuca* extract against *Bacillus subtilis*, *Micrococcus luteus*, and *S. aureus* were 12.5, > 200, and 100 μg mL^{-1}, respectively. In particular, the MIC values observed in the seaweed extract were lower than those of the positive control substance Melittin (> 200 μg mL^{-1}).

Paulert et al. (2007) studied the antimicrobial activity of cell-wall polysaccharides and crude extracts from the seaweed *Ulva fasciata* against bacteria. The antibacterial activity was assessed by agar diffusion assay, and by means of the broth dilution method estimating the minimum inhibitory concentration (MIC). The following human pathogenic bacteria were tested: *Escherichia coli, Staphylococcus aureus*, *Pseudomonas aeruginosa*, *Bacillus cereus*, *Micrococcus luteus*, and two plant pathogens: *Xanthomonas campestris* and *Erwinia carotovora*. Both methanol soluble and methanol insoluble extracts were active against *P. aeruginosa*, *X. campestris*, and *E. carotovora*. The highest activity of extracts was observed against *E. carotovora* (MIC = 1 mg mL^{-1}). In contrast, ulvans did not show any *in vitro* activity towards all test organisms.

The study made by Nair et al. (2007) was done to investigate the antimicrobial potentiality of the marine algae collected from different coastal regions of Gujarat (India) and screened for the same. Several seaweeds species were screened for their potential antibacterial activity against five clinically important bacterial strains, namely *Bacillus cereus*, *Micrococcus flavus*, *Citrobacter freundii*, *Klebsiella pneumoniae*, and *Pseudomonas testosterone*. Acetone and methanol were used for extraction; and the extracted yield was more when the solvent used was methanol. The antibacterial activity was done by both Agar disc diffusion method and Agar ditch method. The five bacterial strains showed varied response towards marine algal extracts. The most susceptible bacterium was *B. cereus* followed by *K. pneumoniae* and *C. freundii*, while the most resistant bacteria were *M. flavus* and *P. testosteroni.* Among the 26 algae screened, *Ulva intestinalis* (formerly *Enteromorpha intestinalis*) was the most potent alga and thus, this alga was selected for further studies. *U. intestinalis* was extracted in petroleum ether, 1,4-dioxan, acetone, methanol, and Dimethylformamide (DMF), and their antibacterial activity was studied against the above-stated five bacterial strains using agar disc method. Maximum extractive value of *U. intestinalis* was in methanol (2.05%), and minimum was in acetone (0.38%). The most susceptible bacterium was *K. pneumoniae*, and maximum antibacterial activity was shown by petroleum ether extract, and minimum was shown by 1,4-dioxan extract. The most resistant bacteria were *M. flavus* and *C. freundii*. The MIC values of *U. intestinalis* extracts ranged from 2500–9.765 μg 0.5 mL^{-1} against *B. cereus* and *K. pneumoniae*. From these results it is concluded that the acetone extract of *U. intestinalis* is the most potent extract and can be used as a lead molecule in drug discovery in inhibiting some of the bacterial strains (Nair et al. 2007).

In the works done by Kandhasamy and Arunachalam (2008), with seaweeds of southeast coast of India, all the crude extracts of algae inhibited the growth of all the pathogens, except 1 or 2 bacterial pathogens. Crude extract of *Ulva lactuca* showed high inhibiting activity against *Staphylococcus aureus* (17 mm); $Gram^+$ bacteria were more sensitive than $Gram^-$ bacteria. Among $Gram^+$ bacteria, *S. aureus* was more susceptible to all the algal extracts. *S. aureus* growth was highly inhibited by the extract of the *U. lactuca* and *Caulerpa racemosa*—the diameter of the inhibition zone of respective algal extracts was 17 and 16 mm, respectively; next to *S. aureus*, *Bacillus subtilis* was susceptible to all the extract of algae used in this study (Kandhasamy and Arunachalam 2008).

Shanmughapriya et al. (2008b) collected 14 seaweeds and tested against 10 human pathogen bacteria and one Human pathogen fungus, using the well diffusion test in the casitone agar medium. In their study, methanol:toluene (3:1) was found to be the best solvent for extracting the antimicrobial principles from fresh algae. However, the ethanolic extract showed no antibacterial activity. *Acrosiphonia orientalis* showed activity against 70% of the tested organisms. Their findings revealed that the tested seaweeds were highly active against $Gram^-$ bacteria than $Gram^+$ bacteria. The antimicrobial principle from seaweed was found to be a lipophilic compound. The compound was stable over a wide range of temperature (30–60°C). The active principles of highly active seaweed *A. orientalis* was bactericidal.

Pressurized (liquid) ethanol extracts from *Haematococcus pluvialis* (microalga) in its red stage possess antimicrobial activity against a $Gram^-$ bacterium, *Escherichia coli*, and a $Gram^+$ bacterium, *Staphylococcus aureus*; this was once again associated with the presence of short-chain fatty acids, namely butanoic and methyl lactic acids (Santoyo et al. 2009).

Ibtissam et al. (2009) evaluated the antibacterial activity of methanolic extracts from 32 macroalgae for the production of antibacterial compounds against *Escherichia coli*, *Staphylococcus aureus*, *Enterococcus faecalis*, *Klebsiella pneumoniae*, and *Enterococcus faecalis*. In all 13 species tested of Chlorophyta, the extract of *Ulva lactuca* has a larger inhibition diameter against *E. coli* (16 mm), *S. aureus* (30 mm), and *E. faecalis* (13 mm). However, the extract of *Ulva compressa* (formerly *Enteromorpha compressa*) is the only one that showed activity against *E. faecalis* (14 mm). Their results indicated that these species of seaweed have the antibacterial activities which makes them interesting for screening for natural products.

Devi et al. (2009) collected some commonly occurring green algae *Codium adhaerens*, *Ulva reticulata*, and *Halimeda tuna* and evaluated its antibacterial activity by agar diffusion method. Seven different solvents, namely acetone, methanol, chloroform, diethyl ether, ethyl acetate, ethanol, and petroleum ether were used for extraction. The ethanol extract showed better results when compared to those for the other extracts. Some extracts were found to be more effective than commercial medicine. The maximum antibacterial activity was noted in ethanol extracts, and showed activity against *Staphylococcus* sp., and the minimum was recorded in methanol extracts against *Escherichia coli*, *Staphylococcus* sp., *Proteus* sp., *Streptococcus* sp., and *Enterococci* sp.

Methanol, ethanol, and acetone extracts of six seaweed species from the southeast coast of India were tested *in vitro* for their antibacterial activities against bacteria (Kolanjinathan and Stella 2009) *Escherichia coli*, *Pseudomonas aeruginosa*, *Staphylococcus aureus*, *Klebsiella pneumonia*, and *Enterococcus faecalis* with the disc diffusion method. Acetone was the best solution for extracting the antimicrobial materials from the algal species used in this experiment, with the exception of *Ulva lactuca*, for which ethanol was the most effective extraction solution. A significant antibacterial activity was not observed between the ethanol and methanol extracts of each alga. In addition, as a result of the comparison of dried and fresh extract of antibacterial activity, it was found that all the test organisms were more sensitive to fresh extracts of the algae. A significant difference in antimicrobial activity was not found between the methanol and ethanol extracts of each alga. For instance, acetone extracts of fresh *Halimeda gracilis* showed effective results against all test organisms; however the acetone extracts of *U. lactuca* were less effective against microorganisms. This result could be related to the presence of bioactive metabolites present in the *Ulva*, which are not soluble in acetone, but can be soluble in methanol.

With a view to explore the finding of new molecules with therapeutic efficacy for human use, the alcoholic extracts of 33 identified species of marine flora, collected from Indian coasts, were prepared and screened for a wide range of biological activities (Lakshmi et al. 2010). From among green algae, the alcoholic extracts from *Caulerpa scalpelliformis*, *Chaetomorpha spiralis* (formerly *Chaetomorpha torta*), and *Ulva reticulata* were active against *Staphylococcus aureus*, *Escherichia coli*, *Mycobacterium tuberculosis*, *Streptococcus faecalis*, *Klebsiella pnuemoniae*, and *Pseudomonas aeruginosa*.

In the study made by Omar et al. (2012), marine algae were collected from the southern coast of Jeddah, Saudi Arabia, and the antibacterial activities of petroleum ether, diethyl ether, ethyl acetate, and methanol extracts of marine algae were studied. The maximum biological activities of the green algae were observed in ethyl acetate extract of *Ulva prolifera* (formerly *Enteromorpha prolifera*) against Methicillin-resistant *Staphylococcus aureus* (inhibition halo: 25 mm), and petroleum ether extracts of *Ulva reticulata* against *Escherichia coli*, *Klebsiella pneumoniae*, and Methicillin-resistant *Staphylococcus aureus* (25, 23, and 20 mm, respectively). At the same time, *Bacillus subtilis* was highly inhibited (17 to 21 mm) by different extracts of the *U. reticulata*. A similar observation was recorded by Thillairajasekar et al. (2009) who found good antimicrobial activity in ethyl acetate extract of algae. Also, they reported that hexane and ethyl acetate extracts of *Ulva* showed the presence of myristic and palmitic acid, linoleic acid, oleic acid, lauric, stearic, and myristic acid, which are known to have potent antibacterial and antifungal agents. Kim et al. (2007b) reported that *Ulva lactuca* exhibited a broad spectrum of antibacterial activity, especially against Methicillin-resistant *Staphylococcus aureus*. A large number of *Ulva* extracts products have been found to have antibacterial activity, many of these structures identified as fatty acids, hydroxyl unsaturated fatty acids, ghycolipids, steroids, phenolics, and terpenoids (Awad 2000). The antimicrobial activity of Ulva organic extract is apparently related to their lipophelic and phenolic contents (Abd El-Baky et al. 2008). Gonzalez et al. (2001) found that the algal extracts such as *Ulva clathrata* (formerly *Enteromorpha ramulosa*) were active against Gram$^+$ and Gram$^-$ bacteria. The results clarified that the minimum inhibition halos (11 to 15, 9 to 15 and 11 to 14 mm) were found in extracts of *U. prolifera* against *B. subtilis*, *S. aureus*, and *K. pneumoniae*, respectively. However, *U. reticulata* extracts recorded the minimum activities (12 to 15 and 13 to 16 mm) against *S. aureus* and *P. aeruginosa*, respectively.

Patra et al. (2009) studied the antibacterial activity of organic solvent extracts of some marine macroalgae, viz., *Chaetomorpha linum* and *Ulva compressa* (formerly *Enteromorpha compressa*) against three Gram$^-$ bacteria (*Shigella flexneri*, *Vibrio cholerae*, and *Escherichia coli*) and two Gram$^+$ bacteria (*Bacillus subtilis* and *Bacillus brevis*). The results revealed that the chloroform and ethyl acetate extracts were active against most of the pathogens, whereas methanol and ethanol extracts were active only against *Shigella flexneri*.

Chromatographic purification of the dichloromethane-soluble fraction of alga, on neutral alumina, using increasing concentrations of ethylacetate/*n*-hexane as eluents, yielded seven labdane diterpenoids as major constituents of green alga *Ulva fasciata* (Chakraborty et al. 2010). Antimicrobial assay showed that the compounds labda-14-ene-3α,8α-diol and labda-14-ene-8α-hydroxy-3-one were inhibitory to the growth of *Vibrio parahaemolyticus* and *Vibrio alginolyticus* with minimum inhibitory concentrations of 30 $\mu g\ mL^{-1}$ by the first, and 40 $\mu g\ mL^{-1}$ by the second, respectively against the former and 30 $\mu g\ mL^{-1}$ by the first, and 80 $\mu g\ mL^{-1}$ by the second compound, respectively, against the latter.

The seaweeds *Ulva compressa* (formerly *Enteromorpha compressa*), *Ulva intestinalis* (formerly *Enteromorpha intestinalis*), *Caulerpa scalpelliformis*, *Caulerpa racemosa*, and *Chaetomorpha linum* were collected from Gulf of Mannar, Tuticorin coast, and six different solvents (hexane, diethyl ether, chloroform, ethyl acetate, ethanol, acetone, and methanol) were used for the extraction (Jebasingh et al. 2011). The antibacterial activity of seaweed extracts was checked against six human pathogens. The acetone extract of *U. compressa* showed higher activity against *Bacillus subtilis* (14.4 mm) and ethyl acetate extract for *Klebsiella pneumoniae* (10.6 mm) showed high activity. In the case of *U. intestinalis*, acetone extract showed maximum activity against *B. subtilis* (19.8 mm), *K. pneumoniae* (17.4 mm), *Escherichia coli* (15.6 mm), and *Pseudomonas aeruginosa* (11.2 mm). *C. racemosa* showed greater activity in both hexane and ethyl acetate extracts for 6 human pathogenic bacteria. Similarly, *C. scalpelliformis* showed maximum activity in acetone extract for all the pathogenic bacteria, mainly against *Enterobacter aerogenes* (Jebasingh et al. 2011).

Antibacterial activity of methanolic extracts from several species of macroalgae collected from Moroccan Mediterranean coasts was evaluated against *Escherichia coli*, *Staphylococcus aureus*, and *Enterococcus faecalis* (Zbakh et al. 2012). The extract of *Caulerpa prolifera* has the larger diameters of inhibition against *S. aureus* (23 mm), *E. coli* (16 mm), and *E. faecalis* (13 mm). However, extracts of *Ulva rigida*, *Ulva compressa* (formerly *Enteromorpha compressa*), and *Caulerpa prolifera* were the only ones to present inhibitory activity against *E. faecalis*. Their respective inhibition diameters are 15 mm, 13 mm, and 12 mm. The results obtained with extracts of *U. rigida* and *C. prolifera* from the Mediterranean Moroccan coasts are similar to those obtained with the same species from the Canary Islands against $Gram^+$ bacteria (Gonzalez et al. 2001). Similar results were also obtained with extracts of ethanol, ethanol/methanol, and hexane of *C. prolifera* from Mexico and the Caribbean (Ballantine et al. 1987, Freile-Pelegrín and Morales 2004).

In the study made by Sujatha et al. (2012), methanol extracts of green seaweeds have been tested for their antibacterial activity against oral bacteria causing dental caries. Different concentrations of the extracts of the four species of seaweeds—*Chaetomorpha antennina*, *Cladophora vagabunda* (as *Cladophora fascicularis*), *Acrosiphonia orientalis* (as *Spongomorpha indica*), and *Ulva fasciata* collected from sea coast of Visakhapatnam (India) have been tested for their antibacterial activity against three oral pathogenic bacteria, *Actinomyces viscosus*, *Streptococcus mitis*, and *Streptococcus mutans*. The antibacterial sensitivity was studied by agar disc diffusion method. Of these, *U. fasciata* has shown greater inhibition on all the three oral bacteria than *C. antennina*, *C. fascicularis*, and *A. orientalis*. The methanol extract of *U. fasciata* showed maximum inhibition on all the oral bacteria, i.e., *S. mutans*, *S. mitis*, and *A. viscosus* at 100 mg mL^{-1} concentration and lowest concentration at 50 mg mL^{-1} on *S. mutans*, *S. mitis*, and *A. viscosus*. *C. fascicularis* has inhibited *S. mitis* and *A. viscosus*, whereas *C. antennina* and *A. orientalis* inhibited only *A. viscosus*. The methanol extract of *C. antennina* showed increased zone of inhibition, maximum at 100 mg mL^{-1} only on *A. viscosus*. It showed no activity on *S. mutans* and *S. mitis*. The methanol extract of *C. fascicularis* showed maximum zone of inhibition on *A. viscosus*, compared to *S. mitis* at 100 mg mL^{-1} concentration, but showed no activity on *S. mutans*. The lowest concentration of *C. fascicularis* extract to inhibit the growth of *A. viscosus* is at 25 mg mL^{-1} concentration, whereas *S. mitis* was inhibited only at 100 mg mL^{-1} concentration. The methanol extract of *A. orientalis* showed maximum inhibition only on *A. viscosus* at 100 mg mL^{-1} concentration, and minimum at 25 mg mL^{-1} concentration. It showed no activity on *S. mutans* and *S. mitis* (Sujatha et al. 2012). The methanol, ethanol, and chloroform extract of *A. orientalis* (formerly *S. indica*) showed a wide varying antibacterial and antifungal activity (Murata et al. 1989).

In a study made by Seenivasan et al. (2010), *in vitro* antibacterial activity of the acetone, methanol, and ethanol extracts of three marine algae from Ennore beach near Chennai (coast of Tamil Nadu) were tested. Among all the marine algae *Ulva fasciata*, *Ulva intestinalis* (formerly *Enteromorpha intestinalis*), and *Chaetomorpha aerea* were collected and tested *in vitro*; *U. fasciata* and *C. aerea* have exhibited average result. *Escherichia coli* in all the solvents have shown significant results for the seaweed *U. fasciata* in selective media. The zone of inhibition was compared with the zone of inhibition produced by the standard antibiotic discs in the antibiotic disc diffusion. Thus the organic (80% ethanolic, methanolic, and acetone) extract of green algae has the ability to inhibit the growth of the $Gram^+$ and $Gram^-$ bacteria (Seenivasan et al. 2010).

Aqueous extracts of some commonly occurring *Cladophora conglomerata*, *Ulva lactuca*, and *Ulva reticulata* were evaluated for antibacterial activity by well diffusion assay (Mansuya et al. 2010). The zone of inhibition ranged between 9–45 mm in aqueous extract and 40 mm in methanolic extract. The maximum activity (45 mm) was recorded from 200 mg of aqueous extract of *U. reticulata* against *Salmonella typhi* and minimum (9 mm) by *U. lactuca* against *Streptococcus pyogenes* at 50 mg level whereas, the methanolic extract showing the maximum activity (40 mm) was recorded from 200 mg of *U. reticulata* against *Escherichia coli*, and *Streptococcus pyogenes* and *C. conglomerata* against *Pseudomonas aeruginosa*.

Antimicrobial assay with the major constituents of green alga *Ulva fasciata* (Chakraborty et al. 2010), showed that the compounds labda-14-ene3α,8α-diol and labda-14-ene-8α-hydroxy-3-one were inhibitory to the growth of *Vibrio parahaemolyticus* and *Vibrio alginolyticus* with minimum inhibitory concentrations of 30 μg mL^{-1}, and 40 μg mL^{-1}, respectively against the former, and 30 μg mL^{-1}, and 80 μg mL^{-1}, respectively, against the latter.

In the study made by Abirami and Kowsalya (2011), the antibacterial activity of seaweed extracts was evaluated. The determination of antibacterial activity of the methanol extracts of *Ulva lactuca* was evaluated against *Bacillus subtilis*, *Bacillus cereus*, *Micrococcus luteus*, *Klebsiella pneumonia*, *Staphylococcus aureus*, and *Escherichia coli*. The most susceptible bacterium was *K. pneumoniae* and *S. aureus* towards organic extracts of *U. lactuca* with highest inhibition zone values (Baky et al. 2008).

In a study made by Dhasarathan and Theriappan (2011), the antibacterial activity of extracts of *Chaetomorpha antennina* and *Ulva fasciata* were evaluated. The results of antimicrobial activity by the well diffusion assay also clearly expressed that *U. fasciata* has a higher concentration of active principles when compared to *C. antennina*. The antimicrobial activity of crude extracts of *U. fasciata* and *C. antennina* against 10 human pathogenic bacterial strains were done, and their zone of inhibition was compared with a standard antibiotic, Tetracycline. The seaweed extracts were shown to possess more active antimicrobial proficiency against *Pseudomonas aeroginosa*, *Staphylococcus aureus*, *Escherichia coli*, *Enterococcus aeroginosa*, and *Salmonella paratyphi* when compared to the standard antibiotic, but little antibacterial activity against *Klebsiella pneumoniae*, *Bacillus subtilis*, *Citrobacter* sp., and *Proteus* sp. *Staphylococcus epidemis* is highly resistant to both test samples as well as standard antibiotics.

Salem et al. (2011) screened the methanolic and ethyl acetate extracts from eight different seaweeds for their antibacterial activities against both $Gram^{+}$ bacteria (*Staphylococcus aureus* and *Bacillus cereus*) and $Gram^{-}$ bacteria (*Escherichia coli*, *Enterococcus faecalis*, *Salmonella* sp., and *Pseudomonas aeruginosa*). The antibacterial activities were expressed as zone of inhibition and minimum inhibitory concentrations. Ethyl acetate extracts of *Caulerpa racemosa*, and *Codium fragile*. *Salmonella* sp., and *P. aeruginosa* were resistant to methanolic extracts of *C. racemosa*. On the other hand, *B. cereus*, *S. aureus*, and *E. coli* were the most sensitive to all seaweed extracts. The susceptibility of $Gram^{+}$ bacteria to the algal extracts was more than those of $Gram^{-}$ bacteria. The activities of ethyl acetate extracts were higher than those of methanolic extracts, and the most powerful inhibitory extract was ethyl acetate extract of *C. racemosa*.

Lavanya and Veerappan (2011) tested the *in vitro* antibacterial activity of six selected marine algae. Extracts of six seaweed samples, namely *Codium decorticatum* and *Caulerpa scalpelliformis*, among others, were selected for antibacterial activity against selected human pathogens, such as species *Vibrio parahaemolyticus*, *Salmonella* sp., *Shewanella* sp., *Escherichia coli*, *Klebsiella pneumoniae*, *Streptococcus pyogenes*, *Staphylococcus aureus*, *Enterococcus faecalis*, *Pseudomonas aeruginosa*, and *Proteus mirabilis*. All the seaweed extracts have shown moderate antibacterial activity < 10 mm of zone of inhibition, out of which only methanolic extract has shown significant activity. The results of their research showed that the minimum antibacterial activity was found in *C. decorticatum*.

Priyadharshini et al. (2011) evaluated the *in vitro* antimicrobial and hemolytic activity of marine macroalgae *Ulva fasciata*. Methanol, butanol, and aqueous extracts were tested against selected fish pathogens, *Aeromonas hydrophila*, *Pseudomonas fluorescens*, *Proteus* sp., *Vibrio alginolyticus*, and *Enterobacter* sp. The extract was subjected to thin layer chromatography (TLC) to determine the presence of peptides and amide groups, and the hemolytic activity was assayed. A maximum of 16 mm inhibition zone was observed against *V. alginolyticus*, and a minimum of 12 mm against *Enterobacter* sp., respectively. Their results showed the use of seaweeds as antimicrobial agents for pharmacology or as a health promoting food for aquaculture. The screening results confirmed that these seaweeds need further study and may be used as possible sources of antimicrobial compounds.

In the study made by Süzgeç-Selçuk et al. (2011), crude extracts derived from 11 marine algae collected from the Marmara, or Turkish coastline of the Black, Aegean, and Mediterranean Seas were screened *in vitro* against the tubercle bacillus (*Mycobacterium tuberculosis*). The algae *Codium bursa* and *Caulerpa racemosa* were found to be quite active (IC_{50} values of 1.38 and 3.12 mg mL^{-1}). The IC_{50} values of the remaining extracts were below 20 mg mL^{-1}. Only marginal activity was obtained towards the tubercle bacillus *M. tuberculosis*.

The *in vitro* antimicrobial activity of the marine green algae *Chaetomorpha aerea* was investigated against $Gram^{+}$ bacteria and $Gram^{-}$ bacteria (Pierre et al. 2011). The water-soluble extract of algae was composed of a sulfated (6.3%) galactan with a molecular weight of 1.160×10^{6} Da, and a global composition close to commercial polysaccharides as dextran sulfate or fucoidan. The polysaccharide was composed of 18% arabinose, 24% glucose, 58% galactose. The re-suspended extracts (methanol, water) exhibited selective antibacterial activities against three Gram$^+$ bacteria, including *Staphylococcus aureus*. Minimum inhibitory concentration and minimum bactericidal concentration tests showed that the sulfated galactan could be a bactericidal agent for this strain (40 mg mL^{-1}).

Several species of common seaweed extracts were tested in laboratory assays for potential industrial applications through evaluation of the antibacterial activity against pathogenic bacteria (five strains) and the antifouling potency against the growth of key species of marine colonizers (seven bacteria, five fungi, and 11 microalgae). The organic extract of *Ulva lactuca* has bacterial antibiosis. The ether extracts were more active in comparison with buthanol extracts against the bacterial strain *Staphylococcus aureus*. The best antifouling results were obtained with *U. lactuca* (0.1–1 µg mL^{-1}) against all strains tested (Águila-Ramírez et al. 2012).

In a previous work, Choi et al. (2012) found that the edible green seaweed *Ulva linza* (formerly *Enteromorpha linza*) displays potent antimicrobial activity against *Prevotella intermedia* and *Porphyromonas gingivalis* without side effects at a moderate dose. Additionally, a mouth rinse containing *U. linza* extract has shown clinical effects against gingivitis, and activity against two bacterial strains (*P. intermedia* and *P. gingivalis*) (Cho et al. 2011). This mouth rinse produced effects similar to those of Listerine®. To discover therapeutic agents against periodontitis from the seaweed with few or no side effects and potent antimicrobial activity, Park et al. (2013) isolated and identified active antimicrobial compounds from *U. linza* extract, and presented data regarding its antimicrobial activity against several oral pathogens. The MIC values of stearidonic acid were 312.50 µg mL^{-1} against *Aggregatibacter actinomycetemcomitans*, 312.50 µg mL^{-1} against *Candida albicans* (yeast), 39.06 µg mL^{-1} against *Fusobacterium nucleatum* subsp. *vincenti*, and 1.250 µg mL^{-1} against *Streptococcus mutans*. The MIC values of gamma-linolenic acid were 78.12 µg mL^{-1} against *A. actinomycetemcomitans*, 78.12 µg mL^{-1} against *C. albicans*, 9.76 µg mL^{-1} against *F. nucleatum* subsp. *vincenti*, and 625 µg mL^{-1} against *S. mutans*. Triclosan and chlorhexidine, which are broad-spectrum antimicrobial agents found in toothpaste, soap, deodorant, detergents, cosmetics, pharmaceuticals, plastics, and fabrics (Nudera et al. 2007, Haraszthy et al. 2006), were used as positive controls. The MIC values of triclosan were not determined against *A. actinomycetemcomitans*, 312.50 µg mL^{-1} against *C. albicans* (yeast), 4.88 µg mL^{-1} against *F. nucleatum* subsp. *vincenti*, not determined against *S. mutans*, 1.250 µg mL^{-1} against *P. intermedia*, and 78.12 µg mL^{-1} against *P. gingivalis*. The MIC values of chlorhexidine were 4.88 µg mL^{-1} against *A. actinomycetemcomitans*, 4.88 µg mL^{-1} against *C. albicans*, 2.44 µg mL^{-1} against *F. nucleatum* subsp. *vincenti*, 9.76 µg mL^{-1} against *S. mutans*, 1.22 µg mL^{-1} against *P. intermedia*, and 1.22 µg mL^{-1} against *P. gingivalis*. The MIC of triclosan was 23 µg mL^{-1} for *P. intermedia* and 6 µg mL^{-1} for *P. gingivalis*, and the MICs of chlorhexidine were 67.5 and 125 µg mL^{-1}, respectively (Park et al. 2013).

In a study made by Selim (2012), two seaweeds were collected from the Red Sea coast. Antimicrobial bioassay against some human pathogenic bacteria and yeast were conducted using disc diffusion method. *Halimeda opuntia* extract exhibited antibacterial activity against six species of microrganisms, with significant inhibition against *Staphylococcus aureus*. Comparative antibacterial studies showed that *Halimeda* extract showed equivalent or better activity as compared with commercial antibiotic when tested against *S. aureus*. Further tests conducted using dilution method showed both extracts as having bacteriostatic mode of action against the tested microorganisms. The largest halos were achieved by the methanol and dimethyformamide extracts of marine algae *H. opuntia* against *Pseudomonas aeruginosa*, *Escherichia coli*, *Serratia marcescens*, *Staphylococcus aureus*, *Enterococcus feacali*, *Bacillus subtilis*, *Bacillus cereus* (bacteria), *Candida albicans*, *Candida utilis*, and *Saccharomyces cerevisiae* (yeasts). Other algae ethanol, chloroform, dimethyl sulfoxide, hexane, or water extracts achieved halos not superior to 10 mm or were not identified (data not shown). Likewise, solvents used alone (controls) produced no halo (data not shown). Methanol and dimethyformamide extracts of marine algae *Halimeda* showed more antimicrobial activity on *S. cerevisiaeis* strain versus other tested microorganisms. Comparative antibacterial studies showed that *Halimeda* extract showed equivalent or better activity as compared with commercial antibiotic when tested against *S. aureus*.

Some seaweeds, such as the green alga, *Caulerpa cylindracea*, have developed immunity to several pathogenic epiphytic *Vibrio* species that live on their thallus surface. A symbiotic allelopathic relationship with the bacteria has been proposed by Rizzo et al. (2013), where the presence of the *Vibrio* actually contributes to successful algal reproduction, allowing the two organisms to function as a holobiont (Ördög et al. 2004). Food-borne illness caused by *Vibrio* and other pathogenic bacteria is an increasing global issue. Isolation of the compounds and the mechanism responsible for algal immunity to these bacteria could produce useful therapeutics and sanitizers. The harnessing and bioengineering of recently characterized allelochemicals represents a potential area of new marine antibacterials.

The antibacterial activity of ethanol, methanol, hexane, and acetone-based extracts of several seaweeds was investigated (Silva 2015). The disc diffusion method was used to evaluate the algae antimicrobial effect against standard strains of *Vibrio parahaemolyticus*, *Escherichia coli*, *Staphylococcus aureus*, *Pseudomonas aeruginosa*, *Salmonella enterica*, and five virulent antibiotic-resistant strains of *Vibrio brasiliensis*, *Vibrio xuii*, and *Vibrio navarrensis* (isolated from the hemolymph of *Litopenaeus vannamei*, the Whiteleg shrimp). The smallest inhibition halos were observed for ethanol extracts of *Ulva fasciata*, when compared to extracts of the other macroalgae. *U. fasciata* was the only algal species with antibacterial activity when methanol was used as solvent. In this case, bioactivity was observed only against *V. navarrensis* (average halo size: 11.3 mm).

Several marine macroalgae were collected from the coast of Gaza strip, Palestine (Elnabris et al. 2013). Crude extracts were prepared using the solvent methanol and evaluated for antibacterial activity by well diffusion method against both Gram$^-$ (*Escherichia coli*, *Pseudomonas aeruginosa*, *Proteus vulgaris*, and *Klebsiella pneumoniae*) and Gram$^+$ bacteria (*Staphylococcus aureus* and *Bacillus subtilis*). The percentage of extracted material (g of crude extract/g of the dry weight of starting material × 100) was obtained from the different algal species using methanol as solvent and 60°C as extraction temperature. Among these algae, *Ulva lactuca* yielded maximum extractable matter (17%), followed by *Ulva compressa* (formerly *Enteromorpha compressa*) (7.2%). The extract of *U. lactuca* exhibited the strongest antibacterial activity against all examined microorganisms, except *E. coli*. In terms of the difference between the averages of diameters of inhibition zones of each seaweed extract and that of the solvent (DMSO), *U. lactuca* produced zone differences of 9.8, 9.3, 5.8, 4.8, and 3.3 mm to *K. pneumoniae*, *S. aureus*, *P. vulgaris*, *B. subtilis*, and *P. aeruginosa*, respectively. *U. compressa* was the second active extract with highest inhibition activity against *K. pneumoniae* followed by *P. aeruginosa*, with zone differences of 7.3 and 3.0 mm, respectively (Elnabris et al. 2013).

Osman et al. (2013) tested macroalgae collected from Egypt against pathogenic Gram$^+$ and Gram$^-$ bacteria, and one clinical yeast strain, *Candida albicans*. The tested species of Chlorophyta were more potent inhibitors than those from Rhodophyta and Phaeophyceae. The extract of *Ulva fasciata* was the most active (phthalate esters derivatives being the active components), followed by *Ulva compressa* (formerly *Enteromorpha compressa*), *Ulva lactuca*, and *Ulva linza* (formerly *Enteromorpha linza*); *Klebsiella pneumoniae* was the most sensitive microorganism. Saritha et al. (2013) reported antibacterial activity of *Ulva lactuca* extracts against *Shigella sonnei*, *B. subtilis*, *E. feacalis*, *S. typhimurium*, *E. coli*, *S. aureus*, *S. pyogenes*, and *Staphylococcus epidermis*. The acetone extract of *U. lactuca* showed broad spectrum of antibacterial activity when compared to other solvent extracts.

Crude methanolic and water extracts of several marine algae species collected from the western coast of Libya were evaluated for antibacterial activity against pathogenic bacteria (four Gram$^+$, four Gram$^-$) (Alghazeer et al. 2013). The extracts showed a significant antibacterial activity against Gram$^+$ (*Staphylococcus aureus*, *Bacillus subtilis*, *Bacillus* spp., and *Staphylococcus epidermidis*) as well as Gram$^-$ bacteria (*Escherichia coli*, *Salmonella typhi*, *Klebsiella* spp., and *Pseudomonas aeruginosa*). The algal aqueous and methanolic extracts displayed different degrees of antimicrobial activities against different bacteria, in some cases methanolic extracts showed higher antibacterial activity than aqueous extracts. Methanol extract of *Caulerpa racemosa* exhibited strong inhibition against *Klebsiella* spp. and *S. typhi* with (16 mm, 16 mm, respectively) which was significantly higher than all other algae extracts. Minimum inhibitory concentration (MIC) values of 25–200 mg mL^{-1} were obtained for the methanol and aqueous extracts in the tests with the bacterial species. MICs of the methanol and aqueous extracts for *Klebsiella* spp. were 25, 200 mg mL^{-1} (*C. racemosa*); whereas, MICs of the methanol and aqueous extracts for *Bacillus* spp. were 50, 200 mg mL^{-1} (Alghazeer et al. 2013). The algal aqueous and methanolic extracts displayed different degrees of antimicrobial activities against different bacteria, whereas some algae were active against all tested bacteria such as *Ulva compressa* (formerly *Enteromorpha compressa*), *Ulva lactuca*, and *Ulva prolifera* (formerly *Enteromorpha prolifera*) which was in agreement with other reports (Val et al. 2001), while others showed no activity against some tested strains. Methanol extract of *Caulerpa racemosa* exhibited strong inhibition against *Salmonella typhi* with 16 mm, which was significantly higher than all other algae extracts. No inhibitory activities have been observed with some of the methanol or aqueous extracts such as *Codium tomentosum* extract that exhibited no activity against *Pseudomonas aeruginosa*, *Klebsiella* spp., and *Escherichia coli* (Alghazeer et al. 2013).

The aim of the study made by Narasimhan et al. (2013) was to investigate the bioactive properties of two green algae samples. *Enteromorpha attenuata* and *Ulva linza* (formerly *Enteromorpha linza*) were collected from the shoreline of Mahabalipuram, Tamil Nadu, India. The antibacterial activities of three seaweeds were assessed against nine food borne pathogens using a disc diffusion assay. The antimicrobial activities were considered to be an indicator of the capacity of seaweeds to synthesize bioactive secondary metabolites. The methanolic extracts (250–1000 μg mL^{-1}) showed average zones of inhibition (10–20 mm) in all tested microorganisms, except *Pseudomonas aeruginosa*, in comparison to the positive control, Streptomycin.

Halimeda tuna was examined for antibacterial activity *in vitro* using the well diffusion method, minimum inhibitory concentration, and minimum bactericidal concentration (Indira et al. 2013). The activity was against 10 bacterial strains (*Staphylococcus aureus*, *Salmonella typhimurium*, *Salmonella paratyphi*, *Klebsiella oxytoca*, *Escherichia coli*, *Proteus mirabillis*, *Lactobacillus vulgaris*, *Pseudomonas* sp., *Klebsiella pneumonia*, and *Vibrio cholerae*). The methanolic extracts in the present study exhibited a broad spectrum of antimicrobial activity compared to the ethanolic and chloroform extracts. Minimum inhibitory concentration (MIC) values of 31.25 to 500 μg mL^{-1} were obtained for the chloroform and ethanol extract; 15.62 to 250 μg mL^{-1} were obtained for the methanol extract in the tests with the bacterial agents. The MIC values obtained in antibacterial assays using aqueous extract were 125 to 500 μg mL^{-1}. The minimum bactericidal concentration (MBC) of the chloroform extracts showed that with the exception of the antibacterial assays against *S. aureus*, *S. typhimurium*, *S. paratyphi*, *K. oxytoca*, and *E. coli*, all the extracts exhibited a MBC at a concentration of 500 μg mL^{-1}, while the aqueous extracts had a MBC value ranging from 250 to 500 μg mL^{-1} (Indira et al. 2013).

Horincar et al. (2014) studied macroalgae from the Romanian Black Sea coast for their volatile compounds content. Seaweed oil extracts had substantial antimicrobial potential against both Gram^{+} (*B. cereus* and *L. monocytogenes*) and Gram^{-} (*E. coli* and *S. enteritidis*) bacteria. The MIC of *Cladophora vagabunda* extracts varied from 1.8 to 3.8 mg mL^{-1}, while the MIC of *Ulva intestinalis* extract was 3.8 mg mL^{-1} for all bacterial strains.

Ulva australis (formerly *Ulva pertusa*) has been reported to show considerable antimicrobial activities against *Bacillus subtilis*, *Staphylococcus aureus*, *Sarcina lutea*, *Escherichia coli*, *Aerobacter aerogenes*, *Proteus vulgaris* (Baik and Kang 1986), *Gardnerella vaginalis* (Ha et al. 2014), *Helicobacter pylori* (Lee et al. 2009b), *Porphyromonas gingivalis*, *Prevotella intermedia* (Choi et al. 2012), and *G. vaginalis* (Choi et al. 2014b). The study made by Choi et al. (2014b) was performed to screen out the extracts of algae and assess the seasonal variation in antimicrobial activity of *U. australis* (formerly *U. pertusa*) against *G. vaginalis*. Seasonal variation in antibacterial activity was observed, with the extracts showing no activity during summer and autumn, and showing antibacterial activity from early winter (December) to middle spring (April). The maximum value of antimicrobial activity (6.5 mm inhibition zone at 5 mg/disc) of *U. australis* against *G. vaginalis* was observed in April. Otherwise, for both chlorophylls *a* and *b*, the highest content (2.87 and 1.37 mg g^{-1}) was observed in March 2009. These results may reflect variation in cellular chemical compositions such as secondary metabolite(s) rather than chlorophyll and biological activities according to season (Choi et al. 2014b). Of the 44 species of seaweed screened for potential anti-*Gardnerella vaginalis* activity, 27 (61.4%) showed antimicrobial activity by the agar disc-diffusion method (Ha et al. 2014). From the 44 species of seaweeds screened, 27 (61.4%) showed antimicrobial activity by disc-diffusion method. The green algae showed the highest activity (83.3%) among the seaweeds screened. The strongest activities against microbial pathogens among Chlorophyta species were exhibited by *Ulva compressa* (formerly *Enteromorpha compressa*), *Ulva linza* (formerly *Enteromorpha linza*), and *U. australis* (formerly *U. pertusa*), which produced a zone of inhibition greater than 5 mm. The MIC values of *U. australis* extracts against *G. vaginalis*, the main cause of vaginosis, were 312 μg mL^{-1}, while the MIC values against both *Candida albicans* (yeast), the main cause of candidiasis, were 2.5 mg mL^{-1}. Against *Lactobacillus gasseri* and *Lactobacillus jensenii*, members of the normal vaginal microflora, no inhibitory effect was seen even at 10 mg mL^{-1}. To identify the primary active compounds, a *U. australis* powder was successively fractionated according to polarity, and the main active agents against *G. vaginalis* were determined to be nitrogenous compounds (156 μg mL^{-1} of the MIC value). According to these results, it was suggested that extracts of the seaweed *U. australis* are valuable for the development of natural therapeutic agents for treating women with bacterial vaginosis (Ha et al. 2014).

Thanigaivel et al. (2014) reported that ethanol extract of the seaweed *Chaetomorpha antennina* was found to be effective against the shrimp pathogen *Vibro parahaemolyticus*. The ethanol extract of *C. antennina* was very effective in controlling *V. parahaemolyticus*, which is resistant to Ampicillin and sensitive to Erythromycin. Ethanol extract of the seaweed was found to be effective against the shrimp pathogen, which showed zone of inhibition towards controlling the growth of *V. parahaemolyticus*. The samples loaded with different concentration from 50, 100, 150, and 200 μL showed zone of inhibition 17, 21, 28, and 36 mm, respectively.

Chandrasekaran et al. (2014) evaluated the antibacterial activity of different extracts of several marine algae against *Enterococcus faecalis* and one clinical isolate of Vancomycin-resistant *E. faecalis*. The maximum antibacterial activity was recorded in the ethyl acetate extracts of *Caulerpa racemosa* than the other extracts. The mean zone of inhibition produced by the extracts in agar diffusion assays against the tested bacterial strains ranged from 7.1 to 14.5 mm. The minimum inhibitory concentration was between 250 and 500 μg mL^{-1}, while the minimum bactericidal concentration was from 500 to 1000 μg mL^{-1}. The ethyl acetate extracts of the seaweeds showed the presence of strong terpenoids, tannins, and phenolic compounds, compared to the other solvent extracts.

Arunachalam et al. (2014) investigated the antimicrobial activity of the marine algae *Ulva lactuca*, *Chaetomorpha linoides*, and *Halimeda macroloba* against six strains of $Gram^+$ bacteria (*Staphylococcus aureus*, *Bacillus subtilis*, and *Lactobacillus acidophilus*) and $Gram^-$ (*Escherichia coli*, *Pseudomonas aeruginosa*, and *Proteus mirabilis*). The maximum activity (7 mm) was observed by the extract of *U. lactuca* against *Proteus mirabilis* by using methanol as a solvent and the lowest activity (2 mm) was recorded by the extract of *U. lactuca* against *L. acidophilus* by using chloroform as a solvent and ethanol extract against *P. aeruginosa*. The lowest activity (2 mm) was seen in the extract of *C. linoides* by using ethanol and methanol as a solvent against *S. aureus*. In *H. macroloba* extract, the lowest activity was recorded against *E. coli* by using chloroform as a solvent. The microbial strains *S. aureus*, *P. aeruginosa*, *B. subtilis*, and *L. acidophilus* were resistant to the chloroform and methanol of all tested seaweed extracts.

Krish and Das (2014) reported that the crude methanol, ethanol, and ethyl acetate extracts of *Cladophora rupestris* collected in a Mediterranean area, showed good antimicrobial activity against *Vibrio harveyii*, *Vibrio parahaemolytical*, and *Vibrio alginolyticus*, which were measured with the agar diffusion method and the zone of inhibition. The fatty acid profile showed palmitic, myristic, oleic, alpha linolenic, palmitoleic, and linoleic acids. In α-amylase inhibitory assay, methanol, ethanol, and ethyl acetate extract showed highest inhibition of 72%, 65%, and 70% at 1000 μg mL^{-1}. In α-glucosidase inhibitory assay, methanol and ethyl acetate showed highest inhibition of 67%, 61%, and 64% at 1000 μg mL^{-1}.

In the study made by Mungmai et al. (2014), *Rhizoclonium hieroglyphicum* from the Nan River in northern Thailand was selected for investigation. The dried alga was extracted by water and 95% (v/v) ethanol to obtain an aqueous extract and ethanolic extract, respectively. Each extract was examined for antimicrobial activity against *Staphylococcus aureus*, Methicillin-resistant *S. aureus*, and *Propionibacterium acnes* by agar well diffusion method. Regarding the application for bacterial infected skin, the antimicrobial activity of the extracts against three $Gram^+$ bacteria, viz., *S. aureus*, Methicillin-resistant *S. aureus*, and *P. acnes* showed that both the 2% and 5% (w/v) extracts could not inhibit the growth of these bacteria.

The European Union also issues directives on food-approved solvent systems and residual limits, which are administered by the European Food Safety Authority. For example, Commission Directive 2010/67/EU sets out the maximum acceptable residues for acetone, hexane, and ethanolic extracts of rosemary. The acceptable concentrations are given as not more than 500 mg kg^{-1} acetone; 25 mg kg^{-1} hexane; and 500 mg kg^{-1} ethanol. The Directive also recommends further purification of the extract with active carbon and/or molecular distillation (EU 2010). Alternatively, solvents that are generally recognized as safe (GRAS) under Sections 201 and 409 of the U.S. Federal Food, Drug, and Cosmetic Act can be utilized, with no issues of toxicity (FDA 2002). For example, in a recent study, Boisvert et al. (2015) used pressurized liquid extraction with approved, food-grade ethanol to extract high yields of antibacterial and antioxidant-rich compounds from the seaweed *Ulva lactuca* (among others), collected at the St. Lawrence Estuary, Canada. Antibacterial activity was tested against three food spoilage bacteria, *Micrococcus luteus*, *Escherichia coli*, and *Brochothrix thermosphacta*, using the microtiter method. The greatest inhibition of *E. coli* (69.5%), *M. luteus* (61.4%), and *B. thermosphacta* (21.4%) was exerted by *U. lactuca* extracts at a concentration of 500 μg mL^{-1}. The greater potency of the *U. lactuca* compared to the other species was thought to be due to a higher mineral concentration of metals such as copper, zinc, silver, and mercury, which have innate antibacterial activity. This choice of food grade solvent and green extraction method exemplifies some of the approaches that can be used in the development of safe algal antibacterials.

The study of Kolsi et al. (2015) was conducted to evaluate the antimicrobial and antifungal activity of hexane, ethyl acetate and methanol extracts of several marine species from Tunisian coastline (Chebba and Sfax). These spices were tested against eight human pathogenic bacteria Gram$^-$ (*Escherichia coli*, *Listeria monocytogene*, *Salmonella enterica*, *Agrobacterium tumefaciens*, *Pseudomonas aerigunosa*), and Gram$^+$ (*Staphylococus aureus*, *Micrococcus luteus*). Methanol extract of *Flabellia petiolata* was moderately active (10 mm) against *S. aureus*. Ethyl acetate and methanol extracts of *Anadyomene stellata* were moderately active (10 mm) only against *S. aureus*. Hexane extract of *Ulva rigida* was moderately active (10 mm) only against *A. tumefaciens*, and the methanolic extract was moderately active (10 mm) against *S. aureus*. Finally, methanol extract of *Codium fragile* was highly active (15 mm) against *S. aureus*.

The study of Manchu et al. (2015) was aimed to investigate the antibacterial activity of *Chaetomorpha antennina* from Rasthacaud coast, Tamil Nadu, India. *C. antennina* extracts were tested against four Gram$^-$ bacteria (*Escherichia coli*, *Klebsiella pneumoniae*, *Proteus vulgaris*, and *Pseudomonas aeruginosa*) and 3 Gram$^+$ bacteria (*Staphylococcus aureus*, *Streptococcus pyogenes*, and *Enterococcus faecalis*) aqueous extract showed activity against *Pseudomonas aeruginosa* (14.33 mm). It has no activity against pathogens like *Escherichia coli*, *Klebsiella pneumoniae*, *Proteus vulgaris*, *Staphylococcus aureus*, *Streptococcus pyogenes*, and *Enterococcus faecalis*. The petroleum ether extract produced a maximum zone of 18.66 mm against *E. faecalis* and a minimum zone of 13.00 mm against *P. vulgaris*. The extract obtained using chloroform showed a maximum activity against *S. pyogenes* (22.00 mm) and minimum activity against *K. pneumoniae* and *S. aureus* (10.66 mm). No antibacterial activity against *P. aeruginosa* was observed. Ethanolic extract didn't provoke zone of inhibition against *K. pneumoniae*, *P. vulgaris*, *P. aeruginosa*, and *S. pyogenes* whereas, the maximum activity (22.66 mm) was recorded against *E. faecalis* and minimum activity (7.66 mm) was against *S. aureus*. Acetone extract pointed out maximum activity against pathogens *S. aureus* (29.33 mm) and *K. pnemoniae* (28.66 mm) and minimum activity against *E. faecalis* (11.66 mm) and *P. aeruginosa* (10.33 mm). Acetone and petroleum ether extracts showed excellent inhibition against bacterial pathogens which were well compared to standard drug Chloramphenicol (30 μg/disc). Chloroform extract showed remarkable antibacterial activity against all pathogens except *P. aeruginosa*. Ethanol extract inhibited *E. coli*, *S. aureus*, and *E. faecalis* whereas, *K. pneumoniae*, *P. vulgaris*, *P. aeruginosa*, and *S. pyogenes* were resistant to ethanolic extract. *E. coli*, *K. pneumoniae*, *P. vulgaris*, *S. aureus*, *S. pyogenes*, and *E. faecalis* were not susceptible to aqueous extract (Manchu et al. 2015).

In the study made by Kosanić et al. (2015), biological activities of two macroalgae from Adriatic coast of Montenegro were evaluated. Extracts of *Ulva lactuca* showed a better antimicrobial activity with minimum inhibitory concentration values ranging from 0.156 to 5 mg mL^{-1}, but it was relatively weak in comparison to standard antibiotics. *Bacillus mycoides* and *Bacillus subtilis* were the most susceptible to the tested extracts. Extracts from *Ulva intestinalis* (formerly *Enteromorpha intestinalis*) also inhibited all of the tested microorganisms, but at slightly higher concentrations.

The study of Johnson and Raja (2015) was aimed to examine the antibacterial potential of the extracts of *Caulerpa corynephora*, *Caulerpa scalpelliformis*, *Chaetomorpha antennina*, *Ulva compressa* (formerly *Enteromorpha compressa*), *Halimeda macroloba*, *Ulva fasciata*, and *Ulva lactuca* from southern east coast of Tamil Nadu (India) against the pathogens *Escherichia coli*, *Klebsiella pneumoniae*, *Pseudomonas aeruginosa*, *Proteus mirabilis*, *Serratia marcescens*, *Salmonella typhi*, *Acinetobacter calcoaceticus*, *Enterobacter cloacae*, *Staphylococcus aureus*, *Enterococcus faecalis*, *Streptococcus pyogenes*, and *Bacillus cereus*. The antibacterial activity of five crude extracts (aqueous, ethanol, chloroform, petroleum ether, and hexane) of *U. lactuca*, *U. fasciata*, *H. macroloba*, *U. compressa*, *C. antennia*, *C. scalpelliformis*, and *C. corynephora* against 12 human bacterial pathogens were determined by paper disc diffusion method. Among the five different crude extracts of *U. lactuca*, antibacterial activities were observed only in hexane, petroleum ether, chloroform, and ethanolic extracts of *U. lactuca*. In *U. lactuca*, among the tested crude extracts, highest antibacterial activity was observed in ethanolic extracts followed by petroleum ether, chloroform, and hexane. Ethanolic extracts not only produced bigger inhibitory zones, but also active against all the pathogens tested, followed by petroleum ether (four), chloroform (three), and hexane (one). Ethanolic and chloroform extracts of *U. lactuca* were active against *S. typhi*, *A. calcoaceticus*, and *S. aureus*, in concordance with the results of Nithya and Dhanalakshmi (2016). Ethanolic and petroleum ether extract of *U. lactuca* were active against *P. aeruginosa*, *S. aureus*, *S. pyogenes*, and *B. cereus*. Ethanolic and hexane extracts produce activity only against *B. cereus*. However, the percentage of inhibition varied with test organisms, highest inhibition was observed against *S. pyogenes* (Gram$^+$, 118.18%), and lowest against *P. mirabilis* (Gram$^-$, 35%). Compared to Amikacin, 50% and higher inhibitory activities were observed in ethanolic extract against *E. coli*, *K. pneumoniae*, *S. typhi*, *A. calcoaceticus*, *S. aureus*, *E. faecalis*, and *S. pyogenes*. The ethanolic extracts showed 92% activity, chloroform 58.33%, and hexane 8.33%. Ethanolic extracts showed 92% activity, chloroform 58.33%, and hexane 8.33%. Ethanolic extract showed 100% activity against Gram$^+$ bacteria, and 87.5% against Gram$^-$ pathogens. In *H. macroloba*, three crude extracts showed antibacterial activity, highest of which was observed in ethanolic extracts, followed by chloroform and hexane. Ethanolic extracts not only produce bigger inhibitory zones, but also active against different pathogens. *P. aeruginosa* and *S. typhi* were inhibited by both ethanolic and chloroform extracts. Ethanolic and hexane extracts of *H. macroloba* were active against *K. pneumoniae* and *B. cereus*. All the Gram$^+$ and Gram$^-$ pathogens were susceptible to the ethanolic extract of the *H. macroloba*. However, the size of the inhibition zone varied. Ethanolic extract of *H. macroloba* was effective against 92% of the tested organisms, highest inhibition was observed against *S. pyogenes* (Gram$^+$, 73%) and lowest against *P. aeruginosa* (Gram$^-$, 29.16%). 50% and more than 50% of inhibition was observed in the ethanolic extract of *H. macroloba* against six different bacterial pathogens namely *E. coli* (Gram$^-$, 65%), *P. mirabilis* (Gram$^-$, 50%), *S. marcescens* (Gram$^-$, 50%), *A. calcoaceticus* (Gram$^-$, 50%), *E. faecalis* (Gram$^+$, 64%), and *S. pyogenes* (Gram$^+$, 73%). Chloroform and hexane extracts were active against 17% of the tested organisms. The size of the inhibition zone was less than 40%. Chloroform extract showed activity only against two different Gram$^-$ pathogens (*P. aeruginosa* and *S. typhi*). Hexane extract showed activity against *K. pneumoniae* (Gram$^-$, 38.09%) and *B. cereus* (Gram$^+$, 26%) bacteria. Petroleum ether, chloroform, and ethanolic extract of *U. compressa* showed antibacterial activity against 11 of the tested 12 bacterial pathogens. The size of the inhibitory zone varied with test organism and type of solvent used for extraction. Compared to control, petroleum ether extracts showed 50% and higher inhibitory activity against *E. coli*, *K. pneumoniae*, *P. aeruginosa*, *S. marcescens*, *A. calcoaceticus*, *E. cloacae*, *S. aureus*, and *S. pyogenes*. Similarly chloroform extract showed 50% and higher activity against *E. coli*, *K. pneumoniae*, *P. aeruginosa*, *P. mirabilis*, *S. typhi*, *A. calcoaceticus*, *S. aureus*, *S. pyogenes*, and *B. cereus*. Ethanolic extract showed 50% and higher inhibitory activity against *P. aeruginosa*, *S. marcescens*, *S. typhi*, *S. aureus*, *S. pyogenes*, and *B. cereus*. An important observation was noted, that 100% of inhibition was exhibited by the ethanolic extract against *S. pyogenes*. Among the five crude extracts of *C. antennia*, the highest antibacterial activity was observed in ethanolic extracts, followed by hexane and chloroform. Ethanolic extracts not only produced bigger inhibitory zones, but were also active against 10 different pathogens, followed by hexane (5) and chloroform (4). *K. pneumoniae* was inhibited by the ethanolic, chloroform, and hexane extracts of *C. antennia*. The size of the inhibitory zone in chloroform extracts were comparatively bigger (9, 8, 7 mm) than those of hexane and ethanolic extracts of *C. antennia*. Ethanolic and chloroform extracts of *C. antennia* were active against *K. pneumoniae*, *S. marcescens*, and *E. cloacae*. Ethanolic and hexane extracts of *C. antennia* were active against *K. pneumoniae*, *P. aeruginosa*, *P. mirabilis*, *A. calcoaceticus*, and *S. aureus*. *S. typhi* was inhibited only by the chloroform extract of *C. antennia*. Similarly, ethanolic extract alone inhibited *E. faecalis*, *S. pyogenes*, and *B. cereus*. Ethanolic extracts of *C. antennia* was effective against 83% of the test organism, except *E. coli* and *S. typhi* (Gram$^-$). However, the percentage of inhibition varied with test organisms. Highest inhibition was observed against *S. pyogenes* (Gram$^+$, 145.45%) and lowest against (33.33%) *K. pneumoniae*, *P. aeruginosa*, and *A. calcoaceticus* (Gram$^-$). 50% and more inhibition was observed against two bacterial pathogens, namely *S. pyogenes* (145.45%) and *E. faecalis* (73%) and three pathogens (Gram$^-$) *P. mirabilis*, *S. marcescens*, and *A. calcoaceticus* 50%, respectively. Chloroform extract was active only against 33.33% of Gram$^+$ pathogens. No inhibitory activity was observed against any of the Gram$^+$ bacteria tested. An observation noted was that the activity of ethanolic extract of *C. antennia* against *S. pyogenes* was 45% more than the control. Hexane extracts of *C. antennia* was effective against 42% of the test organisms. Out of the five different crude extracts of *C. scalpelliformis*, antibacterial activities were observed only in chloroform and ethanolic extracts. However, the size of the inhibition zone varied with test organisms and extracts. Highest activity was observed in ethanolic extracts of *C. scalpelliformis*, followed by chloroform extracts. In *C. scalpelliformis*, ethanolic extracts produced the bigger inhibitory zones, and have also activity against 10 different pathogens. The chloroform extract inhibited only two pathogens. *K. pneumoniae* and *P. aeruginosa* were inhibited by both ethanolic and chloroform extracts of *C. scalpelliformis*. The size of the inhibitory zone in ethanolic extracts was bigger (10.9 mm) than chloroform extract (8.7 mm) against *K. pneumoniae* and *P. aeruginosa*, respectively. Ethanolic extracts of *C. scalpelliformis* were effective against 90% of the test organisms. Except *S. typhi* (Gram$^-$) and *E. faecalis* (Gram$^+$), all the other pathogens were inhibited by ethanolic extracts. Nevertheless, the percentage of inhibition varied with test organisms. Highest inhibition was observed against *S. pyogenes* (Gram$^+$, 82%) and lowest against *S. marcescens* (Gram$^-$, 32%). 50% and more inhibition was observed against two bacterial pathogens *E. coli* (Gram$^-$, 50%) and *S. pyogenes* (Gram$^+$, 82%). Chloroform extract produced less than 50% of inhibition against *K. pneumonie* (38%) and *P. aeruginosa* (29%). In *C. corynephora*, two crude extracts showed antibacterial activity, the higher of which was observed in ethanolic extract, followed by chloroform. Among the 12 bacterial pathogens tested, 10 were susceptible to the ethanolic and chloroform extracts of *C. corynephora*. However, the size of the inhibition zone varied. *S. pyogenes* was inhibited only by the ethanolic extract. *E. faecalis* was not susceptible to the ethanolic and chloroform extract of *C. corynephora*. Ethanolic extracts of *C. corynephora* was effective against 92% of the test organisms. Except *E. faecalis* (Gram$^+$), all the other pathogens were inhibited by the ethanolic extract. Nevertheless, the percentage of inhibition compared to the control varied with test organisms. Highest inhibition was observed in *S. marcescens* and lowest against *S. aureus*. Chloroform extracts of *C. corynephora* were effective against 83.33% of the test organisms, except *E. faecalis* and *S. pyogenes* (Gram$^+$). Compared to the control, the ethanolic extract of *C. corynephora* produced more than 50% of inhibition against 11 different bacterial pathogens. The screened Gram$^-$ were susceptible to the ethanolic extract. Chloroform extracts were comparatively similar to ethanolic extracts, inhibiting 10 different bacterial pathogens (Johnson and Raja 2015).

Cermák et al. (2015) reported the antibacterial long-chain fatty acids in the green microalga *Planktochlorella nurekis* to be significant inhibitors of *Campylobacter jejuni*, *Escherichia coli*, *Salmonella enterica* var. *Enteritidis*, *Salmonella enterica* var. *Infantis*, *Arcobacter butzleri*, and *Lactobacillus johnsonii* using a suspension concentration range of 0.75–6 mg mL^{-1}. The study proposed that green microalgae could be used as an alternative to in-feed antibiotics to prevent disease in livestock and poultry and to maintain the microbial safety of animal products in the human food chain.

Radhika and Mohaideen (2015) used two seaweeds collected from Hare island in the Gulf of Mannar, on Tuticorin coast (India) to evaluate its antibacterial activity. The ethanolic extract fractions of *Ulva lactuca* were tested against different pathogens like *Klebsiella pneumonia*, *Aeromonas hydrophilla*, *Staphylococcus aureus*, *Escherichia coli*, and *Pseudomonas* sp., and the highest zone of inhibition was found in *A. hydrophila*. Similarly, when the crude extract of *U. lactuca* was tested against the same bacterial pathogen, the highest zone of inhibition was found in *E. coli*.

Phytochemical analyses and *in vitro* antibacterial activity of different extracts of hexane, chloroform, ethyl acetate, acetone, and methanol extracts of green algae, *Caulerpa chemnitzia*, against *Bacillus subtilis*, *Streptococcus pyogenes*, *Escherichia coli*, *Klebsiella pneumoniae*, *Proteus mirabilis*, *Proteus vulgaris*, *Pseudomonas aeruginosa*, *Salmonella typhimurium*, *Shigella flexneri*, and *Vibrio cholerae*. The extent of the inhibitory zone, minimum inhibitory concentration (MIC), and minimum bactericidal concentration (MBC) were determined (Raj et al. 2015). The ethyl acetate extract of *C. chemnitzia* showed the presence of phytochemicals, terpenoids, tannins, and phenolic compounds strongly than the other solvent extracts. The mean zone of inhibition produced by the extracts in agar diffusion assays against the tested bacterial strains ranged from 7.1 to 13.6 mm. The MIC was between 125 and 500 μg mL^{-1} while the MBC were between 250 and 1000 μg mL^{-1}. The highest mean zone of inhibition (13.6 mm) and the lowest MIC (125 μg mL^{-1}) and MBC (250 μg mL^{-1}) values were observed in ethyl acetate extract against *B. subtilis*.

In the research work done by Kumar and Padhi (2016), several marine macroalgae were selected for the antimicrobial activities against two bacterial test organisms. The selection of these seaweeds was made on account of their common occurrence in Chilika Lake waters (east coast of India). The two bacterial strains, namely *Pseudomonas aeruginosa* and *Escherichia coli* were selected for the antimicrobial test. Aqueous extracts have no antibacterial activity, but in ethanol and methanol extracts, antimicrobial activities were observed. Methanol extracts exhibited more antimicrobial activity than ethanol extracts. Methanolic extracts of *Chaetomorpha linum* and *Ulva compressa* (formerly *Enteromorpha compressa*) were more effective than ethanolic extracts against the tested bacteria *P. aeruginosa* and *E. coli*. In ethanol extract, zone of inhibition was highest in *U. compressa* (12 mm) against *E. coli*, while lowest with extracts of *C. linum* (6.07 mm). In methanol extracts, highest inhibition effect was shown in *U. compressa* (24 mm) against *E. coli*, followed by *C. linum* (23.5 mm).

Dhanya et al. (2016) reported the antimicrobial activity of extracts from *Ulva reticulata* against several human pathogens. Antimicrobial assay performed for the culture and algal extracts revealed measurable zones of inhibition against human pathogens *Escherichia coli*, *Salmonella typhi*, *Bacillus subitils*, and *Pseudomonas aeruginosa*. The maximum zone of inhibition was obtained with hexane extracts of the effective isolate from *U. reticulata* against *Staphylococcus aureus*, which was 14 mm, followed by *E. coli* (12 mm).

Sulfated polysaccharides isolated from the southwest coast of India seaweeds are commercially and therapeutically interesting due to their various biological activities (Kailas 2015). These sugars were isolated from several seaweed species using water at 100ºC followed by precipitation using ethanol. The antimicrobial activities of the sulfated sugars at a concentration level of 100 g mL^{-1} in water were observed to be appreciable in comparison (> 50%) with the positive controls (concentration of 100 g mL^{-1} in methanol) used against the *Escherichia coli* strains. The sulfated sugars of *Ulva fasciata* exhibited higher extent of broad spectrum antimicrobial activity with respect to both the standards. Similarity of activity was seen in *Ulva prolifera* (formerly *Enteromorpha prolifera*) collected from Kayamkulam location (India). In general, the sulfated sugars exhibited moderate activities against *Staphylococcus aureus* and *Bacillus cereus*, and low activity against *Salmonella abony* with respect to both the standards. Sulfated sugars of *U. fasciata* and *U. prolifera* collected from the Kayamkulam location were observed to be lethal to *S. aureus*. *B. cereus* growth was highly inhibited by the sulfated sugars of *Chaetomorpha antennina*, and *U. linza*. *E. coli* inhibition was seen highest in the *U. fasciata* sulfated sugars (Kailas 2015).

In the study done by Patra and Baek (2016), essential oil from an edible seaweed, *Ulva linza* (formerly *Enteromorpha linza*), was evaluated for its antibacterial activity against foodborne pathogens, along with the mechanism of its antibacterial action. *U. linza* oil at 25 mg/disc was highly active against *Bacillus cereus* (12.3–12.7 mm inhibition zone) and *Staphylococcus aureus* (12.7–13.3 mm inhibition zone). The minimum inhibitory concentration and minimum bactericidal concentration values of *U. linza* oil ranged from 12.5–25 mg mL^{-1}.

In a study done with seaweeds collected from the Persian Gulf, Qeshm Island, Iran, the antibacterial activity of several extracts were evaluated (Mashjoor et al. 2016). All extracts exhibited strong antibacterial activity in $Gram^+$ and $Gram^-$ bacteria. The data showed that *Staphylococcus epidermidis* was the most sensitive strain tested against effects from all macroalgal extracts, specifically ethylacetate extracts minimum inhibition concentration (MIC) = 93 and 187 μg mL^{-1} for the ethyl acetate and methanol extracts, respectively, the inhibition zone ranged from 22 to 28 mm. Of the tested pathogens, *Escherichia coli* for the *Ulva flexuosa* extract (MIC = 93 μg mL^{-1}). Among the bacterial strains tested, *Pseudomonas aeruginosa*, *Enterococcus faecalis*, and *Klebsiella pneumonia* were the most resistant bacteria. The inhibition zone values obtained in the present study for the bacterial strains showed a lower effect (and there was no inhibition zone against *K. pneumonia* and *P. aeruginosa*) compared to the standard antimicrobial agent, Ampicillin, which ranged from 12 to 19 mm overall.

In the work done by Khan (2016), with seaweeds collected from the Karachi coast of Pakistan, MIC of tested green seaweeds extracts ranged from 5 to 30 mg mL^{-1}. Among the ethanol extractions of green seaweeds, the lowest MIC value of 10 mg mL^{-1} was recorded in *Caulerpa scalpelliformis* (from Buleji and Manora beaches), *Ulva compressa* (as *Enteromorpha compressa*) (from Buleji), *Ulva fasciata* (from Manora), *Codium shameelii* and *Ulva lactuca* (from Paradise Point), against *E. coli*, and 5 mg mL^{-1} in *C. scalpelliformis* (Paradise Point) against *Salmonella typhi*, followed by *Udotea indica*, *Ulva lactuca*, and *Valoniopsis pachynema* (from Buleji), *C. shameelii* (from Paradise Point), *Codium indicum* (as *Codium iyengarii*) (from Paradide Point and Manora), with 10 mg mL^{-1}. In acetone extractions, the lowest MIC (5 mg mL^{-1}) was recorded in Buleji species—*C. indicum*, against *E. coli*; in Manora species—*C. indicum*; and in Buleji species—*U. fasciata*, against *S. typhi*. MBC of tested seaweeds extracts ranging from 10 to 40 mg mL^{-1}. Highest MBC value (40 mg mL^{-1}) was recorded in the ethanol extraction of *Udotea indica* and *U. lactuca* (from Buleji), *U. indica* (from Paradise Point), *C. shameelii* and *V. pachynema* (from Manora), against *E. coli* and *S. typhi*. The highest MBC value (40 mg mL^{-1}) recorded in *U. indica* (Buleji), *C. Shameelii* and *Ulva compressa* (Paradise Point), and *C. shameelii* (Manora) (Khan 2016).

The study of the antimicrobial activity of the lipidic extract of the Mediterranean invasive seaweed *Caulerpa cylindracea* was assayed (Stabili et al. 2016). The extract did not show antibacterial activity against *Enterococcus* sp., *Escherichia coli*, *Staphylococcus* sp., and *Streptococcus* sp. By contrast, the degree of inhibition produced by the lipidic extract on some *Vibrio* species was quantified. In particular, using discs with 100 μL of algal extract, *Vibrio fischeri*, *Vibrio inusitatus*, and *Vibrio litoralis* were the most inhibited (diameter of growth inhibition = 0.9 cm). A lower percentage of inhibition was measured on *Vibrio aestuarinus* (0.85 cm), *Vibrio mediterranei* (0.8 cm), and *Vibrio vulnificus* (0.8 cm).

8.5 Antibacterial Activity of Phaeophyceae (Brown Algae)

Methanol extracts obtained from brown algae such as *Saccharina angustata* (formerly *Laminaria angustata*), *Undaria pinnatifida*, and *Sargassum capillare* (as *Sargassum gracile*) found along the coast of Japan inhibited the several kinds of pathogenic bacteria. The extract prepared from *S. gracile* strongly inhibited the growth of *Bacillus mesentericus* (Saito and Nakamura 1951).

Fenical et al. (1973) reported two sesquiterpenes, zonarol and isozonarol from a brown alga, *Dictyopteris undulata* (formerly *Dictyopteris zonarioides*). Neither substances possessed antibacterial properties, but strongly inhibited the growth of 10 species of pathogenic fungi causing diseases in plants (see Chapter 7). Lipid extracts of more than 20 algae found along the coast of Eastern Sicily inhibited the growth of some plant pathogenic bacterium *Xanthomonas malvaciarum* and Tobacco Mosaic Virus (see Chapter 5) under *in vitro* (Caccamese et al. 1980).

Rao and Parekh (1981) reported that the crude extracts of brown seaweeds such as *Dictyota dichotoma* and *Dictyota* sp. were active against the $Gram^{+}$ bacteria such as *Bacillus megatherium* and *Staphylococcus aureus*, but were not active against $Gram^{-}$ bacteria tested. The crude extracts of *Zanardinia typus* (formerly *Zanardinia prototypus*) and *Cystoseira brachycarpa* (formerly *Cystoseira balearica*) exhibited the best antimicrobial and antiviral activities among the seaweeds tested, while extracts obtained from *Lophocladia lallemandii* (Rhodophyta) were not active against the test bacteria but had high antiviral potential. Reichelt and Borowitzka (1984) found that the majority of the algal extracts showed antibacterial activity against $Gram^{+}$ bacteria.

Moreau et al. (1984) found that Mediterranean *Dictyota* species possessed broad activity against different fungi (see Chapter 7), but only *Dictyota dichotoma* inhibited the tested bacteria. A diterpenoid compound isolated from Caribbean *Dictyota bartayresiana* (formerly *Dictyota bartayresii*) was shown to be toxic to assay microorganisms by Norris and Fenical (1982).

In the research work done by Hornsey and Hide (1985), using *Staphylococcus aureus* and *Escherichia coli* as test organisms, various life-cycle phases of 22 species of British marine algae were screened for antimicrobial activity. Out of the 10 species of Phaeophyceae examined, only the growth forms of *Alaria esculenta* showed a significant variation in the levels of antimicrobial activity—the sterile parts of the thallus being active, whilst the reproductive sporophylls were inactive. Consideration of the results obtained with members of the Phaeophyceae reveals slightly different patterns of activity—similar levels of antibiotic production in both haploid and diploid generations as shown by *Dictyota dichotoma*; similar levels of antibiotic activity by both sterile and fertile plants as found in *Laminaria digitata*; and antibiotic production occurring in the sterile regions of the thallus, whilst the reproductive sporophyll areas are devoid of activity as found in *Alaria esculenta* (Hornsey and Hide 1985).

The studies done by Rao and Karmarkar (1986) reveal the poor activity of water extract in all the test plant samples. Among the organic solvent extracts, the butanol, propanol, and acetone extracts, and not those of benzene were highly effective against all post-operative pathogens. Best and higher activity type was found in *Padina tetrastromatica* against *Pseudomonas aeruginosa*, followed by *Padina boergesenii* extract against *Staphylococcus aureus*. Extracts of 12 different species of *Sargassum* were separated into two fractions and tested against nine human pathogenic bacteria. Both fractions of *Sargassum vulgare* showed good antibacterial activity against Gram$^+$ and Gram$^-$ bacteria (Rao et al. 1988). Crude extracts obtained in the diethyl ether from various parts, viz., fronds, stems, and air bladders of *Sargassum johnstonii* screened for their antibacterial potential and found that the extracts of the frond portion showed more bioactivity than the stem and air bladders (Rao 1990). Febles et al. (1995) studied the *in vitro* antibacterial activity of a number of brown seaweeds collected from the littoral of Tenerife (Canary Islands). Three different solvents: *n*-hexane, ethyl acetate, and methanol have been used to obtain extracts from *Sargassum desfontainesii*, *Halopteris scoparia*, and *Stypopodium zonale*. The activity of the extracts was tested using Gram$^+$ and Gram$^-$ bacteria. The methanol extract showed most antibacterial activity, all extracts were mainly active in Gram$^+$ bacteria.

Crude extracts obtained from the whole plant, stem, leafy-portion, receptacle, and vesicle of brown alga *Sargassum swartzii* (formerly *Sargassum wightii*) found along the South east coast of India were tested for their antibacterial activity. Extracts of the leafy-portion and the whole plant exhibited good antibacterial activity (Rao 1990).

Organic extracts of *Lobophora variegata* have shown a broad-spectrum of antibacterial activities (Ballantine et al. 1987, Engel et al. 2006, Morrow et al. 2011, Manilal et al. 2010, 2010b, 2012, Sivakumar 2014, and Gutiérrez-Cepeda et al. 2015). The chloroform-methanolic extract of Caribbean *L. variegata* presented antibacterial activity against *Bacillus subtilis* (Ballantine et al. 1987). Val et al. (2001) did not observed any antimicrobial activity of the methanolic extract of *L. variegata* harvested in Canary Islands (Spain) against a panel of pathogen bacterial strains. Engel et al. (2006) considered two morphotypes of *L. variegata*, crustose and ruffled, which we strongly suspect to be two distinct species. Lipophilic and hydrophilic parts of organic extracts from both morphotypes resulted in growth inhibition of the bacteria *Pseudoalteromonas bacteriolytica*. However, the two morphotypes extracts yielded contrasting IC_{50} values: the lipophilic parts showed volumetric IC_{50s} of 1 and 0.24 (unit less) for the crustose and ruffled types, respectively, and the hydrophilic parts exhibited volumetric IC_{50s} of 0.51 and 0.67, respectively. It would therefore appear that these two different morphotypes have contrasting chemical production. The organic extract of *L. variegata* samples from India showed a strong inhibition against *Salmonella typhi* and *Vibrio cholera*, while being less active against *Klebsilla pneumonia* and *E. coli* (Sivakumar 2014).

Extracts from several southern African seaweeds were screened against 12 bacteria (Vlachos et al. 1997). The methanolic extracts from the brown algae exhibited the largest relative activity, against all the test strains. Of the seaweeds tested, the brown alga *Zonaria subarticulata* showed the highest amount and broadest spectrum of antibacterial activity against the microorganisms tested (inhibition zone diameter > 10 mm), indicating that this seaweed extract had the highest degree and broadest spectrum of antimicrobial activity in the study.

A meroditerpenoid has been isolated from the brown alga *Cystoseira tamariscifolia* and characterized as methoxybifurcarenone, and it possesses antibacterial activity against *Agrobacterium tumefaciens* and

Escherichia coli (Bennamara et al. 1999). In the works of Abourriche et al. (1999), the hexane, diethyl ether, and dichloromethane fractions of the brown alga *C. tamariscifolia* extract showed interesting antibacterial activities. The diethyl ether fraction of *C. tamariscifolia* extract, which was comparable to Tetracyclin against *Agrobacterium tumefaciens* and *Escherichia coli*, showed more potent antibacterial activity than hexane and dichloromethane fractions.

A bioassay-directed was utilized to detect substances with biological activity from several seaweeds collected from the coast of the Gulf of Mexico (Oranday et al. 2004). Antimicrobial activity was found in ether extracts of *Sargassum fluitans* against *Escherichia coli*, *Staphylococcus aureus*, *Candida albicans* (yeast), and *Staphylococcus epidermidis*. The eight fraction of petroleum ether of *S. fluitans* exhibited high activity against *C. albicans* MIC 0.16 $\mu g\ mL^{-1}$. The fact that the tested algae demonstrated antimicrobial activity emphasizes the important role of traditional medicine in the search for antibiotic compounds from natural sources (Oranday et al. 2004).

Ely et al. (2004) tested methanolic extracts of Indian seaweed against clinical bacteria and fungi, using paper disc assays. *Stoechospermum polypodioides* (formerly *Stoechospermum marginatum*) extract was effective only against *Vibrio cholera* and *Klebsiella* sp.

Some fractions obtained from the crude extracts of red, brown, and green seaweeds showed optimum antibacterial activity against the test bacteria, such as *Pseudomonas aeruginosa* and *Proteus vulgaris*. The unsaponifiable part of the lipid extracted in diethyl ether and the saponifiable part of the lipid obtained in benzene from the brown alga *Sargassum johnstonii* exhibited more antibacterial activity against Gram^{+} and Gram^{-} bacteria tested (Parekh et al. 1984). Different concentration of the fractions isolated from the seaweeds showed antibacterial activity against the test bacteria *Staphylococcus aureus* and *Escherichia coli* (Parekh 1998).

The communication made by Lakshmi et al. (2006) deals with the biological activities of the extracts of 48 marine florae. The biological screening includes tests for antibacterial and antifungal properties, among others. From among brown algae, the crude extracts from *Colpomenia sinuosa*, *Polycladia indica* (formerly *Cystoseira indica*), *Dictyopteris woodwardia*, *Sargassum johnstonii*, *Spatoglossum variabile*, *Stoechospermum polypodioides* (formerly *Stoechospermum marginatum*), *Dictyota dichotoma*, *Turbinaria condensata*, *Turbinaria conoides*, and *Turbinaria decurrens* were active against *Staphylococcus aureus*, *Escherichia coli*, *Streptococcus faecalis*, *Klebsiella pnuemoniae*, and *Pseudomonas aeruginosa*.

Tüney et al. (2006) studied the antimicrobial activity of methanol, acetone, diethyl ether, and ethanol extracts of 11 seaweed species against *Candida* sp. (yeast), *Enterococcus faecalis*, *Staphylococcus aureus*, *Streptococcus epidermidis*, *Pseudomonas aeruginosa*, and *Escherichia coli* by disc diffusion method. Diethyl ether extracts of fresh *Cystoseira mediterranea*, *Ectocarpus siliculosus* showed effective results against all tested organisms. However, diethyl ether extracts of some species, such as *Padina pavonica*, *Colpomenia sinuosa*, *Dictyota implexa* (formerly *Dictyota linearis*), and *Dictyopteris polypodioides* (formerly *Dictyopteris membranacea*), gave different results. A significant difference in antimicrobial activity was not observed between the acetone and methanol extracts of each alga.

Atlantic marine algae were screened in growth inhibition assays against the pathogenic bacterium *Pseudoaltermonas bacteriolytica* (Puglisi et al. 2007). The extracts from 50% of all brown algae were active against the tested bacterium; among them, *Hydroclathrus clathratus* was selectively active against *P. bacteriolytica*, further, the hydrophilic extracts from both species of *Sargassum* studied were also active against the pathogenic bacterium (Puglisi et al. 2007).

Methanol, acetone, diethyl ether, and ethanol extracts of 11 seaweed species from the coast of Urla (Turkey) were tested *in vitro* for their antimicrobial activities against *Enterococcus faecalis*, *Staphylococcus aureus*, *Streptococcus epidermidis*, *Pseudomonas aeruginosa*, and *Escherichia coli* with the disc diffusion method. Diethyl ether was the best solution for extracting the effective antibacterial materials from the algae species used in this experiment, with the exception of *Dictyota dichotoma* var. *intricata* (formerly *Dictyota linearis*), for which ethanol was the most effective extraction solution (Tüney et al. 2006). Diethyl ether extracts of fresh *Cystoseira mediterranea* and *Ectocarpus siliculosus* showed effective results against all test organisms. However, diethyl ether extracts of some species, such as *Padina pavonica*, *Colpomenia sinuosa*, *D. dichotoma* var. *intricata*, and *Dictyopteris polypodioides* (formerly *Dictyopteris membranacea*) gave different results. A significant difference in antimicrobial activity was not observed between the acetone and methanol extracts of each alga. In addition, as a result of the comparison of dried and fresh extract antimicrobial activity, it was found that all test organisms were more sensitive to fresh extracts of the algae. Although fresh extracts of *D. dichotoma* var. *intricata* and *E. siliculosus* inhibited the tested bacteria and yeast, their dried extracts had no inhibition activity on either Gram$^-$ or Gram$^+$ bacteria.

In the study executed by Taskin et al. (2007), methanolic extracts of six marine algae from the North Aegean Sea (Turkey) were studied for their antibacterial activity against pathogenic bacteria, three Gram$^+$ (*Staphylococcus aureus*, *Micrococcus luteus*, and *Enterococcus faecalis*) and three Gram$^-$ (*Escherichia coli*, *Enterobacter aerogenes*, and *E. coli*) *in vitro*. The extract of *Cystoseira barbata* had the broadest activity spectrum; *Dictyota dichotoma* and *Halopteris filicina* had the lowest against test microorganisms. The growth of food-borne pathogen *E. coli* was inhibited by only the extracts of *Cladostephus spongiosus* f. *verticillatus* with moderate, and of *C. barbata* with the strong inhibition level (22.33 mm).

Kim et al. (2007c) reported antimicrobial activity of crude extracts and solvent fractions from *Sargassum muticum*. Antimicrobial activities were shown in ethanol, dichloromethane, and *n*-hexane fractions of *S. muticum*. However, butanol, ethyl acetate, and water fractions showed weak antimicrobial activity against the tested microorganisms. Among the five fractions, dichloromethane fraction showed the highest antimicrobial activities against microorganisms tested, such as *Bacillus sublitis*, *Listeria monocytogenes*, *Staphylococcus aureus*, *Escherichia coli*, *Salmonella enteritidis*, and *Pseudomonas aeruginosa*. The polyphenolic compounds from ethanol, *n*-hexane, dichloromethane, ethylacetate, butanol, and water fractions were 63.96, 8.49, 28.11, 172.64, 114.56, and 34.91 mg g^{-1}, respectively. The dichloromethane fraction could be suitable for development as a food preservative.

Several seaweeds were collected from the intertidal zone at Rocky Bay on the east coast of South Africa (Stirk et al. 2007). *Dictyota humifusa* was the only seaweed able to inhibit the Gram$^-$ *Escherichia coli*. Seasonal variation in antibacterial activity was observed, with the extracts generally having no activity in summer and having antibacterial activity in late winter (July collection) and early spring (September and November collections). *D. humifusa* was the most effective seaweed species, having antibacterial activity throughout the year. The most active extract was made from *D. humifusa* collected on 23/05 (IC_{50} 4.75 mg mL^{-1}).

The study done by Jebakumar Solomon and Santhi (2008) had been focused to search for powerful antimicrobial natural products from *Dictyota acutiloba* against human enteric pathogens. Chloroform and acetone extracts of *D. acutiloba* exhibited antimicrobial activity against Methicillin-resistant *Staphylococcus aureus*, Methicillin-susceptible *S. aureus*, *Enterobacter* sp., *Pseudomonas aeruginosa*, *Salmonella typhi*, *Bacillus subtilis*, and *Klebsiella pneumoniae*. Different organic fractions of *D. acutiloba* were assayed for antimicrobial activity by loading 10 μg per sterile disc. Among the hexane, chloroform, acetone and methanol extracts, chloroform and acetone extracts exhibited antimicrobial activity against the tested microorganisms. The highest activity was found in chloroform extract of *D. acutiloba*. The antimicrobial compound present in this marine sea weed has more potent antagonistic effect towards bacterial pathogens and weak antifungal activity. In addition, they showed powerful antimicrobial effect on both Gram$^+$ and Gram$^-$ bacteria. *Enterobacter* sp., *P. aeruginosa*, *B. subtilis*, Methicillin-resistant *S. aureus* and *S. aureus* were highly susceptible to the active principle present in the algal extracts of *D. acutiloba*. On the other hand, *K. pneumoniae* showed moderate susceptibility towards the antimicrobial compound in the marine seaweed. The inhibition zone size of Methicillin *S. aureus* resistant against antibiotic disc containing 5 μg Methicillin was 8 mm, and in case of crude extracts, showed 15 and 14 mm for chloroform and acetone extracts, respectively. Hence the susceptibility of Methicillin-resistant *S. aureus* to the extract of *D. acutiloba* was more pronounced when compared to the antibiotic Methicillin (Jebakumar Solomon and Santhi 2008).

In the work done by Kandhasamy and Arunachalam (2008), with seaweeds of southeast coast of India, bacterial pathogen *Escherichia coli* was resistant to all extracts tested in the present study, except that of *Sargassum tenerrimum* which showed inhibiting zone diameter of 12 mm; *S. tenerrimum* showed lowest inhibiting activity on *Enterobacter faecalis*, *Klebsiella pneumoniae*, and *Enterobacter aerogens* and the diameter of the inhibition zone was 8 mm (Kandhasamy and Arunachalam 2008).

In the study of Patra et al. (2008), the crude extract of *Sargassum* sp. was evaluated for antimicrobial activity against three pathogenic bacteria, namely *Staphylococcus aureus*, *Bacillus subtilis*, and *Escherichia coli*, showed zones of inhibition ranging from 8 to 18 mm. Among the three strains tested, the extract of *Sargassum* sp. was more effective against *B. subtilis* and *E. coli*, showing 18 mm and 16 mm zone of inhibition, respectively at 4000 μg 100 μL^{-1} concentration, in comparison to *S. aureus* showing 10 mm zone of inhibition at the same concentration.

Kumar et al. (2008b) carried out the antimicrobial screening of 12 different seaweeds extracts. The crude extracts were tested against the phytopathogenic bacterium *Pseudomonas syringae*, causing leaf spot disease of the antidiabetic medicinal plant *Gymnema sylvestre*. The methanolic extract of *Sargassum swartzii* (formerly *Sargassum wightii*) showed maximum activity followed by ethyl acetate compared to that of other organic solvent extracts. Thus, their investigation throws fresh light on the appropriate usage of solvent extraction method in preparing potent bio-pesticide.

The study made by Demirel et al. (2009) was conducted to evaluate the antioxidant and antimicrobial activity of methanol, dichloromethane and hexane extracts, as well as the essential oils of brown algae: *Colpomenia sinuosa*, *Dictyota dichotoma*, *Dictyota implexa* (formerly *Dictyota dichotoma* var. *implexa*), *Petalonia fascia*, and *Scytosiphon lomentaria*. Antimicrobial activities of the extracts were assessed against $Gram^+$ and $Gram^-$ bacteria and one yeast strain by the disc diffusion method. According to the results, the dichloromethane extracts generally showed more potent antimicrobial activity than the methanol and hexane extracts at concentrations 1.5 and 1.0 mg/disc. The dichloromethane extracts caused better halo-zones than methanol for all strains. The seaweed extracts are responsible for its activity against $Gram^+$ bacteria, especially *Bacillus subtilis* and *Staphylococcus aureus*. The dichloromethane extracts exhibited a higher degree of activity as compared to the methanol and hexane extracts (Demirel et al. 2009).

In a study by Wang et al. (2009), the bacteriostatic and bactericidal activity of methanol extracted phlorotannins from *Ascophyllum nodosum* harvested in Nova Scotia was compared with hydrolysable and condensed tannins from two trees, Quebracho (*Schinopsis balansaei*) and Chinese sumac (*Rhus semialata*). Four strains of *Escherichia coli* resistant to Ampicillin, Kanamycin, and Nalidixic acid were used to measure inhibition at OD_{600}. At a concentration of only 25 μg mL^{-1}, the phlorotannins from *A. nodosum* exerted bacteriostatic effects on three of the strains for up to 24 hours, after which time two strains resumed growth. Phlorotannins were fully bactericidal to *Escherichia coli* at 50 μg mL^{-1} in the case of two strains, and at 100 μg mL^{-1} in the case of the other two. The hydrolysable and condensed tannins from the three extracts were compared at the same concentrations against two of the *E. coli* strains. However, condensed tannins did not exert a bactericidal effect against either of the two strains, and only had bacteriostatic activity against one strain for 6 hours. Hydrolysable tannins had no bactericidal or bacteriostatic activity against either of the two strains. The treated and untreated *E. coli* cells were examined by transmission electron microscopy. The membrane structure of the untreated cells was smooth and even. The action of the tannins was apparent on the cell walls of treated samples, particularly in the case of *A. nodosum*, where disorganized structures and electron-dense precipitated deposits were visible. The significant differences demonstrated between the marine algal and terrestrial tannins may be attributed to their chemical structure. Although all tannins are composed of cyclic phloroglucinol groups, terrestrial hydrolysable and condensed tannins have a maximum of three rings, whereas algal phlorotannins have up to eight. This means phlorotannins have more hydroxyl groups, enabling them to produce more hydrogen peroxide under aerobic conditions. Hydrogen peroxide is toxic to bacteria (Michalak and Chojnacka 2015). Hydroxyl groups can also form hydrogen bonds with proteins on the bacterial surface, further denaturing the cell.

Methanolic extracts of 17 commonly found seaweeds in the west coast of India were screened for the presence of antimicrobial activity against *Bacillus subtilis*, *Escherichia coli*, *Pseudomonas* sp., *Streptococcus pyrogenes*, *Staphylococcus aureus*, *Proteus vulgaris*, *Klebsiella pneumoneae*, *Serrratia marganii*, and *Candida albicans* (Kotnala et al. 2009). The results of this study indicated that the extracts of *Padina tetrastomatica* were promising against the $Gram^+$ bacteria (*S. pyrogens*, *B. subtilis*, and *S. aureus*). Further purification of the active extracts followed by bioassay indicated that the fraction of ethyl acetate was more active as compared to other fractions. The ethyl acetate fractions of *Padina tetrastromatica* were found to be active against *P. vulgaris*, *S. pyrogens*, and *Pseudomonas* sp.

Cox et al. (2010) found that the antibacterial activity of polyphenolic seaweed extracts was solvent-dependent. Methanol was determined to be the most effective for the brown seaweeds *Himanthalia elongata*, *Laminaria digitata*, and *Saccharina latissima*; acetone and ethanol were more effective for the green and red species. Antibacterial activity was tested on food spoilage (*Enterococcus faecalis* and *Pseudomonas aeruginosa*) and pathogenic (*Listeria monocytogenes* and *Salmonella abony*) bacteria using the microwell plate method. Using methanol only, green and red seaweed extracts had significantly lower antibacterial activity than the brown species, however their potency increased significantly when ethanol and acetone were used as solvents. Methanol extracts of *H. elongata* showed 100% inhibition of *L. monocytogenes*, *E. faecalis*, and *S. abony*, which was, on average, 4.34% greater than that of sodium nitrite and sodium benzoate. *L. digitata* also inhibited *L. monocytogenes* by 100%.

The aim of the study made by Villarreal-Gómez et al. (2010) was to evaluate the antibacterial activity of extracts from the seaweeds *Egregia menziesii*, *Sargassum muticum*, and *Petalonia binghamiae* (formerly *Endarachne binghamiae*) collected from Todos Santos Bay, México. Organic extracts were obtained from bacteria-free algae and from surface-associated bacteria. Pathogen strains of *Staphylococcus aureus*, *Klebsiella pneumoniae*, *Proteus mirabilis*, and *Pseudomonas aeruginosa* were used to test antibacterial activity. 14 bacterial strains and three algae (all except *E. menziesii*) showed antibacterial activity against *P. mirabilis*. None of the extracts from marine algae and bacteria were active against *S. aureus* and *P. aeruginosa*.

Gupta et al. (2010) evaluated the antibacterial activity of three edible Irish brown seaweeds—*Himanthalia elongata*, *Saccharina latissima*, and *Laminaria digitata* in raw and heat processed (95°C) form. Their activity was tested against pathogens, which commonly cause problems in the food industry—*Listeria monocytogenes*, *Salmonella abony*, *Enterococcus faecalis*, and *Pseudomonas aeruginosa*. In microtiter assays, methanol extracts of raw *H. elongata* (60 mg mL^{-1}) inhibited *Listeria monocytogenes* by 98.7%, compared to 96.5% inhibition by the synthetic preservative standard sodium benzoate, and sodium nitrite (96.2%). Raw *H. elongata* was also more potent than the standards against *E. faecalis* and *P. aeruginosa*. Raw *S. latissima* extracts were almost as potent against all four bacteria, followed by *L. digitata*. Heat treatment significantly reduced the antibacterial activity of all seaweeds. These raw seaweed extracts may be useful for incorporation into products such as raw meats and fish, as they would exert their bacterial inhibition during the uncooked, cold storage phase, before inactivation by heat, at which stage the product would be consumed.

Muñoz-Ochoa et al. (2010) screened 60 ethanol extracts of marine flora of Baja California Sur (Mexico) to evaluate the reversing effect of the bacterial resistance to antibiotics in combination with a sub-lethal concentration of Ampicillin or Erythromycin. 35 of the assayed extracts showed inhibitory activity against *Staphylococcus aureus*, 48 were active against *Streptococcus pyogenes*, but none were active against *E. coli*. From the 60 ethanolic extracts, 12 (20%) of them in combination with Ampicillin were able to reverse the resistance of *S. aureus* and eight (13%) with Erythromycin yielded the same reversal with *S. pyogenes*. An extract from *Sargassum horridum* was the only one that reversed the resistance to antibiotics against both *S. aureus* and *S. pyogenes*. The most active extracts were from *Padina mexicana* and *Dictyopteris undulata*, among others.

Rangaiah et al. (2010) investigated the antimicrobial potentiality of the marine algae, two species of brown algae, namely *Sargassum ilicifolium* and *Padina tetrastromatica* by agar well diffusion method. The zone of inhibition was measured for all the different crude algal extracts (chloroform, ethanol, methanol, and water) against six strains of $Gram^{+}$ and $Gram^{-}$ bacterial isolates. Crude extracts revealed a wide range of antimicrobial activity against tested pathogens. Seaweed extracts in different solvents exhibited different antimicrobial activities. In case of *S. ilicifolium*, *P. tetrastromatica*, of the various solvents used for seaweed extractions, maximum inhibition was noticed with ethanol extracts and minimum with chloroform crude extracts.

The methanolic extract of *Iyengaria stellata* showed weak antibacterial activity against *Streptococcus pyogenes* at a concentration of 100 mg 100 mL^{-1} (Khan et al. 2011). Ampicillin, amoxicillin, and cefuroxime were used as standard antibacterial antibiotics (Khan 2000). Saringosterol, loliolide, methyl-4-hydroxybenzoate and propyl-4-hydroxybenzoate were isolated for the first time from the ethyl acetate fraction of *I. stellate* (Khan et al. 2011).

Villarreal-Gómez et al. (2010) evaluated the antibacterial activities of extracts from several seaweeds. Pathogenic strains of *Staphylococcus aureus*, *Klebsiella pneumoniae*, *Proteus mirabilis*, and *Pseudomonas aeruginosa* were used to test antibacterial activity. The extracts from the seaweeds (*Endarachne binghamiae* and *Sargassum muticum*) associated with bacteria inhibited the growth of the Gram$^-$ bacterium *Proteus mirabilis*.

Effective and safe acne vulgaris therapies could be derived from algal extracts. Amiguet et al. (2011) observed that the crude ethylacetate extracts from *Fucus distichus* subsp. *evanescens* showed strong antibacterial activity against *Propionibacterium acnes* (culture collection and clinical isolate), and also against *Hemophilus influenzae*, *Legionella pneumophila*, and *Streptococcus pyogenes*, *Clostridium difficile*, and Methicillin-resistant *S. aureus*, whereas *B. cereus*, *E. coli*, *K. pneumoniae*, and *P. aeruginosa* were not significantly affected. The main active compound was identified as a β-D-galactosyl-*O*-linked glycolipid.

In the work done by Pereira et al. (2011), six meroditerpenoids (epitaondiol, epitaondiol diacetate, epitaondiol monoacetate, stypotriol triacetate, 14-ketostypodiol diacetate, and stypodiol) isolated from the brown alga *Stypopodium flabelliforme*. Antimicrobial activity of the compounds was also evaluated against *Staphylococcus aureus*, *Salmonella typhimurium*, *Proteus mirabilis*, *Bacillus cereus*, *Enterococcus faecalis*, and *Micrococcus luteus*. Compounds exhibiting antimicrobial properties for concentrations lower than 64 μg mL^{-1} are accepted as having notable antimicrobial activity (Gibbons 2004), while those showing activity at concentrations below 10 μg mL^{-1} are considered to be "clinically significant" (Gibbons 2004, Rios and Recio 2005). Epitaondiol monoacetate, stypotriol triacetate, and stypodiol showed some antimicrobial capacity, with the first displaying the major effect against Gram$^+$ and Gram$^-$ bacteria (MIC $\geq$ 114 μg mL^{-1}). *E. faecalis* appeared to be the most sensitive species. As for the remaining compounds, all bacteria were found to be resistant under the tested concentrations.

Subba et al. (2010) reported that the aqueous extract of marine brown alga *Sargassum ilicifolium* showed higher antibacterial activity against the pathogenic bacteria such as *Pseudomona aeruginosa*, *Klebsiella pneumonia*, *Staphylococcus aureus*, and *Bacillus subtilis*. They also reported higher antibacterial activity in the marine brown algae *Padina tetrastromatica* against the following species of *B. subtilis*, *S. aureus*, and *K. pneumonia*. Rosaline et al. (2012) reported 12.67 mm of inhibition in the hexane extract of *Sargassum swartzii* (formerly *Sargassum wightii*) and 10.00 mm of inhibition in *Padina gymnospora* against *S. aureus*. However, no other organism (*P. aeruginosa*, *Salmonella typhi*, *K. pneumonia*, *B. subtilis*) were inhibited by the hexane extract of *S. swartzii* and *P. gymnospora*, except *B. subtilis*—10.67 mm of inhibition in *S. swartzii* (Rosaline et al. 2012).

In the study made by Chiao-Wei et al. (2011), the antibacterial activity of *n*-hexane, dichloromethane, and methanolic extracts of seaweeds, *Sargassum polycystum* and *Padina australis*, were examined using the disc diffusion and broth microdilution methods. Gram$^+$ bacteria, especially *Bacillus cereus* was more susceptible to the seaweed extracts (MIC = 0.130 to 0.065 mg mL^{-1}). Generally, *S. polycystum* extracts exhibited higher bacteriostatic activity (lower MICs) against all the tested bacterial strains when compared with *P. australis*. However, *P. australis* extracts showed a narrow spectrum of bactericidal activity against *B. cereus*. *n*-hexane extracts of *S. polycystum* exhibited promising bacteriostatic agents against *B. cereus* (MIC = 0.065 mg mL^{-1}) with MIC value lower than the standard MIC of potential antimicrobial drug (0.100 mg mL^{-1}).

Vijayabaskar et al. described a sulfated polysaccharide from the brown algae, *Sargassum swartzii* (Vijayabaskar et al. 2012). This sulfated polysaccharide showed a high percentage of carbohydrate (7.40%) and sulfate (5.3%). This sulfated polysaccharide also inhibited both Gram$^+$ and Gram$^-$ bacteria (zone of inhibition: 2–16 mm disc) (Vijayabaskar et al. 2012).

Methanolic and ethyl acetate extracts from eight different seaweeds collected from the Red Sea, Hurghada, Egypt (Salem et al. 2011) were screened for their antibacterial activities against both Gram$^+$ bacteria (*Staphylococcus aureus* and *Bacillus cereus*) and Gram$^-$ bacteria (*Escherichia coli*, *Enterococcus feacalis*, *Salmonella* sp., and *Pseudomonas aeruginosa*). The antibacterial activities were expressed as zone of inhibition and minimum inhibitory concentrations (MIC). Most of the algal extracts exhibited antibacterial activity against all the tested bacterial species. Among Gram$^+$ bacteria, *S. aureus* was the most sensitive to all the seaweed extracts. The higher antibacterial activity (indicated as zone of inhibition) was recorded for ethyl acetate extracts of *Sargassum dentifolium* and *Padina gymnospora* (14.3 and 17.8 mm, respectively); methanolic extracts of *Sargassum hystrix*, *S. dentifolium*, and *Polycladia myrica* (formerly

Cystoseira myrica) (13.8, 18, and 17 mm, respectively), while the inhibition zone of Chloromophenicol was 13.5 mm. Hence, the susceptibility of *S. auerus* to algal extracts was more pronounced when compared to the antibiotic Chloromophenicol. Next to *S. aureus*, *B. cereus* and *E. coli* were very susceptible to all the algal extracts used. Methanolic extracts of *S. hystrix*, *S. dentifolium*, and ethyl acetate extract of *P. myrica* had no antibacterial effect on *Salmonella* sp., while the other extracts showed higher clear zones when compared to the antibiotic Tetracycline. *E. feacalis* was the most resistant bacteria; it did not show any inhibition zones with any extracts from the studied marine algae. MIC of the tested marine algal extracts ranged from 5 to 50 mg mL^{-1}. Lowest MIC value was recorded for the methanolic and ethyl acetate extracts of *S. dentifolium*, *P. gymnospora*, ethyl acetate extracts of *S. hystrix*, *P. myrica*, and methanolic extract of *Sirophysalis trinodis* (formerly *Cystoseira trinodis*) (10 mg mL^{-1}) (Salem et al. 2011).

In the study done by El-Fatimy and Abdel-Moneim (2011), 34 marine algal species were collected and identified. The antibacterial activity of the most dominant species (*Padina pavonica*) was compared with some antibiotics. *P. pavonica* was the most dominant species in all samples, and methanolic crude extract were tested against *Escherichia coli* and *Staphylococcus aureus* bacteria and matched with some famous antibiotics, and all of the treatments affected *E. coli.*

Vijayabaskar and Shiyamala (2011) had investigated the antibacterial activities of methanolic extracts of two seaweeds—*Sargassum swartzii* (formerly *Sargassum wightii*) and *Turbinaria ornata*. The antibacterial activity was tested against various $Gram^{+}$ and $Gram^{-}$ human pathogenic microbes, including nine pathogens such as *Aeromonas hydrophila*, *Bacillus subtilis*, *Enterococcus faecalis*, *Escherichia coli*, *Klebsiella pneumoniae*, *Proteus vulgaris*, *Pseudomonas aeruginosa*, *Shigella flexneri*, and *Staphylococcus aureus*. This activity was due to active components polyphenols and these phenolic compounds affect growth and metabolism of bacteria according to their constitution and concentration. The findings suggest that methanol extracts of *T. ornata* could be utilized as a good source of antimicrobial agent.

A new compound, jolynamine, was isolated from the marine alga *Jolyna laminarioides* collected from the coast of Karachi, Pakistan. Furthermore, the methanolic extracts of both algae showed antimicrobial activities against various bacteria (Khan et al. 2011).

In the study developed by Devi et al. (2012), seaweed *Sargassum swartzii* (formerly *Sargassum wightii*) was screened for the potential bioactive natural substance against human bacterial pathogens. Crude extracts were made using three solvents (acetone, ethanol, and methanol) and then screened for antibacterial activity against human pathogens. The crude extracts were purified by silica gel column chromatography and five fractions obtained from each solvent were collected separately and tested for activity. The second fraction of purified ethanol extract showed maximum activity against seven human bacterial pathogens compared to other fractions of ethanol, methanol, and acetone. This was again subjected for purification by silica gel column chromatography and the three sub-fractions obtained were also tested for the activity. Of the three fractions, the third sub-fraction of ethanol extract showed the highest zone of inhibition against *Escherichia coli* (25.5 mm), followed by *Staphylococcus aureus* (22.85 mm), *Salmonella paratyphi* (19.1 mm), *Salmonella typhi* (18.5 mm), *Pseudomonas aeruginosa* (18.25 mm), *Vibrio cholerae* (17.5 mm), *Klebsiella pneumoniae* (16.15 mm), *Shigella sonnie* (15.2 mm), and the lowest zone of inhibition was observed against *Proteus* sp. (12 mm), and *Klebsiella* sp. (8.5 mm).

Antibacterial activity of methanolic extracts from several species of macroalgae collected from Moroccan Mediterranean coasts was evaluated against *Escherichia coli*, *Staphylococcus aureus*, and *Enterococcus faecalis* (Zbakh et al. 2012). Extract of *Cladostephus spongiosus* showed a broad inhibitory activity against *E. coli* (13 mm), *E. faecalis* (12 mm), and *S. aureus* (24 mm). Negative responses of *Cystoseira compressa* and *Cystoseira humilis* extracts are consistent with those reported by Gonzalez et al. (2001) concerning samples from the Canary Islands.

In the study done by Omar et al. (2012), marine algae were collected from the southern coast of Jeddah, Saudi Arabia, and the antibacterial activities of petroleum ether, diethyl ether, ethyl acetate, and methanol extracts of marine algae were studied. Their crude extracts were tested against different types of $Gram^{+}$ bacteria (*Bacillus subtilis*, Methicillin-resistant *Staphylococcus aureus*) and $Gram^{-}$ bacteria (*Escherichia coli*, *Klebsiella pneumoniae*, and *Pseudomonas aeruginosa*). All marine algae extracts tested exhibited a broad spectrum of antibacterial activity. The maximum inhibition activities were shown for extracts of

Padina pavonica and *Turbinaria triquetra*. The growth inhibitions of bacteria by *Sargassum portierianum* extracts were higher in samples collected during autumn than those investigated in summer. The tested microorganisms that were susceptible to the most effective extracts were further tested for the minimum inhibitory concentration (MIC). The MIC of the tested microorganisms was between 0.5 and 1.25 µg mL^{-1}.

Extraction yield is also impacted by solvent choice, which in turn affects efficacy of the antibacterial. Rajauria et al. (2012) reported different combinations of methanol and water (20–80%) to have a significant effect on the yield of antimicrobial and antioxidant polyphenolic compounds from the Irish brown seaweed *Himanthalia elongata*. The greatest yield (6.8%) was achieved using 60% methanol, compared to the lowest yield (1.2%) using 100% methanol. Disc diffusion and broth dilution tests showed the 60% methanol extract at a concentration of 60 mg mL^{-1} to be the most potent inhibitor of $Gram^+$ *Listeria monocytogenes* and *Enterococcus faecalis*; and $Gram^-$ *Pseudomonas aeruginosa* and *Salmonella abony*. The *H. elongata* extract had greater, or at least equal, zones of inhibition against all bacteria compared to the synthetic food preservatives sodium nitrite and sodium benzoate. Gupta et al. (2012) reported that extracts from *H. elongata* at 6% inhibited the growth of food spoilage microorganisms (*P. aeruginosa* and *E. faecalis*) and food pathogens (*L. monocytogenes* and *S. abony*). Lower concentrations of extract prolonged the lag phase and reduced both the exponential growth rate and final population densities of the culture. Seaweed extract at a concentration of 6% inhibited the growth of all four of the studied organisms.

In the studies done by El Baz et al. (2013), marine algal sulfolipids (SLs) presented a high growth inhibition of the bacterial strains (*B. subtilis* and *E. coli*) at the concentration of 100 µg/well. The highest bacterial growth inhibition was obtained by *Taonia atomaria* sulifolipids (15.0 mm) against *E. coli* and *Dictyota fasciola* (formerly *Dilophus fasciola*) sulfolipids (8 mm) showed the lowest growth inhibition against the same bacterium. Rajauria et al. (2013) report the antimicrobial activity of various methanolic extracts of *Himanthalia elongata* by using the disc diffusion method. Various food spoilage bacteria such as *Enterococcus faecalis* and *Pseudomonas aeruginosa*, and pathogenic bacteria such as *Listeria monocytogenes* and *Salmonella abony* were used to determine the antibacterial activities of brown seaweed aqueous methanolic extracts and synthetic compounds by disc diffusion assay. The zone of inhibition of seaweed extracts was measured by using the reference of the inhibition exhibited by synthetic food antimicrobials.

The antibacterial activity of ethanol, methanol, hexane, and acetone-based extracts of several seaweeds was investigated (Silva 2015). The disc diffusion method was used to evaluate the algae antimicrobial effect against standard strains of *Vibrio parahaemolyticus*, *Escherichia coli*, *Staphylococcus aureus*, *Pseudomonas aeruginosa*, *Salmonella enterica*, and five virulent antibiotic-resistant strains of *Vibrio brasiliensis*, *Vibrio xuii*, and *Vibrio navarrensis* (isolated from the hemolymph of *Litopenaeus vannamei*, the whiteleg shrimp). Only ethanol and methanol extracts were bioactive. The greatest inhibition halos were observed for ethanol extracts of *Padina gymnospora*, ranging from 10.2 (in *E. coli* culture) to 16.7 mm (in *V. brasiliensis* culture). *P. gymnospora* displayed vibrocidal activity, which inhibited the growth of all the *Vibrio* species tested. *S. aureus* and *S. enterica* were resistant to all extracts. The standard strains of *E. coli* and *P. aeruginosa* were susceptible only to the ethanol extract of *P. gymnospora*.

Mohandass et al. (2013) proposed the use of *Sargassum cinereum* extract as a reducing agent in the extracellular synthesis of silver nanoparticles. The MIC against *Staphylococcus aureus* was 2.5 µL (25 µg/disc), and against *Enterobacter aerogenes*, *Salmonella typhi*, and *Proteus vulgaris* were 100 µg/disc.

Crude methanolic and water extracts of several marine algal species collected from the western coast of Libya were evaluated for antibacterial activity against pathogenic bacteria (four $Gram^+$, four $Gram^-$) (Alghazeer et al. 2013). The extracts showed a significant antibacterial activity against $Gram^+$ (*Staphylococcus aureus*, *Bacillus subtilis*, *Bacillus* spp., and *Staphylococcus epidermidis*) as well as $Gram^-$ bacteria (*Escherichia coli*, *Salmonella typhi*, *Klebsiella* spp., and *Pseudomonas aeruginosa*). The algal aqueous and methanolic extracts displayed different degrees of antimicrobial activities against different bacteria; in some cases methanolic extracts showed higher antibacterial activity than aqueous extracts. Among the tested algae, brown algae, namely *Cystoseira crinita* exhibited the highest antibacterial activity. Overall, antibacterial activity of aqueous extracts was higher than that of methanol extracts. In most cases, aqueous extracts of all algae showed profoundly distinct antibacterial activity by having observable inhibition with diameters ranging from 11 to 18 mm on tested bacteria. A remarkable effect was obtained with *C. crinita* extract against *Bacillus* spp. (18 mm) and *Klebsiella* spp. and *S. typhi* (16 mm), compared to all other extracts.

Kim et al. (2013c) confirmed the antimicrobial activity of ethanol extracts of *Saccharina japonica* (formerly *Laminaria japonica*) against oral microbials. The MICs of ethanol extracts were 250 and 62.5 μg mL^{-1} against *Actinomyces naeslundii* and *Actinomyces odontolyticus*, respectively, and 250 and 62.5 μg mL^{-1} for *Fusobacterium nucleatum* and *Porphyromonas gingivalis*, respectively. The MBCs of *A. naeslundii* and *A. odontolyticus* were 500 and 250 μg mL^{-1}, respectively. A dose-dependent effect and a change in the cell surface texture of *Streptococcus mutans*, *A. odontolyticus*, and *P. gingivalis* was observed.

Rajauria and Abu-Ghannam (2013) extracted fucoxanthin from the Irish brown seaweed *Himanthalia elongata* using diethyl ether, n-hexane, and chloroform, and purified the crude extract using preparative thin layer chromatography (TLC). In disc diffusion assays, the purified extract was shown to be a potent inhibitor of *Listeria monocytogenes*, with an inhibition zone of 10.27 mm at a concentration of 1 mg mL^{-1} (25 μg/disc). The extract was 98.4% as effective as an analytical grade fucoxanthin standard (inhibition zone 10.89 mm). Similarly, Deyab and Abou-Dobara (2013) extracted fucoxanthin from the brown seaweed *Turbinaria triquetra* with chloroform and methanol. Extracts were purified by silica column chromatography and identified by nuclear magnetic resonance spectroscopy. *Turbinaria triquetra* showed the greatest bacterial inhibition. Zones of inhibition for *Escherichia coli*, *Bacillus cereus*, *Bacillus subtilis*, *Klebsiella pneumoniae*, *Staphylococcus aureus*, and *Pseudomonasa eruginosa* ranged from 4.0 mm to 7.0 mm (100 μg mL^{-1} extract/disc) and in the case of some $Gram^{-}$ species, were equivalent to antibiotic standards.

Active antibacterial extracts from different brown algae have been found to be made up of saturated and unsaturated fatty acids with a predominance of myristic, palmitic, oleic, and eicosapentaenoic acids in *Sargassum muticum*. Pure palmitic acid showed antibacterial activity at 44 μg mL^{-1} (Bazes et al. 2009). So, the antibacterial activities of the algae tested could be attributed to the type and amount of free fatty acids which have a role in the overall defense against the studied pathogenic $Gram^{+}$ and $Gram^{-}$ bacteria (Benkendorff et al. 2005). Terpenoid content is considered to contribute to antibacterial activity (Fenical and Paul 1984). In the investigation made by Al-Saif et al. (2014), palmitic acid was observed as a major component of the total fatty acids in *Dictyota ciliolata*, and results also indicated that the extracts of this macroalgae was more efficient against the tested bacterial strains.

Al Hazzani et al. (2014) reported higher *in vitro* antimicrobial activity of methanol and acetone extracts from the brown algae (*Saccharina japonica*—formerly *Laminaria japonica*, *Undaria pinnatifida*, and *Ecklonia bicyclis*—formerly *Eisenia bicyclis*) than from the red alga *P. tenera* against $Gram^{+}$ and $Gram^{-}$ bacteria; some were antibiotic-resistant such as Methicillin-resistant *S. aureus* and *P. aeruginosa*, and resistant against yeast, *C. albicans*.

Cox et al. (2014) found that the hydrophilic extract from *Himanthalia elongata* provided up to 100% inhibition of *Salmonella abony* and *Listeria monocytogenes*. Concentration-dependent antibacterial activity was observed in the range of 0.125–8 mg mL^{-1}, but no inhibition was detected under 1 mg mL^{-1}. The extracts showed bactericidal effect in carbohydrate model food systems and bacteriostatic effect in protein model food systems (at 1–10%). Sodium benzoate exhibited greater inhibition at 0.5–2 mg mL^{-1} and showed bacteriostatic effect in carbohydrate medium, whereas the extract had a bactericidal effect.

Plaza et al. (2010) isolated eight compounds from the Pelagophyceae (Ochrophyta) alga *Chrysophaeum taylori* using hexane, chloroform, and methanol. The compounds were of a new chemical structural class which they named chrysophaentins, which consisted of two polyhydroxylated, polyhalogenated ω, ω1-diarylbutene units connected via two ether bonds. The chrysophaentins exerted powerful inhibition *in vitro* of Methicillin-resistant *Staphylococcus aureus* (MIC = 1.5 μg mL^{-1}), Vancomycin-resistant *Enterococcus faecium* (MIC = 2.9 μg mL^{-1}), and multidrug-resistant *Staphylococcus aureus* (MIC = 1.3 μg mL^{-1}). The pharmacological mechanism of action of chrysophaentin is proposed to be unlike any existing antibacterial agent. The functional groups within chrysophaentin act as enzyme inhibitors by binding with guanosine triphosphatase in bacterial cells. This prevents the synthesis of a protein called FtsZ (filamenting temperature-sensitive mutant Z), required for bacterial cell division. The development of antibiotics and other antibacterial products from chrysophaentin continues to be an important area of marine pharmacological investigation (Keffer et al. 2013, Li and Ma 2015).

Moronery et al. (2013) added a spray-dried *Laminaria digitata* extract (9.3% laminarin, 7.8% fucoidan) to minced pork patties at levels of 0.01–0.5% (w/w). After storage in a modified atmosphere for 14 days at 4ºC, at 0.5% the extract exerted the greatest lipid pro-oxidant activity in fresh patties, whereas the lipid oxidation was decreased in cooked patties, but the microbiological status, pH, and water-holding capacity were not influenced.

Marine algae could be a source of therapeutic agents for chronic gastritis and peptic ulceration. Lee et al. (2013) screened 27 Korean species of seaweed for potential anti-*Helicobacter pylori* activity, and seven showed strong inhibitory activity based on the agar diffusion method. The strongest activity was observed for ethanol extracts from *Ishige okamurae*. The inhibition zone of this extract was 9.0 mm at 1 mg/disc, the MIC was 12 μg mL^{-1} based on the broth microdilution assay, and the free urease assay confirmed that the 80% methanol extracts had 75.4% inhibition at 0.1 mg mL^{-1}. Both the *I. okamurae* phenolic compounds and the nitrogen compounds of the extract significantly inhibited *H. pylori*. No toxicity was observed in a study with BALB/c mice at a dose of 5 g kg^{-1} body weight.

Meillisa et al. (2013) proposed subcritical water hydrolysis (200–280ºC, 1.3–6.0 MPa) to extract antibacterial compounds from *Saccharina japonica*, previously de-oiled by SC-CO_2 extraction. Strong antibacterial activity against two $Gram^-$ (*Escherichia coli* and *Salmonella typhimurium*) and two $Gram^+$ bacteria (*S. aureus* and *B. cereus*) was found using acetic acid to aid in the hydrolysis. The MIC values ranged from 1.60 to 3.50 mg mL^{-1}, and the optimal hydrolysis temperatures were 240ºC against *B. cereus* and *S. typhimurium* and 280ºC against *E. coli* and *S. aureus*. Since acetic acid had antibacterial activity at 1%, and only 0.48–0.6% acetic acid remained in the hydrolysate, the addition of acetic acid in the hydrolysis process can improve the extraction of antibacterial substances from the alga.

Oliveira et al. (2014) performed a 40-day experiment consisting of the inclusion of the seaweed *Ascophyllum nodosum* meals at a dose of 20 g kg^{-1} on the diet of Nile tilapia (*Oreochromis niloticus*) inoculated with the bacterium *Aeromonas hydrophila*. The width was greater for the treatment with the algal meal, but there was no influence on the performance parameters of the fingerlings. The occurrence of lesions in animals inoculated with *A. hydrophila* and fed with the alga was lower and declined in a shorter period of time than in the control group; prevention of hepatopancreatic congestion in infected animals was also observed.

Tanniou et al. (2014) tested the antibacterial activity of phenolic products from invasive seaweed *Sargassum muticum*, collected in several European countries, against *Vibrio aestuarianus*, *Vibrio anguillarum*, and *Vibrio parahaemolyticus*. Extracts were active against three marine bacterial strains (*V. aestuarianus*, *V. anguillarum*, and *V. parahaemolyticus*) and three terrestrial strains (*E. coli*, *S. aureus*, and *P. aeruginosa*). In this test, the most active extracts were no longer those from the gradient extremities. Indeed, if we look at the crude extracts, it was especially in those from Ireland and Portugal that the highest activities were found.

In the works of Ismail et al. (2014), the antibacterial activity of brown alga *Zonaria tournefortii* extracts against several multi-resistant *Staphylococcus aureus* were investigated. The aqueous crude extract of *Z. tournefortii* was tested at different concentrations against 13-multi-resistant strains of *S. aureus*; crude extracts of 5 mg mL^{-1} concentration showed strong antibacterial activity against nine strains of *S. aureus* tested with 9–14 mm zone of inhibition. The fraction obtained with 80/20 (water/methanol) elution showed the strongest antibacterial activity especially against *S. aureus* strains 1, 3, and 25, with inhibition zone reaching 25 mm and MIC ranging from 0.25 mg mL^{-1} to 1 mg mL^{-1}. Diameters of inhibition zones with fraction F2 against *S. aureus* strains 1, 21, and 25 were strongest than positive control. Fraction F3, less polar than F2 and obtained by (60:40) (water/methanol), was very inhibitory to *S. aureus* with MIC ranging from 0.5 to 2 mg mL^{-1} (Ismail et al. 2014).

Siahaan et al. (2014) prepared *Saccharina japonica* as a powder (500–900 μm) for the physical adsorption and further release of allyl isothiocyanate, an inhibitor of food-borne bacteria. When the alga was de-oiled, higher adsorption of allyl isothiocyanate was observed in comparison to raw alga. No loss of activity was detected against *E. coli*, *S. typhimurium*, or *B. cereus*, and only a nominal activity against *S. aureus* was observed.

Nogueira et al. (2014) reported modulation of antibiotic activity between the *Padina sanctae-crucis* ethanolic and methanolic extracts and *Escherichia coli* and *Staphylococcus aureus* (bacteria), and a moderate modulatory effect against these microorganisms and *Pseudomonas aeruginosa* (bacterium), *Candida tropicalis*, and *Candida kruseim* (yeasts).

In a study done by Thirunavukkarasu et al. (2014), the main objective was to isolate bioactive molecules from marine seaweed and check the antimicrobial activity against the fish pathogenic bacteria. Based on the disc diffusion method, *Sargassum swartzii* (formerly *Sargassum wightii*) showed a better antimicrobial activity than other seaweed extracts. Chloroform extract showed a minimum zone of inhibition (21.33 mm). The acetone extract of *S. swartzii* produced a maximum zone of inhibition (26 mm) against *Vibrio anguillarum*. Methanol extract of *S. swartzii* showed maximum zone of inhibition (32 mm) against *Vibrio parahaemolyticus*. Ethyl acetate extract showed maximum zone of inhibition against *Vibrio harveyi* (24.66 mm). No zone of inhibition was observed in aqueous extract of all the seaweeds against *Vibrio* sp.

Bioactive extracts from four species of brown seaweeds (*Sargassum plagiophyllum*, *Sargassum fulvellum*, *Padina australis*, and *Sargassum aquifolium*—formerly *Sargassum binderi*) of Malaysia origin were tested *in vitro* for their antibacterial activities against Gram$^+$ bacteria (*Bacillus subtilis* and *Staphylococcus aureus*) and Gram$^-$ bacteria (*Pseudomonas aeruginosa* and *Escherichia coli*) (Jaswir et al. 2014). Four solvents (methanol, acetone, ethyl acetate, and chloroform) were used during the preliminary screening stage to obtain crude extracts. None of the crude extracts displayed antimicrobial activity against the Gram$^-$ bacteria, while the Gram$^+$ bacteria were inhibited. Methanol extract of *S. plagiophyllum* was the best in terms of yield (4.72%) and antimicrobial activity with inhibition zone of 12 mm against *B. subtilis*. Chloroform fraction gave highest inhibition zone of 17 mm against *B. subtilis*.

Moroney et al. (2015) reported an enhancement of pork quality in meat from pigs fed extracts of *Laminaria digitata* containing laminarin and fucoidan for three weeks prior to slaughter. Animals were fed either 450 mg or 900 mg laminarin and fucoidan per kg feed. In both cases, pork meat was found to have an improved fatty acid profile without loss of lipid stability. Saturated fatty acid content was significantly lowered and lipid oxidation was reduced. Aquaculture food products can be a source of food-borne pathogenic and spoilage bacteria. *Vibrio* genera are a common source of food-borne illness in fish products. Seaweeds have been shown to have antibacterial properties against many species that infect farmed fish, which in turn reduces the occurrence of pathogenic bacteria in the final food product (Vatsos and Rebours 2015).

The study conducted by Singh and Raadha (2015) was made to identify marine algae with antimicrobial potency. The powdered *Hydroclathrus clathratus* samples were extracted with a series of solvents of increasing polarity and the crude extracts were screened for their antimicrobial properties. It could either be due to the presence of a compound with a broad spectrum of antibacterial activity, or could be due to the presence of more than one compound, each having its own target of action. It was observed that out of all the solvents, only in methanolic extract of *H. clathratus*, the antibacterial compound had been extracted, and no activity against fungi was observed. It could either be due to the presence of a compound with a broad spectrum of antibacterial activity, or due to the presence of more than one compound, each having its own target of action (Singh and Raadha 2015).

Beaulieu et al. (2015) extracted antibacterial peptides (> 10 kDa mass) from the brown seaweed *Saccharina longicruris* by enzymatic hydrolysis with trypsin. Liquid chromatography-tandem mass spectrometry identified the sub-fractions as peptide precursors to proteins similar to ubiquitin, leucine, histone, and a ribosomal structure, which form part of the innate immune defence of the seaweed. Maximum specific growth rate of the food spoilage bacterium *Staphylococcus aureus* was significantly inhibited by the hydrolysate at concentrations of 0.31 to 2.5 mg mL^{-1} making it a potential agent for food preservation.

Sivagnanam et al. (2015) proposed the supercritical (SC)-CO_2 (45°C, 250 bar) extraction using ethanol as a co-solvent to produce extracts with phenolic content and antioxidant properties higher than those with conventional solvents from brown seaweed (*Saccharina japonica* and *Sargassum horneri*). However, the acetone-methanol extracts exhibited better antimicrobial activities. Among the different seaweeds extracts, the exhibited antimicrobial activity against *Escherichia coli* and *Staphylococcus aureus*, at 28 and 24 mm, respectively, was the highest in the extract of acetone–methanol of *S. horneri*, followed by the acetone–methanol extract of *S. japonica* against *E. coli* and *S. aureus*, at 21 and 20 mm, respectively. No activity was found against *Pseudomonas aeruginosa* and *Salmonella typhimurium*. The SC-CO_2 with ethanol extract also showed good activity similar to the solvent extractions. The SC-CO_2 with ethanol extract of

S. japonica showed good antimicrobial activity against *E. coli* at 18 mm, *Listeria monocytogenes* at 12 mm, *Bacillus cereus* at 10 mm, and *S. aureus* at 18 mm. The SC-CO_2 with ethanol extract of *S. horneri* showed good antimicrobial activity against *E. coli* at 21 mm, *L. monocytogenes* at 14 mm, *B. cereus* at 12 mm, and *S. aureus* at 20 mm. These results clearly show that the *S. horneri* extract had a higher amount of activity than that of *S. japonica*. The acetone–methanol showed more activity than the SC-CO_2–ethanol extracts.

Different solvent extracts of a brown algae, *Sagassum muticum*, collected from Pudumadam, Ramanathapuram, Tamil Nadu, India, were subjected to different bioassays, including antibacterial activity (Moorthi and Balasubramanian 2015). The results revealed that acetone extract had unveiled the maximum of 11 mm zone of inhibition at 40 μL against *Shigella fleschneri*. A similar zone of inhibition (11 mm) was also observed at 50 μL against *Micrococcus* sp. and *S. fleschneri*. Followed by acetone extract, chloroform extract also contributed 11 mm zone of inhibition against *S. fleschneri* and *Salmonella paratyphi* B at 40 and 50 μL, respectively. Besides, methanol extracts revealed meager antibacterial activity (9 mm). Similarly, the antibacterial and synergistic activity of an *Ecklonia cava* extract was investigated against the antibiotic-resistant species *Streptococcus parauberis* (Eom et al. 2015b). For this purpose, a methanolic extract of *E. cava* was prepared, and then tested for its antibacterial activity against *S. parauberis*, showing good growth inhibition. In addition, it has been demonstrated that the *n*-hexane soluble fraction of the *E. cava* methanolic extract showed the highest antibacterial effect with a minimum inhibitory concentration ranging from 256 to 1024 μg mL^{-1}. The authors concluded that *E. cava* represents a great source of antibacterial molecules.

The extracts of *Cystoseira compressa* were tested for their antimicrobial and antioxidant activities in the study made by Güner et al. (2015), with specimens collected from the coast of Urla (Izmir, Turkey). All extracts exhibited moderate antimicrobial activity against tested microorganisms (minimum inhibitory concentration—MIC—ranges are 32–256 μg mL^{-1}).

The study of Kolsi et al. (2015) was conducted to evaluate the antimicrobial activity of hexane, ethyl acetate, and methanol extracts of several marine species from Tunisian coastline (Chebba and Sfax). These spices were tested against eight human pathogenic bacteria—Gram$^-$ (*Escherichia coli*, *Listeria monocytogene*, *Salmonella enterica*, *Agrobacterium tumefaciens*, *Pseudomonas aerigunosa*), and Gram$^+$ (*Staphylococus aureus* and *Micrococcus luteus*). Methanol extract of *Dictyota dichotoma* was moderately active (10 mm) against *S. aureus* and *M. luteus*; Ethyl acetate extract of *D. dichotoma* was moderately active (10 mm) only against *M. luteus*. Ethyl acetate extract of *Cystoseira crinita* was moderately active (10 mm) only against *S. aureus*, and the methanolic extract was highly active (15 mm) against the same bacterium. Methanol extract of *Cystoseira barbata* was highly active (15 mm) against *M. luteus*. Methanol extract of *Dictyopteris polypodioides* (formerly *Dictyopteris membranacea*) was highly active (15 mm) only against *S. aureus*. Finally, methanolic extract of *Sargassum vulgare* was highly active (15 mm) against *M. luteus*.

Similarly, the antimicrobial activity of macroalgae extracts from the Moroccan Atlantic coast was investigated (El Wahidi et al. 2015). 10 marine macroalgae collected from the Moroccan's Atlantic coast were used to prepare dichloromethane and ethanolic extracts, which were tested for their antimicrobial activities against the human pathogenic bacteria: *B. subtilis*, *S. aureus*, *E. coli*, and *P. aeruginosa*, as well as against the two pathogenic yeasts—*Candida albicans* and *Cryptococcus neoformans* (see also Chapter 7). The obtained results showed great and various inhibitory activities depending on the seaweed species and the solvent used. It has been shown that *Cystoseira brachycarpa*, *Cystoseira compressa*, and *Fucus vesiculosus* exhibited the highest antimicrobial activities with a broad spectrum of microbial growth inhibition. The authors concluded about the promising potential of these seaweeds as sources of antimicrobial compounds.

Despite its worldwide distribution, the potential use of *Dictyopteris polypodioides* in pharmacology or as an alginate target remains unexplored due to a lack of research data. Recently however, one of the few reports about bioactivity was performed by Karaki et al. (2013) on antioxidant and anticoagulant activities of some polysaccharides isolated from this species. Otherwise, recent studies on other species belonging to *Dictyopteris* genus have investigated antitumor, antioxidant, and antimicrobial activities (Magalhães et al. 2011, Kim et al. 2014). In this context, *D. polypodioides* was prospected for its antioxidant and biological activities of methanolic extracts against bacteria, yeasts, and fungi with clinical and agricultural relevance as well as for alginate yield and chemical properties. *D. polypodioides* exhibited high activity against the most human pathogenic bacteria tested (*Staphylococcus aureus*, *Bacillus cereus*, *Klebsiella pneumoniae*, *Pseudomonas aeruginosa*, *Escherichia coli*, *Enterococcus faecalis*, *Salmonella* sp.) (Belattmania et al. 2016).

To evaluate the antibacterial activity of various solvent extracts of the marine brown alga *Spatoglossum asperum*, a study was carried out (Pandithurai et al. 2015c). Various organic solvent extracts (aqueous, methanol, chloroform, ethyl acetate, and hexane) of the marine brown alga *S. asperum* were tested against *Bacillus cereus*, *Bacillus subtilis*, *Staphylococcus aureus*, *Klebsiella pneumonia*, *Pseudomonas aeruginosa*, and *Salmonella typhi* by disc diffusion method. The chloroform and methanolic extract showed maximum activity against *S. aureus*—86.88 and 84.94%, respectively—the aqueous extract showed moderate activity against *B. cereus* (54.05%) and ethyl acetate showed minimum activity against *B. cereus* (37.57%), whereas no activity was observed in hexane extract. The study established the methanol extracts of brown seaweed *S. asperum* was highly effective against most of the bacterial pathogens tested. The study showed that the aqueous extract showed moderate antibacterial activity against *B. cereus* 10.27 mm (54.05%), *K. pneumoniae* 8.28 mm (51.75%) and *S. typhi* 10 mm (50%) followed by a lesser activity against *Staphylococcus* 8.05 mm (47.35%), *P. aeruginosa* 9.36 mm (46.80%), and *B. subtilis* 7.12 mm (41.88%). The chloroform extract of *S. asperum* shows highest antibacterial activity among the tested extract against *S. aureus* by showing 14.77 mm (86.88%) followed by moderate activity against *B. subtilis* 10 mm (58.82%) and lesser activity against *B. cereus* 8.17 mm (43%), *S. typhi* 8.05 mm (40.25%), and *P. aeruginosa* 7.17 mm (35.85%). The ethyl acetate extract showed moderate activity against *B. subtilis* 9 mm (52.94%) and *K. pneumoniae* 8 mm (50%). All other species such as *S. typhi* 9.17 mm (45.85%), *S. aureus* 7 mm (41.17%), *P. aeruginosa* 8 mm (41.17%) and *B. cereus* 7.14 mm (37.57%) showed lesser inhibitory activity (Pandithurai et al. 2015c).

In the works of Sivagnanavelmurugan et al. (2015) the polysaccharide fucoidan was extracted from *Sargassum swartzii* (formerly *Sargassum wightii*) and its antibacterial activity was screened by agar well diffusion method. The maximum zone of inhibition observed was 15.66 mm in 20 mg mL^{-1} concentration against *Vibrio parahaemolyticus*. The minimum inhibitory concentration (MIC) of the fucoidan was 12 mg mL^{-1} against *V. parahaemolyticus*.

In the works of Thanigaivel (2015) the bioactive compounds and the toxicity profile of aqueous and ethanol extracts of *Padina gymnospora* and *Sargassum cinereum* species were assessed for the treatment of bacterial (*Pseudomonas aeruginosa*) fish infection. The optimum exposure to seaweed extracts varied with different concentrations, from 100 to 500 mg L^{-1} of dried solvent extracts dispersed in 10 L of water. It was found to be non-toxic up to 250 mg L^{-1}.

In a research work performed by El Shafay et al. (2016) with seaweeds collected from the Red Sea (Egypt), the highest inhibition activity among all extracts was obtained with 100 μL diethyl ether extract *Sargassum fusiforme* against *Staphylococcus aureus*; 2 and 50 μL ethanol extract of *Sargassum vulgare* against *Klebsiella pneumoniae*. Diethyl ether extract of *S. vulgare* showed also inhibition activity against *Shigella flexneri*, *Escherichia coli*, and *Pseudomonas aeruginosa*. However, the methanol extract of *S. vulgare* did not show any noticeable activity against the multidrug resistant bacteria tested, except *K. pneumoniae* and *S. flexneri*. Ethanol extracts of *S. vulgare* show inhibition activity against *P. aeruginosa* and high activity against *K. pneumoniae*, while chloroform extract of *S. vulgare* shows activity against *P. aeruginosa*, *K. pneumoniae*, and *S. aureus* (El Shafay et al. 2016).

Wei et al. (2016) reported that low molecular weight phlorotannins extracted from *Sargassum thunbergii* damaged the cell membrane and cell wall of *Vibrio parahaemolyticus*, causing cytoplasm leakage and deconstruction of membrane permeability. The study suggested that low molecular weight phlorotannins from algae could potentially be used in food safety control and aquacultural drugs. Lee et al. (2014) tested a range of solvent extracts from the brown seaweed, Arame (*Ecklonia bicyclis*—formerly *Eisenia bicyclis*) against antibiotic resistant *Propionibacterium*-related acne. A phlorofucofuroeckol compound (phlorotannin with an alcohol substituent) exhibited the most potent antibacterial activity with an MIC of 32 μg mL^{-1}, while also significantly reversing the resistance of *Propionibacterium* to Erythromycin and Lincomycin. The same research group tested the activity of phlorofucofuroeckol from *E. bicyclis* against Methicillin-resistant *Staphylococcus aureus*. Phlorofucofuroeckol suppressed mecI, mecR1, and mecA gene expression in the resistant *S. aureus* cells. These three genes regulate the expression of Methicillin resistance in bacteria. This resulted in suppression of Penicillin-binding protein 2a production, which is considered the main mechanism by which Methicillin-resistant *S. aureus* strains resist Methicillin (Eom et al. 2014, Lee and Kim 2015). Phlorotannins and their derivatives offer a potentially useful source of natural antibacterial agents for food and medical applications.

A recent study by Dussault et al. (2016) explored the potential of developing several commonly consumed Pacific Island seaweeds into food preservation agents. In broth dilution assays, methanol extracts of the brown species, *Padina* and *Dictyota*, were found to inhibit the growth of Gram$^+$ foodborne pathogens *Listeria monocytogenes*, *Bacillus cereus*, and *Staphylococcus aureus* at low concentrations ($\leq$ 500 µg mL^{-1}). However, the extracts had no activity against Gram$^-$ species, possibly due to the inability of the moderate to low polarity, hydrophobic extracts to breach the hydrophilic lipopolysaccharide, Gram$^-$ bacterial membrane. The extracts are not known to have any toxicity at the concentrations used, making them good candidates for incorporation into foods prone to Gram$^+$ bacterial growth.

In a large study of bioprospecting of several algal species from the Aegean Sea, the antibacterial activity of macroalgal extracts was evaluated (Montalvão et al. 2016). The most significant results were achieved by ethanolic extract of *Cystoseira compressa* with growth inhibition of *Staphylococcus aureus* (50%, at 100 µg mL^{-1}), *Petalonia fascia* extract with 50% of growth inhibition, and *Halopteris scoparia* with 46% of growth inhibition.

In the research work done by Alves et al. (2016b), the antibacterial of *Bifurcaria bifurcata* was evaluated. The largest halo was achieved by the methanolic fraction against *Pseudomonas aeruginosa* (11.3 mm), followed by the dichloromethane fraction against *Escherichia coli* and *P. aeruginosa*, with 8.3 mm for both microorganisms. The methanolic fraction also presented activity against both *E. coli* and *Staphylococcus aureus*. This fraction exhibited stronger and broader activity in comparison to dichloromethane fraction, presenting antimicrobial activity against all studied bacteria. Conversely, dichloromethane extract only showed activity against two Gram$^-$ bacteria (Alves et al. 2016b).

Rodríguez-Martínez et al. (2016) formulated a new, active biodegradable film based on polylactic acid and brown seaweed. An extrusion process was used to incorporate 8% dried *Fucus spiralis* seaweed extract and 0.5% natural sorbic acid into a polylactic acid based biodegradable film. The migration values of the film components into food simulants were found to be generally within the maximum acceptable EU limit. It was concluded that the film was suitable for further development as a packaged food protectant and for shelf-life extension. *F. spiralis* was selected because it showed the greatest inhibition in disc diffusion assays of the foodborne bacteria *Staphylococcus aureus*, *Bacillus cereus*, *Bacillus subtilis*, *Pseudomonas fluorescens*, *Escherichia coli*, *Klebsiella pneumoniae*, *Aeromonas hydrophila*, *Vibrio alginolyticus*, and *Vibrio parahaemolyticus*, when compared to other seaweed genera, *Ascophyllum* and *Bifurcaria*, among others. A concentration of only 10 µg mL^{-1} of *F. spiralis* was required to exert zones of inhibition up to 15 mm (for *B. subtilis*). Although sorbic acid has antimicrobial properties against fungi and bacteria, the addition of *F. spiralis* extract reduces the amount required, and broadens the scope of antibacterial activity. Sorbic acid was first isolated from mountain ash berries, and has antimicrobial action mainly against yeasts and molds, but is selective against bacteria (Lück and Jager 2012). It can also be detected as an acrid taste by some consumers. The addition of small amounts of dried seaweeds, or antibacterial extracts of marine algal compounds to active food-films could minimize the volume of traditional preservatives required. This could reduce costs and bring concomitant nutritional benefits.

In the work done by Belattmania et al. (2016) the main objective was to evaluate antibacterial, antifungal, and antioxidant activities as well as alginate yield and quality of the brown seaweed *Dictyopteris polypodioides*, collected from the Moroccan Atlantic coast. The results showed that *D. polypodioides* exhibited high activity against the most human pathogenic bacteria tested (*Staphylococcus aureus*, *Bacillus cereus*, *Klebsiella pneumoniae*, *Pseudomonas aeruginosa*, *Escherichia coli*, *Enterococcus faecalis*, *Salmonella* sp.). Maximum zone of inhibition was observed against *Bacillus cereus* (36 mm), while minimum activity was noted against *Escherichia coli* (10 mm). In the study performed by Belattmania et al. (2016b), the brown seaweed *Cystoseira humilis* harvested from the Atlantic coast of Morocco has been investigated for fatty acid composition, as well as for antioxidant and antibacterial potentials. The antibacterial activity was limited to *Staphylococcus aureus* and *Bacillus cereus* among all tested pathogenic bacterial strains.

In the works of Le Lann et al. (2016), the extracts of *Halydris saliquosa* showed antibacterial activity at a concentration of 5 mg mL^{-1}, and the ethyl acetate fraction showed a strong bactericidal activity. Indeed, this fraction showed a reduction higher than 5 log of the initial bacterial concentration with *Pseudomonas aeruginosa* and *Escherichia coli* (< 1 UFC mL^{-1}). Concerning *Staphylococcus aureus*, ethyl acetate fraction showed a bactericidal activity with a reduction of 4.5 log of the initial bacterial concentration (75 UFC mL^{-1}) (see Le Lann et al. 2016).

In the work done by Khan (2016), with seaweeds collected from the Karachi coast of Pakistan, MIC of tested brown seaweed extracts ranged from 5 to 30 mg mL^{-1}. Among them, the lowest MIC value (5 mg mL^{-1}) was recorded in *Taonia atomaria* (as *Dictyota ciliata*) sampled from Buleji beach against *E. coli* in the extractions of ethanol, and against *Salmonella typhi* in *Padina pavonica* and *Spatoglossum variabile* sampled from Buleji and Manora beaches, respectively. MBC of tested seaweed extracts ranged from 10 to 40 mg mL^{-1}. Highest MBC value (40 mg mL^{-1}) was recorded in the ethanol extraction of *T. atomaria* (from Mubarak Village) against *E. coli*, and in the same species (Mubarak Village), *Iyengaria stellata* (from Manora beach), *Sargassum boveanum* (from Manora beach), and *Stoechospermum polypodioides* (as *Stoechospermum marginatum*) (Paradise Point and Manora beaches) against *S. typhi*. In acetone extractions the highest MBC (40 mg mL^{-1}) was recorded in *Polycladia indica* (as *Cystoseira indica*) and *Sargassum swartzii* (as *Sargassum wightii*) (Paradise Point), *S. boveanum* (Manora), and *Padina pavonica* (Buleji and Mubarak Village) against *E. coli*. The highest MBC value was recorded in *T. atomaria* (Mubarak Village), *Padina pavonica* (Buleji, Manora, and Mubarak Village), and *S. variabile* (Manora) against *S. typhi* (Khan 2016).

In the study done by Akremi et al. (2017), the Mediterranean brown alga, *Dictyopteris polypodioides* (as *Dictyopteris membranacea*), was investigated. The total phenolic, flavonoid, and tannin contents were quantified, and antibacterial, antifungal, and antitubercular activities of the crude extract and semi-purified ethanol, acetone, and methanol/CH_2Cl_2 fractions were determined. The antimicrobial activity was determined by the agar diffusion method against six strains of $Gram^+$, two strains of $Gram^-$ bacteria, and one yeast strain. Ethanol and acetone fractions exhibited considerable antibacterial activity.

8.6 Antibacterial Activity of Rhodophyta (Red Algae)

In a very early study, Kamimoto (1956) studied the antimicrobial effects of a number of Japanese seaweeds against a small panel of *Mycobacterium* species, including *Mycobacterium tuberculosis*, *Mycobacterium avium*, and *Mycobacterium phlei*. The organic solvent extracts of *Gelidium amansii* and *Pterocladiella capillacea* (formerly *Gelidium capillaceum*) were found to show some inhibition towards *M. avium* and *M. tuberculosis*. *Gelidium filicinum* collected from Chile also exhibited some antimycobacterial effect against *M. tuberculosis* and *Mycobacterium smegmatis* (Maurer 1965).

Chesters and Stott (1956), Roos (1957), and Ikawa et al. (1973) have reported that *Mastocarpus stellatus* (formerly *Gigartina stellata*) and *Ceramium virgatum* (formerly *Ceramium rubrum*) possessed antibacterial properties, and Roos (1957) also reported that *Griffithsia flosculosa* was an antibiotic-producing alga.

Halogenated metabolites from the red alga *Laurencia* species were tested for antibacterial activity against 22 strains of human pathogenic bacteria, including seven strains of antibiotic-resistant bacteria. Laurinterol (Irie et al. 1966), isolaurinterol (Irie et al. 1970), *allo*-laurinterol (Kazlauskas et al. 1976), cupalaurenol (Ichiba and Higa 1986), and 2,3,5,6-tetrbromoindol (Carter et al. 1978) displayed a wide spectrum of antibacterial activity against gram positive bacteria, including Methicillin-resistant *Staphylococcus aureus*, Penicillin-resistant *Streptococus pneumonia*, Vancomycin-resistant *Enterococcus faecalis* and *Enterococcus faecium*. Laurinterol and allo-laurinterol were particularly effective (Vairappan et al. 2004).

Elatol, a halogenated sesquiterepene alcohol from the alga *Coronaphycus elatus* (formerly *Laurencia elata*) (Sims et al. 1974), inhibited six species of human pathogenic bacteria with significant antibacterial activities against *Staphylococcus epidermis*, *Klebsiella pneumonia*, and *Salmonella* sp. (Vairappan 2003). Iso-obtusol from the red alga *Laurencia obtusa* (Gonzalez et al. 1976, 1979) exhibited antibacterial activity against four bacterial species with significant activity against *K. pneumonia* and *Salmonella* sp.

Hornsey and Hide (1974) screened 151 species of seaweed for the production of antibiotics. Of these, *Asparagopsis armata*, *Bonnemaisonia asparagoides*, *Bonnemaisonia hamifera*, *Chondrus crispus*, *Dilsea carnosa*, *Gloiosiphonia capillaris*, *Sphondylothamnion multifidum*, and most members of the family Rhodomelaceae exhibited high antibacterial potential against the test bacteria, such as *Staphylococcus aureus*, *Escherichia coli*, *Bacillus subtilis*, *Streptococcus pyogenes*, and *Proteus morganii*. Although the production of antibiotics would appear to be a characteristic of several families, it has not been possible to establish any major correlation between taxonomy and antibiotic production. In the case of two closely related and morphologically similar species—*C. crispus* and *Mastocarpus stellatus* (formerly *Gigartina stellata*), the former possesses considerable degrees of antimicrobial activity, whilst the latter exhibits no such activity. The results also indicate that the production of antibiotics by the algae is affected by the season of the year (Hornsey and Hide 1974). Cycloeudesmol is an antibiotic cycloproane containing sesquiterpene; it was isolated from the marine alga *Chondria oppositiclada* (Fenical and Sims 1974). Cycloeudesmol was found to be a potent antibiotic against *Staphylococcus aureus* and *Candida albicans* (yeast).

In a study done by Hornsey and Hide (1976), 11 known antibiotic-producing species of British marine algae were quantitatively screened at monthly intervals for the presence of antibacterial compounds. The antibacterial components of the algae were extracted with acetone and then assayed against *Staphylococcus aureus*. The results indicate that seasonal variation in antibiotic production occurs in some marine algae, and also that four main patterns of production can be discerned. These are—(a) the *Polysiphonia* type, in which antibiotic production occurs uniformly throughout the year; (b) the *Chondrus crispus*, *Osmundea pinnatifida* (formerly *Laurencia pinnatifida*) type, which is characterized by maximum antibiotic production occurring during the winter months; (c) the *Dilsea carnosa* type, possessing a summer peak of activity; and (d) the *Codium* type, where there is a spring peak of activity, this is also exhibited by *Halidrys siliquosa*.

According to Ballantine and Almodovar (1977), the degree of activity of the seaweed extracts ranged from slight against a single assay organism, to substantial as in the case of *Asparagopsis taxiformis* which was active against all bioassay microorganisms tested. Aqueous extracts of *A. taxiformis* collected in Puerto Rico have also been shown to be highly antibiotic.

Caccamese et al. (1979) reported the antibacterial potential of *Laurencia decidua* against *Bacillus subtilis* and *Bacillus cereus*. The red alga *Laurencia pacifica* presented activity only against *B. subtilis*.

The red alga *Laurencia mariannensis* afforded a number of new metabolites—the brominated diterpene, 10-hydroxykahukuene B, 2 sesquiterpenes, 9-deoxyelatol and isoda-ctyloxene A, 1 brominated C15-acetogenin, laurenmariallene, and two new naturally occurring halogenated sesquiterpenes that were obtained previously as intermediates in a biomimetic synthetic study of rhodolaureol and rhodolauradiol (Gonzalez et al. 1982). Both 10-hydroxykahukuene B and laurenmariallene had modest antibacterial activity.

The crude extracts obtained from the red seaweeds such as *Falkenbergia rufolanosa* (tetrasporophyte phase of *Asparagopsis armata*) and *Laurencia obtusa* possessed strong inhibitory effects against three pathogenic bacteria tested, whereas the extract of *Hypnea musciformis* showed a very weak activity against the tested bacteria (Pesando and Caram 1984).

In the study done by Hornsey and Hide (1985) using *Staphylococcus aureus* and *Escherichia coli* as test organisms, various life-cycle phases of 22 species of British marine algae were screened for antimicrobial activity. The different growth phases of seven of the 10 species of Rhodophyta screened showed no significant differences in their antibiotic content. However, *Plocamium cartilagineum* and *Gracilariopsis longissima* (formerly *Gracilaria verrucosa*) produced cystocarpic plants with enhanced antibacterial properties, whilst the tetrasporic phase of *Bonnemaisonia hamifera* possessed considerably more activity than the gametophytic generations. Within the Rhodophyta, four main trends can be discerned. These are enhancement during the tetrasporic phase of antimicrobial activity as shown by *B. hamifera*; enhancement during the cystocarpic phase of antimicrobial activity as shown by *P. cartilagineum* and *G. longissima*; possession of similar levels of antimicrobial activity by both cystocarpic and sterile plants as found in *Vertebrata lanosa* (formerly *Polysiphonia lanosa*); and possession of similar levels of antimicrobial activity by both tetrasporic and sterile plants as exhibited by *Chondrus crispus*, *Polysiphonia elongata*, and *V. lanosa* (Hornsey and Hide 1985).

The majority of algal extracts were active against only one or two microorganisms, with most of these (61%) showing activity against the Gram$^+$ *Bacillus subtilus* and (or) *Staphylococcus aureus* (Ballantine et al. 1987). 19% of algal extracts tested showed activity against *Candida albicans* and 15% were inhibitory to the Gram$^-$ *Pseudomonas aeruginosa* and (or) *Escherichia coli*. Reichelt and Borowitzka (1984) similarly found the majority of Australian algal extracts to have activity against Gram$^+$ bacteria. Allen and Dawson (1960), as well as Rao and Parekh (1981), failed to find activity against Gram$^-$ pathogenic bacteria altogether in their algal screening bioassays. Martinez Nadal et al. (1963) have reported antibiotic activity against *E. coli* and *P. aeruginosa* in extracts of *Chondria littoralis*.

Of the 35 seaweeds collected along the coast of Sri Lanka and screened against the human pathogenic bacteria (*Staphylococcous aureus* and *Escherichia coli*) and fungi (see Chapter 7), 26 species exhibited antibacterial and/or antifungal activity (Bandara et al. 1988). The extracts of *Portieria hornemannii* (formerly *Chondrococcus hornemannii*) inhibited the growth of all four microbes tested, and the observed activities were comparable to those of a standard antibiotic Penicillin.

In the study by Soliman et al. (1994), the aqueous extract of *Jania rubens* exerted high antimicrobial activity against *Bacillus subtilis* and slight activity against *Streptococcus aureus*. Gonzalez et al. (2001) showed that the methanol extracts of *J. rubens* and *Jania adhaerescens* did not inhibit the growth of the bacteria *Pseudomonas aeruginosa*, *Serratia marcescens*, *Enterococcus faecium*, *Mycobacterium smegmatis*, *Staphylococcus aureus*, and *Bacillus subtilis*. However, in agreement with the results of Gonzalez et al. (2001), in the study of Karabay-Yavasoglu et al. (2007), none of the tested extracts of *J. rubens* showed activity on the yeast *Candida albicans* (see also Chapter 7).

Fully-grown red seaweed *Gracilaria corticata* showed maximum antibacterial potential against *Staphylococcus aureus* compared to medium and young stages of growth. The middle part of the thalli of fully-grown alga showed maximum bioactive potential compared to the terminal part and the basal part, which hold fast (Vidyavathi and Sridhar 1991).

Extracts from several southern African seaweeds were screened against 12 bacteria (Vlachos et al. 1997). *Polysiphonia virgata* (formerly *Carradoriella virgata*) methanolic extract produced the largest inhibition zones in this division, against the test microorganisms tested in this study. *Portieria hornemannii* and *Plocamium rigidum* showed the broadest spectrum of antimicrobial activity, and inhibited all the Gram$^+$ bacteria tested. *Laurencia complanata* produced the largest zones of inhibition, but inhibition was evident against only five of the six Gram$^+$ bacteria. This indicated that *L. complanata* extract had a high degree of antibacterial activity, but a narrower spectrum of activity than, for example, *P. hamemannii* and *P. rigidum*.

The inhibitory effects of aqueous, ethanolic, and dichloromethane fractions from 16 marine algae from the Atlantic shores of North-East Brittany, France, have been investigated against bacteria frequently associated with immersed surfaces (Hellio et al. 2000). The extracts were tested *in vitro* against isolates of marine bacteria potentially involved at different stages in the formation of biofilms in the sea. The high levels of inhibitory activity of nine extracts against Gram$^+$ bacteria (*Escherichia coli*, *Klebsiella pneumoniae*, and *Pseudomonas aeruginosa*) and their apparent absence of toxicity suggests potential for novel active ingredients. Comparable inhibition zones were only obtained for *Osmundea pinnatifida* (formerly *Laurencia pinnatifida*), *Vertebrata lanosa* (formerly *Polysiphonia lanosa*), which showed strong inhibition of the Gram$^+$ bacteria.

In a study done by Val et al. (2001), extracts from 44 species of seaweed from Canary Islands (Spain) were screened for the production of antifungal and antibacterial compounds against a panel of Gram$^-$ and Gram$^+$ bacteria, mycobacterium, yeasts, and fungi. A total of 28 species displayed antibacterial activity. *Asparagopsis taxiformis* was the species with the strongest activities against the broadest spectrum of target microorganisms. All the species with antibacterial activity were active against Gram$^+$ bacteria (*Staphylococcus aureus* and *Bacillus subtilis*), whereas only two species, *A. taxiformis* and *Omundea hybrida*, were active against mycobacterium. Only one species—*A. taxiformis*—showed activity against the whole panel of six target microorganisms: *Pseudomonas aeruginosa*, *Serratia marcescens*, *Enterococcus faecium*, *Mycobacterium smegmatis*, *Staphylococcus aureus*, and *Bacillus subtilis* (Val et al. 2001).

In a study on several South African seaweeds, where five collections were made over a five-year period, *Osmundaria serrata* had the highest antibacterial activity in winter, while *Dichotomaria diesingiana* (as *Galaxaura diesingiana*) showed little variation in antibacterial activity (Vlachos et al. 2001).

In the work by Yi et al. (2001), three different solvents (ethanol, acetone, and methanol-toluene) were used to extract antibiotics from several marine algae collected from the coast along Fujian, China.

Among them, the ethanol extract showed the strongest activity against the bacteria tested; four species of the Rhodophyta (*Laurencia okamurae*, *Dasya scoparia*, *Grateloupia filicina*, *Gloiopeltis furcata*, and *Plocamium telfairiae*) showed a wide spectrum of antibacterial activity. Every solvent extract from the four species was active against all the bacteria tested. The tested bacterium *Pseudomonas solancearum* was the most sensitive to the extracts of marine algae. Among the actively seaweed extracts, the three solvent extracts of *L. okamurae*, the methanol-toluene and acetone extracts of *D. scoparia*, and the methanol-toluene extracts of *P. telfairiae*, strongly inhibited *Escherichia coli*. Strong antibacterial activities (against *S. aureus* by the ethanol extract of *L. Okamurae*, and against *Pseudomonas solanacearum* by the ethanol and acetone extracts of *L. okamurae*) were observed. None of the extracts from *G. furcata* exhibited any antibacterial activity against the tested bacteria, except *P. solanacearum*. Of the 23 species of algae screened for antibacterial activity, 22 species were active against *P. solanacearum*, 20 species against *E. coli* and *B. subtilis*, and 17 species against *S. aureus*, which may indicate that the bacterium *P. solanacearum* was the most and that *S. areus* was the least sensitive to the extracts of seaweeds. Among the solvents used for the extraction in the present screening test, the ethanol extracts showed the strongest inhibition against the bacteria tested, followed by the acetone extracts, whereas the methanol-toluene extracts showed the weakest inhibition (Yi et al. 2001).

Hexane, chloroform, and ethanol extracts of several marine macroalgae from North Ceará coast (Northeast Brazil) were evaluated for antibacterial activity by the single disc method (Lima-Filho et al. 2002). Best results were shown by the hexane extracts of *Amansia multifida* against enteric Gram$^-$ strains such as *Enterobacter aerogenes*, *Klebsiella pneumoniae*, *Pseudomonas aeruginosa*, *Salmonella typhi*, *Salmonella choleraesuis*, *Serratia marcescens*, *Vibrio cholerae*, and the Gram$^+$ bacteria *Bacillus subtilis* and *Staphylococcus aureus*. Moreover, several different organic solvents have been used to screening algae for antibacterial activity. Earlier, Olessen et al. (1963) related antibacterial activity in the chloroform and acetone extracts of *Asparagopsis taxiformis* (formerly *Falkenbergia hillebrandii*) against *S. aureus*.

According to Ravikumar et al. (2002), propanol extracts of *Gracilaria edulis* showed maximum inhibitory effect against *Klebsiella pneumonia* (9 mm of inhibition zone); the acetone extract of *Palisada cruciata* (formerly *Laurencia cruciata*) gave maximum inhibitory activity (10 mm) against *Pseudomonas aeruginosa*, and the butanol extract of *Hypnea musciformis* gave the maximum inhibitory activity (8 mm) against *Streptococcus pyogens*. In the study done by Kolanjinathan et al. (2009), ethanol extract at 1% concentration of *G. edulis* produced a maximum zone of 13.7 mm against *Staphylococcus aureus* and a minimum zone of 3.1 mm against *Enterobacter aerogenes*.

Vairappan (2003) found that the red algae from the genus *Laurencia* are known to produce a wide range of chemically interesting secondary halogenated metabolites. This investigation delves upon extraction, isolation, structural elucidation, and antibacterial activity of inherently available secondary metabolites of *Laurencia dendroidea* (formerly *Laurencia majuscula*) collected from two locations in water of Sabah, Malaysia. Two major halogenated compounds were identified as elatol and iso-obtusol. Structures of these compounds were determined from their spectroscopic data, such as IR, ^{1}H-NMR, ^{13}C-NMR, and optical rotation. Antibacterial bioassay against human pathogenic bacteria was conducted using disc diffusion method. Elatol inhibited six species of bacteria, with significant antibacterial activities against *Staphylococcus epidermidis*, *Klebsiella pneumoniae*, and *Salmonella* sp., while iso-obtusol exhibited antibacterial activity against four bacterial species, with significant activity against *K. pneumoniae* and *Salmonella* sp. Elatol showed equal and better antibacterial activity compared to tested commercial antibiotics, while iso-obtusol only equaled the potency of commercial antibiotics against *K. pneumoniae* and *Salmonella* sp.

Ethanolic and lipid-soluble extracts from 21 marine algal species (10 Chlorophyta, two Phaeophyceae, and nine Rhodophyta) from the coast of Yucatan, Mexico, were evaluated for antibacterial activity against pathogenic microbes (Freile-Pelegrín and Morales 2004). All species with antibacterial activity (18) were active against the Gram$^+$ bacteria (*Bacillus subtilis*, *Streptococcus faecalis*, and *Micrococcus luteus*), and most of the algal species exhibited activity against *B. subtilis* (89% in ethanolic soluble extracts and 94% in lipid-soluble extracts). The lipid-soluble extract of *Ceramium nitens* exhibited the highest activity among the species tested (Freile-Pelegrín and Morales 2004). Selvin and Lipton (2004) tested the secondary metabolites of two seaweeds for bio-toxicity potential. Both species showed potent activity in antibacterial, brine shrimp cytotoxicity, larvicidal, antifouling, and ichthyotoxicity assays. The red alga *Hypnea musciformis* showed narrow spectrum antibacterial activity.

Atlantic marine algae were screened in growth inhibition assays against the pathogenic bacterium *Pseudoaltermonas bacteriolytica* (Puglisi et al. 2007). Overall, 52% of all red algae yielded extracts active against *P. bacteriolytica*. The extracts from *Amphiroa fragilissima* were only active against this bacterium.

Choudhury et al. (2005) screened the organic solvent extracts of three marine algae, which showed species-specific activity in inhibiting the growth of six virulent strains of bacteria pathogenic to fish (*Edwardsiella tarda*, *Vibrio alginolyticus*, *Pseudomonas fluorescens*, *Pseudomonas aeruginosa*, and *Aeromonas hydrophila*). Three methanol extracts and the chromatographic fractionation of active extracts of *Gracilaria corticata* resulted in enriched fractions with wide spectrum activity and lowered values of minimum inhibitory concentration (MIC).

Prakash et al. (2005) studied the antibacterial activity of 45 extracts of nine algae against the pathogens. The *Otitis media* infected bacterial pathogens were isolated from 25 infected patients. The isolated bacterial species were Gram$^+$ and Gram$^-$ such as *Haemophilus influenza*, *Streptococcus pneumoniae*, *Streptococcus pyogenes*, *Staphylococcus aureus*, and *Moraxella catarrhalis*. *Halymenia floresii* crude extract was found to produce maximum growth inhibition against the bacterial pathogens. Five solvents were used for the extraction of antimicrobials, of which butanol showed the maximum extraction.

The communication done by Lakshmi et al. (2006) deals with the biological activities of the extracts of 48 marine florae. The biological screening includes tests for antibacterial and antifungal (see Chapter 7), among others. From among red algae, the crude extracts from *Chondria dasyphylla*, *Gracilaria corticata*, *Gracilaria canaliculata* (formerly *Gracilaria crassa*), *Halymenia porphyroides*, *Heterosiphonia muelleri*, *Hypnea musciformis*, *Laurencia obtusa*, *Palisada poiteaui* (formerly *Laurencia poiteaui*), *Scinaia moniliformis* (formerly *Scinaia indica*), and *Solieria robusta* were active against *Staphylococcus aureus*, *Escherichia coli*, *Streptococcus faecalis*, *Klebsiella pnuemoniae*, and *Pseudomonas aeruginosa*.

Lanosol enol ether has been shown to be an antibacterial and antifungal component of the red alga *Osmundaria serrata*. The mean bacteriostatic activity of lanosol enol ether was 0.27 mg mL^{-1} and the mean bactericidal activity was 0.69 ± 0.15 mg mL^{-1}. These values are comparable to the estimated concentration of lanosol enol ether in the whole plant of 0.20 mg mL^{-1}. Extracts of *O. serrata* containing lanosol enol ether also caused deformities in some bacteria that were tested (Barreto and Meyer 2006).

Dichloromethane, methanol, and water extracts of 26 species of cultivated seaweeds were screened for their antibacterial activities against five fish pathogenic bacteria strains (*Aeromonas salmonicida* subsp. *salmonicida*, *Aeromonas hydrophila* subsp. *hydrophila*, *Pseudomonas anguilliseptica*, *Vibrio anguillarum*, and *Yersinia ruckeri*). The dichloromethane extracts of *Asparagopsis armata* (and the tetrasporophyte phase, *Falkenbergia rufolanosa*), *Ceramium virgatum* (formerly *Ceramium rubrum*), *Myriogramme minuta* (formerly *Drachiella minuta*), *Gracilaria cornea*, and *Halopithys incurva* showed strong antibacterial activities. *V. anguillarum* and *P. anguilliseptica* were the two most susceptible bacteria strains. The screening results confirm the possible use of seaweeds as a source of antimicrobial compounds or as a health-promoting food for aquaculture (Bansemir et al. 2006). Considerable antimicrobial activities against *V. anguillarum* were presented also by *F. rufolanosa* and *G. cornea*. The MIC values against *V. anguillarum* were < 100 μg mL^{-1} for *A. armata* and < 400 μg mL^{-1} for *C. virgatum*, *F. rufolanosa*, *G. cornea*, and *Halopitys incurva*. The MIC values of the other extracts were > 400 μg mL^{-1}. The MIC value for oxytetracycline against *V. anguillarum* amounted to 0.5 μg mL^{-1}. The growth of *P. anguilliseptica* was strongly inhibited by *A. armata* and *G. cornea* (27 mm). Inhibition zones between 14 mm and 17 mm were measured for *C. virgatum*, *M. minuta*, *F. rufolanosa*, and *H. incurva*. MIC values < 400 μg mL^{-1} for *P. anguilliseptica* were only attained by *G. cornea* and *F. rufolanosa* extracts. The MIC value for oxytetracycline against *P. anguilliseptica* was 0.08 μg mL^{-1}. An influence on the growth of *Aeromonas salmonicida* was detected for eight seaweed species. *A. armata* showed the strongest activity (17 mm). *H. incurva* and *F. rufolanosa* developed inhibition zones between 5 mm and 10 mm. Only three extracts (*A. armata*, *H. incurva*, and *F. rufolanosa*) showed antimicrobial activities against *Aeromonas hydrophila* that are worth mentioning. Antimicrobial effects against *Y. ruckeri* were presented by *A. armata*, *F. rufolanosa*, and *H. incurva*. Among the methanolic extracts, only those of *C. virgatum*, *F. rufolanosa*, and *H. incurva* demonstrated weak inhibiting activities against *V. anguillarum*, *P. anguilliseptica*, and *A. salmonicida*. Summarizing the results, it can be concluded that the most effective seaweeds were *A. armata*, *C. virgatum*, *M. minuta*, *F. rufolanosa*, *G. cornea*, and *H. incurva* (Bansemir et al. 2006).

Carrageenans have proved to have effects against some bacterial strains, such as *Salmonella enteritidis*, *Salmonella typhimurium*, *Vibrio mimicus*, *Aeromonas hydrophila*, *Escherichia coli*, *Listeria monocytogenes*, and *Salmonella aureus*. The growth of all the bacterial strains except *L. monocytogenes* was significantly inhibited by them, particularly by the iota-carrageenan. A growth inhibition experiment using *S. enteritidis* showed that the inhibitory effect of the carrageenans was not bactericidal, but bacteriostatic. Removal of the sulfate residues eliminated the bacteriostatic effect of iota-carrageenan, suggesting that the sulfate residues in carrageenan play an essential role in this effect (Venugopal 2008). In 2014, Sebaaly et al. reported that carrageenans isolated from the red alga *Corallina* sp. exhibited antibacterial activity against *Staphylococcus epidermis*. Infrared spectroscopy (IR) showed that the isolated carrageenan was of lambda-type.

The methanol, dichloromethane, hexane, chloroform, and volatile oil extracts of the red alga *Jania rubens* were tested *in vitro* for their antimicrobial activity (five Gram$^+$, four Gram$^-$ bacteria) (Karabay-Yavasoglu et al. 2007). Gas chromatography-mass-spectrometry (GC-MS) analysis of the volatile components of *J. rubens* identified 40 compounds which constituted 77.53% of the total. The volatile components of *J. rubens* consisted of *n*-docosane (6.35%), *n*-eicosane (5.77%), and *n*-tetratriacontane (5.58%) as major components. The methanol and chloroform extracts (4 mg/disc) showed more potent antimicrobial activity than the hexane and dichloromethane extracts and the volatile oil of *J. rubens*.

In 2007, Salvador et al. evaluated the antifungal and antibacterial activity of 82 Iberian macroalgae (18 Chlorophyta, 25 Phaeophyceae, and 39 Rhodophyta) against three Gram$^+$ bacteria (*Bacillus subtilis*, *Bacillus cereus*, and *Staphylococcus aureus*), two Gram$^-$ bacteria (*Escherichia coli* and *Pseudomonas aeruginosa*), and one yeast (*Candida albicans*) (see also Chapter 7). The bioactivity was analyzed from crude extracts of fresh and lyophilized samples. Of the seaweeds analyzed, 67% were active against at least one of the six test microorganisms. The highest percentage of active *taxa* was found in Phaeophyceae (84%), followed by Rhodophyta (67%), and Chlorophyta (44%). Nevertheless, red algae had both the highest values and the broadest spectrum of bioactivity. In particular, *Bonnemaisonia asparagoides*, *Bonnemaisonia hamifera*, *Asparagopsis armata* (and *Falkenbergia rufolanosa* phase) (Bonnemaisoniales) were the most active *taxa*. In this study, Ceramiales and Gigartinales had noteworthy antimicrobial activity, and Bonnemaisoniales was the order that had the highest bioactivity.

In the study executed by Taskin et al. (2007), methanolic extracts of six marine algae from the North Aegean Sea (Turkey) were studied for their antibacterial activity against pathogenic bacteria, three Gram$^+$ (*Staphylococcus aureus*, *Micrococcus luteus*, and *Enterococcus faecalis*) and three Gram$^-$ (*Enterobacter aerogenes*, and *E. coli*) *in vitro*. Extracts of all the test marine algae, except *Corallina officinalis* showed inhibition against *S. aureus*. On the other hand, highest inhibition activity among all the extracts was shown to *E. aerogenes* by *C. officinalis*.

In the study done by Dubber and Harder (2008), antibacterial effects of hexane and methanol extracts of the macroalgae *Mastocarpus stellatus* and *Ceramium virgatum* (formerly *Ceramium rubrum*) on 12 marine and seven prominent fish pathogenic bacteria at ecologically realistic concentrations of macroalgal secondary metabolites were investigated. Antibacterial activity was determined with a highly sensitive growth inhibition assay that records the fluorescence intensity of stained bacterial DNA. The assay allowed the detection of bacterial growth inhibition at and below the tissue level concentration of algal metabolites. Gram$^+$ marine Bacillaceae were generally more susceptible than Gram$^-$ marine Vibrionaceae. *Listonella anguillarum*, *Pseudomonas anguilliseptica*, and *Aeromonas salmonicida* were the most susceptible fish pathogenic bacteria.

Oh et al. (2008) tested for antimicrobial activity the bromophenols isolated from the red alga *Odonthalia corymbifera* against bacteria and fungi human pathogens, and found that natural products were more active against fungi, and synthetic bromophenols against bacteria. The synthetic bromophenols 3,3′-dibromo-6,6′-dihydroxydiphenylmethane and 3,3′,5,5′-tetrabromo-6,6′-dihydroxydiphenylmethane showed potent antibacterial effect against *Staphylococcus aureus*, *Bacillus subtilis*, *Micrococcus luteus*, *Proteus vulgaris*, and *Salmonella typhimurium*.

Shanmughapriya et al. (2008) showed that extracts from *Gracilaria corticata* were found to be effective against *Pseudomonas aeruginosa* and *Escherichia coli*. It was also effective against *Micrococcus luteus*, *Staphylococcus epidermidis*, and *Enterococcus faecalis*.

The red marine alga *Polysiphonia virgata* was investigated for antimycobacterial activity (Saravanakumar et al. 2008). Oleic acid showed the greatest inhibition of the growth of *Mycobacterium smegmatis*, with a minimum inhibitory quantity (MIQ) of 0.8 µg; linoleic acid and lauric acid had MIQ values of 1.56 and 3.125 µg, respectively. Stearic acid, palmitic acid, and myristic acid did not inhibit the growth of *M. smegmatis*. Using the Bactec-460 radiometric method, oleic acid showed 100% inhibition of the growth of *Mycobacterium tuberculosis*, at a minimum inhibitory concentration (MIC) of 25 µg mL^{-1}; lauric acid, myristic acid, and linoleic acid all showed 100% inhibition at MIC values of 50 µg mL^{-1}. Myristic acid and lauric acid showed 90% and 76% inhibition at 50 µg mL^{-1}. Linoleic acid showed moderate inhibition of the growth of a clinical strain of multidrug-resistant *M. tuberculosis* at 50 µg mL^{-1}.

Saeidnia et al. (2009) evaluated the cytotoxic and antibacterial activity of *Gracilaria salicornia* and *Hypnea flagelliformis*. Ethyl acetate extracts of both algae showed a potent cytotoxic effect against *Artemia salina* (4 µg mL^{-1}). Aqueous methanol (50%) extracts were also effective. None of the methanol and aqueous methanol extracts of the algae showed antibacterial activity against *Staphylococcus aureus* and *Escherichia coli* by the broth dilution method. Only the ethyl acetate extract exhibited antibacterial activity (MIC = 2 µg mL^{-1}) on *S. aureus*.

Manilal et al. (2009b) examined 15 seaweeds for *in vitro* antimicrobial activity against six pathogenic *Vibrio* strains isolated from moribund tiger shrimp (*Penaeus monodon*), six type cultures of prominent shrimp *Vibrio* pathogens, 10 multidrug resistant clinical pathogens, four species of *Candida* obtained from pulmonary tuberculosis (TB) patients, and four species of plant pathogenic fungi to evaluate their potency to be used as natural antibiotics in pharmaceutical and agriculture field. In the preliminary screening process, it was found that out of 15 red algae, the crude methanolic extract of four algae indicate significant antimicrobial activity—one species of genus *Asparagopsis*, two species of genus *Chondrophycus*, and one species of genus *Hypnea*. Bioactivity was analyzed from crude extract of fresh and dried samples prepared from different polar and non-polar solvents. Of these, four species of red algae (*Asparagopsis taxiformis*, *Chondrophycus ceylanicus*—formerly *Laurencia ceylanica*, *Chondrophycus brandenii*—formerly *Laurencia brandenii*, and *Hypnea valentiae*) were found to be highly active. Broadest and highest activity was observed in the crude extract of *Asparagopsis taxiformis*. Among the pathogens tested, shrimp pathogenic *Vibrio* were the most susceptible organisms, while phytopathogens were found to be little resistant.

A new nitrogenous bromophenol, colensolide A, is isolated from the New Zealand red alga *Osmundaria colensoi* together with the known bromophenol lanosol and four of its derivatives (Popplewell and Northcote 2009). Several of the known bromophenols exhibit antibacterial activity and one shows moderate cytotoxicity. Compounds were evaluated for cytotoxicity against the human leukemia cell line HL-60 to 10 µM (see also Chapter 6) and for antibacterial activity against *Mycobacterium smegmatis* to 100 µM. Lanosol butenone exhibited moderate activity against human leukemia cells (IC_{50} 8.0 µM), while lanosol methyl ether, lanosol butenone, and rhodomelol all exhibited antibacterial activity (IC_{50} 7.8, 26.2, and 28.1 µM, respectively).

Vedhagiri et al. (2009) evaluated compounds isolated from *Asparagopsis taxiformis* for their inhibitory action against *Leptospira javanica* isolated from rodent carriers, and found MICs and MBC in the range of 100–1600 µg mL^{-1}, whereas those for Penicillin and Doxycyline were in the range of 25–200 µg mL^{-1}. The gas chromatography-mass-spectrometry (GC-MS) analysis revealed the presence of 4,5-dimethyl-1H-pyrrole-2-carboxylic acid ethyl ester, fatty acids, 14-methyl-pentadecanoicacid methyl ester, octadecanoic acid methyl ester, octadec-9-enoic acid 2,3-dihydroxy-propyl ester, 9-octadecanoic acid, methyl ester, and trace amounts of chlorobenzene.

In the work by Wefky et al. (2009), some algal extracts were tested against fish and human pathogenic bacteria: *Aeromonas hydrophila*, *Vibrio anguillarum*, *Pseudomonas fluorescens*, *Staphylococcus aureus*, *Pseudomonas aeruginosa*, and *Salmonella typhimurium*. The best activities were shown by the methanolic extract of *Pterocladiella capillacea* against *P. fluorescens*, *V. anguillarum*, and *P. aeruginosa*. The strongest antibacterial activities were acheived by the methanolic extract of *P. capillacea* against *P. fluorescens*, *V. anguillraum*, and *P. aeruginosa*, with inhibition zones of 29 mm, 28 mm, and 27 mm, respectively. The ethyl acetate extracts of *P. capillacea* showed antibacterial activities against *A. hydrophila*, *V. anguillaum*, *P. fluorescens*, and *P. aeruginosa*. The tested pathogens were varied in their response to the antibacterial action of the different extracts. The most susceptible organisms were *A. hydrophila* and *P. fluorescens*.

With a view to explore finding the new molecules with therapeutic efficacy for human use, the alcoholic extracts of 33 identified species of marine flora, collected from Indian coasts, were prepared and screened for a wide range of biological activities (Lakshmi et al. 2010). From among red algae, the alcoholic extracts from *Gelidiella acerosa*, *Hypnea musciformis*, and *Portieria hornemannii* (formerly *Chondrococcus hornemanii*) were active against *Staphylococcus aureus*, *Escherichia coli*, *Mycobacterium tuberculosis*, *Streptococcus faecalis*, *Klebsiella pnuemoniae*, and *Pseudomonas aeruginosa*.

Antibacterial activity from chloroform extract of *Gracilaria edulis* was tested against bacterial strains of *Vibrio cholera*, *Staphylococcus aureus*, *Shigella dysenteriae*, *Shigella bodii*, *Salmonella paratyphi*, *Pseudomonas aeruginosa*, and *Klebsiella pneumonia*. Almeida et al. (2011) observed higher activity for *G. edulis* extract against *S. aureus*, in accordance with results obtained by Vallinayagam et al. (2009). Ethanol extract from *Gracilaria debilis* showed antibacterial activity against *S. aureus* but was inactive against *Mycobacterium smegmatis* (Albuquerque et al. 1983). 95% ethanol extract from whole dried *Gracilaria cervicornis* was active against *S. aureus* at a concentration of 5.0 mg mL^{-1} (Perez et al. 1990). Methanol extract from fresh *Gracilaria corticata* was active against *Bacillus subtilis*, *Bacillus megaterium*, *S. aureus*, and *Streptococcus viridians* (Usmanghani et al. 1984). Ethanol extracts from *Gracilaria domingensis* and *Gracilariopsis andersonii* (formerly *Gracilariopsis sjoestedtii*) showed antibacterial activity against *E. coli* and *S. aureus*. Ethanol extracts from *Gracilaria debilis*, *G. domingensis*, and *G. andersonii* were active against *Candida albicans* (yeast) shown by agar plate method (Albuquerque et al. 1983); chloroform, ether, and methanol extracts from *Gracilaria tikvahiae* were inactive (Oranday et al. 2004).

In the study done by Abirami and Kowsalya (2011), the antibacterial activity of seaweed extracts was evaluated. The highest zone was found in the methanolic extract of *Kappaphycus alvarezii* against *Klebsiella pneumonia* at 20 mg/disc concentration (Baky et al. 2008). The maximum zone of inhibition was against *S. aureus* for the methanolic extract of *K. alvarezii*. Wong and Cheung (2002) reported that hexane extract of *Gracilaria* species inhibits only *B. subtilis* in contrast to the result of Rajasulochana et al. 2009, who showed that the methanol extract of *Kappaphycus* species inhibited the bacteria, namely *Pseudomonans fluorescence*, *Staphylococcus aureus*, and *Vibrio cholerae*.

Devi et al. (2008) evaluated 10 edible seaweeds from India for antioxidant and antimicrobial activity against food-borne pathogens. *Gelidiella acerosa* have high phenolic content and exhibited antibacterial activity against *S. aureus*.

The *in vitro* antibacterial activities of *Eucheuma denticulatum* extract showed inhibitory activity only on $Gram^+$ positive organisms tested, including *Staphylococcus aureus* and *Streptococcus pyogenes*, which were expressed in terms of minimum inhibitory concentration and minimum bactericidal concentration test (Al-Haj et al. 2009). Thus, $Gram^-$ pathogens tested, including *Escherichia coli*, *Klebsiella pneumoniae*, and *Pseudomonas aeruginosa*, showed resistant phenotypic pattern to both extracts.

Chloroform extracts, *n*-butanol extracts, water extracts, and volatile compounds from several seaweed species from Black Sea were assessed for antibacterial properties (Kamenarska et al. 2009). Most of the extracts showed pronounced effects against the $Gram^+$ bacterium *Staphylococcus aureus*. Activity towards the $Gram^-$ bacterium *Escherichia coli* was found in only a few cases. The butanol extract from *Jania virgata* (formerly *Haliptilon virgatum*) had considerable activity against *S. aureus*; similarly, the *n*-butanol extract from *Laurencia coronopus* had a significant effect against *S. aureus*, while the *n*-butanol extract from *Palisada perforata* (formerly *Chondrophycus papillosus*) was not active. The activity of the *n*-butanol extract from *L. coronopus* may be attributable to larger quantities and greater variety of N-containing compounds. Esters of phosphoric acid may be responsible for the antibacterial activity of *Gelidium spinosum*. In this alga, as well as in *Ceramium siliquosum* var. *elegans*, the antibacterial activity may be due in part to 1,2-dihydroxy ethane sulfonate, which was identified in the *n* butanol fractions. The *n*-butanol extracts from *Polysiphonia elongata* (formerly *Polysiphonia denudata* f. *fragilis*) and

Polysiphonia denudata possessed a strong activity against the Gram$^+$ bacterium *S. aureus*, but also against the fungus *Candida albicans*. Only the *n*-butanol fraction from *P. elongata* and α-*O*-methyllanosol (one of the main components of the *n*-butanol extract from *P. elongata*) were active against the Gram$^-$ bacterium *E. coli* (250 mg mL^{-1} and 125 mg mL^{-1}, respectively). The observed difference in the activities of these closely related species may be attributed to the fact that *P. denudata* is rich in monobrominated phenols (Kamenarska et al. 2006), while *P. elongata* is rich in dibrominated phenols. Most of the water extracts from the red algae were investigated, except the extracts from the two species from the Corallinaceae (*Ellisolandia elongata*—formerly *Corallina elongata* and *Jania virgata*) and the extract from *P. denudata*, were active against the Gram$^+$ bacterium *S. aureus*, with the minimum inhibitory concentration (MIC) value varying from 250 mg mL^{-1} to 500 mg mL^{-1}. Conversely, the water extract from *P. elongata* possessed low activity against the fungus *C. albicans* (see also Chapter 7) (Kamenarska et al. 2009).

The antimicrobial activities of methanol, methanol–chloroform, diethyl ether, ethyl acetate, and butanol extracts of *Gracilaria changii* were studied (Sasidharan et al. 2009). These extracts were tested against 22 bacteria, six yeasts strains, and 11 fungal isolates by the disc diffusion and broth dilution methods. The results indicated that all of the solvent systems used in the study had equal effectiveness against the tested microorganisms. All of the extract preparations of *G. changii* showed significant activities against seven bacterial and four yeast isolates tested. The minimum inhibitory concentration (MIC) values of *Pseudomonas aeruginosa* and *Bacillus subtilis* were 6.25 mg mL^{-1} and 3.125 mg mL^{-1}, respectively, while the MIC for *Candida albicans* was 3.125 mg mL^{-1}. No fungal isolates showed any susceptibility against the different crude preparations of *G. changii*. The activities were compared to known commercialized antibiotics, such as chloramphenicol and miconazole nitrate.

The antibacterial activities of several seaweeds collected from Pudumadan coastal region (India) were screened against human bacterial pathogens *Staphylococcus aureus*, *Vibrio cholerae*, *Shigella dysentriae*, *Shigella bodii*, *Salmonella paratyphi*, *Pseudomonas aeruginosa*, and *Klebsiella pneumonia* (Vallinayagam et al. 2009). The zone of inhibition ranged between 8.8–1.23 mm. The maximum activity (8.8 mm) was recorded from the extract of *Gracilaria edulis* against *S. aureus*.

Two novel α-pyrone macrolides, neurymenolides, were isolated from the Fijian red alga *Neurymenia fraxinifolia* (Stout et al. 2009). Neurymenolide displayed moderately potent activities against Methicillin-resistant *Staphylococcus aureus* and Vancomycin-resistant *Enterococcus faecium* (IC_{50} of 2.1 µM and 4.5 µM, respectively).

Kappaphycus alvarezii from the coast of Tamil Nadu, India were tested *in vitro* for their antibacterial activities against different types of bacteria using disc diffusion method (Rajasulochana et al. 2009). Methanol extract was used for evaluating the inhibition of different bacteria, such as *Pseudomonas fluorescens*, *Staphylococcus aureus*, *Vibrio cholerae*, and *Proteus mirabilis*. In this study, it is observed that *K. alvarezii* has maximum activity against *P. fluorescens*, *S. aureus*, and less inhibition on *V. cholerae* and *P. mirabilis*. Benzene, *n*-hexane, ethylacetate, methanol, and chloroform/methanol solvents were used for inhibition of *S. aureus* and *Escherichia coli*. It is noted that chloroform: methanol is the best solution for extracting the effective antibacterial materials from the brown algae species. The chloroform: methanol solvent was further used for antibacterial activity against 11 pathogenic bacteria. It is observed from the experiments that the extract residues of algae recorded maximum activity against *S. aureus* with an inhibition zone, compared to other bacteria. The extract residues of brown algae did not show any effect on the growth of *P. vulgaris* and *Pseudomonas aeruginosa* (Rajasulochana et al. 2009).

Bai (2010) screened the marine red alga, *Gracilaria debilis* (formerly *Gracilaria fergusonii*) for preliminary phytochemical analysis (coumerins, phenols, quinones, steroids, and tannins). Coumerins, phenols, quinines, and steroids were present, and tannin was absent in the alga investigated. Different organic solvent extracts—acetone, chloroform, diethyl ether, ethanol, and methanol were evaluated for antibacterial activity, employing Gram$^-$ (*Klebsiella pneumoniae* and *Pseudomonas aeruginosa*) and Gram$^+$ (*Bacillus subtilis* and *Staphylococcus aureus*) bacteria. The ethanol extract of the alga was found to be active against *Pseudomonas aeruginosa* and *Bacillus subtilis*.

Rhimou et al. (2010) screened the antibacterial activity of extracts from 26 seaweeds to assess their potential in the pharmaceutical industry. Their bioactivity was analyzed from crude methanolic extracts of dried samples against three Gram$^+$ bacteria and two Gram$^-$ bacteria using the disc diffusion technique. 96% of the extracts were active against at least one of the five tested microorganisms. *S. aureus* was the most susceptible microorganism. Methanolic extracts of all seaweed extracts tested exhibited a broad spectrum of antibacterial activity, with inhibition diameters ranging from 10 mm to 35 mm. An extract of *Hypnea musciformis* exhibited high antibacterial activity against all the bacteria.

Elsie and Rajan (2010) had investigated the antimicrobial activity of three different solvents extracts of *Gelidiella acerosa* (methanol, ethanol, and acetone) and they have got ethanol extracts showing the presence of a number of biochemical compounds, as compared to acetone and methanol. The antimicrobial activity was determined by well diffusion method and the inhibition of growth of test bacteria such as *Staphylococcus aureus*, *Bacillus cereus*, *Micrococcus luteus*, *Klebsiella pneumonia*, and *Pseudomonas aeruginosa*. *G. acerosa* showed variety in the exploitation of antibacterial activity of zone of inhibition from 5–15 mm against all tested bacteria. The methanol extract showed maximum activity against *B. cereus* (7 mm) and minimum activity in *S. aureus* (5 mm), and it has no activity against the pathogens like *M. luteus*, *K. pneumonia*, *P. aeruginosa*, and *Escherichia coli*. Extracts obtained using ethanol showed a maximum activity against pathogens like *S. aureus* (7 mm), and minimum activity against *M. luteus* (6 mm), *B. cereus* (6 mm), and *K. pneumonia* (5 mm). No activity was shown against the pathogens like *E. coli*, *P. aeruginosa*, and *B. cereus*. Acetone extracts showed a maximum activity against *M. luteus* (15 mm), *B. cereus* (12 mm), and minimum activity in *S. aerues* (10 mm). It has no activity against the pathogens like *M. luteus*, *K. pneumonia*, *P. aeruginosa*, and *E. coli*.

The study done by Al-Haj et al. (2010) was conducted to investigate antibacterial activity of *Gracilaria changii* and *Eucheuma denticulatum* extracts against Methicillin-resistant *Staphylococcus aureus*, *Escherichia coli*, *Streptococcus pyogenes*, *Vibrio cholerae*, *Pseudomonas aeruginosa*, and *Klebsiella pneumoniae*. The minimal inhibitory concentration (MIC) values and minimal bactericidal concentration (MBC) values of methanol extract were used against all assayed bacteria. Results indicated that *G. changii* and *E. denticulatum* extracts must possess major antibacterial components against infectious microorganisms. For isolates of *E. coli*, *P. aeruginosa*, and *K. pneumoniae*, there were no inhibition zones around discs impregnated with methanol extract of *G. changii*. The antibiotic disc which was used as positive control vancomycin showed clear inhibition zones around both Methicillin-resistant *S. aureus* and non-Methicillin-resistant *S. aureus* isolates. MIC and MBC were determined based on the lowest *G. changii* and *E. denticulatum* extract concentrations, reducing colonial growth or killing all bacteria cells. Methanol extracts of *G. changii* and *E. denticulatum* showed strong activity irrespective of Methicillin-resistant *S. aureus* and non-Methicillin-resistant *S. aureus* isolates, whereas the MIC and MBC levels for *G. changii* extract are 25 mg mL^{-1} and 50 mg mL^{-1} for Methicillin-resistant *S. aureus* isolates, and 12.5 mg mL^{-1} and 50 mg mL^{-1} for non-Methicillin-resistant *S. aureus* isolates, respectively, while the MIC and MBC level for *E. denticulatum* extract are 20 mg mL^{-1} and 40 mg mL^{-1} for Methicillin-resistant *S. aureus* isolates and 8.75 mg mL^{-1} and 17.5 mg mL^{-1} for non-Methicillin-resistant *S. aureus* isolates, respectively.

In the studies by Nurul et al. (2010), the antimicrobial activities of eight crude extracts of *Acanthophora spicifera* were evaluated. These extracts were tested *in vitro* against 18 bacteria by disc diffusion method. The results revealed that methanol and ethyl acetate extract from solvent partitioning exhibited broader spectrum activity against tested bacterial strains. These two extracts showed inhibition zones against strains of *Bacillus cereus*, *Bacillus licheniformis*, Methicillin-resistant *Staphylococcus aureus*, *Pseudomonas aeruginosa*, *Yersinia* sp., and *Citrobacter freundii*. While methanol extracts from Soxhlet extraction and butanol from solvent partitioning had no effect against *P. aeruginosa*, the other six extracts exhibited antibacterial activity against the opportunistic strains. Extracts showed a moderate average zone of inhibition ranging from 9–14 mm. *Yersinia* sp. showed a higher inhibition zone of 12 mm.

The antibacterial efficiency of various solvent extracts of algae *Acanthaphora spicifera* was evaluated by Pandian et al. (2011). The methanol extracts showed higher antibacterial activity than petroleum ether and chloroform extracts, and the zone of inhibition ranged between 15.2 mm to 6.2 mm against bacterial strains. From the study, the methanol extract shows promising antibacterial activity against all bacterial strains and it showed maximum activity against *Escherichia coli* (15.2 mm). Similarly, the same extract showed higher activity against *Microsporum gypseum* (15.1 mm). As the concentration increases from

1100 μg mL^{-1}, 500 μg mL^{-1}, 1 mg mL^{-1}, 10 mg mL^{-1}, to 50 mg mL^{-1}, the inhibitory effect also increases the maximum inhibitory effect absorbed with 50 μg mL^{-1} of methanol extract proportionally, as compared to the standard drug. From the study, it's observed that if the crude extract is purified, it could show similar activity when compared to standard antibacterial Ciprofloxacin (Pandian et al. 2011).

The methanolic extract of *Melanothamnus afaqhusainii* showed antibacterial activity against *Staphylococcus aureus*, *Escherichia coli*, *Klebsiella pneumoniae*, *Proteus mirabilis*, and *Pseudomonas aeruginosa* at a concentration of 100 mg 100 mL^{-1} (Khan et al. 2011). Ampicillin, Amoxicillin, and Cefuroxime were used as standard antibacterial antibiotics (Khan 2000).

Asparagopsis taxiformis is a source of halogenated and aromatic volatile organic compounds with strong antimicrobial activity. The most abundant metabolites in *A. taxiformis* are oleic acid (51.33%) and *n*-hexadecanoic acid (42.87%). After Manilal et al. (2010), the red edible alga *A. taxiformis* extracts have antimicrobial activity against *Staphylococcus aureus*, *Pseudomonas aeruginosa*, *Bacillus subtilis*, and *Klebsiella pneumoniae*. Also, ethanol extracts of *A. taxiformis* has inhibitory influences on *Vibrio alginolyticus*, *Vibrio vulnificus*, and *Aeromonas salmonicida* subsp. *salmonicida* (Genovese et al. 2012).

In the research work of Manilal et al. (2012b), the crude extract of the red seaweed, *Asparagopsis* sp. was evaluated for *in vivo* antibacterial activity against the shrimp *Vibrio* pathogens. The algal extract was rationalized with commercial shrimp feed and orally administered for different durations of time, followed by the artificial bacterial challenge experiment. In dose titration experiments, the oral administration of *Asparagopsis* sp. at a dosage of 850 mg kg^{-1} of biomass was highly efficacious in the treatment of natural infestations of Vibriosis in *Penaeus monodon* (Giant tiger prawn or Asian tiger shrimp). The results of the confirmatory dose experiment revealed that the prophylactic treatment with moderate dose of 850 mg kg^{-1} of biomass day^{-1} for four weeks, followed by 14 days of post infection therapy, was highly effective in controlling *Vibrio* spp. infection in shrimps. Moreover, results of the percent survival index and microbiological analysis clearly show that *Asparagopsis* extract-incorporated medicated feed had broad therapeutic potential for managing shrimp Vibriosis.

Lavanya and Veerappan (2011) tested the *in vitro* antibacterial activity of six selected marine algae. Extracts of six seaweed samples, namely *Gracilaria canaliculata* (formerly *Gracilaria crassa*) and *Acanthophora spicifera*, among others, were selected for antibacterial activity against selected human pathogens, such as species *Vibrio parahaemolyticus*, *Salmonella* sp., *Shewanella* sp., *Escherichia coli*, *Klebsiella pneumoniae*, *Streptococcus pyogenes*, *Staphylococcus aureus*, *Enterococcus faecalis*, *Pseudomonas aeruginosa*, and *Proteus mirabilis*. All the seaweed extracts have shown moderate antibacterial activity (< 10 mm of zone of inhibition), out of which only methanolic extract has shown significant activity. The results of their research showed that the highest antibacterial activity was found in *A. spicifera*.

Kolanjinathan and Stella (2011b) screened the methanol crude extracts of *Gracilaria corticata* against human pathogenic bacteria, and compared them to other solvent extracts. The results indicated that the seaweed extracts had traditional claims of effectiveness, and could serve as useful sources of new antimicrobial agents. Antibacterial activity of five different solvents like methanol, acetone, chloroform, hexane, and ethyl acetate extracts of *G. corticata* were evaluated against pathogenic bacteria *Staphylococcus aureus*, *Streptococcus pyogenes*, *Streptococcus epidermis*, *Bacillus subtilis* and *Bacillus cereus*, *Klebsiella pneumoniae*, and *Enterobacter aerogenes*. Among the five solvent extracts tested, the methanol extract showed the greatest zone of inhibition against $Gram^+$ bacteria, as compared to $Gram^-$ bacteria. The methanolic extract of *G. corticata* showed highest mean zone of inhibition against the $Gram^+$ cocci and the minimum inhibitory concentration (MIC) values of *G. corticata* against bacteria was ranged between 1.25 mg and 80 mg.

In an antimicrobial assay done by Manilal et al. (2011), the *Chondrophycus brandenii* (as *Laurencia brandenii*) fraction exhibited a growth inhibition range of 213 mm^2 to 87 mm^2 against the studied bacteria. A high activity range was extended against *B. subtilis* with 213 mm^2, whereas the activity was moderate against *E. coli* at 103 mm^2. Regarding the clinical isolates, *Salmonella typhi* was found to be the most resistant bacteria, producing an inhibition area of 64 mm^2, whereas the fraction was moderately able to inhibit the growth of *Streptococcus pneumoniae* (87 mm^2). The bactericidal potency of the algal fraction was high against the clinical pathogen, *Vibrio cholerae* to the extent of 113 mm^2 at 37ºC. In this study, algal fraction of *C. brandenii* showed maximum antibacterial activity against the Gram$^+$ bacteria, whereas the activity was found to be moderate against Gram$^-$ bacteria. The resistant mechanism of Gram$^-$ bacteria could be due to the permeability provided by the cell wall or to the membrane accumulation tactics (Adwan and Abu-Hasan 1998).

Different solvent extracts of a red alga, *Hypnea flagelliformis*, collected from the Persian Gulf coast were subjected to different bioassays, including antibacterial activity (Jassbi et al. 2013). The water extracts were found to have the most antioxidant activity. The antibacterial minimum inhibitory concentrations (MIC) of the active extracts were determined for the susceptible organisms, *Staphylococcus aureus* and *Bacillus subtilis*, using nutrient-broth micro-dilution (NBMS) bioassay. Among extracts tested for their antibacterial potential, the methanol extract of *H. flagelliformis* was active at 5 mg charged on the paper disc against *B. subtilis* and *S. aureus*, with 10 mm and 9 mm of inhibition, and MICs of 10 mg of extract per mL^{-1} of nutrient-broth.

Several species of common seaweed extracts were tested in laboratory assays for potential industrial applications through evaluation of the antibacterial activity against pathogenic bacteria (five strains) and the antifouling potency against the growth of key species of marine colonizers (seven bacteria, five fungi, and 11 microalgae). The organic extract of *Laurencia johnstonii* has bacterial antibiosis. The ether extracts were more active in comparison to buthanol extracts against the bacterial strain *Staphylococcus aureus*. The best antifouling results were obtained with *L. johnstonii* (0.1–1 µg mL^{-1}) against all strains tested (Águila-Ramírez et al. 2012).

Adaikalaraj et al. (2012) evaluated the antibacterial activity of methanol and aqueous extracts from *Gracilariopsis longissima* (formerly *Gracilaria verrucosa*), *Gracilaria debilis* (formerly *Gracilaria fergusonii*), *Hypnea musciformis*, *Enantiocladia prolifera*, and *Gelidium* spp. against selected bacterial pathogens, and one yeast by using the disc diffusion method. In most of the seaweeds, methanol extract was found to be more effective. *Salmonella typhi*, *S. aureus*, *Bacillus subtilis*, and *Candida albicans* (yeast) were resistant to all aqueous extracts. The highest antibacterial activity (13 mm) was shown by the aqueous extract of *G. longissima* against *Pseudomonas aeruginosa*, and the lowest activity (6 mm) was observed in the methanol extract of *E. prolifera* against *Escherichia coli*. However, in most of the seaweeds, methanol extract was found to be more effective. The microbial strains *S. typhi*, *S. aureus*, *B. subtilis*, and *C. albicans* were resistant to the aqueous extracts of all seaweeds.

In a study done by Selim (2012), two seaweeds were colected from the Red Sea coast. The largest halos were achieved by the methanol and dimethyformamide extracts of marine algae *Sarconema filiforme* against *Pseudomonas aeruginosa*, *Escherichia coli*, *Serratia marcescens*, *Staphylococcus aureus*, *Enterococcus feacali*, *Bacillus subtilis*, *Bacillus cereus* (bacteria), *Candida albicans*, *Candida utilis*, and *Saccharomyces cerevisiae* (yeasts).

Antibacterial activity of methanolic extracts from several species of macroalgae collected from Moroccan Mediterranean coasts was evaluated against *Escherichia coli*, *Staphylococcus aureus*, and *Enterococcus faecalis* (Zbakh et al. 2012). In the case of the extract of *Asparagopsis armata*, the inhibition diameters were smaller than those of *H. musciformis*, and varied between 12 mm and 15 mm. *S. aureus* appeared more sensitive than the other two strains, with larger inhibition zones (Zbakh et al. 2012).

The antibacterial activity of ethanol, methanol, hexane, and acetone-based extracts of several seaweeds was investigated (Silva 2015). The disc diffusion method was used to evaluate the algae antimicrobial effect against standard strains of *Vibrio parahaemolyticus*, *Escherichia coli*, *Staphylococcus aureus*, *Pseudomonas aeruginosa*, *Salmonella enterica*, and five virulent antibiotic-resistant strains of *Vibrio brasiliensis*, *Vibrio xuii*, and *Vibrio navarrensis* (isolated from the hemolymph of *Litopenaeus vannamei*, the whiteleg shrimp). Only ethanol and methanol extracts were bioactive. The observed inhibitory effect of ethanol extracts of *Hypnea musciformis* against virulent antibiotic-resistant bacteria suggests this macroalgal species constitutes a potential source of bioactive compounds.

Lane et al. (2009) extracted bromophycolides (diterpene-benzoate macrolides) from the Fijian red alga *Callophycus serratus* with water, methanol, and dichloromethane. Extracts significantly inhibited Methicillin-resistant *Staphylococcus aureus* and Vancomycin-resistant *Enterococcus faecium*, with maximal inhibitory concentration (IC_{50}) values of 1.4 μM and 5.8 μM, respectively. Their findings suggested that the mechanism of antibacterial action was due to the hydrophobicity and conformational rigidity of the tetrahydropyran structure. Rodrigues et al. (2015b) used dichloromethane to isolate sphaerane bromoditerpenes, including a previously uncharacterized, rare dactylomelane called sphaerodactylomelol, from the red alga *Sphaerococcus coronopifolius*. The extracts were found to inhibit *Escherichia coli*, *Pseudomonas aeruginosa*, *S. aureus*, and *Candida albicans*. The greatest antibacterial was observed against *S. aureus*, with an IC_{50} value of 6.35 μM. Etahiri et al. (2001) isolated bromosphaerone and 12S-hydroxybromosphaerodiol from the same alga, *S. coronopifolius*. Bromosphaerone and 12S-hydroxybromosphaerodiol inhibited *S. aureus* with MIC values of 0.104 μg mL^{-1} and 0.146 μg mL^{-1}, respectively.

Kausalya and Rao (2015) assessed the antimicrobial activity of *Gelidium pusillum* and *Centroceras clavulatum* against $Gram^+$ bacteria *Bacillus subtilis*, *Micrococcus luteus*, *Staphylococcus aureus*, *Staptococcus mutans*, *Streptococcus anginosus*, *Lactobacillus acidophilus*, and $Gram^-$ bacteria *Escherichia coli*, *Enterobacter aerogenes*, *Klebsiella pneumonia*, *Pseudomonas aeuroginosa*, *Erwinia caratovora*, and *Proteus vulgaris*. Among the solvents tested, ethanolic extract of *C. clavulatum* and chloroform extract of *G. pusillum* showed a higher inhibitory activity than other solvents. Ethanolic extract of *C. clavulatum* showed maximum zone of inhibition against *B. subtilis*. The lowest minimum inhibitory concentration (MIC = 8 mg mL^{-1}) value of chloroform extract was observed against bacterial strain *B. subtilis*. Chloroform extract of *G. pusillum* showed maximum zone of inhibition against *K. pneumonia*. The lowest minimum inhibitory concentration (MIC = 12 mg mL^{-1}) value of chloroform extract was observed against *K. pneumonia*.

In the Masters' thesis work of Silva (2015), solvent extracts of wild and IMTA-cultivated *Osmundea pinnatifida* were investigated *in vitro* for their antibacterial activity against *Staphylococcus aureus* ($Gram^+$) and *Escherichia coli* ($Gram^-$). The microorganisms grew in the control wells as in the treatment wells, which means that among the six extracts in study, none gave any inhibition against these clinical isolate strains at the tested concentrations (1562.5–31.25 μg mL^{-1}). Until now, only a few studies aiming the screening of antibacterial capacity of seaweed extracts, through the broth microdilution, have been published (Kamei et al. 2009, Kim et al. 2013c). Among them, the research carried out by Hellio et al. (2000) stands out as the most relevant for this work, since they studied the antibacterial potential of *O. pinnatifida* (formerly *Laurencia pinnatifida*). They found that ethanol and dichloromethane fractions exhibited, respectively, MICs of 96 μg mL^{-1} and 24 μg mL^{-1} against *S. aureus*, and suggested that the latter extract is worthy of further investigation. However, in the particular case of *O. pinnatifida*, studies of this nature are non-existent. Nevertheless, other methodologies have been applied for this purpose. Rizvi and Shameel (2005) studied the antibacterial activity of crude methanol extracts of 14 seaweed species from Pakistan against four $Gram^+$ and five $Gram^-$ bacteria through agar well diffusion technique. At the tested concentration (200 μg 100 $μL^{-1}$ DMSO), the *O. pinnatifida* methanol extract was unable to inhibit the growth of *S. aureus*. Rizvi (2010) used the same methodology to test the antibacterial activity of crude methanol extracts of 26 species of seaweeds (collected in Pakistan) against four $Gram^+$ and seven $Gram^-$ bacteria. Once more, *O. pinnatifida* crude methanol extract (200 μg 100 $μL^{-1}$ DMSO) did not display antibacterial activity against *S. aureus*, and the same was verified against *E. coli*.

The aim of the study done by Cavallo et al. (2013) was to identify seaweeds with antibacterial activity against some pathogenic *Vibrio* species, in order to identify a possible alternative to the commonly used antibiotics in aquaculture. Chloroform/methanol lipidic extracts of several seaweed species were tested for their antibacterial activities against six fish pathogenic *Vibrio* species using the disc diffusion method. *Gracilariopsis longissima* extract had the broadest antibacterial spectrum; indeed, it was active against four *Vibrio* species, although with different inhibition strength. A previous study (Stabili et al. 2012) also demonstrated that the chloroform/methanol lipidic extract of *G. longissima* from the Mar Piccolo of Taranto (Italy) had bacteriostatic activity against *Vibrio alginolyticus* and *Vibrio vulnificus*.

Crude extracts of the red seaweeds *Gellidella acerosa*, *Gracilariopsis longissima* (formerly *Gracilaria verrucosa*), and *Hypnea musciformis* were analyzed for their antimicrobial activity against $Gram^+$ *Salmonella paratyphi*, *Enterococcus aerogenes*, *Staphylococcus epidermidis*, and $Gram^-$ *Salmonella typhi* and *Shigella flexneri* (Varier et al. 2013). Methanol, ethanol, chloroform, and aqueous solvents were used for extraction of seaweeds in their absolute forms. Disc diffusion assay was performed for crude extracts and shade-dried powdered samples of seaweeds. Among the solvent extracts, methanol extract showed best results for both positive and negative strains. Methanol and ethanol extracts of *H. musciformis* showed an inhibition zone of 19 mm and 16 mm of control (15 mm) against *S. typhi*. Chloroform extract of *G. longissima* gave the highest zone of inhibition, measuring 21 mm. Ethanol extract of *G. acerosa* also showed a zone of inhibition of 12 mm. Methanol extract of *G. acerosa* alone gave a zone of 9 mm, compared to control with 19 mm against *E. aerogenes*. Ethanol and chloroform extracts of *G. longissima* gave clearly distinct zone of inhibition measuring 8 mm and 9 mm, with respect to control (25 mm) against *S. epidermidis*. Chloroform extract of *G. longissima* showed similar zone of inhibition while its methanol extract showed 7 mm against *S. typhi*. Methanol extract of *H. musciformis* and chloroform extract of *G. longissima* showed zone of inhibition around 9 mm and 10 mm against *S. flexneri*. None of the aqueous extracts showed antibacterial activity in any of the bacterial cultures.

Six organic and aqueous extracts of 23 red marine algae collected along the Atlantic coast of Morocco were studied for antimicrobial activity (Oumaskour et al. 2013). Antimicrobial activity was tested against 10 Gram$^+$ bacteria *Bacillus cereus*, *Bacillus thuringiensis*, *Bacillus subtilis*, *Clostridium sporogenes*, *Staphylococcus aureus*, *Staphylococcus aureus* subsp. *aureus*, *Mycobacterium smegmatis*, *Enterococcus faecalis*, and *Bacillus* sp., and against two Gram$^-$ bacteria *Escherichia coli* and *Pseudomonas* sp. The positive antibacterial activity with diameter inhibition more than 10 mm was found in methanolic and methanol–dichloromethane (50:50) extract. Also, a Gram$^+$ bacteria presents a sensibility superior to the Gram$^-$, and *S. aureus* subsp. *aureus* was the more sensitive. The antibacterial activity is not uniformly distributed in the various extracts; methanolic extract exhibited greater inhibition against Gram$^+$ bacteria. Methanolic extracts of *Ellisolandia* elongata (formerly *Corallina elongata*), *Halopitys incurva*, *Gracilariopsis longissima* (formerly *Gracilaria verrucosa*), and *Sphaerococcus coronopifolius* inhibited *S. aureus* subsp. *aureus* with a diameter of the inhibition of 15 mm, while methanolic extract of *Pterosiphonia complanata* was active against *B. cereus*, *B. subtilis*, and *S. aureus* subsp. *aureus*. For methanolic extract of *Chondria dasyphylla*, the activity was obtained against *E. faecalis* and *Bacillus* sp. Inhibition zone greater than 15 mm was also observed in dichloromethane: methanol (50:50) extract of *Halopithys incurva* against *B. cereus*. For *Asparagopsis armata*, inhibition was obtained against *B. thuringensis* and *B. subtilis*, while the similar extract of *P. complanata* showed activity toward *B. cereus* and *S. aureus* subsp. *aureus*. *Gelidium corneum* (formerly *Gelidium sesquipedale*) gives a comparable activity, but against *B. cereus*, *S. aureus*, and *E. faecalis* (Oumaskour et al. 2013).

Mendes et al. (2013) characterized the antimicrobial activity of solvent extracts from *Gracilaria vermiculophylla*, *Porphyra dioica*, and *Chondrus crispus*, both from wild and from integrated multi-trophic aquaculture (IMTA). The higher potency in extracts from aquaculture species, when compared to the wild ones may be due to the environmental conditions, such as the presence of larger concentration of compounds from the breeding tanks, the constant water motion and aeration, and to the exposure to higher light intensities during longer periods of time. Aquaculture extracts of *G. vermiculophylla* and *P. dioica* presented a higher content of fatty acids. The ethyl acetate extracts predominated saturated fatty acids, especially palmitic acid, followed by polyunsaturated and monounsaturated fatty acids. Results revealed that test organisms (Gram$^-$ and Gram$^+$ bacteria, as well as one yeast species) were more sensitive to extracts obtained with dried algae, processed continuously at higher temperatures. Results from antimicrobial activity of wild and IMTA seaweed extracts showed stronger antimicrobial activity in extracts of ethyl acetate when compared with those from methanol and diethyl ether; furthermore, among the type of microorganisms tested, there was tendency for inhibition of the Gram$^+$ ones. In general, there appears to be a higher antimicrobial activity for the microorganisms under study in extracts obtained from aquaculture species, when compared to the wild ones. Extracts obtained with dried algae processed at high temperatures (with Soxhlet apparatus) presented wider inhibition zones for some of the tested microorganisms—*Salmonella enteritidis*, *Pseudomonas aeruginosa*, *Listeria innocua*, and both clinical and food isolates of *Staphylococcus aureus*.

Abou Zeid et al. (2014) demonstrated that hot and cold water-extracted polysaccharides from the red seaweed *Pterocladiella capillacea* (formerly *Pterocladia capillacea*) inhibit the growth of Gram$^+$ *Bacillus cereus* and *Staphylococcus aureus*, and Gram$^-$ *Pseudomonas fluorescens* and *Escherichia coli* in disc diffusion assays. In the case of *S. aureus*, cold water-extracted *P. capillacea* had an activity equivalent to 56.8% of the antibiotic standard, Ampicillin.

The Brazilian seaweed assemblage currently comprises 770 *taxa*, distributed along 7,367 km of coastline with highly diverse ecological conditions, suggesting a high biotechnological potential for these species (Fernandes et al. 2014). For example, extracts of *Bryothamnion* spp. show antibacterial effects and lectins from *Bryothamnion seaforthii*, *Bryothamnion triquetrum*, and *Solieria filiformis* act as antibacterials (Holanda et al. 2005, Teixeira et al. 2007b).

The study done by Jiang et al. (2013) was designated to evaluate the antibacterial activity of *Grateloupia livida* extracts *in vitro*. In total, 25 components were identified in the petroleum ether fraction, most of which have known antimicrobial activity. The MIC values against the tested bacteria (*Staphylococcus aureus*, *Escherichia coli*, and *Pseudomonas aeruginosa*) varied between 4 mg mL^{-1} and > 8 mg mL^{-1}.

Ganeshamurthy et al. (2013, 2014) evaluated the antibacterial activity of the seaweed *Portieria hornemannii* (formerly *Chondrococcus hornemanni*) against marine ornamental fish pathogens. The methanolic extracts were found effective, and had more than 80% inhibitory activity against *Aeromonas hydrophila* (20 mm) and *Vibrio parahaemolyticus* (19 mm) at a concentration (25 mg mL^{-1}) showing higher antimicrobial activity.

Several seaweeds collected from Red sea coastal waters of Jeddah, Saudi Arabia, were evaluated for their potential for bioactivity (Al-Saif et al. 2014). Extracts of the algae selected for the study were prepared using ethanol, chloroform, petroleum ether, and water, and assayed for antibacterial activity against *Escherichia coli*, *Pseudomonas aeruginosa*, *Staphylococcus aureus*, and *Enterococcus faecalis*. It was found that chloroform was most effective, followed by ethanol, petroleum ether, and water for the preparation of algal extract with significant antibacterial activities, respectively. Results also indicated that the extracts of red alga *Gracilaria dendroides* were more efficient against the tested bacterial strains.

In the work by Kavita et al. (2014), methanol extract of several seaweed samples were screened against $Gram^+$ (*Staphylococcus aureus* and *Bacillus subtilis*) and $Gram^-$ (*Escherichia coli* and *Pseudomonas aeruginosa*) bacteria. All the methanol seaweed extracts showed activity against at least one bacterium out of four (*E. coli*, *P. aerugenosa*, *S. Aureus*, and *B. subtilis*). Eleven extracts from *Palisada perforata* (formerly *Laurencia papillosa*), *Grateloupia filicina*, *Hypnea valentiae*, *Hypnea pannosa*, *Polysiphonia tuticorinensis*, showed antimicrobial activity against all four bacteria. *P. perforata* gave maximum antimicrobial activity against these bacteria. The zone of growth inhibition was slightly greater in $Gram^+$ bacteria (13–14 mm) than in $Gram^-$ (11–12 mm). This result is in agreement with the previous finding that $Gram^+$ bacteria are generally more susceptible to marine algal extract than $Gram^-$ bacteria (Salem et al. 2011). It was finally tested against four clinical $Gram^-$ isolates (*E. coli*, *P. aerugenosa*, *Klebsiella pneumoniae*, and *Shigella flexineri*) and exhibited antibacterial activity.

García-Bueno et al. (2014) observed by the microplate method that water extracts of *Grateloupia turuturu*, from the French Atlantic coast, were active against the abalone pathogen *Vibrio harveyi* strain. *G. turuturu* showed an antibacterial activity with a maximal growth inhibition in spring of around 16%. However, extracts from *Palmaria palmata* collected from the same coast were inactive.

Cortés et al. (2014) evaluated the antimicrobial activity of *Ceramium virgatum* (formerly *Ceramium rubrum*) from Chile on the bacteria *Yersinia ruckeri* and the oomycete *Saprolegnia parasitica*, causing enteric red mouth disease and saprolegniasis, respectively. The ethanol and dichloromethane extracts were effective against *S. parasitica* using the agar dilution method, and MIC values were determined by the broth dilution method. The whole extract was more active than the individual components, suggesting a synergistic effect among the components.

Methanol, *n*-hexane, and dichloromethane extracts of 12 marine macroalgae from Peniche coast (Portugal) were evaluated for their antibacterial and antifungal activity (Pinteus et al. 2015). The antibacterial activity was evaluated by disc diffusion method against *Bacillus subtilis* ($Gram^+$ bacteria) and *Escherichia coli* ($Gram^-$ bacteria). The high antibacterial activity was obtained by the *Asparagopsis armata* methanolic extract (10 mm–0.1 mg/disc), followed by the *Sphaerococcus coronopifolius n*-hexane extract (8 mm–0.1 mg/disc), and the *A. armata* dichloromethane extract (12 mm–0.3 mg/disc) against *B. subtilis*. There were no positive results against *E. coli*. Positive results were obtained against *B. subtilis* for *S. coronopifolius*, *Ceramium ciliatum*, *Plocamium cartilagineum*, and *Asparagopsis armata*. The most active species were *A. armata* and *P. cartilagineum* and the minimum active concentration against *B. subtilis* of the *n*-hexane fractions was 100 μg/disc for *P. cartilagineum*, and *S. coronopifolius* with 7 mm and 8 mm, respectively. Similarly, the minimum active concentration against *B. subtilis* of the methanolic fraction was 100 μg/disc for *A. armata* and *S. coronopifolius* with 10 mm and 7 mm, respectively. By contrast, the minimum active concentration against *B. subtilis* of the dichloromethane fraction was 300 μg/disc for *C. ciliatum*, *P. cartilagineum*, *A. armata*, and *S. coronopifolius* with 7 mm, 7 mm, 12 mm, and 7 mm, respectively. There were no positive results against *E. coli* (Pinteus et al. 2015).

The study conducted by Singh and Raadha (2015) was made to identify marine algae with antimicrobial potency. The powdered *Gracilaria corticata* var. *cylindrica* samples were extracted with a series of solvents of increasing polarity and the crude extracts were screened for their antimicrobial properties. It could either be due to the presence of a compound with a broad spectrum of antibacterial activity, or could be due to the presence of more than one compound, each having its own target of action. When the extraction of the powdered sample of *G. corticata* var. *cylindrica* using a variety of solvents was performed, petroleum ether, chloroform, and methanol extract contained an antimicrobial compound (Singh and Raadha 2015).

The crude sulfated sugars extracted from the Brazilian seaweed *Gracilaria ornata* was observed to have antimicrobial effects on *Escherichia coli*, *Bacillus subtilis*, *Enterobacter aerogens*, *Salmonella typhi*, *Salmonella choleraesuis*, *Staphylococcus aureus*, and *Pseudomonas aeruginosa* (Amorim et al. 2012). Antibacterial activities against the strains of *Staphylococcus epidermidis*, *S. aureus*, *Enterococcus feacalis*, *E. coli*, and *P. aeruginosa* were exhibited by the sugars obtained from the red algae *Corallina* sp. collected from Lebanese coast (Sebaaly et al. 2014). The literature evidenced the application of sulfated polysaccharides against plant pathogens (Vera et al. 2011) and human pathogens (Kayalvizhi et al. 2012, Pielesz and Machnicka 2014, Choi et al. 2015, Kadam et al. 2015b).

For instance, the antibacterial mode of action of 1,8-dihydroxy-anthraquinone, isolated from the red algae *Pyropia haitanensis* (formerly *Porphyra haitanensis*), against *Staphylococcus aureus* has been investigated (Wei et al. 2015). The reported results showed that the isolated molecule strongly inhibited the cell growth in the logarithmic phase, and that this antibacterial activity is related to its interaction with the cell wall and cell membrane. In fact, an increase of the permeability of the cell envelope was reported, which leads to the leakage of cytoplasm and to cell deconstruction. The authors concluded that 1,8-dihydroxy-anthraquinone represents a natural seaweed product that could be further investigated for its antibacterial activity in food safety control and drugs.

Radhika and Mohaideen (2015) used two seaweeds collected from Hare Island in the Gulf of Mannar, on Tuticorin coast (India) to evaluate its antibacterial activity. The ethanolic fractions of *Gracilaria corticata* were tested against different pathogens like *Klebsiella pneumonia*, *Aeromonas hydrophilla*, *Staphylococcus aureus*, *Escherichia coli*, and *Pseudomonas* sp., and the highest zone of inhibition was found in *K. pneumonia*. Similarly, when the crude extract of *G. corticata* was tested against the same bacterial pathogen, the highest zone of inhibition was found in *Pseudomonas* sp.

Furthermore, laminarin extracted by ultrasound from the Irish brown seaweeds *Ascophyllum nodosum* and *Laminaria hyperborea* was tested for its antimicrobial activity (Kadam et al. 2015b). Ultrasound-assisted extraction of laminarin was carried out using 60% ultrasonic power amplitude and 0.1 M hydrochloric acid for 15 minutes. High concentrations of laminarin (5.82% and 6.24%, dry weight basis, from *A. nodosum* and *L. hyperborea*, respectively) were recorded. After purification, laminarin fractions were tested for their antibacterial activities, showing great bacterial growth inhibition against *S. aureus*, *Listeria monocytogenes*, *E. coli*, and *Salmonella typhimurium*.

Kumar et al. (2015d) evaluated the antimicrobial inhibitory effect of *Champia parvula* of various solvents at the concentration 100 μg mL^{-1}, on pathogenic bacteria like *Klebsiella pneumoniae*, *Proteus vulgaris*, *Bacillus cereus*, *Bacillus subtilius*, *Staphylococcus aureus*, and *Salmonella typhi* by the disc diffusion method. This study revealed a higher zone of inhibition against *S. typhi* (15.4 mm), *B. subtilis* (13.8 mm), *S. aureus* (10.7 mm), and *P. vulgaris* (10.6 mm) in the methanol extract alone, followed by acetone, benzene, chloroform, and ethyl acetate extracts showing moderate activity against most of the pathogens, whereas chloroform extract was inactive only against *B. cereus*.

Lactones are a chemical class of cyclic esters, which includes furanones. The Australian red seaweed *Delisea pulchra* has been studied for its ability to remain free of surface bacterial colonization. Halogenated furanone extracts from *D. pulchra* have been used as effective surface sanitizers in the prevention of *Pseudomonas aeruginosa* biofilm formation. This halogenated furanone also inhibits quorum sensing mechanisms by interfering with bacterial inter-cell communication. In order for bacteria to express specific genes during quorum sensing, signaling molecules called acyl-homoserine lactones (AHLs) are required, as well as luminescence transcriptional activator (LuxR) regulatory proteins. The furanone extract from *D. pulchra* competes with AHL for the LuxR receptor site, thereby inhibiting virulence factor production and pathogenesis in *Pseudomonas aeruginosa* (Hentzer and Givskov 2003, Brameyer and Heermann 2015). Ren et al. (2004) found a similar inhibition of quorum sensing in *Escherichia coli* with a (5*Z*)-4-bromo-5-(bromomethylene)-3-butyl-2(5*H*)-furanone extract from *D. pulchra.* Quorum sensing in *Escherichia coli* was inhibited by blocking *S*-ribosylhomocysteine lyase (LuxS) mediated AI-2 signaling. This influences genes and proteins involved in the normal production of flagellar synthesis, motility, and chemotaxis in the bacterium. Manefield et al. (2001) identified another mechanism of inhibition exerted by a halogenated furanone from *D. pulchra.* The bacterium, *Erwinia carotovora*, produces carbapenem as a virulence factor during quorum sensing. A commercially available 4-bromo-5-(bromomethylene)-3-(1'-hydroxybutyl)-2(5*H*)-furanone was found to inhibit carbapenem production in *Erwinia carotovora* by disrupting the 3-oxo-C6-HSL dependent expression of the *car ABCDEFG* Hoperon. Castillo et al. (2015) also reported that a commercially produced furanone, similar to the *D. pulchra* extract, was effective against Gram$^-$ *Campylobacter jejuni.* When combined with epigallocatechin gallate from green tea and a citric acid extract, AI-2 activity, bacterial motility, and biofilm formation was significantly decreased.

In a large study of bioprospecting of several algal species from the Aegean Sea, the antibacterial activity of macroalgal extracts was evaluated (Montalvão et al. 2016). The most significant results were achieved by ethanolic extract of *Polysiphonia elongata* with growth inhibition of *Staphylococcus aureus* (70% at 100 µg mL^{-1}).

In the study done by Khan (2016), with seaweeds collected from the Karachi coast of Pakistan, MIC of tested red seaweeds extracts ranged from 5 mg mL^{-1} to 20 mg mL^{-1}. In acetone extractions, the highest MBC (40 mg mL^{-1}) was recorded in species collected from Manora aerea (*H. musciformis*) against *E. coli* and the highest value (30 mg mL^{-1}) was recorded in *G. pulvinata* collected from Buleji aerea, against *S. typhi* (see also Table 8.1) (Khan 2016).

Roohi Fatima et al. (2016) enhanced the microbial safety of farmed Mozambique tilapia fish (*Oreochromis mossambicus*) by adding an antibacterial methanol extract of the red seaweed, *Portieria hornemannii*, to housing tanks (1 part per trillion per liter). The instances of *Vibrio parahaemolyticus* infection amongst the fish were significantly decreased compared to the untreated group. Histology tests also showed the same significant decrease of *V. parahaemolyticus* in muscle tissue.

The study done by Kulshreshtha et al. (2016) evaluated the antimicrobial property of red seaweed extracts against *Salmonella enteritidis* using the *Caenorhabditis elegans* (free-living, not parasitic, transparent nematode) infection model. Six red seaweed species were tested for their antimicrobial activity against *S. enteritidis*, and 2, *Sarcodiotheca gaudichaudii* and *Chondrus crispus* were found to exhibit such properties. Spread plate assay revealed that *S. gaudichaudii* and *C. crispus* (1%, w/v) significantly reduced the growth of *S. enteritidis*. Seaweed water extracts of *S. gaudichaudii* and *C. crispus*, at concentrations from 0.4 mg mL^{-1} to 2 mg mL^{-1}, significantly reduced the growth of *S. enteritidis* (log CFU 4.5–5.3 and log 5.7–6.0, respectively). However, methanolic extracts of *C. crispus* and *S. gaudichaudii* did not affect the growth of *S. Enteritidis*. Addition of seaweed water extracts (0.2 mg mL^{-1}, *C. crispus* and *S. gaudichaudii*) significantly decreased biofilm formation and reduced the motility of *S. enteritidis*.

The antibacterial activity of *Jania rubens*, *Ellisolandia elongata* (formerly *Corallina mediterranea*), and *Pterocladiella capillacea* (formerly *Pterocladia capillacea*) were analyzed against human pathogenic bacteria (El-Din and El-Ahwany 2016). Phytochemicals were extracted from the three seaweeds using various solvents, such as methanol, ethanol, acetone, and chloroform. The agar well diffusion method was used to evaluate the antibacterial activity by measuring the zone of inhibition against *Vibrio fluvialis*. Among the three seaweeds screened for their antibacterial activity in this investigation, *P. capillacea* was found to be superior over *E. elongata* and *J. rubens* at controlling the growth of *V. fluvialis*. Among the solvents tested, methanol, ethanol, and acetone extracts of seaweeds exhibited the best activity.

The isolated compounds (Snakeol and Snakediol) extracted from *Laurencia snackeyi* were tested for their antimicrobial activity (Kamada and Vairappan 2017). These compounds showed strong antibacterial activity against *Salmonella typhi* with a MIC/MBC ratio of 2.79, 2.79, 2.79, and 2.72, respectively, indicating a bactericidal antibiosis; they also showed strong antibacterial activity against *Escherichia coli* with a MIC/MBC ratio of 3.02 and 2.76, respectively.

In the work done by Seedevi et al. (2017), the sulfated polysaccharide from *Gracilaria corticata* showed antibacterial activity against 5 clinical bacterial strains, of the 10 pathogens tested. The 100 μg mL^{-1} concentration of the sulfated polysaccharide showed a maximum of 19 mm of inhibition zone against *K. oxytoca*. The lowest inhibition zone of 12 mm was observed against *V. cholera* respectively. In the case of 75 μg mL^{-1} concentration, the sulfated polysaccharide showed the highest activity with 15 mm against *K. oxytoca*, whereas the lowest activity at 12 mm of inhibition zone was observed against *V. cholera*. At 50 μg mL^{-1} concentration, the sulfated polysaccharide showed the maximum inhibition zone of 14 mm against *K. oxytoca* and the lowest of 10 mm inhibition zone were recorded against *S. typhi* and *V. cholera*. At 25 μg mL^{-1} concentration, the sulfated polysaccharide showed 13 mm and 12 mm of inhibition zone against *K. oxytoca*; whereas the lowest of 10 mm inhibition zone was recorded in *S. typhi*, *S. paratyphi*, *S. aureus*, and *V. cholera* respectively. In addition, sulfated polysaccharide did not inhibit the following bacterial pathogens, such as *S. pyogenes*, *E. coli*, *P. mirabilis*, *V. parahaemolyticus*, and *K. pneumoniae* (Seedevi et al. 2017).

CHAPTER 9

Antiparasitic, Insecticidal, and Larvicidal Activities of Seaweeds and their Extracts

9.1 Introduction

Neglected tropical diseases (NTDs) are a diverse group of communicable diseases that prevail in tropical and subtropical conditions in 149 countries and affect more than one billion people, costing developing economies billions of dollars every year. They mainly affect populations living in poverty, without adequate sanitation, and in close contact with infectious vectors and domestic animals and livestock. Effective control against NTDs can be achieved when several public health approaches are combined. Interventions are therefore guided by local epidemiology and availability of appropriate detection, prevention, and control measures that can be delivered locally. Implementation of appropriate measures with high coverage will lead to achieving the WHO NTD Roadmap targets, resulting in the elimination of many diseases and the eradication of at least two by 2020 (WHO 2017).

The World Health Organization (WHO) developed a list of several NTD—Buruli ulcer, Chagas disease, Dengue and Chikungunya, Dracunculiasis (guinea-worm disease), Echinococcosis, Foodborne trematodiases, African trypanosomiasis (sleeping sickness), Leishmaniasis, Leprosy (Hansen's disease), Lymphatic filariasis, Onchocerciasis (river blindness), Rabies, Schistosomiasis, Soil-transmitted helminthiases, Taeniasis/Cysticercosis, Trachoma, and Yaws (endemic treponematoses). On 28 May 2016, the 69th World Health Assembly approved a resolution recognizing Mycetoma as a neglected tropical disease. The resolution also provides for a systematic, technically-driven process for evaluation and potential inclusion of additional diseases among the neglected tropical diseases.

This is in recognition of the fact that there are still many tropical, poverty-related diseases or conditions that remain neglected, and for which advocacy, awareness, and research are required to develop better diagnostic methods, treatments, and control strategies (WHO 2017).

9.2 Neglected Tropical Diseases (NTDs)

9.2.1 African trypanosomiasis

Another NTD with great impact in Africa is Human African trypanosomiasis (HAT), or sleeping sickness. It is transmitted by tsetse flies in sub-Saharan Africa. Following inoculation of infective organisms, the parasites initially proliferate at the site of a fly bite, causing local inflammation (chancre), and subsequently invade the hemolymphatic system. This is known as the early stage of the disease. Later, due to infection, the parasites invade the central nervous system, causing the meningoencephalitic, or late, stage of the disease. Without treatment, the disease progresses to coma and death. Infection models in mice indicate that African trypanosomes trigger potent inflammatory responses, and it has been suggested that survival is determined by the ability of different inbred strains to regulate inflammatory pathology (Namangala et al. 2001).

HAT is caused by the protozoan *Trypanosoma brucei* and is transmitted by insects of the genus *Glossina*, known as tsetse flies, as previously mentioned. Their parasites infect nearly 30,000 people annually, according to official data based on reported cases (Simarro et al. 2011), and another 60 million are living in risk-prone areas. Protozoan hemoflagellates belonging to the complex *T. brucei*. Two subspecies that are morphologically indistinguishable cause distinct disease patterns in humans—*T. b. gambiense* causes West African sleeping sickness and *T. b. rhodesiense* causes East African sleeping sickness. A third member of the complex, *T. b. brucei*, does not infect humans under normal conditions. The WHO reports that *T. b. gambiense* causes 98% of HAT cases (WHO 2017b).

9.2.2 Chagas disease

Chagas disease or American trypanosomiasis is a zoonotic infectious disease affecting humans in Latin America. The disease is caused by the protozoan flagellate *Trypanosoma cruzi* that lives and multiplies within cells from a variety of tissues. The parasite is usually transmitted via the feces of blood-sucking bugs belonging to the subfamily Triatominae (kissing bugs or conenose bugs), with *Triatoma infestans*, *Rhodnius proxilus*, and *Panstrongylus megistus* being the most important vectors. As *T. cruzi* cannot penetrate intact skin, it enters the human body through microlesions that have been introduced and contaminated with feces when individuals scratch the itching vector's bite. In addition, *T. cruzi* is able to break through intact mucous membranes (such as the conjunctiva and the gastric epithelium). Other modes of transmission include blood transfusion, organ transplantation, via breast milk, congenitally via the placenta and by ingestion of contaminated food and drink. Chagas disease is endemic in all South and Central American countries, as well as in Mexico (WHO 2014). In addition, the southern half of the United States contains enzootic cycles of *T. cruzi* and autochthonous vector-borne human infections have been reported in Texas, California, Tennessee, Louisiana, and Mississippi (Bern et al. 2011, Cantey et al. 2012).

9.2.3 Leishmaniasis

Leishmaniasis (or leishmaniosis) is a vector-borne disease caused by obligate protozoan parasites from the genus *Leishmania* (kingdom Protozoa, infrakingdom Excavata, order Trypanosomatida, family Trypanosomatidae), and is transmitted to humans by the bite of infected female phlebotomine (*Nyssomyia* spp., and other species) sandflies (Pinto et al. 2011, Torres et al. 2014). Leishmaniasis is endemic in large areas of the tropics, subtropics, and the Mediterranean basin, including more than 98 countries, where there are a total of 350 million people at risk and 12 million cases of infection. Canine leishmaniasis is a serious problem, and it is estimated that 2.5 million dogs are infected in the Mediterranean basin itself (Moreno and Alvar 2002, WHO 2017).

Leishmaniasis, as previously mentioned, is transmitted by the bite of infected female sandflies, whose hosts are animals such as canids, rodents, marsupials, hyraxes, or human beings. Approximately 53 *Leishmania* species have been described (without considering the synonyms and including all 5 subgenera and complexes: *Leishmania*, *Viannia*, *Sauroleishmania*, *L. enriettii* complex, and *Paraleishmania*); of these, 31 species are known to be parasites of mammals, and 20 species are pathogenic for humans. *Leishmania* parasites cause four main clinical forms of the disease, according to the location of the parasite in mammalian tissues, referred to as visceral, cutaneous, diffuse cutaneous, and mucocutaneous leishmaniasis. The most common form is cutaneous disease, and the 10 countries of Afghanistan, Algeria, Colombia, Brazil, Iran, Syria, Ethiopia, North Sudan, Costa Rica, and Peru together account for 70 to 75% of the global estimated cutaneous leishmaniosis incidences (Alvar et al. 2012, Torres et al. 2014).

9.2.4 Schistosomiasis

Schistosomiasis is the most serious form of parasitism by multicellular organisms, and still remains on the list of the NTDs prioritized by the World Health Organization (WHO 2017). The disease, also known as "bilharzia" or "snail fever", is caused by an infection with blood flukes of the genus *Schistosoma*. It predominantly affects the poor population, representing one of the main public health problems in more than 70 developing countries. Among the *Schistosoma* species, *Schistosoma mansoni* is the most widely spread in Africa and Latin America. The infection occurs when the host's skin is penetrated by the cercaria, the infectious form of the parasite life cycle. Once inside the host, they transform into schistosomula, mature, and form couples in the venous system. The egg is responsible for parasite transmission and is also the main cause of the disease symptoms (Gryssels et al. 2006). The disease affects approximately 240 million people around the world, resulting in an annual mortality rate of 280,000 people.

Current treatment is based only on two drugs—Oxamniquine, which is only effective against the *Schistosoma mansoni* species, and Praziquantel, which is ineffective against young parasites. Therefore, research on new drugs and their targets for the treatment of this disease is urgently needed (Stein et al. 2015).

9.2.5 *Filariasis*

Filariasis is a parasitic disease caused by an infection with Nematoda of the Filarioidea type (CDC 2017). These are spread by blood-feeding black flies and mosquitoes. This disease belongs to the group of diseases called helminthiases.

Eight known filarial nematodes use humans as their definitive hosts. These are divided into three groups according to the niche they occupy in the body—

- Lymphatic filariasis is caused by the worms *Wuchereria bancrofti*, *Brugia malayi*, and *Brugia timori*. These worms occupy the lymphatic system, including the lymph nodes; in chronic cases, these worms lead to the syndrome of elephantiasis;
- Subcutaneous filariasis is caused by *Loa loa* (the eye worm), *Mansonella streptocerca*, and *Onchocerca volvulus*. These worms occupy the subcutaneous layer of the skin, in the fat layer. *L. loa* causes *Loa loa* filariasis, while *O. volvulus* causes river blindness;
- Serous cavity filariasis is caused by the worms *Mansonella perstans* and *Mansonella ozzardi*, which occupy the serous cavity of the abdomen.

9.2.6 *Amoebiasis*

Amoebiasis, also known amoebic dysentery, is an infection caused by any of the amoebas of the Entamoeba group. Symptoms are most common upon infection by *Entamoeba histolytica*. Amoebiasis can be present with no, mild, or severe symptoms. Symptoms may include abdominal pain, mild diarrhea, bloody diarrhea, or severe colitis with tissue death and perforation. This last complication may cause peritonitis. People affected may develop anemia due to loss of blood (Farrar et al. 2013).

9.3 Vector-borne Diseases Control

Mosquito vectors are essential for understanding vector-borne disease transmission dynamics among human populations, because patterns of genetic structure and pathogen transfer through vector populations (Kuno 1997, Service 1997). Dengue is prevalent in more than 100 countries, and threatens the health of approximately 2.5 billion people. Around 80 million people are infected annually at an attack rate of 4% worldwide (Pancharoen et al. 2002). Mosquito control continues to be an important strategy in preventing mosquito-borne diseases (Midega et al. 2010). Diseases that are associated with the transmission of viruses from mosquitoes to humans are an expanding problem in tropical and subtropical regions (Murugan et al. 2012). Currently, most insecticides are non-selective and can be harmful to other organisms and to the environment (Baranitharan 2014). The activity of crude plant extracts is often attributed to the complex mixture of active compounds (Rahuman et al. 2009). Repeated use of synthetic insecticides for mosquito control has disrupted natural biological control systems and led to resurgences in mosquito populations (Das et al. 2007). In view of the residue problems in the environment and development of insect resistance to synthetic insecticides like DDT and chlorinated hydrocarbons (Ali et al. 2014), and the development of resistance to commercial acaricides by parasites has stimulated the search for new control strategies (Kovendan et al. 2012).

9.4 Plant Parasites (nematodes)

Among other plant parasitic nematodes, root-knot nematode (*Meloidogyne* spp.) was one of the most important pests which produced considerable losses to crop plants in different parts of the world (Taylor et al. 1982). Root-knot nematodes (*Meloidogyne* spp.) infected almost all types of plants and caused considerable damage (Adekunle and Akinlua 2007). In those areas where root-knot nematodes were not controlled, average crop yield losses were estimated to be 25%, with damage in individual fields ranging as high as 60% (Sasser et al. 1982). The interaction of nematodes with other pathogens (fungi and bacteria —see Chapters 7 and 8) increased the loss of many field crops (Maiti 1974, Hussain et al. 2013). Pesticides were generally used for the control of pests and diseases. These chemical pesticides caused human health and environmental hazards.

9.5 Seaweed as Sources of Active Compounds against Parasites and Parasite Vectors

Photosynthetic organisms have constituted the basis of traditional medicinal systems for thousands of years, from the first records dating to approximately 2600 B.C. in Mesopotamia. Ancient Egyptian, Chinese, and Indian documents show that medicine in these societies incorporated numerous plant-based remedies and preventives, most of which are still being used today to treat ailments ranging from coughs and colds, to parasitic infections, and inflammation. Today, approximately 80% of the world's population relies on traditional plant-based medicines for primary health care (Gurib-Fakim 2006, Torres et al. 2014).

The traditional use of algae for antiparasitic treatment has gained the attention of several research groups around the world, and marine secondary metabolites are now being evaluated as drug leads for the treatment of neglected diseases, such as leishmaniasis, Chagas disease, and other NTDs. Currently, there are numerous studies aiming to discover antiparasitic natural products from marine organisms. The random exploration of extracts and compounds derived from natural products to identify molecules with antileishmanial and/or trypanocidal activity, requires quantitative, fast, simple, and reproducible bioassays, and conditions that reflect those encountered by the parasite in the host cell (Sereno et al. 2007). Several biological assays involving the manipulation of *Leishmania* promastigotes and amastigotes (Berg et al. 1994), *Trypanosoma cruzi* trypomastigote and amastigotes (Buckner et al. 1996, Romanha et al. 2010), and *T. brucei* bloodstream trypomastigote form (Sykes et al. 2012, Sykes and Avery 2013) are available. Most of these methods allow the evaluation of leishmanicidal or trypanocidal activities a large number of candidate compounds (Canavaci et al. 2010, Bolhassani et al. 2011).

9.5.1 Antiparasitic and insecticidal activity of Chlorophyta (green seaweeds)

Saturated fatty acids such as capric acid, lauric acid, and myristic acid, band monounsaturated fatty acid like palmitoleic acid isolated from green seaweed *Cladophora glomerata* have been tested against mosquito larva *Aedes triseriatus*, and lethal concentration 50 (LC_{50}) values ranging from 3 to 14 ppm have been obtained (LaLonde et al. 1979).

According to Thangam and Kathiresan (1996), *Caulerpa chemnitzia* (as *Caulerpa peltata*), *Caulerpa racemosa*, *Caulerpa scalpelliformis*, *Ulva clathrata* (as *Enteromorpha clathrata*), and *Ulva intestinalis* (as *Enteromorpha intestinalis*) petroleum-ether, and acetone extracts have larvicidal activity against *Aedes aegypti* and *Culex quinquefasciatus*, with $LC_{50} < 200$ mg L^{-1}.

Crude extracts of *Caulerpa racemosa* (IC_{50} = 37.5 µg mL^{-1}), *Ulva fasciata* (IC_{50} = 50 µg mL^{-1}), *Caulerpa faridii* (IC_{50} = 34 µg mL^{-1}), *Codium flabellatum* (IC_{50} = 34 µg mL^{-1}), *Codium indicum* (as *Codium iyengarii*) (IC_{50} = 60.4 µg mL^{-1}), *Ulva reticulata* (IC_{50} = 64.75 µg mL^{-1}), and *Ulva rigida* (IC_{50} = 65.69

μg mL^{-1}) have been documented to exhibit strong activity against the promastigote form of *Leishmania major*, *in vitro* (Sabina et al. 2005).

In the search for therapeutically alternatives to antiprotozoal chemotherapy, Moo-Puc et al. (2008) collected a selection of several tropical seaweeds from the coast of Yucatan (Mexico) in order to undertake ethnopharmacological and chemotaxonomic investigations. 44% of the seaweeds studied had high to moderate anti-trichomonal activity *in vitro*. *Udotea conglutinata* showed the maximal anti-trichomonal (against *Trichomonas vaginalis* trophozoites) activity with IC_{50} values of 1.66 μg mL^{-1}, with good selectivity. So, *U. conglutinata* demonstrated promising anti-trichomonal potential.

13 seaweeds collected from the coast were extracted in methanol:dichloromethane (1:1) and tested for different range of bioactivities, including larvicidal and nematicidal activities (Manilal et al. 2009c). It was found that out of 13 seaweeds, extracts of *Chaetomorpha antennina* and *Caulerpa racemosa* showed lethal effects on root knot nematode *Meloidogyne javanica*, with LD_{50} value above 2 mg mL^{-1}. In the 3rd instar larvicidal assay, secondary metabolities of *Acrosiphonia orientalis* and *C. racemosa* were found to be highly lethal (LD_{50} value 158 and 194 μg mL^{-1}). The present study revealed that *A. orientalis* contained potential mosquito control principles and could be used for the development of biocontrol tactics.

Spavieri et al. (2010b) screened the crude extracts of *Cladophora rupestris*, *Codium fragile* subsp. *fragile* (as *Codium fragile* subsp. *tomentosoides*), *Ulva intestinalis*, and *Ulva lactuca*. The crude extracts showed antiprotozoal activity against *Trypanosoma brucei rhodesiense*, and *C. rupestris* as the most potent, exhibiting an IC_{50} = 3.7 μg mL^{-1}; only *C. rupestris* and *U. lactuta* exhibited moderate trypanocidal activity against *Trypanosoma cruzi* (IC_{50} = 80.8 and 34.9 μg mL^{-1}, respectively). All of the extracts showed leishmanicidal activity when assayed against the axenic amastigotes of *Leishmania donovani*, with IC_{50} values ranging from 12 to 20.2 μg mL^{-1} (Spavieri et al. 2010b).

With a view to explore the new molecules for therapeutic efficacy for human use, the alcoholic extracts of 33 identified species of marine flora, collected from Indian coasts, were prepared and screened for a wide range of biological activities (Lakshmi et al. 2010). Of these, one extract (from *Lyngbya majuscula*, a Cyanobacteria) showed antifilarial activity, and one other (*Chaetomorpha spiralis*—as *Chaetomorpha torta*) showed antiamoebic effect.

13 seaweeds species were collected from Mandapam coastal area and the seaweeds extracts were tested for *in vitro* antiplasmodial activity against *Plasmodium falciparum* (Ravikumar et al. 2011). Among them, *Caulerpa toxifolia* (IC_{50} = 5.06 μg mL^{-1}) showed potential antiplasmodial activity, and it is be comparable to the positive control Artemether (IC_{50} = 4.09 μg mL^{-1}). *Caulerpa chemnitzia* (as *Caulerpa peltata*) (IC_{50} = 16.69 μg mL^{-1}) also exhibited good antiplasmodial activity and the IC_{50} value is lesser than the positive control Chloroquine (IC_{50} = 19.59 μg mL^{-1}). The *in vitro* antiplasmodial activity might be due to the presence of sugars, proteins, and phenols in the ethanolic extracts of seaweeds. It is concluded from the present study that the ethanolic extracts of seaweeds of *C. toxifolia* and *C. chemnitzia* possess lead compounds for development of antiplasmodial drugs. Moreover, the extracts of *Chaetomorpha spiralis* (as *Chaetomorpha torta*), *Chaetomorpha linum* (as *Chaetomorpha crassa*), *Ulva reticulata*, and *Caulerpa racemosa* showed IC_{50} value of more than 100 μg mL^{-1}.

In the studies done by Ravikumar et al. (2011b), a total of eight seaweeds were collected from Kanyakumari district, Tamil Nadu, India. The *in vitro* antiplasmodial activity was performed against *Plasmodium falciparum*. The minimum concentration of inhibitory IC_{50} value was observed with methanolic *Chaetomorpha antennina* extracts (26.37 μg mL^{-1}); the positive controls such as Chloroquine and Artemether showed antiplasmodial activities IC_{50} with 19.10 and 6.03 μg mL^{-1} concentrations, respectively. The antiplasmodial activity of the seaweed extract might be due to the presence of sugars and phenolic compounds. From the present findings, it is concluded that the seaweed extract of *C. antennina* can be further used as a putative antiplasmodial drug in the near future.

Caulerpa veravalensis, *Caulerpa scalpelliformis*, and *Ulva fasciata*, collected from the coastal area of Kanyakumari (Tamil Nadu, India) have nymphicidal and ovicidal activities, whereas *C. veravalensis* has adultoid activity and both *C. veravalensis* and *C. scalpelliformis* ovipositional activity. Hence *C. veravalensis* chloroform extracts could be used for the red cotton bug management (*Dysdercus cingulatus*) in cotton crops (Sahayaraj et al. 2012).

Bianco et al. (2013) evaluated the antiprotozoal activity of several algae species against *Leishmania braziliensis* promastigotes/intracellular amastigotes, and against *Trypanosoma cruzi* epimastigotes/intracellular amastigotes. Extracts from *Anadyomene saldanhae* and *Caulerpa cupressoides* at 50 μg mL^{-1} showed promising results against *L. braziliensis* (87.9 and 51.7% growth inhibition, respectively). Additionally, *A. saldanhae* was effective against *L. brasiliensis* amastigotes (IC_{50}= 24 μg mL^{-1}), and *C. cupressoides* were strongly cytotoxic for bone marrow macrophages (Bianco et al. 2013).

The study of Ali et al. (2013) was made as an attempt to find out the mosquito larvicidal efficacy of ethanolic extracts of seaweeds against the vectors of several tropical diseases, namely dengue, malaria, yellow fever, and filariasis, among others. Among the seaweeds extract, *Caulerpa racemosa* showed toxicity against the 4th instar larvae of *Aedes aegypti*, *Culex quinquefasciatus*, and *Anopheles stephensi* with equivalent LC_{50} value of 0.0556, 0.0675, and 0.0661 μg mL^{-1}, respectively.

In the studies done by Yu et al. (2015), the larvicidal activity, inhibition effect on development, histopathological alteration, and morphological aberration induced by the seaweed extracts derived were evaluated. Among all the solutions tested, the chloroform partition of *Bryopsis pennata* extract exhibited the strongest larvicidal (*Aedes aegypti*) activity (LC_{50} = 82.55 μg mL^{-1}), followed by methanol extract of *B. pennata* (LC_{50} = 160.07 μg mL^{-1}). Mosquito larvae treated with seaweed extract exhibited morphological aberrations, such as damaged anal papillae, distorted body, darkened body, and pale body.

The aims of the study done by Yu et al. (2015b) were to evaluate the mosquitocidal activity of the extracts of seaweed *Bryopsis pennata* against dengue vectors *Aedes aegypti* and *Aedes albopictus*, and to determine the seaweed's toxic effect. Chloroform extract exhibited strong ovicidal activity (with LC_{50} values of 229.3 and 250.5 μg mL^{-1}) and larvicidal activity against *A. aegypti* and *A. albopictus*. The larvicidal potential of chloroform extract was further ascertained when its fraction exhibited strong toxic effect against *A. aegypti* (LC_{50} = 4.7 μg mL^{-1}) and *A. albopictus* (LC_{50} = 5.3 μg mL^{-1}). Methanol extract showed strong repellent effect against female oviposition, along with weak adulticidal activity against mosquito and weak toxicity against brine shrimp (*Artemia* sp.). The mosquitocidal results of *B. pennata* suggest further investigation for the development of effective insecticide.

9.5.2 Antiparasitic and insecticidal activity of Phaeophyceae (brown seaweeds)

Sargassum thunbergii, also known as Hede, is traditionally used as anti-helminthic (Kang 1968). Nara et al. (2005) explored the inhibition potential of extracts from some marine algae against the recombinant *Trypanosoma cruzi* dihydroorotate dehydrogenase (DHOD), an essential enzyme involved in pyrimidine biosynthesis. The extracts from two brown algae, *Fucus distichus* subsp. *evanescens* and *Silvetia babingtonii* (formerly *Pelvetia babingtonii*), showed 59 and 58% decrease in the recombinant DHOD activity, respectively, at 50 μg mL^{-1}, and caused impairment in intracellular amastigotes survival in an *in vitro* *T. cruzi*-HeLa cell infection model at 1 μg mL^{-1}. The data showed that *F. evanescens* and *S. babingtonii* possibly contain inhibitor(s) of *T. cruzi* DHOD activity against the protozoan infection and proliferation in mammalian cells (Nara et al. 2005).

Chemical analysis of the alga *Notheia anomala* collected from the rock platforms along the southern coast of Australia yielded *cis* dihydroxyte-trahydrofuran derivatives. Tetrahydrofuran from *N. anomala* are reported for the first time as potent and selective inhibitors of the larval developments of parasitic nematodes *Haemonchus contortus* and *Trichostrongylus colubriformis* (Capon et al. 1998).

According to Ara et al. (2005), *n*-hexane and methanol of *Spatoglossum asperum* showed nematicidal activity against the plant parasitic nematode, *Meloidogyne javanica*. Both *n*-hexane and methanol fractions of *S. asperum* produced more than 50% mortality of juveniles at 10 mg mL^{-1} after 24 hours. However, after 48 hours, all three fractions (*n*-hexane, chloroform, and methanol) showed more than 50% mortality of juveniles at 10 mg mL^{-1}, whereas the methanol fraction of *S. asperum* also showed more than 50% mortality after 48 hours at 1.0 mg mL^{-1}.

In the studies done by Rizvi and Shameel (2006), methanol extracts of the seaweeds collected in the coastal zones of Karachi (Pakistan), were tested for nematicidal activity against the larvae of *Meloidogyne javanica* root knot nematode. *Stoechospermum polypodioides* appeared to be the most active seaweed, as it caused 80% mortality of the nematode larvae after exposure for 72 hours to its extract. *Jolyna laminarioides* was found to be least active in its nematicidal activity since it caused only 21% mortality after the exposure of its extract for 48 as well as 72 hours.

Sargaquinoic acid, a meroterpenoid isolated from brown seaweed *Sargassum* species, has antimalarial activity against chloroquine-sensitive strain (D10) of *Plasmodium falciparum* (Afolayan et al. 2008).

The brown algae *Bifurcaria bifurcata* is able to synthesize a great number of diterpenes (Ortalo-Magné et al. 2005). The ethyl acetate extract of *B. bifurcata* showed strong trypanocidal activity (IC_{50} = 0.53 μg mL^{-1}) against *Trypanosoma brucei rhodesiense*. Bio-guided fractionation revealed that the isolated diterpene elaganolone presented mild trypanocidal activity against the bloodstream forms of *T. b. rhodesiense* (IC_{50} = 45 μM), compared to the ethyl acetate extract. These data suggest that the trypanocidal activity of the extract may be due to other minor compounds, or to the synergy of several compounds separated during the fractionation process. These extracts have previously shown potent antiprotozoal activities *in vitro* against *Plasmodium falciparum* and *Leishmania donovani*. Four ethyl acetate extracts showed activity under 10 μg mL^{-1}—*B. bifurcata*, *Dictyota dichotoma*, *Pelvetia canaliculata*, and *Sargassum muticum* against *T. brucei rhodesiense* (Galle et al. 2013).

Chinese medicine holds valuable information regarding the use of *Sargassum* seaweed, recorded in ancient manuscripts and summarized in books as the Chinese pharmacopoeia, Compendium of Materia Medica (Liu 2012d). Based on ethnopharmacological knowledge, modern phytochemical studies have recently proved the trypanocidal and leishmanicidal activity of crude extracts of *Sargassum natans* and *Sargassum oligocystum*, respectively (Orhan et al. 2006, Fouladvand et al. 2011).

Orhan et al. (2006) evaluated the *in vitro* antiprotozoal activity of ethanolic extracts of several Turkish marine macroalgae (*Dictyota dichotoma*, *Halopteris scoparia*, and *Sargassum natans*). Although none of the extracts were active against *Trypanosoma cruzi* trypomastigotes, all of the crude extracts elicited a trypanocidal activity against *Trypanosoma brucei rhodesiense* bloodstream form; moreover, the *S. natans* extract was the most active (IC_{50} = 7.4 μg mL^{-1}). Except for the marine algae *H. scoparia*, all of the extracts possessed leishmanicidal potential against axenic amastigote forms (Orhan et al. 2006).

Freile-Pelegrin et al. (2008) analyzed the aqueous and organic extracts of several species of marine algae from the Gulf of Mexico and the Caribbean coast of the Yucatan Peninsula (Mexico). The organic extracts from *Dictyota caribaea*, *Turbinaria turbinata*, and *Lobophora variegata* showed promising results against *Leishmania mexicana* promastigotes *in vitro* (IC_{50} values ranging from 10.9 to 50 μg mL^{-1}) (Freile-Pelegrin et al. 2008).

13 seaweeds collected from the coast were extracted in methanol:dichloromethane (1:1), and tested for different range of bioactivities including larvicidal and nematicidal activities (Manilal et al. 2009c). It was found that out of 13 seaweeds extracts, *Dictyota dichotoma* showed lethal effect on root knot nematode *Meloidogyne javanica*. *D. dichotoma* had lethal effects and killed 50% of nematodes when their LD_{50} value were 1.23 mg mL^{-1}. Of the 13 seaweed extracts studied, *Padina tetrastromatica* were most effective against second instar mosquito larvae, but the activity of *P. tetrastromatica* was not restored in the third instar bioassay. *P. tetrastromatica* extracts were most effective against second instar larvae and having LD_{50} value of 96 μg mL^{-1}.

Spavieri et al. (2010) evaluated the crude extracts of 21 algae against *Trypanosoma brucei rhodesiense*, *Trypanosoma cruzi*, and *Leishmania donovani*. All of the algae extracts showed significant activity against *T. b. rhodesiense*, with *Halidrys siliquosa* and *Bifurcaria bifurcata* (Sargassaceae) being the most potent (IC_{50} = 1.2 and 1.9 μg mL^{-1}, respectively). All the algal extracts also displayed leishmanicidal activity, with *H. siliquosa* and *B. bifurcata* again being the most active (IC_{50} = 6.4 and 8.6 μg mL^{-1}, respectively) (Spavieri et al. 2010).

In 2011, Kar et al. showed that fucoidan administered to BALB/c mice infected with antimony-susceptible or antimony-resistant *Leishmania donovani* strains showed inhibitory effects on the amastigotes of both strains and resulted in a pronounced decrease in parasite burden (200 mg/kg/day; thrice/daily). They further demonstrated that fucoidan induced a protective host cytokine response and significantly increased the ROS and NO levels in infected macrophages, which may have inhibited parasite multiplication (Kar et al. 2011).

In the study of Sultana et al. (2011), *Spatoglossum variabile* showed more or less similar suppressive effect on root knot nematode (*Meloidogyne javanica*) like chemical pesticide and Carbofuran (nematicide).

However, mixed application of *S. variabile* with Carbofuran caused maximum reduction in nematode's penetration in roots and produced greater fresh shoot weight, root length, and maximum yield of tomato under field condition, which is in agreement with the previous study (Sultana et al. 2010), with extracts of *Spatoglossum asperum* and *Sargassum swartzii*.

Vonthron-Sénécheau et al. (2011) screened the hydro-alcoholic and ethyl acetate extracts of several seaweeds from the coast of Normandy, France. The ethyl acetate extracts were more active than the hydro-alcoholic extracts. The most active extract against *Leishmania donovani* axenic amastigotes was the ethyl acetate extract of *Bifurcaria bifurcata*, which had an IC_{50} = 3.9 μg mL^{-1}. Nevertheless, *Dictyopteris polypodioides* (IC_{50} = 10.8 μg mL^{-1}) had higher IC_{50} values than *B. bifurcata*, as they were selective for the parasite than for mammalian cells. The extracts did not show activity against *Trypanosoma cruzi* (Vonthron-Sénécheau et al. 2011).

In the study done by Ravikumar et al. (2011), the larvicidal activity of the seaweed extracts were evaluated against *Aedes aegypti* larvae. The extract of *Dictyota dichotoma* showed minimum level of LC_{50} value of 0.068 μg mL^{-1} and LC_{90} value of 0.140 μg mL^{-1}. The results of the preliminary phytochemical showed the presence of saponin, steroids, terpenoid, phenols, protein, and sugars. So, the ethanolic extracts of *D. dichotoma* possess larvicidal activity.

Dictyota friabilis (as *Dictyota pfaffii*) and *Canistrocarpus cervicornis* have shown antileishmanial activity (Santos et al. 2011, Soares et al. 2012b). Soares and coworkers showed that the diterpene, dolabelladienetriol, obtained from *D. friabilis*, exhibits leishmanicidal activity against intracellular amastigotes (IC_{50} = 44 μM) and anti-HIV-1 activity (see also Chapter 5). These data are outstanding because HIV-1 is known to exacerbate the *Leishmania* load in macrophage infection; therefore, the leishmanicidal and anti-HIV-1 activities of dolabelladienetriol make it a promising candidate for leishmaniasis chemotherapy, either in isolated cases or in cases associated with HIV-1 (Soares et al. 2012).

Bianco et al. (2013) evaluated the antiprotozoal activity of several algae species against *Leishmania braziliensis* promastigotes/intracellular amastigotes, and against *Trypanosoma cruzi* epimastigotes/intracellular amastigotes. Extracts from *Canistrocarpus cervicornis*, *Dictyota* sp., and *Padina* sp. at 50 μg mL^{-1} showed promising results against *L. braziliensis* (85.9, 93.3, and 80.9% growth inhibition, respectively). Only *Dictyota* sp. was effective against *T. cruzi* (60.4% growth inhibition). Additionally, *Padina* sp. was effective against *L. brasiliensis* amastigotes (IC_{50}= 40 μg mL^{-1}), and *C. cervicornis* was strongly cytotoxic for bone marrow macrophages (Bianco et al. 2013).

A tropical disease was widely distributed with 44 million people affected by common chronic manifestation and 120 millions of people infected worldwide by the mosquito *Culex quinquefasciatus*, a vector of lymphatic filariasis. Different alternative methods are being tested in different parts of the world to reduce the problems associated with commercial insecticides. Use of synthetic pesticides develops resistance in mosquito's enzyme systems, which results in sequestration of the insecticide. The study done by Nisha and Poonguzhali (2013) was undertaken to investigate the larvicidal potential of *Sargassum swartzii* (as *Sargassum wightii*) against the mosquito vector *C. quinquefasciatus*. Acetone extract of *S. swartzii* was found to be effective. This may be due to the presence of sulfated polysaccharides, saponins, dioctyl phthalate, tannins, and other cytotoxic bioactive compounds present in the seaweed. The larvicidal activity of *S. swartzii* (as *S. wightii*) extracts was confirmed by the works of Nisha and Poonguzhali (2013), Achary et al. (2014), and Kumar et al. (2012), against *Aedes aegypti* and *Anopheles sundaicus* vectors, respectively.

In the study done by Karthick et al. (2014), the seaweed extracts were tested, *in vitro*, into two different concentrations (2 and 4 mg mL^{-1}). *Stoechospermum polypodioides* showed high nematicidal activity in both the concentrations of methanol extracts of tested seaweeds. In other studies, the aqueous and ethanol extracts of *S. polypodioides* were also found to display strong nematicidal activities against the larvae of *Meloidogyne javanica* (Abid et al. 1993, Sultana et al. 2000). When compared to green and red seaweeds, the marine brown algae from the coast of Pakistan as well as those of other coastal areas have always revealed much stronger nematicidal activities against juveniles (Whapham et al. 1994, Ara et al. 1996, Sultana et al. 2000, Noreen et al. 2002, Zaki et al. 2005).

Lobophora variegata extracts presented antiprotozoal activities against six protozoan parasites, namely *Trichomonas vaginalis* (a common and worldwide parasite which infects the urogenital tract of men and women), *Entamoeba histolytica* (parasite infecting humans and other primates), *Giardia intestinalis* (responsible for enteric protozoan infections), *Schizochytrium aggregatum* (Labyrinthulomycota), *Leishmania mexicana* (one of the causative species of leishmaniasis), and *Trypanosoma cruzi* (causative species of trypanomiasis) (Vieira et al. 2015). Sulfoquinovosyldiacylglycerols (SQDGs) extracted from *L. vareigata* were shown to exhibit an *in vitro* antiprotozoal activity against *Entamoeba histolytica* with an IC_{50} of 3.9 µg mL^{-1}, and a moderate activity against *T. vaginalis* trophozoites with an IC_{50} of 8 µg mL^{-1}. Engel et al. (2006) observed differences in the antiprotozoal activities of both *Lobophora* studied earlier. While both hydrophilic and lipophilic parts of the organic extract of the crustose type inhibited the growth of *Schizochytrium aggregatum*, only the lipophilic part of the ruffled type showed a significant inhibition (Engel et al. 2006). The organic extract exhibited anti-trichomonal activity with an IC_{50} of 1.39 µg mL^{-1} (Moo-Puc et al. 2008), an IC_{50} of 3.2 µg mL^{-1} against *Trichomonas vaginalis* (Cantillo-Ciau et al. 2010). The same extract exhibited a moderate *in vitro* antiprotozoal activity against *Trypanosoma cruzi* with an IC_{50} of 9.72 µg mL^{-1} (León-Deniz et al. 2009). Cantillo-Ciau et al. (2010) identified three sulfoquinovosyldiacylglycerols (SQDGs; 1-*O*-palmitoyl-2-*O*-myristoyl-3-*O*-(6'''-sulfo-α-*D*-quinovopyranosyl) glycerol, 1,2-di-*O*-palmitoyl-3-O-(6'''-sulfo-α-*D*-quinovopyranosyl) glycerol, and 1-Opalmitoyl-2-*O*-oleoyl-3-*O*-(6'''-sulfo-α-*D*-quinovopyranosyl) glycerol with antiprotozoal activity from a lipophilic fraction.

Malaria leads to severe anemia and cycles of fever as the parasite takes hold. Fucoidan is one agent that shows some promise as an inhibitory agent for *Plasmodium* infection *in vitro* and in mouse models (Chen et al. 2009). *Undaria pinnatifida* fucoidan significantly inhibited the invasion of erythrocytes by *Plasmodium falciparum* merozoites. Four-day suppressive testing in *Plasmodium berghei*-infected mice (a model system) with fucoidan resulted in a 37% suppressive effect versus the control group and a significant delay in the deaths from anemia.

Alcoholic extracts of eight different types of seaweeds from Iran's Persian Gulf were tested for their antimalarial and acetylcholinesterase enzyme (AChE) inhibitory activities (Ghannadi et al. 2013). The extract of *Sargassum boveanum* showed the highest AChE inhibitory activity (IC_{50} = 1 mg mL^{-1}), while *Polycladia indica* (as *Cystoseira indica*) exhibited the least activity (IC_{50} = 11 mg mL^{-1}).

In the study done by Bantoto and Dy (2013), the crude extracts of Philippine *Padina minor* and *Dictyota dichotoma* var. *intricata* (as *Dictyota linearis*) were compared for their larvicidal activity and established their LC_{50} and LC_{90} so that we can further compare the values with other similar studies. If a potent larvicidal activity is detected, it may serve as a suitable alternative to expensive synthetic insecticides. At 20 mg mL^{-1}, no dead larvae were observed in *D. dichotoma* var. *intricata* extract, while in *P. minor*, 8% of the larvae died. From 40 to 80 mg mL^{-1}, the number of dead larvae increased with the *P. minor* extract having a consistently higher percentage of larval mortality than the *D. dichotoma* var. *intricata* extract. At 100 mg mL^{-1}, both the extracts showed 100% larval mortality. In the positive and negative controls, no dead larva was observed after 24 hours of exposure. The LC_{50} of *D. dichotoma* var. *intricata* was 60 mg mL^{-1}. *P. minor* has lower LC_{50} than *D. dichotoma* var. *intricata* at 50.8 mg mL^{-1}. Likewise, *P. minor* extract has a lower LC_{90} (84 mg mL^{-1}) than *D. dichotoma* var. *intricata* (91.6 mg mL^{-1}). *P. minor* has significantly lower LC_{50} than *D. dichotoma* var. *intricata* but at LC_{90}, their effectiveness was comparable. Hence, the extract of *P. minor* is more potent than *D. dichotoma* var. *intricata*.

In the study done by Khan et al. (2015), 32 seaweeds from Karachi coast (Pakistan) were evaluated to determine nematicidal activity against *Meloidogyne javanica* (egg hatching and larval mortality tests) *in vitro*. It is observed that *Sargassum tenerrimum* and *Padina tetrastromatica* showed maximum egg hatching (96%) and larval mortality (99%) and (100%), respectively, in water and methanol extract at 10% concentration, after exposure for 72 hours.

Stein et al. (2015) screened *in vitro* the extracts of several seaweeds from Brazil against *Schistosoma mansoni*. The extracts displaying activity at 500 µg mL^{-1} exposure were considered to be active and subjected to further testing at a lower concentration (100 µg mL^{-1}). The three *Dictyota* species showed activity at the two concentrations tested. The supercritical extract of *D. menstrualis* kills very effectively both males and females after exposure for 24 hours. Additionally, *D. dichotoma* submitted to supercritical extraction kills the *Schistosoma* worms after 72 hours. Surprisingly, *D. mertensii* extracted with dichloromethane/methanol kills males after 72 hours and females after 120 hours of incubation. *Padina gymnospora* extracted with dichloromethane/methanol did not show anthelmintic activity (Stein et al. 2015).

The study done by Valentina et al. (2015) was undertaken to investigate the larvicidal and pupicidal potential of the seaweeds of *Turbinaria conoides* against the mosquito species *Aedes aegypti*, *Anopheles stephensi*, and *Culex quinquefasciatus*. By using three different solvent seaweed extracts (aqueous, acetone, and ethanol), LC_{50} and LC_{90} values were found, and the pupae mortality was also analyzed. The values of larvicidal activity against *A. aegypti* were recorded and LC_{50} were found to be 18.74 mg L^{-1} of aqueous extract followed by 6427 mg L^{-1} of ethanol extract and 100.07 mg L^{-1} of acetone extract of *T. conoides*, against *A. aegypti*. The LC_{90} values of *T. conoides* aqueous and acetone extracts were 164.59 and 269.76 mg L^{-1}, respectively. The order of hierarchy of bio-larvicidal activity of the three different extracts of *T. conoides* against *A. aegypti* was found to be aqueous > ethanol > acetone at the LC_{50} level. The results showed that the aqueous extract of seaweed exhibited a high degree of blocking the development by induction of great mortality of larvae and pupae.

9.5.3 Antiparasitic and insecticidal activity of Rhodophyta (red seaweeds)

α-Kainic acid, an aminoacid isolated from the red seaweed *Digenea simplex*, exhibits antihelmintic activity against roundworm *Ascaris lumbricoides* (Rim et al. 1974), and also exhibits killing effects towards American cockroach *Periplaneta americana* (Maeda et al. 1984).

Sesquiterpenes consisting of three isoprene units are known for their bioactivity. Laurepinnacin and Isolaurepinnacin, which are acetylinic sesquiterpene ethers isolated from red seaweed *Laurencia pinnata*, are potent toward Azuki bean beetle *Callosobruchus chinensis* (Fukuzawa and Masamune 1981).

Terpenes are secondary metabolites made up of isoprene units. Monoterpenes are terpenes that consist of two isoprene units. The halogenated monoterpenes isolated from *Plocamium cartilagineum*, such as violacene, mertensene, dibromomertensene, dihydromertensene, and 1,4,6-trichloro-3-(2′-chlorovinyl)-1,3-dimethylcycohexane exhibit insecticidal potential against various insects (San-Martin et al. 1991, Argandoña et al. 2000).

The pharmacological potential of marine algae as sources of new treatments for parasitic disease is proven by the Kainic acid, an amino acid content isolated from the tropical species *Digenea simplex* (Rhodophyta, Ceramiales). This species has been known for its anti-helminthic and insecticidal properties in East Asian countries for more than 1,000 years. *Laurencia microcladia* and *Jania capillacea* from the Gulf of Mexico have been used for their anti-helminthic and antiprotozoal properties (Nitta et al. 1958, Moo-Puc et al. 2008).

In the study done by Naqvi et al. (1992), 25 species of marine algae of Karachi coast were tested for nematicidal activity, but positive results were obtained with only some of them. Among the red algae positive extracts, *Centroceras clavulatum* reduced the number of active nematodes by 95%, followed by *Cystoclonium purpureum* (78%), and *Scinaia fascicularis* (44%).

Chondriamide, a new bis(indole) amide and 3-indolacrylamide have been isolated from the red algae *Chondria atropurpurea*, and showed anthelmintic activity against *Nippostrongylus brasiliensis* (Nematoda) (Davyt et al. 1998).

11 sesquiterpenes and one long chain aldehyde have been isolated from the dichloromethane extract of the red alga *Laurencia dendroidea* (as *Laurencia scoparia*) (Davyt et al. 2001). Four of them are new natural products. Scopariol is a new natural product with an unusual rearranged chamigrane-type structure. The *in vitro* activity of compounds against the parasitant stage of *Nippostrongylus brasiliensis* (nematode that infects rodents, primarily rats) was studied. Some of the seaweed extracts demonstrated *in vitro* moderate anthelmintic activity (lower than 100 μM), and none of the compounds assayed were more active than the reference drugs (e.g., albendazole, EC_{50} = 0.34 μM).

Diterpenes with four isoprene units can be isolated from various seaweeds. Seven brominated diterpenes of the parguerene and isoparguerene series derived from red seaweed *Jania rubens* exhibit marked anthelmintic effects against earthworm *Allolobophora caliginosa* (Awad 2004).

Crude extracts of *Osmundea pinnatifida* (formerly *Laurencia pinnatifida*) (IC_{50} = 6.25 μg mL^{-1}), *Melanothamnus afaqhusainii* (IC_{50} = 32.6 μg mL^{-1}), *Gracilaria corticata* (IC_{50} = 38 μg mL^{-1}), *Scinaia hatei* (IC_{50} = 14.1 μg mL^{-1}), *Scinaia moniliformis* (as *Scinaia indica*) (IC_{50} = 59.6 μg mL^{-1}), *Centroceras clavulatum* (IC_{50} = 57.89 μg mL^{-1}), and *Botryocladia leptopoda* (IC_{50} = 60.81 μg mL^{-1}) have been documented to exhibit strong activity against the promastigote form of *Leishmania major*, *in vitro* (Sabina et al. 2005).

A lectin from the marine red alga *Gracilaria ornata* was purified and characterized (Leite et al. 2005). *G. ornata* lectin significantly affected the development of *Callosobruchus maculatus* larvae (Cowpea weevil or Cowpea seed beetle), indicating the possibility of using this lectin in a biotechnological strategy for insect management of stored Cowpea seeds.

Aqueous and organic extracts of several species of marine algae collected from the Gulf of Mexico and Caribbean coast of the Yucatan Peninsula (Mexico) were evaluated for their antileishmanial *in vitro* activity against *Leishmania mexicana* promastigote forms (Freile-Pelegrin et al. 2008). Organic extracts from *Laurencia microcladia* possessed promising *in vitro* activity against *L. mexicana* promastigotes (LC_{50} values ranging from 10.9 to 49.9 μg mL^{-1}).

The sesquiterpene (8*R*)-8-bromo-10-*epi*-β-snyderol, isolated from the chloroform-methanol extract of *Laurencia obtusa*, showed antimalarial activity, with IC_{50} values of 2700 and 4000 ng mL^{-1} against the D6 and W2 clones of *Plasmodium falciparum*, respectively (Topcu et al. 2003).

Bromophycolides were isolated from extracts of the Fijian red alga *Callophycus serratus* (Lane et al. 2009), and among these bromophycolides, several exhibited activities in the low micromolar range against the human malaria parasite, *Plasmodium falciparum* (malaria tropica), prompting evaluation of antimalarial activities for previously reported bromophycolides A-I and debromophycolide A. Bromophycolides A, D, E, H, and M, representing both 15- and 16-membered lactone frameworks, exhibited potent antimalarial activity with $IC_{50's}$ of 0.3–0.9 μM, suggesting that neither mode of lactonization confers an inherent bioactivity advantage. Furthermore, a macrolide motif appears to be essential for antimalarial activity, considering that non-macrocyclic callophycoic acids and callophycols also isolated from *C. serratus* were less active against *P. falciparum* (Lane et al. 2009).

Crude extracts of *Boergeseniella fruticulosa*, *Calliblepharis jubata*, *Ceramium virgatum*, *Chylocladia verticillata*, *Claviclonium ovatum*, *Corallina officinalis*, *Cryptopleura ramosa*, *Cystoclonium purpureum*, *Dumontia contorta* (as *Dumontia incrassata*), *Furcellaria lumbricalis*, *Gelidium pulchellum*, *Gracilaria gracilis*, *Halopithys incurva*, *Halurus equisetifolius*, *Jania rubens*, *Lomentaria articulata*, *Mastocarpus stellatus*, *Osmundea hybrida*, *Osmundea pinnatifida*, *Plocamium cartilagineum*, *Polyides rotundus*, *Pyropia leucosticta* (as *Porphyra leucosticta*) and *Porphyra linearis* were evaluated for biological activity against *Trypanosoma brucei rhodesiense*, *Trypanosoma cruzi* trypomastigote, and *Leishmania donovani* axenic amastigotes. All the algal extracts showed activity against the *T. brucei rhodesiense* bloodstream form, with *C. officinalis* and *C. virgatum* being the most potent (IC_{50} values of 4.8 and 5.5 μg mL^{-1}, respectively). Except for *P. leucosticta*, the extracts from all the seaweeds elicit leishmanicidal activity with IC_{50} values ranging from 16.5 to 85.6 μg mL^{-1}. None of the algal extracts inhibited the growth of *T. cruzi* (Allmendinger et al. 2010).

Elatol, a halogenated sesquiterpene isolated from red seaweed *Laurencia dendroidea* and *Laurencia microcladia*, exhibits potent larvicidal effects against mosquito *Aedes aegypti*; the compound also has antileishmanial, antitumor, acaricidal, and repellent activities (Santos et al. 2010, Born et al. 2012, Campos et al. 2012). Another sesquiterpene isolated from red seaweed *Laurencia nipponica*, deoxyprepacifenol, exhibits larvicidal activity against mosquito *Culex pipiens* (Watanabe et al. 1989).

Felício et al. (2010) reported that the *n*-hexane and dichloromethane fractions of *Bostrychia tenella* from the Sao Paulo Coast, Brazil, showed activity against *Trypanosoma cruzi* trypomastigotes and *Leishmania amazonensis* promastigotes. In a trypanocidal assay, the *n*-hexane and dichloromethane fractions showed IC_{50} values of 16.8 and 19.1 μg mL^{-1}, respectively. For the leishmanicidal assay, the *n*-hexane and dichloromethane sub-fractions (obtained by chromatographic methods) were active against *L. amazonensis* promastigotes, exhibiting IC_{50} values of 1.5, 2.7, 4.4, and 4.3 μg mL^{-1}, respectively (Felício et al. 2010).

The sesquiterpenes elatol and triquinane obtained from the Brazilian red algae *Laurencia dendroidea*, showed antiprotozoal activity but no cytotoxicity to mammalian cells. Elatol is the major constituent of *L. dendroidea* and showed trypanocidal activity against the trypomastigotes (IC_{50} = 1.38 μM) and amastigotes (IC_{50} = 1.01 μM) of *Trypanosoma cruzi* (Veiga-Santos et al. 2010). This compound also

Table 9.1 Seaweed extracts with larvicidal activity (LC_{50} < 100 mg L^{-1}) against mosquito larvae (vector), or antiparasitic activity.

Species	Solvent	Parasite or vector species	LC_{50} or IC_{50} values	References
Chlorophyta (Green seaweed)				
Acrosiphonia orientalis	Dichloromethane: methanol (1:1) Methanol	*Culex quinquefasciatus* *Aedes aegypti* *C. quinquefasciatus*	100 μg mL^{-1}, mortality 62.6% 2nd instar, 86.13 μg mL^{-1} 2nd instar, 94.42 μg mL^{-1}	Manilal et al. 2009c Manilal et al. 2011b
Anadyomene saldanhae	*n*-hexane	*Leishmania braziliensis*	Amastigotes, 24 μg mL^{-1}	Bianco et al. 2013
Bryopsis pennata	Chloroform Methanol	*A. aegypti*	82.55 μg mL^{-1} 160.07 μg mL^{-1}	Yu et al. 2015
	Chloroform	*Aedes albopictus* *A. aegypti*	250.5 μg mL^{-1} 229.3 μg mL^{-1}	Yu et al. 2015b
Caulerpa chemnitzia (as *Caulerpa peltata*)	Petroleum-ether, acetone Ethanol	*A. aegypti, C. quinquefasciatus* *Plasmodium falciparum*	< 100 mg L^{-1} 16.69 μg mL^{-1}	Thangam and Kathiresan 1996 Ravikumar et al. 2011
C. cupressoides	*n*-hexane	*Leishmania braziliensis*	50 μg mL^{-1}	Bianco et al. 2013
C. faridii	Crude extracts	*Leishmania major*	34 μg mL^{-1}	Sabina et al. 2005
C. racemosa	Petroleum-ether, acetone Crude extracts Methanol: dichloromethane (1:1)	*A. aegypti, C. quinquefasciatus* *Leishmania major* *Meloidogyne javanica*	< 100 mg L^{-1} 37.5 μg mL^{-1} 2 mg mL^{-1}	Thangam and Kathiresan 1996 Sabina et al. 2005 Manilal et al. 2009c
C. racemosa	Methanol: dichloromethane (1:1)	*C. quinquefasciatus*	3rd instar 194 μg mL^{-1}	Manilal et al. 2009c
C. racemosa	Caulerpinic acid	*Culex pipiens*	2nd instar, 1.42 ppm	Alarif et al. 2010
C. racemosa	Caulerpin	*C. pipiens*	2nd instar, 1.42 ppm 3rd instar, 1.81 ppm 4th instar, 1.99 ppm	Alarif et al. 2010
C. racemosa	Ethanol: water (3:1)	*A. aegypti* *C. quinquefasciatus* *Anopheles stephensi*	0.0556 μg mL^{-1} 0.0675 μg mL^{-1} 0.0661 μg mL^{-1}	Ali et al. 2013

Table 9.1 contd....

Table 9.1 contd....

Species	Solvent	Parasite or vector species	LC_{50} or IC_{50} values	References
C. scalpelliformis	Acetone Methanol: acetone (1:9) Methanol: acetone (2:8) Methanol-acetone (3:7)	*A. aegypti*	53.7 mg L^{-1} 31.62 mg L^{-1} 15.85 mg L^{-1} 23.44 mg L^{-1}	Thangam and Kathiresan 1991
C. scalpelliformis	Petroleum-ether, acetone	*A. aegypti, C. quinquefasciatus*	< 100 mg L^{-1}	Thiangam and Kathiresan 1996
C. scalpelliformis	Chloroform	*Dysdercus cingulatus*	3rd instar nymphs, 108.23 mg L^{-1}	Sahayaraj et al. 2012
C. scalpelliformis	Ethanol: water (3:1)	*A. aegypti* *C. quinquefasciatus* *A. stephensi*	0.070 µg mL^{-1} 0.067 µg mL^{-1} 0.066 µg mL^{-1}	Ali et al. 2013
C. toxifolia	Ethanol	*Plasmodium falciparum*	5.06 µg mL^{-1}	Ravikumar et al. 2011
C. veravalensis	Chloroform	*D. cingulatus*	3rd instar nymphs, 59.37 mg L^{-1}	Sahayaraj et al. 2012
Chaetomorpha antennina	Methanol: dichloromethane (1:1) Methanol	*M. javanica* *P. falciparum*	> 2 mg mL^{-1} 26.37 g mL^{-1}	Manilal et al. 2009c Ravikumar et al. 2011b
C. linum (as *C. crassa*)	Ethanol	*P. falciparum*	> 100 µg mL^{-1}	Ravikumar et al. 2011
C. spiralis (as *C. torta*)	Ethanol	*P. falciparum*	> 100 µg mL^{-1}	Ravikumar et al. 2011
*Cladophora glomerata**	Methanol	*Aedes triseriatus*	LC_{50} of capric [10:0], lauric [12:0], myristic [14:0] and palmitoleic [16:1] are 14, 7, 4 and 3 ppm, respectively	LaLonde et al. 1979
C. rupestris	Crude extracts	*Leishmania donovani* *Trypanosoma brucei rhodesiense* *T. cruzi*	12–20.2 µg mL^{-1} 3.7 µg mL^{-1} 80.8 µg mL^{-1}	Spavieri et al. 2010b
C. rupestris	Chloroform: methanol (3:1)	*P. falciparum* blood stage *P. berghei* liver stage	11.9 µg mL^{-1} 37.3 µg mL^{-1}	Spavieri et al. 2013
Codium flabellatum	Crude extracts	*L. major*	Promastigote, 34 µg mL^{-1}	Sabina et al. 2005
C. fragile subsp. *fragile* (as *C. fragile* subsp. *tomentosoides*)	Crude extracts	*L. donovani*	12–20.2 µg mL^{-1}	Spavieri et al. 2010b
C. fragile subsp. *fragile* (as *C. fragile* subsp. *tomentosoides*)	Chloroform: methanol (3:1)	*P. falciparum* blood stage *P. berghei* liver stage	11.8 µg mL^{-1} 34.6 µg mL^{-1}	Spavieri et al. 2013
C. indicum (as *C. iyengarii*)	Crude extracts	*L. major*	Promastigote, 60.4 µg mL^{-1}	Sabina et al. 2005

Microdictyon pseudohapteron	Petroleum ether fraction	*C. quinquefasciatus*	50 mg L^{-1}	Devi et al. 1997
Udotea conglutinata	Dichloromethane: methanol (7:3)	*Trichomonas vaginalis*	1.66 μg mL^{-1}	Moo-Puc et al. 2008
Ulva clathrata (as *Enteromorpha clathrata*)	Petroleum-ether, acetone	*A. aegypti* *C. quinquefasciatus*	< 200 mg L^{-1}	Thangam and Kathiresan 1996
U. intestinalis (as *E. intestinalis*)	Petroleum-ether, acetone	*A. aegypti* *C. quinquefasciatus*	< 200 mg L^{-1}	Thangam and Kathiresan 1996
U. intestinalis	Chloroform: methanol (3:1)	*P. falciparum* blood stage	18.2 μg mL^{-1}	Spavieri et al. 2013
U. lactuta	Crude extracts	*L. donovani* *T. cruzi*	Amastigotes, 12–20.2 μg mL^{-1} 34.9 μg mL^{-1}	Spavieri et al. 2010b
U. lactuca	Chloroform: methanol (3:1)	*P. falciparum* blood stage *P. berghei* liver stage	3.8 μg mL^{-1} 14.9 μg mL^{-1}	Spavieri et al. 2013
U. linza (as *U. fasciata*)	Crude extracts Chloroform	*L. major* *Dysdercus cingulatus*	50 μg mL^{-1} Nymphicidal and ovicidal activities	Sabina et al. 2005 Sahayaraj et al. 2012
U. intestinalis	Crude extracts	*L. donovani*	Amastigotes, 12–20.2 μg mL^{-1}	Spavieri et al. 2010b
Ulva reticulata	Crude extracts Methanol	*L. major* *P. falciparum*	64.75 μg mL^{-1} > 100 μg mL^{-1}	Sabina et al. 2005 Ravikumar et al. 2011
U. rigida	Crude extracts	*L. major*	65.69 μg mL^{-1}	Sabina et al. 2005
Phaeophyceae (Brown seaweed)				
Bifurcaria bifurcata	Ethyl acetate	*Trypanosoma brucei rhodesiense*	0.53 μg mL^{-1}	Ortalo-Magné et al. 2005
B. bifurcata	Crude extracts	*T. b. rhodesiense*	1.9 μg mL^{-1}	Spavieri et al. 2010
B. bifurcata	Ethyl acetate	*Leishmania donovani*	3.9 μg mL^{-1}	Vonthron-Sénécheau et al. 2011
B. bifurcata	Ethyl acetate	*L. donovani, Plasmodium falciparum*	10 μg mL^{-1}	Galle et al. 2013
Canistrocarpus cervicornis	Organic extracts	*Leishmania braziliensis*	50 μg mL^{-1}	Bianco et al. 2013
Cystoseira baccata	Chloroform: methanol (3:1)	*P. falciparum* blood stage *P. berghei* liver stage	3.4 μg mL^{-1} 32.6 μg mL^{-1}	Spavieri et al. 2013
C. tamariscifolia	Chloroform: methanol (3:1)	*P. falciparum* blood stage *P. berghei* liver stage	3.3 μg mL^{-1} 49.4 μg mL^{-1}	Spavieri et al. 2013
Dictyopteris polypodioides	Ethyl acetate	*L. donovani*	10.8 μg mL^{-1}	Vonthron-Sénécheau et al. 2011

Table 9.1 contd....

Table 9.1 contd....

Species	Solvent	Parasite or vector species	LC_{50} or IC_{50} values	References
Dictyota caribaea	Organic extracts	*Leishmania mexicana*	Promastigotes, 10.9–50 μg mL^{-1}	Freile-Pelegrin et al. 2008
D. dichotoma	Acetone Methanol: acetone (1:9) Methanol: acetone (2:8)	*A. aegypti*	61.65 mg L^{-1} 25.70 mg L^{-1} 28.21 mg L^{-1}	Thangam and Kathiresan 1991
D. dichotoma	Ethanol	*T. cruzei* *L. denovani* *Potamogeton falciparum*	17.1 μg mL^{-1} 52.0 μg mL^{-1} 33.9 μg mL^{-1}	Orhan et al. 2006
D. dichotoma	Methanol: dichloromethane (1:1)	*M. javanica*	1.23 mg mL^{-1}	Manilal et al. 2009c
D. dichotoma	Ethanol	*A. aegypti*	Larva, 0.140 μg mL^{-1}	Ravikumar et al. 2011
D. dichotoma	Ethyl acetate	*T. brucei rhodesiense*	10 μg mL^{-1}	Galle et al. 2013
D. dichotoma	Supercritical extraction	*Schistosoma mansoni*	100 mg mL^{-1}, 24 h	Stein et al. 2015
D. dichotoma var. *intricata* (as *D. linearis*)	Ethanol	*A. aegypti*	100 mg mL^{-1}	Bantoto and Dy 2013
D. friabilis (as *D. pfaffii*)	Dolabelladienetriol	*Leishmania* sp.	44 μM	Soares et al. 2012
D. menstrualis	Supercritical extraction	*S. mansoni*	100 mg mL^{-1}, 24 h	Stein et al. 2015
D. mertensii	Dichloromethane, methanol	*S. mansoni*	100 mg mL^{-1}, 72–120 h	Stein et al. 2015
Dictyota sp.	Dichloromethane: methanol (2:1)	*L. braziliensis* *T. cruzi*	50 μg mL^{-1}, 93.3% growth inhibition 50 μg mL^{-1}, 60.4% growth inhibition	Bianco et al. 2013
F. vesiculosus	Chloroform: methanol (3:1)	*P. falciparum* blood stage	15.7 μg mL^{-1}	Spavieri et al. 2013
Laminaria digitata	Chloroform: methanol (3:1)	*P. falciparum* blood stage	17.6 μg mL^{-1}	Spavieri et al. 2013
Lobophora variegata	Organic extracts	*Leishmania mexicana*	Promastigotes, 10.9–50 μg mL^{-1}	Freile-Pelegrin et al. 2008
L. variegata	Methanol	*A. aegupty* *C. quinquefasciatus*	2nd instar, 70.38 μg mL^{-1} 3rd instar, 95.52 μg mL^{-1} 2nd instar, 79.43 μg mL^{-1}	Manilal et al. 2011
L. variegata	Sulfoquinovosyldiacylglycerols	*Entamoeba histolytica* *T. vaginalis*	3.9 μg mL^{-1} Trophozoites, 8 μg mL^{-1}	Vieira et al. 2015
F. vesiculosus	Chloroform: methanol (3:1)	*P. falciparum* blood stage	15.7 μg mL^{-1}	Spavieri et al. 2013
Padina minor	95% Ethanol	*A. aegypti*	50.8 mg mL^{-1}	Bantoto and Dy 2013
Padina sp.	Methanol	*L. braziliensis*	50 μg mL^{-1}, 80.9% growth inhibition	Bianco et al. 2013

P. tetrastromatica	Methanol: dichloromethane (1:1)	*C. quinquefasiatus*	2nd instar, 96 μg mL^{-1}	Manilal et al. 2009c
P. tetrastromatica	Methanol	*A. aegypti* *C. quinquefasciatus*	2nd instar, 97.41 μg mL^{-1} 2nd instar, 97.94 μg mL^{-1}	Manilal et al. 2011
P. tetrastromatica	Water and methanol extract 10% concentration, after 72 h exposure	*M. javanica*	Egg hatching (96%) and larval mortality (100%)	Khan et al. 2015
Polycladia indica (as *Cystoseira indica*)	Methanol	*Plasmodium falciparum*	Acetylcholinesterase enzyme (AChE) inhibitory activity: 11 mg mL^{-1}	Ghannadi et al. 2013
Saccorhiza polyschides	Chloroform: methanol (3:1)	*P. falciparum* blood stage	16.1 μg mL^{-1}	Spavieri et al. 2013
Sargassum boveanum	Methanol	*P. falciparum*	Acetylcholinesterase enzyme (AChE) inhibitory activity: 1 mg mL^{-1}	Ghannadi et al. 2013
S. incisifolium (as *S. heterophyllum*)	Diclomethane and metanol extracts	*P. falciparum* *In vitro*	IC_{50}= 2.8 mg mL^{-1}	Lategan et al. 2009
S. muticum	Chloroform: methanol (3:1)	*P. falciparum* blood stage	18.2 μg mL^{-1}	Spavieri et al. 2013
S. natans	Ethanol	*Trypanosoma brucei rhodesiense*	7.4 μg mL^{-1}	Orhan et al. 2006
S. oligocystum	Methanol	*P. falciparum*	Acetylcholinesterase enzyme (AChE) inhibitory activity: 2.5 mg mL^{-1}	Ghannadi et al. 2013
S. polycystum (as *S. myriocystum*)	Ethanol: water (3:1)	*A. aegypti* *C. quinquefasciatus*	0.086 μg mL^{-1} 0.098 μg mL^{-1}	Ali et al. 2013
S. swartzii	Ethyl acetate fraction	*A. stephensi*	11.75 ppm	Khanavi et al. 2011
S. swartzii (as *S. wightii*)	Methanol	*A. aegypti* *C. quinquefasciatus*	2nd instar, 84.82 μg mL^{-1} 3rd instar, 97.28 μg mL^{-1} 2nd instar, 87.09 μg mL^{-1}	Manilal et al. 2011
S. tenerrimum	Water and methanol extract 10% concentration, after 72 h exposure	*M. javanica*	Egg hatching (96%) and larval mortality (99%)	Khan et al. 2015
Spatoglossum asperum	*n*-Hexane, methanol	*M. javanica*	10 mg mL^{-1}, 24 h	Ara et al. 2005
S. asperum	Methanol	*A. aegypti* *C. quinquefasciatus*	2nd instar, 81.23 μg mL^{-1} 3rd instar, 96.13 μg mL^{-1} 2nd instar, 83.17 μg mL^{-1} 3rd instar, 97.71 μg mL^{-1}	Manilal et al. 2011

Table 9.1 contd....

Table 9.1 contd.…

Species	Solvent	Parasite or vector species	LC_{50} or IC_{50} values	References
Stoechospermum polypodioides	Methanol	*M. javanica*	2nd instar, 80% mortality, 100 µg mL^{-1}, 72 h exposure	Rizvi and Shameel 2006
S. polypodioides	Methanol	*A. aegypti* *C. quinquefasciatus*	2nd instar, 82.95 µg mL^{-1} 3rd instar, 97.83 µg mL^{-1} 2nd instar 85.11 mL^{-1} 3rd instar, 98.59 mL^{-1}	Manilal et al. 2011
Turbinaria conoides	Water Acetone Ethanol	*A. aegypti*	18.74 mg L^{-1} 100.07 mg L^{-1} 6427 mg L^{-1}	Valentina et al. 2015
T. decurrens	Ethanol: water (3:1)	*A. aegypti* *A. stephensi*	0.079 µg mL^{-1} 0.099 µg mL^{-1}	Ali et al. 2013
T. turbinata	Organic extracts	*Leishmania mexicana*	Promastigotes, 10.9–50 µg mL^{-1}	Freile-Pelegrin et al. 2008
Rhodophyta (Red seaweed)				
Acanthophora muscoides	Petroleum ether fraction	*C. quinquefasciatus*	62.5 mg L^{-1}	Devi 1997
Bostrychia tenella	*n*-hexane, dichloromethane	*T. cruzi* *Leishmania amazonensis*	16.8 µg mL^{-1} 19.1 µg mL^{-1}	Felício et al. 2010
Botryocladia leptopoda	Crude extracts	*Leishmania major*	60.81 µg mL^{-1}	Sabina et al. 2005
Callophycus serratus	Bromophycolides	*Plasmodium falciparum*	0.3–0.9 µM	Lane et al. 2009
Centroceras clavulatum	Crude extracts	*L. major*	Promastigote, 57.89 µg mL^{-1}	Sabina et al. 2005
C. clavulatum	Dichloromethane: methanol (1:1)	*C. quinquefasciatus*	LC 58.3 µg mL^{-1}, mortality 100%	Manilal et al. 2009
C. clavulatum	Methanol	*A. aegypti* *C. quinquefasciatus*	2nd instar, 91.54 µg mL^{-1} 2nd instar, 97.72 µg mL^{-1}	Manilal et al 2011
Ceramium virgatum	Crude extracts	*T. brucei rhodesiense* bloodstream form	5.5 µg mL^{-1}	Allmendinger et al. 2010
C. virgatum	Chloroform: methanol (3:1)	*P. falciparum* blood stage *P. berghei* liver stage	13.6 µg mL^{-1} 26.4 mL^{-1}	Spavieri et al. 2013

Chondria armata	Domoic acid	*Periplaneta Americana* (American cockroach) *Musca domestica* (House fly) *Blatella germanica* (German cockroach)	Adulticidal, LC 1 µg g^{-1}, 100% mortality Adulticidal, LC 0.1 µg/insect, 40% mortality Adulticidal, LC 0.6 µg/insect, 60% mortality	Maeda et al. 1984
C. armata	Isodomoic acid A Isodomoic acid B Isodomoic acid C	*Periplaneta americana* (American cockroach)	Adulticidal, LC 3.2×10−8 mol Adulticidal, LC 3.2×10−8 mol Adulticidal, LC 6.4×10−8 mol	Maeda et al. 1984
C. atropurpurea	Acetone extract fractioned with *n*-hexane, ethyl acetate, and methanol: Chondriamide	*Nippostrongylus brasiliensis*	EC_{80} = 0.09 mM	Davyt et al. 1998
C. dasyphylla	Ethyl acetate fraction	*A. stephensi*	10.62 ppm	Khanavi et al. 2011
Chondrus crispus	Hydro-alcoholic extract	*L. donovani*	9.7 µg mL^{-1}, 95% inhibition	Vonthron-Sénécheau et al. 2011
Corallina officinalis	Crude extracts	*T. brucei rhodesiense* bloodstream form	4.8 µg mL^{-1}	Allmendinger et al. 2010
C. officinalis	Chloroform: methanol (3:1)	*P. falciparum* blood stage	8.6 µg mL^{-1}	Spavieri et al. 2013
Dasya baillouviana (as *Dasya pedicellata*)	Methanol	*T. brucei rhodesiense* *L. donovani* *P. falciparum*	Bloodstream form, 0.37 µg mL^{-1} Trypomastigotes, 62.02 µg mL^{-1} 16.76–69.98 µg mL^{-1} 0.38 µg mL^{-1}	Süzgeç-Selçuk et al. 2011
Digenea simplex	α-Kainic acid	*Periplaneta americana* (American cockroach)	Adulticidal, LC 100 µg g^{-1}, 100% mortality	Maeda et al. 1984
Gracilaria corticata	Crude extracts	*L. major*	38 µg mL^{-1}	Sabina et al. 2005
G. canaliculata (as *G. crassa*)	Crude extracts	*A. aegypti*	52.2 mg mL^{-1}	Anandhan and Sorna Kumari 2011
Gracilariopsis longissima (as *Gracilaria verrucosa*)	Crude extract	*A. aegypti* *A. stephensi* *C. quinquefasciatus*	0.125 ppm 0.131 ppm 0.135 ppm	Sundar et al. 2013
Grateloupia lithophila	Methanol Acetone Benzene	*Culex quinquefasciatus*	431.90 mg L^{-1} 349.74 mg L^{-1} 425.42 mg L^{-1}	Poonguzhali et al. 2012

Table 9.1 contd....

Table 9.1 contd....

Species	Solvent	Parasite or vector species	LC_{50} or IC_{50} values	References
Halopithys incurva	Chloroform: methanol (3:1)	*P. berghei* liver stage	28.8 mL^{-1}	Spavieri et al. 2013
Hypnea musciformis	Dichloromethane: methanol (1:1)	*Culex* sp.	2nd instar, LC 6 mg mL^{-1}, mortality 61.2%	Selvin and Lipton 2004
H. valentia	Crude extracts	*A. aegypti*	53.4 mg mL^{-1}	Anandhan and Sorna Kumari 2011
Laurencia dendroidea (as *Laurencia scoparia*)	Scopariol	*Nippostrongylus brasiliensis*	< 100 μM	Davyt et al. 2001
L. dendroidea	Elatol	*Trypanosoma cruzi* *Leishmania amazonensis*	1.01–1.38 μM 0.45–4 μM	Veiga-Santos et al. 2010
L. dendroidea	Dichloromethane: methanol (2:1)	*A. aegypti*	LC 100 ppm, mortality 100%	Bianco et al. 2013b
L. microcladia	Organic extracts	*Leishmania mexicana*	Promastigotes, 10.9–49.9 μg mL^{-1}	Freile-Pelegrin et al. 2008
L. nipponica	Deoxyprepacifenol Z-Laureatin Z-isolaureatin	*C. pipiens*	6.83 ppm 2.86 ppm 6.14 ppm	Watanabe et al. 1989
L. obtusa	Chloroform-methanol	*P. falciparum*	2700–4000 ng mL^{-1}	Topcu et al. 2003
L. pinnata	Laurepinnacin	*Mamestra brassicae* (Cabbage armyworm)	Larvicidal, 100% mortality	Fukuzawa and Masamune 1981
Mastocarpus stellatus	Ethyl acetate	*P. falciparum, L. donovani, T. cruzi*	2.8 μg mL^{-1}, SI > 30	Vonthron-Sénécheau et al. 2011
Melanothamnus afaqhusainii	Crude extracts	*L. major*	32.6 μg mL^{-1}	Sabina et al. 2005
Osmundea pinnatifida (formerly *L. pinnatifida*)	Crude extracts	*L. major*	6.25 μg mL^{-1}	Sabina et al. 2005
O. pinnatifida	Chloroform: methanol (3:1)	*P. falciparum* blood stage *P. berghei* liver stage	14.5 μg mL^{-1} 52.9 mL^{-1}	Spavieri et al. 2013
Palisada perforata (as *Laurencia papillosa*)	Non-polar crude extract	*Culex pipiens*	2nd instar, 57.6 ppm 3rd instar, 64.5 ppm 4th instar, 70.1 ppm	Abou-Elnaga et al. 2011
P. perforata (as *L. papillosa*)	(12E)-*cis* maneonene-E	*Tribolium confusum* (Confused flour beetle)	LD_{50} values were 0.21 and 0.16% after 6 and 12 days of treatment	Abou-Elnaga et al. 2011
Plocamium telfairiae	Aplysiaterpenoid A	*C. pipiens* *Anopheles gambiae*	LC 10 ppm, 100% mortality 0.1 ppm	Watanabe et al. 1989 Watanabe et al. 1990

P. telfairiae	Telfairine	*C. pipiens* *A. gambiae*	LC 10 ppm, 100% mortality 0.1 ppm	Watanabe et al. 1989 Watanabe et al. 1990
Pyropia leucosticta (as *Porphyra leucosticta*)	Crude extracts	*L. donovani*	16.5 to 85.6 μg mL^{-1}	Allmendinger et al. 2010
Scinaia hatei	Crude extracts	*L. major*	14.1 μg mL^{-1}	Sabina et al. 2005
S. moniliformis (as *S. indica)*	Crude extracts	*L. major*	59.6 μg mL^{-1}	Sabina et al. 2005

*Fresh water algae

proved to have antileishmanial activity when tested against the promastigote (IC_{50} = 4 μM) and intracellular amastigote (IC_{50} = 0.45 μM) forms of *Leishmania amazonensis*.

The individual fractions of *Chondrophycus brandenii* (as *Laurencia brandenii*) were evaluated by the insecticidal activity on *Sitophilus oryzae* (insect Rice weevil), maggoticidal activity against 2nd instar larvae of *Sarcophaga* sp. (insect), and termiticidal activity against *Microtermes obesi*. It was found that the fraction eluted using petroleum ether:chloroform (6:4) exhibited broader biological activities (Manilal et al. 2011). The maggoticidal activity of *C. brandenii* was studied by feeding the maggots on an artificial diet containing different concentrations of the active fraction of the seaweed. Maximum mortality (100%) was recorded at 60 mg g^{-1}, whereas at 20 mg g^{-1} extract showed a decrease in activity. The results indicate that the *C. brandenii* fraction contains compounds that are toxic to maggots, since the LD_{50} value is moderately low. The termiticidal assay represents a rapid, inexpensive, and simple bioassay for testing seaweed bioactivity. In the present study, a fatal effect was observed when 3 mg mL^{-1} concentration of algal fraction of *C. brandenii* was applied to the Petri dish, causing 90% mortality at exposure for 6 hours. The mortality rate for the positive control experiment was much lower (< 10), while that of negative control gave a mortality of 15%. The study showed that the termite population declined drastically relative to the concentration and time of exposure.

Süzgeç-Selçuk et al. (2011) showed that methanolic extracts of *Dasya baillouviana* (as *Dasya pedicellata*) and *Gelidium crinale* were active against *Trypanosoma brucei rhodesiense* bloodstream forms, against which *D. baillouviana* extract was the most potent (IC_{50} = 0.37 μg mL^{-1}). The same extract also impaired the survival of *Trypanosoma cruzi* trypomastigotes (IC_{50} = 62.02 μg mL^{-1}). All of the extracts showed leishmanicidal activity (IC_{50} values ranging from 16.76 to 69.98 μg mL^{-1}) (Süzgeç-Selçuk et al. 2011).

In the study done by Anandhan and Sorna Kumari (2011), the bio-restraining potentials of marine macroalgae collected from Rameshwaram, Tamil Nadu, were evaluated. Larvicidal bioassay (against mosquito larva of *Aedes aegypti*) was carried out with the two algae—*Gracilaria canaliculata* (as *Gracilaria crassa*) and *Hypnea valentiae*. The LC_{50} determined by the *G. caniculata* and *H. valentia* was noted as 52.2 and 53.4 mg mL^{-1}, respectively.

Vonthron-Sénécheau et al. (2011) tested the hydro-alcoholic and ethyl acetate extracts of 20 species of seaweeds collected in the Normandy coast (France), *in vitro*, against the protozoa responsible for the three major endemic parasitic diseases—*Plasmodium falciparum*, *Leishmania donovani*, and *Trypanosoma cruzi*. Almost half the extracts showed good activity against *P. falciparum*. The ethyl acetate extract of *Mastocarpus stellatus* showed the best antiplasmodial activity as well as the best selectivity index (IC_{50} = 2.8 μg mL^{-1}, SI > 30). The active extracts were almost entirely (97%) ethyl acetate extracts, while the hydro-alcoholic extracts were mainly inactive. This finding suggests that the active antiprotozoal compounds were relatively nonpolar, except for the hydro-alcoholic extract of *Chondrus crispus*, which was quite active against *L. donovani* (95% inhibition of parasite growth at 9.7 μg mL^{-1}). Moreover, Vonthron-Sénécheau et al. (2011) showed, for the first time, *in vitro* antiplasmodial activity by eight French species of red algae: *Calliblepharis jubata*, *C. crispus*, *Dilsea carnosa*, *Gelidium spinosum* (as *G. latifolium*), *Gracilaria gracilis*, *Grateloupia turuturu*, *Halurus flosculosus*, and *M. stellatus*.

The study done by Poonguzhali and Nisha (2012) focuses on larvicidal activity of methanol, acetone, and benzene extract of *Grateloupia lithophila* against *Culex quinquefasciatus* larva (mosquito larva). *G. lithophila* was found to effective against the larva of *Culex* in all the three extracts tested. LC_{50} value of the methanol extract was 431.90, acetone extract was 349.74, and benzene extract was 425.42 mg L^{-1}, while for *G. lithophila* LC_{90} value of the methanol extract was 860.60, acetone extract was 724.54, and benzene extract was 793.92 mg L^{-1}.

In an attempt to find natural products that could be employed in the effective control of the dengue mosquito, the larvicidal activities of crude extracts obtained from 15 seaweed species collected from northeastern Brazil have been determined (Bianco et al. 2013b). Extracts of *Laurencia obtusa* and *Hypnea musciformis* at concentrations of 300 ppm showed ≥ 50% mortality against the fourth instar larvae of *Aedes aegyptis*. The strongest larvicidal activities (> 91% mortality at 50 ppm) were obtained with extracts of *L. obtusa*. A halogenated sesquiterpene, identified as elatol, was isolated by sequential fractionation of the *n*-hexane extract of this species of seaweed, and the isolated compound exhibited potent larvicidal activity with an LC_{50} value of 10.7 ppm. The isolation of elatol from seaweed could represent an interesting prospect for a novel agent against the dengue mosquito.

The study done by Poonguzhali and Nisha (2013) was undertaken to investigate the larvicidal potential of *Gracilara corticata* against the medically important species of malaria vector *Anopheles*. *G. corticata* was found to effective against the larva *A. stephensi*. The *G. corticata* LC_{50} value of the benzene extract was 297.40, followed by the methanol extract 189.69, and the acetone extract 145.38 mg L^{-1}.

Stein et al. (2015) screened *in vitro* the extracts of several seaweeds from Brazil against *Schistosoma mansoni*. The extracts displaying activity at 500 µg mL^{-1} exposure were considered to be active and subjected to further testing at a lower concentration (100 µg mL^{-1}). The supercritical extracts of the red algae *Plocamium brasiliense*, *Chondria littoralis*, and *Spyridia hypnoides* were active against *S. mansoni* only at the higher concentration (500 µg mL^{-1}) tested. Interestingly, *Laurencia dendroidea* and *Laurencia catarinensis* submitted to hexane extraction display distinct results against schistosoma worms. *L. catarinensis* hexane extracts kill the worms in 24 hours, and *L. dendroidea* killed the worms in females in 24 hours and males in 72 hours. In addition, chloroform extract of *L. dendroidea* was also effective against males after 120 hours in incubation. *Gracilaria ornata* submitted to the dichloromethane/methanol extraction displays anthelmintic activity (100 µg mL^{-1}) only against male worms after 120 hours. The macroalgae *Pterocladiella capillacea*, *Ochtodes secundiramea*, and *L. dendroidea* extracted with dichloromethane/methanol (2:1), *L. dendroidea* extracted with methanol, and *L. catarinensis* extracted with chloroform did not present anthelmintic activity (Stein et al. 2015).

In a study made by Ishii et al. (2017), an ethyl acetate extract of *Laurencia nidifica* was found to have toxic effect against brine shrimp larvae (*Artemia salina*). Among isolated compounds, only laurinterol showed strong toxicity against *A. salina*. Further experiments revealed that laurinterol also exhibited repellent activity against the maize weevil *Sitophilus zeamais*, insecticidal activity against the termite *Reticulitermes speratus*, and acetylcholinesterase inhibitory effect. Insecticidal activity was evaluated by a topical application method using adult termites of *R. speratus*. Even though laurinterol exhibited lower activity than all of the commercial insecticides including rotenone, fenitrothion, and pyrethrins, it showed good activity against termites with LD_{50} 2.2 µg/insect.

CHAPTER 10

Anti-inflammatory, Anti-allergic, Antipyretic, Antinociceptive, Antithrombotic, and Anti-coagulant Activities of Seaweeds and their Extracts

10.1 Introduction

Inflammation has been linked with pathogenesis of many diseases like cancer, atherosclerosis, neurodegenerative diseases, diabetes mellitus, obesity, arthritis, cardiovascular diseases, Alzheimer's and Parkinson's diseases, and other deadly illnesses (Lee and Weinblantt 2001, Firestein 2006, Klegeris et al. 2007, Filippin et al. 2008). Inflammation can result in genetic defects, immunoregulation, and mechanism defects which lead to tissue damage. Biological sources for active compounds that have medical importance are on the increase in recent times (Kaboli et al. 2001). Some of the common allergic reactions correspond to the residual side effects of synthetic compounds that form bulk of the materials used for the production of pharmaceutical products. Aspirin can cause stomach bleeding, acetaminophen can cause liver damage. Cox-2 inhibitor Vioxx® and Celebrex® can cause heart problems, and non-steroidal anti-inflammatory drugs (NSAID's) were reported to contribute to numerous deaths yearly (Singh 1998, Wolf et al. 1999, Clegg et al. 2006, Pramitha and Kumari 2016). Furthermore, pathogenic organisms are becoming resistant to drugs over the last decades. However, living organisms have an inherent mechanism to withstand biotic and abiotic factors, therefore they serve as a reservoir for various active compounds which can trigger some immunological responses in the human beings (Pomponi 2001).

Drugs which are in use presently for the management of pain and inflammatory conditions are either narcotics (Opiods) or nonnarcotics (Salicylates, Corticosteroids and Hydrocortisone), and all of the drugs possess well-known toxic and side effects (Kasim et al. 2010).

10.2 Anti-inflammatory Activity and Effects on the Immune Response

Inflammation occurs frequently in living tissues, and is responsible for numerous deaths, and is a precursor to some deadly diseases. Anti-inflammatory compounds derived from seaweeds, which have attracted interest, are a promising replacement of current anti-inflammatory drugs. Macroalgae have both pro- and anti-inflammatory compounds. The latter include sulfated polysaccharides (fucoidans) from brown seaweeds, alkaloids (Caulerpin I, II and III) (see Fig. 10.3) from red and green seaweeds, polyunsaturated fatty acids (Docosahexaenoic acid: EDA, Eicosapentaenoic acid: EPA, Stearidonic acid: SA, and Eicosatrienoic acid: ETA) (see Fig. 10.4), carotenoids (fucoxanthin and astaxanthin) (see Fig. 2.6, Chapter 2), Pheophytin A, and Vidalols A and B. Anti-inflammatory assays include edema, erythema, tumor necrosis factor (TNF-α), interleukin (IL 1β, 6, 8), Nitric oxide (NO), inducible Nitric oxide synthase (iNOS), Prostaglandin E (PGE 2 and 3), Cyclooxygenase (COX-2), transcription factor (NF-κB), and leukotrienes (LB 3 and 4) (Jaswir and Monsur 2011).

Seaweeds are generally rich in 20 carbon atoms of polyunsaturated fatty acids, mainly eicosapentaenoic and docosahexanoic, also known as PUFAs, and capable of metabolizing various PUFAs via oxidative pathways (Stefanov et al. 1988, Gerwick and Bernart 1993). The metabolized products of PUFAs, called oxylipins, resemble human eicosanoid hormones, which carry out a range of physiologically important functions. The anomalous production of these compounds underlies several diseases related to inflammation, and thus eicosanoids and their derivatives have received wide attention in the search for anti-inflammatory drugs (Smit 2004, Serhan 2005).

10.3 Allergic Diseases and Anti-allergic Therapeutics

Allergic diseases are one of the major public health problems in the developed world. It was estimated that approximately one third of the general population and one-fifth of the population in western countries was affected by allergic diseases (Ono 2000, Arthur 2007). Allergic rhinitis, asthma, and atopic eczema are among the most common causes of chronic ill-health. The prevalence, severity, and complexity of these allergic diseases are rapidly rising and considerably adding to the burden of health-care costs (Kay 2000). Therefore, the knowledge about the pathophysiology of allergic diseases has increased, offering new opportunities for therapeutic intervention. Substantially, allergy is caused by an exaggerated reaction of the immune system to harmless environmental substances, such as animal dander, house dust mites, foods, pollen, insects, and chemical agents (Milián and Díaz 2004, Arshad 2010).

The initial event responsible for the development of allergic diseases is the generation of allergen-specific CD4+ Th2 cells. Once generated, effector Th2 cells produce IL-4, IL-5, IL-9, and IL-13 which cause the production of allergen-specific IgE by B cells (Akdis et al. 2005). Subsequently, allergic reactions are induced upon binding of allergen to IgE, which is tethered to the high affinity IgE receptor on the surface of mast cells and basophils (see Fig. 10.1). Following the aggregation of cell-surface receptors is a cascade of intracellular events, including the increase of intracellular Ca^{2+} level, the release of preformed inflammatory mediators from secretory granules such as histamine and β-hexosaminidase, and the generation

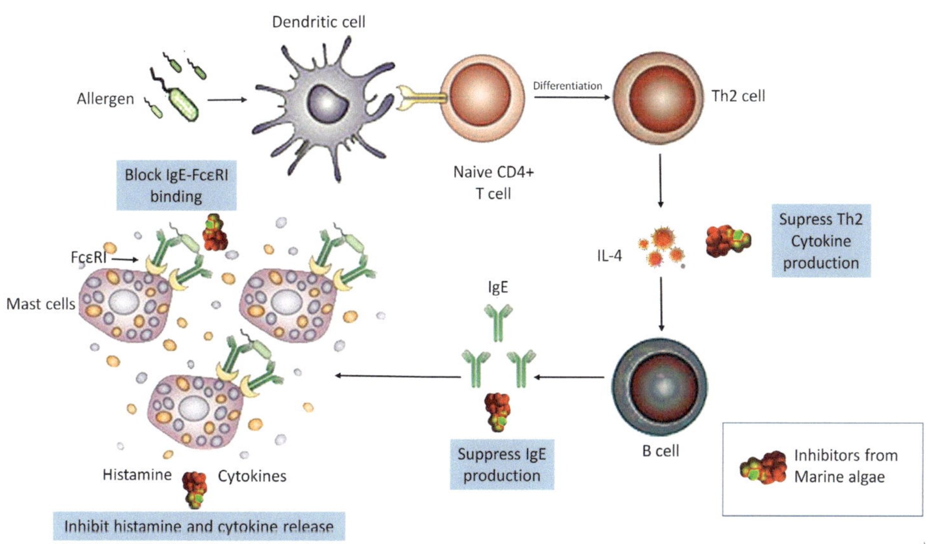

Figure 10.1 Targets for therapeutic intervention in allergy.

and secretion of the newly synthesized substances such as leukotrienes, prostaglandins, and cytokines. These mediators cause allergic inflammatory responses due to airway constriction, mucous production, and recruitment of inflammatory cells (Galli et al. 2008, Vo et al. 2012). According to this mechanism, the control of Th2-type cytokine expression, IgE levels, and inflammatory mediator production are especially important for the regulation of type I allergic reaction, thus allergic diseases may be managed. The current drugs that are used to treat allergies, such as antihistamines or corticosteroids, ameliorate symptoms but do not stop progression (Uermösi et al. 2010). There are also concerns regarding the side-effects from chronic use of current drugs, particularly by children (Li and Brown 2009). Thus, the search for potential drug candidates containing higher anti-allergy activity is increasing in the pharmaceutical industry. In this regard, natural bioactive compounds and their derivatives are great sources for the development of new generation anti-allergic therapeutics which are more effective with fewer side-effects (Vo et al. 2012b).

Allergy is a growing health issue that greatly impacts the professional and social life of patients. Finding the safe and efficient therapeutics for prevention and treatment of allergic diseases are always needed. Herein, a large number of anti-allergic agents from marine algae have been identified based on the specific assay system or screening approaches. The extensive studies of marine algae due to their anti-allergic activity will contribute to the development of novel anti-allergic therapeutics. Thus, it is believed that marine algae play an important role in the pharmaceutical industry to develop novel drugs against allergic disorders (Vo et al. 2012b).

10.4 Antithrombotic and Anticoagulant Activities

The maintenance of blood fluidity is controlled by the inner lining of the circulatory system, a thin layer of cells—the endothelium. Endothelial cells cover the entire vasculature of vertebrates, and thus, form a hemo-compatible blood container with a large surface, which has been estimated to amount to 3,000–6,000 m^2 (Gimbrone 1987). Although on many places not more than 0.2 μm thick, endothelial cells form a diffuse tissue of about 720 g in an adult person. Most of these cells are microvascular endothelial cells, and over 600 g of them cover the surface of capillaries (Wolinsky 1980). In the capillaries, exchange of nutrients and hormones occurs, facilitated by the exposure of a large endothelial surface area to a relatively small volume of blood. In contrast, in arteries and veins the ratio between surface and blood volume is 100- to 500-fold less. Furthermore, the shear forces evoked by the flowing blood vary in different types of blood vessels. Consequently, considerable variation exists in the regulation of general and specific functions between arteries, veins, arterioles, post capillary venules, and capillaries. In addition, specific tissue demands also underlie a large variation between microvascular endothelial cells from various tissues (Aird 2007).

In relation to their role as a gatekeeper between blood and tissues, endothelial cells are equipped with several general main functions, while in addition many specific functions are attributed to their specific location and tissue. Living endothelial cells are required to prevent thrombus formation. Specific functions and adaptations of endothelial cells aim at maintaining blood fluidity and preventing thrombus formation. This is only possible because the endothelium also actively controls the extravasation of fluid, solutes, hormones, and macromolecules (Mehta and Malik 2006), as well as that of platelets and blood cells. This guarantees the availability of appropriate amounts of clotting factors and platelets. Furthermore, the endothelium, together with the smooth muscle cells, plays an important role in the control of perfusion of specific tissues and the blood tension by local vasoregulation. This is further facilitated by the responsiveness of endothelial cells to vasoactive agents, and—in particular in the lung—by involvement of the endothelium in the conversion and catabolism of vasoactive agents. Another general feature of endothelial cells is that they can become activated by inflammatory cytokines, triggers of the innate and acquired immune response, and by other stresses such as hypoxia and metabolic stress (Pober and Sessa 2007, Bakker et al. 2009, Fraisl et al. 2009, Semenza 2010). Such activations induce the transcription of new genes, which enable endothelial cells to offer a new repertoire of activities and receptors. Endothelial cells control the recruitment of leukocytes in areas where these inflammatory cells are needed, in particular by expressing specific proteins cell adhesion molecules. They adapt their metabolism according to oxygen tension and metabolic needs. In inflammation, various functions of the endothelium are pivotal and well recognized from features such as tissue swelling due to a reduced barrier function, vasodilation, which also causes the red color, increased leukocyte infiltration, and alterations in the control of coagulation and thrombus formation into a more procoagulant direction. Furthermore, endothelial cells are involved in the healing process after wounding or inflammation by being the vector of angiogenesis, the formation of new blood vessels, which is essential for proper formation of granulation tissue and tissue repair, as well as for re-canalization of mural and obstructing fibrin clots.

Macromolecule recognition processes are common in cells and their specificity is their most important characteristic. Many research programs exploit recognition events and these have become focus areas of research in biology, chemistry, medicine, and pharmacology. Biological reactions that involve recognition events include processes such as cell agglutination and coagulation, the stimulation of cell migration, and fertilization (Smit 2004).

Lectins, sometimes referred to as hemagglutinins or agglutinins, are glycoproteins with an ability to agglutinate red blood cells (Boyd and Reguera 1949). Various polysaccharides are present on cell surfaces, and as a result many cells including microbes and yeasts (Bird et al. 1992, Cisar et al. 1995), tumor cells (Hori et al. 1986), and erythrocytes are selectively agglutinated by lectins (Chen et al. 1995). Lectins are inhibited by sugars of the same type as those on the surface of the cells being agglutinated (Sharma and Sahni 1993). They are useful in exploring properties of biological structures and processes, and have found applications in biology, cytology, biochemistry, medicine, and food science and technology. Lectins from *Codium* spp. have been developed into commercially available reagents and are routinely used in biochemical studies.

10.4.1 The blood coagulation system

The human blood coagulation system is the process leading to the arrest of bleeding (hemostasis) and includes the transformation of liquid blood into a solid state to reduce the loss of blood from injured blood vessels. This process covers three mechanisms such as formation of prothrombinase, conversion of prothrombin into the thrombin which is a key protein of coagulating cascade, where thrombin activates a series of coagulant factors, and finally, conversion of soluble fibrinogen into insoluble fibrin (see Fig. 10.2) (Mišurcová et al. 2015).

The blood coagulation system consists of cellular elements (blood platelets, white cells, to some extent red cells, and microvascular remnants or microparticles), coagulation enzymes, proteins cofactors, and several anticoagulant proteins (Spronk et al. 2003). The mechanism of blood coagulation is based on the enzyme cascade divided in the intrinsic, extrinsic, and common pathway, where a series of coagulation factors promote the formation of the end-product fibrin (Spronk et al. 2003, Wijesekara et al. 2011). As

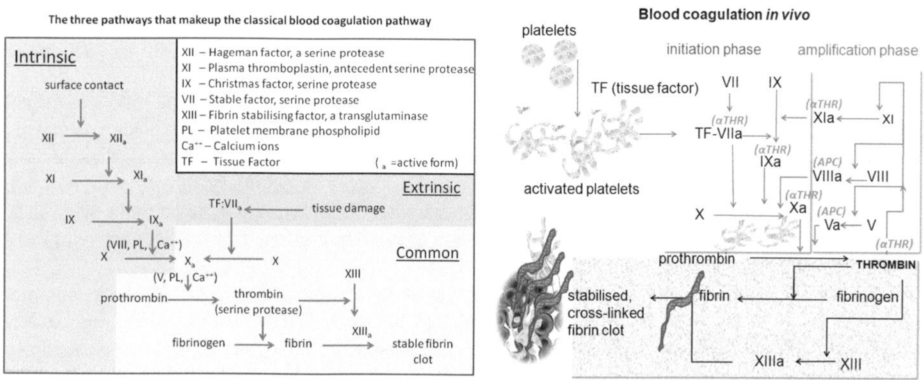

Figure 10.2 Blood coagulation pathways *in vivo*; aTHR (Antithrombin); APC (Activated protein C) (adapted from Beards 2012).

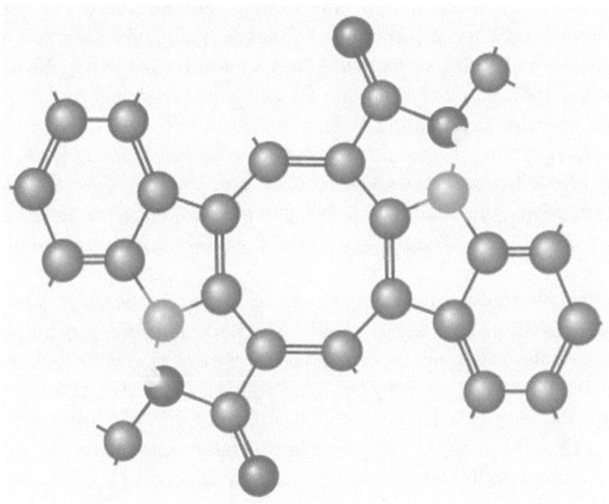

Figure 10.3 3D conformer of Caulerpin (after PubChem 2017).

it can be concluded from Fig. 10.2, during the intrinsic pathway activated Stuart-Prower factor (X) can also be activated by the extrinsic pathway. Firstly, the intrinsic cascade begins with the formation of primary complex of collagen by high molecular weight kininogen (HMWK), prekallikrein, and Hageman factor (XII). During the activation, the single-chain protein of the native Hageman factor is divided into two chains of different molecular weights (28 kDa and 58 kDa). However, both chains remain linked by a disulfide bond. The 28 kDa light chain contains the active site, and this molecule is called activated Hageman factor (XIIa), which can activate plasma thromboplastin antecedent (PTA) or antihemophilic factor—C (XI). Further, HMWK, known as Fitzgerald factor, binds to the factor XI, and in the presence of Ca^{2+} ions, it facilitates the activation process of factor XIa. This factor XIa activates Christmas factor, plasma thromboplastin component (PTC), or antihemophilic B factor (factor IX) in the reaction requiring Ca^{2+} ions, factor VIII, and phospholipids. Antihemophilic factor VIII is obviously an essential factor for this step of coagulation cascade, and its deficiency is associated with hemophilia A, while the deficiency of factor IX relates to hemophilia B (Adelson et al. 1963). Activated IXa factor further activates Stuart-Prower factor (X) to factor Xa; and factor X is the first molecule of the common pathway of coagulation cascade. The extrinsic pathway could be considered as an alternative way of the activation of factor X in the cooperation with two main components—tissue factor (TF) and factor VII. Blood coagulation factor VII, formerly known as proconvertin, belongs to the serine protease enzyme class, and its main role in extrinsic pathway is to initiate the coagulation process in conjunction with TF.

TF is constitutively present on cell membranes within and around the vessels and serves as the cell surface receptor for serine protease factor VIIa. Carboxylated GLA domain of factor VIIa binds to negatively charged phospholipids in the presence of calcium. Binding of VIIa to negatively charged phospholipids greatly enhances the protein-protein binding of VIIa to TF. Upon a vessel injury, tissue factor, normally found outside of blood vessels, is exposed to the blood where it forms a catalytic complex with factor VIIa activating factor IX and catalyzing the conversion of inactive protease factor X into active protease factor Xa (Spronk et al. 2003, Mirzaahmadi et al. 2011).

Both intrinsic and extrinsic pathways lead to the activation of factor X into factor Xa (the common pathway), which combines with its cofactor—activated proaccelerin (factor Va)—in the presence of calcium and phospholipid to produce thrombin for the conversion of fibrinogen to fibrin. Fibrin monomers spontaneously polymerize and form an insoluble gel (clots) which is held together by noncovalent and electrostatic forces and is stabilized by fibrin-stabilizing factor XIII, catalyzing the formation of peptide bonds between fibrin monomers. Clots together with aggregated platelets (thrombi) block damaged blood vessels and prevent further bleeding (Chatterjee et al. 2010).

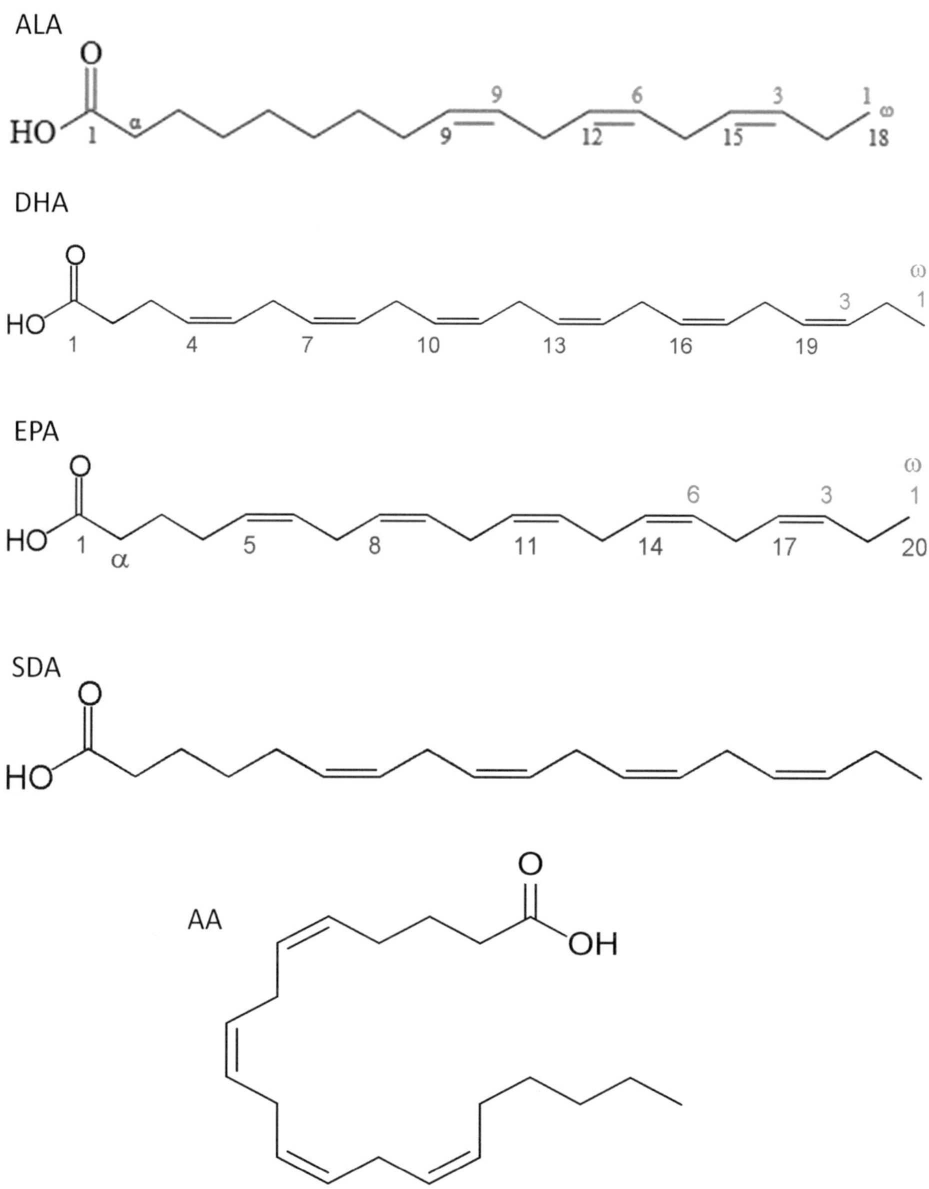

Figure 10.4 Polyunsaturated fatty acids (PUFAs): chemical structure of alpha-linolenic acid (ALA), docosahexaenoic acid (DHA), eicosapentaenoic acid (EPA), stearidonic acid (SDA), and arachidonic acid (AA) (after Edgar 2017).

Hemostatic abnormalities can lead to serious health problems, such as excessive bleeding or thrombosis. Thus, the human coagulation mechanism should be strictly regulated by the inactivation of procoagulant enzymes, fibrinolysis, and hepatic clearance of activated clotting factors (Kalafatis et al. 1997) via tissue pathway inhibitor (TFPI), heparin-antithrombin pathway, and protein C pathway (Esmon 2005). The first inhibitor process TFPI inactivates factor VIIa bound to tissue factor (TF). The second antithrombin

(ATIII)-heparin mechanism inactivates factor Xa, thrombin, factor IXa, and factor VIIa bound to cell surface tissue factor (Rao et al. 1995). The latter protein C pathway is based on the activation of protein C by the thrombin-thrombomodulin complex on endothelium. This natural anticoagulant system exerts its anticoagulant effect by regulating an activity of factors VIIIa and Va, cofactors in tenase, and prothrombinase complexes, respectively (Dahlbäck and Villoutreix 2005, Esmon 2003).

Heparin, a highly sulfated glycosaminoglycan, is naturally produced by basophils and mast cells; in medicine, it is principally used as an anticoagulant to treat and prevent blood clots in the veins and arteries. Heparin molecule possesses a specific local structure, and it is composed of penta-saccharide sequence with a specific pattern of sugar residues along with a sulfation pattern required to induce a conformational activation of antithrombin. Heparin also has an additional anticoagulant mechanism in which polysaccharide brings antithrombin and thrombin in a ternary complex, in which both the inhibitor and proteinase are bound to the same polysaccharide chain (Streusand et al. 1995, Pereira et al. 2002). The heparin usage is limited due to several side effects, i.e., a serious side effect resulting in degradation of platelets causing thrombocytopenia (Bick and Frenkel 1999, Castelli et al. 2007).

Fucoidans (see Fig. 2.1, Chapter 2) have *in vivo* and *in vitro* heparin-like antithrombotic and anticoagulant activities that are mediated by blood coagulation inhibitors such as heparin cofactor II or antithrombin III (Church et al. 1989, Colliec et al. 1991, Matou et al. 2002). The anticoagulation activity is the result of direct fucan-thrombin interaction (Graufel et al. 1989), and it usually increases with the amount of sulfation (Nishino and Nagumo 1991, 1992). Sulfated fucans from *Fucus vesiculosus* and *Ascophyllum nodosum* have been patented as anticoagulant substances. The work was motivated by the need to find a potential replacement for cattle derived heparin and the fear of the transmission of bovine spongiform encephalitis (BSE) through the use of bovine-derived products (Trento et al. 2001). Sulfated fucoidan has several advantages over heparin. It shows concentration-dependent inhibition of thrombin generation from platelets; it exhibits concentration dependent inhibition of thrombin-induced platelet aggregation; it lacks the hypotensive effect found in thrombin; it reduces the sticking of polymorph nucleated leukocytes to rabbit aorta; and it shows a dose dependent inhibition of thrombin-induced thrombosis (Trento et al. 2001). Some older literature reports laminaran (see Fig. 2.2, Chapter 2) as having anticoagulant properties (Chapman 1980), but it is possible that this activity comes from fucoidan, which is often present in the same extracted fraction as laminaran (Smit 2004).

10.5 The Potential of Seaweeds for Alternatives to Conventional Medication

The search for alternatives to conventional medication has led to a search for various living organisms for the purpose of obtaining active compounds of medicinal values in them. Algae form part of the few living organisms that have been identified as sustainable sources of bioactive compounds (Veena et al. 2007).

Seaweeds are generally grown in an aquatic environment and have the ability to withstand fluctuation in salinity around them, strong tidal current, variation in light intensity, and constant fluctuation in temperature (Ehrlich 2010). These adaptive properties influence its physical (morphology) and biochemical constituents. Analyses of microalgae indicate the presence of various bioactive compounds including anti-inflammatory, antithrombotic, anti-allergic, anti-coagulant, and other activities (Bhakuni and Rawat 2005, Pramitha and Kumari 2016). However, findings from these studies are numerous in various scientific journals which occlude the extent of scientific researchers on antioxidant, anti-inflammatory potential of seaweed and make it difficult to know the direction and predict future trend.

Marine organisms are currently undergoing detailed investigations with the objective of isolating biologically active molecules along with the search for new compounds. Among marine organisms, macroalgae are rich sources of structurally diverse bioactive compounds with different bioactivity spectra and biomedical value (Pramitha and Kumaro 2016). The development of ways to obtain large quantities of the natural metabolites is currently the most important quest (Flora and Rani 2013). Macroalgal substances serve as the viable source of drugs for the world population and several algae-based drugs are in extensive clinical use.

10.5.1 Chlorophyta (green seaweeds)

Pheophytin A is a derivative of chlorophyll, extracted from *Ulva prolifera* (as *Enteromorpha prolifera*), and was reported to possess anti-inflammatory activities. It suppressed production of Formyl-Met-Leu-Phe (FMLP) of human polymorphonuclear leukocytes, superoxide in mouse, and edema formation in BALB/c mouse ear, respectively, induced by chemotaxis (PMNs) in Boyden's chamber experiment, by 12-*O*-tetradecanoylphorbol-13-acetate (TPA) using the cytochrome C reduction method, and by TPA-induced inflammation reaction (Okai and Okai 1997).

Codium latum from Japanese waters yielded sulfated arabinan α(1→5) as an active molecule which gave 12.6 times more anti-thrombin activity when standard heparin was taken as one (Uehara et al. 1992). 13 species of green algae belonging to the family Codiaceae were screened from Indian coasts (Siddhanta et al. 1999, Shanmugam et al. 2001). It was found that *Codium dwarkense* and *Codium tomentosum* showed promising activity. Bioassay guided purification of both the species yielded sulfated arabinan as the most active component and sulfated arabinogalactan as the relatively less active component (Shanmugam 1998, Siddhanta et al. 1999, Shanmugam et al. 2001). It was found that arabinose and sulfate play an important role in eliciting anticoagulant activity (Shanmugam 1998, Siddhanta et al. 1999). Two active components have been identified in fractions obtained by molecular exclusion from *Codium fragile* subsp. *fragile* (as *Codium fragile* subsp. *tomentosoides*) (Rogers et al. 1990b). The high molecular weight component (approximately 1700 kDa) exhibited greater anticoagulant activity by prolonging PT (prolongation of prothrombin time), aPTT (activated partial thromboplastin time), and TT (thrombin time assays).

A high MW proteoglycan (1.8 × 106) from *Codium fragile* subsp. *atlanticum* exhibited potent blood anticoagulant activity. It has been observed that the activity is directly proportional to the carbohydrate and sulfate contents of proteoglycan and inversely proportional to the protein content of proteoglycans (Rogers et al. 1990). High molecular weight sulfated proteoglycan and low molecular weight sulfated polysaccharides from the same species were reported to have strong activity (Jurd et al. 1995). aPTT, PT, and TT assays were applied and it was found that an increase in anticoagulant activity was demonstrated with increasing concentration and sulfate content of algal products. The proteoglycans (18.4% sulfate) possess the greatest anticoagulant activity, followed by sulfated polysaccharides with 10.2% and 7.5% sulfate, respectively. It has also been reported that polysaccharides isolated by proteolytic digestion from green seaweed *Caulerpa cupressoides* have a low anticoagulant potential in relation with polysaccharide extract obtained by ion-exchange chromatography (Rodrigues et al. 2011).

According to Fernández et al. (2013), a highly sulfated 3-linked β-arabinan (Ab1) with arabinose in the pyranose form was obtained from green seaweed *Codium vermilara*. It comprised major amounts of units sulfated on C-2 and C-4, and constituted the first polysaccharide of this type isolated in the pure form and fully characterized. Ab1 showed anticoagulant activity by global coagulation tests. Less sulfated arabinans obtained from the same seaweed have less or no activity. Ab1 exerts its activity through direct and indirect (antithrombin- and heparin cofactor II-mediated) inhibition of thrombin.

The sulfated polysaccharides from 23 green algae species have been tested for anticoagulant activity and have discovered a high rhamnose-containing sulfated polysaccharide from *Monostroma nitidum*, the purified version of which was more significant than standard heparin (Hayakawa et al. 2000).

Highly sulfated galactan (13.1% of sulfate) containing mainly galactose with a small amount of glucose from green algae *Codium cylindricum* has been found as an anticoagulant with the similar activity compared to heparin, but weaker than heparin (Matsubara et al. 2001). Anticoagulant activity of sulfated polysaccharides isolated from green seaweed *Monostroma nitidum* has also been described. Among two sulfated polysaccharides of *M. nitidum* with different structure of their molecules, different extent of anticoagulant activities has been evaluated. Evidently, both polysaccharide fractions exhibit high anticoagulant activities; however, differences between them are directly due to their structural feature discrepancy, and sulfated polysaccharide with a lower molecular size and higher sulfate content show notably higher anticoagulant activity (Mao et al. 2008). Further, sulfated polysaccharide composed mainly of (1,2)-linked L-rhamnose residues with sulfate groups substituted at positions C-3 and/or C-4 isolated from marine green algae *M. latissimum* show high anticoagulant activities proved by APTT and TT assays.

Anticoagulant property of this sulfate polysaccharide has been mainly attributed to powerful potentiation of thrombin by HP II (Mao et al. 2009).

Green seaweed *Ulva conglobata* collected from three various locations in China (coasts of Quigdao, Yantai, and Rizhao) exhibited different polysaccharide compositions with major representation of rhamnose 71.90 mol%, 72.26 mol%, and 63.77 mol%, respectively; variable contents of glucose and fucose; and trace amounts of xylose, galactose, and mannose. Their sulfate ester contents were 35.20%, 23.04%, and 28.06%, respectively. The highest anticoagulant activity was established in hot water extract from *U. conglobata* collected from the coast of Quigdao in comparison to extracts from the other locations. Polysaccharide extracted from the species collected in the coast of Rizhao possessed higher content of sulfated ester and lower content of rhamnose than the extract from the species collected in the coast of Yantai; thus, the first one show a lower anticoagulant activity than the latter (Mao et al. 2006). The extent of anticoagulant activity of algal polysaccharides varied in dependence on different sugar residues and their proportion, as well as on sulfation content and structural features.

The influence of different concentrations of extraction solutions (0.50 M NaCl, 0.75 M NaCl, and 1.00 M NaCl) used for the elution of sulfated polysaccharides of three species—*Caulerpa cupressoides*, *Caulerpa racemosa*, and *Caulerpa prolifera*—has been confirmed. Dissimilarity in the presence of sulfate groups in chemical structures of their sulfated polysaccharides obtained by ion-exchange chromatography has been observed. Lower content of sulfate groups has been established in *C. cupressoides*, followed by *C. racemosa*, and higher content has been determined in *C. prolifera*, which is in accordance with the extent of their anticoagulant activities of 17.37 IU mg^{-1}, 22.17 IU mg^{-1}, and 25.64 IU mg^{-1}, respectively, in comparison to heparin activity of 193 IU mg^{-1} (Rodrigues et al. 2012).

So far, *Ulva compressa* (as *Enteromorpha compressa*), an edible green alga, has been widely used due to its nutritional and medicinal importance. This alga has been identified as a potential anti-anaphylactic agent (Raman et al. 2004). The administration of *U. compressa* extract along with ovalbumin (OVA) caused a lowering of serum IgE level in different murine models of Balb/C, C57BL/6, and SWR/J mice. It indicated that *U. compressa* showed the same potency and consistency in its action as an anti-allergic alga.

Water-soluble acidic polysaccharides from the cell walls of *Ulva rigida* are mainly composed of disaccharides that contain glucuronic acid and sulfated rhamnose. The structure of disaccharides resembles that of glycosaminoglycans (GAGs) as they both contain glucuronic acid and sulfated sugars. Certain types of GAGs can even activate macrophages and therefore the acidic polysaccharides from *U. rigida* probably modulate macrophage activity. In the present study, we evaluated the effects of *U. rigida* polysaccharides on several RAW264.7 murine macrophage activities, including expression of inflammatory cytokines and receptors, nitric oxide and prostaglandin E2 (PGE2) production, and nitric oxide synthase 2 (NOS-2) and cyclooxygenase-2 (COX-2) gene expression. *U. rigida* acidic polysaccharides induced a more than two-fold increase in the expression of several chemokines (chemokine (C motif) ligand 1, chemokine (C-X-C motif) ligand 12, chemokine (C-C motif) ligand 22 and chemokine (C-X-C motif) ligand 14 (Cxcl14)) and in the expression of IL6 signal transducer and IL12 receptor beta 1. Incubation of macrophages with *U. rigida* polysaccharides also induced an increase in nitrite production, although this effect decreased considerably after desulphation of polysaccharides, suggesting that the sulphate group is important for the stimulatory capacity of these molecules. *U. rigida* polysaccharides also stimulated macrophage secretion of PGE2 and induced an increase in COX-2 and NOS-2 expression. The results indicate that *U. rigida* acid polysaccharide can be used as an experimental immunostimulant for analysing inflammatory responses related to macrophage functions. In addition, these polysaccharides may also be of clinical interest for modifying certain macrophage activities in diseases where macrophage function is impaired or needs to be boosted (Leiro et al. 2007).

In the work done by Khan et al. (2008), 37 species of common seaweeds from the coast of Korea were screened for anti-inflammatory activity. Methanol extracts of the seaweeds were tested against mouse ear edema and erythema induced by phorbol myristate acetate. *Ulva linza* extracts strongly reduced ear edema/erythema by 84%/70%. *U. linza* extract has IC_{50} of 20 mg mL^{-1}, 26 mg mL^{-1}, and 31 mg mL^{-1}, respectively for edema, erythema, and blood flow (Khan et al. 2008). According to Kimiya et al. (2008), various extracts of *Codium fragile* and *Umbraulva japonica* (as *Ulva japonica*) have been found to inhibit more than 50% of β-hexosaminidase released from RBL-2H3 cells at concentrations of 100 μg mL^{-1} and 200 μg mL^{-1}.

Caulerpin is the only reported alkaloid from seaweed with anti-inflammatory activities. Caulerpin is a bisindole alkaloid because it contains two indole groups (benzylpyrrole derived from tryptophan) linked together by eight carbons' cyclic ring with two carboxyl groups (Fig. 10.3) (Kasim et al. 2010). Caulerpin has been isolated from mainly green algae. Isolation of Caulerpin (CLP) from seaweed (*Caulerpa* spp.) was first conducted as far back as 1970 and tagged as CLP I (Aguilar-Santos 1970). Reduction of leukocytes in mice was observed after Caulerpin was injected in carrageenan induced cells. Caulerpin significantly reduced 48.3% of the number of leukocytes, but it was lower than indomethacin, which caused 72.1% reduction. Furthermore, Caulerpin was reported to greatly reduce neutrophils recruitment compared to other leukocytes. Structurally, Caulerpin resembles indomethacin, and its anti-inflammatory potential was probably due to the presence of indole group as an active site (Éverton et al. 2009).

Anti-inflammatory ingredients of algal extracts can be applied in the treatment of orodental diseases (stomatitis, gingivitis, periodontal diseases, and dental plaque) (Sirbu et al. 2008). Literature describes the addition of algal extracts (5% and 10%) together with chlorhexidine to fibrillary collagen matrixes to form stomatological gels. Hydro-alcoholic extracts from Black Sea algae *Ulva lactuca* (and other seaweeds) have been used. Therapeutic effect at nanostructure level has been achieved in dental medicine in the treatment of periodontal diseases. In the study on dental gels it is significant to evaluate their rheological properties, in particular pseudo-plastic rheological behavior and decrease of an apparent viscosity with the increase of shearing (Sirbu et al. 2008). The mechanisms and compounds responsible for anti-inflammatory activity have not been identified yet (Margret et al. 2009).

Inflammatory bowel diseases, which include Crohn's disease and ulcerative colitis, are characterized by chronic and relapsed gut inflammation. *Caulerpa mexicana*, found in tropical areas, such as the Brazilian coastland, exhibit *in vitro* and *in vivo* anti-inflammatory properties, such as the ability to reduce both cell migration to different sites and edema formation induced by chemical irritants (Bitencourt et al. 2015). Acute experimental colitis was induced in BALB/c mice by treatment with 3% dextran sodium sulfate orally for 14 days. During this 14-day period, *C. mexicana* methanolic extract (2 mg/kg/day) was given intravenously on alternate days. Treatment with the methanolic extract significantly attenuated body weight loss and severe clinical symptoms. This was associated with a remarkable amelioration of colonic architecture disruption and a significant reduction in pro-inflammatory cytokine production.

In the works of Montalvão et al. (2016), the anti-inflammatory potential of several algae specimens collected from the Aegean Sea were screened. Extracts from *Chaetomorpha aerea* and *Caulerpa racemosa* clearly showed a reduction in the release of IL-β, IL-6, TNF-α, and IFN-ϒ pro-inflammatory cytokines, when compared to positive controls.

The seaweed *Monostroma angicava* grows in upper part of intertidal zone and has been widely used as food for centuries in China. A novel rhamnan-type sulfated polysaccharide was isolated from *M. angicava* collected from the coast of the Yellow Sea of China. The structure, anticoagulant activity, and inhibition mechanism of the sulfated polysaccharide (see Table 10.1) were investigated by Li et al. (2017). The investigation revealed that one of the fractions is a novel sulfated rhamnan that differs from previously described sulfated polysaccharides from green seaweed and may be a potential anticoagulant polysaccharide.

10.5.2 Phaeophyceae (brown seaweeds)

Although marine algae were believed to be safe and efficient agents for anti-inflammatory, anti-allergic, antithrombotic, and anticoagulant treatment, they have not been as extensively studied as terrestrial plants. Several brown macroalgae have been determined for their capability against allergic responses.

The presence of anticoagulant activity in brown algae was first reported in 1941, where *Saccharina japonica* (as *Laminaria japonica*) showed anticoagulant effect, its active compound being located in the holdfasts (Kimura 1941). Laminarin shows anticoagulant activity only after structural modifications such as sulfation, reduction, or oxidation. The anticoagulant activity is improved chemically by increasing the degree of sulfation (Shanmugam and Mody 2000).

The most widely recognized and studied bioactivity in marine sulfated polysaccharides is the heparin-like anticoagulant activity exhibited by fucoidans and other fucans of Phaeophyceae. This was first reported from *Fucus vesiculosus* fucoidan isolated by Springer and colleagues, who observed inhibition of fibrin clot formation and antithrombin activity (Springer et al. 1957, Bernardi and Springer 1962). The most active fucoidan fractions predominantly consisted of sulfated fucose residues. Thrombin inhibition activity of these fractions exceeded that of heparin (Springer et al. 1957). Bernardi and Springer (1962) further characterized these active fucoidan fractions and demonstrated that the material was essentially homogeneous, with molecular weight 7.4 × 104 Daltons, and possessed 60–80% of the activity of heparin in the re-calcification time tests, and 15–18% heparin activity in the whole human blood. Adams and Thorpe (1957) found that fucoidan fractions F13 and A showed activity of 8.9 heparin units mg^{-1} and 9 heparin units mg^{-1}, respectively.

Some structural features of fucoidan are most likely required for certain specific activities. Earlier, Nishino and colleagues found that in sulfated polysaccharide fractions from *Ecklonia kurome*, a higher content of fucose and sulfate groups coincided with higher anticoagulant activities (Nishino et al. 1989, Nishino and Nagumo 1991).

Leite et al. (1998) investigated the anticoagulant activity of acidic polysaccharide extracts from *Spatoglossum schroederi*. Polysaccharide extracts were obtained by ion-exchange chromatography using different salt molarities (0.15–3.0 M NaCl) and lower concentrations from the range of 0.5–0.7 M and 1.0–1.5 M of salt, resulting in higher content of alginic acid and xylose in obtained extract, respectively, while the extracts enriched with sulfated xylofucan were obtained with using higher concentrations (2.5–3.0 M) of salt. Anticoagulant activities of all polysaccharide fractions were low (0–22.4 IU mg^{-1}) in comparison to heparin activity of 150.0 IU mg^{-1}. However, all acidic polysaccharide fractions showed the ability to stimulate the synthesis of antithrombotic heparin sulfate produced by the rabbit aorta endothelial cells in culture to the same amount of heparin (Leite et al. 1998).

It has been reported that fucoidans (pure fucans) from *Ecklonia bicyclis* (as *Eisenia bicyclis*) (Usui et al. 1980), *Sargassum fusiforme* (as *Hizikia fusiformis*) (Dobashi et al. 1989), *Saccharina longissima* (as *Laminaria angustata* var. *longissima*) (Kitamura et al. 1991), and *Silvetia babingtonii* (as *Pelvetia wrightii*) (Anno et al. 1966) showed considerably high anticoagulant activity. *Sargassum cinctum* from Indian waters was studied for its sulfated polysaccharides (SPS) content and anticoagulant activity. It was observed that all the fractions contained mainly galactose and a trace of fucose with sulfate (29.5%), and hence activity was also poor (Mody et al. 1997). 16 species of British marine algae were screened, of which four species, i.e., *Laminaria digitata*, *Laminaria hyperborea*, *Saccharima latissima* (as *Laminaria saccharina*), and *Fucus spiralis* showed potent anticoagulant activity (Deacon-Smith et al. 1985). Sargassan is composed of glucuronic acid, galactose, mannose, xylose and fucose in the molar ratio of 4.57:8.40:1.00:2.48:2.53, and of sulfate (18%) (Abdel-Fattah and Hussein 1973, Abdel-Fattah et al. 1974). Although sargassan has high uronic acid content, the polysaccharide was shown to have much higher anticoagulant activity than heparin (Abdel-Fattah et al. 1974). Similar high anticoagulant activity has been reported for purified fucan sulfates from *Padina pavonica* (xylofucomannoglucuronan) (Hussein et al. 1980), *Padina tetrastromatica* (xylofucogalactomannoglucuronan) (Rao et al. 1984), and *Dictyota dichotoma* (glucuron-oxylomannogalactofucan) (Abdel-Fattah et al. 1978). These results indicate that the relationship between the sugar composition and the anticoagulant activity of fucan sulfates is very complex (Shanmugam and Mody 2000).

The sulfated fucan obtained from brown seaweed *Laminaria brasiliensis* shows a strong anticoagulant activity of 30.0 IU mg^{-1}; however, desulfation of this fucan totally abolishes its anticoagulant activity (Pereira et al. 1999). According to Ushakova et al. (2009), fucoidans originated from brown seaweeds *Saccharina latissima* (formerly *Laminaria saccharina*), *Laminaria digitata*, and *Fucus distichus* contained the highest amount of sulfate in the range of 27.5–34.8% and the lowest amount of uronic acid and showed the highest anticoagulant activity by APTT assay in the amounts of 33.0 IU mg^{-1}, 24.2 IU mg^{-1}, and 26.9 IU mg^{-1}, respectively, and by PT assay in the amounts of 40.8 s, 33.2 s, and 33.0 s, respectively; and, finally, anticoagulant activities performed by TT assay were in the amounts of 72.8 s, 36.0 s, and 29.0 s, respectively. In accordance with the previous author, in brown seaweeds, *S. latissima*, *L. digitata*, and *F. distichus* have been established as the highest anticoagulant activities of 33.0 IU mg^{-1}, 24.2 IU mg^{-1}, and 26.9 IU mg^{-1}, respectively. In other brown seaweeds, such as *Fucus serratus*, *Fucus distichus* subsp. *evanescens*, *Fucus spiralis*, and *Ascophyllum nodosum*, lower anticoagulant activities have been determined in the amounts

Table 10.1 Bioactivities of seaweeds.

Species	Extract/Compound	Bioactivity	References
Chlorophyta (green seaweed)			
Auxenochlorella pyrenoidosa (as *Chlorella pyrenoidosa*) *	Crude extract	Anti-allergic	Chen et al. 2015b
Bryopsis maxima	Sulfated polysaccharide	Anticoagulant	Ciancia et al. 2010, Arata et al. 2015
Capsosiphon fulvescens	Methanol extract	Anti-edema and anti-erythema	Khan et al. 2008
Caulerpa cupressoides	Sulfated polysaccharide	Anticoagulant	Costa et al. 2010
C. cupressoides	Lectin, sulfated polysaccharide	Anti-inflammatory	Vanderlei et al. 2010
C. cupressoides	Sulfated polysaccharide	Antinociceptive, anti-inflammatory and anticoagulant	Rodrigues et al. 2011, 2012
C. cupressoides var. *lycopodium*	Sulfated polysaccharide	Anticoagulant	Rodrigues et al. 2013
C. cupressoides var. *lycopodium*	Sulfated polysaccharide	Antinociceptive	Rodrigues et al. 2013b
C. mexicana	Aqueous and methanol extracts	Antinociceptive and anti-inflammatory	Bitencourt et al. 2011, Da Matta et al. 2011
Caulerpa mexicana	Methanolic extract	Anti-inflammatory	Bitencourt et al. 2015
C. okamurae	Sulfated polysaccharide	Anticoagulant	Ciancia et al. 2010, Arata et al. 2015
C. prolifera	Sesquiterpene (Caulerpyne)	Anti-inflammatory and anti-allergic	Mayer et al. 1993, Vo et al. 2012b, c
C. prolifera	Sulfated polysaccharide	Anticoagulant	Costa et al. 2010
C. racemosa	Crude extracts	Antinociceptive and anti-inflammatory	Souza et al. 2009
C. racemosa	Caulerpin	Antinociceptive and anti-inflammatory	De Souza et al. 2009
C. racemosa	Sulfated polysaccharide	Anticoagulant	Rodrigues et al. 2012
C. racemosa	Methanol extract	Anti-inflammatory	Radhika et al. 2013
C. racemosa	Sulfated polysaccharide	Antinociceptive and anti-inflammatory	Ribeiro et al. 2014
C. racemosa	Crude extracts	Anti-inflammatory	Montalvão et al. 2016
C. sertularioides	Sulfated polysaccharide	Anticoagulant	Costa et al. 2010
Caulerpa sp.	Caulerpin	Anti-inflammatory	Éverton et al. 2009
Chaetomorpha antennina	Sulfated polysacharide	Anticoagulant	Ganesh et al. 2009

Table 10.1 contd....

Table 10.1 contd....

Species	Extract/Compound	Bioactivity	References
*Chlorella marina**	Lycopene	Anti-inflammatory	Renju et al. 2013
Cladophora falklandica	Sulfated xylogalactoarabinans	Anticoagulant	Arata et al. 2016
C. vagabunda (as *Cladophora fascicularis*)	2-(20,40-dibromophenoxy) -4,6-dibromoanisol	Anti-inflammatory	Kuniyoshi et al. 1985
Codium cylindricum	Sulfated polysaccharide (Ulvan-like)	Anticoagulant	Matsubara et al. 2001
C. dwarkense	Sulfated arabinogalactan	Anticoagulant	Shanmugam et al. 2001, 2002
C. fragile	Sulfated arabinans and galactans (and/or sulfated arabinogalactans)	Anticoagulant	Ciancia et al. 2007
C. fragile	Methanol extract	Anti-edema and anti-erythema	Khan et al. 2008
C. fragile	Dichloromethane extract	Anti-inflammatory, antipyretic and analgesic	Kang et al. 2012b
C. indicum	Sulfated arabinogalactan	Anticoagulant	Shanmugam et al. 2002
C. isthmocladum	Sulfated polysaccharide	Anticoagulant	Costa et al. 2010
C. isthmocladum	Sulfated polysaccharide	Anticoagulant	Ciancia et al. 2010, Arata et al. 2015
C. latum	Sulfated polysaccharide	Anticoagulant	Ciancia et al. 2010, Arata et al. 2015
Codium spp.	Lectins	Anticoagulant	Sharma and Sahni 1993
C. subtubulosum (as *C. divaricatum*)	Sulfated polysaccharide	Anticoagulant	Ciancia et al. 2010, Arata et al. 2015
C. subtubulosum (as *C. divaricatum*)	Sulfated polysaccharide	Anticoagulant	Li et al. 2015
C. vermilara	Sulfated arabinan and galactan (and/or sulfated arabinogalactan)	Anticoagulant	Ciancia et al. 2007, Fernández et al. 2013
Cymopolia barbata	Bromohydroquinones (Cymopol, Cyclocymopol)	Anti-inflammatory and anti-allergic	Mayer et al. 1993, Vo et al. 2012b, c
Dunaliella salina (as *Dunaliella bardawil*)*	Beta-carotene	Anti-inflammatory	Lavy et al. 2003
*Dunaliella tertiolecta**	Phytosterols	Anti-inflammatory	Caroprese et al. 2012
Gayralia oxysperma (as *Monostroma oxyspermum*)	Sulfated polysaccharide	Anticoagulant	Seedevi et al. 2015
Monostroma angicava	Rhamnan-type sulfated polysaccharide	Anticoagulant	Li et al. 2017

M. nitidum	Sulfated polysaccharide (Rhamnose)	Anticoagulant	Mao et al. 2008 Hayakawa et al. 2000
M. latissimum	Sulfated polysaccharide (Rhamnose)	Anticoagulant	Mao et al. 2009
Penicillus capitatus	Highly pyruvylated sulfated galactan	Anticoagulant	Arata et al. 2015
Prasiola crispa	Dichloromethane, ethyl acetate, *n*-hexane, methanol extracts	Anti-snake venom	Da Silva et al. 2017
Rhipocephalus phoenix	Sesquiterpene (Rhipocephalin)	Anti-inflammatory and anti-allergic	Tan et al. 2000, Vo et al. 2012b, c
Ulva australis (as *Ulva pertusa*)	Polyunsaturated fatty acids (PUFAs)	Anti-allergic	Ishihara et al. 1998
U. australis (as *U. pertusa*)	Methanolic extract	Anti-inflammatory	Khan et al. 2008
U. australis (as *U. pertusa*)	3-Hydroxy-4,7-megastigmadien-9-one	Anti-inflammatory	Ali et al. 2017
U. clathata	Sulfated polysaccharide	Anticoagulant	Qi et al. 2012
U. compressa	Extract	Anti-allergic	Raman et al. 2004
U. compressa	Methanol extract	Anti-edema and anti-erythema	Khan et al. 2008
U. conglobata	Methanol extract	Neuroprotective (see also Chapter 11) and anti-inflammatory	Jin et al. 2006
U. conglobata	Sulfated polysaccharide (Rhamnose)	Anticoagulant	Mao et al. 2006
U. lactuca	3-0-β-D-glucopyranosy-lstigmasta-5,25-diene	Anti-inflammatory	Awad 2000
U. lactuca	Methanol extract (Sulfated polysaccharide)	Anti-inflammatory	Sirbu et al. 2008 Margret et al. 2009
U. lactuca	Crude extracts (enzymatic digestion)	Analgesic, anti-inflammatory	De Araujo et al. 2016
U. linza	Methanolic extract	Anti-inflammatory	Khan et al. 2008
U. linza	Sulfated polysaccharide	Anticoagulant	Wang et al. 2013b
U. prolifera	Pheophytin A	Anti-inflammatory	Okai and Okai 1997
U. reticulata	Mathanolic extract	Anti-inflammatory	Hong et al. 2011
U. rigida	Acidic polysaccharide	Anti-inflammatory	Leiro et al. 2007
U. rigida	Methanol extract	Antipyretic	Sheeba and Paula 2014

Table 10.1 contd....

Table 10.1 contd....

Species	Extract/Compound	Bioactivity	References
Phaeophyceae (brown seaweed)			
Ascophyllum nodosum	Fucoidan	Anticoagulant	Chevolot et al. 2001
A. nodosum	Fucoidan	Antithrombotic and anticoagulant	Trento et al. 2001
A. nodosum	Sulfated fucan	Anticoagulant	Pomim et al. 2005
A. nodosum	Fucoidan	Anti-inflammatory, antiangiogenic, anticoagulant, antiadhesive	Cumashi et al. 2007
Canistrocarpus cervicornis (as *Dictyota cervicornis*)	Sulfated polysaccharide	Anticoagulant	Costa et al. 2010
C. cervicornis	Fucans	Anticoagulant	Camara et al. 2011
Cladosiphon okamuranus	Fucoidan	Anti-inflammatory	Matsumoto et al. 2004
C. okamuranus	Sulfated polysaccharide	Anticoagulant	Cumashi et al. 2007
Colpomenia bullosa	Methanol extract	Anti-edema and anti-erythema	Khan et al. 2008
C. sinuosa	Methanol extract	Anti-edema and anti-erythema	Khan et al. 2008
Costaria costata	Methanol extract	Anti-edema and anti-erythema	Khan et al. 2008
C. costata	Crude extracts (serine protease)	Anti-thrombotic	Kim et al. 2013d
Cystoseira compressa	Water extract	Anti-inflammatory	Mhadhebi et al. 2014
C. compressa	Fucoidan	Anti-inflammatory, Antiulcerogenic	Ammar et al. 2015
C. crinita	Water extract	Anti-inflammatory	Mhadhebi et al. 2014
C. crinita	Fucoidan	Anti-inflammatory, Antiulcerogenic	Ammar et al. 2015
C. crinita	Extracts	Anti-inflammatory	Montalvão et al. 2016
C. sedoides	Water extract	Anti-inflammatory	Mhadhebi et al. 2014
C. sedoides	Fucoidan	Anti-inflammatory, Antiulcerogenic	Ammar et al. 2015
Dictyopteris delicatula	Sulfated polysaccharide	Anticoagulant	Costa et al. 2010
D. delicatula	Sulfated polysaccharide	Anticoagulant	Magalhães et al. 2011
D. polypodioides	Water-soluble polysaccharides, such as fucoidan, laminaran, and mannuronan	Anticoagulant	Karaki et al. 2013
Dictyota menstrualis	Sulfated polysaccharide	Anticoagulant	Costa et al. 2010
D. menstrualis	Sulfated heterofucans	Anticoagulant	Albuquerque et al. 2004

D. menstrualis	Heterofucan	Antinociceptive and anti-inflammatory	Albuquerque et al. 2013
D. dichotoma	Glucuron-oxylomannogalactofucan	Anticoagulant	Abdel-Fattah et al. 1978
D. dichotoma	Methanol extract	Anti-edema and anti-erythema	Khan et al. 2008
D. fasciola	Extracts	Anti-inflammatory	Montalvão et al. 2016
D. mertensii	Sulfated polysaccharide	Anticoagulant	Costa et al. 2010
Ecklonia arborea	Phlorotannins	Anti-allergic	Shibata et al. 2002 Sugiura et al. 2009
E. arborea	Extracts	Anti-allergic	Sugiura et al. 2008
E. arborea (as *Eisenia arborea*)	Methanol/chloroform extract (Phlorotannins)	Anti-allergic and anti-inflammatory	Sugiura et al. 2011
E. bicyclis	Fucoidan	Anticoagulant	Usui et al. 1980
E. bicyclis (as *Eisenia bicyclis*)	Phlorotannins	Anti-inflammatory and anti-allergic	Shibata et al. 2002, 2003, Vo et al. 2012b, c
E. cava	Sulfated polysaccharide	Anticoagulant	Athukorala et al. 2006
E. cava	Phlorotannins	Anti-inflammatory	Shin et al. 2006
E. cava	Sulfated polysaccharide	Anticoagulant	Jung et al. 2007
E. cava	Enzymatic extract	Anti-inflammatory and anti-allergic	Ahn et al. 2008
E. cava	Methanol extract	Anti-edema and anti-erythema	Khan et al. 2008
E. cava	Ethanol extracts	Anti-inflammatory and anti-allergic	Kim et al. 2008b
E. cava	Phlorotannins	Anti-allergic	Li et al. 2008b, Le et al. 2009
E. cava	Phlorotannins	Anti-inflammatory and anti-allergic	Ryu et al. 2008
E. cava	Extracts	Anti-allergic	Sugiura et al. 2008
E. cava	Methanol extract	Anti-edema and anti-erythema	Jung et al. 2009b
E. cava	Phlorotannins	Anti-inflammatory and anti-allergic	Le et al. 2009
E. cava	Methanol extract	Anti-inflammatory and anti-allergic	Shim et al. 2009
E. cava	Organic solvent extracts	Anti-inflammatory and anti-allergic	Kim and Bae 2010
E. cava	Fucoidan	Anti-inflammatory and anti-allergic	Kang et al. 2011
E. cava	Sulfated polysaccharide	Anticoagulant	Wijesinghe et al. 2011
E. cava	Fucoidan	Anti-inflammatory	Lee et al. 2012d

Table 10.1 contd....

Table 10.1 contd....

Species	Extract/Compound	Bioactivity	References
E. kurome	Sulfated polysaccharide (Fucoidan)	Anticoagulant	Nishino et al. 1989, Nishino and Nagumo 1991
E. kurome	Sulfated polysaccharide	Anticoagulant	Kang et al. 2011
E. kurome	Polyphenol phlorotannins	Anti-inflammatory	Kim and Himaya 2011
E. stolonifera	Methanol extract	Anti-edema and anti-erythema	Khan et al. 2008
Fucus distichus	Fucoidan	Anti-inflammatory, antiangiogenic, anticoagulant, and antiadhesive	Cumashi et al. 2007
F. distichus	Fucoidan	Anticoagulant	Ushakova et al. 2009
F. evanescens	Fucoidan	Anti-inflammatory, antiangiogenic, anticoagulant, and antiadhesive	Cumashi et al. 2007
F. evanescens	Fucoidan	Anticoagulant	Ushakova et al. 2009
F. serratus	Fucoidan	Anti-inflammatory, antiangiogenic, anticoagulant, and antiadhesive	Cumashi et al. 2007
F. serratus	Fucoidan	Anticoagulant	Ushakova et al. 2009
F. spiralis	Sulfated polysaccharides	Anticoagulant	Deacon-Smith et al. 1985
F. spiralis	Fucoidan	Anti-inflammatory, antiangiogenic, anticoagulant, and antiadhesive	Cumashi et al. 2007
F. spiralis	Fucoidan	Anticoagulant	Ushakova et al. 2009
F. vesiculosus	Fucoidan	Inhibition of fibrin clot formation and antithrombin	Springer et al. 1957
F. vesiculosus	Fucan	Anticoagulant	Bernardi and Springer 1962
Fucus vesiculosus	Fucoidan	Anticoagulant	Dürig et al. 1997
F. vesiculosus	Sulfated fucans	Anticoagulant	Trento et al. 2001
F. vesiculosus	Fucoidan	Anti-inflammatory	Park et al. 2011
F. vesiculosus	Fucoidan	Antifibrotic and anti-inflammatory	Lim et al. 2015
Ishige foliacea	Methanol extracts	Anti-allergic	Sugiura et al. 2008
I. okamurae	Methoxylated fatty acid (7-methoxy-9-methylhexadeca-4,8- dienoic)	Anti-inflammatory and anti-allergic	Cho et al. 2008, Vo et al. 2012b, c
I. okamurae	Methanol extract	Anti-edema and anti-erythema	Khan et al. 2008

I. okamurae	Methanol extracts	Anti-allergic	Sugiura et al. 2008
I. okamurae	Ethanolic extract	Anti-inflammatory	Kim et al. 2009b
I. sinicola	Methanol extract	Anti-edema and anti-erythema	Khan et al. 2008
Laminaria brasiliensis	Sulfated fucan	Anticoagulant	Pereira et al. 1999
L. digitata	Fucoidan	Anticoagulant	Ushakova et al. 2009
L. hyperborea	Sulfated polysaccharide	Anticoagulant	Deacon-Smith et al. 1985
Lessonia flavicans	Fucoidan	Anticoagulant	Chandía and Matsuhiro 2008
L. flavicans	Fucoidan	Anticoagulant	Li et al. 2008
L. nigrescens	Alginate	Anti-allergic	Asada et al. 1997
Lobophora variegata	Macrolides, lobophorins A and B	Anti-inflammatory	Jiang et al. 1999
L. variegata	Sulfated fucans (galactofucan)	Anti-inflammatory	Cumashi et al. 2007 Paiva et al. 2011
L. variegata	Sulfated polysaccharides (fucans and fucoidans)	Anticoagulant and anti-inflammatory	Medeiros et al. 2008
L. variegata	Sulfated polysaccharide	Anti-inflammatory	Siqueira et al. 2011
Macrocystis pyrifera	Alginate	Anti-allergic	Jeong et al. 2006
Myagropsis myagroides	Fucoxanthin	Anti-inflammatory	Heo et al. 2010
Padina boergesenii	Extract	Anti-inflammatory, antivenom	Vasanthi et al. 2003
P. gymnospora	Sulfated fucan	Anticoagulant	Silva et al. 2005
P. minor	Water extract	Anti-inflammatory	Peerapornpisal et al. 2010
P. pavonica	Fucan sulfates (xylofucomannoglucuronan)	Anticoagulant	Hussein et al. 1980
Padina sp.	Water extract	Anti-inflammatory	Krishnankartha and Chakaborty 2013
P. tetrastromatica	Sulfated fucan (xylofucogalactomannoglucuronan)	Anticoagulant	Rao et al. 1984
P. tetrastromatica	Sulfated polysaccharides	Anticoagulant	Manoj et al. 2013
P. tetrastromatica	Methanol extract	Anti-inflammatory	Radhika et al. 2013
Petalonia binghamiae	Extracts	Anti-inflammatory, anti-allergic	Kimiya et al. 2008

Table 10.1 contd....

Table 10.1 contd....

Species	Extract/Compound	Bioactivity	References
P. binghamiae	Hexane, ethyl acetate and dichloromethane extracts	Anti-inflammatory	Yoon et al. 2009, Yang et al. 2010, 2010c
Saccharina cichorioides (as *Laminaria cichorioides*)	Sulfated fucan	Anticoagulant	Yoon et al. 2007
S. japonica	Laminarin	Anticoagulant	Kimura 1941
S. japonica	Methanol extract	Anti-edema and anti-erythema	Khan et al. 2008
S. japonica	Ethanol extract	Anti-allergic	Samee et al. 2009, 2009b
S. japonica (as *Laminaria japonica*)	Crude extract	Anti-allergic	Chen et al. 2015b
S. latissima	Sulfated polysaccharide	Anticoagulant	Deacon-Smith et al. 1985
S. latissima	Fucoidan	Anticoagulant	Ushakova et al. 2009
S. longissima	Fucoidan	Anticoagulant	Kitamura et al. 1991
Sargassum cervicorne	Ethanol extract	Anti-allergic	Samee et al. 2009, 2009b
S. cinctum	Sulfated polysaccharides	Anticoagulant	Mody et al. 1997
S. confusum	Methanol extract	Anti-edema and anti-erythema	Khan et al. 2008
S. filipendula	Sulfated polysaccharide	Anticoagulant	Costa et al. 2010
Sargassum fulvellum	Water and dichloromethane extracts	Antipyretic, analgesic, anti-inflammatory, anti-edema	Kang et al. 2008b
S. fulvellum	Methanol extract	Anti-edema and anti-erythema	Khan et al. 2008
S. fulvellum	Hexane extract	Anti-inflammatory	Gwon et al. 2013
S. fusiforme	Fucoidan	Anticoagulant	Dobashi et al. 1989
S. fusiforme	Methanol extract	Anti-edema and anti-erythema	Khan et al. 2008
Sargassum graminifolium	Ethanol extract	Anti-allergic	Samee et al. 2009, 2009b
S. hemiphyllum	Methanol extract	Anti-allergic	Na et al. 2005b
S. hemiphyllum	Water (100°C) extract	Anti-inflammatory	Hwang et al. 2011
S. hemiphyllum	Sulfated polysaccharide	Anti-inflammatory	Hwang et al. 2015b
S. horneri	Crude polysaccharide fraction	Anticoagulant	Athukorala et al. 2007
S. horneri	95% ethanol extract	Anti-inflammatory	Kang et al. 2008b

S. horneri	Methanol extract	Anti-edema and anti-erythema	Khan et al. 2008
S. horneri	Ethanol/water xtract	Anti-allergic	Yoshioka et al. 2014
S. horneri	Enzymatic digestion (sulfated polysaccharide)	Anti-inflammatory	Sanjeewa et al. 2017
S. ilicifolium	Mathanolic extract	Anti-inflammatory and analgesic	Simpi et al. 2013
S. micracanthum	Extract	Anti-allergic	Sugiura et al. 2008
S. micracanthum	Phlorotanin	Anti-inflammatory	Yoon et al. 2009
S. polyceratium	Ethanol extract	Antinociceptive	Santos et al. 2015
S. ringgoldianum	Methanol extract	Anti-edema and anti-erythema	Khan et al. 2008
S. ringgoldianum	Methanol extract	Anti-allergic	Sugiura et al. 2008
S. sagamianum	Methanol extract	Anti-edema and anti-erythema	Khan et al. 2008
S. siliquastrum	Fucoxanthin	Cytoprotective effect on H_2O_2 mediated cell damage and UV-B induced cell damage	Heo et al. 2008, Heo and Jeon 2009
S. siliquastrum	Sargachromanol G	Anti-inflammatory	Yoon et al. 2012
S. siliquosum	Ethanol extract	Hepatoprotective	Vasquez et al. 2012
S. swartzii (as *S. wightii*)	Hexane, methanol and butanol extracts	Anti-inflammatory	Dar et al. 2007
S. swartzii	Mathanolic extract	Anti-inflammatory	Hong et al. 2011
S. swartzii (as *Sargassum wightii*)	Sulfated polysaccharides	Anticoagulant	Manoj et al. 2013
S. swartzii (as *S. wightii*)	Methanol extract	Anti-inflammatory	Radhika et al. 2013
S. swartzii (as *S. wightii*)	Alginic acid	Anti-inflammatory	Sarithakumari et al. 2013
S. swartzii (as *S. wightii*)	Phloroglucinol	Anticoagulant	Karthik et al. 2016
S. swartzii (as *S. wightii*)	Sulfated polysaccharide	Anti-inflammatory and antinociceptive	Neelakandan and Venkatesan 2016
S. swartzii (as *S. wightii*)	Ethyl acetate extract	Anti-inflammatory	Pramitha and Kumari 2016
S. tenerrimum	Ethanol extract	Anti-allergic	Samee et al. 2009, 2009b
S. tenerrimum	Sulfated polysaccharides	Anticoagulant	Manoj et al. 2013
S. tenerrimum	Phloroglucinol	Anticoagulant	Karthik et al. 2016
S. thunbergii	Acetone/dichloromethane and methanol extracts	Anti-allergic	Na et al. 2004
S. thunbergii	Dichloromethane and methanol extracts	Anti-allergic	Lee et al. 2006

Table 10.1 contd....

Table 10.1 contd....

Species	Extract/Compound	Bioactivity	References
S. thunbergii	Methanol extract	Anti-edema and anti-erythema	Khan et al. 2008
S. thunbergii	Extract	Anti-allergic	Sugiura et al. 2008
S. thunbergii	Water and dichloromethane extracts	Antipyretic, analgesic, anti-inflammatory	Kang et al. 2008b
S. thunbergii	Ethanol extract	Anti-allergic	Samee et al. 2009, 2009b
S. vulgare	Fucan	Anticoagulant, antithrombotic, antioxidant and anti-inflammatory	Dore et al. 2013
S. yezoense	95% ethanol extract	Anti-inflammatory	Kang et al. 2008b
Scytosiphon lomentaria	Methanol extract	Anti-edema and anti-erythema	Khan et al. 2008
S. lomentaria	Extracts	Anti-inflammatory, anti-allergic	Kimiya et al. 2008
Scytosiphon sp.	Crude extract	Anti-allergic	Chen et al. 2015b
Silvetia babingtonii	Fucoidan	Anticoagulant	Anno et al. 1966
Spatoglossum schroederi	Sulfated xylofucan	Anticoagulant	Leite et al. 1998
S. schroederi	Sulfated galactofucan	Anticoagulant, antithrombotic	Rocha et al. 2005
S. schroederi	Sulfated polysaccharide	Anticoagulant	Costa et al. 2010
S. schroederi	Sulfated polysaccharide	Antinociceptive and anti-inflammatory	Farias et al. 2011
S. schroederi	Methanol extract	Anti-inflammatory	Júnior et al. 2015
Stypopodium flabelliforme	Diterpene (Epitaondiol)	Anti-inflammatory and anti-allergic	Terracciano et al. 2006, Vo et al. 2012b, c
S. flabelliforme	Terpenoid (Isoepitaondiol)	Anti-inflammatory	Areche et al. 2010, D'Orazio et al. 2012
Taonia atomaria	Sargaquinone	Anti-inflammatory	Ishitsuka et al. 1979, Tziveleka et al. 2005
Turbinaria conoides	Water extract	Anti-inflammatory	Yoon et al. 2009
T. conoides	Water extract	Anti-inflammatory	Boonchum et al. 2011
T. conoides	Sulfated polysaccharides	Anticoagulant	Manoj et al. 2013
T. conoides	Phloroglucinol	Anticoagulant	Karthik et al. 2016
T. ornata	Water-soluble crude polysaccharide	Anti-inflammatory	Ananthi et al. 2010
T. ornata	Water extract	Anti-inflammatory	Arivuselvan et al. 2011
T. ornata	Sulfated polysaccharides	Anticoagulant	Manoj et al. 2013
T. tricostata	Fucoidan	Hepatoprotective	Chale-Dzul et al. 2015

Undaria pinnatifida	Polyunsaturated fatty acids (PUFAs)	Anti-inflammatory and anti-allergic	Ishihara et al. 1998
U. pinnatifida	Fucoidan	Anti-allergic	Maruyama et al. 2005
U. pinnatifida	Polyunsaturated fatty acids (PUFAs)	Anti-inflammatory	Khan et al. 2007
U. pinnatifida	Methanol extract	Anti-edema and anti-erythema	Khan et al. 2008
U. pinnatifida	Extracts	Anti-inflammatory, anti-allergic	Kimiya et al. 2008
U. pinnatifida	Fatty acids (FA) and Eicosapentaenoic acids (EPA)	Anti-inflammatory	Simopoulos 2002, Khan et al. 2009
U. pinnatifida	Fucoidan	Anticoagulant	Irhimeh et al. 2009
U. pinnatifida	Fucoidan	Anti-inflammatory	Hong et al. 2012
U. pinnatifida	Sulfated polysaccharide	Anticoagulant	Faggio et al. 2015
Rhodophyta (red seaweed)			
Acanthophora muscoides	Sulfated polysaccharide	Antinociceptive and anti-inflammatory	Quindere et al. 2013
A. muscoides	Sulfated polysaccharide	Anticoagulant	Gurgel-Rodrigues et al. 2016, Gurgel-Rodrigues et al. 2016b
Acrosorium ciliolatum	Dichloromethane/methanol (50:50) extract	Anti-inflammatory	Oumaskour et al. 2013
Agardhiella ramosissima	Sulfated polysaccharide	Antinociceptive and anti-inflammatory	Batista et al. 2014
Ahnfeltiopsis flabelliformis	Methanol extract	Anti-edema and anti-erythema	Khan et al. 2008
Asparagopsis armata	Dichloromethane/methanol (50:50) extract	Anti-inflammatory	Oumaskour et al. 2013
A. taxiformis	Sulfated polysaccharides	Anticoagulant	Manilal et al. 2012c
Botryocladia occidentalis	Sulfated galactan	Anticoagulant, antithrombotic	Farias et al. 2000, Mourão 2015
Bryothamnion seaforthii	Sulfated polysaccharides	Antinociceptive	Viana et al. 2002, Vieira et al. 2004
B. triquetrum	Sulfated polysaccharides Methanol extract	Anti-inflammatory and antinociceptive	Viana et al. 2002, Almeida et al. 2011, Cavalcante-Silva et al. 2012
Ceratodictyon spongiosum	Peptide (*trans, trans*-ceratospongamide)	Anti-inflammatory and anti-allergic	Tan et al. 2000, Vo et al. 2012b, c
Chondracanthus acicularis	Carrageenan	Anticoagulant	Houck et al. 1957
C. acicularis (as *Gigartina acicularis*)	Dichloromethane/methanol (50:50) extract	Anti-inflammatory	Oumaskour et al. 2013
C. tenellus	Methanol extract	Anti-edema and anti-erythema	Khan et al. 2008
Chondria dasyphylla	Dichloromethane/methanol (50:50) extract	Anti-inflammatory	Oumaskour et al. 2013

Table 10.1 contd....

Table 10.1 contd....

Species	Extract/Compound	Bioactivity	References
Chondrophycus undulates (as *Laurencia undulata*)	Extract	Anti-allergic	Kim et al. 2008b, Jung et al. 2009
Chondrus crispus	Lambda-carrageenan	Anticoagulant	Hawkins et al. 1962, Hawkins and Leonard 1963, Kindness et al. 1979
C. crispus	Dichloromethane/methanol (50:50) extract	Anti-inflammatory	Oumaskour et al. 2013
C. ocellatus	Carrageenan	Anticoagulant	Parish 1987
C. ocellatus	Methanol extract	Anti-edema and anti-erythema	Khan et al. 2008
C. pinnulatus	Iota and kappa/iota hybrid carrageenans	Anticoagulant	Yermak et al. 2006
Coccotylus brodiei	Lambda-carrageenan	Anticoagulant	Efinmov et al. 1983
Corallina pilulifera	Methanol extract	Anti-edema and anti-erythema	Khan et al. 2008
Corallina sp.	Sulfated galactan, carrageenan	Anticoagulant	Sebaaly et al. 2014
Chrysymenia wrightii	Extract	Anti-allergic	Kimiya et al. 2008
Delesseria sanguinea	Sulfated polysaccharide	Anticoagulant	Elsener 1938
D. sanguinea	Sulfated polysaccharide	Anti-inflammatory	Grunewald et al. 2009, Almeida et al. 2011
Dichotomaria marginata (as *Galaxaura marginata)*	Ethanol/acetic acid extract	Anti-inflammatory	Rozas and Freitas 2007
D. obtusata	Water extract	Analgesic and anti-inflammatory	Vázquez et al. 2011
D. obtusata	Methanol extract	Anti-inflammatory and antinociceptive	García Delgado et al. 2013
Digenea simplex	Sulfated polysaccharide	Anti-inflammatory	Pereira et al. 2014
Ellisolandia elongata (as *Corallina elongata*)	Dichloromethane/methanol (50:50) extract	Anti-inflammatory	Oumaskour et al. 2013
Eucheuma denticulatum	Carrageenan	Anticoagulant	Anderson 1969
Fushitsunagia catenata (as *Lomentaria catenata*)	Methanol extract	Anti-edema and anti-erythema	Khan et al. 2008
F. catenata (as *L. catenata*)	Sulfated galactan	Anticoagulant	Pushpamali et al. 2008
Gelidium corneum (as *Gelidium sesquipedale*)	Dichloromethane/methanol (50:50) extract	Anti-inflammatory	Oumaskour et al. 2013

G. crinale	Sulfated galactan	Anticoagulant	Pereira et al. 2005
G. crinale	Galactan	Anti-inflammatory and antinociceptive	Almeida et al. 2011, De Sousa et al. 2013
Gigartina pistillata	Dichloromethane/methanol (50:50) extract	Anti-inflammatory	Oumaskour et al. 2013
G. skottsbergii	Carrageenan	Anticoagulant	Carlucci et al. 1997a
Gracilaria birdiae	NaOH and ultrasound-enhanced extract (sulfated polysaccharides)	Anticoagulant	Khan et al. 2008
G. birdiae	Sulfated polysaccharide	Anti-inflammatory	Vanderlei et al. 2011
G. caudata	Sulfated polysaccharide	Anticoagulant	Costa et al. 2010
G. caudata	Sulfated polysaccharide	Anti-inflammatory and antinociceptive	Almeida et al. 2011, Chaves et al. 2013
G. cornea	Sulfated polysaccharide	Antinociceptive and anti-inflammatory	Coura et al. 2012
G. cornea	Sulfated polysaccharides	Anxiolytic	Monteiro et al. 2016
G. cornea	Sulfated agaran	Anti-inflammatory, antinociceptive	Souza et al. 2017
G. edulis	Methanol extract	Anti-inflammatory	Radhika et al. 2013
G. edulis	Water and methanol extracts	Anti-inflammatory	Vijayalakshmi 2015
G. tenuistipitata	Water extract	Anti-inflammatory	Almeida et al. 2011, Chen et al. 2013
G. textorii	Water extract	Anti-inflammatory	Okada et al. 1994, Almeida et al. 2011
Gracilariopsis longissima	Methanol extract	Anti-edema and anti-erythema	Khan et al. 2008
G. longissima (as *Gracilaria verrucosa*)	Fatty acids	Anti-inflammatory	Lee et al. 2009c, Almeida et al. 2011
G. longissima (as *Gracilaria verrucosa*)	Polysaccharide fraction	Anti-inflammatory	Yoshizawa et al. 1996, Almeida et al. 2011
Grateloupia cornea	Methanol extract	Anti-edema and anti-erythema	Khan et al. 2008
G. elliptica (as *Pachymeniopsis elliptica*)	Methanol extract	Anti-edema and anti-erythema	Khan et al. 2008
G. filicina	Sulfated polysaccharide	Anticoagulant	Muruganantham 2001
G. indica	Sulfated galactan	Anticoagulant	Sen et al. 1994
G. livida	Sulfated polysaccharide	Anticoagulant	Tang et al. 2017
G. turuturu	Carrageenan	Anticoagulant	Efinmov et al. 1983
Halopithys incurva	Dichloromethane/methanol (50:50) extract	Anti-inflammatory	Oumaskour et al. 2013
Halymenia floresia	Sulfated polysaccharides	Anticoagulant	Rodrigues et al. 2011b

Table 10.1 contd....

Table 10.1 contd....

Species	Extract/Compound	Bioactivity	References
Helminthocladia australis	Methanol extract	Anti-edema and anti-erythema	Khan et al. 2008
Hypnea cervicornis	Mucin-binding agglutinin	Anti-inflammatory and antinociceptive	Bitencourt et al. 2008, Almeida et al. 2011
H. cervicornis	Lectin	Anti-inflammatory	Figueiredo et al. 2010
H. charoides	Methanol extract	Anti-edema and anti-erythema	Khan et al. 2008
H. musciformis	Dichloromethane/methanol (50:50) extract	Anti-inflammatory	Oumaskour et al. 2013
H. musciformis	Sulfated galactan	Anticoagulant	Alves et al. 2016c
H. valentiae	Extract	Anti-inflammatory, antivenom	Vasanthi et al. 2003
Iridaea spp.	Sulfated galactan	Anticoagulant	Chargaff et al. 1936, Elsner 1938
Kappaphycus alvarezii	Water extract	Anti-inflammatory	Ranganayaki et al. 2014
Laurencia aldingensis	Sulfated agaran	Anti-inflammatory, antivenom	Da Silva et al. 2016
L. claviformis	Terpenoid (Pacifenol)	Anti-inflammatory	D'Orazio et al. 2012
L. filiformis	Terpenoid (Pacifenol)	Anti-inflammatory	D'Orazio et al. 2012
L. glandulifera	Diterpenoid (Neorogioltriol)	Anti-inflammatory	Almeida et al. 2011, Chatter et al. 2011
L. obtusa	Acetogenins: (12Z)-*cis*-maneonene-D and (12E)-*cis*-maneonene-E	Anti-inflammatory	Almeida et al. 2011, Ayyad et al. 2011
L. obtusa	Methanol and etahnol extracts	Analgesic, anti-inflammatory	Lajili et al. 2016, 2016b
L. snackeyi	Crude extract	Anti-inflammatory	Vairappan et al. 2013
L. tasmanica	Terpenoid (Pacifenol)	Anti-inflammatory	D'Orazio et al. 2012
Lithothamnion corallioides	Multi-mineral aquamin	Anti-inflammatory	Almeida et al. 2011, Ryan et al. 2011
Mazzaella laminarioides	Sulfated polysaccharide	Anticoagulant	Chargaff et al. 1936
Meristotheca papulosa	Methanol extract	Anti-edema and anti-erythema	Khan et al. 2008
Neorhodomela aculeata	Methanol extract	Anti-inflammatory	Lim et al. 2006, Almeida et al. 2011
Nothogenia fastigiata	Xylomannan	Anticoagulant	Kolender et al. 1997
Osmundaria obtusiloba (as *Vidalia obtusiloba*)	Bromophenols (Vidalols A and B)	Anti-inflammatory	Wiemer et al. 1991
Osmundea pinnatifida (as *Laurencia pinnatifida*)	Dichloromethane/methanol (50:50) extract	Anti-inflammatory	Oumaskour et al. 2013
Pachymeniopsis lanceolata	Methanol extract	Anti-edema and anti-erythema	Khan et al. 2008

Palmaria palmata	Dichloromethane/methanol (50:50) extract	Anti-inflammatory	Oumaskour et al. 2013
Phacelocarpus peperocarpos	Extract	Anti-venom	Mayer et al. 1993, Song et al. 2003
Plocamium brasiliense	*n*-Hexane, dichloromethane, ethyl acetate and hydroalcoholic solution extracts	Anti-inflammatory, anti-snake venom	Silva et al. 2015b
Polyopes affinis	Methanol extract	Anti-allergic	Na et al. 2005
P. affinis	Ethanolic extract	Anti-inflammatory	Almeida et al. 2011, Lee et al. 2011c
Pterocladiella capillacea	Lectins	Anti-inflammatory and antinociceptive	Silva et al. 2010
Pterosiphonia complanata	Dichloromethane/methanol (50:50) extract	Anti-inflammatory	Oumaskour et al. 2013
Pyropia dentata	Extract	Anti-allergic	Kimiya et al. 2008
P. dentata	Methanol extract	Anti-inflammatory	Kazlowska et al. 2010
P. dentata (as *Porphyra dentata*)	Methanol extract	Antiallergic, anti-inflammatory	Kazlowska et al. 2013
P. haitanensis	Porphyran	Anticoagulant	Zhang et al. 2010b
P. tenera	Porphyran	Anti-allergic	Ishihara et al. 2005
P. vietnamensis (as *Porphyra vietnamensis*)	Water extract	Anti-inflammatory, analgesic and antiulcer	Bhatia et al. 2015
P. yezoensis	Porphyran	Anti-allergic	Ishihara et al. 2005
P. yezoensis	Methanol extract	Anti-edema and anti-erythema	Khan et al. 2008
P. yezoensis (as *Porphyra yezoensis*)	Glycoprotein	Anti-inflammatory	Almeida et al. 2011, Shin et al. 2011
Pyropia sp. (*Porphyra* sp.)	Crude extract	Anti-allergic	Chen et al. 2015b
Schizymenia binderi	Sulfated galactan	Anticoagulant	Zúniga et al. 2006
Solieria filiformis	Iota-carrageenan	Anti-inflammatory, Anticoagulant	Araújo et al. 2012
Sphaerococcus coronopifolius	Extract	Anti-venom	Mayer et al. 1993, Song et al. 2003
Stenogramma interruptum	Carrageenan	Anticoagulant	Cáceres et al. 2000
Symphyocladia latiuscula	Methanol extract	Anti-edema and anti-erythema	Khan et al. 2008

*Marine microalgae

of 19.1 IU mg^{-1}, 15.1 IU mg^{-1}, 16.6 IU mg^{-1}, and 13.4 IU mg^{-1}, respectively. On the other hand, sulfated polysaccharide of *Cladosiphon okamuranus* with the highest content of 23.4% of uronic acids and the lowest content of sulfate groups of 15.1% has showed almost no anticoagulant effect (Cumashi et al. 2007). Analyses of anticoagulant activities of fucoidans have demonstrated that native fucoidans interact with ATIII like heparin and promoted thrombin activation followed by the formation of a complex with thrombin and ATIII. However, in contrast to heparin, they may directly inhibit thrombin (Ushakova et al. 2009). Molecular weight is also important for the anticoagulant activity of fucoidans because of long enough sugar chain, and suitable conformation is needed to bind thrombin. This is because a light decrease in molecular weight of sulfated fucan dramatically reduces its effect on thrombin inactivation mediated by HCII. Although sulfated fucan with 45 tetra-saccharide repeating units could bind to HCII, it is unable to link efficiently the plasma inhibitor and thrombin, because for this action chains with 100 or more tetra-saccharide repeating units are necessary. The template mechanism may predominate over the allosteric effect in the case of linear sulfated fucan inactivation of thrombin heparin cofactor II being present. Linear sulfated fucan requires significantly longer chains than mammalian glycosaminoglycans to achieve the anticoagulant activity (Pomin et al. 2005, Li et al. 2008). Molecular size of the most active fucoidans has been approximately between 50 Da and 100.000 Da, whereas fractions with higher molecular weight have exceeded 850.000 Da inclined to lower the anticoagulant activity (Shanmugam and Mody 2000).

Anticoagulant activity of 75% fucoidan from *Undaria pinnatifida* consisted of 24.7% fucose, 20.35% galactose, no mannose, 29.07% sulfate, 2.19% protein, and 7% bond ions evaluated by aPTT, TT, PT, and ATIII assays. It was established that this fucoidan exhibited a very strong hemostasis effect. It prolonged aPTT time from 38.8 s for control to 172.5 s at 63 mg L^{-1}; TT time was also prolonged at a higher rate in which it was 15.2 s at baseline and went up to 240.1 s at 15.6 mg L^{-1}. Further, ATIII decreased with the fucoidan treatment from 108% for control to 89% at 10,000 mg L^{-1}; and finally, low concentration of fucoidan had no effect on PT until the concentration of 125 mg L^{-1} of PT, when it began to increase (Irhimeh et al. 2009). Also, crude polysaccharide fraction of brown seaweed *Sargassum horneri* consisted of 97% of polysaccharide and 2% of protein showed a strong anticoagulant effect exceeding 300 s by aPTT assay, whereas anticoagulant active compounds were mainly concentrated in the fraction with molecular weight higher than 30 kDa (Athukorala et al. 2007). Brown seaweed *Dictyopteris polypodioides* growing on the Lebanese coast was collected in two different seasons—in May and July. It was rich in polysaccharides whose main component was alginic acid (11%) as well as water-soluble polysaccharides, such as fucoidan, laminaran, and mannuronan, whose amounts differed depending on the season between 3.75% in May and 5.8% in July. The anticoagulant activities were established in different fractions of polysaccharides composed from fucose and laminaran residues, mixture of fucose, laminaran and mannuronan, and mannuronan. The highest anticoagulant activities of 42.5 s, 43.1 s, and 42.1 s by aPTT assay in the application doses of 2.5 mg of different fractions were determined, respectively (Karaki et al. 2013).

Polyunsaturated fatty acids (Fig. 10.4) have attracted a great deal of attention because of their antioxidant, anti-atherosclerotic, anti-inflammatory, and immunoregulatory activities (Wojenski et al. 1991, Calder 1998, Kim et al. 2010c). Especially, their potential inhibitory effects on allergic reactions have been evidenced recently. In particular, two polyunsaturated fatty acids of 18:4*n*–3 and 16:4*n*–3 purified from the marine alga *Undaria pinnatifida* exhibit effective inhibition on the production of leukotriene B4, leukotriene C4, and 5-hydroxyeicosatetraenoic acid in MC/9 mouse mast cells (Ishihara et al. 1998). Moreover, α-linolenic acid (18:3 (*n*–3)) induced ameliorative changes in metabolism of omega-3/omega-6 polyunsaturated fatty acids, histamine content, and histamine release from RBL-2H3 cells. Namely, the concentration of α-linolenic acid and docosahexenoic acid (DHA, 22:6 (*n*–3)) was increased, while linolenic acid (18:2 (*n*–6)) was slightly and arachidonic acid (20:4 (*n*–6)) was markedly decreased in mast cells (Kawasaki et al. 1994). Also, histamine content and release was remarkably lowered in the α-linolenic acid-treated RBL-2H3 cells. Likewise, PGE_2 production and histamine release were diminished in the canine mastocytoma cell line C2 treated with α-linolenic acid (Gueck et al. 2003), α-linolenic acid, and docosahexaenoic acid (Gueck et al. 2004). Thus, the anti-allergic effect of these polyunsaturated fatty acids was suggested either by the decrease in histamine content or by inhibition of the release of chemical mediator resulting from changes in the fatty acid composition.

PUFAs was reported not to cause significant difference in apoptosis and reduction of iNOS compared to control when tested for modulation of inflammation and necrosis (Caplan and Jilling 2001). Fatty acids (FA) and eicosapentaenoic acids (EPA) extracted from brown seaweed *Undaria pinnatifida* exhibited anti-inflammatory against mouse ear edema, erythema, and blood flow induced by phorbol myristate acetate (PMA) (Simopoulos 2002, Khan et al. 2009).

A role of polysaccharides from marine algae as anti-allergic agents has been suggested. Alginic acid (see Fig. 3.2, Chapter 3), a naturally occurring hydrophilic colloidal polysaccharide obtained from the several species of brown seaweeds, exhibited different effects against hyaluronidase activity and histamine release from mast cells (Asada et al. 1997). In the *in vivo* conditions, alginic acid inhibited compound 48/80-induced systemic anaphylaxis with doses of 0.25–1 g kg^{-1} and significantly inhibited passive cutaneous anaphylaxis by 54.8% at 1 g kg^{-1} for 1 hour pretreatment (Jeong et al. 2006). Besides, alginic acid was found to have a maximum suppression rate (60.8%) on histamine release from rat peritoneal mast cells at a concentration of 0.01 μg mL^{-1}. Furthermore, the anti-allergic activities of alginic acid were also observed due to its suppressive effects on activity and expression of histidine decarboxylase, production of IL-1β and TNF-α, and protein level of nuclear factor (NF)-κB/*Rel A* in PMA plus A23187-stimulated HMC-1 cells (Jeong et al. 2006). Noticeably, alginic acid oligosaccharide, a lyase lysate of alginic acid, has been revealed to be able to reduce IgE production in the serum of BALB/c mice immunized with β-lactoglobulin (Yoshida et al. 2004, Uno et al. 2006). Moreover, antigen-induced Th2 development was blocked by alginic acid oligosaccharide treatment via enhancing the production of IFN-ϒ and IL-12, and down-regulating IL-4 production in splenocytes of mice (Yoshida et al. 2004).

The molecular weight of sulfated polysaccharides and their anticoagulant activity relationship were considered by Pomim and co-workers (2005), who observed that linear, sulfated fucan required significant longer chains than mammalian glycosaminoglycans to achieve anticoagulant activity. It has been shown that selective cleavage to reduce the molecular size of the fucan by only a small amount dramatically reduced its effect on thrombin inactivation mediated by heparin cofactor II. This is because lower molecular weight fucans appear to bind to heparin cofactor II but, unlike the native (full length) fucan, are unable to effectively facilitate the heparin cofactor II interaction with thrombin (Pomim et al. 2005). The importance of sulfate group location on the sugar residues for anticoagulant activity has been reported by Chevolot et al. (2001). Their studies established the role of 2-*O*-sulfated and 2,3-*O* di-sulfated fucose residues for the anticoagulant activity of fucoidan from *Ascophyllum nodosum*. Good anticoagulant activity was shown by the native fucoidan from *Lessonia flavicans* (as *Lessonia vadosa*), with a molecular weight of 320 kDa compared to a smaller depolymerized fraction with a molecular weight of 32 kDa, which presented weaker anticoagulant activity (Li et al. 2008).

Secretion of interleukin (IL)-6, and its mRNA, by LPS induced murine colonic epithelial cell line was reportedly lowered by fucoidan obtained from *Cladosiphon okamuranus* (Matsumoto et al. 2004). Reduction of mRNA of MCP-1 was also achieved by application of fucoidan, consequently, leading to reduction in MCP-1 production, which exhibit concentration dependent pattern of fucoidan (Park et al. 2011). Suppression of phosphorylation of p38 and extracellular kinase ERK was reportedly caused by fucoidan treatment of cell (Cui et al. 2010, Park et al. 2011). Furthermore, fucoidan successfully inhibit activation of other pathways known to aid occurrence of inflammation. These pathways include NF-κB, Akt, and JNK (Park et al. 2011). However, c-Jun N-terminal kinase (JNK) pathway was not affected by fucoidan, according to Cui et al. (2010).

Fucoxanthin (see Fig. 2.6, Chapter 2) extracted from *Ishige okamurae* have been reported to induce apoptosis in human leukemia cells HL60 by generating reactive oxygen species (ROS) and induced breaking of caspases-3 and -7, poly-ADP-ribose, and reducing B-cell lymphoma-extra-large (Bcl-xL) level (Kim et al. 2010d). Fucoxanthin extracted from *Sargassum siliquastrum* provides cytoprotective effect on H_2O_2 mediated cell damage and UV-B induced cell damage, thus, identified as being able to offer potential therapeutic application towards several diseases (Heo et al. 2008, Heo and Jeon 2009) in which inflammation was not an exception (Shiratori et al. 2005, Heo et al. 2010, Kim et al. 2010e). Inhibition of activities of both iNOS and COX-2, which resulted in reduced NO and PGE_2 produced, was reportedly observed when LPS induced RAW 246.7. Reverse transcriptase-polymerase chain reaction (RT-PCR) analysis revealed that mRNA of iNOS and COX-2 were affected by fucoxanthin. Expression of cytokines IL-6, TNF-α, and IL-1β were also suppressed by fucoxanthin along with their respective mRNA (Shiratori et al. 2005, Heo et al. 2010, Kim et al. 2010e). Fucoxanthin reduced infiltration of cells and protein concentration in aqueous humor of LPS induced rat eye in a dose dependent manner (Shiratori et al. 2005). It also reduced nuclear translocation of P50 and P65 protein and cytoplasmic degradation of inhibitor of B (IκB)-α, which lead to low level of nuclear factor (NF)κB transactivation, and inhibit phosphorylation of mitogen-activated protein kinases (MAPKs, JNK, ERK and p38) (Heo et al. 2010).

Fucoidan (see Fig. 2.1, Chapter 2) from *Undaria pinnatifida* reduced the concentrations of both IL-4 and IL-13 in bronchoalveolar lavage fluid (BAL), and inhibited the increase of antigen-specific IgE in OVA-induced mouse airway hypersensitivity (Maruyama et al. 2005). In a recent study, Yanase et al. (2009) have reported that the peritoneal injection of fucoidan caused an alleviative effect of plasma IgE level by suppressing a number of IgE-expressing and IgE-secreting B cells from OVA-sensitized mice. On the other hand, the inhibitory effect of fucoidan on IgE production was determined by preventing Cε germline transcription and NF-κB p52 translocation in B cells (Oomizu et al. 2006). Yet, the inhibitory activity of fucoidan has not been observed if B cells were pre-stimulated with IL-4 and anti-CD40 antibody before the administration of fucoidan. Thus, it suggested that fucoidan may not prevent a further increase of IgE in patients who have already developed allergic diseases and high levels of serum IgE. However, Iwamoto et al. (2011) have recently determined that fucoidan effectively reduced IgE production in both peripheral blood mononuclear cells from atopic dermatitis patients and healthy donors. These findings indicated that fucoidan suppresses IgE production by inhibiting immunoglobulin class switching to IgE in human B cells, even after the onset of atopic dermatitis.

In addition, brown algae of *Ecklonia bicyclis* (as *Eisenia bicyclis*) and *Ecklonia arborea* (as *Eisenia arborea*) were also identified to contain many anti-allergic phlorotannins. Phloroglucinol, phlorofucofuroeckol A, dieckol, and 8,8'-bieckol from *E. bicyclis* considerably inhibited hyaluronidase enzyme with IC_{50} values of 280 μM, 140 μM, 120 μM, and 40 μM, respectively (Shibata et al. 2002). The effect of these phlorotannins against hyaluronidase enzyme is stronger than well-known inhibitors such as catechins (IC_{50} = 620 μM) and sodium cromoglycate (IC_{50} = 270 μM). Notably, 8,8'-bieckol, the strongest hyaluronidase inhibitor among the tested phlorotannins, acted as a competitive inhibitor with an inhibition constant of 35 μM. Moreover, these phlorotannins caused the inactivation of enzyme secretory phospholipase A2s, lipoxygenases, and cyclooxygenase. Herein, 8,8'-bieckol showed the pronounced inhibitory effects on soybean lipoxygenases and 5-lipoxygenases with IC_{50} values of 38 μM and 24 μM, respectively. Meanwhile, dieckol presented a significant inhibition of COX-1 with inhibition rate of 74.7% (Shibata et al. 2003). Likewise, several phlorotannins of eckol, 6,6'-bieckol, 6,8'-bieckol, 8,8'-bieckol, PFF-A, and PFF-B from *E. arborea* were also confirmed as strong inhibitors of hyaluronidase, phospholipase A2, cyclooxygenase, and lipoxygenases (Shibata et al. 2008, Sugiura et al. 2009), which correlated to suppression in synthesis and release of leukotoriene and prostaglandin from RBL cells (Sugiura et al. 2009). Among phlorotannins obtained from *E. arborea*, PFF-B exposed the strongest activity against histamine and β-hexosaminidase release with IC_{50} value of 7.8 μM (Sugiura et al. 2006b, 2007). Obviously, PFF-B had a 2.8–6.0 times greater inhibitory activity than those of epigallocatechin gallate (IC_{50} = 22.0 μM) or Tranilast (IC_{50} = 46.6 μM), a clinically used anti-allergic drug (Matsubara et al. 2004).

It was also reported that a sulfated galactofucan from the seaweed *Spatoglossum schroederi* has significant antithrombotic activity in a rat model of venous thrombosis (Rocha et al. 2005). The *in vivo* action of this sulfated galactofucan progresses slowly, showing maximal effectiveness within 8 hours post-injection, unlike heparin, which produces a rapid but transient antithrombotic effect. *In vitro* tests on endothelial cells have shown that the galactofucan stimulates the production of heparin sulfate, leading to the hypothesis that its delayed action *in vivo* is tied to the need for an accumulation of the heparin sulfate on blood vessel surfaces. The galactofucan, despite its high sulfation level, lacks significant anticoagulation activity, making it a suitable candidate as an antithrombotic agent (Rocha et al. 2005).

As expected, extracts from brown algae *Ecklonia arborea* (as *Eisenia arborea*), *Ecklonia cava*, *Ishige foliacea*, *Ishige okamurae*, *Sargassum micracanthum*, *Sargassum ringgoldianum*, and *Sargassum thunbergii* could suppress histamine release from RBL-2H3 cells. Noticeably, *E. arborea* and *S. thunbergii* exhibited inhibitory effect without significant cytotoxicity. In a further study, *E. arborea* was examined for its anti-allergic effect *in vivo*, using Brown Norway rats as an allergy model animal (Sugiura et al. 2008). The rats were immunized with ovalbumin (OVA) by oral administration and fed with *E. arborea* powder (1–5 g/rat) daily. Interestingly, both the OVA-specific and total serum IgE levels were suppressed in the rats fed with a diet of dried *E. arborea* powder. Moreover, the histamine levels of the test groups were significantly reduced compared to that of the control, and were similar to that of the basal diet group. Specially, the diet with *E. arborea* induced a change in the Th1/Th2 balance due to suppressing the release of Th2 cytokines, IL-4, and IL-10, and enhancing the expression of Th1 cytokine IFN-ϒ from the spleen and mesenteric lymph nodes. Herein, the extract of *S. tenerrimum* exhibited the most active suppression of passive cutaneous anaphylaxis (PCA) and active cutaneous anaphylaxis (ACA), which is comparable to anti-allergic drug disodiumcromoglycate. In another study, brown alga *Sargassum hemiphyllum*, a Korean folk medicine for the therapeutic treatment of various allergic diseases, has been determined to inhibit atopic allergic reaction via regulation of inflammatory mediators in mast cells (Na et al. 2005b). In particular, methanol extract of *S. hemiphyllum* effectively inhibited the release of histamine, β-hexosaminidase, IL-8, and TNF-α from the activated mast cells. When orally administered for 1 hour, *S. hemiphyllum* produced a marked inhibitory effect on PCA reaction with inhibition rate of 49.71% at the dose of 0.1 g kg^{-1}. These evidences indicate that *S. hemiphyllum* has an inhibitory effect on the allergic reaction and thus it may be useful in the treatment of allergic inflammatory diseases, such as atopic dermatitis.

Some studies have been conducted *in vivo* in order to investigate the anti-inflammatory potential of fucoidans using gamma/LPS induced RAW 264.7 macrophage (Nakamura et al. 2006, Yang et al. 2006, Kang et al. 2011), C6 glioma cells (Do et al. 2010), microglial cell purified from whole brains of neonatal (1 day old) Sprague-Dawley rats (Cui et al. 2010), and BV2 microglia cells (Park et al. 2011). Fucoidan, when applied to gamma and LPS induced inflammations, acted as anti-inflammation by hindering expression of NO secretion (Nakamura et al. 2006, Yang et al. 2006, Do et al. 2010, Guangling et al. 2011). Likewise, LPS alter the cells' shape to amoeboid, which was reportedly inhibited by 62.5 μg mL^{-1} of fucoidan (Cui et al. 2010). Fucoidan significantly reduced NO and COX2 at 50 μg mL^{-1} and 100 μg mL^{-1} of fucoidan due to reduction of iNOS and COX-2 (Park et al. 2011). Cui et al. (2010) stated 125 μg mL^{-1} of fucoidan caused reduction of NO (75%) and iNOS' mRNA and protein (50%). Also, Kang et al. (2011) reported that purified sulfated polysaccharide from *Ecklonia cava*, fucoidan, reduced NO and PGE_2, in a dose dependent manner, by inhibiting iNOS and COX-2, respectively in LPS induced inflammation of RAW 264.7.

Reduction in degraded carrageenan induced edema was exhibited by hexane, methanol, and butanol extracts of *Sargassum swartzii* (as *Sargassum wightii*), with butanol having maximum inhibitory effect. Seasonal variation of anti-inflammatory potential of *S. swartzii* was possibly due to the reduction in nutrient availability during spring compared to winter. All extracts from *S. swartzii* harvested in winter caused significant reduction of edema, compared to the spring harvest (Dar et al. 2007).

In the study conducted by Khan et al. (2008), out of methanol extracts of 37 seaweeds examined, *Undaria pinnatifida* extracts strongly reduced ear edema/erythema by 85%/78%. *U. pinnatifida* extract has IC_{50} of 10 mg mL^{-1}, 15 mg mL^{-1}, and 18 mg mL^{-1}, respectively for edema, erythema, and blood flow (Khan et al. 2008). According to Kimiya et al. (2008), various extracts of *Ecklonia cava*, *Petalonia binghamiae*, *Scytosiphon lomentaria*, and *Undaria pinnatifida* have been found to inhibit more than 50% of β-hexosaminidase release from RBL-2H3 cells at concentrations of 100 μg mL^{-1} and 200 μg mL^{-1}. Among them, *P. binghamiae* appeared to be most effective against degranulation of both RBL-2H3 cells and mouse eosinophils.

On the other hand, several brown seaweeds including *Sargassum tenerrimum* and *Sargassum cervicorne* from Pakistan, and *Sargassum graminifolium*, *Sargassum thunbergii*, and *Saccharina japonica* (as *Laminaria japonica*) from China were evaluated for their activity against hyaluronidase, which is known to be involved in permeability of the vascular system and allergic reaction (Samee et al. 2009). *S. tennerimum* was revealed to be a strong inhibitor of hyaluronidase enzyme with the 50% inhibitory concentrations (IC_{50}) of 21 μg mL^{-1}. Meanwhile, disodium cromoglycate, an anti-allergic drug, had an IC_{50} value of 39 μg mL^{-1}, and catechin, a natural inhibitor of hyaluronidase, had an IC_{50} value of 20 μg mL^{-1}. Other seaweeds exhibited inhibitory effect with IC_{50} range of 109.5–269 μg mL^{-1}. Furthermore, intraperitoneal administration of *S. tenerrimum*, *S. cervicorne*, and *S. graminifolium* resulted in the suppression of both passive cutaneous anaphylaxis (PCA) and active cutaneous anaphylaxis (ACA) in female BALB/c mice triggered by ovalbumin (OVA) and shrimp allergen (Samee et al. 2009b).

Crude-ethanol extract of *Sargassum micracanthum* was partitioned into hexane, dichloromethane, ethyl acetate, butanol, and water fractions. 20 μg mL^{-1} of each of dichloromethane and hexane fractions that contained phlorotanin exhibited 85.6% and 85.2% reduction of NO production, respectively; PGE_2 production was also reduced in lipopolysaccharides (LPS) induced RAW 246.7 macrophage (Yoon et al. 2009).

Ecklonia cava has been known as an inhibitor of FcεRI activity, a high-affinity receptor for IgE on the cell surface of mast cell and basophils. Shim et al. (2009) have evidenced that the methanol extract of *E. cava* caused a reduction in the cell surface expression of FcεRI, blockade of the IgE-FcεRI interaction, and suppression of the mRNA expression of total cellular FcεRI α-chain in human basophilic KU812F cells. Thus, *E. cava* extract inhibited the degranulation via attenuating the FcεRI-mediated release of histamine in KU812F cells.

Moreover, the ethanol extract of *Ecklonia cava* was observed to inhibit OVA-induced asthmatic reactions in a mouse asthma model (Kim et al. 2008b). The asthmatic reactions in OVA-induced mice were characterized by an increase in the number of eosinophils in bronchoalveolar lavage fluid, a marked influx of inflammatory cells into the lung around blood vessels and airway luminal narrowing, the development of airway hyper-responsiveness, the production of tumor necrosis factor-alpha (TNF-α) and Th2 cytokines, including IL-4 and IL-5 in the bronchoalveolar lavage (BAL) fluid, and the detection of allergen-specific immunoglobulin E (IgE) in the serum. However, the administration of *E. cava* extract prior to the final airway OVA challenge resulted in a significant inhibition of all asthmatic reactions. Also, several contributors for the development of asthmatic reactions such as eosinophil peroxidase (EPO), matrix metalloproteinase-9 (MMP-9), and suppressor of cytokine signaling-3 (SOCS-3) were down-regulated by *E. cava* extract treatment. Accordingly, the extracts of marine brown algae have been indicated to be a useful therapeutic approach for the treatment of allergic diseases.

Several studies report the presence of anti-inflammatory compounds in seaweeds. The anti-inflammatory effect of *Sargassum thunbergii* and *Sargassum fulvellum* against taphorbol myriste acetate induced ear edema, erythema, and blood flow (Kang et al. 2008b), and *Turbinaria conoides* against degraded carrageenan-induced hind paw edema has been reported (Boonchum et al. 2011).

Likewise, ethyl acetate crude extract of *Petalonia binghamiae* reduced NO and PGE_2 production, with IC_{50} of 38.8 μg mL^{-1} and 9.3 μg mL^{-1}, respectively, in lipopolysaccharides (LPS) induced RAW 246.7 (Yang et al. 2010c). Both reports agreed that extracts modulated iNOS and COX-2 expression. Although all five fractions used by Yoon et al. (2009) modulated iNOS and COX-2 proteins expression in the macrophages, hexane and dichloromethane caused better modulation in a dose dependent manner. Suppression of iNOS and COX-2 was as a consequence of reduction of their respective messenger ribonucleic acid (mRNA) by hexane and dichloromethane fractions (Yoon et al. 2009). Also, the two works reported that extracts caused reduction in level of TNF-α, IL-6 (Yoon et al. 2009, Yang et al. 2010c). The two potent fractions in Yoon et al. (2009) exhibited dose dependent decrease in IL-1β protein and mRNA of TNF-α, IL-6, and IL-1β (Yoon et al. 2009). IC_{50} of ethyl acetate extract of *P. binghamiae* was 19.4 μg mL^{-1} for IL-6 (Yang et al. 2010c). Extracts were proved to be non-toxic by not affecting mRNA of β-actin (a house keeping beneficial protein) (Yoon et al. 2009, Yang et al. 2010), or less toxic because concentration of ethyl acetate extract of *P. binghamiae* below 50 μg mL^{-1} did not affect cell viability (Yang et al. 2010c).

Complete blockage of NO production and PGE_2 was achieved at 100 μg mL^{-1} and 200 μg mL^{-1} of fucoidan, respectively (Kang et al. 2011). Although fucoidan was reported to induce inducible iNOS in RAW 264.7 macrophage cells (Nakamura et al. 2006), induction of iNOS was not significant in BV2 (Park et al. 2011) toward NO production. Since NO has both pro- and anti-inflammatory potential (Vane et al. 1994), slight increase of NO caused by fucoidan can be linked to anti-inflammatory potential. Impact of fucoidans alone on PGE_2 production has not been reported. Slight increase of PGE_2 production is an act of anti-inflammatory because PGE_2 have both pro- and anti-inflammatory activities (Calder 2009) which have not been reportedly associated with fucoidans. However, anti-inflammatory threshold limit of concentration of NO and PGE2 production needs to be investigated and established.

Fucoidans and sulfated fucans (galactofucan) extracted from brown seaweed *Lobophora variegata* have been reported to possess anti-inflammatory potential because they inhibit leucocyte migration to inflammation site (Cumashi et al. 2007). Sulfated fucans from *Saccharina latissima* (as *Laminaria saccharina*) were also reported to inhibit adhesion of neutrophil to platelets under flow and also actively inhibit recruitment of leukocytes to inflammation site. Heterofucan extracted from *L. variegata* possessed anti-inflammatory activity in acute zymosan-induced arthritis in rat by reducing cell infiltration and NO level in intra-articular lavage fluid, edema injured knee, and serum TNF-α (Paiva et al. 2011). Fucoidan extracted blocked induction of acute peritonitis by interfering with L- and P-selectins (lectin found on the surface of leukocytes) which support reduction in polymorphonuclear neutrophils (PMN) transmission/transmigration to the abdominal cavity (Cumashi et al. 2007, Croci et al. 2011). Fucoidan, behaving like heparin or heparan sulfate, possibly got attached to L- and P-selectins since both selectins are known to interact with sulfated polysaccharide.

Methanolic extracts of *Sargassum swartzii* at 175 mg kg^{-1} and 350 mg kg^{-1} per body weight, applied for 24 hours, showed 52.12% and 45.85% anti-inflammatory activities, respectively against degrade carrageenan induced hind paw edema of acute inflammatory model. Leucocytes and granulomas were significantly reduced by dosages of extract compared to 25 mg indomethacin kg^{-1} body weight and 5 mg prednisone kg^{-1} body weight, respectively (Hong et al. 2011).

In the study performed by Simpi et al. (2013), the methanolic extract of *Sargassum ilicifolium* was used to evaluate its analgesic and anti-inflammatory activity. Extract of *S. ilicifolium* (1–100 mg kg^{-1}) shows significant analgesic activity, in a dose dependent manner, by reducing significantly the number of acetic acid-induced writhes. The dose of the methanolic extract (50–100 mg kg^{-1}) produced a dose dependent response, i.e., % inhibition of edema, the effect increases over a 3 hour period. The greatest effect was observed at a dose of 50 mg kg^{-1} at 3 hours of 87.5% inhibitory effect elicited 3 hours after injection of degraded carrageenan 0.1 l (1%) compared to standard Indomethacin (10 g kg^{-1}), which produced 100% inhibition of edema.

Sargassum fulvellum has been used to treat various inflammatory diseases, including lump, dropsy, swollen and painful scrotum, and urination problems for several centuries with no side effects. This study aims to investigate the anti-inflammatory effect of the hexane fraction of *S. fulvellum* in lipopolysaccharide (LPS)-stimulated RAW 264.7 cells and phorbol 12-myristate 13-acetate (PMA)-induced mouse-ear edema. The anti-inflammatory activity of hexane fraction of *S. fulvellum* in LPS-stimulated RAW 264.7 cells was investigated by assessing the inhibition of nitric oxide (NO) and pro-inflammatory cytokine production during *Griess* reaction and enzyme-linked immunosorbent assay (ELISA), respectively (Gwon et al. 2013).

Sulfated polysaccharides (fucan) from the algae *Sargassum vulgare* were extracted, fractionated in acetone, and examined with respect to chemical composition, anticoagulant, anti-inflammatory, antithrombotic effects, and cellular proliferation (Dore et al. 2013). These sulfated polysaccharides prolonged activated partial thromboplastin time (aPTT) and exhibited high antithrombotic action *in vivo*, with a concentration 10 times higher than heparin activity; and it displays strong anti-inflammatory action at all concentrations tested in the degraded carrageenan-induced paw edema model, demonstrated by reduced edema and cellular infiltration.

In the study by Montalvão et al. (2016), the anti-inflammatory potential of several algae specimens collected from the Aegean Sea were screened. Extracts from *Cystoseira crinita* and *Dictyota fasciola* (as *Dilophus fasciola*) clearly showed a reduction in the release of IL-β, IL-6, TNF-α, and IFN-ϒ pro-inflammatory cytokines, when compared to positive controls.

10.5.3 Rhodophyta (red seaweeds)

Carrageenophytes like *Furcellaria* lumbricalis (as *Furcellaria fastigiata*), *Eucheuma denticulatum* (as *Eucheuma spinosum*), *Chondracanthus acicularis* (as *Gigartina acicularis*), *Gigartina pistillata*, *Sarcothalia radula* (as *Gigartina radula*), and *Iridaea* have been studied and it was found that only *C. acicularis* showed very potent anticoagulant activity (see also Chapter 3) (Houck et al. 1957). However, other studies on *Iridaea* (Chargaff et al. 1936, Elsner 1938) have confirmed the presence of anticoagulant activity. Similarly, carrageenan of *E. denticulatum* (as *E. spinosum*) was reported to have anticoagulant activity (Anderson 1969). Polysaccharide extract of *C. acicularis* was checked for its *in vivo* effect on dogs. When the extract was administered intravenously at a dose of 1 mg kg^{-1} and at a concentration of 2 mg mL^{-1}, it showed maximal anticoagulant activity after 4 hours, which was monitored by whole blood clotting time estimations (Houck et al. 1957). Effect of carrageenan on human platelets has also been studied (McMillan et al. 1979, Kindness et al. 1979, 1980, Güven et al. 1991). It has been reported in *ex vivo* study that carrageenan had anticoagulant effect and inhibited platelet aggregation.

Anticoagulant activity of algal polysaccharides was first reported in *Mazzaella laminarioides* (as *Iridaea laminarioides*) (Chargaff et al. 1936). This extract, a "galactan sulfuric acid ester", was shown to possess 30 U mg^{-1} of heparin equivalence. Subsequent studies described similar anticoagulant properties in agar and carrageenan. A group of 19 species belonging to red, brown, and green algae were screened by Elsener in 1938, but anticoagulant activity was found only in *Delesseria sanguinea*. Since these studies, there have been many reports relating to the anti-hemostatic properties of extracts and purified fractions from red, brown, and green algae (Shanmugam and Mody 2000).

The principal basis of the anticoagulant activity of carrageenan appeared to be an anti-thrombotic property (Hawkins and Leonard 1962, 1963). Lambda-carrageenan showed greater anti-thrombotic activity than kappa-carrageenan, probably due to its higher sulfate content, whereas the activity of the unfractionated material remained between the two. Lambda-carrageenan consistently prolonged the clotting time and was more toxic than kappa-carrageenan. The difference in sulfate content between the two carrageenans did not correspond directly to differences in anticoagulation action and toxicity (Anderson and Duncan 1965). It was found that the toxicity of carrageenans depended on the molecular weight (Anderson 1967). Similar results were obtained with lambda-carrageenan of *Coccotylus brodiei* (as *Phyllophora brodiaei*), which gave the highest blood anticoagulant activity (Efinmov et al. 1983). Carrageenan of *Grateloupia turuturu* also showed anticoagulant activity (Efinmov et al. 1983).

Carrageenans exhibit anticoagulant activity and inhibit platelet aggregation, as previously described (Hawkins and Leonard 1962, 1963, Kindness et al. 1979). Among the various carrageenan types, lambda carrageenan (primarily from *Chondrus crispus*) has approximately twice the activity of non-fractioned carrageenan and four times the activity of iota and kappa carrageenan extracted from *Eucheuma denticulatum* (as *Eucheuma spinosum*) and *Kappaphycus alvarezii* (as *Eucheuma cottonii*), respectively. The most active carrageenan has approximately 1/15 activity of heparin (Hawkins and Leonard 1962), but the sulfated galactan from *Grateloupia indica* collected from Indian waters exhibited activity as significant as heparin (Sen et al. 1994). The principal basis of the anticoagulant activity of carrageenan was found to be an antithrombotic property. Because of the higher sulfate content, lambda carrageenan showed greater antithrombotic activity than kappa carrageenan, whereas the activity of the non-fractioned material remained between the two. Similar results were obtained with lambda carrageenan of *Coccotylus brodiei* (as *Phyllophora brodiaei*), which gave the highest blood anticoagulant activity (Sen et al. 1994).

Vidalols A and B are bromophenols successively extracted from Caribbean alga *Osmundaria obtusiloba* (as *Vidalia obtusiloba*) and possess anti-inflammatory activities. The two compounds were able to inhibit bee venom-derived phospholipase A_2 (96%) and reduce phorbol ester induced mouse ear edema (58% to 82%). However, due to their susceptibility to oxidation during bioassay, it was not clear if the vidalols or their oxidative form ortho-quinnoes were responsible for their anti-inflammatory potentials (Wiemer et al. 1991).

Many sulfated polysaccharides from Rhodophyta macroalgae have also yielded potential anticoagulant, and other activities (Cáceres 2000, Farias et al. 2000, Pereira et al. 2005, Glauser et al. 2009). Polysaccharides from tetrasporic plants (lambda-carrageenan) of *Stenogramma interruptum* showed higher activity than those isolated from cystocarpic (kappa and iota-carrageenans) fronds in TT test (Cáceres et al. 2000). Studies on a sulfated galactan from the red seaweed *Botryocladia occidentalis* are particularly expressive. This species produces 2,3-di-*O*-sulfated D-galactan, a compound that exhibits anticoagulant activity, comparable to heparin, which appeared because of the inhibition of thrombin and factor X. Its activity was more significant than similar sulfated galactans, from invertebrate sources, which had only single sulfate per galactose residue (Farias et al. 2000). In the same way, a polysaccharide chain but with lower amounts of 2, 3-di-*O*-sulfated D-galactose in *Gelidium crinale* was less potential in a clotting time assay when compared to that from *B. occidentalis* (Pereira et al. 2005). The two sulfated polysaccharides did not differ in thrombin inhibition mediated by antithrombin; however, assays that replaced antithrombin with heparin cofactor II, the sulfated galactan from *B. occidentalis*, was more inhibitory than that from *G. crinale*. Nevertheless, when factor Xa was the target protease, the sulfated galactan from *G. crinale* was a more efficient anticoagulant than that from *B. occidentalis*. These observations unambiguously demonstrated that the distribution of 2,3-di-sulfated galactose along the polysaccharide chain modulates the interaction of the polysaccharides with specific proteases in the coagulation system (Vavilala and D'Souza 2015).

The extent of anticoagulant activity of carrageenans seemed to be influenced also by a different composition of polysaccharides extracted from red seaweed in various stages of their life cycle. Anticoagulant activity of carrageenans of iota-type and the mixture of kappa/iota-type extracted from cystocarpic (reproductive) and sterile plants of *Chondrus pinnulatus* harvested on the Russian Pacific coast had extremely high values exceeding 600 s, established by aPTT assay in contrary to that from sterile plants. Since molecular weights of cystocarpic plant carrageenans were almost two times higher (420 kDa and 389 kDa) than molecular weights of carrageenans from sterile plants (290 kDa and 220 kDa), high anticoagulant activity may be caused by high molecular weight of this polysaccharide (Yermak et al. 2006).

Some species of red macroalgae are also known to be useful additional materials for control of allergic responses. The red algal *Polyopes affinis* (as *Carpopeltis affinis*) has been confirmed to be effective against atopic allergic reaction *in vitro* (Na et al. 2005). It was observed that *P. affinis* (10 $\mu g\ mL^{-1}$) significantly alleviated the release of histamine, β-hexosaminidase, IL-8, and TNF-α from mast cells with inhibition rates of 58.7%, 86.6%, 36.5%, and 59.4%, respectively.

Porphyran (see Fig. 2.3, Chapter 2), a sulfated polysaccharide isolated from red seaweeds, has been recognized to be effective against different allergic responses. According to Ishihara et al. (2005), Porphyran of red algae *Pyropia tenera* (as *Porphyra tenera*) and *Pyropia yezoensis* (as *Porphyra yezoensis*) were capable of inhibiting the contact hypersensitivity reaction induced by 2,4,6-trinitrochlorobenzene via decreasing the serum level of IgE in Balb/c mice.

Chondrus ocellatus is distributed spontaneously in intertidal zone of the southeast seaside of China (Zhou et al. 2006). In ancient times the alga was not only edible, but was also used as a medicine to cure chronic constipation and bone fracture (Fengwu and Yushun 1993). *C. ocellatus* is an important economic original alga that can produce carrageenan. Antitumor, antiviral, anticoagulant, and immunomodulatory activities of carrageenan have been found in recent years (Parish and Snowden 1985, Schaeffer and Krylov 2000). Many evidences indicated that lambda-carrageenan, which contained 1,4-linked D-galactose 2,6-disulfate units, and highly sulfated group, has more bioactivities than other carrageenan types (Anderson and Duncan 1965, Güven et al. 1991, Zhou et al. 2006).

Ethanol/acetic acid extract (1 mg/ear) of *Dichotomaria marginata* (as *Galaxaura marginata*) was reported to cause 95% reduction of mouse ear edema with its four hexane-thin layer chromatography (TLC) fractions inhibited 55%, 75%, 100%, and 100%. EC_{50} of ethanol:acetic extract was reached at 0.31 mg/ear and its anti-inflammatory activity was associated to inhibition of the bee-venom phospholipase A_2 (PLA_2) (Rozas and Freitas 2007). Other compounds isolated from red algae *Phacelocarpus peperocarpos* (as *Phacelocarpus labillardieri*) and *Sphaerococcus coronopifolius* also showed bee-venom PLA_2 inhibitory activity (Mayer et al. 1993, Song et al. 2003). Extracts from *Hypnea valentiae* also inhibited the action by *Naja nigricollis* venom, when inoculated into mice and when *in vitro* assayed (Vasanthi et al. 2003).

According to Kimiya et al. (2008) various extracts of *Chrysymenia wrightii* and *Pyropia dentata* (as *Porphyra dentata*) have been found to inhibit more than 50% of β-hexosaminidase release from RBL-2H3 cells at concentrations of 100 $\mu g\ mL^{-1}$ and 200 $\mu g\ mL^{-1}$. Moreover, the protective effect of an edible red alga *Chondrophycus undulatus* (as *Laurencia undulata*) against murine allergic airway reactions was elucidated *in vivo* (Jung et al. 2009). In a similar experimental model as examined by Kim et al. (2008b), the ethanol extract of *C. undulata* also exhibited a significant inhibition of all asthmatic reactions in ovalbumin (OVA)-induced mice.

Pyropia dentata (as *Porphyra dentata*), a red edible seaweed, has long been used worldwide in folk medicine for the treatment of inflammatory diseases such as hypersensitivity, lymphadenitis, bronchitis, etc. The aims of the study done by Kazlowska et al. (2010) was to clarify the anti-inflammatory role of *P. dentata* crude extract and its identified phenolic compounds by investigating their effect on the nitric oxide (NO)/inducible nitric oxide synthase (iNOS) transcription pathway in macrophage RAW 264.7 cells. Methanolic extract of *P. dentata* and its phenolic compounds (catechol and rutin, but not hesperidin) was also reported to inhibit secretion of NO and suppress iNOS and NF-κB (Kazlowska et al. 2010).

Porphyran derivate obtained from red seaweed *Pyropia haitanensis* (as *Porphyra haitanensis*) with fully sulfated modification showed the highest anticoagulant activities by aPTT, TT, and PT assays 396.47 s, 311.70 s, and 298.03 s, respectively, in a dose of 100 mg mL^{-1}. Besides the degree of sulfation, the distribution of sulfated group it is also an important factor influencing the anticoagulant activity of porphyran, that have a linear backbone of alternating 3-linked β-D-galactosyl units and 4-linked α-L-galactosyl 6-sulfate or 3,6-anhydro-α-L-galactosyl units. In the case of alkali treatment of porphyran, 6-*o*-desulphatation in (1,4)-linked residue was carried out, almost without any change of anticoagulant activity. Moreover, 6-*o*-sulfated derivate showed a lower anticoagulant effect than 2,2',4-*o*-sulfated derivate. From these results, it could be concluded that sulfate groups at C-6 were not necessary for the anticoagulant activity, whereas an increase in the overall molecular weight and position of sulfate groups at C-2, C-3, and C-4 was a dominant factor for the anticoagulant effect of porphyran (Zhang et al. 2010b).

Sulfated galactans and carrageenan were extracted from red algae *Corallina* collected at the Lebanese coast in the yields of 2.5% and 10%, and their anticoagulant activities were determined by ATPP assay. Carrageenans exhibited more powerful anticoagulant activity from 78.4 s to 120 s at doses of 0.05 mg and 0.5 mg, respectively, than sulfated galactans that reached 104.3 s at their highest dose of 5 mg, but only iota-carrageenan was used for analyses (Sebaaly et al. 2014). The results obtained by Tang et al. (2017), with sulfated polysaccharides extracted from *Grateloupia livida* can serve, soon, as readily available alternative natural sources of anticoagulant and antioxidant agents (see Table 10.1).

CHAPTER 11

Neurological Activities of Seaweeds and their Extracts

11.1 Introduction

Hominization (the history of the human lineage) is reported to have its primitive beginnings at least 4–6 million years ago, with the most significant changes in brain development occurring over the past 2–2.5 million years (Aiello and Wheeler 1995, Hawks et al. 2000, Glazko and Nei 2003). Thus, modern-day humans are left with an organ that is the source of all the qualities that define humanity. It is the epicenter of function and intelligence, the interpreter of the senses, the initiator of body movement, and the controller of behavior (Hynes 2016, Cornish et al. 2017).

11.1.1 Did seaweed make us who we are today?

Seaweeds are part of the cultural heritage of Asian countries, much more so than in western countries. Nevertheless, according to a recent archaeological study, cooked and partially-eaten seaweeds were found at a 14,000-year-old site in southern Chile, suggesting seaweeds have been a part of human diet for a very long time and in other parts of the world as well (Dillehay et al. 2008).

Millions of years ago something happened, allowing early *Homo sapiens* to branch out from the primitive hominoid family tree. Was this crucial turn in human evolution partly driven by seaweed and its content of essential nutrients? Over the past 2–2.5 million years, human brains have gone through the most significant development, and thus modern-day humans are left with an organ that is the source of all the qualities that define humanity.

Our ancestors needed lots of energy-rich foods just to get by, and for this impressive, significant brain development, they also needed certain essential nutrients. Without nutrients like magnesium and zinc, modern brains cannot function, and according to a number of scientific studies, it is likely that the access to certain essential nutrients influenced the evolution of the human brain so that it could become the brain we have today (Hynes 2016, Cornish et al. 2017).

11.1.2 From primitive ancestor to modern human

Nutrients needed for this transition from a primitive ancestor to modern *Homo sapiens* were (and still are) available in seaweeds. Seaweeds could be found and harvested in abundance on shores, and for a foraging lifestyle, a rich coastal environment would be a significant source of a consistent supply of these nutrients.

Cornish et al. (2017) note that the human lineage is estimated to have diverged from our closest living relatives, the chimpanzees, around 5–7 million years ago. However, the changing patterns of resource distribution associated with the extensive drying and expansion of the African savannahs between 2 and 2.5 million years ago have been the impetus for a shift in foraging behavior among early members of the genus *Homo*. Foraging over longer distances for food would have contributed to bipedalism and a different body stature, as increasingly larger ranges had to be traversed, and in the case of our primitive ancestors, this would undoubtedly lead to significant changes in diet. Coastal areas may very well have attracted early hominoids in search of food.

Our ancestors would find foods like fish, crustaceans, snails, seaweeds, bird eggs, and perhaps occasional dead marine vertebrates. However, they probably did not have the necessary rudimental understanding of seasonal tidal cycles and their influence on shellfish availability. Seaweeds of different types, on the other hand, could be found across the intertidal zone from the high-water mark to the subtidal regions, and could be readily and repeatedly harvested for food by all family members, including women and children. The nutrients in seaweed did not just benefit our ancestors. Seaweed is just as healthy and nutritious for humans today as it was millions of years ago (Svennevig 2017).

11.2 Essential Nutrients for Brain Development and Preservation

It has long been suspected that the relative abundance of specific nutrients can affect cognitive processes and emotions. Newly described influences of dietary factors on neuronal function and synaptic plasticity have revealed some of the vital mechanisms that are responsible for the action of diet on brain health and mental function. Several gut hormones that can enter the brain, or that are produced in the brain itself, influence cognitive ability. In addition, well-established regulators of synaptic plasticity, such as brain-derived neurotrophic factor, can function as metabolic modulators, responding to peripheral signals such as food intake. Understanding the molecular basis of the effects of food on cognition will help us to determine how best to manipulate diet to increase the resistance of neurons to insults and promote mental fitness (Gómez-Pinilla 2008).

11.2.1 List of the nutrients (see also Chapter 2)

Astaxanthin: Astaxanthin, a unique member of the xanthophylls, is a deep-red-colored phytonutrient that can be synthesized by a green microalga *Haematococcus lacustris* (formerly *Haematococcus pluvialis*). Distinct from other xanthophylls, astaxanthin has 2 hydroxyl groups (Ambati et al. 2014, Regnier et al. 2015). Astaxanthin spans the bi-lipid layer and is long enough that the two hydroxyl groups jut into the fluid phase near the membrane, and that when electrons are extracted from these hyroxyl groups by free radicals, the molecule is resonance stabilized. Astaxanthin can dramatically decrease the risk of cardiovascular disease (Fassett and Coombes 2011). A diet supplemented with astaxanthin (75 or 200 mg kg^{-1} body weight per day) for 8 weeks has been shown to improve endothelium-dependent vasodilatation in resistance vessels, reduce systolic blood pressure, and improve cardiovascular remodeling in spontaneously hypertensive rats. In addition, astaxanthin (100 and 500 mg 100 g^{-1}) for 60 days protects against serum protein oxidation in hyper-cholesterolemic rabbits (Monroy-Ruiz et al. 2011, Augusti et al. 2012). Studies have also demonstrated that astaxanthin can easily cross the blood-brain barrier (BBB) to protect the brain from acute injury and chronic neurodegeneration (Shen et al. 2009, Yuan et al. 2011). The neuroprotective properties of this molecule involve anti-oxidation, anti-inflammation, and anti-apoptosis (Zhang et al. 2014b, c).

Iron: Micronutrient deficiencies, especially those related to iodine and iron, are linked to different cognitive impairments, as well as to potential long-term behavioral changes. Among the cognitive impairments caused by iron deficiency, those referring to attention span, intelligence, and sensory perception functions are mainly cited, as well as those associated with emotions and behavior, often directly related to the presence of iron deficiency anemia. In addition, iron deficiency without anemia may cause cognitive disturbances. At present, the prevalence of iron deficiency and iron deficiency anemia is 2–6% among European children. The importance of iron deficiency relative to proper cognitive development and the alterations that can persist through adulthood because of this deficiency is thus clear (Jáuregui-Lobera 2014).

Copper: There are several ways in which copper loss might inflict neuronal damage. As a cofactor of cytochrome c oxidase, which forms a part of the mitochondrial electron transport chain driving ATP production, copper is critical for energy generation in all cells. This function is especially important for energy-hungry neurons. Loss of copper could lead to an energy deficiency that dramatically affects neurons' abilities to maintain electrical potential and transport signals up and down long axons. Loss of copper could also increase oxidative stress, which is thought to be the primary driver of cell death in neurodegenerative diseases (White 2014).

Taurine can be found in red algae (Rhodophyta), marine fish, shellfish, and meat of mammals. It is present in large amounts in the central nervous system and in the retina. The highest concentrations occur in the developing brain. Its levels in adults are approximately 1/3 of those in newborns.

Magnesium: It can be found in legumes, pumpkin and squash seeds, nuts, and macroalgae. It plays an important role in neuroprotection and cognition. It is important due to its ability to store new information in neural networks.

Zinc can be found in many foods, but is particularly plentiful in various cuts of meat, especially liver. It is extremely abundant in oysters. Crustaceans and most seaweeds are also robust sources of zinc. It plays an important role in learning, development, and memory.

Vitamin B_{12} is found exclusively in animal products such as meat, eggs, fish, and milk—with one exception–it is also confirmed in *Porphyra*/*Pyropia* species of seaweeds, and it is quite likely to be present in others that have yet to be adequately analyzed. B_{12} is important for blood flow in the brain and cognitive functions like language.

Iodine is found abundantly in seaweeds, especially in brown seaweeds. It is a necessary element for the synthesis of thyroid hormones, which are essential for the development of the central nervous system.

Poly-unsaturated fatty-acids (PUFAs): The original sources of PUFAs are not, as often thought, fish and shellfish, but microalgae and seaweed (see Fig. 10.4, Chapter 10).

11.3 Neuroprotective Effect of Marine Algae

A comprehensive summary of macroalgae-derived pure compounds with different neuroprotection activities is presented in Table 11.1.

Typically afflicting adults in mid-life, neurodegenerative diseases are characterized by motor or cognitive symptoms that get progressively worse with age, and that usually reduce life expectancy. Human neurodegenerative disease can result from a variety of environmental and genetic causes. Genetic factors have been instrumental in developing our understanding of the etiology and progression of such diseases, and can range from mutations that increase the risk for a disorder, to mutations that are the sole, direct cause of a disease. As with another collection of related diseases—cancer—neurodegeneration can result from dominant or recessive mutations (Lessing and Bonini 2009).

Among the series of identified disease conditions that relate to extensive loss of function and quantity of neurons—Parkinson's, Alzheimer's, and Huntington's, are well-defined neurodegenerative diseases. However, the etiology of those diseases has not yet been identified clearly, even though some of them were reported centuries before (Wood-Kaczmar et al. 2006).

The involvement of reactive oxygen species, such as hydrogen peroxide (H_2O_2) has been suggested in neurodegenerative disorders such as Alzheimer's, Parkinson's, and Minamata diseases (Simonian and Coyle 1996). The neurotoxicity of H_2O_2 is exerted mainly by the formation of the highly reactive hydroxil ($OH^{\bullet}$) radical, although depletion of glutathione (GSH) levels, and the secondary rupture of calcium homeostasis, can also contribute to the toxic effects of H_2O_2 (Fig. 11.1) (Farber et al. 1990, Rimpler et al. 1999).

A matter of concern is the high prevalence of dementia in elderly people. Dementia is chronic and progressive and affects several brain functions, including memory, thinking, orientation, calculation, learning capacity, language, and judgment (APA 2017). The deficits in cognitive function are commonly accompanied, and occasionally preceded, by deterioration in emotional control, social behavior, or motivation (Starr 2010). The most common cause of dementia is Alzheimer's disease, accounting for 60–70% of the cases. People living with dementia have poor access to appropriate healthcare, even in most high-income country settings, where only around 50% of the people living with dementia receive a diagnosis. In low and middle income countries, less than 10% of the cases are diagnosed. As populations age, due to increasing life expectancy, the number of people with dementia is increasing. We estimate that there were 46.8 million people worldwide living with dementia in 2015, and this number will reach 131.5 million in 2050 (World Alzheimer's Report 2016). Although Alzheimer's disease is a form of dementia, there are other syndromes that have similar symptoms—such as depression, hallucinations, and memory loss—syndromes that include dementia of Lewy bodies, vascular dementia, frontotemporal dementia, etc. (Farlow 2010). However, and notwithstanding the fact that the ethical background has some similarities, this article will focus exclusively on Alzheimer's disease.

Also, recent scientific findings determined that mild cognitive impairment (MCI) can be detected more than 10 years before full diagnosis, and that amyloid-β peptide deposits can be detected by amyloid imaging even earlier (Jack et al. 2011). Because the changes caused by MCI are not severe enough to affect daily life, the patient does not meet diagnostic guidelines for dementia.

Although significant research has already been performed with regard to the diagnosis and treatment of currently incurable neurodegenerative dementias, such as Alzheimer's, for the time being, it is still considered an incurable disease. The discovery of genes responsible for early-onset Alzheimer's dementia will not only make early diagnosis and treatment of the disease possible before brain damage occurs, but can also lead to the prediction of the disease through genetic technology (Nordgren 2010).

In accordance, neuropathological studies demonstrated that Alzheimer's was associated with deficiency in the brain neurotransmitter acetylcholine (ACh) (Tabet 2006). The inhibition of acetylcholinesterase (AChE) enzyme, which catalyzes the breakdown of ACh, may be one of the most realistic approaches to the symptomatic treatment of Alzheimer's (Pangestuti and Kim 2010). Presently, a variety of plants has been reported to possess AChE inhibitory activity. *Huperzia serrata*, a Chinese terrestrial herb has been demonstrated to be a potent AChE inhibitor (Cheng et al. 1996). Several studies have shown AChE inhibitory activity of several marine algae species. A list of marine algae reported to have significant AChE inhibitory activity can be seen in Table 11.1.

Neurodegenerative diseases are estimated to surpass cancer as the second most common cause of death among elderly by the 2040s (Bjarkam et al. 2001, Ansari et al. 2010). For this reason, a great deal of attention has been expressed by scientists regarding safe and effective neuroprotective agents. Several categories of natural and synthetic neuroprotective agents have been reported. However, synthetic neuroprotective agents are believed to have certain side effects, such as dry mouth, tiredness, drowsiness, sleepiness, anxiety or nervousness, difficulty to balance, etc. (Narang et al. 2008). Hence, nowadays researchers have an interest in studying natural bioactive compounds that can act as neuroprotective agents. Marine algae represent one potential candidate as neuroprotective agents. Nevertheless, development of marine algae as neuroprotective agents still faces several challenges. The rationale for marine algal neuroprotective effects treatment in the CNS is based on established observations and experiments *in vitro* or in animal models only. Up till now, none of the marine algal neuroprotective effects have been examined in human subjects. Therefore, small clinical studies and further large-scale controlled studies are needed. Another important challenge in the development of marine algae as neuroprotective agents is that many drugs failed to provide real neuroprotection in practice. Potential reasons for this failure include inappropriate use of specific neuroprotection/s for a given disease or stage of disease progression, or the use of suboptimal doses (Gilgun-Sherki et al. 2001). Hereafter, future studies are needed focusing on the synergistic benefits of consuming different marine algae species, recommended doses and timing of intake, and preparation methods for marine algal bioactive compounds to maximize the desired protective effect in the prevention of neurodegenerative diseases (Pangestuti and Kim 2010).

11.4 Seaweeds and Neurophysiological Activities

Macroalgae-derived compounds with neuroprotective activity may provide the main nutrients for the prevention and treatment of neurodegenerative diseases such as Alzheimer's, Huntington's, Parkinson's, and other neurodegenerative diseases. Much of these bioactive compounds are derived from Phaeophyceae

Table 11.1 Bioactive properties of some compounds extracted from seaweeds.

Species	Extract/Compound	Activity	References
Chlorophyta (green seaweed)			
Caulerpa racemosa	Methanolic extract	Inhibiting AChE $IC_{50} = 5.5$ mg mL^{-1}	Stirk et al. 2007
C. racemosa	Alkaloid: Caulerpin	Inhibition of nociception 100 µM kg^{-1} in Swiss albino mice	De Souza et al. 2009
C. racemosa	Bisindole alkaloid (racemosin A)	Increase 5.5% of cell viability in SH-SY5Y cells (neuroblast from neural tissue) $IC_{50} = 10$ µM	Liu et al. 2013c
C. racemosa	Bisindole alkaloid (racemosin B)	Increase 14.6% of cell viability in SH-SY5Y cells (neuroblast from neural tissue) $IC_{50} = 10$ µM	Liu et al. 2013c
C. racemosa	Hexane, dichloromethane and methanol extracts	Inhibiting AChE $IC_{50} = 0.086$, 0.089 and 0.095 mg mL^{-1}, respectively Inhibiting BuChE $IC_{50} = 0.156$, > 0.2 and 0.118 mg mL^{-1}, respectively	Gany et al. 2014
C. racemosa	Terpenoid (α-tocospirone)	13.85% increases in cell viability in SH-SY5Y cells $IC_{50} = 10$ µM	Yang et al. 2015
C. racemosa	Sterol (23E)-3β-hydroxystigmasta-5,23-dien28-one	11.31% increases in cell viability in SH-SY5Y cells $IC_{50} = 10$ µM	Yang et al. 2015
C. racemosa	Sterol (22E)-3β-hydroxycholesta-5,22-dien24-one	15.98% increases in cell viability in SH-SY5Y cells $IC_{50} = 10$ µM	Yang et al. 2015
Cladophora vagabunda (as *Cladophora fascicularis*)	Methanolic extract	Inhibiting AChE $IC_{50} = 2$ mg mL^{-1}	Natarajan et al. 2009
Codium capitatum	Methanolic extract	Inhibiting AChE $IC_{50} = 7.8$ mg mL^{-1}	Stirk et al. 2007
C. capitatum	50% Aqueous methanol extract	Inhibiting AChE $IC_{50} = 0.11$ mg mL^{-1}	Rengasamy et al. 2015
C. duthieae	50% Aqueous methanol extract	Inhibiting AChE $IC_{50} = 0.14$ mg mL^{-1}	Rengasamy et al. 2015
C. fragile	Sterol: Clerosterol	Exhibit reducing activity to COX-2, iNOS and TNF-α $IC_{50} = 3$ µg mL^{-1}	Lee et al. 2013b

Table 11.1 contd....

Table 11.1 contd....

Species	Extract/Compound	Activity	References
Halimeda incrassata	Water extracts	Neuroprotective and antioxidant properties	Fallarero et al. 2003
H. cuneata	Methanolic extract	Inhibiting AChE $IC_{50} = 5.7$ mg mL^{-1}	Stirk et al. 2007
H. cuneata	50% Aqueous methanol extract	Inhibiting AChE $IC_{50} = 0.07$ mg mL^{-1}	Rengasamy et al. 2015
Ulva australis (as *Ulva pertusa*)	Sulfated polysaccharide (ulvan)	Scavenging activity for superoxide radicals	Qi et al. 2005
U. fasciata	Methanolic extract	Inhibiting AChE $IC_{50} = 4.8$ mg mL^{-1}	Stirk et al. 2007
U. fasciata	50% Aqueous methanol extract	Inhibiting AChE $IC_{50} = 0.13$ mg mL^{-1}	Rengasamy et al. 2015
U. prolifera (as *Enteromorpha prolifera*)	Pheophorbide A	Antioxidant activity $IC_{50} = 71.9$ μM	Cho et al. 2011b
U. reticulata	Methanolic extract	Inhibiting AChE $IC_{50} = 10$ mg mL^{-1}	Natarajan et al. 2009
U. reticulata	Methanolic extract	Inhibiting AChE $IC_{50} = 10$ mg mL^{-1}	Suganthy et al. 2010
Phaeophyceae (brown seaweed)			
Agarum clathratum subsp. *yakishiriense*	Crude, ethyl acetate, *n*-butanol extracts	Neuronal protection from ischemic injury	Kim et al. 2014f
Alaria esculenta	Methanol/water (1:1) at 50°C extract	Extract fractions inhibit the formation of amyloid fibrils by α-Synuclein	Giffin et al. 2017
Cystoseira humilis	Methanolic extract	AChE inhibitory capacity: 50% 10 mg mL^{-1}	Custódio et al. 2016
C. nodicaulis	Methanolic extract	AChE inhibitory capacity: 64.4% BuChE inhibitory capacity: 110% 10 mg mL^{-1}	Custódio et al. 2016
C. tamariscifolia	Methanolic extract	AChE inhibitory capacity: 85% BuChE inhibitory capacity: 86% 10 mg mL^{-1}	Custódio et al. 2016
C. usneoides	Methanolic extract	AChE inhibitory capacity: 47% 10 mg mL^{-1}	Custódio et al. 2016

Dictyopteris undulata	Sesquiterpene: Zonarol	Activates the Nrf2/ARE pathway, induces phase-2 enzymes, and protects neuronal cells from oxidative stress	Shimizu et al. 2015
Dictyota humifusa	Methanolic extract	Inhibiting AChE $IC_{50} = 4.8$ mg mL^{-1}	Stirk et al. 2007
Ecklonia bicyclis	Phlorotannins	Suppression of BACE-1 enzyme activity $IC_{50} = 5.35$ µM	Jung et al. 2010
E. bicyclis (as *Eisenia bicyclis*)	Phlorotannins from ethyl acetate fraction	Decreased Aβ-induced cell death $IC_{50} = 800$ µM	Ahn et al. 2012
E. bicyclis (as *Eisenia bicyclis*)	Phlorotannins	Protection from retinal neuronal death	Kim et al. 2012c
E. bicyclis (as *Eisenia bicyclis*)	Ethanolic extract	Inhibitory properties against AChE, BChE and total ROS with inhibition percentages (%) of 68.01, 95.72, and 73.20 at concentrations of 25 µg mL^{-1}, respectively	Choi et al. 2015c
E. cava	Phlorotannins: Dieckol and phlorofucofuroeckol	Improvement of memory and possible involvement of the AChE inhibition	Myung et al. 2005
E. cava	Phlorotannin: Triphlorethol-A	Anti-oxidative activity: Scavenging activity against ROS and DPPH via activation of ERK protein	Kang et al. 2005
E. cava	Phlorotnnins	Scavenging activity against hydroxyl, superoxide and peroxyl radicals $IC_{50} = 392.5$, 115.2 and 128.9 µM, respectively	Li et al. 2009
E. cava	Enzymatic extract: Phlorotannins	The phlorotannin-rich fraction significantly potentiated the pentobarbital-induced sleep at > 50 mg Kg^{-1}	Cho et al. 2012
E. cova	Phlorotannins	Neuroprotective effects against H_2O_2-induced oxidative stress in murine hippocampal HT22 cells $IC_{50} = 50$ µM	Kang et al. 2012e
E. cava	Phloroglucinol	Reduce the toxicity ROS induced by hydrogen peroxide $IC_{50} = 10$ µg mL^{-1}	Kim et al. 2012d
E. cava	Phlorotannin: 8,8'-Bieckol	Phlorotannin: 8,8'-Bieckol Reduce COX-2, NO and prostaglandin E2 (PGE2) $IC_{50} = 100$ µM	Yang et al. 2014
E. cava	Ethanolic extract	Extracts have potential analgesic effects in the case of postoperative pain and neuropathic pain	Kim et al. 2014g

Table 11.1 contd....

Table 11.1 contd....

Species	Extract/Compound	Activity	References
E. cava	Phlorotannin (eckol)	Inhibiting BuChE $IC_{50} = 29\ \mu M$	Choi et al. 2015b
E. cava	Phlorotannin (7-phloroeckol)	Inhibiting BuChE $IC_{50} = 0.95\ \mu M$	Choi et al. 2015b
E. kurome	Acidic oligosaccharide sugar chain (AOSC)	Blocking the fibril formation of Aβ $IC_{50} = 100\ \mu g\ mL^{-1}$	Hu et al. 2004b
E. maxima	Crude extract was sequentially extracted with *n*-hexane, dichloromethane, ethyl acetate, and finally *n*-butanol	IC_{50} values for the solvent fractions ranged from 62.61 to 150.8 $\mu g\ mL^{-1}$, with the ethyl acetate fraction having the best inhibitory activity	Kannan et al. 2013
E. stolonifera	Phlorotannin (dieckol)	Inhibiting AChE 17.11 μM	Yoon et al. 2008
E. stolonifera	Phlorotannin (eckstolonol)	Inhibiting AChE and BuChE $IC_{50} = 42.66$ and $230.27\ \mu M$	Yoon et al. 2008
E. stolonifera	Phlorotannin (eckol)	Inhibiting AChE $IC_{50} = 20.56\ \mu M$	Yoon et al. 2008
E. stolonifera	Phlorotannin (2-phloroeckol)	Inhibiting AChE $IC_{50} = 38.13\ \mu M$	Yoon et al. 2008
E. stolonifera	Phlorotannin (7-phloroeckol)	Inhibiting AChE and BuChE $IC_{50} = 4.89$ and $136.71\ \mu M$	Yoon et al. 2008
E. stolonifera	Phlorotannin (phlorofucofuroeckol A)	Inhibiting AChE and BuChE $IC_{50} = 4.89$ and $136.71\ \mu M$	Yoon et al. 2008
E. stolonifera	Sterol (fucosterol)	Inhibiting AChE $IC_{50} = 421.72\ \mu M$	Yoon et al. 2008
E. stolonifera	Methanolic extract	Inhibiting AChE $IC_{50} = 108.11\ mg\ mL^{-1}$	Yoon et al. 2009c
Fucus vesiculosus	Fucoidan (Sigma)	Fucoidan completely blocks microglial uptake of fDNA at only 40 ng mL^{-1}	Li et al. 2004
F. vesiculosus	Fucoidan	Anti-oxidative activity: Inhibit superoxide radicals, hydroxyl radicals and lipid peroxidation $IC_{50} = 0.058$, 0.157 and 1.250 mg mL^{-1}, respectively	De Souza et al. 2007

F. vesiculosus	Fucoidan (Sigma)	Fucoidan has protective effect via inducible nitric oxide synthase (iNOS)	Lee et al. 2007
F. vesiculosis	Fucoidan (Sigma)	Fucoidan inhibits TNF-alpha- and IFN-gamma-stimulated NO production via p38 MAPK, AP-1, JAK/STAT and IRF-1	Do et al. 2009
F. vesiculosus	Fucoidan (Sigma)	Fucoidan inhibits beta amyloid induced microglial clustering at 10 μM	Huang et al. 2009
F. vesiculosus	Fucoidan	Stimulate cAMP-response elementbinding protein (CREB) and BDNF expressions in brain IC_{50} = 50 mg kg^{-1} in male Sprague–Dawley rats	Lee et al. 2012e
F. vesiculosus	Phlorotnnins	Suppressing the overproduction of intracellular ROS induced by hydrogen peroxide IC_{50} = 0.068 mg mL^{-1}	Liu and Gu 2012
Ishige okamurae	Phlorotannin: 6,6′-Bieckol	Inhibiting AChE IC_{50} = 46.42 μM	Yoon et al. 2009c
I. okamurae	Phlorotannin: Diphlorethohydroxycarmalol (DPHC)	Neuroprotection against hydrogen peroxide (H_2O_2)-induced oxidative stress in murine hippocampal neuronal cells IC_{50} = 50 μM	Heo et al. 2012
Marginariella boryana	Sulfated fucans	Prevents the accumulation of Aβ	Wozniak et al. 2015
Padina australis	Dichloromethane extract	Inhibiting AChE IC_{50} = 0.149 mg mL^{-1}	Gany et al. 2014
P. gymnospora	Methanolic extract	Inhibiting AChE IC_{50} = 3.5 mg mL^{-1}	Yoon et al. 2008
P. gymnospora	Methanolic extract	Inhibiting AChE IC_{50} = 3.5 mg mL^{-1}	Natarajan et al. 2009
P. gymnospora	Acetone extracts	IC_{50} value < 10 μg mL^{-1} for both AChE and BuChE	Shanmuganathan et al. 2015
P. tetrastromatica	Fucoxanthin	Anti-oxidative activity: Reduce lipid peroxidation in rats IC_{50} = 0.83 μM	Sangeetha et al. 2009
P. tetrastromatica	Chloroform and ethanol extracts	Chloroform extract at 600 mg Kg^{-1} showed significant anticonvulsant activity	Yende et al. 2016
Papenfussiella lutea	Sesquiterpenes	Inhibiting AChE IC_{50} = 48–65 μM	Ryu et al. 2003

Table 11.1 contd....

Table 11.1 contd....

Species	Extract/Compound	Activity	References
Saccharina japonica	Fucoidan	Protective effect in MPTP-induced neurotoxicity. In addition, reduce behavioural deficits and cell death and increase dopamine $IC_{50} = 25$ mg kg^{-1}, once per day in mice	Luo et al. 2009
S. japonica	Fucoidan	Inhibiting microglia which inhibits LPS-induced NO production via suppression of p38 and ERK phosphorylation $IC_{50} = 125$ μg mL^{-1}	Cui et al. 2010
S. japonica	Fucoidan	Anti-oxidative activity: Reduce the toxicity of H_2O_2 in PC12 cells via activation of PI3K/Akt pathway $IC_{50} = 60$ μg mL^{-1}	Gao et al. 2012
S. japonica	Ethanolic extract	Promoted neurite outgrowth in a dose-dependent manner with optimal concentrations of 15 μg mL^{-1}	Hannan et al. 2014, 2014b
S. japonica	Fucoidan	Reduced 6-hydroxydopamine (6-OHDA) and reduced the loss of dopaminergic in neurons $IC_{50} = 20$ mg kg^{-1} in rats	Zhang et al. 2014d
Sargassum fulvellum	Pigment: Pheophytin A	Produce neurite outgrowth (from 20 to 100% in the present of 10 ng mL^{-1} of NGF) and activate $IC_{50} = 3.9$ μg mL^{-1} in PC12 cells	Ina et al. 2007
S. macrocarpum	Carotenoids	Promote neurite outgrowth activity to 0.4 in PC12 cells $IC_{50} = 6.25$ μg mL^{-1}	Tsang et al. 2001
S. macrocarpum	Meroterpenoid: Sargaquinoic acid	Signalling pathway of TrkA-MAP kinase pathway $IC_{50} = 3$ μg mL^{-1}	Kamei and Tsang 2003
S. macrocarpum	Meroterpene: Sargachromenol	Promote survival of PC-12 cells and neurite outgrowth through activation of cAMP and MAP kinase pathways $IC_{50} = 9$ μM	Tsang et al. 2005
S. micracanthum	Plastoquinones	Anti-oxidative activity: Lipid peroxidation $IC_{50} = 0.95–44.3$ μg mL^{-1} DPPH $IC_{50} = 3–52.6$ μg mL^{-1}	Mori et al. 2005b
S. fulvellum	Ethanolic extract	Promoted neurite outgrowth in a dose-dependent manner with optimal concentrations of 5 μg mL^{-1}	Hannan et al. 2012

S. fusiforme (as *Hijikia fusiformis*)	Fucoxanthin	Anti-oxidative activity: DPPH radical scavenging	Yan et al. 1999
S. fusiforme	Fucoidan	Ameliorating learning and memory deficiencies, and otential ingredient on treatment of Alzheimer's disease	Hu et al. 2016
S. horneri	Total sterols and β-sitosterol	Antidepressant effect	Zhao et al. 2016
S. polycystum	Hexane, dichloromethane and methanol extracts	Inhibiting AChE IC_{50} = 0.115, 0.180 and 0.162 mg mL^{-1}, respectively	Gany et al. 2014
S. sagamianum	Sesquiterpenes	Inhibiting AChE IC_{50} = 48–65 μM	Ryu et al. 2003
S. sagamianum	Plastoquinones: Sargaquinoic acid and sargachromenol	Inhibiting AChE IC_{50} = 23.2 and 32.7 μM, respectively Inhibiting BuChE IC_{50} = 26 μM (for sargaquinoic)	Choi et al. 2007
S. siliquastrum	Fucoxanthin	Anti-oxidative activity: Inhibit hydrogen peroxide in Vero cells IC_{50} = 100 uM	Heo et al. 2008
S. siliquastrum	Meroditerpenoids	These compounds exhibited moderate to significant radical-scavenging activity as well as weak inhibitory activities against sortase A and isocitrate lyase	Jung et al. 2008
Sargassum sp.	Methanolic extract	Inhibiting AChE IC_{50} = 1 mg mL^{-1}	Natarajan et al. 2009
S. swartzii (as *Sargassum wightii*)	Alginic acid	Polysaccharide inhibition activities to COX-2, lipoxygenase (5-LOX), xanthine oxidase (XO) and myeloperoxidase (MPO) in type II collagen induced arthritic rats IC_{50} = 100 mg kg^{-1}	Sarithakumari and Kurup 2013
S. swartzii (as *S. wightii*)	Petroleum ether, hexane, benzene and dichloromethane extracts	Inhibiting AChE IC_{50} = 19.33, 46.81, 27.24, 50.56 μg mL^{-1}, respectively Inhibiting BuChE IC_{50} = 17.91, 32.75, 12.98, 36.16 μg mL^{-1}, respectively	Syad et al. 2013
S. vulgare	Methanolic extract	Inhibiting AChE IC_{50} = 3.5 mg mL^{-1}	Custódio et al. 2016
Scytothamnus australis	Sulfated fucans	Prevents the accumulation of Aβ	Wozniak et al. 2015
Splachnidium rugosum	Sulfated fucans	Prevents the accumulation of Aβ	Wozniak et al. 2015
Turbinaria decurrens	Fucoidan	Potential neuroprotective effect in Parkinson's deasese	Meenakshi et al. 2016

Table 11.1 contd....

Table 11.1 contd....

Species	Extract/Compound	Activity	References
Undaria pinnatifida	Ethanolic extract	Promoted neurite outgrowth in a dose-dependent manner with optimal concentrations of 5 μg mL^{-1}	Hannan et al. 2014, 2014b
U. pinnatifida	Ethanolic extract	Neurogenesis, neuroprotection, anti-inflammatory and anti-Alzheimer's	Bhuiyan et al. 2015
U. pinnatifida	Glycoprotein	Neurogenesis, neuroprotection, anti-inflammatory and anti-Alzheimer's Showed predominantly AChE, BChE, and BACE1 inhibitory activities with IC_{50} values of 63.56, 99.03 and 73.35 μg mL^{-1}, respectively	Rafiquzzaman et al. 2015
U. pinnatifida	Sulfated fucans	Prevents the accumulation of Aβ	Wozniak et al. 2015
Zonaria spiralis	Phloroglucinol: Spiralisone A and Chromone 6	Kinases inhibitory to CDK5/p25, CK1δ and GSK3β IC_{50} = 10.0, < 10 and < 10 μM, respectively	Zhang et al. 2012c
Rhodophyta (red seaweed)			
Amphiroa beauvoisii	50% Aqueous methanol extract	Inhibiting AChE IC_{50} = 0.12 mg mL^{-1}	Rengasamy et al. 2015
A. bowerbankii	Methanolic extract	Inhibiting AChE IC_{50} = 5.3 mg mL^{-1}	Stirk et al. 2007
A. ephedraea	Methanolic extract	Inhibiting AChE IC_{50} = 5.1 mg mL^{-1}	Stirk et al. 2007
Asparagopsis armata	Methanolic extract	AChE inhibitory capacity: 58.4% BuChE inhibitory capacity: 81.4% 10 mg mL^{-1}	Custódio et al. 2016
Bryothamnion triquetrum	Water extract	Protect GT1–7 cells death produced by severe (180 min.) chemical hypoxia/aglycemia insult	Fallarero et al. 2003, 2006
Chondracanthus acicularis	Lambda-carrageenan	Anti-oxidative activity: Inhibit superoxide radicals, hydroxyl radicals and lipid peroxidation IC_{50} = 0.046, 0.357 and 2.267 mg mL^{-1}, respectively	De Souza et al. 2007

Chondrophycus undulatus (as *Laurencia undulata*)	Glycerol glycosides: Floridoside	Suppress pro-inflammatory responses in microglia by markedly inhibiting the production of nitric oxide (NO) and reactive oxygen species (ROS) $IC_{50} = 10\ \mu M$	Kim et al. 2013g
Chondrus crispus	Methanolic extract	Extract-mediated protection against Parkinson's disease pathology	Liu et al. 2015
Eucheuma denticulatum	Iota-carrageenan	Anti-oxidative activity: Inhibit superoxide radicals, hydroxyl radicals and lipid peroxidation $IC_{50} = 0.332$, 0.281 and 0.830 mg mL^{-1}, respectively	De Souza et al. 2007
Gelidiella acerosa	Petroleum ether, hexane, benzene, dichloromethane, chloroform, ethyl acetate, acetone, methanol, and water extracts	Inhibiting AChE Benzene extract, $IC_{50} = 434.61\ \mu g\ mL^{-1}$ Ethyl acetate, $IC_{50} = 444.44\ \mu g\ mL^{-1}$ Inhibiting BuChE Benzene extract, $IC_{50} = 163.01\ \mu g\ mL^{-1}$ Chloroform extract, $IC_{50} = 375\ \mu g\ mL^{-1}$	Syad et al. 2012
G. acerosa	Phytol	*In vitro* and *in vivo* antioxidant activities (25–125 μg mL^{-1}) with an IC_{50} value of 95.27 μg mL^{-1} and cholinesterase inhibitory potential (5–25 μg mL^{-1}) with IC_{50} values of 2.704 and 5.798 μg mL^{-1} for AChE and BuChE, respectively	Syad et al. 2016
Gelidium amansii	Ethanol extract	Neurogenesis (synaptogenesis promotion)	Hannan et al. 2013, 2014
G. foliaceum	50% Aqueous methanol extract	Inhibiting AChE $IC_{50} = 0.16$ mg mL^{-1}	Rengasamy et al. 2015
Gloiopeltis furcata	2-(3-Hydroxy-5-oxotetrahydrofuran-3-yl) acetic acid, glutaric acid, succinic acid, nicotinic acid, (E)-4-hydroxyhex-2-enoic acid, cholesterol, 7-hydroxycholesterol, uridine, glycerol, phlorotannin, fatty acids	Inhibiting AChE 1.4–12.50 μg mL^{-1} Inhibiting BuChE 6.56–75.25 μg mL^{-1}	Fang et al. 2010
Gracilaria edulis	Methanolic extract	Inhibiting AChE $IC_{50} = 3$ mg mL^{-1}	Natarajan et al. 2009
G. edulis	Methanolic extract	Inhibiting AChE $IC_{50} = 3$ mg mL^{-1}	Yoon et al. 2008
G. gracilis	Methanolic extract	Inhibiting AChE $IC_{50} = 1.5$ mg mL^{-1}	Natarajan et al. 2009

Table 11.1 contd....

Table 11.1 contd....

Species	Extract/Compound	Activity	References
Gracilariopsis chorda	Ethanolic extract	Neuronal cell viability and cell cytotoxicity testing revealed that the ethanol extract afforded the most neuroprotection at a concentration of 15 μmL^{-1}, at which *G. chorda* extract significantly increased cell viability to 119.0%–3.2%, and decreased cell death to 80.5%–10.3%	Mohibbullah et al. 2015
G. chorda	Ethanolic extract	Extract concentration-dependently increased neurite outgrowth, with an optimal concentration of 30 mu g mL^{-1}	Mohibbullah et al. 2016b
Hypnea valentine	Methanolic extract	Inhibiting AChE $IC_{50} = 2.6$ mg mL^{-1}	Natarajan et al. 2009
H. valentiae	Methanolic extract	Inhibiting AChE $IC_{50} = 2.6$ mg mL^{-1}	Suganthy et al. 2010
Kappaphycus alvarezii	Kappa-carrageenan	Anti-oxidative activity: Inhibit superoxide radicals, hydroxyl radicals and lipid peroxidation $IC_{50} = 0.112$, 0.335 and 0.323 mg mL^{-1}, respectively	De Souza et al. 2007
K. alvarezii	Ethanolic extract	Promotes neurite outgrowth in hippocampal neurons	Tirtawijaya et al. 2016
Ochtodes secundiramea	Dichloromethane/methanol extract: Halogenated monoterpenes	Extract showed 48% AChE inhibition at 400 µg mL^{-1}	Machado et al. 2015
Porphyra/Pyropia sp. (Korean purple laver)	Phycoerythrobilin	Antioxidant activity $IC_{50} = 0.048$ mmol g^{-1}	Yabuta et al. 2010
Pyropia yezoensis (as *Porphyra yezoensis*)	Ethanolic extract	Increased neurite outgrowth at an optimal concentration of 15 µg mL^{-1}	Mohibbullah et al. 2016
Rhodomela confervoides	Bromophenols	Antioxidant activity $IC_{50} = 5.22$–23.60 µM	Li et al. 2012
Rhodomelopsis africana	50% Aqueous methanol extract	Inhibiting AChE $IC_{50} = 0.12$ mg mL^{-1}	Rengasamy et al. 2015

*Marine microalgae

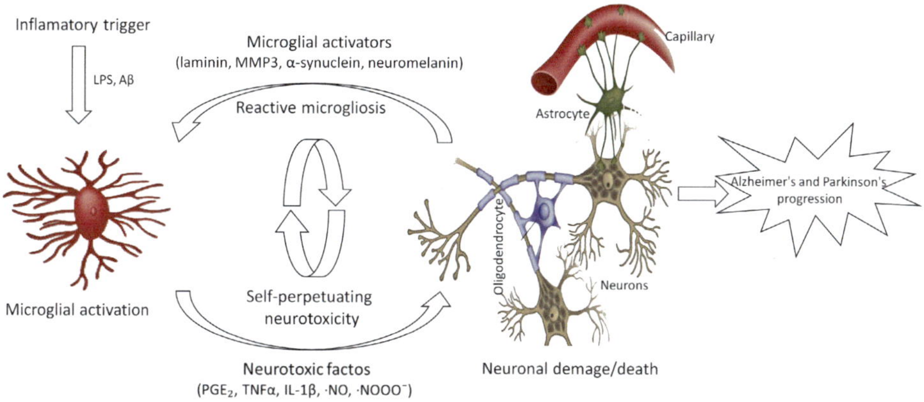

Figure 11.1 Microglia-mediated neurotoxicity in Alzheimer's and Parkinson's diseases.

(57.6%), followed by Rhodophyta (28.3%), and Chlorophyta (14.1%) (see Table 11.1) (Alghazwi et al. 2016).

11.4.1 Chlorophyta (green seaweed)

Several studies demonstrated that NO generated by iNOS causes injury and cell death of neuron and oligodendrocytes in the CNS, hence NO is implicated in pathogenesis of various neurodegenerative disease (Heales et al. 1999, Lee et al. 2000). Anti-neuroinflammatory activity of another marine algae species, *Ulva conglobata* has been reported. *U. conglobata* methanolic extracts could suppress the expression of pro-inflammatory enzymes, iNOS and cyclooxygenase-2 (COX-2), which accounted for the large production of NO and PGE2, respectively (Salvemini et al. 1995, Jin et al. 2006). Among other mediators released by microglia, NO and PGE2 are the main cytotoxic mediators participating in the innate response in the CNS (Vane and Botting 1995, Boscá et al. 2005). Pro-inflammatory mediators have been found to be elevated in the brain of early Alzheimer's disease (Blasko et al. 2004). For this reason, agents that inhibit the production of pro-inflammatory mediators have been previously considered to be potential candidates for the treatment of neurodegenerative diseases.

Fallarero et al. (2003) showed that *Halimeda incrassata* are an effective ROS scavenger in hypothalamic (GT1–7) cells of mice. Eight compounds have been found from macroalgae with neuroprotective activity against beta amyloid protein (Aβ) (see Table 11.1). Five compounds are reported from the green alga *Caulerpa racemosa* (Alghazwi et al. 2016).

11.4.2 Phaeophyceae (brown seaweed)

Yan et al. (1999) demonstrated that carotenoids have a strong radical scavenging activity and are found as a major antioxidant in marine algae (Nomura et al. 1997). Young and Lowe (2001) indicated that structure, physical form, location or site of action, potential interaction with another antioxidant, concentration, and partial pressure to oxygen may affect the antioxidant activities of carotenoids in biological systems. Fucoxanthin obtained from *Padina tetrastromatica* has shown higher potential to be used as an antioxidant than β-carotene in modulating antioxidant enzyme in plasma and liver of retinol deficient rats (Kumar et al. 2008c, Sangeetha et al. 2009). Nevertheless, the exact mechanisms of action as to how fucoxanthin exerts antioxidative effects in rats induced by retinol deficiency are not yet completely understood. Moreover, the cytoprotective effect of fucoxanthin against ROS formation induced by H_2O_2 in monkey kidney fibroblast (Vero) cells has been observed (Heo et al. 2008).

Parkinson's disease is usually characterized by the loss of dopaminergic neurons, and the presence of 1-methyl-4-phenyl-1,2,3,6-tetrahydropyridine (MPTP) can induce Parkinsonism (Langston et al. 1999). Administration of MPTP can result in motor dysfunction, like that which occurs in Parkinson's disease, which makes it a favorable experimental model for this disease (Gerlach et al. 1991). One compound (Fucoidan) has been found to attenuate the neurotoxicity of MPTP activity. This sulfated polysaccharide, derived from *Saccharina japonica*, has been demonstrated in mice models to be effective at a dosage of 25 mg kg^{-1} in protecting the cells from MPTP-induced neurotoxicity by reducing the behavioral deficits and cell death and increasing the level of dopamine (Luo et al. 2009).

Tau is a microtubule-associated protein (MAP) found in axons, and this protein is responsible for regulating the stability of microtubules (Goedert et al. 1988, Drechsel et al. 1992, Hirokawa et al. 1996). Hyper-phosphorylation of tau results in its dissociation from microtubules and aggregation in the form of neurofibrillary tangles (Avila et al. 2006). Hyper-phosphorylated tau protein is a major component in neurofibrillary tangles, which is a hallmark of Alzheimer's disease, and dysregulation of kinases and phosphatases has been found to increase tau hyper-phosphorylation levels (Selkoe 1997, Hanger et al. 2009). Only three compounds (Spiralisone A, B, and Chromone 6) from seaweed have kinase inhibitory activity, and these compounds have been isolated from brown algae (Phaeophyceae). Besides, these compounds were isolated from a single species, *Zonaria spiralis*, collected in Australia, and all of them are phloroglucinols. The most active compound is spiralisone B, inhibiting the kinases-cyclin-dependent kinase 5 (CDK5/p25), casein kinase 1 (CK1δ), and glycogen synthase kinase 3 beta (GSK3β), with IC_{50} values of 3, 5, and 5.4 μM, respectively (Zhang et al. 2012c).

A common pathological hallmark of various neurodegenerative diseases is the loss of subsets of neurons (Mattson 2000). Neurodegeneration of these neural subsets may be a consequence of various forms of neural cell death, including necrosis and apoptosis (Bains and Shaw 1997). A study carried out by Jhamandas et al. (2005) successfully showed that fucoidan (see Fig. 2.1, Chapter 2) isolated from *Fucus vesiculosus*, could protect cholinergic neuronal death in rats induced by Aβ1–42. Fucoidan pretreatment blocked the activation of caspase-9 and caspase-3. Caspase-9 and caspase-3 have been suggested to mediate the terminal stages of neuronal apoptosis (Cowan et al. 2001). Caspase-9 and caspase-3 are two of several central components of the machinery responsible for apoptosis. Therefore, the ability of fucoidan to block the activation of caspase-9 and caspase-3 suggest that inhibition of neuronal death by fucoidan mainly occurs through apoptotic inhibition. In neurodegenerative diseases, apoptosis might be pathogenic, and targeting this process might mitigate neurodegenerative diseases (Vila and Przedborski 2003).

Neurite outgrowth is a fundamental neuronal feature and plays an important role in neuronal development during embryogenesis and in the adult brain (Khodosevich and Monyer 2010). *Sargassum macrocarpum* and its two active components, sargaquinoic acid and sargachromechanol, have been shown to promote neurite outgrowth in pheochromocytoma (PC12) cells in rats (Kamei and Sagara 2002, Tsang and Kamei 2004, Tsang et al. 2005). Structure of sargaquinoic acid and neurite outgrowth promoting relationship has been reported by Tsang et al. (2001). They reported that quinone is the structural moiety of the sargaquinoic acid molecule, which is responsible for the neurite outgrowth-promoting activity. Notably, the hydroxyl group bonded to quinone had a significant effect on neuritogenic activity. In addition, pheophytin a, a chlorophyll-related compound and its analog, vitamin B_{12} derived from *Sargassum fulvellum*, also has potential neurite outgrowth-promoting activity (Ina and Kamei 2006, Ina et al. 2007).

Brown algae are a rich source of phlorotannins, which are a well-known group of bioactive tannins. In the wide range of bioactivities, the effectiveness of phlorotannins as an inhibitor of neuroinflammation has been considerably researched, and reports indicate that use of brown algae as food or food ingredients might have highly positive results on brain health. Ethanolic extract of *Ecklonia cava* (Phaeophyceae, Laminariales) is reported to inhibit lipopolysaccharide-induced cyclooxygenase-2 (COX-2) and iNOS expression in BV-2 microglia cells (Jung et al. 2009b). Detailed analysis of the extract and its constituents revealed that dieckol is the effector material responsible for the suppression of endotoxin-stimulated proinflammatory enzymes iNOS and COX-2 production in the murine BV-2 microglia (Jung et al. 2009c). In addition to dieckol, several other phlorotannins present in brown algae, such as phloroglucinol, eckol, 7-phloroeckol, phlorofucofuroeckol A, and dioxinodehydroeckol (see Table 11.1) have also shown activity against neuroinflammation (Jung et al. 2013).

Another study conducted by Cui et al. (2010) provide the first evidence that fucoidan isolated from *Saccharina japonica* (as *Laminaria japonica*) has a potent inhibitory effect against LPS-induced NO production in BV2 cells. In their study, the average molecular weight of fucoidan was 7000 Dalton, consisting of 48% total sugar (including 28% fucose) and 29% sulfate. Fucoidan at a concentration of 125 µg mL^{-1} significantly inhibited NO production to 75%. NO is a cytotoxic, short-lived, highly diffusible signaling molecule (Heales et al. 1999).

The results obtained by Custódio et al. (2016) suggest that the brown seaweeds *Cystoseira tamariscifolia* and *Cystoseira nodicaulis* contain molecules appropriate for nutraceuticals, and may constitute functional foods aiming for the improvement of cognitive functions through their capacity to scavenge free radicals, chelate metals, protect neuronal cells from oxidative stress, and inhibit enzymes involved in the degradation of the neurotransmitter ACh. Assays are already in progress aiming to isolate and identify the bioactive compounds in those extracts.

11.4.3 Rhodophyta (red seaweed)

The amino acid (α-alkokainic acid) isolated from the red alga *Digenea simplex* showed a potent neurophysiological activity in mammals (Biscoe et al. 1975, Ferkany and Coyle 1983). 5-Iodo-5'-deoxy-tubercidin was isolated from the red alga *Hypnea valentiae*, which causes pronounced relaxation of muscles and hypothermia in mice, and blocks polysynaptic and monosynaptic reflexes. This compound is one of the most interesting algal metabolites discovered by using a bioassay directed isolation procedure (Kazlauskas et al. 1983).

Bromophenols are an interesting class of bioactive compound commonly present in red algae. Among algae-derived bromophenols, anti-inflammatory effects were observed for vidalols A and B, which were obtained from the red alga *Osmundaria obtusiloba* (as *Vidalia obtusiloba*). Wiemer et al. (1991) reported that bioactivity-guided experiments showed that crude extract of alga *O. obtusiloba* potentially inhibited (6–9% inactivation at 1.6 mg mL^{-1}) bee venom-derived phospholipase A_2 (PLA_2) activity. In the brain, strict regulation of PLA_2 activity is of crucial importance, as PLA_2 regulation accounts for a balance between arachidonic acid's conversion into proinflammatory mediators and their reincorporation into the membrane. A disruption of this well-balanced system (increased level of PLA_2 activity) results in induced oxidative stress and neuroinflammation, which may cause several neurological diseases. Therefore, specific inhibitors of PLA_2 regulation could be a pharmaceutical target. Despite the potency of the compounds vidalols A and B in the inflammatory activity of the brain, their potential to be used as inhibitors of neuroinflammation has not yet been evaluated.

Novoa et al. (2001) reported that the antioxidant and ROS scavenging activity of *Bryothamnion triquetrum* are related to their high phenolic contents. Fallarero et al. (2003) showed that *B. triquetrum* are a potent ROS scavenger in the hypothalamic (GT1–7) cells of mice. Red algae extracts and purified materials have shown potential against neuroinflammation. *Neorhodomela aculeata*, which is a popular red alga used in Chinese medicines, is reported as potential inhibitor of inflammation in neuronal and microglial cells. Furthermore, Lim et al. (2006) demonstrated that *N. aculeata* could scavenge DPPH with an IC_{50} = 90 µg mL^{-1} and at a concentration of 20 µg mL^{-1} completely suppressed H_2O_2 induced lipid peroxidation in brain homogenate of rats.

Aqueous extracts of *Bryothamnion triquetrum* have been demonstrated to protect GT1–7 cell deaths produced by severe (180 min) chemical hypoxia/aglycemia insult, which further reduced the cytotoxicity and early production of free radicals. The protection exerted by *B. triquetrum* extract seems to be linked to its ability to reduce free radical generation. The authors suggest that the protective effects of *B. triquetrum* extract are partially related to the presence of ferulic acid (Fallarero et al. 2003, 2006).

In the study made by Najam et al. (2010), the administration of methanolic extract of *Hypnea musciformis* significantly increased the level of dopamine on rats and mice. The possible effect of *H. musciformis* on dopamine and other brain biogenic amines indicate that *H. musciformis* probably have psychotropic and anxiolytic profile. The increased level of dopamine could also be beneficial, keeping in view the etiology of Parkinson's disease. In this study, the level of serotonin was found to be decreased after the administration of *H. musciformis*. The regular use of seaweeds as a diet will relieve the symptoms of anxiety, because the known anxiolytics also manifest their effect by decreasing the concentration of serotonin (Najam et al. 2010).

Gloiopeltis furcata is the only species of red algae that provided all 15 compounds, which highlights the potential of discovering neuroprotective compounds from another Rhodophyta species. Of interest from all these compounds studied by far, 2-(3-hydroxy-5-oxotetrahydrofuran-3-yl) acetic acid isolated from this species has multiple inhibition activities, with IC_{50} values of 1.4 and 12.6 μg mL^{-1} for AChE and BuChE, respectively. For comparison, the known cholinesterase inhibitor nicotinic acid has an IC_{50} value of 1.14 μg mL^{-1} against AChE, and 20.9 μg mL^{-1} against BuChE (see Table 11.1) (Fang et al. 2010).

Mohibbullah et al. (2015) collected 23 edible seaweeds from Korean and Indonesian coasts to screen for marine seaweeds with potent neuroprotective activity. Hippocampal neurons of rats (DIV 9) were cultured in the presence of ethanol extracts and the cultures were treated with three different concentrations of seaweed extract: 5, 15, and 30 μg mL^{-1}. About 1/3 of the tested seaweeds exhibited neuroprotective activity. Cell viability and cell cytotoxicity testing revealed that the ethanol *Gracilariopsis chorda* extract afforded the most neuroprotection at a concentration of 15 μg m^{-1}, at which *G. chorda* significantly increased cell viability to 3.2–119.0%, and decreased cell death to 10.3–80.5%. *Undaria pinnatifida* (brown algae) had almost the same level of neuroprotection as *G. chorda*, and others like *Sargassum fulvellum*, *Sargassum nigrifolium* (brown algae), *Pyropia yezoensis* (as *Porphyra yezoensis*), *Gracilaria coronopifolia, Gracilaria tenuistipitata* (red algae), *Ecklonia bicyclis* (as *Eisenia bicyclis*) (brown alga), and *Carpopeltis cornea* (red alga) also exhibited moderate neuroprotective effects (Mohibbullah et al. 2015).

Liu et al. (2015) demonstrated that dietary supplementation of the worms with an extract from the cultivated red seaweed *Chondrus crispus* decreased the accumulation of α-synulein, and protected the worms from the neuronal toxin-, 6-OHDA, and the induced dopaminergic neurodegeneration. These effects were associated with a corrected slowness of movement. We also showed that the enhancement of oxidative stress tolerance and an up-regulation of the stress response genes, sod-3 and skn-1, may have served as the molecular mechanism for the *C. crispus*-extract-mediated protection against Parkinson's disease pathology. Altogether, apart from its potential as a functional food, the tested red seaweed, *C. crispus*, might find promising pharmaceutical applications for the development of potential novel anti-neurodegenerative drugs for humans (Liu et al. 2015).

CHAPTER 12

Thalassotherapy and Marine Cosmeceuticals

12.1 Introduction

Riverine, estuarine, or coastal populations have always used algae in the development of home remedies that were then used to treat various health problems. These applications are the product of empirical knowledge of many generations, and in most cases their mechanism of action is unknown, that is, few scientific studies would have been described beyond simple collection and ethnographic recording. Nevertheless, recent investigations carried out with the purpose of analyzing the components and causes that alter the functioning and the balance of our organism, are already giving their first results. Thus, we know today that the good results obtained in the treatment of goiter based on the use of laminarian and fucoid algae are due to the fact that the origin of this disease is directly related to a diet low in iodine, which is thus enriched by the ingestion of these algae (see Chapters 2 and 3), in which iodine is present in very significant amounts (Pereira 2011, Pereira and Correia 2015).

Humanity is progressively confronted with growing problems of an ecological order and increasingly underlines the need to return to nature. This desire will influence the research work in this domain. Having this goal in mind, the research begins with seawater itself, which shows extremely important qualities in the field of cosmetics and "well-being" products.

12.1.1 Seawater

Water, and especially seawater, is considered to be essential to life on our planet. It contains all the substances necessary and conducive to the development of the living being (minerals, catalysts, vitamins, amino acids, etc.). More than 70% of the earth is covered by ocean, which is home to up to 90% of the planet's organisms. The ocean provides many unique environments and rich resources, and there are numerous marine organisms with great potential to produce bioactive compounds that can be used as pharmaceuticals, nutraceuticals, and cosmeceuticals. Therefore, more research is necessary to explore, identify, understand, and eventually make use of the organisms living in the ocean (Gomez et al. 2009).

In the 19th century, the French biologist René Quinton could prove experimentally the remarkable similarity between seawater and human plasma—"Our organism is a kind of aquarium whose internal environment is constituted by seawater" (Quinton 1904). Quinton's book "L'eau de mer, milieu organique" became the *vademecum* of thalassotherapy, buttressed by the work of Dr. Louis Bagot, who created a center at Roscoff (Brittany). With the 20th century came the first Rockroum marine medical center, with the hydrotherapy as a guiding principle. Publicity for kine-balneotherapy came from a car accident of a champion bicycle racer—Louison Bobet. He was saved by Dr. Bagot and, in 1964, opened the first modern thalassotherapy center at Quiberon (Brittany). Although it was beneficial in bringing this therapy to the attention of the public at large, it carried a damaging impact for the reputation of medical and thermal cures. In the 1980s the term and concepts were misused and some institutes imported seawater, sea mud, and algae at distances far from the ocean, contrary to true therapy, which must be practiced at the shore, preferably in facilities not farther away from the water's edge than a good 100 m. Close to 50 procedures are part of thalassotherapy (Charlier and Chaineux 2009).

Seawater has a recognized action in the treatment of eczemas, dermatoses, psoriasis, nasopharyngeal inflammations, conjunctivitis, vaginitis, and other infections of the external genital organs. In cosmetology, it has also been used with good results in moisturizing the skin, and making it more firm. It allows for a more uniform tan, and regularizes the sebaceous production at the level of the hair follicles, avoiding the consequent formation of scaling of the scalp (dandruff). It will be interesting to mention the important role of seawater in the absorption of saline and metallic ions, favoring the excretion of toxic residues and a certain oxygenation of tissues (Barata 1995).

As might be expected, these characteristics of seawater have led to the development of bio-cosmetics based on seawater and all its substances or organisms: algae, sea salts, organic and mineral elements, etc.

12.2 Cosmeceuticals

Today's society is showing a growing interest in everything that can help improve the quality of life, increasingly believing that many of the solutions are found in nature itself and in the chemical synthesis of drugs (usually with various contraindications, physiological and/or metabolic, or undesirable side effects/changes). Accompanying this trend, the demand for products made from natural raw materials has increased (preferably those of vegetable origin), allowing for the appearance of a new niche market (natural and/or biological), in which algae have flourished, thanks to its role in human well-being, which is increasingly being recognized (Pereira and Correia 2015). For example, in Europe, consumption of fresh algae, in a cumulative perspective (for direct feeding and/or therapeutic and cosmetic uses), mainly of *Ascophyllum*, *Fucus*, *Laminaria* (brown algae), *Gelidium* and *Chondrus* (red algae), is around 70,000 tons/year (Mesnildrey et al. 2012, Netalgae 2017).

The term cosmeceuticals was created three decades back by Albert Kligman to define the cosmetic products with biologically active substances carrying either medicinals or drug-like benefits (Gao et al. 2008, Manela-Azulay and Bagatin 2009). In general, cosmeceuticals are often used in the field of dermatology to improve skin tone/whitening, increase skin radiance, decrease the appearance of skin wrinkling, and provide antiaging benefits (Kligman 2000, Thornfeldt 2005, Draelos 2008, Gao et al. 2008, Manela-Azulay and Bagatin 2009). Cosmetic products associated with cosmeceutical compounds have become one of the major market niche with significant annual growth within recent years (Gao et al. 2008, Anunciato and da Rocha Filho 2012). Parallel to the development of cosmeceutical industry, also in research and development fields, thousands of chemical substances obtained either naturally or synthetically have been tested and/or investigated to potentially use as active compounds in cosmeceutical products (Gao et al. 2008).

Cosmetics is a field in which seaweed has been consolidating its presence, and it is possible to find a wide variety of products from topical application—from slimming creams to perfumes, shampoos, sunscreens, and bath salts, which center their activity in the properties and in the actions that they induce to the human organism (Pereira and Correia 2015).

Some of the compounds in algae can be potentially used in cosmeceuticals (see Table 12.1), for example, phlorotannins, sulfated polysaccharides, and tyrosinase inhibitors. The term "cosmeceuticals" is derived from "cosmetics" with the potential of "pharmaceuticals", and it refers to specific products containing active ingredients (Thomas and Kim 2013).

Algae produce both primary metabolites, which are directly involved in normal growth, development, or reproduction conditions to perform physiological functions, and secondary metabolites, which are performed under different stress conditions such as UV radiation exposure, salinity, temperature changes, or environmental pollutants. Primary metabolites in algae include polysaccharides, proteins, amino acids, and fatty acids (Fig. 10.4). Secondary metabolites produced in algae tissues are pigments, phenolic

Table 12.1 Bioactive properties of some compounds extracted from seaweeds.

Species	Extract/Compound	Cosmetics properties and/or products	References
Chlorophyta (green seaweed)			
Codium tomentosum	Extract[3]	Moisturizing	Majmudar 2012
C. tomentosum	Extract	Skin moisturization and protection	Wang et al. 2013c, 2013d
*Chlorella vulgaris**	Extracts	Antistretch marks creams, body lotions, eye creams, face masks, shower gels	Thalgo 2017, La-Mer 2017
Cladophora glomerata	Chlorophylls: a, b, c, and d	Antibacterial, antioxidant, coloring, antibacterial, deodorizing, tissue growth stimulating agents	Goldberg 1943, Spears 1988, Lanfer-Marquez et al. 2005
*Dunaliella salina**	Carotenoids: β-carotene, astaxanthin, lutein, and fucoxanthin	Antioxidant, anti-inflammatory, antiaging, antiphotoaging agents, radical scavengers, colorants, and tyrosinase inhibitors	Spears 1988, Borowitzka 2013, Christaki et al. 2013
Haematococcus lacustris (as *Haematococcus pluvialis*) *	Carotenoids: β-carotene, astaxanthin, lutein, and fucoxanthin	Antioxidant, anti-inflammatory, antiaging, antiphotoaging agents, radical scavengers, colorants, and tyrosinase inhibitors	Spears 1988, Borowitzka 2013, Christaki et al. 2013
Ulva australis (as *Ulva pertusa*)	Proteins (amino acids)	Moisturizers, antioxidants, and natural sunscreens	Houston 2005, Borowitzka 2013
U. compressa (as *Enteromorpha compressa*)	Micronlzed algae	Antiaging and smoothing face creams, body scrubs, firming body lotions, face peelings	Thalgo 2017, Ziaja 2017
U. compressa (as *E. compressa*)	Extracts	Body lotions, cleansing gels, day and night face creams, eye creams, face masks, fluids, tonics, hair shampoos	Thalgo 2017, La-Mer 2017
U. lactuca	Chlorophylls: a, b, c, and d	Antibacterial, antioxidant, coloring, antibacterial, deodorizing, tissue growth stimulating agents	Goldberg 1943, Spears 1988, Lanfer-Marquez et al. 2005
U. lactuca	Seaweed lipopeptide mixed with clay	Anti-elastase, collagen synthesis stimulation	Delaunay and Volle 2011
U. lactuca	Seaweed polysaccharide mixed with clay[1]	Antioxydant activity, anti-elastase, collagen synthesis stimulation, anti-aging	Demais et al. 2007, Olmix 2017
U. lactuca	Sulfated polysaccharide (ulvan)	Antioxidative, chelating, gelling, moisturizing, and protective agents	Lahaye and Robic 2007, Robic et al. 2009
U. lactuca	Tripeptide: arginine, glycine, aspartic acid	Stimulation of collagen production via TGF-β, elastine, increase in the biosynthesis of collagen I	Guglielmo and Montanari 2008
U. lactuca	Carotenoids: β-carotene, astaxanthin, lutein, and fucoxanthin	Antioxidant, anti-inflammatory, antiaging, antiphotoaging agents, radical scavengers, colorants, and tyrosinase inhibitors	Spears 1988, Borowitzka 2013, Christaki et al. 2013

Table 12.1 contd....

Table 12.1 contd....

Species	Extract/Compound	Cosmetics properties and/or products	References
U. lactuca	Fatty acids	Antioxidant, cytoprotective Nrf2-ARE pathway	Wang et al. 2013e
U. lactuca	Extracts	Exfoliating gel, body mask, bath salts, moisturizing cream (components of the talassotherapy kit[2])	Lusalgae 2017
U. rigida (as *U. armoricana*)	Sulfated polysaccharide (ulvan)	Antioxidative, chelating, gelling, moisturizing, and protective agents	Lahaye and Robic 2007, Robic et al. 2009
U. rigida	Sulfated polysaccharide (ulvan)	Antioxidative, chelating, gelling, moisturizing, and protective agents	Lahaye and Robic 2007, Robic et al. 2009
U. rotundata	Sulfated polysaccharide (ulvan)	Antioxidative, chelating, gelling, moisturizing, and protective agents	Lahaye and Robic 2007, Robic et al. 2009
Ulva sp.	Lectins	Antiadhesive agents, antibacterial, anti-inflammatory, antiviral	Samarakoon and Jeon 2012
Phaeophyceae (brown seaweed)			
Alaria esculenta	Extract	Skin anti-ageing	Verdy et al. 2011
Ascophyllum nodosum	Sulfated polysaccharide (fucoidan)	Antioxidant, anticellulite, antiviral, anti-inflammatory, anti-aging, antiphotoaging agents, elastase, tyrosinase inhibitors	Chizhov et al. 1999, Wijesinghea and Jeona 2012
A. nodosum	Phlorotannins: eckols, fucols, fucophlorethols, fuhalols, phlorethols	Antiaging, antiphotoaging, anti-inflammatory, antiallergic, chelating agents, antioxidants, natural UV screens, histamine, tyrosinase, hyaluronidase inhibitors	Ryu et al. 2008, Stengel et al. 2011, Borowitzka 2013, Thomas and Kim 2013
A. nodosum	Extract	Skin moisturization and protection	Wang et al. 2013c, 2013d
A. nodosum	Extract	Anti-free-radical, tyrosinase inhibiting[4]	Codif 2017
A. nodosum	Extract	Skin conditioning regerating and sebum regulating agent[5]	Codif 2017
A. nodosum	Micronlzed algae	Anticellulite body masks, face creams, slimming body creams, serum	Dermika 2017, Thalgo 2017
Bifurcaria bifurcata	Extracts	Exfoliating gel, body mask, bath salts, moisturizing cream (components of the talassotherapy kit[2])	Lusalgae 2017
Cladosiphon okamuranus	Extract	Skin moisturization and protection	Wang et al. 2013c, 2013d
Cystoseira nodicaulis	Phlorotannins (bieckol, fucophloroethol, 7-phloroeckol and phlorofucofuroeckol)	Anti-inflammatory, antioxidant, anti-skin aging, anti-wrinkling (hyaluronidase Inhibition), lipid peroxidation inhibition	Ferreres et al. 2012

C. tamariscifolia	Phlorotannins (bieckol, fucophloroethol, 7-phloroeckol and phlorofucofuroeckol)	Anti-inflammatory, antioxidant, anti-skin aging, anti-wrinkling (hyaluronidase Inhibition), lipid peroxidation inhibition	Ferreres et al. 2012
C. usneoides	Phlorotannins (bieckol, fucophloroethol, 7-phloroeckol and phlorofucofuroeckol)	Anti-inflammatory, antioxidant, anti-skin aging, anti-wrinkling (hyaluronidase Inhibition), lipid peroxidation inhibition	Ferreres et al. 2012
Durvillaea antarctica	Extract	Skin moisturization and protection	Wang et al. 2013c, 2013d
Durvillea spp.	Alginates	Emulsion stabilizers, chelating agents, colloids, gelling, immunostimulating agents, moisturizing, protective colloids	Tonnesen and Karlsen 2002, Mafinowska 2011
Ecklonia bicyclis	Phlorotannins: dieckol, eckol, phloroglucinol, phlorofucofuroeckol A and 8.8′-bieckol	Hyaluronidase inhibition, anti-wrinkling	Shibata et al. 2002
E. cava	Proteins (amino acids)	Radical scavengers, antioxidant, chelating agents	Heo et al. 2003, Ahn et al. 2004, Freurence 2004
E. cava	Phlorotannins: dieckol and phloroglucinol	Histamine release, anti-allergic	Le et al. 2009
E. cava	Phlorotannins: 6,6′-bieckol and dioxinodehydroeckol	Matrix metalloproteinases (MMPs) inhibition, anti-wrinkling	Zhang et al. 2012, Kong et al. 2011b
E. cava	Sulfated polysaccharide (fucoidan)	Antioxidant, anticellulite, antiviral, anti-inflammatory, antiaging, antiphotoaging agents, elastase, tyrosinase inhibitors	Chizhov et al. 1999, Wijesinghea and Jeona 2012
E. cava	Phlorotannins: dieckol	Hair growth	Kang et al. 2012c
E. cava	Phlorotannins: eckols, fucols, fucophlorethols, fuhalols, phlorethols	Antiaging, antiphotoaging, anti-inflammatory, antiallergic, chelating agents, antioxidants, natural UV screens, histamine, tyrosinase, hyaluronidase inhibitors	Ryu et al. 2008, Stengel et al. 2011, Borowitzka 2013, Thomas and Kim 2013
E. cava	Phlorotannins: dieckol, eckol, eckstolonol, fucodiphlorethol G, phloroglucinol, triphlorethol A	Antioxidant, UV protection	Heo et al. 2009b, Ko et al. 2011, Kim et al. 2014e
E. cava	Phlorotannins: dioxinodehy-droeckol	Hair growth	Bak et al. 2013
E. cava	Phlorotannins: dieckol	Adipogenesis inhibitory effect	Ko et al. 2013
E. cava	Phlorotannins: dieckol, dioxinodehydroeckol, eckol, phloroglucinol, phloroglucinol and 7-phloroeckol	Tyrosinase inhibition, whitening effect	Heo et al. 2009b, Yoon et al. 2009b, Lee et al. 2015c

Table 12.1 contd....

Table 12.1 contd....

Species	Extract/Compound	Cosmetics properties and/or products	References
E. kurome	Phlorotannins: dieckol, eckol, phloroglucinol, phlorofucofuroeckol A and 8.8′-bieckol	Hyaluronidase inhibition, anti-wrinkling	Shibata et al. 2002
E. stolonifera	Phlorotannins: dieckol, eckol, eckstolonol, phloroglucinol, phlorofucofuroeckol A	Tyrosinase inhibition, whitening effect	Kang et al. 2004
E. stolonifera	Phlorotannins: eckol and dieckol	Matrix metalloproteinases (MMPs) inhibition, anti-wrinkling	Joe et al. 2006
E. stolonifera	Phlorotannins: phlorofucofuroeckol A	Anti-inflammatory	Kim et al. 2011c
Fucus spiralis	Phlorotannins (bieckol, fucophloroethol, 7-phloroeckol and phlorofucofuroeckol)	Anti-inflammatory, antioxidant, anti-skin aging, anti-wrinkling (hyaluronidase Inhibition), lipid peroxidation inhibition	Ferreres et al. 2012
F. vesiculosus	Sulfated polysaccharide (fucoidan)	Antioxidant, anticellulite, antiviral, anti-inflammatory, antiaging, antiphotoaging agents, elastase, tyrosinase inhibitors	Chizhov et al. 1999, Wijesinghea and Jeona 2012
F. vesiculosus	Sulfated polysaccharide (fucoidan)	Anticoagulant, antioxidant, skin fibroblast stimulation	Rupérez et al. 2002
F. vesiculosus	Phlorotannins: eckols, fucols, fucophlorethols, fuhalols, phlorethols	Antiaging, antiphotoaging, anti-inflammatory, antiallergic, chelating agents, antioxidants, natural UV screens, histamine, tyrosinase, hyaluronidase inhibitors	Ryu et al. 2008, Stengel et al. 2011, Borowitzka 2013, Thomas and Kim 2013
F. vesiculosus	Fucoidan and alginate	Antioxidative properties, prevent skin aging and cutaneous disorders	Je et al. 2009
F. vesiculosus	Micronlzed algae	Topical cosmetic compositions for treating or preventing cellulite	Al-Bader et al. 2012, 2013, Oriflame 2017
F. vesiculosus	Extract	Emollient, humectant, masking, oral care, skin conditioning	Jones 2015
F. vesiculosus	Micronlzed algae	Antiaging creams, anticellulite body masks, body scrubs, slimming body creams	Bielenda 2017, Dermika 2017, Thalgo 2017
Himanthalia elongata	Fatty acids and volatile compounds	Antioxidant and antimicrobial activity	Plaza et al. 2010b
Ishige foliacea	Phlorotannins: octaphlorethol A	Tyrosinase inhibition, whitening effect	Kim et al. 2013f
I. okamurae	Phlorotannins: diphlorethohydroxycarmalol	Antioxidant, UV protection	Heo et al. 2010b
Laminaria digitata	Carbohydrates (69%), minerals (20%), proteins (11%)	Lipolytic[5]	Gedouin et al. 2006, Codif 2017

L. digitata	Micronlzed algae	Antiacne creams, antiaging creams, anticellulite body lotions, peelings, slimming body masks moisturizing face creams	Bielenda 2017, Dermika 2017, Thalgo 2017
L. hyperborea	Extracts	Antiacne and antiaging face creams, body lotions, face masks, tonics, fluids, moisturizing, tonics	Thalgo 2017, La-Mer 2017
L. ochroleuca	Extracts	Antiacne creams, antiaging creams and serums, cleansing gels, day and night face creams, fluids, tonics, hair shampoos and conditioners, sun protection creams	Thalgo 2017, La-Mer 2017
Laminaria sp.	Laminarans	Anti-inflammatory, antioxidant, anti-inflammatory, and anticellulite agents	Stengel et al. 2011
Laminaria sp.	Alginates	Emulsion stabilizers, chelating agents, colloids, gelling, immunostimulating agents, moisturizing, protective colloids	Tonnesen and Karlsen 2002, Mafinowska 2011
Lessonia sp.	Alginates	Emulsion stabilizers, chelating agents, colloids, gelling, immunostimulating agents, moisturizing, protective colloids	Tonnesen and Karlsen 2002, Mafinowska 2011
Macrocystis sp.	Alginates	Emulsion stabilizers, chelating agents, colloids, gelling, immunostimulating agents, moisturizing, protective colloids	Tonnesen and Karlsen 2002, Mafinowska 2011
Padina boergesenii	Sulfated polysaccharides	Collagen formation and epidermal regeneration	Kordjazi et al. 2013
P. pavorica	Methanolic extract	Can kill fungi and bad bacteria, maintaining skin flora in state of balance	Saidani et al. 2012
P. pavorica	Extract	Keratinocytes differentiation, protein synthesis activation	Gutiérrez 1995, Texinfine 2017
P. tetrastromatica	Sulfated polysaccharides	Collagen formation and epidermal regeneration	Kordjazi et al. 2013
Pelvetia canaliculata	Ethanol extract: alginic acid, aminoacids, flavonoids, fucoidans, polyols	Antioxidant, collagen synthesis stimulation, proteoglycans synthesis stimulation. Antiobesity effects	Hupel et al. 2011, Jang and Choung 2013
Saccharina japonica (as *Laminara japonica*)	Carotenoids: Astaxanthin, β-carotene, fucoxanthin, lutein	Antioxidant, anticellulite, antiviral, anti-inflammatory, antiaging, antiphotoaging agents, elastase, tyrosinase inhibitors	Chizhov et al. 1999, Wijesinghea and Jeona 2012
S. japonica	Polysaccharides	Skin moisturization and protection	Wang et al. 2013c, 2013d
S. longicruris	Galactofucan (638 and 1529 kDa)	Synthesis of matrix metalloproteinase and collagen-I, fibroblasts growth rate	Rioux et al. 2013

Table 12.1 contd....

Table 12.1 contd....

Species	Extract/Compound	Cosmetics properties and/or products	References
S. sculpera (as *Kjellmaniella crassifolia*)	Fucoidan	Antiaging, antiwrinkle	Mizutani et al. 2010, TaKaRa 2017
Sargassum fusiforme (as *Hijikia fusiformis*)	Fucoxanthin	*In vivo* inducer of the Nrf2-ARE	Liu et al. 2011c
S. fusiforme (as *H. fusiformis*)	Phlorotannins: 4-hydroxyphenethyl alcohol	Tyrosinase inhibition, whitening effect	Jang et al. 2014
S. macrocarpum	Sargafuran	Skin care cosmetics to prevent or treat acne	Kamei et al. 2009
S. polycystum	Flavonoids, tannins, terpenoids, phenols, saponins	Antimelanogenisis or skin-whitening effect	Song et al. 2009, Chan et al. 2011
S. siliquastrum	Water (20°C) extract	Skin-whitening agent	Cha et al. 2011
Scytosiphon lomentaria	Proteins (amino acids)	Radical scavengers, antioxidant, chelating agents	Heo et al. 2003, Ahn et al. 2004, Freurence 2004
Silvetia babingtonii (as *Pelvetia wrightii*)	Polysaccharides	Anticellulite	Rozkin et al. 1991
Turbinaria conoides	Fucoidan and alginate	Antioxidative properties, prevent skin aging and cutaneous disorders	Je et al. 2009
Undaria pinnatifida	Sulfated polysaccharide (fucoidan)	Antioxidant, anticellulite, antiviral, anti-inflammatory, antiaging, antiphotoaging agents, elastase, tyrosinase inhibitors	Chizhov et al. 1999, Wijesinghea and Jeona 2012
Undaria pinnatifida	Extract	Skin moisturization and protection	Wang et al. 2013c, 2013d
Rhodophyta (red seaweed)			
Acanthophora muscoides	Sulfated polysaccharide (carrageenan)	Anticoagulant, antinociceptive and anti-inflammatory, gelling agents	Prud'homme van Reine and Trono Jr. 2001, Quindere et al. 2013, Gurgel-Rodrigues et al. 2016, Gurgel-Rodrigues et al. 2016b
A. nayadiformis (as *A. delilei*)	Proteins (amino acids)	Radical scavengers, antioxidant, chelating agents	Heo et al. 2003, Ahn et al. 2004, Freurence 2004
Chondria armata	Galactoglycerolipids	Antimicrobial activity (see also Chapters 7 and 8)	Al-Fadhli et al. 2006, Fabrowska et al. 2015

Chondrus crispus	Fatty acids: AA, ALA, DHA, EPA, GLA, LA, oleic acid, palmitic acid	Antiallergic, antiaging, anti-inflammatory, antiwrinlde, antioxidant, antiaging, antimicrobial, emollients, regenerating agents, used in eczema and psoriasis treatment	Sanghvi and Lo 2010, Stengel et al. 2011
C. crispus	Sulfated polysaccharide (carrageenan)	Gelling agents, protective colloids, thickeners	Stengel et al. 2011
C. crispus	Polysaccharides	Skin moisturization and protection	Wang et al. 2013c, 2013d
C. crispus	Extracts	Exfoliating gel, body mask, bath salts, moisturizing cream (components of the talassotherapy kit[2])	Lusalgae 2017
C. crispus	Extracts	Body lotions, face creams, fluids, make-up removers, shampoos, hair conditioners	Thalgo 2017, La-Mer 2017
Corallina officinalis	Sulfated polysaccharides	Antioxidant	Yang et al. 2011
C. pilulifera	Phlorotannins: eckols, fucols, fucophlorethols, fuhalols, phlorethols	Antiaging, antiphotoaging, anti-inflammatory, antiallergic, chelating agents, antioxidants, natural UV screens, histamine, tyrosinase, hyaluronidase inhibitors	Ryu et al. 2009, Stengel et al. 2011, Borowitzka 2013, Thomas and Kim 2013
Eucheuma serra	Lectins	Antiadhesive agents, antibacterial, anti-inflammatory, antiviral	Samarakoon and Jeon 2012
Furcellaria lumbricalis	Micronlzed algae	Topical cosmetic compositions for treating or preventing cellulite	Al-Bader et al. 2012, 2013, Oriflame 2017
Gelidium sp.	Agar	Emulsion stabilizers, gelling agents, thickeners	Iberagar 2010
Gracilaria sp.	Agar	Emulsion stabilizers, gelling agents, thickeners	Iberagar 2010
Grateloupia elliptica	Extract	Prevention of hair loss	Kang et al. 2012d
Kappaphycus alvarezii (as *Eucheuma cottonii*)	Ethanolic and aqueous extracts	Hair growth	Fard et al. 2011
Laurencia pacifica	Laurinterol	Can kill bad bacteria (*Staphylococcus aureus*), maintaining skin flora in state of balance	Fenical 1976
Laurencia sp.	Bromophenols	Antioxidant, antimicrobial, antithrombotic agents	Liu et al. 2011b
Palmaria palmata	MAAs	Anti-UV	Yuan et al. 2009
P. palmata	Proteins (amino acids)	Moisturizers, antioxidants, natural sunscreens	Houston 2005, Borowitzka 2013
Porphyra umbilicalis	MAAs	Anti-UVA	Carreto and Carigan 2001, Zhang et al. 2004

Table 12.1 contd....

Table 12.1 contd....

Species	Extract/Compound	Cosmetics properties and/or products	References
P. umbilicalis	Fatty acids: AA, ALA, DHA, EPA, GLA, LA, oleic acid, palmitic acid	Antiallergic, antiaging, anti-inflammatory, antiwrinlde, antioxidant, antiaging, antimicrobial, emollients, regenerating agents, used in eczema and psoriasis treatment	Sanghvi and Lo 2010, Stengel et al. 2011
P. umbilicalis	Proteins (amino acids)	Moisturizers, antioxidants, and natural sunscreens	Houston 2005, Borowitzka 2013
P. umbilicalis	Extract	Skin conditioning	Jones 2015
P. umbilicalis	Extract	Sunscreen formulation with red algae extract; photoprotective formulation with anti-aging properties	Mercurio et al. 2015
Portieria spp.	Phycobiliproteins: Allophycocyanin, phycoerythrin, phycocyanin	Antioxidant, anti-inflammatory, colorants, radical scavenging agents	Stengel et al. 2011, Borowitzka 2013
Pyropia dentata (as *Porphyra dentata*)	Phytosterols: brassicasterol, ergosterol, fucosterol, ergosterol, 3-sitosterol	Antiallergic, anti-inflammatory agents, antioxidants, radical scavengers	Stengel et al. 2011, Borowitzka 2013, Kazlowska et al. 2013
P. haitanensis (as *Porphyra haitanensis*)	Sulfated galactans	*In vivo* antioxidant activity	Zhang et al. 2004
Rhodella spp.*	Phycobiliproteins: Allophycocyanin, phycoerythrin, phycocyanin	Antioxidant, anti-inflammatory, colorants, radical scavenging agents	Stengel et al. 2011, Borowitzka 2013
Rhodomela confervoides	Methanolic extracts	Can kill fungi and bad bacteria, maintaining skin flora in state of balance	Saidani et al. 2012
Vertebrata lanosa (as *Polysiphonia lanosa*)	Extract	Skin moisturization and protection	Wang et al. 2013c, 2013d

*Marine microalgae, [1]Revertime™, [2]Sealgae™ (see Fig. 12.2), [3]Codiavelane®, [4]Algowhite, [5]Pheofiltrat compounds, sterols, vitamins, and other bioactive agents. Table 12.1 summarizes the main algal active constituents with properties that have direct relevance in cosmetics. Due to such large amounts of high-value chemicals, algae perform multidirectional action on skin, and are used in many kinds of cosmetic products (Mafinowska 2011, Stengel et al. 2011, Borowitzka 2013).

12.2.1 Industrial processes used in the elaboration of cosmetic products

Once harvested, the seaweeds are subject to several transformations for their incorporation into various cosmetics. It is essential that these processes do not alter the bioactive molecules, so that they can be fully operational in the final products. These techniques and the quality of the methods have evolved in the last years and there are many options and the selection of some over the others, depending essentially on the end to be achieved (obtaining the active principle, functional stability, and type of product to be manufactured) and the available technical resources. The commonly used techniques are:

Drying in air or in industrial ventilated ovens or chambers, for the removal of water from algae tissues (dehydration), is promoted by a controlled environment process where the temperature must not exceed 40°C to avoid the destruction of the active elements (for example, protein molecules, such as enzymes, which are thus not denatured) present in algae. This is, of all the methods mentioned here, not only the simplest, but the most cost-effective. On the other hand, it is the least direct, and often incorporates other methods as a process step.

After the drying process, the purification of it is needed; algae biomass is micronized using milling, bashing, grinding, cutting, crushing, and other techniques to reduce the particle size. This process demands the separation of the bioactive compounds from the algal biomass because of the physical disruption of their cells. Usually, there are different kinds of mills and homogenizers used to disrupt algal cells (Olaizola 2003). The micronization process significantly alters the surface area and, consequently, the functional properties of the algal particles. This is related to the structural changes in the particles that have functional groups (hydroxyl, carboxyl), which exhibit higher biological activity of micronized algae (Huang et al. 2010, Dyminiska et al. 2012). In addition, the penetration of active ingredients could occur into the deeper layers of the skin with the maximum possible fragmentation of the material. Finally, algal biomass may be applied in cosmetic products in the micronized form—the dried powder form (Fabrowska et al. 2015).

In the extraction by liquid phase, the compounds are extracted with different solvents, aqueous or organic (water, glycerin, ethyl alcohol, etc.) that allow the separation and isolation of the metabolites with different bioactive functions, depending on their chemical affinity with the solvent, in different phases, some of which are discarded successively for purification and concentration. Depending on the solvents used, the duration of the extraction, temperature, pH, and other conditions, different extraction efficiencies can be achieved. In case of water and other polar solvents, extracts rich in proteins, polysaccharides, and other water-soluble bioactives are obtained. On the contrary, if nonpolar solvents are used, extracts rich in fatty acids, phospholipids steroids, and other lipophilic compounds are obtained (Fabrowska et al. 2015, Pereira and Correia 2015).

As far as the novel extraction methods are concerned, accelerated solvent extraction (ASE), enzyme assisted extraction (EAE), microwave assisted extraction (MAE), supercritical fluid extraction (SFE), and ultrasound assisted extraction (UAE) (Herrero et al. 2010, Wang et al. 2010, Kim et al. 2013e, Rodrigues et al. 2015c) should be stated. The ASE, also known as enhanced solvent extraction (ESE), pressurized fluid extraction (PFE), or high pressure solvent extraction (HPSE), uses conventional solvents, such as water, ethanol, hexane, petroleum ether, but the extraction process is conducted in extraction cells using high temperature and pressure. With increasing temperature, the solvent pressure also increases in the closed extraction cell. So, the solvent may remain in the liquid state at temperatures above its normal boiling point because of the high pressure. The combination of high pressure and temperature in ASE technique enables faster extraction process. Moreover, this method is more efficient than conventional ones, because the higher temperature increases solubility and the rate of mass transfer, and decreases the viscosity and surface tension of solvents. Moreover, the ASE method requires small amounts of solvents, which makes this technique more attractive than, for example, Soxhlet extraction (Azmir et al. 2013). ASE is widely used as an efficient, fast, and environment-friendly method for obtaining bioactive agents from algae. Using this extraction method, carotenoids, fatty acids, lipids, and different types of antioxidants can be obtained from algae (Breithaupt 2004, Chen et al. 2013b, Rodrigues et al. 2015c).

12.3 Skin: Anti-wrinkling and Whitening

As the biggest organ in the human body, the skin has an important role in many physical functions (Wang et al. 2013c, Wang et al. 2013d). Skin is composed of the epidermis, dermis, and hypodermis. Epidermis is subdivided into five separate strata: basal, spinous, granular, lucid, and corneum.

In the epidermis, the predominant keratinocyte cells repair skin damage, while melanocytes contain melanin, thus determining skin color and protecting the skin from UV light (Brenner and Hearing 2007). Langerhans cells, a type of dendritic cell, provide a degree of immunity (Wollenberg et al. 1996), as they take up microbial antigens in the skin and transform microbial antigens into antigen presenting cells by interacting with T cells. Sebaceous glands produce an oily substance known as sebum that lubricates the skin, although the occlusion and infection of these can trigger acne (Zouboulis 2004). Receptors in the skin detect various environmental stimuli and respond accordingly, with mechanoreceptors detecting sensations and thermoreceptors detecting heat. These receptors can cause sweat glands to produce sweat, thus maintaining temperature—homeostasis, as well as getting rid of waste (Denda et al. 2007). Insulation is also provided in the subcutaneous layer of the skin, where fat is stored. Skin absorbs not only oxygen and water, but also certain drugs, such as topical steroids. In addition, skin is rich in 7-dehydrocholesterol, and when it is exposed to UV light, this substance is converted into vitamin D (cholecalciferol) (see also Chapter 2), which is otherwise obtained by dietary means, such as by ingesting dairy products (Chen et al. 2007).

The immune function of the skin prevents damage from UV light using pigmentation (Wickett and Visscher 2006). Under the epidermis is the dermis, which is mainly composed of connective tissues, including blood vessels, sweat glands, nerves, fibroblasts, collagen, and elastin. Collagen and elastin, which are cross-linked, provide support for the skin. Hyaluronic acid (HA) is also a major component of the dermis, where it is involved in tissue repair. HA is of fundamental importance in water retention, and can absorb water about 1000 times its own volume. Nevertheless, collagen and HA break down with aging, causing wrinkles to appear and the skin to lose firmness. The third layer, the hypodermis, is composed mainly of fat and a layer of loose connective tissue. It provides insulation to the body, stores energy, and mechanically allows the attenuation and dispersion of externally applied pressure (Benbow 2009).

12.3.1 Skin aging

Skin aging can be defined as a complex biological process that is a catastrophic outcome of either intrinsic or genetically programmed aging that occurs with time and exposure of skin to environmental factors such as UV radiation. However, the aging process reduces the skin thickness, elasticity of the skin, and curling of elastic fibers in the skin, and gives rise to wrinkles in the skin (Tsukahara et al. 2001, McCullough et al. 2006, Ryu et al. 2014). Wrinkles, especially developed in forehead, corners of the eyes, and cheeks are considered to be one of the catastrophic cosmetic problems, as wrinkles can provide some indication of the age of a person (Nagashima et al. 1999, Magnenat-Thalmann et al. 2002). The increase in the aging population because of increasing longevity, declining fertility rates, and the psychosocial impact of skin aging all around the globe has created a demand and good market for cosmeceutical products which contain anti-wrinkling properties (Anderson and Hussey 2000, McCullough et al. 2006). In contrast, compounds capable of increasing matrix metalloproteinase (MMP) inhibitory activities, hyaluronidase inhibitory activity, expression of collagen, and elastase inhibitory activity, may have potential to be used as active ingredients in new anti-wrinkle cosmetic products (Tsuji et al. 2001, Kim et al. 2006, Kim et al. 2008c, Ferreres et al. 2012, Thomas and Kim 2013).

Three genera of seaweeds are routinely used in cosmetics—*Laminaria*, *Fucus*, and *Chondrus*—for their ability to nourish and rehydrate the skin (De Roeck-Holtzhauer 1991). Topical application of fucoidan has been shown to have antiaging activity by increasing the moisture and elasticity of the cells (Fujimura et al. 2002). Fucoidan is known to stimulate the production of HGF—Heparin Growth Factor, which promotes growth in a range of cells and tissues, and this is exploited commercially by the Takara-Bio Company in Japan (TaKaRa 2017).

12.3.2 Skin whitening

Asian women generally prefer white skin to brown skin. In Asia, therefore, skin whitening products appear to be the largest and continually growing segment in the skin care market (Boonme et al. 2009). The melanin produced in mammalian skin plays an important role in determining skin color, protecting human skin from the harmful effects of ultraviolet (UV) radiation, and scavenging toxic drugs and chemicals. Besides, melanin has taken much interest among scientific communities, as there's an inverse relation of melanin with the development of melanoma skin cancers (Gilchrest and Eller 1999, Taylor 2002). In melanin biosynthesis process, first tyrosinase acts on *L*-tyrosin to produce 3–4-dihydroxy-phenylalanine (*L*-dopa), and then Dopa oxidase acts on *L*-dopa to produce melanin (Martínez-Esparza et al. 1998, Solano et al. 2006). In contrast, whitening effects can be achieved in cosmeceutical products by incorporating tyrosinase inhibitors, which are responsible for suppressing melanin bio-synthesis in the skin (Park 2006).

12.3.3 Wound healing

Kappaphycus alvarezii (as *Eucheuma cottonii*) (a red alga) ethanolic extract accelerated wound healing and hair growth significantly in mammals (Fard et al. 2011, Ha and Park 2013). A silver-loaded seaweed-based cellulosic fiber greatly improved epidermal skin physiology, barrier function (trans-epidermal water loss), *Stratum corneum* hydration, and surface pH in atopic dermatitis and eczema patients over time in a controlled, randomized single blinded human study (Zikeli 2006, Fluhr et al. 2010).

Fucoidans (see Fig. 2.1, Chapter 2) stimulate fibroblast and epithelial cell growth and increase TGF-β1 secretion that accelerate wound healing, and modulate growth factor dependent pathways in tissue repairs (Shakespeare 2001, Matou et al. 2002, Leoni et al. 2015). Fucoidan/chitosan hydrogels effectively contract and heal dermal burns. Various soluble seaweed-based dressings prevent interference with the healing granulation tissues. They do not cause pain, hence are advantageous for venous, neuropathic, and ischemic ulcers (Sezer et al. 2008, Sasikala et al. 2013).

Alginic acid (see Fig. 3.2, Chapter 3) are highly viscous bio-absorbable guluronic and mannuronic acid polysaccharides from brown seaweed, that can inhibit scar formation by functioning as physical barriers to invading fibroblasts and help accelerate wound healing. Alginates are used in tissue engineering and clinical applications, especially with the addition of collagen (Dae et al. 2007, Gong et al. 2007).

12.4 Thalassotherapy

Thalassotherapy is a term that emerged at the beginning of the 18th century, originating in the Greek "thalasso" (sea) and "therapia" (cure). The pioneers of its use were the French in the middle of the last century and, thanks to the results obtained and published on the well-being and quality of life obtained by those who used them, spread rapidly throughout the world. Historically, the Chinese, Egyptians, and Romans already knew some of their properties, having produced several panaceas from algae (Pereira 2010b, Pereira and Correia 2015).

Many treatments have received the thalassotherapy label; all do not deserve it. Due to its recent success, the term has been usurped to indicate so-called "spa" treatments. It would be best to limit its use to seaside treatments calling upon climate factors, seawater bathing, impact of waves, marine (e.g., algal) products poultices, seawater medication; the approach is very like thermalism, but is geared to the role of the sea and seawater (Charlier and Chaineux 2009).

The reality is that the components of the sea (be they minerals or organic ones) have always been part of our lives. Water and, in particular, seawater are considered essential to the maintenance of life on our planet, since it contains all substances or elements necessary and conducive to the development of the living being (minerals, catalysts, vitamins, amino acids, etc.). In therapeutic terms, seawater has a recognized action in the treatment of diseases that afflict skin tissues, whether cutaneous (skin) and/or mucous membranes, such as eczema, dermatosis, psoriasis, nasopharyngeal inflammation and some forms of conjunctivitis, vaginitis and other infections and affections of the external genital organs.

Some benefits associated with Thalassotherapy are:

- Increased skin permeability;
- their ionic characteristics, when interacting with the skin, facilitate the penetration of the cosmetic compounds during or after the immediate to these treatments; regulates the organic functions through the neuro-endocrine system, as enzymatic cofactor;
- relaxes tight and tight muscles, giving a rested appearance to the skin;
- tends to normalize seborrheic secretion;
- reduces hyperhidrosis (excessive perspiration, including feet);
- promotes organic remineralization at the cutaneous tissue level;
- activates the cutaneous metabolism and locally stimulates the blood circulation, indirectly promoting the oxygenation and nutrition of these tissues.

12.4.1 Therapeutic uses of thalassotherapy

Thalassotherapy, often assisted by compounds of algal origin, is thus indicated for noninvasive treatments of various affections:

- rheumatic problems, by reducing pain and interrupting the evolution of other disorders associated with this clinical stage. In the case of arthritis and arthrosis that are not very advanced, the action of seawater can interrupt its progress in the long term, especially if the therapy is accompanied by adequate food and better life habits;
- healthy bones and joints. Seaweeds are naturally high in minerals that are essential for healthy bones and joints. They are also high in vitamins and antioxidants—important components of cosmetic preparations (Luop 2001). Calcium can comprise up to 7% of dry weight (Pengzhan et al. 2003) and is also found in association with polysaccharides, such as alginate and carrageenans. The calcareous red seaweeds such as *Lithothamnion* spp. can contain as high 35% calcium. An Irish Product Aquamin™ is a multi-mineral preparation from *Lithothamnion* spp. It has been shown to be high in bioavailable calcium, more so than the standard calcium carbonate that is used in supplements, and is effective against mobilization of bone calcium through parathyroid hormonal activity (Aqumin 2017). Similar activity has been observed with a mix of oyster shell and the brown seaweed *Sargassum fusiforme* (as *Cystophyllum fusiforme*) (Fujita 2005);
- problems of blood circulation (especially in the lower limbs, more subject to premature wear thanks to our bipedal posture), in which sea water, as a toning and anti-inflammatory, stimulates circulation and locally eliminates the sensation of fatigue;
- skin diseases in which the disinfecting power of iodine dissolved in sea water and extracted from some algae destroys organisms that promote skin diseases, benefiting mainly patients suffering from acnes, boils, and other skin conditions caused or aggravated by Bacterial action;
- stress and fatigue, in which aquatic gymnastics in their own tanks, with relaxation and balanced feeding, help the release of stress, and through exercise, promote the release of toxins and facilitate skin absorption and consequent remineralization. In addition, it can also contribute to the weight loss process and actively collaborates in the reduction of cellulite, when accompanied by a balanced diet and physical activity;

- respiratory affections, as is known, the action of dissolved marine salts in the sterilized sea water, aspirated through the nasal passages, decongest the nasal mucosa, and relieve the symptoms of allergic rhinitis and sinusitis.

12.4.2 Sea bath and seaweed at home—some practical advice

To enjoy the benefits of thalassotherapy at home, place a handful of dried seaweed, for example *Fucus* (brown alga), inside a bath glove and immerse it in a hot tub (35–37°C), along with a handful of bath salts. Massage the body with the glove (Fig. 12.1), as this will favor the penetration of the active substances through the skin. It is important to keep the temperature constant, so it is recommended to periodically add hot water. The treatment should be done for 5 or 6 weeks (2 to 3 baths per week), with an average duration of 10 minutes each, so that their effects can be felt.

For the penetration of mineral substances into the body to take place immediately after the first baths, it is advisable to wrap in a towel (without adding creams or other substances) and rest for a few minutes. In this way, while facilitating the absorption process, the heart rate and body temperature are restored (Pereira 2010b).

12.4.3 Algae in clinical thalassotherapy

A rather wide range of marine algae enter in the packs used in thalassotherapy, and algal flours or algal salts are sold for use in home bath therapy. Poultices of *Fucus*, *Laminaria*, *Ulva*, and *Ascophyllum* (with or without *Lithothamnium* powder) are heated to 40ºC, or even 50ºC and applied, at thalassotherapy clinics, to limbs to relieve pains due to rheumatism and arthritis, but also to shoulders, shoulder blades, and lumbar regions to treat chronic back pains. Of the more than 20,000 identified species of algae, some 60 are in use in pharmacy, food, and in cosmetology. Most commonly used in these areas are *Porphyra, Eucheuma* (red algae), *Laminaria*, and *Undaria* (brown algae) (Charlier 2002, Charlier and Chaineux 2009).

Phymatolithon calcareum (as *Lithothamnium calcareum*) is a calcareous marine alga whose thallus is used in the treatment of decalcification, osteoporosis, painful joints, chronic tiredness, painful stages of rheumatism, gingivitis, stomach pains, and in anti-acid cures. *Lithothamnium* and *Fucus* enter frequently

Figure 12.1 To enjoy the thalassotherapy benefits at home, place a handful of dried seaweed, *Fucus* for example, inside a glove bath and immerse it in a bathtub full of hot water (35/37°C), with a handful of bath salts. Massage the body with the glove; it will thereby promote penetration of active substances through the skin (adapted from Pereira 2010b).

Figure 12.2 Thalassotherapy Sealgae™ kit made by the Lusalgae® company.

in the composition of sacks used in thalassotherapy. Among other marine macroalgae, rockweed (*Fucus vesiculosus*) has also proved effective against excess weight for individuals with a normal appetite. Rockweeds have furthermore an anti-seborrheic effect on greasy hair (Charlier 1975, 2002).

12.5 Bioactive Compounds used in Cosmetics

Due to the chemical diversity and unique properties, algae have been the subject of many studies and are widely used in the cosmetics industry. Algae contain different biochemical compounds including polysaccharides, proteins, lipids, phenolic compounds, pigments, vitamins, and other bioactives, as well as macro and microelements (Stengel et al. 2011, Samarakoon and Jeon 2012, Borowitzka 2013, Christaki et al. 2013).

Seaweeds produce both primary metabolites, which are directly involved in normal growth, development, or reproduction conditions to perform physiological functions, and secondary metabolites, which are performed under different stress conditions, such as UV radiation exposure, salinity, temperature changes, or environmental poisons. Primary metabolites in algae include polysaccharides, proteins, amino acids, and fatty acids. Secondary metabolites produced in seaweed tissues are pigments, phenolic compounds, sterols, vitamins, and other bioactive agents (see Table 12.1) (Carvalho and Pereira 2015).

12.5.1 Lipids

The word "lipid" comes from the Greek word *lipos*, meaning fat and greasy to the touch. Lipids are a diverse group of organic compounds found in plants, animals, and micro-organisms. They comprise one of the three large classes of foods and, with proteins and carbohydrates, are components of all living cells (Alvarez and Rodríguez 2000). Lipids represent a group of chemical compounds that have a lipophilic character as a common feature Lipids present in algae include fatty acids, glyco- and phospholipids, sterols, and others (Sanghvi and Lo 2010, Stengel et al. 2011, Borowitzka 2013, Kazlowska et al. 2013). Currently, most of the studies are focused on fatty acids, derived from macro- and microalgae, with direct applications in the production of biodiesel (Singh and Gu 2010, Pereira 2017). However, algal fatty acids and other lipophilic compounds have anti-inflammatory, antiallergic, and antioxidant activities (Mafinowska 2011,

Stengel et al. 2011). In addition, they can act as emollients—the softening and smoothing compounds, which protect the skin against the loss of water (Mafinowska 2011).

12.5.2 Phenolic compounds

Phenolic compounds are one of the most interesting water soluble plant substances naturally present in cell vacuole, which carry out numerous biological activities (Harborne 1998, Manach et al. 2005). Chemical structures of phenols can be defined as an aromatic ring of carbon, which bear one or more hydroxyl substituents.

Within the last few decades, thousands of phenolic compounds have been isolated and identified from the kingdom Plantae, including both terrestrial and marine plants with their structures (Tückmantel et al. 1999, Urquiaga and Leighton 2000). However, there are marked differences that exist in structures of terrestrial polyphenol compounds compared to marine polyphenols (phlorotannins), as flavonoids and Gallic acid are the building blocks of terrestrial polyphenols, but phlorotannins (reported only in brown algae) are chains of 1,3,5-trihydroxybenzene formed in the acetate-malonate pathway with a wide range of molecular weights from 126 to 65,000 Da (Liu and Gu 2012e). The marine brown seaweeds play a significant role of the bioactive compounds production, and they are the only organisms on earth producing phlorotannins, which are polyphenols exhibiting important biological activities (Thomas and Kim 2001, Li et al. 2011).

Phlorotannins are phloroglucinol-based compounds, biosynthesized by the acetate-malonate pathway, highly hydrophilic, and with a wide range of molecular sizes (Target and Arnold 2001). Phlorotannins are reported as anti-allergic, whitening, anti-wrinkling, and potential anti-aging agents, owing to their strong inhibitory effect on the activation of hyaluronidase (HAase) (Table 12.1) (Ferreres et al. 2012, Sanjeewa et al. 2016).

12.5.3 Pigments

Colorants from natural sources are gaining importance mainly due to health and environmental issues. Algae contain a wide range of photosynthetic pigments. Three major classes of photosynthetic pigments are chlorophylls, carotenoids (carotenes and xanthophylls), and phycobilins. Phycocyanin and phycoerythrin belong to the major class of phycobilins photosynthetic pigment, while fucoxanthin and peridinin belong to carotenoid group of photosynthetic pigment. The table below elucidates the different types of algae and the major pigment they contain (Pereira 2015).

Each taxonomic group of algae contain different types of pigments: the green, brown, red, and blue-green algae. The green algae (Chlorophyta) include predominantly chlorophyll a and b, the brown algae (Phaeophyceae) are characterized by a high content of fucoxanthin, and the red (Rhodophyta) and the blue-green (Cyanobacteria) algae by a high content of phycoerythrin and phycocyanin. Algal pigments are divided into chlorophylls, carotenoids, xanthophylls, and phycobilins. Their biological role concerns the photoprotection of the photosynthetic apparatus from excess light dissipation by scavenging ROS (Pereira 2009, 2015).

Generally, chlorophylls occur in algae in the highest concentrations among other pigments. These green photosynthetic pigments with antioxidant and antimutagenic properties are widely used in cosmetics, pharmaceutical, and food industries (Hosikian et al. 2010). Chlorophylls can absorb light in the red and blue regions and thus, emit a green color. Therefore, these compounds are used as natural coloring agents (Spears 1988). Moreover, chlorophylls have antibacterial and deodorizing properties (Goldberg 1943). The strong antioxidant activity and their ability to stimulate tissue growth qualifies these pigments as valuable raw materials in cosmetics (Horwitz 1951, Lanfer-Marquez et al. 2005).

Currently, most of the studies involving algal pigments are focused on carotenoids as commercially relevant bioactive compounds. Carotenoids are lipophilic isoprenoid molecules, which are classified as carotenes (β-carotene, a-carotene, lycopene) and xanthophylls (lutein, astaxanthin, zeaxanthin, violaxanthin, loraxanthin, fucoxanthin) (Christaki et al. 2013). They are widely used as natural colorants and strong antioxidants with additional anti-inflammatory, anticancer, and radical scavenging properties (Stengel et al. 2011, Borowitzka 2013, Christaki et al. 2013). β-carotene was the first bioactive agent commercially produced among other carotenoids (Vílchez et al. 2011, Christaki et al. 2013, Mata-Gómez et al. 2014).

12.5.4 Polysaccharides

Polysaccharides (see also Chapters 2, 3, and Table 12.1) are the largest group of all the active metabolites occurring in algae (about 60% of all bioactive) (Pereira 2016d). These compounds are composed from various building blocks and form long carbohydrate molecules of monosaccharide units joined together by glyosidic bonds. They are hydrophilic, water soluble, and they have a rather regular structure. Algal polysaccharides are structural components of cell walls and act as energy storage units. There are many kinds of biologically active polysaccharides in algae tissues. Generally, these compounds are moisturizing and antioxidative ingredients in cosmetics. They are also widely used as stabilizers in emulsions or gelling agents (Stengel et al. 2011, Pereira 2016d).

Polysaccharides, including those of algal origin, have entered, or undoubtedly will enter, production of vaccines, drug delivery, anticancer, antithrombogenic, antiadhesive drugs, and diagnostics (Table 12.1) (Colwell 1985).

Agars (see Fig. 3.1, Chapter 3) are also well-known polysaccharides extracted from red algae, such as *Gracilaria* and *Gelidium* species. They can create stiff and brittle gels, and they are applied as thickeners and emulsion stabilizers (Pereira and Ribeiro-Claro 2015). Compared to polysaccharides produced by microalgae, they are less popular in the cosmetics market than macroalgal polysaccharides (Borowitzka 2013). However, many cyanobacteria (De Philippis et al. 2001) as well as unicellular red algae such as *Porphyridium* and *Rhodella* (Arad and Levy-Ontman 2010) contain these compounds.

The main polysaccharide found in the brown seaweeds (Phaeophyceae) is alginate (see Fig. 3.2, Chapter 3), a linear copolymer of mannuronic (M) and guluronic acid (G). The extraction process is based on the conversion of an insoluble mixture of alginic acid salts of the cell wall in a soluble salt (alginate) which is appropriate for the water extraction (Lahaye 2001, Pereira 2013b). This polysaccharide is derived from several genera of brown algae (e.g., mixed Fucales and Laminariales) that are utilized as raw materials by commercial alginate producers (Pereira 2013b). Because of their chelating properties, alginates are widely used as gelling agents in cosmetics and as thickeners, protective colloids, or emulsion stabilizers (Mafinowska 2011). With their water-binding capacity, they moisturize the skin and have a unique immunostimulating activity (Table 12.1) (Fabrowska et al. 2015).

Carrageenans (see Fig. 3.3, Chapter 3), extracted from several algae, i.e., *Chondrus crispus*, *Mastocarpus stellatus*, *Kappaphycus alvarezii*, *Eucheuma denticulatum*, Gigartinaceae and Phyllophoraceae, enter cough medicines, toothpastes, lotions, sun ray filterers, shaving creams, shampoos, hair conditioners, and deodorants. More than 20% of carrageenan production is used in pharmacy and cosmetology. They are said to be invaluable for the manufacture of wash-removable creams and ointments. Excipients of algal origin are used in vanishing creams—the rapid evaporation of the emulsion's aqueous phase on the skin leaves a thin protective medicated oily microfilm. Dried and pulverized *Lithothamnium* and *Phymatolithon* are used to make absorbent face and/or beauty masks (Charlier and Chaineux 2009, Pereira 2010b).

One of the best studied fucans from brown algae is fucoidan, which was first isolated by Kylin in 1913. The fucoidan (see Fig. 2.1, Chapter 2) from *Fucus vesiculosus* has been available commercially for decades (Sigma-Aldrich Chemical Company, St. Louis, MO, USA). Early work on its structure showed that it contained primarily (1→2) linked 4-O-sulfated fucopyranose residues. However, 3-linked fucose with 4-sulfated groups were subsequently reported to be present on some of the fucose residues (Pereira et al. 2013b). Moon et al. (2011) found that fucoidan treatment increased type I procollagen and inhibited UVB (ultraviolet B radiation)-induced matrix metalloproteinase expression. Consequently, it is suggested that fucoidans may be potentially used as therapeutic agents to prevent skin photoaging. Some studies showed that fucoidans can minimize human leukocyte elastase activity and thus, protect the elastic fibers of the skin (Senni et al. 2006). Fucoidans also act as tyrosinase inhibitors and consequently, may reduce skin pigmentation while used in skin whitening products (Table 12.1) (Wijesinghea and Jeona 2012, Thomas and Kim 2013).

Laminaran (see Fig. 2.2, Chapter 2) is a small glucan present in either soluble or insoluble form. The first form is characterized by complete solubility in cold water, while the other is only soluble in hot water (Kylin 1913, Chevolot et al. 2001). This polysaccharide is composed of D-glucose with β-(1,3) linkages, with β-(1,6) intra-chain branching (Pereira and Ribeiro-Claro 2015). Laminarans are derived mainly from brown seaweed *Laminaria*, which are also another biologically active polysaccharide. These compounds are reported to have antioxidant, anticoagulant, anti-inflammatory, antiviral, and antitumoral properties (Stengel et al. 2011). As far as the application of laminarans is concerned, they are generally used in anticellulite cosmetics products (Pereira et al. 2013b, Fabrowska et al. 2015).

Ulvan (see Fig. 2.4, Chapter 2) represents 8–29% of the algae dry weight and is produced by species belonging to the phylum Chlorophyta (green algae), mostly belonging to the class Ulvophyceae (Robic et al. 2009). It is mainly made up of disaccharide repeating sequences composed of sulfated rhamnose and glucuronic acid, iduronic acid, or xylose, and have gelling properties (Pereira and Ribeiro-Claro 2015). The mechanism of gel formation by ulvans is complex and involves the formation of spherically shaped structures of ulvans in the presence of boric acid and calcium ions (Robic and Lahaye 2007, Robic et al. 2009). Apart from their ability to create gels, ulvans have moisturizing, protective, and antioxidative properties (Fabrowska et al. 2015). The unique chemical and physicochemical properties of ulvan make this family of polysaccharides attractive candidates for novel functional and biologically active polymers for the food/feed, pharmaceutical, chemical aquaculture, and agriculture domains (Table 12.1) (Lahaye and Robic 2007).

12.5.5 Proteins and amino acids

Proteins are present in algae in different forms, for example, as simple or conjugated proteins (conjugated proteins are composed of simple proteins bounded to non-proteinaceous substance). Moreover, algae may contain protein derivatives, such as enzymes or peptides, as well as free amino acids (Stengel et al. 2011). Proteins and their derivatives are characterized by their diverse structure, their cellular location, and their functions. Most of these components have antioxidant, antiaging, anticancer, anti-inflammatory, and protective activities. Furthermore, algal proteins are applied as moisturizing agents on skin and hair (Samarakoon and Jeon 2012). Therefore, proteins may be successfully used in functional cosmetics and cosmeceuticals (Sekar and Chandramohan 2008).

Amino acids generally act as moisturizing agents in cosmetics products because many of them are constituents of the natural moisturizing factor (NMF) in human skin (Fabrowska et al. 2015). Algae may contain both endogenous amino acids, such as alanine, serine, proline, and exogenous amino acids, for example, histidine, tyrosine, tryptophan (Stengel et al. 2011, Samarakoon and Jeon 2012). Some algae species are good sources of the essential amino acids; for example, *Ulva australis* contains histidine and taurine, which have antioxidative and antihypertensive properties (Zhang et al. 2004, Houston 2005). Other studies showed that the red alga *Palmaria palmata* (dulse) and the brown alga *Himanthalia elongata* (sea spaghetti) contain high concentrations of glutamic acid, serine, and alanine (Galland-Irmouli et al. 1999, Pereira 2016).

Mycosporine-like amino acids (MAAs) are a family of secondary metabolites whose production is directly or indirectly related to the absorption of solar energy, and which protect marine organisms exposed to high UV radiation. MAAs have been detected in diverse organisms and especially in red macroalgae (Rhodophyta)—*Chondrus crispus*, *Palmaria palmata*, *Gelidium* spp., *Porphyra/Pyropia* spp., *Gracilaria cornea*, *Asparagopsis armata*, *Solieria chordalis*, *Grateloupia lanceola*, and *Curdiea racovitzae* (Bedoux et al. 2014). The content of MAAs is higher in summer and at a moderate depth (0–1 m) (Reef et al. 2009). MAAs may be potentially used in cosmetics and toiletries as UV protectors and activators of cell proliferation (Bedoux et al. 2014).

12.5.6 Vitamins and minerals

Algae are a source of different vitamins (see also Chapter 2), such as vitamin C (ascorbic acid), vitamin A (β-carotene and other carotenoids), vitamin E (α-tocopherol), and B vitamins (B_1, B_2, B_3, B_5, B_6, and B_{12}) (Pereira 2011, Fabrowska et al. 2015). Also, thanks to the freely permeable algal cell walls to low molecular weight constituents, such as water, ions, and gases, algae may include macro- and microelements ions (Wang and Chen 2009). Depending on their natural habitat, algae contain various kinds of macroelements, such as Ca, Na, K, Cl, and microelements, such as Zn, Mg, Cu, I, Fe, Mn, which are essential for the proper functioning of the skin (Wang and Chen 2009, Pereira 2011, Fabrowska et al. 2015). To sum up, algae represent a very attractive cosmetic raw material because of their natural origin and rich bioactive phytochemicals with multidirectional effects on skin.

References

Abatis, D., C. Vagias, D. Galanakis, J.N. Norris, D. Moreau, C. Roussakis and V. Roussis. 2005. Atomarianones A and B: two cytotoxic meroditerpenes from the brown alga *Taonia atomaria*. Tetrahedron Letters 46(49): 8525–8529.

Abbott, I.A. and G.J. Hollenberg. 1976. Marine Algae of California. Stanford University Press, Redwood City, CA, 844 pp.

Abbott, I.A. 1999. Marine red algae of the Hawaiian Islands. Bishop Museum Press, Honolulu, Hawai'I, 465 pp.

Abd El-Baky, H.H., F.K. El Baz and G.S. El Baroty. 2008. Evaluation of marine alga *Ulva lactuca* as a source of natural preservative ingredient. American-Eurasian Journal of Agricultural & Environmental Sciences 3(3): 434–444.

Abdel-Azeem, A. 2017. *Botryotrichum piluliferum*. Atlas of Living Australia. Avaible online at: http://bie.ala.org.au/species/23f93440-e8cf-4a3e-85ff-dde67c4adf71.

Abdel-Fattah, A.F. and M.M. Hussein. 1973. Isolation of water insoluble laminaran-like polysaccharide from *Sargassum linifolium*. Qualitas Plantarum-Plant Foods for Human Nutrition 22(2): 181–187.

Abdel-Fattah, A.F., M.M. Hussein and H.M. Salem. 1974. Studies of the purification and some properties of sargassan, a sulphated heteropolysaccharide from *Sargassum linifolium*. Carbohydrate Research 33: 9–17.

Abdel-Fattah, A.F., M.M.-D. Hussein and S.T. Fouad. 1978. Carbohydrates of the brown seaweed *Dictyota dichotoma*. Phytochemistry 17: 741–743.

Abd-Ellatef, G.E.F., O.M. Ahmed, E.S. Abdel-Reheim and A.H.Z. Abdel-Hamid. 2017. *Ulva lactuca* polysaccharides prevent Wistar rat breast carcinogenesis through the augmentation of apoptosis, enhancement of antioxidant defense system, and suppression of inflammation. Breast Cancer-Targets and Therapy 9: 67–83.

Abdelfat, Af. and M.M. Hussein. 1973. Isolation of water insoluble Laminaran-like polysaccharide from *Sargassum linifolium*. Qualitas Plantarum-Plant Foods for Human Nutrition 22: 181–187.

Abdullah, N.S., S. Muhamad, I.C. Omar and H. Abdullah. 2013. Determination of antioxidant and cytotoxic activities of red seaweed, (*Gracilaria manilaensis*) against different cancer cell lines. Journal of Food Science and Engineering 3: 616–624.

Abid, M., M.J. Zaki and M.A. Maqbool. 1993. Use of the brown alga *Stoechospermum marginatum* for the control of root-knot nematode in brinjal. Marine Research 2: 39–44.

Abirami, R.G. and S. Kowsalya. 2011. Nutrient and Nutraceutical Potentials of Seaweed Biomass *Ulva lactuca* and *Kappaphycus alvarezii*. Journal of Agricultural Science and Technology 5(1): 109–115.

Abou-Elnaga, Z.S., W.M. Alarif and S.S. Al-lihaibi. 2011. New larvicidal acetogenin from the red alga *Laurencia papillosa*. CLEAN-Soil Air Water 39(8): 787–794.

Abou Zeid, A.H., E.A. Aboutabl, A.A. Sleem and H.M. El-Rafie. 2014. Water soluble polysaccharides extracted *from Pterocladia capillacea* and *Dictyopteris membranacea* and their biological activities. Carbohydr. Polym. 113: 62–66.

Abourriche, A., M. Charrouf, M. Berrada, A. Bennamara, N. Chaib and C. Francisco. 1999. Antimicrobial activities and cytotoxicity of the brown alga *Cystoseira tamariscifolia*. Fitoterapia 70: 611–614.

Abrantes, J.L., J. Barbosa, D. Cavalcanti, R.C. Pereira, C.L. Frederico Fontes, V.L. Teixeira, T.L. Moreno Souza and I.C. Paixao. 2010. The effects of the diterpenes isolated from the Brazilian brown algae *Dictyota pfaffii* and *Dictyota menstrualis* against the *Herpes simplex* type-1 replicative cycle. Planta Medica 76: 339–344.

Abu, R., Z. Jiang, M. Ueno, S. Isaka, S. Nakazono, T. Okimura, K. Cho, K. Yamaguchi, D. Kim and T. Oda. 2015. Anti-metastatic effects of the sulfated polysaccharide ascophyllan ssolated from *Ascophyllum nodosum* on B16 melanoma. Biochemical and Biophysical Research Communications 458(4): 727–732.

Achary, A., K. Muthalagu and M.S. Guru. 2014. Identification of phytochemicals from *Sargassum wightii* against *Aedes aegypti*. International Journal of Pharmaceutical Sciences Review and Research 29(1): 314–319.

Adachi, M. and S. Watanabe. 2007. Evaluation of combined deactivators-supplemented agar medium (CDSAM) for recovery of dermatophytes from patients with tinea pedis. Medical Mycology 45: 347–349.

Adaikalaraj, G., R.D. Patric, M. Johnson, N. Janakiraman and A. Babu. 2012. Antibacterial potential of selected red seaweeds from Manapad coastal areas, Thoothukudi, Tamil Nadu, India. Asian Pacific Journal of Tropical Biomedicine 2(2): S1077–S80.

Adams, S.S. and H.M. Thorpe. 1957. The anticoagulant activity and toxicity of laminarin sulphate K. Journal of Pharmacy and Pharmacology 9: 459–463.

Adekunle, O.K. and A. Akinlua. 2007. Nematicidal effects of *Leucaena leucocephala* and *Gliricidia sepium* extracts on *Meloidogyne incognita* infecting okra. Journal of Agricultural Sciences 52: 53–63.

Adelson, E., J.J. Rheingold, O. Parker, M. Steiner and J.C. Kirby. 1963. The survival of factor VIII (antihemophilic globulin) and factor IX (plasma thromboplastin component) in normal humans. Journal of Clinical Investigation 42: 1040–1047.

Adhikari, U., C.G. Mateu, K. Chattopadhyay, C.A. Pujol, E.B. Damonte and B. Ray. 2006. Structure and antiviral activity of sulfated fucans from *Stoechospermum marginatum*. Phytochemistry 67: 2474–2482.

Adwan, K. and N. Abu-Hasan. 1998. Gentamicin resistance in clinical strains of Enterobacteriaceae associated with reduced gentamicin uptake. Folia Microbiol 43(4): 438–440.

Afolayan, A.F., J.J. Bolton, C.A. Lategan, P.J. Smith and D.R. Beukes. 2008. Fucoxanthin, tetraprenylated toluquinone and toluhydroquinone metabolites from *Sargassum heterophyllum* inhibit the *in vitro* growth of the malaria parasite *Plasmodium falciparum*. Zeitschrift für Naturforschung C 63: 848–852.

Águila-Ramírez, R.N., A. Arenas-González, C.J. Hernández-Guerrero, B. González-Acosta, J.M. Borges-Souza, B. Veron, J. Pope and C. Hellio. 2012. Antimicrobial and antifouling activities achieved by extracts of seaweeds from Gulf of California, Mexico. Hidrobiologica 22: 8–15.

Aguilar-Briseño, J.A., L.E. Cruz-Suarez, J.-F. Sassi, D. Ricque-Marie, P. Zapata-Benavides, E. Mendoza-Gamboa, C. Rodríguez-Padilla and L.M. Trejo-Avila. 2015. Sulphated polysaccharides from *Ulva clathrata* and *Cladosiphon okamuranus* seaweeds both inhibit viral attachment/entry and cell-cell fusion, in NDV infection. Marine Drugs 13(2): 697–712.

Aguilar-Santos, G.A. and M.S. Doty. 1968. Chemical studies on three species of marine algal genus Caulerpa. pp. 173–176. *In*: Froedenthal, H.D. (ed.). Drugs from the Sea. Marine Technology Society, University of Houston.

Aguilar-Santos, G. 1970. Caulerpin, a new red pigment from green algae of the genus *Caulerpa*. Journal of the Chemical Society C: Organic 6: 842–843.

Ahmadi, A., S. Zorofchian Moghadamtousi, S. Abubakar and K. Zandi. 2015. Antiviral potential of algae polysaccharides isolated from marine sources: A review. BioMed Research International 825203: 10 pp.

Ahmed, H.H., M.M. Hegazi, H.I. Abd-Alla, E.F. Eskander and M.S. Ellithey. 2011. Antitumour and antioxidant activity of some red sea seaweeds in Ehrlich ascites carcinoma *in vivo*. Zeitschrift für Naturforschung C 66(7-8): 367–376.

Ahn, B.R., H.E. Moon, H.R. Kim, H.A. Jung and J.S. Choi. 2012. Neuroprotective effect of edible brown alga *Eisenia bicyclis* on amyloid beta peptideinduced toxicity in PC12 cells. Archives of Pharmacal Research 35: 1989–1998.

Ahn, C.B., Y.J. Jeon, D.S. Kang T.S. Shin and B.M. Jung. 2004b. Free radical scavenging activity of enzymatic extracts from a brown seaweed *Scytosiphon lomentaria* by electron spin resonance spectrometry. Food Research International 37: 253–258.

Ahn, G., E. Park, D.S. Kim, Y.J. Jeon, T. Shin, J.W. Park, H.C. Woo, K.W. Lee and Y. Jee. 2008. Anti-inflammatory effects of enzymatic extract from *Ecklonia cava* on TPA-induced ear skin edema. Food Science and Biotechnology 17: 745–750.

Ahn, M.J., K.D. Yoon, C.Y. Kim, S.Y. Min, Y.U. Kim, H.J. Kim, J.H. Kim, C.-G. Shin, C.-K. Lee, T.G. Kim, S.H. Kim, H. Huh and J. Kim. 2002. Inhibition of HIV-1 reverse transcriptase and HIV-1 integrase and antiviral activity of Korean seaweed extracts. Journal of Applied Phycology 14: 325–329.

Ahn, M.J., K.D. Yoon, S.Y. Min, J.S. Lee, J.H. Kim, T.G. Kim, S.H. Kim, N.G. Kim, H. Huh and J. Kim. 2004. Inhibition of HIV-1 reverse transcriptase and protease by phlorotannins from the brown alga *Ecklonia cava*. Biological and Pharmaceutical Bulletin 27: 544–547.

Ahn, M.J., K.D. Yoon, C.Y. Kim, J.H. Kim, C.G. Shin and J. Kim. 2006. Inhibitory activity on HIV-1 reverse transcriptase and integrase of a carmalol derivative from a brown alga, *Ishige okamurae*. Phytotherapy Research 20: 711–713.

Aiello, L.C. and P. Wheeler. 1995. The expensive-tissue hypothesis: the brain and the digestive system in human and primate evolution. Current Anthropology 36: 199–221.

Airanthi, M., N. Sasaki, S. Iwasaki, N. Baba, M. Abe, M. Hosokawa and K. Miyashita. 2011. Effect of brown seaweed lipids on fatty acid composition and lipid hydroperoxide levels of mouse liver. Journal of Agricultural and Food Chemistry 59: 4156–4163.

Aird, W.C. 2007. Phenotypic heterogeneity of the endothelium I. Structure, function, and mechanisms. Circulation Research 100: 158–173.

Aisa, Y., Y. Miyakawa, T. Nakazato, H. Shibata, K. Saito, Y. Ikeda and M. Kizaki. 2005. Fucoidan induces apoptosis of human HS-Sultan cells accompanied by activation of caspase-3 and down-regulation of Erk pathways. American Journal of Hematology 78(1): 7–14.

Ajisaka, T. 1991. *Cladosiphon novae-caledoniae* Kylin (Phaeophyceae, Chordariales) from New Caledonia. South Pacific Study 12(1): 1–6.

Akamatsu, E., M. Shimanaga and Y. Kamei. 2003. Isolation of an anti-influenza virus substance, MC26 from a marine brown alga *Sargassum piluliferum* and its antiviral activity against influenza virus. Coastal Bioenvironment 1: 29–34.

Akdis, M., K. Blaser and C.A. Akdis. 2005. T regulatory cells in allergy: novel concepts in the pathogenesis, prevention, and treatment of allergic diseases. Journal of Allergy and Clinical Immunology 116: 961–968.

Akhtar, P. and V. Sultana. 2002. Biochemical studies of some seaweed species from Karachi coast. Records Zoological Survey of Pakistan 14: 1–4.

Aki, T., M. Yamamoto, T. Takahashi, K. Tomita, R. Toyoura, K. Iwashita, S. Kawamoto, M. Hosokawa, K. Miyashita and K. Ono. 2014. Regulation of polyunsaturated fatty acid biosynthesis by seaweed fucoxanthin and its metabolite in cultured hepatocytes. Lipids 49: 133–141.

Akiyama, T., R. Ueoka, R.W. van Soest and S. Matsunaga. 2009. Ceratodictyols, 1-glyceryl ethers from the red alga-sponge association *Ceratodictyon spongiosum/Haliclona cymaeformis*. Journal of Natural Products 72(8): 1552–1554.

Akremi, N., D. Cappoen, R. Anthonissen, L. Verschaeve and A. Bouraoui. 2017. Phytochemical and *in vitro* antimicrobial and genotoxic activity in the brown algae *Dictyopteris membranacea*. South African Journal of Botany 108: 308–314.

Al Hazzani, A.A., A.I. Shehata, N.M.S. Moubayed and H.J. Al Houri. 2014. Antimicrobial and biochemical properties of selected edible brown and red marine macroalgae. Journal of Pure and Applied Microbiology 8: 1275–1282.

Alagan, V.T., R.N. Valsala and K.D. Rajesh. 2017. Bioactive chemical constituent analysis, *in vitro* antioxidant and antimicrobial activity of whole plant methanol extracts of *Ulva lactuca* Linn. British Journal of Pharmaceutical Research 15(1).

Alam, K., T. Agua, R. Maven, R. Taie, K.S. Rao, I. Burrows, M.E. Huber and T. Ral. 1994. Preliminary screening of seaweeds, seagrass and lemongrass oil from Papupa New Guinea for antimicrobial and antifungal activity. International Journal of Pharmacognosy 32(4): 396–399.

Alarif, W.M., Z.S. Abou-Elnaga, S.E. Ayyad and S.S. Al-lihaibi. 2010. Insecticidal metabolites from the green alga *Caulerpa racemosa*. CLEAN-Soil Air Water 38(5-6): 548–557.

Al-Bader, T., A. Byrne, J. Gillbro, A. Mitarotonda, A. Metois, F. Vial, A.V. Rawlings and A. Laloeuf. 2012. Effect of cosmetic ingredients as anticellulite agents: synergistic action of actives with *in vitro* and *in vivo* efficacy. Journal of Cosmetic Dermatolo 11: 17–26.

Al-Bader, T., M. Davis, A. Laloeuf and A.V. Rawlings. 2013. Topical cosmetic compositions for treating or preventing cellulite. EP 2414047 B1.

Albuquerque, I.R.L., K.C.S. Queiroz, L.G. Alves, E.A. Santos, E.L. Leite and H.A.O. Rocha. 2004. Heterofucans from Dictyota menstrualis have anticoagulant activity. Brazilian Journal of Medical and Biological Research 37: 167–171.

Albuquerque, I.R.L., S.L. Cordeiro, D.L. Gomes, J.L. Dreyfuss, L.G.A. Filgueira, E.L. Leite, H.B. Nader and H.A.O. Rocha. 2013. Evaluation of anti-nociceptive and anti-inflammatory activities of a heterofucan from *Dictyota menstrualis*. Marine Drugs 11(8): 2722–2740.

Albuquerque, M.R., Campos-Takaki and M.L. Koening. 1983. Detection of antimicrobial activity in marine seaweeds. Revista do Instituto de Antibioticos, Universidade Federal de Pernambuco 21: 127–138.

Alderkamp, A.-C., M. Van Rijssel and H. Bolhuis. 2007. Characterization of marine bacteria and the activity of their enzyme systems involved in degradation of the algal storage glucan laminarin. FEMS Microbiology Ecology 59: 108–117.

Ale, M.T., H. Maruyama, H. Tamauchi, J.D. Mikkelsen and A.S. Meyer. 2011a. Fucoidan from *Sargassum* sp. and *Fucus vesiculosus* reduces cell viability of lung carcinoma and melanoma cells *in vitro* and activates natural killer cells in mice *in vivo*. International Journal of Biological Macromolecules 49(3): 331–336.

Ale, M.T., H. Maruyama, H. Tamauchi, J.D. Mikkelsen and A.S. Meyer. 2011b. Fucose-containing sulfated polysaccharides from brown seaweeds inhibit proliferation of melanoma cells and induce apoptosis by activation of caspase-3 *in vitro*. Marine Drugs 9(12): 2605–2621.

Ale, M.T. 2012. Fucose-containing sulfated polysaccharides from brown seaweed: Extraction technology and biological activity assessment, PhD Thesis, DTU Chemical Engineering.

Ale, M.T. and A.S. Meyer. 2013. Fucoidans from brown seaweeds: an update on structures, extraction techniques and use of enzymes as tools for structural elucidation. RSC Advances 3(22): 8131–8141.

Alekseyenko, T.V., S.Y. Zhanayeva, A.A. Venediktova, T.N. Zvyagintseva, T.A. Kuznetsova, N.N. Besednova and T.A. Korolenko. 2007. Antitumor and antimetastatic activity of fucoidan, a sulfated polysaccharide isolated from the Okhotsk Sea *Fucus evanescens* brown alga. Bulletin of Experimental Biology and Medicine 143(6): 730–732.

Al-Fadhli, A., S. Wahidulla and L. D'Souza. 2006. Glycolipids from the red alga *Chondria armata*. Glycobiology 16: 902–915.

Alghazeer, R., F. Whida, E. Abduelrhman, F. Gammoudi and S. Azwai. 2013. Screening of antibacterial activity in marine green, red and brown macroalgae from the western coast of Libya. Natural Science 5: 7–14.

Alghazwi, M., Y.Q. Kan, W. Zhang, W.P. Gai, M.J. Garson and S. Smid. 2016. Neuroprotective activities of natural products from marine macroalgae during 1999–2015. Journal of Applied Phycology 28(6): 3599–3616.

Al-Haj, N.A., N.I. Mashan, M.N. Shamsudin, H. Mohamad, C.S. Vairappan and Z. Sekawi. 2009. Antibacterial activity in marine algae *Eucheuma denticulatum* against *Staphylococcus aureus* and *Streptococcus pyogenes*. Research Journal of Biological Sciences 4: 519–524.

Al-Haj, N.A., N.I. Mashan, M.N. Shamsudin, H. Mohamad, C.S. Vairappan and Z. Sekawi. 2010. Antibacterial activity of marine source xtracts against multidrug resistance organisms. American Journal of Pharmacology and Toxicology 5(2): 95–102.

Ali, H., P.J. Houghton and A. Soumyanath. 2006. α-amylase inhibitory activity of some Malaysian plants used to treat diabetes; with particular reference to *Phyllanthus amarus*. Journal of Ethnopharmacology 107: 449–455.

Ali, I., Z. Manzoor, J.E. Koo, J.E. Kim, S.H. Byeon, E.S. Yoo, H.K. Kang et al. 2017. 3-Hydroxy-4,7-megastigmadien-9-One, isolated from *Ulva pertusa*, attenuates TLR9-mediated inflammatory response by down-regulating mitogen-activated protein kinase and NF-κ B pathways. Pharmaceutical Biology 55(1): 435–440.

Ali, M.S., F. Mazhar, M. Saleem, M. Jahangir, K. Pervez, K. Usmanghani and V.U. Ahmad. 2000. Chemistry and biology of algae from sea coasts of Karachi, pp. 33–44. *In*: Ahmad, V.U. (ed.). Proceedings Nationall ONR Symposium on Arabian Sea as a Resource of Biological Diversity. HEJ Research Institute of Chemistry, Karachi University, Karachi.

Ali, M.S., M. Saleem, R. Yammdagni and M.A. Ali. 2002. Steroid and antibacterial glycosides from marine green alga *Codium iyengarii* Borgesen. Natural Product Letters 16: 407–413.

Ali, M.S., S. Ravikumar, J.M. Beula, V. Anuradha and N. Yogananth. 2014. Insecticidal compounds from Rhizophoraceae mangrove plants for the management of dengue vector *Aedes aegypti*. Journal of Vector Borne Diseases 51(2): 106–114.

Ali, M.Y.S., S. Ravikumar and J.M. Beula. 2013. Mosquito larvicidal activity of seaweeds extracts against *Anopheles stephensi, Aedes aegypti* and *Culex quinquefasciatus*. Asian Pacific Journal of Tropical Disease 3(3): 196–201.

Al-Judaibi, A. 2014. Antibacterial effects of extracts of two types of red sea algae. Journal of Biosciences and Medicines 2: 74–82.

Allen, M.B. and E.Y. Dawson. 1960. Production of antibacterial substances by benthic tropical marine algae. Journal of Bacteriology 79: 459–460.

Allmendinger, A., J. Spavieri, M. Kaiser, R. Casey, S. Hingley-Wilson, A. Lalvani, M. Guiry, G. Blunden and D. Tasdemir. 2010. Antiprotozoal, antimycobacterial and cytotoxic potential of twenty-three British and Irish red algae. Phytotherapy Research 24: 1099–1103.

Almeida, C.L.F., H.S. Falcão, G.R.M. Lima, C.A. Montenegro, N.S. Lira, P.F. Athayde-Filho, L.C. Rodrigues, M.F.V. de Souza, J.M. Barbosa-Filho and L.M. Batista. 2011. Bioactivities from marine algae of the genus *Gracilaria*. International Journal of Molecular Sciences 12: 4550–4573.

Almodovar, L.R. 1964. Ecological aspects of some antibiotic algae in Puerto Rico. Botanica Marina 6: 143–146.

Al-Saif, S.S.A.-l., N. Abdel-Raouf, H.A. El-Wazanani and I.A. Aref. 2014. Antibacterial substances from marine algae isolated from Jeddah coast of Red sea, Saudi Arabia. Saudi Journal of Biological Sciences 21: 57–64.

Alvar, J., I.D. Vélez, C. Bern, M. Herrero, P. Desjeux, J. Cano, J. Jannin and M. den Boer. 2012. Leishmaniasis worldwide and global estimates of its incidence. PLoSONE 7(5): e35671.

Alvarez, A.M.R. and M.L.G. Rodríguez. 2000. Lipids in pharmaceutical and cosmetic preparations. Grasas y Aceites 51(1-2): 74–96.

Alves, Anabela, Rui A. Sousa and Rui L. Reis. 2013a. Processing of degradable ulvan 3D porous structures for biomedical applications. Journal of Biomedical Materials Research Part A 101(4): 998–1006.

Alves, A., R.A. Sousa and R.L. Reis. 2013b. *In vitro* cytotoxicity assessment of ulvan, a polysaccharide extracted from green algae. Phytotherapy Research 27: 1143–1148.

Alves, C., S. Pinteus, A. Horta and R. Pedrosa. 2016. High cytotoxicity and anti-proliferative activity of algae extracts on an *in vitro* model of human hepatocellular carcinoma. SpringerPlus 5(1339): 1–13.

Alves, C., S. Pinteus, T. Simoes, A. Horta, J. Silva, C. Tecelao and R. Pedrosa. 2016b. *Bifurcaria fifurcata*: A key macro-alga as a source of bioactive compounds and functional ingredients. International Journal of Food Science and Technology 51(7): 1638–1646.

Alves, M., J. Almeida-Lima, A.A.O. Paiva, E.L. Leite and H.A.O. Rocha. 2016c. Extraction process optimization of sulfated galactan-rich fractions from *Hypnea musciformis* in order to obtain antioxidant, anticoagulant, or immunomodulatory polysaccharides. Journal of Applied Phycology 28(3): 1931–1942.

Alwarsamy, M., R. Gooneratne and R. Ravichandran. 2016. Effect of fucoidan from *Turbinaria conoides* on human lung adenocarcinoma epithelial (A549) cells. Carbohydrate Polymers 152: 207–213.

Alzahrani, A.S., S. Aldasouqi, S.A. Salam, A. Sultan. 2005. Autoimmune thyroid disease with fluctuating thyroid function. PLoS medicine 2(5): e89.

Amado-Filho, G.M., L.T. Salgado, M.F. Rebelo, C.E. Rezende, C.S. Karez and W.C. Pfeiffer. 2008. Heavy metals in benthic organisms from Todos os Santos Bay, Brazil. Brazilian Journal of Biology 68(1): 95–100.

Amano, H., M. Kakinuma, D.A. Coury, H. Ohno and T. Hara. 2005. Effect of a seaweed mixture on serum lipid level and platelet aggregation in rats. Fisheries Science 71(5): 1160–1166.

Ambati, R.R., P. Siew Moi, S. Ravi and R.G. Aswathanarayana. 2014. Astaxanthin: Sources, extraction, stability, biological activities and its commercial applications—a review. Marine Drugs 12: 128–152.

Ambika, S. and K. Sujatha. 2014. Comparative studies on brown, red and green alga seaweed extracts for their antifungal activity against *Fusarium oxysporum* f. sp. *udum* in Pigeon pea var. CO (Rg)7 (*Cajanus cajan* (L.) Mills.). Journal of Biopesticides 7(2): 167–176.

Ambika, S. and K. Sujatha. 2015. Antifungal activity of brown, red and green alga seaweed extracts against *Macrophomina phaseolina* (Tassi) Goid., in pigeonpea var. CO (Rg) 7. International Journal of Agricultural Sciences 11(2): 210–216.

American Cancer Society 2015. Cancer facts and figures. Atlanta. Available online at: http://www.cancer.org/acs/groups/content/@editorial/documents/document/acspc044552.pdf.

Amico, V., G. Oriente, M. Piattelli, C. Tringali, E. Fattorusso, S. Magno and L. Mayol. 1978. Caulerpenyne, an unusual sequiterpenoid from the green alga *Caulerpa prolifera*. Tetrahedron Letters 19(38): 3593–3596.

Amiguet, V.T., L.E. Jewell, H. Mao, M. Sharma, J.B. Hudson, T. Durst, M. Allard, G. Rochefort and J.T. Arnason. 2011. Antibacterial properties of a glycolipid-rich extract and active principle from Nunavik collections of the macroalgae *Fucus evanescens* C. Agardh (Fucaceae). Canadian Journal of Microbiology 57: 745–749.

Ammar, H., S. Lajili, R.B. Said, D. Le Cerf, A. Bouraoui and H. Majdoub. 2015. Physico-chemical characterization and pharmacological evaluation of sulfated polysaccharides from three species of Mediterranean brown algae of the genus *Cystoseira*. DARU Journal of Pharmaceutical Sciences 23(1): 1–8.

Amorim, R.N.S., J.A.G. Rodrigues, M.L. Holanda, A.L.G. Quinderé, R.C.M. de Paula, V.M.M. Melo and N.M.B. Benevides. 2012. Antimicrobial effect of a crude sulfated polysaccharide from the red seaweed *Gracilaria ornata*. Brazilian Archives of Biology and Technology 55: 171–181.

Anandhan, S. and H. Sorna Kumari. 2011. Biorestraining potentials of marine macroalgae collected from Rameshwaram, Tamil nadu. Journal of Research in Biology 5: 385–392.

Ananthi, S., H.R. Raghavendran, A.G. Sunil, V. Gayathri, G. Ramakrishnan and H.R. Vasanthi. 2010. *In vitro* antioxidant and *in vivo* anti-inflammatory potential of crude polysaccharide from *Turbinaria ornata* (marine brown alga). Food and Chemical Toxicology 48: 187–192.

Anastyuk, S.D., N.M. Shevchenko, S.P. Ermakova, O.S. Vishchuk, E.L. Nazarenko, P.S. Dmitrenok and T.N. Zvyagintseva. 2012. Anticancer activity *in vitro* of a fucoidan from the brown alga *Fucus evanescens* and its low-molecular fragments, structurally characterized by tandem mass-spectrometry. Carbohydrate Polymers 87(1): 186–194.

Anderson, G.F. and P.S. Hussey. 2000. Population aging: a comparison among industrialized countries. Health Affairs 19: 191–203.

Anderson, R.J., H. Stegenga and J.J. Bolton. 2016. *Codium capitatum*—Seaweeds of the South African South Coast. University of Cape Town. Available online at: http://southafrseaweeds.uct.ac.za/descriptions/green/codium_capitatum.php.

Anderson, W. and J.G.C. Duncan. 1965. The anticoagulant activity of carrageenan. Journal of Pharmacy and Pharmacology 17: 647–654.

Anderson, W. 1967. Carrageenan—Structure and biological activity. Canadian Journal of Pharmaceutical Sciences 2: 81–90.

Anderson, W. 1969. Proceedings of the International Seaweed Symposium 6: 627–635.

Ando, K., S. Suzuki, K. Suzuki, K. Kodama, G. Tamura and K. Arima. 1969. Isolation of fatty acids with antitumor activity from fungal mycelia. The Journal of Antibiotics (Tokyo) 22: 18–22.

Ando, K., K. Kodama, A. Kato, G. Tamura and K. Arima. 1972. Antitumor activity of glyceryl ethers. Cancer Research 32: 125–129.

Andrade-Sorcia, G., R. Riosmena-Rodríguez, R. Muñiz-Salazar, J.M. López-Vivas, G.H. Boo, K.M. Lee and S.M. Boo. 2014. Morphological reassessment and molecular assessment of *Sargassum* (Fucales: Phaeophyceae) species from the Gulf of California, Mexico. Phytotaxa 183(4): 201–223.

Andriole, V.T. 1993. Infections with *Aspergillus* species. Clinical Infectious Diseases 17: S481–S486.

Anno, K., H. Terahata, Y. Hayashi and N. Seno. 1966. Isolation and purification of fucoidin from brown seaweed *Pelvetia wrightii*. Agricultural and Biological Chemistry 30(5): 495–499.

Ansari, J., A. Siraj and N. Inamdar. 2010. Pharmacotherapeutic approaches of Parkinson's disease. International Journal of Pharmacology 6: 584–590.

Anthoni, J.F. 2007. The intertidal rocky shore—identifying red seaweeds of the intertidal rocky shore. *Vidalia colensoi* (formerly *Osmundaria colensoi*). Available online at: http://www.seafriends.org.nz/enviro/habitat/rsred.htm.

Antunes, E.M., A.F. Afolayan, M.T. Chiwakata, J. Fakee, M.G. Knott, C.E. Whibley, D.T. Hendricks, J.J. Bolton and D.R. Beukes. 2011. Identification and *in vitro* anti-esophageal cancer activity of a series of halogenated monoterpenes isolated from the South African seaweeds *Plocamium suhrii* and *Plocamium cornutum*. Phytochemistry 72(8): 769–772.

Anunciato, T.P. and P.A. da Rocha-Filho. 2012. Carotenoids and polyphenols in nutricosmetics, nutraceuticals, and cosmeceuticals. Journal of Cosmetic Dermatology 11: 51–54.

Anuradha, V., K. Byju, K. Jacob, A. Gopinath, K.J. Prashob Peter, T.R. Gireesh Kumar, G. Vasundhara, E. Rosamine, H.S. Harishankar, N.C. Kumar and S.M. Nair. 2015. Evaluation of biochemical and nutritional potential of seaweeds from Lakshadweep Archipelago. Journal of Aquatic Food Product Technology 24(7): 639–647.

APA. 2017. American psychiatric association. Available online at: https://www.psychiatry.org/.

Apostolidis, E. and C.M. Lee. 2010. *In vitro* potential of *Ascophyllum nodosum* phenolic antioxidant mediated α-glucosidase and α-amylase inhibition. Journal of Food Science 75: 97–102.

Applegate, R.D. and P.B. Gray. 1995. Nutritional value of seaweed to ruminants. Rangifer 15: 15–18.

Aqumin. 2017. Marine minerals for health. Available online at: http://aquamin.com/.

Ara, J., S. Ehteshamul-Haque, V. Sultana, R. Qasim and A. Gaffar. 1996. Nematicidal activity of seaweeds against *Meloidogyne javanica* root knot nematode. Pakistan Journal of Nematology 14: 129–131.

Ara, J., V. Sultana, S. Ehteshamul-Haque, S.A. Qureshi and V.U. Ahmad. 1998. Bioactivity of seaweeds against soil home plant pathogens. Phytologia 85: 292–299.

Ara, J., V. Sultana and V.U. Qasim. 2002. Hypolipidemic activity of seaweed from Karachi coast. Phytotherapy Research 16: 479–483.

Ara, J., V. Sultana, R. Qasim, S. Ehteshamul-Haque and V.U. Ahmad. 2005. Biological activity of *Spatoglossum asperum*: A brown alga. Phytotherapy Research 19: 618–623.

Arad, S. and O. Levy-Ontman. 2010. Red microalgal cell-wall polysaccharides: biotecnological aspects. Current Opinion in Biotechnology 21: 358–364.

Aragão, M.J. 2009. Plantas e algas medicinais. Livros Horizonte, Lisboa, pp. 153–174.

Araki, C.H. 1937. Acetylation of agar like substance of *Gelidium amansii*. Journal of the Chemical Society 58: 1338–1350.

Arasaki, S. and T. Arasaki. 1983. Low calorie, high nutrition vegetables from the sea. To help you look and feel better. Japan Publications Inc., Tokyo, 196 pp.

Arata, P.X., I. Quintana, D.J. Canelon, B.E. Vera, R.S. Compagnone and M. Ciancia. 2015. Chemical structure and anticoagulant activity of highly pyruvylated sulfated galactans from tropical green seaweeds of the order bryopsidales. Carbohydrate Polymers 122: 376–386.

Arata, P.X., I. Quintana, M.P. Raffo and M. Ciancia. 2016. Novel sulfated xylogalactoarabinans from green seaweed *Cladophora falklandica*: Chemical structure and action on the fibrin network. Carbohydrate Polymers 154: 139–150.

Araújo, I.W.F., E.S.O. Vanderlei, A.R.G. Rodrigues, C.O. Coura, A.L.G. Quinderé, B.P. Fontes, I.N.L. Queiroz, R.J.B. Jorge, M.M. Bezerra, A.A.R. Silva, H.V. Chaves, H.S.A. Monteiro, R.C.M. Paula and N.M.B. Benevides. 2011. Effects of a sulfated polysaccharide isolated from the red seaweed *Solieria filiformis* on models of nociception and inflammation. Carbohydrate Polymers 86: 1207–1215.

Araújo, I.W.F., J.A.G. Rodrigues, E.S.O. Vanderlei, G.A. de Paula, T.B. Lima and N.M.B. Benevides. 2012. Iota-carrageenans from *Solieria filiformis* (Rhodophyta) and their effects in the inflammation and coagulation. Acta Scientiarum 34(2): 127–135.

Aravindan, S., C.R. Delma, S.S. Thirugnanasambandan, T.S. Herman and N. Aravindan. 2013. Anti-pancreatic cancer deliverables from sea: first-hand evidence on the efficacy, molecular targets and mode of action for multifarious polyphenols from five different brown-algae. PLoS One 8(4): 1–15.

Archer, G.L. 1998. *Staphylococcus aureus*: a well-armed pathogen. Clinical Infectious Diseases 26: 1179–1181.

Areche, C., A. San-Martín, J. Rovirosa, M.A. Muñoz, A. Hernández-Barragán, M.A. Bucio and P. Joseph-Nathan. 2010. Stereostructure reassignment and absolute configuration of isoepitaondiol, a meroditerpenoid from *Stypopodium flabelliforme*. Journal of Natural Products 73(1): 79–82.

Argandoña, V., T. Del Pozo, A. San-Martín and J. Rovirosa. 2000. Insecticidal activity of *Plocamium cartilagineum* monoterpenes. Boletin de la Sociedad Chilena de Quimica 45(3): 371–376.

Arivuselvan, N., M. Radhiga and P. Anantharaman. 2011. *In vitro* antioxidant and anticoagulant activities of sulphated polysaccharides from brown seaweed (*Turbinaria ornata*) (Turner) J. Agardh. Asian Journal of Pharmaceutical and Biological Research 1: 232–238.

Armisen, R. and F. Galatas. 2000. Agar. pp. 21–40. *In*: Phillips, G. and P. Williams (eds.). Handbook of Hydrocolloids. CRC Press, Taylor & Francis Group, Boca Raton, FL.

Arshad, S.H. 2010. Does exposure to indoor allergens contribute to the development of asthma and allergy? Current Allergy and Asthma Reports 10: 49–55.

Artan, M., Y. Li, F. Karadeniz, S.H. Lee, M.M. Kim and S.K. Kim. 2008. Anti-HIV-1 activity of phloroglucinol derivative, 6,6-bieckol, from *Ecklonia cava*. Bioorganic & Medicinal Chemistry 16: 7921–7926.

Arthur, C.O. 2007. Antiallergic effects of probiotics. Journal of Nutrition 137: 794S–797S.

Arulvasu, C., S.S. Shakthi, V. Rangasamy and M. Hemavathi. 2014. *In-vitro* antioxidant and anticancer activity of selected seaweeds of Kovalam coast, Tamilnadu. (IJIRSE) International Journal of Innovative Research in Science & Engineering 14(143): 1–7.

Aruna, P., P. Mansuya, S. Sridhar, J.S. Kumar and S. Babu. 2010. Pharmacognostical and antifungal activity of selected seaweeds from Gulf of Mannar region. Recent Research in Science and Technology 2(1): 115–119.

Aruna, P., P. Mansuya, S. Sridhar, J.S. Kumar and S. Babu. 2011. Pharmacognostical and antifungal activity of selected seaweeds from Gulf of Mannar region. Recent Research in Science and Technology 2: 115–119.

Arunachalam, P., R. Uthandakalai and R. Rajsmaill. 2014. Evaluation of antibacterial activity of some selected green seaweed extracts from Muttam coastal areas, Kanyakumari, Tamil Nadu, India. Journal of Coastal Life Medicine 2(2): 112–115.

Asada, M., M. Sugie, M. Inoue, K. Nakagomi, S. Hongo, K. Murata, S. Irie, T. Takeuchi, N. Tomizuka and S. Oka. 1997. Inhibitory effect of alginic acids on hyaluronidase and on histamine release from mast cells. Bioscience, Biotechnology, and Biochemistry 61: 1030–1032.

Asai, A., T. Sugawara, H. Ono and A. Nagao. 2004. Biotransformation of fucoxanthinol into amarouciaxanthin A in mice and HepG2 Cells: Formation and cytotoxicity of fucoxanthin metabolites. Drug Metabolism & Disposition 32(2): 205–211.

Aschner, J.L. and M. Aschner. 2005. Nutritional aspects of manganese homeostasis. Molecular Aspects of Medicine 26: 353–362.

ASSC. 2016. History of salt. America's Sea Salt Company—Salt works webpage. Available online at: http://www.saltworks.us/salt_info/si_HistoryOfSalt.asp.

Astorga-España, M.S. and A. Mansilla. 2014. Sub-antarctic macroalgae: opportunities for gastronomic tourism and local fisheries in the region of magallanes and Chilean Antarctic territory. Journal of Applied Phycology 26: 973–978.

Astorga-España, M.S., B. Rodríguez Galdón, E.M. Rodríguez Rodríguez and C. Díaz Romero. 2015. Mineral and trace element concentrations in seaweeds from the sub-Antarctic ecoregion of Magallanes (Chile). Journal of Food Composition and Analysis 39: 69–76.

Aswathi, E.M. and P. Jamila. 2014. Antifungal phytochemical screening and repellent effect in seaweed *G. edulis*. International Journal of Recent Scientific Research 5(11): 2037–2042.

Athukorala, Y., K.N. Kim and Y.J. Jeon. 2006. Antiproliferative and antioxidant properties of an enzymatic hydrolysate from brown alga, *Ecklonia cava*. Food and Chemical Toxicology 44(7): 1065–1074.

Athukorala, Y., K.W. Lee, S.K. Kim and Y.J. Jeon. 2007. Anticoagulant activity of marine green and brown algae collected from Jeju Island in Korea. Bioresour Technol 98:1711–1716.

Athukorala, Y., S. Trang, C. Kwok and Y.V. Yuan. 2016. Antiproliferative and antioxidant activities and mycosporine-like amino acid profiles of wild-harvested and cultivated edible canadian marine red macroalgae. Molecules 21(1): 1–17.

Atmar, R.L. 2010. Noroviruses: state of the art. Food and Environmental Virology 2: 117–126.

Augusti, P.R., A. Quatrin, S. Somacal, G.M.M. Conterato, R. Sobieski, A.R. Ruviaro, L.H. Maurer, M.M.F. Duarte, M. Roehrs and T. Emanuelli. 2012. Astaxanthin prevents changes in the activities of thioredoxin reductase and paraoxonase in hypercholesterolemic rabbits. Journal of Clinical Biochemistry and Nutrition 51: 42–49.

Avila, J., I. Santa-Maria, M. Perez, F. Hernandez and F. Moreno. 2006. Tau phosphorylation, aggregation, and cell toxicity. Journal of Biomedicine and Biotechnology 2006: 1–5.

Awad, N.E. 2000. Biologically active steroid from the green alga *Ulva lactuca*. Phytotherapy Research 14: 641–643.

Awad, N.E. 2004. Bioactive brominated diterpenes from the marine red alga *Jania rubens* (L.) Lamx. Phytotherapy Research 18: 275–279.

Awang, A.N., J.L. Ng, P. Matanjun, M.R. Sulaiman, T.S. Tan and Y.B.H. Ooi. 2014. Anti-obesity property of the brown seaweed, *Sargassum polycystum* using an *in vivo* animal model. Journal of Applied Phycology 26: 1043–1048.

Aylagas, E., M. Machín-Sánchez, C. Gil-Rodríguez. 2010. Revision of the *Laurencia* genus in the Canary Islands. I. *Laurencia microcladia* (Ceramiales, Rhodomelaceae). Vieraea 39: 53–64.

Ayyad, S.E., M.O. Slama, A.H. MoKhtar and A.F. Anter. 2001. Cytotoxic bicyclic diterpene from the brown alga *Sargassum crispum*. Bollettino Chimico Farmaceutico 140(3): 155–159.

Ayyad, S.E., K.O. Al-Footy, W.M. Alarif, T.R. Sobahi, S.A. Bassaif, M.S. Makki, A.M. Asiri, A.Y. Al Halwani, A.F. Badria and F.A. Badria. 2011. Bioactive C15 acetogenins from the red alga *Laurencia obtusa*. Chemical and Pharmaceutical Bulletin (Tokyo) 59(10): 1294–1298.

Ayyad, S.E.N., O.B. Abdel-Halim, W.T. Shier and T.R. Hoye. 2003. Cytotoxic hydroazulene diterpenes from the brown alga *Cystoseira myrica*. Zeitschrift Fur Naturforschung C-a Journal of Biosciences 58(1-2): 33–38.

Azmir, J., I.S.M. Zaidul, M.M. Rahman, K.M. Sharif, A. Mohamed, F. Sahena, M.H.A. Jahurul, K. Ghafoor, N.A.N. Norulaini and A.K.M. Omar. 2013. Techniques for extraction of bioactive compounds from plant materials: A review. Journal of Food Engineering 117(4): 426–436.

Azuma, K., T. Ishihara, H. Nakamoto, T. Amaha, T. Osaki, T. Tsuka, T. Imagawa, S. Minami, O. Takashima, S. Ifuku, M. Morimoto, H. Saimoto, H. Kawamoto and Y. Okamoto. 2012. Effects of oral administration of fucoidan extracted from *Cladosiphon okamuranus* on tumor growth and survival time in a tumor-bearing mouse model. Marine Drugs 10: 2337–2348.

Baba, M., M. Nakajima, D. Schols, R. Pauwels, J. Balzarini and E. De Clercq. 1988a. Pentosan polysulfate, a sulfated oligosaccharide, is a potent and selective anti-HIV agent *in vitro*. Antiviral Research 9: 335–343.

Baba, M., R. Snoeck, R. Pauwels and E. de Clercq. 1988b. Sulfated polysaccharides are potent and selective inhibitors of various enveloped viruses, including Herpes simplex virus, cytomegalovirus, vesicular stomatitis virus, and human immunodeficiency virus. Antimicrobial Agents and Chemotherapy 32: 1742–1745.

Babio, N., S.M. Bullo, J. Basora, N. Ibarrola-Jurado, J. Fernandez-Ballart, M.A. Martinez-Gonzalez, L. Serra-Majem, R. Gonzalez-Perez and J. Salas-Salvado. 2012. Association between red meat consumption and metabolic syndrome in a Mediterranean population at high cardiovascular risk: cross-sectional and 1-year follow-up assessment. Nutrition, Metabolism, and Cardiovascular Disease 22: 200–207.

Baddley, J.W., P.G. Pappas, A.C. Smith and S.A. Moser. 2003. Epidemiology of *Aspergillus terreus* at a university hospital. Journal of Clinical Microbiology 41: 5525–5529.

Bae, S.J. and Y.H. Choi. 2007. Methanol extract of the seaweed *Gloiopeltis furcata* induces G2/M arrest and inhibits cyclooxygenase-2 activity in human hepatocarcinoma Hepg2 cells. Phytotherapy Research 21(1): 52–57.

Baghel, R.S., P. Kumari, C.R.K. Reddy and B. Jha. 2014. Growth, Pigments, and biochemical composition of marine red alga *Gracilaria crassa*. Journal of Applied Phycology 26(5): 2143–2150.

Bai, N.R. 2010. Evaluation of *Gracilaria fergusonii* for phytochemical analysis and antibacterial activity. Plant Archives 10(2): 711–713.

Baik, S.E. and J.W. Kang. 1986. Antimicrobial activity of the volatile and lipid fractions of marine algae. Korean Journal of Phycology 1: 293–310.

Bains, J.S. and C.A. Shaw. 1997. Neurodegenerative disorders in humans: The role of glutathione in oxidative stress-mediated neuronal death. Brain Research Reviews 25: 335–358.

Bak, S.S., B.N. Ahn, J.A. Kim, S.H. Shin, J.C. Kim, M.K. Kim, Y.K. Sung and S.K. Kim. 2013. *Ecklonia cava* promotes hair growth. Clinical and Experimental Dermatology 38(8): 904–910.

Bakker, W., E.C. Eringa, P. Sipkema and V.W.M. Hinsbergh. 2009. Endothelial dysfunction and diabetes: roles of hyperglycemia, impaired insulin signaling and obesity. Cell and Tissue Research 335: 165–189.

Baky, H.H.A., K.F. El Baz and S.G.E. Baroty. 2008. Evaluation of marine alga *Ulva lactuca* as a source of natural preservative ingredient. Journal of Agriculture and Environmental Science 3: 434–444.

Balasubramaniam, V., J.C. Lee, M.F. Noh, S. Ahmad, I.A. Brownlee and A. Ismail. 2016. Alpha-amylase, antioxidant, and anti-inflammatory activities of *Eucheuma denticulatum* (N.L. Burman) F.S. Collins and Hervey. Journal of Applied Phycology 28(3): 1965–1974.

Ballantine, D.L. and L.R. Almodovar. 1977. Epiphytism and antibiotic activity of extracts of tropical marine algae. pp. 23–30. *In:* Progress in Marine Research in the Caribbean and Adjacent Regions II Symposium, Caracas, Venezuela.

Ballantine, D.L., W.H. Gerwick, S.M. Velez, E. Alexander and P. Guevara. 1987. Antibiotic activity of lipid-soluble extracts from Caribbean marine algae. Hydrobiologia 151/152: 463–469.

Ballesteros, E., D. Martin and M.J. Uriz. 1992. Biologicalactivity of extracts from some mediterranean macrophytes. Botanica Marina 35(6): 481–485.

Balunas, M.J. and A.D. Kinghorn. 2005. Drug discovery frommedicinal plants. Life Science Journal 78: 431–441.

Bandara, B.M.R., A.A.I. Gunatilaka, N.S. Kumar, W.R. Wimalasiri, N.K.B. Adikaram and S. Balasubramaniam. 1988. Antimicrobial activity of some marine algae of Sri Lanka. Journal of the National Science Foundation of Sri Lanka 16: 209–221.

Bandyopadhyay, S.S., Mojdeh H. Navid, T. Ghosh, P. Schnitzler and B. Ray. 2011. Structural features and *in vitro* antiviral activities of sulfated polysaccharides from *Sphacelaria indica*. Phytochemistry 72(2-3): 276–283.

Bansemir, A., M. Blume, S. Schröder and U. Lindequist. 2006. Screening of cultivated seaweeds for antibacterial activity against fish pathogenic bacteria. Aquaculture 252(1): 79–84.

Banskota, A.H., R. Stefanova, S. Sperker, S. Lall, J.S. Craigie and J.T. Hafting. 2013. Lipids isolated from the cultivated red alga *Chondrus crispus* inhibit nitric oxide production. Journal of Applied Phycology 26: 1565–1571.

Bantoto, V. and D. Dy. 2013. The larvicidal activity of brown algae *Padina minor* (Yamada 1925) and *Dicyota linearis* (Greville 1830) against the dengue vector, *Aedes aegypti* (Linn 1762) (Diptera: Culicidae). Journal of Vector Borne Diseases 50: 68–70.

Baranitharan, M. 2014. Mosquitocidal efficacies of medicinal plant of *Coleus aromaticus* Benth (Lamiaceae) leaf extracts chikungunya vector, *Aedes aegypti* (Linn.) (Diptera: Culicidae). International Journal of Current Research in Chemistry and Pharmaceutical Sciences 1: 61–67.

Barata, E.A.F. 1995. A cosmetologia—Princípios básicos. Tecnopress, São Paulo, Brasil, 176 pp.

Barbarino, E. and S.O. Lourenço. 2005. An evaluation of methods for extraction and quantification of protein from marine macro and microalgae. Journal of Applied Phycology 17(5): 447–460.

Barbier, P., S. Guise, P. Huitorel, P. Amade, D. Pesando, C. Briand and V. Peyrot. 2001. Caulerpenyne from *Caulerpa taxifolia* has an antiproliferative activity on tumor cell line SK-N-SH and modifies the microtubule network. Life Sciences 70(4): 415–429.

Barbosa, J.P., R.C. Pereira, J. Abrantes, C.C. Dos Santos, M.A. Rebello, I. de Palmer Paixao Frugulhetti and V.L. Teixeira. 2004. *In vitro* antiviral diterpenes from the Brazilian brown alga *Dictyota pfaffii*. Planta Medica 70: 856–860.

Barbosa, M., P. Valentao and P.B. Andrade. 2014. Bioactive compounds from macroalgae in the new millennium: Implications for neurodegenerative diseases. Marine Drugs 12(9): 4934–4972.

Barnes, M. 2008. *Chaetomorpha linum*. Spaghetti algae. Marine Life Information Network: Biology and Sensitivity Key Information Sub-programme. Plymouth: Marine Biological Association of the United Kingdom. Available online at: http://www.marlin.ac.uk/species/detail/138.

Barona, J., J.J. Jones, R.E. Kopec, M. Comperatore, C. Andersen, S.J. Schwartz, R.H. Lerman and M.L. Fernandez. 2012. A Mediterranean-style low-glycemic-load diet increases plasma carotenoids and decreases LDL oxidation in women with metabolic syndrome. Journal of Nutritional Biochemistry 23: 609–615.

Barreto, M., A.T. Critchley and C. Straker. 2002. Extracts from seaweeds can promote fungal growth. Journal of Basic Microbiology 42(5): 302–310.

Barreto, M. and J.J.M. Meyer. 2006. Isolation and antimicrobial activity of a Lanosol derivative from *Osmundaria serrata* (Rhodophyta) and a visual exploration of its biofilm covering. South African Journal of Botany 72(4): 521–528.

Barrow, C. and F. Shahidi (eds.). 2008. Marine Nutraceuticals and Functional Foods. CRC Press, Taylor & Francis Group, Boca Raton, FL.

Barton, C., J.C. Kouokam, A.B. Lasnik, O. Foreman, A. Cambon, G. Brock, D.C. Montefiori, F. Vojdani, A.A. McCormick, B.R. O'Keefe and K.E. Palmer. 2014. Activity of and effect of subcutaneous treatment with the broad-spectrum antiviral lectin Griffithsin in two laboratory rodent models. Antimicrobial Agents and Chemotherapy 58(1): 120–127.

Baserga, R., C. Sell and P. Porcu. 1994. The role of the IGF-I receptor in the growth and transformation of mammalian cells. Cell Proliferation 27: 63–71.

Baserga, R. 1999. The IGF-I receptor in cancer research. Experimental Cell Research 253: 1–6.

Batista, J.A., E.G.N. Dias, T.V. Brito, R.S. Prudêncio, R.O. Silva, R.A. Ribeiro, M.H.L.P. Souza, R.C.M. de Paulad and J.P.A. Feitosa. 2014. Polysaccharide Isolated from *Agardhiella Ramosissima*: Chemical Structure and Anti-Inflammation Activity. Carbohydrate Polymers 99: 59–67.

Bauer, A.W.W.M., M. Kirby, J.C. Sherris and M. Turck. 1966. Antibiotic susceptibility testing by a standardized single disk method. American Journal of Clinical Pathology 45: 493–496.

Bazes, A., A. Silkina, P. Douzenel, F. Fay, N. Kervarec, D. Morin, J.-P. Berge and N. Bourgougnon. 2009. Investigation of the antifouling constituents from the brown alga *Sargassum muticum* (Yendo) Fensholt. Journal of Applied Phycology 21(4): 395–403.

Beards, G. 2012. Blood coagulation pathways *in vivo* and the classical blood coagulation pathway. Available online at: https://commons.wikimedia.org/wiki/File:Coagulation_*in_vivo*.png and https://commons.wikimedia.org/wiki/File:Classical_blood_coagulation_pathway.png.

Beaulieu, L., S. Bondu, K. Doiron, L.-E. Rioux and S.L. Turgeon. 2015. Characterization of antibacterial activity from protein hydrolysates of the macroalga *Saccharina longicruris* and identification of peptides implied in bioactivity. Journal of Functional Foods 17: 685–697.

Bechelli, J., M. Coppage, K. Rosell and J. Liesveld. 2011. Cytotoxicity of algae extracts on normal and malignant cells. Leukemia Research and Treatment Vol. 2011, Article ID 373519: 7 pp.

Beck-Sagué, C.M. and W.R. Jarvis. 1993. National nosocomial infections surveillance system. Secular trends in the epidemiology of nosocomial fungal infections in the United States, 1980–1990. The Journal of Infectious Diseases 167: 1247–1251.

Bedoux, G., K. Hardouin, A.S. Burlot and N. Bourgougnon. 2014. Chapter 12—Bioactive components from seaweeds: Cosmetic applications and future development. pp. 345–379. *In*: Curry, B.E. (ed.). Advances in Biological Research 71. Academic Press, Elsevier, London.

Belattmania, Z., A. Reani, M. Barakate, R. Zrid, S. Elatouani, M. Hassouani, A. Eddaoui, F. Bentiss and B. Sabour. 2016. Antimicrobial, antioxidant and alginate potentials of *Dictyopteris polypodioides* (Dictyotales, Phaeophyceae) from the Moroccan Atlantic coast. Der Pharma Chemica 8(2): 216–226.

Belattmania, Z., A.H. Engelen, H. Pereira, Ester A. Serrão, M. Barakate, S. Elatouani, R. Zrid, F. Bentiss, N. Chahboun, A. Reani and B. Sabour. 2016b. Potential uses of the brown seaweed *Cystoseira humilis* biomass: 2—Fatty acid composition, antioxidant and antibacterial activities. Journal of Materials and Environmental Science 7(6): 2074–2081.

Belliot, G., A. Lavaux, D. Souihel, D. Agnello and P. Pothier. 2008. Use of Murine Norovirus as a surrogate to evaluate resistance of Human Norovirus to disinfectants. Applied and Environmental Microbiology 74: 3315–3318.

Benbow, M. 2009. Maintaining skin integrity and preventing pressure damage. Nursing and Residential Care 11: 44–450.

Benkendorff, K., A.R. Davis, C.N. Rogers and J.B. Bremner. 2005. Free fatty acids and steroids in the benthic spawn of aquatic molluscs, and their associated antimicrobial properties. Journal of Experimental Marine Biology and Ecology 316(1): 29–44.

Bennamara, A., A. Abourriche, M. Berrada, M. Charrouf, N. Chaib, M. Boundouma and X.F. Garjeall. 1999. Methoxybifurcarenone: an antifungal and antibacterial meroditerpenoid from the brown alga *Cystoseira tamariscifolia*. Phytochemistry 52: 37–40.

Bennet, J.W. and M. Klich. 2003. Mycotoxins. Clinical Microbiology Reviews 16: 497–516.

Bensch, K. 2016. *Fusarium udum*—Mycobank. Available online at: http://www.mycobank.org/BioloMICS.aspx?TableKey=14682616000000063&Rec=14899&Fields=All.

Béress, A., O. Wassermann, T. Bruhn, L. Béress, E.N. Kraiselburd, L.V. Gonzalez, G.E. de Motta and P.I. Chavez. 1993. A new procedure for the isolation of anti-HIV compounds (polysaccharides and polyphenols) from the marine alga *Fucus vesiculosus*. Journal of Natural Products 56: 478–488.

Berg, K., L. Zhai, M. Chen, A. Kharazmi and T.C. Owen. 1994. The use of a water-soluble formazam complex to quantitate the cell number and mitochrondrial-function of *Leishmania* major promastigotes. Parasitology Research 80: 235–239.

Bern, C., S. Kjos, M.J. Yabsley and S.P. Montgomery. 2011. *Trypanosoma cruzi* and Chagas' disease in the United States. Clinical Microbiology Reviews 24: 655–681.

Bernal, J. 2015. Thyroid hormones in brain development and function. TDF—Thyroid Disease Manager: 1–62. Available online at: http://www.thyroidmanager.org/chapter/thyroid-hormones-in-brain-development-and-function/.

Bernardi, G. and G.F. Springer. 1962. Properties of highly purified fucan. The Journal of Biological Chemistry 237: 75–80.

Besednova, N.N., I.D. Makarenkova, T.N. Zvyagintseva, T.I. Imbs, L.M. Somova and T.S. Zaporozhets. 2016. Antiviral activity and pathogenetic targets for seaweed sulfated polysaccharides in Herpesvirus infections. Biochemistry (Moscow) Supplement Series B: Biomedical Chemistry 10(1): 31–42.

Bhakta, T., P.K. Mukherjee, K. Mukherjee, S. Banerjee, C.M. Subhash, T.K. Maity, M. Pal and B.P. Saha. 1999. Evaluation of hepatoprotective activity of Cassia fistula leaf extract. Journal of Ethnopharmacology 66: 277–282.

Bhakuni, D.S. and D.S. Rawat. 2005. Bioactive Marine Natural Products. Springer, New Delhi, India, 354 pp.

Bhatia, S., A. Sharma, K. Sharma, M. Kavale, B.B. Chaugule, K. Dhalwal, A.G. Namdeo and K.R. Mahadik. 2008. Novel algal polysaccharides from marine source: Porphyran. Pharmacognosy Reviews 2(4): 271–276.

Bhatia, S., K. Sharma, A. Sharma, K. Nagpa and T. Bera. 2015. Anti-inflammatory, Analgesic and Antiulcer properties of Porphyra vietnamensis. Avicenna Journal of Phytonedicine 5(1): 69–77.

Bhuiyan, M.M.H., M. Mohibbullah, M.A. Hannan, Y.K. Hong, J.S. Choi, I.S. Choi and I.S. Moon. 2015. *Undaria pinnatifida* promotes spinogenesis and synaptogenesis and potentiates functional presynaptic plasticity in hippocampal neurons. The American Journal of Chinese Medicine 43: 529–542.

Bianco, A.C. and B.W. Kim. 2006. Deiodinases: implications of the local control of thyroid hormone action. Journal of Clinical Investigation 116(10): 2571–2579.

Bianco, E.M., S.Q. de Oliveira, C. Rigotto, M.L. Tonini, G.T. da Rosa, F. Bittencourt, L.P. Gouvêa, C. Aresi, M.T. de Almeida, M.I. Moritz, C.D. Martins, F. Scherner, J.L. Carraro, P.A. Horta, F.H. Reginatto, M. Steindel, C.M. Simões, E.P. Schenkel. 2013. Anti-infective potential of marine invertebrates and seaweeds from the Brazilian coast. Molecules 18: 5761–5778.

Bianco, Everson M., Luana Pires, Geanne K.N. Santos, Kamilla A. Dutra, Thiago N.V. Reis, Edson R.T.P.P. Vasconcelos, Adilma L.M. Cocentino and Daniela M.A.F. Navarro. 2013b. Larvicidal activity of seaweeds from northeastern Brazil and of a halogenated sesquiterpene against the dengue mosquito (*Aedes aegypti*). Industrial Crops and Products: 270–275.

Biard, J.F., J.F. Verbist, Y. Letourneux and R. Floch. 1980. Cétols diterpeniques à activité antimicrobienne de *Bifurcaria bifurcata*. Planta Medica 40(1): 288–294.

Bibbins-Domingo, K., G.M. Chertow, P.G. Coxson, A. Moran, J.M. Lightwood, M.J. Pletcher and L. Goldman. 2010. Projected effect of dietary salt reductions on future cardiovascular disease. New England Journal of Medicine 362: 590–599.

Bibiana, M.A., K. Nithya, M.S. Manikandan, P. Selvamani and S. Latha. 2012. Antimicrobial evaluation of the organic extracts of *Sargassum wigthii* (brown algae) and *Kappaphycus alvarezii* (red algae) collected from the coast of Meemesal, Tamilnadu. International Journal of Pharmaceutical, Chemical and Biological Sciences 2(4): 439–446.

Bick, R.L. and E.P. Frenkel. 1999. Clinical aspects of heparin-induced thrombocytopenia and thrombosis and other side effects of heparin therapy. Clinical and Applied Thrombosis/Hemostasis 5: S7–S15.

Bielenda. 2017. Institute Bielenda Professional. Available online at: http://bielenda.pl/en/series/professional-formula-series.

Billakanti, J.M., O.J. Catchpole, T.A. Fenton, K.A. Mitchell and A.D. MacKenzie. 2013. Enzyme-assisted extraction of fucoxanthin and lipids containing polyunsaturated fatty acids *from Undaria pinnatifida* using dimethyl ether and ethanol. Process Biochemistry 48(12): 1999–2008.

Birch, L.L. and J.O. Fisher. 1998. Development of eating behaviors among children and adolescents. Pediatrics 101: 539–549.

Bird, K.T., T.C. Chiles, R.E. Longley, A.F. Kendrick and M.D. Kinkema. 1992. Agglutinins from marine macroalgae of the southeastern United States. Journal of Applied Phycology 5: 213–218.

Biscoe, T.J., R.H. Evans, P.M. Headley, M. Martin and J.C. Watkins. 1975. Domic and quisqualic acids as potent amino acids excitants of frog and rat spinal neurons. Nature 255: 166–167.

Bitencourt, F.S., J.G. Figueiredo, M.R. Mota, C.C. Bezerra, P.P. Silvestre, M.R. Vale, K.S. Nascimento, A.H. Sampaio, C.S. Nagano, S. Saker-Sampaio, C.S. Nagano, S. Saker-Sampaio, W.R. Farias, B.S. Cavada, A.M. Assreuy and N.M. de Alencar. 2008. Antinociceptive and anti-inflammatory effects of a mucin-binding agglutinin isolated from the red marine *alga Hypnea cervicornis*. Naunyn-Schmiedeberg's Archives of Pharmacology 377(2): 139–148.

Bitencourt, M.A., G.R. Dantas, D.P. Lira, J.M. Barbosa-Filho, G.E. de Miranda, B.V. Santos and J.T. Souto. 2011. Aqueous and methanolic extracts of *Caulerpa mexicana* suppress cell migration and ear edema induced by inflammatory agents. Marine Drugs 9(8): 1332–1345.

Bitencourt, M.A.O., H.M.D. Silva, G.M.F. Abílio, George E.C. Miranda, A.M.A. Moura, J.X. de Araújo-Júnior, E.J.D. Silveira, B.V.O. Santos and J.T. Souto. 2015. Anti-inflammatory effects of methanolic extract of green algae *Caulerpa mexicana* in a murine model of ulcerative colitis. Revista Brasileira de Farmacognosia 25: 677–682.

Bixler, H.J. 1996. Recent developments in manufacturing and marketing carrageenan. Hydrobiologia 326/327: 35–57.

Bixler, H.J. and H. Porse. 2011. A decade of change in the seaweed hydrocolloid industry. Journal of Applied Phycology 23: 321–335.

Bjarkam, C.R., J.C. Sørensen, N.Å. Sunde, F.A. Geneser and K. Østergaard. 2001. New strategies for the treatment of Parkinson's disease hold considerable promise for the future management of neurodegenerative disorders. Biogerontology 2: 193–207.

Blackman, A.J. and R.J. Wells. 1978. Flexilin and trifarin, terpene 1,4-diaectoxybuta I ,3-dienes from the *Caulerpa* species (Chlorophyta). Tetrahedron Letters 3063–3064.

Blackmold. 2016. *Asperigillus*. Available on line at: http://blackmold.awardspace.com/aspergillus.html.

Blasko, I., M. Stampfer-Kountchev, P. Robatscher, R. Veerhuis, P. Eikelenboom and B. Grubeck-Loebenstein. 2004. How chronic inflammation can affect the brain and support the development of Alzheimer's disease in old age: The role of microglia and astrocytes. Aging Cell 3: 169–176.

Bodey, G.P. and S. Vartivarian. 1989. Aspergillosis. European Journal of Clinical Microbiology & Infectious Diseases 8: 413–437.

Bogentoff, C.B. 1981. Antacid composition for neutralizing stomach acid. Germany Patent 2,722,484.

Bohn, J.A. and J.N. BeMiller. 1995. (1→3)-β-d-Glucans as biological response modifiers: a review of structure-functional activity relationships. Carbohydrate Polymers 28: 3–14.

Boisvert, C., L. Beaulieu, C. Bonnet and É. Pelletier. 2015. Assessment of the antioxidant and antibacterial activities of three species of edible seaweeds. Journal of Food Biochemistry 39: 377–387.

Bolhassani, A., T. Taheri, Y. Taslimi, S. Zamanilui, F. Zahedifard, N. Seyed, F. Torkashvand, B. Vaziri and S. Rafati. 2011. Fluorescent *Leishmania* species: Development of stable GFP expression and its application for *in vitro* and *in vivo* studies. Experimental Parasitology 127: 637–645.

Bondu, S., C. Bonnet, J. Gaubert, É. Deslandes, S.L. Turgeon and L. Beaulieu. 2014. Bioassay-guided fractionation approach for determination of protein precursors of proteolytic bioactive metabolites from macroalgae. Journal of Applied Phycology (27): 2059–2074.

Boo, H.-J., J.H. Hyun, S.C. Kim, J.I. Kang, M.K. Kim, S.Y. Kim, H. Cho, E.-S. Yoo and H.-K. Kang. 2011. Fucoidan from *Undaria pinnatifida* induces apoptosis in A549 human lung carcinoma cells. Phytotherapy Research 25: 1082–1086.

Boo, H.-J., J.-Y. Hong, S.-C. Kim, J.-I. Kang, M.-K. Kim, E.-J. Kim, J.-W. Hyun, Y.-S. Koh, E.-S. Yoo, J.-M. Kwon and H.-K. Kang. 2013. The anticancer effect of fucoidan in PC-3 prostate cancer cells. Marine Drugs 11(8): 2982–2999.

Boonchum, W., Y. Peerapornpisal, D. Kanjanapothi, J. Pekkoh, D. Amornlerdpison and C. Pumas. 2011. Antimicrobial and anti-inflammatory properties of various seaweeds from the Gulf of Thailand. International Journal of Agriculture and Biology 13(1): 100–104.

Boonme, P., V.B. Junyaprasert, N. Suksawad and S. Songkro. 2009. Microemulsions and nanoemulsions: novel vehicles for whitening cosmeceuticals. Journal of Biomedical Nanotechnology 5: 373–383.

Borai, I.H., M.K. Ezz, M.Z. Rizk, A.A. Matloub, H.F. Aly, A. El, R. Farrag and G.I. Fouad. 2015. Hypolipidemic and anti-atherogenic effect of sulphated polysaccharides from the green alga *Ulva fasciata*. International Journal of Pharmaceutical Sciences Review and Research 31: 1–12.

Born, F.S., E.M. Bianco and C.A. da Camara. 2012. Acaricidal and repellent activity of terpenoids from seaweeds collected in Pernambuco, Brazil. Natural Product Communications 7(4): 463–466.

Borowitzka, M.A. 2013. High-value products from microalgae—their development and commercialization. Journal of Applied Phycology 25: 743–756.

Boscá, L., M. Zeini, P. Través and S. Hortelano. 2005. Nitric oxide and cell viability in inflammatory cells: A role for NO in macrophage function and fate. Toxicology 208: 249–258.

Bouhadir, K.H., K.Y. Lee, E. Alsberg, K.L. Damm, K.W. Anderson and D.J. Mooney. 2001. Degradation of partially oxidized alginate and its potential application for tissue engineering. Biotechnology Progress 17(5): 945–950.

Bouhlal, R., H. Riadi and N. Bourgougnon. 2010. Antiviral activity of the extracts of Rhodophyceae from Morocco. African Journal of Biotechnology 9: 7968–7975.

Bouhlal, R., H. Riadi, J. Martínez and N. Bourgougnon. 2010b. The antibacterial potential of the seaweeds (Rhodophyceae) of the Strait of Gibraltar and the Mediterranean coast of Morocco. African Journal of Biotechnology 9: 6365–6372.

Bouhlal, R., C. Haslin, J.-C. Chermann, S. Colliec-Jouault, C. Sinquin, G. Simon, S. Cerantola, H. Riadi and N. Bourgougnon. 2011. Antiviral activities of sulfated polysaccharides isolated from *Sphaerococcus coronopifolius* (Rhodophyta, Gigartinales) and *Boergeseniella thuyoides* (Rhodophyta, Ceramiales). Marine Drugs 9: 1187–1209.

Bourgougnon, N., M. Lahaye, J.-C. Chermann and J.-M. Kornprobst. 1993. Composition and antiviral activities of a sulfated polysaccharide from *Schizymenia dubyi* (Rhodophyta, Gigartinales). Bioorganic & Medicinal Chemistry Letters 3(6): 1141–1146.

Bourgougnon, N., C. Roussakis, J.-M. Kornprobs and M. Lahaye. 1994. Effects *in vitro* of sulfated polysaccharide from *Schizymenia dubyi* (Rhodophyta, Gigartinales) on a non-small-cell bronchopulmonary carcinoma line (NSCLC-N6). Cancer Letters 85: 87–92.

Bourgougnon, N., M. Lahaye, B. Quemener, J.C. Chermann, M. Rimbert, M. Cormaci, G. Furnari and J.M. Komprobst. 1996. Annual variation in composition and *in vitro* anti-HIV-1 activity of the sulfated glucuronogalactan from *Schizymenia dubyi* (Rhodophyta, Gigartinales). Journal of Applied Phycology 8: 155–161.

Box, A., A. Sureda, J. Terrados, A. Pons and S. Deudero. 2008. Antioxidant response and caulerpenyne production of the alien *Caulerpa taxifolia* (Vahl) epiphytized by the invasive algae Lophodadia lallemandii (Montagne). Journal of Experimental Marine Biology and Ecology 364: 24–28.

Boyd, W.C. and R.M. Reguera. 1949. Hemagglutinating substances in various plants. The Journal of Immunology 62: 333–339.

Brailski, Kh. and B. Dimitrov. 1987. Treatment of duodenal ulcer with the Bulgarian preparation Gastralgin. Vutreshni Bolesti 26(2): 79–82.

Brameyer, S. and R. Heermann. 2015. Specificity of signal-binding *via* non-AHL LuxR-type receptors. PLoS ONE 10(4): e0124093.

Brannon-Peppas, L. and J.O. Blanchette. 2004. Nanoparticle and targeted systems for cancer therapy. Advanced Drug Delivery Reviews 56(11): 1649–5169.

Braune, W. and M. Guiry. 2011. Seaweeds—A colour guide to common benthic green. *In*: A.R.G. Gatner (ed.). Brown and Red Algae of the World's Oceans. K.G. Verlag, Ruggell, 601 pp.

Breithaupt, D.E. 2004. Simultaneous HPLC determination of carotenoids used as food coloring additives: Applicability of accelerated solvent extraction. Food Chemistry 86(3): 449–456.

Brenner, M. and V.J. Hearing. 2007. The protective role of melanin against UV damage in human skin. Photochem. Photobiol. 84(3): 539–549.

Bu, T., M. Liu, L. Zheng, Y. Guo and X. Lin. 2010. Alpha-clucosidase inhibition and the *in vivo* hypoglycemic effect of butyl-isobutyl-phthalate derived from the *Laminaria japonica* rhizoid. Phytotherapy Research 24(11): 1588–1591.

Buck, C.B., C.D. Thompson, J.N. Roberts, M. Muller, D.R. Lowy and J.T. Schiller. 2006. Carrageenan is a potent inhibitor of papillomavirus infection. PLoS Pathogens 2: 671–680.

Buckner, F.S., C.L. Verlinde, A.C. LaFlamme and W.C. VanVoorhis. 1996. Efficient technique for screening drugs for activity against *Trypanosoma cruzi* using parasites expressing beta-galactosidase. Antimicrobial Agents and Chemotherapy 40: 2592–2597.

Buggeln, J.S. and R.G. Carige. 1973. The physiology and biochemistry of *Chondrus crispus* Stackhouse. pp. 81–102. *In*: Harvey, M.J. and J. MacLachlan (eds.). *Chondrus crispus*. Nova Scotian Institute of Science, Halifax, Nova Scotia.

Bulliyya, G. 2000. Key role of dietary fats in coronary heart disease under progressive urbanization and nutritional transition. Asia Pacific Journal of Clinical Nutrition 9: 289–297.

Burgi, H. 2010. Iodine Excess. Best Practice & Research Clinical Endocrinology & Metabolism 24(1): 107–15.

Burkholder, P.R., L.M. Burkholder and L.R. Almodovar. 1960. Antibiotic activity of some marine algae of Puerto Rico. Botanica Marina 2: 149–156.

Burkholder, P.R. 1973. The ecology of marine antibiotics and coral reefs, pp. 117–182. *In*: Jones, O.A. and R. Endean (eds.). Biology and Geology of Coral Reefs. Academic Press, New York.

Burtin, P. 2003. Nutritional value of seaweeds. Electronic Journal of Environmental, Agricultural and Food Chemistry 2: 498–503.

Bushing, B. 2014. The Invasion of the "Devil Weed": *Sargassum horneri* invading California waters. California Diver. Available online at: http://californiadiver.com/the-invasion-of-the-devil-weed-sargassum-horneri-invading-california-waters/.

Cabioc'h, J., J.-Y. Floc'h, A. Le Toquin, J.-F. Boudouresque, A. Meinesz and M. Verlaque. 2014. Guide des algues des mers d'Europe. Edition Delachaux et Niestlé, Paris, 272 pp.

Caccamese, S., L.P. Hager, K.l. Ribehart and R.B. Setzer. 1979. Characterization of *Laureneia* species by gas chromatography mass spectrometry. Botanica Marina 22: 41–46.

Caccamese, S., R. Azzolina, G. Furnari, M. Cormaci and S. Grasso. 1980. Antimicrobial and antiviral activities of extracts from Mediterranean algae. Botanica Marina 23: 285–288.

Caccamese, S., R. Azzolina, G. Furnari, M. Cormaci and S. Grasso. 1981. Antimicrobial and antiviral activities of some marine algae from eastern Sicily. Botanica Marina 24(7): 365–368.

Cáceres, P.J., M.J. Carlucci, E.B. Damonte, B. Matsuhiro and E.A. Zúñiga. 2000. Carrageenans from Chilean samples of *Stenogramme interrupta* (Phyllophoraceae): Structural analysis and biological cctivity. Phytochemistry 53(1): 81–86.

Cai, C., C. Li, S. Wu, Q. Wang, Z. Guo and P. He. 2012. Large scale preparation of phycobiliproteins from *Porphyra yezoensis* using co-precipitation with ammonium sulfate. Natural Science 4(8): 536–543.

Calder, P.C. 1998. Immunoregulatory and anti-inflammatory effects of *n*–3 polyunsaturated fatty acids. Brazilian Journal of Medical and Biological Research 31: 467–490.

Calder, P.C. 2009. Polyunsaturated fatty acids and inflammatory processes: New twists in an old tale. Biochimie 91: 791–795.

Calvano, T.P., P.J. Blatz, T.J. Vento, B.L. Wickes, D.A. Sutton, E.H. Thompson, C.E. White, E.M. Renz and D.R. Hospenthal. 2011. *Pythium aphanidermatum* infection following combat trauma. Journal of Clinical Microbiology 49(10): 3710–3713.

Camara, B.G.R., L.S. Costa, G.P. Fidelis, L.T.D.B. Nobre, N. Dantas-Santos, S.L. Cordeiro, M.S.S.P. Costa, L.G. Alves and H.A.O. Rocha. 2011. Heterofucans from the brown seaweed *Canistrocarpus cervicornis* with anticoagulant and antioxidant activities. Marine Drugs 9(1): 124–138.

Campo, V.L., D.F. Kawano, D.B. da Silva, Jr. and I. Carvalho. 2009. Carrageenans: Biological properties, chemical modifications and structural analysis—a review. Carbohydrate Polymers 77(2): 167–180.

Campos, A., C.B. Souza, C. Lhullier, M. Falkenberg, E.P. Schenkel, R.M. Ribeiro-do-Valle and J.M. Siqueira. 2012. Antitumour effects of elatol, a marine derivative compound obtained from red algae *Laurencia microcladia*. Journal of Pharmacy and Pharmacology 64(8): 1146–1154.

Canavaci, A.M.C., J.M. Bustamante, A.M. Padilla, C.M.P. Brandan, D. Xu, C.L. Boehlke and R.L. Tarleton. 2010. *In vitro* and *in vivo* high-throughput assays for the testing of anti-*Trypanosoma cruzi* compounds. PLOS Neglected Tropical Diseases 4: e740.

Canducci, F., M. Debiaggi, M. Sampaolo, M.C. Marinozzi, S. Berre, C. Terulla, G. Gargantini, P. Cambieri, E. Romero and M. Clementi. 2008. Two-year prospective study of single infections and co-infections by respiratory syncytial virus and viruses identified recently in infants with acute respiratory disease. Journal of Medical Virology 80(4): 716–723.

Cannon, J.L., E. Papafragkou, G.W. Park, J. Osborne, L.A. Jaykus and J. Vinjé. 2006. Surrogates for the study of norovirus stability and inactivation in the environment: a comparison of murine norovirus and feline calicivirus. Journal of Food Protection 69: 2761–2765.

Cannon, R.D., E. Lamping, A.R. Holmes, K. Niimi, K. Tanabe, M. Niimi and B.C. Monk. 2007. *Candida albicans* drug resistance another way to cope with stress. Microbiology 153(10): 3211–3217.

Cantey, P.T., S.L. Stramer, R.L. Townsend, H. Kamel, K. Ofafa, C.W. Todd, M. Currier, S. Hand, W. Varnado, E. Dotson, C. Hall, P.L. Jett and S.P. Montgomery. 2012. The United States *Trypanosoma cruzi* infection study: evidence for vector-borne transmission of the parasite that causes Chagas disease among United States blood donors. Transfusion 52: 1922–1930.

Cantillo-Ciau, Z., R. Moo-Puc, L. Quijano and Y. Freile-Pelegrín. 2010. The tropical brown alga *Lobophora variegata*: a source of antiprotozoal compounds. Marine Drugs 8: 1292–1304.

Caper, W. 2011. Typhoid fever: Dirty food, dirty water! Beaport Publishing—High-Interest Nonfiction for Grades K-8 Company, New York, 32 pp.

Caplan, M.S. and T. Jilling. 2001. The role of polyunsaturated fatty acid supplementation in intestinal inflammation and neonatal necrotizing enterocolitis. Lipids 36(9): 1053–1057.

Capon, R.J., E.L. Ghisalberti and P.R. Jefries. 1981. New sesquiterpenes from *Caulerpa flesilis* var. *muelleri*. Australian Journal of Chemistry 34: 1775–1778.

Capon, R.I., R.A. Barrow, S. Rochfort, M. Jobling, C. Skene, E. Larcey, I.H. Gill, T. Friedel and D. Wadsworth. 1998. Marine Nematodes: Tetrahydrofuran from a southern Australian brown alga, *Notheia anomala*. Tetrahedron 54: 2227–2242.

Cardoso, S.M., L.G. Carvalho, P.J. Silva, M.S. Rodrigues, O.R. Pereira and L. Pereira. 2014. Bioproducts from seaweeds: a review with special focus on the Iberian Peninsula. Current Organic Chemistry 18(7): 896–917.

Cardoso, S.M., O.R. Pereira, A.M. Seca, D. Pinto and A.M. Silva. 2015. Seaweeds as preventive agents for cardiovascular diseases: From nutrients to functional Foods. Marine Drugs 13: 6838–6865.

Carlucci, M.J., C.A. Pujol, M. Ciancia, M.D. Noseda, M.C. Matulewicz, E.B. Damonte and A.S. Cerezo. 1997a. Antiherpetic and anticoagulant properties of carrageenans from the red seaweed *Gigartina skottsbergii* and their cyclized derivatives: correlation between structure and biological activity. International Journal of Biological Macromolecules 20(2): 97–105.

Carlucci, M.J., L.A. Scolaro, M.I. Errea, M.C. Matulewicz and E.B. Damonte. 1997b. Antiviral activity of natural sulphated galactans on *Herpes* virus multiplication in cell culture. Planta Medica 63: 429–432.

Carlucci, M.J., L.A. Scolaro and E.B. Damonte. 1999. Inhibitory action of natural carrageenans on *Herpes simplex* virus infection of mouse astrocytes. Chemotherapy 45: 429–436.

Carlucci, M.J., L.A. Scolaro and E.B. Damonte. 2002. *Herpes simplex* virus type 1 variants arising after selection with an antiviral carrageenan: lack of correlation between drug susceptibility and synphenotype. Journal of Medical Virology 68(1): 92–98.

Carlucci, M.J., L.A. Scolaro, M.D. Noseda, A.S. Cerezo and E.B. Damonte. 2004. Protective effect of a natural carrageenan on genital *Herpes simplex* virus infection in mice. Antiviral Research 64(2): 137–141.

Caroprese, M., M. Albenzio, M.G. Ciliberti, M. Francavilla and A. Sevi. 2012. A mixture of phytosterols from *Dunaliella tertiolecta* affects proliferation of peripheral blood mononuclear cells and cytokine production in sheep. Veterinary Immunology and Immunopathology 150(1-2): 27–35.

Carreto, J.I., M.O. Carignan and N.G. Montoya. 2001. Comparative studies on mycosporine-like amino acids, paralytic shellfish toxins and pigment profiles of the toxic dinoflagellates *Alexandrium tamarense, A. catenella* and *A. minutum*. Marine Ecology Progress Series 223: 49–60.

Carroll, K.C. and J.A. Hobden. 2016. Normal human microbiota. *In*: Weitz, M. and B. Kearns (eds.). Jawetz, Melnick & Adelberg's Medical Microbiology (27th ed.). The McGraw-Hill Education, LLC.

Carter, G.T., Jr. Rinehart, L.H. Li and S.L. Kuentzel. 1978. Brominated indoles from *Laurencia brongniartii*. Tetrahedron Letters 19: 4479–4482.

Cartlidge, P. 2000. The epidermal barrier. Seminars in Neonatology 5: 273–280.

Carvalho, D. and M.M. Lopes. 2016a. Atlas de Micologia (Atlas of Mycology). Available online at: http://atlasmicologia.blogspot.pt/2013/06/microsporum-gypseum.html.

Carvalho, L.G. 2013. Avaliação do potencial biotecnológico de micro e macroalgas da flora portuguesa. Dissertação de Mestrado em Biodiversidade e Biotecnologia Vegetal, Departamento de Ciências da Vida, Universidade de Coimbra, Portugal, 104 pp. Available online at: https://estudogeral.sib.uc.pt/handle/10316/24928.

Carvalho, L.G. and L. Pereira. 2015. Review of marine algae as source of bioactive metabolites, Chapter 8. pp. 192–224. *In*: Pereira, L. and J.M. Neto (eds.). Marine Algae—Biodiversity, Taxonomy, Environmental Assessment and Biotechnology. Science Publishers, an Imprint of CRC Press/Taylor & Francis Group, Boca Raton, FL.

Casas, M.P., V. Rodríguez-Hermida, P. Pérez-Larrán, E. Conde, M.T. Liveri, D. Ribeiro, E. Fernandes and H. Domínguez. 2016. *In vitro* bioactive properties of phlorotannins recovered from hydrothermal treatment of *Sargassum muticum*. Separation and Purification Technology 167: 117–126.

Cassano, V., Y. Metti, A.J.K. Millar, M.C.G. Rodríguez, A. Sentíes, J. Díaz-Larrea, M.C. Oliveira and M.T. Fujii. 2012. Redefining the taxonomic status of *Laurencia dendroidea* (Ceramiales, Rhodophyta) from Brazil and the Canary Islands. European Journal of Phycology 47(1): 67–81.

Cassolato, J.E., M.D. Noseda, C.A. Pujol, F.M. Pellizzari, E.B. Damonte and M.E. Duarte. 2008. Chemical structure and antiviral activity of the sulfated heterorhamnan isolated from the green seaweed *Gayralia oxysperma*. Carbohydrate Research 343(18): 3085–3095.

Castelli, R., E. Cassinerio, M.D. Cappellini, F. Porro, G. Graziadei and F. Fabris. 2007. Heparin induced thrombocytopenia: pathogenic, clinical, diagnostic and therapeutic aspects. Cardiovascular & Hematological Disorders-Drug Targets 7: 153–162.

Castillo, S., N. Heredia and S. García. 2015. 2 (5*H*)-Furanone, epigallocatechin gallate, and a citric-based disinfectant disturb quorum-sensing activity and reduce motility and biofilm formation of *Campylobacter jejuni*. Folia Microbiologica 60: 89–95.

Castro-Gonzalez, M.I., F.P.G. Romo, S. Perez-Estrella and S. Carrillo-Dominguez. 1996. Chemical composition of the green alga *Ulva lactuca*. Ciencias Marinas 22(2): 205–213.

Cavaco, T.M. 2017. Fatty acid composition and bioactive compounds of edible seaweeds. Potential as functional food ingredients. PhD thesis proposal, Universidade do Algarve, Faro, Portugal, 138 pp.

Cavalcante-Silva, L.H., C.B. da Matta, M.V. de Araujo, J.M. Barbosa-Filho, D.P. de Lira, B.V. de Oliveira Santos, G.E. de Miranda and M.S. Alexandre-Moreira. 2012. Antinociceptive and anti-inflammatory activities of crude methanolic extract of red alga *Bryothamnion triquetrum*. Marine Drugs 10(9): 1977–1992.

Cavallo, A.R., M.I. Acquaviva, L. Stabili, E. Cecere, A. Petrocelli and M. Narracci. 2013. Antibacterial activity of marine macroalgae against fish pathogenic *Vibrio* species. Central European Journal of Biology 8(7): 646–653.

Cavas, L., Y. Baskin, K. Yurdakoc and N. Olgun. 2006. Antiproliferative and newly attributed apoptotic activities from an invasive marine alga: *Caulerpa racemosa* var. *cylindracea*. Journal of Experimental Marine Biology and Ecology 339(1): 111–119.

CDC—Center for Disease Control and Prevention. 2011. Updated Norovirus outbreak management and disease prevention guidelines. MMWR Recommendations and Reports 60: 1–18.

CDC—Center for Disease Control and Prevention. 2016. LDL and HDL: bad and good cholesterol. Available online at: http://www.cdc.gov/cholesterol/ldl_hdl.htm.

CDC—Center for Disease Control and Prevention. 2017. Parasites—Lymphatic Filariasis. Available online at: https://www.cdc.gov/parasites/lymphaticfilariasis/index.html.

Celikler, S., S. Tas, O. Vatan, S. Ziyanok-Ayvalik, G. Yildiz and R. Bilaloglu. 2009. Anti-hyperglycemic and antigenotoxic potential of *Ulva rigida* ethanolic extract in the experimental diabetes mellitus. Food Chem Toxicol 47: 1837–1840.

Cercaci, L., A. Conchillo, M.T. Rodriguez-Estrada, A.D. Ansorena, I. Astiasaran and G. Lerker. 2006. Preliminary study on health-related lipid components of bakery products. Journal of Food Protection 69: 1393–1401.

Ceresini, P. 1999. *Thanatephorus cucumeris* (formerly *Rhizoctonia solani*)—Pathogen profile. Available online at: https://projects.ncsu.edu/cals/course/pp728/Rhizoctonia/Rhizoctonia.html.

Cermák, L., Š. Pražáková, M. Marounek, M. Skrivan and E. Skrivanová. 2015. Effect of green alga *Planktochlorella nurekis* on selected bacteria revealed antibacterial activity *in vitro*. Czech Journal of Animal Science 60: 427–435.

Cerna, M. 2011. Seaweed proteins and amino acids as nutraceuticals. Advances in Food and Nutrition Research 64: 297–312.

Cersosimo, M.G. and W.C. Koller. 2006. The diagnosis of manganese-induced parkinsonism. Neurotoxicology 27: 340–346.

Cha, S.-H., G.N. Ahn, S.J. Heo, K.N. Kim, K.-W. Lee, C.B. Song, S.K. Cho and Y.-J. Jeon. 2006a. Screening of extracts from marine green and brown algae in Jeju for potential marine Angiotensin-I converting enzyme (ACE) inhibitory activity. Journal of the Korean Society of Food Science and Nutrition 35(3): 307–314.

Cha, S.-H., K.-W. Lee and Y.-J. Jeon. 2006b. Screening of extracts from red algae in Jeju for potentials marine Angiotensin-I converting enzyme (ACE) inhibitory activity. Algae 21: 343–348.

Cha, S.H., S.C. Ko, D. Kim and Y.J. Jeon. 2011. Screening of marine algae for potential tyrosinase inhibitor: those inhibitors reduced tyrosinase activity and melanin synthesis in zebra fish. Journal of Dermatology 38: 354–363.

Chakraborty, K., A.P. Lipton, R.P. Raj and K.K. Vijayan. 2010. Antibacterial labdane diterpenoids of *Ulva fasciata* Delile from Southwestern coast of the Indian Peninsula. Food Chemistry 119(4): 1399–1408.

Chakraborty, S. and S.C. Santra. 2008. Biochemical composition of eight benthic algae collected from Sunderban. Indian Journal of Marine Sciences 37(3): 329–332.

Chale-Dzul, J., R. Moo-Puc, D. Robledo and Y. Freile-Pelegrín. 2015. Hepatoprotective effect of the fucoidan from the brown seaweed *Turbinaria tricostata*. Journal of Applied Phycology 27: 2123–2135.

Chambers, L.D., C. Hellio, K.R. Stokes, S.P. Dennington, L.R. Goodes, R.J.K. Wood and F.C. Walsh. 2011. Investigation of *Chondrus crispus* as a potential source of new antifouling agents. International Biodeterioration & Biodegradation 65: 939–946.

Chan, P.T., P. Matanjun, S.M. Yasir and T.S. Tan. 2014. Antioxidant and hypolipidaemic properties of red seaweed, *Gracilaria changii*. Journal of Applied Phycology 26(2): 987–997.

Chan, Y.Y., K.H. Kim and S.H. Cheah. 2011. Inhibitory ects of *Sargassum polycystum* on tyrosinase activity and melanin formation in B16F1O murine melanoma cells. Journal of Ethnopharmacology 137(3): 1183–1188.

Chanda, S., R. Dave, M. Kaneria and K. Nagani. 2010. Seaweeds: A novel, untapped source of drugs from sea to combat Infectious diseases. pp. 473–480. *In*: Mendez-Vilas, A. (ed.). Current Research, Technology and Education Topics in Applied Microbiology and Microbial Biotechnology. FORMATEX Research Center, Badajoz, Spain.

Chandía, N.P. and B. Matsuhiro. 2008. Characterization of a fucoidan from *Lessonia vadosa* (Phaeophyta) and its anticoagulant and elicitor properties. International Journal of Biological Macromolecules 42: 235–240.

Chandrasekaran, M., V. Venkatesalu and G.A. Raj. 2014. Antibacterial activity of selected marine macro algae against vancomycin resistant *Enterococcus faecalis*. Journal of Coastal Life Medicine 2(12): 940–946.

Chandy, T., D.L. Mooradian and G.H.R. Rao. 1998. Chitosan/polyethylene glycol-alginate microcapsules for oral delivery of hirudin. Journal of Applied Polymer Science 70: 2143–2153.

Chang, C.-J., Y.-H. Yang, Y.-C. Liang, C.-J. Chiu, K.-H. Chu, H.-N. Chou and B.-L. Chiang. 2011. A novel Phycobiliprotein alleviates allergic airway inflammation by modulating immune responses. American Journal of Respiratory and Critical Care Medicine 183(1): 15–25.

Chapman, V.J. and D.J. Chapman. 1980. Seaweeds and their uses. Chapman and Hall, New York, pp. 1–334.

Chargaff, E., F.W. Bancroft and M.S. Brown. 1936. Studies on the chemistry of coagulation II. On the inhibition of blood clotting by substance oh high molecular weight. The Journal of Biological Chemistry 115: 155–161.

Chari, R.V. 2008. Targeted cancer therapy: conferring specificity to cytotoxic drugs. Accounts of Chemical Research 41(1): 98–107.

Charlier, R.H. 1975. Thalassotherapy. pp. 807–813. *In*: Fairbridge, R.W. and Schwartz, M. (eds.). Encyclopedia of Coasts and Beaches. Englewood Cliffs, NJ.

Charlier, R.H. 2002. Thalassotherapy. Proceedings of the Romanian Seminar on Advances in Marine Research, Constanta.

Charlier, R.H. and M.-C.P. Chaineux 2009. The healing sea: A sustainable coastal ocean resource: Thalassotherapy. Journal of Coastal Research 25(4): 838–856.

Chatter, R., R. Ben Othman, S. Rabhi, M. Kladi, S. Tarhouni, C. Vagias, V. Roussis, L. Guizani-Tabbane and R. Kharrat. 2011. *In vivo* and *in vitro* anti-inflammatory activity of neorogioltriol, a new diterpene extracted from the red algae *Laurencia glandulifera*. Marine Drugs 9(7): 1293–1306.

Chatterjee, M.S., W.S. Denney, H. Jing and S.L. Diamond. 2010. Systems biology of coagulation initiation: kinetics of thrombin generation in resting and activated human blood. PLOS Computational Biology 6: 1–24.

Chattopadhyay, K., C.G. Mateu, P. Mandal, C.A. Pujol, E.B. Damonte and B. Ray. 2007. Galactan sulfate of *Grateloupia indica*: Isolation, structural features and antiviral activity. Phytochemistry 68(10): 1428–1435.

Chausmer, A.B. 1998. Zinc, insulin and diabetes. The Journal of the American College of Nutrition 17: 109–115.

Chaves, L.S., L.A. Nicolau, R.O. Silva, F.C. Barros, A.L. Freitas, K.S. Aragao, A.R. Ribeiro, M.H. Souza, A.L. Barbosa and J.V. Medeiros. 2013. Antiinflammatory and antinociceptive effects in mice of a sulfated polysaccharide fraction extracted from the marine red algae *Gracilaria caudata*. Immunopharmacology and Immunotoxicology 35(1): 93–100.

Chen, H., P. Sun and Q. Song. 1995. Lectins, a powerful defensive weapon of aquatic animals. Journal of Oceanography of Huanghai & Bohai Seas 13: 61–71.

Chen, H., Qifei Cong, Z. Du, W. Liao, L. Zhang, Y. Yao and K. Ding. 2016. Sulfated fucoidan FP08S2 inhibits lung cancer cell growth *in vivo* by disrupting angiogenesis *via* targeting VEGFR2/VEGF and blocking VEGFR2/Erk/VEGF signaling. Cancer Letters 382: 44–52.

Chen, H.-M., L. Zheng and X.-J. Yan. 2005. The preparation and bioactivity research of agaro-oligosaccharides. Food Technology and Biotechnology 43(1): 29–36.

Chen, J.H., J.D. Lim, E.H. Sohn, Y.S. Choi and E.T. Han. 2009. Growth-inhibitory effect of a fucoidan from brown seaweed *Undaria pinnatifida* on *Plasmodium parasites*. Parasitology Research 104: 245–250.

Chen, J.P., I.M. Chu, M.Y. Shiao, B.R.S. Hsu and S.H. Fu. 1998. Microencapsulation of islets in PEG-amine modified alginate-poly(l-lysine)-alginate microcapsules for constructing bioartificial pancreas. Journal of Fermentation and Bioengineering 86: 185–190.

Chen, K.J., C.K. Tseng, F.R. Chang, J.I. Yang, C.C. Yeh, W.C. Chen, S.F. Wu, H.W. Chang and J.C. Lee. 2013. Aqueous extract of the edible *Gracilaria tenuistipitata* inhibits Hepatitis C viral replication via cyclooxygenase-2 suppression and reduces virus-induced inflammation. PLoS One 8(2): 28.

Chen, K.-T., C.-H. Cheng, Y.-H. Wu, W.-C. Lu, Y.-H. Lin and H.-T. Lee. 2013b. Continuous lipid extraction of microalgae using high-pressure carbon dioxide. Bioresource Technology 146: 23–26.

Chen, M.-C., W.-L. Hsu, P.-A. Hwang and T.-C. Chou. 2015. Low molecular weight fucoidan inhibits tumor angiogenesis through downregulation of HIF-1/VEGF signaling under hypoxia. Marine Drugs 13(7): 4436–4451.

Chen, M.-Z., J. Yu, Z.-H. Liao and X. Wang. 2008. Study on the antitumor activity and antioxidation of polysaccharide from *Gracilaria lemaneiformis* in sarcoma 180 bearing mice. Chinese Journal of Marine Drugs 27(2): 46–49.

Chen, S., Y. Zhao, Y. Zhang and D. Zhang. 2014. Fucoidan induces cancer cell apoptosis by modulating the endoplasmic reticulum stress cascades. PLoS One 9: e108157.

Chen, T.C., F. Chimeh, Z. Lu, J. Mathieu, K.S. Person, A. Zhang, N. Kohn, S. Martinello, R. Berkowitz and M.F. Holick. 2007. Factors that influence the cutaneous synthesis and dietary sources of vitamin D. Archives of Biochemistry and Biophysics 460(2): 213–217.

Chen, Y., H. Lin, Z. Li and Q. Mou. 2015b. The anti-allergic activity of polyphenol extracted from five marine algae. Journal of Ocean University of China 14(4): 681–684.

Cheng, D.H., H. Ren, X.C. Tang and A. Huperzine. 1996. A novel promising acetylcholinesterase inhibitor. Neuroreport 8: 97–101.

Chesters, G.C.G. and J.A. Stott. 1956. Production of antibiotic substances by seaweeds. pp. 49–54. Proceedings of the 2nd International Seaweed Symposium, Pergamon Press, New York.

Chevolot, L., B. Mulloy, J. Ratiskol, A. Foucault and S. Colliec-Jouault. 2001. A disaccharide repeat unit is the major structure in fucoidans from two species of brown algae. Carbohydrate Research 330(4): 529–535.

Chia, Y.Y., M.S. Kanthimathi, J. Rajarajeswaran, K.S. Khoo and H.M. Cheng. 2015. Antioxidant, antiproliferative, genotoxic and Cytoprotective effects of the methanolic extract of *Padina tetrastromatica* on human breast adenocarcinoma and embryonic fibroblast cell lines. Frontiers in Life Science 8(4): 411–418.

Chiao-Wei, C., H. Siew-Ling and W. Ching-Lee. 2011. Antibacterial activity of *Sargassum polycystum* C. Agardh and *Padina australis* Hauck (Phaeophyceae). African Journal of Biotechnology 10(64): 14125–14131.

Chiellini, F. and A. Morelli. 2011. Ulvan: A versatile platform of biomaterials from renewable resources. *In*: R. Pignatello (ed.). Biomaterials—Physics and Chemistry. INTECH Open Access Publisher. Available online at: http://www.intechopen.com/books/biomaterials-physics-and-chemistry/ulvan-aversatile-platform-of-biomaterials-from-renewable-resources.

Chignell, C.F. and R.H. Sik. 2003. A photochemical study of cells loaded with 2',7'-dichlorofluorescin: Implications for the detection of reactive oxygen species generated during UVA irradiation. Free Radical Biology & Medicine 34(8): 1029–1034.

Chin, Y.X., P.E. Lim, C.A. Maggs, S.M. Phang, Y. Sharifuddin and B.D. Green. 2015. Anti-diabetic potential of selected Malaysian seaweeds. Journal of Applied Phycology 27(5): 2137–2148.

Chiu, Y.-H., Y.-L. Chan, L.W. Tsai, T.-L. Li and C.-J. Wu. 2012. Prevention of human enterovirus 71 infection by kappa carrageenan. Antiviral Research 95: 128–134.

Chiu, Y.-H., Y.-L. Chan, T.-L. Li and C.-J. Wu. 2012b. Inhibition of Japanese encephalitis virus infection by the sulfated polysaccharide extracts from *Ulva lactuca*. Marine Biotechnology 14(4): 468–478.

Chizhov, A.O., A. Dell, H.R. Morris, S.M. Haslam, R.A. McDowell, A.S. Shashkov, N.E. Nifantev, E.A. Khatuntseva and A.I. Usov. 1999. A study of fucoidan from the brown seaweed *Chorda filum*. Carbohydrat Research 320: 108–119.

Cho, H.B., L.H. Lee, O.H. Lee, H.S. Choi, J.S. Choi and B.Y. Lee. 2011. Clinical and microbial evaluation of the effects on gingivitis of a mouth rinse containing an *Enteromorpha linza* (*E. linza*) extract. Journal of Medicinal Food 14(12): 1670–1676.

Cho, J., Y.P. Gyawali, S. Ahn, M.N.A. Khan, I. Kong and Y. Hong. 2008. A methoxylated fatty acid isolated from the brown seaweed *Ishige okamurae* inhibits bacterial phospholipase A2. Phytotherapy Research 22(8): 1070–1074.

Cho, J.C., R.S. Sharma, D.A. Sutton, N.P. Wiederhold, C. Sanders, B.L. Wickes and S.J. Estrada. 2015. Fungal arthritis secondary to *Colletotrichum gloeosporioides*. JMM Case Reports 2(1): 1–4.

Cho, M., K.-S. Lee, I.-J. Kang, M.-H. Won and S. You. 2011b. Antioxidant properties of extract and fractions from *Enteromorpha prolifera*, a type of green seaweed. Food Chemistry 127: 999–1006.

Cho, M., G.M. Park, S.N. Kim, T. Amna, S. Lee and W.S. Shin. 2014. Glioblastoma-specific anticancer activity of pheophorbide A from the edible red seaweed *Grateloupia elliptica*. Journal of Microbiology and Biotechnology 24(3): 346–353.

Cho, M.L., B.Y. Lee and S.G. You. 2010. Relationship between oversulfation and conformation of low and high molecular weight fucoidans and evaluation of their *in vitro* anticancer activity. Molecules 16(1): 291–297.

Cho, S., D. Han, S.B. Kim, M. Yoon, H. Yang, Y.H. Jin, J. Jo, H. Yong, S.H. Lee, Y.J. Jeon and M. Shimizu. 2012. Depressive effects on the central nervous system and underlying mechanism of the enzymatic extract and its phlorotannin-rich fraction from *Ecklonia cava* edible brown seaweed. Bioscience, Biotechnology, and Biochemistry 76(1): 163–168.

Choi, B.W., G. Rhu, S.H. Park, E.S., Kim, J. Shin, S. Roh, H.C. Shin and B.H. Lee. 2007. Anticholinesterase activity of plastoquinones from *Sargassum sagamianum*, lead compounds for Alzheimer's disease therapy. Phytotherapy Research 21: 423e426.

Choi, B.W., H.S. Lee, H.C. Shin and B.H. Lee. 2015b. Multifunctional activity of polyphenolic compounds associated with a potential for Alzheimer's disease therapy from *Ecklonia cava*. Phytotherapy Research 29: 549–553.

Choi, E.M., A. Kim, Y.O. Kim and J.K. Hwang. 2005. Immunomodulating activity of arabinogalactan and fucoidan *in vitro*. Journal of Medicinal Food 8: 446–453.

Choi, J.-I., H.-J. Kim and J.-W. Lee. 2011. Structural feature and antioxidant activity of low molecular weight laminarin degraded by gamma irradiation. Food Chemistry 129: 520–523.

Choi, J.S., Y.M. Ha, C.U. Joo, K.K. Cho, S.J. Kim and I.S. Choi. 2012. Inhibition of oral pathogens and collagenase activity by seaweed extracts. Journal of Environmental Biology 33: 115–121.

Choi, J.-S., Y.-M. Ha, B.-B. Lee, H.E. Moon, K.K. Cho and I.S. Choi. 2013. Seasonal variation of antibacterial activities in the green alga *Ulva pertusa* Kjellman. Journal of Environmental Biology 35: 341–344.

Choi, J.-S., Y.-M. Ha, B.-B. Lee, H.E. Moon, K.K. Cho and I.S. Choi. 2014b. Seasonal variation of antibacterial activities in the green alga *Ulva pertusa* Kjellman. Journal of Environmental Biology 35: 341–344.

Choi, J.S., S. Haulader, S. Karki, H.J. Jung, H.R. Kim and H.A. Jung. 2015c. Acetyl- and butyryl-cholinesterase inhibitory activities of the edible brown alga *Eisenia bicyclis*. Archives of Pharmacal Research 38: 1477–1487.

Choi, S.M., E.J. Jang and J.D. Cha. 2015. Synergistic effect between fucoidan andantibiotics against clinic methicillin-resistant *Staphylococcus aureus*. Advancesin Bioscience and Biotechnology 6: 275–285.

Choi, Y., E. Kim, S. Moon, J.-D. Choi, M.-S. Lee and Y.-M. Kim. 2014. Phaeophyta extracts exhibit antiviral activity against feline calicivirus. Fisheries and Aquatic Sciences 17: 155–158.

Chojnacka, K. and S.-K. Kim. 2015. Introduction of marine algae extracts. pp. 1–14. *In*: Kim, S.-K. and K. Chojnacka (eds.). Marine Algae Extracts, Wiley-VCH Verlag GmbH & Co. KGaA, Weinheim, Germany.

Choudhury, S., A. Sree, S.C. Mukherjee, P. Pattnaik and M. Bapuji. 2005. *In vitro* antibacterial activity of extracts of selected marine algae and mangroves against fish pathogens. Asian Fisheries Science 18: 285–294.

Chowdhary, A., S. Kathuria, J. Xu, J. and J.F. Meis. 2013. Emergence of azole-resistant *Aspergillus fumigatus* strains due to agricultural azole use creates an increasing threat to human health. PLoS Pathogens 9(10): 1–5.

Chowdhary, A., J.F. Meis, J. Guarro, G.S. de Hoog, S. Kathuria, M.C. Arendrup, S. Arikan-Akdagli, M. Akova, T. Boekhout, M. Caira, J. Guinea, A. Chakrabarti, E. Dannaoui, A. van Diepeningen, T. Freiberger, A.H. Groll, W.W. Hope, E. Johnson, M. Lackner, K. Lagrou, F. Lanternier, C. Lass-Flörl, O. Lortholary, J. Meletiadis, P. Muñoz, L. Pagano, G. Petrikkos, M.D. Richardson, E. Roilides, A. Skiada, A.M. Tortorano, A.J. Ullmann, P.E. Verweij3, O.A. Cornely and M. Cuenca-Estrella. 2014. ESCMID and ECMM joint clinical guidelines for the diagnosis and management of systemic phaeohyphomycosis: diseases caused by black fungi. Clinical Microbiology and Infection 20(Suppl 3): 47–75.

Chowdhury, M.M.H., K. Kubra, M.B. Hossain, M.G. Mustafa, T. Jainab, M.R. Karim and M.E. Mehedy. 2015. Screening of antibacterial and antifungal activity of freshwater and marine algae as a prominent natural antibiotic available in Bangladesh. International Journal of Pharmacology 11(7): 828–833.

Christaki, E., E. Bonos, I. Giannenas and P. Florou-Paneri. 2013. Functional properties of carotenoids origznatmg from algae. Journal of the Science of Food and Agriculture 93(1): 5–11.

Chung, J.Y., J.H. Choo, M.H. Lee and J.K. Hwang. 2006. Anticarcinogenic activity of macelignan isolated from *Myristica fragrans* (Mitmeg) against *Streptococcus mutans*. Phytomedicine 13: 261–266.

Church, F.C., J.B. Meade, R.E. Treanor and H.C. Whinna. 1989. Antithrombin activity of fucoidan. The Journal of Biological Chemistry 264: 3618–3623.

Cian, R.E., O. Martínez-Augustin and S.R. Drago. 2012. Bioactive properties of peptides obtained by enzymatic hydrolysis from protein byproducts of *Porphyra columbina*. Food Research International (49): 364–372.

Cian, R.E., M. Alaiz, J. Vioque and S.R. Drago. 2013. Enzyme proteolysis enhanced extraction of ACE inhibitory and antioxidant compounds (peptides and polyphenols) from *Porphyra columbina* residual cake. Journal of Applied Phycology 25(4): 1197–1206.

Cian, R.E., M.A. Fajardo, M. Alaiz, J. Vioque, R.J. Gonzalez and S.R. 2014. Drago. Chemical composition, nutritional and antioxidant properties of the red edible seaweed *Porphyra columbina.* International Journal of Food Sciences and Nutrition 65(3): 299–305.

Cian, R.E., M.S. Caballero, N. Sabbag, R.J. González and S.R. Drago. 2014b. Bio-accessibility of bioactive compounds (ACE inhibitors and antioxidants) from extruded Maize products added with a red seaweed *Porphyra columbina.* LWT—Food Science and Technology 55(1): 51–58.

Cian, R.E., A.G. Garzón, D.B. Ancona, L.C. Guerrero and S.R. Drago. 2015. Hydrolyzates from *Pyropia columbina* seaweed have antiplatelet aggregation, antioxidant and ACE I inhibitory peptides which maintain bioactivity after simulated gastrointestinal digestion. LWT—Food Science and Technology 64: 881–888.

Ciancia, M., I. Quintana, M.I. Vizcargüénaga, L. Kasulin, A. de Dios, J.M. Estevez and A.S. Cerezo. 2007. Polysaccharides from the green seaweeds *Codium fragile* and *C. vermilara* with controversial effects on hemostasis. International Journal of Biological Macromolecules 41(5): 641–649.

Ciancia, M., I. Quintana and A.S. Cerezo. 2010. Overview of anticoagulant activity of sulfated polysaccharides from seaweeds in relation to their structures, focusing on those of green seaweeds. Current Medicinal Chemistry 1723: 2503–2529.

Cirne-Santos, C.C., T.M.L. Souza, V.L. Teixeira, C.F.L. Fontes, M.A. Rebello, L.R.R. Castello-Branco, C.M. Abreu, A. Tanuri, I.C. Frugulhetti and D.C. Bou-Habib. 2008. The dolabellane diterpene Dolabelladienetriol is a typical noncompetitive inhibitor of HIV-1 reverse transcriptase enzyme. Antiviral Research 77: 64–71.

Cisar, J.O., A.L. Sandberg, C. Abeygunawardana, G.P. Reddy and C.A. Bush. 1995. Lectin recognition of host-like saccharide motifs in streptococcal cell wall polysaccharides. Glycobiology 5: 655–662.

Clegg, D., D. Reda, C. Harris, M. Klein, J. O'Dell, M. Hooper, J. Bradley, C. Bingham, M. Weisman, C. Jackson, N. Lane, J. Cush, L. Moreland, H. Schumaker, C. Oddis, F. Wolfe, J. Molitor, D. Yokum, T. Schnitzer, D. Furst, A. Sawitzke, H. Shi, K. Brandt, R. Moskowitz and H. Williams. 2006. Glucosamine, chondroitin sulfate and the two in combination for painful knee osteoarthritis. The New England Journal of Medicine 354(8): 795–808.

Cleland, J.G.F., K. Swedberg and P.A. Poole-Wilson. 1998. Successes and failure of current treatment of heart failure. Lancet: 352: SI19–SI28.

Clements, J.E. and M.C. Zink. 1996. Molecular biology and pathogenesis of animal lentivirus infections. Clinical Microbiology Reviews 9: 100–117.

CLSI. 2015. Methods for dilution antimicrobial susceptibility tests for bacteria that grow aerobically; approved standard—10^{ed}, M07-A10. Clinical and Laboratory Standards Institute, Wayne, USA, 87 pp.

Codif. 2017. Technologie naturelle. Available online at: http://www.codif-tn.com/.

Colliec, S., A.M. Fischer, J. Tapon-Bretaudiere, C. Boisson, P. Durand and J. Jozefonvicz. 1991. Anticoaggulant properties of a fucoidan fraction. Thrombosis Research 64: 143–154.

Collins, P.L. and A.K. Ruth. 2013. Respiratory syncytial virus and metapneumovirus. pp. 1086–1123. *In*: Knipe, D.M. and P.M. Howley (eds.). Fields Virology, vol. I. Lippincott-Raven, Philadelphia.

Colwell, R.R. 1985. Marine polysaccharides for pharmaceutical and microbiological applications. pp. 363–376. *In*: Colwell, R.R., E.R. Pariser and J. Sinskey (eds.). Biotechnology of Marine Polysaccharides. McGraw-Hill, New York.

Coombe, D.R., C.R. Parish, I.A. Ramshaw and J.M. Snowden. 1987. Analysis of the inhibition of tumour metastasis by sulphated polysaccharides. International Journal of Cancer 39(1): 82–88.

Cordeiro, R.A., V.M. Gomes, A.F.U. Carvalho and V.M.M. Melo. 2006. Effect of proteins from red seaweed *Hypnea musciformis* (Wulfen) Lamouroux on the growth of human pathogen yeasts. Brazilian Archives of Biology and Technology 49(6): 915–924.

Cornish, M.L., A.T. Critchley and O.G. Mouritsen. 2015. A role for dietary macroalgae in the amelioration of certain risk factors associated with cardiovascular disease. Phycologia 54(6): 649–666.

Cornish, M.L., A.T. Critchley and O.G. Mouritsen. 2017. Consumption of seaweeds and the human brain. Journal of Applied Phycology 29(5): 2377–2398.

Cortés, Y., E. Hormazábal, H. Leal, A. Urzúa, A. Mutis, L. Parra and A. Quiroz. 2014. Novel antimicrobial activity of a dichloromethane extract obtained from red seaweed *Ceramium rubrum* (Hudson) (Rhodophyta: Florideophyceae) against *Yersinia ruckeri* and *Saprolegnia parasitica*, agents that cause diseases in salmonids. Electronic Journal of Biotechnology 17(3): 126–31.

Cosoveanu, A., O. Axine and B. Iacomi. 2010. Antifungal activity of macroalgae extracts. Scientific Papers—UASVM Bucharest Series A (Vol. LIII): 442–447.

Costa, L.S., G.P. Fidelis, S.L. Cordeiro, R.M. Oliveira, D.A. Sabry, R.B. Camara, L.T. Nobre, M.S. Costa, J. Almeida-Lima, E.H. Farias, E.L. Leite and H. A. Rocha. 2010. Biological activities of sulfated polysaccharides from tropical seaweeds. Biomedicine & Pharmacotherapy 64(1): 21–28.

Costa, L.S., G.P. Fidelis, C.B. Telles, N. Dantas-Santos, R.B. Camara, S.L. Cordeiro, M.S. Costa, J. Almeida-Lima, R.F. Melo-Silveira, R.M. Oliveira, I.R. Albuquerque, G.P. Andrade and H.A. Rocha. 2011a. Antioxidant and antiproliferative activities of heterofucans from the seaweed *Sargassum filipendula*. Marine Drugs 9(6): 952–966.

Costa, L.S., C.B.S. Telles, R.M. Oliveira, L.T.D.B. Nobre, N. Dantas-Santos, R.B.G. Camara, M.S.S.P. Costa, J. Almeida-Lima, R.F. Melo-Silveira, I.R.L. Albuquerque, E.L. Leite and H.A.O. Rocha. 2011b. Heterofucan from *Sargassum filipendula* induces apoptosis in HeLa cells. Marine Drugs 9(4): 603–614.

Coura, C.O., I.W.F. de Araujo, E.S.O. Vanderlei, J.A.G. Rodrigues, A.L.G. Quindere, B.P. Fontes, I.N.L. de Queiroz, D.B. de Menezes, M.M. Bezerra, A.A.R.E. Silva, H.V. Chaves, R.J.B. Evangelista and N.M.B. Benevides. 2012. Antinociceptive and anti-inflammatory activities of sulphated polysaccharides from the red seaweed *Gracilaria cornea*. Basic & Clinical Pharmacology & Toxicology 110(4): 335–341.

Cowan, C.M., J. Thai, S. Krajewski, J.C. Reed, D.W. Nicholson, S.H. Kaufmann and A.J. Roskams. 2001. Caspases 3 and 9 send a pro-apoptotic signal from synapse to cell body in olfactory receptor neurons. Journal of Neuroscience 21: 7099–7109.

Cowan, M.M. 1999. Plant products as antimicrobial agents. Clinical Microbiology Reviews 12: 564–582.

Cox, R.A. and M.R. Garcia-Palmieri. 1990. Cholesterol, triglycerides. and associated lipoproteins. pp. 153–160. *In*: H.K. Walker, W.D. Hall and J.W. Hurst (eds.). Clinical Methods: The History, Physical, and Laboratory Examinations, 3rd edition. Butterworths, a division of Reed Publishing, Boston.

Cox, S., N. Abu-Ghannam and S. Gupta. 2010. An assessment of the antioxidant and antimicrobial activity of six species of edible Irish seaweeds. International Food Research Journal 17: 205–220.

Cox, S., G.H. Turley, G. Rajauria, N. Abu-Ghannam and A.K. Jaiswal. 2014. Antioxidant potential and antimicrobial efficacy of seaweed (*Himanthalia elongata*) extract in model food systems. Journal of Applied Phycology 26(4): 1823–1831.

Critchley, A.T. and M. Ohno. 1998. Seaweed resources of the world. Japan International Cooperation Agency, Yokosuka, Japan, 431 pp.

Croci, D.O., A. Cumashi, N.A. Ushakova, M.E. Preobrazhenskaya, A. Piccoli, L. Totani, N.E. Ustyuzhanina, M.I. Bilan, A.I. Usov, A.A. Grachev, G.E. Morozevich, A.E. Berman, C.J. Sanderson, M. Kelly, P. Di Gregorio, C. Rossi, N. Tinari, S. Iacobelli, G.A. Rabinovich and N.E. Nifantiev. 2011. Fucans, but not fucomannoglucuronans, determine the biological activities of sulfated polysaccharides from *Laminaria saccharina* brown seaweed. PLoS One 6(2): 0017283.

Cui, Y.Q., L.J. Zhang, T. Zhang, D.Z. Luo, Y.J. Jia, Z.X. Guo, Q.B. Zhang, X. Wang and X.M. Wang. 2010. Inhibitory effect of fucoidan on nitric oxide production in lipopolysaccharide activated primary microglia. Clinical Experimental Pharmacalogy and Physiology 37: 422–428.

Cumashi, A., N.A. Ushakova, M.E. Preobrazhenskaya, A. D'Incecco, A. Piccoli, L. Totani, N. Tinari, G.E. Morozevich, A.E. Berman, M.I. Bilan, A.I. Usov, N.E. Ustyuzhanina, A.A. Grachev, C.J. Sanderson, M. Kelly, G.A. Rabinovich, S. Iacobelli and N.E. Nifantiev. 2007. A comparative study of the anti-inflammatory, anticoagulant, antiangiogenic, and antiadhesive activities of nine different fucoidans from brown seaweeds. Glycobiology 17(5): 541–552.

Cuong, H.D., T.T. Thuy, T.T. Huong, B.M. Ly and T.T.T. Van. 2015. Structure and hypolipidaemic activity of fucoidan extracted from brown seaweed *Sargassum henslowianum*. Natural Product Research 29(5): 411–415.

Custódio, L., L. Silvestre, M.I. Rocha, M.J. Rodrigues, C. Vizetto-Duarte, H. Pereira, L. Barreira and J. Varela. 2016. Methanol extracts from *Cystoseira tamariscifolia* and *Cystoseira nodicaulis* are able to inhibit cholinesterases and protect a Human dopaminergic cell line from hydrogen peroxide-induced cytotoxicity. Pharmaceutical Biology 54(9): 1687–1696.

Cyclamen. 2016. Fungi—*Botrytis cinerea*. Available online at: http://www.cyclamen.com/en/professional/diseases/8/20.

D'Orazio, N., M.A. Gammone, E. Gemello, M. De Girolamo, S. Cusenza and G. Riccioni. 2012. Marine bioactives: pharmacological properties and potential applications against inflammatory diseases. Marine Drugs 10: 812–33.

Da Matta, C.B., E.T. de Souza, A.C. de Queiroz, D.P. de Lira, M.V. de Araujo, L.H. Cavalcante-Silva, G.E. de Miranda, J.X. de Araujo-Junior, J.M. Barbosa-Filho, B.V. de Oliveira Santos and M.S. Alexandre-Moreira. 2011. Antinociceptive and anti-inflammatory activity from algae of the genus *Caulerpa*. Marine Drugs 9(3): 307–318.

Da Silva, A.C.R., L.G. Ferreira, M.E.R. Duarte, M.T. Fujii, E.F. Sanchez, M.D. Noseda and A.L. Fuly. 2016. Protective effect of the sulfated agaran isolated from the red seaweed *Laurencia aldingensis* against toxic effects of the venom of the snake, *Lachesis muta*. Marine Biotechnology 18(6): 619–629.

Da Silva, A.C.R., A.M.G. Pires, C.J.B. Ramos, E.F. Sanchez, D.N. Cavalcanti, V.L. Teixeira and A.L. Fuly. 2017. The seaweed *Prasiola crispa* (Chlorophyta) neutralizes toxic effects of Bothrops jararacussu snake venom. Journal of Applied Phycology 29(2): 781–788.

Dae, H.P., S.C. Won, H.Y. Sean, S.S. Jung H.S. Chul. 2007. A developmental study of artificial skin using the alginate dermal substrate. Key Engineering Materials 342–343.

Dahlback, B. and B.O. Villoutreix. 2005. The anticoagulant protein C pathway. FEBS Letters 579: 3310–3316.

Dallman, P.R. 1986. Biochemical basis for the manifestations of Iron deficiency. Annual Review of Nutrition 6: 13–40.

Dame, M.K., I. Veerapaneni, N. Bhagavathula, M. Naik and J. Varani. 2011. Human colon tissue in organ culture: Calcium and multi-mineral-induced mucosal differentiation. *In Vitro* Cellular & Developmental Biology. Animal 47(1): 32–38.

Damonte, E.B., J. Neyts, C.A. Pujol, R. Snoeck, G. Andrei, S. Ikeda, M. Witvrouw, D. Reymen, H. Haines, M.C. Matulewicz, A. Cerezo, C.E. Coto and E. De Clercq. 1994. Antiviral activity of a sulphated polysaccharide from the red seaweed *Nothogenia fastigiata*. Biochemical Pharmacology 47(12): 2187–2192.

Damonte, E.B., M.C. Matulewicz and A.S. Cerezo. 2004. Sulfated seaweed polysaccharides as antiviral agents. Current Medicinal Chemistry 11: 2399–2419.

Dancey, J. and E.A. Sausville. 2003. Issues and progress with protein kinase inhibitors for cancer treatment. Nature Reviews Drug Discovery 2(4): 296–313.

Dar, A., H.S. Baig, S.M. Saifullah, V.U. Ahmad, S. Yasmeen and M. Nizamuddin. 2007. Effect of seasonal variation on the anti-inflammatory activity of *Sargassum wightii* growing on the N. Arabian Sea coast of Pakistan. Journal of Experimental Marine Biology and Ecology 351: 1–9.

Darah, I., S.H. Lim. 2015. *In vitro* antimicrobial activities of methanolic extract from marine alga *Enteromorpha intestinalis*. Asian Pacific Journal of Tropical Biomedicine 5(9): 785–788.

Das, N.G., D. Goswami and B. Rabha. 2007. Preliminary evaluation of mosquito larvicidal efficacy of plant extracts. Journal of Vector Borne Diseases 44: 145–148.

Davies, J. and D. Davies. 2010. Origins and evolution of antibiotic resistance. Microbiology and Molecular Biology Reviews 74(3): 417–433.

Davis, R., S. Zivanovic, D.H. D'Souza and P.M. Davidson. 2012. Effectiveness of chitosan on the inactivation of enteric viral surrogates. Food Microbiology 32: 57–62.

Davis, T.A., F. Llanes, B. Volesky and A. Mucci. 2003. Metal selectivity of *Sargassum* spp. and their alginates in relation to their α-L-guluronic acid content and conformation. Environmental Science and Technology 37: 261–267.

Davyt, D., W. Entz, R. Fernandez, R. Mariezcurrena, A.W. Mombru, I. Saldana, L. Dominguez, J. Coil and E. Manta. 1998. A new indole derivative from the red alga *Chondria atropurpurea* isolation, structure determination, and anthelmintic activity. Journal of Natural Products 61: 1560–1563.

Davyt, D., R. Fernandez, L. Suescun, A.W. Mombru, J. Saldana, L. Dominguez, J. Coll, M.T. Fujii and E. Manta. 2001. New sesquiterpene derivatives from the red alga *Laurencia scoparia*. Isolation, structure determination, and anthelmintic activity. Journal of Natural Products 64(12): 1552–1555.

Dawczynski, C., R. Schubert and G. Jahreis. 2007. Amino acids, fatty acids, and dietary fibre in edible seaweed products. Food Chemistry 103: 891–899.

De Araujo, I.W.F., J.A.G. Rodrigues, A.L.G. Quindere, J.D.T. Silva, G.D. Maciel, N.A. Ribeiro, E.D.O. Vanderlei, K.A. Ribeiro, H.V. Chaves, K.M.A. Pereira, M.M. Bezerra and N.M.B. Benevides. 2016. Analgesic and anti-inflammatory actions on bradykinin route of a polysulfated fraction from alga *Ulva lactuca*. International Journal of Biological Macromolecules 92: 820–830.

De Benoist, B., M. Andersson, I. Egli, B. Takkouche and H. Allen. 2004. Iodine status worldwide—WHO Global Database on Iodine Deficiency. Department of Nutrition for Health and Development. World Health Organization, Geneva, 49 pp.

De Hoog, G.S., J. Guarro and M.J. Figueras. 2001. Atlas of clinical fungi, 2 ed. Amer Society for Microbiology Press, Utrecht, Netherlands, 1126 pp.

De Ines, C., V.H. Argandona, J. Rovirosa, A. San-Martin, A.R. Diaz-Marrero, M. Cueto and A. Gonzalez-Coloma. 2004. Cytotoxic activity of halogenated monoterpenes from Plocamium cartilagineum. Zeitschrift für Naturforschung C 59(5-6): 339–344.

De Jesus Raposo, M.F., A. de Morais and R. de Morais. 2015. Marine polysaccharides from algae with potential biomedical applications. Marine Drugs 13(5): 2967–3028.

De Kluijver, M., G. Gijswijt, R. de Leon and I. da Cunda. 2017. *Stypopodium zonale*—Interactive Guide to Caribbean Diving. Marine Species Identification Portal. Available online at: http://species-identification.org/species.php?species_group=caribbean_diving_guide&id=502.

De La Mare, J.A., J.C. Lawson, M.T. Chiwakata, D.R. Beukes, A.L. Edkins and G.L. Blatch. 2012. Quinones and halogenated monoterpenes of algal origin show anti-proliferative effects against breast cancer cells *in vitro*. Invest New Drugs 30(6): 2187–2200.

De Oliveira, V.P., N.M. de Figueiredo, Y.Y. Valentin and R.C. Garla. 2009. Four new additions to the marine flora of Fernando de Noronha Archipelago, Tropical western South Atlantic Ocean. Check List 5(2): 210–215.

De Philippis, R., C. Sili, R. Paperi and M. Vincenzini. 2001. Exopolysaccharide-producing cyanobacteria and their possible exploitation: a review. Journal of Applied Phycology 13: 293–299.

De Roeck-Holtzhauer, Y. 1991. Uses of seaweeds in cosmetics. pp. 83–94. *In*: Guiry, M. and G. Blunden (eds.). Seaweed Resources in Europe: Uses and Potential. John Wiley & Sons, Chichester.

De SF-Tischer, P.C., L.B. Talarico, M.D. Noseda, S.M.P.B. Guimarães, E.B. Damonte and M.E.R. Duarte. 2006. Chemical structure and antiviral activity of carrageenans from *Meristiella gelidium* against *Herpes simplex* and Dengue virus. Carbohydrate Polymers 63(4): 459–465.

De Sousa, A.A., N.M. Benevides, A. de Freitas Pires, F.P. Fiuza, M.G. Queiroz, T.M. Morais, M.G. Pereira and A.M. Assreuy. 2013. A report of a galactan from marine alga *Gelidium crinale* with *in vivo* anti-inflammatory and antinociceptive effects. Fundamental & Clinical Pharmacology 27(2): 173–180.

De Souza, E.T., D.P. de Lira, A.C. de Queiroz, D.J.C. da Silva, A.B. de Aquino, E.A.C. Mella, V.P. Lorenzo, G.E.C. de Miranda, J.X. de Araújo-Júnior, M.C.O. Chaves, J.M. Barbosa-Filho, P.F. de Athayde-Filho, B.V.O. Santos and M.S. Alexandre-Moreira. 2009. The antinociceptive and anti-inflammatory activities of caulerpin, a bisindole alkaloid isolated from seaweeds of the genus *Caulerpa*. Marine Drugs 7: 689–704.

De Souza, L.M., G.L. Sassaki, M.T. Romanos and E. Barreto-Bergter. 2012 Structural characterization and anti-HSV-1 and HSV-2 activity of glycolipids from the marine algae *Osmundaria obtusiloba* isolated from Southeastern Brazilian coast. Marine Drugs 10: 918–931.

De Souza, M.C.R., C.T. Marques, C.M.G. Dore, F.R.F. da Silva, H.A.O. Rocha and E.L. Leite. 2007. Antioxidant activities of sulfated polysaccharides from brown and red seaweeds. Journal of Applied Phycology 19: 153–160.

De Souza, R.J., A. Mente, A. Maroleanu, A.I. Cozma, V. Ha, T. Kishibe, E. Uleryk, P. Budylowski, H. Schunemann, J. Beyene and S.S. Anand. 2015. Intake of saturated and trans unsaturated fatty acids and risk of all cause mortality, cardiovascular disease, and type 2 diabetes: systematic review and meta-analysis of observational studies. BMJ 11(351): 1–16.

De Souza-Barros, C., V.L. Teixeira and I.C.N.P. Paixão. 2015. Seaweeds with anti-*Herpes simplex* virus type 1 activity. Journal of Applied Phycology 27(4): 1623–1637.

De Souza-Barros, C., V. Garrido, V. Melchiades, R. Gomes, M.W.L. Gomes, V.L. Teixeira and I.C.N.P. Paixão. 2017. Therapeutic efficacy in Balb/c mice of extract from marine alga *Canistrocarpus cervicornis* (Phaeophyceae) against *Herpes simplex* virus type 1. Journal of Applied Phycology 29(2): 769–773.

Deacon-Smith, R.A., J.P. Lee-Potterand and D.J. Rogers. 1985. Anticoagulant activity in extracts of British marine algae. Botanica Marina 28: 333–338.

Deegan, H., H.M. Bates and L.J. McCargar. 2005. Assessment of iron status in adolescents: dietary, biochemical and lifestyle determinants. Journal of Adolescent Health 37(1): 75 pp.

Deffrasnes, C., M.E. Hamelin and G. Boivin. 2007. Human metapneumovirus. Seminars in Respiratory and Critical Care Medicine 28: 213–221.

Deig, E.F., D.W. Ehresmann, M.T. Hatch and D.J. Riedlinger. 1974. Inhibition of herpesvirus replication by marine algae extracts. Antimicrobial Agents and Chemotherapy 6(4): 524–525.

Delage, N., A. d'Harlingue, B.C. Ceccaldi and G. Bompeix. 2003. Occurrence of mycotoxins in fruit juices and wine. Food Control 14(4): 225–227.

Delaunay, D. and I. Voile. 2011. Composition dermatologique et/ou cosmétique atilisée pour la régénation de la peau. WO 2011051591 A1.

Delépine, R., B. Delesalle and C. Lambert. 1979. Sur l'existence dun tétrasporophyte dans le cycle de la Rhodophycée *Chaetangium fastigiatum* (Bory) J. Ag. aux îles Kerguelen. Comptes Rendus de l'Académie des Sciences, Paris 289(D): 595–598.

Dellai, A., S. Laajili, V.L. Morvan, J. Robert and A. Bouraoui. 2013. Antiproliferative activity and phenolics of the mediterranean seaweed *Laurencia obusta*. Industrial Crops and Products 47: 252–255.

Delma, C.R., S.T. Somasundaram, G.P. Srinivasan, M. Khursheed, M.D. Bashyam and N. Aravindan. 2015. Fucoidan from *Turbinaria conoides*: A multifaceted 'deliverable' to combat pancreatic cancer progression. International Journal of Biological Macromolecules 74: 447–457.

Demais, H., J. Brendle, H. Le Deit, L. Laza Anca, L. Lurton and D. Brault. 2007. Argiles intercalés. EP 1786862 A1.

Dembitsky, V.M. and T. Maoka. 2007. Allenic and cumulenic lipids. Progress in Lipid Research 46(6): 328–375.

Demirel, Z., F.F. Yilmaz-Koz, U.N. Karabay-Yavasoglu, G. Ozdemir and A. Sukatar. 2009. Antimicrobial and antioxidant activity of brown algae from the Aegean Sea. Journal of Serbian Chemical Society 74: 619–628.

Denda, M., T. Sokabe, T. Fukumi-Tominaga and M. Tominaga. 2007. Effects of skin surface temperature on epidermal permeability barrier homeostasis Journal of Investigative Dermatology 127(3): 654–659.

Denning, D.W. 1998. Invasive aspergillosis. Clinical Infectious Diseases 26: 781–805.

Denning, D.W., A. Marianas and J. Cohen. 1998. An EORTC multicenter prospective survey of invasive aspergillosis in haematological patients: diagnosis and therapeutic outcome. Journal of Infection 37: 173–180.

Depix, M.S., J. Martinez, F. Santibanez, J. Rovirosa, A. San Martin and R.B. Maccioni. 1998. The compound 14-Keto-Stypodiol Diacetate from the algae *Stypopodium flabelliforme* inhibits microtubules and cell proliferation in DU-145 human prostatic cells. Molecular and Cellular Biochemistry 187(1-2): 191–199.

Dermika. 2017. Awakens your beauty. Available online at: http://dermika.pl/en/.

Dethier, M.N. and S.L. Williams. 2009. Seasonal stresses shift optimal intertidal algal habitats. Marine Biology 156: 555–567.

Devery, R., A. Miller and C. Stanton. 2001. Conjugated linoleic acid and oxidative behaviour in cancer cells. Biochemical Society Transactions 29: 341–344.

Devi, K., G.K. Thirumaran, K. Manivannan and P. Anantharaman. 2009. Element composition of certain seaweeds from Gulf of Mannar marine biosphere reserve; Southeast Coast of India. World Journal of Dairy and Food Sciences 4(1): 46–55.

Devi, K.N., T.T.A. Kumar, K.V. Dhaneesh, T. Marudhupandi and T. Balasubramanian. 2012. Evaluation of antibacterial and antioxidant properties from brown seaweed, *Sargassum wightii* (Greville, 1848) against Human bacterial pathogenes. International Journal of Pharmacy and Pharmaceutical Sciences 4(3): 143–149.

Devi, K.P., N. Suganthy, P. Kesika and S.K. Pandian. 2008. Bioprotective properties of seaweeds: *in vitro* evaluation of antioxidant activity and antimicrobial activity against food borne bacteria in relation to polyphenolic content. BMC Complementary and Alternative Medicine 8: 3–38.

Devi, P., W. Solimabi, L. D'Souza and S.Y. Kamat. 1997. Toxic effects of coastal and marine plant extracts on mosquito larvae. Botanica Marina 40: 533–535.

Devi, V.D., A. Minhajdeen, V. Pavithra, S. Brindha, S.S. Jaya, R.S. Saranya, S. Varun, P. Senthilkumar and S. Sudha. 2014. Antioxidant and antifungal activities of methanolic extract of *Chaetomorpha linus* from Indian southeast coast. International Journal of Current Research and Review 06(07): 11–16.

Devillé, C., J. Damas, P. Forget, G. Dandrifosse and O. Peulen. 2004. Laminarin in the dietary fibre concept. Journal of the Science of Food and Agriculture 84: 1030–1038.

Deyab, M.A. and M.I. Abou-Dobara. 2013. Antibacterial activity of some marine algal extracts against most nosocomial bacterial infections. The Egyptian Journal of Experimental Biology (Botany) 9: 281–286.

Dhanya, K.I., V.I. Swati, K.S. Vanka and W.J. Osborne. 2016. Antimicrobial activity of *Ulva reticulata* and its endophytes. Journal of Ocean University of China 15(2): 363–369.

Dhargalkar, V.K. and N. Pereira. 2005. Seaweeds: promising plants of the millennium. Science and Culture 71: 60–66.

Dhasarathan, P and P. Theriappan. 2011. Phytochemical characterization and antimicrobial efficiency of seaweed samples, Ulva *fasciata* and *Chaetomorpha antennina*. International Journal of Pharma and Bio Sciences 2(1): 288–293.

Dhemla, S. and K. Varma. 2015. Health risks of excessive sodium and the role of value chain in India. Current Research in Nutrition and Food Science 3(2): 165–176.

Di Guardia, S., R. Valls, V. Mesguiche, J.M. Brunel and G. Culioli. 1999. Enantioselectivesynthesisofbifurcadiol: A natural antitumor marine product. Tetrahedron Letters 40(48): 8359–8360.

Dias, P.F., J.M. Siqueira, L.F. Vendruscolo, T.D. Neiva, A.R. Gagliardi, M. Maraschin and R.M. Ribeiro-Do-Valle. 2005. Antiangiogenic and antitumoral properties of a polysaccharide isolated from the seaweed *Sargassum stenophyllum*. Cancer Chemotherapy and Pharmacology 56(4): 436–446.

Díaz-Rubio, M.E., J. Perez-Jimenez and F. Saura-Calixto. 2009. Dietary fiber and antioxidant capacity in *Fucus vesiculosus* products. International Journal of Food Sciences and Nutrition 2: 23–34.

Dickman, M.B. 1993. *Collettotrichum gloeosporioides*—Crop Knowledge Master. Available online at: http://www.extento.hawaii.edu/kbase/crop/type/c_gloeo.htm.

Didier, C., I. Kerblat, C. Drouet, A. Favier, J.C. Beani and M.J. Richard. 2001. Induction of thioredoxin by ultraviolet-a radiation prevents oxidative-mediated cell death in human skin fibroblasts. Free Radical Biology & Medicine 31(5): 585–598.

Diekema, D.J., S.A. Messer, R.J. Hollis, R.N. Jones and M.A. Pfaller. 2003. Activities of caspofungin, itraconazole, posaconazole, ravuconazole, voriconazole, and amphotericin B against 448 recent clinical isolates of filamentous fungi. Journal of Clinical Microbiology 41: 3623–3626.

Dillard, H.R. 1988. Vegetable crops—Bean anthracnose. Available online at: http://vegetablemdonline.ppath.cornell.edu/factsheets/Beans_Anthracnose.htm.

Dillehay, T.D., C. Ramrez, M. Pino, M.B. Collins, J. Rossen and J.D. Pino-Navarro. 2008. Monte Verde: Seaweed, food, medicine, and the peopling of South America. Science 320: 784–786.

Dinesha, S., T. Menona, L.E. Hannab, V. Sureshc, M. Sathuvand and M. Manikannan. 2016. *In vitro* anti-HIV-1 activity of fucoidan from *Sargassum swartzii*. Internationl Journal of Biological Macromolecules 82: 83–88.

Diogo, J.V., S.G. Novo, Sabrina Galdo, M.J. González, M. Ciancia and A.C. Bratanich. 2015. Antiviral activity of lambda-carrageenan prepared from red seaweed (*Gigartina skottsbergii*) against BoHV-1 and SuHV-1. Research in Veterinary Science C98: 142–144.

Dixon, D.D. and T.J. Walsh. 1992. Human pathogenesis. pp. 249–267. *In*: Bennett, J.W. and M.A. Klich (eds.). *Aspergillus*, Biology and Industrial Application. Butterworth-Heinemann, Boston.

Dixon, D.M., M.M. McNeil, M.L. Cohen, B.G. Gellin and J.R. La Montagne. 1996. Fungal infections: a growing threat. Public Health Reports 111: 226–236.

Do, H., S. Pyo and E.H. Sohn. 2009. Suppression of iNOS expression by fucoidan is mediated by regulation of p38 MAPK, JAK/STAT, AP-1 and IRF-1, and depends on up-regulation of scavenger receptor B1 expression in TNF-alpha- and IFN-gamma-stimulated C6 glioma cells. The Journal of Nutritional Biochemistry 21: 671–679.

Do, H., S. Pyo and E.H. Sohn. 2010. Suppression of iNOS expression by fucoidan is mediated by regulation of p38 MAPK, JAK/STAT, AP-1 and IRF-1, and depends on up-regulation of scavenger receptor B1 expression in TNF-alpha- and IFN-gamma-stimulated C6 glioma cells. The Journal of Nutritional Biochemistry 21(8): 671–679.

Dobashi, K., T. Nishino, M. Fujihara and T. Nagumo. 1989. Isolation and preliminary characterization of fucose-containing sulfated polysaccharides with blood-anticoagulant activity from the brown seaweed *Hizikia fusiforme*. Carbohydrate Research 194: 315–320.

Domingo, A.C. and J.A. Corrales. 2001. Inventory and distribution of edible seaweeds in Ilocos Sur. University of Northern Philippines, Tamag, Vigan City, 12 pp.

Dore, C.M.P.G., M.G.C.F. Alves, L.S.E.P. Will, T.G. Costa, D.A. Sabry, L.A.H.A.O. Rocha, L.G.A. Filgueira and E.L. Leite. 2013. A sulfated polysaccharide, fucans, isolated from brown algae *Sargassum vulgare* with anticoagulant, antithrombotic, antioxidant and anti-inflammatory effects. Carbohydrate Polymers 91: 467–475.

Draelos, Z.D. 2008. The cosmeceutical realm. Clinics in Dermatology 26: 627–632.

Drechsel, D.N., A. Hyman, N.H. Cobb and M. Kirschner. 1992. Modulation of the dynamic instability of tubulin assembly by the microtubuleas-sociated protein tau. Molecular Biology of the Cell 3: 1141–1154.

Drum, R. 2013. Sea vegetables for food and medicine. Well Being Journal, July/August. Available online at: http://www.eidon.com/SeaweedArticle.pdf.

Drum, R. 2016. Medicinal uses of seaweeds. Available online at: http://www.ryandrum.com/seaweeds.htm.

Duarte, M., D. Noseda, M. Noseda, S. Tulio, C. Pujol and E. Damonte. 2001. Inhibitory effect of sulfated galactans from the marine alga *Bostrychia montagnei* on *Herpes simplex* virus replication *in vitro*. Phytomedicine 8: 53–58.

Duarte, M.E., J.P. Cauduro, N.G. Noseda, M.D. Noseda, A.G. Gonçalves, C.A. Pujol, E.B. Damonte and A.S. Cerezo. 2004. The structure of the agaran sulfate from *Acanthophora spicifera* (Rhodomelaceae, Ceramiales) and its antiviral activity. Relation between structure and antiviral activity in agarans. Carbohydrate Research 339: 35–347.

Dubber, D. and T. Harder. 2008. Extracts of *Ceramium rubrum, Mastocarpus st*ellatus and *Laminaria digitata* inhibit growth of marine and fish pathogenic bacteria at ecologically realistic concentrations. Aquaculture 274: 196–200.

Dubois, D., M. Pihet, C.L. Clec'h, A. Croue, H. Beguin, J.P. Bouchara and D. Chabasse. 2005. Cutaneous phaeohyphomycosis due to *Alternaria infectoria*. Mycopathologia 160(2): 117–123.

Dumelod, B.D., R.P. Ramirez, C.L. Tiangson, E.B. Barrios and L.N. Panlasigui. 1999. Carbohydrate availability of *arroz caldo* with lambda-carrageenan. International Journal of Food Sciences and Nutrition 50: 283–289.

Dumitriu, S. 2004. Polysaccharides: Structural Diversity and Functional Versatility, 2nd edition. CRC Press, Taylor & Francis Group, Boca Raton, FL, 1224 pp.

Dürig, J., T. Bruhn, K.H. Zurborn, K. Gutensohn, H.D. Bruhn and L. Béress. 1997. Anticoagulant fucoidan fractions from *Fucus vesiculosus* induce platelet activation *in vitro*. Thrombosis Research 85: 479–491.

Dussault, D., K.D. Vu, T. Vansach, F.D. Horgen and M. Lacroix. 2016. Antimicrobial effects of marine algal extracts and cyanobacterial pure compounds against five foodborne pathogens. Food Chemistry 199: 114–118.

Dyminiska, L., M. Szatkowski, M. Wróbel-Kwiatkowska, A. Kurzawa, W. Syska, A. Ggot, M. Zawadzki, M. Ptak, M. Maczka, J. Hanuza and J. Szopa. 2012. Improved properties of micronized genetically modified flax fibers. Journal of Biotechnology 164: 292–299.

Eccles, R., C. Meier, M. Jawad, R. Weinmullner, A. Grassauer and E. Prieschl-Grassauer. 2010. Efficacy and safety of an antiviral iota-carrageenan nasal spray: A randomized, double-blind, placebo-controlled exploratory study in volunteers with early symptoms of the common cold. Respiratory Research 11(108): 1465–9921.

Eccles, R., B. Winther, S.L. Johnston, P. Robinson, M. Trampisch and S. Koelsch. 2015. Efficacy and safety of iota-carrageenan nasal spray versus placebo in early treatment of the common cold in adults: The ICICC trial. Respiratory Research 16(1): 1–11.

Edgar, G. 2008. *Codium duthieae*. *In*: Baldock, R.N. (ed.). Algae Revealed. Australian Marine Life (2nd ed.). S Australian State Herbarium. Available online at: http://www.flora.sa.gov.au/efsa/algae_revealed/pdf/Codium_duthieae.pdf.

Edgar, G. 2009. *Ulva australis*. *In*: Baldock, R.N. (ed.). Algae Revealed. Australian Marine Life (2nd ed.). S Australian State Herbarium. Available online at: http://www.flora.sa.gov.au/efsa/algae_revealed/pdf/Ulva_australis.pdf.

Edgar, G. 2009b. *Ulva flexuosa*. *In*: Baldock, R.N. (ed.). Algae Revealed. Australian Marine Life (2nd ed.). S Australian State Herbarium. Available online at: http://www.flora.sa.gov.au/efsa/algae_revealed/pdf/Enteromorpha_flexuosa.pdf.

Edgar, G. 2016. Laminaran/laminarin—chemical structure of β-1,3-1,6-glucan (image in public domain). Available online at: https://commons.wikimedia.org/w/index.php?curid=5274129.

Edgar, G. 2017. Polyunsaturated fatty acids (PUFAs) (images in public domain). Available online at: https://en.wikipedia.org/wiki/Polyunsaturated_fatty_acid.

Efinmov, V.S., A.I. Usov, T.S. Ol'skaya, A. Baliunis and M.Y. Rozkin. 1983. Comparative study of the anticoagulant activity of sulfated polysaccharides from red marine algae. Farmakologiia I Toksikologiia (Moscow) 46(3): 61–67 (in Russian).

Egan, S., T. Harder, C. Burke, P. Steinberg, S. Kjelleberg and T. Thomas. 2013. The seaweed holobiont: Understanding seaweed-bacteria interactions. FEMS Microbiology Reviews 37: 462–476.

Ehresmann, D.W., E.F. Dieg, M.T. Hatch, L.H. DiSalvo and N.A. Vedros. 1977. Antiviral substances from California marine algae. Journal of Phycology 13: 37–40.

Ehrlich, H. 2010. Biological materials of marine origin: Invertebrates. Biologically-Inspered Systems, Volume 1. Springer, London, 572 pp.

Eitsuka, T., K. Nakagawa, M. Igarashi and T. Miyazawa. 2004. Telomerase inhibition by sulfoquinovosyldiacylglycerol from edible purple laver (*Porphyra yezoensis*). Cancer Letters 212(1): 15–20.

Eklof, J. 2012. Chemical structure of Porphyran (image in public domain). Available online at: https://www.cazypedia.org/index.php/File:Porphyran.png.

El Atouani, S., S. Tahiri, A. Reani, F. Bentiss, M.E. Krati, A. Sahibed-Dine, A. Schamel, A. Aarfane and B. Sabour. 2016. Hexavalent chromium uptake from aqueous solutions using raw biomass of the invasive brown seaweed *Sargassum muticum* from the Moroccan shorelines: kineticas and isotherms. European Scientific Institute, ESI 12(30): 243–262.

El Baz, F.K., G.S. El Baroty, H.H. Abd El Baky, O.I. Abd El-Salam and E.A. Ibrahim. 2013. Structural characterization and biological activity of sulfolipids from selected marine algae. Grasas y Aceites 64(5): 561–571.

El Din, N.G.S. and Z.M. El-Sherif. 2012. Nutritional value of some Algae from the North-Western Mediterranean Coast of Egypt. Journal of Applied Phycology 24(3): 613–626.

El Khoury, D., H.D. Goff, S. Berengut, R. Kubant and G.H. Anderson. 2014. Effect of sodium alginate addition to chocolate milk on glycemia, insulin, appetite and food intake in healthy adult men. European Journal of Clinical Nutrition 68(5): 613–618.

El Khoury, D., H.D. Goff and G.H. Anderson. 2015. The role of alginates in regulation of food intake and glycemia: a gastroenterological perspective. Critical Reviews in Food Science and Nutrition 55(10): 1406–1424.

El Shafay, S.M., S.S. Ali and M.M. El-Sheekh. 2016. Antimicrobial activity of some seaweeds species from Red sea, against multidrug resistant bacteria. The Egyptian Journal of Aquatic Research 42(1): 65–74.

El Wahidi, M., B. El Amraoui, M. El Amraoui and T. Bamhaoud. 2015. Screening of antimicrobial activity of macroalgae extracts from the Moroccan Atlantic coast. Annales Pharmaceutiques Françaises 73(3): 190–196.

Elbanna, S.M. and M.M. Hegazi. 2011. Screening of some seaweeds species from South Sinai, Red Sea as potential bioinsecticides against mosquito larvae: *Culex pipiens*, Egypt. Egyptian Academic Journal of Biological Sciences 4(2): 21–30.

El-Din, S.M.M. and A.M.D. El-Ahwany. 2016. Bioactivity and phytochemical constituents of marine red seaweeds (*Jania rubens*, *Corallina mediterranea* and *Pterocladia capillacea*). Journal of Taibah University for Science 10: 471–484.

El-Fatimy, E.S. and A.S. Abdel-Moneim. 2011. Antibacterial activity of methanolic extract of dominant marine alga (*Padina pavonia*) of Tolmeta coasts, Libya. Journal of American Science 7(4): 745–751.

Elizondo-Gonzalez, R., L.E. Cruz-Suarez, D. Ricque-Marie, E. Mendoza-Gamboa, C. Rodriguez-Padilla and L.M. Trejo-Avila. 2012. *In vitro* characterization of the antiviral activity of fucoidan from *Cladosiphon okamuranus* against Newcastle disease virus. Virology Journal 9(1): 1–9.

El-Kamel, A., M. Sokar, V. Naggar and S. Al Gamal. 2002. Chitosan and sodium alginate-based bioadhesive vaginal tablets. AAPS PharmSciTech 4: E44.

Ellis, D. 2016a. *Candida glabrata*—Mycology online, The University of Adelaide. Available online at: http://www.mycology.adelaide.edu.au/Fungal_Descriptions/Yeasts/Candida/Candida_glabrata.html.

Ellis, D. 2016b. *Trichophyton mentagrophytes*—Mycology online, The University of Adelaide. Available online at: http://www.mycology.adelaide.edu.au/Fungal_Descriptions/Dermatophytes/Trichophyton/mentagrophytes.html.

Ellis, D. 2016c. *Microsporum canis*—Mycology online, The University of Adelaide. Available online at: http://www.mycology.adelaide.edu.au/Fungal_Descriptions/Dermatophytes/Microsporum/Microsporum_canis.html.

Ellis, D. 2016d. *Aspergillus flavus*—Mycology online, The University of Adelaide. Available online at: http://www.mycology.adelaide.edu.au/Fungal_Descriptions/Hyphomycetes_(hyaline)/Aspergillus/flavus.html.

Ellis, D. 2016e. *Fusarium solani*—Mycology online, The University of Adelaide. Available online at: http://www.mycology.adelaide.edu.au/Fungal_Descriptions/Hyphomycetes_(hyaline)/Fusarium/solani.html.

Ellis, D. 2016f. *Candida krusei*—Mycology online, The University of Adelaide. Available online at: http://www.mycology.adelaide.edu.au/Fungal_Descriptions/Yeasts/Candida/Candida_krusei.html.

Ellis, D. 2016g. *Epidermophyton floccosum*—Mycology online, The University of Adelaide. Available online at: http://www.mycology.adelaide.edu.au/Fungal_Descriptions/Dermatophytes/Epidermophyton/.

Ellis, D. 2016h. *Trichophyton rubrum*—Mycology online, The University of Adelaide. Available online at: http://www.mycology.adelaide.edu.au/Fungal_Descriptions/Dermatophytes/Trichophyton/rubrum.html.

Ellis, D. 2016i. *Penicillium*—Mycology online, The University of Adelaide. Available online at: http://www.mycology.adelaide.edu.au/Fungal_Descriptions/Hyphomycetes_(hyaline)/Penicillium/.

Ellis, D. 2016j. *Trichophyton schoenleinii*—Mycology online, The University of Adelaide. Available online at: http://www.mycology.adelaide.edu.au/Fungal_Descriptions/Dermatophytes/Trichophyton/schoenlenii.html.

Ellis, D. 2016k. *Trichophyton simii*—Mycology online, The University of Adelaide. Available online at: http://www.mycology.adelaide.edu.au/Fungal_Descriptions/Dermatophytes/Trichophyton/.

Ellis, D. 2016l. *Epicoccum nigrum*—Mycology online, The University of Adelaide. Available online at: http://www.mycology.adelaide.edu.au/Fungal_Descriptions/Hyphomycetes_(dematiaceous)/Epicoccum/.

Ellis, D. 2016m. *Cladosporium cladosporioides*—Mycology online, The University of Adelaide. Available online at: http://www.mycology.adelaide.edu.au/Fungal_Descriptions/Hyphomycetes_(dematiaceous)/Cladosporium/.

Ellis, D. 2016n. *Trichophyton tonsurans*—Mycology online, The University of Adelaide. Available online at: http://www.mycology.adelaide.edu.au/Fungal_Descriptions/Dermatophytes/Trichophyton/tonsurans.html.

Ellis, D. 2016o. *Candida tropicalis*—Mycology online, The University of Adelaide. Available online at: http://www.mycology.adelaide.edu.au/Fungal_Descriptions/Yeasts/Candida/Candida_tropicalis.html.

Ellis, D. 2016p. *Geotrichum*—Mycology online, The University of Adelaide. Available online at: http://www.mycology.adelaide.edu.au/Fungal_Descriptions/Hyphomycetes_(hyaline)/Geotrichum/.

Ellis, D. 2016q. *Candida parapsilosis*—Mycology online, The University of Adelaide. Available online at: http://www.mycology.adelaide.edu.au/Fungal_Descriptions/Yeasts/Candida/Candida_parapsilosis.html.

Ellis, D. 2016r. *Microsporum gypseum*—Mycology online, The University of Adelaide. Available online at: http://www.mycology.adelaide.edu.au/Fungal_Descriptions/Dermatophytes/Microsporum/Microsporum_gypseum.html.

Ellis, D. 2016s. *Candida kefyreum*—Mycology online, The University of Adelaide. Available online at: http://www.mycology.adelaide.edu.au/Fungal_Descriptions/Yeasts/Candida/Candida_keyfyr.html.

Ellis, D. 2016t. *Cunninghamella bertholletiae*—Mycology online, The University of Adelaide. Available online at: http://www.mycology.adelaide.edu.au/Fungal_Descriptions/Zygomycetes/Cunninghamella/.

Ellis, D. 2016u. *Trichophyton violaceum*—Mycology online, The University of Adelaide. Available online at: http://www.mycology.adelaide.edu.au/Fungal_Descriptions/Dermatophytes/Trichophyton/violaceum.html.

Ellis, D. 2016v. *Trichophyton verrucosum*—Mycology online, The University of Adelaide. Available online at: http://www.mycology.adelaide.edu.au/Fungal_Descriptions/Dermatophytes/Trichophyton/verrucosum.html.

Ellof, J.N. 1998. Which extractant should be used for the screening and isolation of antimicrobial components from plants? Journal of Ethnopharmacology 60(1): 1–8.

Ellouali, M., C. Boisson-Vidal, P. Durand and J. Jozefonvicz. 1993. Antitumor activity of low molecular weight fucans extracted from brown seaweed *Ascophyllum nodosum*. Anticancer Research 13(6A): 2011–2019.

Elnabris, K.J., A.A. Elmanama and W.N. Chihadeh. 2013. Antibacterial activity of four marine seaweeds collected from the coast of Gaza Strip, Palestine. Mesopotamian Journal of Marine Science 28(1): 81–92.

El-Raheem, A.E.-H. and Y. El-Said. 1988. Preparation and characterization of agar beads containing phenobarbitone sodium. Journal of Microencapsulation 5: 159–163.

El-Said, G.F. and A. El-Sikaily. 2013. Chemical composition of some seaweed from Mediterranean Sea coast, Egypt. Environmental Monitoring and Assessment 185(7): 6089–6099.

El-Sarraf, W. and G. El-Shaarawy. 1994. Chemical composition of some marine macroalgae from the Mediterranean Sea of Alexandria, Egypt. The Bulletin of the High Institute of Public Health 24: 523–534.

El-Sheekh, M.M., M.M. Gharieb, S.M. El-Sabbagh and W.T. Hamza. 2014. Antimicrobial efficacy of some marine macroalgae of Red Sea. International Journal of Microbiology and Immunology Research 3(3): 021–028.

Elsie, B.H. and M.S.D.J. Rajan. 2010. Evaluation of antimicrobial activity and phytochemical screening of *Gelidium acerosa*. Journal of Pharmaceutical Science & Research 2(11): 704–707.

Elsie, B.H., M.S. Dhanarajan and P.N. Sudha. 2011. *In vitro* screening of secondary metabolites and antimicrobial activities of ethanol and acetone extracts from red seaweed *Gelidium acerosa*. International Journal of Chemistry Research 2(2): 27–29.

Elsner, H. 1938. Hoppe-Seyler's Zeitschrift fur Physiologische Chemie 252: 196–200.

Ely, R., T. Supriya and C.G. Naik. 2004. Antimicrobial activity of marine organisms collected off the coast of South East India. Journal of Experimental Marine Biology and Ecology 309(1): 121–127.

Encyclopaedia Britannica. 1998. Wilt plant disease. Available online at: https://www.britannica.com/science/wilt#ref165245.

Engel, S., M.P. Puglisi, P.R. Jensen and W. Fenical. 2006. Antimicrobial activities of extracts from tropical Atlantic marine plants against marine pathogens and saprophytes. Marine Biology 149(5): 991–1002.

Englert, G., T. Bjornland and S. Liaaenjensen. 1990. 1D and 2D NMR-study of some allenic carotenoids of the fucoxanthin series. Magnetic Resonance in Chemistry 28(6): 519–528.

Enoki, T., H. Sagawa, T. Tominaga, E. Nishiyama, N. Komyama, T. Sakai, F.-G. Yu, K. Ikai and I. Kato. 2003. Drugs, foods or drinks with the use of algae-derived physiologically active substances. Patent US6475990 B1.

Ensembl Fungi. 2016. *Fusarium verticillioides*—Release 34. Available online at: http://fungi.ensembl.org/Fusarium_verticillioides/Info/Index.

Eom, S.H., S.H. Lee, N.Y. Yoon, W.K. Jung, Y.J. Jeon, S.K. Kim, M.S. Lee and K.M. Kim. 2012. α-Glucosidase and α-amylase-inhibitory activities of phlorotannins from *Eisenia bicyclis*. Journal of the Science of Food and Agriculture 92: 2084–2090.

Eom, S.-H., D.-S. Lee, Y.-J. Jung, J.-H. Park, J.-I. Choi, M.-J. Yim, J.-M. Jeon, H.-W. Kim, K.-T. Son, J.-Y. Je, M.S. Lee and Y.M. Kim. 2014. The mechanism of antibacterial activity of phlorofucofuroeckol-A against methicillin-resistant *Staphylococcus aureus*. Applied Microbiology and Biotchnology 98: 9795–9804.

Eom, S.-H., S.-Y. Moon, D.-S. Lee, H.-J. Kim, K. Park, E.-W. Lee, T.H. Kim, Y.-H. Chung, M.-S. Lee and Y.-M. Kim. 2015. *In vitro* antiviral activity of Dieckol and Phlorofucofuroeckol-A isolated from edible brown alga *Eisenia bicyclis* against murine norovirus. ALGAE 30(3): 241–246.

Eom, S.-H., J.A. Santos, J.-H. Kim, W.-K. Jung, D.-H. Kim and Y.-K. Kim. 2015b. *In vitro* antibacterial and synergistic activity of an *Ecklonia cava* extract against antibiotic-resistant *Streptococcus parauberis*. Fisheries and Aquatic Sciences 18(3): 241–247.

Erdmann, K., B.W.Y. Cheung and H. Schröder. 2008. The possible roles of food-derived bioactive peptides in reducing the risk of cardiovascular disease. The Journal of Nutritional Biochemistry 19: 643–654.

Erenler, R., K. Pabuccu, A.S. Yaglioglu, I. Demirtas and F. Gul. 2016. Chemical constituents and antiproliferative effects of cultured *Mougeotia nummuloides* and *Spirulina major* against cancerous cell lines. Zeitschrift Fur Naturforschung Section C-a Journal of Biosciences 71(3-4): 87–92.

Erickson, K.L., J.A. Beutler, G.N. Gray, J.H. Cardellina and M.R. Boyd. 1995. Majapolene A, a cytotoxic peroxide, and related sesquiterpenes from the red alga *Laurencia majuscula*. Journal of Natural Products 58(12): 1848–1860.

Ermakova, S., R. Sokolova, S.M. Kim, B.H. Um, V. Isakov and T. Zvyagintseva. 2011. Fucoidans from brown seaweeds *Sargassum hornery*, *Ecklonia cava*, *Costaria costata*: Structural characteristics and anticancer activity. Applied Biochemistry and Biotechnology 164: 841–850.

Ertürk, Ö. and B. Taş. 2011. Antibacterial and antifungal effects of some marine algae. Kafkas Üniversitesi Veteriner Fakültesi Dergisi 17 (Suppl A): S121–S124.

Esmon, C.T. 2005. The interactions between inflammation and coagulation. British Journal of Haematology 131: 417–430.

Espeche, M.E., E.R. Fraile and A.M.S. Mayer. 1984. Screening of Argentine marine algae for antimicrobial activity. Proceedings of the International Seaweed Symposium 11: 525–528.

Etahiri, S., V. Bultel-Poncé, C. Caux and M. Guyot. 2001. New bromoditerpenes from the red alga *Sphaerococcus coronopifolius*. Journal of Natural Products 64: 1024–1027.

EU. Commission Directive. 2010. 2010/67/EU of 20 October 2010 amending Directive 2008/84/EC laying down specific purity criteria on food additives other than colours and sweeteners. Official Journal of the European Union—EUR-Lex 277: 17–26.

Éverton, T.S., P.L. Daysianne, C.Q. Aline, J.C.S. Diogo, B.A. Anansa, A.C.M. Eliane, P.L. Vitor, E.C.M. George, X.A.J. João, C.O.C. Maria, M.B.F. José, F.A. Petrônio, V.O.S. Bárbara and S.A. Magna. 2009. The antinociceptive and anti-inflammatory activities of Caulerpin, a bisindole alkaloid isolated from seaweeds of the genus *Caulerpa*. Marine Drugs 7: 689–704.

Fabregas, J., D. García, M. Fernandez-Alonso b, A.I. Rocha, P. Gómez-Puertas, J.M. Escribano, A. Otero and J.M. Coll. 1999. *In vitro* inhibition of the replication of haemorrhagic septicaemia virus (VHSV) and African swine fever virus (ASFV) by extracts from marine microalgae. Antiviral Research 44: 67–73.

Fabrowska, J., B. Łęska, G. Schroeder, B. Messyasz and M. Pikosz. 2015. Biomass and extracts of algae as material for cosmetics, pp. 681–706. *In*: Se-Kwon Kim and K. Chojnacka (eds.). Marine Algae Extracts. Wiley-VCH Verlag GmbH & Co. KGaA, Weinheim.

Faggio, C., M. Morabito, S.A. Minicante, G.L. Piano, M. Pagano and G. Genovese. 2015. Potential use of polysaccharides from the brown alga *Undaria pinnatifida* as anticoagulants. Brazilian Archives of Biology and Technology 58(5): 798–804.

Fallarero, A., J.J. Loikkanen, P.T. Männistö, O. Castañeda and A. Vidal. 2003. Effects of aqueous extracts of *Halimeda incrassata* (Ellis) Lamouroux and *Bryothamnion triquetrum* (S.G. Gmelim) Howe on hydrogen peroxide and methyl mercury-induced oxidative stress in GT1–7 mouse hypothalamic immortalized cells. Phytomedicine 10: 39–47.

Fallarero, A., A. Peltoketo, J. Loikkanen, P. Tammela, A. Vidal and P. Vuorela. 2006. Effects of the aqueous extract of *Bryothamnion* triquetrum on chemical hypoxia and aglycemia-induced damage in GT1-7 mouse hypothalamic immortalized cells Phytomedicine 13: 240–245.

Fang, Z., S.Y. Jeong, H.A. Jung, J.S. Choi, B.S. Min and M.H. Woo. 2010. Anticholinesterase and antioxidant constituents from *Gloiopeltis furcata*. Chemical and Pharmaceutical Bulletin (Tokyo) 58: 1236–1239.

Faral-Tello, P., S. Mirazo, C. Dutra, A. Perez, L. Geis-Asteggiante, S. Frabasile, E. Koncke, D. Davyt, L. Cavallaro, H. Heinzen and J. Arbiza. 2012. Cytotoxic, virucidal, and antiviral activity of South American plant and algae extracts. Scientific World Journal 2012: 174837.

Farasat, M., R.A. Khavari-Nejad, S.M.B. Nabavi and F. Namjooyan. 2014. Antioxidant activity, total phenolics and flavonoid contents of some edible green seaweeds from northern coasts of the Persian Gulf. Iranian Journal of Pharmaceutical Research 13: 163–170.

Farber, J.L., M.E. Kyle and J.B. Coleman. 1990. Mechanisms of cell injury by activated oxygen species. Laboratory Investigation 62: 670–679.

Fard, S.G., F.T. Shamsabadi, M. Emadi, G.Y. Meng, K. Muhammad and S. Mohamed. 2011. Ethanolic extract of *Eucheuma cottonii* promotes *in vivo* hair growth and wound healing. Journal of Animal and Veterinary Advances 10(5): 601–605.

Farias, E.H.C., V.H. Pomim, A.P. Valente, H.B.O. Nader, H.A. Rocha and P.A.S. Morão. 2008. A preponderantly 4-sulfated, 3-linked galactan from the green alga *Codium isthmocladum*. Glicobiology 18: 250–259.

Farias, W.R., P.C. Lima, N.V. Rodrigues, R.C. Siqueira, R.M. Amorim, M.G. Pereira and A.M. Assreuy. 2011. A novel antinociceptive sulphated polysaccharide of the brown marine alga *Spatoglossum schroederi*. Natural Product Communications 6(6): 863–866.

Farias, W.R.L., A.P. Valente, M.S. Pereira and P.A.S. Morão. 2000. Structure and anticoagulante activity of sulfated galactans. Isolation of a unique sulfated galactan from the red algae *Botryocladia occidentalis* and comparison of its anticoagulant action with that of sulfated galactans from invertebrates. The Journal of Biological Chemistry 275: 29299–29307.

Farlow, M. 2010. Alzheimer disease. *In*: Fillit, H.M., K. Rockwood, K. Woodhouse (eds.). Textbook of Geriatric Medicine and Gerontology–Brocklehurst's. Saunders/Elservier, Philadelphia, 1119 pp.

Farooqi, A.A., G. Butt and Z. Razzaq. 2012. Algae extractsand methyl jasmonate anti-cancer activities in prostate cancer: choreographers of 'the dance macabre'. Cancer Cell International 12(1): 1475–2867.

Farrar, J., P. Hotez, T. Junghanss, G. Kang, D. Lalloo, J.N. White. 2013. Manson's tropical diseases, 23rd Ed, pp. 664–671. Elsevier Health Sciences, London.

Farrar, J.J., B.M. Pryor, R.M. Davis. 2004. *Alternaria* diseases of carrot. Plant Disease 88: 776–784.

Fassett, R.G. and J.S. Coombes. 2011. Astaxanthin: A potential therapeutic agent in cardiovascular disease. Marine Drugs 9: 447–465.

FDA. 2002. Federal Food, Drug, and Cosmetic Act (FD & C Act) 2002. United States Code, Title 21; The Food and Drug Administration: Silver Spring, MD, USA.

Febles, C.I., A. Arias, M.C. Gil-Rodriguez, A. Hardisson and A. Sierra Lopez. 1995. *In vitro* study of antimicrobial activity in algae (Chlorophyta, Phaeophyta and Rhodophyta) collected from the coast of Tenerife (Spanish). Anuario del Instituto de Estudios Canarios 34: 181–192.

Feldman, S., S. Reynaldi, C. Stortz, A. Cerezo and E. Damonte. 1999. Antiviral properties of fucoidan fractions from *Leathesia difformis*. Phytomedicine 6: 335–340.

Felício, R., S. Albuquerque, M.C.M. Young, N.S. Yokoyad and H.M. Debonsia. 2010. Trypanocidal, leishmanicidal and antifungal potential from marine red alga *Bostrychia tenella* J. Agardh (Rhodomelaceae, Ceramiales). Journal of Pharmaceutical and Biomedical Analysis 52: 763–769.

Fengwu, J. and Z. Yushun. 1993. The dictionary of Chinese oceanic medicine. Ocean Press, 160 pp.

Fenical, W., J.J. Sim, D. Squatrito, R.M. Wing, P. Radlick. 1973. Zonarol and isozonarol, fungitoxic hydroquinones from the brown seaweed *Dictyopteris zonarioides*. The Journal of Organic Chemistry 38: 2383–2385.

Fenical, W. and J.J. Sims. 1974. Cycloeudesmol, an antibiotic cyclopropane conatinnin sequiterpene from the marine alga, *Chondria oppositiclada* Dawson. Tetrahedron Letters 13: 1137–1140.

Fenical, W. 1976. Chemical variation in a new bromochamigrene derivative from the red seaweed *Laurencia pacifica*. Phytochemistry 15(4): 511–512.

Fenical, W. and V.J. Paul. 1984. Antimicrobial and cytotoxic terpenoids from tropical green algae of the family Udoteaceae. Hydrobiologia 117: 137–140.

Ferkany, J.W. and J.T. Coyle. 1983. Kainic acid selectively stimulates the release of endogenous excitatory acidic amino acids. Journal of Pharmacology and Experimental Therapeutics 225: 399–406.

Ferlay, J., E. Steliarova-Foucher, J. Lortet-Tieulent, S. Rosso, J.W. Coebergh, H. Comber, D. Forman and F. Bray. 2013. Cancer incidence and mortality patterns in Europe: estimates for 40 countries in 2012. European Journal of Cancer 49(6): 1374–1403.

Ferlay, J., I. Soerjomataram, R. Dikshit, S. Eser, C. Mathers, M. Rebelo, D. M. Parkin, D. Forman and F. Bray. 2015. Cancer incidence and mortality worldwide: sources, methods and major patterns in Globocan 2012. International Journal of Cancer 136(5): 9.

Fernandes, D.R.P., V.P. de Oliveira and Y.Y. Valentin. 2014. Seaweed biotechnology in Brazil: Six decades of studies on natural products and their antibiotic and other biological activities. Journal of Applied Phycology 26(5): 1923–1937.

Fernandez, L.E., O.G. Valiente, V. Mainardi, J.L. Bello, H. Velez and A. Rosado. 1989. Isolation and characterization of an antitumor active agar-type polysaccharide of *Gracilaria dominguensis*. Carbohydrate Research 190(1): 77–83.

Fernández, P.V., I. Quintana, A.S. Cerezo, J.J. Caramelo, L. Pol-Fachin, H. Verli, J.M. Estevez and M. Ciancia. 2013. Anticoagulant activity of a unique sulfated pyranosic (1→3)-β-L-Arabinan through direct interaction with thrombin. The Journal of Biological Chemistry 288(1): 223–233.

Ferramosca, A., A. Conte, F. Guerra, S. Felline, M.G. Rimoli, E. Mollo, V. Zara and A. Terlizzi. 2016. Metabolites from invasive pests inhibit mitochondrial complex II: A potential strategy for the treatment of human ovarian carcinoma? Biochemical and Biophysical Research Communications 473(4): 1133–1138.

Ferreres, F., G. Lopes, A. Gil-Izquierdo, P.B. Andrade, C. Sousa, T. Mouga and P. Valentão. 2012. Phlorotannin extracts from fucales characterized by HPLC-DAD-ESI-MSn: approaches to hyaluronidase inhibitory capacity and antioxidant properties. Marine Drugs 10: 2766–2781.

Fidelis, G.P., R.B. Camara, M.F. Queiroz, M.S. Santos Pereira Costa, P.C. Santos, H.A. Rocha and L.S. Costa. 2014. Proteolysis, NaOH and ultrasound-enhanced extraction of anticoagulant and antioxidant sulfated polysaccharides from the edible seaweed, *Gracilaria birdiae*. Molecules 19(11): 18511–1826.

Figueiredo, J.G., F.S. Bitencourt, T.M. Cunha, P.B. Luz, K.S. Nascimento, M.R. Mota, A.H. Sampaio, B.S. Cavada, F.Q. Cunha and N.M. Alencar. 2010. Agglutinin isolated from the red marine alga *Hypnea cervicornis* J. Agardh reduces inflammatory hypernociception: involvement of nitric oxide. Pharmacology Biochemistry and Behavior 96: 371–377.

Filippin, L.I., R. Vercelino, N.P. Marroni and R.M. Xavier. 2008. Redox signalling and the inflammatory response in rheumatoid arthritis. Clinical & Experimental Immunology 153(3): 415–422.

Firestein, G.S. 2006. Inhibiting inflammation in rheumatoid arthritis. The New England Journal of Medicine 354: 80–82.

Fischel, J.L., R. Lemee, P. Formento, C. Caldani, J.L. Moll, D. Pesando, A. Meinesz, P. Grelier, P. Pietra and A. Guerriero 1995. Cell growth inhibitory effects of caulerpenyne, a sesquiterpenoid from the marine algae *Caulerpa taxifolia*. Anticancer Research 15(5B): 2155–2160.

Fitton, J.H. 2011. Therapies from Fucoidan; multifunctional marine polymers. Marine Drugs 9(10): 1731–1760.

Fitton, J.H., D.N. Stringer and S.S. Karpiniec. 2015. Therapies from Fucoidan: an update. Marine Drugs 13(9): 5920–5946.

Fitzgerald, C., L. Mora-Soler, E. Gallagher, P. O'Connor, J. Prieto, A. Soler-Vila and M. Hayes. 2012. Isolation and characterization of bioactive pro-peptides with *in vitro* renin inhibitory activities from the macroalga *Palmaria palmata*. Journal of Agricultural and Food Chemistry 60: 7421–7427.

Flemming, R.V., T.J. Walsh and E.J. Anaissie. 2002. Emerging and less common fungal pathogens. Infectious Disease Clinics of North America 16: 915–933.

Fleurence, J. 1999. Seaweed proteins: biochemical, nutritional aspects and potential uses. Trends in Food Science and Technology 10: 25–28.

Fleury, B.G., A. Kelecom, R.C. Pereira and V.L. Teixeira. 1994. Polyphenols, terpenes and sterols in Brazilian Dictyotales and Fucales (Phaeophyta). Botanica Marina 37: 457–462.

Flora, G. and M.V.S. Rani. 2013. GC-MS analysis of *Acanthophora spicifera*. International Journal of Pharma and Bio Sciences 4(1): 649–653.

Fluhr, J.W., M. Breternitz, D. Kowatzki, A. Bauer, J. Bossert, P. Elsner and U.C. Hipler. 2010. Silver-loaded seaweed-based cellulosic fiber improves epidermal skin physiology in atopic dermatitis: Safety assessment, mode of action and controlled, randomized single-blinded exploratory *in vivo* study. Experimental Dermatol 19(8): 1600–0625.

Foley, S.A., E. Szegezdi, B. Mulloy, A. Samali and M.G. Tuohy. 2011. An unfractionated fucoidan from *Ascophyllum nodosum*: Extraction, characterization, and apoptotic effects *in vitro*. Journal of Natural Products 74(9): 1851–1861.

Fontecave, M. and J.L. Pierre 1993. Iron: Metabolism, toxicity and therapy. Biochimie 75: 767–773.

Forstermann, U. 2010. Nitric oxide and oxidative stress in vascular disease. Pflugers Archiv European Journal of Physiology 459: 923–939.

Foster, G.G., A.N. Hodgson. 1998. Consummation and apparent dry matter digestibility of six intertidal macroalgae by *Turbo sarmaticus* (Mollusca: Vetigastropoda: Turbinidae). Aquaculture 167: 211–227.

Fouladvand, M., A. Barazesh, F. Farokhzad, H. Malekizadeh and K. Sartavi. 2011. Evaluation of *in vitro* anti-leishmanial activity of some brown, green and red algae from the Persian Gulf. European Review for Medical and Pharmacological Sciences 15: 597–600.

Fraisl, P., M. Mazzone, T. Schmidt and P. Carmeliet. 2009. Regulation of angiogenesis by oxygen and metabolism. Developmental Cell 16: 167–179.

Francisco, C., B. Banaigs, R. Valls and L. Codomier. 1985. Mediterraneol a, a novel rearranged diterpenoid-hydroquinone from the marine alga *Cystoseira mediterranea*. Tetrahedron Letters 26(22): 2629–2632.

Francisco, C., B. Banaigs, J. Teste and A. Cave. 1986. Mediterraneols: a novel biologically active class of rearranged diterpenoid metabolites from *Cystoseira mediterranea* (Pheophyta). The Journal of Organic Chemistry 51(7): 1115–1120.

Freile-Pelegrín, Y. and J.L. Morales. 2004. Antibacterial activity in marine algae from the coast of Yucatan, Mexico. Botanica Marina 47: 140–146.

Freile-Pelegrin, Y., D. Robledo, M.J. Chan-Bacab and B.O. Ortega-Morales. 2008. Antileishmanial properties of tropical marine algae extracts. Fitoterapia 79: 374–377.

Freurence, J. 2004. Seaweed proteins, Chapter 9. *In*: Yada, R.Y. (ed.). Proteins in Food Processing. Woodhead Publishing Limited and CRC Press LLC, Boca Raton.

Frick, M.H., O. Elo, K. Haapa, O.P. Heinonen, P. Heinsalmi, P. Helo, J.K. Huttunen, P. Kaitaniemi, P. Koskinen, V. Manninen, H. Mäenpää, M. Mälkönen, M. Mänttäri, S. Norola, A. Pasternack, J. Pikkarainen, M. Romo, T. Sjöblom and E.A. Nikkilä. 1987. Helsinki heart study: Primary-prevention trial with Gemfibrozil in middle-aged men with dyslipidemia. Safety of treatment, changes in risk factors, and incidence of coronary heart disease. The New England Journal of Medicine 317(20): 1237–1245.

Fujimura, T., Y. Shibuya, S. Moriwaki, K. Tsukahara, T. Kitahara, T. Sano, Y. Nishizawa and Y. Takema. 2000. Fucoidan is the active component of *Fucus vesiculosus* that promotes contraction of fibroblast-populated collagen cells. Biological and Pharmaceutical Bulletin 23(10): 1180–1184.

Fujimura, T., K. Tsukahara, S. Moriwaki, T. Kitahara, T. Sano and Y. Takema. 2002. Treatment of human skin with an extract of *Fucus vesiculosus* changes its thickness and mechanical properties. Journal of Cosmetic Science 53(1): 1–9.

Fujita, T. 2005. Active absorbable algal calcium (AAACa) changes calcium paradigm. Clinical Calcium 15: 87–93.

Fukahori, S., H. Yano, J. Akiba, S. Ogasawara, S. Momosaki, S. Sanada, K. Kuratomi, Y. Ishizaki, F. Moriya, M. Yagi and M. Kojiro. 2008. Fucoidan, a major component of brown seaweed, prohibits the growth of human cancer cell lines *in vitro*. Molecular Medicine Reports 1: 537–542.

Fukuda, Y., T. Sugahara, M. Ueno, Y. Fukuta, Y. Ochi, K. Akiyama, T. Miyazaki, S. Masuda, A. Kawakubo and K. Kato. 2006. The anti-tumor effect of *Euchema serra* agglutinin on colon cance rcells *in vitro* and *in vivo*. Anticancer Drugs 17(8): 943–947.

Fukuzawa, A. and T. Masamune. 1981. Laurepinnacin and isolaurepinnacin, new acetylenic cyclic ethers from the marine red alga *Laurencia pinnata* Yamada. Tetrahedron Letters 22(41): 4081–4084.

Fuller, R.W., J.H. Cardellina, Y. Kato, L.S. Brinen, J. Clardy, K.M. Snader and M.R. Boyd. 1992. A pentahalogenated monoterpene from the red alga *Portieria hornemannii* produces a novel cytotoxicity profile against a diverse panel of human tumor cell lines. Journal of Medicinal Chemistry 35(16): 3007–3011.

Fuller, R.W., J.H. Cardellina, J. Jurek, P.J. Scheuer, B. Alvarado-Lindner, M. McGuire, G.N. Gray, J.R. Steiner, J. Clardy, E. Menez, R.H. Shoemaker, D.J. Newman, K.M. Snader and J. Clardy. 1994. Isolation and structure/activity features of halomon-related antitumor monoterpenes from the red alga *Portieria hornemannii*. Journal of Medicinal Chemistry 37(5): 4407–4411.

Funahashi, H., T. Imai, Y. Tanaka, K. Tsukamura, Y. Hayakawa, T. Kikumori, T. Mase, T. Itoh, M. Nishikawa, H. Hayashi, A. Shibata, Y. Hibi, M. Takahashi and T. Narita. 1999. Wakame seaweed suppresses the proliferation of 7,12-dimethylbenz(a)-anthracene-induced mammary tumors in rats. GANN Japanese Journal of Cancer Research 90(9): 922–927.

Funahashi, H., T. Imai, T. Mase, M. Sekiya, K. Yokoi, H. Hayashi, A. Shibata, T. Hayashi, M. Nishikawa, N. Suda, Y. Hibi, Y. Mizuno, K. Tsukamura, A. Hayakawa and S. Tanuma. 2001. Seaweed prevents breast cancer? Japanese Journal of Cancer Research 92(5): 483–487.

Funaki, M., M. Nishizawa, T. Sawaya, S. Inoue and T. Yamagishi. 2001. Mineral composition in the holdfast of three brown algae of the genus *Laminaria*. Fisheries Science 67: 295–300.

Fung, A., N. Hamid and J. Lu. 2013. Fucoxanthin content and antioxidant properties of *Undaria pinnatifida*. Food Chemistry 136(2): 1055–1062.

Galal, H.R.M., W.M. Salem and F.N. El-Deen. 2011. Biological control of some pathogenic fungi using marine algae extracts. Research Journal of Microbiology 6: 645–657.

Galland-Irmouli, A.V., J. Fleurence, R. Lamghari, M. Lun, C. Rouxel, O. Barbaroux, J.-P. Bronowicki, C. Villaume and J.-L. Guéant. 1999. Nutritional value of proteins from edible seaweed, *Palmaria palmata* (dulse). The Journal of Biological Chemistry 10: 353–359.

Galle, J.B., B. Attioua, M. Kaiser, A.M. Rusig, A. Lobstein and C. Vonthron-Senecheau. 2013. Eleganolone, a diterpene from the French marine alga *Bifurcaria bifurcata* inhibits growth of the human pathogens *Trypanosoma brucei* and *Plasmodium falciparum*. Marine Drugs 11: 599–610.

Galli, S.J., M. Tsai and A.M. Piliponsky. 2008. The development of allergic inflammation. Nature 454: 445–454.

Gamal-Eldeen, A.M., E.F. Ahmed and M.A. Abo-Zeid. 2009. *In vitro* cancer chemopreventive properties of polysaccharide extract from the brown alga, *Sargassum latifolium*. Food and Chemical Toxicology 47(6): 1378–1384.

Gambato, G., É.G. Baroni, C.S.C. Garcia, R. Frassini, C.O.S. Frozza, S. Moura, C.M.P. Pereira, M.T. Fujii, P. Colepicolo, A.P.F. Lambert, J.A.P. Henriques and M. Roesch-Ely. 2014. Brown algae *Himantothallus grandifolius* (Desmarestiales, Phaeophyceae) suppresses proliferation and promotes apoptosis-mediated cell death in tumor cells. Advances in Biological Chemistry 4: 98–108.

Gammone, M.A. and N. D'Orazio. 2015. Anti-obesity activity of the marine carotenoid fucoxanthin. Marine Drugs 13(4): 2196–2214.

Gammone, M.A., G. Riccioni and N. D'Orazio. 2015. Carotenoids: Potential allies of cardiovascular health? Food & Nutrition Research 59: 26762.

Ganesan, P., K. Noda, Y. Manabe, T. Ohkubo, Y. Tanaka, T. Maoka, T. Sugawara and T. Hirata. 2011. Siphonaxanthin, a marine carotenoid from green algae, effectively induces apoptosis in human leukemia (Hl-60) cells. Biochimica et Biophysica Acta (BBA) 5: 497–503.

Ganesh, A.E., S. Das, G. Arun, S. Balamurugan and R.R. Raj. 2009. Heparin like compound from green alga *Chaetomorpha antennina*—As potential anticoagulant agent. Asian Journal of Medical Sciences 1: 114–116.

Ganeshamurthy, R., T.T. Ajith Kumar, N.B. Dhayanithi and K. Tissera. 2013. Evaluation of antibacterial activity of bioactive compounds obtained from the seaweed *Chondrococcus hornemanni* on ichthyopathogenic bacteria affecting marine ornamental fish. Journal of Coastal Life Medicine 1(1): 34–38.

Ganeshamurthy, R., N.B. Dhayanithi, T.T.A. Kumar and S. Kumaresan. 2014. Evaluation of antibacterial activity and immunostimulant of red seaweed *Chondrococcus hornemanni* (Kuetzing, 1847) against marine ornamental fish pathogens. Journal of Coastal Life Medicine 2(1): 64–69.

Ganguly, R. and G.N. Pierce. 2012. Trans fat involvement in cardiovascular disease. Molecular Nutrition and Food Research 56: 1090–1096.

Gany, S.A., S.C. Tan and S.Y. Gan. 2014. Antioxidative, anticholinesterase and anti-neuroinflammatory properties of Malaysian brown and green seaweeds. International Journal of Biological, Biomolecular, Agricultural, Food and Biotechnological Engineering 8(11): 1251–1257.

Gao, J., C. Caballero-George, B. Wang, K.V. Rao, A.G. Shilabin and M.T. Hamann. 2009. 5-OHKF and NorKA, depsipeptides from a Hawaiian collection of *Bryopsis pennata*: binding properties for NorKA to the human neuropeptide Y Y1 receptor. Journal of Natural Products 72(12): 2172–2176.

Gao, X.-H., L. Zhang, H. Wei, H.-D. Chen. 2008. Efficacy and safety of innovative cosmeceuticals. Clinics in Dermatology 26: 367–374.

Gao, Y., C. Dong, J. Yin, J. Shen, J. Tian and C. Li. 2012. Neuroprotective effect of fucoidan on H2O2-induced apoptosis in PC12 cells via activation of PI3K/Akt pathway. Cellular and Molecular Neurobiology 32: 523–529.

García-Bueno, N., P. Decottignies, V. Turpin, J. Dumay, C. Paillard, V. Stiger-Pouvreau, N. Kervarec, Y.F. Pouchus, A.A. Marín-Atucha and J. Fleurence. 2014. Seasonal antibacterial activity of two red seaweeds, *Palmaria palmata* and *Grateloupia turuturu*, on European abalone pathogen *Vibrio harveyi*. Aquatic Living Resources 27: 83–89.

García Delgado, N., A.I.F. Vázquez, H.C. Sánchez, R.M.S. del Valle, Y.S. Gómez and A.M.S. Alfonso. 2013. Anti-inflammatory and antinociceptive activities of methanolic extract from red seaweed *Dichotomaria obtusata*. Brazilian Journal of Pharmaceutical Sciences 49: 65–74.

García, I., R. Castroviejo and C. Neira. 1993. Las algas en Galicia: alimentación y otros usos. Consellería de Pesca, Marisqueo e Acuicultura—Xunta de Galícia, La Coruña, 231 pp.

Garg, H.S. 1993. Bioactive substances in marine algae. pp. 1–8. *In*: Marine Biotechnology. Plenum press, New York.

Garrido, V., G.A. Teixeira, V.L. Teixeira, P. Ocampo, W.J. Ferreira, D.N. Cavalcanti, S. Campos, P. Md, P. Olaya, C.C. dos Santos, V. Giongo and I.C.P. Paixão. 2011. Evaluation of the acute toxicity of dolabelladienotriol, a potential antiviral from the brown alga *Dictyota pfaffii*, in BALB/c mice. Revista Brasileira de Farmacognosia 21: 209–215.

Gavini, E., V. Sanna, C. Juliano, M.C. Bonferoni and P. Giunchedi. 2002. Mucoadhesive vaginal tablets as veterinary delivery system for the controlled release of an antimicrobial drug, Acriflavine. AAPS PharmSciTech 3(3): E20.

Gedouin, A., R. Valle and P.Y. Morvan. 2006. Use of algae extract to stimulate the oxygen uptake by the cells having lipolytic effect to produce ATP molecules. FR 2879098 A1.

Geisen, U., M. Zenthoefer, M. Peipp, J. Kerber, J. Plenge, A. Managò, M. Fuhrmann, R. Geyer, S. Hennig, D. Adam, L. Piker, G. Rimbach and H. Kalthoff. 2015. Molecular mechanisms by which a *Fucus vesiculosus* extract mediates cell cycle inhibition and cell death in pancreatic cancer cells. Marine Drugs 13(7): 4470–4491.

Gelderblom, H.R. 1991. Assembly and morphology of HIV: potential effect of structure on viral function: AIDS 5(6): 617–637.

Gelderblom, H.R. 1996. Structure and classification of viruses. In: Baron, S. (ed.). Medical Microbiology. The University of Texas Medical Branch at Galveston, Galveston.

Geng, M.Y., F.C. Li, X.L. Xin, J. Li, Z.W. Yan and H.S. Guan. 2003. The potential molecular targets of marine sulfated polymannuroguluronate interfering with HIV-1 entry interaction between SPMG and HIV-1 gp120 and CD4 molecule. Antiviral Research 59: 127–135.

Genovese, G., L. Tedone, M.T. Hamann and M. Morabito. 2009. The mediterranean red alga *Asparagopsis*: A source of compounds against Leishmania. Marine Drugs 7(3): 361–366.

Genovese, G., C. Faggio, C. Gugliandolo, A. Torre, A. Spano, M. Morabitoa and T.L. Maugeri. 2012. *In vitro* evaluation of antibacterial activity of *Asparagopsis taxiformis* from the Straits of Messina against pathogens relevant in aquaculture. Marine environmental research 73: 1–6.

Genovese, G., S. Leitner, S.A. Minicante and C. Lass-Florl. 2013. The Mediterranean red alga *Asparagopsis taxiformis* has antifungal activity against *Aspergillus* species. Mycoses 56: 516–519.

Georg, J.M., M. Kristensen and A. Astrup. 2012. Effect of alginate supplementation on weight loss in obese subjects completing a 12-Wk energy-restricted diet: a randomized controlled trial. The American Journal of Clinical Nutrition 96(1): 5–13.

Gerber, P., J.D. Dutcher, E.V. Adams and J.H. Sherman. 1958. Protective effect of seaweed extracts for chicken embryos infected with Influenza B or Mumps. Proceedings of the Society for Experimental Biology and Medicine 99: 590–593.

Gerlach, M., P. Riederer, H. Przuntek and M.B. Youdim. 1991. MPTP mechanisms of neurotoxicity and their implications for Parkinson's disease. European Journal of Pharmacology: Molecular Pharmacology 208: 273–286.

Gerwick, W.H. and M.W. Bernart. 1993. Eicosanoids and related compounds from marine algae. pp. 101–152. *In*: Attaway, D.H. and O.R. Zaborsky (eds.). Marine Biotechnology—Pharmaceutical and Bioactive Natural Products. Springer, Boston, US.

Ghannadi, A., A. Plubrukarn, K. Zandi, K. Sartavi and A. Yegdaneh. 2013. Screening for antimalarial and acetylcholinesterase inhibitory activities of some Iranian seaweeds. Research in Pharmaceutical Sciences 8(2): 113–118.

Ghislain, M., D.H. Kwak, W.O. Brice, C.A. Maranguy, D.B. Sylvatrie, D.H. Lee, G.M. Pissibanganga, K. Kisung, J. Seo and K.C. Young. 2014. Anticancer effects of different seaweeds on human colon and breast cancers. Maraine Drugs 12: 4898–4911.

Ghosh, P., U. Adhikari, P.K. Ghosal, C.A. Pujol, M.J. Carlucci, E.B. Damonte and B. Ray. 2004. *In vitro* anti-herpetic activity of sulfated polysaccharide fractions from *Caulerpa racemosa*. Phytochemistry 65: 3151–3157.

Ghosh, T., C.A. Pujol, E.B. Damonte, S. Sinha and B. Ray. 2009. Sulfated xylomannans from the red seaweed *Sebdenia polydactyla*: structural features, chemical modification and antiviral activity. Antiviral Chemistry & Chemotherapy 19: 235–242.

Gibbons, S. 2004. Anti-staphylococcal plant natural products. Natural Product Reports 21: 263–277.

Giffin, J.C., R.C. Richards, C. Craft, N. Jahan, C. Leggiadro, T. Chopin, M. Szemerda, S.L. MacKinnon and K.V. Ewart. 2017. An extract of the marine alga *Alaria esculenta* modulates α-synuclein folding and amyloid formation. Neuroscience Letters 644: 87–93.

Gilaberte, M., R. Bartralot, J.M. Torres, F.S. Reus, V. Rodríguez, A. Alomar and R.M. Pujol. 2005. Cutaneous alternariosis in transplant recipients: clinicopathologic review of 9 cases. Journal of the American Academy of Dermatology 52: 653–659.

Gilchrest, B.A. and M.S. Eller. 1999. DNA photodamage stimulates melanogenesis and other photoprotective responses. Journal of Investigative Dermatology 4: 35–40.

Gilgun-Sherki, Y., E. Melamed and D. Offen. 2001. Oxidative stress induced-neurodegenerative diseases: The need for antioxidants that penetrate the blood brain barrier. Neuropharmacology 40: 959–975.

Gimbrone, M.A. 1987. Vascular endothelium in hemostasis and thrombosis. Churchill Livingstone, New York, 264 pp.

Girond S., J.M. Crance, H. Van Cuyck-Gandre, J. Renaudet and R. Deloince. 1991. Antiviral activity of carrageenan on hepatitis A virus replication in cell culture. Research in Virology 142: 261–270.

Glauser, B.F., R.M. Rezende, F.R. Melo, M.S. Pereira, I.M.B. Francischetti, R.Q. Monteiro and P.A.S. Mourão. 2009. Anticoagulant activity of a sulfated galactan: serpin-independent effect and specific interaction with factor Xa. Thrombosis and Haemostasis 102: 1183–1193.

Glazko, G.V. and M. Nei. 2003. Estimation of divergence times for major lineages of primate species. Molecular Biology and Evolution 20: 424–434.

Glombitza, K.-W., M. Forster and W.F. Farnham. 1982. Antibiotics from algae, 25: Polyhydroxyphenyl ethers from the brown alga *Sargassum muticum* (Yendo) Fensholt, Part 2. Botanica Marina 25(9): 449–453.

Glombitza, K.-W. and S.-M. Li. 1991. Hydroxyphlorethols from the brown alga *Carpophyllum Maschalocarpum*. Phytochemistry 30(8): 2741–2745.

Go, H., H.-J. Hwang and T.-J. Nam. 2009. Glycoprotein extraction from *Laminaria japonica* promotes IEC-6 cell proliferation. International Journal of Molecular Medicine 24: 819–824.

Go, H., H.-J. Hwang and T.-J. Nam. 2010. A glycoprotein from *Laminaria japonica* induces apoptosis in HT-29 colon cancer cells. Toxicology *in vitro* 24: 1546–1553.

Goedert, M., C. Wischik, R. Crowther, J. Walker and A. Klug. 1988. Cloning and sequencing of the cDNA encoding a core protein of the paired helical filament of Alzheimer disease: identification as the microtubule-associated protein tau. Proceedings of the National Academy of Sciences 85: 4051–4055.

Goldberg, S.L. 1943. The use of water soluble chlorophyll in oral sepsis: an experimental study of 300 cases. The American Journal of Surgery 62: 117–123.

Goldmann, D.A. 2001. Epidemiology and prevention of pediatric viral respiratory infections in health-care institutions. Emerging Infectious Diseases 7: 249–253.

Gomaa, H.H.A. and G.A. Elshoubaky. 2015. Antiviral activity of sulfated polysaccharides carrageenan from some marine seaweeds. International Journal of Current Pharmaceutical Review and Research 7(1): 34–42.

Gomes, D., C. Telles, M. Costa, J. Almeida-Lima, L. Costa, T. Keesen and H. Rocha. 2015. Methanolic extracts from brown seaweeds *Dictyota cilliolata* and *Dictyota menstrualis* induce apoptosis in human cervical adenocarcinoma Hela cells. Molecules 20(4): 6573–6591.

Gómez-Alpízar, L. 2001. *Verticillium dahliae*. Available online at: https://projects.ncsu.edu/cals/course/pp728/Verticillium/Vertifin.htm.

Gomez, C.G., M.V.P. Lambrecht, J.E. Lozano, M. Rinaudo and M.A. Villar. 2009. Influence of the extraction-purification conditions on final properties of alginates obtained from brown algae (*Macrocystis pyrifera*). International Journal of Biological Macromolecules 44 (4): 365–371.

Gómez-Ordóñez, E., A. Jiménez-Escrig and A.P. Rupérez. 2010. Dietary fibre and physicochemical properties of several edible seaweeds from the northwestern Spanish coast. Food Research International 43: 2289–2294.

Gómez-Ordóñez, E., A. Jiménez-Escrig and A.P. Rupérez. 2012. Effect of the red seaweed *Mastocarpus stellatus* intake on lipid metabolism and antioxidant status in healthy Wistar rats. Food Chemistry 135: 806–811.

Gómez-Pinilla, F. 2008. Brain foods: the effects of nutrients on brain function. Nature Reviews Neuroscience 9(7): 568–578.

Gong, T.F., C.Y. Fang, W. Chen and P.N. Wang. 2007. Physiochemical properties of alginate in tissue engineering research and its clinical application. Journal of Clinical Rehabilitative Tissue Engineering Research 11(18): 3613–3616.

Gonzalez, A.G., J. Darias, A. Diaz J.D. Fournero, J.D. Martin and C. Perez. 1976. Evidence for the biogenesis of halogenated chamigrenes from the red alga *Laurencia obtusa*. Tetrahedron Letters 17(35): 3051–3054.

Gonzalez, A.G., M.J. Delgado, V.S. Martin, M. Martinez-Ripoll and J. Fayos. 1979. X-ray study of sesquiterpene constituents of the alga *Laurencia obtusa* leads to structure revision. Tetrahedron Letters 29: 2717–2718.

Gonzalez, A.G., J.D. Martin, V.S. Martıin, M. Norte and R. Perez. 1982. Biomimetic approach to the synthesis of rhodolaureol and rhodolauradiol. Tetrahedron Letters 23: 2395–2398.

Gonzalez, d.V., A., G. Platas, A. Basilio, A. Cabello, J. Gorrochategui, I. Suay, F. Vicente, E. Portillo, M. Jiménez del Río, G.G. Reina and F. Peláez. 2001. Screening of antimicrobial activities in red, green and brown macroalgae from Gran Canaria (Canary Islands, Spain). International Microbiology 4(1): 35–40.

Gonzalez, M.E., B. Alarcon and L. Carrasco. 1987. Polysaccharides as antiviral agents: Antiviral activity of carrageenan. Antimicrobial Agents and Chemotherapy 31: 1388–1393.

Gonzalez, R., S. Rodriguez, C. Romay, O. Ancheta, A. Gonzalez, J. Armesto, D. Remirez and N. Merino. 1999. Anti-inflammatory activity of Phycocyanin extract in acetic acid-induced colitis in rats. Pharmacological Research 39(1): 55–59.

Gorham, J. and S.A. Lewey. 1984. Seasonal changes in the chemical composition of *Sargassum muticum*. Marine Biology 80: 103–107.

Govindan, M., S.A. Abbas, F.J. Schmitz, R.H. Lee, J.S. Papkoff and D.L. Slate. 1994. New cycloartanol sulfates from the alga *Tydemania expeditionis*: inhibitors of the protein tyrosine kinase pp60v-src. Journal of Natural Products 57(1): 74–78.

Grassauer, A., R. Weinmuellner, C. Meier, A. Pretsch, E. Prieschl-Grassauer and H. Unger. 2008. Iota-carrageenan is a potent inhibitor of rhinovirus infection. Virology Journal 5(107): 2008.

Graufel, V., B. Kloareg, S. Mabeau, P. Durand and J. Jozefonvicz. 1989. New natural polysaccharides with potent antithrombic activity: fucans from brown algae. Biomaterials 10: 363–368.

Greig, N.D., K. Lahiri and K. Sambamurti. 2002. Butyrylcholinesterase: An important new target in Alzheimer's disease therapy. International Psychogeriatrics 14: 77–91.

Groll, A.H., P.M. Shah, C. Mentzel, M. Schneider, G. Just-Nuebling and K. Huebner. 1996. Trends in the postmortem epidemiology of invasive fungal infections at a university hospital. Journal of Infection 33: 23–32.

Gross, H., D.E. Goeger, P. Hills, S.L. Mooberry, D.L. Ballantine, T.F. Murray, F.A. Valeriote and W.H. Gerwick. 2006. Lophocladines, bioactive alkaloids from the red alga *Lophocladia* sp. Journal of Natural Products 69(4): 640–644.

Grunewald, N., I. Groth and S. Alban. 2009. Evaluation of seasonal variations of the structure and anti-inflammatory activity of sulfated polysaccharides extracted from the red alga *Delesseria sanguinea* (Hudson) Lamouroux (Ceramiales, Delesseriaceae). Biomacromolecules 10(5): 1155–1162.

Gryssels, B., K. Polman, J. Clerinx and L. Kestens. 2006. Human schistosomiasis. Lancet 368: 1106–1118.

Guandalini, S. and H. Vaziri (eds.). 2011. Diarrhea—Clinical gastrology diagnostic and therapeutic advances. Springer, Humana Press, LLC, 500 pp.

Guangling, J., Y. Guangli, Z. Junzeng and H.E. Stephen. 2011. Chemical structures and bioactivities of sulfated polysaccharides from marine algae. Marine Drugs 9: 196–223.

Guardia, S.D., R. Valls, V. Mesguiche, J.-M. Brunei and G. Gulioli. 1999. Enantioselective synthesis of bifurcadiol: a natural antitumor marine product. Tetrahedron Letters 40: 8359–8360.

Guarro, J. 2013. Fusariosis, a complex infection caused by a high diversity of fungal species refractory to treatment. European Journal of Clinical Microbiology & Infectious Diseases 32: 1491–1500.

Gubler, D.J. 2002. Epidemic Dengue/Dengue hemorrhagic fever as a public health, social and economic problem in the 21st century. Trends in Microbiology 10: 100–103.

Gueck, T., A. Seidel and H. Fuhrmann. 2003. Effects of essential fatty acids on mediators of mast cells in culture. Prostaglandins Leukot Essent Fatty Acids 68: 317–322.

Gueck, T., A. Seidel, D. Baumann, A. Meister and H. Fuhrmann. 2004. Alterations of mast cell mediator production and release by gamma-linolenic and docosahexaenoic acid. Veterinary Dermatology 15: 309–314.

Guedes, E.A.C., T.G. da Silva, J.S. Aguiar, L.D. de Barros, L.M. Pinotti and A.E.G. Sant'Ana. 2013. Cytotoxic activity of marine algae against cancerous cells. Revista Brasileira de Farmacognosia 23: 668–673.

Guedes, E.M.C., M.A.S. Araújo, A.K.P. Souza, L.I.S. Souza, L.D. Barros, F.C.A. Maranhão and G.A.E. Sant'Ana. 2012. Antifungal activities of different extracts of marine macroalgae against dermatophytes and *Candida* species. Mycopathologia 174(3): 223–232.

Guerra-Rivas, G., C.M. Gómez-Gutiérrez, G. Alarcón-Arteaga, I.E. Soria-Mercado and N.E. Ayala-Sánchez. 2011. Screening for anticoagulant activity in marine algae from the Northwest Mexican Pacific coast. Journal of Applied Phycology 23(3): 495–810.

Guglielmo, M. and D. Montanari. 2008. Cosmetic composition with a lifting effect for sustaining relaxed tissues. Patent WO 2008146116 A2.

Guillon, F. and M. Champ. 2000. Structural and physical properties of dietary fibres, and consequences of processing on human physiology. Food Research International 33: 233–245.

Guimarães, S.M., M.R.A. Braga, M. Cordeiro-Marino and A.G. Pedrini. 1986. Morphology and taxonomy of *Jolyna laminarioides*, a new member of the Scytosiphonales (Phaeophyceae) from Brazil. Phycologia 25(1): 99–108.

Guiry, M.D. 2016. Seaweed site: information on marine algae. Available online at: http://www.seaweed.ie/uses_general/humanfood.php.

Guiry, M.D. and G.M. Guiry. 2017. *Sargassum oligocystum*—AlgaeBase. World-wide electronic publication (National University of Ireland, Galway. Available online at: http://www.algaebase.org/search/species/detail/?species_id=4111.

Guix, S., M. Asanaka, K. Katayama, S.E. Crawford, F.H. Neill, R.L. Atmar and M.K. Estes. 2007. Norwalk virus RNA is infectious in mammalian cells. Journal of Virology 81: 12238–12248.

Güner, A., C. Köksal, Ş.B. Erel, H. Kayalar, A. Nalbantsoy, A. Sukatar and N.Ü. Karabay Yavaşoğlu. 2015. Antimicrobial and antioxidant activities with acute toxicity, cytotoxicity and mutagenicity of *Cystoseira compressa* (Esper) Gerloff & Nizamuddin from the coast of Urla (Izmir, Turkey). Cytotechnology 67: 135–143.

Gunji, S., J. Santoso, Y. Yoshie-Stark and T. Suzuki. 2007. Effects of extracts from tropical seaweeds on DPPH radicals and Caco-2 Cells treated with hydrogen peroxide. Food Science and Technology Research 13(3): 275–279.

Gupta, S., G. Rajauria and N. Abu-Ghannam. 2010. Study of the microbial diversity and antimicrobial properties of Irish edible brown seaweeds. International Journal of Food Science & Technology 45: 482–489.

Gupta, S. and N. Abu-Ghannam. 2011. Bioactive potential and possible health effects of edible brown seaweeds. Trends in Food Science and Technology 22: 315–326.

Gupta, S., S. Cox, G. Rajauria, A.K. Jaiswal and N. Abu-Ghannam. 2012. Growth inhibition of common food spoilage and pathogenic microorganisms in the presence of brown seaweed extracts. Food and Bioprocess Technology 5(5): 1907–1916.

Gurgel Rodrigues, J.A., A.M. Freire Tovar, A.L. Gomes Quinderé, P.A. de Souza Mourão, I. Nilo Lino de Queiroz and N.M. Barros Benevides. 2016. Extraction and structural properties of *Acanthophora muscoides* (Rhodophyceae) extracellular matrix sulfated polysaccharides and their effects on coagulation. Acta Scientiarum. Technology 38: 273–282.

Gurgel Rodrigues, J.A., I.N.L. de Queiroz, A.L.G. Quinderé, N.M.B. Benevides, A.M.F. Tovar and P.A. de Souza Mourão. 2016b. Mild-acid hydrolysis of a native polysulfated fraction from *Acanthophora muscoides* generates sulfated oligosaccharides displaying *in vitro* thrombin generation inhibition. Acta Scientiarum. Biological Sciences 38(1): 7–15.

Gurib-Fakim, A. 2006. Medicinal plants: traditions of yesterday and drugs of tomorrow. Molecular Aspects of Medicine 27: 1–93.

Gutiérrez, G. 1995. Compositions of *Padina* algae or their extracts, and their pharmaceutical, food compositions, or use for the culture of molluscs or arthropods. EP 0655250 Al.

Gutiérrez-Cepeda, A., J.J. Fernández, M. Norte, S. Montalvao, P. Tammela and M.L. Souto. 2015. Acetate-derived metabolites from the brown alga *Lobophora variegata*. Journal of Natural Products 78: 1716–1722.

Güven, K., Y. Özsoy and O. Ulutin. 1991. Anticoagulant, fibrinolytic and antiaggregant activity of carrageenans and alginic acid. Botanica Marina 34(5): 429–432.

Gwon, W.G., M.S. Lee, J.S. Kim, J.I. Kim, C.W. Lim, N.G. Kim and H.R. Kim. 2013. Hexane fraction from *Sargassum fulvellum* inhibits lipopolysaccharide-induced inducible nitric oxide synthase expression in RAW 264.7 cells *via* NF-κB pathways. The American Journal of Chinese Medicine 41(3): 565–484.

Ha, W.H. and D.H. Park. 2013. Effect of seaweed extract on hair growth promotion in experimental study of C57BL/6 mice. Archives of Craniofacial Surgery 14(1): 1–10.

Ha, Y.-M., J.-S. Choi, B.-B. Lee, H.E. Moon, K.K. Cho and I.S. Choi. 2014. Inhibitory effects of seaweed extracts on the growth of the vaginal bacterium *Gardnerella vaginalis*. Journal of Environmental Biology 35: 537–542.

Häggström, M. 2014. Medical gallery of Mikael Häggström. Wiki Journal of Medicine 1(2): 2002–4436 (In Public Domain). Available online at: https://en.wikipedia.org/wiki/Virus#/media/File:Viral_infections_and_involved_species.png.

Haines, J. 1995. *Aspergillus* in compost: straw man or fatal flaw. Biocycle 6: 32–35.

Hajjeh, R.A., A.N. Sofair, L.H. Harrison, G.M. Lyon, B.A. Arthington-Skaggs, S.A. Mirza, M. Phelan, J. Morgan, M.A. Lee-Yang, L.E. Ciblak, M. Phelan, J. Morgan, W. Lee-Yang, M.A. Ciblak, L.E. Benjamin, L.T. Sanza, S. Huie, S.F. Yeo, M.E. Brandt and D.W. Warnock. 2004. Incidence of bloodstream infections due to *Candida* species and *in vitro* susceptibilities of isolates collected from 1998 to 2000 in a population-based active surveillance program. Journal of Clinical Microbiology 42: 1519–152.

Haldimann, M., A. Alt, A. Blanc and K. Blondeau. 2005. Iodine content of food groups. Journal of Food Composition and Analysis 18: 461–471.

Hamann, M.T. and P.J. Scheuer. 1993. Kahalalide F: a bioactive depsipeptide from the sacoglossan mollusk *Elysia rufescens* and the green alga *Bryopsis* sp. Journal of the American Chemical Society 115(13): 5825–5826.

Hamdy, A.H., E.A. Aboutabl, S. Sameer, A.A. Hussein, A.R. Diaz-Marrero, J. Darias and M. Cueto. 2009. 3-keto-22-epi-28-nor-cathasterone, a brassinosteroid-related metabolite from *Cystoseira myrica*. Steroids 74(12): 927–930.

Hammer, K.A., C.F. Carson and T.V. Riley. 2003. Antifungal activity of the components of *Melaleuca alternifolia* (tea tree) oil. Journal of Applied Microbiology 95: 853–860.

Han, D.K., K.D. Park, K.D. Ahn, S.Y. Jeong and Y.H. Kim. 1989. Preparation and surface characterization of PEO-grafted and heparin-immobilized polyurethanes. Journal of Biomedical Materials Research 23: 87–104.

Han, Y.S., J.H. Lee and S.H. Lee. 2015a. Fucoidan inhibits the migration and proliferation of human colon cancer HT-29 cells via the phosphoinositide-3kinase/Akt/mechanistic target of rapamycin pathways. Molecular Medicine Reports 12: 3446–3452.

Han, Y.S., J.H. Lee and S.H. Lee. 2015b. Antitumor effects of fucoidan on humancolon cancer cells via activation of Akt signaling. Biomolecules & Therapeutics 23: 225–232.

Hanahan, D. and R.A. Weinberg. 2011. Hallmarks of cancer: the next generation. Cell 144(5): 646–764.

Haneji, K., T. Matsuda, M. Tomita, H. Kawakami, K. Ohshiro, J.-N. Uchihara, M. Masuda, N. Takasu, Y. Tanaka, T. Ohta and N. Mori. 2005. Fucoidan extracted from *Cladosiphon okamuranus* Tokida induces apoptosis of human T-cell leukemia virus type 1-infected T-cell lines and primary adult T-cell leukemia cells. Nutrition and Cancer 52(2): 189–201.

Hanger, D.P., A. Seereeram and W. Noble. 2009. Mediators of tau phosphorylation in the pathogenesis of Alzheimer's disease. Expert Review of Neurotherapeutics 9: 1647–1666.

Hannan, M.A., J.Y. Kang, Y.K. Hong, H.S. Lee, M.T.H. Chowdhury, J.-S. Choi, I.S. Choi and I.S. Moon. 2012. A brown alga *Sargassum fulvellum* facilitates neuronal maturation and synaptogenesis. *In Vitro* Cellular & Developmental Biology—Animal 48: 535–544.

Hannan, M.A., J.-Y. Kang, Y.-K. Hong, H.S. Lee, J.-S. Choi, I.S. Choi and I.S. Moon. 2013. The Marine alga *Gelidium amansii* promotes the development and complexity of neuronal cytoarchitecture. Phytotherapy Research 27(1): 21-29.

Hannan, M.A., M. Mohibbullah, Y.-K. Hong, J.H. Nam and I.S. Moon. 2014. *Gelidium amansii* promotes dendritic spine morphology and synaptogenesis, and modulates NMDA receptor-mediated postsynaptic current. *In Vitro* Cellular & Developmental Biology—Animal 50(5): 445–452.

Hannan, M.A., M. Mohibbullah, S.Y. Hwang, K. Lee, Y.C. Kim, Y.K. Hong and I.S. Moon. 2014b. Differential neuritogenic activities of two edible brown macroalgae, *Undaria pinnatifida* and *Saccharina japonica*. The American Journal of Chinese Medicine 42: 1371–1384.

Hans, R.G. 1996. Chapter 41—Structure and classification of viruses. *In*: Baron, S. (ed.). Medical Microbiology, 4th edition. The University of Texas Medical Branch at Galveston.

Hansman, G.S., X.J. Jiang and K.Y. Green (eds.). 2010. Caliciviruses: Molecular and cellular virology. Caister Academic Press, Norfolk, UK, 248 pp.

Harada, H. and Y. Kamei. 1997. Selective cytotoxicity of marine algae extracts to several human leukemic cell lines. Cytotechnology 25: 213–219.

Harada, H., T. Noro and Y. Kamei. 1997. Selective antitumor activity *in vitro* from marine algae from Japan coasts. Biological and Pharmaceutical Bulletin 20(5): 541–546.

Harada, H., U. Yamashita, H. Kurihara, E. Fukushi, J. Kawabata and Y. Kamei. 2002. Antitumor activity of palmitic acid found as a selective cytotoxic substance in a marine red alga. Anticancer Research 22(5): 2587–2590.

Haraszthy, V.I., H.S. Reynolds, P.K. Sreenivasan, R. Subramanyam, D. Cummins and J.J. Zambon. 2006. Media- and method-dependent variations in minimal inhibitory concentrations of antiplaque agents on oral bacteria. Letters in Applied Microbiology 43: 256–261.

Harborne, J.B. 1998. Phytochemical methods a guide to modern techniques of plant analysis. Springer Science & Business Media, Springer Netherlands, 310 pp.

Harden, C.J., J.C. Richardson, P.W. Dettmar, B.M. Corfe and J.R. Paxman. 2012. An ionic-gelling alginate drink attenuates postprandial glycaemia in males. Journal of Functional Foods 4(1): 122–128.

Harden, E.A., R. Falshaw, S.M. Carnachan, E.R. Kern and M.N. Prichard. 2009. Virucidal activity of polysaccharide extracts from four algal species against *Herpes simplex* virus. Antiviral Research 83: 282–289.

Hardoko, T. Siratantri, Eveline, M. Yogabuana and S. Olivia. 2014. An *in vitro* study of antidiabetic activity of *Sargassum duplicatum* and *Turbinaria decurens* seaweed. International Journal of Pharmaceutical Science Invention 3(2): 13–18.

Hardouin, K., A.-S. Burlot, A. Umami, A. Tanniou, V. Stiger-Pouvreau, I. Widowati, G. Bedoux and N. Bourgougnon. 2014. Biochemical and antiviral activities of enzymatic hydrolysates from different invasive French seaweeds. Journal of Applied Phycology 26: 1029–1042.

Hardouin, K., G. Bedoux, A.-S. Burlot, C. Donnay-Moreno, J.-P. Bergé, P. Nyvall-Collén and N. Bourgougnon. 2016. Enzyme-assisted extraction (EAE) for the production of antiviral and antioxidant extracts from the green seaweed *Ulva armoricana* (Ulvales, Ulvophyceae). Algal Research 16: 233–239.

Haroun-Bouhedja, F., F. Lindenmeyer, H. Lu, C. Soria, J. Jozefonvicz and C. Boisson-Vidal. 2002. *In vitro* effects of fucans on MDA-MB231 tumor cell adhesion and invasion. Anticancer Research 22(4): 2285–2292.

Harvey, W.H. 1849. Phycologia britannica, or, a history of British sea-weeds: containing coloured figures, generic and specific characters, synonymes, and descriptions of all the species of algae inhabiting the shores of the British Islands. Reeve & Benham, London, pp. 217–270.

Haslin, C., M. Lahaye, M. Pellegrini and J.C. Chermann. 2001. *In vitro* anti-HIV activity of sulfated cell-wall polysaccharides from gametic, carposporic and tetrasporic stages of the mediterranean red alga *Asparagopsis armata*. Planta Medica 67: 301–305.

Hassan, S., S.A. El-Twab, M. Hetta and B. Mahmoud. 2011. Improvement of lipid profile and antioxidant of hypercholesterolemic albino rats by polysaccharides extracted from the green alga *Ulva lactuca* Linnaeus. Saudi Journal of Biological Sciences 18: 333–340.

Haug, A. and A. Jensen. 1954. Seasonal variation in the chemical composition of *Alaria esculenta, Laminaria saccharina, Laminaria hyperborea* and *Laminaria digitata* from Northern Norway, pp. 1–14. *In*: Akademisk Trykningssentral. Norwegian Institute of Seaweed Research Blindern, Norway.

Hawkins, W.W. and V.G. Leonard. 1962. Antipeptic and antithrombic properties of carrageenin. Journal of Laboratory and Clinical Medicine 60: 641–648.

Hawkins, W.W. and V.G. Leonard. 1963. Antithrombotic activity of carrageenan in human blood. Canadian Journal of Biochemistry and Physiology 41(5): 1325–1327.

Hawks, J., K. Hunley, S.-H. Lee and M. Wolpoff. 2000. Population bottlenecks and pleistocene human evolution. Molecular Biology and Evolution 17(1): 2–22.

Hayakawa, Y., T. Hayashi, J.-B. Lee, P. Srisomporn, M. Maeda, T. Ozawa and N. Sakuragawa. 2000. Inhibition of thrombin by sulfated polysaccharides isolated from green algae. Biochimica et Biophysica Acta (BBA)—Protein Structure and Molecular Enzymology 1543(1): 86–94.

Hayakawa, K. and T. Nagamine. 2009. Effect of fucoidan on the biotinidase kinetics in human hepatocellular carcinoma. Anticancer Research 29: 1211–1218.

Hayashi, K., T. Nakano, M. Hashimoto, K. Kanekiyo and T. Hayashi. 2008. Defensive effects of a fucoidan from brown alga *Undaria pinnatifida* against *Herpes simplex* virus infection. International Immunopharmacology 8: 109–116.

Hayashi, K., J.B. Lee, T. Nakano and T. Hayashi. 2013. Anti-influenza A virus characteristics of a fucoidan from sporophyll of *Undaria pinnatifida* in mice with normal and compromised immunity. Microbes & Infection Institut Pasteur 15: 302–309.

He, F.J. and G.A. MacGregor. 2010. Reducing population salt intake worldwide: from evidence to implementation. Progress in Cardiovascular Diseases 42: 363–382.

Heales, S., J. Bolaños, V. Stewart, P. Brookes, J. Land and J. Clark. 1999. Nitric oxide, mitochondria and neurological disease. Biochimica et Biophysica Acta 1410: 215–228.

Healy, G.N., C.E. Matthews, D.W. Dunstan, E.A.H. Winkler and N. Owen. 2011. Sedentary time and cardio-metabolic biomarkers in US adults: NHANES 2003–06. European Heart Journal 32: 590–597.

Hedayati, M.T., A.C. Pasqualotto, P.A. Warn, P. Bowyer and D.W. Denning. 2007. *Aspergillus flavus*: human pathogen, allergen and mycotoxin producer. Microbiology 153: 1677–1692.

Hellio, C., G. Bremer, A.M. Pons, Y. Le Gal and N. Bourgougnon. 2000. Inhibition of the development of microorganisms (bacteria and fungi) by extracts of marine algae from Brittany, France. Applied Microbiology and Biotechnology 54(4): 543–549.

Hellio, C., D. De La Broise, L. Dufossé, Y. Le Gal and N. Bourgougnon. 2001. Inhibition of marine bacteria by extract of macroalgae: Potential use for environmentally friendly paints. Marine Environmental Research 52: 231–247.

Hellio, C., J.P. Berge, C. Beaupoil, Y. Le Gal and N. Bourgougnon. 2002. Screening of marine algal extracts for anti-settlement activities against microalgae and macroalga. Biofouling 18(3): 205–215.

Hemmingson, J., R. Falshaw, R. Furneaux and K. Thompson. 2006. Structure and antiviral activity of the galactofucan sulfates extracted from *Undaria pinnatifida* (Phaeophyta). Journal of Applied Phycology 18: 185–193.

Hennequart, F. 2007. Seaweed applications in human health/nutrition: The example of algal extracts as functional ingredients in novel beverages. 4th International Symposium Health and Sea, Granville, France.

Hentzer, M. and M. Givskov. 2003. Pharmacological inhibition of quorum sensing for the treatment of chronic bacterial infections. Journal of Clinical Investigation 112: 1300–1307.

Heo, 5.J., K.W. Lee, C.B. Song and Y.J. Jean. 2003. Antioxidant activity of enzymatic extracts from brown seaweeds. Algae 18: 71–81.

Heo, S.J., S.C. Ko, S.M. Kang, H.S. Kang, J.P. Kim, S.H. Kim, K.W. Lee, M.G. Cho and Y.J. Jeon. 2008. Cytoprotective effect of fucoxanthin isolated from brown algae *Sargassum siliquastrum* against H_2O_2-induced cell damage. European Food Research and Technology 228: 145–151.

Heo, S.J. and Y.J. Jeon. 2009. Protective effect of fucoxanthin isolated from *Sargassum siliquastrum* on UV-B induced cell damage. Journal of Photochemistry and Photobiology B: Biology 95: 101–107.

Heo, S.J., J.Y. Hwang, J.I. Choi, J.S. Han, H.J. Kim and Y.J. Jeon. 2009. Diphlorethohydroxycarmalol isolated from *Ishige Okamurae*, a brown alga, a potent alpha-glucosidase and alpha-amylase inhibitor, alleviates postprandial hyperglycemia in diabetic Mice. European Journal of Pharmacology 615(1-3): 252–256.

Heo, S.-J., S.-C. Ko, S.-H. Cha, D.-H. Kang, H.-S. Park, Y.-U. Choi, D. Kim, W.-K. Jung and Y.-J. Jeon. 2009b. Effect of phlorotannins isolated fromEcklonia cava on melanogenesis and their protective effect against photo-oxidative stress induced by UV-B radiation. Toxicol. *in vitro* 23: 1123–1130.

Heo, S.J., W.J. Yoon, K.N. Kim, G.N. Ahn, S.M. Kang, D.H. Kang, Abu affan, C. Oh, W.K. Jung and Y.J. Jeon. 2010. Evaluation of anti-inflammatory effect of fucoxanthin isolated from brown algae in lipopolysaccharidestimulated RAW 264.7 macrophages. Food and Chemical Toxicology 48(8-9): 2045–2051.

Heo, S.-J., S.-C. Ko, S.-M. Kang, S.-H. Cha, S.-H. Lee, D.-H. Kang, W.-K. Jung, A. Affan, C. Oh and Y.-J. Jeon. 2010b. Inhibitory effect of diphlorethohydroxycarmalol on melanogenesis and its protective effect against UV-B radiation-induced cell damage. Food and Chemical Toxicology 48: 1355–1361.

Heo, S.J., K.N. Kim, W.J. Yoon, C. Oh, Y.U. Choi, A. Affan, Y.J. Lee, H.S. Lee and D.H. Kang. 2011. Chromene induces apoptosis via caspase-3 activation in human leukemia HL-60 cells. Food and Chemical Toxicology 49(9): 1998–2004.

Heo, S.J., S.H. Cha, K.N. Kim, S.H. Lee, G. Ahn, D.H. Kang, C. Oh, Y.U. Choi, A. Affan, D. Kim and Y.J. Jeon. 2012. Neuroprotective effect of phlorotannin isolated from *Ishige okamurae* against H2O2-induced oxidative stress in murine hippocampal neuronal cells, HT22. Applied Biochemistry and Biotechnology 166: 1520–1532.

Hernández-Mireles, T. and M. Rito-Palomares. 2006. Improved recovery of B-phycoerythrin produced by the red microalga *Porphyridium cruentum*. Journal of Chemical Technology and Biotechnology 6: 989–996.

Herrero, M., J.A. Mendiola, A. Cifuentes and E. Ibáñez. 2010. Supercritical fluid extraction: Recent advances and applications. Journal of Chromatography A 1217(16): 2495–2511.

Hetta, M., S. Hassan, S. Abdel-Tawab, M. Bastawy and B. Mahmoud. 2009. Hypolipidaemic effect of *Acanthophora spicifera* (red alga) and *Cystoseira trinode* (brown alga) on albino rats. Iranian Journal of Science and Technology Transaction a-Science 33: 287–297.

Hewitt, W. and S. Vincent. 1989. Theory and application of microbiological assay. Academic Press, London, 323 pp.

Hibbett, D.S., M. Binder, J.F. Bischoff, M. Blackwell, P.F. Cannon, O.E. Eriksson, S. Huhndorf, T. James, P.M. Kirk, R. Lucking, H. Thorsten Lumbsch, F. Lutzoni, P.B. Matheny, D.J. McLaughlin, M.J. Powell, S. Redhead, C.L. Schoch, J.W. Spatafora, J.A. Stalpers, R. Vilgalys, M.C. Aime, A. Aptroot, R. Bauer, D. Begerow, G.L. Benny, L.A. Castlebury, P.W. Crous, Y.C. Dai, W. Gams, D.M. Geiser, G.W. Griffith, C. Gueidan, D.L. Hawksworth, G. Hestmark, K. Hosaka, R.A. Humber, K.D. Hyde, J.E. Ironside, U. Koljalg, C.P. Kurtzman, K.H. Larsson, R. Lichtwardt, J. Longcore, J. Miadlikowska, A. Miller, J.M. Moncalvo, S. Mozley-Standridge, F. Oberwinkler, E. Parmasto, V. Reeb, J.D. Rogers, C. Roux, L. Ryvarden, J.P. Sampaio, A. Schussler, J. Sugiyama, R.G. Thorn, L. Tibell, W.A. Untereiner, C. Walker, Z. Wang, A. Weir, M. Weiss, M.M. White, K. Winka, Y.J. Yao and N. Zhang. 2007. A higher-level phylogenetic classification of the Fungi. Mycological Research 111: 509–547.

Hidari, K.I.P.J., N. Takahashi, N. Arihara, M. Nagaoka, K. Morita and T. Suzuki. 2008. Structure and anti-Dengue virus activity of sulfated polysaccharide from a marine alga. Biochemical and Biophysical Research Communications 376: 91–95.

Higashi-Okai, K., S. Otani and Y. Okai. 1999. Potent suppressive effect of a Japanese edible seaweed, *Enteromorpha prolifera* (Sujiao-Nori) on initiation and promotion phases of chemically induced mouse skin tumorigenesis. Cancer Letters 140(1-2): 21–25.

Hirayama, M., H. Shibata, K. Imamura, T. Sakaguchi and K. Hori. 2016. High-mannose specific lectin and its recombinants from a carrageenophyta *Kappaphycus alvarezii* represent a potent anti-HIV activity through high-affinity binding to the viral envelope glycoprotein Gp120. Marine Biotechnology 18(2): 215–231.

Hiroishi, S., K. Sugie, T. Yoshida, J. Morimoto, Y. Taniguchi, S. Imai and J. Kurebayashi. 2001. Antitumor effects of *Marginisporum crassissimum* (Rhodophyceae), a marine red alga. Cancer Letters 167(2): 145–150.

Hirokawa, N., T. Funakoshi, R. Sato-Harada and Y. Kanai. 1996. Selective stabilization of tau in axons and microtubule-associated protein 2C in cell bodies and dendrites contributes to polarized localization of cytoskeletal proteins in mature neurons. The Journal of Cell Biology 132: 667–679.

Hirsch, M.S., H.F. Günthard, J.M. Schapiro, F.B. Vézinet, B. Clotet, S.M. Hammer, V.A. Johnson, D.R. Kuritzkes, J.W. Mellors, D. Pillay, P.G. Yeni, D.M. Jacobsen and D.D. Richman. 2008. Antiretroviral drug resistance testing in adult HIV-1 infection: 2008 recommendations of an international AIDS 4, society-USA panel. Clinical Infectious Diseases 47: 266–285.

Hoang, M.H., J.-Y. Kim, J.H. Lee, S.G. You and S.-J. Lee. 2015. Antioxidative, hypolipidemic, and anti-inflammatory activities of sulfated polysaccharides from *Monostroma nitidum*. Food Science and Biotechnology 24: 199–205.

Holanda, M.L., V.M.M. Melo, L.M.C.M. Silva, R.C.N. Amorim, M.G. Pereira and N.M.B. Benevides. 2005. Differential activity of a lectin from *Solieria filiformis* against human pathogenic bacteria. Brazilian Journal of Medical and Biological Research 38: 1769–1773.

Holdt, S.L. and S. Kraan. 2011. Bioactive compounds in seaweed: functional food applications and legislation. Journal of Applied Phycology 23(3): 543–597.

Holmes, E.M. 1895. Marine algae from Japan. Journal of the Linnean Society of London, Botany 31: 248–260.

Holtkamp, A.D., S. Kelly, R. Ulber and S. Lang. 2009. Fucoidans and fucoidanases-focus on techniques for molecular structure elucidation and modification of marine polysaccharides. Applied Microbiology and Biotechnology 82(1): 1–11.

Homsey, I. and D. Hide. 1974. The production of antimicrobial compounds by British marine algae I. Antibiotic-producing marine algae. British Phycological Journal 9: 353–361.

Hong, D.D., H.M. Hien and H.T.L. Anh 2011. Studies on the analgesic and anti-inflammatory activities of *Sargassum swartzii* (Turner) C. Agardh (Phaeophyta) and *Ulva reticulata* Forsskal (Chlorophyta) in experiment animal models. African Journal of Biotechnology 10(12): 2308–2314.

Hong, S.W., H.S. Lee, K.H. Jung, H. Lee and S.S. Hong. 2012. Protective effect of fucoidan against acetaminophen-induced liver injury. Archives of Pharmacal Research 35: 1099–1105.

Hoppe, H.A. 1979. Marine algae and their products and constituents in pharmacy. pp. 25–119. *In*: H.A. Hoppe, T. Levring and Y. Tanaka (eds.). Marine Algae in Pharmaceutical Science. Walter de Gruyter, New York.

Horgen, F.D., D.B. delos Santos, G. Goetz, B. Sakamoto, Y. Kan, H. Nagai and P.J. Scheuer. 2000. A new depsipeptide from the sacoglossan mollusk *Elysia ornata* and the green alga *Bryopsis* species. Journal of Natural Products 63(1): 152–154.

Hori, K., K. Miyazawa, N. Fusetani, K. Hashimoto and K. Ito. 1986. Hypnins, low-molecular weight peptidic agglutinins isolated from a marine red alga, *Hypnea japonica*. Biochimica et Biophysica Acta 873: 228–236.

Horincar, V.B., G. Parfene, A.K. Tyagi, D. Gottardi, R. Dinica, M.E. Guerzoni and G. Bahrim. 2014. Extraction and characterization of volatile compounds and fatty acids from red and green macroalgae from the Romanian Black Sea in order to obtain valuable bioadditives and biopreservatives. Journal of Applied Phycology 26: 551–559.

Hornsey, L.S. and D. Hide. 1974. The production of antimicrobial compounds by British marine algae. British Phycological Journal 9(4): 353–361.

Hornsey, I.S. and D. Hide 1976. The production of antimicrobial compounds by British marine algae II. Seasonal variation in production of antibiotics, British Phycological Journal 11(1): 63–67.

Hornsey, I.S. and D. Hide. 1985. The production of antimicrobial compounds by British marine algae. IV. Variation of antimicrobial activity with algal generation. British Phycological Journal 20(1): 21–25.

Horwitz, B. 1951. Role of chlorophyll in proctology. The American Journal of Surgery 81(1): 81–84.

Hoshino, T., T. Hayashi, K. Hayashi, J. Hamada, J.-B. Lee and U. Sankawa. 1998. An antivirally active sulfated polysaccharide from *Sargassum horneri* (Turner) C. Agardh. Biological and Pharmaceutical Bulletin 21: 730–734.

Hosikian, A., S. Lim, R. Halim and M.K. Danquah. 2010. Chlorophyll extraction from microalgae: A review on the process engineering aspects. International Journal of Chemical Engineering vol. 2010, Article ID 391632, 11 pages.

Hosokawa, M., M. Kudo, H. Maeda, H. Kohno, T. Tanaka and K. Miyashita. 2004. Fucoxanthin induces apoptosis and enhances the antiproliferative effect of the PPARgamma ligand, troglitazone, on colon cancer cells. Biochimica et Biophysica Acta (BBA) 18: 1–3.

Hosokawa, M., T. Okada, N. Mikami, I. Konishi and K. Miyashita. 2009. Bio-Functions of Marine Carotenoids. Food Science and Biotechnology 18(1): 1–11.

Hossain, M.M., F. Sultana, M. Kubota, H. Koyama and M. Hyakumachi. 2007. The plant growth-promoting fungus *Penicillium simplicissimum* GP17-2 induces resistance in *Arabidopsis thaliana* by activation of multiple defense signals. Plant and Cell Physiology 48(12): 1724–1736.

Hou, L.L., C. Gao, L. Chen, G.Q. Hu and S.Q. Xie. 2013. Essential role of autophagy in fucoxanthin-induced cytotoxicity to human epithelial cervical cancer Hela cells. Acta Pharmacologica Sinica 34(11): 1403–1410.

Hou, X. and X. Yan. 1998. Study on the concentration and seasonal variation of inorganic elements in 35 species of marine algae. Science of the Total Environment 222: 141–156.

Houbraken, J.A.M.P., J.C. Frisvad and R.A. Samson. 2010. Taxonomy of *Penicillium citrinum* and related species. Fungal Diversity 44: 117–133.

Houck, J.C., R.K. Morris and E.J. Lazaro. 1957. Anticoagulant, lipemia clearing and other effects of anionic polysaccharides extracted from seaweed. Proceedings of the Society for Experimental Biology and Medicine 96: 528–530.

Houghton, D., M.D. Wilcox, P.I. Chater, I.A. Brownlee, C.J. Seal and J.P. Pearson. 2015. Biological activity of alginate and its effect on pancreatic lipase inhibition as a potential treatment for obesity. Food Hydrocolloids 49: 18–24.

Houston, M.C. 2005. Nutraceuticals, vitamins, antioxidants, and minerals in the prevention and treatment of hypertension. Progress in Cardiovascular Diseases 47(6): 396–449.

Howard, S.J., D. Cerar, M.J. Anderson, A. Albarrag, M.C. Fisher, A.C. Pasqualotto, M. Laverdiere, M.C. Arendrup, D.S. Perlin and D.W. Denning. 2009. Frequency and evolution of azole resistance in *Aspergillus fumigatus* associated with treatment failure. Emerging Infectious Diseases 15(7): 1068–1076.

Hsu, H.Y., T.Y. Lin, P.A. Hwang, L.M. Tseng, R.H. Chen, S.M. Tsao and J. Hsu. 2013. Fucoidan induces changes in the epithelial to mesenchymal transition and decreases metastasis by enhancing ubiquitin-dependent TGF beta receptor degradation in breast cancer. Carcinogenesis 34: 874–884.

Hsu, H.Y., T.Y. Lin, Y.C. Wu, S.M. Tsao, P.A. Hwang, Y.W. Shih and J. Hsu. 2014. Fucoidan inhibition of lung cancer *in vivo* and *in vitro*: Role of the Smurf2-dependent ubiquitin proteasome pathway in TGFβ-receptor degradation. Oncotarget 5: 7870–7885.

Hu, J., M. Geng, J. Li, X. Xin, J. Wang, M. Tang, J. Zhang, X. Zhang and J. Ding. 2004b. Acidic oligosaccharide sugar chain, a marine-derived acidic oligosaccharide, inhibits the cytotoxicity and aggregation of amyloid beta protein. Journal of Pharmaceutical Sciences 95: 248–255.

Hu, P., Z. Li, M. Chen, Z. Sun, Y. Ling, J. Jiang and C. Huang. 2016. Structural elucidation and protective role of a polysaccharide from *Sargassum fusiforme* on ameliorating learning and memory deficiencies in mice. Carbohydrate Polymers 139: 150–158.

Hu, X., X. Jiang, H. Hwang, S. Liu and H. Guan. 2004. Antitumour activities of alginate-derived oligosaccharides and their sulphated substitution derivatives. European Journal of Phycology 39(1): 67–71.

Huang, C.C., Y.F. Chen and C.C.R. Wang. 2010. Effects of micronization on the physicochemical properties of peels of three root and tuber crops. Journal of the Science of Food and Agriculture 90: 759–763.

Huang, H.L., S.L. Wu, H.F. Liao, C.M. Jiang, R.L. Huang, Y.Y. Chen, Y.C. Yang and Y.J. Chen. 2005. Induction of apoptosis by three marine algae through generation of reactive oxygen species in human leukemic cell lines. Journal of Agricultural and Food Chemistry 53(5): 1776–1781.

Huang, S.-F. 2017. *Valonia aegagropila*—BiotaTaiwanica, Algae of Taiwan. Available online at: http://algae.biota.biodiv.tw/pages/1848.

Huang, S.H., Y.S. Lin, C.W. Wu and C.J. Wu. 2014. Assessment of the inhibition of Dengue virus infection by carrageenan via real-time monitoring of cellular oxygen consumption rates within a microfluidic device. Biomicrofluidics 8 (024110): 1–14.

Huang, W.C., F.C. Yen, Y.J. Shiao, F.S. Shie, J.L. Chan, C.N. Yang, Y.L. Sung, F.L. Huang and H.J. Tsay. 2009. Enlargement of Abeta aggregates through chemokine-dependent microglial clustering. Neuroscience Research 63: 280–287.

Hudson, J.B., J.H. Kim, M.K. Lee, R.E. DeWreede and Y.K. Hong. 1998. Antiviral compounds in extracts of Korean seaweeds: evidence for multiple activities. Journal of Applied Phycology 10: 427–434.

Hudson, J.B., J.H. Kim, M.K. Lee, R.E. DeWreede and Y.K. Hong. 1999. Antiviral compounds in extracts of Korean seaweeds: Evidence for multiple activities. Journal of Applied Phycology 10: 427–434.

Hughes, J. 2017. *Cryptopleura ramosa*—MarLIN, The Marine Life Information Network. Available online at: http://www.marlin.ac.uk/species/detail/1687.

Huheihel, M., V. Ishanu, J. Tal and S.M. Arad. 2002. Activity of *Porphyridium* sp. polysaccharide against *Herpes simplex* viruses *in vitro* and *in vivo*. Journal of Biochemical and Biophysical Methods 50(2-3): 189–200.

Hui, B., J. Li, M.Y. Geng. 2008. Sulfated polymannuroguluronate, a novel anti-acquired immune deficiency syndrome drug candidate, decreased vulnerability of PC12 cells to human immunodeficiency virus tat protein through attenuating calcium overload. Journal of Neuroscience Research 86(5): 1169–1177.

Huisman, J. and C. Parker. 2016. *Gracilaria canaliculata*—FloraBase. Available online at: https://florabase.dpaw.wa.gov.au/browse/profile/35899.

Humphrey, A.M. 2004. Chlorophyll as a color and functional ingredient. Journal of Food Science 69(5): C422–C425.

Hupel, M., C. Lecointre, A. Meudec, N. Poupart and E. Ar Gall. 2011. Comparison of photoprotective responses to UV radiation in the brown seaweed *Pelvetia canaliculata* and the marine angiosperm *Salicornia ramosissima*. Journal of Experimental Marine Biology and Ecology 401: 36–47.

Hurtado, A.Q., P.-E. Lim, J. Tan, S.-M. Phang, I.C. Neish and A.T. Critchley. 2016. Chapter 3—Biodiversity and biogeography of commercial tropical carrageenophytes in the Southeast Asian region. pp. 51–74. *In*: Pereira, L. (ed.). Carrageenans: Sources and Extraction Methods, Molecular Structure, Bioactive Properties and Health Effects. Nova Science Publisher, N.Y.

Husni, A., R. Wijayanti and Ustadi. 2014. Inhibitory activity of α-amylase and α-glucosidase by *Padina pavonica* extracts. Journal of Biological Sciences 14: 515–520.

Hussain, F., S.S. Shaukat, M. Abid, F. Usman and M. Akbar. 2013. Control of *Meloidogyne javanica* and *Fusarium solani* in chilli (*Capsicum annuum* L.) with the application of chitin. Pakistan Journal of Nematology 31: 165–170.

Hussein, M.M.D., A. Abdelaziz and H.M. Salem. 1980. Sulfated heteropolysaccharides from *Padina pavonia*. Phytochemistry 19(10): 2131–2132.

Huynh, T.G., S.T. Yeh, Y.C. Lin, J.F. Shyu, L.L. Chen and J.C. Chen. 2011. White shrimp *Litopenaeus vannamei* immersed in seawater containing *Sargassum hemiphyllum* var. *chinense* powder and its extract showed increased immunity and resistance against *Vibrio alginolyticus* and white spot syndrome virus. Fish and Shellfish Immunology 31: 286–293.

Hwang, E.K., H. Amano and C.S. Park. 2008. Assessment of the nutritional value of *Capsosiphon fulvescens* (Chlorophyta): developing a new species of marine macroalgae for cultivation in Korea. Journal of Applied Phycology 20: 147–151.

Hwang, H., T. Chen, R.G. Nines, H.C. Shin and G.D. Stoner. 2006. Photochemoprevention of UVB-induced skin carcinogenesis in SKH-1 Mice by brown algae polyphenols. International Journal of Cancer 119(12): 2742–2749.

Hwang, P.A., S.Y. Chien, Y.L. Chan, M.K. Lu, C.H. Wu, Z.L. Kong and C.J. Wu. 2011. Inhibition of lipopolysaccharide (LPS)-induced inflammatory response by *Sargassum hemiphyllum* sulfated polysaccharide extract in RAW 264.7 macrophage cells. Journal of Agricultural and Food Chemistry 59: 2062–2068.

Hwang, P.A., Y.L. Hung, Y.K. Tsai, S.Y. Chien and Z.L. Kong. 2015. The brown seaweed *Sargassum hemiphyllum* exhibits α-amylase and α-glucosidase inhibitory activity and enhances insulin release *in vitro*. Cytotechnology 67: 653–660.

Hwang, P.-A., Y.-L. Hung and S.-Y. Chien. 2015b. Inhibitory activity of *Sargassum hemiphyllum* sulfated polysaccharide in arachidonic acidinduced animal models of inflammation. Journal of Food and Drug Analysis 23(1): 49–56.

Hynes, T. 2016. Anatomy of the brian. Mayfield Brain & Spine. Available online at: http://www.mayfieldclinic.com/PE-AnatBrain.htm.

Hyun, J.H., S.C. Kim, J.I. Kang, M.K. Kim, H.J. Boo, J.M. Kwon, Y.S. Koh, J.W. Hyun, D.B. Park, E.S. Yoo and H.K. Kang. 2009. Apoptosis inducing activity of fucoidan in HCT-15 colon carcinoma cells. Biological and Pharmaceutical Bulletin 32(10): 1760–1764.

Iantorno, M., U. Campia, N. Di Daniele, S. Nistico, G.B. Forleo, C. Cardillo and M. Tesauro. 2014. Obesity, inflammation and endothelial dysfunction. Journal of Biological Regulators & Homeostatic Agents 28: 160–176.

Iberagar. 2010. Agar-agar, poderoso espessante e gelificante natural. Food Ingredients Brasil 14: 49–50. Available online at: http://www.revista-fi.com/materias/146.pdf.

Ibraheem, I.B.M., N. Abdel-Raouf, M.S. Abdel-Hameed and K. El-Yamany. 2012. Antibacterial and antiviral activities against Newcastle disease from marine algae isolated from Qusier and Marsa-Alam seashore (Red Sea), Egypt. African Journal of Biotechnology 11(33): 8332–8340.

Ibrahim, A.M.M., M.H. Mostafa, M.H. El-Masry and M.M.A. El-Naggar. 2005. Active biological materials inhibiting tumor initiation extracted from marine algae. Egyptian Journal of Aquatic Research 31: 146–155.

Ibtissam, C., R. Hassane, J. Martinez-Lopez, D.S.J. Francisco, G.V.J. Antonio, B. Hassan and K. Mohamed. 2009. Screening of antibacterial activity in marine green and brown macroalgae from the coast of Morocco. African Journal of Biotechnology 8(7): 1258–1262.

Ichiba, T. and T. Higa. 1986. New cuparene-derived sesquiterpenes with unprecedented oxygenation pattern from the sea hare *Aplysia dactylomela*. The Journal of Organic Chemistry 51: 3364–3366.

Iizima-Mizui, N., M. Fujihara, J. Himeno, K. Komiyama, I. Umezawa and T. Nagumo. 1985. Antitumor activity of polysaccharide fractions from the brown seaweed *Sargassum kjellmanianum*. The Kitasato Archives of Experimental Medicine 58(3): 59–71.

Ikawa, M., V.M. Thomas, L.J. Buckley and J.J. Uebel. 1973. Sulfur and the toxicity of the red alga *Ceramium rubrum* to *Bacillus subtilis*. Journal of Phycology 9: 302–304.

Imbs, T.I., N.P. Krasovskaya, S.P. Ermakova, T.N. Makarieva, N.M. Shevchenko and T.N. Zvyagintseva. 2009. Comparative study of chemical composition and antitumor activity of aqueous-ethanol extracts of brown algae *Laminaria cichorioides*, *Costaria costata*, and *Fucus evanescens*. Russian Journal of Marine Biology 35(2): 164–170.

Imbs, T.I., S.P. Ermakova, O.S. Malyarenko, V.V. Isakov and T.N. Zvyagintseva. 2016. Structural elucidation of polysaccharide fractions from the brown alga *Coccophora langsdorfii* and *in vitro* investigation of their anticancer activity. Carbohydrate Polymers 135: 162–168.

Ina, A. and Y. Kamei. 2006. Vitamin B_{12}, a chlorophyll-related analog to pheophytin a from marine brown algae, promotes neurite outgrowth and stimulates differentiation in PC12 cells. Cytotechnology 52: 181–187.

Ina, A., K. Hayashi, H. Nozaki and Y. Kamei. 2007. Pheophytin a, a low molecular weight compound found in the marine brown alga *Sargassum fulvellum,* promotes the differentiation of PC12 cells. International Journal of Developmental Neuroscience 25: 63–68.

Indergaard, M. and J. Minsaas. 1991. Animal and human nutrition. pp. 21-64. *In*: M.D. Guiry and G. Blunden (eds.). Seaweed Resources in Europe: Uses and Potential. John Wiley & Sons, Chichester.

Indira, K., S. Balakrishnan, M. Srinivasan, S. Bragadeeswaran and T. Balasubramanian. 2013. Evaluation of *in vitro* antimicrobial property of seaweed (*Halimeda tuna*) from Tuticorin coast, Tamil Nadu, Southeast coast of India. African Journal of Biotechnology 12(3): 284–289.

Inés, C., V.H. Argandona, J. Rovirosa, A. San-Martin, A.R. Diaz-Marrero, M. Cueto and A. Gonzalez-Coloma. 2004. Cytotoxic activity of halogenated monoterpenes from *Plocamium cartilagineum*. Zeitschrift für Naturforschung C 59(5-6): 339–344.

International Diabetes Federation. 2013. IDF Diabetes Atlas, 6th edition. Brussels, Belgium. Available online at: http://www.idf.org/diabetesatlas.

Irhimeh, M.R., J.H. Fitton and R.M. Lowenthal. 2009. Pilot clinical study to evaluate the anticoagulant activity of fucoidan. Blood Coagulation & Fibrinolysis 20: 607–610.

Irie, T., M. Suzuki, E. Kurosawa and T. Masamune. 1966. Laurinterol and debromolaurinterol, constituents from *Laurencia intermedia*. Tetrahedron Letters 1837–1840.

Irie, T., M. Suzuki, E. Kurosawa and T. Masamune. 1970. Laurinterol, debromolaurinterol and isolaurinterol, constituents of *Laurencia intermedia* Yamada. Tetrahedron Letters 26: 3271–3277.

Isbary, G. and W. Stolz. 2012. Common healthcare challenges. pp. 117–155. *In*: Laroussi, M., M.G. Kong, G. Morfill and W. Stolz (eds.). Plasma Medicine: Applications of Low-temperature Gas Plasmas in Medicine and Biology (1st ed.). Cambridge University Press, New York.

Ishihara, K., M. Murata, M. Kaneniwa, H. Saito, K. Shinohara and M. Maeda-Yamamoto. 1998. Inhibition of icosanoid production in MC/9 mouse mast cells by *n*–3 polyunsaturated fatty acids isolated from edible marine algae. Bioscience, Biotechnology, and Biochemistry 62: 1412–1415.

Ishihara, K., S. Oyamada, R. Matsushima, M. Murata and T. Muraoka. 2005. Inhibitory effect of porphyran, prepared from dried Nori, on contact hypersensitivity in mice. Bioscience, Biotechnology, and Biochemistry 69(10): 1824–1830.

Ishii, T., T. Nagamine, B.C.Q. Nguyen and S. Tawata. 2017. Insecticidal and repellent activities of Laurinterol from the Okinawan red alga *Laurencia nidifica*. Records of Natural Products 11(1): 63–68.

Ishitsuka, M., T. Kusumi, Y. Nomura, T. Konno and H. Kakisawa. 1979. New geranylgeranybenzoquinone derivatives from *Sargassum tortile*. Chemical Letters: 1269–1272.

Islam, M.N., I.J. Ishita, S.E. Jin, R.J. Choi, C.M. Lee, Y.S. Kim, H.A. Jung and J.S. Choi. 2013. Anti-inflammatory activity of edible brown alga *Saccharina japonica* and its constituents pheophorbide A and pheophytin A in LPS-stimulate RAW 264.7 macrophage cells. Food Chemistry and Toxicology 55: 541–548.

Islam, M.N., S.H. Choi, H.E. Moon, H.A. Park, H.C. Woo and J.S. Choi. 2014. The inhibitory activities of the edible green alga *Capsosiphon fulvescens* on rat lens aldose reductase and advanced and advanced glycation end products formation. European Journal of Nutrition 53: 233–242.

Ismail, A., K.B.H. Salah, M. Ahmed, M. Mastouri, A. Bouraoui and M. Aouni. 2014. Antibacterial and antifungal activities of brown alga *Zonaria tournefortii* (JV Lamouroux). Allelopathy Journal 34(1): 143–154.

Ito, H. and M. Sugiura. 1976. Antitumor polysaccharide fraction from *Sargassum thunbergii*. Chemical and Pharmaceutical Bulletin 24: 1114–1115.

Ito, H., K. Kasama, S. Naruse and K. Shimura. 1982. Antitumor effect of palmitoleic acid on Ehrlich ascites tumor. Cancer Letters 17: 197–203.

Ito, K. and Y. Tsuchida. 1972. The effect of algal polysaccharides on depressing of plasma cholesterol level in rats. Proceedings of International Seaweed Symposium 7: 451–55.

Itoh, H., H. Noda, H. Amano, C. Zhuaug, T. Mizuno and H. Ito. 1993. Antitumor activity and immunological properties of marine algal polysaccharides, especially fucoidan, prepared from *Sargassum thunbergii* of Phaeophyceae. Anticancer Research 13(6A): 2045–2052.

Itoh, H., H. Noda, H. Amano and H. Ito. 1995. Immunological analysis of inhibition of lung metastases by fucoidan (Giv-a) prepared from brown seaweed *Sargassum thunbergii*. Anticancer Research 15(5B): 1937–1947.

Ivanova, E., R. Toshkova, V. Ivanova, R. Rouseva and M. Koarova. 1994b. Stimulating effect of *Ulva lactuca* hetero-polysaccharide on the immune response in mice. Doklady Bolgarskoi Akademii Nauk 47: 109–112.

Ivanova, V., R. Rouseva, M. Kolarova, J. Serkedjieva, R. Rachev, N. Manolova. 1994. Isolation of a polysaccharide with antiviral effect from *Ulva lactuca*. Preparative Biochemistry 24: 83–97.

Iwai, K. 2008. Antidiabetic and antioxidant effects of polyphenols in brown alga *Ecklonia stolonifera* in genetically diabetic KK-Ay mice. Plant Foods for Human Nutrition 63: 163–169.

Iwamoto, K., T. Hiragun, S. Takahagi, Y. Yanase, S. Morioke, S. Mihara, Y. Kameyoshi and M. Hide. 2011. Fucoidan suppresses IgE production in peripheral blood mononuclear cells from patients with atopic dermatitis. Archives of Dermatological Research 303: 425–431.

Iwashima, M., J. Mori, X. Ting, T. Matsunaga, K. Hayashi, D. Shinoda, H. Saito, U. Sankawa and T. Hayashi. 2005. Antioxidant and antiviral activities of plastoquinones from the brown alga *Sargassum micracanthum*, and a new chromene derivative converted from the plastoquinones. Biological and Pharmaceutical Bulletin 28: 374–377.

Iwen, P.C., M.E. Rupp, A.N. Langnas, E.C. Reed and S.H. Hinrichs. 998. Invasive pulmonary aspergillosis due to *Aspergillus terreus*: 12-year experience and review of the literature. Clinical Infectious Diseases 26: 1092–1097.

Jack, C.R., M.S. Albert, D.S. Knopman, G.M. McKhann, R.A. Sperling, M.C. Carrillo, B. Thies and C.H. Phelps. 2011. Introduction to the recommendations from the National Institute on Aging-Alzheimer's Association workgroups on diagnostic guidelines for Alzheimer's disease. Alzheimer's and Dementia 7: 257–262.

Jain, R., S. Sonawane and N. Mandrekar. 2008. Marine organisms: potential source for drug discovery. Current Science 94(3): 292.

Jakobsen, S.E. 2016. Can we use seaweed instead of salt? Science Nordic, July 2016. Available online at: http://sciencenordic.com/can-we-use-seaweed-instead-salt.

James, T.Y., P.M. Letcher, J.E. Longcore, S.E. Mozley-Standridge, D. Porter, M.J. Powell, G.W. Griffith and R. Vilgalys. 2006. A molecular phylogeny of the flagellated fungi (Chytridiomycota) and description of a new phylum (Blastocladiomycota). Mycologia 98(6): 860–871.

Jang, M.S., H.-Y. Park and K.-H. Nam. 2014. Whitening effects of 4-hydroxyphenethyl alcohol isolated from water boiled with *Hizikia fusiformis*. Food Science and Biotechnology 23: 555–560.

Jang, W.S. and S.Y. Choung. 2013. Antiobesity effects of the ethanol extract of *Laminaria japonica* Areshoung in high-fat-diet-induced obese rat. Evidence-based Complementary and Alternative Medicine 492807(10): 10. Available online at: https://www.ncbi.nlm.nih.gov/pubmed/23365609.

Jassbi, A.R., M. Mohabati, S. Eslami, J. Sohrabipour and R. Miri. 2013. Biological activity and chemical constituents of red and brown algae from the Persian Gulf. Iranian Journal of Pharmaceutical Research 12: 339–348.

Jaswir, I. and H.A. Monsur. 2011. Anti-inflammatory compounds of macro algae origin: A review. Journal of Medicinal Plants Research 5(33): 7146–7154.

Jaswir, I., D. Noviendri, H.M. Salleh, M. Taher and K. Miyashita. 2011. Isolation of fucoxanthin and fatty acids analysis of *Padina australis* and cytotoxic effect of fucoxanthin on human lung cancer (H1299) cell lines. African Journal of Biotechnology 10(81): 18855–18862.

Jaswir, I., A.H.T. Tope, R.A. Raus, H.A. Monsur and N. Ramli. 2014. Study on anti-bacterial potentials of some Malaysian brown seaweeds. Food Hydrocolloids 42: 275–279.

Jáuregui-Lobera, I. 2014. Iron deficiency and cognitive functions. Neuropsychiatric Disease and Treatment 10: 2087–2095.

Jayant, R.D., M.J. McShane and R. Srivastava. 2009. Polyelectrolyte-coated alginate microspheres as drug delivery carriers for dexamethasone release. Drug Delivery 16: 331–340.

Jayasooriya R.G.P.T., T.-J. Nam, G.-Y. Kim and H. Choi. 2012. Methanol extract of *Polyopes lancifolius* suppresses tumor necrosis factor-α-induced matrix metalloproteinase-9 expression in t24 bladder carcinoma cells. Tropical Journal of Pharmaceutical Research 13: 405–412.

Je, J.-Y., P.-J. Park, E.-K. Kim, J.-S. Park, H.-D. Yoon, K.-R. Kim and C.-B. Ahn. 2009. Antioxidant activity of enzymatic extracts from the brown seaweed *Undaria pinnatifida* by electron spin resonance spectroscopy. Lwt-Food Science and Technology 42(4): 874–878.

Jebakumar Solomon, S.R.D. and V.S. Santhi. 2008. Purification of bioactive natural product against Human microbial pathogens from marine sea weed *Dictyota acutiloba* J. Ag. World Journal of Microbiology and Biotechnology 24(9): 1747–1752.

Jebasingh, S.E.J., S. Rosemary, S. Elaiyaraja, K. Sivaraman, M. Lakshmikandan, A. Murugan and P. Raja. 2011. Potential antibacterial activity of selected green and red seaweeds. Journal of Pharmaceutical and Biomedical Sciences 05(05): 1–7.

Jeon, Y.-J., Y. Athukorala and J. Lee. 2005. Characterization of agarose product from agar using DMSO. Algae 20: 61–67.

Jeong, H.J., S.A. Lee, P.D. Moon, H.J. Na, R.K. Park, J.Y. Um, H.M. Kim and S.H. Hong. 2006. Alginic acid has anti-anaphylactic effects and inhibits inflammatory cytokine expression *via* suppression of nuclear factor-κB activation. Clinical & Experimental Allergy 36: 785–794.

Jeong B.-E., E.-J. Ko and H.-G. Joo. 2012. Cytoprotective effects of fucoidan, an algae-derived polysaccharide on 5-fluorouracil-treated dendritic cells. Food and Chemical Toxicology 50: 1480–1484.

Jespersen, L., L.D. Stromdahl, K. Olsen and L.H. Skibsted. 2005. Heat and light stability of three natural blue colorants for use in confectionery and beverages. European Food Research and Technology 220(3-4): 261–266.

Jhamandas, J.H., M.B. Wie, K. Harris, D. MacTavish and S. Kar. 2005. Fucoidan inhibits cellular and neurotoxic effects of β-amyloid (Aβ) in rat cholinergic basal forebrain neurons. European Journal of Neuroscience 21: 2649–2659.

Ji, H., H. Shao, C. Zhang, P. Hong and H. Xiong. 2008b. Separation of the polysaccharides in *Caulerpa racemosa* and their chemical composition and antitumor activity. Journal of Applied Polymer Science 110(3): 1435–1440.

Ji, N.-Y., X.-M. Li, H. Xie, J. Ding, K. Li, L.-P. Ding and B.-G. Wang. 2008. Highly oxygenated triterpenoids from the marine red alga *Laurencia mariannensis* (Rhodomelaceae). Helvetica Chimica Acta 91(10): 1940–1946.

Jiang, B., X. Xu, L. Li and W. Yuan. 2003. Study on '911' anti-HBV effect in HepG2.2.15 cells culture. Modern Preventive Medicine 30: 517–518.

Jiang, R.W., M.E. Hay, C.R. Fairchild, J. Prudhomme, K.L. Roch, W. Aalbersberg and J. Kubanek. 2008. Antineoplastic unsaturated fatty acids from Fijian macroalgae. Phytochemistry 69(13): 2495–500.

Jiang, Z., T. Okimura, K. Yamaguchi and T. Oda. 2011. The potent activity of sulfated polysaccharide, ascophyllan, isolated from *Ascophyllum nodosum* to induce nitric oxide and cytokine production from mouse macrophage RAW264.7 cells: Comparison between ascophyllan and fucoidan. Nitric Oxide 25: 407–415.

Jiang, Z., Y. Chen, F. Yao, W. Chen, S. Zhong, F. Zheng and G. Shi. 2013. Antioxidant, antibacterial and antischistosomal activities of extracts from *Grateloupia livida* (Harv) Yamada. PLOS ONE 8(11): 1–10.

Jiang, Z., R. Abu, S. Isaka, S. Nakazono, M. Ueno, T. Okimura, K. Yamaguchi and T. Oda. 2014. Inhibitory effect of orally-administered sulfated polysaccharide ascophyllan isolated from *Ascophyllum nodosum* on the growth of sarcoma-180 solid tumor in mice. Anticancer Research 34(4): 1663–1671.

Jiang, Z.-D., P.R. Jensen and W. Fenical. 1999. Lobophorins A and B, new antiinflammatory macrolides produced by a tropical marine bacterium. Bioorganic & Medicinal Chemistry Letters 9: 2003–2006.

Jiao, G.L., G.L. Yu, J.Z. Zhang and S.H. Ewart. 2011. Chemical structures and bioactivities of sulfated polysaccharides from marine algae. Marine Drugs 9(2): 196–223.

Jiao, G.L., G.L. Yu, W. Wang, X.L. Zhao, J.Z. Zhang and S.H. Ewart. 2012. Properties of polysaccharides in several seaweeds from Atlantic Canada and their potential anti-Influenza viral activities. Journal of Ocean University of China 11(2): 205–212.

Jiao, L., X. Li, T. Li, P. Jiang, L. Zhang, M. Wu and L. Zhang. 2009. Characterization and anti-tumor activity of alkali-extracted polysaccharide from *Enteromorpha intestinalis*. International Immunopharmacology 9(3): 324–329.

Jiménez-Escrig, A. and I.G. Cambrodón. 1999. Evaluación nutricional y efectos fisiológicos de macroalgas marinas comestibles. Archivos Latinoamericanos de Nutrición 2: 114–120.

Jimenez-Escrig, A. and F.J. Sanchez-Muniz. 2000. Dietary fi bre from edible seaweeds: chemical structure, physicochemical properties and effects on cholesterol metabolism. Nutrition Research 20: 585–598.

Jiménez-Escrig, A., M. Alaiz, J. Vioque and P. Rupérez. 2010. Health-promoting activities of ultra-filtered okara protein hydrolysates released by *in vitro* gastrointestinal digestion: Identification of active peptide from soybean lipoxygenase. European Food Research and Technology 230: 655–663.

Jiménez-Escrig, A., E. Gómez-Ordóñez and P. Rupérez. 2011. Seaweed as a source of novel nutraceuticals: Sulfated polysaccharides and peptides. pp. 326–337. *In*: Se-Kwon Kim (ed.). Advances in Food and Nutrition Research, Volume 64, Marine Medicinal Foods: Implications and Application, Macro and Microalgae. Academic Press, an imprint of Elsevier, Waltham, USA.

Jin, D.Q., C.S. Lim, J.Y. Sung, H.G. Choi, I. Ha and J.S. Han. 2006. *Ulva conglobata*, a marine algae, has neuroprotective and anti-inflammatory effects in murine hippocampal and microglial cells. Neuroscience Letters 402(1-2): 154–158.

Jin, J.O., M.G. Song, Y.N. Kim, J.I. Park and J.Y. Kwak. 2010. The mechanism of fucoidan-induced apoptosis in leukemic cells: Involvement of ERK1/2, JNK, glutathione, and nitric oxide. Molecular Carcinogenesis 49: 771–782.

Joe, M.-J., S.-N. Kim, H.-Y. Choi, W.-S. Shin, G.-M. Park, D.-W. Kang and Y.K. Kim. 2006. The inhibitory effects of eckol and dieckol from *Ecklonia stolonifera* on the expression of matrix metalloproteinase-1 in human dermal fibroblasts. Biological and Pharmaceutical Bulletin 29: 1735–1739.

Johnson B.A.M. and P.D. Raja. 2015. Bio-efficacy of green seaweeds from South East coast of Tamil Nadu, India. Journal of Microbiology & Experimentation 2(2): 1–7.

Johnston, H.W. 1966. The biological and economic importance of algae, 2. Tuatara 14(1): 30–62.

Jolles, B., M. Remington and P. S. Andrews. 1962. Effects of sulfated degraded laminarin on experimental tumor growth. British Journal of Cancer 16: 109–115.

Jolles, B., M. Remington and P.S. Andrews. 1963. Effects of sulphated degraded laminarin on experimental tumor growth. British Journal of Cancer 17(1): 109–115.

Jones, A. 2015. CBI product factsheet: Marine extracts in Europe. CBI Market Intelligence, The Hague, The Netherlands, 14 pp.

Joyce, J.A. 2005. Therapeutic targeting of the tumor microenvironment. Cancer Cell 7(6): 513–520.

Jung, H.A., S.K. Hyun, H.R. Kim and J.S. Choi. 2006. Angiotensin-converting enzyme I inhibitory activity of phlorotannins from *Ecklonia stolonifera*. Fisheries Science 6: 1292–1299.

Jung, H.A., S.H. Oh and J.S. Choi. 2010. Molecular docking studies of phlorotannins from *Eisenia bicyclis* with BACE1 inhibitory activity. Bioorganic & Medicinal Chemistry Letters 20: 3211–3215.

Jung, H.A., S.E. Jin, B.R. Ahn, C.M. Lee and J.S. Choi. 2013. Anti-inflammatory activity of edible brown alga Eisenia bicyclis and its constituents fucosterol and phlorotannins in LPS-stimulated Raw264.7 macrophages. Food and Chemical Toxicology 59: 199–206.

Jung, M., K.H. Jang, B. Kim, B.H. Lee, B.W. Choi, K.-B. Oh and J. Shin. 2008. Meroditerpenoids from the brown alga *Sargassum siliquastrum*. Journal of Natural Products 71: 1714–1719.

Jung, W.K., Y. Athukorala, Y.J. Lee, S.H. Cha, C.H. Lee, T. Vasanthan, K.S. Choi, S.H. Yoo, S.K. Kim and Y.J. Jeon. 2007. Sulfated polysaccharide purified from *Ecklonia cava* accelerates antithrombin III-mediated plasma proteinase inhibition. Journal of Applied Phycology 19: 425–430.

Jung, W.K., I. Choi, S. Oh, S.G. Park, S.K. Seo, S.W. Lee, D.S. Lee, S.J. Heo, Y.J. Jeon, J.Y. Je, C.B. Ahn, J.S. Kim, K.S. Oh, Y.M. Kim, C. Moon and I.W. Choi. 2009. Anti-asthmatic effect of marine red alga (*Laurencia undulata*) polyphenolic extracts in a murine model of asthma. Food and Chemical Toxicology 47: 293–297.

Jung, W.K., Y.W. Ahn, S.H. Lee, Y.H. Choi, S.K. Kim, S.S. Yea, I. Choi, S.G. Park, S.K. Seo, S.W. Lee and I.W. Cho. 2009b. *Ecklonia cava* ethanolic extracts inhibit lipopolysaccharide-induced cyclooxygrnase-2 and inducible nitric oxide synthase expression in BV2 microglia via the MMP kinase and NF-kB pathways. Food and Chemical Toxicology 47: 410–417.

Jung, W.K., S.J. Heo, Y.J. Jeon, C.M. Lee, Y.M. Park, H.G. Byun, Y.H. Choi, S.G. Park and I.W. Choi. 2009c. Inhibitory effects and molecular mechanism of dieckol isolated from marine brown alga on COX-2 and iNOS in microglial cells. Journal of Agricultural and Food Chemistry 57(10): 4439–4446.

Júnior, S. Q., V.H.A. Carneiro, T.P.C. Fontenelle, L.S. Chaves, J.X. Mesquita, T.V. de Brito, R.S. Prudêncio, J.S. de Oliveira, Jand-Venes Rolim Medeiros, K.S. Aragão, R.A. Ribeiro, A.L.R. Barbosa and A.L.P. Freitas. 2015. Antioxidant and anti-inflammatory activities of methanol extract and its fractions from the brown seaweed *Spatoglossum schroederi*. Journal of Applied Phycology 27(6): 2367–2376.

Jurd, K.M., D.J. Rogers, G. Blunden and D.S. McLellan. 1995. Anticoagulant properties of sulfated polysaccharides and a proteoglycan from *Codium fragile* ssp *atlanticum*. Journal of Applied Phycology 7(4): 339–345.

Kaboli, P.J., B.N. Doebbeling, K.G. Saag and G.E. Rosenthal. 2001. Use of complementary and alternative medicine by older patients with arthritis: a population-based study. Arthritis & Rheumatology 45: 398–403.

Kadam, S.U., B.K. Tiwari and C.P. O'Donnell. 2013. Application of novel extraction technologies for bioactives from marine algae. Journal of Agricultural and Food Chemistry 61(20): 4667–4675.

Kadam, S.U., Brijesh K. Tiwari and C.P. O'Donnell. 2015a. Extraction, structure and biofunctional activities of Laminarin from brown algae. International Journal of Food Science & Technology 50(1): 24–31.

Kadam, S.U., C.P. O'Donnell, D.K. Rai, M.B. Hossain, C.M. Burgess, D. Walsh and B.K. Tiwari. 2015b. Laminarin from Irish brown seaweeds *Ascophyllum nodosum* and *Laminaria hyperborea*: ultrasound assisted extraction, characterization and bioactivity. Marine Drugs 13(7): 4270–4280.

Kadir, A., S. Latifah, H. Yaakob and R.M. Zulkifli. 2013. Potential anti-Dengue medicinal plants: a review. Journal of Natural Medicines 67(4): 677–689.

Kaeffer, B., C. Benard, M. Lahaye, H.M. Blottiere and C. Cherbut. 1999. Biological properties of ulvan, a new source of green seaweed sulphated polysaccharides, on cultured normal and cancerous colonic epithelial cells, Planta Medica 65(6): 527–531.

Kahn, J.S. 2007. Newly discovered respiratory viruses: significance and implications. Current Opinion in Pharmacology 7: 478–483.

Kailas, A.P. 2015. Characterization and quantification of major biochemical components and secondary metabolites of some green and red seaweeds from the Kerala coast. PhD Thesis, Cochin University of Science and Technology, Kochi, 468 pp.

Kalafatis, M., J.O. Egan, C. van't Veer, K.M. Cawthern and K.G. Mann. 1997. The regulation of clotting factors. Critical Reviews in Eukaryotic Gene Expression 7: 241–280.

Kaliaperumal, N., S. Kalimuthu and J.R. Ramalingam. 1995. Economically important seaweeds. pp. 1–35. *In*: M. Devaraj (ed.). Special Publication 62. Central Marine Fisheries Research Institute, Indian Council of Agricultural Research, Cochin, India.

Kalitnik, A.A., A.O.B. Barabanova, V.P. Nagorskaya, A.V. Reunov, V.P. Glazunov, T.F. Solov'eva and I.M. Yermak. 2013. Low molecular weight derivatives of different carrageenan types and their antiviral activity. Journal of Applied Phycology 25(1): 65–72.

Kamada, T. and C.S. Vairappan. 2017. Non-halogenated new sesquiterpenes from Bornean *Laurencia snackeyi*. Natural Product Research 31(3): 333–340.

Kamat, S.Y., S. Wahidulla, L. D'Souza, C.G. Naik, V. Ambiye, D.S. Bhakuni, A.K. Goel, H.S. Garg and R.C. Srimal. 1992. Bioactivity of marine organisms. VI. Antiviral evaluation of marine algal extracts from the Indian coast. Botanica Marina 35(2): 161–164.

Kamei, Y. and A. Sagara. 2002. Neurite outgrowth promoting activity of marine algae from Japan against rat adrenal medulla pheochromocytoma cell line, PC12D. Cytotechnology 40: 99–106.

Kamei, Y. and C.K. Tsang. 2003. Sargaquinoic acid promotes neurite outgrowth via protein kinase A and MAP kinases-mediated signaling pathways in PC12D cells. International Journal of Developmental Neuroscience 21: 255–262.

Kamei, Y., M. Sueyoshi, K. Hayashi, R. Terada and H. Nozaki. 2009. The novel anti-*Propionibacterium* acnes compound, Sargafuran, found in the marine brown alga *Sargassum macrocarpum*. The Journal of Antibiotics 62(5): 259–263.

Kamenarska, Z., A. Ivanova, R. Stancheva, K. Stefanov, S. Dimitrova-Konaklieva and S. Popov. 2006. Polar constituents of some Black Sea red and brown algae and their application into chemotaxonomy and chemoevolution. Algological Studies 119: 139–154.

Kamenarska, Z., J. Serkedjieva, H. Najdenski, K. Stefanov, I. Tsvetkova, S. Dimitrova-Konaklieva and S. Popov. 2009. Antibacterial, antiviral, and cytotoxic activities of some red and brown seaweeds from the Black Sea. Botanica Marina 52(1): 80–86.

Kamimoto, K. 1956. Antibacterial substances extracted from seaweeds. Effects of extracts from seaweeds on the growth of some acid-fast bacteria. Nippon Saikingaku Zasshi 11: 307–313.

Kan, Y., T. Fujita, B. Sakamoto, Y. Hokama and H. Nagai. 1999. Kahalalide K: a new cyclic depsipeptide from the Hawaiian green alga *Bryopsis* species. Journal of Natural Products 62(8): 1169–1172.

Kanagarajjeevitha, J. Damahe, S. Das, T.R. Chowdhury and S.S. Khora. 2014. *In vitro* antioxidant and cytotoxic activity of brown alga *Padina boergesenii*. International Journal of Drug Development and Research 6(2): 110–119.

Kanazawa, K., Y. Ozaki, T. Hashimoto, S.K. Das, S. Matsushita, M. Hirano, T. Okada, A. Komoto, N. Mori and M. Nakatsuka. 2008. Commercial-scale preparation of biofunctional fucoxanthin from waste parts of brown sea algae *Laminalia japonica*. Food Science and Technology Research 14(6): 573–582.

Kandel, E.S. and N. Hay. 1999. The regulation and activities of the multifunctional serine/threonine kinase Akt/PKB. Resultados da procura experimental. Cell Research 253: 210–29.

Kandhasamy, M. and K.D. Arunachalam. 2008. Evaluation of *in vitro* antibacterial property of seaweeds of southeast coast of India. African Journal of Biotechnology 7(12): 1958–1961.

Kanegawa, K., H. Harada, H. Myouga, Y. Katakura, S. Shirahata and Y. Kamei. 2000. Telomerase inhibiting activity *in vitro* from natural resources, marine algae extracts. Cytotechnology 33(1-3): 221–227.

Kang, H., H. Kim, D. Byun, B. Son, T. Nam and J. Choi. 2004. Tyrosinase inhibitors isolated from the edible brown alga *Ecklonia stolonifera*. Archives of Pharmacal Research 27: 1226–1232.

Kang, J.-I., S.-C. Kim, M.-K. Kim, H.-J. Boo, Y.-J. Jeon, Y.-S. Koh, E.-S. Yoo, S.-M. Kang and H.-K. Kang. 2012c. Effect of dieckol, a component of *Ecklonia cava*, on the promotion of hair growth. International Journal of Molecular Sciences 13: 6407–6423.

Kang, J.I., S.C. Kim, S.C. Han, H.J. Hong, Y.J. Jeon, B. Kim, Y.S. Koh, E.S. Yoo and H.K. Kang. 2012d. Hair-loss preventing effect of *Grateloupia elliptica*. Biomolecules & Therapeutics 20(1): 118–124.

Kang, J.W. 1968. Illustrated encyclopedia of fauna and flora of Korea: Marine algae. Samhwa Press, Seoul, 465 pp.

Kang, J.Y., M.N.A. Khan, N.H. Park, J.Y. Cho, M.C. Lee, H. Fujii and Y.K. Hong. 2008b. Antipyretic, analgesic, and anti-inflammatory activities of the seaweed *Sargassum fulvellum* and *Sargassum thunbergii* in mice. Journal of Ethnopharmacology 116: 187–190.

Kang, J.-Y., Q.-H. Luyen, M.N.A. Khan, J.-S. Choi, I.S. Choi and Y.-K. Hong. 2012b. *In vivo* Anti-inflammatory, antipyretic, and analgesic activities of the aquaculturable green seaweed *Codium fragile* extracts in mice. Journal of Life Science 22(6): 852–856.

Kang, K., Y. Park, H.J. Hwang, S.H. Kim, J.G. Lee and H.-C. Shin. 2003. Antioxidative properties of brown algae polyphenolics and their perspectives as chemopreventive agents against vascular risk factors. Archives of Pharmacological Research 26: 286–293.

Kang, K.A., K.H. Lee, S. Chae, Y.S. Koh, B.S. Yoo, J.H. Kim, Y.M. Ham, J.S. Baik, N.H. Lee and J.W. Hyun. 2005. Triphlorethol-A from *Ecklonia cava* protects V79-4 lung fibroblast against hydrogen peroxide induced cell damage. Free Radical Research 39: 883–892.

Kang, S.I., Y.J. Jin, H.C. Ko, S.Y. Choi, J.H. Hwang, I. Whang, M.H. Kim, H.S. Shin, H.B. Jeong and S.J. Kim. 2008. *Petalonia* improves glucose homeostasis in streptozotocin-induced diabetic mice. Biochemical and Biophysical Research Communications 373(2): 265–269.

Kang, S.-I., H.-S. Shin, H.-M. Kim, S.-A. Yoon, S.-W. Kang, H.-C. Ko and S.-J. Kim. 2012. *Callophyiis japonica* extract improves high-fat diet-induced obesity and inhibits adipogenesis in 3T3-L1 cells. Animal Cells and Systems 16: 447–454.

Kang, S.M., K.N. Kim, S.H. Lee, G. Ahn, S.H. Cha, A.D. Kim, X.D. Yang, M.C. Kang an Y.J. Jeon. 2011. Anti-inflammatory activity of polysaccharide purified from AMG-assistant extract of *Ecklonia cava* in LPS-stimulated RAW 264.7 macrophages. Carbohydrate Polymers 85: 80–85.

Kang, S.M., S.H. Cha, J.Y. Ko, M.C. Kang, D. Kim, S.J. Heo, J.S. Kim, M.S. Heu, Y.T. Kim, W.K. Jung and Y.J. Jeon. 2012e. Neuroprotective effects of phlorotannins isolated from a brown alga, *Ecklonia cava*, against H_2O_2-induced oxidative stress in murine hippocampal HT22 cells. Environmental Toxicology and Pharmacology 34: 96–105.

Kang, Y., J. Lu, X. Xu, X. Sun, N. Xu and X. Zhao. 2015. Effect of cell proliferation activity and morphology in lung cancer cells by polysaccharide of *Gracilaria lemaneiformis*. Journal of Fisheries of China 39(9): 015.

Kang, Y.N., H. Li, J. Wu, X.T. Xu, X. Sun, X.D. Zhao and N.J. Xu. 2016. Transcriptome profiling reveals the antitumor mechanism of polysaccharide from marine algae *Gracilariopsis lemaneiformis*. PLoS One 11(6): 1–16.

Kannan, R.R.R., M.A. Aderogba, A.R. Ndhlala, W.A. Stirk and J.V. Staden. 2013. Acetylcholinesterase inhibitory activity of phlorotannins isolated from the brown alga, *Ecklonia maxima* (Osbeck) Papenfuss. Food Research International 54(1): 1250–1254.

Kaper, J.B., J.P. Nataro and H.L.T. Mobley. 2004. Pathogenic *Escherichia coli*. Nature Reviews Microbiology 2(2): 123–140.

Kapil, U. 2007. Health consequences of iodine deficiency. Sultan Qaboos University Medical Journal 7(3): 267–272.

Kar, S., G. Sharma and P.K. Das. 2011. Fucoidan cures infection with both antimony-susceptible and -resistant strains of *Leishmania donovani* through Th1 response and macrophage-derived oxidants. Journal of Antimicrobial Chemotherapy 66: 618–625.

Karabay-Yavasoglu, N.U., A. Sukatar, G. Ozdemir and Z. Horzum. 2007. Antimicrobial activity of volatile components and various extracts of the red alga *Jania rubens*. Phytotherapy Research 21(2): 153–156.

Karaki, N., C. Sebaaly, N. Chahine, T. Faour, A. Zinchenko, S. Rachid and H. Kanaan. 2013. The antioxidant and anticoagulant activities of polysaccharides isolated from the brown algae *Dictyopteris polypodioides* growing on the Lebanese coast. Journal of Applied Pharmaceutical Science 3(2): 43–51.

Karen, B. 2014. *Sargassum* spp. seaweeds. The Bermudian Magazine. Available online at: http://environment.bm/sargassum-seaweed/.

Karmakar, P., C.A. Pujol, E.B. Damonte, T. Ghosh and B. Ray. 2010. Polysaccharides from *Padina tetrastromatica*: Structural features, chemical modification and antiviral activity. Carbohydrate Polymers 80(2): 513–520.

Karnjanapratum, S., M. Tabarsa, M. Cho and S. You. 2012. Characterisation and immunomodulatory activities of sulfated polysaccharides from *Capsosiphon fulvescens*. International Journal of Biological Macromolecules 51: 720–729.

Karthick, N. 2014. Biopotentials of organic fertilizers produced from three different seaweeds on soil biota. Ph.D. Thesis, Manonmaniam Sundaranar University, Tirunelveli, India.

Karthick, N., V.P. Kumar and S. Umamaheswari. 2014. *In vitro* nematicidal activity of different seaweed extracts against *Meloidogyne javanica*. International Journal of Development Research 4(9): 1841–1843.

Karthik, R., V. Manigandan, R. Sheeba, R. Saravanan and P.R. Rajesh. 2016. Structural characterization and comparative biomedical properties of phloroglucinol from Indian brown seaweeds. Journal of Applied Phycology 28(6): 3561–3573.

Karthikaidevi, G., K. Manivannan, G. Thirumaran, P. Anantharaman and T. Balasubaramanian. 2009. Antibacterial properties of selected green seaweeds from Vedalai coastal waters; Gulf of Mannar marine biosphere reserve. Global Journal of Pharmacology 3(2): 107–112.

Kasai, Y., K. Komatsu, H. Shigemori, M. Tsuda, Y. Mikami and J. Kobayashi. 2005 Cladionol A, a ploketide glycoside from marine-derived fungus *Gliocladoium* species. Journal of Natural Products 68: 777–779.

Kasai, A., S. Arafuka, N. Koshiba, D. Takahashi and K. Toshima. 2015. Systematic synthesis of low-molecular weight fucoidan derivatives and their effect oncancer cells. Organic & Biomolecular Chemistry 13: 10556–10568.

Kasim, C.G., P. Aline and S. Ekrem. 2010. Alkaloids in marine algae. Marine Drugs 8: 269–284.

Kaufman, H.L., S. Wadler and K. Antman. 2008. Molecular targeting in oncology. Humana Press, Springer, Berlin, 748 pp.

Kausalya, M. and G.M.N. Rao. 2015. Antimicrobial activity of *Gelidium pusillum* and *Centroceras clavulatum* from Visakhapatnam Coast, India. Journal of Algal Biomass Utilization 6(2): 40–48.

Kavita, K., V.K. Singh and B. Jha. 2014. 24-branched delta5 sterols from *Laurencia papillosa* red seaweed with antibacterial activity against Human pathogenic bacteria. Microbiological Research 169(4): 301–306.

Kawamura-Konishi, Y., N. Watanabe, M. Saito, N. Nakajima, T. Sasaki, T. Katayama and T. Enomoto. 2012. Isolation of a new phlorotannin, a potent inhibitor of carbohydrate-hydrolyzing enzymes, from the brown alga *Sargassum patens*. Journal of Agricultural and Food Chemistry 60: 5565–5570.

Kawasaki, M., M. Toyoda, R. Teshima, J. Sawada and Y. Saito. 1994. Effect of alphalinolenic acid on the metabolism of omega-3 and omega-6 polyunsaturated fatty acids and histamine release in RBL-2H3 cells. Biological and Pharmaceutical Bulletin 17: 1321–1325.

Kay, AB. 2000. Overview of 'allergy and allergic diseases: with a view in the future'. British Medical Bulletin 56: 843–64.

Kayalvizhi, K., V. Subramanian, P. Anantharaman and K. Kathiresan. 2012. Antimicrobial activity of seaweeds from the Gulf of Mannar. International Journal of Pharmaceutical Applications 3: 306–314.

Kazlauskas, R., P.T. Murphy, R.J. Quinn and R.J. Wells. 1976. New laurene derivatives from *Laurencia filiformis*. Australian Journal of Chemistry 29: 2533–2539.

Kazlauskas, R., P.T. Murphy, R.J. Wells, J.A. Baired-Lambert and D.D. Jamieson. 1983. Halogenated pyrrolo (2, 3-d) pyrimidine nucclosides from marine organisms. Australian Journal of Chemistry 83: 165–170.

Kazlowska, K., T. Hsu, C.C. Hou, W.C. Yang and G.J. Tsai. 2010. Antiinflammatory properties of phenolic compounds and crude extract from *Porphyra dentata*. Journal of Ethnopharmacology 128: 123–130.

Kazlowska, K., H.T. Lin, S.H. Chang and G.J. Tsai. 2013. *In vitro* and *in vivo* anticancer effects of sterol fraction from red algae *Porphyra dentata*. Evidence-Based Complementary and Alternative Medicine 493869(10): 26.

Kazutosi, N. 2002. Seaweeds Kaiso, bountiful harvest from the sea: sustenance for health and well-being. Japan Seaweed Association, 106 pp.

Keffer, J.L., S. Huecas, J.T. Hammill, P. Wipf, J.M. Andreu and C.A. Bewley. 2013. Chrysophaentins are competitive inhibitors of FtsZ and inhibit Z-ring formation in live bacteria. Bioorganic & Medicinal Chemistry 21: 5673–5678.

Keivan, Z., S. Tajbakhsh, I. Nabipour, Z. Rastian, F. Yousefi, S. Sharafian and K. Sartavi. 2010. *In vitro* antitumor activity of *Gracilaria corticata* (a red alga) against Jurkat and molt-4 human cancer cell lines. African Journal of Biotechnology 9: 6787–6790.

Kellogg, J., M.H. Grace and M.A. Lila. 2014. Phlorotannins from Alaskan seaweed inhibit carbolytic enzyme activity. Marine Drugs 12: 5277–5294.

Kelly, G.S. 1999. Larch arabinogalactan: clinical relevance of a novel immune enhancing polysaccharide. Alternative Medicine Review 4: 96–103.

Kevadiya, B.D., G.V. Joshi, H.A. Patel, P.G. Ingole, H.M. Mody and H.C. Bajaj. 2010. Montmorillonite-alginate nanocomposites as a drug delivery system: Intercalation and *in vitro* release of vitamin B_1 and vitamin B_6. Journal of Biomaterials Applications 25: 161–177.

Khairy, H.M. and M.A. El-Sheikh. 2015. Antioxidant activity and mineral composition of three Mediterranean common seaweeds from Abu-Qir bay, Egypt. Saudi Journal of Biological Sciences 22(5): 623–630.

Khaled, N., M. Hiba and C. Asma. 2012. Antioxidant and antifungal activities of *Padina pavonica* and *Sargassum vulgare* from the Lebanese Mediterranean coast. Advances in Environmental Biology 6(1): 42–48.

Khalifa, K.S., R.A. Hamouda, D. Hanafy and A. Hamza. 2016. *In vitro* antitumor activity of silver nanoparticles biosynthesized by marine algae. Digest Journal of Nanomaterials and Biostructures 11(1): 213–221.

Khallil, A.M., I.M. Daghman and A.A. Fady. 2015. Antifungal potential in crude extracts of five selected brown seaweeds collected from the western Libya coast. Journal of Microbiology and Modern Techniques 1(1): 1–8.

Khan, A.M. 2000. Phytochemical and structural studies on the chemical constituents of *Taxus wallichiana, Tanacetum gracile, Jolyna laminarioides* and other marine algae. Doctoral dissertation. University of Karachi, Karachi, Pakistan. Available online at: http://eprints.hec.gov.pk/979/1/711.html.

Khan, A.M., S. Noreen, Z.P. Imran, Atta-ur-Rahman and M.I. Choudhary. 2011. A new compound, jolynamine, from marine brown alga *Jolyna laminarioides*. Natural Product Research 25(9): 898–904.

Khan, A.R. 2016. Studies on seaweeds as therapeutic agents against diarrhea causing organisms. PhD Thesis, Centre of Excellence in Marine Biology, University of Karachi, Karachi, Pakistan, 382 pp.

Khan, M.N., S.J. Yoon, J.S. Choi, N.G. Park, H.H. Lee, J.Y. Cho and Y.K. Hong. 2009. Anti-edema effects of brown seaweed (*Undaria pinnatifida*) extract on phorbol 12-myristate 13-acetate-induced mouse ear inflammation. The American Journal of Chinese Medicine 37(2): 373–381.

Khan, M.N.A., J.S. Choi, M.C. Lee, E.K. Kim, T.J. Nam, H.F. Fujii and Y.K. Hong. 2008. Anti-inflammatory activities of methanol extracts from various seaweed species. Journal of Environmental Biology 29(4): 465–469.

Khan, M.N.J., Y. Cho, M.C. Lee, J.Y. Kang, N.G. Park, H. Fujii and Y.K. Hong. 2007. Isolation of two anti-inflammatory and one proinflammatory polyunsaturated fatty acids from the brown seaweed *Undaria pinnatifida*. Journal of Agricultural and Food Chemistry 55: 6984–6988

Khan, S.N. 2007. *Macrophomina phaseolina* as causal agent for charcoal rot of sunflower. Mycopath 5(2): 111–118.

Khan, S.A., M. Abid and F. Hussain. 2015. Nematicidal activity of seaweeds against *Meloidogyne javanica*. Pakistan Journal of Nematology 33(2): 195–203.

Khanavi, M., M. Nabavi, N. Sadati, M. Shams Ardekani, J. Sohrabipour, S.M. Nabavi, P. Ghaeli and S.N. Ostad. 2010. Cytotoxic activity of some marine brown algae against cancer cell lines. Biological Research 43(1): 31–37.

Khanavi, M., P.B. Toulabi, M.R. Abai, N. Sadati, F. Hadjiakhoondi, A. Hadjiakhoondi and H. Vatandoost. 2011. Larvicidal activity ofmarine algae, *Sargassum swartzii* and *Chondria dasyphylla*, against malaria vector *Anopheles stephensi*. Journal of Vector Borne Diseases 48(4): 241–244.

Khanavi, M., R. Gheidarloo, N. Sadati, M.R. Ardekani, S.M. Nabavi, S. Tavajohi and S.N. Ostad. 2012. Cytotoxicity of fucosterol containing fraction of marine algae against breast and colon carcinoma cell line. Pharmacognosy Magazine 8(29): 60–64.

Khanzada, A.K., W. Shaikh, T. Kazi and S. Soofia. 2007. Antifungal activity, elemental analysis and determination of total protein of seaweed, *Solieria robusta* (Greville) Kylin from the coast of Karachi. Pakistan Journal of Botany 39(3): 931–937.

Khodosevich, K. and H. Monyer. 2010. Signaling involved in neurite outgrowth of postnatally born subventricular zone neurons *in vitro*. BMC Neuroscience 11(18): 1471–2202.

Khotimchenko, Y.S., V.V. Kovalev, O.V. Savchenko and O.A. Ziganshina. 2001. Physicalchemical properties, physiological activity, and usage of alginates, the polysaccharides of brown algae. Russian Journal of Marine Biology 27(1): 53–64.

Kim, A.D., K.A. Kang, M.J. Piao, K.C. Kim, J. Zheng, C.W. Yao, J.W. Cha, C.L Hyun, S.J. Boo, N.H. Lee, Y.S. Na, J.W. Hyun. 2014. *Dictyopteris undulata* extract induces apoptosis in human colon cancer cells. Biotechnology and Bioprocess Engineering 19: 419–425.

Kim, A.R., M.-S. Lee, T.-S. Shin, H. Hua, B.-C. Jang, J.-S. Choi, D.-S. Byun, T. Utsuki, D. Ingram and H.-R. Kim. 2011c. Phlorofucofuroeckol A inhibits the LPS-stimulated iNOS and COX-2 expressions in macrophages via inhibition of NF-κB, Akt, and p38 MAPK. Toxicology *in vitro* 25: 1789–1795.

Kim, D.-H., S.-H. Eom, T.H. Kim, B.-Y. Kim, Y.-M. Kim and S.-B. Kim. 2013. Deodorizing effects of Phlorotannins from edible brown alga *Eisenia bicyclis* on methyl mercaptan. Journal of Agricultural Science 5(1): 1–9.

Kim, D.-W., K. Sapkota, J.-H. Choi, Y.-S. Kim, S. Kimc and S.-J. Kim. 2013d. Direct acting anti-thrombotic serine protease from brown seaweed *Costaria costata*. Process Biochemistry 48: 340–350.

Kim, E.J., S.Y. Park, J.Y. Lee and J.H. Park. 2010a. Fucoidan present in brown algae induces apoptosis of human colon cancer cells. BMC Gastroenterology 10(96): 10–96.

Kim, E.Y., D.G. Kim, Y.R., Kim, H.J. Hwang, T.J. Nam and I.S. Kong. 2011b. An improved method of protein isolation and proteome analysis with *Saccharina japonica* (Laminariales) incubated under different pH conditions. Journal of Applied Phycology 23: 123–130.

Kim, H.J., W.J. Kim, B.-W. Koo, D.-W. Kim, J.H. Lee and W.S.K. Nugroho. 2016. Anticancer activity of sulfated polysaccharides isolated from the antarctic red seaweed *Iridaea cordata*. Ocean and Polar Research 38(2): 129–137.

Kim, H.S., K. Lee, K.A. Kang, N.H. Lee, J.W. Hyun and H.-S. Kim. 2012d. Phloroglucinol exerts protective effects against oxidative stress-induced cell damage in SH-SY5Y cells. Journal of Pharmacological Sciences 119: 186–192.

Kim, I.H., D.-G. Lee, S.H. Lee, J.-M. Ha, S.K. Kim and J.-H. Lee. 2007b. Antibacterial activity of *Ulva lactuca* against methicillin-resistant *Staphylococcus aureus* (MRSA). Biotechnology and Bioprocess Engineering 12: 579–582.

Kim, I.H. and J.H. Lee. 2008. Antimicrobial activities against methicillin resistant *Staphylococcus aureus* from macroalgae. Journal of Industrial Engineering Chemistry 14: 568–572.

Kim, I.H., K.Y. Yoo, J.H. Park, B.C. Yan, J.H. Ahn, J.C. Lee, H.M. Kwon, J.D. Kim, Y.M. Kim, S.G. You, I.J. Kang and M.H. Won. 2014f. Comparison of neuroprotective effects of extract and fractions from *Agarum clathratum* against experimentally induced transient cerebral ischemic damage. Pharmaceutical Biology 52: 335–343.

Kim, J., J. Kim, M. Lee, Y. Hong and R. DeWreede. 1999. Multiple antiviral activities in extracts of seaweeds from British Columbia. Pharmaceutical Biology 37: 300–306.

Kim, J.A., C.S. Kong and S.K. Kim. 2010c. Effect of *Sargassum thunbergii* on ROS mediated oxidative damage and identification of polyunsaturated fatty acid components. Food and Chemical Toxicology 48: 1243–1249.

Kim, J.G., D.W. Lim, S. Cho, D. Han and Y.T. Kim 2014g. The edible brown seaweed *Ecklonia cava* reduces hypersensitivity in postoperative and neuropathic pain models in rats. Molecules 19: 7669–7678.

Kim, J.H., A.M. Huang, K. Bannister, T.J. Choi, G.H.N. Towers, R.E. DeWreede, J.B. Hudson, H. Jin and Y.K. Hong. 1997. Biological activities of seaweed extracts from British Columbia, Canada, and Korea. I. Antiviral activity. Canadian Journal of Botany 75: 1656–1660.

Kim, J.M., S. Araki, D.J. Kim, C.B. Park, N. Takasuka, H. Baba-Toriyama, T. Ota, Z. Nir, F. Khachik, N. Shimidzu, Y. Tanaka, T. Osawa, T. Uraji, M. Murakoshi, H. Nishino and H. Tsuda. 1998. Chemopreventive effects of carotenoids and curcumins on mouse colon carcinogenesis after 1,2-dimethylhydrazine initiation. Carcinogenesis 19(1): 81–85.

Kim, J.Y., M.Y. Yoon, M.R. Cha, J.H. Hwang, E. Park, S.U. Choi, H.R. Park and Y.I. Hwang. 2007. Methanolic extracts of *Plocamium telfairiae* induce cytotoxicity and caspase-dependent apoptosis in HT-29 human colon carcinoma cells. Journal of Medicinal Food 10(4): 587–593.

Kim, J.-Y., J.-A. Lee, K.-N. Kim, W.-J. Yoon, W.J. Lee and S.-Y. Park. 2007c. Antioxidative and antimicrobial activities of *Sargassum muticum* extracts. Journal of the Korean Society of Food Science and Nutrition 36(6): 663–669.

Kim, K.A., S.M. Kim, S.W. Kang, S.I. Jeon, B.H. Um and S.H. Jung. 2012c. Edible seaweed, *Eisenia bicyclis*, protects retinal ganglion cells death caused by oxidative stress. Marine Biotechnology 14: 383–395.

Kim, K.C., M.J. Piao, J. Zheng, C.W. Yao, J.W. Cha, M.H.S.R. Kumara, X. Han, H.K. Kang, N.H. Lee and J.W. Hyun. 2014e. Fucodiphlorethol G purified from *Ecklonia cava* suppresses ultraviolet B radiation-induced oxidative stress and cellular damage. Biomolecules & Therapeutics 22: 301–307.

Kim, K.-H., S.-H. Eom, H.-J. Kim, D.-S. Lee, O. Nshimiyumukiza, D. Kim, Y.-M. Kim and L. Myung-Suk. 2014d. Antifungal and synergistic effects of an ethyl acetate extract of the edible brown seaweed *Eisenia bicyclis* against *Candida* species. Fisheries and Aquatic Sciences 17(2): 209–214.

Kim, K.-J. and B.-Y. Lee. 2012. Fucoidan from the sporophyll of *Undaria pinnatifida* suppresses adipocyte differentiation by inhibition of inflammation-related cytokines in 3T3-L1 cells. Nutrition Research 32: 439–447.

Kim, K.N., Y.M. Ham, J.Y. Moon, M.J. Kim, D.S. Kim, W.J. Lee, N.H. Lee and C.G. Hyun. 2009. *In vitro* cytotoxic activity of *Sargassum thunbergii* and *Dictyopteris divaricata* (Jeju Seaweeds) on the HL-60 tumour cell line. International Journal of Pharmacology 5: 298–306.

Kim, K.N., S.J. Heo, S.M. Kang, G. Ahn and Y.J. Jeon. 2010d. Fucoxanthin induces apoptosis in human leukemia HL-60 cells through a ROS-mediated Bcl-xL pathway. Toxicology *in vitro* 24: 1648–1654.

Kim, K.N., S.J. Heo, W.J. Yoon, S.M. Kang, G. Ahn, T.H. Yi and Y.J. Jeon. 2010e. Fucoxanthin inhibits the inflammatory response by suppressing the activation of NF-κB and MAPKs in lipopolysaccharide-induced RAW 264.7 macrophages. European Journal of Pharmacology 649: 369–375.

Kim, K.-N., M.-S. Yang, G.-O. Kim, W.J. Lee, N.H. Lee and C.-G. Hyun. 2012a. *Hydroclathrus clathratus* induces apoptosis in HL-60 leukaemia cells via caspase activation, upregulation of pro-apoptotic Bax/Bcl-2 ratio and ROS production. Journal of Medicinal Plants Research 6(9): 1497–1504.

Kim, K.-N., H.-M. Yang, S.-M. Kang, D. Kim, G. Ahn and Y.-J. Jeon. 2013f. Octaphlorethol A isolated from *Ishige foliacea* inhibits α-MSH-stimulated induced melanogenesis *via* ERK pathway in B16F10 melanoma cells. Food and Chemical Toxicology 59: 521–526.

Kim, K.-T., L.-E. Rioux and S.L. Turgeon. 2014c. Alpha-amylase and alpha-glucosidase inhibition is differentially modulated by fucoidan Ootained from *Fucus vesiculosus* and *Ascophyllum nodosum*. Phytochemistry 98: 27–33.

Kim, K.Y., K.A. Nam, H. Kurihara and S.M. Kim. 2008. Potent alpha-glucosidase inhibitors purified from the red alga *Grateloupia elliptica*. Phytochemistry 69(16): 2820–2825.

Kim, K.Y., T.H. Nguyen, H. Kurihara and S.M. Kim. 2010b. α-Glucosidase inhibitory activity of bromophenol purified from the red alga *Polyopes lancifolia*. Journal of Food Science 75: 145–150.

Kim, M., Y.X. Li, P. Dewapriya, B.M. Ryu and S.K. Kim. 2013g. Floridoside suppresses pro-inflammatory responses by blocking MAPK signaling in activated microglia. BMB Reports 46: 398–403.

Kim, M.J. and H.K. Kim. 2012. Insulinotrophic and hypolipidemic effects of *Ecklonia cava* in streptozotocin-induced diabetic mice. Asian Pacific Journal of Tropical Medicine 5: 374–379.

Kim, M.J., J. Jeon and J.S. Lee. 2014b. Fucoidan prevents high-fat diet-induced obesity in animals by suppression of fat accumulation. Phytotherapy Research 28: 137–143.

Kim, M.M., Q.V. Ta, E. Mendis, N. Rajapakse, W.K. Jung, H.G. Byun, Y.J. Jeon and S.K. Kim. 2006. Phlorotannins in *Ecklonia cava* extract inhibit matrix metalloproteinase activity. Life Sciences 79(15): 1436–1443.

Kim, M.M., N. Rajapakse and S.K. Kim. 2009b. Anti-inflammatory effect of *Ishige okamurae* ethanolic extract *via* inhibition of NF-κB transcription factor in RAW 264.7 cells. Phytotherapy Research 23(5): 628–634.

Kim, S.K., D.Y. Lee, W.K. Jung, J.H. Kim, I. Choi, S.G. Park, S.K. Seo, S.W. Lee, C.M. Lee, S.S. Yea, Y.H. Choi and I.W. Choi. 2008b. Effects of *Ecklonia cava* ethanolic extracts on airway hyperresponsiveness and inflammation in a murine asthma model: role of suppressor of cytokine signaling. Biomedicine & Pharmacotherapy 62: 289–296.

Kim, S.K. and I. Wijesekara. 2010. Development and biological activities of marine derived bioactive peptides: A review. Journal of Functional Foods 2: 1–9.

Kim, S.K. and S.W. Himaya. 2011. Medicinal effects of phlorotannins from marine brown algae. Advances in Food and Nutrition Research 64: 97–109.

Kim, S.M., Y.F. Shang and B.H. Um. 2011. A preparative method for isolation of Fucoxanthin from *Eisenia bicyclis* by centrifugal partition chromatography. Phytochemical Analysis 22(4): 322–329.

Kim, T.H. and J.S. Bae. 2010. *Ecklonia cava* extracts inhibit lipopolysaccharide induced inflammatory responses in human endothelial cells. Food and Chemical Toxicology 48: 1682–1687.

Kim, Y.H., C.B. Chung, J.G. Kim, K.I. Ko, S.H. Park, J.-H. Kim, S.Y. Eom, Y.S. Kim, Y.-I. Hwang, K.H. Kim. 2008c. Anti-wrinkle activity of Ziyuglycoside I isolated from a *Sanguisorba officinalis* root extract and its application as a cosmeceutical ingredient. Bioscience, Biotechnology, and Biochemistry 72: 303–311.

Kim, Y.-H., J.H. Kim, H.-J. Jin and S.Y. Lee. 2013c. Antimicrobial activity of ethanol extracts of *Laminaria japonica* against oral microorganisms. Anaerobe 21: 34–38.

Kim, Y.-H., S. Park, M.H. Kim, Y.-K. Choi, Y.-H. Yang, H.J. Kim, H. Kim, K.-G. Songd and S.H. Lee. 2013e. Ultrasound-assisted extraction of lipids from Chlorella vulgaris using [Bmim][MeSO4]. Biomass and Bioenergy 56: 99–103.

Kim, Y.M., I.H. Kim and T.J. Nam. 2012b. Induction of apoptosis signaling by glycoprotein of *Capsosiphon fulvescens* in human gastric cancer (Ags) cells. Nutrition and Cancer 64(5): 761–769.

Kim, Y.M., I.H. Kim and T.J. Nam. 2013b. Inhibition of AGS human gastric cancer cell invasion and proliferation by *Capsosiphon fulvescens* glycoprotein. Molecular Medicine Reports 8(1): 11–16.

Kimiya, T., K. Ohtani, S. Satoh, Y. Abe, Y. Ogita, H. Kawakita, H. Hamada, Y. Konishi and S. Kubota. 2008. Inhibitory effects of edible marine algae extracts on degranulation of RBL-2H3 cells and mouse eosinophils. Fisheries Science 74: 1157–1165.

Kimura, J. 1941. Studies on the substances of antibloodcoagulation activity in marine algae, IV. in *Laminaria japonica*. Hokkaido Igaku Zasshi (Acta Medica Hokkaidonensia), 19: 427–436 (in Japanese).

Kimura, Y., K. Watanabe and H. Okuda. 1996. Effects of soluble sodium alginate on cholesterol excretion and glucose tolerance in rats. Journal of Ethnopharmacology 54: 47–54.

Kindness, G., W.F. Long and F.B. Williamson. 1980. Anticoagulant effects of sulphated polysaccharides in normal and antithrombin III-deficient plasmas. British Journal of Pharmacology 69(4): 675–677.

Kindness, G., W.F. Long, F.B. Williamson and J. Boyd. 1979. Effects of carrageenans on the aggregation of human blood platelets. Thrombosis Research 15(1-2): 3–15.

King, R.J.B. and M.W. Robins. 2006. Cancer biology. Pearson, Prentice Hall, Harlow, 292 pp.

Kintzios, S.E. 2004. What do we know about cancer and its therapy? pp. 1–14. *In*: Kintzios, S.E. and G.M. Barberaki (eds.). Plants That Fight Cancer. CRC Press, Taylor and Francis Group, LLC.

Kiriyama, S., Y. Okazaki and A. Yoshida. 1968. Hypocholesterolemic effect of polysaccharides and polysaccharide foodstuffs in cholesterol fed rats. Journal of Nutrition 97: 382–88.

Kitamura, K., M. Matsuo and T. Yasui. 1991. Fucoidan from brown seaweed *Laminaria angustata* var *longissima*. Agricultural and Biological Chemistry 55(2): 615–616.

Kladi, M., H. Xenaki, C. Vagias, P. Papazafiri and V. Roussis. 2006. New cytotoxic sesquiterpenes from the red algae *Laurencia obtusa* and *Laurencia microcladia*. Tetrahedron 62(1): 182–189.

Kladi, M., C. Vagias, P. Papazafiri, G. Furnari, D. Serio and V. Roussis. 2007. New sesquiterpenes from the red alga *Laurencia microcladia.* Tetrahedron 63(32): 7606–7611.

Kladi, M., C. Vagias, P. Papazafiri, S. Brogi, A. Tafi and V. Roussis. 2009. Tetrahydrofuran acetogenins from Laurencia glandulifera. Journal of Natural Products 72(2): 190–193.

Klegeris, A., E.G. McGeer and P.L. McGeer. 2007. Therapeutic approaches to inflammation in neurodegenerative disease. Current Opinion in Neurology 20: 351–357.

Kligman, A. 2005. The future of cosmeceuticals: An interview with Albert Kligman, Md, Phd. Interview by Zoe Diana Draelos. Dermatologic Surgery 31(7 Pt 2): 890–891.

Kligman, A.M. 2000. Cosmetics: a dermatologist looks to the future: promises and problems. Dermatologic Clinics 18: 699–709.

Kloareg, B., M. Demarty and S. Mabeau. 1986. Polyanionic characteristics of purified sulphated homofucans from brown algae. International Journal of Biological Macromolecules 8: 380–386.

Kluytmans, J., A. Van Belkum and H. Verbrugh. 1997. Nasal carriage of *Staphylococcus aureus*: epidemiology, underlying mechanisms, and associated risks. Clinical Microbiology Reviews 10(3): 505–520.

Knott, M.G., H. Mkwananzi, C.E. Arendse, D.T. Hendricks, J.J. Bolton and D.R. Beukes. 2005. Plocoralides A–C, polyhalogenated monoterpenes from the marine alga *Plocamium corallorhiza.* Phytochemistry 66(10): 1108–1112.

Knox, R. and R. Woodroffe. 1957. Semi-solid agar media for rapid drug sensitivity tests on cultures of *Mycobacterium tuberculosis*. Journal of General Microbiology 16: 647–659.

Knutsen, S., D. Myslabodski, B. Larsen and A. Usov. 1994. A modified system of nomenclature for red algal galactans. Botanica Marina 37: 163–170.

Ko, S.-C., S.-H. Cha, S.-J. Heo, S.-H. Lee, S.-M. Kang, Y.-J. Jeon. 2011. Protective effect of *Ecklonia cava on* UVB-induced oxidative stress: *in vitro* and *in vivo* zebrafish model, Journal Applied Phycology 23: 697–708.

Ko, S.C., L. Myoungsook, L. Ji-Hyeok, L. Seung-Hong, L. Yunsook and Y.J. Jeon. 2013. Dieckol, a phlorotannin isolated from a brown seaweed, Ecklonia cava, inhibits adipogenesis through AMP-activated protein kinase (AMPK) activation in 3T3-LI preadipocytes. Environmental Toxicology and Pharmacology 36: 1253–1260.

Koehn, F.E., S.P. Gunasekera, D.N. Niel and S.S. Cross. 1991. Halitunal, an unusual diterpene aldehyde from the marine alga *Halimeda tuna.* Tetrahedron Letters 32(2): 169–172.

Koenighofer, M., T. Lion, A. Bodenteich, E. Prieschl-Grassauer, A. Grassauer, H. Unger, C.A. Mueller and T. Fazekas. 2014. Carrageenan nasal spray in virus confirmed common cold: individual patient data analysis of two randomized controlled trials. Multidisciplinary Respiratory Medicine 9(57): 1–12.

Kohanski, M.A., D.J. Dwyer and J.J. Collins. 2010. How antibiotics kill bacteria: from targets to networks. Nature Reviews Microbiology 8(6): 423–435.

Koishi, A.C., P.R. Zanello, É.M. Bianco, J. Bordignon and C.N.D. dos Santos. 2012. Screening of Dengue virus antiviral activity of marine seaweeds by an *in situ* enzyme-linked immunosorbent assay. PLoS ONE 7(12): e51089.

Kojima, T., M. Hashida, S. Muranishi and H. Sezaki. 1978. Antitumor activity of timed-release derivative of mitomycin C, agarose bead conjugate. Chemical and Pharmaceutical Bulletin 26: 1818–1824.

Kokilam, G. and S. Vasuki. 2014. Biochemical and phytochemical analysis on *Ulva fasciata* and *Caulerpa taxifolia*. International Journal of Pharmacy and Pharmaceutical Science Research 4(1): 7–11.

Kolanjinathan, K. and D. Stella. 2009. Antibacterial activity of marine macro algae against Human pathogens. Recent Research in Science and Technology 1: 20–22.

Kolanjinathan, K., P. Ganesh and M. Govindarajan. 2009. Antibacterial activity of ethanol extracts of seaweeds against fish bacterial pathogens. European Review for Medical and Pharmacological Sciences 13: 173–177.

Kolanjinathan, K. and D. Stella. 2011. Comparitive studies on antimicrobial activity of *Ulva reticulata* and *Ulva lactuca* against human pathogens. International Journal of Pharmaceutical & Biological Archives 2(6): 1738–1744.

Kolanjinathan, K. and D. Stella. 2011b. Pharmacological effect of *Gracilaria corticata* solvent extracts against human pathogenic bacteria and fungi. International Journal of Pharmaceutical & Biological Archives 2(6): 1722–1728.

Kolb, N., L. Vallorani and V. Stocchi. 1999. Chemical composition and evaluation of protein quality by amino acid score method of edible brown marine algae Arame (*Eisenia bicyclis*) and Hijiki (*Hijikia fusiforme*). Acta Alimentaria 28: 213–222.

Kolb, N., L. Vallorani, N. Milanovi and V. Stocchi. 2004. Evaluation of marine algae Wakame (*Undaria pinnatifida*) and Kombu (*Laminaria digitata*, *L. japonica*) as food supplements. Food Technology and Biotechnology 42: 57–61.

Kolender, A.A., C.A., Pujol, E.B. Damonte, M.C. Matulewicz and A.S. Cerezo. 1997. The system of sulfated α-(1→3)-linked D-mannans from the red seaweed *Nothogenia fastigiata*: structures, antiherpetic and anticoagulant properties. Carbohydrate Research 304: 53–60.

Kolsi, R.B.A., D. Frikha, I. Jribi, A. Hamza, L. Feki and K. Belghith. 2015. Screening of antibacterial and antifongical activity in marine macroalgae and magnoliophytea from the coast of Tunisia. International Journal of Pharmacy and Pharmaceutical Sciences 7(3): 47–51.

Kong, C.-S., J.-A. Kim, N.-Y. Yoon and S.-K. Kim. 2009. Induction of apoptosis by phloroglucinol derivative from *Ecklonia cava* in MCF-7 Human breast cancer cells. Food and Chemical Toxicology 47(7): 1653–1658.

Kong, C.-S., J.-A. Kim, B.-N. Ahn and S.-K. Kim. 2011b. Potential effect of phloroglucinol derivatives from *Ecklonia cava* on matrix metalloproteinase expression and the inflammatory profile in lipopolysaccharide-stimulated human THP-1 macrophages. Fisheries Science 77: 867–873.

Kong, F., Y. Mao, F. Cui, X.Z. Zhang and Z. Gao. 2011. Morphology and molecular identification of *Ulva* forming green tides in Qingdao China. Journal of Ocean University of China 10: 73–79.

Kordjazi, M., B. Shabanpour, E. Zabihi, M.A. Faramarzi, F. Feizi, H.A. Gavlighi, M.A. Feghhi and S.A. Hosseini. 2013. Sulfated polysaccharides purified from two species of *Padina* improve collagen and epidermis formation in the Rat. International Journal of Molecular and Cellular Medicine 2(4): 156–163.

Korukanti, V.P., H. Ponnam and B.R. Akondi. 2013. Evaluation of antiobesity activity of *Fucus vesiculosus*. Indian Journal of Research in Homoeopathy 7(3): 126–132.

Kosanić, M., B. Ranković and T. Stanojković. 2015. Biological activities of two macroalgae from Adriatic coast of Montenegro. Saudi Journal of Biological Sciences 22(4): 390–397.

Kosovel, V., A. Avanzini, V. Scarcia and A. Furlani. 1988. Algae as possible sources of antitumoural agents. Preliminary evaluation of the "*in vitro*" cytostatic activity of crude extracts. Pharmacological Research Communications (5): 27–31.

Kotake-Nara, E., M. Kushiro, H. Zhang, T. Sugawara, K. Miyashita and A. Nagao. 2001. Carotenoids affect proliferation of human prostate cancer cells. Journal of Nutrition 131(12): 3303–3306.

Kotake-Nara, E., M. Terasaki and A. Nagao. 2005. Characterization of apoptosis induced by fucoxanthin in human promyelocytic leukemia cells. Bioscience, Biotechnology, and Biochemistry (BBB) 69(1): 224–227.

Kotnala, S., A. Garg and A. Chatterji. 2009. Screening for the presence of antimicrobial activity in few Indian seaweeds. Pertanika Journal of Tropical Agricultural Science 32: 69–75.

Kovendan, K., K. Murugan, S. Vincent and D.R. Barnard. 2012. Mosquito larvicidal properties of *Orthosiphon thymiflorus* (Roth) Sleesen. (Family: Labiatae) against mosquito vectors, *Anopheles stephensi, Culex quinquefasciatus* and *Aedes aegypti* (Diptera: Culicidae). Asian Pacific Journal of Tropical Medicine 5(4): 299–305.

Koyanagi, S., N. Tanigawa, H. Nakagawa, S. Soeda and H. Shimeno. 2003. Oversulfation of fucoidan enhances its anti-angiogenic and antitumor activities. Biochemical Pharmacology 65(2): 173–179.

Kraan, S. 2008. *Sargassum muticum* (Yendo) Fensholt in Ireland: an invasive species on the move. Journal of Applied Phycology 20: 825–832.

Kraan, S. 2012. Algal polysaccharides, novel applications and outlook. pp. 489–532. *In*: Chuan-Fa Chang (ed.). Biochemistry, Genetics and Molecular Biology—Carbohydrates—Comprehensive Studies on Glycobiology and Glycotechnology. INTECH, Science, Technology and Medicine open access publisher. Available online at: http://www.intechopen.com/books/carbohydrates-comprehensive-studies-on-glycobiology-and-glycotechnology/algal-polysaccharides-novel-applications-and-outlook.

Kraft, G.T. 2009. Algae of Australia: marine benthic algae of lord howe island and the southern great barrier reef, 2—Brown algae. CSIRO Publishing/Australian Biological Resources, Australia, 370 pp.

Kreitman, R.J. 2006. Immunotoxins for targeted cancer therapy. The AAPS Journal 8(3). E532–E551.

Kremer, B.P. 1975. Physiologisch-chemische charakteristik verschiedener thallusbereiche von *Fucus serratus*. Helgolander Meeresun 127: 115–127.

Krish, S. and A. Das. 2014. *In vitro* bioactivity of marine seaweed, *Cladophora rupestris*. International Journal of Pharma and Bio Sciences 5(1): 898–908.

Krishnaiah, D., R. Sarbatly, D.M.R. Prasad and A. Bono. 2008. Mineral content of some seaweeds from Sabah's south China sea. Asian Journal of Scientific Research 1: 166–170.

Krishnankartha, P. and K. Chakaborty. 2013. Antioxidant and antiinflammatory potential of the aqueous extract and polysaccharide fraction from brown marine macroalgae *Padina* sp. from Gulf of Mannar of Peninsular India. Journal of Coastal Life Medicine 1: 19–20.

Krotkiewski, M., M. Aurell, G. Holm, G. Grimby and J. Szczepanik. 1991. Effects of a sodium-potassium ion-exchanging seaweed preparation in mild hypertension. American Journal of Hypertension 4(6): 483–488.

Ktari, L. and M. Guyot. 1999. A cytotoxic oxysterol from the marine alga *Padina pavonica* (L.) Thivy. Journal of Applied Phycology 11: 511–513.

Ktari, L., A. Blond and M. Guyot. 2000. 16β-hydroxy-5α-cholestane-3,6-dione, a novel cytotoxic oxysterol from the red alga *Jania rubens*. Bioorganic & Medicinal Chemistry Letters 10(22): 2563–2565.

Kubanek, J., A.C. Prusak, T.W. Snell, R.A. Giese, K.I. Hardcastle, C.R. Fairchild, W. Aalbersberg, C. Raventos-Suarez and M.E. Hay. 2005. Antineoplastic diterpene-benzoate macrolides from the Fijian red alga *Callophycus serratus*. Organic Letters 7(23): 5261–5264.

Kubanek, J., A.C. Prusak, T.W. Snell, R.A. Giese, C.R. Fairchild, W. Aalbersberg and M.E. Hay. 2006. Bromophycolides C-I from the Fijian red alga *Callophycus serratus*. Journal of Natural Products 69(5): 731–735.

Kulik, M.M. 1995. The potential for using cyanobacteria (blue-green algae) and algae in the biological control of plant pathogenic bacteria and fungi. European Journal of Plant Pathology 101: 585–599.

Kulshreshtha, G., A.-S. Burlot, C. Marty, A. Critchley, J. Hafting, G. Bedoux, N. Bourgougnon and B. Prithiviraj. 2015. Enzyme-assisted extraction of bioactive material from *Chondrus crispus* and *Codium fragile* and Its effect on *Herpes simplex* virus (HSV-1). Marine Drugs 13: 558–580.

Kulshreshtha, G., T. Borza, B. Rathgeber, G.S. Stratton, N.A. Thomas, A.T. Critchley, J. Hafting and B. Prithiviraj. 2016. Red Seaweeds *Sarcodiotheca gaudichaudii* and *Chondrus crispus* down regulate virulence factors of *Salmonella enteritidis* and induce immune responses in *Caenorhabditis elegans*. Frontiers in Microbiology 7(421): 1–12.

Kumar, C.S., D.V.L. Sarada and V.L. Rengasamy. 2008b. Seaweed extracts control the leaf spot disease of the medicinal plant *Gymnema sylvestre*. Indian Journal of Science and Technology 1(3): 1–5.

Kumar, J.I.N., R.N. Kumar, K. Patel, S. Viyol and R. Bhoi. 2009. Nutrient composition and calorific value of some seaweeds from Bet Dwarka, West Coast of Gujarat, India. Our Nature 7: 18–25.

Kumar, J.I.N., M. Barotand and R.N. Kumar. 2014. Phytochemical analysis and antifungal activity of selected seaweeds from Okha coast, Gujarat, India. Life Sciences Leaflets 52: 57–70.

Kumar, K.P., K. Murugana, K. Kovendana, A.N. Kumar, J.-S. Hwang and D.R. Barnard. 2012. Combined effect of seaweed (*Sargassum wightii*) and *Bacillus thuringiensis* var. *israelensis* on the coastal mosquito, *Anopheles sundaicus*, in Tamil Nadu, India. Sciencie Asia 38: 141–146.

Kumar, K.S., K. Ganesan and P.V.S. Rao. 2015a. Seasonal variation in nutritional composition of *Kappaphycus alvarezii* (Doty) Doty—an edible seaweed. Journal of Food Science and Technology-Mysore 52(5): 2751–2760.

Kumar, M., V. Gupta, P. Kumari, C.R.K. Reddy and B. Jha. 2011a. Assessment of nutrient composition and antioxidant potential of Caulerpaceae seaweeds. Journal of Food Composition and Analysis 24: 270–278.

Kumar, M., P. Kumari, N. Trivedi, M.K. Shukla, V. Gupta, C.R.K. Reddy and B. Jha. 2011b. Minerals, Pufas and antioxidant properties of some tropical seaweeds from Saurashtra coast of India. Journal of Applied Phycology 23(5): 797–810.

Kumar, M.R. and S.B. Padhi. 2016. Antibacterial activity of some marine macro algae from the coastal environments of Southern Odisha. Journal of Pharmacy 6(9): 14–18.

Kumar, R., S. Singh and O.V. Singh. 2008. Bioconversion of lignocellulosic biomass: biochemical and molecular perspectives. Journal of Industrial Microbiology and Biotechnology 35(5): 377–391.

Kumar, R.V., S. Murugesan, S. Bhuvaneshwari and S. Thennarasan. 2015d. *In vitro* antibacterial effects of red alga *Champia parvula* (C. Agardh) of various solvents against human pathogenic bacteria. International Journal of Advances in Pharmaceutics 4(6): 111–116.

Kumar, S.A., M. Magnusson, L.C. Ward, N.A. Paul and L. Brown. 2015b. Seaweed supplements normalise metabolic, cardiovascular and liver responses in high-carbohydrate, high-fat fed rats. Marine Drugs 13(2): 788–805.

Kumar, S.D., B. Narayan and B. Vallikannan. 2008c. Fucoxanthin restrains oxidative stress induced by retinol deficiency through modulation of Na^+ Ka^+-ATPase and antioxidant enzyme activities in rats. European Journal of Nutrition 47: 432–441.

Kumar, T.V., S. Lakshmanasenthil, D. Geetharamani, T. Marudhupandi, G. Suja and P. Suganya. 2015c. Fucoidan—a A-D-Glucosidase inhibitor from *Sargassum wightii* with relevance to type 2 diabetes mellitus therapy. International Journal of Biological Macromolecules 72: 1044–1047.

Kung'u, J. 2016. *Alternaria*: a well recognized allergy causing fungus—Mold & Bacteria Consulting Services. Available online at: https://www.moldbacteria.com/mold/alternaria.html.

Kung'u, J. 2016b. *Cladosporium sphaerospermum*—Mold & Bacteria Consulting Services. Available online at: http://www.moldbacteriaconsulting.com/fungi/cladosporium-sphaerospermum.html.

Kuniyoshi, M., K. Yamada and T. Higa. 1985. A biologically active diphenyl ether from the green alga *Cladophora fascicularis*. Experientia 41: 523–524.

Kuno, G. 1997. Factors influencing the transmission of dengue viruses. *In*: Gubler, D.J. and G. Kuno (eds.). Dengue and Dengue Hemorrhagic Fever. CAB International, Wallingford, Oxon, UK.

Kurihara, H., T. Mitani, J. Kawabata and K. Takahashi. 1999. Inhibitory potencies of bromophenols from Rhodomelaceae algae against α-glucosidase activity. Fisheries Science 65: 300–303.

Kuritzkes, D.R. 2007. HIV resistance: frequency, testing, mechanisms. Topics in HIV Medicine 15: 150–154.

Kuznetsova, T.A., N.N. Besednova, A.N. Mamaev, A.P. Momot, N.M. Shevchenko and T.N. Zvyagintseva. 2003. Anticoagulant activity of fucoidan from brown algae *Fucus evanescens* of the Okhotsk Sea. Bulletin of Experimental Biology and Medicine 136(5): 471–473.

Kwiatek, M.A., S. Roman, A. Fareeduddin, J.E. Pandolfino and P.J. Kahrilas. 2011. An alginate-antacid formulation (Gaviscon Double Action Liquid®) can eliminate the postprandial "acid pocket" in symptomatic GERD patients. Alimentary Pharmacology & Therapeutics 34(1): 59–66.

Kwon, M.-J. and T.-J. Nam. 2006. Porphyran induces apoptosis related signal pathway in AGS gastric cancer cell lines. Life Sciences 79(20): 1956–1962.

Kwon-Chung, K.J. and J.E. Bennett. 1992. Medical mycology. Lea & Febiger, Philadelphia, 866 pp.

Kyaw, S.P.P., M.K. Wai, T. Nyunt and U. Soe-Htun. 2009. The morphology and distribution of *Stoechospermum polypodioides* (Lamouroux) J. Agardh (Dictyotales, Phaeophyta) from Myanmar. Journal of Myanmar Academy of Arts and Science. 7(5): 239–251.

Kylin, H. 1913. Biochemistry of sea algae. Zeitschrift für Physikalische Chemie 83: 171–197.

La Rosa, G., M. Fratini, S.D. Libera, M. Iaconelli and M. Muscillo. 2013. Viral infections acquired indoors through airborne, droplet or contact transmission. Annali dell'Istituto Superiore di Sanita 49: 124–132.

Lahaye, M. 1991. Marine algae as sources of fibers: determination of soluble and insoluble dietary fiber contents in some, sea vegetables. Journal of the Science of Food and Agriculture 54: 587–594.

Lahaye, M. 2001. Chemistry and physico-chemistry of phycocolloids. Cahiers de Biologie Marine 42(1-2): 137–157.

Lahaye, M. and A. Robic. 2007. Structure and functional properties of Ulvan, a polysaccharide from green seaweeds. Biomacromolecules 8(6): 1765–1774.

Lajili, S., S.A. Azouaou, M. Turki, C.D. Muller and A. Bouraoui. 2016. Anti-inflammatory, analgesic activities and gastro-protective effects of the phenolic contents of the red alga, *Laurencia obtusa*. European Journal of Integrative Medicine 8(3): 298–306.

Lajili, S., M. Deghrigue, H.B.H. Amor, C.D. Muller and A. Bouraoui. 2016b. *In vitro* immunomodulatory activity and *in vivo* anti-inflammatory and analgesic potential with gastroprotective effect of the mediterranean red alga *Laurencia obtusa*. Pharmaceutical Biology 54(11): 2486–2495.

Lakmal, H.H.C., K.W. Samarakoon, W.W. Lee, J.-H. Lee, D.T.U. Abeytunga, H.-S. Lee and Y.-J. Jeon. 2014. Anticancer and antioxidant effects of selected Sri Lankan marine algae. Journal of the National Science Foundation of Sri Lanka 42(4): 315–323.

Lakshmana-Senthil, S., T.V. Kumar, D. Geetharamani, G. Suja, R. Yesudas and A. Chacko. 2015. Fucoidan—An α-amylase inhibitor from *Sargassum wightii* with relevance to NIDDM. International Journal of Biological Macromolecules 81: 644–647.

Lakshmi, V., A.K. Goel, M.N. Srivastava and D.K. Kulshreshtha. 2006. Raghubir. Bioactivity of marine organisms: Part IX—Screening of some marine flora from Indian coasts. Indian Journal of Experimental Biology 44: 137–141.

Lakshmi, V., A.K. Goel, M.N. Srivastava and R. Raghubir. 2010. Bioactivity of marine organisms: Part XI—Screening of some marine flora from the Indian coasts. Bangladesh Journal of Scientific and Industrial Research 45(3): 271–276.

LaLonde, C.D., C.F. Wong, L.C. Gardener, D.J. Eckert, D.R. King and R.H. Zimmerman. 1979. Response of *Aedes triseriatus* larvae to fatty acids of *Cladophora*. Journal of Chemical Ecology 5(3): 371–381.

Lamela, M., J. Anca, R. Villar, J. Otero and J.M. Calleja. 1989. Hypoglycemic activity of several seaweed extracts. Journal of Ethnopharmacology 27: 35–43.

Lamela, M., J.M. Anca, M.J. Vazquez-Freire, A. Gato and J.M. Calleja. 1993. Hypoglycemic activity of a fucan from *Hinaanthalia elongata*. *In*: Actes du 2e Colloque Europien d'Ethnophmacologie et de la 11' Conference Internationale d'Ethnomédecine, Heidelberg, 24–27 mars 1993.

La-Mer. 2017. My skin—and what it needs. Available online at: https://www.la-mer.com/en.

Lampe, M.F., L.M. Ballweber, C.E. Isaacs, D.L. Patton and W.E. Stamm. 1998. Killing of *Chlamydia trachomatis* by novel antimicrobial lipids adapted from compounds in human breast milk. Antimicrobial Agents and Chemotherapy 45: 1239–1244.

Lane, A.L., E.P. Stout, M.E. Hay, A.C. Prusak, K. Hardcastle, C.R. Fairchild, S.G. Franzblau, K.L. Roch, J. Prudhomme, W. Aalbersberg and J. Kubanek. 2007. Callophycoic acids and callophycols from the Fijian red alga *Callophycus serratus*. The Journal of Organic Chemistry 72: 7343–7351.

Lane, A.L., E.O. Stout, A. Lin, J. Prudhomme, K. Le, C.R. Fairchild, S.G. Franzblau, M.E. Hay, W. Aalbersberg and J. Kubanek. 2009. Antimalarial bromophycolides J–Q from the Fijian red alga *Callophycus serratus*. The Journal of Organic Chemistry 74: 2736–2742.

Lane, A.L., L. Nyadong, A.S. Galhena, T.L. Shearer, E.P. Stout, R.M. Parry, M. Kwasnik, M.D. Wang, M.E. Hay, F.M. Fernandez and J. Kubanek. 2009b. Desorption electrospray ionization mass spectrometry reveals surface-mediated antifungal chemical defense of a tropical seaweed. PNAS 106(18): 7314–7319.

Lanfer-Marquez, U.M., R.M.C. Barros and P. Sinnecker. 2005. Antioxidant activity of chlorophylls and their derivatives. Food Research International 38: 885–891.

Langston, J., L. Forno, J. Tetrud, A. Reeves, J. Kaplan and D. Karluk. 1999. Evidence of active nerve cell degeneration in the *substantia nigra* of humans years after 1-methyl-4-phenyl-1,2,3,6-tetrahydropyridine exposure. Annals of Neurology 46: 598–605.

Larsen, B., A. Haug and T. Painter. 1970. Sulphated polysaccharides in brown algae. 3. The native state of fucoidan in *Ascophyllum nodosum* and *Fucus vesiculosus*. Acta Chemica Scandinavica 24(3): 3339–3352.

Larsen, B., D.M. Salem, M.A. Sallam, M.M. Mishrikey and A.I. Beltagy. 2003. Characterization of the alginates from algae harvested at the Egyptian Red Sea coast. Carbohydrate Research 338: 2325–36.

Lategan, C., T. Kellerman, A.F. Afolayan, M.G. Mann, E.M. Antunes, P.J. Smith, J.J. Bolton and D.R. Beukes. 2009. Antiplasmodial and antimicrobial activities of South African marine algal extracts. Pharmaceutical Biology 47: 408–413.

Laurberg, P., C. Cerqueira, L. Ovesen, L.B. Rasmussen, H. Perrild, S. Andersen, I.B. Pedersen and A. Carle. 2010. Iodine intake as a determinant of thyroid disorders in populations. Best Practice & Research Clinical Endocrinology & Metabolism 24(1): 13–27.

Laurienzo, P. 2010. Marine polysaccharides in pharmaceutical applications: an overview. Marine Drugs 8(9): 2435–2465.

Lavakumar, V., K.F.H.N. Ahamed and V. Ravichandiran. 2012. Anticancer and antioxidant effect of *Acanthophora spicifera* against EAC induced carcinoma in mice. Journal of Pharmacy Research 5(3): 1503–1507.

Lavanya, R. and N. Veerapan, 2011. Antibacterial potential of six seaweeds collected from Gulf of Mannar of Southeast Coast of India. Advances in Biological Research 5(1): 38–44.

Lavanya, R. and N. Veerappan. 2012. Pharmaceutical properties of marine macro algal communities from Gulf of mannnar against Human pathogens. Asian Pacific Journal of Tropical Diseases 2(1): S320–S323.

Lavy, A., Y. Naveh, R. Coleman, S. Mokady and M.J. Werman. 2003. Dietary *Dunaliella bardawil*, a beta-carotene-rich alga, protects against acetic acid-induced small bowel inflammation in rats. Inflammatory Bowel Diseases 9(6): 372–379.

Lay, M.K., R.L. Atmar, S. Guix, U. Bharadwaj, H. He, F.H. Neill, K.J. Sastry, Q. Yao and M.K. Estes. 2010. Norwalk virus does not replicate in human macrophages or dendritic cells derived from the peripheral blood of susceptible humans. Virology 406: 1–11.

Le Lann, K., C. Jegou and V. Stiger-Pouvreau. 2008. Effect of different conditioning treatments on total phenolic content and antioxidant activities in two Sargassacean species: Comparison of the frondose *Sargassum muticum* (Yendo) Fensholt and the cylindrical *Bifurcaria bifurcata* R. Ross. Phycological Research 56(4): 238–245.

Le Lann, K., G. Surget, C. Couteau, L. Coiffard, S. Cérantola, F. Gaillard, M. Larnicol, M. Zubia, F. Guérard, N. Poupart and V. Stiger-Pouvreau. 2016. Sunscreen, antioxidant, and bactericide capacities of phlorotannins from the brown macroalga *Halidrys siliquosa*. Journal of Applied Phycology 28(6): 3547–3559.

Le, Q.T., Y. Li, Z.J. Qian, M.M. Kim and S.K. Kim. 2009. Inhibitory effects of polyphenols isolated from marine algae *Ecklonia cava* on histamine release. Process Biochem 44: 168–176.

Leal, D., B. Matsuhiro, M. Rossi and F. Caruso. 2008. FT-IR spectra of alginic acid block fractions in three species of brown seaweeds. Carbohydrate Research 343: 308–316.

Lee, B., B. Sur, J. Park, H. Shin, S. Kwon, M. Yeom, S.J. Kim, K. Kim, I. Shim and C.S. Yin. 2012e. Fucoidan ameliorates scopolamine-induced neuronal impairment and memory dysfunction in rats *via* activation of cholinergic system and regulation of cAMP-response element-binding protein and brain-derived neurotrophic factor expressions. Journal of the Korean Society for Applied Biological Chemistry 55: 711–720.

Lee, B.B., J.S. Choi, Y.M. Ha, K.H. Do and I.S. Choi. 2009b. Inhibitory effects of seaweed extracts on *Helicobacter pylori* growth and urease activity. pp. 144. *In*: 18th Annual Meeting of North Pacific Marine Science Organization (PICES), Oct 23-Nov 1, Jeju, Korea.

Lee, B.-B., J.-S. Choi, H.E. Moon, Y.-M. Ha, M.S. Kim, K.K. Cho and I.S. Choi. 2013. Inhibition of growth and urease of *Helicobacter pylori* by Korean edible seaweed extracts. Botanical Sciences: 515–222.

Lee, C., G.H. Park, E.M. Ahn, B. Kim, C.-I. Park and J.H. Jang. 2013b. Protective effect of *Codium fragile* against UVB-induced pro-inflammatory and oxidative damages in HaCaT cells and BALB/c mice. Fitoterapia 86: 54–63.

Lee, C.W. and J.S. Han. 2012. Hypoglycemic effect of *Sargassum ringgoldianum* extract in STZ-induced diabetic mice. Prev. Nutrition & Food Science 17: 8–13.

Lee, D.-G., J.W. Hyun, K.-A. Kang, J.-O. Lee, S.-H. Lee, B.-J. Ha, J.-M. Ha, E.Y. Lee and J.-H. Lee. 2004b. *Ulva lactuca*: a potential seaweed for tumor treatment and immune stimulation. Biotechnology and Bioprocess Engineering 9(3): 236–238.

Lee, D.M. and M.E. Weinblatt. 2001. Rheumatoid arthritis. Lancet 358: 903–911.

Lee, D.-S., W.S. Park, S.-J. Heo, S.-H. Cha, D. Kim, Y.-J. Jeon, S.-G. Park, E.B. Shim, I.-W. Choi and W.-K. Jung. 2011c. *Polyopes affinis* alleviates airway inflammation in a murine model of allergic asthma. Journal of Biosciences 36(5): 869–77.

Lee, E.-J. and M.-K. Sung. 2003. Chemoprevention of azoxymethane-induced rat colon carcinogenesis by seatangle, a fiber-rich seaweed. Plant Foods for Human Nutrition 58(3): 1–8.

Lee, H., C. Kang, E.-S. Jung, J.-S. Kim and E. Kim. 2011b. Antimetastatic activity of polyphenol-rich extract of *Ecklonia cava* through the inhibition of the Akt pathway in A549 human lung cancer cells. Food Chemistry 127(3): 1229–1236.

Lee, H., J.S. Kim and E. Kim. 2012c. Fucoidan from seaweed *Fucus vesiculosus* inhibits migration and invasion of human lung cancer cell via PI3K-Akt-mTOR pathways. PLoS One 7: e50624.

Lee, H.J., Y.A. Kim, J.W. Ahn, H.J. Na, H.M. Kim and Y. Seo. 2006. Screening of Korean marine plants for their inhibitory effect on histamine release from RPMC *in vitro*. Biotechnology and Bioprocess Engineering 11: 80–83.

Lee, H.J., H.T. Dang, G.J. Kang, E.J. Yang, S.S. Park, W.J. Yoon, J.H. Jung, H.K. Kang and E.S. Yoo. 2009c. Two enone fatty acids isolated from *Gracilaria verrucosa* suppress the production of inflammatory mediators by down-regulating NFkappaB and STAT1 activity in lipopolysaccharide-stimulated RAW 264.7 cells. Archives of Pharmacal Research 32(3): 453–462.

Lee, H.-R., H. Do, S.-R. Lee, E.-S. Sohn, S. Pyo and E. Son. 2007. Effects of fucoidan on neuronal cell proliferation: Association with NO production through the iNOS pathway. Journal of Food and Nutrition Sciences 12: 74–78.

Lee, J., M. Grabb, G. Zipfel and D. Choi. 2000. Brain tissue responses to ischemia. Journal of Clinical Investigation 106: 723–731.

Lee, J.-B., K. Hayashi, M. Maeda and T. Hayashi. 2004. Antiherpetic activities of sulfated polysaccharides from green algae. Planta Medica 70: 813–817.

Lee, J.-B., A. Takeshita, K. Hayashi and T. Hayashi. 2011. Structures and antiviral activities of polysaccharides from *Sargassum trichophyllum*. Carbohydrate Polymers 86(2): 995–999.

Lee, J.-H., S.-H. Eom, E.-H. Lee, Y.-J. Jung, H.-J. Kim, M.-R. Jo, K.-T. Son, H.J. Lee, J.H. Kim, and M.-R. Lee. 2014. *In vitro* antibacterial and synergistic effect of phlorotannins isolated from edible brown seaweed *Eisenia bicyclis* against acne-related bacteria. Algae 29: 47–55.

Lee, J.W., J.H. Wang, K.M. Ng, C.H. Tan, P. Rabina and S.S. Teo, SS. 2015. *In-vitro* anticancer activity of *Eucheuma cottonii* extracts against HeLa cell line, human lung carcinoma cell line (SK-LU-1), human colon carcinoma cell line (HCT-116), and fibroblast. International Journal of Current Multidisciplinary Studies 1(2): 69–73.

Lee, J.Y., Y.J. Kim, H.J. Kim, Y.S. Kim and W. Park. 2012. Immunostimulatory effect of laminarin on RAW 264.7 mouse macrophages. Molecules 17: 5404–5411.

Lee, K.Y. and D.J. Mooney. 2012. Alginate: Properties and biomedical applications. Progress in Polymer Science 37: 106–126.

Lee, N.Y., S.P. Ermakova, H.K. Choi, M.I. Kusaykin, N.M. Shevchenko, T.N. Zvyagintseva and H.S. Choi. 2008. Fucoidan from *Laminaria cichorioides* inhibits AP-1 transactivation and cell transformation in the mouse epidermal JB6 cells. Molecular Carcinogenesis 47(8): 629–637.

Lee, N.Y., S.P. Ermakova, T.N. Zvyagintseva, K.W. Kang, Z. Dong and H.S. Choi. 2008b. Inhibitory effects of fucoidan on activation of epidermal growth factor receptor and cell transformation in Jb6 Cl41 cells. Food and Chemical Toxicology 46(5): 1793–1800.

Lee, O.-K., K.-Y. Yoon, K.-J. Kim, S.-G. You and B.-Y. Lee. 2011a. Seaweed extracts as a potential tool for the attenuation of oxidative damage in obesity-related pathologies. Journal of Phycology 47: 548–556.

Lee, S.H., Y. Li, F. Karadeniz, M.M. Kim and S.K. Kim. 2009. α-Glucosidase and α-amylase inhibitory activities of phloroglucinol derivatives from edible marine brown alga, *Ecklonia cava*. Journal of the Science of Food and Agriculture 89: 1552–1558.

Lee, S.H., M.H. Park, S.J. Heo, S.M. Kang, S.C. Ko, S.J. Han and Y.J. Jeon. 2010. Dieckol isolated from *Ecklonia cava* inhibits α-glucosidase and α-amylase *in vitro* and alleviates post-prandial hyperglycemia in streptozotocin-induced diabetic mice. Food and Chemical Toxicology 48: 2633–2637.

Lee, S.H., K.H. Min, J.S. Han, D.H. Lee, D.B. Park, W.K. Jung, P.J. Park, B.T. Jeon, S.K. Kim and Y.J. Jeon. 2012b. Effects of brown alga, *Ecklonia cava* on glucose and lipid metabolism in C57BL/KsJ-db/db mice, a model of type 2 diabetes mellitus. Food and Chemical Toxicology 50: 575–582.

Lee, S.-H. C.-I. Koa, G. Ahn, S. You, J.-S. Kim, M.S. Heu, J. Kim, Y. Jee and Y.-J. Jeon. 2012d. Molecular characteristics and anti-inflammatory activity of the fucoidan extracted from *Ecklonia cava*. Carbohydrate Polymers 89: 599–606.

Lee, S.H. and S.-K. Kim. 2015b. Biological phlorotannins of *Eisenia bicyclis*. pp. 453–464 pp. *In*: Kim, S.-K. and K. Chojnacka (eds.). Marine Algae Extracts, Wiley-VCH Verlag GmbH & Co. KGaA, Weinheim, Germany.

Lee, S.-H., S.-M. Kang, C.H. Sok, J.T. Hong, J.-Y. Oh, Y.-J. Jeon. 2015c. Cellular activities and docking studies of eckol isolated from *Ecklonia cava* (Laminariales, Phaeophyceae) as potential tyrosinase inhibitor. Algae 30: 163–170.

Lee, S.-G., H.-G. Cho and S.-Y. Paik. 2015d. Molecular epidemiology of norovirus in South Korea. BMB Reports 48: 61–67.

Leibbrandt, A., C. Meier, M. Konig-Schuster, R. Weinmullner, D. Kalthoff, B. Pflugfelder, P. Graf, B. Frank-Gehrke, M. Beer, T. Fazekas, H. Unger, E. Prieschl-Grassauer and A. Grassauer. 2010. Iota-carrageenan is a potent inhibitor of Influenza A Virus Infection. Plos One 5(12): 0014320.

Leiro, J.M., R. Castro, J.A. Arranz and J. Lamas. 2007. Immunomodulating activities of acidic sulphated polysaccharides obtained from the seaweed *Ulva rigida* C. Agardh. International Immunopharmacology 7: 879–888.

Leite, E.L., M.G.L. Medeiros, H.A.O. Rocha, G.G.M. Farias, L.F. da Silva, S.F. Chavante, L.D. de Abreu, C.P. Dietrich and H.B. Nader. 1998. Structure and pharmacological activities of a sulfated xylofucoglucuronan from alga *Spatoglossum schroederi*. Plant Science 132: 215–228.

Leite, Y.F.M.M., L.M.C.M. Silva, R.C.N. Amorim, E.A. Freire, D.M.M. Jorge, T.B. Grangeiro and N.M.B. Benevides. 2005. Purification of a lectin from the marine red alga *Gracilaria ornata* and its effect on the development of the Cowpea Weevil *Callosobruchus maculatus* (Coleoptera: Bruchidae). Biochimica et Biophysica Acta (BBA)—General Subjects 1724(1-2): 137–145.

Lemée, R., D. Pesando, M. Durand-Clément, A. Dubreuil, A. Meinesz, A. Guerriero and F. Pietra. 1993. Preliminary survey of toxicity of the green alga *Caulerpa taxifolia* introduced into the Mediterranean. Journal of Applied Phycology 5(5): 485–493.

León-Deniz, L.V., E. Dumonteil, R. Moo-Puc and Y. Freile-Pelegrin. 2009. Antitrypanosomal *in vitro* activity of tropical marine algae extracts. Pharmaceutical Biology 47(9): 864-871.

Leoni, G., P.-A. Neumann, R. Sumagin, T. Denning and A. Nusrat. 2015. Wound repair: role of immune–epithelial interactions. Mucosal Immunology 8(5): 959–968.

LeRoith, D., H. Werner, D.B. Johnson and J. Robert. 1995. Molecular and cellular aspects of the insulin-like growth factor I receptor. Endocrine Reviews 16: 143–163.

Lessing, D. and N.M. Bonini. 2009. Maintaining the brain: Insight into human neurodegeneration from *Drosophila mutants*. Nature Reviews. Genetics 10(6): 1–26.

Li, B., F. Lu, X.J. Wei and R.X. Zhao. 2008. Fucoidan: structure and bioactivity. Molecules 13(8): 1671–1695.

Li, D.M. and G.S. de Hoog. 2009. Cerebral phaeohyphomycosis—a cure at what lengths? The Lancet Infectious Diseases 9: 376–383.

Li, K., X.-M. Li, J.B. Gloer and B.-G. Wang. 2012. New nitrogen-containing bromophenols from the marine red alga *Rhodomela confervoides* and their radical scavenging activity. Food Chemistry 135: 868–872.

Li, L., R. Ni, Y. Shao and S. Mao. 2014. Carrageenan, and its applications in drug delivery. Carbohydrate Polymers 103: 1–11.

Li, N., W. Mao, M. Yan, X. Liu, Z. Xia, S. Wang, B. Xiao, C. Chen, L. Zhang and S. CaoKey. 2015. Structural characterization and anticoagulant activity of a sulfated polysaccharide from the green alga *Codium divaricatum*. Carbohydrate Polymers 121: 175–182.

Li, N., X. Liu, X.X. He, S.Y. Wang, S.J. Cao, Z. Xia, H.L. Xian, L. Qin, and W.J. Mao. 2017. Structure and anticoagulant property of a sulfated polysaccharide isolated from the green seaweed *Monostroma angicava*. Carbohydrate Polymers 159: 195–206.

Li, X. and S. Ma. 2015. Advances in the discovery of novel antimicrobials targeting the assembly of bacterial cell division protein FtsZ. European Journal of Medicinal Chemistry 95: 1–15.

Li, X.M. and L. Brown. 2009. Efficacy and mechanisms of action of traditional Chinese medicines for treating asthma and allergy. Journal of Allergy and Clinical Immunology 123: 297–308.

Li, Y., L. Liu, D. Liu, S. Woodward, S.W. Barger, R.E. Mrak and W.S. Griffin. 2004. Microglial activation by uptake of fDNA via a scavenger receptor. Journal of Neuroimmunology 147: 50–55.

Li, Y., S.H. Lee, Q.T. Le, M.M. Kim and S.K. Kim. 2008b. Anti-allergic effects of phlorotannins on histamine release via binding inhibition between IgE and FcεRI. Journal of Agricultural and Food Chemistry 56: 12073–12080.

Li, Y., Z.-J. Qian, B. Ryu, S.-H. Lee, M.-M. Kim and S.-K. Kim. 2009. Chemical components and its antioxidant properties *in vitro*: an edible marine brown alga, *Ecklonia cava*. Bioorganic & Medicinal Chemistry 17: 1963–1973.

Li, Y., Z.-J. Qian, M.-M. Kim and S.-K. Kim. 2011. Cytotoxic activities of phlorethol and fucophlorethol derivatives isolated from laminariaceae *Ecklonia cava*. Journal of Food Biochemistry 35(2): 357–369.

Li, Y.X., I. Wijesekara, Y. Li and S.K. Kim. 2011b. Phlorotannins as bioactive agents from brown algae. Process Biochemistry 46: 2219–2224.

Liao, W.-R., J.-Y. Lin, W.-Y. Shieh, W.-L. Jeng and R. Huang. 2003. Antibiotic activity of lectins from marine algae against marine vibrios. Journal of Industrial Microbiology & Biotechnology 30: 433–439.

Lieu, P.T., M. Heiskala, P.A. Peterson and Y. Yang. 2001. The roles of iron in health and disease. Molecular Aspects of Medicine 22: 1–87.

Lim, C.S., D.Q. Jin, J.Y. Sung, J.H. Lee, H.G. Choi, I. Ha and J.S. Han. 2006. Antioxidant and anti-inflammatory activities of the methanolic extract of *Neorhodomela aculeata* in hippocampal and microglial cells. Biological and Pharmaceutical Bulletin 29(6): 1212–1216.

Lim, J.D., S.R. Lee, T. Kim, S.-A. Jang, S.C. Kang, H.J. Koo, E. Sohn, J.P. Bak, S. Namkoong, H.K. Kim, I.S. Song, N. Kim, E.-H. Sohn and J. Han. 2015. Fucoidan from *Fucus vesiculosus* protects against alcohol-induced liver damage by modulating inflammatory mediators in mice and HepG2 cells. Marine Drugs 13: 1051–1067.

Lima-Filho, J.V.M., A.F.F.U. Carvalho, S.M. Freitas and V.M.M. Melo. 2002. Antibacterial activity of extracts of six macroalgae from the North Eastern Brazilian coast. Brazilian Journal of Microbiology 33: 311–313.

Lin, A.S., E.P. Stout, J. Prudhomme, K. Le Roch, C.R. Fairchild, S.G. Franzblau, W. Aalbersberg, M.E. Hay and J. Kubanek. 2010. Bioactive bromophycolides R-U from the Fijian red alga *Callophycus serratus*. Journal of Natural Products 73(2): 275–278.

Lin, H.C., S.T. Chou, M.Y. Chuang, T.Y. Liao, W.S. Tsai and T.H. Chiu. 2012. The effects of *Caulerpa microphysa* enzyme-digested extracts on ACE-inhibitory activity and *in vitro* anti-tumour properties. Food Chemistry 134: 2235–2241.

Lin, L., K.L. Chow and Y. Leng. 2009. Study of hydroxyapatite osteoinductivity with an osteogenic differentiation of mesenchymal stem cells. Journal of Biomedical Materials Research Part A 89: 326–335.

Lin, S.-J., J. Schranz and S.M. Teutsch. 2001. Aspergillosis case-fatality rate: systematic review of the literature. Clinical Infectious Diseases 32(3): 358–366.

Lindberg, M. and S. Lindstrom. 2017. *Gloiopeltis furcata*—Seaweeds of Alaska. Available online at: http://www.seaweedsofalaska.com/species.asp?SeaweedID=74.

Lindenbach, B.D. and C.M. Rice. 2001. Flaviviridae: the viruses and their replication. pp. 589–639. *In*: Knipe, D.M. and P.M. Howley (eds.). Fundamental Virology. Lippincott Williams & Wilkins, Philadelphia.

Lins, K.O., D.P. Bezerra, A.P. Alves, N.M. Alencar, M.W. Lima, V.M. Torres, W.R. Farias, W.R. Farias, C. Pessoa, M.O. de Morais and L.V. Costa-Lotufo. 2008. Antitumor properties of a sulfated polysaccharide from the red seaweed *Champia feldmannii* (Diaz-Pifferer). Journal of Applied Toxicology 29(1): 20–26.

Littler, D., M. Littler, K. Bucher and J.N. Norris. 1989. Marine plants of the Caribbean, a Field guide from Florida to Brazil. Smithsonian Institution Press, Washington D.C., 272 pp.

Liu, A.-H., D.-Q. Liu, T.-J. Liang, X.-Q. Yu, M.-T. Feng, L.-G. Yao, Y. Fang, B. Wang, L.-H. Feng, M.-X. Zhang and S.-C. Mao. 2013b. Caulerprenylols A and B, two rare antifungal prenylated para-xylenes from the green alga *Caulerpa racemosa*. Bioorganic & Medicinal Chemistry Letters 23: 2491–2494.

Liu, B., L.A. Wasilwa, T.E. Morelock, N.R. O'Neill and J.C. Correll. 2007. Comparison of *Colletotrichum orbiculare* and several allied *Colletotrichum* spp. for mtDNA RFLPs, intron RFLP and sequence variation, vegetative compatibility, and host specificity. Phytopathology 97(10): 1305–1314.

Liu, C.L., Y.T. Chiu and M.L. Hu. 2011c. Fucoxanthin enhances HO-1 and NQOI expression in murine hepatic BNL CL2 cells through activation of the Nrf2/ARE system partially by its pro-oxidant activity Journal of Agriculture and Food Chemistry 59: 11344–11351.

Liu, C.L., Y.P. Lim and M.L. Hu. Liu, C.L., Y.P. Lim and M.L. Hu. 2013. Fucoxanthin enhances cisplatin-induced cytotoxicity via nfkappab-mediated pathway and downregulates DNA repair gene expression in human hepatoma HepG2 cells. Marine Drugs 11(1): 50–66.

Liu, D.-Q., S.-C. Mao, H.-Y. Zhang, X.-Q. Yu, M.-T. Feng, B. Wang, L.-H. Feng and Y.-W. Guo. 2013c. Racemosins A and B, two novel bisindole alkaloids from the green alga *Caulerpa racemosa*. Fitoterapia 91: 15–20.

Liu, F., J. Wang, A.K. Chang, B. Liu, L. Yang, Q. Li, P. Wang and X. Zou. 2012a. Fucoidan extract derived from *Undaria pinnatifida* inhibits angiogenesis by human umbilical vein endothelial cells. Phytomedicine 19(8-9): 797–803.

Liu, F., J. Wang, A.K. Chang, B.L., L. Yang, Q. Li, P. Wang and X. Zou. 2012c. Fucoidan extract derived from *Undaria pinnatifida* inhibits angiogenesis by human umbilical vein endothelial cells. Phytomedicine 19(8–9): 797–803.

Liu, H. and L. Gu. 2012. Phlorotannins from brown algae (*Fucus vesiculosus*) inhibited the formation of advanced glycation endproducts by scavenging reactive carbonyls. Journal of Agricultural and Food Chemistry 60: 1326–1334.

Liu, H. and L. Gu. 2012e. Phlorotannins from brown algae (*Fucus vesiculosus*) inhibited the formation of advanced glycation endproducts by scavenging reactive carbonyls. Journal of Agricultural and Food Chemistry 60: 1326–1334.

Liu, H.Y., M.Y. Geng, X.L. Xin, F.C. Li, H.X. Chen, H.S. Guan and J. Ding. 2005. Multiple and multivalent interactions of novel anti-AIDS drug candidates, sulfated polymannuronate (SPMG)-derived oligosaccharides, with gp120 and their anti-HIV activities. Glycobiology 15: 501–510.

Liu, J., A.H. Banskota, A.T. Critchley, J. Hafting and B. Prithiviraj. 2015. Neuroprotective effects of the cultivated *Chondrus crispus* in a *C. elegans* model of Parkinson's disease. Marine Drugs 13: 2250–2266.

Liu, J.C., F.L. Hsu, J.C. Tsai, P. Chan, J.Y.H. Liu, G.N. Thomas, B. Tomlinson, M.Y. Lo and J.Y. Lin. 2003. Antihypertensive effects of tannins isolated from traditional Chinese herbs as non-specific inhibitors of angiotensin converting enzyme. Life Sciences 73: 1543–1555.

Liu, L., M. Heinrich, S. Myers and S.A. Dworjanyn. 2012d. Towards a better understanding of medicinal uses of the brown seaweed *Sargassum* in traditional Chinese medicine: A phytochemical and pharmacological review. Journal of Ethnopharmacology 142: 591–619.

Liu, M., P.L. Hansen and X. Lin. 2011b. Bromophenols in marine algae and their bioactivifies, Marine Drugs: 9: 1273–1292.

Liu, M., W. Zhang, J. Wei, L. Qiu and X. Lin. 2012b. Marine bromophenol bis(2,3-dibromo-4,5-dihydroxybenzyl) ether, induces mitochondrial apoptosis in K562 cells and inhibits topoisomerase I *in vitro*. Toxicology Letters 211(2): 126–134.

Liu, X., X. Li, L. Gao, C. Cui, C. Li, J. Li and B. Wang. 2011. Extraction and PTP1B inhibitory activity of bromophenols from the marine red alga *Symphyocladia latiuscula*. Chinese Journal of Oceanology and Limnology 29: 686–690.

Liu, Y., J.B. Morgan, V. Coothankandaswamy, R. Liu, M.B. Jekabsons, F. Mahdi, D.G. Nagle and Y.-D. Zhou. 2009b. The *Caulerpa* pigment caulerpin Inhibits HIF-1 activation and mitochondrial respiration. Journal of Natural Products 72(12): 2104–2109.

Liu, Z., I. Romero-Canelón, B. Qamar, J.M. Hearn, A. Habtemariam, N.P.E. Barry, A.M. Pizarro, G.J. Clarkson and P.J. Sadler. 2014. The potent oxidant anticancer activity of organoiridium catalysts. Angewandte Chemie International Edition 53(15): 3941–3946.

Lloréns, J.L.P., I.H. Carero, J.J.V. Oñate, F.G.B. Murillo and A. León. 2016. Las algas se comen: un periplo por la biologia, la historia, las curiosidades y la gastronomia. Universidade de Cádiz, Editorial UCA, Cádiz, 336 pp.

Lo, W.M. and E.C. Li-Chan. 2005. Angiotensin I converting enzyme inhibitory peptides from *in vitro* pepsin-pancreatin digestion of soy protein. Journal of Agricultural and Food Chemistry 53: 3369–3376.

Lobban, C.S., D.J. Chapman and B.P. Kremer. 1988. Experimental phycology: a laboratory manual. Phycological Society of America. Cambridge University Press, Cambridge, UK, 320 pp.

Lopes, G., P.B. Andrade and P. Valentão. 2015. Screening of a marine algal extract for antifungal activities. pp. 411–420. *In*: Stengel, D.B. and S. Connan (eds.). Natural Products from Marine Algae: Methods and Protocols, Methods in Molecular Biology, vol. 1308, Springer Science+Business Media, New York.

López-López, I., S. Bastida, C. Ruiz-Capillas, L. Bravo, M.T. Larrea, F. Sánchez-Muniz, S. Cofrades and F. Jiménez-Colmenero. 2009. Composition and antioxidant capacity of low salt meat emulsion model systems containing edible seaweeds. Meat Science 83: 492–498.

Lorbeer, A.J., R. Tham and W. Zhang. 2013. Potential products from the highly diverse and endemic macroalgae of southern Australia and pathways for their sustainable production. Journal of Applied Phycology 25(3): 717–732.

Lordan, S., R.P. Ross and C. Stanton. 2011. Marine bioactives as functional food ingredients: potential to reduce the incidence of chronic diseases. Marine Drugs 9(6): 1056–1100.

Lordan, S., T.J. Smyth, A. Soler-Vila, C. Stanton and R.P. Ross. 2013. The α-amylase and α-glucosidase inhibitory effects of Irish seaweed extracts. Food Chemistry 141: 2170–2176.

Lourenço, R. and M.E. Camilo. 2002. Taurine: A conditionally essential amino acid in humans: An overview in health and disease. Nutrición Hospitalaria 17: 262–270.

Loya, S., M. Bakhanashvili, Y. Kashman and A. Hizi. 1995. Peyssonols A and B, two novel inhibitors of the reverse transcriptases of human immunodeficiency virus types 1 and 2. Archives of Biochemistry and Biophysics 316(2): 789–796.

Lozano, I., J.M. Wacyk, J. Carrasco and M.A.C.-S. Martín. 2016. Red macroalgae *Pyropia columbina* and *Gracilaria chilensis*: sustainable feed additive in the *Salmo salar* diet and the evaluation of potential antiviral activity against infectious Salmon anemia virus. Journal of Applied Phycology 28(2): 1343–1351.

LTS. 2005. Health and food technology. Home economics, learning and teaching, resource management, Scotland. Available online at: http://www.educationscotland.gov.uk/Images/HFTResourceMgtH_tcm4-277559.pdf.

Lu, C.X., J. Li, Y.X. Sun, X. Qi, Q.J. Wang, X.L. Xin and M.Y. Geng. 2007. Sulfated polymannuroguluronate, a novel anti-AIDS drug candidate, inhibits HIV-1 Tat induced angiogenesis in Kaposi's sarcoma cells. Biochemical Pharmacology 74: 1330–1339.

Lück, E. and M. Jager. 2012. Antimicrobial food additives: Characteristics—uses—effects. Springer Science & Business Media, Berlin, Germany, 262 pp.

Lundsteen, S. and R. Nielsen. 2015. Coccotylus brodiei, *C. truncatus* and other Phyllophoraceae (Rhodophyta) in Danish waters. Forum om marin bundfauna og–flora. *Coccotylus* and other Phyllophoraceae, 20 pp.

Luo, D., Q. Zhang, H. Wang, Y. Cui, Z. Sun, J. Yang, Y. Zheng, J. Jia, F. Yu and X. Wang. 2009. Fucoidan protects against dopaminergic neuron death *in vivo* and *in vitro*. European Journal of Pharmacology 617: 33–40.

Luo, M., B. Shao, W. Nie, X.W. Wei, Y.L. Li, B.L. Wang, Z.Y. He, X. Liang, T.H. Ye and Y.Q. Wei. 2015. Antitumor and adjuvant activity of lambda-carrageenan by stimulating immune response in cancer immunotherapy. Scientific Reports 5(11062).

Luo, Z., D. Tian, M. Zhou, W. Xiao, Y. Zhang, M. Li, B. Sui, W. Wang, H. Guan, H. Chen, Z.F. Fu and L. Zhao. 2015. λ-carrageenan P32 is a potent inhibitor of Rabies Virus Infection. PLoS ONE 10(10): e0140586.

Luop, M.P. 2001. Antioxidants and vitamins in cosmetics. Clinics in Dermatology 19: 467–473.

Lusalgae. 2017. Research activities, wholesale and retail trade of cosmetic and hygiene products. Available online at: https://www.facebook.com/Lusalgae/.

Lüscher-Mattli, M. 2000. Polyanions—a lost chance in the fight against HIV and other virus diseases? Antiviral Chemistry and Chemotherapy 11: 249–259.

Ly, B.M., N.Q. Buu, N.D. Nhut, P.D. Thinh, T. Thi and T. Van. 2005. Studies on fucoidan and its production from Vietnamese brown sea weeds. ASEAN Journal Science and Technology Development 22: 371–380.

Lycke, E., M. Johansson, B. Svennerholm and U. Lindahl. 1991. Binding of *Herpes simplex* virus to cellular heparan sulphate, an initial step in the adsorption process. Journal of General Virology 72: 1131–1137.

Lynch, G., L. Low, S. Li, A. Sloane, S. Adams, C. Parish, B.L. Kemp and A.L. Cunningham. 1994. Sulfated polyanions prevent HIV infection of lymphocytes by disruption of the CD4-gp120 interaction, but do not inhibit monocyte infection. Journal of Leukocyte Biology 56: 266–272.

Ma, C.Y., W.S. Liu, K.C. Kwok and F. Kwok. 1996. Isolation and characterization of proteins from soymilk residue (okara). Food Research International 8: 799–805.

Ma, J.-W. and Tan W.-C. 1984. Screening for antimicrobial activities in marine algae from the Qingdao coast, China. Proceedings of the International Seaweed Symposium 11: 517–520.

Ma, M., J. Zhao, S. Wang, S. Li, Y. Yang, J. Shi, X. Fan and L. He. 2006. Bromophenols coupled with methyl gamma-ureidobutyrate and bromophenol sulfates from the red alga *Rhodomela confervoides*. Journal of Natural Products 69(2): 206–210.

MacArtain, P., I. Gill, R. Ch, M. Brooks, R. Campbell and I.R. Rowland. 2007. Nutritional value of edible seaweed. Nutrition Reviews 65: 535–543.

MacCaughey, V. 1918. Algae of the Hawaiian archipelago II. The Botanical Gazette 65(2): 121–149.

Macedo, N.R.P.V., N.S.R. Ribeiro, R.C. Villaça, W. Ferreira, A.M. Pinto, V.L. Teixeira, C. Cirne-Santos, I.C.N.P. Paixão and V. Giongo. 2012. Caulerpin as potencial antiviral drug against herpes simplex vírus type 1. Brazilian Journal of Pharmacognosy 22(4): 861–867.

Machado, L.P., W.M.S. Bispo, S.T. Matsumoto, F.O. Reis, R.B. Santos and L.F.G. Oliveira, Jr. 2011. Triagem de macroalgas com potencial antifúngico no controle *in vitro* da antracnose do mamoeiro (*Carica papaya* L.). Revista Brasileira de Agrociência 17(4): 463–467.

Machado, L.P., S.T. Matsumoto, C.M. Jamal, M.B. Silva, D.C. Centeno, P.C. Neto, L.R. Carvalho and N.S. Yokoya. 2014. Chemical analysis and toxicity of seaweed extracts with inhibitory activity against tropical fruit anthracnose fungi. Journal of the Science of Food and Agriculture 94: 1739–1744.

Machado, L.P., S.T. Matsumoto, G.R.F. Cuzzuol and L.F.G. Oliveira, Jr. 2014b. Influence of laboratory cultivation on species of Rhodophyta physiological evaluations and antifungal activity against phytopathogens. Revista Ciência Agronômica 45(1): 52–61.

Machado, L.P., L.R. Carvalho, M.C.M. Young, E.M. Cardoso-Lopes, D.C. Centeno, L. Zambotti-Villela, P. Colepicolo, N.S. Yokoya. 2015. Evaluation of acetylcholinesterase inhibitory activity of Brazilian red macroalgae organic extracts. Revista Brasileira de Farmacognosia 25: 657–662.

Mackay, I.M., K.C. Jacob, D. Woolhouse, K. Walker, M.W. Syrmis and D.M. Whiley. 2003. Molecular assays for detection of human metapneumovirus. Journal of Clinical Microbiology 41: 100–105.

Madigan, M.T., J.M. Martinko, K.S. Bender, D.H. Buckley, D.A. Stahl and T. Brock. 2014. Brock biology of microorganisms (14th ed.). Pearson, San Francisco, CA, 1032 pp.

Maeda, M., T. Kodama, T. Tanaka, Y. Ohfune, K. Nomoto, K. Nishimura and T. Fujita. 1984. Insecticidal and neuromuscular activities of domoic acid and its related compounds. Journal of Pesticide Science 9(1): 27–32.

Maeda, H., M. Hosokawa, T. Sashima, K. Murakami-Funayama and K. Miyashita. 2009. Anti-obesity and anti-diabetic effects of fucoxanthin on diet-induced obesity conditions in a murine model. Molecular Medicine Reports 2: 897–902.

Maehre, H.K., M.K. Malde, K.E. Eilertsen and E.O. Elvevoll. 2014. Characterization of protein, lipid and mineral contents in common Norwegian seaweeds and evaluation of their potential as food and feed. Journal of the Science of Food and Agriculture 94(15): 3281–3290.

Mafinowska, P. 2011. Algae extracts as active cosmetic ingredients. Zeszyty Naukowe, Poznań University of Economics Press 212: 123–129.

Magalhães, K.D., L.S. Costa, G.P. Fidelis, R.M. Oliveira, L.T.D. Barreto Nobre, N. Dantas-Santos, R.F.G. Camara, I.R.L. Albuquerque, S.L. Cordeiro, D.A. Sabry, M.S.S. Pereira Costa, L.G. Alves and H.A.O. Rocha. 2011. Anticoagulant, antioxidant and antitumor activities of heterofucans from the seaweed *Dictyopteris delicatula*. International Journal of Molecular Sciences 12(5): 3352–3365.

Mageswaran, R., V. Balakrishnan and S. Balasubramaniam. 1985. Boron content of marine algae from the Mandaitivu and Kirinda coasts and mineral content of nine species of algae from the Kirinda coast. Journal of the National Science Council of Sri Lanka 13: 131–140.

Magill, C.L., B.F. Gilmore and C.A. Maggs. 2009. An antiviral product from *Dumontia contorta* (S). Phycologist 76: 18.

Magnenat-Thalmann, N., P. Kalra, J.L. Lévêque, R. Bazin, D. Batisse and B. Querleux. 2002. A computational skin model: fold and wrinkle formation. IEEE transactions on information technology in biomedicine 6: 317–323.

Maiti, I.B. 1974. Purification and mode of action of phytase from *Phaseolus aureus*. Phytochemistry 13: 1047–1051.

Majmudar, G. 2012. Compositions of marine botanicals to provide nutrition to aging and environmentally damaged skin. US 8318178 B2.

Makela, M.J., T. Puhakka, O. Ruuskanen, M. Leinonen, P. Saikku, M. Kimpimaki, S. Blomqvist, T. Hyypia and P. Arstila. 1998. Viruses and bacteria in the etiology of the common cold. Journal of Clinical Microbiology 36: 539–542.

Makkar, F. and K. Chakraborty. 2016. Antidiabetic and anti-inflammatory potential of sulphated polygalactans from red seaweeds *Kappaphycus alvarezii* and *Gracilaria opuntia*. International Journal of Food Properties, DOI: http://dx.doi.org/10.1080/10942912.2016.1209216.

Malagolia, B.G., F.T.G.S. Cardozo, J.H.S. Gomesa, V.P. Ferraz, C.M.O. Simões and F.C. Braga. 2014.Chemical characterization and antiherpes activity of sulfated polysaccharides from *Lithothamnion muelleri*. International Journal of Biological Macromolecules 66: 332–337.

Mamatha, B.S., K.K. Namitha, A. Senthil, J. Smitha and G.A. Ravishankar. 2007. Studies on use of *Enteromorpha* in snack food. Food Chemistry 101: 1707–1713.

Manach, C., G. Williamson, C. Morand, A. Scalbert and C. Rémésy. 2005. Bioavailability and bioefficacy of polyphenols in humans. I. Review of 97 bioavailability studies. The American Journal of Clinical Nutrition 81: 230S–242S.

Manchu, N., Y. Melpha, S. Jeeva and J.E. James. 2015. Phytochemical analysis and antibacterial activity of *Chaetomorphs antennina* from Rasthacaud coast, Tamil Nadu, India. World Journal of Pharmaceutical Research 4(3): 1337–1351.

Mandal, P., C.G. Mateu, K. Chattopadhyay, C.A. Pujol, E.B. Damonte and B. Ray. 2007. Structural features and antiviral activity of sulphated fucans from the brown seaweed *Cystoseira indica*. Antiviral Chemistry and Chemotherapy 18: 153–162.

Mandal, P., C.A. Pujol, M.J. Carlucci, K. Chattopadhyay, E.B. Damonte and B. Ray. 2008. Anti-herpetic activity of a sulfated xylomannan from *Scinaia hatei*. Phytochemistry 69: 2193–2199.

Manefield, M., M. Welch, M. Givskov, G.P.C. Salmond and S. Kjelleberg. 2001. Halogenated furanones from the red alga, *Delisea pulchra*, inhibit carbapenem antibiotic synthesis and exoenzyme virulence factor production in the phytopathogen Erwinia carotovora. FEMS Microbiology Letters 205: 131–138.

Manela-Azulay, M., E. Bagatin. 2009. Cosmeceuticals vitamins. Clinics in Dermatology 27: 469–474.

Manhanzva, M.T., J. Mutsvangwa, I.A. Beck, L.M. Frenke, M. Tshabalala, J. Manasa, A.T. Murahwa and L. Gwanzura. 2015. The burden of HIV-associated drug resistance mutations in an early infant diagnosis program: A glance through the pediatric window of Zimbabwe. Journal of Infectious Diseases & Therapy 3: 1–6.

Manigandan, M. and K. Kolanjinathan. 2014. *In vitro* antifungal activity of certain seaweeds collected from Mandapam coast, Tamilnadu, India. Indian Journal of Applied Research 4(10): 557–559.

Manilal, A., S. Sujith, J. Selvin, G.S. Kiran and C. Shakir. 2009. *In vivo* antiviral activity of polysaccharide from the Indian green alga, *Acrosiphonia orientalis* (J. Agardh): potential implication in shrimp disease management. World Journal of Fish and Marine Sciences 1(4): 278–282.

Manilal, A., S. Sujith, G.S. Kiran, J. Selvin, C. Shakir, R.G. and A.P. Lipton. 2009b. Antimicrobial potential and seasonality of red algae collected from the Southwest coast of India tested against shrimp, human and phytopathogens. Annals of Microbiology 59(2): 207–219.

Manilal, A., S. Sujith, G.S. Kiran, J. Selvin, C. Shakir, R. Gandhimathi and M.V.N. Panikkar. 2009c. Biopotentials of seaweeds collected from southwest coast of India. Journal of Marine Science and Technology 17(1): 67–73.

Manilal, A., S. Sujith, J. Selvin, G.S. Kiran, C. Shakir and A.P. Lipton. 2010. Antimicrobial potential of marine organisms collected from the southwest coast of India against multiresistant human and shrimp pathogens. Scientia Marina 74: 287–296.

Manilal, A., S. Sujith, B. Sabarathnam, G.S. Kiran, J. Selvin, C. Shakir and A.P. Lipton. 2010b. Antifouling potentials of seaweeds collected from the southwest coast of India. World Journal of Agricultural Sciences 6: 243–248.

Manilal, A., S. Sujith, B. Sabarathnam, G.S. Kiran, J. Selvin, C. Shakir and A.P. Lipton. 2011. Biological activity of red alga *Laurencia brandenii*. Acta Botanica Croatica 70(1): 81–90.

Manilal, A., N. Thajuddin, J. Selvin, A. Idhayadhulla, R.S. Kumar and S. Sujith. 2011b. *In vitro* mosquito larvicidal activity of marine algae against the human vectors, *Culex quinquefasciatus* (Say) and *Aedes aegypti* (Linnaeus) (Diptera: Culicidae). International Journal of Zoology Research 7(3): 272–278.

Manilal, A., J. Selvin, N. Thajuddin, S. Sujith, M. Panikkar, A. Idhayadhulla R.S. Kumar. 2012. Biopotentials of marine alga, *Lobophora variegata* collected from the south Indian littoral. Thalassas 28(1): 47–54.

Manilal, A., J. Selvin and S. George. 2012b. *In vivo* therapeutic potentiality of red seaweed, *Asparagopsis* (Bonnemaisoniales, Rhodophyta) in the treatment of vibriosis in *Penaeus monodon* Fabricius. Saudi Journal of Biological Sciences 19: 165–175.

Manilal, A., S. Sujith, J. Selvin, M.V.N. Panikkar and S. George. 2012c. Anticoagulant potential of polysaccharide isolated from Indian red alga, *Asparagopsis taxiformis* (Delile) Trevisan. Thalassas: An International Journal of Marine Sciences 28: 9–15.

Manivannan, K., G.D. Karthikai, G. Thirumaran and P. Anantharaman. 2008. Mineral composition of marine macroalgae from Mandapam coastal regions; southeast coast of India. American Eurasian Journal of Botany 1(2): 58–67.

Manivannan, K., G.D. Karthikai, P. Anantharaman and T. Balasubramanian. 2011. Antimicrobial potential of selected brown seaweeds from Vedalai coastal waters, Gulf of Mannar. Asian Pacific Journal of Tropical Biomedicine 1(2): 114–120.

Mann, G.V. 1994. Metabolic consequences of dietary trans fatty acids. Lancet 343: 1268–1271.

Manoj, S.G.M., P.S.M. Kumar, M. Vasanthi and A. Achary 2013. Anticoagulant property of sulphated polysaccharides extracted from marine brown algae collected from Mandapam Island, India. African Journal of Biotechnology 12(16): 1937–1945.

Manojkumar, K.S. 2013. *In vitro* cancer chemopreventive properties of polysaccharide extract from the brown alga, *Sargassum wightii*. IOSR Journal of Pharmacy and Biological Sciences 8(3): 06–11.

Mansuya, P., P. Aruna, S. Sridhar, J.S. Kumar and S. Babu. 2010. Antibacterial activity and qualitative phytochemical analysis of selected seaweeds from Gulf of Mannar region. Journal of Experimental Sciences 1(8): 23–26.

Mao, W., B. Li, Q. Gu, Y. Fang H. and Xing. 2004. Preliminary studies on the chemical characterization and antihyperlipidemic activity of polysaccharide from the brown alga *Sargassum fusiforme*. Hydrobiologia 512: 263–266.

Mao. W., X. Zang and Y. Li H. Zhang. 2006. Sulfated polysaccharides from marine algae *Ulva conglobata* and their anticoagulant activity. Journal of Applied Phycology 18: 9–14.

Mao, W., H. Li, Y. Li, H. Zhang, X. Qi and H. Sun. 2009. Chemical characteristic and anticoagulant activity of the sulfated polysaccharide isolated from *Monostroma latissimum* (Chlorophyta). International Journal of Biological Macromolecules 44: 70–74.

Mao, W.J., F. Fang, H.Y. Li, X.H. Qi, H.H. Sun, Y. Chen and S.D. Guo. 2008. Heparinoid-active two sulfated polysaccharides isolated from marine green algae *Monostroma nitidum*. Carbohydrate Polymers 74: 834–839.

Marchetti, M., S. Pisani, V. Pietropaolo, L. Seganti, R. Nicoletti and N. Orsi. 1995. Inhibition of *Herpes simplex* virus infection by negatively charged and neutral carbohydrate polymers. Journal of Chemotherapy 7: 90–96.

Maréchal, J.-P., G. Culioli, C. Hellio, H. Thomas-Guyon, M.E. Callow, A.S. Clare and A. Ortalo-Magné. 2004. Seasonal variation in antifouling activity of crude extracts of the brown alga *Bifurcaria bifurcata* (Cystoseiraceae) against

Cyprids of *Balanus amphitrite* and the marine bacteria *Cobetia marina* and *Pseudoalteromonas haloplanktis*. Journal of Experimental Marine Biology and Ecology 313(1): 47–62.

Margret, R.J., S. Kumaresan and S. Ravikumar. 2009. A preliminary study on the anti-inflammatory activity of methanol extract of *Ulva lactuca* in rat. Journal of Environmental Biology 30: 899–902.

Margulis, L. and M.J. Chapman. 2009. Kingdoms and domains—An illustrated guide to the phyla of life on earth. Academic Press, London, 659 pp.

Marinho-Soriano, E., P.C. Fonseca, M.A.A. Carneiro and W.S.C. Moreira. 2006. Seasonal variation in the chemical composition of two tropical seaweeds. Bioresource Technology 97: 2402–2406.

Mariya, V. and V.S. Ravindran. 2013. Biomedical and pharmacological significance of marine macro algae-review. Indian Journal of Geo-Marine Sciences 42(5): 527–537.

Marlett, J.A., M.I. McBurney and J.L. Slavin. 2002. Position of the American Dietetic Association: health implications of dietary fiber. Journal of American Diabetes Association 102: 993–1000.

Marques, J.F. and L. Pereira. 2016. Chapter 6—A revision on the red alga *Dilsea carnosa*. pp. 101–145. *In*: Pereira, L. (ed.). Carrageenans: Sources and Extraction Methods, Molecular Structure, Bioactive Properties and Health Effects. Nova Science Publishers, N.Y.

Marrazzo, J.M., C. del Rio, D.R. Holtgrave, M.S. Cohen, S.C. Kalichman, K.H. Mayer, J.S. Montaner, D.P. Wheeler, R.M. Grant, B. Grinsztejn, N. Kumarasamy, S. Shoptaw, R.P. Walensky, F. Dabis, J. Sugarman and C.A. Benson. 2014. International antiviral society-USA panel, 2014. HIV prevention in clinical care settings: 2014 recommendations of the International Antiviral Society-USA Panel. JAMA 312: 390–409.

Marsh, J. and J.A. Goode. 2007. Ciba foundation symposium 186—antimicroabial peptides. Novartis Foundation Symposia, John Wiley & Sons, Chichester, 283 pp.

Marti, R., M.J. Uriz and X. Turon. 2004. Seasonal and spatial variation of species toxicity in Mediterranean seaweed communities: Correlation to biotic and abiotic factors. Marine Ecology Progress Series 282: 73–85.

Martín-Algarra, S., E. Espinosa, J. Rubio, J.J. Lopez Lopez, J.L. Manzano, L.A. Carrión, A. Plazaola, A. Tanovic and L. Paz-Ares. 2009. Phase II study of weekly Kahalalide F in patients with advanced malignant melanoma. European Journal of Cancer 45: 732–735.

Martínez-Esparza, M., C. Jiménez-Cervantes, F. Solano, J.A. Lozano and J.C. García-Borrón. 1998. Mechanisms of melanogenesis inhibition by tumor necrosis factor-α in B16/F10 mouse melanoma cells. European Journal of Biochemistry 255: 139–146.

Martinez Nadal, N.G., L.V. Rodriguez and C. Casillas. 1963. Sarganin and chonalgin, new antibiotic substances from marine algae from Puerto Rico. Antimicrobial Agents Chemotherapy 3: 68–72.

Martínez-Navarrete, N., M.M. Camacho, J. Martínez-Lahuerta, J. Martínez-Monzó and P. Fito. 2002. Iron deficiency and iron fortified foods—a review. Food Research International 35: 225–231.

Martin, K.C. 2016. *Pythium aphanidermatum*—Soilborne plant pathogens class project. Available online at: https://projects.ncsu.edu/cals/course/pp728/Pythium/Pythium_aphanidermatum.html.

Marudhupandi, T., T.T.A. Kumar, S. Lakshmanasenthil, G. Suja and T. Vinothkumar. 2015. *In vitro* anticancer activity of fucoidan from *Turbinaria conoides* against A549 cell lines. International Journal of Biological Macromolecules 72: 919–923.

Maruyama, H., H. Tamauchi, M. Hashimoto and T. Nakano. 2003. Antitumor activity and immune response of Mekabu fucoidan extracted from sporophyll of *Undaria pinnatifida*. *In Vivo* 17(3): 245–249.

Maruyama, H., H. Tamauchi, M. Hashimoto and T. Nakano. 2005. Suppression of Th2 immune responses by Mekabu fucoidan from *Undaria pinnatifida* sporophylls. International Archives of Allergy and Immunology 137: 289–294.

Maruyama, H., H. Tamauchi, M. Iizuka and T. Nakano. 2006. The role of Nk cells in antitumor activity of dietary fucoidan from *Undaria pinnatifida* sporophylls (Mekabu). Planta Medica 72(15): 1415–1417.

Mashjoor, S., M. Yousefzadi, M.A. Esmaeili and R. Rafiee. 2016. Cytotoxicity and antimicrobial activity of marine macro algae (Dictyotaceae and Ulvaceae) from the Persian Gulf. Cytotechnology 68(5): 1717–1726.

Mata-Gómez, L.C., J.C. Montañez, A. Méndez-Zavala and C.N. Aguilar. 2014. Biotechnological production of carotenoids by yeasts: an overview. Microbial Cell Factories 13: 12 pp.

Matanjun, P., S. Mohamed, N. Mustapha and K. Muhammad. 2009. Nutrient content of tropical edible seaweeds, *Eucheuma cottonii*, *Caulerpa lentillifera* and *Sargassum polycystum*. Journal of Applied Phycology 21: 75–80.

Mathew, M. and R.S. Verma. 2009. Humanized immunotoxins: a new generation of immunotoxins for targeted cancer therapy. Cancer Science 100(8): 1359–1365.

Matou, S., D. Helley, D. Chabut, A. Bros and A.-M. Fischer. 2002. Effect of fucoidan on fibroblast growth factor-2-induced angiogenesis *in vitro*. Thrombosis Research 106: 213–221.

Matsubara, K., Y. Matsuura, A. Bacic, M. Liao, K. Hori and K. Miyazawa. 2001. Anticoagulant properties of a sulfated galactan preparation from a marine green alga, *Codium cylindricum*. International Journal of Biological Macromolecules 28: 395–399.

Matsubara, K., Y. Matsuura, A. Bacic, M.L. Liao, K. Hori and K. Miyazawa. 2001. Anticoagulant properties of a sulfated galactan preparation from a marine green alga, *Codium cylindricum*. International Journal of Biological Macromolecules 28: 395–399.

Matsubara, K., C. Xue, X. Zhao, M. Mori, T. Suqawara and T. Hirata. 2005. Effect of middle molecular weight fucoidans on *in vitro* and *ex vivo* angiogenesis ofendothelial cells. International Journal of Molecular Medicine 15: 695–699.

Matsubara, M., S. Masaki, K. Ohmori, A. Karasawa and K. Hasegawa. 2004. Differential regulation of IL-4 expression and degranulation by anti-allergic olopatadine in rat basophilic leukemia (RBL-2H3) cells. Biochemical Pharmacology 67: 1315–1326.

Matsuda, T., J. Sasaki, H. Kurihara, M. Hatano and K. Takahashi. 2005. Anticancer benefit of *Sargassum horneri* extract. Hokkaido Daigaku Suisan Kagaku Kenkyu Iho 56: 75–86.

Matsuhiro, B., A.F. Conte, E.B. Damonte, A.A. Kolender, M.C. Matulewicz, E.G. Mejias, C.A. Pujol and E.A. Zuniga. 2005. Structural analysis and antiviral activity of a sulfated galactan from the red seaweed *Schizymenia binderi* (Gigartinales, Rhodophyta). Carbohydrate Research 340: 2392–2402.

Matsumoto, M., M. Hosokawa, N. Matsukawa, M. Hagio, A. Shinoki, M. Nishimukai, K. Miyashita, T. Yajima and H. Hara. 2010. Suppressive effects of the marine carotenoids, fucoxanthin and fucoxanthinol on triglyceride absorption in lymph duct-cannulated rats. European Journal of Nutrition 49: 243–249.

Matsumoto, S., M. Nagaoka, T. Hara, I. Kimura-Takagi, K. Mistuyama and S. Ueyama. 2004. Fucoidan derived from *Cladosiphon okamuranus* Tokida ameliorates murine chronic colitis through the down-regulation of interleukin-6 production on colonic epithelial cells. Clinical & Experimental Immunology 136: 432–439.

Matsumoto, T., A. Kitano, N. Oshitani, A. Obata, M. Hiki, H. Hashimura, K. Okawa, H. Nagura and K. Kobayashi. 1988. Immunoglobulin-containing cells in the colonic mucosa of rabbits with carrageenan-induced colitis. Diseases of the Colon & Rectum 31(9): 723–729.

Mattos, B.B., M.T. Romanos, L.M. de Souza, G. Sassaki and E. Barreto-Bergter. 2011. Glycolipids from macroalgae: potential biomolecules for marine biotechnologi? Revista Brasileira de Farmacognosia 21: 244–247.

Mattson, M.P. 2000. Apoptosis in neurodegenerative disorders. Nature Reviews Molecular Cell Biology 1: 120–130.

Maurer, C.C. 1965. Antibacterial substances in Chilean marine algae. Anal Facult Quim Farm 16: 114–121.

Mayakrishnan, V., P. Kannappan, N. Abdullah and A.B.A. Ahmed. 2013. Cardioprotective activity of polysaccharides derived from marine algae: An overview. Trends in Food Science & Technology 30(2): 98–104.

Mayer, A., A. Rodríguez, O. Taglialatela-Scafati and N. Fusetani. 2013. Marine pharmacology in 2009–2011: Marine compounds with antibacterial, antidiabetic, antifungal, anti-inflammatory, antiprotozoal, antituberculosis, and antiviral activities; affecting the immune and nervous systems, and other miscellaneous mechanisms of action. Marine Drugs 11(7): 2510–2573.

Mayer, A.M., V.J. Paul, W. Fenical, J.N. Norris, M.S. Carvalho and R.S. Jacobs. 1993. Phospholipase A_2 inhibitors from marine algae. Hydrobiologia 260/261: 521–529.

Mazumder, S., P.K. Ghosal, C.A. Pujol, M.J. Carlucci, E.B. Damonte and B. Ray. 2002. Isolation, chemical investigation and antiviral activity of polysaccharides from *Gracilaria corticata* (Gracilariaceae, Rhodophyta). International Journal of Biological Macromolecules 31: 87–95.

McCullough, J.L. and K.M. Kelly. 2006. Prevention and treatment of skin aging. Annals of the New York Academy of Sciences 1067: 323–331.

McFarlane, S.I., A. Kumar and J.R. Sowers. 2003. Mechanisms by which angiotensin-converting enzyme inhibitors prevent diabetes and cardiovascular disease. American Journal of Cardiology 91 (Suppl.): 30H–37H.

McHugh, D.J. 2003. A Guide to the Seaweed Industry. FAO, Fisheries Technical Paper 441: 73–90.

McKeown, S.R., R.L. Cowen and K.J. Williams. 2007. Bioreductive drugs: from concept to clinic. Clinical Oncology 19(6): 427–442.

McLean, B.J., J. Hobson-Peters, C.E. Webb, D. Watterson, N.A. Prow, H.D. Nguyen, S. Hall-Mendelin, D. Warrilow, C.A. Johansen, C.C. Jansen, A.F. van den Hurk, N.W. Beebe, E. Schnettler, R.T. Barnard and A.R. Hall. 2015. A novel insect-specific flavivirus replicates only in Aedes-derived cells and persists at high prevalence in wild Aedes vigilax populations in Sydney, Australia. Virology 486: 272–283.

McMillan, R.M., D.E. McIntyre and J.L. Gordon. 1979. Stimulation of human platelets by carrageenans. Journal of Pharmacy and Pharmacology 31: 148–152.

Medeiros, V.P., K.C.S. Queiroz, M.L. Cardoso, G.R.G. Monteiro, F.W. Oliveira, S.F. Chavante, L.A. Guimaraes, H.A.O. Rocha and E.L. Leite. 2008. Sulfated galactofucan from *Lobophora variegata*: Anticoagulant and anti-inflammatory properties. Biochemistry (Moscow) 73(9): 1018–1024.

Meenakshi, S., S. Umayaparvathi, R. Saravanan, T. Manivasagam and T. Balasubramanian. 2016. Neuroprotective effect of fucoidan from *Turbinaria decurrens* in MPTP intoxicated Parkinsonic mice. International Journal of Biological Macromolecules 86: 425–433.

Mehdinezhad, N., A. Ghannadi and A. Yegdaneh. 2016. Phytochemical and biological evaluation of some *Sargassum* species from Persian Gulf. Research in Pharmaceutical Sciences 11(3): 243–249.

Mehta, D. and A.B. Malik. 2006. Signaling mechanisms regulating endothelial permeability. Physiological Reviews 86: 279–367.

Meillisa, A., E.A. Siahaan, J.N. Park, H.C. Woo and B.S. Chun. 2013. Effect of subcritical water hydrolysate in the brown seaweed *Saccharina japonica* as a potential antibacterial agent on food-borne pathogens. Journal of Applied Phycology 25: 763–769.

Meiyu, G., L. Fuchuan, X. Xianliang, L. Jing, Y. Zuowei and G. Huashi. 2003. The potential molecular targets of marine sulfated polymannuroguluronate interfering with HIV-1 entry: interaction between SPMG and HIV-1 rgp120 and CD4 molecule. Antiviral Research 59(2): 127–135.

Melnick, J.L. 1996. Current status of poliovirus infections. Clinical Microbiology Reviews 9: 293–300.

Mendes, G.S., A.R. Soares, F.O. Martins, M.C. Albuquerque, S.S. Costa, Y. Yoneshigue-Valentin, L.M. Gestinari, N. Santos and M.T. Romanos. 2010. Antiviral activity of the green marine alga *Ulva fasciata* on the replication of human Metapneumovirus. Revista do Instituto de Medicina Tropical de São Paulo 52(1): 3–10.

Mendes, G.S., A.R. Soares, L. Sigiliano, F. Machado, C. Kaiser, N. Romeiro, L. Gestinari, N. Santos and M.T.V. Romanos. 2011. *In vitro* anti-HMPV activity of meroditerpenoids from marine alga *Stypopodium zonale* (Dictyotales). Molecules 16: 8437–8450.

Mendes, G.S., I.C. Bravin, Y. Yoneshigue-Valentin, N.S. Yokoya and M.T.V. Romanos. 2012. Anti-HSV activity of *Hypnea musciformis* cultured with different phytohormones. Revista Brasileira de Farmacognosia 22: 789–794.

Mendes, G.S., M.E.R. Duarte, F.G. Colodi, M.D. Noseda, L.G. Ferreira, S.D. Berte, J.F. Cavalcanti, N. Santos and M.T.V. Romanos. 2014. Structure and anti-metapneumovirus activity of sulfated galactans from the red seaweed *Cryptonemia seminervis*. Carbohydrate Polymers 101: 313–323.

Mendes, M., R. Pereira, I. Sousa Pinto, A.P. Carvalho and A.M. Gomes. 2013. Antimicrobial activity and lipid profile of seaweed extracts from the North Portuguese coast. International Food Research Journal 20(6): 3337–3345.

Mendis, E. and S.-K. Kim. 2011. Present and future prospects of seaweeds in developing functional foods. pp. 1–15. *In*: Se-Kwon Kim (ed.). Advances in Food and Nutrition Research, Volume 64, Marine Medicinal Foods: Implications and Application, Macro and Microalgae. Academic Press, an imprint of Elsevier, Waltham, USA.

Mendoza, L., R. Vilela, K. Voelz, A.S. Ibrahim, K. Voigt and S.C. Lee. 2015. Human fungal pathogens of Mucorales and Entomophthorales. Cold Spring Harbor Perspectives in Medicine 5: a019562.

Mercurio, D.G., T.A.L. Wagemaker, V.M. Alves, C.G. Benevenuto, L.R. Gaspar and P. M.B.G. Maia Campos. 2015. *In vivo* photoprotective effects of cosmetic formulations containing UV filters, vitamins, *Ginkgo biloba* and red algae extracts. Journal of Photochemistry and Photobiology B: Biology 153: 121–126.

Mergler, D. 1999. Neurotoxic effects of low level exposure to manganese in human populations. Environmental Research 80: 99–102.

Merrill, E.W. and E.W. Salzman. 1983. Polyethylene oxide as a biomaterial. American Society for Artificial Internal Organs Journal 6: 60–64.

Mertin, J. and R. Hunt. 1976. Influence of polyunsaturated fatty acids on survival of skin allografts and tumor incidence in mice. Proceedings of the National Academy of Sciences U.S.A. 73: 928–931.

Mesnildrey, L., C. Jacob, K. Frangoudes, M. Reunavot and M. Lesueur. 2012. Seaweed industry in France. Report. Interreg program NETALGAE. Les publications du pôle halieutique Agrocampus Ouest 9: 34 pp.

Metti, D.Y. 2012. The morphology and molecular phylogenies of the *Laurencia* complex Genera with a description of species from Eastern Australia. PhD Thesis, UNSW, Sydney, 355 pp.

Meyers, S.P. and L. Hoyo. 1966. Observations on the growth of the marine hyphomycete *Varicosporina ramulosa*. Canadian Journal of Botany 44(9): 1133–1140.

Mhadhebi, L., A. Dellai, A. Clary-Laroche, R.B. Said, J. Robert and A. Bouraoui. 2012. Anti-inflammatory and antiproliferative activities of organic fractions from the Mediterranean brown seaweed, *Cystoseira compressa*. Drug Development Research 73(2): 82–89.

Mhadhebi, L., A. Mhadhebi, J. Robert and A. Bouraoui. 2014. Antioxidant, anti-inflammatory and antiproliferative effects of aqueous extracts of three mediterranean brown seaweeds of the genus *Cystoseira*. Iranian Journal of Pharmaceutical Research 13(1): 207–220.

Mhadhebi, L., K. Chaiebb and A. Bouraoui. 2012b. Evaluation of antimicrobial activity of organic fractions of six marine algae from Tunisian Mediterranean coasts. International Journal of Pharmacy and Pharmaceutical Sciences 4(1): 534–537.

Miao, B.C., M.Y. Geng, J. Li, F. Li H. Chen, H.S. Guan and J. Ding. 2004. Sulfated polymannuroguluronate, a novel anti-acquired immune deficiency syndrome (AIDS) drug candidate, targeting CD4 in lymphocytes. Biochemical Pharmacology 68: 641–649.

Michalak, I. and K. Chojnacka. 2015. Algae as production systems of bioactive compounds. Engineering in Life Sciences 15: 160–176.

Midega, J.T., E.J. Muturi, F.N. Baliraine C.M. Mbogo J. Githure, J.C. Beier and G. Yan. 2010. Population structure of *Anopheles gambiae* along the Kenyan coast. Acta Tropica 114: 103–108.

Milchakova, N. 2011. Marine plants of the Black Sea. An illustrated field guide. Digit Print Press, Sevastopol, Ukraine, 144 pp.

Milián, E. and A.M. Díaz. 2004. Allergy to house dust mites and asthma. Puerto Rico Health Sciences Journal 23: 47–57.

Mills, G. 2012. Seaweed: a rich source of vitamins and bioactive compounds. Available online at: http://www.inshore-ireland.com/News/seaweed-a-rich-source-of-vitamins-and-bioactive-compounds.html.

Min, E.Y., I.H. Kim, J. Lee, E.Y. Kim, Y.H. Choi and T.J. Nam. 2014. The effects of fucodian on senescence are controlled by the p16INK4a-pRb and p14Arf-p53 pathways in hepatocellular carcinoma and hepatic cell lines. International Journal of Oncology 45: 47–56.

Minghou, J. 1990. Processing and extraction of phycocolloids. Regional Workshop on the Culture and Utilization of Seaweeds, Volume II. FAO Technical Resource Papers. Cebu City, Philippines.

Mirza, S.A., M. Phelan, D. Rimland, E. Graviss, R. Hamill, M.E. Brandt, T. Gardner, M. Sattah, G. Ponce de Leon, W. Baughman and R.A. Hajjeh. 2003. The changing epidemiology of cryptococcosis: an update from population-based active surveillance in 2 large metropolitan areas, 1992–2000. Clinical Infectious Diseases 36: 789–794.

Mirzaahmadi, S., G. Asaadi-Tehrani, M. Bandehpour, N. Davouri, L. Tahmasbi, N. Hosseinzadeh, H. Mirzahoseini, K. Parivar and B. Kazemi. 2011. Expression of recombinant human coagulation factor VII by the Lizard *Leishmania* expression system. Journal of Biomedicine and Biotechnology 2011: 1–8.

Mise, T. and T. Yasumoto. 2011. Simultaneous treatment of cancer cells lines with the anticancer drug cisplatin and the antioxidant fucoxanthin. British Journal of Pharmacology and Toxicology 2(3): 127–131.

Mišurcová, L., S. Kráčmar, B. Klejdus and J. Vacek. 2010. Nitrogen content, dietary fiber, and digestibility in algal food products. Czech Journal of Food Sciences 28: 27–35.

Mišurcová, L., L. Machů and J. Orsavová. 2011a. Seaweed minerals as nutraceuticals. pp. 371–390. *In*: Se-Kwon Kim (ed.). Advances in Food and Nutrition Research, Vol. 64. Academic Press, Burlington.

Mišurcová, L., J. Ambrožová and D. Samek. 2011b. Seaweed lipids as nutraceuticals. pp. 339–355. *In*: Se-Kwon Kim (ed.). Advances in Food and Nutrition Research, Vol. 64. Academic Press, Burlington.

Mišurcová, L., J. Orsavová and J.V. Ambrožová. 2015. Algal polysaccharides and health, pp. 109–144. *In*: Kishan Gopal Ramawat and Jean-Michel Mérillon (eds.). Polysaccharides—Bioactivity and Biotechnology. Springer International Publishing AG, part of Springer Nature.

Mitchell, J. 1744. Account of the Yellow fever which prevailed in Virginia in the years 1737, 1741, and 1742, in a letter to the late Cadwallader Colden, Esq. of New York, from the late John Mitchell, M.D.F.R.S. of Virginia. American Medical and Philosophical Register 181–215.

Mitchell, K. 2000. Keith Michell's practically macrobiotic cookbook. Healing Arts Press, Rochester, Vermont, 248 pp.

Mithlesh, P., H.S. Garg and D.S. Bhakuni. 1988. Chemical constituents of *Ulva fasciata* from indian coast. Journal of the Indian Chemical Society 65: 675–677.

Miyashita, K. and M. Hosokawa. 2008. Beneficial health effects of seaweed carotenoid, fucoxanthin in marine nutraceuticals and functional foods. pp. 297–308. *In*: Shahid, F. and C. Barrow (eds.). Marine Nutraceuticals and Functional Foods. CRC Press, Taylor & Francis Group, Boca Raton, FL.

Miyashita, K., B. Narayan, T. Tsukui, H. Kamogawa, M. Abe and M. Hosokawa. 2011. Brown seaweed lipids as potential source of omega-3 PUFA in biological systems. pp. 329–339. *In*: Kim, S.-K. (ed.). Handbook of Marine Macroalgae: Biotechnology and Applied Phycology. John Wiley & Sons, West Sussex, UK.

Miyashita, K., S. Nishikawa, F. Beppu, T. Tsukui, M. Abe and M. Hosokawa. 2011. The allenic carotenoid fucoxanthin, a novel marine nutraceutical from brown seaweeds. Journal of the Science of Food and Agriculture 91: 1166–1174.

Miyata, M., T. Koyama, T. Kamitani, T. Toda and K. Yazawa. 2009. Anti-obesity effect on rodents of the traditional Japanese food, Tororokombu, shaved *Laminaria.* Bioscience, Biotechnology, and Biochemistry 73(10): 2326–2328.

Miyazaki, S., A. Nakayama, M. Oda, M. Takada and D. Attwood. 1994. Chitosan and sodium alginate based bioadhesive tablets for intraoral drug delivery. Biological and Pharmaceutical Bulletin 17: 745–747.

Mizushina, Y., Y. Sugiyama, H. Yoshida, S. Hanashima, T. Yamazaki, S. Kamisuki, K. Ohta, M. Takemura, T. Yamaguchi, A. Matsukage, S. Yoshida, M. Saneyoshi, F. Sugawara and K. Sakagauchi. 2001. Galactosyldiacylglycerol, a mammalian DNA polymerase alpha-specific inhibitor from a sea alga, *Petalonia bingbamiae*. Biological and Pharmaceutical Bulletin 24(9): 982–987.

Mizutani, S., S. Deguchi, E. Kobayashi, E. Nishiyama, H. Sagawa and I. Kato. 2010. Fucoidan containing cosmetics. US 7678368 B2.

Mochizuki, H., J. Takido, H. Oda and H. Yokogoshi. 1999. Improving effect of dietary taurine on marked hypercholesterolemia induced by a high-cholesterol diet in streptozotocin-induced diabetic rats. Bioscience, Biotechnology, and Biochemistry 63: 1984–1987.

Mody, K.H., A. Saikrishna Murthy, A.K. Siddhanta, B.K. Ramavat and O.P. Mairh. 1997. Investigation on the sulphated polysaccharides of *Sargassum cinctum* of Saurashtra coast. pp. 83–90. *In*: Subba Rangaiah, G. (ed.). Recent Trends in Algal Research. Andhra University, India.

Moen, L.K. and G.F. Clark. 1993. A novel reverse transcriptase inhibitor from *Fucus vesiculosus*. International AIDS Conference 9: 145–161.

Mohamed, S.F. and F.A. Agili. 2013. Antiviral sulphated polysaccharide from brown algae *Padina pavonia* characterization and structure elucidation. International Journal of ChemTech Research 5(4): 1469–1476.

Mohamed, S.F., T.M. Slamy and S.Y. Shaeiri. 2015. Characterization of sulphated polysaccharide with antiviral activity from marine brown alga *Cystoseira myrica* collected from Jazan coasts, KSA. International Journal of PharmTech Research 8(10): 198–203.

Mohandass, C., A.S. Vijayaraj, R. Rajasabapathy, S. Satheeshbabu, S.V. Rao, C. Shiva and L. De-Mello. 2013. Biosynthesis of silver nanoparticles from marine seaweed *Sargassum cinereum* and their antibacterial activity. Indian Journal of Pharmaceutical Sciences 75: 606–610.

Mohapatra, L., S.K. Bhattamishra, R. Panigrahy, S. Parida and P. Pati. 2016. Antidiabetic effect of *Sargassum wightii* and *Ulva fasciata* in high fat diet and multi low dose Streptozotocin induced type 2 diabetic mice. UK Journal of Pharmaceutical and Biosciences 4(2): 13–23.

Mohibbullah, M., M.A. Hannan, J.Y. Choi, M.M.H. Bhuiyan, Y.K. Hong, J.S. Choi, I.S. Choi and I.S. Moon. 2015. The edible marine alga *Gracilariopsis chorda* alleviates hypoxia/reoxygenation-induced oxidative stress in cultured hippocampal neurons. Journal of Medicinal Food 18: 960–971.

Mohibbullah, M., M.A. Hannan, I.S. Park, I.S. Moon and Y.K. Hong. 2016. The edible red seaweed *Gracilariopsis chorda* promotes axodendritic architectural complexity in hippocampal neurons. 2016. Journal of Medicinal Food 19(7): 638–644.

Mohibbullah, M., M.M.H. Bhuiyan, M.A. Hannan, P. Getachew, Y.K. Hong, J.S. Choi and I.S. Choi. 2016. The edible red alga *Porphyra yezoensis* promotes neuronal survival and cytoarchitecture in primary hippocampal neurons. Cellular and Molecular Neurobiology 36: 669–682.

Mohsen, M.S.A., S.F. Mohamed, F.M. Ali and O.H. El-Sayed. 2007. Chemical structure and antiviral activity of water-soluble sulfated polysaccharides from *Sargassum latifolium*. Journal of Applied Sciences Research 3: 1178–1185.

Mohsin, S., R. Mahadevan and G. Muraleedhara Kurup. 2014. Free-radical-scavenging activity and antioxidant effect of Ascophyllan from marine brown algae *Padina tetrastromatica*. Biomedicine & Preventive Nutrition 4(1): 75–79.

Monroy-Ruiz, J., M.A. Sevilla, R. Carron and M.J. Montero. 2011. Astaxanthin-enriched-diet reduces blood pressure and improves cardiovascular parameters in spontaneously hypertensive rats. Pharmacological Research 63: 44–50.

Montalvão, S., Z. Demirel, P. Devi, V. Lombardi, V. Hongisto, M. Perala, J. Hattara, E. Imamoglu, S.S. Tilvi, G. Turan, M.C. Dalay and P. Tammela. 2016. Large-scale bioprospecting of cyanobacteria, micro- and macroalgae from the Aegean Sea. New Biotechnology 33(3): 399–406.

Montanha, J.A., N. Bourgougnon, J. Boustie and M. Amoros. 2009. Antiviral activity of carrageenans from marine red algae. Latin American Journal of Pharmacy 28: 443–448.

Monteiro, R. and I. Azevedo. 2010. Chronic inflammation in obesity and the metabolic syndrome. Mediators of Inflammation 289645(10): 14 pp.

Monteiro, V.S., F.B. Teles, C.O. Coura, R.B. Souza, C.N.D. Lima, D.V.D. Costa, E.R. Honorio, S.D. Escudeiro, E.M.C. Chaves and S.M.M. Vasconcelos. 2016. Involvement of the gabaergic system in the anxiolytic effect of sulfated polysaccharides from the red seaweed *Gracilaria cornea*. Journal of Applied Phycology 28(3): 1997–2004.

Montero, D., G. Walther, A. Perez-Martin, E. Roche and A. Vinet. 2011. Endothelial dysfunction, inflammation, and oxidative stress in obese children and adolescents: markers and effect of lifestyle intervention. Obesity Reviews 13: 441–445.

Monto, A.S. and K.M. Sullivan. 1993. Acute respiratory illness in the community. Frequency of illness and the agents involved. Epidemiology and Infection 110: 145–160.

Monto, A.S., A.M. Fendrick and M.W. Sarnes. 2001. Respiratory illness caused by picornavirus infection: a review of clinical outcomes. Clinical Therapeutics 23: 1615–1627.

Monto, A.S. 2002. Epidemiology of viral respiratory infections. American Journal of Medicine 112: 4S–12S.

Moon, H.E., N. Islam, B.R. Ahr, S.S. Chowdhury, H.S. Sohn, H.A. Jung and J.S. Choi. 2011. Protein tyrosine phosphatase 1B and α-glucosidase inhibitory phlorotannins from edible brown algae *Ecklonia stolonifera* and *Eisenia bicyclis*. Bioscience, Biotechnology, and Biochemistry 75: 1472–1480.

Moon, H.J., S.H. Lee, M.J. Ku, B.C. Yu, M.J. Jeon, S.H. Jeong, V.A. Stonik, T.N. Zvyagintseva, S.P. Ermakova and Y.H. Lee. 2009. Fucoidan inhibits UVB-induced MMP-1 promoter expression and down regulation of type I procollagen synthesis in human skin fibroblasts. European Journal of Dermatology 19(2): 129–134.

Moo-Puc, R., D. Robledo and Y. Freile-Pelegrin. 2008. Evaluation of selected tropical seaweeds for *in vitro* anti-trichomonal activity. Journal of Ethnopharmacology 120: 92–97.

Moo-Puc, R., D. Robledo and Y. Freile-Pelegrín. 2009. *In vitro* cytotoxic and antiproliferative activities of marine macroalgae from Yucatán, Mexico (Actividad citotóxica y antiproliferativa *in vitro* de macroalgas marinas de Yucatán, México). Ciencias Marinas 35(4): 345–358.

Moo-Puc, R., D. Robledo and Y. Freile-Pelegrín. 2011a. Enhanced antitumoral activity of extracts derived from cultured *Udotea flabellum* (Chlorophyta). Evidence-Based Complementary and Alternative Medicine Vol. 2011, Article ID 969275, 7 pp.

Moo-Puc, R.D. Robledo and Y. Freile-Pelegrín. 2011b. Improved antitumoral activity of extracts derived from cultured *Penicillus dumetosus*. Evidence-Based Complementary and Alternative Medicine Vol. 2011, Article ID 969275: 7 pp.

Moorthi, P.V. and C. Balasubramanian. 2015. Antimicrobial properties of marine seaweed, *Sargassum muticum* against human pathogens. Journal of Coastal Life Medicine 3(2): 122–125.

Morales, J.L., Z.O. Cantillo-Ciau, I. Sanchez-Molina and G.J. Mena-Rejon. 2006. Screening of antibacterial and antifungal activities of six marine macroalgae from coasts of Yucatan peninsula. Pharmaceutical Biology 44: 632–635.

Morán-Santibañez, K., L.E. Cruz-Suárez, D. Ricque-Marie, D. Robledo, Y. Freile-Pelegrín, M.A. Peña-Hernández, C. Rodríguez-Padilla and L.M. Trejo-Avila. 2016. Synergistic effects of sulfated polysaccharides from Mexican seaweeds against Measles virus. BioMed Research International vol 2016, Article ID 8502123: 11 pp.

Moreau, D., H. Thomas-Guyon, C. Jacquot, M. Jugé, G. Culioli, A. Ortalo-Magné, L. Piovetti and C. Roussakis. 2006. An extract from the brown alga *Bifurcaria bifurcata* induces irreversible arrest of cell proliferation in a non-small-cell bronchopulmonary carcinoma line. Journal of Applied Phycology 18(1): 87–93.

Moreau, J., D. Pesando and B. Caram 1984. Antifungal and antibacterial screening of Dictyotales from the French Mediterranean coast. Proceedings of International Seaweed Symposium 11: 521–524.

Moreira, A.R.S., J. Benedi, L. Gonzalez-Torres, R. Olivero-David, S. Bastida, M.I. Sánchez-Reus, M.J. González-Muñoz and F.J. Sánchez-Muniz. 2011. Effects of diet enriched with restructured meats, containing *Himanthalia elongata* on hypercholester-olaemic induction, CYP7A1 expression and antioxidant enzyme activity and expression in growing rats. Food Chemistry 129: 1623–1630.

Moreira, A.S., L. Gonzalez-Torres, R. Olivero-David, S. Bastida, J. Benedi and F.J. Sanchez-Muniz. 2010. Wakame and Nori in restructured meats included in cholesterol-enriched diets affect the antioxidant enzyme gene expressions and activities in Wistar rats. Plant Foods in Human Nutrition 65: 290–298.

Moreno, J. and J. Alvar. 2002. Canine leishmaniasis: epidemiological risk and the experimental model. Trends in Parasitology 18: 399–405.

Mori, J., M. Iwashima, H. Wakasugi, H. Saito, T. Matsunaga, M. Ogasawara, S. Takahashi, H. Suzuki and T. Hayashi. 2005b. New plastoquinones isolated from the brown alga, *Sargassum micracanthum*. Chemical and Pharmaceutical Bulletin 53(9): 1159–1163.

Mori, K., T. Ooi, M. Hiraoka, N. Oka, H. Hamada, M. Tamura and T. Kusumi. 2004. Fucoxanthin and its metabolites in edible brown algae cultivated in deep seawater. Marine Drugs 2(2): 63–72.

Mori, T., B.R. O'Keefe, R.C. Sowder II, S. Bringans, R. Gardella, S. Berg, P. Cochran, J.A. Turpin, W. Buckheit, Jr., J.B. McMahon and M.R. Boyd. 2005. Isolation and characterization of griffithsin, a novel HIV-inactivating protein, from the red alga *Griffithsia* sp. The Journal of Biological Chemistry 280: 9345–9353.

Morokutti-Kurz, M., M. Konig-Schuster, C. Koller, C. Graf, P. Graf, N. Kirchoff, B. Reutterer, J.M. Seifert, H. Unger, A. Grassauer, E. Prieschl-Grassauer and S. Nakowitsch. 2015. The intranasal application of Zanamivir and carrageenan is synergistically active against influenza A virus in the murine model. PLoS One 10(6): 1–16.

Moroney, N.C., M.N. O'Grady, J.V. O'Doherty and J.P. Kerry. 2013. The impact and mode of action of phenolic compounds extracted from brown seaweed on mixed anaerobic microbial cultures. Journal of Applied Microbiology 114: 964–973.

Moroney, N.C., M.N. O'Grady, R.C. Robertson, C. Stanton, J.V. O'Doherty and J.P. Kerry. 2015. Influence of level and duration of feeding polysaccharide (laminarin and fucoidan) extracts from brown seaweed (*Laminaria digitata*) on quality indices of fresh pork. Meat Science 99: 132–141.

Morris, C.J. 2003. Carrageenan-induced paw edema in the rat and mouse, pp. 115–121. *In*: Paul G. Winyard and Derek A. Willoughby (eds.). Inflammation Protocols. Humana, Totowa, NJ.

Morrissey, J., S. Kraan and M.D. Guiry. 2001. A guide to commercially important seaweeds on the Irish coast. Bord Iascaigh Mhara/Irish Fisheries Board, Dublin, pp. 5–40.

Morrow, K., V. Paul, M. Liles and N. Chadwick. 2011. Allelochemicals produced by Caribbean macroalgae and cyanobacteria have species-specific effects on reef coral microorganisms. Coral Reefs 30: 309–320.

Morse, S.A. and T.A. Meitzner. 2016. The science of microbiology. *In*: Weitz, M. and B. Kearns (eds.). Jawetz, Melnick & Adelberg's Medical Microbiology (27th ed.). The McGraw-Hill Education, LLC.

Mosaddegh, M., B.M. Gharanjik, F. Naghibi, S. Esmaeili, A. Pirani, B.E. Tehrani, B. Keramatian and A. Hassanpour. 2014. A survey of cytotoxic effects of some marine algae in the Chabahar coast of Oman Sea. Research Journal of Pharmacognosy (RJP) 1: 27–31.

Moshaverinia, A., C. Chen, K. Akiyama, S. Ansari, X. Xu, W.W. Chee, S.R. Schricker and S. Shi. 2012. Alginate hydrogel as a promising scaffold for dental-derived stem cells: An *in vitro* study. Journal of Materials Science: Materials in Medicine 23: 3041–3051.

Motshakeri, M., M. Ebrahimi, Y.M. Goh, H.H. Othman, M. Hair-Bejo and S. Mohamed. 2014. Effects of brown seaweed (*Sargassum polycystum*) extracts on kidney, liver, and pancreas of type 2 diabetic rat model. Evidence-Based Complementary and Alternative Medicine 2014: 11.

Mouradi-Givernaud, A., T. Givernaud, H. Morvan and J. Cosson. 1992. Agar from *Gelidium latifolium* (Rhodophyceae, Gelidiales)—biochemical composition and seasonal variations. Botanica Marina 35: 153–159.

Mourão, P.A.S. 2015. Perspective on the use of sulfated polysaccharides from marine organisms as a source of new antithrombotic drugs. Marine Drugs 13: 2770–2784.

Mouritsen, O.G. 2013. Seaweeds: Edible, Available, and Sustainable. University of Chicago Press, Chicago, USA, 304 pp.

Mtolera, M.S.P. and A.K. Semesi. 1996. Antimicrobial activity of extracts from six green algae from Tanzania. Current Trends in Marine Botanical Research in East African Region 2(5): 211–217.

Mumper, R.J., A.S. Hoffman, P.A. Puolakkainen, L.S. Bouchard and W.R. Gombotz. 1994. Calcium-alginate beads for the oral delivery of transforming growth factor-β_1 (TGF-β_1): stabilization of TGF-β_1 by the addition of polyacrylic acid within acid-treated beads. Journal of Controlled Release 30: 241–251.

Mungmai, L., S. Jiranusornkul, Y. Peerapornpisal, B. Sirithunyalug and P. Leelapornpisid. 2014. Extraction, characterization and biological activities of extracts from freshwater macroalga [*Rhizoclonium hieroglyphicum* (C. Agardh) K tzing] cultivated in Northern Thailand. Chiang Mai Journal of Science 41(1): 14–26.

Muñoz-Ochoa, M., J.I. Murillo-Alvarez, L.A. Zermeno-Cervantes, S. Martinez-Diaz and R. Rodriguez-Riosmena. 2010. Screening of extracts of algae from Baja California Sur, Mexico as reversers of the antibiotic resistance of some pathogenic bacteria. European Review for Medical and Pharmacological Sciences 14(9): 739–147.

Murad, H., A. Ghannam, M. Al-Ktaifani, A. Abbas and M. Hawat. 2015. Algal sulfated carrageenan inhibits proliferation of MDA-MB-231 cells via apoptosis regulatory genes. Molecular Medicine Reports 11(3): 2153–8258.

Murata, M., K. Ishihara and H. Saito. 1999. Hepatic fatty acid oxidation enzyme activities are stimulated in rats fed the brown seaweed, *Undaria pinnatifida* (Wakame). Journal of Nutrition 129(1): 146–151.

Murata, M. and J. Nakazoe. 2001. Production and use of marine algae in Japan. Japan Agricultural Research Quartery 35: 281–290.

Murata, Y., K. Mita, E. Miyamoto and M. Ueda. 1989. Antimicrobial activity of N, N-Dialkyl polymethylenediamines against some plaque bacteria. Antimicrobial Agents Chemotherapy 33(9): 1636–1638.

Murphy, C., S. Hotchkiss, J. Worthington and S.R. McKeown. 2014. The potential of seaweed as a source of drugs for use in cancer chemotherapy. Journal of Applied Phycology 26(5): 2211–2264.

Murugan, K., P. Mahesh Kumar, K. Kovendan, D. Amerasan, J. Subrmaniam and J.S. Hwang. 2012. Larvicidal, pupicidal, repellent and adulticidal activity of *Citrus sinensis* orange peel extract against *Anopheles stephensi, Aedes Aegypti* and *Culex quinquefasciatus* (Diptera: Culicidae). Parasitology Research 111(4): 1757–1769.

Murugan, K. and V.V. Iyer. 2014. Antioxidant and antiproliferative activities of extracts of selected red and brown seaweeds from the Mandapam coast of Tamil Nadu. Journal of Food Biochemistry 38(1): 92–101.

Muruganantham, A. 2001. Heparin from seaweeds: A search in *Grateloupia filicina* (Wulf) J. Ag. Ph.D. Thesis Annamalai University, Chidambaram, Tamil Nadu, India, 146 pp.

Muto, S., K. Niimura, M. Oohara, Y. Oguchi, K. Matsunaga, K. Hirose, J. Kakuchi, N. Sugita and T. Furusho. 1988. Polysaccharides from marine algae and antiviral drugs containing the same as active ingredients. EP295956. Eur. Patent.

Myung, C.S., H.C. Shin, H.Y. Bao, S.J. Yeo, B.H. Lee and J.S. Kang. 2005. Improvement of memory by dieckol and phlorofucofuroeckol in ethanol-treated mice: possible involvement of the inhibition of acetylcholinesterase. Archives of Pharmacal Research 28(6): 691–698.

Na, H.J., Y. Seo, E.H. Lee, H.M. Kim and S.H. Hong. 2004. Regulatory effect of Th-2 cytokine production in mast cells by 02PS15. Journal of Applied Pharmacology 12: 79–84.

Na, H.J., P.D. Moon, H.J. Lee, H.R. Kim, H.J. Chae, T. Shin, Y. Seo, S.H. Hong and H.M. Kim. 2005. Regulatory effect of atopic allergic reaction by *Carpopeltis affinis*. Journal of Ethnopharmacology 101: 43–48.

Na, H.J., P.D. Moon, S.G. Ko, H.J. Lee, H.A. Jung, S.H. Hong, Y. Seo, J.M. Oh, B.H. Lee, B.W. Choi and H.M. Kim. 2005b. *Sargassum hemiphyllum* inhibits atopic allergic reaction *via* the regulation of inflammatory mediators. Journal of Pharmacological Sciences 97: 219–226.

Nacife, V.P., M. Soeiro, T.C. Araújo-Jorge, H. Neto and M. Meirelles. 2000. Ultrastructural, immunocytochemical and flow cytometry study of mouse peritoneal cells stimulated with carrageenan. Cell Structure and Function 25: 337–350.

Nacife, V.P., M. Soeiro, R.N. Gomes, H. D'Avila, H. Neto and M. Meirelles. 2004. Morphological and biochemical characterization of macrophages activated by carrageenan and lipopolysaccharide *in vivo*. Cell Structure and Function 29: 27–34.

Nagai, K., N. Yotsukura, H. Ikegami, H. Kimura and K. Morimoto. 2008. Protein extraction for 2 DE from the lamina of *Ecklonia kurome* (Laminariales): Recalcitrant tissue containing high levels of viscous polysaccharides. Electrophoresis 3: 672–681.

Nagamine, T., K. Hayakawa, T. Kusakabe, H. Takada, K. Nakazato, E. Hisanaga and M. Iha. 2009. Inhibitory effect of fucoidan on Huh7 hepatoma cells through down regulation of CXCL12. Nutrition and Cancer 61: 340–347.

Nagappan, T. and C.S. Vairappan. 2014. Nutritional and bioactive properties of three edible species of green algae, genus *Caulerpa* (Caulerpaceae). Journal Applied Phycology 26: 1019–1027.

Nagashima, H., K. Hanada and I. Hashimoto. 1999. Correlation of skin phototype with facial wrinkle formation, Photodermatology. Photoimmunology & Photomedicine 15: 2–6.

Nagorskaia, V.P., A.V. Reunov, L.A. Lapshina, I.M. Ermak and A.O. Barabanova. 2008. Influence of kappa/beta-carrageenan from red alga *Tichocarpus crinitus* on development of local infection induced by Tobacco mosaic virus in xanthi-nc tobacco leaves. Izv Akad Nauk Ser Biol 3: 360–364.

Nair, R., R. Chabhadiya and S. Chanda. 2007. Marine algae: Screening for a potent antibacterial agent. Journal of Herbal Pharmacotherapy 7(1): 73–86.

Najam, R., S.P. Ahmed and I. Azhar. 2010. Pharmacological activities of *Hypnea musciformis*. African Journal of Biomedical Research 13: 69–74.

Nakajima, K., A. Yokoyama and Y. Nakajima. 2009. Anticancer effects of a tertiary sulfonium compound, dimethylsulfoniopropionate, in green sea algae on Ehrlich Ascites Carcinoma-Bearing Mice. Journal of Nutritional Science and Vitaminology 55(5): 434–438.

Nakamura, T., H. Suzuki, Y. Wada, T. Kodama and T. Doi. 2006. Fucoidan induces nitric oxide production *via* p38 mitogen-activated protein kinase and NF-κB-dependent signalling pathways through macrophage scavenger receptors. Biochemical and Biophysical Research Communications 343: 286–294.

Nakano, M., Y. Nakamur, K. Takikawa, M. Kouketsu and I. Arita. 1979. Sustained release of sulfamethizole from agar beads. Journal of Pharmacy and Pharmacology 31: 869–872.

Nakano, T., T. Noro and Y. Kamei. 1997. *In vitro* promoting activity of human interferon β production by extracts of marine algae from Japan. Cytotechnology 25: 239–241.

Nakano, T., H. Kidaka, J. Uchida, K. Nakajima and Y. Hata. 1998. Hypotensive effects of wakame. Journal of Japan Society of Nutrition and Food Sciences 20: 92.

Nakano, T. and Y. Kamei. 2005. *In vivo* antiviral activity of an interferon β production-promoting fraction from the marine brown alga, *Sargassum hemiphyllum* (Turner) C. Agardh. Botanica Marina 48: 84–86.

Nakashima, H., Y. Kido, N. Kobayashi, Y. Motoki, M. Neushul and N. Yamamoto. 1987a. Antiretroviral activity in a marine red alga; reverse transcriptase inhibition by an aqueous extract of *Schizymenia pacifica*. Journal of Cancer Research and Clinical Oncology 113: 413–16.

Nakashima, H., Y. Kido, N. Kobayashi, Y. Motoki, M. Neushul and N. Yamamoto. 1987b. Purification and characterization of an avian myeloblastosis and Human immunodeficiency virus reverse transcriptase inhibitor, sulfated polysaccharides extracted from sea algae. Antimicrobial Agents and Chemotherapy 31: 1524–1528.

Nakatsu, T., B.N. Ravi and D.J. Faulkner 1981. Antimicrobial constituents of Udotea flabellum. The Journal of Organic Chemistry 46: 2435–2438.

Nakayasu, S., R. Soegima, K. Yamaguchi and T. Oda. 2009. Biological activities of fucose-containing polysaccharide ascophyllan isolated from the brown alga *Ascophyllum nodosum*. Bioscience, Biotechnology, and Biochemistry 73: 961–964.

Nakazawa, S., H. Kuroda, F. Abe, T. Nishino, M. Otsuki and I. Umezaki. 1974. Antitumor effect of water-extracts from marine algae (I). Chemotherapy 22: 1435–1422.

Nakazawa, S., F. Abe, H. Kuroda, K. Kohno, T. Higashi and I. Umezaki. 1976. Antitumor effect of water-extracts from marine algae(II). *Sargassum horneri* (Turner) C. Agardh. Chemotherapy 24: 443–447.

Nakazawa, Y., T. Sashima, M. Hosokawa and K. Miyashita. 2009. Comparative evaluation of growth inhibitory effect of stereoisomers of Fucoxanthin in human cancer cell lines. Journal of Functional Foods 1(1): 88–97.

Nalamothu, N., A. Potluriand M.B. Muppalla. 2014. Review on marine alginates and its applications. Indo American Journal of Pharmaceutical Research 4(10): 4006–4015.

Nam, K.W. 2007. Validation of the generic name *Palisada* (Rhodomelaceae, Rhodophyta). Algae 22: 53–55.

Namangala, B., W. Noel, P. De Baetselier, L. Brys and A. Beschin. 2001. Relative contribution of interferon-gamma and interleukin-10 to resistance to murine African trypanosomosis. The Journal of Infectious Diseases 183: 1794–1800.

Namvar, F., S. Mohamed, S.G. Fard, J. Behravan, N.M. Mustapha, N.B.M. Alitheen and F. Othman. 2012. Polyphenol-rich seaweed (*Eucheuma cottonii*) extract suppresses breast tumour via hormone modulation and apoptosis induction. Food Chemistry 130(2): 376–382.

Namvar, F., R. Mohamad, J. Baharara, S. Zafar-Balanejad, F. Fargahi and H.S. Rahman. 2013. Antioxidant, antiproliferative, and antiangiogenesis effects of polyphenol-rich seaweed (*Sargassum muticum*). BioMed Research International vol. 2013, Article ID 604787: 9 pp.

Naqvi, B.S., A. Khan, D.A. Shaikh and M.R. Shaikh. 1992. Nematicidal properties of selected marine algae from Karachi coast. Journal of Islamic Academy of Sciences 5(3): 171–172.

Naqvi, S., W.A. Solimabi, S.Y. Kamat, L. Fernandes, C.V.G. Reddy, D.S. Bhakuni and B.N. Dhawan. 1980. Screening of some marine plants from the Indian coast for biological activity. Botanica Marina 24: 51–55.

Nara, T., Y. Kamei, A. Tsubouchi, T. Annoura, K. Hirota, K. Iizumi, Y. Dohmoto, T. Ono and T. Aoki. 2005. Inhibitory action of marine algae extracts on the *Trypanosoma cruzi* dihydroorotate dehydrogenase activity and on the protozoan growth in mammalian cells. Parasitology International 54: 59–64.

Narang, S., D. Gibson, A.D. Wasan, E.L. Ross, E. Michna, S.S. Nedeljkovic and R.N. Jamison. 2008. Efficacy of dronabinol as an adjuvant treatment for chronic pain patients on opioid therapy. The Clinical Journal of Pain 9: 254–264.

Narasimhan, M.K., S.K. Pavithra, V. Krishnan and M. Chandrasekaran. 2013. *In vitro* analysis of antioxidant, antimicrobial and antiproliferative activity of *Enteromorpha attenuata*, *Enteromorpha linza* and *Gracilaria corticata* extracts. Jundishapur Journal of Natural Pharmaceutical Products 8(4): 151–159.

Nascentes, C.C., M.A.Z. Arruda, A.R.A. Nogueira and J.A. Nobrega. 2004. Direct determination of copper and zinc in fruit juices and bovine Milk by thermospray fame furnace atomic absorption spectrometry. Talanta 64: 912–917.

Natarajan, S., K.P. Shanmugiahthevar and P.D. Kasi. 2009. Cholinesterase inhibitors from *Sargassum* and *Gracilaria gracilis*: Seaweeds inhabiting South Indian coastal areas (Hare Island, Gulf of Mannar). Natural Product Research 23(4): 355–369.

Nayaka, S.C., A.C.U. Shankar, S.R. Niranjana, H.S. Prakash and C.N. Mortensen. 2009. Anthracnose disease of chilli pepper. University of Mysore, Asian Seed Health Centre & Department of Studies in Applied Botany and Biotechnology Technical Bulletin: 1–14.

NCGC (National Clinical Guideline Centre). 2011. Hypertension The clinical management of primary hypertension in adults. Commissioned by the National Institute for Health and Clinical Excellence. Published by the National Clinical Guideline Centre at The Royal College of Physicians, 11 St Andrews Place, Regents Park, London, 861 pp.

Neelakandan, Y. and A. Venkatesan. 2016. Antinociceptive and anti-inflammatory effect of sulfated polysaccharide fractions from *Sargassum wightii* and *Halophila ovalis* in male Wistar rats. Indian Journal of Pharmacology 48(5): 562–570.

Nelson, T.E. and B.A. Lewis. 1974. Separation and characterization of the soluble and insoluble components of insoluble laminaran. Carbohydrate Research 33: 63–74.

Netalgae. 2017. Inter-regional network to promote sustainable development in the marine algal industry. Available online at: http://www.netalgae.eu/index-en.php.

Neto, A.I., I. Titley and P.M. Raposeiro. 2005. Rocky shore marine flora of the Azores. Secrataria Regional do Ambiente e do Mar, Azores, Portugal, 159 pp.

Nicholson, K.G., T. Mcnally, M. Silverman, P. Simons, J.D. Stockton and M.C. Zambon. 2006. Rates of hospitalization for influenza, respiratory syncytial virus and human metapneumovirus among infants and young children. Vaccine 24: 102–8.

Nicolai, J.J., P. Speelman, G.N. Tytgat and J. van der Stadt. 1981. Comparison of the combination of cholestyramine/alginates with placebo in the treatment of postgastrectomy biliary reflux gastritis. European Journal of Clinical Pharmacology 21(3): 189–194.

Nicoletti, E., F. Della Pieta, V. Calderone, P. Bandecchi, M. Pistello, I. Morelli and F. Cinelli. 1999. Antiviral properties of a crude extract from a green alga *Caulerpa taxifolia* (Vahl) C. Agardh. Phytotherapy Research 13(3): 245–247.

NIH (National Cancer Institute). 2016. Causes and prevention. Available online at: https://www.cancer.gov/about-cancer/causes-prevention.

Nisha, L.L.J.L. and T.V. Poonguzhali. 2013. Effect of two seaweeds *Chaetomorpha antennina* (Bory de Saint-Vincent) Kützing and *Sargassum wightii* Greville as larvicide against mosquito vector *Culex quinquefasciatus*. International Journal of Biology, Pharmacy and Allied Sciences (IJBPAS) 2(3): 705–711.

Nisha L.L.J.L., I. Vincent and T.V. Poonguzhali. 2013b. Larvicidal activity of two seaweeds, *Cheatomorpha antenina* (Bory de Saint-Vincent) Kutzing and *Sargassum wightii* Greville against mosquito vector, *Aedes aegypti*. Asian Journal of Biochemical and Pharmaceutical Research 2(3): 1–7.

Nishibori, N., M. Itoh, M. Kashiwagi, H. Arimochi and K. Morita. 2012. *In vitro* cytotoxic effect of ethanol extract prepared from sporophyll of *Undaria pinnatifida* on human colorectal cancer cells. Phytotherapy Research 26(2): 191–196.

Nishide, E., H. Anzai and H. Uchida. 1993. Effects of alginates on the ingestion and excretion of cholesterol in the rat. Journal of Applied Phycology 5: 207–211.

Nishide, E. and H. Uchida. 2003. Effects of *Ulva* powder on the ingestion and excretion of cholesterol in rats. pp. 165–168. *In*: A.R. Chapman, R.J. Anderson, V.J. Vreeland and I.R. Davison (eds.). Proceedings of the 17th International Seaweed Symposium, Oxford University Press, Oxford.

Nishiguchi, T., Z. Jiang, M. Ueno, S. Takeshita, K. Cho, S. Roh, K.H. Kang, K. Yrnaguchi, D. Kim and T. Oda. 2014. Reevaluation of bactericidal, cytotoxic, and macrophage-stimulating activities of commercially available *Fucus vesiculosus* fucoidan. Algae 29(3): 237–247.

Nishikawa, Y., M. Okabe, K. Yoshimoto, G. Kurono and F. Fukuoka. 1976. Chemical and biochemical studies on carbohydrate esters. II. Antitumor activity of saturated fatty acids and their ester derivatives against Ehrlich Ascites carcinoma. ChemicalL & Pharmaceutical Bulletin 24: 387–393.

Nishino, T., G. Yokoyama, K. Dobashi, M. Fujihara and T. Nagumo. 1989. Isolation, purification and characterization of fucose-containing sulfated polysaccharides from the brown seaweed *Ecklonia kurome* and their blood-anticoagulant activities. Carbohydrate Research 186: 119–129.

Nishino, T. and T. Nagumo. 1991. The sulfate-content dependence of the anticoagulant activity of a fucan sulfate from the brown seaweed *Ecklonia kurome*. Carbohydrate Research 214: 193–197.

Nishino, T. and T. Nagumo. 1992. Anticoagulant and antithrombin activities of oversulfated fucans. Carbohydrate Research 229: 355–362.

Nisizawa, K. 2006. Seaweeds Kaiso—Bountiful harvest from the seas. *In*: Critchley, A., M. Ohno and D. Largo (eds.). World Seaweed Resources—An Authoritative Reference System. ETI Information Services Ltd., pp. Hybrid Windows and Mac DVD-ROM; ISBN: 90-75000-75080-75004.

Nithya, P. and B. Dhanalakshmi. 2016. Antibacterial activity of methanol extracts from selected seaweed of south east coast of India. International Journal of Applied Research 2(9): 714–718.

Nitta, I., H. Watase and Y. Tomiie. 1958. Structure of kainic acid and its isomer, allokainic acid. Nature 181: 761–762.

Nizamuddin, M. 1963. Studies on the green alga, *Udotea indica*. Pacific Science XVII: 243–245.

Nkwenkeu, S.-F., G. Kennedy, S. Philippe and J. Zayed. 2002. Oral manganese intake estimated with dietary records and with direct chemical analysis. Science of the Total Environment 287: 147–153.

Noda H., H. Amano, K. Arashima, S. Hashimoto and K. Nisizawa. 1989. Antitumour activity of polysaccharides and lipids from marine algae. Nippon Suisan Gakkaishi 55: 1265–271.

Noda, H., H. Amano, K. Arashima and K. Nisizawa. 1990. Antitumor activity of marine algae. Hydrobiologia 204(1): 577–584.

Noda, H. 1993. Health benefits and nutritional properties of nori. Journal of Applied Phycology 5: 255–258.

Nogueira, C.C.R., I.C.N.P. Paixão, C.C. Cirne-Santos, P.R.S. Stephens, R.C. Villaça, H.S. Pereira and V.L. Teixeira. 2016. Anti-HIV-1 activity in human primary cells and Anti-HIV1 RT inhibitory activity of extracts from the red seaweed *Acanthophora spicifera*. Journal of Medicinal Plants Research 10(35): 621–625.

Nogueira, L.F., E.C. Morais, M.A. Brito, B.S. Santos, D.L. Vale, B.F. Lucena, F.G. Figueredo, G.M. Guedes, S.R. Tintino, C.E. Souza, R.B.S.S. Nogueira, E.F.F. Matias, M.F.B. Morais-Braga, E.V.L. Cunha, M.A. Lima and H.D.M. Coutinho. 2014. Evaluation of antibacterial, antifungal and modulatory activity of methanol and ethanol extracts of *Padina sanctae-crucis*. African Health Sciences 14: 372–376.

Nomura, M., H. Kamogawa, E. Susanto, C. Kawagoe, H. Yasui, N. Saga, M. Hosokawa and K. Miyashita. 2013. Seasonal variations of total lipids, fatty acid composition, and fucoxanthin contents of *Sargassum horneri* (Turner) and *Cystoseira hakodatensis* (Yendo) from the northern seashore of Japan. Journal of Applied Phycology 25(4): 1159–1169.

Nomura, T., M. Kikuchi, A. Kubodera and Y. Kawakami. 1997. Proton-donative antioxidant activity of fucoxanthin with 1, 1-diphenyl-2-picrylhydrazyl (DPPH). IUBMB Life 42: 361–370.

Nordgren, A. 2010. Responsible genetics: The moral responsibility of geneticists for the consequences of human... Springer, Dordrecht, 276 pp.

Noreen, A., M. Amer-Zareen, J. Zaki and M. Abid. 2002. Effect of seaweeds on population of *Verticillium chlamydosporium* and control of root-knot nematode on egg plant. Proceeding of the National Symposium on Nematology 103–109.

Norris, J.N. and W. Fenical. 1982. Chemical defense in tropical marine algae. pp. 417–431. *In*: Rutzler, K. and I.G. MacIntyre (eds.). The Atlantic Barrier Reef Ecosystem at Carrie Bow Cay, Belize I. Structure and Communities. Smithsonian Contributions to the Marine Sciences 12. Smithsonian Institution Press, Washington, DC.

Norris, J.N. and W.H. Fenical. 1985. Natural products chemistry: uses in ecology and systematics. pp. 121–145. *In*: M.M. Littler and D.S. Littler (eds.). Handbook of Phycological Methods. Ecological Field Methods: Macroalgae. Cambridge University Press, New York.

Norziah, M.H. and C.Y. Ching. 2000. Nutritional composition of edible seaweed *Gracilaria changgi*. Food Chemistry 68(1): 69–76.

Novaczek, I. 2001. A guide to the common and edible and medicinal sea plants of the Pacific Islands. University of the South Pacific, 40 pp.

Novoa, A.V., M. Motidome, J. Mancini Filho, A. Fallarero Linares, M. Tanae, L. Torres and A. Lapa. 2001. Actividad antioxidante y ácidos fenólicos del alga marina *Bryothamnion triquetrum* (SG Gmelim) Howe; Antioxidant activity related to phenolic acids in the aqueous extract of the marine seaweed *Bryothamnin triquetrum* (SG Gmelim) Howe. Revista Brasileira de Ciências Farmacêuticas/Brazilian Journal of Pharmaceutical Sciences 37: 373–382.

Nowicki, M., M. Nowakowska, A. Niezgoda and E. Kozik. 2012. *Alternaria* black spot of crucifers: Symptoms, importance of disease, and perspectives of resistance breeding. Vegetable Crops Research Bulletin 76: 5–19.

Nudera, W.J., M.I. Fayad, B.R. Johnson, M. Zhu, C.S. Wenckus, E.A. Begole and C.D. Wu. 2007. Antimicrobial effect of triclosan and triclosan with Gantrez on five common endodontic pathogens. Journal of Endodontics 33: 1239–1242.

Numata, A., S. Kanbara, C. Takahashi, R. Fujiki, M. Yoneda, E. Fujita and Y. Nabeshima. 1991. Cytotoxic activity of marine algae and a cytotoxic principle of the brown alga *Sargassum tortile*. Chemical and Pharmaceutical Bulletin 39(8): 2129–2131.

Numata, A., C. Takahashi, Y. Ito, T. Takeda, K. Kawai, Y. Osami, E. Matsumura, M. Imachi, T. Ito and T. Hasegawa. 1993. Communesins, cytotoxic metabolites of a fungus isolated from a marine alga. Tetrahedron Letters 34: 2355–2358.

Nurul, Z.A., I. Darah, S.F. Shaida and S.A. Nor. 2010. Screening for antimicrobial activity of various extracts of *Acanthophora spicera* (Rhodomelaceae, Ceramiales) from Malaysian waters. Research Journal of Biological Sciences 5(5): 368–375.

Nussinovitich, A. 1997. Hydrocolloid applications: Gum technology in the food and other Industries. pp. 1–18. Blackie Academic & Professional, London.

Nwosu, F., J. Morris, V.A. Lund, D. Stewart, H.A. Ross and G.J. McDougall. 2011. Anti-proliferative and potential anti-diabetic effects of phenolic-rich extracts from marine edible algae. Food Chemistry 126: 1006–1012.

N'Yeurt, A.D.R. 1999. A preliminary illustrated field guide to the common marine algae of the Cook Islands (Rarotonga and Aitutaki). World Wide Fund for Nature, 74 pp.

O'Dell, B.L. 2000. Role of zinc in plasma membrane function. Journal of Nutrition 130: 1432S–1436S.

O'neill, A.N. 1955a. Derivatives of 4-O-Beta-D-galactopyranosyl-3,6-anhydro-D-galactose from κ-carrageenin. Journal of the American Chemical Society 77(23): 6324–6326.

O'neill, A.N. 1955b. 3,6-anhydro-D-galactose as a constituent of κ-carrageenin. Journal of the American Chemical Society 77(10): 2837–2839.

Odunsi, S.T., M.I. Vázquez-Roque, M. Camilleri, A. Papathanasopoulos, M.M. lark, L. Wodrich, M. Lempke, S. McKinzie, M. Ryks, D. Burton and A.R. Zinsmeister. 2010. Effect of alginate on satiation, appetite, gastric function, and selected gut satiety hormones in overweight and obesity. Obesity (Silver Spring) 18: 1579–1584.

Ogata, M., T. Matsui, T. Kita and A. Shigematsu. 1999. Carrageenan primes leukocytes to enhance lipopolysaccharide-induced tumor necrosis factor alpha production. Infection and Immunity 67: 3284–3289.

Oh, K.-B., J.H. Lee, S.-C. Chung, J. Shin, H.J. Shin, H.-K. Kim and H.-S. Lee. 2008. Antimicrobial activities of the bromophenols from the red alga *Odonthalia corymbifera* and some synthetic derivatives. Bioorganic & Medicinal Chemistry Letters 18(1): 104–108.

Oh, S.-M., C.G. Park, J.H. Kang, E.-J. Kim, H.Y. Chee, B.H. Lee and K.B. Lee. 2011. Dieckol inhibits 12-*O*-tetradecanoylphorbol-13-acetateinduced SK-Hep1 human hepatoma cell motility through suppression of matrix metalloproteinase-9 activity. Journal of the Korean Society for Applied Biological Chemistry 54(3): 376–381.

Ohigashi, H., Y. Sakai, K. Yamaguchi, I. Umezaki and K. Koshimizu. 1992. Possible anti-tumor promoting properties of marine algae and *in vivo* activity of wakame seaweed extract. Bioscience, Biotechnology, and Biochemistry 56(6): 994–995.

Ohta, K., Y. Mizushina, N. Hirata, M. Takemura, F. Sugawara, A. Matsukage, S. Yoshida and K. Sakaguchi. 1998. Sulfoquinovosyldiacylglycerol, KM043, a new potent inhibitor of eukaryotic DNA polymerases and HIV-reverse transcriptase type 1 from a marine red alga, *Gigartina tenella*. Chemical and Pharmaceutical Bulletin 46(4): 684–686.

Ohta, Y., J.-B. Lee, K. Hayashi and T. Hayashi. 2009. Isolation of sulfated galactan from *Codium fragile* and its antiviral effect. Biological and Pharmaceutical Bulletin 32: 892–898.

Okada, Y., N. Miyauch, K. Suzuki, T. Kobayashi, C. Tsutsui, K. Mayuzumi and T. Okuyama. 1994. Search for naturally occurring substances for prevention against the complications of diabetes; inhibitory effect on aldose reductase and platelet aggregation. Nature Medicine 48: 324–329.

Okada, Y., A. Ishimaru, R. Suzukia and T. Okuyama. 2004. A new phloroglucinol derivative from the brown alga *Eisenia bicyclis*: Potential for the effective treatment of diabetic complications. Journal of Natural Products 67: 103–105.

Okai, Y. and K.H. Okai. 1997. Potent anti-inflammatory activity of pheophytin a derived from edible green alga, *Enteromorpha prolifera* (Sujiaonori). International Journal of Immunopharmacology 19: 355–358.

Olaizola, M. 2003. Commercial development of microalgal biotechnology: from the test tube to the marketplace. Biomolecular Engineering 20: 459–466.

Olessen, P.E., A. Maretzki and L.A. Almodovar. 1963. An investigation of antimicrobial substances from marine algae. Botanica Marina 6: 226–232.

Oliveira, S.T.L., G. Veneroni-Gouveia, A.C. Santos, S.M.N. Sousa, M.L. Veiga, C.C. Krewer and M.M. Costa. 2014. *Ascophyllum nodosum* in the diet of tilapia (*Oreochromis niloticus*) and its effect after inoculation of *Aeromonas hydrophila*. Pesquisa Veterinária Brasileira 34: 403–408.

Olmix. 2017. Specialist of natural algae-based solutions. Available online at: https://www.olmix.com/.

Omar, H.H., H.M. Shiekh, N.M. Gumgumjee, M.M. El-Kazan and A.M. El-Gendy. 2012. Antibacterial activity of extracts of marine algae from the Red Sea of Jeddah, Saudi Arabia. African Journal of Biotechnology 11(71): 13576–13585.

Ondetti, M.A. and D.W. Cushman. 1982. Enzymes of the renin-angiotensin system and their inhibitors. Annual Review of Biochemistry 51: 283–308.

Onianwa, P.C., I.G. Adethola, C.M.A. Iwebue, M.F. Ojo and O.O. Tella. 1999. Trace heavy metals composition of some Nigerian beverages and fruit drinks. Food Chemistry 66: 275–279.

Onianwa, P.C., A.O. Adeemo, O.E. Idowu and E.E. Ogabiela. 2000. Copper and zinc contents of Nigerian foods and estimates of the adult dietary intakes. Food Chemistry 72: 89–95.

Ono, S.J. 2000. Molecular genetics of allergic diseases. Annual Review of Immunology 18: 347–366.

Onsoyen, E. 1997. Alginates. pp. 22–44. *In*: A. Imeson (ed.). Thickening and Gelling Agents for Food. Blackie Academic and Professional, London, UK.

Oomizu, S., Y. Yanase, H. Suzuki, Y. Kameyoshi and M. Hide. 2006. Fucoidan prevents Cε germline transcription and NF-κB p52 translocation for IgE production in B cells. Biochemical and Biophysical Research Communications 350: 501–507.

Oranday, M.A., M.J. Verde, S.J. Martínez-Lozano and N.H. Waksman. 2004. Active fractions from four species of marine algae. International Journal of Experimental Botany 73: 165–170.

Ördög, V., W. Stirk, R. Lenobel, M. Bancírová, M. Strnad, J. van Staden, J. Szigeti and L. Németh. 2004. Screening microalgae for some potentially useful agricultural and pharmaceutical secondary metabolites. Journal of Applied Phycology 16: 309–314.

Ordoñez-Librado, J.L., V. Anaya-Martinez, A.L. Gutierrez-Valdez, E. Montiel-Flores, D.R. Corona, D. Martinez-Fong and M.R. Avila-Costa. 2010. L-DOPA treatment reverses the motor alterations induced by manganese exposure as a Parkinson disease experimental model. Neuroscience Letters 471: 79–82.

Orhan, I., B. Sener, T. Atici, R. Brun, R. Perozzo and D. Tasdemir. 2006. Turkish freshwater and marine macrophyte extracts show *in vitro* antiprotozoal activity and inhibit FabI, a key enzyme of *Plasmodium falciparum* fatty acid biosynthesis. Phytomedicine 13: 388–393.

Oriflame. 2017. Oriflame Cosmetics SA. Available online at: https://oriflame.com/.

Ortalo-Magné, A., G. Culioli, R. Valls, B. Pucci and L. Piovetti. 2005. Polar acyclic diterpenoids from *Bifurcaria bifurcata* (Fucales, Phaeophyta). Phytochemistry 66: 2316–2323.

Ortiz, J., E. Uquiche, P. Robert, N. Romero, V. Quitral and C. Llantén. 2009. Functional and nutritional value of the Chilean seaweeds *Codium fragile, Gracilaria chilensis* and *Macrocystis pyrifera*. European Journal of Lipid Science and Technology 111: 320–327.

Osman, M.E.H., A.M. Abushady and M.E. Elshobary. 2010. *In vitro* screening of antimicrobial activity of extracts of some macroalgae collected from abu-Qir bay Alexandria. African Journal of Biotechnology 9(42): 7203–7208.

Osman, M.E.H., A.M. Aboshadya and M.E. Elshobary. 2013. Production and characterization of antimicrobial active substance from some macroalgae collected from AbuQir bay (Alexandria) Egypt. African Journal of Biotechnology 12: 6847–6858.

Ostrosky-Zeichner, L., J.H. Rex, P.G. Pappas, R.J. Hamill, R.A. Larsen, H.W. Horowitz, W.J. Powderly, N. Hyslop, C.A. Kauffman, J. Cleary, J.E. Mangino and J. Lee. 2003. Antifungal susceptibility survey of 2,000 bloodstream *Candida* isolates in the United States. Antimicrobial Agents and Chemotherapy 47: 3149–3154.

O'Sullivan, A.M., Y.C. O'Callaghan, M.N. O'Grady, B. Queguineur, D. Hanniffy, D.J. Troy, J.P. Kerry and N.M. O'Brien. 2011. *In vitro* and cellular antioxidant activities of seaweed extracts prepared from five brown seaweeds harvested in spring from the west coast of Ireland. Food Chemistry 126: 1064–1070.

Oumaskour, K., N. Boujaber, S. Etahiri and O. Assobhei. 2012. Screening of antibacterial and antifungal activities in green and brown algae from the coast of Sidi Bouzid (El Jadida, Morocco). African Journal of Biotechnology 11(104): 16831–16837.

Oumaskour, K., N. Boujaber, S. Etahiri and O. Assobhei. 2013. Anti-inflammatory and antimicrobial activities of twenty-tree marine red algae from the coast of Sidi Bouzid (El Jadida-Morocco). International Journal of Pharmacy and Pharmaceutical Sciences 5(3): 145–149.

Pabayo, R., J.C. Spence, N. Cutumisu, L. Casey and K. Story. 2012. Sociodemographic, behavioural and environmental correlates of sweetened beverage consumption among pre-school children. Public Health Nutrition 15: 1338–1346.

Pacheco, F.C., Janny A. Villa-Pulgarin, F. Mollinedo, M.N. Martín, J.J. Fernández and A.H. Daranas. 2011. New polyether triterpenoids from *Laurencia viridis* and their biological evaluation. Marine Drugs 9(11): 2220–2235.

Padmakumar, K. and K. Ayyakkannu. 1997. Seasonal variation of antibacterial and antifungal activities of the extracts of marine algae from Southern coasts of India. Botanica Marina 40: 507–515.

Pádua, M., P. Fontoura and A. Mathias. 2004. Chemical Composition of *Ulvaria oxysperma* (Kützing) Bliding, *Ulva lactuca* (Linnaeus) and *Ulva fascita* (Delile). Brazilian Archives of Biology and Technology 47(1): 49–55.

Paiva, A.A.O., A.J.G. Castro, M.S. Nascimento, L.S.E.P. Will, N.E. Santos, R.M. Araújo, C.A.C. Xavier, F.A. Rocha E.L. Leite. 2011. Antioxidant and antiinflammatory effect of polysaccharides from *Lobophora variegata* on zymosan-induced arthritis in rats. International Immunopharmacology 11(9): 1241–1250.

Paiva, L., E. Lima, R.F. Patarra, A.I. Neto and J. Baptista. 2014. Edible Azorean macroalgae as source of rich nutrients with Impact on human health. Food Chemistry 164: 128–135.

Palermo, J.A., P.B. Flower and A.M. Seldes. 1992. Chondriamides A and B, new indolic metabolites from the red alga *Chondria* sp. Tetrahedron Letters 33(22): 3097–3100.

Pallansch, M.A. and R.P. Roos. 2001. Enteroviruses: polioviruses, coxsackieviruses, echoviruses and newer enteroviruses. pp. 723–775. *In*: Fields, B.N., D. Knipe and P. Howley (ed.). Virology, 4th ed. Lippincott Williams & Wilkins, Philadelphia.

Pancharoen, C., W. Kulwichi, T. Tantawichien, U. Thisyakorn and C. Thisyakorn. 2002. Dengue infection: a global concern. Journal of the Medical Association of Thailand 85(Suppl 1): S25–S33.

Pande, B.N. and A.B. Gupta. 1977. Antibiotic properties in *Chlorococcum humicolum* (Naeg) Raben H. (Chlorophyceae). Phycologia 16: 439–441.

Pandian, P., S. Selvamuthukumar, R. Manavalan and V. Parthasarathy. 2011. Screening of antibacterial and antifungal activities of red marine algae *Acanthaphora spicifera* (Rhodophyceae). Journal of Biomedical Sciences and Research 3(3): 444–448.

Pandithurai, M., S. Murugesan, S. Bhuvaneswari and S. Thennarasan. 2015. *In vitro* α-amylase and α-glucosidase inhibition activity of methanolic extract of marine brown alga *Spatoglossum asperum*. International Journal of Advances in Pharmaceutics 4(5): 83–87.

Pandithurai, M., M. Subbiah, S. Vajiravelu and N.T. Selvan. 2015b. Antifungal activity of various solvent extracts of marine brown alga *Spatoglossum asperum*. International Journal of Pharmaceutical Chemistry 05(08): 277–280.

Pandithurai, M., S. Murugesan and V. Sivamurugan. 2015c. Antibacterial activity of various solvent extracts of marine brown alga Spatoglossum asperum. International Journal of Pharmacological Research 5(6): 133–137.

Pandurangan, A., M. Periasamy, S. Sekaran, S.K. Jebamalai and B. Sarangam. 2010. Pharmacognostical and antifungal activity of selected seaweeds from Gulf of Mannar region. Recent Research in Science and Technology 2(1): 115–119.

Pangestuti, R. and S.K. Kim. 2010. Neuroprotective properties of chitosan and its derivatives. Marine Drugs 8: 2117–2128.

Panlasigui, L.N., C.L.P. Tiangson, A.M.M. Quijano, H.T. Arugay and B.D. Dumelod. 1997. Blood glucose response of normal subjects to a carrageenan-incorporated Filipino food. J. Nutr. Diet. Assoc. Phil. 11: 87–92.

Panlasigui, L.N., B.D. Dumelod, L. Alonde, J.B. Torres, A.L. Basilio, M.T. Concepcion, L. Mercado and C. Olermo. 1998. Metabolic effects of local seaweed extracts on I. Blood glucose levels. Terminal Report.

Panlasigui, L.N., O.Q. Baello, J.M. Dimatangal and B.D. Dumelod. 2003. Blood cholesterol and lipid lowering effects of carrageenan on human volunteers. Asia Pacific Journal of Clinical Nutrition 12: 209–14.

Pappas, P.G., J.H. Rex, J.D. Sobel, S.G. Filler, W.E. Dismukes, T.J. Walsh and J.E. Edwards. 2004. Guidelines for treatment of candidiasis. Clinical Infectious Diseases 38(2): 161–189.

Pappas, P.G., C.A. Kauffman, D. Andes, D.K. Benjamin, Jr., T.F. Calandra, J.E. Edwards, Jr., S.G. Filler, J.F. Fisher, B.J. Kullberg, L. Ostrosky-Zeichner, A.C. Reboli, J.H. Rex, T.J. Walsh and J.D. Sobel. 2009. Clinical practice guidelines for the management of candidiasis: 2009 update by the Infectious Diseases Society of America. Clinical Infectious Diseases 48(5): 503–535.

Parati, G., J.E. Ochoa and G. Bilo. 2012. Blood pressure variability, cardiovascular risk, and risk for renal disease progression. Current Hypertension Reports 14: 421–131.

Pardo, B., L. Paz-Ares, J. Tabernero, E. Ciruelos, M. Garcia, R. Salazar, A. Lopez, M. Blanco, A. Nieto, J. Jimeno, M.A. Izquierdo and J.M. Trigo. 2008. Phase I clinical and pharmacokinetic study of Kahalalide F administered weekly as a 1-hour infusion to patients with advanced solid tumors. Clinical Cancer Research 14(4): 1116–1123.

Parekh, K.S., H. Parekh and P.S. Rao. 1984. Antibacterial activity of Indian seaweeds. Phykos 23: 216–221.

Parekh, K.S. 1985. Effect of Antibacterial substances from seaweeds on the growth of bacterial substances from seaweeds on the growth of bacteria. Proceedings of the All India Symposium in Marine Plants, pp. 175–178.

Parish, C.R. and J.M. Snowden. 1985. Lymphocytes express a diverse array of specific receptors for sulfated polysaccharides. Cellular Immunology 91(1): 201–214.

Parish, C.R., D.R. Coombe, K.B. Jakobsen, F.A. Bennett and P.A. Underwood. 1987. Evidence that sulphated polysaccharides inhibit tumour metastasis by blocking tumour-cell-derived heparanases. International Journal of Cancer 40(4): 511–518.

Park, E.Y., E.H. Kim, M.H. Kim, Y.W. Seo, J.I. Lee and H.S. Jun. 2012. Polyphenol-rich fraction of brown alga *Ecklonia cava* collected front Gijang, Korea, reduces obesity and glucose levels in high-fat diet-induced obese mice. Evidence-Based Complementary and Alternative Medicine Article ID 418912. 11 pp.

Park, H., H. Hwang, G.-Y. Kim, H.-J. Cha, W.-J. Kim, N. Kim, Y. Yoo and Y. Choi. 2013. Induction of apoptosis by fucoidan in human leukemia U937 cells through activation of P38 MAPK and modulation of Bcl-2 family. Marine Drugs 11(7): 2347–2364.

Park, H.J., M. Kurokawa, K. Shiraki, N. Nakamura, J.S. Choi and M. Hattori. 2005. Antiviral activity of the marine alga *Symphyocladia latiuscula* against Herpes simplex virus (HSV-1) *in vitro* and its therapeutic efficacy against HSV-1 infection in mice. Biological and Pharmaceutical Bulletin 28: 2258–2262.

Park, H.Y., H.H. Min, P. Cheol, J. Cheng-Yun, K. Gi-Young, C. Il-Whan, D.K. Nam, N. Taek-Jeong, K.K. Taeg and H.C. Yung. 2011. Anti-inflammatory effects of fucoidan through inhibition of NF-κB, MAPK and Akt activation in lipopolysaccharide-induced BV2 microglia cells. Food and Chemical Toxicology 49: 1745–1752.

Park, M.H. and J.S. Han. 2012. Hypoglycemic effect of *Padina arborescens* extract in Streptozotocin-induced diabetic mice. Preventive Nutrition and Food Science 17(4): 239–244.

Park, N.H., J.S. Choi, S.Y. Hwang, Y.C. Kim, Y.K. Hong, K.K. Cho and I.S. Choi. 2013b. Antimicrobial activities of stearidonic and gamma-linolenic acids from the green seaweed *Enteromorpha linza* against several oral pathogenic bacteria. Botanical Studies 54: 39–47.

Park, S.H., J.H. Song, T. Kim, W.S. Shin, G.M. Park, S. Lee, Y.J. Kim, P. Choi, H. Kim, H.S. Kim, D.H. Kwon, H.J. Choi and J. Ham. 2012b. Anti-human rhinoviral activity of polybromocatechol compounds isolated from the Rhodophyta, *Neorhodomela aculeata*. Marine Drugs 10(10): 2222–2233.

Park, Y.I. 2006. 4.4. Skin whitening effect. New Perspectives on Aloe: 127.

Parker, K.C. 2017. *Pythium aphanidermatum*. PP728 Soilborne plant pathogens class project. Available online at: https://projects.ncsu.edu/cals/course/pp728/Pythium/Pythium_aphanidermatum.html.

Parrizas, M., A.R. Saltiel and D. LeRoith. 1997. Insulin like growth factor I inhibits apoptosis using the PI3-kinase and MAPK pathway. Journal of Biological Chemistry 272: 154–161.

Pasquet, V., J.-R. Cherouvrier, F. Farhat, V. Thiery, J.-M. Piot, J.-B Berard, R. Kaas, B. Serive, T. Patrice, J.-P. Cadoret and L. Picot. 2011. Study on the microalgal pigments extraction process: Performance of microwave assisted extraction. Process Biochemistry 46(1): 59–67.

Pasquier, B., M. Armand, C. Castelian, F. Guillon, P. Borel, H. Lafont and D. Lairon. 1996. Emulsification and lipolysis of triacylglycerols are altered by viscous soluble dietary fibres in acidic gastric medium *in vitro*. Biochemical Journal 314: 269–275.

Patarra, R.F., L. Paiva, A.I. Neto, E. Lima and J. Baptista. 2011. Nutritional value of selected macroalgae. Journal of Applied Phycology 23: 205–208.

Patel, S. 2012. Therapeutic Importance of Sulfated Polysaccharides from Seaweed: Updating the Recent Findings. 3 Biotech 2: 171–185.

Paterson, D.L. 2006. Resistance in gram-negative bacteria: Enterobacteriaceae. American Journal of Infection Control 34: S64–73.

Patra, J.K., S.K. Rath, K. Jena, V.K. Rathod and H. Thatoi. 2008. Evaluation of antioxidant and antimicrobial activity of seaweed (*Sargassum* sp.) extract: A study on inhibition of Glutathione-S-transferase activity. Turkish Journal of Biology 32: 119–125.

Patra, J.K., A.P. Patra, N.K. Mahapatra, H.N. Thatoi, S.I. Das, R.K. Sahu and G.C. Swain. 2009. Antimicrobial activity of organic solvent extracts of three marine macroalgae from Chilika Lake, Orissa, India. Malaysian Journal of Microbiology 5(2): 128–131.

Patra, J.K., G. Das and K.H. Baek. 2015. Antibacterial mechanism of the action of *Enteromorpha linza* L. essential oil against *Escherichia coli* and *Salmonella typhimurium*. Botanical Studies 56: 13–21.

Patra, J.K., G. Das and K.H. Baek. 2015. Chemical composition and antioxidant and antibacterial activities of an essential oil extracted from an edible seaweed, *Laminaria japonica* L. Molecules 20: 12093–12113.

Patra, J.K. and K.-H. Baek. 2016. Antibacterial activity and action mechanism of the essential oil from *Enteromorpha linza* L. against foodborne pathogenic bacteria. Molecules 21(388): 1–11.

Pattama, R.A. and A. Chirapart. 2006. Nutritional evaluation of tropical green seaweeds *Caulerpa lentillifera* and *Ulva reticulata*. Kasetsart Journal—Natural Science 40: 75–83.

Paul, N.A., R. de Nys and P.D. Steinberg. 2005. Chemical defence against bacteria in the red alga *Asparagopsis armata*; linking structure with function. Phycologia 44(4): 78–78.

Paul, N.A., R. de Nys and P.D. Steinberg. 2006. Chemical defence against bacteria in the red alga *Asparagopsis armata*: Linking structure with function. Marine Ecology Progress Series 306: 87–101.

Paul, S. and R. Kundu. 2013. Antiproliferative activity of methanolic extracts from two green algae, *Enteromorpha intestinalis* and *Rizoclonium riparium* on HeLa cells. DARU Journal of Pharmaceutical Sciences 21(1): 1–12.

Paul, V.J., M.M. Littler, D.S. Littler and W. Fenical. 1987. Evidence for chemical defence in tropical green alga, *Caulerpa ashmeadii* (caulerpeae: chlorophyll). Isolation of new bioactive sesquiterpenoids. Journal of Chemical Ecology 13(5): 1171–1185.

Paulert, R.J., M.J.S. Stadnik and G.C. Pizzolatti. 2007. Antimicrobial properties of extracts from the green seaweed *Ulva fasciata* Delile against pathogenic bacteria and fungi. Algological Studies 123: 123–130.

Pavliga, S.N., G.G. Kompanets and V.Y. Tsygankov. 2016. The experimental research (*in vitro*) of carrageenans and fucoidans to decrease activity of Hantavirus. Food and Environmental Virology 8(2): 120–124.

Paxman, J.R., J.C. Richardson, P.W. Dettmar and B.M. Corfe. 2008. Alginate reduces the increased uptake of cholesterol and glucose in overweight male subjects: A pilot study. Nutrition Research 28: 501–505.

Payghami, N., S. Jamili, A. Rustaiyan, S. Saeidnia, M. Nikan and A.R. Gohari. 2015. Alpha-amylase inhibitory activity and sterol composition of the marine algae, *Sargassum glaucescens*. Pharmacognosy Research 7(4): 314–321.

Pearce-Pratt, R. and D.M. Phillips. 1996. Sulfated polysaccharides inhibit lymphocyte-to-epithelial transmission of human immunodeficiency virus-1. Biology of Reproduction 54: 173–182.

Peerapornpisal, Y., D. Amornlerdpison, U. Jamjai, T. Taesotikul, Y. Pongpaibul, M. Nualchareon and D. Kanja. 2010. Antioxidant and anti-inflammatory activities of brown marine alga, *Padina minor* Yamada. Chiang Mai Journal of Science 37(3): 507–516.

Peña-Rodríguez, A., T.P. Mawhinney, D. Ricque-Marie and L.E. Cruz-Suárez. 2011. Chemical Composition of Cultivated Seaweed *Ulva clathrata* (Roth) C. Agardh. Food Chemistry 129: 491–498.

Peng, Z., M. Liu, Z. Fang and Q. Zhang. 2012. *In vitro* antioxidant effects and cytotoxicity of polysaccharides extracted from *Laminaria japonica*. International Journal of Biological Macromolecules 50: 1254–1259.

Peng, Y., E. Xie, K. Zheng, M. Fredimoses, X. Yang, X. Zhou, Y. Wang, B. Yang, X. Lin and J. Liu. 2013. Nutritional and chemical composition and antiviral activity of cultivated seaweed *Sargassum naozhouense* Tseng et Lu. Marine Drugs 11: 20–32.

Peng, Y., J. Hu, B. Yang, X.P. Lin, X.F. Zhou, X.W. Yang and Y.H. Liu. 2015. *In*: Brijesh K. Tiwari and Declan Troy (eds.). Sustainability: Food and Non-Food Applications. Elsevier Science, Academic Press, London, 472 pp.

Pengzhan, Y., P.Z. Quanbin, L. Ning, X. Zuhong, W. Yanmei and L. Zhi'en. 2003. Polysaccharides from *Ulva pertusa* (Chlorophyta) and preliminary studies on their antihyperlipidemia activity. Journal of Applied Phycology 15: 21–27.

Pennington, J.E. 1994. Respiratory infections: Diagnosis and management. Thorax 49 (12): 1279–1279.

Percival, E.J.V. and R.H. McDowell. 1967. Chemistry and enzymology of marine algal polysaccharides. Academic Press, London, New York, 219 pp.

Pereira, D.M., J. Cheel, C. Areche, A. San-Martin, J. Rovirosa, L.R. Silva, P. Valentao and Paula B. Andrade. 2011. Anti-proliferative activity of meroditerpenoids isolated from the brown alga *Stypopodium flabelliforme* against several cancer cell lines. Marine Drugs 9(5): 852–862.

Pereira, H.S., L.R. Leao-Ferreira, N. Moussatche, V.L. Teixeira, D.N. Cavalcanti, L.J. Costa, R. Diaz and I.C. Frugulhetti. 2004. Antiviral activity of diterpenes isolated from the Brazilian marine alga *Dictyota menstrualis* against human immunodeficiency virus type 1 (HIV-1). Antiviral Research 64(1): 69–76.

Pereira, J.G., J.X. Mesquita, K.S. Aragão, Á.X. Franco, M.H.L.P. Souza, T.V. Brito, J.M. Dias, R.O. Silva and J.-V.R. Medeiros. 2014. Polysaccharides isolated from *Digenea simplex* Inhibit Inflammatory and Nociceptive Responses. Carbohydrate Polymers 108: 17–25.

Pereira, L., A. Sousa, H. Coelho, A.M. Amado and P.J.A. Ribeiro-Claro. 2003. Use of FTIR, FT-Raman and ^{13}C-NMR spectroscopy for identification of some seaweed phycocolloids. Biomolecular Engineering 20: 223–228.

Pereira, L. 2004. Estudos em macroalgas carragenófitas (Gigartinales, Rhodophyceae) da costa portuguesa—aspectos ecológicos, bioquímicos e citológicos. PhD Thesis, Universidade de Coimbra, Coimbra, Portugal, 293 pp.

Pereira, L. 2008. As algas marinhas e respectivas utilidades. Monografias. Available online at: http://br.monografias.com/trabalhos913/algas-marinhas-utilidades/algas-marinhas-utilidades.pdf.

Pereira, L. 2009. Guia ilustrado das macroalgas—conhecer e reconhecer algumas espécies da flora Portuguesa (Illustrated guide of macroalgae—to know and recognize some species of portuguese flora). Imprensa da Universidade de Coimbra, Coimbra, Portugal, 90 pp.

Pereira, L., A.M. Amado, A.T. Critchley, F. van de Velde and P.J.A. Ribeiro-Claro. 2009. Identification of selected seaweed polysaccharides (Phycocolloids) by vibrational spectroscopy (FTIR-ATR and FT-Raman). Food Hydrocolloids 23: 1903–1909.

Pereira, L., A.T. Critchley, A.M. Amado and P.J.A. Ribeiro-Claro. 2009b. A comparative analysis of phycocolloids produced by underutilized *versus* industrially utilized carrageenophytes (Gigartinales, Rhodophyta). Journal of Applied Phycology 21(5): 599–605. Pereira, L. 2010a. Littoral of Viana do Castelo—ALGAE (Bilingual). Câmara Municipal de Viana do Castelo, Portugal, 68 pp.

Pereira, L. 2010b. Littoral of Viana do Castelo—ALGAE (Bilingual). Uses in agriculture, gastronomy and food industry (Bilingual). Câmara Municipal de Viana do Castelo, Portugal, 68 pp.

Pereira, L. 2011. A review of the nutrient composition of selected edible seaweeds. pp. 15–47. *In*: Vitor H. Pomin (ed.). Seaweed: Ecology, Nutrient Composition and Medicinal Uses. Nova Science Publishers Inc., New York.

Pereira, L. and F. van de Velde. 2011. Portuguese carrageenophytes: carrageenan composition and geographic distribution of eight species (Gigartinales, Rhodophyta). Carbohydrate Polymers 84: 614–623.

Pereira, L. 2011b. Microfungos—Chaves simples para identificação de alguns géneros. Universidade de Coimbra, Coimbra, 23 pp.

Pereira, L. 2013. Population studies and carrageenan properties in eight Gigartinales (Rhodophyta) from western coast of Portugal. The Scientific World Journal, Vol. 2013: 11 pp.

Pereira, L., F.S. Gheda and P.J.A. Ribeiro-Claro. 2013b. Analysis by vibrational spectroscopy of seaweed polysaccharides with potential use in food, pharmaceutical and cosmetic industries. International Journal of Carbohydrate Chemistry, vol. 2013, Article ID 537202: 7 pp.

Pereira, L. 2013c. Cytological and cytochemical aspects in selected carrageenophytes (Gigartinales, Rhodophyta). pp. 81–104. *In*: Heimann, K. and C. Katsaros (eds.). Advances in Algal Cell Biology. De Gruyter, Berlin.

Pereira, L. 2015. Seaweed flora of the European North Atlantic and Mediterranean. pp. 65–178. *In*: Se-Kwon Kim (ed.). Springer Handbook of Marine Biotechnology. Springer, Berlin Heidelberg.

Pereira, L. and F. Correia. 2015. Algas marinhas da costa Portuguesa—ecologia, biodiversidade e utilizações. Nota de Rodapé Edições, Paris, 340 pp.

Pereira, L. and P.J.A. Ribeiro-Claro. 2015. Analysis by vibrational spectroscopy of seaweed with potential use in food, pharmaceutical and cosmetic. pp. 225–247. *In*: Leonel Pereira and João M. Neto (eds.). Marine Algae—Biodiversity, Taxonomy, Environmental Assessment and Biotechnology. Science Publishers, an Imprint of CRC Press/Taylor & Francis Group, Boca Raton, FL.

Pereira, L. 2016. Edible seaweeds of the world. Science Publishers, an Imprint of CRC Press/Taylor & Francis Group, Boca Raton, FL, 453 pp.

Pereira, L. 2016b. *Fucus vesiculosus*—Portuguese Seaweeds Website (MACOI). Available online at: http://macoi.ci.uc.pt/spec_list_detail.php?spec_id=78.

Pereira, L. 2016c. *Sargassum muticum*—Portuguese Seaweeds Website (MACOI). Available online at: http://macoi.ci.uc.pt/spec_list_detail.php?spec_id=157.

Pereira, L. (ed.). 2016d. Carrageenans—sources and extraction methods, molecular structure, bioactive properties and health effects. Nova Science Publishers, NY, 268 pp.

Pereira, L., F. Soares, A.C. Freitas, A.C. Duarte and P. Ribeiro-Claro. 2017. Extraction, characterization and use of carrageenans. *In*: P.N. Sudha (ed.). Industrial Applications of Marine Biopolymers. CRC Press, Taylor & Francis Group, Boca Raton, FL, 472 pp.

Pereira, L. 2017a. *Bryopsis hypnoides*—Portuguese Seaweeds Website (MACOI). Available online at: http://macoi.ci.uc.pt/spec_list_detail.php?spec_id=503.

Pereira, L. 2017b. *Bryopsis pennata*—Portuguese Seaweeds Website (MACOI). Available online at: http://macoi.ci.uc.pt/spec_list_detail.php?spec_id=445.

Pereira, L. 2017c. *Caulerpa cylindracea*—Portuguese Seaweeds Website (MACOI). Available online at: http://macoi.ci.uc.pt/spec_list_detail.php?spec_id=551.

Pereira, L. 2017d. *Caulerpa cupressoides*—Portuguese Seaweeds Website (MACOI). Available online at: http://macoi.ci.uc.pt/spec_list_detail.php?spec_id=487.

Pereira, L. 2017e. *Caulerpa mexicana*—Portuguese Seaweeds Website (MACOI). Available online at: http://macoi.ci.uc.pt/spec_list_detail.php?spec_id=390.

Pereira, L. 2017f. *Caulerpa taxifolia*—Portuguese Seaweeds Website (MACOI). Available online at: http://macoi.ci.uc.pt/spec_list_detail.php?spec_id=491.

Pereira, L. 2017g. *Halimeda macroloba*—Portuguese Seaweeds Website (MACOI). Available online at: http://macoi.ci.uc.pt/spec_list_detail.php?spec_id=370.

Pereira, L. 2017h. *Alaria esculenta*—Portuguese Seaweeds Website (MACOI). Available online at: http://macoi.ci.uc.pt/spec_list_detail.php?spec_id=241.

Pereira, L. 2017i. *Canistrocarpus cervicornis*—Portuguese Seaweeds Website (MACOI). Available online at: http://macoi.ci.uc.pt/spec_list_detail.php?spec_id=300.

Pereira, L. 2017j. *Cystoseira balearica*—Portuguese Seaweeds Website (MACOI). Available online at: http://macoi.ci.uc.pt/spec_list_detail.php?spec_id=484.

Pereira, L. 2017k. *Cystoseira compressa*—Portuguese Seaweeds Website (MACOI). Available online at: http://macoi.ci.uc.pt/spec_list_detail.php?spec_id=464.

Pereira, L. 2017l. *Cystoseira humilis*—Portuguese Seaweeds Website (MACOI). Available online at: http://macoi.ci.uc.pt/spec_list_detail.php?spec_id=188.

Pereira, L. 2017m. *Cystoseira mediterranea*—Portuguese Seaweeds Website (MACOI). Available online at: http://macoi.ci.uc.pt/spec_list_detail.php?spec_id=485.

Pereira, L. 2017n. *Sargassum filipendula*—Portuguese Seaweeds Website (MACOI). Available online at: http://macoi.ci.uc.pt/spec_list_detail.php?spec_id=450.

Pereira, L. 2017o. *Sargassum vulgare*—Portuguese Seaweeds Website (MACOI). Available online at: http://macoi.ci.uc.pt/spec_list_detail.php?spec_id=242.

Pereira, L. 2017p. *Taonia atomaria*—Portuguese Seaweeds Website (MACOI). Available online at: http://macoi.ci.uc.pt/spec_list_detail.php?spec_id=165.

Pereira, L. 2017q. *Asparagopsis armata*—Portuguese Seaweeds Website (MACOI). Available online at: http://macoi.ci.uc.pt/spec_list_detail.php?spec_id=7.

Pereira, L. 2017r. *Gelidiella acerosa*—Portuguese Seaweeds Website (MACOI). Available online at: http://macoi.ci.uc.pt/spec_list_detail.php?spec_id=440.

Pereira, L. 2017s. *Palisada perforata*—Portuguese Seaweeds Website (MACOI). Available online at: http://macoi.ci.uc.pt/spec_list_detail.php?spec_id=372.

Pereira, L. 2017t. *Polysiphonia denudata*—Portuguese Seaweeds Website (MACOI). Available online at: http://macoi.ci.uc.pt/spec_list_detail.php?spec_id=135.

Pereira, M.G., N.M.B. Benevides, M.R.S. Melo, A.P. Valente, F.R. Melo and P.A.S. Morão. 2005. Structure and anticoagulante activity of a sulfated galactan from the red alga, *Gelidium crinale*. Is there a specific structural requirement for the anticoagulant action? Carbohydrate research 340: 2015–2023.

Pereira, M.S., B. Mulloy and A.S. Mourao. 1999. Structure and anticoagulant activity of sulfated fucans. Journal of Biological Chemistry 274: 7656–7667.

Pereira, M.S., F.R. Melo and P.A.S. Mourão. 2002. Is there a correlation between structure and anticoagulant action of sulfated galactans and sulfated fucans? Glycobiology 12: 573–580.

Peres, J.C.F., L.R. de Carvalho, E. Goncalez, L.O.S. Berian and J.D. Felicio. 2012. Evaluation of antifungal activity of seaweed extracts. Ciência e Agrotecnologia 36(3): 294–299.

Pérez, M.J., E. Falque and H. Dominguez. 2016. Antimicrobial action of compounds from marine seaweed. Marine Drugs 14(52): 1–38.

Perez, R.M., J.G., Avila, S. Perez, A. Martinez and G. Martinez. 1990. Antimicrobial activity of some American algae. Journal of Ethnopharmacology 9: 111–116.

Pérez-Riverol, A., A.P. Ramos, L.F.M. Díaz, C.B.A. Herrera, OV. Iglesias and G.d.B. Alonso. 2014. *In vitro* antiviral activity of aqueous extract from red seaweed *Tricleocarpa fragilis* to Influenza A virus. Revista Cubana de Farmacia 48(2): 316–328.

Perl, A., R. Gonzalez-Cabello and P. Gergely. 1983. Stimulation of lectin-dependent cell-mediated cytotoxicity against adherent Hep-2 cells by Carrageenan. Clinical & Experimental Immunology 54(2): 567–572.

Perrone, C., E. Cecere and G. Furnari. 2006. Growth pattern assessment in the genus *Acanthophora* (Rhodophyta, Ceramiales). Phycologia 45: 37–43.

Perry, N.B., J.W. Blunt and M.H. Munro. 1991. A cytotoxic and antifungal 1,4 naphthoquinone and related compounds from a New Zealand brown alga, *Landsburgia quercifolia*. Journal of Natural Products 54(4): 978–985.

Pesando, D. and B. Caram. 1984. Screening of marine algae from French Mediterranean coast for antibacterial and antifungal activity. Botanica Marina 27: 381–386.

Pfaller, M.A. and D.J. Diekema. 2004. Rare and emerging opportunistic fungal pathogens: concern for resistance beyond *Candida albicans* and *Aspergillus fumigatus*. Journal of Clinical Microbiology 42(10): 4419–4431.

Phang, S.-M. and K. Lewmanomont. 2016. *Gracilaria changii*—PROSEA (Panta Resources of South-East Asia). Available noline at: http://uses.plantnet-project.org/en/Gracilaria_changii_(PROSEA).

Phang, S.-M. and K. Lewmanomont. 2016b. *Gracilaria edulis*—PROSEA (Panta Resources of South-East Asia). Available noline at: http://uses.plantnet-project.org/en/Gracilaria_edulis_(PROSEA).

Philchenkov, A., M. Zavelevich, T. Imbs, T. Zvyagintseva and T. Zaporozhets. 2007. Sensitization of human malignant lymphoid cells to etoposide by fucoidan, a brown seaweed polysaccharide. Experimental Oncology 29(3): 181–185.

Philchenkov, A.A., M.P. Zavelevich, N.N. Khranovskaya, T.S. Zaporozhets, T.I. Imbs, T.N. Zvyagintseva et al. 2006. Fucoidan from Far-Eastern seaweeds modulate *in vitro* apoptosis in human MT-4 leukemic cells. Russian Journal of Biotherapy 4: 30–37.

Piazza, V., V. Roussis, F. Garaventa, G. Greco, V. Smyrniotopoulos, C. Vagias and M. Faimali. 2011. Terpenes from the red alga *Sphaerococcus coronopifolius* inhibit the settlement of barnacles. Marine Biotechnology 13(4): 764–772.

Pielesz, A. and A. Machnicka. 2014. Antibacterial activity and cellulose acetate electrophoresis in monitoring collagen hydrogels modified with saccharides. African Journal of Pharmacy and Pharmacology 8: 175–184.

Pierre, G., V. Sopena, C. Juin, A. Mastouri, M. Graber and T. Maugard. 2011. Antibacterial activity of a sulfated galactan extracted from the marine alga *Chaetomorpha aerea* against *Staphylococcus aureus*. Biotechnology and Bioprocess Engineering 16[5]: 937–945.

Pietrangelo, A. 2002. Physiology of iron transport and the hemochromatosis gene. American Journal of Physiology—Gastrointestinal and Liver Physiology 282: G403–G414.

Pihlanto, A., S. Akkanen and H.J. Korhonen. 2008. ACE-inhibitory and antioxidant properties of potato (*Solanum tuberosum*). Food Chemistry 1: 104–112.

Pinteus, S., C. Alves, H. Monteiro, E. Araújo, A. Horta and R. Pedrosa. 2015. *Asparagopsis armata* and *Sphaerococcus coronopifolius* as a natural source of antimicrobial compounds. World Journal of Microbiology and Biotechnology 31(3): 445–451.

Pinto, A.M.V., J.P.G. Leite, W.J. Ferreira, D.N. Cavalcanti, R.C. Villaça, V. Giongo, V.L. Teixeira and I.C.N.P. Paixão. 2012. Marine natural seaweed products as potential antiviral drugs against Bovine viral diarrhea virus. Revista Brasileira de Farmacognosia Brazilian Journal of Pharmacognosy 22(4): 813–817.

Pinto, A.M.V., J.P.G. Leite, C.J.B. Ramos, R.R. da Fonseca, V.L. Teixeira and I. Paixo. 2014. Anti-herpesvirus bovine type 5 activities of extracts obtained from *Plocamium brasiliense*. Journal of Applied Phycology 26(5): 2021–2027.

Pinto, M.C., K. Barbieri, M.C.E. Silva, M.A.S. Graminha, C. Casanova, A.J. Andrade and A.E. Eiras. 2011. Octenol as attractant to *Nyssomyia neivai* (Diptera: Psychodidae: Phlebotominae) in the field. Journal of Medical Entomology 48: 39–44.

Piovan, A., R. Seraglia, B. Bresin, R. Caniato and R. Filippini. 2013. Fucoxanthin from *Undaria pinnatifida*: Photostability and coextractive effects. Molecules 18(6): 6298–6310.

Pires, C.R.C. 2016. Identificação e análise do potencial antifúngico dos polissacarídeos das algas *Saccharina latissima* e *Laminaria ochroleuca*. Tese de Mestrado em Biodiversidade e Biotecnologia Vegetal, Departamento de Ciências da Vida, Universidade de Coimbra, Portugal, 88 pp. Available online at: https://estudogeral.sib.uc.pt/handle/10316/31319.

Pise, N.M. and A.B. Sabale. 2010. Biochemical composition of seaweeds along central west coast of India, Pharmacognosy Journal 2(7): 148–150.

Pitt, J.I. 1994. The current role of *Aspergillus* and *Penicillium* in human and animal health. Journal of Medical and Veterinary Mycology S1: 17–32.

Pitt, J.I. and A.D. Hocking. 1999. Fungi and food spoilage, 2nd edition. A Chapman and Hall Food Science Book, Aspen Publication, Gaithersburg, Marylan, 326–327.

Plaza, A., J.L. Keffer, G. Bifulco, J.R. Lloyd and C.A. Bewley. 2010. Chrysophaentins A–H, antibacterial bisdiarylbutene macrocycles that inhibit the bacterial cell division protein FtsZ. Journal of the American Chemical Society 132: 9069–9077.

Plaza, M., S. Santoyo, L. Jaime, R.G. Garcia-Blairsy, M. Herrero, F.J. Senorans and E. Ibãnez. 2010b. Screening for bioactive compounds from algae. Journal of Pharmaceutical and Biomedical Analysis 51: 450–455.

Plouguerné E., L.M. de Souza, G.L. Sassaki, J.F. Cavalcanti, M.T. Villela Romanos, B.A. da Gama, R.C. Pereira and E. Barreto-Bergter. 2013. Antiviral sulfoquinovosyldiacylglycerols (SQDGs) from the Brazilian brown seaweed *Sargassum vulgare*. Marine Drugs 11: 4628–4640.

Plouguerné, E., C. Hellio, E. Deslandes, B. Veron, V. Stiger-Pouvreau. 2008. Anti-microfouling activities in extracts of two invasive algae: *Grateloupia turuturu* and *Sargassum muticum*. Botanica Marina 51(3): 202–208.

Plouguerné, E., E. Ioannou, P. Georgantea, C. Vagias, V. Roussis, C. Hellio, E. Kraffe, V. Stiger-Pouvreau. 2010. Anti-microfouling activity of lipidic metabolites from the invasive brown alga *Sargassum muticum* (Yendo) Fensholt. Marine Biotechnology 12(1): 52–61.

Pober, J.S. and W.C. Sessa. 2007. Evolving functions of endothelial cells in inflammation. Nature Reviews Immunology 7: 803–815.

Polat, S. and Y. Ozogul. 2009. Fatty acid, mineral and proximate composition of some seaweeds from the northeastern Mediterranean coast. Italian Journal of Food Science 21(3): 317–324.

Pomin, V.H., M.S. Pereira, A.P. Valente, D.M. Tollefsen, M.S.G. Pavão and P.A.S. Mourão. 2005. Selective cleavage and antcoagulant activity of a sulfated fucan: stereospecific removal of a 2-sulfate ester from the polysaccharide by mild acid hydrolysis, preparation of oligosaccharides, and heparin cofactor II-dependent anticoagulant activity. Glycobiology 15: 369–381.

Pomponi, S.A. 1999. The bioprocess technological potential of the sea. Journal of Biotechnology 70: 5–13.

Pomponi, S.A. 2001. The oceans and human health: the discovery and development of marine-derived drugs. Oceonogrophy 14: 78–87.

Ponce, N., C.A. Pujol, E.B. Damonte, M.L. Flores and C.A. Stortz. 2003. Fucoidans from the brown seaweed *Adenocystis utricularis*: extraction methods, antiviral activity and structural studies. Carbohydrate Research 338: 153–165.

Ponnanikajamideen, C. Malarkodi, M. Malinil and S.R. Kumar. 2014. Explore the antimicrobial potential from organic solventes of brown seaweed extracts (*Sargassum longifolium*) alleviating to pharmaceuticals. International Journal of Pharmacognosy 1(1): 82–89.

Poonguzhali, T.V. and L.L.J.L. Nisha. 2012. Larvicidal activity of two seaweeds, *Ulva fasciata* and *Grateloupia lithophila* against mosquito vector, *Culex quinquefasciatus*. International Journal of Current Science 4: 163–168.

Poonguzhali, T.V. and J.L.L.L. Nisha. 2013. Larvicidal activity of two seaweeds, *Enteromorpha flexuosa* J. Agardh and *Gracilara corticata* J. Agardh against mosquito vector, *Anopheles stephensi*. Biosciences Biotechnology Research Asia 10(1): 333–337.

Popplewell, W.L. and P.T. Northcote. 2009. Colensolide A: a new nitrogenous bromophenol from the New Zealand marine red alga *Osmundaria colensoi*. Tetrahedron Letters 50(49): 6814–6817.

Population Council. 2016. Microbicides program. Online Population Council News. Available online at: http://www.popcouncil.org/research/results?projects=yes&experts=yes&resources=yes&keywords=carrageenan.

Prabhakar, K., L.S. Kumar, S. Rajendran, M. Chandrasekaran, K. Bhaskar and K.S. Khan. 2008. Antifungal activity of plant extracts against *Candida* species from oral lesions. Indian Journal of Pharmaceutical Science 70(6): 801–803.

Prakash, S., S.A.J. Firthous and B.V. Bhimba. 2005. Biomedical potential of seaweed against *Otitis media* infected bacterial pathogens. Seaweed Research and Utilization 27: 105–109.

Pramitha, V.S. and N.S. Kumari. 2016. Anti-inflammatory, anti-oxidant, phytochemical and gc-ms analysis of marine brown macroalga, *Sargassum wighti*. International Journal of Pharmaceutical, Cemical and Biological Sciences 6(1): 7–15.

Pratt, R., T.C. Daniels, J.J. Eiler, J.B. Gunnison, W.D. Kumler, J.F. Oneto, L.A. Strait, H.A. Spoehr, G.J. Hardin, H.W. Milner, J.H. Smith and H.H. Strain. 1944. Chlorellin, an antibacterial substance from *Chlorella*. Science 99: 351–352.

Preeprame, S., K. Hayashi, J.-B. Lee, U. Sankawa and T. Hayashi. 2001. A novel antivirally active fucan sulfate derived from an edible brown alga, *Sargassum horneri*. Chemical and Pharmaceutical Bulletin 49. 484–485.

Premnathan, M., K. Chandra, S.K. Bajpai and K. Kathiresan. 1992. A survey of some Indian marine plants for antiviral activity. Botanica Marina 35: 321–324.

Preskitt, L. 2001. *Dictyosphaeria cavernosa*—Marine Algae of Hawai`i. Botany, University of Hawaii at Manoa. Avalable online at: https://www.hawaii.edu/reefalgae/invasive_algae/pdf.

Preskitt, L. 2001b. *Dictyosphaeria versluysii*—Marine Algae of Hawai`i. Botany, University of Hawaii at Manoa. Avalable online at: https://www.hawaii.edu/reefalgae/invasive_algae/pdf%20files/dictyosphaeria_versluyii.pdf.

Priyadharshini, S., S. Bragadeeswaran, K. Prabhu and S. Sophia Rani. 2011. Antimicrobial and hemolytic activity of seaweed extracts *Ulva fasciata* (Delile 1813) from Mandapam, Southeast coast of India. Asian Pacific Journal of Tropical Biomedicine 1(1): S37–S39.

Prud'homme van Reine, H.F. and G.C. Trono, Jr. (eds.). 2001. Cryptogams: Algae—Plant Resources of South-East Asia 15(1). Backhuys Publishers, Leiden, the Netherlands.

PubChem. 2017. 3D conformer of Caulerpin. Available online at: https://pubchem.ncbi.nlm.nih.gov/compound/5326018#section=3D-Conformer.

Puglisi, M.P., S. Engel, P.R. Jensen and W. Fenical. 2007. Antimicrobial activities of extracts from Indo-Pacific marine plants against marine pathogens and saprophytes. Marine Biology 150(4): 531–540.

Pujol, C.A., M.I. Errea, M.C. Matulewicz and E.B. Damonte. 1996. Antiherpetic activity of S1, an algal derived sulphated galactan. Phytotherapy Research 10: 410–413.

Pujol, C.A., J.M. Estevez, M.J. Carlucci, M. Ciancia, A.S. Cerezo and E.B. Damonte. 2002. Novel DL-galactan hybrids from the red seaweed *Gymnogongrus torulosus* are potent inhibitors of Herpes simplex virus and Dengue virus. Antiviral Chemistry and Chemotherapy 13(2): 83–89.

Puntarulo, S. 2005. Iron, oxidative stress and human health. Molecular Aspects of Medicine 26: 299–312.

Pushpamali, W.A., C. Nikapitiya, M. De Zoysa, I. Whang I, S.J. Kim and J. Lee. 2008. Isolation and purification of an anticoagulant from fermented red seaweed *Lomentaria catenata*. Carbohydrate Polymers 73: 274–279.

Puspita, M., M. Déniel, I. Widowati, O.K. Radjasa, P. Douzenel, C. Marty, L. Vandanjon, G. Bedoux and N. Bourgougnon. 2017. Total phenolic content and biological activities of enzymatic extracts from *Sargassum muticum* (Yendo) Fensholt. Journal of Applied Phycology: 1–17.

Qi, H., Q. Zhang, T. Zhao, R. Chen, H. Zhang and X. Niu. 2005. Antioxidant activity of different sulphate content derivatives of polysaccharide extracted from *Ulva pertusa* (Chlorophyta) *in vitro*. International Journal of Biological Macromolecules 37(4): 195–199.

Qi, H., Q. Zhang, T. Zhao, R. Hu, K. Zhang and Z. Li. 2006. *In vitro* antioxidant activity of acetylated and benzoylated derivatives of polysaccharide extracted from *Ulva pertusa* (Chlorophyta). Bioorganic Medicinal Chemistry Letters 16(9): 2441–2445.

Qi, H., D. Li, J. Zhang, L. Liu and Q. Zhang. 2008. Study on extraction of agaropectin from *Gelidium amansii* and its anticoagulant activity. Chinese Journal of Oceanology and Limnology 26(2): 186–189.

Qi, M., Y. Morch, I. Lacik, K. Formo, E. Marchese, Y. Wang, K.K. Danielson, K. Kinzer, S. Wang, B. Barbaro, G. Kollarikova, D. Chorvat, Jr., D. Hunkeler, G. Skjak-Braek, J. Oberholzer and B.L. Strand. 2012. Survival of human islets in microbeads containing high guluronic acid alginate crosslinked with Ca^{2+} and Ba^{2+}. Xenotransplantation 19(6): 355–364.

Qin, J., H. Su, Y. Zhang, J. Gao, L. Zhu, X. Wu, H. Pan and X. Li. 2010. Highly brominated metabolites from marine red alga *Laurencia similis* inhibit protein tyrosine phosphatase 1B. Bioorganic & Medicinal Chemistry Letters 20: 7152–7154.

Qu, W., H. Ma, Z. Pan, L. Luo, Z. Wang and R. He. 2010. Preparation and antihypertensive activity of peptides from *Porphyra yezoensis*. Food Chemistry 123: 14–20.

Queiroz, K.C.S., C.F. Assis, V.P. Medeiros, H.A. Rocha, H. Aoyama, C.V. Ferreira and E.L. Leite. 2006. Cytotoxicity effect of algal polysaccharides on Hl60 cells. Biochemistry 71: 1312–1315.

Queiroz, K.C.S., V.P. Medeiros, L.S. Queiroz, L.R.D. Abreu, H.A.O. Rocha, C.V. Ferreira, M.B. Juca, H. Aoyama and E.L. Leite. 2008. Inhibition of reverse transcriptase activity of HIV by polysaccharides of brown algae. Biomedicine & Pharmacotherapy 62: 303–307.

Quemener, B., M. Lahaye and C. Bobin-Dubigeon. 1997. Sugar determination in ulvans by a chemical-enzymatic method coupled to high performance anion exchange chromatography. Journal of Applied Phycology 9: 179–188.

Quindere, A.L.G., B.P. Fontes, E.D.O. Vanderlei, I.N.L. de Queiroz, J.A.G. Rodrigues, I.W.F. de Araujo, R.J.B. Jorge, D.B. de Menezes, A.A.R. e Silva, H.V. Chaves, J.S.A.M. Evangelista, M.M. Bezerra and N.M.B. Benevides. 2013. Peripheral antinociception and anti-edematogenic effect of a sulfated polysaccharide from *Acanthophora muscoides*. Pharmacological Reports 65(3): 600–613.

Quinton, R. 1904. L'eau de mer—Milieu organique. Masson et Cie, Éditeurs, Paris, 497 pp.

Quirós, A.R.B., C. Ron, J. López-Hernández and M.A. Lage-Yusty. 2004. Determination of folates in seaweeds by high-performance liquid chromatography. Journal of Chromatography A 1032: 135–139.

Qurrat-ul-Ain, S. Sharma, G.K. Khuller and S.K. Garg. 2003. Alginate-based oral drug delivery system for tuberculosis: pharmacokinetics and therapeutic Eefects. Journal of Antimicrobial Chemotherapy 51(4): 931–938.

Rabanal, Melissa, Nora M.A. Ponce, Diego A. Navarro, Ricardo M. Gómez and Carlos A. Stortz. 2014. The system of fucoidans from the brown seaweed *Dictyota dichotoma*: Chemical analysis and antiviral activity. Carbohydrate Polymers 101: 804–811.

Radhika, D., C. Veerabahu and R. Priya. 2013. Anti-inflammatory activities of some seaweed collected from the gulf of Mannar Coast, Tuticorin, South India. International Journal of Pharma and Bio Sciences 4(1): 39–44.

Radhika, D. and A. Mohaideen. 2015. Fourier transform infrared analysis of *Ulva lactuca* and *Gracilaria corticata* and their effect on antibacterial acivity. Asian Journal of Pharmaceutical and Clinical Research 8(2): 209–212.

Radhika, D. and R. Priya. 2016. Antifungal activity of *Acanthophora spicifera*, *Padina tetrastromatica* and *Caulerpa scalpelliformis* against some fungal pathogenes in crude and fractionated form. World Journal of Pharmacy and Pharmaceutical Sciences 5(3): 1155–1162.

Rafiquzzaman, S.M., E.Y. Kim, J.M. Lee, M. Mohibbullah, M.B. Alam, I.S. Moon, J.M. Kim and I.S. Kong. 2015. Anti-Alzheimers and antiinflammatory activities of a glycoprotein purified from the edible brown alga *Undaria pinnatifida*. Food Research International 77: 118–124.

Ragan, M.A. 1984. Bioactivities in marine genera of Atlantic Canada: the unexplored potential. Proceedings of the Nova Scotian Institute of Science 34: 81–132.

Rahuman, A.A., A. Bagavan, C. Kamaraj, M. Vadivelu, A.A. Zahir, G. Elango and G. Pandiyan. 2009. Evaluation of indigenous plant extracts against larvae of *Culex quinquefasciatus* Say (Diptera: Culicidae). Parasitology Research 104(3): 637–643.

Rai, M.K. and P.D. Bridge. 2009. Applied mycology. Wallingford, Cambridge, UK, 336 pp.

Raj, G.A., M. Chandrasekaran, S. Krishnamoorthy and V. Venkatesalu. 2015. Antibacterial activity of diff erent solvent extracts of *Caulerpa chemnitzia* (Esper) J.V. Lamououx, from Mandapam, Gulf of Mannar Southeast Coast, Tamil Nadu, India. Journal of Medicinal Herbs and Ethnomedicine 1(1): 24–31.

Rajapakse, N. and S.-K. Kim. 2011. Nutritional and digestive health benefits of seaweed. pp. 16–28. *In*: Se-Kwon Kim (ed.). Advances in Food and Nutrition Research, Volume 64, Marine Medicinal Foods: Implications and Application, Macro and Microalgae. Academic Press, an imprint of Elsevier, Waltham, USA.

Rajarajan, R. and S. Selvaraju. 2014. Antifungal activity of *Caulerpa racemosa* against *Fusarium oxysporum* of *Cucumis sativus*. International Journal of Advanced Research in Biological Sciences 1(8): 16–19.

Rajasulochana, P., R. Dhamotharan, P. Krishnamoorthy and S. Murugesan. 2009. Antibacterial activity of the extracts of marine red and brown algae. Journal of American Science 5(3): 20–25.

Rajauria, G., A.K. Jaiswal, N. Abu-Ghannam and S. Gupta. 2012. Antimicrobial, antioxidant and free radical-scavenging capacity of brown seaweed *Himanthalia elongata* from western coast of Ireland. Journal of Food Biochemistry 37: 322–335.

Rajauria, G. and N. Abu-Ghannam. 2013. Isolation and partial characterization of bioactive Fucoxanthin from *Himanthalia elongata* brown seaweed: a TLC-based approach. International Journal of Analytical Chemistry 802573(10): 6 pp.

Rajauria, G., A.K. Jaiswal, N. Abu-Gannam and S. Gupta. 2013. Antimicrobial, antioxidant and free radical-scavenging capacity of brown seaweed *Himanthalia elongata* from western coast of Ireland. Journal of Food Biochemistry 37: 322–335.

Raman, B.V., D.N. Rao and T.M. Radhakrishnan. 2004. *Enteromorpha compressa* (L.) Greville and edible green alga as a source of anti-allergic principle(s). Indian Journal of Clinical Biochemistry 19: 105–109.

Ramirez-Higuera, A., L. Quevedo-Corona, N. Paniagua-Castro, G. Chamorro-Ceballos, A. Milliar-Garcia and M.E. Jaramillo-Flores. 2014. Antioxidant enzymes gene expression and antihypertensive effects of seaweeds *Ulva linza* and *Lessonia trabeculata* in rats fed a high-fat and high-sucrose diet. Journal of Applied Phycology 26(1): 597–605.

Ramnani, P., R. Chitarrari, K. Tuohy, J. Grant, S. Hotchkiss, K. Philp, R. Campbell, C. Gill and I. Rowland. 2012. *In vitro* fermentation and prebiotic potential of novel low molecular weight polysaccharides derived from agar and alginate seaweeds. Anaerobe 18(1): 1–6.

Rang, H.P., J.M. Ritter, R.J. Flower and G. Henderson. 2010. Rang & Dale´s Pharmacology. Elsevier—Churchill Livingstone, London, 776 pp.

Rangaiah, G.S., P.A. Lakshmi and K. Sruthikeerthia. 2010. The antimicrobial activity of the crude extracts of Chlorophycean seaweeds *Ulva, Caulerpa* and *Spongomorpha* spp. against clinical and phytopathogens. Drug Invention Today 2: 311–314.

Ranganayaki, P., S. Susmitha and R. Vijayaraghavan. 2014. Study on metabolic compounds of *Kappaphycus alvarezii* and its *in-vitro* analysis of anti-inflammatory activity. Internationl Journal of Current Research and Academic Review 2(10): 157–166.

Rani, R.J., S.K. Sundar, B. Parthipan and J.M. Antonisamy. 2013. *In vitro* cytotoxicity studies on selected seaweeds from South East coast of India. Indo American Journal of Pharmaceutical Research 3(4): 3182–3191.

Rao, L.V., O. Nordfang, A.D. Hoang and U.R. Pendurthi. 1995. Mechanism of antithrombin III inhibition of factor VIIa/tissue factor activity on cell surfaces: comparison with tissue factor pathway inhibitor/factor Xa-induced inhibition of factor VIIa/tissue factor activity. Blood 85: 121–129.

Rao, N.V.S.A.V.P., K.V. Sastry and E.V. Rao. 1984. Studies on Indian polysaccharides. Part 2. Carbohydrates of *Padina tetrastromatica*. Phytochemistry 23: 2531–2533.

Rao, O.S., K.G. Girijavallaban, S. Muthusamy, V. Chandrika, C.P. Gopinathan, S. Kalimuthu and M. Najmuddin. 1991. Bioactivity in marine algae, pp. 373–377. *In*: Thompson, M.F., R. Sarojini and R. Nagabhushanam (eds.). Bioactive Compounds from Marine Organisms with Emphasis on the Indian Ocean: An Indo-United States Symposium, Oxford and IBH Pub. Co., New Delhi.

Rao, P. and S.M. Karmarkar. 1986. Biological investigation of genus *Sargassum*—antifungal activity of fractions of different species of *Sargassum*. Phykos 25. 6–11.

Rao, P.P.S., R.S. Rao and S.M. Karmarkar. 1988. Antibacterial activity from Indian species of *Sargassum*. Botanica Mararina 31: 295–298.

Rao, P.S. and K.S. Parekh. 1981. Antibacterial activity of Indian seaweed extracts. Botanica Marina 24: 577–582.

Rao, P.S. 1990. Biological investigation of Indian Phaeophyceae XI. Screening of different parts of *Sargassum wighti* Greville for this antibacterial activity. Indian Bot. Rep. 9: 52–55.

Rao, P.V.S., V.A. Mantri and K. Ganesan. 2007. Mineral composition of edible seaweed *Porphyra vietnamensis*. Food Chemistry 102: 215–218.

Raper, K.B. and D.I. Fennell. 1965. The genus *Aspergillus*. William & Wilkins Co., Baltimore, 686 pp.

Ratana-Arporn, P. and A. Chirapart. 2006. Nutritional evaluation of tropical green seaweeds *Caulerpa lentillifera* and *Ulva reticulata*. Kasetsart Journal—Natural Science 40: 75–83.

Ratnaike, R.N. and T.E. Jones. 1998. Mechanisms of drug-induced diarrhoea in the elderly. Drugs & Aging 13(3): 245–253.

Ravikumar, S., L. Anburajan, G. Ramanathan and N. Kaliaperumal. 2002. Screening of seaweed extracts against antibiotic resistant post operative infectious pathogens. Seaweed Research and Utilization 24(1): 95–99.

Ravikumar, S., M.S. Ali and J.M. Beula. 2013. Mosquito larvicidal efficacy of seaweed extracts against dengue vector of *Aedes aegypti*. Asian Pacific Journal of Tropical Biomedicine: S143–S146.

Ravikumar, S., S.J. Inbaneson, P. Suganthi, R. Gokulakrishnan and M. Venkatesan. 2011. *In vitro* antiplasmodial activity of ethanolic extracts of seaweed macroalgae against *Plasmodium falciparum*. Parasitology Research. 108: 1411–1416.

Ravikumar, S., G. Ramanathan, M. Gnanadesigan, A. Ramu and V. Vijayakumar. 2011b. *In vitro* antiplasmodial activity of methanolic extracts from seaweeds of South West Coast of India. Asian Pacific Journal of Tropical Medicine: 862–865.

Ray, K. 2002. Typhoid fever. Rosen Publishing, New Ork, 62 pp.

Ray, S., C.A. Pujol, E.B. Damonte and B. Ray. 2015. Additionally sulfated xylomannan sulfates from *Scinaia hatei* and their antiviral activities. Carbohydrate Polymers 131: 315–321.

Razina, T.G., Y.P. Zueva, Y.N. Amosova et al. 2010. Effect of fucoidan from the marine brown alga *Laminaria japonica* on the progress of Ehrlich's adenocarcinoma and Lewis lung carcinoma and on cyclophosphan efficacy in mice. Tikhookean. Med. Zhurnal 2: 36–39.

Recalde, M.P., M.J. Carlucci, M.D. Noseda and M.C. Matulewicz. 2012. Chemical modifications of algal mannans and xylomannans: effects on antiviral activity. Phytochemistry 73(1): 57–64.

Recalde, P.M.P., M.D. Noseda, C.A. Pujol, M.J. Carlucci and M.C. Matulewicz. 2009. Sulfated mannans from the red seaweed *Nemalion Helminthoides* of the south atlantic. Phytochemistry 7(8): 1062–1068.

Reddy, B.S., S. Numoto and C.I. Choi. 1985. Effect of dietary *Laminaria angustata* (brown seaweed) on azoxymethane-induced intestinal carcinogenesis in male F344 rats. Nutrition and Cancer 7(1-2): 59–64.

Reddy, K.S., S.E. Pawar and C.R. Bhatia. 1994. Inheritance of Powdery mildew (*Erysiphe polygoni*) resistance in Mungbean (*Vigna radiata*). Theoretical and Applied Genetics 88(8): 945–948.

Reddy, P. and S. Urban. 2009. Meroditerpenoids from the southern Australian marine brown alga *Sargassum fallax*. Phytochemistry 70(2): 250–255.

Reed, M. 1906. The economic seaweeds of Hawaii and their food value. Hawaii Agricultural Experiment Station, Honolulu, Hawaii, 88 pp.

Reef, R., P. Kaniewska and O. Hoegh-Guldberg. 2009. Coral skeletons defend against ultraviolet radiation. PloS One 4: e7995.

Rees, J.R., R.W. Pinner, R.A. Hajjeh, M.R. Brandt and A.L. Reingold. 1998. The epidemiologic features of invasive mycotic infection in the San Francisco bay area, 1992–1993: results of a population-based laboratory active surveillance. Clinical Infectious Diseases 27: 1138–1147.

Regnier, P., J. Bastias, V. Rodriguez-Ruiz, N. Caballero-Casero, C. Caballo, D. Sicilia, A. Fuentes, M. Maire, M. Crepin, D. Letourneur, V. Gueguen, S. Rubio and G. Pavon-Djavid. 2015. Astaxanthin from *Haematococcus pluvialis* prevents oxidative stress on human endothelial cells without toxicity. Marine Drugs 13: 2857–2874.

Reichelt, J.L. and M.A. Borowitzka. 1984. Antimicrobial activity from marine algae: Results of a large-scale screening programme. Proceedings of the International Seaweed Symposium 11: 158–168.

Ren, D., L.A. Bedzyk, R.W. Ye, S.M. Thomas and T.K. Wood. 2004. Differential gene expression shows natural brominated furanones interfere with the autoinducer-2 bacterial signaling system of *Escherichia coli*. Biotechnology and Bioengineering 88: 630–642.

Renaud, SM. and J.T. Luong-Van. 2006. Seasonal variation in the chemical composition of tropical Australian marine macroalgae. Journal of Applied Phycology 18: 381–387.

Rengasamy, K.R., M.A. Aderogba, S.O. Amoo, W.A. Stirk and J. van Staden. 2013. Potential antiradical and α-glucosidase inhibitors from *Ecklonia maxima* (Osbeck) Papenfuss. Food Chemistry 2141: 1412–1415.

Rengasamy, K.R.R., S.O. Amoo, A.O. Aremu, W.A. Stirk, J. Gruz, M. Subrtova, K. Dolezal and J. Van Staden. 2015. Phenolic profiles, antioxidant capacity, and acetylcholinesterase inhibitory activity of eight South African seaweeds. Journal of Applied Phycology 27(4): 1599–1605.

Renju, G.L., G. Muraleedhara Kurup and C.H. Saritha Kumari. 2013. Anti-inflammatory activity of lycopene isolated from *Chlorella marina* on type II collagen induced arthritis in Sprague Dawley rats. Immunopharmacology and Immunotoxicology 35(2): 282–291.

Renn, D.W., H. Noda, H. Amano, T. Nishino and K. Nishizana.1994a. Antihypertensive and antihyperlipidemic effects of funoran. Fisheries Science 60: 423–427.

Renn, D.W., H. Noda, H. Amano, T. Nishino and K. Nishizana. 1994b. Study on hypertensive and antihyperlipidemic effect of marine algae. Fisheries Science 60: 83–88.

Rhimou, B., R. Hassane, M. Jose and Bourgougnon Nathalie. 2010. The antibacterial potential of the seaweeds (Rhodophyceae) of the Strait of Gibraltar and the Mediterranean Coast of Morocco. African Journal of Biotechnology 9(38): 6365–6372.

Riaz, B., R. Najam, H. Anser and M.S. Ali. 2014. Evaluation of *Iyengaria stellata* for its hypolipidemic and hepatoprotective activity. Pakistan Journal of Pharmaceutical Sciences 27: 1775–1779.

Ribeiro, N.A., T.M. Abreu, H.V. Chaves, M.M. Bezerra, H.S.A. Monteiro, R.J.B. Jorge and N.M.B. Benevides. 2014. Sulfated polysaccharides isolated from the green seaweed *Caulerpa racemosa* plays antinociceptive and anti-inflammatory activities in a way dependent on HO-1 pathway activation. Inflammation Research 63: 569–580.

Ribier, J. and J.C. Godineau. 1984. Les algues. La Maison Rustique, Flammarion, France, pp. 15–26.

Riccioni, G., N. D'Orazio, S. Franceschelli and L. Speranza. 2011. Marine carotenoids and cardiovascular risk markers. Marine Drugs 9: 1166–1175.

Richards, J.T., E.R. Kern, L.A. Glasgow, J.C. Overall, Jr., E.F. Deig and M.T. Hatch. 1978. Antiviral activity of extracts from marine algae. Antimicrobial Agents and Chemotherapy 14: 24–30.

Richardson, M.D. and D.W. Warnock. 2012. Fungal infection: Diagnosis and management, 4th edition. Wiley-Blackwell, Hoboken, New Jersey, 476 pp.

Rim, H.J., J.H. Ha, J.S. Lee, I. Hyun and K.B. Uh. 1974. Pyrantel embonate in mass treatment of ascariasis and comparison with piperazine adipate and santonin-kainic acid complex. The Korean Journal of Parasitology 12(2): 141–146.

Rimpler, M.M., U. Rauen, T. Schmidt, T. Möröy and H. De Groot. 1999. Protection against hydrogen peroxide cytotoxicity in Rat-1 fibroblasts provided by the oncoprotein BCL-2: maintenance of calcium homoeostasis is secondary to the effect of BCL-2 on cellular glutahione. Biochemical Journal 340: 291–297.

Rink, L. and H. Haase. 2007. Zinc homeostasis and immunity. Trends in Immunology 28: 1–4.

Rios, J.L. and M.C. Recio. 2005. Medicinal plants and antimicrobial activity. Journal of Ethnopharmacology 100: 80–84.

Riou, D., S. Colliec-Jouault, D. Pinczon du Sel, S. Bosch, S. Siavoshian, V. Le Bert, C. Tomasoni, C. Sinquim, P. Durand and C. Roussakis. 1996. Antitumor and antiproliferative effects of a fucan extracted from *Ascophyllum nodosum* against a non-small-cell bronchopulmonary carcinoma line. Anticancer Research 16(3A): 1213–1218.

Rioux, L.E., S.L. Turgeon and M. Beaulieu. 2007. Characterization of Polysaccharides Extracted from Brown Seaweeds. Carbohydrate Polymers 69(3): 530–537.

Rioux, L.E., S.L. Turgeon and M. Beaulieu. 2010. Structural characterization of laminaran and galactofucan extracted from the brown seaweed *Saccharina longicruris*. Phytochemistry 71: 1586–1595.

Rioux, L.E., V. Moulin, M. Beaulieu and S.L. Turgeon. 2013. Human skin fibroblast response is differentially regulated by galactofucan and low molecular weight galactofucan. Bioactive Carbohydrates and Dietary Fibre 1: 105–110.

Rissanen, T.H., S. Voutilainen, J.K. Virtanen, B. Venho, M. Vanharanta, J. Mursu and J.T. Salonen. 2003. Low intake of fruits, berries and vegetables is associated with excess mortality in men: The Kuopio Ischemic Heart Disease Risk Factor (KIHD) study. Journal of Nutrition 133: 199–204.

Rizk, M.Z., H.F. Aly A.A. Matloub and G.I. Fouad. 2016. The anti-hypercholesterolemic effect of ulvan polysaccharide extracted from the green alga *Ulva fasciata* on aged hypercholesterolemic rats. Asian Journal of Pharmaceutical and Clinical Research 9(3): 11 pp.

Rizvi, M.A. and M. Shameel. 2001. Distribution of elements in marine algae of Karachi coast. Pakistan Journal of Botany 33: 357–363.

Rizvi, M.A. and M. Shameel. 2004. Studies on the bioactivity and elementology of marine algae from the coast of Karachi, Pakistan. Phytotherapy Research 18: 865–872.

Rizvi, M.A. and M. Shameel. 2005. Pharmaceutical biology of seaweeds from the Karachi coast of Pakistan. Pharmaceutical Biology 43(2): 97–107.

Rizvi, M.A. and M. Shameel. 2006. *In vitro* nematicidal activities of sea weed extracts from Karachi coast. Pakistan Journal of Botany 38(4): 1245–1248.

Rizvi, M.A. 2010. Comparative antibacterial activities of seaweed extracts from Karachi coast, Pakistan. Pakistan Journal of Pharmacology 27(2): 53–57.

Rizzo, L., C. Manaia, C. Merlin, T. Schwartz, C. Dagot, M.C. Ploy, I. Michael and D. Fatta-Kassinos. 2013. Urban wastewater treatment plants as hotspots for antibiotic resistant bacteria and genes spread into the environment: A review. Science of the Total Environment 447: 345–360.

Rizzo, L., S. Fraschetti, P. Alifano, M.S. Tredici and L. Stabili. 2016. Association of *Vibrio* community with the Atlantic Mediterranean invasive alga *Caulerpa cylindracea*. Journal of Experimental Marine Biology and Ecology 475: 129–136.

Roberts, M.A. and B. Quemenery. 1999. Measurement of carrageenans in food: Challenges, progress, and trends in analysis. Trends in Food Science & Technology 10: 169–181.

Roberts, J.N., C.B. Buck, C.D. Thompson, R. Kines, M. Bernardo, P.L. Choyke, D.R. Lowy and J.T. Schiller. 2007. Genital transmission of HPV in a mouse model is potentiated by nonoxynol-9 and inhibited by carrageenan. Nature Medecine 13: 857–861.

Robertson, R.C., F. Guiheneuf, B. Bahar, M. Schmid, D.B. Stengel, G.F. Fitzgerald, R.P. Ross and C. Stanton. 2015. The antiinflammatory effect of algae-derived lipid extracts on lipopoly-saccharide (LPS)-stimulated human THP-1 macrophages. Marine Drugs 13: 5402–5424.

Robic, A. and M. Lahaye. 2007 Structure and funcfional properties of ulvan, a polysaccharide from green seaweeds. Biomacromolecules 6: 1765–1774.

Robic, A., C. Rondeau-Mouro, J.F. Sassi, Y. Lerat and M. Lahaye. 2009. Structure and interactions of ulvan in the cell wall of the marine green algae *Ulva rotundata* (Ulvales, Chlorophyceae). Carbohydrate Polymers 77(2): 206–216.

Robledo, D. and Y.F. Pelegrin. 1997. Chemical and mineral composition of six potentially edible seaweed species of Yucatan. Botanica Marina 40(4): 301–306.

Rocha, F.D., A.R. Soares, P.J. Houghton, R.C. Pereira, M.A. Kaplan and V.L. Teixeira. 2007. Potential cytotoxic activity of some Brazilian seaweeds on human melanoma cells. Phytotherapy Research 21(2): 170–175.

Rocha, H.A.O., F.A. Moraes, E.S. Trindade, C.R.C. Franco, R.J.S. Torquato, S.S. Veiga, A.P. Valente, P.A.S. Morão, E.L. Leite, H.B. Nade rand C.P. Dietrich. 2005. Structural and hemostatic activities of a sulfated galactofucan from the brown alga *Spatoglossum schroederi*: an ideal antithrombotic agent? The Journal of Biological Chemistry 280: 41278–41288.

Rodrigues, D., A.C. Freitas, L. Pereira, T.A. P. Rocha-Santos, M.W. Vasconcelos, M. Roriz, L.M. Rodríguez-Alcalá, A.M.P. Gomes and A.C. Duarte. 2015. Chemical composition of red, brown and green macroalgae from Buarcos bay in central west coast of Portugal. Food Chemistry 183(2): 197–207.

Rodrigues, D., C. Alves, A. Horta, S. Pinteus, J. Silva, G. Culioli, O.P. Thomas and R. Pedrosa. 2015b. Antitumor and antimicrobial potential of bromoditerpenes isolated from the red alga, *Sphaerococcus coronopifolius*. Marine Drugs 13: 713–726.

Rodrigues, D., S. Sousa, A. Silva, M. Amorim, L. Pereira, T.A. Rocha-Santos, A.M. Gomes, A.C. Duarte and A.C. Freitas. 2015c. Impact of enzyme and ultrasound assisted extraction methods on biological properties of red, brown and green seaweeds from the Central West Coast of Portugal. Journal of Agricultural and Food Chemistry 10: 10.

Rodrigues, J.A.G., E.S.O. Vanderlei, É.F. Bessa, F.A. Magalhaes, R.C.M. de Paula, V. Lima N.M.B. Benevides. 2011. Anticoagulant activity of a sulfated polysaccharide isolated from the green seaweed *Caulerpa cupressoides*. Brazilian Archives of Biology and Technology 54: 691–700.

Rodrigues, J.A.G., I.N.L. de Queiroz, E.F. Bessa, C.O. Coura, R.C.N. Amorim and N.M.B. Benevides. 2011b. Anticoagulant activity of sulfated polysaccharides fractions from an aqueous extract obtained from the red seaweed *Halymenia floresia* (Clemente) C. Agardh. Acta Scientiarum. Technology 33: 371–378.

Rodrigues, J.A.G., A.L.G. Quinderé, I.N.L. de Queiroz, C.O. Coura and N.M.B. Benevides. 2012. Comparative study of sulfated polysaccharides from *Caulerpa* spp. (Chlorophyceae). Biotechnological tool for species identification? Maringá 34: 381–389.

Rodrigues, J.A.G., E.M. Neto, L.A.C. Teixeira, R.C.M. de Paula, P.A.S. Mourao and N.M.B. Benevides. 2013. Structural features and inactivation of coagulation proteases of a sulfated polysaccharidic fraction from *Caulerpa cupressoides* var. *lycopodium* (Caulerpaceae, Chlorophyta). Acta Scientiarum. Technology 35: 611–619.

Rodrigues, J.A.G., E.D.O. Vanderlei, A.L.G. Quindere, V.S. Monteiro, S.M.M. de Vasconcelos and N.M.B. Benevides. 2013b. Antinociceptive activity and acute toxicological study of a novel sulfated polysaccharide from *Caulerpa cupressoides* var. *lycopodium* (Chlorophyta) in Swiss mice. Acta Scientiarum-Technology 35(3): 417–425.

Rodríguez, A., K. Kleinbeck, O. Mizenina, L. Kizima, K. Levendosky, N. Jean-Pierre, G. Villegas, B.E. Ford, M.L. Cooney, N. Teleshova, M. Robbiani, B.C. Herold, T. Zydowsky and J. A. Fernandez Romero. 2014. *In vitro* and *in vivo* evaluation of two carrageenan-based formulations to prevent HPV acquisition. Antiviral Research 108: 88–93.

Rodríguez, M.C., E.R. Merino, C.A. Pujol, E.B. Damonte, A.S. Cerezo and M.C. Matulewicz. 2005. Galactans from cystocarpic plants of the red seaweed *Callophyllis variegata* (Kallymeniaceae, Gigartinales). Carbohydrate Research 340: 2742–2751.

Rodríguez-Martínez, A.V., R., Sendón, M.J. Abad, M.V. González-Rodríguez, J. Barros-Velázquez, S.P. Aubourg, P. Paseiro-Losada and A. Rodríguez-Bernaldo de Quirós. 2016. Migration kinetics of sorbic acid from polylactic acid and seaweed based films into food simulants. LWT—Food Science and Technology 65: 630–636.

Rodríguez-Moreno, F., E. Gonzáles-Reimers, F. Santolaria-Fernandez, L. Galindo-Martin, O. Hernandez-Torres, N. Batista-Lopez and M. Molina-Perez. 1997. Zinc, copper, manganese, and iron in chronic alcoholic liver disease. Alcohol 14: 39–44.

Roemer, T. and D.J. Krysan. 2014. Antifungal drug development: challenges, unmet clinical needs, and new approaches. Cold Spring Harbor Perspectives in Medicine 4(5): 1–15.

Rogers, D.J., K.M. Jurd, G. Bluendan, S. Paoletti and F. Zanetti. 1990. Anticoagulant activity of a proteoglycan in extracts of *Codium fragile* ssp. atlanticum. Journal of Applied Phycology 2: 357–361.

Rogers, D.J., K.M. Jurd, D.S. McLellan and G. Bluendan. 1990b. 4th Colloque sur la valorization des produits de la mer 13: 201–206.

Rogers, K. (ed.). 2011. Bacteria and Viruses. Rosen Education Service, Britannica Educational Publishing, New York, 216 pp.

Rohani-Ghadikolaei, K., E. Abdulalian and W.K. Ng. 2012. Evaluation of the proximate, fatty acid and mineral composition of representative green, brown and red seaweeds from the Persian Gulf of Iran as potential food and feed resources. Journal of Food Science and Technology-Mysore 49(6): 774–780.

Romanha, A.J., S.L. Castro, Md.N. Soeiro, J. Lannes-Vieira, I. Ribeiro, A. Talvani, B. Bourdin, B. Blum. B. Olivieri, C. Zani, C. Spadafora, E. Chiari, E. Chatelain, G. Chaves, J.E. Calzada, J.M. Bustamante, L.H. Freitas-Junior, L.I. Romero, M.T. Bahia, M. Lotrowska, M. Soares, S.G. Andrade, T. Armstrong, W. Degrave and Z.D.A. Andrade. 2010. *In vitro* and *in vivo* experimental models for drug screening and development for Chagas disease. Memórias do Instituto Oswaldo Cruz 105: 233–238.

Romanos, M., M.J. Andrada-Serpa, S. Dos, A. Ribeiro, Y. Yoneshigue-Valentin, S.S. Costa and M.D. Wigg. 2002. Inhibitory effect of extracts of Brazilian marine algae on human T-cell lymphotropic virus type 1 (HTLV-1) induced syncytium formation *in vitro*. Cancer Investigation 20: 46–54.

Romarís-Hortas, V., C. García-Sartal, M.C. Barciela-Alonso, R. Domínguez-González, A. Moreda-Piñeiro and P. Bermejo-Barrera. 2011. Bioavailability study using an *in vitro* method of iodine and bromine in edible seaweed. Food Chemistry 124: 1747–1752.

Romay, C., R. Gonzalez, N. Ledon, D. Remirez and V. Rimbau. 2003. C-phycocyanin: A biliprotein with antioxidant, anti-inflammatory and neuroprotective effects. Current Protein and Peptide Science 4(3): 207–216.

Roohi Fatima, M., S. Dinesh, T. Mekata, T. Itami and R. Sudhakaran. 2016. Therapeutic efficiency of *Portieria hornemannii* (Rhodophyta) against *Vibrio parahaemolyticus* in experimentally infected *Oreochromis mossambicus*. Aquaculture 450: 369–374.

Roos, H. 1957. Untersuchungen fiber das Vorkommen antimikrobieller Substanzen in Meeresalgen. Kieler Meeresforsch 13: 41–58.

Ropellato, J., M.M. Carvalho, L.G. Ferreira, M.D. Noseda, C.R. Zuconelli, A.G. Goncalves, D.R. Ducatti, J.C. Kenski, P.L. Nasato, S.M. Winnischofer and M.E. Duarte. 2015. Sulfated heterorhamnans from the green seaweed *Gayralia oxysperma*: partial depolymerization, chemical structure and antitumor activity. Carbohydrate Polymers 117: 476–485.

Rosaline, X.D., S. Sakthivelkumar, K. Rajendran and S. Janarthanan. 2012. Screening of selected marine algae from the coastal Tamil Nadu, South India for antibacterial activity. Asian Pacific Journal of Tropical Biomedicine 2(1): S140–S146.

Rossi, F. and M. van Griensven. 2014. Polymer functionalization as a powerful tool to improve scaffold performances. Tissue Engineering Part A 20(15-16): 2043–2051.

Roszkowski, W., J. Beuth, H.L. Ko, G. Uhlenbruck and G. Pulverer. 1989. Blocking of lectin-like adhesion molecules on pulmonary cells inhibits lung sarcoma L-1 colonization in Balb/C-Mice. Experientia 45(6): 584–588.

Rowley, S. 2008. *Heterosiphonia plumosa*—MarLIN, The Marine Life Information Network. Available online at: http://www.marlin.ac.uk/species/detail/11.

Rozas, E. and J.C. Freitas. 2007. Anti-inflammatory activity of the apolar extract from the seaweed *Galaxaura marginata* (Rhodophyta, Nemaliales). Journal of Venomous Animals and Toxins including Tropical Diseases 13(2): 544-548.

Rozkin, M., M.N. Levina, V.S. Efimov and A.I. Usov. 1991. The anticoagulant and lipolysis-stimulating activity of polysaccharides from marine brown algae. Farmakology Toksicology 54: 40–42.

Rudtanatip, T., S. Asuvapongpatana, B. Withyachumnarnkul and K. Wongprasert. 2014. Sulfated galactans isolated from the red seaweed *Gracilaria fisheri* target the envelope proteins of white spot syndrome virus and protect against viral infection in shrimp haemocytes. Journal of General Virology 95: 1126–1134.

Rueckert, R.R. 1996. Picornaviridae: the viruses and their replication. pp. 609–654. *In*: Fields, B.N., D.M. Knipe and P.M. Howley (eds.). Fields Virology, 3rd ed. Lippincott-Raven Publishers, Philadelphia, Pa.

Rupérez, P. and F. Saura-Calixto. 2001. Dietary fibre and physicochemical properties of edible Spanish seaweeds. European Food Research and Technology 212: 349–354.

Rupérez, P. 2002. Mineral content of edible marine seaweeds. Food Chemistry 79: 23–26.

Rupérez, P., O. Ahrazem and J.A. Leal. 2002. Potential antioxydant capacity of sulfated polysaccharides from the edible marine brown seaweed *Fucus vesiculosus*. Journal of Agriculture and Food Chemistry 50: 840–845.

Ruqqia, K., V. Sultana, J. Ara, S. Ehteshamul-Haque and M. Athar. 2015. Hypolipidaemic potential of seaweeds in normal, triton-induced and high-fat diet-induced hyperlipidaemic rats. Journal of Applied Phycology 27: 571–579.

Russell, D.J. and G.H. Balazs. 2000. Identification manual for dietary vegetation of the Hawaiian green turtle, *Chelonia mydas*. NOAA TM-NMFS-SWFSC-294. 49 pp.

Russo, R., M. Malinconico and G. Santagata. 2007. Effect on cross-linking with calcium ions on the physical properties of alginate films. Biomacromolecules 8: 3193–3197.

Ryan, S., D.M. O'Gorman and Y.M. Nolan. 2011. Evidence that the marine-derived multimineral Aquamin has anti-inflammatory effects on cortical glial-enriched cultures. Phytotherapy Research 25(5): 765–767.

Ryu, B., Y. Li, Z.J. Qian, M.M. Kim and S.K. Kim. 2008. Exhibitory effects of compounds from brown alga *Ecklonia cava* on the human osteoblasts. Journal of Biotechnology 136: 577–588.

Ryu, B., Z.J. Qian, N.M. Kim, K.W. Nam and S.K. Kim. 2009. Anti-photoaging activity and inhibition of matrix metalloproteinase (MMP) by marine red alga, *Corallina pilulifera* methanol extract. Radiation Physics and Chemistry 78: 98–105.

Ryu, G., S.H. Park, E.S. Kim, B.W. Choi, S.Y. Ryu and B.H. Lee. 2003. Cholinesterase inhibitory activity of two farnesylacetone derivatives from the brown alga *Sargassum sagamianum*. Archives of Pharmacal Research 26: 796–799.

Ryu, J., S.-J. Park, I.-H. Kim, Y.H. Choi and T.-J. Nam. 2014. Protective effect of *Porphyra*-334 on UVA-induced photoaging in human skin fibroblasts. International Journal of Molecular Medicine 34: 796–803.

Sáa, C.F. 2002. Atlantic seavegetables—nutrition and health: properties, recipes, description. Algamar, Redondela-Pontevedra, Spain, 300 pp.

Sabina, H., S. Tasneem, Y. Sambreen Kausar, M.I. Choudhary and R. Aliya. 2005. Antileishmanial activity in the crude extract of various seaweed from the coast of Karachi, Pakistan. Pakistan Journal of Botany 37: 163–168.

Sabry, O.M., S. Andrews, K.L. McPhail, D.E. Goeger, A. Yokochi, K.T. LePage, T.F. Murray and W.H. Gerwick. 2005. Neurotoxic meroditerpenoids from the tropical marine brown alga *Stypopodium flabelliforme*. Journal of Natural Products 68(7): 1022–1030.

Sachindra, N.M., E. Sato, H. Maeda, M. Hosokawa, Y. Niwano, M. Kohno and K. Miyashita. 2007. Radical scavenging and singlet oxygen quenching activity of marine carotenoid fucoxanthin and its metabolites. Journal of Agricultural and Food Chemistry 55(21): 8516–8522.

Saeidnia, S., A.R. Gohari, A.R. Shahverdi, P. Permeh, M. Nasiri, K. Mollazadeh and F. Farahani. 2009. Biological activity of two red algae, *Gracilaria salicornia* and *Hypnea flagelliformis* from Persian Gulf. Pharmacognosy Research 1(6): 428–430.

Sahayaraj, K., S. Rajesh, A. Asha and J.M. Rathi. 2012. Marine algae for the cotton pest and disease management. Recent trends in Agriculture, Water and Environment Research: 48–60.

Sahnouni, F., Z. Benattouche, A. Matallah-Boutiba, M. Benchohra, W.M. Chentouf, D. Bouhadi and Z. Boutiba. 2016. Antimicrobial activity of two marine algae *Ulva rigida* and *Ulva intestinalis* collected from Arzew gulf (Western Algeria). Journal of Applied Environmental and Biological Sciences 6(1): 242–248.

Sahoo, D. 2010. Common seaweeds of India. I.K. International Pvt Ltd, New Delhi, India, 196 pp.

Saidani, K., F. Bedjou, F. Benabdesselam and N. Touati. 2012. Antifungal activity of methanolic extracts of four Algerian marine algae species. African Journal of Biotechnology 39: 9496 9500.

Saito, K. and Y. Nakamura. 1951. Sarganin and related phenols from marine alga and medical function. Journal of the Chemical Society of Japan. Pure Chemistry Section 72: 992–993.

Sakthivel, R., S. Muniasamy, G. Archunan and K.P. Devi. 2016. *Gracilaria edulis* exhibit antiproliferative activity against human lung adenocarcinoma cell line A549 without causing adverse toxic effect *in vitro* and *in vivo*. Food & Function 7(2): 1155–1165.

Salazar, R., H. Cortes-Funes, E. Casado, B. Pardo, A. Lopez-Martin, C. Cuadra, J. Tabernero, C. Coronado, M. Garcia, A. Soto Matos-Pita, B. Miguel-Lillo, M. Cullell-Young, J.L. Iglesias Dios and L. Paz-Ares. 2013. Phase I study of weekly kahalalide F as prolonged infusion in patients with advanced solid tumors. Cancer Chemotherapy and Pharmacology 72(1): 75–83.

Salem, T.A. and A.M. Ibrahim. 2011. Anticancer activity of egyptian marine alga *Ulva rigida*. International Journal of Health Sciences 5 2(1): 6–8.

Salem, W.M., H. Galal and F.N. El-deen. 2011. Screening for antibacterial activities in some marine algae from the red sea (Hurghada, Egypt). African Journal of Microbiology Research 5(15): 2160–2167.

Salgueiro, M.J., M.B. Zubillaga, A.E. Lysionek, M.I. Sarabia, R.A. Caro, T. De Paoli, A. Hager, R. Weill and J.R. Boccio. 2000. Zinc as an essential micronutrient: A review. Nutrition Research 20: 737–755.

Salgueiro, M.J., M.B. Zubillaga, A.E. Lysionek, R.A. Caro, R. Weill and J.R. Boccio. 2002. The role of zinc in the growth and development of children. Nutrition 18: 510–519.

Salta, M., J.A., Wharton, S.P. Dennington, P. Stoodley and K.R. Stokes. 2013. Anti-biofilm performance of three natural products against initial bacterial attachment. International Journal of Molecular Sciences 14: 21757–21780.

Salter, A.M. 2013. Dietary fatty acids and cardiovascular disease. Animal 7(S1): 163–171.

Salvador, N., A.G. Goméz, L. Lavelli and M.A. Ribera. 2007. Antimicrobial activity of Iberian macroalgae. Scientia Marina 71: 101–113.

Salvemini, D., P. Manning, B. Zweifel, K. Seibert, J. Connor, M. Currie, P. Needleman and J. Masferrer. 1995. Dual inhibition of nitric oxide and prostaglandin production contributes to the antiinflammatory properties of nitric oxide synthase inhibitors. Journal of Clinical Investigation 96: 301–308.

Samarakoon, K. and Y.J. Jeon. 2012. Bio-functionalities of proteins derived from marine algae—A review. Food Research International 48: 948–960.

Samee, H., Z.X. Li, H. Lin, J. Khalid and Y.C. Guo. 2009. Anti-allergic effects of ethanol extracts from brown seaweeds. Journal of Zhejiang University Science B 10: 147–153.

Samee, H., Z.X. Li, H. Lin, J. Khalid and B.P. Wang. 2009b. *In vivo* study of antiallergenicity of ethanol extracts from *Sargassum tenerrimum, Sargassum cervicorne* and *Sargassum graminifolium* turn. European Food Research and Technology 229: 435–441.

Samuels, G.J. 2006. *Trichoderma*: Systematics, the sexual state, and ecology. Phytopathology 96(2): 195–206.

Sandsdalen, E., T. Haug, K. Stensva and O.B. Styrvold. 2003. The antibacterial effect of a polyhydroxylated fucophlorethol from the marine brown alga, *Fucus vesiculosus*. World Journal of Microbiology and Biotechnology 19: 777–782.

Sangeetha, R., N. Bhaskar and V. Baskaran. 2009. Comparative effects of β-carotene and fucoxanthin on retinol deficiency induced oxidative stress in rats. Molecular and Cellular Biochemistry 331: 59–67.

Sangha, J.S., D. Fan, A.H. Banskota, R. Stefanova, W. Khan, J. Hafting, J. Craigie, A.T. Critchley and B. Prithiviraj. 2013. Bioactive components of the edible strain of red alga, *Chondrus crispus*, enhance oxidative stress tolerance in *Caenorhabditis elegans*. Journal of Functional Foods 5: 1180–1190.

Sanghvi, A.M. and Y.M. Lo. 2010. Present and potential industrial applications of macro- and microalgae. Recent Patents on Food, Nutrition & Agriculture 2: 187–194.

Sanjeewa, K.K.A., E.-A. Kim, K.-T. Son and Y.J. Jeon. 2016. Bioactive properties and potentials cosmeceutical applications of phlorotannins isolated from brown seaweeds: A review. Journal of Photochemistry & Photobiology, B: Biology 162: 100–105.

Sanjeewa, K.K.A., I.P.S. Fernando, E.A. Kim, G. Ahn, Y. Jee and Y.J. Jeon. 2017. Anti-inflammatory activity of a sulfated polysaccharide isolated from an enzymatic digest of brown seaweed *Sargassum horneri* in Raw 264.7 cells. Nutrition Research and Practice 11(1): 3–10.

San-Martin, A., R. Negrete and J. Rovirosa. 1991. Insecticide and acaricide activities of polyhalogenated monoterpenes from Chilean *Plocamium cartilagineum*. Phytochemistry 30(7): 2165–2169.

Sansom, C.E., L. Larsen, N.B. Perry, M.V. Berridge, E.W. Chia, J.L. Harper and V.L. Webb. 2007. An antiproliferative bis-prenylated quinone from the New Zealand brown alga *Perithalia capillaris*. Journal of Natural Products 70(12): 2042–2044.

Santhanam, R. 2015. Nutritional marine life. CRC Press, Taylor & Francis Group, Boca Raton, FL, 290 pp.

Santos, A.K.F.S, D.V. da Fonseca, P.R.R. Salgado, V.M. Muniz, P.A. Torres, N.S. Lira, C. da Silva Dias, L.C.M. Pordeus, J.M. Barbosa-Filho and R.N. de Almeida. 2015. Antinociceptive activity of *Sargassum polyceratium* and the isolation of its chemical components. Santos, A.K.F.S, D.V. da Fonseca, P.R.R. Salgado, V.M. Muniz, P.A. Torres, N.S. Lira, C. da Silva Dias, L.C.M. Pordeus, J.M. Barbosa-Filho and R.N. de Almeida. Revista Brasileira de Farmacognosia 25: 683–689.

Santos, A.O., P. Veiga-Santos, T. Ueda-Nakamura, D.B. Sudatti, É.M. Bianco, R.C. Pereira and C.V. Nakamura. 2010. Effect of elatol, isolated from red seaweed *Laurencia dendroidea*, on *Leishmania amazonensis*. Marine Drugs 8: 2733–2743.

Santos, A.O., E.A. Britta, E.M. Bianco, T. Ueda-Nakamura, B.P. Filho, R.C. Pereira, C.V. Nakamura. 2011. 4-Acetoxydolastane diterpene from the Brazilian brown alga *Canistrocarpus cervicornis* as antileishmanial agent. Marine Drugs 9: 2369–2383.

Santos, M.G.M., M.H.C. Lagrota, M.M.F.S. Miranda, Y. Yoneshigue-Valentin and M.D. Wigg. 1999. A screening for the antiviral effect of extracts from Brazilian marine algae against acyclovir resistant *Herpes simplex* virus type 1. Botanica Marina 42: 227–230.

Santoso, J., Y. Yoshie-Stark and T. Suzuki. 2006. Comparative contents of minerals and dietary fibres in several tropical seaweeds. Bulletin Teknologi Hasil Perikanan 9: 1–11.

Santoyo, S., I. Rodríguez-Meizoso, A. Cifuentes, L. Jaime, G. García-Blairsy Reina, F.J. Señorans and E. Ibáñez. 2009. Green processes based on the extraction with pressurized fluids to obtain potent antimicrobials from *Haematococcus pluvialis* microalgae. LWT–Food Science and Technology 42: 1213–1218.

Saravanakumar, D.E.M., P.I. Folb, B.W. Campbell and P. Smith. 2008. Antimycobacterial activity of the red alga *Polysiphonia virgata*. Pharmaceutical Biology 46(4): 254–260.

Saritha, K., A.E. Mani, M. Priyalaxmi and J. Patterson. 2013. Antibacterial activity and biochemical constituents of seaweed *Ulva lactuca*. Global Journal of Pharmacology 7: 276–282.

Sarithakumari, C.H., G.L. Renju and G.M. Kurup. 2013. Anti-inflammatory and antioxidant potential of alginic acid isolated from the marine algae, *Sargassum wightii* on adjuvant-induced arthritic rats. Inflammopharmacology 21(3): 261–268.

Sasidharan, S., I. Darah and M.K.M.J. Noordin. 2009. Screening antimicrobial activity of various extracts of *Gracilaria changii*. Pharmaceutical Biology 47: 72–76.

Sasikala, L., B. Durai and R. Rathinamoorthy. 2013. Manuka honey loaded chitosan hydrogel films for wound dressing applications. International Journal of PharmTech Research 5(4): 1774–1785.

Sasser, J.N., T.L. Kirkpatrick and R.A. Dybas. 1982. Efficacy of avermectins for root-knot control in tobacco. Plant Disease 66: 691–693.

Sastry, V.M.V.S. and G.R.K. Rao. 1994. Antibacterial substances from marine algae: Successive extraction using benzene, chloroform and methanol. Botanica Marina 37: 357–360.

Sato, M., T. Hosokawa, T. Yamaguchi, T. Nakano, K. Muramoto, T. Kahara, K. Funayama, A. Kobayashi and T. Nakano. 2002a. Angiotensin I-converting enzyme inhibitory peptides derived from Wakame (*Undaria pinnatifida*) and their antihypertensive effect in spontaneously hypertensive rats. Journal of Agricultural and Food Chemistry 21: 6245–6252.

Sato, M., T. Oba, T. Yamaguchi, T. Nakano, T. Kahara, K. Funayama, A. Kobayashi and T. Nakano. 2002b. Antihypertensive effects of hydrolysates of Wakame (*Undaria pinnatifida*) and their Angiotensin-I-converting enzyme inhibitory activity. Annals of Nutrition and Metabolism 6: 259–267.

Sato, M., T. Hosokawa, T. Yamaguchi, T. Nakano, K. Muramoto, T. Kahara, K. Funayama, A. Kobayashi and T. Nakano. 2005. Angiotensin I-converting enzyme inhibitory peptides derived from Wakame (*Undaria pinnatifida*) and their antihypertensive effect in spontaneously hypertensive rats. Journal of Agricultural and Food Chemistry 50: 6245–6252.

Sato, Y., K. Morimoto, T. Kubo, T. Sakaguchi, A. Nishizono, M. Hirayama and K. Hori. 2015. Entry inhibition of Influenza viruses with high mannose binding lectin ESA-2 from the red alga *Eucheuma Serra* through the recognition of viral hemagglutinin. Marine Drugs 13(6): 3454–3465.

Schaeffer, D.J. and V.S. Krylov. 2000. Anti-HIV Activity of extracts and compounds from algae and cyanobacteria. Carbohydrate Polymers 41: 185–190.

Schepetkin, I.A. and M.T. Quinn. 2006. Botanical polysaccharides: macrophage immunomodulation and therapeutic potential. International Immunopharmacology 6: 317–333.

Schiener, P., K.D. Black, M.S. Stanley and D.H. Green. 2015. The seasonal variation in the chemical composition of the Kelp species *Laminaria digitata, Laminaria hyperborea, Saccharina latissima* and *Alaria esculenta*. Journal of Applied Phycology 27(1): 363–373.

Schildgen, O., T. Glatzel, T. Geikowski, B. Scheibner, B. Matz, L. Bindl, M. Born, S. Viazov, A. Wilkesmann, G. Knöpfle, M. Roggendorf and A. Simon. 2005. Human metapneumovirus RNA in encephalitis patient. Emerging Infectious Diseases 11: 467–470.

Schmiedeberg, J.E.O. 1885. Gesellschaft deutscher naturforscher und arzte: Leipzig. Tageblatt der Versammlung 58: 427.

Schmitt, S. and N. Mamoozadeh. 2014. *Dictyopteris delicatula*. Tropical Field Phycology Field Book, STRI, Bocas Del Toro, Smithsonian Tropical Research Institute. Available online at: http://biogeodb.stri.si.edu/bocas_database/search/species/3527.

Schneider, T., K. Ehrig, I. Liewert and S. Alban. 2015. Interference with the Cxcl12/Cxcr4 axis as potential antitumor strategy: Superiority of a sulfated galactofucan from the brown alga *Saccharina latissima* and fucoidan over heparins. Glycobiology 25(8): 812–824.

Schoch, C.L., G.-H. Sung, F. López-Giráldez, J.P. Townsend, J. Miadlikowska, V. Hofstetter, B. Robbertse, P.B. Matheny, F. Kauff, Z. Wang, C. Gueidan, R.M. Andrie, K. Trippe, L.M. Ciufetti, A. Wynns, E. Fraker, B.P. Hodkinson, G. Bonito, J.Z. Groenewald, M. Arzanlou, G.S. de Hoog, P.W. Crous, D. Hewitt, D.H. Pfister, K. Peterson, M. Gryzenhout, M.J. Wingfield, A. Aptroot, S.O. Suh, M. Blackwell, D.M. Hillis, G.W. Griffith, L.A. Castlebury, A.Y. Rossman, H.T. Lumbsch, R. Lücking, B. Büdel, A. Rauhut, P. Diederich, D. Ertz, D.M. Geiser, K. Hosaka, P. Inderbitzin, J. Kohlmeyer, B. Volkmann-Kohlmeyer, L. Mostert, K. O'Donnell, H. Sipman, J.D. Rogers, R.A. Shoemaker, J. Sugiyama, R.C. Summerbell, W. Untereiner, P.R. Johnston, S. Stenroos, A. Zuccaro, P.S. Dyer, P.D. Crittenden, M.S. Cole, K. Hansen, J.M. Trappe, R. Yahr, F. Lutzoni and J.H. Spatafora. 2009. The Ascomycota tree of life: a phylum-wide phylogeny clarifies the origin and evolution of fundamental reproductive and ecological traits. Systematic Biology 58(2): 224–239.

Schoenaker, D.A.J.M., M. Toeller, N. Chaturvedi, J.H. Fuller and S.S. Soedamah-Muthu. 2012. Dietary saturated fat and fibre and risk of cardiovascular disease and all-cause mortality among type 1 diabetic patients: the EURODIAB Prospective Complications Study. Diabetologia 55: 2131–2141.

Sebaaly, C., S. Kassem, E. Grishina, H. Kanaan, A. Sweidan, M.S. Chmit and H.M Kanaan. 2014. Anticoagulant and antibacterial activities of polysaccharides of red algae *Corallina* collected from Lebanese coast. Journal of Applied Pharmaceutical Science 4(4): 30–37.

Seedevi, P., M. Moovendhan, S. Sudharsan, S. Vasanthkumar, A. Srinivasan, S. Vairamani and A. Shanmugam. 2015. Structural characterization and bioactivities of sulfatedpolysaccharide from *Monostroma oxyspermum*. International Journal of Biological Macromolecules 72: 1459–1465.

Seedevi, P., M. Moovendhan, S. Viramani and A. Shanmugam. 2017. Bioactive potential and structural chracterization of sulfated polysaccharide from seaweed (*Gracilaria corticata*). Carbohydrate Polymers 155: 516–524.

Seenivasan, R., H. Indu, G. Archana and S. Geetha. 2010. The antibacterial activity of some marine algae from south east coast of India. American-Eurasian Journal of Agricultural & Environmental Sciences 9(5): 480–489.

Sekar, D. and K. Kolanjinathan. 2015. Screning of antifungal activity of selected macroalgae against *Candida albicans* and *Candida glabrata*. Indo–Asian Journal of Multidisciplinary Research (IAJMR) 1(4): 315–319.

Sekar, S. and M. Chandramohan. 2008. Phycobiliproteins as a commodity: trends in applied research, patents and commercialization. Journal of Applied Phycology 20(2): 113–136.

Sekiya, M., H.F.K. Tsukamura, T. Imai, A.H.T. Kiuchi and A. Nakao. 2005. Intracellular signaling in the induction of apoptosis in a human breast cancer cell line by water extract of Mekabu. International Journal of Clinical Oncology 10(2): 122–126.

Selim, R.E., S.M. Ahmed, S.R. El-Zemity, S.S. Ramses and Y.T.A. Moustafa. 2015. Antifungal activity and seasonal variation of green alga (*Ulva lactuca*) extracts. Asian Journal of Agriculture and Food Sciences 5(05): 419–247.

Selim, S.A. 2012. Antimicrobial, antiplasmid and cytotoxicity potentials of marine algae *Halimeda opuntia* and *Sarconema filiforme* collected from Red Sea coast. International Journal of Medical, Health, Biomedical, Bioengineering and Pharmaceutical Engineering 6(1): 24–29.

Selkoe, D.J. 1997. Alzheimer's disease: genotypes, phenotypes, and treatments. Science 275: 630–631.

Selvaraj. R., M. Ayyappan, R. Shiva and S. Kabilan. 2006. Screening for antifungal activity of macro algae. Seaweed Res. Util 28(1): 89–98.

Selvin, J. and A.P. Lipton. 2004. Biopotentials of *Ulva fasciata* and *Hypnea musciformis* collected from the peninsular coast of India. Journal of Marine Science and Technology 12(1): 1–6.

Semenza, G.L. 2010. Vascular responses to hypoxia and ischemia. Arteriosclerosis, Thrombosis, and Vascular Biology 30: 648–652.

Sen, A.K., A.K. Das, N. Banerji, A.K. Siddhanta, K.H. Mody, B.K. Ramavat, V.D. Chauhan, J.R. Vedasiromoni and D.K. Ganguly. 1994. A new sulfated polysaccharide with potent blood anti-coagulant activity from the red seaweed *Grateloupia indica*. International Journal of Biological Macromolecules 16: 279–280.

Senni, K., F. Gueniche, A. Foucault-Bertaud, S. Igondjo-Tchen, F. Fioretti, S. Colliec-Jouault, P. Durand, J. Guezennec, G. Godeau and D. Letourneur. 2006. Fucoidan a sulfated polysaccharide from brown algae is a potent modulator of connective tissue proteolysis. Archives of Biochemistry and Biophysics 445: 56–64.

Senthilkumar, D. and S. Jayanthi. 2016. Partial characterization and anticancer activities of purified glycoprotein extracted from green seaweed *Codium decorticatum*. Journal of Functional Foods 25: 323–332.

SenthilKumar, P. and S. Sudha. 2012. Evaluation of alpha-amylase and alpha-glucosidase inhibitory properties of selected seaweeds from gulf o Mannar. International Research Journal of Pharmacy 3(8): 128–130.

Senthilkumar, P., S. Sudha and S. Prakash. 2014. Antidiabetic activity of aqueous extract of *Padina boergesenii* in streptozotocin-induced diabetic rats. International Journal of Pharmacy and Pharmaceutical Sciences 6(5): 418–422.

Sereno, D., A.C. da Silva, F. Mathieu-Daude and A. Ouaissi. 2007. Advances and perspectives in *Leishmania* cell based drug-screening procedures. Parasitology International 56: 3–7.

Serhan, C.N. 2005. Novel ω-3-derived local mediators in anti-inflammation and resolution. Pharmacology & Therapeutics 105: 7–21.

Serkedjieva, J., M. Konaklieva, S. Dimitrova-Konaklieva, V. Ivanova, K. Stefanov and S. Popov. 2000. Antiinfluenza virus effect of extracts from marine algae and invertebrates. Zeitschrift Fur Naturforschung C-A Journal of Biosciences 551(2): 87–93.

Serkedjieva, J. 2000b. Antiherpes virus effect of the red marine alga *Polysiphonia denudata*. Zeitschrift für Naturforschung C 55: 830–835.

Serkedjieva, J. 2004. Antiviral activity of the red marine alga *Ceramium rubrum*. Phytotherapy Research 18: 480–483.

Service, M.W. 1997. Mosquito (Diptera: Culicidae) dispersal: the long and short of it. Journal of Medical Entomology 34: 579–588.

Sethi, P. 2015. *Caulerpa taxifolia* (M. Vahl) C. Agardh against Tobacco Necrotic Virus (TNV). International Journal of Virology and Molecular Biology 4(1): 1–3.

Sezer, A.D., C. Erdal, H.P. Fatih, O. Zeki, L.B. Ahmet and A. Julide. 2008. Preparation of fucoidanechitosan hydrogel and its application as burn healing accelerator on rabbits. Biological and Pharmaceutical Bulletin 31(12): 2326–2333.

SF-Tischer, P.C., L.B. Talarico, M.D. Noseda, S.M.P.B. Guimarães, E.B. Damonte and M.E.R. Duarte. 2006. Chemical structure and antiviral activity of carrageenans from *Meristiella gelidium* against herpes simplex and dengue virus. Carbohydrate Polymers 63(4): 459–465.

Shakespeare, P. 2001. Burn wound healing and skin substitutes. Burns 27: 517–522.

Shalaby, E.A. 2011. Algae as promising organisms for environment and health. Plant Signaling & Behavior 6(9): 1338–1350.

Shamsabadi, F.T., A. Khoddami, S.G. Fard, R. Abdullah, H.H. Othman and S. Mohamed. 2013. Comparison of tamoxifen with edible seaweed (*Eucheuma cottonii* L.) extract in suppressing breast tumor. Nutrition and Cancer 65(2): 255–262.

Shamsian, B.S., N. Rezaei, M.T. Arzanian, S. Alavi, O. Khojasteh and A. Eghbali. 2009. Severe hypochromic microcytic anemia in a patient with congenital atransferrinemia. Pediatric Hematology and Oncology 26(5): 356–362.

Shanmugaasokan, L., V. Thirumalairaj, D. Geetharamani and T. Maruthupandi. 2013. Screening of seaweeds collected from southeast coastal area of India for α-amylase inhibitory activity, antioxidant activity and biocompatibility. International Journal of Pharmacy and Pharmaceutical Sciences 5: 240–244.

Shanmugam, A. and C. Palpandi. 2008. Biochemical composition and fatty acid profile of the green alga *Ulva reticulata*. Asian Journal of Biochemistry 3: 26–31.

Shanmugam, M. 1998. Studies on the biomedically and industrially important metabolites of Indian marine algae. PhD thesis, Bhavnagar University, 1998.

Shanmugam, M., K.H. Mody, R.M. Oza and B.K. Ramavat. 2001. Blood anticoagulant activity of a green marine alga *Codium dwarkense* (Codiaceae, Chlorophyta) in relation to its growth stages. Indian Journal of Marine Sciences 30(1): 49–52.

Shanmugam, M., K.H. Mody, B.K. Ramavat, A.S.K. Murthy and A.K. Siddhanta. 2002. Screening of Codiacean algae (Chlorophyta) of the Indian coasts for blood anticoagulant activity. Indian Journal of Geo-Marine Sciences 31: 33–38.

Shanmugam, S. and K.H. Mody. 2000. Heparinoid-active sulphated polysaccharides from marine algae as potential blood anticoagulant agents. Current Science 79: 1672–1683.

Shanmuganathan, B., D.S. Malar, S. Sathya and K.P. Devi. 2015. Antiaggregation potential of Padina gymnospora against the toxic Alzheimer's beta-amyloid peptide 25–35 and cholinesterase inhibitory property of its bioactive compounds. PLOS ONE: 1–14.

Shanmughapriya, S., A. Manilal, S. Sujith, J. Selvin, G.S. Kiran and K.N. Seenivasan. 2008b. Antimicrobial activity of seaweeds extracts against multiresistant pathogens. Annals of Microbiology 58(3): 535-541.

Shanmughapriya, S., J., Krishnaveni, J. Selvin, R. Gandhimathi, M. Arunkumar, T. Thangavelu, G.S. Kiran and K. Natarajaseenivasan. 2008. Optimization of extracellular thermotolerant alkaline protease produced by marine *Roseobacter* sp. (MMD040). Bioprocess and Biosystems Engineering 31: 427–433.

Shaoo, D. 2010. Common seaweeds of India. I.K. International Publishing House Pvt. Ltd. New Delhi, India, 196 pp.

Sharif, N., N. Munir, F. Saleem, F. Aslam and S. Naz. 2014. Prolific anticancer bioactivity of algal extracts (Review). American Journal of Drug Delivery and Therapeutics 1(2): 060–072.

Sharma, G.M. and M.K. Sahni. 1993. Marine proteins and clinical chemistry. pp. 153–180. *In*: Attaway, D.H. and O.R. Zaborsky (eds.). Marine Biotechnology, Vol. 1. Pharmaceutical and Bioactive Natural Products. Plenum Press, NY.

Sheeba, M.M. and J.J.P. Paula. 2014. Evaluation of antipyretic activity of methanol extract of *Ulva rigida* c. ag (green seaweed) in Hare island, Thoothukudu, Tamil Nadu, India. International Journal of Experimental Pharmacology 4(2): 105–108.

Shen, H., C.C. Kuo, J. Chou, A. Delvolve, S.N. Jackson, J. Post, A.S. Woods, B.J. Hoffer, Y. Wang and B.K. Harvey. 2009. Astaxanthin reduces ischemic brain injury in adult rats. The FASEB Journal 23: 1958–1968.

Sherman, W.C., C.A. Elveajem and E.B. Hart. 1934. Further Studies on the Availability of Iron in Biological Materials. The Journal of Biological Chemistry 107: 383–394.

Sheth, B.B. 1967. Alginate containing antacid compositions. US Patetent 3,326,755.

Sheu, J.H., S.Y. Huang, G.H. Wang and C.Y. Duh. 1997. Study on cytotoxic oxygenated desmosterols isolated from the red alga *Galaxaura marginata*. Journal of Natural Products 60(9): 900–903.

Sheu, J.H., G.H. Wang, P.J. Sung and C.Y. Duh. 1999. New cytotoxic oxygenated fucosterols from the brown alga *Turbinaria conoides*. Journal of Natural Products 62(2): 224–227.

Shevchenko, N.M., S.D. Anastyuk, R.V. Menshova, O.S. Vishchuk, V.I. Isakov, P.A. Zadorozhny, T.V. Sikorskaya and T.N. Zvyagintseva. 2015. Further studies on structure of fucoidan from brown alga *Saccharina gurjanovae*. Carbohydrate Polymers 121: 207–216.

Shi, D., F. Xu, J. He, J. Li, X. Fan and J. Han. 2008. Inhibition of bromophenols against PTP1B and anti-hyperglycemic effect of *Rhodomela confervoides* extract in diabetic rats. Chinese Science Bulletin 53: 2476–2479.

Shi, D., J. Li, S. Guo, H. Su and X. Fan. 2009. The antitumor effect of bromophenol derivatives *in vitro* and *Leathesia nana* extract *in vivo*. Chinese Journal of Oceanology and Limnology 27(2): 277–282.

Shi, P.-Y. (ed.). 2012. Molecular virology and control of flaviviruses. Caister Academic Press, Norfolk, UK, 358 pp.

Shi, Z.Y., Y.Z. Guo and Z. Wang. 2000. Pharmacological activity of fucoidan from *Laminaria japonica*. Journal of Shanghai Fisheries University 9: 268–271.

Shibata, T., K. Fujimoto, K. Nagayama, K. Yamaguchi and T. Nakamura. 2002. Inhibitory activity of brown algal phlorotannins against hyaluronidase. International Journal of Food Science & Technology 37:703–709.

Shibata, T., K. Nagayama, R. Tanaka, K. Yamaguchi and T. Nakamura. 2003. Inhibitory effects of brown algal phlorotannins on phospholipase A2s, lipoxygenases and cyclooxygenases. Journal of Applied Phycology 15: 61–66.

Shibata, T., K. Nagayama, R. Tanaka, K. Yamaguchi, Y. Sugiura, K. Matsuda, Y. Yamada, K. Imai, M. Kakinuma and H. Amano. 2008. Radical scavenging and hyarulonidase inhibitory activities of phlorotannins from the edible brown alga *Eisenia arborea*. Food Science and Technology Research 14: 595–598.

Shieh, M.-T., D. WuDunn, R.I. Montgomery, J.D. Esko and P.G. Spear. 1992. Cell surface receptors for *Herpes simplex* virus are heparan sulfate proteoglycans. The Journal of Cell Biology 116: 1273–1281.

Shim, S.Y., Q.T. Le, S.H. Lee and S.K. Kim. 2009. *Ecklonia cava* extract suppresses the high-affinity IgE receptor, FcεRI expression. Food and Chemical Toxicology 47(3): 555–560.

Shimamura, N. 2010. Agar. *In*: Rediscovering the treasures of food. The Tokyo Foundation. Available online at: http://www.tokyofoundation.org/en/topics/japanese-traditionalfoods/vol.-4-agar

Shimizu, H., T. Koyama, S. Yamada, S.A. Lipton and T. Satoh. 2015. Zonarol, a sesquiterpene from the brown algae *Dictyopteris undulata*, provides neuroprotection by activating the Nrf2/ARE pathway. Biochemical and Biophysical Research Communications 457(4): 718–722.

Shin, E.S., H.J. Hwang, I.H. Kim and T.J. Nam. 2011. A glycoprotein from *Porphyra yezoensis* produces anti-inflammatory effects in liposaccharidestimulated macrophages via the TLR4 signaling pathway. International Journal of Molecular Medicine 28(5): 809–815.

Shin, H.-C., H.J. Hwang, K.J. Kang and B.H. Lee. 2006. An antioxidative and antiinflammatory agent for potential treatment of osteoarthritis from Ecklonia cava. Archives Pharmacan Research 29(2): 165–171.

Shin, H.J., S.J. Oh, S.I. Kim, H. Won Kim and J.-H. Son. 2009. Conformational characteristics of β-Glucan in Laminarin probed by terahertz spectroscopy. Applied Physics Letters 94(11): 111911–111911-3.

Shin, S.S. and M. Yoon. 2012. The herbal composition GGExl8 from *Laminaria japonica*, *Rheum palmatum*, and *Ephedra sinica* inhibits high-fat diet-induced hepatic steatosis via hepatic PPARα activation. Pharmaceutical Biology 50: 1261–1268.

Shiratori, K., K. Ohgami, I. Ilieva, X.H. Jin, Y. Koyama, K. Miyashita, K. Yoshida, S. Kase and S. Ohno. 2005. Effects of fucoxanthin on lipopolysaccharide-induced inflammation *in vitro* and *in vivo*. Experimental Eye Research 81: 422–428.

Shoeib, N.A., M.C. Bibby, G. Blunden, P.A. Linley, D.J. Swaine, R.T. Wheelhouse and C.W. Wright. 2004. *In-vitro* cytotoxic activities of the major bromophenols of the red alga *Polysiphonia lanosa* and some novel synthetic isomers. Journal of Natural Products 67(9): 1445–1449.

Shojaei, A.H., J. Paulson and S. Honary. 2000. Evaluation of poly(acrylic acid-co-ethylhexyl acrylate) films for mucoadhesive transbuccal drug delivery: factors affecting the force of mucoadhesion. Journal of Controlled Release 67: 223–232.

Shum, D., J.L. Smith, A.J. Hirsch, B. Bhinder, C. Radu, D.A. Stein, J.A. Nelson, K. Fruh and H. Djaballah. 2010. High-content assay to identify inhibitors of dengue virus infection. ASSAY and Drug Development Technologies 8: 553–570.

Siahaan, E.A., P. Pendleton, H.-C. Woo and B.-S. Chun. 2014. Brown seaweed (*Saccharina japonica*) as an edible natural delivery matrix for allyl isothiocyanate inhibiting food-borne bacteria. Food Chemistry 152: 11–17.

Siamopoulou, P., A. Bimplakis, D. Iliopoulou, C. Vagias, P. Cos, D. Vande Berghe and V. Roussis. 2004. Diterpenes from the brown algae *Dictyota dichotoma* and *Dictyota linearis*. Phytochemistry 65: 2025–2030.

Siddhanta, A.K., M. Shanmugam, K.H. Mody, A.M. Goswami and B.K. Ramavat. 1999. Sulphated polysaccharides of *Codium dwarkense* Boergs. from the west coast of India: Chemical composition and blood anticoagulant activity. International Journal of Biological Macromolecules 26: 151–154.

Sieburth, J.M. 1964. Antibacterial substances produced by marine algae. Development in Industrial Microbiology 5: 124–134.

Silkina, A., A. Bazes, J.-L. Mouget and N. Bourgougnon. 2012. Comparative efficiency of macroalgal extracts and booster biocides as antifouling agents to control growth of three diatom species. Marine Pollution Bulletin 64(10): 2039–2046.

Silva, G.A., T.F.S. Domingos, R.R. Fonseca, E.F. Sanchez, V.L. Teixeira and A.L. Fuly. 2015b. The red seaweed *Plocamium brasiliense* shows anti-snake venom toxic effects. Journal of Venomous Animals and Toxins including Tropical Diseases 21(2): 1–9.

Silva, L.M., V. Lima, M.L. Holanda, P.G. Pinheiro, J.A. Rodrigues, M.E. Lima and N.M. Benevides. 2010. Antinociceptive and antiinflammatory activities of lectin from marine red alga *Pterocladiella capillacea*. Biological and Pharmaceutical Bulletin 33: 830–835.

Silva, P.J. 2015. Screening of biotechnological potential of *Osmundea pinnatifida*: cultivation trials and biological activities. Master Dissertation in Vegetal Biotechnology, Department of Life Sciences, Faculty of Sciences and Technology, University of Coimbra, 92 pp.

Silva, R.G. 2014. Estudos em *Bifurcaria bifurcata*: aquacultura, atividade antioxidante e antifúngica. Dissertação de Mestrado em Biologia, Departamento de Ciências da Vida, Universidade de Coimbra, Portugal, 83 pp. Available online at: https://estudogeral.sib.uc.pt/handle/10316/33257.

Silva, T.M.A., L.G. Alves, K.C.S. de Queiroz, M.G.L. Santos, C.T. Marques, S.F. Chavante, H.A.O. Rocha and E.L. Leite. 2005. Partial characterization and anticoagulant activity of a heterofucan from the brown seaweed *Padina gymnospora*. Brazilian Journal of Medical and Biological Research 38: 523–533.

Simarro, P.P., A. Diarra, J.A. Ruiz Postigo, J.R. Franco and J.G. Jannin. 2011. The Human African trypanosomiasis control and surveillance programme of the World Health Organization 2000–2009: The Way Forward. PLOS Neglected Tropical Diseases 5(2): 0001007.

Simonian, N.A. and J.T. Coyle. 1996. Oxidative stress in neurodegenerative disease. Annual Review of Pharmacology and Toxicology 36: 83–106.

Simopoulos, A.P. 2002. Omega-3 fatty acids in inflammation and autoimmune diseases. The Journal of the American College of Nutrition 21(6): 495–505.

Simpi, C.C., C.V. Nagathan, S.R. Karajgi and N.V. Kalyane. 2013. Evaluation of marine brown algae *Sargassum ilicifolium* extract for analgesic and anti-inflammatory activity. Pharmacognosy Research 5(3): 146–149.

Sims, J.J., G.H. Lin and R.M. Wing. 1974. Marine natural products X. elatol, a halogenated sesquiterpene alcohol from the red alga *Laurencia elata*. Tetrahedron Letters 15: 3487–3490.

Singer, M. 1981. The fanatic's ecstatic aromatic guide to onion, garlic shallots and leeks. Prentice-Hall, Englewood Cliffs, NJ, 262 pp.

Singh, G. 1998. Recent considerations in nonsteroidal antiinflammatory drug gastropathy. American Journal of Medicine 105(1B): 315–385.

Singh, I.P. and S.B. Bharate. 2006. Phloroglucinol compounds of natural origin. Natural Product Reports 23(4): 558–591.

Singh, J. and S. Gu. 2010. Commercialization potential of microalgae for biofuels production. Renewable & Sustainable Energy Reviews 14: 2596–2610.

Singh, M.J. and C.K. Raadha. 2015. Studies on the antimicrobial potency of the marine algae—*Gracilaria corticata* var. *cylindrica* and *Hydroclathrus clathratus*. Advances in Life Sciences and Health 2(1): 50–55.

Singh, V., S. Kaul, R. Chander and N.K. Kapoor. 1990. Stimulation of low density lipoprotein receptor activity in liver membrane of Guggulsterone treated rats. Pharmacological Research 22(1): 37–44.

Singh, V., N.K. Kapoor and B.N. Dhawan. 1992a. Effect of Picroliv on protein and nucleic acid synthesis. Indian Journal of Experimental Biology 30(1): 68–69.

Singh, V., P.K. Visen, G.K. Patnaik, N.K. Kapoor and B.N. Dhawan. 1992b. Effect of Picroliv on low density lipoprotein receptor binding of rat hepatocytes in hepatic damage induced by paracetamol. Indian Journal of Biochemistry and Biophysics 29(5): 428–432.

Siqueira, R.C., M.S. da Silva, D.B. de Alencar, F. Pires Ade, N.M. de Alencar, M.G. Pereira, B.S. Cavada, A.H. Sampaio, W.R. Farias and A.M. Assreuy. 2011. *In vivo* antiinflammatory effect of a sulfated polysaccharide isolated from the marine brown algae *Lobophora variegata*. Pharmaceutical Biology 49(2): 167–174.

Sirbu, R., T. Negreanu-Pirjol, A. Bechir, M. Maris and D. Maris. 2008. Rheological characterization of collagen gels from marine resources of Black Sea and chlorhexidine salt for using in dental medicine. World Academy of Science, Engineering and Technology 19: 604–609.

Siri-Tarino, P.W., Q. Sun, F.B. Hu and R.M. Krauss. 2010. Meta-analysis of prospective cohort studies evaluating the association of saturated fat with cardiovascular disease. American Journal of Clinical Nutrition 91: 535–546.

Sithranga, B.N. and K. Kathiresan. 2010. Anticancer drugs from marine flora: an overview. Journal of Oncology, Article ID 214186: 1–18.

Sivagnanam, S.P., S. Yin, J.H. Choi, Y.B. Park, H.C. Woo and B.S. Chun. 2015. Biological properties of fucoxanthin in oil recovered from two brown seaweeds using supercritical CO_2 extraction. Marine Drugs 13: 3422–3442.

Sivagnanavelmurugan, M., G. Karthik Ramnath, B. Jude Thaddaeus, A. Palavesam, and G. Immanuel. 2015. Effect of *Sargassum wightii* fucoidan on growth and disease resistance to *Vibrio parahaemolyticus* in *Penaeus monodon* post larvae. Aquaculture Nutrition 21(6): 960–969.

Sivakumar, S. 2014. *In vitro* activity of seaweed extracts collected from Gulf of Mannar coast islands Tamilnadu on clinical isolates. International Journal of Pharmaceutical and Biological Archive 5: 99–103.

Sivaperumal, P., T.V. Sankar and P.G. Viswanathan. 2007. Heavy Metal Concentration in Fish, Shell Fish and Fish Products from Internal Markets of India Vis-a-Vis International Standards. Food Chemistry 102: 612–620.

Skoler-Karpoff, S., G. Ramjee, K. Ahmed, L. Altini, M.G. Plagianos, B. Friedland, S. Govender, A. De Kock, N. Cassim, T. Palanee, G. Dozier, R. Maguire and P. Lahteenmaki. 2008. Efficacy of Carraguard for prevention of HIV infection in women in South Africa: a randomised, double-blind, placebo-controlled trial. The Lancet 372(9654): 1977–1987.

Škrovánková, S. 2011. Seaweed vitamins as nutraceuticals. Advances in Food and Nutrition Research 64: 357–369.

Sloots, T.P., D.M. Whiley, S.B. Lambert and M.D. Nissen. 2008. Emerging respiratory agents: new viruses for old diseases. Journal of Clinical Virology 42: 233–243.

Smit, J.A. 2004. Medicinal and pharmaceutical uses of seaweed natural products: A review. Journal of Applied Phycology 16: 245–262.

Smith, D.B. and H.W. Cook. 1954. Physical studies on carrageenin and carrageenin fractions. Archives of Biochemistry and Biophysics 45: 23–233.

Smith, D.B., A.N. Oneill and A.S. Perlin. 1955. Studies on the heterogeneity of carrageenin. Canadian Journal of Chemistry 33(8): 1352–1360.

Smith, G.S. and O.N. Carvil. 1977. Field screening of commercial and experimental soybean cultivars for their reaction to *Macrophomina phaseolina*. Plant Disease 81: 363–368.

Smith, H.M. 1904. The seaweed industries of Japan. Bulletin of the Bureau of Fisheries (USA) 24: 133–181.

Smith, J.L., G. Summers and R. Wong. 2010. Nutrient and heavy metal content of edible seaweeds in New Zealand. New Zealand Journal of Crop and Horticultural Science 38: 19–28.

Smyrniotopoulos, V., L.-A.D. Abatis, C. Tziveleka, V. Tsitsimpikou, A. Loukis Roussis and C. Vagias. 2003. Acetylene sesquiterpenoid esters from the green alga *Caulerpa prolifera*. Journal of Natural Products 66: 21–25.

Smyrniotopoulos, V., A. Quesada, C. Vagias, D. Moreau, C. Roussakis and V. Roussis. 2008. Cytotoxic bromoditerpenes from the red alga *Sphaerococcus coronopifolius*. Tetrahedron 64: 5184–5190.

Smyrniotopoulos, V., C. Vagiasa and V. Roussis. 2009. Sphaeroane and neodolabellane diterpenes from the red alga *Sphaerococcus coronopifolius*. Marine Drugs 7: 184–195.

Smyrniotopoulos, V., C. Vagias, C. Bruyère, D. Lamoral-Theys, R. Kiss and V. Roussis. 2010. Structure and *in vitro* antitumor activity evaluation of brominated diterpenes from the red alga *Sphaerococcus coronopifolius*. Bioorganic & Medicinal Chemistry 18(3): 1321–1330.

Snelders, E., H.A. van der Lee, J. Kuijpers, A.J. Rijs, J. Varga, R.A. Samson, E. Mellado, A.R.T. Donders, W.J.G. Melchers and P.E. Verweij. 2008. Emergence of Azole Resistance in Aspergillus Fumigatus and Spread of a Single Resistance Mechanism. PLoS. Medicine 5(11): 1629–1637.

Soares, A.R., J.L. Abrantes, T.M.L. Souza, C.F.L. Fontes, R.C. Pereira, I.C. de Palmer Paixao Frugulhetti and V.L. Teixeira. 2007. *In vitro* antiviral effect of meroditerpenes isolated from the Brazilian seaweed *Stypopodium zonale* (Dictyotales). Planta Medica 73: 1221–1224.

Soares, A.R., M. Robaina, G.S. Mendes, T.S. Silva, L. Gestinari, O.S. Pamplona, Y. Yoneshigue-Valentin, C.R. Kaiser and M.T.V. Romanos. 2012. Antiviral activity of extracts from Brazilian seaweeds against *Herpes simplex* virus. Revista Brasileira de Farmacognosia 22: 714–723.

Soares, D.C., T.C. Calegari-Silva, U.G. Lopes, V.L. Teixeira, I.C.N. de Palmer Paixão, C. Cirne-Santos, D.C. Bou-Habib and E.M. Saraiva. 2012b. Dolabelladienetriol, a compound from *Dictyota pfaffii* algae, inhibits the infection by *Leishmania amazonensis*. PLOS Neglected Tropical Diseases 6: e1787.

Soares, F. 2015. Antifungal, antibacterial and antiviral activity of *Chondracathus teedei* var. *lusitanicus* (Gigartinaceae, Rhodophyta). MsC Thesis, University of Coimbra, Coimbra, Portugal, 154 pp.

Soares, F., C. Fernandes, P. Silva, L. Pereira and T. Gonçalves. 2016. Antifungal activity of carrageenan extracts from the red alga *Chondracanthus teedei* var. *lusitanicus*. Journal of Applied Phycology (2016): 1–8.

Soeda, S., S. Ishida, H. Shimeno and A. Nagamatsu. 1994. Inhibitory effect of oversulfated cucoidan on invasion through reconstituted basement membrane by murine Lewis lung carcinoma. GANN Japanese Journal of Cancer Research 85(11): 1144–1150.

Soetan, K.O., C.O. Olaiya and O.E. Oyewole. 2010. The importance of mineral elements for humans, domestic animals and plants: a review. African Journal of Food Science 4(5): 200–222.

Sokolova, E.V., A.O. Barabanova, R.N. Bogdanovich, V.A. Khomenko, T.F. Solov'eva and I.M. Yermak. 2011. *In vitro* antioxidant properties of red algal polysaccharides. Biomedicine & Preventive Nutrition 1(3): 161–167.

Sokolova, E.V., L.N. Bogdanovich, T.B. Ivanova, A.O. Byankina, S.P. Kryzhanovskiy and I.M. Yermak. 2014. Effect of carrageenan food supplement on patients with cardiovascular disease results in normalization of lipid profile and moderate modulation of immunity system markers. PharmaNutrition 2: 33–37.

Solano, F., S. Briganti, M. Picardo and G. Ghanem. 2006. Hypopigmenting agents: an updated review on biological, chemical and clinical aspects. Pigment Cell Research 19: 550–571.

Soliman, F.M., S.F. El-Tohamy, M.M. Fathy, A. Ramdan, N.A. Afify and O.A. Sanad. 1994. Phytochemical and biological investigation of *Jania rubens* (L.) Lamx. Amino acids, proteins, nitrogenous bases and biological screening. Journal of Drug Research of Egypt 21: 155–164.

Somasundarama, S.N., S. Shanmugama, B. Subramaniana and R. Jaganathan. 2016. Cytotoxic effect of fucoidan extracted from *Sargassum cinereum* on colon cancer cell line HCT-15. International Journal of Biological Macromolecules 91: 1215–1223.

Son, Y.K., S.E. Jin, H.R. Kim, C.W. Woo, H.A. Jung and J.S. Choi. 2011. Inhibitory activities of the edible brown alga *Laminaria japonica* on glucose-mediated protein damage and rat lens aldose reductase. Fisheries Science 77: 1069–1079.

Song, D., G. Blond and A. Furstner. 2003. Study towards bioactive pyrone derivatives from the marine red alga *Phacelocarpus labillardieri*. Tetrahedron 59: 6899–6904.

Song, T.Y., C. Chen, N.C. Yang, C.S. Fu, Y.T. Chang and C.L. Chen. 2009. The correlation of *in vitro* mushroom tyrosinase activity with cellular tyrosinase activity and melanin formation in melanoma cells A2058. Journal of Food and Drug Analysis 17: 156–162.

Songa, A. 1995. Variation and survival of *Macrophomina phaseolina* in relation to screening common bean (*Phaseolus vulgaris* L.) for resistance. PhD thesis, University of Reading, UK, pp. 7–32, 89–90.

Souza, E.T., A.C. Queiroz, G.E.C. Miranda, V.P. Lorenzo, E.F. Silva, T.L.M. Freire-Dias, Y.K. Cupertino-Silva, G.M.A. Melo, B.V.O. Santos, M.C.O. Chaves and M.S. Alexandre-Moreira. 2009. Antinociceptive activities of crude methanolic extract and phases, *n*-butanolic, chloroformic and ethyl acetate from *Caulerpa racemosa* (Caulerpaceae). Revista Brasileira de Farmacognosia 19: 115–120.

Souza, R.B., A.F. Frota, R.S. Sousa, N.A. Cezario, T.B. Santos, L.M.F. Souza, C.O. Coura, V.S. Monteiro, G. Cristino Filho, S.M. Vasconcelos, R.M. da Cunha, L.M. Aguiar and N.M. Benevides. 2017. Neuroprotective effects of sulphated agaran from marine alga *Gracilaria cornea* in rat 6-hydroxydopamine Parkinson's disease model: Behavioural, neurochemical and transcriptional alterations. Basic & Clinical Pharmacology & Toxicology 120(2): 159–170.

Spavieri, J., A. Allmendinger, M. Kaiser, R. Casey, S. Hingley-Wilson, A. Lalvani, M.D. Guiry, G. Blunden and D. Tasdemir. 2010. Antimycobacterial, antiprotozoal and cytotoxic potential of twenty-one brown algae (Phaeophyceae) from British and Irish waters. Phytotherapy Research 24(11): 1724–1729.

Spavieri, J., M. Kaiser, R. Casey, S. Hingley-Wilson, A. Lalvani, G. Blunden and D. Tasdemir. 2010b. Antiprotozoal, antimycobacterial and cytotoxic potential of some british green algae. Phytotherapy Research 24: 1095–1098.

Spavieri, J., A. Allmendinger, M. Kaiser, M.A. Itoe, G. Blunden, M.M. Mota and D. Tasdemir. 2013. Assessment of dual life stage antiplasmodial activity of British seaweeds. Marine Drugs 11(10): 4019–4034.

Spears, K. 1988. Developments in food colorings: the natural alternatives. Trends in Biotechnology 6: 283–288.

Spicer, W.J. 2007. Clinical microbiology and infectious diseases: An illustrated colour text. Churchill Livingstone, Elsevier, Edinburgh, 264 pp.

Spilman, C.H., K.L. Hart, D.M. Dinh and T.J. Vidmar. 1992. Spontaneous hypercholesterolemia in Cynomolgus Monkeys: evidence for defective low-density lipoprotein catabolism. Biochimica et Biophysica Acta 22(1): 26–34.

Spoden, G., L. Kühling, N. Cordes, B. Frenzel, M. Sapp, K. Boller, L. Florin and M. Schelhaas. 2013. Human papillomavirus types 16, 18, and 31 share similar endocytic requirements for entry. Journal of Virology 87(13): 7765–7773.

Springer, G.F., H.A. Wurzel, G.M.J. McNeal, N.J. Ansel and M.F. Doughty. 1957. Isolation of anticoagulant fractions from crude fucoidan. Proceedings of the Society for Experimental Biology and Medicine 94: 4004–409.

Spronk, H.M.H., J.W.P. Govers-Riemslag and H. Cate. 2003. The blood coagulation system as a molecular machine. Bioessays 256: 1220–1228.

Stabili, L., M.I. Acquaviva, F. Biandolino, R.A. Cavallo, S.A. De Pascali, F.P. Fanizzi, M. Narracci, A. Petrocelli, E. Cecere. 2012. The lipidic extract of the seaweed *Gracilariopsis longissima* (Rhodophyta, Gracilariales): a potential resource for biotechnological purposes? New Biotechnology 29(3): 443–450.

Stabili, L., M.I. Acquaviva, F. Biandolino, R.A. Cavallo, S.A. De Pascali, F.C. Fanizzi, M. Narracci, E. Cecere and A. Petrocelli. 2014. Biotechnological potential of the seaweed *Cladophora rupestris* (Chlorophyta, Cladophorales) lipidic extract. New Biotechnology 31: 436–444.

Stabili, L., S. Fraschetti, M.I. Acquaviva, R.A. Cavallo, S.A. Pascali, F.P. Fanizzi, C. Gerardi, M. Narracci and L. Rizzo. 2016. Antibacterial and antioxidant activity of the Mediterranean invasive alga *Caulerpa cylindracea*: can they transform the invasion into a gain? Marine Drugs 14: 1–4.

Stanley, N. 1987. Production, properties and uses of carrageenan. FAO Fisheries Technical Paper, pp. 116–146.

Stanojković, T.P., K. Šavikin, G. Zdunić, Z. Kljajić, N. Grozdanić and J. Antić. 2013. *In vitro* antitumoral activities of *Padina pavonia* on human cervix and breast cancer cell lines. Journal of Medicinal Plants Research 7(8): 419–424.

Starr, J. 2010. The older adult with intellectual disability. *In*: Fillit, H.M., K. Rockwood, K. Woodhouse (eds.). Textbook of geriatric medicine and gerontology–Brocklehurst's. Saunders/Elservier, Philadelphia, 1119 pp.

Stefanov, K., M. Konaklieva, E.Y. Brechany and W.W. Christie. 1988. Fatty acid composition of some algae from the Black Sea. Phytochemistry 27: 3495–3497.

Stein, E.M., D.X. Andreguetti, C.S. Rocha, M.T. Fujii, M.S. Baptista, P. Colepicolo and G.L. Indig. 2011. Search for cytotoxic agents in multiple *Laurencia* complex seaweed species (Ceramiales, Rhodophyta) harvested from the Atlantic Ocean with emphasis on the Brazilian state of Espírito Santo. Revista Brasileira de Farmacognosia 21: 239–243.

Stein, E.M., P. Colepicolo, F.A.K. Afonso and M.T. Fujii. 2011b. Screening for antifungal activities of extracts of the Brazilian seaweed genus *Laurencia* (Ceramiales, Rhodophyta). Brazilian Journal of Pharmacognosy 21(2): 290–295.

Stein, E.M., L.P. Machado, H.K. Roffato, P.A. Miyasato, E. Nakano, P. Colepicolo and D.X. Andreguetti. 2015. Antischistosomal activity from Brazilian marine algae. Revista Brasileira de Farmacognosia 25: 663–667.

Stengel, D.B., S. Connan and Z.A. Popper. 2011. Algal chemodiversity and bioactivity: Sources of natural variability and implications for commercial application. Biotechnology Advances 29(5): 483–501.

Stephanie, B., D. Eric, F.M. Sophie, B. Christian and G.L. Yu. 2010. Carrageenan from *Solieria chordalis* (Gigartinales): Structural analysis and immunological activities of the low molecular weight fractions. Carbohydrate Polymers 81: 448–460.

Stephens, P.R.S., C.C. Cirne-Santos, C. de Souza-Barros, V.L. Teixeira, L.A.D. Carneiro, L.S.C. Amorim, J.S.P. Ocampo, L.R.R. Castello-Branco and I.C.N.P. Paixão. 2017. Diterpene from marine brown alga *Dictyota fri*abilis as a potential microbicide against HIV-1 in tissue explants. Journal of Applied Phycology 29(2): 775–780.

Stevan, F.R., M.B. Oliveira, D.F. Bucchi, Noseda, M. Iacomini and M.E. Duarte. 2001. Cytotoxic effects against HeLa cells of polysaccharides from seaweeds. Journal of Submicroscopic Cytology and Pathology 33(4): 477–484.

Stirk, W.A., D.L. Reinecke and J.V. Staden. 2007. Seasonal variation in antifungal, antibacterial and acetylcholinesterase activity in seven South African seaweeds. Journal of Applied Phycology 19(3): 271–276.

Stout, E.P., A.P. Hasemeyer, A.L. Lane, T.M. Davenport, S. Engel, M.E. Hay, C.R. Fairchild, J. Prudhomme, K. Le Roch, W. Aalbersberg and J. Kubanek. 2009. Antibacterial neurymenolides from the Fijian red alga *Neurymenia fraxinifolia*. Organic Letters 11(1): 225–228.

Strand, B.L., O. Gaserod, B. Kulseng, T. Espevik and G. Skjak-Braek. 2002. Alginate-polylisine-alginate microcapsules: Effect of size reduction on capsule properties. Journal of Microencapsulation 19: 612–630.

Streusand, V.J., I. Björk, P.G.W. Gettins, M. Petitou and S.T. Olson. 1995. Mechanism of acceleration of antithrombin-proteinase reactions by low affinity heparin. The Journal of Biological Chemistry 270: 9043–9051.

Su, X., S. Zivanovic and D.H. D'Souza. 2009. Effect of chitosan on the infectivity of murine norovirus, feline calicivirus, and bacteriophage MS2. Journal of Food Protection 72: 2623–2628.

Subba, R.G., P. Lakshmi and E. Manjula. 2010. Antimicrobial activity of seaweeds *Gracilaria*, *Padina* and *Sargassum* spp. on clinical and phytopathogens. International Journal of Chemical and Analytical Science (6): 114–117.

Suetsuna, K. and T. Nakano. 2000. Identification of an antihypertensive peptide from peptic digest of Wakame (*Undaria pinnatifida*). The Journal of Nutritional Biochemistry 9: 450–454.

Suetsuna, K., K. Maekawa and J.R. Chen. 2004. Antihypertensive effects of *Undaria pinnatifida* (Wakame) peptide on blood pressure in spontaneously hypertensive rats. The Journal of Nutritional Biochemistry 15: 267–272.

Sugahara, T., Y. Ohama, A. Fukuda, M. Hayashi, A. Kawakubo and K. Kato. 2001. The cytotoxic effect of *Eucheuma serra* agglutinin (ESA) on cancer cells and its application to molecular probe for drug delivery system using lipid vesicles. Cytotechnology 36(1-3): 93–99.

Suganthy, N., S.K. Pandian and K.P. Devi. 2010. Neuroprotective effect of seaweeds inhabiting South Indian coastal area (Hare Island, Gulf of Mannar marine biosphere reserve): Cholinesterase inhibitory effect of *Hypnea valentiae* and *Ulva reticulata*. Neuroscience Letters 468: 216–219.

Sugawa-Katayama, Y. and M. Katayama. 2009. Release of minerals from dried Hijiki, *Sargassum fusiforme* (Harvey) Setchell, during water-soaking. Trace Nutrients Research 24: 106–109.

Sugawara, I., W. Itoh, S. Kimura, S. Mori and K. Shimada. 1989. Further characterization of sulfated homopolysaccharides as anti-HIV agents. Experientia 45: 996–998.

Sugawara, T., V. Baskaran, W. Tsuzuki and A. Nagao. 2002. Brown algae fucoxanthin is hydrolyzed to fucoxanthinol during absorption by Caco-2 human intestinal cells and mice. Journal of Nutrition 132(5): 946–951.

Sugawara, T., K. Matsubara, R. Akagi, M. Mori and T. Hirata. 2006. Antiangiogenic activity of brown algae fucoxanthin and its deacetylated product, Fucoxanthinol. Journal of Agricultural and Food Chemistry 54(26): 9805–9810.

Sugiura, Y., Y. Takeuchi, M. Kakinuma and H. Amano. 2006. Inhibitory effects of seaweeds on histamine release from rat basophile leukemia cells (RBL-2H3). Fisheries Science 72(6): 1286–1291.

Sugiura, Y., K. Matsuda, Y. Yamada, M. Nishikawa, K. Shioya, H. Katsuzaki, K. Imai and H. Amano. 2006b. Isolation of a new anti-allergic phlorotannin, phlorofucofuroeckol-B, from an edible brown alga, *Eisenia arborea*. Bioscience, Biotechnology, and Biochemistry 70: 2807–2811.

Sugiura, Y., K. Matsuda, Y. Yamada, M. Nishikawa, K. Shioya, H. Katsuzaki, K. Imai and H. Amano. 2007. Anti-allergic phlorotannins from the edible brown alga, *Eisenia arborea*. Food Science and Technology Research 13: 54–60.

Sugiura, Y., K. Matsuda, T. Okamoto, M. Kakinuma and H. Amano. 2008. Anti-allergic effects of the brown alga *Eisenia arborea* on Brown Norway rats. Fisheries Science 74: 180–186.

Sugiura, Y., K. Matsuda, T. Okamoto, Y. Yamada, K. Imai, T. Ito, M. Kakinuma and H. Amano. 2009. The inhibitory effects of components from a brown alga, *Eisenia arborea*, on degranulation of mast cells and eicosanoid synthesis. Journal of Functional Foods 1: 387–393.

Sugiura, Y., K. Imai and H. Amano. 2011. The Anti-allergic and anti-inflammatory effects of seaweed polyphenol (phlorotannin). Foods & Food Ingredients Jornal of Japan 216(1).

Sujatha, L., T. Lalitha Govardhan and G.S. Rangaiah. 2012. Antibacterial activity of green seaweeds on oral bacteria. Indian Journal of Natural Products and Resources 3(3): 328–333.

Sukatar, A., N.U. Karabay-Yavaşsoglu, G. Ozdemir and Z. Horzum. 2006. Antimicrobial activity of volatile component and various extracts of *Enteromorpha linza* (Linnaeus) J. Agardh from the Coast of Izmir, Turkey. Annals of Microbiology 56(3): 275–279.

Sukumaran, P. and R. Thevanathan. 2014. Antifungal activity of *Pithophora oedogonia* against plant pathogens. Report and Opinion 6(8): 67–71.

Sultana, V., S. Ehteshamul-Haque, J. Ara and V.U. Ahmad. 2000. Utilization of seaweeds for the control of root diseases of tomato. pp. 193–206. *In*: Ahmad, V.U (ed.). Proceedings of National ONR Symposium on Arabian Sea as a Resource of Biological Diversity. HEJRIC, University of Karachi, Pakistan.

Sultana, V., S.E. Haque, J. Ara and M. Athar. 2005. Comparative efficacy of brown, green and red seaweed in the control of root infecting fungi in okra. International Journal of Environmental Science and Technology 2(2): 129–132.

Sultana, V., S. Ehteshamul-Haque, J. Ara and M. Athar. 2010. Effect of brown seaweeds and pesticides on root rotting fungi and root-knot nematode infecting tomato roots. Journal of Applied Botany and Food Quality 83: 50–53.

Sultana, V., G.N. Baloch, J. Ara, S. Ehteshamul-Haque, R.M. Tariq and M. Athar. 2011. Seaweeds as an alternative to chemical pesticides for the management of root diseases of sunflower and tomato. Journal of Applied Botany and Food Quality 84: 162–168.

Sun, Y., Y. Xu, K. Liu, H. Hua, H. Zhu and Y. Pei. 2006. Gracilarioside and gracilamides from the red *alga Gracilaria asiatica*. Journal of Natural Products 69(10): 1488–1491.

Sundar, R., V. Sundar and G. Sakkarapani. 2013. Toxicity of *Gracilaria verrucosa* methonal extract against selected medically important vector mosquitoes (diptera: culicidae) of Puducherry, India. International Journal of Current Research 5(11): 3388–3392.

Susanto, A., P.S. Sudharto and R.Y. Purba. 2005. Enhancing biological control of basal stem rot disease (*Ganoderma*) in oil palm plantations. Mycopathologia 159(1): 153–157.

Süzgeç-Selçuk, S., A.H. Mericli, K.C. Guven, M. Kaiser, R. Casey, S. Hingley-Wilson, A. Lalvani and D. Tasdemir. 2011. Evaluation of Turkish seaweeds for antiprotozoal, antimycobacterial and cytotoxic activities. Phytotherapy Research 25(5): 778–783.

Suzuki, Y., I. Yamamoto and I. Umezawa. 1980. Antitumor affect of seaweed. Partial purification and the antitumor effect of polysaccharides from *Laminaria angustata* Kjellman var. *longissima* Miyabe. Chemotherapy 28: 165–170.

Svennevig, B. 2017. Did seaweed make us who we are today? University of Southern Denmark. Available online at: www.sciencedaily.com/releases/2017/02/170228131040.htm.

Sweeney, N.J., P. Klemm, B.A. McCormick, E. Moller-Nielsen, M. Utley, M.A Schembri, D.C. Laux and P. Cohen. 1996. The *Escherichia coli* K-12 gntP gene allows *E. coli* F18 to occupy a distinct nutritional niche in the streptomycin-treated mouse large intestine. Infection and Immunity 64(9): 3497–3503.

Syad, A.N., K.P. Shunmugiah and P.D. Kasi. 2012. Assessment of anticholinesterase activity of *Gelidiella acerosa*: implications for its therapeutic potential against Alzheimer's disease. Evidence-Based Complementary and Alternative Medicine vol. 2012, Article ID 497242, 8 pp.

Syad, A.N., K.P. Shunmugiah and P.D. Kasi. 2013. Antioxidant and anti-cholinesterase activity of *Sargassum wightii*. Pharmaceutical Biology 51(11): 1401–1410.

Syad, A.N., B.S. Rajamohamed, K.P. Shunmugaiah and P.D. Kasi. 2016. Neuroprotective effect of the marine macroalga *Gelidiella acerosa*: identification of active compounds through bioactivity-guided fractionation. Pharmaceutical Research 54: 2073–2081.

Sykes, M.L., J.B. Baell, M. Kaiser, E. Chatelain, S.R. Moawad, D. Ganame, J.R. Ioset and V.M. Avery. 2012. Identification of compounds with anti-proliferative activity against *Trypanosoma brucei* strain 427 by a whole cell viability based HTS campaign. PLOS Neglected Tropical Diseases 6: e1896.

Sykes, M.L. and V.M. Avery. 2013. Approaches to protozoan drug discovery: phenotypic screening. Journal of Medicinal Chemistry 56: 7727–7740.

Synytsya, A., W.-J. Kim, S.-M. Kim, R. Pohl, A. Synytsya, F. Kvasnicka, J. Copíková and Y.I. Park 2010. Structure and antitumour activity of fucoidan isolated from sporophyll of Korean brown seaweed *Undaria pinnatifida*. Carbohydrate Polymers 81: 41–48.

Synytsya, A., R. Bleha, R. Pohl, K. Hayashi, K. Yoshinaga, T. Nakano and T. Hayashi. 2014. Mekabu fucoidan: structural complexity and defensive effects against avian Influenza A viruses. Carbohydrate Polymers 111: 633–644.

Tabarsa, M., M. Rezaei, Z. Ramezanpour and J.R. Waaland. 2012a. Chemical compositions of the marine algae *Gracilaria salicornia* (Rhodophyta) and *Ulva lactuca* (Chlorophyta) as a potential food source. Journal of the Science of Food and Agriculture 92(12): 2500–2506.

Tabarsa, M., M. Rezaei, Z. Ramezanpour, J.R. Waaland and R. Rabiei. 2012b. Fatty acids, amino acids, mineral contents, and proximate composition of some brown seaweeds. Journal of Phycology 48(2): 285–292.

Tabet, N. 2006. Acetylcholinesterase inhibitors for Alzheimer's disease: Anti-inflammatories in acetylcholine clothing! Age Ageing 35: 336–338.

Taboada, C., R. Millan and I. Miguez. 2009. Evaluation of the marine alga *Ulva rigida* as a food supplement: Effect of intake on intestinal, hepatic, and renal enzyme activities in rats. Journal of Medicinal Food 14: 161–166.

Taboada, C., R. Millán and I. Míguez. 2010. Composition, nutritional aspects and effect on serum parameters of marine algae *Ulva rigida*. Journal of the Science of Food and Agriculture 90: 445–449.

Tada, R., J.-P. Latgé and V. Aimanianda. 2013. Undressing the fungal cell wall/cell membrane-the antifungal drug targets. Current Pharmaceutical Design 19(20): 3738–3747.

Taherpour, A., B. Archangi, S. Ghaemmaghami, H. Zolgharnein and K. Ghanemi. 2016. Screening of marine algae (*Padina* sp.) from the Lengeh Port, Persian Gulf for antibacterial and antifungal activities. Journal of Coastal Life Medicine 4(9): 698–702.

Takada, N., R. Watanabe, K. Suenaga, K. Yamada and D. Uemura. 2001. Isolation and structures of hedaols A, B, and C, new bisnorditerpenes from a Japanese brown alga. Journal of Natural Products 64(5): 653–655.

Takahashi, M. 1983. Studies on the mechanisms of host mediated antitumor action of crude fucoidan from a brown alga *Eisenia bicyclis*. Journal of the Japan Society of the Reticuloendothelial System 22: 269–283.

TaKaRa. 2017. Takara Bio Inc. Available online at: http://www.takara-bio.com/.

Takeda, A. 2000. Movement of zinc and its functional significance in the brain. Brain Research Reviews 34: 137–148.

Takeda, A. 2003. Manganese action in brain function. Brain Research Reviews 41: 79–87.

Takeda, K., K. Tomimori, R. Kimura, C. Ishikawa, T.K. Nowling and N. Mori. 2012. Anti-tumor activity of fucoidan is mediated by nitric oxide released from macrophages. International Journal of Oncology 40: 251–260.

Talarico, L.B., R.G. Zibetti, P. Faria, L.A. Scolaro, M.E. Duarte, M.D. Noseda, C.A. Pujol and E.B. Damonte. 2004. Anti-*Herpes simplex* virus activity of sulfated galactans from the red seaweeds *Gymnogongrus griffithsiae* and *Cryptonemia crenulata*. International Journal of Biological Macromolecules 34: 63–71.

Talarico, L.B., C.A. Pujol, R.G. Zibetti, P.C. Faraí, M.D. Noseda, M.E, Duarte and E.B. Damonte. 2005. The antiviral activity of sulfated polysaccharides against Dengue virus is dependent on virus serotype and host cell. Antiviral Research 66: 103–110.

Talarico, L.B. and E.B. Damonte. 2007. Interference in Dengue virus adsorption and uncoating by carrageenans. Virology 363(2): 473–485.

Talarico, L.B., M.E.R. Duarte, R.G.M. Zibetti, M.D. Noseda and E.B. Damonte. 2007b. An algal-derived DL-galactan hybrid is an efficient preventing agent for *in vitro* Dengue virus infection. Planta Medica 73: 1464–1468.

Talarico, L.B., M.D. Noseda, D.R.B. Ducatti, M.E. Duarte and E.B. Damonte. 2011. Differential inhibition of dengue virus infection in mammalian and mosquito cells by iota-carrageenan. Journal of General Virology 92(6): 1332–1342.

Talarico, L.B., C.A. Pujol and E.B. Damonte. 2016. Biological activities of carrageenans against human infectious diseases, chapter 11. pp. 245–268. *In*: Pereira, L. (ed.). Carrageenans: Sources and Extraction Methods, Molecular Structure, Bioactive Properties and Health Effects. Nova Science Publishers, New York.

Talbot, N.J. 2003. On the trail of a cereal killer: Exploring the biology of *Magnaporthe grisea*. Annual Review of Microbiology 57: 177–202.

Tan, L.T., R.T. Williamson, W.H. Gerwick, K.H. Watts, K. McGough and R. Jacobs. 2000. *Cis, cis* and *trans*, trans-ceratospongamide, new bioactive cyclic heptapeptides from the Indonesian red alga *Ceratodictyon spongiosum* and symbiotic spong *Sigmadocia symbiotica*. The Journal of Organic Chemistry 65: 419–425.

Tan, S.P., McLoughlin, L. O'Sullivan, M.L. Prieto, G.E. Gardiner, P.G. Lawlor and H. Hughes. 2013. Development of a novel antimicrobial seaweed extract-based hydrogel wound dressing. International Journal of Pharmaceutics 456: 10–20.

Tang, F., F. Chen and F. Li. 2013. Preparation and potential *in vivo* anti-influenza virus activity of low molecular-weight κ-carrageenans and their derivatives. Journal of Applied Polymer Science 127(3): 2110–2115.

Tang, H.F., Y. Yang-Hua, X.S. Yao, Q.Z. Xu, S.Y. Zhang and H.W. Lin. 2002. Bioactive steroids from the brown alga *Sargassum carpophyllum*. Journal of Asian Natural Products Research 4(2): 95–101.

Tang, L., Y.C. Chen, Z.B. Jiang, S.P. Zhong, W. Chen, F.C. Zheng and G.G. Shi. 2017. Purification, partial characterization and bioactivity of sulfated polysaccharides from *Grateloupia livida*. International Journal of Biological Macromolecules 94: 642–652.

Tanniou, A., L. Vandanjon, M. Incera, E. Serrano Leon, V. Husa, J. Le Grand, J.-L. Nicolas et al. 2014. Assessment of the spatial variability of phenolic contents and associated bioactivities in the invasive alga *Sargassum muticum* sampled along its European range from Norway to Portugal. Journal of Applied Phycology 26(2): 1215–1230.

Tapia, C., Z. Escobar, E. Costa, J. Sapag-Hagar, F. Valenzuela, C. Basualto, M.N. Gai and M. Yazdani-Pedram. 2004. Comparative studies on polyelectrolyte complexes and mixtures of chitosan-alginate and chitosan-carrageenan as prolonged diltiazem clorohydrate release systems. European Journal of Pharmaceutics and Biopharmaceutics 57: 65–75.

Tapiero, H. and K.D. Tew. 2003. Trace elements in human physiology and pathology: Zinc and metallothioneins. Biomedicine & Pharmacotherapy 57: 399–411.

Target, N.M. and T.M. Arnold. 2001. Effects of secondary metabolites on digestion in marine herbivores. pp. 391–411. *In*: McClintock, J.B. and B.J. Baker (eds.). Marine Chemical Ecology. CRC Press, Boca Raton, FL, USA.

Tariq, V.N. 1991. Antifungal activity in crude extracts of marine red algae. Mycological Research 95: 1433–1440.

Taskin, E., M. Ozturk, E. Taskin and O. Kurt. 2007. Antibacterial activities of some marine algae from the Aegean Sea (Turkey). African Journal of Biotechnology 6: 2746–2751.

Taskin, E., Z. Caki, M. Ozturk and E. Taskin. 2010. Assessment of *in vitro* antitumoral and antimicrobial activities of marine algae harvested from the eastern Mediterranean Sea. African Journal of Biotechnology 9: 4272–4277.

Taylor, A.L., J.N. Sasser and L.A. Nelson. 1982. Relationship of climate and characteristics to geographical distribution of Meloidogyne species in agricultural soil. Cooperative Publication, Department of Plant Pathology, North Carolina State University and the United States Agency for International Development, Raleigh, NC, 65 pp.

Taylor, S.C. 2002. Skin of color: biology, structure, function, and implications for dermatologic disease. Journal of the American Academy of Dermatology 46: S41–S62.

Teas, J., M.L. Harbison and R.S. Gelman. 1984. Dietary seaweed (*Laminaria*) and mammary carcinogenesis in rats. Cancer Research 44(7): 2758–2761.

Teas, J., M.R. Irhimeh, S. Druker, T.G. Hurley, J.R. Hébert, T.M. Savarese and M.S. Kurzer. 2011. Serum IGF-1 concentrations change with soy and seaweed supplements in healthy postmenopausal American women. Nutrition and Cancer 63: 743–748.

Teas, J., S. Vena, D.L. Cone and M. Irhimeh. 2013. The consumption of seaweed as a protective factor in the etiology of breast cancer: Proof of principle. Journal of Applied Phycology 25(3): 771–779.

Teixeira, E.H., M.H. Napimoga, V.A. Carneiro, T.M. de Oliveira, K.S. Nascimento, C.S. Nagano, J.B. Souza, A. Havt, V.P. Pinto, R.B. Goncalves, W.R. Farias, S. Saker-Sampaio, A.H. Sampaio and B.S. Cavada. 2007b. *In vitro* inhibition of oral *Streptococci* binding to the acquired pellicle by algal lectins. Journal of Applied Microbiology 103(4): 1001–1006.

Teixeira, V.L., T. Tomassini, B.G. Fleury and A. Kelecom. 1986a. Dolastane and secodolastane diterpenes from the marine brown alga *Dictyota cervicornis*. Journal of Natural Products 49: 570–575.

Teixeira, V.L., T. Tomassini and A. Kelecom. 1986b. Cervicol, a further secodolastane diterpene from the marine brown alga *Dictyota cervicornis* Kützing. Bulletin des Sociétés Chimiques Belges 95: 263–268.

Teixeira, V.L. and A.A. Kelecom. 1988. Chemotaxonomic study of diterpenes from marine brown algae of the genus *Dictyota*. Science of the Total Environment 75: 271–283.

Teixeira, V.L., F.D. Rocha, P.J. Houghton, M.A.C. Kaplan and R.C. Pereira. 2007. α-amylase inhibitors from Brazilian seaweeds and their hypoglycemic potential. Fitoterapia 78(1): 35–36.

Teixeira, V.L. 2012. Produtos naturais de algas marinhas bentônicas. Revista Virtual de Química 5: 343–362. Available online at: http://rvq.sbq.org.br/index.php/rvq/article/viewDownloadInterstitial/346/275.

Temple, W.J. and K.K. Gladwin. 2003. Fruits, vegetables and the prevention of cancer: Research challenges. Nutrition 19: 467–470.

Teng, H., Y. Yang, H. Wei, Z. Liu, Z. Liu, Y. Ma, Z. Gao, L. Hou and X. Zou. 2015. Fucoidan suppresses hypoxia-induced lymphangiogenesis and lymphatic metastasis in mouse hepatocarcinoma. Marine Drugs 13(6): 3514–3530.

Terada, A., H. Hara and T. Mitsuoka. 1995. Effect of dietary alginate on the faecal microbiota and faecal metabolic activity in humans. Microbial Ecology in Health and Disease 8(6): 259–266.

Terasaki, M., A. Hirose, B. Narayan, Y. Baba, C. Kawagoe, H. Yasui, N. Saga, M. Hosokawa and K. Miyashita. 2009. Evaluation of recoverable functional lipid components of several brown seaweeds (Phaeophyta) from Japan with special reference to fucoxanthin and fucosterol contents. Journal of Phycology 45(4): 974–980.

Terracciano, S., M. Aquino, M. Rodriquez, M.C. Monti, A. Casapullo, R. Riccio and L. Gomez-Paloma. 2006. Chemistry and biology of anti-inflammatory marine natural products: molecules interfering with cyclooxygenase, NF-κB and other unidentified targets. Current Medicinal Chemistry 13: 1947–1969.

Teruya, T., T. Konishi, S. Uechi, H. Tamaki and M. Tako. 2007. Anti-proliferative activity of oversulfated fucoidan from commercially cultured *Cladosiphon okamuranus* Tokida in U937 cells. International Journal of Biological Macromolecules 41(3): 221–226.

Teruya, T., H. Tatemoto, T. Konishi and M. Tako. 2009. Structural characteristics and in vitro macrophage activation of acetyl fucoidan from *Cladosiphon okamuranus*. Glycoconjugate Journal 26(8): 1019–1028.

Texinfine. 2017. Laboratoires ICP-Texinfine. Available online at: http://www.icp-texinfine.com/.

Thaisomboonsuk, B.K., E.T. Clayson, S. Pantuwatana, D.W. Vaughn and T.P. Endy. 2005. Characterization of Dengue-2 virus binding to surfaces of mammalian and insect cells. American Journal of Tropical Medicine and Hygiene 72: 375–383.

Thalgo. 2017. La Beaute marine. Available online at: http://www.thalgo.com/.

Thangam, T.S. and K. Kathiresan. 1991. Mosquito larvicidal effect of seaweed extracts. Botanica Marina 34: 433–435.

Thangam, T.S. and K. Kathiresan. 1996. Marine plants for mosquito control. 1–5 pp. *In*: Wildey, K.B. (ed.). Proceedings of the Second International Conference on Urban Pests, 7–10 July, Edinburgh, Scotland.

Thanh-Sang, V. and S.-K. Kim. 2013. Fucoidans as a natural bioactive ingredient for functional foods. Journal of Functional Foods 5(1): 16–27.

Thanigaivel, S., N. Chandrasekaran, A. Mukherjee and J. Thomas. 2015. Investigation of seaweed extracts as a source of treatment against bacterial fish pathogen. Aquaculture 448: 82–86.

Thanigaivel, S., S. Vijayakumar, A. Mukherjee, N. Chandrasekaran and J. Thomas. 2014. Antioxidant and antibacterial activity of *Chaetomorpha antennina* against shrimp pathogen *Vibrio parahaemolyticus*. Aquaculture 433: 467–475.

Thennozhi, K., T. Selvarani and B.K. Prabhu. 2013. Anti-viral activity of aqueous extract from brown seaweed, *Sargassum ilicifolium* (Turner) on three different types Herpes simplex virus. Advance Research in Pharmaceuticals and Biologicals 3(1): 370–373.

Thenmozhi, K., T. Selvarani and B.K. Prabhu. 2013. Estimation of anti-fungal activity of aqueous extracts of brown seaweed *Sargassum ilicifolium*. Advance Research in Pharmaceuticals and Biologicals 3(I): 374–377.

Thennarasan, S. and Murugesan. 2015. Biochecmial composition of marine brown alga *Lobophora variegata* from Mandapam in the south east coast of Tamil Nadu. International Journal of Phytopharmacy 5(3): 25–29.

Therkelsen, G.H. 1993. Carrageenan. pp. 145–180. *In*: Whistler, R.L. and J.N. Bemiller (eds.). Industrial Gums: Polysaccharides and Their Derivatives. Academic Press, San Diego.

Thillairajasekar, K., V. Duraipandiyan, P. Perumal and S. Ignacimuthu. 2009. Antimicrobial activity of *Trichodesmium erythraeum* (Ehr) (microalaga) from South East Coast of Tamil Nadu, India. International Journal of Integrative Biology 5: 167–170.

Thinakaran, T. and K. Sivakumar. 2013. Antifungal activity of certain seaweeds from Phuthumadam coast. International Journal of Research and Reviews in Pharmacy and Applied Science 3(3): 341–350.

Thinh, P., R. Menshova, S. Ermakova, S. Anastyuk, B. Ly and T. Zvyagintseva. 2013. Structural characteristics and anticancer activity of fucoidan from the brown alga *Sargassum mcclurei*. Marine Drugs 11(5): 1456–1476.

Thirunavukkarasu, R., P. Pandiyan, K. Subaramaniyan, D. Balaraman, S. Manikkam, B. Sadaiyappan and G.E.G. Jothi. 2014. Screening of marine seaweeds for bioactive compound against fish pathogenic bacteria and active fraction analysed by gas chromatography–mass spectrometry. Journal of Coastal Life Medicine 2(5): 367–375.

Thivy, F. and V.D. Chauhan. 1963. *Caulerpa veravalensis*, a new species from Gujarat, India. Botanica Marina 5(4): 97–100.

Thomas, N.V. and S.K. Kim. 2001. Potential pharmacological applications of polyphenolic derivatives from marine brown algae. Environmental Toxicology and Pharmacology 32: 325–335.

Thomas, N.V. and S.K. Kim. 2013. Beneficial effects of marine algal compounds in cosmaceuticals. Marine Drugs 11: 146–164.

Thomesa, P.B., R. Murugan, P. Balu and R. Ramasamy. 2010. Cardioprotective activity of *Cladosiphon okamuranus* fucoidan against isoproterenol induced myocardial infarction in rats. Phytomedicine 18: 52–57.

Thompson, K.D. and C. Dragar. 2004. Antiviral activity of *Undaria pinnatifida* against *Herpes simplex* virus. Phytotherapy Research 18: 551–555.

Thornfeldt, C. 2005. Cosmeceuticals containing herbs: fact, fiction, and future. Dermatologic Surgery 31: 873–881.

Thuy, T.T., B.M. Ly, T.T. Van, N.V. Quang, H.C. Tu, Y. Zheng, C. Seguin-Devaux, B. Mi and U. Ai. 2015. Anti-HIV activity of fucoidans from three brown seaweed species. Carbohydrate Polymers 115: 122–128.

Tierney, M.S., A.K. Croft and M. Hayes. 2010. A review of antihypertensive and antioxidant activities in macroalgae. Botanica Marina 53: 387–408.

Tiker, N. 2016. Phycoerythrobilin is the typical chromophore in phycoerythrin (image in public domain). Available online at: https://upload.wikimedia.org/wikipedia/commons/thumb/7/71/Phycoerythrobilin2.svg/2000px-Phycoerythrobilin2.svg.png.

Tirtawijaya, G., M. Mohibbullah, M.D.N. Meinita, I.S. Moon and Y.K. Hong. 2016. The ethanol extract of the rhodophyte *Kappaphycus alvarezii* promotes neurite outgrowth in hippocampal neurons. Journal of Applied Phycology 28: 2515–2522.

Tobacman, J.K. 2001. Review of harmful gastrointestinal effects of carrageenan in animal experiments. Environmental Health Perspectives 109(10): 983–994.

Tokida, J. 1954. The marine algae of south Saghalien. Mem. Fac. Fish., Hokkaido Univ. 2(1): 1–264.

Tolnai, S. and J.F. Morgan. 1962. Studies on the *in vitro* antitumor activity of fatty acids V. unsaturated acids. Journal of Physiology and Biochemistry 40: 869–875.

Tolnai, S. and J.F. Morgan. 1966. Studies on the *in vitro* antitumor activity of fatty acids VII. effect of amino acid-fatty acids salts. Canadian Journal of Biochemistry 44: 979–981.

Tonnesen, H.H. and J. Karlsen. 2002. Alginate in drug delivery systems. Drug Development and Industrial Pharmacy 2(6): 621–630.

Topcu, G., Z. Aydogmus, S. Imre, A.C. Goren, J.M. Pezzuto, J.A. Clement and D.G.I. Kingston. 2003. Brominated sesquiterpenes from the red alga *Laurencia obtusa*. Journal of Natural Products 66: 1505–1508.

Torres, F.A.E., T.G. Passalacqua, A.M.A. Velásqueza, R.A. de Souzab, P. Colepicolod and M.A.S. Graminha. 2014. New drugs with antiprotozoal activity from marine algae: a review. Revista Brasileira Farmacognosia 24: 265–276.

Torres, M.R., A.P. Sousa, E.A. Silva Filho, D.F. Melo, J.P. Feitosa, R.C. de Paula and M.G. Lima. 2007. Extraction and physicochemical characterization of *Sargassum vulgare* alginate from Brazil. Carbohydrate Research 342: 2067–2074.

Torsdottir, I., M. Alpsten, G. Holm, A.S. Sandberg and J. Tolli. 1991. A small dose of soluble alginate-fiber affects postprandial glycemia and gastric-emptying in humans with diabetes. Journal of Nutrition 121(6): 795–799.

Toskas, G., S. Heinemann, C. Heinemann, C. Cherif, R.-D. Hund, V. Roussis and T. Hanke. 2012. Ulvan and ulvan/chitosan polyelectrolyte nanofibrous membranes as a potential substrate material for the cultivation of osteoblasts. Carbohydrate Polymers 89(3): 997–1002.

Townsend, G.F., J.F. Morgan, S. Tolnai, B. Hazlett, H.J. Morton and R.W. Shuel. 1960. Studies on the *in vitro* antitumor activity of fatty acids I. 10-hydroxy-2-decenoic acid from royal jelly. Cancer Research 20: 503–510.

Trease, G.E. and W.C. Evanes. 2009. Pharmacognosy, 16th ed. Elsevier Health Sciences, London, 616 pp.

Trejo-Avila, L.M. M.E. Morales-Martinez, D. Ricque-Marie, L.E. Cruz-Suarez, P. Zapata Benavides, K. Moran-Santibanez and C. Rodriguez-Padilla. 2014. *In vitro* anti-canine distemper virus activity of fucoidan extracted from the brown alga *Cladosiphon okamuranus*. VirusDisease 25: 474–480.

Trento, F., F. Cattaneo, R. Pescador, R. Porta and R. Ferro. 2001. Antithrombin activity of an algal polysaccharide. Thrombosis Research 102: 457–465.

Trick, W.E., S.K. Fridkin, J.R. Edwards, R.A. Hajjeh and R.P. Gaynes. 2002. The national nosocomial infections surveillance system hospitals. Secular trend of hospitalacquired candidemia among intensive care unit patients in the United States during 1989–1999. Clinical Infectious Diseases 35: 627–630.

Trinchero, J., N.M.A. Ponce, Q.L. Cordoba, M.L. Flores, S. Pampuro, C.A. Stortz, H. Salomon and G. Turk. 2009. Antiretroviral activity of fucoidans extracted from the brown seaweed *Adenocystis utricularis*. Phytotherapy Research 23: 707–712.

Trono, G.C. 2016. *Portieria hornemannii*—PROSEA, Plant Resources of South-East Asia. Available online at: http://uses.plantnet-project.org/en/Portieria_hornemannii_(PROSEA).

Truus, K., M. Vaher and I. Taure. 2001. Algal biomass from *Fucus vesiculosus* (Phaeophyta): investigation of the mineral and alginate components. Proceedings of the Estonian Academy of Sciences-Chemistry 50: 95–103.

Tsang, C., A. Sagara and Y. Kamei. 2001. Structure-activity relationship of a neurite outgrowth-promoting substance purified from the brown alga, *Sargassum macrocarpum*, and its analogues on PC12D cells. Journal of Applied Phycology 13: 349-357.

Tsang, C. and Y. Kamei. 2004. Sargaquinoic acid supports the survival of neuronal PC12D cells in a nerve growth factor-independent manner. European Journal of Pharmacology 488: 11–18.

Tsang, C., A. Ina, T. Goto and Y. Kamei. 2005. Sargachromenol, a novel nerve growth factor-potentiating substance isolated from *Sargassum macrocarpum*, promotes neurite outgrowth and survival *via* distinct signaling pathways in PC12D cells. Neuroscience 132: 633–643.

Tseng, C.K. 1945. The terminology of seaweed colloids. Science 101: 597–602.

Tsi, D., N.P. Das and B.K. Tan. 1995. Effects of aqueous celery (*Apium graveolens*) extract on lipid parameters of rats fed a high fat diet. Planta Medica 61(1): 18–21.

Tsuchiya, Y. 1950. Physiological studies on the vitamin C content of marine algae. Tohoku Journal of Agricultural Research 1: 97–102.

Tsuge, K., M. Okabe, T. Yoshimura, T. Sumi, H. Tachibana and K. Yamada. 2004. Dietary effects of Porphyran from *Porphyra yezoensis* on growth and lipid metabolism of Sprague Dawley rats. Food Science and Technology Research 10: 147–51.

Tsuji, N., S. Moriwaki, Y. Suzuki, Y. Takema and G. Imokawa. 2001. The role of Elastases secreted by fibroblasts in wrinkle formation: implication through selective inhibition of elastase activity. Photochemistry and Photobiology 74: 283–290.

Tsukahara, K., Y. Takema, S. Moriwaki, N. Tsuji, Y. Suzuki, T. Fujimura and G. Imokawa. 2001. Selective inhibition of skin fibroblast elastase elicits a concentration-dependent prevention of ultraviolet B-induced wrinkle formation. Journal of Investigative Dermatology 117: 671–677.

Tsukidate, J.I. 1984. On the most suitable growth conditions of young thalli of *Sargassum patens* C. Agardh and *Sargassum tortile* C. Agardh. Bull. Nansei. Reg. Fish. Res. Lab. 16: 1–9.

Tsukui, T., K. Konno, M. Hosokawa, H. Maeda, T. Sashima and K. Miyashita. 2007. Fucoxanthin and fucoxanthinol enhance the amount of docosahexaenoic acid in the liver of KKAy obese/diabetic mice. Journal of Agricultural and Food Chemistry 55: 5025–5029.

Tsukui, T., N. Baba, M. Hosokawa, T. Sashima and K. Miyashita. 2009. Enhancement of hepatic docosahexaenoic acid and arachidonic acid contents in C57BL/6J mice by dietary fucoxanthin. Fisheries Science 75: 261–263.

Tu, R., C.L. Lu, K. Thyagarajan, E. Wang, H. Nguyen, S. Shen, C. Hata and R.C. Quijano. 1993. Kinetic study of collagen fixation with polyepoxy fixatives. Journal of Biomedical Materials Research 27: 3–9.

Tückmantel, W., A.P. Kozikowski and L.J. Romanczyk. 1999. Studies in polyphenol chemistry and bioactivity. 1. Preparation of building blocks from (+)-catechin. Procyanidin formation. Synthesis of the cancer cell growth inhibitor, 3-O-galloyl-(2R,3R)-epicatechin-4β,8-[3-O-galloyl-(2R,3R)-epicatechin]. Journal of the American Chemical Society 121: 12073–12081.

Tüney, I., B.H. Cadirci, D. Ünal and A. Sukatar. 2006. Antimicrobial activities of the extracts of marine algae from the coast of Urla (Izmir, Turkey). Turkish Journal of Biology 30: 171–175.

Turkington, T.K., A. Petran, T. Yonow and D.J. Kriticos. 2014. *Fusarium graminearum*. HarvestChoice Pest Geography, InSTePP-HarvestChoice, Background, St. Paul, MN, 7 pp.

Tyler-Walters, H. 2017. *Delesseria sanguinea*—MarLIN, The Marine Life Information Network. Available online at: http://www.marlin.ac.uk/species/detail/1338.

Tziveleka, L.-A., D. Abatis, K. Paulus, R. Bauer, C. Vigias and V. Roussis. 2005. Marine polyprenylated hydroquinones, quinones and chromenols with inhibitory effects on leukotriene formation. Chemistry & Biology 2: 901–909.

Uchida, J.Y. 2016. *Phytophthora infestans*—Crop knowledge master. Available online at: http://www.extento.hawaii.edu/Kbase/crop/type/p_infest.htm.

Uehara, T., M. Takeshita and M. Maeda. 1992. Studies on anticoagulant-active arabinan sulfates from the green-alga, *Codium latum*. Carbohydrate Research 235: 309–311.

Ueno, H., S. Horie, Y. Nishi, H. Shogawa, M. Kawasaki, S. Suzuki, T. Hayashi, M. Arisawa, M. Shimizu, M. Yoshizaki and N. Morita. 1998. Chemical and pharmaceutical studies on medicinal plants in Paraguay. Geraniin, an angiotensin-converting enzyme inhibitor from "Paraparai mi" *Phyllanthus niruri*. Journal of Natural Products 51: 357–359.

Uermösi, C., R.R. Beerli, M. Bauer, V. Manolova, K. Dietmeier, R.B. Buser, T.M. Kündig, P. Saudan and M.F. Bachmann. 2010. Mechanisms of allergen-specific desensitization. Journal of Allergy and Clinical Immunology 126: 375–383.

Underwood, E.J. 1977. Trace elements in human nutrition (4th ed.). Academic Press, New York, 545 pp.

Unnikrishnan, P.S., K. Suthindhiran and M.A. Jayasri. 2015. Antidiabetic potential of marine algae by inhibiting key metabolic enzymes. Frontiers in Life Science 8(2): 148–159.

Unnikrishnan, P.S., K. Suthindhiran and M.A. Jayasri. 2016. Alpha-amylase inhibition and antioxidant activity of marine green algae and its possible role in diabetes management. Pharmacognosy Magazine 11(S4): 511–515.

Uno, T., M. Hattori and T. Yoshida. 2006. Oral administration of alginic acid oligosaccharide suppresses IgE production and inhibits the induction of oral tolerance. Bioscience, Biotechnology, and Biochemistry 70: 3054–3057.

Urones, J.G., M.E.M. Araujo, F.M.S.B. Palma, P. Basabe, I.S. Marcos, R.F. Moro, A.M. Lithgow and J. Pineda. 1992. Meroterpenes from *Cystoseira usneoides* II. Phytochemistry 31(6): 2105–2109.

Urquiaga, I. and F. Leighton. 2000. Plant polyphenol antioxidants and oxidative stress. Biological Research 33(2): 55–64.

Ushakova, N.A., G.E. Morozevich, N.E. Ustyuzhanina, M.I. Bilan, A.I. Usov, N.E. Nifantiev and M.E. Preobrazhenskaya. 2009. Anticoagulant activity of fucoidans from brown algae. Biochemistry (Moscow) 3: 77–83.

Usmanghani, K., M. Shameel, M. Sualeh, K.H. Khan and Z.A. Mahmood. 1984. Antibacterial and antifungal activities of marine algae from Karachi seashore of Pakistan. Fitoterapia 55: 73–77.

Usol'tseva, R.V., P. Zhao, M.I. Kusaikin, A. Jia, W. Yuan, M. Zhang, C. Liu and S.P. Ermakova. 2017. Structural characteristics and antitumor activity of fucoidans from the brown alga *Sargassum muticum*. Chemistry of Natural Compounds 53(2): 219–223.

Usui, T., K. Asari and T. Mizuno. 1980. Isolation of highly purified fucoidan from *Eisenia bicyclis* and its anticoagulant and antitumor activities. Agricultural and Biological Chemistry 44: 1965–1966.

Vairappan, C.S. 2003. Potent antibacterial activity of halogenated metabolites from Malaysian Red algae *Laurencia majuscula* (Rhodomelaceae, Ceramiales). Biomolecular Engineering 20(4-6): 255–259.

Vairappan, C.S., T. Kawamoto, H. Miwa and M. Suzuki. 2004. Potent antibacterial activity of halogenated compounds against antibiotic-resistant bacteria. Planta Medica 70: 1087–1090.

Vairappan, C.S., T. Kamada, W.-W. Lee and Y.-J. Jeon. 2013. Anti-inflammatory activity of halogenated secondary metabolites of *Laurencia snackeyi* (Weber-van Bosse) Masuda in LPS-stimulated RAW 264.7 macrophages. Journal of Applied Phycology 25: 1805–1813.

Val, A., G. Platas, A. Basilio, A. Cabello, J. Gorrochategui, I. Suay, F. Vicente, E. Portillo, M.J. del Rio, R.G. Garcia and F. Peláez. 2001. Screening of antimicrobial activities in red, green and organisms VI. Antiviral evaluation of marine algal brown macroalgae from Gran Canaria (Canary Islands, Spain). International Microbiology 4(1): 35–40.

Valente, J., T. Valente, P. Alves, P. Ferreira and A. Silva and I.J. Correia. 2012. Alginate based scaffolds for bone tissue engineering. Materials Science and Engineering C32: 2596–2603.

Valentina, J., T.V. Poonguzhali and L.L.J.L. Nisha. 2015. Mosquito larvicidal and pupicidal activity of seaweed extracts against *Aedes aegypti*, *Anopheles stephensi* and *Culex quinquefasciatus*. International Journal of Mosquito Research 2(4): 54–59.

Valkonen, J.P.T. and H. Koponen. 1990. The seed-borne fungi of Chinese cabbage (*Brassica pekinensis*), their pathogenicity and control. Plant Pathology 39: 510–516.

Vallim, M.A., J.E. Barbosa, D.N. Cavalcanti, J.C. De-Paula, V.A.G.G. Silva, V.L. Teixeira and I.C.N.P. Paixão. 2010. *In vitro* antiviral activity of diterpenes isolated from the Brazilian brown alga *Canistrocarpus cervicornis*. Journal of Medicinal Plants Research 4: 2379–2382.

Vallinayagam, K., R. Arumugam, R.R.R. Kannan, G. Thirumaran and P. Anantharaman. 2009. Antibacterial activity of some selected seaweeds from Pudumadam coastal regions. Global Journal of Pharmacology 3: 50–52.

van de Velde F. and G.A. de Ruiter. 2002. Carrageenan. pp. 245–274. *In*: Vandamme, E.J., S.D. Baets and A. Steinbèuchel (eds.). Biopolymers V6, Polysaccharides II, Polysaccharides from Eukaryotes. Wiley, Weinheim.

van de Velde, F., L. Pereira and H.S. Rollema. 2004. The revised NMR chemical shift data of carrageenans. Carbohydrate Research 339: 2309–2313.

Van Den Hoogen, B.G., G.J. Van Doornum, J.C. Fockens, J.J. Cornelisse, W.E. Beyer, R. de Groot, A.D. Osterhaus and R.A. Fouchier. 2003. Prevalence and clinical symptoms of human metapneumovirus infection in hospitalized patients. The Journal of Infectious Diseases 188: 1571–1577.

Van Den Hoogen, B.G., J.C. De Jong, J. Groen, T. Kuiken, R. de Groot, R.A. Fouchier and A.D. Osterhaus. 2001. A newly discovered human pneumovirus isolated from young children with respiratory tract disease. Nature Medicine 7: 719–724.

Van der Linden, J.W.M., M.C. Arendrup, A. Warris, K. Lagrou, H. Pelloux, P. Hauser, E. Chryssanthou, E. Mellado, S.E. Kidd, A.M. Tortorano, E. Dannaoui, P. Gaustad, J.W. Baddley, A. Uekötter, C. Lass-Flörl, N. Klimko, C.B. Moore, D.W. Denning, A.C. Pasqualotto, C. Kibbler, S. Arikan-Akdagli, D. Andes, J. Meletiadis, L. Naumiuk, M. Nucci, W.J.G. Melchers and P.E. Verweij. 2015. Prospective multicenter international surveillance of azole resistance in *Aspergillus fumigatus*. Emerging Infectious Diseases 21(6): 1041–1044.

Van Rooijen, N. and A. Sanders. 1997. Elimination, blocking, and activation of macrophages: three of a kind? Journal of Leukocyte Biology 62: 702–709.

van West, P. 2006. *Saprolegnia parasitica*, an Oomycete pathogen with a fishy appetite: New challenges for an old problem. Mycologist 20(3): 99–104.

Vanden Bossche, H., P. Marichal, H. Geerts and P.A.J. Janssen. 1988. The molecular basis for itraconazole's activity against *Aspergillus fumigatus*. pp. 171–197. *In*: Vanden Bossche, H., D.W.R. Mackenzie and G. Cauwenbergh (eds.). *Aspergillus* and *Aspergillosis*. Springer, Boston.

Vanderlei, E.S., K.K. Patoilo, N.A. Lima, A.P. Lima, J.A. Rodrigues, L.M. Silva, M.E. Lima, V. Lima and N.M. Benevides. 2010. Antinociceptive and anti-inflammatory activities of lectin from the marine green alga *Caulerpa cupressoides*. International Immunopharmacology 10: 1113–1118.

Vanderlei, E.S.O., I.W.F. de Araújo, A.L.G. Quinderé, B.P. Fontes, Y.R.G. Eloy, J.A.G. Rodrigues, A.A.R. Silva, H.V. Chaves, R.J.B. Jorge, D.B. de Menezes, J.S.A.M. Evangelista, M.M. Bezerra and N.M.B. Benevides. 2011. The involvement of the HO-1 pathway in the anti-inflammatory action of a sulfated polysaccharide isolated from the red seaweed *Gracilaria birdiae*. Inflamm. Res. 60:1121–1130.

Vane, J. and R. Botting. 1995. New insights into the mode of action of anti-inflammatory drugs. Inflammation Research 44: 1–10.

Vane, J.R., J.A. Mitchell, I. Appleton, A. Tomlinson, B.D. Bishop, J. Croxtall and D.A. Willoughby. 1994. Inducible isoforms of cyclooxygenase and nitric-oxide synthase in inflammation. Proceedings of the National Academy of Sciences U.S.A. 91: 2046–2050.

Varier, K.M., M.C.J. Milton, C. Arulvasu and B. Gajendran. 2013. Evaluation of antibacterial properties of selected red seaweeds from Rameshwaram, Tamil Nadu, India. Journal of Academia and Industrial Research 1(11): 667–670.

Vasanthi, H., A. Jaswanth, V. Krishnaraj, G. Rajamanickam and A. Saraswathy. 2003. *In vitro* snake venom detoxifying action of some marine algae of Goulf of Mannar, South East of India. Phytotherapy Research 17: 1217–1219.

Vasquez, R, J. Aposto, J.D. Ramos, M. Morale, K. Padiernos, C. Pangilinan, C. Payuran and J. Princesa. 2012. Hepatoprotective effects of aqueous sulfated polysaccharide extract from *Sargassum siliquosum* J.G. Agardh on paracetamol-induced oxidative liver toxicity and antioxidant properties. International Journal of Pharmaceutical Frontier Research 2: 15–27.

Vatsos, I.N. and C. Rebours. 2015. Seaweed extracts as antimicrobial agents in aquaculture. Journal of Applied Phycology 27: 2017–2035.

Vaugelade, P., C. Hoebler, F. Bernard, F. Guillon, M. Lahaye, P.H. Duee and B. Darcy-Vrillon. 2000. Non-starch polysaccharides extracted from seaweed can modulate intestinal absorption of glucose and insulin response in the pig. Reproduction Nutrition Development 40: 33–47.

Vavilala, S.L. and J.S. D'Souza. 2015. Algal polysaccharides and their biological applications. pp. 411–452. *In*: S.-K. Kim and K. Chojnacka (eds.). Marine Algae Extracts. Wiley-VCH Verlag GmbH & Co., Weinheim, Germany.

Vázquez, A.I.F., C.M.D. Sánchez, N.G. Delgado, A.M.S. Alfonso, Y.S. Ortega and H.C. Sánchez. 2011. Anti-inflammatory and analgesic activities of red seaweed *Dichotomaria obtusata*. Brazilian Journal of Pharmaceutical Sciences 47(1): 111–118.

Vazquez-Freire, M.J., M. Lamela, J. Anca and J.M. Calleja. 1993. Neuropharrnacological activity of *Cystoseira usneoides*. Medicaments et Aliments: Lapproche Ethnopharmacologique 7: 295–297.

Vazquez-Freire, M.J., M. Lamela and J.M. Calleja. 1996. Hypolipidaemic activity of a polysaccharide extract from *Fucus vesiculosus* L. Phytotherapy Research 10(8): 647–650.

Vedhagiri, K., A. Manilal, T. Valliyammai, S. Shanmughapriya, S. Sujith, J. Selvin and K. Natarajaseenivasan. 2009. Antimicrobial potential of a marine seaweed *Asparagopsis taxiformis* against *Leptospira javanica* isolates of rodent reservoirs. Annals of Microbiology 59(3): 431–437.

Veena, C.K., A. Josephine, S.P. Preetha and P. Varalakshmi. 2007. Beneficial role of sulfated polysaccharides from edible seaweed *Fucus vesiculosus* in experimental hyperoxaluria. Food Chemistry 100: 1552–1559.

Veiga-Santos, P., K.J. Pelizzaro-Rocha, A.O. Santos, T. UedaNakamura, B.P. Dias Filho, S.O. Silva, D.B. Sudatti, E.M. Bianco, R.C. Pereira and C.V. Nakamura. 2010. *In vitro* antitrypanosomal activity of elatol isolated from red seaweed *Laurencia dendroidea*. Parasitology 137: 1661–1670.

Velmurugan, S., N. Jerin, M.M. Babu, F. Bindhu, S.A. Dhas and T. Citarasu. 2015. Screening and characterization of antiviral compounds from *Enteromorpha flexuosa* against white spot syndrome virus (WSSV) and its *in vivo* influence on Indian white shrimp *Fenneropenaeus indicus*. Aquaculture International 23(1): 65–80.

Vengadesan, P. and S.A. John. 2014. Study on antifungal activity of *Sargassum wightii* from the Mandapam coast Tamil Nadu, India. Asian Journal of Pharmaceutical Science & Technology 4(1): 37–38.

Venkatesalu, V., P. Sundaramoorthy, M. Anantharaj, M. Chandrasekaran and A. Senthilkumar. 2012. Seasonal variation on fatty acid composition of some marine marco algae from Gulf of Mannar Biosphere Reserve, Southeast coast of India. Indian Journal of Geo-Marine Sciences 41(5): 442–450.

Venkatesan, J., B. Lowe, S. Anil, P. Manivasagan, A.A. Al Kheraif, K.H. Kang and S.K. Kim. 2015a. Seaweed polysaccharides and their potential biomedical applications. Starch-Starke 67(5-6): 381–390.

Venkatesan, J., I. Bhatnagar, P. Manivasagan, K.H. Kang and S-K. Kim. 2015b. Alginate composites for bone tissue engineering: A review. International Journal of Biological Macromolecules 72: 269–281.

Venkateswaran, P.S., I. Millman and B.S. Blumberg. 1989. Interaction of fucoidan from *Pelvetia fastigiata* with surface antigens of hepatitis B and woodchuck hepatitis viruses. Planta Medica 55: 265–270.

Venugopal, V. (ed.). 2008. Marine Products for Healthcare: Functional and bioactive Nutraceutical Compounds from the Ocean. CRC Press, Taylor & Francis Group, Boca Raton, F.L. 528 pp.

Vera, J., J. Castro, A. Gonzalez and A. Moenne. 2011. Seaweed polysaccharides and derived oligosaccharides stimulate defense responses and protection against pathogens in plants. Marine Drugs 9: 2514.

Verdy, C., J.-E. Branka and N. Mekideche. 2011. Quantitative assessment of lactate and progerin production in normal human cutaneous cells during normal ageing: effect of an *Alaria esculenta* extract. International Journal of Cosmetic Science 33(5): 462–466.

Vereecken, C. and L. Maes. 2010. Young children's dietary habits and associations with the mothers' nutritional knowledge and attitudes. Appetite 54: 44–51.

Viana, G.S., A.L.P. Freitas, M.M.L. Lima, L.A.P. Vieira, M.C.H. Andrade and N.M.B. Benevides. 2002. Antinociceptive activity of sulfated carbohydrates from the red algae *Bryothamnion seaforthii* (Turner) Kutz. and *B. triquetrum* (S.G. Gmel.) M. Howe. Brazilian Journal of Medical and Biological Research 35(6): 713–722.

Vidyavathl, N. and K.R. Sridhar. 1991. Seasonal and geographical variation in the antimicrobial activity of seaweeds from the Mangalore coast of India. Botanica Marina 34: 279–284.

Vieira, C., J. Gaubert, O. De Clerck, C. Payri, G. Culioli and O.P. Thomas. 2015. Biological activities associated to the chemodiversity of the brown algae belonging to genus *Lobophora* (Dictyotales, Phaeophyceae). Phytochemistry Reviews: 1–17.

Vieira, L.A.P., A.L.P. Freitas, J.P.A. Feitosa, D.C. Silva and G.S.B. Viana. 2004. The alga *Bryothamnion seaforthii* contains carbohydrates with antinociceptive activity. Brazilian Journal of Medical and Biological Research 37(7): 1071–1079.

Vijayabaskar, P. and V. Shiyamala. 2011. Antibacterial activities of brown marine algae (*Sargassum wightii* and *Turbinaria ornata*) from the Gulf of Mannar biosphere reserve. Advances in Biological Research 5(2): 99– 102.

Vijayabaskar, P., N. Vaseela and G. Thirumaran. 2012. Potential antibacterial and antioxidant properties of a sulfated polysaccharide from the brown marine algae *Sargassum swartzii*. Chinese Journal of Natural Medicines 10: 421–428.

Vijayalakshmi, S. 2015. Screening and anti-inflammatory activity of methanolic and queous extracts of seaweed *Gracilaria edulis*. International Journal of Modern Chemistry and Applied Science 2(4): 248–250.

Vila, M. and S. Przedborski. 2003. Targeting programmed cell death in neurodegenerative diseases. Nature Reviews Neuroscience 4: 365–375.

Vílchez, C., E. Forján, M. Cuaresma, F. Bédmar, I. Garbayo and J.M. Vega. 2011. Marine carotenoids: Biological functions and commercial applications. Marine Drugs 9(3): 319–333.

Villanueva, M.J., M. Morcillo, M.D. Tenorio, I. Mateos-Aparicio, V. Andrés and A. Redondo-Cuenca. 2014. Health-promoting effects in the gut and influence on lipid metabolism of *Himanthalia elongata* and *Gigartina pistillata* in hypercholesterolaemic Wistar rats. European Food Research and Technology 238: 409–416.

Villarreal-Gómez, L.J., I.E. Soria-Mercado, G. Guerra-Rivas and N.E. Ayala-Sánchez. 2010. Antibacterial and anticancer activity of seaweeds and bacteria associated with their surface. Revista de Biología Marina y Oceanografía 45(2): 267–275.

Villarroel L.H. and A.B. Zanlungo. 1981. Structural studies on the Porphyran from *Porphyra columbina*. Carbohydrate Research 88: 139–145.

Vinayak, R., S. Puttananjaiah, A. Chatterji and B. Salimath. 2014. Anti-proliferative and angio-suppressive effect of *Stoechospermum marginatum* (C. Agardh) Kutzing extract using various experimental models. Nutrition Research and Practice 8(4): 377–385.

Vishchuk, O.S., S.P. Ermakova, D.T. Pham, N.M. Shevchenko, B.M. Ly and T.N. Zvyagintseva. 2009. Anticancer activity of brown algae-derived fucoidans. Pacific Medical Journal 3: 86–89.

Vishchuk, O.S., D.V. Tarbeeva, S.P. Ermakova and T.N. Zvyagintseva. 2012. Structural characteristics and biological activity of fucoidans from the brown algae *Alaria* sp. and *Saccharina japonica* of different reproductive status. Chemistry and Biodiversity 9(4): 817–828.

Vishchuk, O.S., S.P. Ermakova and T.N. Zvyagintseva. 2013. The fucoidans from brown algae of far-eastern seas: Anti-tumor activity. Food Chemistry 141: 1211–1217.

Vishchuk, O.S., S.P. Ermakova and T.N. Zvyagintseva. 2013b. The effect of sulfated (1-->3)-Alpha-L-fucan from the brown alga *Saccharina cichorioides* Miyabe on resveratrol-induced apoptosis in colon carcinoma cells. Marine Drugs 11(1): 194–212.

Vishchuk, O.S., H.M. Sun, Z. Wang, S.P. Ermakova, J.J. Xiao, T. Lu, P.P. Xue, T.N. Zvyagintseva, H. Xiong, C. Shao, W. Yan, Q. Duan and F. Zhu. 2016. Pdz-binding kinase/T-lak cell-originated protein kinase is a target of the fucoidan from brown alga *Fucus evanescens* in the prevention of EGF-induced neoplastic cell transformation and colon cancer growth. Oncotarget 7(14): 18763–1873.

Viswanathan, S., C. Ebciba and S.R.T. Nallamuthu. 2014. Phytochemical screening and in vitro antibacterial, antioxidant and anticancer activity of *Amphiroa fragilissima* (Linneaus) JV Lamoroux. International Journal of Innovative Research in Science, Engineering and Technology 3(5): 12933–129948.

Vivek, M., P.S. Kumar, S. Steffi and S. Sudha. 2011. Biogenic silver nanoparticles by *Gelidiella acerosa* extract and their antifungal effects. Avicenna Journal of Medical Biotechnology 3(3): 143–148.

Vlachos, V., A.T. Critchley and A. von Holy. 1997. Antimicrobial activity of extracts from selected Southern African marine macro-algae. South African Journal of Science 93: 328–332.

Vlachos, V., A.T. Critchley and A. von Holy. 2001. Effect of postcollection storage time and season on the antibacterial activity of selected southern African marine macroalgae. pp. 207–213. *In*: Chen, F. and Y. Jiang (eds.). Algae and their Biotechnological Potential. Kluwer Academic Publishers, The Netherlands.

Vlieghe, P., T. Clerc, C. Pannecouque, M. Witvrouw, E. De Clercq, J. Salles and J.L. Kraus. 2002. Synthesis of new covalently bound kappa-carrageenan–AZT conjugates with improved anti-HIV activities. Journal of Medicinal Chemistry 45: 1275–1283.

Vo, T.-S. and S.-K. Kim. 2010. Potential anti-HIV agents from marine resources: an overview. Marine Drugs 8: 2871–2892.

Vo, T.-S., J.A. Kim, D.H. Ngo, C.S. Kong and S.-K. Kim. 2012. Protective effect of chitosan oligosaccharides against FcεRI-mediated RBL-2H3 mast cell activation. Process Biochemmistry 47: 327–330.

Vo, T.-S., D.-H. Ngo and S.-K. Kim. 2012b. Marine algae as a potential pharmaceutical source for anti-allergic therapeutics. Process Biochemistry 47: 386–394.

Vo, T.-S., D.-H. Ngo and S.-K. Kim. 2012c. Potential targets for anti-inflammatory and anti-allergic activities of marine algae: An overview. Inflammation & Allergy—Drug Targets 11: 90–101.

Volk, T. 2010. *Saccharomyces cerevisiae*. Available online at: http://bioweb.uwlax.edu/bio203/s2007/nelson_andr/.

Volk, T. 2010b. *Trichoderma viride*. Available online at: http://botit.botany.wisc.edu/toms_fungi/nov2004.html.

Vonthron-Sénécheau, C., M. Kaiser, I. Devambez, A. Vastel, I. Mussio and A.M. Rusig. 2011. Antiprotozoal activities of organic extracts from French marine seaweeds. Marine Drugs 9: 922–933.

Wainwright, M. 1991. Miracle cure: The story of Penicillin and the golden age of antibiotics. Basil Blackwell, Oxford, 288 pp.

Walker, A.R.P. 1998. The remedying of iron deficiency: What priority should it have? British Journal of Nutrition 79: 227–235.

Walker, G.M. 1998b. Yeast physiology and biotechnology. John Wiley & Sons, Chichester, 362 pp.

Wallentinus, I. and C.D. Nyberg. 2007. Introduced marine organisms as habitat modifiers. Marine Pollution Bulletin 55: 323–332.

Walsh, C. 2003. Antibiotics: actions, origins, resistance. American Society of Microbiology, Washington, DC, 335 pp.

Walsh, T.J., A. Groll, J. Hiemenz, R. Flemming, E. Roilides and E. Anaissie. 2004. Infections due to emerging and uncommon medically important fungal pathogens. Clinical Microbiology and Infection 10 (Suppl. 1): 48–66.

Walsh, T.J., E.J. Anaissie, D.W. Denning, R. Herbrecht, D.P. Kontoyiannis, K.A. Marr, V.A. Morrison, B.H. Segal, W.J. Steinbach, D.A. Stevens, J.A. van Burik, J.R. Wingard and T.F. Patterson. 2008. Treatment of aspergillosis: clinical practice guidelines of the Infectious Diseases Society of America. Clinical Infectious Diseases 46(3): 327–360.

Wang, C.-Y., T.-C. Wu, S.-L. Hsieh, Y.-H. Tsai, C.-W. Yeh and C.-Y. Huang. 2015. Antioxidant activity and growth inhibition of human colon cancer cells by crude and purified fucoidan preparations extracted from *Sargassum cristaefolium*. Journal of Food and Drug Analysis 23(4): 766–777.

Wang, G.C. 2002. Isolation and purification of phycoerythrin from red alga *Garcilaria verrucosa* by expanded-bed-adsorption and ion-exchange chromatography. Chromatographia 56(7-8): 509–513.

Wang, H., E.V. Ooi and P.O. Ang, Jr. 2008. Antiviral activities of extracts from Hong Kong seaweeds. Journal of Zhejiang University-SCIENCE B 9: 969–976.

Wang, H.-M., Y.-T. Chou, Z.-H. Wen, C.-Z. Wang, C.-H. and M.-L. Ho. 2013c. Novel biodegradable porous scaffold applied to skin regeneration. PLoS ONE 8(6): e56330.

Wang, J. and C. Chen. 2009. Biosorbents for heavy metals removal and their future. Biotechnology Advances 27: 195–226.

Wang, J., W. Jin, Y. Hou, X. Niu, H. Zhang and Q. Zhang. 2013d. Chemical composition and moisture-absorption/retention ability of polysaccharides extracted from five algae. International Journal of Biological Macromolecules 57: 26–29.

Wang, L., X. Wang, H. Wu and R. Liu. 2014. Overview on biological activities and molecular characteristics of sulfated polysaccharides from marine green algae in recent years. Marine Drugs 12(9): 4984–5020.

Wang, R., V.J. Paul and H. Luesch. 2013e. Seaweed extracts and unsaturated fatty acid constituents from the green alga *Ulva lactuca* as activators of the cytoprotective Nrf2-ARE pathway. Free Radical Biology and Medecine 57: 141–153.

Wang, S.C., S.W.A. Bligh, S.S. Shi, Z. Wang, Z.B. Hu, J. Crowder, C. Branford-White and C. Vella. 2007. Structural features and anti-HIV-1 activity of novel polysaccharides from red algae *Grateloupia longifolia* and *Grateloupia filicina*. International Journal of Biological Macromolecules 41: 369–375.

Wang, T., R. Jónsdóttir, H.G. Kristinsson, G.O. Hreggvidsson, J.Ó. Jónsson, G. Thorkelsson and G. Ólafsdóttir. 2010. Enzyme-enhanced extraction of antioxidant ingredients from red algae *Palmaria palmata*. LWT—Food Science and Technology 43(9): 1387–1393.

Wang, W., S.-X. Wang and H.-S. Guan. 2012a. The antiviral activities and mechanisms of marine polysaccharides: an overview. Marine Drugs 10(12): 2795–2816.

Wang, W., P. Zhang, G.L. Yu, C.X. Li, C. Hao, X. Qi, L.J. Zhang and H.S. Guan. 2012b. Preparation and anti-influenza A virus activity of κ-carrageenan oligosaccharide and its sulphated derivatives. Food Chemistry 133: 880–888.

Wang, W., J.D. Wu, X.S. Zhang, C. Hao, X.L. Zhao, G.L. Jiao, X.D. Shan, W.J. Tai and G.L. Yu. 2017. Inhibition of influenza A virus infection by fucoidan targeting viral neuraminidase and cellular EGFR pathway. Scientific Reports 7: 40760.

Wang, W.-L. 2014. *Padina australis*—BiotaTaiwanica, Algae of Taiwan. Availabe online at: http://algae.biota.biodiv.tw/pages/922.

Wang, X., Z. Zhang, Z. Yao, M. Zhao and H. Qi. 2013b. Sulfation, anticoagulant and antioxidant activities of polysaccharide from green algae *Enteromorpha linza*. International Journal of Biological Macromolecules 58: 225–230.

Wang, Y., F. Han, B. Hu, J.B. Li and W.G. Yu. 2006. *In vivo* prebiotic properties of alginate oligosaccharides prepared through enzymatic hydrolysis of alginate. Nutrition Research 26(11): 597–603.

Wang, Y., Z. Xu, S. Bach and T. McAllister. 2009. Sensitivity of *Escherichia coli* to seaweed (*Ascophyllum nodosum*) phlorotannins and terrestrial tannins. Asian-Australas. Journal of Animal Science 22: 238–245.

Wang, Y., Z. Jiang, D. Kim, M. Ueno, T. Okimura, K. Yamaguchi and T. Oda. 2013. Stimulatory effect of the sulfated polysaccharide Ascophyllan on the respiratory burst in Raw264.7 macrophages. International Journal of Biological Macromolecules 52: 164–169.

Washington, N. and G. Denton. 1995. Effect of alginate and alginate-cimetidine combination therapy on stimulated postprandial gastro-oesophageal reflux. Journal of Pharmacy and Pharmacology 47(11): 879–882.

Watanabe, F., S. Takenaka, H. Katsura, E. Miyamoto, K. Abe, Y. Tamura, T. Nakatsuka and Y. Nakano. 2000. Characterization of a vitamin B_{12} compound in the edible purple laver, *Porphyra yezoensis*. Bioscience, Biotechnology, and Biochemistry 64: 2712–2715.

Watanabe, K., K. Umeda and M. Miyakado. 1989. Isolation and identification of three insecticidal principles from the red alga, *Laurencia nipponica*. Agricultural and Biological Chemistry 53: 2513–2515.

Watanabe, K., K. Umeda, Y. Kurita, C. Takayama and M. Miyakado. 1990. Two insecticidal monoterpenes, telfairine and aplysiaterpenoid A, from the red alga *Plocamium telfairiae*: Structure elucidation, biological activity, and molecular topographical consideration by a semiempirical molecular orbital study. Pesticide Biochemistry and Physiology 37(3): 275–286.

Wefky, S., A.H. Shobier, S.A.A. Ghani and E.S.H. El Ashry. 2009. Screening of the antibacterial activity of some marine algae against fish and human pathogens. Bulletin of High Institute of Public Health 39(1): 112–124.

Wei, Y., Q. Liu, C. Xu, J. Yu, L. Zhao and Q. Guo. 2016. Damage to the membrane permeability and cell death of *Vibrio parahaemolyticus* caused by phlorotannins with low molecular weight from *Sargassum thunbergii*. Journal of Aquatic Food Product Technology 25(3): 323–333.

Wei, Y., Q. Liu, J. Yu, Q. Feng, L. Zhao, H. Song and W. Wang. 2015. Antibacterial mode of action of 1,8-dihydroxy-anthraquinone from *Porphyra haitanensis* against *Staphylococcus aureus*. Natural Product Research 29(10): 976–979.

Weigl, J.A.I., W. Puppe, C.U. Meyer, R. Berner, J. Foster, H.J. Schmitt and F. Zepp. 2007. Ten years' experience with year-round active surveillance of up to 19 respiratory pathogens in children. European Journal of Pediatrics 166: 957–966.

Weinberg, R.A. 2013. The biology of cancer (2nd edition). Garland Science, London, 960 pp.

Welch, A.M. 1962. Preliminary survey of fungistatic properties of marine algae. Journal of Bacteriology 83: 97–99.

Wells, J.C.K., A.A. Marphatia, T.J. Cole and D. McCoy. 2012. Associations of economic and gender inequality with global obesity prevalence: understanding the female excess. Social Science & Medicine 75: 482–490.

Wen, X., C. Peng, H. Zhou, Z. Lin, G. Lin, S. Chen and P. Li. 2006. Nutritional composition and assessment of *Gracilaria lemaneiformis* Bory. Journal of Integrative Plant Biology 48: 1047–1053.

Whapham, C.A., T. Jenkins, G. Blunden and S.D. Hankins. 1994. The role of seaweed extracts *Ascophyllum nodosum* in the reduction in fecundity of *Meloidogyne javanica*. Fundamental & Applied Nematology 17: 181–183.

White, A. 2014. Metals on our mind. The Scientist. Available online at: http://www.the-scientist.com/?articles.view/articleNo/39517/title/Metals-on-our-Mind/.

WHO. 2001. Iron deficiency anaemia: assessment, prevention and control. *In*: A Guide for Programme Managers. World Health Organization, Geneva.

WHO. 2002. Biological and social aspects of cancer. pp. 3–7. *In*: National Cancer Control Programmes: Policies and Managerial Guidelines, 2nd ed. World Health Organization, Geneva. Available online at: http://apps.who.int/iris/bitstream/10665/42494/1/9241545577.pdf.

WHO. 2014. Chagas disease (American trypanosomiasis). World Health Organ Fact Sheet 340. Available online at: http://www.who.int/mediacentre/factsheets/fs340/en/.

WHO. 2015. Global action plan on antimicrobial resistance. World Health Organization, Geneva. Available online at: http://www.wpro.who.int/entity/drug_resistance/resources/global_action_plan_eng.pdf.

WHO. 2017. Neglected tropical diseases. Available online at: http://www.who.int/neglected_diseases/diseases/en/.

WHO. 2017b. Trypanosomiasis, human African (sleeping sickness). Available online at: http://www.who.int/mediacentre/factsheets/fs259/en/.

WHO. 2017c. Top 10 causes of death. Available online at: http://www.who.int/gho/mortality_burden_disease/causes_death/top_10/en/.

Wickett, R.R. and M.O. Visscher. 2006. Structure and function of the epidermal barrier. American Journal of Infection Control 34: S98–S110.

Wiederhold, N.P. and R.E. Lewis. 2009. Antifungal activity against *Scedosporium* species and novel assays to assess antifungal pharmacodynamics against filamentous fungi. Medical mycology 47(4): 422–432.

Wiemer, D.F., D.D. Idler and W. Fenical. 1991. Vidalols A and B, new antiinflammatory bromophenols from the Caribbean marine red alga *Vidalia obtusiloba*. Experientia 47(91): 851–853.

Wijesekara, I., R. Pangestuti and S.-K. Kim. 2011. Biological activities and potential health benefits of sulfated polysaccharides derived from marine algae. Carbohydrate Polymers 84: 14–21.

Wijesinghe, W.A., S.C. Ko and Y.J. Jeon. 2011. Effect of phlorotannins isolated from *Ecklonia cava* on angiotensin I-converting enzyme (ACE) inhibitory activity. Nutrition Research and Practice 5: 93–100.

Wijesinghea, W.A.J.P. and Y.-J. Jeona. 2012. Biological activities and potential industrial applications of fucose rich sulfated polysaccfzarides and fucoidans isolated from brown seaweeds: a review. Carbohydrate Polymers 88: 13–20.

Williams, J.V., P.A. Harris, S.J. Tollesfson, L.L. Halbarnt-Rush, J.M. Pingsterhaus, K.M. Edwards, P.F. Wright, J.E. Crowe, Jr. 2004. Human metapneumovirus and lower respiratory tract disease in otherwise healthy infants and children. The New England Journal of Medicine 350: 443–450.

Willis, R.J. 2007. The history of allelopathy. Springer, Berlin, Germany, 299 pp.

Willstatter, R. and H.J. Page. 1914. Tests on chlorophyll, XXIV The pigments of brown algae. Justus Liebigs Annalen der Chemie 404(1/3): 237–271.

Winberg, P.C., D. Skropeta and A. Ullrich. 2011. Seaweed cultivation pilot trials—towards culture systems and marketable products. Australian Government Rural Industries Research and Development Corporation. RIRDC Publication No. 10/184. PRJ - 000162. Available online at: http://ro.uow.edu.au/cgi/viewcontent.cgi?article=1001&context=smfc.

Witvrouw, M. and E. De Clercq. 1997. Sulfated polysaccharides extracted from sea algae as potential antiviral drugs. General Pharmacology 29: 497–511.

Witvrouw, M., J.A. Este, M.Q. Mateu, D. Reymen, G. Andrei, R. Snoeck, S. Ikeda, R. Pauwels, N.V. Bianchini, J. Desmyter and E. De Clercq. 1994. Activity of sulfated polysaccharide extracted from the red seaweed *Aghardhiela tenera* against Human-immunodeficiency-virus and other enveloped viruses. Antiviral Chemistry and Chemotherapy 5: 297–303.

Wojenski, C.M., M.J. Silver and J. Waker. 1991. Eicosapentaenoic acid ethyl ester as an antithrombotic agent: comparison to an extract of fish oil. Biochimica et Biophysica Acta 1081: 33–38.

Wolf, M., D. Lichtensteein and G. Singh. 1999. Gastrointestinal toxicity on nonsteroidal anti-inflammatory drugs. The New England Journal of Medicine 340(24): 1888–1889.

Wolff, J. 1989. Excess iodide inhibits the thyroid by multiple mechanisms. Advances in Experimental Medicine and Biology 261: 211–244.

Wolinsky, H. 1980. A proposal linking clearance of circulating lipoproteins to tissue metabolic-activity as a basis for understanding atherogenesis. Circulation Research 47: 301–311.

Wollenberg, A., S. Kraft, D. Hanau and T. Bieber. 1996. Immunomorphological and ultrastructural characterization of langerhans cells and a novel, inflammatory dendritic epidermal cell (IDEC) population in lesional skin of atopic eczema. Journal of Investigative Dermatology 106: 446–453.

Womersley, H.B.S. 1987. The marine benthic flora of southern Australia. Part II. South Australian Government Printing Division, Adelaide, Australia, 481 pp.

Womersley, H.B.S. 2003. Marine benthic flora of southern Australia. Available online at: http://www.flora.sa.gov.au/algae_flora/The_Marine_Benthic_Flora_of_SA_static_index.shtml.

Wong, K.H. and P.C.K. Cheung. 2000. Nutritional evaluation of some subtropical red and green seaweeds. Part I. Proximate composition, amino acid profile and some physico-chemical properties. Food Chemistry 71: 475–482.

Wong, K.H. and P. Cheung. 2002. Nutritional evaluation of some subtropical red and green seaweeds part-1, proximate composition, amino acid profiles and some physico-chemical properties. Journal of Food Chemistry 71: 475–482.

Wong, M.-Y. 2003. *Fusarium oxysporum* f. sp. *lycopersici* (Sacc.) W.C. Snyder and H.N. Hans. Available online at: https://projects.ncsu.edu/cals/course/pp728/Fusarium/Fusarium_oxysporum.htm.

Wood-Kaczmar, A., S. Gandhi and N. Wood. 2006. Understanding the molecular causes of Parkinsons disease. Trends in Molecular Medicine 12: 521–528.

World Alzheimer Report. 2016. Improving healthcare for people living with dementia. Available online at: https://www.alz.co.uk/research/worldalzheimerreport2016sheet.pdf.

Wozniak, M., T. Bell, Á. Dénes, R. Falshaw and R. Itzhaki. 2015. Anti-HSV1 activity of brown algal polysaccharides and possible relevance to the treatment of Alzheimer's disease. International Journal of Biological Macromolecules 74: 530–540.

Wu, D., S.N. Han, R.T. Bronson, D.E. Smith and S.N. Meydani. 1998. Dietary supplementation with mushroom-derived protein-bound glucan does not enhance immune function in young and old mice. Journal of Nutrition 128: 193–197.

Wu, J. and X. Ding 2002. Characterization of inhibition and stability of soy-protein-derived angiotensin I-converting enzyme inhibitory peptides. Food Research International 3: 367–375.

Wu, J.M. and C.A. Staton. 2012. Anti-angiogenic drug discovery: lessons from the past and thoughts for the future. Expert Opin. Drug Discov. 7(8): 723–743.

Wu, L., J. Sun, X. Su, Q. Yu, Q. Yu and P. Zhang. 2016. A review about the development of fucoidan in antitumor activity: progress and challenges. Carbohydrate Polymers 154: 96–111.

WuDunn, D. and P.G. Spear. 1989. Initial interaction of *Herpes simplex* virus with cells is binding to heparan sulfate. Journal of Virology 63: 52–58.

Xianliang, X., G. Meiyu, G. Huashi and L. Zelin. 2000. Study on the mechanismof inhibitory actionof 911 onreplication of HIV-1 *in vitro*. Chinese Journal of Marine Drugs 19(4): 15–22.

Xin, X., M. Geng, H. Guan and Z. Li. 1999. Study on the mechanism of inhibitory action of 911 on replication of HIV-1 *in vitro*. Chinese Journal of Marine Drugs 19(4): 15–18.

Xin, X.L., H. Ding, M.Y. Geng, P.F. Liang, Y.X. Li and H.S. Guan. 2000a. Studies of the anti-AIDS effects of marine polysaccharide drug 911 and its related mechanisms of action. Chinese Journal of Marine Drugs 6: 4–8.

Xin, X.L., M.Y. Geng, H.S. Guan and Z.L. Li. 2000b. Study on the mechanism of inhibitory action of 911 on replication of HIV-1 *in vitro*. Chinese Journal of Marine Drugs 19: 15–18.

Xu, N., X. Fan, L. Han, W. Lin and C. Zeng. 2002. Study on antibacterial activity of ethanol extract of seaweed. Haiyang Yu Huzhao 33: 265–270.

Xu, N., X. Fan, X. Yan and C.K. Tseng. 2004b. Screening marine algae from China for their antitumor activities. Journal of Applied Phycology 16(6): 451–456.

Xu, X., F. Song, S. Wang, S. Li, F. Xiao, J. Zhao, Y. Yang, S. Shang, L. Yang and J. Shi. 2004. Dibenzyl bromophenols with diverse dimerization patterns from the brown alga *Leathesia nana*. Journal of Natural Products 67(10): 1661–1666.

Xu, W.-J., X.-J. Liao, S.-H. Xu, J.-Z. Diao, B. Du, X.-L. Zhou and S.-S. Pan. 2008. Isolation, structure determination, and synthesis of Galaxamide, A rare cytotoxic cyclic pentapeptide from a marine algae *Galaxaura filamentosa*. Organic Letters 10: 4569–4572.

Xue, M., Y. Ge, J. Zhang, Q. Wang, L. Hou, Y. Liu, L. Sun and Q. Li. 2012. Anticancer properties and mechanisms of fucoidan on mouse breast cancer *in vitro* and *in vivo*. PLoS One 7: e43483.

Xue, M., Y. Ge, J. Zhang, Y. Liu, Q. Wang, L. Hou and Z. Zheng. 2013. Fucoidan inhibited 4T1 mouse breast cancer cell growth *in vivo* and *in vitro* via downregulation ofWnt/β-catenin signaling. Nutrition and Cancer 65: 460–468.

Yabur, R., Y. Bashan and G. Hernandez-Carmona. 2007. Alginate from the macroalgae *Sargassum sinicola* as a novel source for microbial immobilization material in wastewater treatment and plant growth promotion. Journal of Applied Phycology 19: 43-53.

Yabuta, Y., H. Fujimura, C.S. Kwak, T. Enomoto and F. Watanabe. 2010. Antioxidant activity of the phycoerythrobilin compound formed from a dried Korean purple laver (*Porphyra* sp.) during *in vitro* digestion. Food Science and Technology Research 16: 347–352.

Yamada, T., A. Ogamo, T. Saito, J. Watanabe, H. Uchiyama and Y. Nakagawa. 1997. Preparation and anti-HIV activity of low-molecular-weight carrageenans and their sulfated derivatives. Carbohydrate Polymers 32: 51–55.

Yamada, T., A. Ogamo, T. Saito, H. Uchiyama and Y. Nakagawa. 2000. Preparation of O-acylated low-molecular-weight carrageenans with potent anti-HIV activity and low anticoagulant effect. Carbohydrate Polymers 41: 115–120.

Yamada, T., A. Ogamo, T. Saito, J. Watanabe, H. Uchiyama and Y. Nakagawa. 2000a. Preparation and anti-HIV activity of low-molecular-weight carrageenans and their sulfated derivatives. Carbohydrate Polymers 32: 51–55.

Yamada, T., A. Ogamo, T. Saito, H. Uchiyama and Y. Nakagawa. 2000b. Preparation of O-acylated low-molecular-weight carrageenans with potent anti-HIV activity and low anticoagulant effect. Carbohydrate Polymers 41: 115–120.

Yamada, Y., T. Miyoshi, S. Tanada and M. Imaki. 1991. Digestibility and energy availability of Wakame (*Undaria pinnatifida*) seaweed (in Japanese). Nippon Eiseigaku Zasshi 46: 788–794.

Yamamoto, I., T. Nagumo, K. Yagi, H. Tominaga and M. Aoki. 1974. Antitumor effect of seaweeds I. Antitumor effect of extracts from *Sargassum* and *Laminaria*. The Japanese Journal of Experimental Medicine 44: 543–546.

Yamamoto, I., T. Nagumo, M. Fujihara, M. Takahashi, Y. Ando, M. Okada and K. Kawai. 1977. Antitumor effect of seaweeds II. Fractionation and partial characterization of the polysaccharide with antitumor activity from *Sargassum fulvellum*. The Japanese Journal of Experimental Medicine 47: 133–140.

Yamamoto, I., T. Nagumo, M. Takahashi, M. Fujihara, Y. Suzuki and N. Iizima, 1981. Antitumor effect of seaweeds III. Antitumor effect of an extract from *Sargassum kjellmanianum*. The Japanese Journal of Experimental Medicine 51: 187–189.

Yamamoto, I., M. Takahashi, E. Tamura and H. Maruyama. 1982. Antitumor activity of crude extracts from edible marine algae against L-1210 leukemia. Botanica Marina 25: 455–457.

Yamamoto, I., M. Takahashi, E. Tamura, H. Maruyama and H. Mori. 1984. Antitumor activity of edible marine algae: effect of crude fucoidan fractions prepared from edible brown seaweeds against L-1210 leukemia. pp. 145–148. *In*: Bird, C.J. and M.A. Raga (eds.). Eleventh International Seaweed Symposium: Proceedings of the Eleventh International Seaweed Symposium, held in Qingdao, People's Republic of China, June 19–25, 1983. Springer Netherlands, Dordrecht.

Yamamoto, I., M. Takahashi, E. Tamura, H. Maruyama and H. Mori. 1984a. Antitumor activity of edible marine algae: Effect of fucoidan fractions prepared from edible seaweeds against L-1210 leukemia. Proceedings of International Seaweed Symposium 11: 145–148.

Yamamoto, I., M. Takahashi, T. Suzuki, H. Seino and H. Mori. 1984b. Antitumor effect of seaweeds IV. Enhancement of antitumor activity by sulfation of a crude fucoidan fraction from *Sargassum kjellmanianum*. The Japanese Journal of Experimental Medicine 54: 143–151.

Yamamoto, I. and H. Maruyama. 1985. Effect of dietary seaweed preparations on 1, 2-dimethylhydrazine induced intestinal carcinogenesis in rats. Cancer Letters 26: 241–51.

Yamamoto, I., H. Maruyama and M. Moriguchi. 1987. The effect of dietary seaweeds on 7,12-dimethylbenz[a]anthracene-induced mammary tumorigenesis in rats. Cancer Letters 35: 109–118.

Yamamoto, I., H. Maruyama, M. Takahashi and K. Komiyama. 1986. The effect of dietary or intraperitoneally injected seaweed preparations on the growth of sarcoma-180 cells subcutaneously implanted into mice. Cancer Letters 30(2): 125–131.

Yamamoto, K., C. Ishikawa, H. Katano, T. Yasumoto and N. Mori. 2011. Fucoxanthin and its deacetylated product, fucoxanthinol, induce apoptosis of primary effusion lymphomas. Cancer Letters 300(2): 225–234.

Yan, M.-D., C.-J. Yao, J.-M. Chow, C.-L. Chang, P.-A. Hwang, S.-E. Chuang, J. Whang-Peng and G.-M. Lai. 2015. Fucoidan elevates microRNA-29b to regulate DNMT3b-MTSS1 axis and inhibit EMT in human hepatocellular carcinoma cells. Marine Drugs 13(10): 6099–6116.

Yan, X.J., Y. Chuda, M. Suzuki and T. Nagata. 1999. Fucoxanthin as the major antioxidant in *Hijikia fusiformis*, a common edible seaweed. Bioscience Biotechnology and Biochemistry 63(3): 605–607.

Yanase, Y., T. Hiragun, K. Uchida, K. Ishii, S. Oomizu, H. Suzuki, S. Mihara, K. Iwamoto, H. Matsuo, N. Onishi, Y. Kameyoshi and M. Hide. 2009. Peritoneal injection of fucoidan suppresses the increase of plasma IgE induced by OVA-sensitization. Biochemical and Biophysical Research Communications 387: 435–439.

Yang, C., D. Chung, I.-S. Shin, H.Y. Lee, J.C. Kim, Y.J. Lee and S.G. You. 2008. Effects of molecular weight and hydrolysis conditions on anticancer activity of fucoidans from sporophyll of *Undaria pinnatifida*. International Journal of Biological Macromolecules 43(5): 433–437.

Yang, E.-J., J.-Y. Moon, M.-J. Kim, D.S. Kim, C.-S. Kim, W.J. Lee, N.H. Lee and C.-G. Hyun. 2010. Inhibitory effect of Jeju endemic seaweeds on the production of pro-inflammatory mediators in mouse macrophage cell line RAW 264.7. Journal of Zhejiang University-Science B (Biomedicine & Biotechnology) 11: 315–322.

Yang, E.J., J.Y. Moon, M.J. Kim, D.S. Kim, W.J. Lee, N.H. Lee and C.G. Hyun. 2010c. Anti-inflammatory effect of *Petalonia binghamiae* in LPSinduced macrophages is mediated by suppression of iNOS and COX2. International Journal of Agriculture and Biology 12: 754–758.

Yang, H., M. Zeng, S. Dong, Z. Liu and R. Li. 2010b. Anti-proliferative activity of phlorotannin extracts from brown algae *Laminaria japonica* Aresch. Chinese Journal of Oceanology and Limnology 28(1): 122–130.

Yang, H.-N. 2014a. *Sargassum hemiphyllum*—BiotaTaiwanica, Algae of Taiwan. Available at: http://algae.biota.biodiv.tw/pages/1598.

Yang, J.I., C.C. Yeh, J.C. Lee, S.C. Yi, H.W. Huang, C.N. Tseng and H.W. Chang. 2012. Aqueous extracts of the edible *Gracilaria tenuistipitata* are protective against H_2O_2-induced DNA damage, growth inhibition, and cell cycle arrest. Molecules 17(6): 7241–7254.

Yang, J.W., S.Y. Yoon, S.J. Oh, S.K. Kim and K.W. Kang. 2006. Bifunctional effects of fucoidan on the expression of inducible nitric oxide synthase. Biochemical and Biophysical Research Communications 346: 345–350.

Yang, L., P. Wang, H. Wang, Q. Li, H. Teng, Z. Liu, W. Yang, L. Hou and X. Zou. 2013. Fucoidan derived from *Undaria pinnatifida* induces apoptosis in human hepatocellular carcinoma SMMC-7721 cells via the ROS-mediated mitochondrial pathway. Marine Drugs 11(6): 1961–1976.

Yang, P., D.-Q. Liu, T.-J. Liang, J. Li, H.-Y. Zhang, A.-H. Liu, Y.-W. Guo and S.C. Mao. 2015. Bioactive constituents from the green alga *Caulerpa racemosa*. Bioorganic & Medicinal Chemistry 23: 38–45.

Yang, Y., D. Liu, J. Wu, Y. Chen and S. Wang. 2011. *In vitro* antioxidant activities of sulfated polysaccharide fiactions extracted from *Corallina officinalis*. International Journal of Biological Macromolecules 49: 1031–1037.

Yang, Y.-I., S.-H. Jung, K.-T. Lee and J.-H. Choi. 2014. 8, 8′-Bieckol, isolated from edible brown algae, exerts its anti-inflammatory effects through inhibition of NF-kB signaling and ROS production in LPS-stimulated macrophages. International Immunopharmacology 23: 460–468.

Yang, Y.J., S.J. Nam, G. Kong and M.K. Kim. 2010a. A case-control study on seaweed consumption and the risk of breast cancer. British Journal of Nutrition 103(9): 1345–1353.

Yasuhara-Bell, J. and Y. Lu. 2010. Marine compounds and their antiviral activities. Antiviral Research 86(3): 231–240.

Yazdani, D., Y.H. Tan, M.A. Zainal-Abidin and I.B. Jaganath. 2011. A review on bioactive compounds isolated from plants against plant pathogenic fungi. Journal of Medicinal Plant Research 5(16): 6584–6589.

Ye, G., Q. Lu, W. Zhao, D. Du, L. Jin and Y. Liu. 2014. Fucoxanthin induces apoptosis in human cervical cancer cell line HeLa via PI3K/Akt pathway. Tumor Biology 35(11): 11261–11267.

Ye, J., Y. Li, K. Teruya, Y. Katakura, A. Ichikawa, H. Eto, M. Hosoi, S. Nishimoto and S. Shirahata. 2005. Enzyme-digested fucoidan extracts derived from seaweed mozuku of *Cladosiphon novae-caledoniae* Kylin inhibit invasion and angiogenesis of tumor cells. Cytotechnology 47(1-3): 117–126.

Yeh, C.-C., J.-I. Yang, J.-C. Lee, C.-N. Tseng, Y.-C. Chan, Y.-C. Hseu, J.-Y. Tang, L.-Y. Chuang, H.-W. Huang, F.-R. Chang and H.-W. Chang. 2012. Antiproliferative effect of methanolic extract of *Gracilaria tenuistipitata* on oral cancer cells involves apoptosis, DNA damage, andoxidative stress. BMC Complementary and Alternative Medicine 12(1): 1–9.

Yende, S.R., U.N. Harle and A.M. Ittadwar. 2016. Insignificant anticonvulsant activity of *Padina tetrastromatica* (brown macroalgae) in mice. Journal of Pharmaceutical Negative Results 7: 33–36.

Yendo, K. 1902. Uses of marine algæ in Japan. The Year Book of the Minnesota Seaside Station, pp. 1–18.

Yendo, K. 1907. The Fucaceae of Japan. Journal of the College of Science, Tokyo Imperial University 21(Article 12): 1–174.

Yermak, I.M., A.O. Barabanova, V.P. Glazunov, V.V. Isakov, K.Y. Hwan, S.K. Soon, T.V. Titlynova and T.F. Solov'eva. 2006. Carrageenans from cystocarpic and sterile plants of *Chondrus pinnulatus* (Gigartinaceae, Rhodophyta) collected from the Russian Pacific coast. Journal of Applied Phycology 18: 361–368.

Yew, W.W., S. Tonb, K.S. Lui, S.K. Leung, C.H. Chau and E.P. Wang. 2001. Comparison of MB/BacT system and agar proportion method in drug susceptibility testing of *Mycobacterium tuberculosis*. Diagnostic Microbiology and Infectious Disease 39: 229–232.

Yi, Z., Chen Y.-S. and Lu H.-S. 2001. Screening for antibacterial and antifungal activities in some marine algae from the Fujian coast of China with three different solvents. Chinese Journal of Oceanology and Limnology 19(4): 327–331.

Yikrazuul. 2009. Fucoxanthin molecule (image into the public domain). Available online at: https://commons.wikimedia.org/wiki/File:Fucoxanthin.svg.

Ying, C.J., F. Zhang, X.Y. Zhou, X.T. Hu, J. Chen, X.R. Wen, Y. Sun, K.Y. Zheng, R.X. Tang and Y.J. Song. 2015. Anti-inflammatory effect of astaxanthin on the sickness behavior induced by diabetes mellitus. Cellular and Molecular Neurobiology 35(7): 1027–1037.

Yip, W.H., L.S. Joe, W.A.W. Mustapha, M.Y. Maskat and M. Said. 2014. Characterization and stability of pigments extracted from *Sargassum binderi* obtained from Semporna, Sabah. Sains Malaysiana 43(9): 1345–1354.

Yoon, N.Y., H.Y. Chung, H.R. Kim and J.E. Choi. 2008. Acetyl- and butyrylcholinesterase inhibitory activities of sterols and phlorotannins from *Ecklonia stolonifera*. Fisheries Science 74: 200–207.

Yoon, N.Y., T.-K. Eom, M.-M. Kim and S.-K. Kim. 2009b. Inhibitory effect of phlorotannins isolated from *Ecklonia cava* on mushroom tyrosinase activity and melanin formation in mouse B16F10 melanoma cells. Journal of Agricultural and Food Chemistry 57: 4124–4129.

Yoon, N.Y., S.-H. Lee and S.-K. Kim. 2009c. Phlorotannins from *Ishige okamurae* and their acetyl- and butyrylcholinesterase inhibitory effects. J. Funct. Foods 1: 331–335.

Yoon, S.J., Y.R. Pyun, J.K. Hwang and P.A.A. Mourao. 2007. Sulphated fucan from the brown alga *Laminaria cichorioides* has mainly heparin cofactor II-dependant anticoagulant activity. Carbohydrate Research 342: 2326–2330.

Yoon, W.J., Y.M. Ham, S.S. Kim, B.S. Yoo, J.Y. Moon, J.S. Baik, N.H. Lee and C.G. Hyun. 2009. Suppression of pro-inflammatory cytokines, iNOS, and COX-2 expression by brown algae *Sargassum micracanthum* in RAW 264.7 macrophages. EurAsian Journal of BioSciences 3(17): 130–143.

Yoon, W.J., S.J. Heo, S.C. Han, H.J. Lee, G.J. Kang, H.K. Kang, J.W. Hyun, Y.S. Koh and E.S. Yoo. 2012. Anti-inflammatory effect of sargachromanol G isolated from *Sargassum siliquastrum* in RAW 264.7 cells. Arch. Pharm. Res. 35(8): 1421–1430.

Yoshida, T., A. Hirano, H. Wada, K. Takahashi and M. Hattori. 2004. Alginic acid oligosaccharide suppresses Th2 development and IgE production by inducing IL-12 production. International Archives of Allergy and Immunology 133: 239–247.

Yoshimoto, M., K. Higaki, E. Nanba and M. Ikequchi. 2015. Anti-proliferationactivity of fucoidan in MKN45 gastric cancer cells and downregulation of phosphorylated ASK1, a cell cycle-regulated kinase. Yonago Acta Medica 58: 1–7.

Yoshioka, H., M. Ishida, K. Nishi, H. Oda, H. Toyohara and T. Sugahara. 2014. Studies on anti-allergic activity of *Sargassum horneri* extract. Journal of Functional Foods 10: 154–60.

Yoshizawa, Y., J. Tsunehiro, K. Nomura, M. Itoh, F. Fukui, A. Ametani and S. Kaminogawa. 1996. *In vivo* macrophage-stimulation activity of the enzyme-degraded water-soluble polysaccharide fraction from a marine alga (*Gracilaria verrucosa*). Bioscience, Biotechnology, and Biochemistry 60: 1667–1671.

Young, A.J. and G.M. Lowe. 2001. Antioxidant and prooxidant properties of carotenoids. Archives of Biochemistry and Biophysics 385: 20–27.

Yu, K.-X., C.-L. Wong, R. Ahmad and I. Jantan. 2015. Larvicidal activity, inhibition effect on development, histopathological alteration and morphological aberration induced by seaweed extracts in *Aedes aegypti* (Diptera: Culicidae). Asian Pacific Journal of Tropical Medicine 8(12): 1006–1012.

Yu, K.-X., C.-L. Wong, R. Ahmad and I. Jantan. 2015b. Mosquitocidal and oviposition repellent activities of the extracts of seaweed *Bryopsis pennata* on *Aedes aegypti* and *Aedes albopictus*. Molecules 20: 14082–14102.

Yu, P., N. Li, X. Liu, G. Zhou, Q. Zhang and P. Li. 2003a. Antihyperlipidemic effects of different molecular weight sulfated polysaccharides from *Ulva pertusa* (Chlorophyta). Pharmacological Research 48: 543–549.

Yu, P., Z. Quanbin, L. Ning, X. Zuhong, W. Yanmei and L. Zhi'en. 2003b. Polysaccharides from *Ulva pertusa* (Chlorophyta) and preliminary studies on their antihyperlipidemia activity. Journal of Applied Phycology 15: 21–27.

Yu, R.X., X.M. Hu, S.Q. Xu, Z.J. Jiang and W. Yang. 2011. Effects of fucoxanthin on proliferation and apoptosis in human gastric adenocarcinoma MGC-803 cells via JAK/STAT signal pathway. European Journal of Pharmacology 657(1-3): 10–19.

Yuan, Y.V., M.F. Carrington and N.A. Walsh. 2005. Extracts from dulse (*Palmaria palmata*) are effective antioxidants and inhibitors of cell proliferation *in vitro*. Food and Chemical Toxicology 43(7): 1073–1081.

Yuan, Y.V. and N.A. Walsh. 2006. Antioxidant and antiproliferative activities of extracts from a variety of edible seaweeds. Food and Chemical Toxicology 44(7): 1144–1150.

Yuan, Y.V., N.D. Westcott, C. Hu and D.D. Kitts. 2009. Mycosporine-like amino acid composition of edible red alga *Palmaria palmata* (Dulse) harvested from the west and east coasts of Grand Manan Island, New Brunswick. Food Chemistry 12: 321–328.

Yuan, H.M., J.M. Song, W.W. Zhang, X.G. Li, N. Li and X.L. Gao. 2006. Antioxidant activity and cytoprotective effect of kappa-carrageenan oligosaccharides and their different derivatives. Bioorganic & Medicinal Chemistry Letters 16: 1329–1334.

Yuan, J.P., J. Peng, K. Yin and J.H. Wang. 2011. Potential health-promoting effects of astaxanthin: a high-value carotenoid mostly from microalgae. Molecular Nutrition & Food Research 55: 150–165.

Yuvaraj, N., P. Kanmani, R. Satishkumar, K.A. Paari, V. Pattukumar and V. Arul. 2011. Extraction, purification and partial characterization of *Cladophora glomerata* against multidrug resistant human pathogen *Acinetobacter baumannii* and fish pathogens. World Journal of Fish and Marine Sciences 3: 51–57.

Zacharopoulos, V.R. and D.M. Phillips. 1997. Vaginal formulations of carrageenan protect mice from herpes simplex virus infection. Clinical and Diagnostic Laboratory Immunology 4(4): 465–468.

Zakaria, L, S. Sahak, M. Zakaria and B. Salleh. 2009. Characterisation of *Colletotrichum* species associated with anthracnose of banana. Tropical Life Sciences Research 20: 119–125.

Zaki, M., J. Amer-Zareen, A. Sattar and M.Q. Khan. 2005. Effect of seaweeds on the efficacy of *Pasteuria penetrans* in the control of root-knot nematode, *Meloidogyne javanica* in egg plant. International Journal of Phycology and Phycochemistry 1: 65–72.

Zandi, K., M. Fouladvand, P. Pakdel and K. Sartavi. 2007. Evaluation of *in vitro* antiviral activity of a brown alga (*Cystoseira myrica*) from the Persian Gulf against Herpes simplex virus type 1. African Journal of Biotechnology 6 (22): 2511–2514.

Zandi, K., S. Ahmadzadeh, S. Tajbakhsh, Z. Rastian, F. Yousefi, F. Farshadpour and K. Sartavi. 2010. Anticancer activity of *Sargassum oligocystum* water extract against human cancer cell lines. European Review for Medical and Pharmacological Sciences 14(8): 669–673.

Zandi, K., S. Tajbakhsh, I. Nabipour, Z. Rastian, F. Yousefi, S. Sharafian and K. Sartavi. 2010b. *In vitro* antitumor activity of *Gracilaria corticata* (a red alga) against Jurkat and Molt-4 Human cancer cell Lines. African Journal of Biotechnology 9(40): 6787–6790.

Zayed, A., K. Muffler, T. Hahn, S. Rupp, D. Finkelmeier, A. Burger-Kentischer and R. Ulber. 2016. Physicochemical and biological characterization of fucoidan from *Fucus vesiculosus* purified by dye affinity chromatography. Marine Drugs 14(79): 1–15.

Zbakh, H., H. Chiheb, H. Bouziane, V.M. Sánchez and H. Riadi. 2012. Antibacterial activity of benthic marine algae extracts from the mediterranean coast of Morocco. Journal of Microbiology, Biotechnology and Food Sciences 2(1): 219–228.

Zee, S. 1991. Body weight loss with the aid of alginic acid. Archives of Internal Medicine 3-4: 113–114.

Zeitlin, L., K.J. Whaley, T.A. Hegarty, T.R. Moench and R.A. Cone. 1997. Tests of vaginal microbicides in the mouse genital herpes model. Contraception 56(5): 329–335.

Zhai, B. and X. Lin. 2011. Recent progress on antifungal drug development. Current Pharmaceutical Biotechnology 12(8): 1255–1262.

Zhang, C., Y. Li, X. Shi and S.K. Kim. 2010. Inhibition of the expression on MMP-2, 9 and morphological changes via human fibrosarcoma cell line by 6,6'-Bieckol from marine alga *Ecklonia cava*. BMB Reports 43(1): 62–68.

Zhang, F.L., Y. He, Y. Zheng, W.J. Zhang, Q. Wang, Y.J. Jia, H.L. Song, H.T. An, H.B. Zhang and Y.J. Qian. 2014d. Therapeutic effects of fucoidan in 6-hydroxydopamine-lesioned rat model of Parkinson's disease: role of NADPH oxidase-1. CNS Neuroscience & Therapeutics 20: 1036–1044.

Zhang, H., X. Xiao, M.M. Conte, Z. Khalil and R.J. Capon. 2012c. Spiralisones A–D: acylphloroglucinol hemiketals from an Australian marine brown alga, *Zonaria spiralis*. Organic & Biomolecular Chemistry 10: 9671–9676.

Zhang, H.J., W.J. Mao, F. Fang, H.Y. Li, H.H. Sun, Y. Gehen and X.H. Qi. 2008. Chemical characteristics and anticoagulant activities of a sulphated polysaccharide and its fragments from *Monostroma latissimum*. Carbohydrate Polymers 71(3): 428–434.

Zhang, J., C. Tiller, J. Shen, C. Wang, G.S. Girouard, D. Dennis, C.J. Barrow, M. Miao and H.S. Ewart. 2007. Antidiabetic properties of polysaccharide- and polyphenolic-enriched fractions from the brown seaweed *Ascophyllum nodosum*. Canadian Journal of Physiology and Pharmacology 85(11): 1116–11123.

Zhang, M., L.F. Bi, J.H. Fang, X.L. Su, G.L. Da, T. Kuwamori and S. Kagamimori. 2004. Beneficial Effects of taurine on serum lipids in overweight or obese non-diabetic subjects. Amino Acids 26(3): 267–271.

Zhang, Q., N. Li, X. Lin, Z. Zhao, Z. Li and Z. Xu. 2004. The structure of a sulfated galactan from *Porphyra haitanensis* and its *in vivo* antioxidant activity. Carbohydrate Research 339: 105–111.

Zhang, Q.B., H.M. Qi, T.T. Zhao, E. Deslandes, N.M. Ismaeli, F. Molloy and A.T. Critchley. 2005. chemical characteristics of a polysaccharide from *Porphyra capensis* (Rhodophyta). Carbohydrate Research 340(15): 2447–2450.

Zhang, R.L., W.D. Luo, T.N. Bi and S.K. Zhou. 2012b. Evaluation of antioxidant and immunity-enhancing activities of *Sargassum pallidum* aqueous extract in gastric cancer rats. Molecules 17(7): 8419–8429.

Zhang, W. J.-Y. Du, Z. Jiang, T. Okimura, T. Oda, Q. Yu and J.-O. Jin. 2014. Ascophyllan purified from *Ascophyllum nodosum* induces Th1 and Tc1 immune responses by promoting dendritic cell maturation. Marine Drugs 12: 4148–4416.

Zhang, W., T. Oda, Q. Yu and J.-O. Jin. 2015. Fucoidan from *Macrocystis pyrifera* has powerful immune-modulatory effects compared to three other fucoidans. Marine Drugs 13(3): 1084–1104.

Zhang, X.-F., Y.-C. Dai, W. Zhong, M. Tan, Z.-P. Lv, Y.-C. Zhou and X. Jiang. 2012. Tannic acid inhibited norovirus binding to HBGA receptors, a study of 50 Chinese medicinal herbs. Bioorganic & Medicinal Chemistry 20:1616–1623.

Zhang, X.S., X. Zhang, M.L. Zhou, X.M. Zhou, N. Li, W. Li, Z.X. Cong, Q. Sun, Z. Zhuang, C.X. Wang and J.X. Shi. 2014b. Amelioration of oxidative stress and protection against early brain injury by astaxanthin after experimental subarachnoid hemorrhage. Journal of Neurosurgery 121: 42–54.

Zhang, X.S., X. Zhang, Q. Wu, W. Li, C.X. Wang, G.B. Xie, X.M. Zhou, J.X. Shi and M.L. Zhou. 2014c. Astaxanthin offers neuroprotection and reduces neuroinflammation in experimental subarachnoid hemorrhage. Journal of Surgical Research 192: 206–213.

Zhang, Y., J.Y. Han, J. Mu and J. Feng. 2013b. Bioactivity and constituents of several common seaweeds. Chinese Scince Bulletin 58: 2282–2289.

Zhang, Z., Q. Zhang, J. Wang, H. Song, H. Zhang and X. Niu. 2010b. Regioselective syntheses of sulfated porphyrans from *Porphyra haitanensis* and their antioxidant and anticoagulant activities *in vitro*. Carbohydrate Polymers 79: 1124–1129

Zhang, Z., K. Teruya, H. Eto and S. Shirahata. 2011. Fucoidan extract induces apoptosis in MCF-7 cells via a mechanism involving the ROS-dependent JNK activation and mitochondria-mediated pathways. PLoS One 6(11): 1–14.

Zhang, Z., K. Teruya, T. Yoshida, H. Eto and S. Shirahata. 2013. Fucoidan extract enhances the anti-cancer activity of chemotherapeutic agents in Mda-Mb-231 and Mcf-7 breast cancer cells. Marine Drugs 11(1): 81–98.

Zhao, D., L. Zheng, L. Qi, S. Wang, L. Guan, Y. Xia and J. Cai. 2016. Structural features and potent antidepressant effects of total sterols and β-sitosterol extracted from *Sargassum horneri*. Marine Drugs 14(7): 123.

Zhao, J., X. Fan, S. Wang, S. Li, S. Shang, Y. Yang, N. Xu, Y. Lu and J. Shi. 2004. Bromophenol derivatives from the red alga *Rhodomela confervoides*. Journal of Natural Products 67(6): 1032–1035.

Zhao, J., M. Ma, S. Wang, S. Li, P. Cao, Y. Yang, Y. Lu, J. Shi, N. Xu, X. Fan and L. He. 2005. Bromophenols coupled with derivatives of amino acids and nucleosides from the red alga *Rhodomela confervoides*. Journal of Natural Products 68: 691–694.

Zhao, L., M.D. Weir and H.H. Xu. 2010. An injectable calcium phosphate-alginate hydrogel-umbilical cord mesenchymal stem cell paste for bone tissue engineering. Biomaterials 31: 6502–6510.

Zhou, G., Y. Sun, H. Xin, Y. Zhang, Z. Li and Z. Xu. 2004. *In vivo* antitumor and immunomodulation activities of different molecular weight lambda-carrageenans from *Chondrus ocellatus*. Pharmacological Research 50: 47–53.

Zhou, G., W. Sheng, W. Yao and C. Wang. 2006. Effect of low molecular lambda-carrageenan from *Chondrus ocellatus* on antitumor H-22 activity of 5-Fu. Pharmacological Research 53(2): 129–134.

Zhu, C., R. Cao, S.-X. Zhang, Y.-N. Man and X.-Z. Wu. 2013. Fucoidan inhibits the growth of hepatocellular carcinoma independent of angiogenesis. Evidence-Based Complementary and Alternative Medicine Vol. 2013, Article ID 692549, 7 pp.

Zhu, W., V.E.C. Ooi, P.K.S. Chan and P.O. Ang. 2003a. Inhibitory effect of extracts of marine algae from Hong Kong against *Herpes simplex* viruses. pp. 159–164. *In*: Chapman, A.R.O., R.J. Anderson, V.J. Vreeland and I.R. Davison (eds.). Proceedings of the 17th International Seaweed Symposium, Oxford University Press, Oxford.

Zhu, W., V.E.C Ooi, P.K.S. Chan and P.O. Ang. 2003b. Inhibitory effect of extracts of marine algae from Hong Kong against *Herpes simplex* viruses. pp. 164–166. *In:* Chapman, A.R.O., R.J. Anderson, V.J. Vreeland and I.R. Davison (eds.). Proceedings of the 17th International Seaweed Symposium, Oxford University Press, Oxford.

Zhu, W., L. Chiu, V. Ooi, P. Chan and P. Ang, Jr. 2006. Antiviral property and mechanisms of a sulphated polysaccharide from the brown alga *Sargassum patens* against *Herpes simplex* virus type 1. Phytomedicine 13: 695–670.

Zhuang, C., H. Itoh, T. Mizuno and H. Ito. 1995. Antitumor active fucoidan from the brown seaweed, Umitoranoo (*Sargassum thunbergii*). Bioscience, Biotechnology, and Biochemistry 59(4): 563–567.

Ziaja. 2017. Focus on skin. Available online at: http://ziaja.co.uk/.

Zikeli, S. 2006. Production process of a new cellulosic fiber with antimicrobial properties. Current Problems in Dermatology 33: 110–126.

Zimmermann, M.B. and C.M. Crill. 2010. Iodine in enteral and parenteral nutrition. Best Practice & Research Clinical Endocrinology & Metabolism 24(1): 143–158.

Zouboulis, C.C. 2004. Acne and sebaceous gland function. Clinics in Dermatology 22: 360–366.

Zubia, M., M.-S. Fabre, V. Kerjean and E. Deslandes. 2009a. Antioxidant and cytotoxic activities of some red algae (Rhodophyta) from Brittany coasts (France). Botanica Marina 52(3): 268–277.

Zubia, M., M.S. Fabre, V. Kerjean, K.L. Lann, V. Stiger-Pouvreau, M. Fauchon and E. Deslandes. 2009b. Antioxidant and antitumoural activities of some Phaeophyta from Brittany coasts. Food Chemistry 116(3): 693–701.

Zúniga, E.A., B. Matsuhiro and E. Mejías. 2006. Preparation of a low-molecular weight fraction by free radical depolymerization of the sulfated galactan from *Schizymenia binderi* (Gigartinales, Rhodophyta) and its anticoagulant activity. Carbohydrate Polymers 66: 208–215.

Index
of Algae Species

A

Acanthopeltis japonica Okamura 67, 272
Acanthophora muscoides (Linnaeus) Bory 444, 471, 510
Acanthophora spicifera (M. Vahl) Børgesen 43, 119, 139, 146, 179, 182, 193, 210, 211, 214, 273, 310, 340, 343, 345, 419, 420
Acrocladus herpesticus (Montagne) Boedeker (as *Cladophoropsis zollingeri*) 351
Acrosiphonia coalita (Ruprecht) Scagel, Garbary, Golden and M.W. Hawkes 180, 183
Acrosiphonia orientalis (J. Agardh) P.C. Silva 353, 383, 385, 432, 439
Acrosiphonia orientalis (J. Agardh) P.C. Silva (as *Spongomorpha indica*) 353, 385
Acrosorium ciliolatum (Zanardini) Kylin 471
Acrosorium flabellatum Yamada 273
Acrosorium Zanardini ex Kützing 194, 273, 471
Actinotrichia fragilis (Forsskål) Børgesen 273
Adenocystis utricularis (Bory de Saint-Vincent) Skottsberg 18, 67, 100, 102, 186, 204
Agardhiella ramosissima (Harvey) Kylin 471
Agardhiella subulata (C. Agardh) Kraft and M.J. Wynne (as *Agardhiella tenera*) 67, 194
Agarum clathratum subsp. *yakishiriense* Yamada ex G.H. Boo and P.C. Silva 490
Aglaothamnion pseudobyssoides (P. Crouan and H. Crouan) Halos 273
Ahnfeltiopsis flabelliformis (Harvey) Masuda 166, 169, 273, 471
Ahnfeltiopsis flabelliformis (Harvey) Masuda (as *Gymnogongrus flabelliformis*) 273
Ahnfeltiopsis paradoxa (Suringar) Masuda (as *Ahnfeltia paradoxa*) 67, 274
Alaria esculenta (Linnaeus) Greville 18, 19, 67, 104, 109, 118, 122, 151, 226, 306, 332, 396, 490, 506
Alaria Greville 18, 19, 67, 104, 109, 118, 122, 151, 152, 226, 227, 306, 332, 396, 490, 506
Alaria marginata Postels and Ruprecht 67, 152
Alsidium corallinum C. Agardh 194, 370
Amphiroa anceps (Lamarck) Decaisne (as *Amphiroa dilatata*) 274
Amphiroa beauvoisii J.V. Lamouroux (as *Amphiroa zonata*) 496
Amphiroa bowerbankii Harvey 496
Amphiroa ephedraea (Lamarck) Decaisne 370
Amphiroa fragilissima (Linnaeus) J.V. Lamouroux 43, 340, 414
Anadyomene saldanhae A.B. Joly and E.C. Oliveira 433, 439
Anadyomene stellata (Wulfen) C. Agardh 2, 3, 67, 353, 391
Analipus japonicus (Harvey) M.J. Wynne 67, 181, 186
Anthophycus longifolius (as *Sargassum longifollium*) 335, 363
Arthrocardia carinata (Kützing) Johansen 370
Ascophyllum nodosum (Linnaeus) Le Jolis 18, 19, 66, 68, 100–102, 104, 109, 118, 122, 127, 145, 151, 152, 159, 162, 164, 186, 227, 305, 306, 334, 363, 399, 405, 425, 456, 460, 464, 477, 506
Ascophyllum Stackhouse 18, 19, 35, 66, 68, 100–102, 104, 109, 118, 122, 127, 131, 145, 151, 152, 159, 162, 164, 186, 227, 305, 306, 334, 363, 399, 405, 409, 425, 456, 460, 464, 477, 504, 506, 517
Asparagopsis armata Harvey 43, 44, 68, 194, 209, 274, 309, 310, 341, 345, 370, 411, 414, 415, 421, 423, 424, 471, 496, 521
Asparagopsis armata Harvey (and *Falkenbergia rufolanosa* phase) 341, 371, 411, 414, 415
Asparagopsis Montagne 43, 44, 68, 194, 209, 274, 309, 310, 339–341, 343, 345, 370, 371, 411–416, 420, 421, 423, 424, 471, 496, 521
Asparagopsis taxiformis (Delile) Trevisan 43, 44, 68, 310, 340, 341, 343, 411–413, 416, 420
Asparagopsis taxiformis (Delile) Trevisan (and *Falkenbergia hillebrandii* phase) 371, 413
Avrainvillea digitata D.S. Littler and Littler 215
Avrainvillea elliottii A. Gepp and E.S. Gepp 180, 183
Avrainvillea obscura (C. Agardh) J. Agardh 326, 352, 353

B

Bangia fuscopurpurea (Dillwyn) Lyngbye 68, 194
Betaphycus gelatinus (Esper) Doty ex P.C. Silva 68, 138, 214, 275, 308
Betaphycus gelatinus (Esper) Doty ex P.C. Silva (as *Eucheuma gelatinae*) 214, 275, 308
Bifurcaria bifurcata R. Ross 19, 20, 68, 100, 118, 227, 305, 306, 308, 335, 363, 409, 434, 435, 441, 506
Bifurcariopsis capensis (Areschoug) Papenfuss 363

Boergeseniella fruticulosa (Wulfen) Kylin 438
Boergeseniella thuyoides (Harvey) Kylin 44, 45, 68, 194, 209, 210
Bonnemaisonia asparagoides (Woodward) C. Agardh 341, 371, 411, 415
Bonnemaisonia C. Agardh 45, 166, 169, 275, 339, 341, 371, 411, 415
Bonnemaisonia hamifera Hariot 45, 166, 169, 275, 341, 411, 415
Boodlea composita (Harvey) F. Brand 68, 326, 353
Bossiella P.C. Silva 194
Bostrychia montagnei Harvey 194, 207, 211, 275
Bostrychia radicans (Montagne) Montagne 68, 209
Bostrychia tenella (J.V. Lamouroux) J. Agardh 68, 341, 438, 444
Botryocladia leptopoda (J. Agardh) Kylin 340, 438, 444
Botryocladia occidentalis (Børgesen) Kylin 275, 310, 471, 482
Brongniartella byssoides (Goodenough and Woodward) F. Schmitz 45, 275, 310
Brown algae (Phaeophyceae) 99, 214, 500, 519
Bryopsis hypnoides J.V. Lamouroux 4, 180, 325, 353, 382
Bryopsis J.V. Lamouroux 3, 4, 180, 181, 183, 215, 216, 325, 326, 353, 382, 433, 439, 461
Bryopsis maxima Okamura ex Segawa 461
Bryopsis myosuroides Kützing 353
Bryopsis pennata J.V. Lamouroux 4, 215, 325, 326, 433, 439
Bryopsis plumosa (Hudson) C. Agardh 3, 4, 180, 325, 382
Bryothamnion seaforthii (Turner) Kützing 424, 471
Bryothamnion triquetrum (S.G. Gmelin) M. Howe 45, 68, 275, 309, 424, 496, 501

C

Calliblepharis fimbriata (Greville) Kützing 371
Calliblepharis jubata (Goodenough and Woodward) Kützing 44, 46, 47, 275, 438, 448
Callithamnion tetragonum (Withering) S.F. Gray 275
Callophycus serratus (Harvey ex Kützing) 46, 68, 276, 341, 371, 421, 438, 444
Callophyllis crispata Okamura 276
Callophyllis japonica Okamura 146
Callophyllis laciniata (Hudson) Kutzing 44, 46, 68
Callophyllis variegata (Bory) Kützing 46, 68, 194, 208, 209, 211
Caloglossa leprieurii (Montagne) G. Martens 68, 206
Canistrocarpus cervicornis (Kützing) De Paula and De Clerck 19, 20, 102, 186, 203, 228, 305, 435, 441, 464
Canistrocarpus cervicornis (Kützing) De Paula and De Clerck (as *Dictyota cervicornis*) 20, 102, 118, 186, 203, 228, 305, 464
Capsosiphon fulvescens (C. Agardh) Setchell and N.L. Gardner 4, 68, 108, 116, 117, 122, 159, 167, 173, 216, 461
Carpopeltis cornea (Okamura) Okamura 502
Catenella caespitosa (Withering) L.M. Irvine (as *Catenella repens*) 68, 116, 119
Caulacanthus ustulatus (Mertens ex Turner) Kützing 277
Caulerpa ambigua Okamura 326
Caulerpa ashmeadii Harvey 352, 353
Caulerpa brachypus Harvey 69, 181, 183
Caulerpa chemnitzia (Esper) J.V. Lamouroux 5, 69, 216, 301, 302, 325, 394, 431, 432, 439
Caulerpa chemnitzia (Esper) J.V. Lamouroux (as *Caulerpa peltata*) 5, 302, 431, 432, 439
Caulerpa chemnitzia (Esper) J.V. Lamouroux (as *Caulerpa racemosa* var. *laetevirens*) 216, 301
Caulerpa corynephora Montagne 69, 392
Caulerpa cupressoides (Vahl) C. Agardh 5, 69, 301, 302, 326, 352, 433, 457, 458, 461
Caulerpa cupressoides var. *lycopodium* Weber-van Bosse 69
Caulerpa cylindracea Sonder 3, 5, 388, 395
Caulerpa cylindracea Sonder (as *Caulerpa racemosa* var. *cylindracea*) 5
Caulerpa faridii Nizamuddin 431
Caulerpa J.V. Lamouroux 3, 5–7, 69, 108, 116, 117, 122, 144, 147, 152, 170, 173, 179–181, 183, 214, 216–218, 301, 302, 325–330, 351–353, 382–386, 388–390, 392, 394, 395, 431–433, 439, 457–459, 461, 489, 499
Caulerpa lentillifera J. Agardh 69, 108, 116, 117, 122
Caulerpa mexicana Sonder ex Kützing 5, 325, 352, 459, 461
Caulerpa microphysa (Sonder ex Kützing) G. Murray 69, 170, 173
Caulerpa okamurae f. *oligophylla* Okamura 183
Caulerpa okamurae Weber-van Bosse 69, 181
Caulerpa paspaloides (Bory) Greville 352
Caulerpa prolifera (Forsskål) J.V. Lamouroux 3, 6, 301, 302, 328, 351, 352, 385, 458
Caulerpa racemosa (Forsskål) J. Agardh 3, 5, 6, 69, 144, 147, 179, 180, 216, 301, 302, 325, 326, 328, 329, 382–384, 386, 388–390, 431–433, 458, 459, 489, 499
Caulerpa scalpelliformis (R. Brown ex Turner) C. Agardh 6, 69, 181, 327, 328, 330, 351, 384, 386, 392, 395, 431, 432
Caulerpa serrulata (Forsskål) J. Agardh 69, 326
Caulerpa serrulata f. *lata* (Weber-van Bosse) C.K. Tseng 217
Caulerpa sertularioides (S.G. Gmelin) M. Howe 3, 7, 69, 302
Caulerpa taxifolia (M. Vahl) C. Agardh 7, 69, 214, 301, 302, 351
Caulerpa veravalensis Thivy and V.D. Chauhan 7, 69, 180, 325, 382, 432
Caulerpa zeyheri Kützing 355
Centroceras clavulatum (C. Agardh) Montagne 46, 70, 209, 309, 344, 371, 422, 437, 438, 444
Ceramium boydenii E.S. Gepp 70, 111, 277
Ceramium ciliatum (J. Ellis) Ducluzeau 371, 424
Ceramium deslongchampsii Chauvin ex Duby 70
Ceramium japonicum Okamura 277
Ceramium kondoi Yendo 70
Ceramium nitens (C. Agardh) J. Agardh 413
Ceramium siliquosum var. *elegans* (Roth) *G. Furnari* 417
Ceramium virgatum Roth 44, 47, 70, 194, 207, 209, 211, 344, 410, 414, 415, 424, 438, 444

Ceramium virgatum Roth (as *Ceramium rubrum*) 47, 194, 207, 344, 410, 414, 415, 424
Ceratodictyon spongiosum Zanardini 277, 471
Chaetomorpha aerea (Dillwyn) Kützing 3, 7, 70, 152, 330, 355, 385, 386, 459
Chaetomorpha antennina (Bory) Kützing 8, 70, 180, 183, 219, 385, 386, 390–392, 394, 432, 440, 461
Chaetomorpha Kützing 3, 7–9, 70, 116, 117, 152, 167, 180, 181, 183, 219, 302, 325, 327, 329, 330, 352, 355, 356, 382, 384–386, 390–392, 394, 432, 440, 459, 461
Chaetomorpha ligustica (Kützing) Kützing (as *Lola capillaris*) 116, 117
Chaetomorpha linoides Kützing 8, 70, 352, 390
Chaetomorpha linum (O.F. Müller) Kützing 3, 8, 70, 167, 181, 302, 329, 384, 394, 432
Chaetomorpha linum (O.F. Müller) Kützing (as *Chaetomorpha crassa*) 70, 181, 302, 432
Chaetomorpha spiralis Okamura 9, 181, 327, 384, 432
Chaetomorpha spiralis Okamura (as *Chaetomorpha torta*) 9, 327, 384, 432
Champia compressa Harvey 70, 340
Champia feldmannii Díaz-Piferrer 70, 277
Champia parvula (C. Agardh) Harvey 44, 47, 70, 425
Chlorodesmis W.H. Harvey and J.W. Bailey 152
Chnoospora implexa J. Agardh 70, 363
Chnoospora minima (Hering) Papenfuss 70, 109
Chondracanthus acicularis (Roth) Fredericq 47, 70, 195, 207, 209, 211, 372, 471, 481, 496
Chondracanthus acicularis (Roth) Fredericq (as *Gigartina acicularis*) 47, 195, 207, 471, 481
Chondracanthus canaliculatus (Harvey) Guiry 372
Chondracanthus intermedius (Suringar) Hommersand (as *Gigartina intermedia*) 70, 277
Chondracanthus teedei var. *lusitanicus* (Rodrigues) Bárbara and Cremades 44, 47, 346
Chondracanthus tenellus (Harvey) Hommersand (as *Gigartina tenella*) 70, 169, 173, 195
Chondria armata (Kützing) Okamura 195, 206, 372, 445, 510
Chondria C. Agardh 44, 48, 70, 71, 169, 173, 195, 206, 207, 214, 277, 340, 372, 411, 412, 414, 423, 437, 445, 449, 471, 510
Chondria capillaris (Hudson) M.J. Wynne (as *Chondria tenuissima*) 70, 277
Chondria crassicaulis Harvey 71, 169, 173
Chondria dasyphylla (Woodward) C. Agardh 44, 48, 71, 207, 340, 414, 423, 471
Chondria littoralis Harvey 412, 449
Chondria oppositiclada E.Y. Dawson 411
Chondriopsis dasyphylla f. pyrifera J. Agardh (as *Laurencia intricata*) 372
Chondrophycus brandenii (Saito and Womersley) Nam (as *Laurencia brandenii*) 48, 341, 372, 416, 420, 448
Chondrophycus ceylanicus (J. Agardh) M.J. Wynne 278, 341, 416
Chondrophycus ceylanicus (J. Agardh) M.J. Wynne (as *Laurencia ceylanica*) 341, 416
Chondrophycus intermedius (Yamada) Garbary and J.T. Harper (as *Laurencia intermedia*) 146
Chondrophycus undulatus (Yamada) Garbary and Harper (as *Laurencia undulata*) 71, 169, 483, 497
Chondrus crispus Stackhouse 44, 48, 71, 111, 119, 123, 137, 138, 146, 169, 211, 278, 372, 411, 423, 426, 445, 448, 472, 482, 497, 502, 511, 520, 521
Chondrus ocellatus f. *crispus* Okamura 279
Chondrus ocellatus Holmes 48, 71, 279, 483
Chondrus pinnulatus (Harvey) Okamura 71, 483
Chorda filum (Linnaeus) Stackhouse 71, 100, 228
Chordaria flagelliformis (O.F. Müller) C. Agardh 71, 228
Chrysophaeum taylori I.F. Lewis and H.F. Bryan 404
Chrysymenia wrightii (Harvey) Yamada 279, 472, 483
Chylocladia verticillata (Lightfoot) Bliding 438
Cladophora conglomerata Pilger 385
Cladophora falklandica (J.D. Hooker and Harvey) J.D. Hooker and Harve 462
Cladophora fascicularis (Mertens ex C. Agardh) Kützing 9, 385, 489
Cladophora glomerata (Linnaeus) Kützing 8, 71, 327, 431, 440, 505
Cladophora herpestica (Montagne) Kützing (as *Cladophoropsis zollingeri*) 219
Cladophora Kützing 3, 8, 9, 71, 108, 152, 165, 180, 184, 219, 325, 327, 356, 385, 389, 390, 431, 432, 440, 462, 489, 505
Cladophora prolifera (Roth) Kützing 3, 8, 71, 180, 184
Cladophora rupestris (Linnaeus) Kützing 9, 71, 108, 152, 165, 390, 432
Cladophora vagabunda (Linnaeus) Hoek (as *Cladophora fascicularis*) 9, 71, 385, 389, 489
Cladophoropsis vaucheriaeformis (Areschoug) Papenfuss 301
Cladosiphon novae-caledoniae Kylin 20, 228, 306
Cladosiphon okamuranus Tokida 21, 71, 100, 102, 148, 182, 186, 464, 476, 477, 506
Cladostephus spongiosus f. *verticillatus* (Lightfoot) Prud homme van Reine 398
Claviclonium ovatum (J.V. Lamouroux) Kraft and Min-Thein 438
Coccophora langsdorfii (Turner) Greville 230
Coccotylus brodiei (Turner) Kützing (as *Phyllophora brodiaei*) 49, 139, 472, 482
Codium adhaerens C. Agardh 9, 71, 181, 184, 219, 330, 383
Codium amplivesiculatum Setchell and N.L. Gardner 356
Codium bursa (Linnaeus) C. Agardh 3, 10, 72, 386
Codium capitatum P.C. Silva 10, 326, 352, 489
Codium coactum Okamura (as *Codium coarctatum*) 220
Codium contractum Kjellman 167
Codium cuneatum Setchell and N.L. Gardner (as *Codium simulans*) 357
Codium cylindricum Holmes 72, 457, 462
Codium decorticatum (Woodward) M. Howe 10, 72, 180, 181, 302, 325, 328, 382, 386
Codium decorticatum (Woodward) M. Howe (as *Codium elongatum*) 180, 302, 325, 382
Codium duthieae P.C. Silva 10, 72, 352
Codium dwarkense Børgesen 72, 180, 325, 382, 457
Codium flabellatum P.C. Silva ex M. Nizamuddin 431, 440

Codium fragile (Suringar) Hariot 10, 72, 108, 180, 181, 301, 328, 330, 386, 391, 432, 457, 458
Codium fragile subsp. *atlanticum* (A.D. Cotton) P.C. Silva 457
Codium fragile subsp. *fragile* (Suringar) Hariot (as *Codium fragile* subsp. *tomentosoides*) 432, 457
Codium geppiorum O.C. Schmidt 72, 326
Codium indicum S.C. Dixit (as *Codium iyengarii*) 11, 72, 325, 352, 395, 431
Codium intricatum Okamura 72
Codium isthmocladum Vickers 11, 72, 301, 302
Codium latum Suringar 181, 457
Codium platylobium Areschoug 72, 352
Codium pugniforme Okamura 221
Codium shameelii M. Nizamuddin 395
Codium spongiosum Harvey 180
Codium subtubulosum Okamura (as *Codium divaricatum*) 72
Codium tenue (Kützing) Kützing 72, 352
Codium tomentosum Stackhouse 3, 11, 72, 151, 389, 457, 505
Codium vermilara (Olivi) Delle Chiaje 73, 457
Colpomenia (Endlicher) Derbès and Solier 19, 21, 30, 73, 109, 144, 181, 187, 204, 205, 214, 231, 305, 331, 334, 337, 363, 397, 399, 464
Colpomenia bullosa (De A. Saunders) Yamada 181, 187, 231, 464
Colpomenia sinuosa (Mertens ex Roth) Derbès and Solier 19, 21, 73, 109, 144, 181, 204, 205, 214, 305, 331, 334, 337, 363, 397, 399
Colpomenia tuberculata D.A. Saunders 363
Constantinea simplex Setchell 195, 206
Corallina crassisima (Yendo) K. Hind and G.W. Saunders (as *Marginisporum crassissimum*) 279
Corallina Linnaeus 49, 52, 90, 111, 125, 139, 195, 196, 207, 209, 279, 280, 372, 373, 415, 418, 423, 425, 426, 438, 445, 472, 484, 511
Corallina officinalis Linnaeus 49, 52, 195, 415, 438, 445, 511
Corallina panizzoi R. Schnetter and U. Richter 209
Corallina pilulifera Postels and Ruprecht 207, 472
Corallina vancouveriensis Yendo 207
Coronaphycus elatus (C. Agardh) Metti (as *Laurencia elata*) 49, 209, 373, 410
Corynomorpha prismatica (J. Agardh) J. Agardh 111
Costaria costata (C. Agardh) De A. Saunders 23, 73, 231, 464
Cryptonemia crenulata (J. Agardh) J. Agardh 49, 73, 195, 209, 211
Cryptonemia seminervis (C. Agardh) J. Agardh 209
Cryptopleura ramosa (Hudson) L. Newton 44, 49, 196, 207, 211, 438
Cryptosiphonia woodii (J. Agardh) J. Agardh 196
Curdiea racovitzae Hariot 521
Cymopolia barbata (Linnaeus) J.V. Lamouroux 325, 462
Cystoclonium purpureum (Hudson) Batters 50, 51, 73, 279, 437, 438
Cystophora moniliformis (Esper) Womersley and Nizamuddin 231
Cystoseira abies-marina (S.G. Gmelin) 19, 21, 73, 116
Cystoseira amentacea var. stricta Montagne (as *Cystoseira stricta*) 232
Cystoseira baccata (S.G. Gmelin) P.C. Silva 73, 441
Cystoseira barbata (Stackhouse) C. Agardh 21, 73, 187, 337, 363, 398, 407
Cystoseira brachycarpa (as *Cystoseira balearica* Sauvageau) 22, 395
Cystoseira brachycarpa J. Agardh 19, 22, 395, 407
Cystoseira compressa (Esper) Gerloff and Nizamuddin 22, 334, 402, 407, 409, 464
Cystoseira corniculata (Turner) Zanardini 109
Cystoseira crinita Duby 22, 73, 334, 403, 407, 481
Cystoseira humilis Schousboe ex Kützing 19, 22, 402, 409, 490
Cystoseira indica (Thivy and Doshi) Mairh 102, 189, 204, 331, 367, 397, 410, 436, 443
Cystoseira mediterranea Sauvageau 19, 23, 305, 397
Cystoseira nodicaulis (Withering) M. Roberts 23, 73, 501, 506
Cystoseira sedoides (Desfontaines) C. Agardh 334
Cystoseira tamariscifolia (Hudson) Papenfuss 19, 23, 73, 306, 331, 333, 396, 501
Cystoseira usneoides (Linnaeus) M. Roberts 233, 490, 507

D

Dasya baillouviana (S.G. Gmelin) Montagne (as *Dasya pedicellata*) 280, 445, 448
Dasya scoparia Harvey 373, 413
Dasyphila plumarioides Yendo 340
Dasysiphonia japonica (Yendo) H.-S. Kim (as *Heterosiphonia japonica*) 280
Delesseria sanguinea (Hudson) J.V. Lamouroux 50, 51, 472, 482
Delisea pulchra (Greville) Montagne 425
Derbesia tenuissima (Moris and De Notaris) P. Crouan and H. Crouan 145, 147
Desmarestia ligulata (Stackhouse) J.V. Lamouroux 233, 306
Desmarestia viridis (O.F. Müller) J.V. Lamouroux 233
Dichotomaria diesingiana (Zanardini) Huisman, J.T. Harper and G.W. Saunders (as *Galaxaura diesingiana*) 373, 412
Dichotomaria falcata (Kjellman) Kurihara and Masuda (as *Galaxaura falcata*) 280
Dichotomaria marginata (J. Ellis and Solander) Lamarck (as *Galaxaura marginata*) 472, 483
Dichotomaria obtusata (J. Ellis and Solander) Lamarck (as Galaxaura robusta) 280, 472
Dictyopteris delicatula J.V. Lamouroux 23, 73, 102, 187, 203, 205, 234, 305, 308, 364, 464
Dictyopteris divaricata (Okamura) Okamura 304, 305
Dictyopteris longifolia Papenfuss 364
Dictyopteris polypodioides (A.P. De Candolle) J.V. Lamouroux 24, 73, 308, 334, 336, 338, 397, 407, 409, 410, 435, 441, 476
Dictyopteris polypodioides (A.P. De Candolle) J.V. Lamouroux (as *Dictyopteris membranacea*) 24, 308, 334, 338, 397, 407, 410

Dictyopteris prolifera (Okamura) Okamura 305
Dictyopteris undulata Holmes 145, 331, 395, 400, 491
Dictyopteris undulata Holmes (as *Dictyopteris zonarioides*) 331, 395
Dictyopteris woodwardia (R. Brown ex Turner) C. Agardh 204, 331, 397
Dictyosphaeria cavernosa (Forsskål) Børgesen 11, 180, 184, 325, 357, 382
Dictyosphaeria versluysii Weber-van Bosse 12, 326, 352
Dictyota acutiloba J. Agardh 73, 398
Dictyota bartayresiana J.V. Lamouroux 19, 24, 73, 332, 364, 395
Dictyota bartayresiana J.V. Lamouroux (as *Dictyota bartayresii*) 24, 364, 395
Dictyota caribaea Hörnig and Schnetter 24, 74, 304, 434, 442
Dictyota cervicornis Kützing (as *Dictyota indica*) 20, 102, 118, 186, 203, 228, 305, 464
Dictyota ceylanica Kützing 116, 118
Dictyota ciliolata Sonder ex Kützing 74, 235, 404
Dictyota ciliolata Sonder ex Kützing (as *Dictyota ciliata*) 41, 410
Dictyota coriacea (Holmes) I.K. Wang, Hy.S. Kim and W.J. Lee (as *Pachydictyon coriaceum*) 235
Dictyota dichotoma (Hudson) J.V. Lamouroux 19, 24, 25, 74, 109, 181, 182, 187, 204, 205, 306, 331, 337, 395–399, 407, 434–436, 460
Dictyota dichotoma var. *intricata* (as *Dictyota linearis*) 25, 74, 182, 397, 436
Dictyota fasciola (Roth) J.V. Lamouroux (as *Dilophus fasciola*) 403, 481
Dictyota fastigiata Sonder (as *Dilophus fasciculatus*) 187
Dictyota flabellata (Collins) Setchell and N.L. Gardner 364
Dictyota friabilis Setchell 74, 182, 435
Dictyota friabilis Setchell (as *Dictyota pfaffii*) 182, 435
Dictyota humifusa Hörnig, Schnetter and Coppejans 332, 398, 491
Dictyota implexa (Desfontaines) J.V. Lamouroux (as *Dictyota dichotoma* var. *implexa*) 25, 397, 399
Dictyota implexa (Desfontaines) J.V. Lamouroux (as *Dictyota linearis*) 25, 182, 397, 436
Dictyota J.V. Lamouroux 19, 20, 24, 25, 41, 73, 74, 100, 102, 109, 116, 118, 126, 181, 182, 186, 187, 203–205, 228, 235, 304–306, 308, 331, 332, 337, 364, 395–399, 403, 404, 407, 409, 410, 434–436, 442, 460, 464, 481, 491
Dictyota menstrualis (Hoyt) Schnetter, Hörning and Weber-Peukert 74, 100, 203, 205, 305, 308, 464
Dictyota mertensii (C. Martius) Kützing 19, 25, 74, 102, 305, 308
Dictyota spiralis Montagne 74, 364
Digenea simplex (Wulfen) C. Agardh 74, 196, 280, 437, 445, 472, 501
Dilsea carnosa Schmidel) Kuntze 50, 51, 74, 280, 339, 373, 411, 478
Dumontia contorta (S.G. Gmelin) Ruprecht (as *Dumontia incrassata*) 51, 52, 74, 196, 280, 438
Dumontia simplex A.D. Cotton 196
Durvillaea antarctica (Chamisso) Hariot 74, 109, 507
Durvillaea Bory 74, 109, 131, 507

E

Ecklonia arborea (Areschoug) M.D. Rothman, Mattio and J.J. Bolton (as *Eisenia arborea*) 25, 74, 188, 236, 465, 478
Ecklonia bicyclis Kjellman (as *Eisenia bicyclis*) 25, 74, 118, 145, 147, 153, 159, 204, 303–305, 336, 365, 404, 408, 460, 478, 491, 502, 507
Ecklonia cava Kjellman 26, 74, 147, 159, 167, 171, 173, 205, 407, 478–480
Ecklonia Hornemann 25, 26, 74, 100, 109, 118, 131, 145, 147, 153, 159, 160, 167, 171, 173, 188, 204, 205, 236, 303–305, 332, 336, 365, 404, 407, 408, 460, 465, 478–480, 491, 502, 507
Ecklonia kurome Okamura 74, 100, 460
Ecklonia maxima (Osbeck) Papenfuss 74, 332
Ecklonia radiata (C. Agardh) J. Agardh 74, 109
Ecklonia stolonifera Okamura 74, 160, 188
Ectocarpus confervoides Le Jolis 238
Ectocarpus siliculosus (Dillwyn) Lyngbye 26, 27, 365, 397
Egregia menziesii (Turner) Areschoug 74, 400
Ellisolandia elongata (J. Ellis and Solander) K.R. Hind and G.W. Sauders (as *Corallina elongata*) 51, 52, 111, 125, 196, 280, 373, 418, 423, 426, 472
Ellisolandia elongata (J. Ellis and Solander) K.R. Hind and G.W. Saunders (as *Corallina mediterranea*) 52, 111, 426
Enantiocladia prolifera Falkenberg 373, 421
Endarachne binghamiae J. Agardh 168, 400, 401
Enteromorpha attenuata (C. Agardh) Greville 221, 357, 389
Euchema denticulatum (N.L. Burman) Collins and Hervey 52, 196, 207
Eucheuma amakusaense Okamura 74, 281
Eucheuma denticulatum (N.L. Burman) Collins and Hervey (as *Eucheuma spinosum*) 75, 137, 138, 157, 208, 308, 373, 417, 419, 472, 481, 482, 497, 520
Eucheuma serra (J. Agardh) J. Agardh 75, 511

F

Farlowia mollis Harvey and Bailey) Farlow and Setchell 196, 206
Flabellia petiolata (Turra) Nizamuddin 184, 351, 357, 391
Fucus ceranoides Linnaeus 26, 27, 75, 238, 306
Fucus distichus Linnaeus 75, 100, 104, 155, 306, 460, 466
Fucus distichus subsp. *evanescens* (C. Agardh) H.T. Powell 75, 188, 306, 307, 365, 401, 433, 460
Fucus guiryi G.I. Zardi, K.R. Nicastro, E.S. Serrão and G.A. Pearson (as *Fucus spiralis* var. *platycarpus*) 365
Fucus Linnaeus 26–28, 35, 66, 75, 100, 101, 102, 104, 109, 116, 118, 123, 126, 127, 131, 151, 155, 162, 188, 206, 238, 306, 307, 332, 334, 365, 401, 407, 409, 433, 456, 459, 460, 466, 492, 500, 504, 508, 515, 517, 518, 520
Fucus serratus Linnaeus 26, 27, 75, 101, 102, 306, 460
Fucus spiralis Linnaeus 27, 28, 75, 109, 116, 118, 334, 409, 460, 508
Fucus vesiculosus Linnaeus 28, 66, 75, 101, 102, 123, 127, 151, 162, 206, 306, 332, 407, 456, 459, 466, 492, 500, 518, 520

Furcellaria lumbricalis (Hudson) J.V. Lamouroux 51, 52, 75, 196, 281, 438, 481, 511
Furcellaria lumbricalis (Hudson) J.V. Lamouroux (as *Furcellaria fastigiata*) 52, 281, 481
Fushitsunagia catenata (Harvey) Filloramo and G.W. Saunders (as *Lomentaria catenata*) 166, 169, 472

G

Galaxaura filamentosa R.C.Y. Chou 281
Galaxaura rugosa (J. Ellis and Solander) J.V. Lamouroux (as *Galaxaura elongata*) 196
Ganonema farinosum (J.V. Lamouroux) K.C. Fan and Yung C. Wang (as *Liagora farinosa*) 75, 196, 281, 373
Ganonema farinosum J.V. Lamouroux) K.C. Fan and Yung C. Wang (as *Liagora farinosa*) 196, 281
Gastroclonium ovatum (Hudson) Papenfuss 281
Gayralia oxysperma (Kützing) K.L. Vinogradova ex Scagel et al. 75, 117, 184, 221, 302, 462
Gayralia oxysperma (Kützing) K.L. Vinogradova ex Scagel et al. (as *Monostroma oxyspermum*) 462
Gelidiella acerosa (Forsskål) Feldmann and G. Hamel 52, 75, 111, 116, 119, 196, 282, 341, 342, 417, 419, 497
Gelidium abbottiorum R.E. Norris 373
Gelidium amansii (J.V. Lamouroux) J.V. Lamouroux 75, 111, 128, 169, 173, 282, 373, 410, 497
Gelidium attenuatum (Turner) Thuret 75, 196
Gelidium corneum (Hudson) J.V. Lamouroux (as *Gelidium sesquipedale*) 76, 146, 423, 472
Gelidium crinale (Hare ex Turner) Gaillon 448, 482
Gelidium filicinum Bory 76, 410
Gelidium foliaceum (Okamura) E.M. Tronchin 497
Gelidium J.V. Lamouroux 51, 53, 63, 75, 76, 111, 116, 119, 128, 131, 146, 169, 173, 178, 196, 199, 205, 208, 282, 343, 344, 373, 374, 380, 410, 417, 421–423, 438, 448, 472, 482, 497, 504, 511, 520, 521
Gelidium japonicum (Harvey) Okamura 76
Gelidium microdon Kützing 116, 119
Gelidium microdon Kützing (as *Gelidium spinulosum*) 196, 373
Gelidium micropterum Kützing 76
Gelidium pristoides (Turner) Kützing 119
Gelidium pulchellum (Turner) Kützing 51, 53, 438
Gelidium pusillum (Stackhouse) Le Jolis 76, 344, 422
Gelidium robustum (N.L. Gardner) Hollenberg and I.A. Abbott (as *Gelidium cartilagenium*) 178
Gelidium spinosum (S.G. Gmelin) P.C. Silva (as *Gelidium latifolium*) 51, 53, 76, 343, 417, 448
Gibsmithia hawaiiensis Doty 340
Gigartina pistillata (S.G. Gmelin) Stackhouse 51, 53, 76, 147, 197, 473, 481
Gigartina skottsbergii Setchell and N.L. Gardner 53, 76, 140, 179, 210, 211
Gloiopeltis furcata (Postels and Ruprecht) J. Agardh 54, 76, 146, 169, 197, 282, 310, 374, 413, 497, 502
Gloiosiphonia capillaris (Hudson) Carmichael 374, 411
Gracilaria birdiae E.M. Plastino and E.C. Oliveira 76, 473
Gracilaria bursa-pastoris (S.G. Gmelin) P.C. Silva 76
Gracilaria canaliculata Sonder (as *Gracilaria crassa*) 54, 76, 207, 340, 343, 374, 414, 420, 448
Gracilaria caudata J. Agardh 283, 310
Gracilaria cervicornis (Turner) J. Agardh 309, 417
Gracilaria changii (B.M. Xia and I.A. Abbott) I.A. Abbott, J. Zhang and B.M. Xia 54, 76, 147, 418, 419
Gracilaria chilensis C.J. Bird, McLachlan and E.C. Oliveira 77
Gracilaria cornea J. Agardh 77, 414, 521
Gracilaria cornea J. Agardh (as *Hydropuntia cornea*) 80, 283
Gracilaria coronopifolia J. Agardh 77, 502
Gracilaria corticata (J. Agardh) J. Agardh 77, 157, 197, 207, 214, 310, 340, 343, 412, 414, 416, 417, 420, 425, 427, 437, 445
Gracilaria corticata var. *cylindrica* Umamaheswara Rao 425
Gracilaria debilis (Forsskål) Børgesen 77, 417, 418, 421
Gracilaria debilis (Forsskål) Børgesen (as *Gracilaria fergusonii*) 418, 421
Gracilaria dendroides Gargiulo, De Masi and Tripodi 424
Gracilaria domingensis (Kützing) Sonder ex Dickie 77, 209, 211, 417
Gracilaria dura (C. Agardh) J. Agardh 112, 375
Gracilaria edulis (S.G. Gmelin) P.C. Silva 54, 77, 344, 413, 417, 418, 497
Gracilaria fergusonii J. Agardh 418, 421
Gracilaria foliifera (Forsskål) Børgesen 77
Gracilaria gracilis (Stackhouse) Steentoft, L.M. Irvine and Farnham 51, 55, 77, 125, 438, 448
Gracilaria Greville 51, 54, 55, 76–78, 111, 112, 119, 123, 125, 129, 146, 147, 157, 166, 169, 197, 207, 209, 211, 214, 283, 285, 309, 310, 340, 343, 344, 374–376, 411–414, 416–425, 427, 437, 438, 445, 448, 449, 473, 497, 502, 511, 520, 521
Gracilaria lemaneiformis (Bory) Greville 284
Gracilaria manilaensis Yamamoto and Trono 284
Gracilaria opuntia Durairatnam, nom. inval. 157
Gracilaria ornata Areschoug 425, 438, 449
Gracilaria pacifica I.A. Abbott 77
Gracilaria pulvinata Skottsberg (as *Gracilaria pygmaea*) 375
Gracilaria salicornia (C. Agardh) E.Y. Dawson 77, 416
Gracilaria subsecundata Setchell and N.L. Gardner 375
Gracilaria tenuistipitata C.F. Chang and B.M. Xia 77, 211, 502
Gracilaria textorii (Suringar) De Toni 77, 169
Gracilaria vermiculophylla (Ohmi) Papenfuss 55, 78, 166, 423
Gracilariopsis andersonii (Grunow) E.Y. Dawson (as *Gracilariopsis sjoestedtii*) 78, 285, 417
Gracilariopsis chorda (Holmes) Ohmi 498, 502
Gracilariopsis longissima (S.G. Gmelin) M. Steentoft, L.M. Irvine and W.F. Farnham 55, 78, 112, 169, 375, 411, 421–423, 445, 473
Gracilariopsis longissima (S.G. Gmelin) M. Steentoft, L.M. Irvine and W.F. Farnham (as *Gracilaria verrucosa*) 169, 285, 376, 411, 421–423, 445, 473
Grateloupia acuminata Holmes 285
Grateloupia cornea Okamura (as *Prionitis cornea*) 169, 473

Grateloupia elliptica Holmes 78, 146, 157, 511
Grateloupia elliptica Holmes (as *Pachymeniopsis elliptica*) 473
Grateloupia filicina (J.V. Lamouroux) C. Agardh 51, 55, 78, 166, 376, 413, 424
Grateloupia imbricata Holmes 286
Grateloupia indica Børgesen 78, 139, 197, 209, 482
Grateloupia lanceola (J. Agardh) J. Agardh 78, 166, 170, 294, 521
Grateloupia lithophila Børgesen 445, 448
Grateloupia livida (Harvey) Yamada 78, 424, 484
Grateloupia longifolia Kylin 197
Grateloupia sparsa (Okamura) Chiang 286
Grateloupia turuturu Yamada 55, 56, 78, 112, 125, 424, 448, 482
Green algae (Chlorophyta) 2, 13, 17, 70, 99, 106, 121, 148, 173, 180, 181, 214, 301, 302, 304, 325–327, 330, 334, 351, 352, 382–386, 389, 390, 394, 457, 459, 482, 519, 521
Griffithsia C. Agardh 78, 198, 206, 410
Griffithsia flosculosa (J. Ellis) Batters 410
Gymnogongrus griffithsiae (Turner) C. Martius 56, 57, 79, 198, 208–211
Gymnogongrus torulosus (J.D. Hooker and Harvey) F. Schmitz 210

H

Halicoryne wrightii Harvey 221
Halidrys siliquosa (Linnaeus) Lyngbye 29, 242, 306, 411, 434
Halimeda copiosa Goreau and E.A. Graham 326
Halimeda cuneata Hering 326, 357
Halimeda discoidea Decaisne 79, 221, 301
Halimeda gracilis Harvey ex J. Agardh 383
Halimeda incrassata (J. Ellis) J.V. Lamouroux 79, 181, 490, 499
Halimeda macroloba Decaisne 12, 152, 164, 301, 325, 326, 352, 390, 392
Halimeda opuntia (Linnaeus) J.V. Lamouroux 12, 79, 180, 184, 325, 326, 352, 387
Halimeda tuna (J. Ellis and Solander) J.V. Lamouroux 3, 12, 180, 301, 325, 328, 351, 352, 383, 389
Halopithys incurva (Hudson) Batters 57, 198, 414, 423, 438, 446, 473
Halopteris filicina (Grateloup) Kützing 365, 398
Halopteris scoparia (Linnaeus) Sauvageau 27, 29, 79, 396, 409, 434
Halurus equisetifolius (Lightfoot) Kützing 56, 57, 79, 343, 438
Halurus flosculosus (J. Ellis) Maggs and Hommersand 79, 448
Halymenia dilatata Zanardini 79, 170
Halymenia floresii (Clemente) C. Agardh 414
Halymenia porphyroides Børgesen 207, 340, 376, 414
Hapalospongidion macrocarpum (Feldmann) León-Álvarez and González-González 365
Helminthocladia australis Harvey 474
Heterosiphonia gibbesii (Harvey) Falkenberg 287
Heterosiphonia muelleri (Sonder) De Toni 207, 340, 376, 414
Heterosiphonia plumosa (J. Ellis) Batters 56, 57, 310
Himanthalia elongata (Linnaeus) S.F. Gray 27, 29, 80, 100, 110, 118, 123, 127, 145, 147, 151, 160, 350, 365, 399, 400, 403, 404, 508, 521
Himantothallus grandifolius (A. Gepp and E.S. Gepp) Zinova 243
Hormophysa cuneiformis (J.F. Gmelin) P.C. Silva (as Hormophysa triquetra) 80, 243
Hormosira banksii (Turner) Decaisne 80, 110
Hyalosiphonia caespitosa Okamura 287
Hydroclathrus clathratus (C. Agardh) M.A. Howe 30, 80, 181, 188, 205, 214, 243, 331, 366, 397, 406
Hydrolithon boergesenii (Foslie) Foslie (as *Hydrolithon reinboldii*) 287
Hypnea cervicornis J. Agardh 119, 198, 474
Hypnea charoides J.V. Lamouroux 80, 287
Hypnea flagelliformis Greville ex J. Agardh 343, 416, 421
Hypnea japonica Tanaka 81, 287
Hypnea musciformis (Wulfen) J.V. Lamouroux 56, 57, 81, 112, 147, 206, 207, 209, 211, 310, 340, 342, 344, 376, 411, 413, 414, 417, 419, 421, 422, 446, 448, 501
Hypnea pannosa J. Agardh 81, 340, 424
Hypnea saidana Holmes 81
Hypnea spinella (C. Agardh) Kützing 81, 209
Hypnea valentiae (Turner) Montagne 58, 340, 341, 343, 416, 424, 448, 483, 501

I

Iridaea Bory de Saint-Vincent 81, 138, 288, 474, 481, 482
Iridaea cordata (Turner) Bory de Saint-Vincent 81, 288
Ishige foliacea Okamura 81, 167, 243, 466, 478, 508
Ishige okamurae Yendo 30, 81, 145, 155, 160, 188, 205, 405, 477, 478, 493
Ishige sinicola (Setchell and N.L. Gardner) Chihara 81
Iyengaria stellata (Børgesen) Børgesen 30, 144, 147, 366, 400, 410

J

Jania adhaerens J.V. Lamouroux 56, 58, 209, 288, 310
Jania capillacea Harvey 437
Jania crassa J.V. Lamouroux 198, 209, 211
Jania rubens (Linnaeus) J.V. Lamouroux 56, 58, 112, 377, 412, 415, 426, 437, 438
Jania sagittata (J.V. Lamouroux) Blainville (as *Cheilosporum sagittatum*) 377
Jania spectabilis (Harvey ex Grunow) J.H. Kim, Guiry and H.-G. Choi (as *Cheilosporum spectabile*) 199
Jania virgata (Zanardini) Montagne 417, 418
Jania virgata (Zanardini) Montagne (as *Haliptilon virgatum*) 417
Jolyna laminarioides S.M. Guimarães 30, 144, 402, 433

K

Kallymenia perforata J. Agardh 288
Kallymenia requienii (J. Agardh) J. Agardh 199

Kappaphycus alvarezii (Doty) Doty ex P.C. Silva 59, 82, 116, 137–139, 158, 199, 289, 309, 341, 377, 417, 418, 474, 482, 498, 511, 515, 520
Kappaphycus alvarezii (Doty) Doty ex P.C. Silva (as *Eucheuma cottonii*) 116, 199, 289, 309, 482, 511, 515
Kappaphycus cottonii (Weber-van Bosse) Doty ex P.C. Silv 82, 199, 289

L

Laminaria abyssalis A.B. Joly and E.C. Oliveira 82, 181, 188
Laminaria brasiliensis A.B. Joly and E.C. Oliveira 244, 460, 467
Laminaria digitata J.V. Lamouroux 31, 82, 101, 102, 104, 110, 118, 123, 127, 163, 366, 396, 399, 400, 405, 406, 442, 460, 508
Laminaria hyperborea (Gunnerus) Foslie 27, 31, 82, 100, 334, 425, 460
Laminaria J.V. Lamouroux 1, 27, 31, 34, 35, 82, 100–104, 110, 118, 121, 123, 126, 127, 131, 145, 149, 151, 161, 163, 168, 181, 188, 190, 214, 244, 250, 252, 253, 303, 304, 306, 307, 334, 339, 366, 367, 395, 396, 399, 400, 404–406, 425, 442, 459, 460, 467, 468, 479, 481, 501, 504, 508, 509, 515, 517, 521
Laminaria ochroleuca Bachelot de la Pylaie 27, 31, 82, 151, 334, 339
Laminaria setchellii P.C. Silva 82
Landsburgia quercifolia Harvey 245
Laurencia aldingensis Saito and Womersley 342, 474
Laurencia catarinensis Cordeiro-Marino and Fujii 290, 342, 449
Laurencia claviformis Børgesen 474
Laurencia complanata (Suhr) Kützing 377, 412
Laurencia coronopus J. Agardh 82, 199, 417
Laurencia dendroidea J. Agardh 59, 206, 209, 211, 342, 344, 413, 437, 438, 446, 449
Laurencia dendroidea J. Agardh (as *Laurencia majuscula*) 59, 206, 413
Laurencia dendroidea J. Agardh (as *Laurencia scoparia*) 59, 437, 446
Laurencia filiformis (C. Agardh) Montagne 377, 474
Laurencia glandulifera (Kützing) Kützing 290, 474
Laurencia gracilis J.D. Hooker and Harvey (as *Laurencia filiformis*) 377, 474
Laurencia intricata J.V. Lamouroux 377, 474
Laurencia J.V. Lamouroux 48, 49, 56, 59–61, 82, 83, 112, 119, 141, 146, 161, 170, 199, 200, 206, 207, 209, 211, 290, 292, 294, 309, 310, 339–342, 344, 345, 372, 373, 377–379, 410–414, 416, 417, 420, 421, 422, 424, 427, 437, 438, 446, 448, 449, 472, 474, 483, 497, 511
Laurencia johnstonii Setchell and Gardner 421
Laurencia mariannensis Yamada 411
Laurencia microcladia Kützing 59, 309, 437, 438
Laurencia nidifica J. Agardh 83, 449
Laurencia nipponica Yamada 438
Laurencia obtusa (Hudson) J.V. Lamouroux 56, 59, 83, 207, 339, 340, 410, 411, 414, 438, 448
Laurencia okamurae Yamada 112, 146, 170, 413
Laurencia pacifica Kylin 411, 511
Laurencia pinnata Yamada 83, 437
Laurencia similis K.W. Nam and Y. Saito 161
Laurencia snackeyi (Weber-van Bosse) M. Masuda 427
Laurencia tasmanica J.D. Hooker and Harvey 474
Laurencia translucida Fujii and Cordeiro-Marina 342
Laurencia viridis Gil-Rodríguez and Haroun 83
Leathesia marina (Lyngbye) Decaisne (as *Leathesia difformis*) 27, 32, 168, 188, 204, 245
Leathesia marina (Lyngbye) Decaisne (as *Leathesia nana*) 245
Lessonia Bory 83, 100, 131, 145, 214, 245, 467, 477, 509
Lessonia flavicans Bory (as *Lessonia vadosa*) 83, 100, 467, 477
Lessonia nigrescens Bory 214, 245
Lessonia trabeculata Villouta and Santelices 145
Lithophyllum kotschyanum Unger 340
Lithophyllum okamurae Foslie 166, 170
Lithophyllum pygmaeum (Heydrich) Heydrich (as *Lithophyllum moluccense*) 340
Lithothamnion corallioides (P. Crouan and H. Crouan) P. Crouan and H. Crouan 474
Lithothamnion Heydrich 107, 199, 295, 474, 516
Lithothamnion muelleri Lenormand ex Rosanoff 199
Lobophora J. Agardh 32, 83, 102, 118, 123, 181, 188, 203, 205, 245, 304, 332, 366, 396, 434–436, 442, 467, 481
Lobophora variegata (J.V. Lamouroux) Womersley ex E.C. Oliveira 32, 83, 102, 118, 123, 181, 188, 203, 205, 245, 304, 366, 396, 434, 435, 442, 467, 481
Lomentaria articulata (Hudson) Lyngbye 292, 438
Lomentaria catenata Harvey 83, 166, 169, 472
Lomentaria clavellosa (Lightfoot ex Turner) Gaillon 292
Lomentaria hakodatensis Yendo 293
Lophocladia (J. Agardh) F. Schmitz 293, 395
Lophocladia lallemandii (Montagne) F. Schmitz 395

M

Macrocystis C. Agardh 83, 100, 110, 131, 181, 189, 246, 467, 509
Macrocystis pyrifera (Linnaeus) C. Agardh 83, 100, 110, 181, 189, 246, 467
Macrocystis pyrifera (Linnaeus) C. Agardh (as *Macrocystis integrifolia*) 246
Marginariella boryana (A. Richard) Tandy 189, 493
Martensia denticulata Harvey 170
Mastocarpus stellatus (Stackhouse) Guiry 56, 59, 84, 137, 138, 146, 293, 378, 410, 411, 415, 438, 446, 448, 520
Mastocarpus stellatus (Stackhouse) Guiry (as *Gigartina stellata*) 378, 410, 411
Mastophora rosea (C. Agardh) Setchell 340
Mazzaella capensis (J. Agardh) Fredericq 378
Mazzaella japonica (Mikami) Hommersand 84, 293
Mazzaella laminarioides (Bory de Saint-Vincent) Fredericq 84, 474, 482
Mazzaella laminarioides (Bory de Saint-Vincent) Fredericq (as *Iridaea laminarioides*) 482
Mazzaella parksii (Setchell and N.L. Gardner) Hughey, P.C. Silva and Hommersand (as *Mazzaella cornucopiae*) 199, 207

Melanothamnus afaqhusainii M. Shameel 144, 378, 420, 437, 446
Meristotheca coacta Okamura 293
Meristotheca gelidium (J. Agardh) E.J. Faye and M. Masuda (as *Meristiella gelidium*) 199, 208
Meristotheca papulosa (Montagne) J. Agardh 84, 474
Mesophyllum expansum (Philippi) Cabioch and M.L. Mendoza (as *Lithophyllum expansum*) 199
Microdictyon pseudohapteron A. Gepp and E.S. Gepp 441
Microdictyon umbilicatum (Velley) Zanardini (as *Microdictyon agardhianum*) 325
Monostroma angicava Kjellman 84, 459, 462
Monostroma grevillei (Thuret) Wittrock 84, 185
Monostroma latissimum Wittrock 84
Monostroma nitidum Wittrock 13, 84, 149, 167, 181, 222, 457
Myagropsis myagroides (Mertens ex Turner) Fensholt 168, 246, 467
Myelophycus simplex (Harvey) Papenfuss 168, 189, 246
Myriogramme minuta Kylin (as *Drachiella minuta*) 378, 414

N

Nemacystus decipiens (Suringar) Kuckuck 85
Nemalion elminthoides (Velley) Batters 85, 199
Neomeris annulata Dickie 326
Neorhodomela aculeata (Perestenko) Masuda 199, 474, 501
Neorhodomela larix (Turner) Masuda 85, 378
Neorhodomela munita (Perestenko) Masuda 199
Neosiphonia japonica (Harvey) M.-S. Kim and I.K. Lee 170
Nereocystis luetkeana (K. Mertens) Postels and Ruprecht 85, 181, 189, 246
Neurymenia fraxinifolia (Mertens ex Turner) J. Agardh 293, 378, 418
Notheia anomala Harvey and J.W. Bailey 433
Nothogenia fastigiata (Bory) P.G. Parkinson 60, 200, 206, 207, 211, 474

O

Ochtodes secundiramea (Montagne) M. Howe 344, 378, 449, 498
Odonthalia corymbifera (S.G. Gmelin) Greville 85, 158, 378, 415
Odonthalia dentata (Linnaeus) Lyngbye 339
Osmundaria colensoi (J.D. Hooker and Harvey) R.E. Norris 64, 300, 378, 416
Osmundaria obtusiloba (C. Agardh) R.E. Norris 200, 207, 209, 474, 482, 501
Osmundaria obtusiloba (C. Agardh) R.E. Norris (as *Vidalia obtusiloba*) 200, 203, 207, 474, 482, 501
Osmundaria serrata (Suhr) R.E. Norris 345, 412, 414
Osmundea hybrida (A.P. de Candolle) K.W. Nam 56, 60, 340, 438
Osmundea pinnatifida (Hudson) Stackhouse 56, 60, 85, 112, 116, 214, 294, 339, 345, 378, 411, 412, 422, 437, 438, 446, 474
Osmundea pinnatifida (Hudson) Stackhouse (as *Laurencia pinnatifida*) 60, 339, 345, 378, 411, 412, 422, 437, 474
Osmundea truncata (Kützing) K.W. Nam and Maggs 85, 379
Otohimella japonica (Yamada) Mas. Suzuki, T. Segawa, Hi. Mori and H. Nozaki (as *Liagora japonica*) 294

P

Pachymeniopsis lanceolata (K. Okamura) Y. Yamada ex S. Kawabata 166, 170, 294, 474
Pachymeniopsis lanceolata (K. Okamura) Y. Yamada ex S. Kawabata (as *Grateloupia lanceolata*) 78, 166, 170, 294
Padina Adanson 27, 32, 33, 85, 86, 100, 102, 110, 118, 155, 160, 164, 168, 181, 189, 203, 205, 247, 332–336, 338, 339, 366, 367, 396, 397, 399–403, 405, 406, 408–410, 434–437, 442, 460, 467, 493, 499, 509
Padina antillarum (Kützing) Piccone 85, 247, 339
Padina arborescens Holmes 85, 155, 160, 168
Padina australis Hauck 32, 86, 181, 189, 205, 366, 401, 406, 493
Padina boergesenii Allender and Kraft 32, 86, 164, 396, 467, 509
Padina boryana Thivy (as *Padina tenuis*) 86, 332
Padina concrescens Thivy 366
Padina gymnospora (Kützing) Sonder 27, 33, 102, 110, 181, 203, 205, 333, 335, 336, 338, 401, 403, 408, 437
Padina jonesii Tsuda 332
Padina minor Yamada 436, 442
Padina pavonica (Linnaeus) Thivy 27, 33, 100, 118, 333, 334, 339, 397, 402, 403, 410, 460
Padina sanctae-crucis Borgesen 405
Padina sulcata Ni-Ni-Win, S.G.A. Draisma and H. Kawai 164
Padina tetrastomatica Hauck 338, 399
Palisada cruciata (Harvey) K.W. Nam (as *Laurencia cruciata*) 379, 413
Palisada flagellifera (J. Agardh) K.W. Nam 344
Palisada perforata (Bory) K.W. Nam 56, 60, 86, 112, 119, 141, 200, 294, 310, 417, 424, 446
Palisada perforata (Bory) K.W. Nam (as *Chondrophycus papillosus*) 60, 379, 417
Palisada perforata (Bory) K.W. Nam (as *Laurencia papillosa*) 60, 112, 119, 141, 200, 294, 310, 379, 424, 446
Palisada poiteaui (J.V. Lamouroux) K.W. Nam (as *Laurencia poiteaui*) 61, 207, 340, 414
Palisada yamadana (M. Howe) K.W. Nam (as *Laurencia yamadana*) 294
Palmaria palmata (Linnaeus) F. Weber and D. Mohr 56, 61, 86, 113, 120, 121, 124, 125, 151, 171, 173, 200, 214, 295, 424, 475, 511, 521
Palmophyllum crassum (Naccari) Rabenhorst 185
Papenfussiella lutea Kylin 189, 493
Pelvetia canaliculata (Linnaeus) Decaisne and Thuret 27, 33, 87, 110, 249, 334, 434, 509
Penicillus capitatus Lamarck 13, 180, 185, 326, 463
Penicillus dumetosus (J.V. Lamouroux) Blainville 222

Perithalia capillaris J. Agardh 249
Petalonia binghamiae (J. Agardh) K.L. Vinogradova 87, 160, 163, 168, 249, 400, 467, 479, 480
Petalonia binghamiae (J. Agardh) K.L. Vinogradova (as *Endarachne binghamiae*) 168, 400, 401
Petalonia fascia (O.F. Müller) Kuntze 27, 33, 87, 367, 399, 409
Petrospongium rugosum (Okamura) Setchell and N.L. Gardner 168
Peyssonelia Decaisne 200
Peyssonnelia rubra (Greville) J. Agardh 379
Phacelocarpus Endlicher and Diesing 170, 475, 483
Phacelocarpus peperocarpos (Poiret) M.J. Wynne (as *Phacelocarpus labillardieri*) 475, 483
Phyllophora crispa (Hudson) P.S. Dixon (as *Phyllophora nervosa*) 200
Phyllospora comosa (Labillardière) C. Agardh 249
Phymatolithon calcareum (Pallas) W.H. Adey and D.L. McKibbin ex Woelkering and L.M. Irvine (as *Lithothamnium calcareum*) 87, 295, 517
Pithophora roettleri (Roth) Wittrock (as *Pithophora oedogonia*) 87, 329
Plocamium brasiliense (Greville) M. Howe and W.R. Taylor 200, 209, 211, 449, 475
Plocamium cartilagineum (Linnaeus) P.S. Dixon 56, 61, 87, 205, 209, 295, 309, 345, 379, 411, 424, 437, 438
Plocamium cartilagineum (Linnaeus) P.S. Dixon (as *Gelidium cartilagineum*) 205
Plocamium corallorhiza (Turner) J.D. Hooker and Harvey 295
Plocamium cornutum (Turner) Harvey 296
Plocamium rigidum Bory 412
Plocamium suhrii Kützing 296
Plocamium telfairiae (W.J. Hooker and Harvey) Harvey ex Kützing 413, 446
Polycladia indica (Thivy and Doshi) Draisma, Ballesteros, F. Rousseau and T. Thibaut (as *Cystoseira indica*) 189, 204, 331, 367, 397, 410, 436, 443
Polycladia myrica (S.G. Gmelin) Draima, Ballesteros, F. Rousseau and T. Thibaut (as *Cystoseira myrica*) 214, 249, 305, 367, 401
Polyides rotundus (Hudson) Gaillon 438
Polyneura bonnemaisonii (C. Agardh) Maggs and Hommersand 296
Polyopes affinis (Harvey) Kawaguchi and Wang (as *Carpopeltis affinis*) 87, 170, 201, 296, 475, 483
Polyopes lancifolius (Harvey) Kawaguchi and Wang 87, 158, 166
Polyopes lancifolius (Harvey) Kawaguchi and Wang (as *Sinkoraena lancifolia*) 166, 170
Polysiphonia denudata (Dillwyn) Greville ex Harvey 61, 62, 201, 207, 211, 417, 418
Polysiphonia elongata (Hudson) Sprengel 411, 417, 426
Polysiphonia elongata (Hudson) Sprengel (as *Polysiphonia denudata* f. *fragilis*) 417
Polysiphonia Greville 61, 62, 87, 88, 116, 120, 125, 170, 201, 203, 207, 211, 297, 300, 339, 379, 381, 411, 412, 416, 417, 424, 426, 512
Polysiphonia mollis J.D. Hooker and Harvey 62, 88, 116, 120
Polysiphonia stricta (Dillwyn) Greville 125, 297
Polysiphonia stricta (Dillwyn) Greville (as *Polysiphonia urceolata*) 297
Polysiphonia tuticorinensis Børgesen 379, 424
Polysiphonia virgata (C. Agardh) Sprengel 412, 416
Polysiphonia virgata (C. Agardh) Sprengel (as *Carradoriella virgata*) 379, 412
Porphyra C. Agardh 62, 63, 88, 89, 104, 113, 116, 120, 121, 124, 125, 146, 147, 170, 171, 173, 214, 297, 298, 308, 379, 380, 423, 425, 438, 447, 475, 483, 484, 487, 498, 502, 511, 512, 517, 521
Porphyra dioica J. Brodie and L.M. Irvine 146, 379, 423
Porphyra linearis Greville 62, 88, 438
Porphyra purpurea (Roth) C. Agardh 88, 297
Porphyra umbilicalis Kützing 88, 113, 120, 124, 511
Portieria hornemannii (Lyngbye) P.C. Silva 63, 89, 158, 201, 297, 340, 342, 379, 412, 417, 424, 426
Portieria hornemannii (Lyngbye) P.C. Silva (as *Chondrococcus hornemannii*) 63, 158, 201, 297, 412
Portieria Zanardini 63, 89, 158, 201, 297, 340, 342, 379, 412, 417, 424, 426, 512
Postelsia palmaeformis Ruprecht 89, 181, 190
Prasiola crispa (Lightfoot) Kützing 463
Pseudocodium devriesii Weber-van Bosse 358
Pterocladiella capillacea (S.G. Gmelin) Santelices and Hommersand 62, 63, 89, 113, 116, 120, 170, 201, 207, 209, 211, 379, 410, 416, 423, 426, 449, 475
Pterocladiella capillacea (S.G. Gmelin) Santelices and Hommersand (as *Gelidium capillaceum*) 63, 410
Pterocladiella capillacea (S.G. Gmelin) Santelices and Hommersand (as *Pterocladia capillacea*) 63, 201, 207, 380, 423, 426
Pterosiphonia complanata (Clemente) Falkenberg 201, 209, 380, 423, 475
Ptilonia (Harvey) J. Agardh 339
Ptilophora pinnatifida J. Agardh (as *Beckerella pinnatifida*) 380
Punctaria latifolia Greville 190, 250
Punctaria plantaginea (Roth) Greville 250
Pyropia columbina (Montagne) W.A. Nelson 89, 113, 171, 173, 201, 298
Pyropia columbina (Montagne) W.A. Nelson (as *Porphyra columbina*) 171, 298
Pyropia dentata (Kjellman) N. Kikuchi and M. Miyata (as *Porphyra dentata*) 89, 475, 483, 512
Pyropia haitanensis (Kjellman) N. Kikuchi and M. Miyata (as *Porphyra haitanensis*) 89, 380, 425, 484
Pyropia J. Agardh 62, 63, 89, 90, 104, 113, 116, 120, 121, 124, 125, 147, 170, 171, 173, 201, 214, 297, 298, 308, 380, 425, 438, 447, 475, 483, 484, 487, 498, 502, 512, 521
Pyropia leucosticta (Thuret) Neefus and J. Brodie (as *Porphyra leucosticta*) 62, 63, 89, 438, 447
Pyropia spiralis (E.C. Oliveira and Coll) M.C. Oliveira, D. Milstein and E.C. Oliveira (as *Porphyra spiralis*) 90, 120

Pyropia tenera (Kjellman) (as *Porphyra tenera*) 90, 121, 170, 173, 483
Pyropia vietnamensis (Tak.Tanaka and Pham-Hoàng Ho) J.E. Sutherland and Monotilla (as *Porphyra vietnamensis*) 90, 113
Pyropia yezoensi (Ueda) M.S. Hwang and H.G. Choi (as *Porphyra yezoensis*) 90, 124, 147, 171, 173, 214, 308, 483, 498, 502

R

Red algae (Rhodophyta) 99, 128, 345
Rhipocephalus phoenix (J. Ellis and Solander) Kützing 222, 463
Rhizoclonium hieroglyphicum (C. Agardh) Kützing 358, 390
Rhizoclonium riparium (Roth) Harvey (as *Rhizoclonium implexum*) 14, 116, 117, 222
Rhodomela confervoides (Hudson) P.C. Silva 162, 299, 343, 498, 512
Rhodomelopsis africana M.A. Pocock 498
Rhodymenia californica Kylin 380
Rhodymenia divaricata E.Y. Dawson 340
Rosenvingea intricata (J. Agardh) Børgesen 91, 367
Rugulopteryx okamurae E.Y. Dawson) I.K. Hwang, W.J. Lee and H.S. Kim (as *Dilophus okamurae*) 250
Rytiphlaea tinctoria (Clemente) C. Agardh 380

S

Saccharina angustata (Kjellman) C.E. Lane, C. Mayes, Druehl and G.W. Saunders (as *Laminaria angustata*) 34, 91, 100, 250, 304, 367, 395
Saccharina cichorioides (Miyabe) C.E. Lane, C. Mayes, Druehl and G.W. Saunders 91, 190, 307, 468
Saccharina cichorioides (Miyabe) C.E. Lane, C. Mayes, Druehl and G.W. Saunders (as *Laminaria cichorioides*) 190, 307, 468
Saccharina gurjanovae (A.D. Zinova) Selivanova, Zhigadlova and G.I. Hansen (as *Laminaria gurjanovae*) 307
Saccharina japonica (Areschoug) C.E. Lane, C. Mayes, Druehl and G.W. Saunders 34, 91, 102, 110, 118, 123, 127, 145, 155, 161, 163, 168, 171, 172, 214, 252, 306, 404–406, 459, 479, 494, 500, 501, 509
Saccharina japonica (Areschoug) C.E. Lane, C. Mayes, Druehl and G.W. Saunders (as *Laminaria japonica*) 34, 131, 145, 161, 163, 214, 252, 306, 404, 459, 468, 479, 501
Saccharina japonica (Areschoug) C.E. Lane, C. Mayes, Druehl and G.W. Saunders (as *Laminaria japonica* var. *ochotensis*) 252
Saccharina japonica (Areschoug) C.E. Lane, C. Mayes, Druehl and G.W. Saunders (as *Laminaria ochotensis*) 168
Saccharina latissima (Linnaeus) C.E. Lane, C. Mayes, Druehl and G.W. Saunders 27, 34, 91, 104, 306, 339, 399, 400, 460, 481
Saccharina latissima (Linnaeus) C.E. Lane, C. Mayes, Druehl and G.W. Saunders (as *Laminaria saccharina*) 306, 460, 481
Saccharina longicruris (Bachelot de la Pylaie) Kuntze 34, 92, 104, 406
Saccharina longissima (Miyabe) C.E. Lane, C. Mayes, Druehl and G.W. Saunders 35, 92, 101, 303, 460
Saccharina longissima (Miyabe) C.E. Lane, C. Mayes, Druehl and G.W. Saunders (as *Laminaria angustata* var. *longissima*) 35, 303, 460
Saccharina religiosa (Miyabe) C.E. Lane, C. Mayes, Druehl and G.W. Saunders 92
Saccharina religiosa (Miyabe) C.E. Lane, C. Mayes, Druehl and G.W. Saunders (as *Laminaria religiosa*) 253
Saccharina sculpera C.E. Lane, C. Mayes, Druehl and G.W. Saunders 92, 190
Saccharina sculpera C.E. Lane, C. Mayes, Druehl and G.W. Saunders (as *Kjellmaniella crassifolia*) 190, 254, 510
Saccorhiza polyschides (Lightfoot) Batters 27, 31, 35, 92, 110, 151, 254, 334, 443
Sarcodia ceylanica Harvey ex Kützing 113
Sarcodiotheca furcata (Setchell and N.L. Gardner) Kylin 380
Sarconema filiforme (Sonder) Kylin 92, 113, 340, 380, 421
Sarconema filiforme (Sonder) Kylin (as *Sarconema furcellatum*) 340
Sarconema Zanardini 92, 113, 340, 380, 421
Sarcothalia atropurpurea (J. Agardh) Hommersand (as *Gigartina atropurpurea*) 201
Sarcothalia crispata (Bory) Leister 92
Sarcothalia radula (Esper) Edyvane and Womersley (as *Gigartina radula*) 481
Sargassum acinarium (Linnaeus) Setchell 100, 104
Sargassum angustifolium C. Agardh 254
Sargassum aquifolium (Turner) (as *Sargassum crassifolium*) 35, 92, 144, 155, 161, 164, 368, 406
Sargassum aquifolium (Turner) C. Agardh (as *Sargassum binderi*) 35, 155, 161, 368, 406
Sargassum boveanum J. Agardh 335, 410, 436, 443
Sargassum C. Agardh 35–40, 92–94, 100, 102, 104, 110, 116, 119, 127, 132, 144, 145, 147, 150, 151, 155, 161, 164, 168, 171, 173, 181, 182, 190, 203–206, 214, 254, 262, 263, 303–305, 307, 308, 331–338, 368, 395–408, 410, 433–436, 443, 460, 468, 469, 476–481, 494, 495, 500, 502, 510, 516
Sargassum capillare Kützing (as *Sargassum gracile*) 395
Sargassum carpophyllum J. Agardh 255
Sargassum cervicorne Greville 468, 479
Sargassum cinctum J. Agardh 92, 460
Sargassum cinereum J. Agardh 403, 408
Sargassum confusum C. Agardh 92, 190
Sargassum coreanum J. Agardh 92, 168
Sargassum cymosum C. Agardh 93, 203, 205
Sargassum dentifolium (Turner) C. Agardh 401
Sargassum desfontainesii (Turner) C. Agardh 396
Sargassum elegans Suhr 368
Sargassum fallax Sonder 255
Sargassum filipendula C. Agardh 35, 36, 102, 119, 305, 308, 334
Sargassum fluitans (Børgesen) Børgesen 397

Sargassum fulvellum (Turner) C. Agardh 93, 406, 468, 480, 481, 494, 500, 502
Sargassum fusiforme (Harvey) Setchell 36, 93, 110, 145, 147, 173, 263, 305, 408, 460, 510, 516
Sargassum fusiforme (Harvey) Setchell (as *Cystophyllum fusiforme*) 36, 516
Sargassum fusiforme (Harvey) Setchell (as *Hizikia fusiformis*) 36, 145, 147, 168, 173, 256, 263, 305, 460
Sargassum glaucescens J. Agardh 156
Sargassum graminifolium C. Agardh 468, 479
Sargassum hemiphyllum (Turner) C. Agardh 36, 93, 181, 205, 479
Sargassum hemiphyllum var. *chinense* J. Agardh 191
Sargassum henslowianum C. Agardh 93, 150
Sargassum horneri (Turner) C. Agardh 37, 93, 151, 171, 204, 406, 476
Sargassum horridum Setchell and N.L. Gardner 400
Sargassum hystrix J. Agardh 401
Sargassum ilicifolium (Turner) C. Agardh 37, 164, 332, 335, 400, 401, 481
Sargassum ilicifolium (Turner) C. Agardh (as *Sargassum cristaefolium*) 332
Sargassum ilicifolium (Turner) C. Agardh (as *Sargasum duplicatum*) 156, 164
Sargassum incisifolium (Turner) C. Agardh 368, 443
Sargassum incisifolium (Turner) C. Agardh (as *Sargassum heterophyllum*) 443
Sargassum johnstonii Setchell and N.L. Gardner 37, 204, 331, 396, 397
Sargassum latifolium (Turner) C. Agardh 206
Sargassum macrocarpum C. Agardh 500
Sargassum marginatum (C. Agardh) J. Agardh 308
Sargassum mcclurei Setchell 205, 307
Sargassum micracanthum (Kützing) Endlicher 204, 478, 480
Sargassum miyabei Yendo (as *Sargassum kjellmanianum*) 37, 214, 303, 305
Sargassum muticum (Yendo) Fensholt 36, 38, 93, 334, 398, 400, 401, 404, 405, 434
Sargassum naozhouense C.K. Tseng and Lu Baoren 204
Sargassum natans (Linnaeus) Gaillon 38, 434
Sargassum nigrifolium Yendo 502
Sargassum notarisii Zanardini (as *Sargassum crispum*) 214
Sargassum oligocystum Montagne 39, 93, 307, 434
Sargassum pallidum (Turner) C. Agardh 93
Sargassum patens C. Agardh 204
Sargassum piluliferum (Turner) C. Agardh 168
Sargassum plagiophyllum C. Agardh 406
Sargassum polyceratium Montagne 93, 203
Sargassum polycystum C. Agardh 39, 93, 161, 164, 214, 332, 336, 401
Sargassum polycystum C. Agardh (as *Sargassum myriocystum*) 39, 336
Sargassum portierianum Zanardini 403
Sargassum ramifolium Kützing 369
Sargassum ringgoldianum Harvey 161, 214, 304, 478
Sargassum siliquastrum (as *Sargassum tortile*) 39, 94, 127, 214, 305, 477
Sargassum siliquastrum (Mertens ex Turner) C. Agardh 39
Sargassum stenophyllum C. Martius 305, 307, 334
Sargassum subrepandum (Forsskål) C. Agardh 262
Sargassum swartzii C. Agardh 40, 94, 164, 214, 307, 332, 334–338, 396, 399, 401, 402, 406, 408, 410, 435, 479, 481
Sargassum swartzii C. Agardh (as *Sargassum wightii*) 40, 332, 334–336, 338, 396, 399, 401, 402, 406, 408, 410, 435, 469, 479, 495
Sargassum tenerrimum J. Agardh 94, 333, 335, 398, 436, 479
Sargassum thunbergii (Mertens ex Roth) Kuntze 40, 94, 181, 214, 303, 305, 408, 433, 478–480
Sargassum vulgare C. Agardh 36, 40, 94, 116, 181, 203, 205, 334, 336, 337, 396, 407, 408, 481
Sargassum vulgare var. *nanum* E. de Paula 192
Sargassum yendoi Okamura and Yamada 264
Sargassum yezoense (Yamada) Yoshida and T. Konn 470
Schizymenia binderi (J. Agardh ex Kützing) J. Agardh 94, 201, 209, 211, 475
Schizymenia dubyi (Chauvin ex Duby) J. Agardh 170, 206, 299
Schizymenia pacifica (Kylin) Kylin 206
Scinaia fascicularis (Børgesen) Huisman 437
Scinaia hatei Børgesen 94, 202, 380, 447
Scinaia moniliformis J. Agardh (formerly *Scinaia indica*) 94, 207, 299, 340, 414, 438
Scytosiphon lomentaria (Lyngbye) Link 40, 94, 168, 181, 192, 214, 264, 304, 369, 399, 470, 479, 510
Scytothamnus australis (J. Agardh) J.D. Hooker and Harvey 192, 495
Sebdenia flabellata (J. Agardh) P.G. Parkinson (as *Sebdenia polydactyla*) 94, 202
Sebdenia polydactyla (Børgesen) M. Balakrishnan 202, 206
Silvetia babingtonii (Harvey) E.A. Serrão, T.O. Cho, S.M. Boo and Brawley (as *Pelvetia wrightii*) 94, 100, 433, 460, 470, 510
Silvetia babingtonii (Harvey) E.A. Serrão, T.O. Cho, S.M. Boo and Brawley (as *Pelvetia babingtonii*) 433
Silvetia compressa (J. Agardh) E. Serrão, T.O. Cho (formerly *Pelvetia compressa*) 192
Silvetia siliquosa (C.K. Tseng and C.F. Chang) E.A. Serrão, T.O. Cho, S.M. Boo and S.H. Brawley (as *Pelvetia siliquosa*) 95, 170, 173
Sirophysalis trinodis (Forsskål) Kützing (as *Cystoseira trinodis*) 95, 146, 369, 402
Solieria chordalis (C. Agardh) J. Agardh 62, 63, 171, 202, 521
Solieria filiformis (Kützing) P.W. Gabrielson 95, 424, 475
Solieria robusta (Greville) Kylin 95, 144, 147, 207, 340, 341, 380, 414
Spatoglossum asperum J. Agardh 40, 144, 147, 156, 265, 308, 331, 332, 336, 408, 433, 435, 443
Spatoglossum schroederi (C. Agardh) Kützing 41, 95, 102, 156, 192, 305, 308, 460, 470, 478
Spatoglossum stipitatum (Tanaka and K. Nozawa) Bittner et al. (as *zonaria stipitata*) 265
Spatoglossum variabile Figari and De Notaris 204, 308, 331, 369, 397, 410, 434

Sphacelaria indica Reinke 95, 192
Sphaerococcus coronopifolius Stackhouse 62, 64, 95, 116, 120, 202, 209, 299, 309, 345, 380, 422–424, 475, 483
Sphondylothamnion multifidum (Hudson) Nägeli 381, 411
Splachnidium rugosum (Linnaeus) Greville 95, 192, 495
Spyridia clavata Kützing 202, 209, 211
Spyridia filamentosa (Wulfen) Harvey 95, 214, 299, 381
Spyridia fusiformis Børgesen 95
Spyridia hypnoides (Bory) (as *Spyridia insignis*) 202
Spyridia hypnoides (Bory) Papenfuss 449
Stenogramma interruptum (C. Agardh) Montagne (as *Stenogramme interrupta*) 202, 475, 482
Stephanocystis hakodatensis (Yendo) Draisma, Ballesteros, F. Rousseau and T. Thibaut (as *Cystoseira hakodatensis*) 95, 151
Stephanocystis osmundacea (Turner) Trevisan (as *Cystoseira osmundacea*) 96, 369
Stilophora tenella (Esper) P.C. Silva 193
Stoechospermum polypodioides (J.V. Lamouroux) J. Agardh 41, 193, 204, 265, 331, 332, 369, 397, 410, 433, 435, 444
Stoechospermum polypodioides (J.V. Lamouroux) J. Agardh (as *Stoechospermum marginatum*) 41, 102, 193, 204, 265, 331, 332, 397, 410
Stypopodium flabelliforme Weber-van Bosse 265, 401, 470
Stypopodium zonale (J.V. Lamouroux) Papenfuss 41, 182, 193, 203, 205, 334, 369, 396
Symphyocladia latiuscula (Harvey) Yamada 158, 162, 202, 207, 299, 373, 475
Symphyocladia marchantioides (Harvey) Falkenberg 207

T

Taonia atomaria (Woodward) 36, 41, 96, 193, 266, 403, 410, 470
Taonia atomaria (Woodward) J. Agardh (as *Dictyota ciliata*) 41, 410
Tichocarpus crinitus (S.G. Gmelin) Ruprecht 203
Tolypiocladia glomerulata (C. Agardh) F. Schmitz 340
Trichogloeopsis mucosissima (Yamada) I.A. Abbott and Doty 299
Tricleocarpa cylindrica (J. Ellis and Solander) Huisman and Borowitzka 203, 209, 300
Tricleocarpa cylindrica (J. Ellis and Solander) Huisman and Borowitzka (as *Galaxaura cylindrica*) 203, 300
Tricleocarpa cylindrica (J. Ellis and Solander) Huisman and Borowi (as *Galaxaura fastigiata*) 300
Tricleocarpa fragilis (Linnaeus) Huisman and R.A. Townsend 203, 300
Tricleocarpa fragilis (Linnaeus) Huisman and R.A. Townsend (as *Galaxaura oblongata*) 300
Turbinaria condensata Sonder 204, 331, 369, 397
Turbinaria conoides (J. Agardh) Kützing 42, 96, 157, 161, 164, 204, 214, 266, 331–333, 335, 336, 338, 397, 437, 444, 470, 480, 510
Turbinaria decurrens Bory 42, 96, 164, 204, 331, 397, 495
Turbinaria ornata (Turner) J. Agardh 96, 193, 205, 332, 335, 402
Turbinaria tricostata E.S. Barton 470
Turbinaria triquetra (J. Agardh) Kützing 403, 404
Turbinaria turbinata (Linnaeus) Kuntze 96, 304, 308, 434
Tydemania expeditionis Weber-van Bosse 222, 326, 352, 358

U

U. flexuosa Wulfen (formerly *Enteromorpha flexuosa*) 185, 223, 224, 303, 360
Udotea argentea Zanardini 326
Udotea conglutinata (J. Ellis and Solander) J.V. Lamouroux 223, 432, 441
Udotea flabellum (J. Ellis and Solander) M. Howe 180, 185
Udotea indica A. Gepp and E.S. Gepp 13, 96, 180, 325, 359, 382, 395
Ulva australis Areschoug (as *Ulva pertusa*) 13, 96, 167, 180, 185, 223, 325, 352, 359, 389, 463, 490, 505, 521
Ulva clathrata (as *Enteromorpha muscoides*) 3, 14, 96, 180, 325, 326, 382, 384, 431, 441
Ulva clathrata (formerly *Enteromorpha clathrata*) 14, 180, 325, 326, 382, 431, 441
Ulva compressa Linnaeus 3, 14, 97, 116, 117, 181, 326, 330, 351, 383–385, 388–390, 392, 394, 395, 458
Ulva compressa Linnaeus (as *Enteromorpha compressa*) 14, 167, 181, 185, 326, 351, 383–385, 388–390, 392, 394, 395, 458
Ulva conglobata Kjellman 97, 458, 499
Ulva fasciata Delile 14, 97, 108, 122, 149, 164, 180, 302, 325, 326, 352, 382, 384–386, 388, 392, 394, 395, 431, 432
Ulva flexuosa Wulfen 15, 97, 303, 330, 395
Ulva intestinalis Linnaeus 15, 16, 97, 152, 301, 330, 351, 382, 384, 385, 389, 391, 431, 432
Ulva intestinalis Linnaeus (as *Enteromorpha intestinalis*) 15, 117, 167, 224, 301, 330, 351, 382, 384, 385, 391, 431
Ulva lactuca Linnaeus 15, 16, 97, 148, 180, 214, 301, 326, 327, 329, 330, 382–393, 395, 432, 459
Ulva Linnaeus 3, 13–17, 96–98, 108, 116, 117, 121, 122, 145, 148, 149, 152, 159, 164, 167, 172, 180, 181, 185, 214, 223, 226, 301–303, 325–327, 329, 330, 351, 352, 359, 362, 382–395, 431, 432, 441, 457–459, 463, 490, 499, 505, 506, 517, 521
Ulva Linnaeus (as *Enteromorpha* Link) 14–16, 71, 117, 167, 180, 181, 185, 214, 221, 224, 225, 301, 325, 326, 330, 351, 352, 357, 359, 362, 382–385, 387–392, 394, 395, 431, 441, 457, 458, 490, 505
Ulva linza Linnaeus 15, 16, 97, 145, 180, 302, 352, 387–390, 394, 458
Ulva linza Linnaeus (as *Enteromorpha linza*) 15, 167, 180, 185, 225, 352, 387–390, 394
Ulva ohnoi M. Hiraoka and S. Shimada 145
Ulva prolifera O.F. Müller 16, 97, 214, 301, 330, 384, 389, 394, 457
Ulva prolifera O.F. Müller (as *Enteromorpha prolifera*) 15, 16, 214, 225, 301, 330, 384, 389, 394, 457, 490
Ulva reticulata Forsskål 16, 98, 326, 327, 383–385, 394, 431, 432, 441

Ulva rigida C. Agardh 16, 17, 98, 148, 159, 226, 301, 330, 385, 391, 431, 458
Ulva rigida C. Agardh (as *Ulva armoricana*) 186, 506
Ulva stenophylla Setchell and N.L. Gardner 98
Ulva taeniata (Setchell) Setchell and N.L. Gardner (as *Ulva dactylifera*) 362
Umbraulva japonica (Holmes) Bae and I.K. Lee (as *Ulva japonica*) 98, 458
Undaria pinnatifida (Harvey) Suringar 36, 42, 98, 100, 102, 104, 111, 119, 123, 126, 127, 145, 147, 151, 161, 169, 171–173, 181, 182, 193, 204, 205, 267, 304, 306, 307, 395, 404, 436, 471, 476–479, 496, 502, 510

V

Valonia aegagropila C. Agardh 17, 98, 325, 352, 362
Valonia utricularis (Roth) C. Agardh 16, 17, 99, 186
Valoniopsis pachynema (G. Martens) Børgesen 17, 180, 325, 363, 382, 395
Vertebrata lanosa (Linnaeus) T.A. Christensen (as *Polysiphonia lanosa*) 99, 203, 300, 339, 381, 411, 412, 512
Vidalia colensoi (J.D. Hooker and Harvey) J. Agardh (as *Osmundaria colensoi*) 64, 300
Vidalia obtusiloba (Mertens ex C. Agardh) J. Agardh 200, 203, 207, 474, 482, 501

Y

Yuzurua poiteaui (J.V. Lamouroux) Martin-Lescanne (as *Laurencia poitei*, formerly *Palisada poitei*) 339

Z

Zanardinia typus (Nardo) P.C. Silva (as *Zanardinia prototypus*) 193, 370, 395
Zonaria spiralis (J. Agardh) Papenfuss 496, 500
Zonaria subarticulata (J.V. Lamouroux) Papenfuss 370, 396
Zonaria tournefortii (J.V. Lamouroux) Montagne 36, 42, 335, 405

Made in United States
Cleveland, OH
21 January 2025

13714316R00402